THE GENETIC CODE

The three-letter and one-letter abbreviations of the amino acids. "Ter" denotes termination codon. The initiation codon, AUG, is shaded green, and the three termination codons—UAA, UAG, and UGA—are shaded red.

second base

first base of codon (5′ end)

third base of codon (3′ end)

	U	C	A	G	
U	UUU ⎤ Phe UUC ⎦ F UUA ⎤ Leu UUG ⎦ L	UCU ⎤ UCC ⎬ Ser UCA ⎪ S UCG ⎦	UAU ⎤ Tyr UAC ⎦ Y UAA ⎤ UAG ⎦ Ter	UGU ⎤ Cys UGC ⎦ C UGA Ter UGG Trp W	U C A G
C	CUU ⎤ CUC ⎬ Leu CUA ⎪ L CUG ⎦	CCU ⎤ CCC ⎬ Pro CCA ⎪ P CCG ⎦	CAU ⎤ His CAC ⎦ H CAA ⎤ Gln CAG ⎦ Q	CGU ⎤ CGC ⎬ Arg CGA ⎪ R CGG ⎦	U C A G
A	AUU ⎤ AUC ⎬ Ile AUA ⎦ I AUG Met M	ACU ⎤ ACC ⎬ Thr ACA ⎪ T ACG ⎦	AAU ⎤ Asn AAC ⎦ N AAA ⎤ Lys AAG ⎦ K	AGU ⎤ Ser AGC ⎦ S AGA ⎤ Arg AGG ⎦ R	U C A G
G	GUU ⎤ GUC ⎬ Val GUA ⎪ V GUG ⎦	GCU ⎤ GCC ⎬ Ala GCA ⎪ A GCG ⎦	GAU ⎤ Asp GAC ⎦ D GAA ⎤ Glu GAG ⎦ E	GGU ⎤ GGC ⎬ Gly GGA ⎪ G GGG ⎦	U C A G

Amino Acids and Their Symbols			Codons					
Alanine	Ala	A	GCA	GCC	GCG	GCU		
Arginine	Arg	R	AGA	AGG	CGA	CGC	CGG	CGU
Asparagine	Asn	N	AAC	AAU				
Aspartic acid	Asp	D	GAC	GAU				
Cysteine	Cys	C	UGC	UGU				
Glutamic acid	Glu	E	GAA	GAG				
Glutamine	Gln	Q	CAA	CAG				
Glycine	Gly	G	GGA	GGC	GGG	GGU		
Histidine	His	H	CAC	CAU				
Isoleucine	Ile	I	AUA	AUC	AUU			
Leucine	Leu	L	UUA	UUG	CUA	CUC	CUG	CUU
Lysine	Lys	K	AAA	AAG				
Methionine	Met	M	AUG					
Phenylalanine	Phe	F	UUC	UUU				
Proline	Pro	P	CCA	CCC	CCG	CCU		
Serine	Ser	S	AGC	AGU	UCA	UCC	UCG	UCU
Threonine	Thr	T	ACA	ACC	ACG	ACU		
Tryptophan	Trp	W	UGG					
Tyrosine	Tyr	Y	UAC	UAU				
Valine	Val	V	GUA	GUC	GUG	GUU		

GENETICS THE CONTINUITY OF LIFE

Daniel J. Fairbanks

W. Ralph Andersen

GENETICS THE CONTINUITY OF LIFE

Brooks/Cole Publishing Company • Wadsworth Publishing Company

I(T)P® an International Thomson Publishing Company

Pacific Grove, CA • Albany • Belmont, CA • Boston • Cincinnati • Johannesburg • London • Madrid • Melbourne • Mexico City
New York • Scottsdale, AZ • Singapore • Tokyo • Toronto

Biology Editor: *Gary Carlson*
Developmental Editor: *Mary Arbogast*
Developmental Consultants: *Elmarie Hutchinson, Brian Jones*
Assistant Editor: *Marie Carigma-Sambilay*
Editorial Assistant: *Larisa Lieberman*
Marketing Manager: *Tami Cueny*
Project Editor: *John Walker*
Print Buyer: *Karen Hunt*
Permissions Editor: *Bob Kauser*
Production: *Jonathan Peck, Dovetail Publishing Services*
Interior Design: *Rob Hugel*
Cover Design: *John Walker*
Copy Editor: *Brian Jones*
Art Editor: *Dovetail Publishing Services*
Illustrator: *Precision Graphics, Marcus Alan Vincent*
Photo Researcher: *Stuart Kenter Associates*
Indexer: *Pilar Wyman*
Compositor: *New England Typographic Service*
Printer: *World Color*

Printed in the United States of America
1 2 3 4 5 6 7 8 9 10

Cover Images:
Atomic scale image of DNA using scanning tunneling
microscopy courtesy of John D. Baldeschwieler, California
Institute of Technology.
Pea flower courtesy of Daniel J. Fairbanks

Interior images:
Page 2, Science Source/Photo Researchers; pages 18 and 20, Ken
Eward/Photo Researchers; page 56: Oscar Miller/Photo
Researchers; page 88, E. Kiselva/D. Fawcett/Visuals Unlimited;
page 156, Ken Eward/Photo Researchers; page 176, Daniel J.
Fairbanks; page 178, K. G. Murti/Visuals Unlimited;
page 294, Science Source/Photo Researchers;
page 320, D. M. Phillips/Visuals Unlimited; page 344, Berenice
Quinzani Jordão; pages 346 and 380, Daniel J. Fairbanks;
page 418, Dan Sudia/Photo Researchers; page 444, Oliver
Meckes/Photo Researchers; page 480, Science Source/Photo
Researchers; page 512, H. C. Stutz; pages 548, 580, 606, and 632,
Science Source/Photo Researchers; page 658, Marcus Alan Vincent;
page 660, Daniel J. Fairbanks; page 682, Oliver Meckes/Photo
Researchers; page 700, Science Source/Photo Researchers;
page 720, Robert Becker/Custom Medical Stock Photo;
page 742, J. L. Carson/Custom Medical Stock Photo;
page 762, G. Bottner/Photo Researchers;
page 786, Photo Researchers

For more information, contact Brooks/Cole Publishing Company,
511 Forest Lodge Road, Pacific Grove, CA 93940, or electronically
at http://www.brookscole.com

International Thomson Publishing Europe
Berkshire House
168-173 High Holborn
London, WC1V 7AA, United Kingdom

Nelson ITP, Australia
102 Dodds Street
South Melbourne
Victoria 3205 Australia

Nelson Canada
1120 Birchmount Road
Scarborough, Ontario
Canada M1K 5G4

International Thomson Publishing Southern Africa
Building 18, Constantia Square
138 Sixteenth Road, P.O. Box 2459
Halfway House, 1685 South Africa

International Thomson Editores
Seneca, 53
Colonia Polanco
11560 México D.F. México

International Thomson Publishing Asia
60 Albert Street
#15-01 Albert Complex
Singapore 189969

International Thomson Publishing Japan
Hirakawa-cho Kyowa Building, 3F
2-2-1 Hirakawa-cho, Chiyoda-ku
Tokyo 102 Japan

Library of Congress Cataloging-in-Publication Data

Fairbanks, Daniel J.
    Genetics: the continuity of life/ Daniel J. Fairbanks,
  W. Ralph Andersen.
        p.   cm.
    Includes bibliographical references and index.
    ISBN 0-534-25272-9
    1. Genetics.   I. Anderson, W. Ralph.   II. Title
  QH430.F35   1999
  576.5—dc21   98-32298

BRIEF CONTENTS

DETAILED CONTENTS

PREFACE

Which should a genetics professor teach first, Mendelian genetics or molecular genetics? Most textbook authors have chosen to begin with transmission genetics, an organization that parallels the historical development of the discipline. With this approach students are often left with the impression that the field of transmission genetics is finished, and that molecular genetics is the frontier. However, the application of molecular tools to genetic analysis has brought transmission genetics firmly into the molecular arena. To understand transmission genetics in today's world, students need a strong molecular background.

Our book begins with genetics at its most fundamental level: the molecular information encoded in DNA. It takes readers through the transfer of genetic information from DNA to RNA to protein and finally to the outward characteristics that we observe in organisms. It proceeds in a logical fashion from molecules to cells to organisms to populations.

This organization offers several advantages for students and professors. Because students enter the genetics course with basic knowledge of DNA and genes from earlier biology courses, they can begin to forge connections between molecular and Mendelian concepts from the first week of class. They are exhilarated to encounter modern molecular concepts and techniques from the start rather than feeling that the course is replaying ideas they already know. As the term goes on, this organization gives professors the power to explain in a multidimensional way how genetics actually works. And with

this approach, they can teach their students the way most geneticists think when tackling real genetic research.

OBJECTIVES

In writing *Genetics: The Continuity of Life*, we address four major objectives:

1. *Integrate modern molecular biology with all other areas of genetics.* Because of its unique organization, our book teaches the entire realm of genetics in a modern context, like no other book does. For example, as students confront Mendelian genetics, they do so with a molecular foundation and can relate Mendelian principles to the underlying molecular causes. The linkage between genotype at the DNA level and phenotype at the level of observation is a key theme of every chapter—a theme that can work only if we begin with molecular genetics.

2. *Teach students how to conduct genetic analysis.* After each major concept, we provide students with *real genetic data* that have been published in scientific journals, and show them how to analyze and interpret the data step by step, applying the concepts they just learned. We highlight key publications in the history of genetics as well as recent research. The advantage of such an approach is twofold: First, students learn how scientists acquire, organize, analyze, and interpret real genetic data. They learn about specific problems that scientists have faced and see how the scientists confronted those problems. Second, by

analyzing key experiments that helped to build the science of genetics, students become acquainted with the history and development of genetics. Rather than simply reading about the history, they analyze the work of scientists who made genetics what it now is. No other book delves as deeply into analysis of real experiments as ours does.

3. *Present genetics using a clear, friendly style that conveys the excitement of scientific discovery, stimulates imagination, and invites students to learn more.* The writing style of our book is conversational. We want students to feel that they are discussing genetics, not just reading about it. The book is illustrated with excellent drawings and photographs that visually portray the concepts discussed in the text. The artwork is integrated with the text so that students have the chance to work step by step through a diagram until they understand the concept. We have also interwoven the worked examples with the conceptual part of the text so that they fit together coherently.

Throughout, we have tried to convey the passion for scientific discovery that motivates most geneticists. Genetics is a dynamic science with important advances and applications appearing in the news almost weekly. As students sense how exciting genetics can be, they begin to imagine how they can contribute to the science of genetics as researchers. We hope that they will go beyond the book and seek information from both current and historical sources. To help them, we provide carefully chosen suggested readings and free online access to articles through *InfoTrac College Edition.*

4. *Provide students with a resource for study that assists them in their careers long after they have completed the course.* We view a genetics textbook as a resource not only for a genetics course, but also for a scientific career. Most students will not learn all of the information in our book in a single-semester genetics course. We hope, however, that during their course, they will become so familiar with their genetics textbook that they can turn to it for easily accessible additional information years after completing the course. Although the rapid progress of genetic discovery ensures that many of the details in any genetics textbook will become outdated shortly after its publication, the core principles and experimental examples that constitute the central messages of our book will remain informative and valuable for years to come.

ORGANIZATION

Genetics: The Continuity of Life includes 28 chapters: an introductory chapter followed by six major parts. The first chapter introduces students to the importance of genetics and genetic analysis, the purpose of model organisms, and the concept of the gene. The first four parts provide core principles, beginning with the molecular foundations

of genetics and moving sequentially through the genetics of cells, organisms, and populations. The fifth part focuses on four special topics in genetics: transposable elements, development, cancer, and immunity. The final part discusses applications of genetics in medicine, forensics, agriculture, and industry and concludes by examining of the legal and ethical issues that such applications raise.

Part I: Molecular Foundations of Genetics

Part I provides students with the molecular background to understand the material in the rest of the book. Chapter 2 introduces students to DNA structure and replication. After providing fundamental information about gene expression, Chapter 3 focuses on transcription and RNA processing in both prokaryotes and eukaryotes. Chapter 4 covers translation and the processing and functions of proteins. Chapter 5 discusses mutation at the molecular level. Chapter 6 ties the concepts from the previous four chapters together to show how genetic variation at the DNA level is expressed as outward phenotypic variation, setting the stage for the chapters that follow.

Part II: Genetics of Cells

Part II covers genetics at the cellular level in prokaryotes and eukaryotes. Chapter 7, on the genetics and genome organization of bacteria, includes recent studies on whole genome sequencing as well as classical genetic mapping. In this chapter students also learn about bacterial culture techniques and bacterial viruses, information that is important for understanding recombinant DNA methodology. Chapter 8 reviews gene regulation in both prokaryotes and eukaryotes with a focus on how it influences the phenotype. Chapter 9 introduces the fundamental techniques of molecular biology, such as DNA cloning, genetic engineering of eukaryotic genes for expression in bacteria, the polymerase chain reaction, gel electrophoresis of nucleic acids and proteins, DNA sequencing, and computerized use of DNA sequence databases. This chapter provides the foundation for understanding the applications of molecular biology in eukaryotic genetics. Chapter 10 introduces students to the organization of the eukaryotic genome, culminating with an analysis of the complete DNA sequence of the *Saccharomyces cerevisiae* genome as an example of whole genome organization in eukaryotes. The review of mitosis, meiosis, and life cycles in Chapter 11 sets the stage for the next part.

Part III: Genetics of Organisms

In Part III the advantages of a molecular foundation become apparent. Chapter 12 discusses Mendelian genetics. Most students will have studied Mendelian genetics be-

fore, both in high school and in an introductory biology course in college. Rather than review the same material they have already encountered (but may not fully understand), this chapter presents Mendelian genetics in an entirely new light. It begins with a discussion of Mendel's experiments as he perceived them, and shows how he derived the fundamental principles of inheritance from the results of his experiments. Then the chapter brings Mendel's work into a modern context, showing how an understanding of gene expression and mutation at the molecular level can explain Mendelian principles at the phenotypic level. The chapter also introduces students to probability analysis and teaches them to use the binomial distribution and chi-square analysis. The chapter concludes with the story of the origin of genetics and applies statistical analysis to questions raised about Mendel's research. Chapter 13 teaches such important genetic concepts as incomplete dominance, codominance, multiple alleles, pleiotropy, penetrance, variable expressivity, and epistasis, all in a molecular context. With a strong background in the molecular basis of inheritance, students can grasp these concepts with ease. This chapter also introduces students to DNA markers and their use in genetics research. Chapter 14 describes the molecular basis of sex determination in humans and *Drosophila*, and the inheritance of X-linked alleles. Chapter 15 discusses chromosome mapping, starting with the classical approaches in model organisms, then moving to DNA marker-based chromosome mapping in humans and other species. The chapter ends with an explanation of how mapped DNA markers are used to clone important genes through chromosome walking and jumping. Chapter 16 discusses the fine structure of the gene. It shows how classical studies in fungal genetics led researchers to develop models of recombination and gene conversion at the molecular level. Then it discusses intragenic recombination and fine-structure mapping of mutations. Chapter 17 introduces students to the fundamentals of cytogenetics, including alterations in chromosome number and structure, and ends by showing how molecular analysis has played a role in human cytogenetics. Chapter 18 discusses extranuclear inheritance, focusing on mitochondrial and plastidial genomes and their inheritance.

Part IV: Genetics of Populations

In Part IV Chapter 19 introduces population genetics and provides detailed discussions of selection, inbreeding, and random genetic drift. Chapter 20 outlines quantitative genetics with a focus on the genetic basis for continuous variation and the relationship between heritability and the effectiveness of selection. Chapter 21 introduces students to the genetic basis of evolution, highlighting the contribution of molecular biology to an understanding of human origins.

Part V: Gene Expression and the Organism

Part V treats four important topics in current research on gene expression. Transposable elements were introduced in Chapter 10, but Chapter 22 discusses them in more detail, especially the transposable elements of maize and *Drosophila*. Chapter 23 covers developmental genetics, highlighting studies of *Caenorhabditis elegans*, *Drosophila melanogaster*, and mammals. The chapter concludes with a discussion of the emerging field of plant developmental genetics. Chapter 24 introduces students to the genetic basis of cancer. They learn how genes regulate the cell cycle and how mutations in those genes can cause cancer to develop. The chapter also provides practical information on cancer prevention and treatment. Chapter 25 discusses the genetic basis of immunity with a focus on human immunity. It also introduces students to immune system dysfunction including autoimmune disorders and HIV infection.

Part VI: Applications of Genetics

Throughout the book we discuss many applications of genetics, but Part VI shows students how genetics is used in applied research. Chapter 26 presents applications to medicine and forensics. Topics include the Human Genome Project and its implications, genetic testing and screening, DNA fingerprinting, genetic pharmacology, gene therapy, and clinical genetics. We introduce new examples and refer students to examples from previous chapters. Chapter 27 reviews the applications of genetics in agriculture and industry from prehistory to modern times. It tells the genetic origin of domesticated plants and animals and shows how genetic erosion threatens our food supply. This discussion leads into modern breeding and biotechnology of agricultural species. Chapter 28 addresses the legal and ethical issues that arise from applications of genetics. After a historical discussion of the eugenics movement, we discuss current legal and ethical issues, such as protection of intellectual property, confidentiality and informed consent, admissibility of DNA evidence in court, and safety issues associated with recombinant DNA research. The chapter concludes with a discussion of the importance of science education as a means of encouraging people in all walks of life to properly inform themselves about scientific issues.

PEDAGOGICAL FEATURES

A single goal guided the development of pedagogical features in this book: Provide students with a straightforward, accessible presentation that integrates genetic concepts and analysis. Genetics is a rigorous and challenging discipline; accordingly, our book is designed to help students focus on learning genetic concepts and analysis with as little distraction as possible. We have

integrated the artwork and photographs with the text to enliven and clarify the discussion, and we have avoided using sidebars and other features that can distract students from an intensive study of core principles.

Focus on Analysis: Examples and Problems

Nine out of ten instructors state that analyzing and solving problems is the greatest challenge for students in a genetics course. Students *learn* genetics by *doing* genetic analysis, and for this reason every genetics textbook provides students with questions and problems to help them practice genetic analysis. Most also provide worked examples at the end of each chapter to show students step by step how to interpret and analyze data. In many of these texts, the worked examples and end-of-chapter problems use fabricated data that are easy to analyze mathematically.

Our approach to teaching genetic analysis departs from this standard method. We require students to work as much as possible with real data from published research, which are not as mathematically simple as fabricated data. With such an approach, students, like professional researchers, must wrestle with ambiguities, sampling error, alternative interpretations of data, and tentative conclusions. Although this approach may challenge students, it also prepares them to understand genetics in the real world.

In addition, our worked examples are woven into the text so that genetic concepts and analysis are presented in a unified way. We have carefully chosen the worked examples from published data that clearly illustrate each major concept.

The end-of-chapter problems range from easy conceptual exercises to analyses that require students to synthesize information from various parts of the chapter. Like the worked examples, many of the end-of-chapter problems are based on published data.

Our website (http://www.brookscole.com/biology) and *InfoTrac* provide online resources that supplement the book's focus on analysis. Our website contains additional questions and problems that we regularly update to include recent research. Many of the questions and problems at the website ask students to consult published research that is available in *InfoTrac College Edition*, an online resource with recent original articles from leading journals such as *Science, Annual Review of Genetics, Annual Review of Microbiology,* and *Evolution.*

Study Aids

Each chapter begins with a set of key concepts that give students an overview of what they are about to study. Following each major topic in the chapter is a boldfaced concept review, which reinforces the key ideas of that topic. We highlight key terms in boldface and define the most important key terms in a glossary at the end of the book. Each chapter concludes with a summary of the major principles discussed in the chapter.

Design and Layout

The design and layout of our book are also intended to capture interest and facilitate learning. The illustrations and text are placed to complement one another visually and to help students avoid turning pages as they compare the illustrations with the text. The judicious use of colors in the text and illustrations help students focus on concepts and organize their thoughts as they study.

Suggested Readings

Each chapter ends with an annotated list of suggested readings. Rather than simply list a long set of references, we have chosen readings that will guide students to excellent sources of further information. Some of the readings are review articles that provide more detailed information than that found in our book. Others are important original research that helped scientists develop the concepts discussed in the chapter. In some cases, we refer students to books or collections of articles. Besides the suggested readings at the end of the chapter, we provide references for all data in the worked examples and end-of-chapter problems so that students can consult the original sources to enhance their study. We also give full references for reprinted illustrations and photographs in the figure captions, rather than at the end of the book, so that students can easily consult these sources as well.

Technology-Based Resources and Supplements

A host of valuable resources accompany this book. The website for this book is available free to students and provides online study aids, reviews of current articles, specific information about each chapter in the text, supplemental questions and problems, career tips, and links to genetics-related sites. **InfoTrac College Edition** delivers online access to full-length articles from more than 700 periodicals. Students can log on to this online library with their personal ID and perform searches, read assigned articles, use articles as a resource to analyze assigned problems from the website, and print articles for their personal use. The Brooks/Cole **BioLink** is a lecture presentation tool and image bank that allows instructors to assemble images and related text (captions and titles) for electronic presentation in lectures and to post lecture notes and URLs to a course website. BioLink includes over 2000 images, animations, and Quicktime® movies, including more than 100 images from this book. **Genetics Updates**, an online genetics newsletter, features concise essays on current topics and issues in the field of genetics with links to related sites. The **Instructor's Resource**

Manual contains sample syllabi, chapter outlines, lists of main concepts, solutions to all questions and problems from the book, teaching hints, hints on how to use media and web materials, and a conversion guide to illustrate how to move from a Mendel-first approach to a molecular-first approach using *Genetics: The Continuity of Life*. The **Student Companion to Genetics: The Continuity of Life** contains detailed solutions to all end-of-chapter questions and problems. *Current Perspectives in Genetics*, by Shelly Cummings, features some 50 selected articles in molecular, transmission, population, and human genetics as a complement to genetics courses. **Transparencies** with over 100 color reproductions of art from the book are available to instructors. *Genetics on the Web*, by Daniel J. Kurland, gives students exercises to become proficient with the Internet as they study genetics; this book includes a list of URL sites and online materials for each topic in genetics.

ACKNOWLEDGMENTS

The writing, assembly, publication, and distribution of this textbook has been a major cooperative effort on the part of many dedicated professionals who have worked together for the past 6 years. The idea for this book began in 1992 when Cecie Starr and Daniel Fairbanks had a delightful telephone conversation about biology. Cecie suggested that her publisher, Jack Carey, approach Daniel about the possibility of writing a genetics textbook. Daniel and Jack agreed that the time had come to develop an excellent genetics textbook with a molecular-first approach. Ralph Andersen soon joined on as coauthor. Daniel wrote most of the text while Ralph did most of the background library research and prepared many of the worked examples and end-of-chapter problems. Ralph also managed the day-to-day activities of our joint laboratory, allowing us to continue to conduct and publish our research during the time we were working on the textbook.

Numerous reviewers from universities and colleges throughout the country generously provided comments and criticisms at each stage of manuscript development. Their comments have proved invaluable in helping us decide how to organize the book and what to emphasize, and in ensuring accuracy. We are grateful to the reviewers for pointing out errors in the manuscript, but accept full responsibility ourselves for any errors that remain. The names of the reviewers are listed on the following page. We also thank our student assistants at BYU—Michelle Aliff, Carol Gregory, Gordon Harkness, Amy Allgaier, Adam Scrumm, Matthew Tonioli, Paula Randall, Angela Hawkes, Michelle Blauer, and Jeremy Beard—for their contributions in assisting us with the research and clerical work for this book.

The experienced team of professionals at Brooks/Cole and Wadsworth Publishing companies have generously shared their diverse skills to make this book possible. Publishers Jack Carey and Gary Carlson have worked tirelessly to coordinate all aspects of the text and ancillary development. Kristin Milotich, Kerri Abdinoor, Marie Carigma-Sambilay, and Larisa Lieberman have managed the many details associated with the development of the book and its ancillaries. Elmarie Hutchinson, with her skilled developmental editing, taught us how to write clearly and how to reach students with the interweaving of text, examples, and illustrations. Mary Arbogast oversaw the editing and assembly of the final manuscript. She spent many hours on the telephone helping us through the unfamiliar process of the final stages of text development. John Walker supervised a group of professionals who developed the captivating design of the book and its cover. The marketing team of Halee Dinsey, Marlene Veach, and Rita Frumkin enthusiastically developed the most detailed and informative marketing brochure ever produced for a genetics textbook.

Jonathan Peck of Dovetail Publishing Services brilliantly orchestrated the efforts of artists, copyeditor, compositor, proofreader, and photo researcher. He put countless hours into the production of this book and, in the process, became a true friend. Brian Jones is a remarkable copyeditor who meticulously scrutinized every sentence and every mathematical exercise. His queries demonstrated to us that his fascination with genetics, and his devotion to improving the quality of our book, went well beyond his duties as a copyeditor. Our special thanks go to the artists of Precision Graphics, and to Marcus Alan Vincent whose artwork makes this book so understandable and visually appealing. We also thank the many authors, artists, photographers, and publishers who so generously gave permission to adapt their drawings and reprint their photographs. We especially thank Cecie Starr and Stephen Wolfe for allowing us to adapt many drawings from their books. Our photo researcher, Stuart Kenter, acquired many photographs to highlight the chapter opening pages and to illustrate concepts throughout the book. We are grateful to Stephanie Gintowt, Cindy Marschat, and their colleagues at New England Typographic Services for their skillful typesetting and layout.

We ultimately owe our ability to write this book to those who taught us genetics, and we express our heartfelt gratitude to them for shaping our careers. Their devotion to teaching is reflected in the content of this book. We also thank the thousands of students who completed our genetics courses and collectively taught us how to teach genetics. Finally, we cannot find appropriate words to express our love and appreciation to our families, who supported us so completely in the long and demanding process of writing a textbook.

September 1998

Daniel J. Fairbanks
W. Ralph Andersen

REVIEWERS

GENETICS THE CONTINUITY OF LIFE

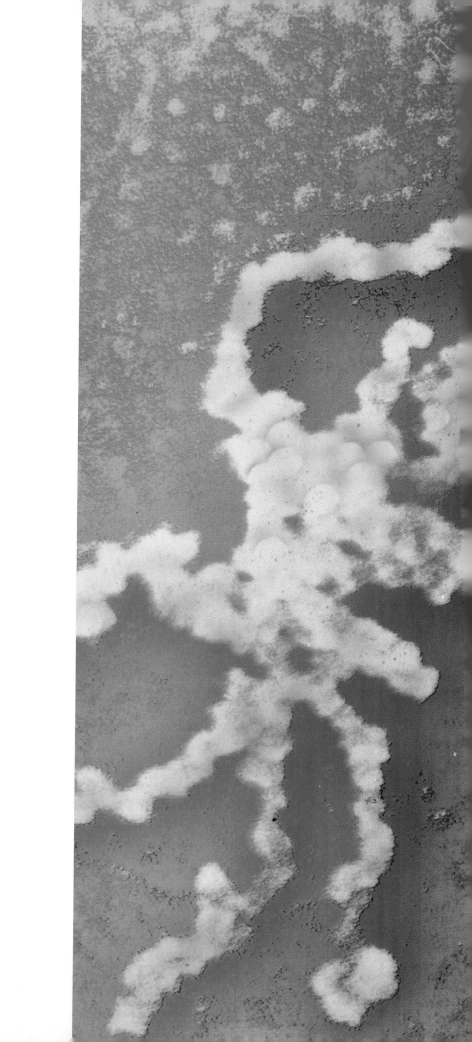

CHAPTER 1

KEY CONCEPTS

Genetics, one of the most rapidly progressing branches of biology, has applications that affect all of humanity.

~

Many of the principles of genetics are universal among the vast diversity of species on earth, making it possible to apply concepts discovered in one species to other species.

~

The concept of the gene is central to the study of genetics.

~

Genetics, an analytical science, is best understood through a study of how genetic experiments are conducted and their results interpreted.

INTRODUCTION

For more than two billion years, living organisms have procreated, faithfully transmitting their hereditary information from one generation to the next. **Genetics** is the study of how this hereditary information is organized, expressed, and inherited. You are about to begin a fascinating and challenging study that will greatly expand your perception of the world. Genetics is a central theme of modern biology. It is a rapidly progressing science that touches all of humanity through its applications in medicine and agriculture. Each week, exciting discoveries in genetics appear in news reports, generating both hope and fear among the public. Hope that scientists will develop more effective treatments for genetic disorders, diseases, and cancer, and discover ways to produce more food for an ever increasing world population. Fear that some people might misuse the powerful tools of genetic research, or that the benefits of genetic research may fail to reach those who need them the most. Such hopes and fears can be properly addressed only with a correct understanding of the principles of genetics and their applications.

This chapter sets the stage for your study of genetics. It provides fundamental concepts that will guide you through the more detailed chapters that follow. We begin first with the role of genetics in human society. We will then discuss the universality of genetic principles and

how this universality guides experimentation and choice of organisms in genetic research. Next, we will lay out the organization for your study of genetics. The final sections of this chapter explore the concept of the gene and genetic analysis as common themes throughout all areas of genetics.

1.1 GENETICS IN HUMAN SOCIETY

The science of genetics emerged at the start of the twentieth century. Gregor Mendel first explained the foundation principles of genetics in 1865, based on his meticulous experiments with pea plants. However, no one grasped the importance of his work at the time, and genetics remained in obscurity until 1900, when three botanists independently rediscovered Mendel's principles. In 1905, William Bateson coined the term *genetics* to describe the study of inherited characteristics. By that time, genetics was already recognized as an important branch of biology.

Some aspects of genetics, however, are among the oldest of sciences. Humans recognized long ago that all forms of life bear offspring "after their kind" (to use a biblical phrase). Offspring tend to resemble their parents,

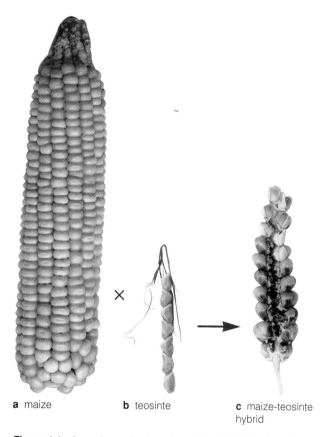

a maize   **b** teosinte   **c** maize-teosinte hybrid

Figure 1.1 An early application of genetics. **(a)** Maize (corn) cannot survive beyond one generation without human assistance, because it is incapable of dispersing its seeds. The seeds remain on the cob until humans remove them. **(b)** Teosinte, a wild relative of maize, is thought by many scientists to be the ancestor of maize. Young plants appear similar to maize, but each plant produces only a few small triangular-shaped seeds on a small spike near the top of the plant. At maturity, the seeds fall from the plant and disperse naturally. Teosinte lacks the high productivity, palatability, and nutritional characteristics of maize. **(c)** A maize-teosinte hybrid. In spite of the substantial differences between the two, maize and teosinte are fully interfertile. When maize and teosinte are hybridized, they produce a hybrid plant with characteristics that are intermediate between the two parents. These hybrid plants are fully fertile, and when the seeds from a hybrid are grown, they produce a wide variety of plant types.

usually having a mixture of traits from both of them. It is therefore reasonable to expect that a vigorous and productive plant or animal would produce vigorous and productive offspring. Plant and animal breeding are probably the earliest applications of scientific principles, and are among the most important. People who cultivated their food saved the best individuals for producing future generations. Repeated selection of individuals for breeding eventually resulted in new species of domesticated plants and animals, such as corn, wheat, rice, cattle, and chickens, species that in many cases have been changed to such an extent that they are no longer capable of surviving without human care (Figure 1.1).

For millennia, the application of genetic principles has played a central role in shaping human history and civilization. Selection for desirable inherited traits led to the domestication of food plants and animals, which made agriculture possible. Because agriculture is a prerequisite for a civilization to rise and flourish, it is no surprise that genetic improvement of agricultural species has continued from prehistory to the present day.

In the early part of the twentieth century, scientists began to apply experimental genetics in a concerted scientific effort to improve domesticated plant and animal species. Astounding gains in agricultural production resulted. While improved cultural practices such as disease control, improved feeding regimes, irrigation, pest and weed control, and fertilization contributed significantly to these increases, the largest single contributor to improved agricultural production is genetic improvement. Genetic improvement of wheat, rice, and a number of other important food crops in less developed countries, coupled with implementation of modern cultural practices, resulted in doubled, tripled, or even quadrupled crop yields over a period of only twenty years. So rapid and influential were these production gains that this international effort became known as the green revolution. But with these gains in agricultural productivity have come social, economic, and environmental concerns, some of which are being addressed by reevaluating priorities in agricultural breeding and biotechnology programs.

Our present understanding of genetics has revolutionized modern medical research. For example, scientists have engineered genes that produce human insulin in bacteria for treatment of people afflicted with diabetes (Figure 1.2). Before genetically engineered human insulin was developed, diabetic patients received insulin extracted from the pancreatic glands of cattle and pigs. However, some patients developed allergies to nonhuman insulins, which differ slightly from human insulin. Although the genes that encode genetically engineered human insulin were constructed by scientists, the insulin is identical to natural human insulin and does not cause allergic reactions. Several other pharmaceutical products produced in genetically altered bacteria are now on the market as well.

In other medical applications, newborn babies are routinely screened for genetic disorders for which early medical intervention can prevent permanent damage. Certain genetic disorders can be identified in human fetuses before birth. Our understanding of the genetic basis of cancer has grown tremendously in the past few years, allowing for development of methods that permit earlier and more accurate detection of cancer. Early detection of cancer makes it possible to begin treatments during the initial stages of tumor development when treatments are most effective.

The Human Genome Project is now at the center of important and controversial applications of genetics to medicine. This is a large international project aimed at determining the entire human DNA sequence and study-

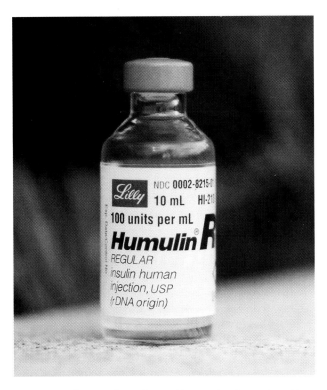

Figure 1.2 Genetically engineered human insulin produced in bacteria.

Figure 1.3 Strip mine spoils revegetated with shrubs that have been genetically developed to grow on poor soils. (Photo courtesy of Howard C. Stutz)

ing the chromosomal locations and DNA sequence variations for all genes in humans. Given that the human DNA sequence consists of approximately 3 billion nucleotide subunits of DNA, and that many portions of the DNA must be resequenced using DNA from different individuals to study the extent of sequence variation, this project is no small task. Nonetheless, technological improvements and coordination of efforts are contributing to rapid accumulation of sequence data. The target date for completion of the entire human DNA sequence is 2005.

The potential benefits of the Human Genome Project are great. Already some genes, such as the one responsible for cystic fibrosis, have been identified and their DNA sequence determined. Such information allows scientists to develop effective treatments for genetic disorders. Many other important genes are now being analyzed as the project progresses. Most opponents of the Human Genome Project do not dispute its potential benefits. Instead, they question whether the enormous expense required to complete the project is justified when many other important and beneficial projects could be supported with the funds appropriated to finance it.

Genetics has entered the world of industry. Enzymes often have industrial applications. For example, enzymes that degrade proteins are added to detergents, and some of these enzymes have been genetically improved to be more effective as detergent additives. Genetically altered bacteria and fungi are being used for waste treatment and hazardous-waste cleanup. Mining companies use plants

that have been genetically selected to grow on poor soils to revegetate mine spoils that might otherwise lay barren (Figure 1.3). Even such items as pigments for cosmetics are produced in genetically selected plant cell cultures.

In recent years, genetics has often entered the courtroom. DNA fingerprinting permits law enforcement officials to identify criminals from small amounts of DNA obtained at crime scenes. Not only can DNA fingerprinting assist in identifying a criminal, it can also exonerate innocent suspects. DNA fingerprinting is also used in paternity testing and in reuniting lost or kidnapped children with their true parents (Figure 1.4).

With the enormous financial benefits of genetic research, courtrooms are filled with disputes over the ownership of genetically altered organisms—and even of the genes themselves. Patents have been granted to protect a company's ownership of genetically engineered bacteria, genetically improved plants and animals, and genes that have been constructed in the laboratory or transferred from one species to another. Even some laboratory techniques have been patented, legally restricting their use and providing royalties to companies that own the rights. As a result, biotechnology patent law is now among the most active and profitable areas of the legal profession.

Genetics is a foundation discipline for all areas of biology. As the molecular foundations of genetics became clear, the relationship of genetics to other major fields of biology likewise clarified. Many of the fundamental principles of cell biology were discovered through genetic analysis of mutations. Cancer is caused by mutations in genes that regulate the cell cycle. Mammals have immunity because certain genes rearrange themselves to produce antibodies and other proteins for the immune system. Ecology and evolutionary biology rely on principles of population genetics, and molecular analysis has become an important tool in studying ecological and evolutionary mechanisms. Taxonomy has traditionally been based on analysis of structural similarities and differences; these analyses are now augmented by

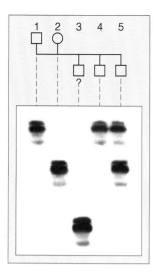

Figure 1.4 Paternity testing with human DNA fingerprints. Each lane represents a separate individual, as diagrammed in the pedigree above the lanes. Lanes 1 and 2 represent husband and wife, respectively. Lanes 3, 4, and 5 represent three male children. The DNA fingerprints in lanes 4 and 5 match those of at least one of the parents in lanes 1 and 2. The DNA fingerprint in lane 3, however, cannot be matched with either lane 1 or lane 2. This means that at least one of the people represented by lanes 1 and 2 is not the biological parent of the child represented in lane 3. (From Tautz, D. 1990 Genomic fingerprinting goes simple. *BioEssays* 12:44–46. Copyright 1990. Reprinted by permission of John Wiley & Sons, Inc.)

comparisons of similarities and differences in DNA sequence. Clearly, a good understanding of genetics is essential for every student of biology.

The techniques of genetic analysis have revolutionized several areas of scholarly research extending beyond the boundaries of traditional biology. DNA analysis is now a routine part of some archeological and historical research. For example, reconstruction of Egyptian royal family lineages by analysis of DNA obtained from mummies will provide significant insights about Egyptian history. And long before there was DNA analysis, human pedigrees derived from information in genealogical libraries helped us understand more about our history, and allowed geneticists to make significant progress in understanding certain inherited traits in humans.

Genetic research has affected the lives of all human beings on the planet through applications in medicine, agriculture, industry, and forensics. There are associated economic and legal concerns that also affect us all. Genetic analysis has become an integral part of nearly every field of the biological sciences. There is no doubt that it is among the most important and rapidly progressing areas in science.

~

Genetics has played a significant role in shaping human society both historically and in modern times. Modern application of genetics has been particularly important in medicine, agriculture, industry, and forensics.

1.2 THE UNIVERSALITY OF GENETIC PRINCIPLES

Among the most remarkable aspects of genetic principles is their universality. DNA is the hereditary material of all cellular organisms, which use the same genetic code to translate the information contained in DNA into proteins. Viruses (noncellular infectious particles) have either DNA or RNA, a molecule very similar to DNA, as their hereditary material. The structure of genes in all organisms is also remarkably similar. Because of this universality, patterns observed in one species often apply to many others, making genetics a constant theme within the diversity of life.

It is understandable that early researchers in genetics did not realize how widely applicable their discoveries were. In 1865, Gregor Mendel (Figure 1.5) reported his now famous studies on the inheritance of characteristics in peas. It is easy to imagine Mendel in a dramatic setting in which he struggled to convince others of the revolutionary significance of his discoveries, but was so far ahead of his time that no one could understand his work or recognize its importance. In reality, Mendel himself was reluctant to conclude that the principles he had discovered were universal. In his original paper, he stated:

> That, so far, no generally applicable law governing the formation and development of hybrids has been successfully formulated can hardly be wondered at by anyone who is acquainted with the extent of the task, and can appreciate the difficulties with which experiments of this class have to contend. A final decision can only be arrived at when we shall have before us the results of detailed experiments made on plants belonging to the most diverse orders.

To his credit as a scientist, Mendel was unwilling to assume that his results applied to organisms other than pea plants until more information had been collected. Although Mendel began to collect such information, a general understanding of the near universality of his principles had to wait 35 years, until the turn of the century. By then, Mendel had died, never knowing that he would become known as the founder of genetics. Likewise, although DNA was described at about the time when Mendel reported his experiments and was hypothesized to be hereditary material as early as the 1890s, its universality as the hereditary material was not recognized until the 1950s.

~

Underlying the discussions in the chapters that follow are two major principles: (1) DNA is the inherited material, and (2) its inheritance is manifest as traits that appear in consistent patterns in parents and offspring. Throughout our study of genetics, we will repeatedly encounter discoveries made in one organism that are widely—even universally—applicable throughout all of life.

Figure 1.5 Gregor Mendel, an Austrian monk whose methodical experiments with garden peas led him to devise a set of rules that are now considered the basic principles of inheritance. (Portrait by Marcus Alan Vincent.)

1.3 MODEL ORGANISMS

Because of the universality of many principles in genetics, we can use model organisms to help us understand how these principles apply generally. A **model organism** is a species that is preeminently suited to the study at hand. Model organisms have characteristics that permit efficient genetic analysis. People unacquainted with genetics often wonder why so much research is devoted to the common fruit fly, *Drosophila melanogaster* (Figure 1.6a). These flies are a seemingly unimportant organism in that they inflict no harm on humans and are not an agricultural pest.

To a geneticist, however, the fruit fly is a wonderful organism for study. Thousands of them may be raised in small bottles on food made of inexpensive cornmeal, yeast, agar, and molasses. A single mating pair may produce several hundred offspring. They complete their life cycle in about 2 weeks at room temperature. They may be handled easily for study when anesthetized, and then revived to continue reproducing. Males and females can be readily distinguished, even with the unaided eye, although most researchers observe the flies under a dissecting microscope. Many genetic variants are readily available, and the flies can easily be shipped through the mail by individ-

ual researchers or centralized stocking facilities. Fruit fly larvae have enormous chromosomes in the cells of their salivary glands that can be studied easily and in detail, unlike the tiny chromosomes of most species.

Fruit flies are found almost everywhere on earth and may easily be captured by setting out a bottle with an old banana in it, making them an ideal organism for studying the genetics of populations on a local, regional, or worldwide basis. *Drosophila melanogaster* has numerous relatives within the genus *Drosophila* that are also easy to raise and study, but some have much more limited geographical ranges. This makes the genus a model system for the study of speciation and evolutionary genetics. The pioneering work of Thomas Hunt Morgan and his students during the first three decades of the twentieth century brought the fruit fly to the forefront of genetic research. Fruit flies have been studied intensively ever since, resulting in a very large body of research information that is now used for comparative purposes. In short, model organisms, such as *Drosophila melanogaster*, provide the most efficient means to discover the basic principles of genetics.

To summarize, some of the characteristics of a model organism are as follows:

- It can be raised easily and inexpensively.
- It produces large numbers of progeny.
- Its generation time is short.
- Genetic variants within the species are readily available for study.
- It has been the subject of previous studies that have produced relevant background information.

Along with *D. melanogaster*, there are several other widely used model organisms (Figure 1.6). *Escherichia coli*, a bacterium that inhabits mammalian intestines, is the most widely used bacterial species. Much of what we now understand about how DNA replicates and mutates was first discovered in *E. coli*. The mold *Neurospora crassa* has been particularly advantageous for certain genetic studies because all four nuclei produced by each meiosis are readily available for study. Studies of *Neurospora crassa* contributed substantially to the elucidation of biochemical pathways and to understanding how mutations affect those pathways. Important work on mitochondrial genetics and some of the most extensive DNA sequencing has been done using brewer's yeast, *Saccharomyces cerevisiae*. Maize, *Zea mays* (often called *corn* in the United States), was the model organism for some of the earliest studies on the chromosomal basis of inheritance. Studies of maize also produced the discovery of transposable elements, segments of DNA that move from one position to another on chromosomes. *Arabidopsis thaliana*, a tiny plant that produces hundreds of seeds and has a life cycle of only 4–6 weeks, has provided much information about

a *Drosophila melanogaster*

b *Arabidopsis thaliana*

c *Zea mays*

d *Mus musculus domesticus*

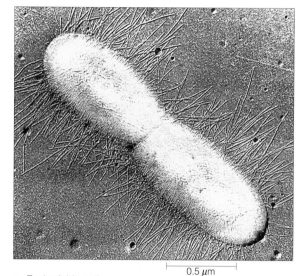

e *Escherichia coli*

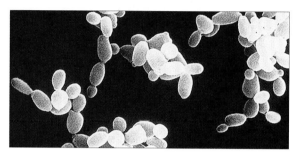

f *Saccharomyces cerevisiae*

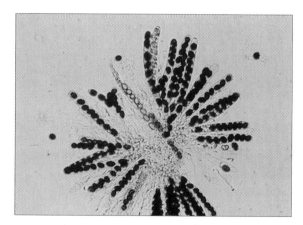

g *Neurospora crassa*

Figure 1.6 Several model organisms used in genetics research. **(a)** *Drosophila melanogaster*, an insect, **(b)** *Arabidopsis thaliana*, a plant, **(c)** *Zea mays*, another plant, **(d)** *Mus musculus domesticus*, the laboratory mouse, **(e)** *Escherichia coli*, a bacterium, **(f)** *Saccharomyces cerevisiae*, a fungus, and **(g)** *Neurospora crassa*, another fungus. (Photos courtesy of (a) Carolina Biological Supply, (e) CNRI/SPL Photo Researchers, (f) David M. Phillips, Visuals Unlimited, and (g) James W. Richardson, Visuals Unlimited.)

the genetics of plant development. Among mammals, the laboratory mouse, *Mus musculus domesticus*, has proven to be a good model organism. Much of the work elucidating the genetic bases of cancer and immunity has used the mouse and its cells.

Once a principle has been described using a model organism, it may be tested much more efficiently in organisms of more practical importance, such as humans, disease-causing organisms, and agricultural species. As you read this book, you will learn about landmark stud-

ies conducted on these and other model organisms that have confirmed the universality of many principles in genetics.

Can humans be considered a model organism? Humans clearly do not have all the characteristics of a model organism for genetics. We have relatively long generation times, we produce few progeny per mating pair, and selective mating for research purposes is ethically inappropriate. Whereas the science of genetics grew rapidly during the first eight decades of the twentieth century, progress in human genetics lagged well behind the advances obtained using model organisms in many areas. However, this situation has changed dramatically in recent years. DNA analysis can now be applied routinely and in large scale to study human genetics. Researchers can compensate for the limitations of more traditional genetic analysis in humans through the use of DNA analysis. We can now say that humans are genetically one of the best characterized species, a statement that would have been difficult to justify a few years ago. As you read this book, you will encounter many examples of recent advances in human genetics.

~

Model organisms are species selected for genetic study because they possess characteristics that allow important research questions to be addressed efficiently. Information learned from model organisms may be applied to humans or other species of economic or social importance that cannot be so efficiently studied. Recent DNA analysis has made humans one of the best characterized species genetically.

Throughout this book, after presenting new information, we will apply it in examples in which we confront a problem that we then solve. As our first example, we consider one of the model organisms.

Example 1.1   The laboratory mouse, *Mus musculus domesticus*, as a model organism.

Problem: The laboratory mouse has been used since 1902 as a model organism in genetics. Describe how the laboratory mouse fits the criteria for a model organism. Are there any characteristics about the mouse that might make it a better candidate than other model organisms for certain studies?

Solution: The laboratory mouse meets all the criteria described above for a model organism. It is a small animal that can be raised conveniently and inexpensively in a small space indoors, typically in a laboratory vivarium. After 18–22 days of gestation, a female gives birth to 5–10 pups. She may become pregnant again only 28 hours after giving birth. (Most researchers wait until female mice have had sufficient time to wean their litters before allowing

the female mice to remate, however.) Mice may mate for the first time when they are 10 weeks old, and researchers schedule subsequent matings to produce as many as five generations per year. Because many genetic variants of mouse have been discovered and are used to found reproductive lineages, different genetic stocks of mice are readily available from supply houses and university research centers. The research literature from many studies conducted over about a century covers all topics of mouse biology. Because the biology of the mouse is similar to that of other mammals, it is the best model organism for studies that address questions about humans and other mammals that cannot themselves be as efficiently studied. For example, mice often serve as the subjects for studies of cancer and immunity; in many cases, information acquired from such studies may then be applied to humans.

1.4   ORGANIZING THE STUDY OF GENETICS

Genetics is among the most logical and integrated branches of biology. The structures and functions of DNA, RNA, and proteins dictate how parents transmit their traits to their offspring, and the transmission of traits between generations dictates how variation in traits is distributed within and between populations over time. Once the whole picture is in view, each individual detail makes sense. Unfortunately, the whole scope of genetics cannot be studied at once. Thus, to introduce students to the study of genetics, teachers usually divide it into three parts: molecular genetics, transmission genetics, and the genetics of populations. The order of these parts in this book parallels levels of organization in biology: from smallest to largest: molecules, cells, organisms, populations, and species.

Molecular genetics is the study of the principal molecules of heredity (DNA, RNA, and protein), how they are organized within the cell, and how they interact within the cell. This is the most fundamental level of genetics. Although many of the principles of genetics were understood long before molecular genetics even became a field of study, the molecular basis of most of these principles is now well known. Genes contained within DNA and expressed through RNA and proteins determine the traits that are studied at other levels of genetics. Part I of this book, Molecular Foundations of Genetics, introduces the molecules of heredity, and Part II, Genetics of Cells, examines these molecules as parts of cells.

Transmission genetics is the study of how genetic material, and the traits encoded by that material, are transmitted from parents to offspring. Transmission genetics is often called classical genetics because it dominated the

field of genetics during the first half of the twentieth century. In recent years, transmission genetics has become fully integrated with molecular genetics. Many of the molecular foundations of transmission genetics are now well understood and are an active part of modern genetic research. Part III of this book, Genetics of Organisms, presents transmission genetics in a modern context, fully integrated with molecular genetics.

In Part IV, Genetics of Populations, we examine the third major level, the genetics of populations. Within this level there are three areas: population genetics, quantitative genetics, and evolutionary genetics. **Population genetics** is the study of inheritance in populations and how inheritance is affected by external forces, such as natural selection. **Quantitative genetics** is the study of how multiple genes interact with one another and the environment. **Evolutionary genetics** is the study of all levels of genetics as they affect evolutionary processes.

Part V, Gene Expression and the Organism, explores gene expression with all three levels in view, including transposable elements, development, cancer, and immunity. We conclude in Part VI, Applications of Genetics, with some of the applications of genetics in medicine, agriculture, industry, and forensics.

~

A foundation in molecular genetics serves as the basis for understanding transmission genetics, which, in turn, underlies the genetics of populations.

1.5 THE CONCEPT OF THE GENE

The concept of the gene is central to the study of genetics. It is possible to describe a gene as a segment of DNA, a place on a chromosome, something that is passed from parent to offspring, and something that is present among individuals in a population. Although these various descriptions of a gene seem quite different, in each case we are actually referring to the same thing, but at different levels. This makes it difficult to state a simple yet comprehensive definition. To understand the concept of the gene, let's consider how that concept developed.

Before the twentieth century, most biologists viewed the substance of heredity as something fluid, such as blood, semen, or some substance within animal blood or plant sap. They envisioned this substance as something that could be blended and reblended with each generation. Many nineteenth-century biologists also viewed the substance of heredity as changeable rather than constant, and under the influence of the environment.

Gregor Mendel was the first to present solid evidence of the constant inherited units that we now call genes. In Mendel's day, the word *gene* had yet to be invented. Mendel used the German words *Merkmal* (characteristic, mark, or sign), *Anlage* (plan, design, or predisposition), and *Element* (element) to identify what we now call a gene. He understood genes to be individual hereditary factors that carry the potential for the outward appearance, or phenotype, of an organism. He concluded that genes remain constant from one generation to the next without being changed by the environment. He also concluded that genes are shuffled during formation of the germ cells and at fertilization, permitting all possible combinations of genes in predictable ratios among progeny, a concept that explains genetic variation among individuals. When Wilhelm Johannsen coined the word *gene* in 1909, the concept of genes as constant hereditary units that adhere to Mendel's principles of inheritance was already well established.

In Mendel's interpretation of his experiments, a single gene corresponds to variation for a single trait. For example, a single gene determines whether the flowers on a pea plant are purple or white. A second gene determines whether a plant is tall or dwarf. This type of phenotypic variation in which contrasting characters fall into discrete classes is known as *discontinuous variation*. However, among all species there are many traits that do not fall into discrete phenotypic classes. Instead, individuals differ from each other over a continuous range without any discrete phenotypes. This type of phenotypic variation is called *continuous variation*, exemplified by the distribution for human height illustrated in Figure 1.7. After Mendel's principles were rediscovered in 1900, some biologists assumed that, although discontinuous variation was due to individual genes, continuous variation was due to some form of blending inheritance. William Bateson rejected this explanation and proposed in 1902 that continuous variation derives from the influence of multiple genes on the same trait, and that each of these genes is inherited according to Mendel's principles, a phenomenon called *polygenic inheritance*. A few years later, several researchers showed that continuous variation is indeed due to polygenic inheritance combined with the effect of environmental variation.

Shortly after Mendel's principles were rediscovered in 1900, Archibald Garrod recognized that the absence of a functional enzyme causes certain inherited disorders in humans. An enzyme is a protein that catalyzes a specific biochemical reaction in the cell. By 1911, Bateson recognized the link between genes and enzymes. He proposed that enzymes themselves are not inherited, but genes are, and somehow a gene determines whether or not a functional enzyme is present in an individual. When a functional enzyme is present, a particular phenotype appears, but when the enzyme is absent, a different phenotype appears. George Beadle and Edward Tatum further demonstrated the relationship between genes and enzymes in the 1940s and hypothesized that each gene corresponds to an enzyme. How a gene could produce an enzyme, however, remained a mystery.

The constancy of genes was called into question shortly after the rediscovery of Mendel's principles.

Figure 1.7 Continuous variation for height in humans. Students from a biology class have lined up by height. Height is manifest in continuous gradations.

Bateson, at one point, assumed that genes could not be changed but could be lost, resulting in the loss of the enzyme produced by the gene. However, studies with maize and fruit flies revealed that new forms of existing genes occasionally arise by changes called *mutations*. Constancy of genes appeared to be the rule, but it became evident that genes can mutate and that mutations can be inherited, which creates new genetic variation.

In 1902, Walter Sutton and Theodor Boveri proposed that genes are located on *chromosomes* and that each chromosome contains many genes. Supporting this proposal, Thomas Hunt Morgan showed in 1910 that a gene responsible for eye color in *Drosophila melanogaster* is located on the X chromosome. The following year, Alfred Sturtevant, who was an undergraduate studying under Morgan, determined the relative positions of several *D. melanogaster* genes on the X chromosome, beginning the practice of *gene mapping*. Within a few years, Sturtevant, Morgan, and Calvin Bridges, along with their associates, had mapped numerous genes to their chromosomal locations in *D. melanogaster*. By then, gene mapping in other model organisms was well under way.

Even as the concept of the gene as part of a chromosome became accepted, there was recognition that chromosomes in the nucleus of eukaryotic cells are not the exclusive location of genes. During the first few decades of the twentieth century, researchers found that a few inherited traits in eukaryotes were controlled by genes in the cell cytoplasm rather than genes in the nucleus. Later, these cytoplasmic genes were found to be located on circular DNA molecules in mitochondria or chloroplasts.

And other locations for genes were discovered. Prokaryotes (bacteria) lack nuclei and do not have linear chromosomes like eukaryotes (organisms with a cell nucleus). Prokaryotic genes are found in a circular DNA molecule within the cell. Genes were also found in prokaryotic and eukaryotic viruses.

An aspect of the concept of the gene that developed slowly was the understanding of its chemical composition. For the first half of the twentieth century, most scientists hypothesized that protein was the genetic material. The first clues about the actual chemical composition of genes surfaced in 1926 when Frederick Griffith discovered that a hereditary substance (which he called the "transforming principle") could be transferred from dead bacteria into living bacteria. In 1944, Oswald Avery, Colin MacLeod, and Maclyn McCarty demonstrated that Griffith's transforming principle is DNA. In 1952, Alfred Hershey and Martha Chase showed that the hereditary material of bacterial viruses is also DNA.

James Watson and Francis Crick found themselves in a fortunate situation in the early 1950s. They suspected that DNA is the genetic material and sought to determine its chemical structure. They had access to unpublished photographs taken by Rosalind Franklin and Maurice Wilkins of X-ray diffraction patterns of DNA that revealed much about its structure. Watson and Crick used information about DNA chemistry, X-ray diffraction images, and data on DNA composition published by Erwin Chargaff to deduce the chemical structure of DNA in 1953. According to Watson and Crick's model, a single DNA molecule consists of two linear chains of nucleotide

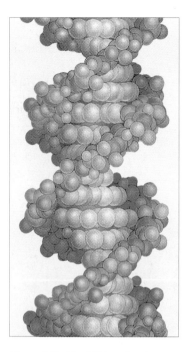

Figure 1.8 The DNA molecule.

subunits with the chains wound around each other to form a double helix (Figure 1.8). The structure of DNA immediately suggested how DNA could be replicated and soon led to the discovery of what a gene is at its most basic level.

By the late 1950s, there was little doubt that DNA is the universal genetic material and that genes are segments of a DNA molecule. Although proteins are not the genetic material, they are the substance of enzymes and are, therefore, related to genes. Like DNA, proteins are chainlike molecules composed of subunits, but the subunits of proteins are amino acids rather than nucleotides. In the early 1960s, Charles Yanofsky showed that there is a linear relationship between mutations in the nucleotide sequence of a gene and changes in the amino acid sequence of the protein encoded by the gene. These results suggested that genes and proteins are colinear, which means that the linear sequence of nucleotides in DNA corresponds to the linear sequence of amino acids in proteins.

It was also obvious in eukaryotic cells that proteins are not synthesized directly from genes in DNA because DNA remains in the cell nucleus while proteins are synthesized on ribosomes in the cytoplasm. There had to be an intermediate between genes and protein. As demonstrated by Severo Ochoa, Sydney Brenner, François Jacob, and Matthew Meselson, the intermediate is RNA, a molecule very similar in structure to DNA. The information in a gene is copied, through a process called transcription, into an RNA molecule. Then the RNA moves to the cytoplasm, where its information is transferred, through a process called translation, into the linear chain of amino acids constituting a protein.

Transcription and translation of genes became collectively known as gene expression.

Direct biochemical studies, as well as theoretical studies based on math and statistics, demonstrated that each amino acid in a protein corresponds to three nucleotides in a gene. During the late 1950s and early 1960s, a group of scientists, including Marshall Nirenberg, Heinrich Matthaei, Philip Leder, and Har Gobind Khorana, deciphered the corresponding triplet nucleotide codes for each of the twenty amino acids. These relationships turned out to be the same in all organisms and are now known collectively as the genetic code.

Subsequent research has revealed detailed information about genes at the molecular level. Much current research is focused on gene regulation, the ways by which a cell determines which genes should be expressed, when they should be expressed, and how much of the gene's product should be produced.

As the concept of the gene developed from the time of Mendel to the present, several fundamental aspects of the gene emerged:

- A gene is a linear segment of nucleotides in a DNA molecule.

- A gene in DNA is transcribed into an RNA molecule. In most cases, the nucleotide sequence information in the RNA molecule is translated into a linear sequence of amino acids called a polypeptide. The polypeptide is then processed and folded into a protein. Many proteins function as enzymes that regulate chemical reactions in the cell.

- A gene may mutate. Mutations are rare changes in the nucleotide sequence of DNA. By altering the amino acid sequence of the protein encoded by a mutated gene, mutations may alter or eliminate the function of a protein. Mutations in genes are the underlying source of observable genetic variation.

- Genes are found in chromosomes in cell nuclei, in mitochondria and chloroplasts, in bacteria, and in viruses.

- Single genes may determine the phenotypes of discontinuously varying traits. For such traits, as Mendel demonstrated, the ratios of phenotypes in the offspring are predictable from information about the traits in the parents.

- The phenotypes of continuously varying traits are determined by many genes acting together plus environmental effects.

Our examination of how the concept of the gene developed provides an overview of the topics we will discuss in Parts I–III. As we proceed through the book, we will often make connections among these fundamental aspects. Let's now take a look in the example that follows at how we can connect the gene as DNA that encodes a protein to the phenotypes that Mendel observed.

Mendel hybridized pea plants that had purple flowers with pea plants that had white flowers and observed that all of the first-generation progeny plants had purple flowers. However, in the second and subsequent generations, he observed purple-flowered plants and white-flowered plants among the progeny in a predictable pattern that we will review momentarily. Mendel concluded that an underlying constant factor (which we now call a gene) determines whether a plant will have purple or white flowers, and that this factor is transmitted unchanged from one generation to the next.

Problem: Mendel's results indicate that a single gene determines whether a plant will have purple or white flowers. Describe how a gene could cause this variation.

Solution: A purple-colored pigment (called anthocyanin) causes pea flowers to be purple. White flowers lack this pigment. The purple pigment is synthesized from a colorless precursor through reactions catalyzed by enzymes that are encoded by genes. As long as all the genes produce functional enzymes, the purple pigment is produced and the flowers are purple, as illustrated in Figure 1.9a. However, if a mutation changes one of the genes so that it can no longer produce a functional enzyme, the pigment cannot form and the flower will be white, as illustrated in Figure 1.9b.

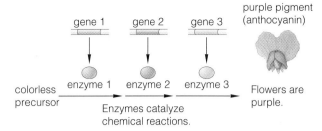

a The presence of functional enzymes promotes synthesis of purple pigment, resulting in purple flowers.

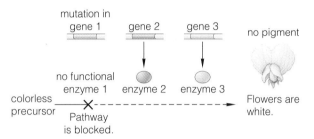

b The absence of a functional enzyme prevents synthesis of purple pigment, resulting in white flowers.

Figure 1.9 Pigment synthesis in pea flowers.

1.6 GENETIC ANALYSIS

Science may be seen as a body of theories and facts, but this is a very narrow view. We see the true excitement of science when we look at the process of discovery as well as the information discovered. Genetics is an analytical science, and the process of discovery unfolds before us when we examine how genetic experiments are conducted and their results interpreted.

The experiments of Gregor Mendel, performed in the 1850s, and his interpretation of his experimental results are a superb example of how genetic analysis can be used to discover and confirm underlying principles. Mendel's approach of applying designed experimentation with mathematical analysis to answer biological questions was unique in his day. Perhaps for that reason, no one recognized the importance of his experiments during his lifetime. However, once Mendel's work was rediscovered in 1900, his approach to experimentation and analysis set the pattern for much of the genetic research of the twentieth century. Let's take a brief look at some of his work as an example of the investigative nature of genetics.

Mendel was a contemporary of Darwin, and the evolutionary development of new species was a hotly debated topic at the time. Mendel studied at the University of Vienna, where the local clergy vigorously criticized Mendel's botany professor, Franz Unger, for teaching the mutability of species. Mendel was a devout clergyman himself, but took a keen interest in evolution. He read much of the scientific literature about plant evolution, speciation, and hybridization. From his reading he concluded that while many experiments on plant hybridization had been conducted, none of them were designed in such a way that they could fully answer the question of how hybridization could produce new species. Mendel took it upon himself to conduct a large, meticulous, 8-year study of pea plants to address this question.

One of his experiments, in which he used tall plants and dwarf plants, serves to illustrate the logic of his design, analysis, and interpretation. Mendel hybridized tall pea plants and dwarf pea plants. The hybrid offspring from these crosses were all tall, a result that didn't surprise Mendel, for many of his predecessors had seen similar results. He then allowed these hybrid plants to self-fertilize. In the second-generation progeny, he counted 787 tall plants and 277 dwarf plants, approximating very closely a ratio of three tall plants to one dwarf plant. Mendel performed similar experiments with six other traits, and each time the same pattern appeared: all of the first-generation hybrid progeny had only one of

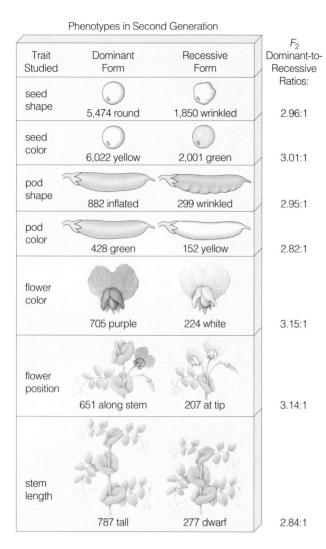

Phenotypes in Second Generation

Trait Studied	Dominant Form	Recessive Form	F_2 Dominant-to-Recessive Ratios:
seed shape	5,474 round	1,850 wrinkled	2.96:1
seed color	6,022 yellow	2,001 green	3.01:1
pod shape	882 inflated	299 wrinkled	2.95:1
pod color	428 green	152 yellow	2.82:1
flower color	705 purple	224 white	3.15:1
flower position	651 along stem	207 at tip	3.14:1
stem length	787 tall	277 dwarf	2.84:1

Figure 1.10 The results of some of Mendel's experiments. After hybridizing the two phenotypes, Mendel observed ratios close to 3:1 among the self-fertilized second generation for all seven traits he studied.

the two contrasting characters, but the other character reappeared in the second generation in approximately one-quarter of the individuals (see Figure 1.10 for Mendel's actual results).

Mendel developed a theoretical interpretation that explained his results. He concluded that each plant carried hereditary factors (which we now call genes) that influenced the traits he was studying. When plants with contrasting characters were crossed, each parent plant contributed a factor to the offspring so that the hybrid offspring carried both factors simultaneously. Mendel concluded that both factors were present in the hybrid, and that one of these factors was dominant and the other recessive. Rather than speaking of factors, we now say that a hybrid for plant height has a dominant allele and a recessive allele of the gene controlling plant height. Different alleles are defined as different versions of the same gene. Mendel represented the dominant allele with

an uppercase A and the recessive allele with a lowercase a. He then represented the first-generation hybrid plants as Aa because they carried both alleles. Only the character associated with the dominant allele appeared in the phenotype of the hybrid plants. In the plant height example, the hybrid plants simultaneously carried the allele for tall stature and the allele for dwarf stature, but the tall-stature allele was dominant, so all the first-generation progeny were tall.

Mendel further developed his interpretation in light of the cell theory of his day. He concluded that during formation of the germ cells (pollen and egg cells) in the hybrid plant, the dominant and recessive alleles segregated from each other so that half of the pollen grains received the dominant allele A and the other half received the recessive allele a. The same was true for the egg cells; half received A and half a. Under these conditions, random union of pollen cells and egg cells at fertilization should result in a mathematical series of all possible combinations of alleles in the second-generation progeny. Mendel represented this series as

$$\frac{A}{A} + \frac{A}{a} + \frac{a}{A} + \frac{a}{a}$$

where the symbols in the numerator of each fraction denote the alleles from the pollen cells and the symbols in the denominator of each fraction denote the alleles from the egg cells. With random union of pollen and egg cells at fertilization, each of these four possible combinations has an equal probability of appearing in the second-generation offspring of a hybrid plant.

Three of the four combinations above have at least one dominant A allele, which causes tall plants. One of the four combinations (a/a) has only a, which causes dwarf plants. Under this theory, about three-fourths of the progeny should be tall, while the remaining one-fourth should be dwarf. This explained the 3:1 ratio that Mendel observed in the second-generation progeny for each of the seven traits he studied. Having developed a theoretical interpretation to explain his results, Mendel designed additional experiments to test his interpretation. The results of these additional experiments supported his interpretation, lending it additional credibility.

Mendel also discovered through his experiments that the inheritance of each set of contrasting characters was independent of the inheritance of the other sets of contrasting characters. For example, the inheritance of plant height was independent of the inheritance of flower color. He interpreted this to mean that all possible combinations of characters could be obtained in the second-generation offspring from hybrids.

It is enlightening to see what Mendel did and did not conclude from his study. He concluded that hereditary factors in hybrids segregated into germ cells in a 1:1 ratio and that the random union of these cells resulted in

second-generation phenotypes in a 3:1 ratio due to dominance. In deriving this conclusion, he interpreted his results in light of the cell theory that was common in the scientific literature of his day. Mendel also concluded that inheritance followed predictable mathematical patterns, a concept that was completely novel in his day but was clearly demonstrated by his results. He made no attempt to explain what the hereditary factors were or to speculate about their composition. He also did not attempt to explain how an underlying factor could cause an observable phenotype, or how the cells could carry and express such factors. This lack of speculation makes sense because his results, and the results of others, provided no information about these phenomena. Mendel concluded his paper by attempting to reconcile the observations of other plant hybridists in light of his theory and to explain how hybridization and the subsequent new combinations of genes could result in new species of plants.

Mendel's approach illustrates some of the fundamental aspects of genetic analysis that are as applicable today as they were for him well over a century ago. He chose an appropriate organism for the experiments he had in mind. The seven traits he chose to analyze could be identified easily and unambiguously and were not influenced by environmental variation. This allowed Mendel to focus on heredity alone, without the confounding effects of environmental influences, a point that he stressed in his paper as crucial to the success of his experiments. He recognized that in order to obtain valid information, he needed to analyze large numbers of individuals. He designed his experiments to do this, and appropriately applied mathematical analysis to his results. He then developed an interpretation based on his results and on the scientific theories of his day, and tested his interpretation by further experimentation. He concluded his report by placing his experimental results and their interpretation in the context of previous experimentation.

In this discussion, we have only briefly looked at part of Mendel's work as an example of genetic analysis and discovery. We will examine his work in detail in Chapter 12. To peruse the original German version and an English translation of his paper on the World Wide Web, see the links at http://www.brookscole.com/biology.

Let's return to Mendel's experiment with purple-flowered and white-flowered pea plants that we used in Example 1.2 to see how the concept of the gene at the DNA level fits with Mendel's analysis.

Example 1.3 Applying further analysis to Mendel's results: Developing a model to explain dominance.

When Mendel crossed purple-flowered plants with white-flowered plants, all of the first-generation progeny had purple flowers. As Figure 1.10 shows, 705 of the second-generation progeny had purple flowers, while 224 had white flowers, very close to a 3:1 ratio. These results show clearly that the purple-flower characteristic is dominant to the white-flower characteristic.

Problem: Based on this information and on the information from Example 1.2, propose a model that explains (a) why the allele conferring purple flowers is dominant to the allele conferring white flowers and (b) why Mendel observed a 3:1 segregation ratio for this trait in the second-generation progeny.

Solution: (a) As we learned in Example 1.2, the allele that causes purple flowers encodes a functional enzyme that is essential for formation of purple pigment. The allele responsible for white flowers fails to encode this functional enzyme. Mendel's results tell us that the allele responsible for purple flowers is dominant and that the allele responsible for white flowers is recessive. Using Mendel's terminology, let's call the dominant allele A and the recessive allele a. When Mendel crossed a purple-flowered plant with a white-flowered one, the first-generation offspring inherited both alleles (one from each parent), and can be represented as Aa. The A allele encodes a functional enzyme for purple-pigment formation, so the enzyme is present in the developing flowers of the first-generation progeny. The a allele, which is also present in these plants, fails to encode this functional enzyme, but because the A allele is present, functional enzyme is produced, pigment is formed, and purple flowers develop (Figure 1.11). The A allele compensates for the failure of the a allele to produce functional enzyme, so the A allele is said to be dominant to the a allele.
(b) According to Mendel's model, three-fourths of the second-generation progeny (on average) should have at least one dominant A allele, as indicated in the following mathematical series of all possible combinations of alleles in the second-generation progeny:

$$\frac{A}{A} + \frac{A}{a} + \frac{a}{A} + \frac{a}{a}$$

So functional enzyme is present in about three-fourths of the second-generation progeny, causing the flowers to be purple. About one-fourth of the second-generation progeny have only an a allele and, therefore, do not have functional enzyme. Because no pigment is produced in the flowers of these individuals, the flowers are white (Figure 1.11b).

Investigation, discovery, analysis, and interpretation are what science is about. Throughout this book, you will read about actual experimental results from published studies and learn how these results can be analyzed and

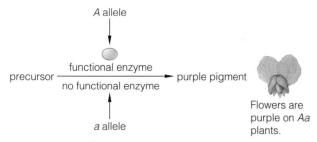

a The dominant *A* allele produces a functional enzyme that promotes purple pigment synthesis. This compensates for the recessive *a* allele, which fails to produce this functional enzyme.

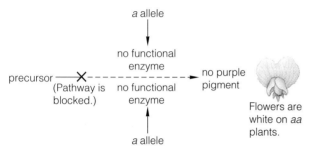

b When only the *a* allele is present, no functional enzyme is produced, and, therefore, no purple pigment is produced.

Figure 1.11 The effect of dominance.

interpreted. Among these experiments, you will encounter elegant logic and brilliant insights. In many cases, the experimental evidence confirming a principle of genetics is as informative as the principle itself.

There is one final aspect to keep in mind before embarking on your study. Genetics is one of the most dynamic fields of scientific research—a field in which new discoveries are emerging at an ever increasing rate. Most of the fundamental principles of genetics have been strengthened as additional evidence supporting them has been obtained. However, many of the details have changed in recent years. For that reason, while fundamental principles tend to remain in place, some of the detailed information in any genetics textbook is already out of date by the time the book is printed.

PROBLEMS AND QUESTIONS

1. Why can genetics be considered simultaneously to be a twentieth-century science and among the oldest fields of science?

2. Describe some of the ways in which genetics has been applied in agriculture, industry, and medicine. Why do these applications raise legal questions?

3. Find a recently published general biology textbook for college students. Choose an area of biology, such as cellular biology, plant or animal diversity, physiology, ecology, or evolution, and browse through the chapters on this area. Which genetic topics are mentioned in these chapters? Can you think of additional genetic topics related to this area that are not mentioned in the textbook?

4. In what ways are the principles of genetics universal or nearly universal?

5. Describe some of the characteristics of a model organism. Why have some model organisms with little economic value been selected for genetic research?

6. What characteristics of a model organism do humans lack? How have these limitations been overcome in human genetic analysis?

7. Prior to the 1980s, the plant species *Arabidopsis thaliana* had received little research attention because it had little economic importance. However, since that time it has become an important model organism for plant genetics. G. P. Redei, in an article entitled *"Arabidopsis* as a genetic tool" (*Annual Review of Genetics* 9:111–127), discussed the use of *Arabidopsis thaliana* as a model organism. Look up this article in a library and list the characteristics of *A. thaliana* that make it a good choice as a model organism. What disadvantages does *A. thaliana* have as a model organism?

8. What are some of the distinguishing features of prokaryotes and eukaryotes? Why are these two groups often separated when discussing the principles of genetics?

9. What are the distinguishing features of the major divisions of genetics? How are these areas interrelated?

10. Describe some of the most fundamental aspects of a gene.

11. Briefly indicate how the following people and groups contributed to the development of our current concept of the gene:

> Gregor Mendel
>
> William Bateson
>
> Archibald Garrod
>
> Walter Sutton and Theodor Boveri
>
> Thomas Hunt Morgan
>
> Alfred Sturtevant and Calvin Bridges
>
> George Beadle and Edward Tatum
>
> Oswald Avery, Colin MacLeod, and Maclyn McCarty

Alfred Hershey and Martha Chase

Rosalind Franklin and Maurice Wilkins

Erwin Chargaff

James Watson and Francis Crick

Charles Yanofsky

Severo Ochoa, Sydney Brenner, François Jacob, and Matthew Meselson

Marshall Nirenberg, Heinrich Matthaei, Philip Leder, and Har Gobind Khorana

12. Why should we study the experiments that revealed genetic principles along with the principles themselves?

13. Briefly describe how Mendel's results led him to develop a model that explained his results. What aspects of Mendel's approach allowed him to discover the basic principles of inheritance?

14. Go to a library and find some recent issues of the weekly journals *Science* and *Nature*. In the first several pages of each issue are news reports describing recent research in the sciences. Which of the news items describe research that involves genetics? Are any political, legal, or ethical issues that touch on genetics mentioned? If so, what are they and why is there concern over these issues?

FOR FURTHER READING

A very well written book about the investigative nature of biology is **Moore, J. A. 1993.** *Science as a Way of Knowing: The Foundations of Modern Biology.* **Cambridge, Mass.: Harvard University Press**. A short and easily readable book on the history of genetics that describes many of the events leading to our modern concept of a gene is **Sturtevant, A. H. 1965.** *A History of Genetics.* **New York: Harper & Row**. A book detailing the development of the human genome initiative is **Cook-Deegan, R. M. 1994.** *The Gene Wars: Science, Politics, and the Human Genome.* **New York: Norton**. A book that examines the prelude to Mendel's work and Mendel's contributions to genetics is **Olby, R. C. 1985.** *Origins of Mendelism, 2nd ed.* **Chicago: University of Chicago Press**. A thorough and well documented biography of Mendel that focuses on his experiments and their impact is **Orel, V. 1996.** *Gregor Mendel: The First Geneticist.* **Oxford: Oxford University Press**.

For additional reading, go to InfoTrac College Edition, your online research library at: http: www.infotrac-college.com/brookscole

PART I

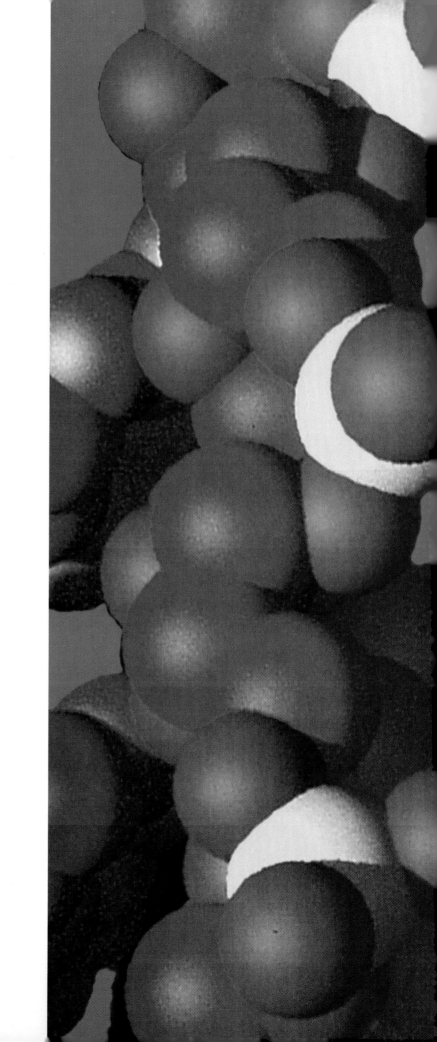

MOLECULAR FOUNDATIONS OF GENETICS

CHAPTER 2

DNA STRUCTURE AND REPLICATION

Since the dawn of recorded history, people have proposed various substances as the hereditary material. Early speculations focused on semen, vapors, blood, and particles in the blood. The idea that blood is the genetic material lingers in our language with such terms as *blood relatives, bloodlines, royal blood,* and even *consanguineous,* which means literally "of the same blood." Establishing that DNA, rather than blood or any other substance, is the genetic material of all cellular organisms ranks as one of the greatest accomplishments in science.

In this chapter we will examine some of the pioneering experiments that confirmed DNA as the genetic material. We will then look at the structure of the DNA molecule and how that structure was discovered. Among the most important characteristics of the genetic material is its ability to replicate. In the final sections of this chapter, we will see how the structure of the DNA molecule provides a means of replication, and we will examine several ways that DNA molecules replicate.

2.1 THE GENETIC MATERIAL

We usually think of the discovery of DNA as something recent, and indeed DNA's structure, mode of replication, and function were not known until the 1950s. However, the actual discovery of DNA as a substance dates back to the 1860s, about the same time as Gregor Mendel's discoveries. With refinements in microscopy, cells and their components were under intensive study in the 1860s. The most obvious cellular component was the nucleus, and for this reason a number of scientists focused their research efforts on it. In 1866, Ernst Haeckel proposed that the nucleus contains the factors necessary for heredity, setting the stage for determining what substances are present in the nucleus.

In the late 1860s, Friedrich Miescher (Figure 2.1) began physiological and chemical studies of pus cells. Pus cells are simple human cells that were easily available because antiseptic medicine was not often practiced and postsurgical infections were common. Miescher intended to characterize substances called *proteins* (from the Greek term *proteios,* meaning "of first importance"). Proteins had been discovered about 30 years earlier and were thought to be the most important material in the cell. When Miescher separated the nuclei from the rest of the cell components, he discovered that the nuclei had large quantities of a substance that did not behave chemically like protein. This substance was eventually named **deoxyribonucleic acid**, which is usually abbreviated to **DNA**.

Miescher thought that DNA was a reservoir of cellular phosphorus made available on demand by

Figure 2.1 Friedrich Miescher, who described some of the chemical properties of DNA in 1871. (Portrait by Marcus Alan Vincent.)

digestion. He also thought protein must be the hereditary material, as is obvious from a statement he made that reflected the prevailing view for the next 80 years:

> The many asymmetric carbon atoms [of protein] allow such a colossal number of stereoisomers that the richness and variety of hereditary transmission may find their expression in it.

Miescher studied DNA from salmon sperm and reported an empirical formula for it ($C_{29}H_{49}N_9P_3O_{22}$) that is fairly close to the actual proportional composition of these elements in DNA. Miescher died in 1895 of tuberculosis at the age of 51. Although 26 years had passed since his discovery of DNA, its function remained unknown.

Late in the nineteenth century, Edmund Wilson used evidence from the staining properties of DNA to conclude that

> there is considerable ground for the hypothesis that in a chemical sense this substance is the most essential nuclear element handed on from cell to cell, whether by cell division or fertilization.

Thus, before the rediscovery of Mendel's principles of inheritance in 1900, DNA had already been postulated as the genetic material. However, Wilson later reversed his view, stating that it was nuclear protein that persists through cell division and that DNA appeared and disappeared during the various stages of a cell's life.

With the discovery and characterization of chromosomes, it became increasingly clear that chromosomes are inherited and that they are composed of chromatin, a combination of both protein and DNA. The view that the protein in chromatin, rather than the DNA, is the hereditary material persisted until the middle of the twentieth century.

The Transforming Principle

In 1928, Frederick Griffith reported the now classic experiments that eventually stimulated certain other researchers to identify the genetic material in bacteria. Griffith was a U.S. Army medical officer attempting to develop a vaccine against pneumonia. He studied the infection patterns of the pneumonia-causing bacterium *Streptococcus pneumoniae* using mice as the experimental host organisms. Griffith studied two types of the bacterium, one with a polysaccharide on its capsular surface that prevents it from being destroyed by the host's immune system, and one that is unencapsulated and lacks the polysaccharide. The host's immune system easily destroys the unencapsulated cells. The bacteria with the polysaccharide have a smooth appearance when grown as colonies on a petri plate and are called S for "smooth," whereas the bacteria lacking the polysaccharide form colonies with a rough appearance and are called R for "rough." The S-type bacteria are virulent (disease causing), and the R-type bacteria are avirulent (harmless).

When Griffith injected live S-type bacteria into a mouse, the mouse died from pneumonia. But if he heat-killed the S-type bacteria before injecting them, then the mouse did not die, presumably because the dead S-type bacteria could no longer reproduce. When Griffith injected live R-type bacteria into a mouse, the mouse did not die because the mouse's immune system destroyed the R-type bacteria. However, when Griffith injected heat-killed S-type bacteria *together with* live R-type bacteria into a mouse, the mouse died and he found live S-type bacteria in the mouse's blood (Figure 2.2).

Griffith concluded that something in the dead S-type bacteria is not destroyed by the heat treatment and is transmitted to the live R-type bacteria, transforming them into S-type bacteria. Moreover, whatever is transmitted must be inherited because he observed that the descendants of the cells transformed from R into S were also S. Griffith called the transmitted substance the "transforming principle."

DNA as the Transforming Principle

In 1944, Oswald Avery, Colin MacLeod, and Maclyn McCarty reported a follow-up to Griffith's experiments in which they sought to identify Griffith's transforming

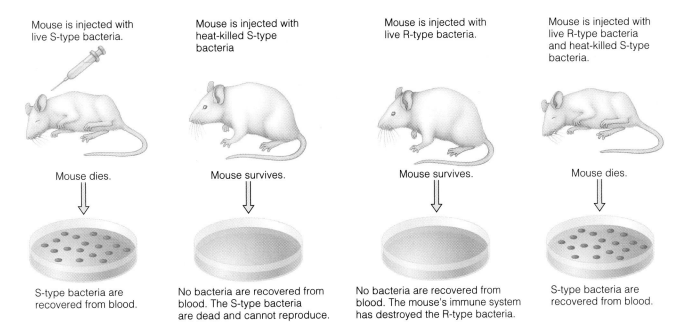

Mouse is injected with live S-type bacteria.

Mouse is injected with heat-killed S-type bacteria

Mouse is injected with live R-type bacteria.

Mouse is injected with live R-type bacteria and heat-killed S-type bacteria.

Mouse dies.

Mouse survives.

Mouse survives.

Mouse dies.

S-type bacteria are recovered from blood.

No bacteria are recovered from blood. The S-type bacteria are dead and cannot reproduce.

No bacteria are recovered from blood. The mouse's immune system has destroyed the R-type bacteria.

S-type bacteria are recovered from blood.

Figure 2.2 Frederick Griffith's experiments with rough (R) and smooth (S) strains of *Streptococcus pneumoniae* in mice.

principle. They heat-killed S cells, then treated the cells chemically to purify a viscous substance that transformed R cells into S cells. They specifically treated the substance to remove protein and RNA. Their chemical analysis of this substance revealed that its composition matched DNA. There remained, however, the possibility that the substance was contaminated with small amounts of protein or RNA, and that the contaminant, rather than the DNA, was responsible for the transformation. When they treated the substance with deoxyribonucleases (DNA-degrading enzymes), they discovered that it no longer transformed R cells into S cells. However, if they treated the substance with proteases (protein-degrading enzymes) or ribonucleases (RNA-degrading enzymes), the substance retained its transforming ability (Figure 2.3). These experiments ruled out the possibility that a protein or RNA contaminant of the DNA preparation might actually be the transforming principle. Avery, MacLeod, and McCarty determined that

> within the limits of the methods, the active fraction contains no demonstrable protein, unbound lipid, or serologically reactive polysaccharide and consists principally, if not solely, of a highly polymerized, viscous form of deoxyribonucleic acid [DNA].

Avery, MacLeod, and McCarty's experiments demonstrated quite conclusively that Griffith's transforming principle is DNA. In spite of these results, however, many scientists still regarded protein as the best candidate for the hereditary material. Let's now review some of the data that Avery, MacLeod, and McCarty reported in their classic paper.

Example 2.1 Chemical composition of the transforming principle.

Avery, MacLeod, and McCarty, as reported in their 1944 paper (*Journal of Experimental Medicine* 79:137–158), purified a substance capable of transforming avirulent live R cells into virulent live S cells. When they analyzed the chemical composition of the purified substance from two preparations, they obtained the following results:

Chemical Composition

Preparation No.	Carbon	Hydrogen	Nitrogen	Phosphorus
37	34.27%	3.89%	14.21%	8.57%
42	35.50%	3.76%	15.36%	9.04%
Theoretical proportions for DNA	34.20%	3.21%	15.32%	9.05%

Problem: What aspects of these results suggest that the substance is DNA?

Solution: The observed proportions of all four elements match closely the theoretical proportions for DNA. Had the substance contained significant amounts of proteins, polysaccharides, or lipids, the observed percentages (especially for phosphorus and nitrogen) should have differed substantially from the theoretical proportions for DNA.

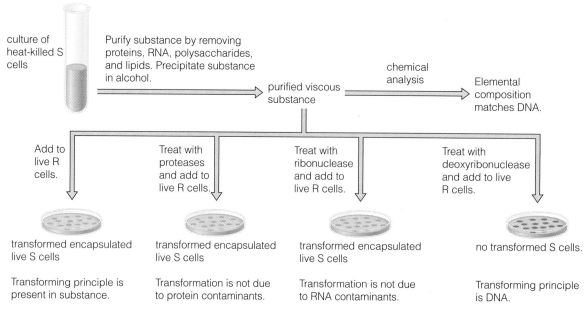

Figure 2.3 Avery, MacLeod, and McCarty's experiments demonstrating that Griffith's transforming principle is DNA.

The Genetic Material in Bacterial Viruses

In 1952, Alfred Hershey and Martha Chase published the results of experiments designed to determine whether protein or DNA is the hereditary material of the bacteriophage T2 (a virus that infects *E. coli* cells). Like many other viruses, this bacterial virus consists only of DNA and protein. When a virus of this type infects a bacterial cell, much of the virus remains on the outer surface of the cell, but some part of the virus capable of producing progeny is injected into the cell. Eventually these newly formed progeny viruses cause the cell to burst, releasing viruses to infect other cells.

The virus protein contains sulfur but no phosphorus, and the virus DNA contains phosphorus but no sulfur. Hershey and Chase used viruses to infect bacteria that had been provided either radioactive sulfur (^{35}S) or radioactive phosphorus (^{32}P). As the new viruses synthesized their DNA and proteins, they incorporated the radioactive label that was present in the host bacterial cells. Those exposed to radioactive sulfur incorporated it into their protein. Those exposed to radioactive phosphorus incorporated it into their DNA. Hershey and Chase then added the labeled viruses to bacterial cells that had been grown with no radioactive labels. These bacteria were agitated in blenders to knock the viral particles off the surfaces of the cells. After agitation, the bacterial cells were separated from the surrounding material by centrifugation.

Most of the ^{32}P label was found within the bacterial cells, whereas most of the ^{35}S label had remained outside of the cells, indicating that the substance transferred into the bacteria to form progeny viruses was DNA (Figure 2.4). Because a small amount of protein also entered the

bacterial cells, Hershey and Chase recognized that their experiments did not provide unequivocal proof that DNA was the genetic material of the T2 phage, but their results combined with those of other experimenters provided strong collective evidence of the importance of DNA in heredity.

A letter from Hershey to James Watson, who was an American postdoctoral fellow in England at the time, supported Watson's belief that DNA is the hereditary material and provided him with additional motivation to search out the structure of DNA. In Watson's words:

> Al Hershey had sent me a long letter from Cold Spring Harbor summarizing the recently completed experiments by which he and Martha Chase established that a key feature of the infection of a bacterium by a phage was the injection of the viral DNA into the host bacterium. Most important, very little protein entered the bacterium. Their experiment was thus a powerful new proof that DNA is the primary genetic material.... Nonetheless, almost no one in the audience of over four hundred microbiologists [at a scientific meeting] seemed interested as I read long sections of Hershey's letter.

However, any lack of interest in DNA was short-lived. Just one year later, in 1953, the structure of DNA was discovered, and geneticists soon recognized the now overwhelming evidence that DNA is indeed the genetic material.

Why did it take so long for scientists to recognize DNA as the genetic material? Early in the twentieth century, chromosomes were known to consist of proteins and DNA, and it was known that chromosomes carried the

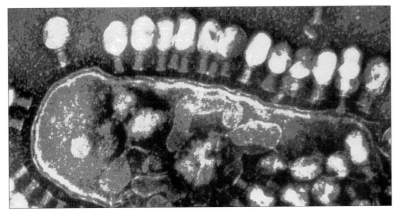

a The micrograph shows T2 bacteriophages infecting an E. coli cell.

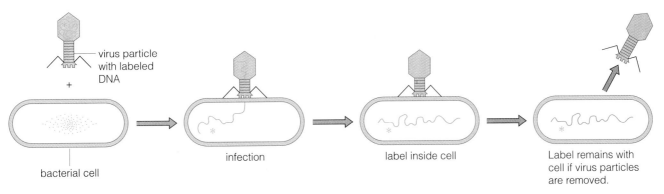

virus particle with labeled DNA

+

bacterial cell

infection

label inside cell

Label remains with cell if virus particles are removed.

b When viruses labeled with radioactive phosphorus infected the cells, the radioactive label was found in the cells, indicating that the viruses injected DNA into the cells.

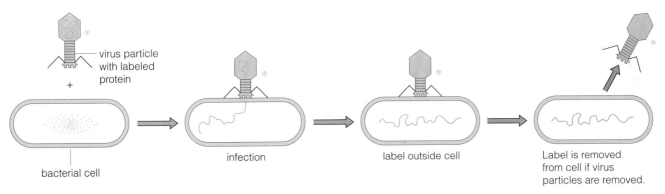

virus particle with labeled protein

+

bacterial cell

infection

label outside cell

Label is removed from cell if virus particles are removed.

c When viruses labeled with radioactive sulfur infected the cells, the radioactive label was found outside the cells, indicating that protein was not injected into the cells.

Figure 2.4 Hersey and Chase's experiments with T2 bacteriophage in *E. coli*.

genes. Proteins are composed of amino acid subunits, which in varying combinations could specify innumerable traits. DNA is composed of nucleotide subunits. There are 20 amino acids found in proteins, but only 4 nucleotides found in DNA. Protein thus seemed the better choice as the material that in various combinations could specify innumerable traits.

~

Several key experiments demonstrated that DNA, rather than protein, is the substance of heredity.

2.2 NUCLEIC ACID STRUCTURE

DNA is classified as a **nucleic acid**, as is **ribonucleic acid**, or **RNA**. They are called *nucleic* because much of the material found in the cell nucleus is DNA and RNA. And they are referred to as *acids* because they tend to be acidic when dissolved in aqueous solution. As we just discussed, DNA is the genetic material of all cellular organisms. RNA plays crucial roles in gene expression and in DNA replication in all organisms, and many viruses

have RNA instead of DNA as their genetic material. Let's look first at DNA.

The DNA Molecule

DNA is a remarkable molecule. Compared to most other molecules, its size is enormous. Although we measure the size of most molecules in angstroms (one angstrom, Å, equals one ten-billionth of a meter, 10^{-10} m), we measure certain DNA molecules in millimeters. You have 46 separate DNA molecules in the nucleus of each of your cells. These DNA molecules, if stretched out end-to-end, are on average about 50 mm long, about the length of your little finger. If all the DNA molecules in the nucleus of one of your cells were lined up end-to-end, the total length would be over 2 meters.

The amount of information contained in this DNA is also enormous. The DNA in the nucleus of one diploid human cell consists of about 6 billion pairs of subunits. If a single letter the size of the letters in this sentence represented each subunit pair, and if 6 billion of those letters were written out in a single line, it would extend for over 6600 km. Consider the distance between Los Angeles and New York City, about 4000 km, and you can imagine the enormous amount of genetic information in a single human cell. It is no wonder that the effort currently under way to determine the entire molecular sequence of the human genome is an enormous task.

Stood upright, a model of the DNA molecule looks like a spiral staircase, as shown in Figure 2.5. The light blue "steps," which represent nitrogenous bases, ascend between the dark blue stairway supports, which represent the two sugar-phosphate backbones of the molecule. Like a spiral staircase, the two backbones wind around each other in a regular and repeating conformation to form a double-helical shape.

Nucleotide Structure

The DNA molecule is composed of four subunits called **nucleotides**. Each nucleotide consists of a five-carbon sugar, deoxyribose, with a nitrogenous base attached to one end and a phosphate group attached to the other.

In the five-carbon deoxyribose sugar, the first four of the five carbon atoms, together with an oxygen atom, form a five-cornered ring, with the fifth carbon extending out

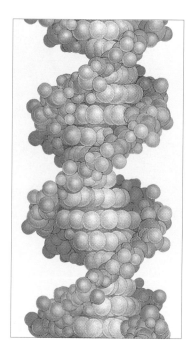

Figure 2.5 A space-filling model of DNA.

from the ring. The five carbon atoms are designated $1'$, $2'$, $3'$, $4'$, and $5'$ (pronounced "one-prime," "two-prime," "three-prime," etc.). The primes distinguish the carbons of the sugar from the numbered carbon and nitrogens in the nitrogenous base. The nitrogenous base is attached to the $1'$ carbon of the sugar, and a phosphate group is attached to the $5'$ carbon. An oxygen atom that is normally present on the $2'$ carbon in the ribose sugar is absent in DNA, hence the prefix *deoxy* in the term *deoxyribonucleic acid*.

The four different nucleotides in DNA are identical in structure except for the composition of the nitrogenous bases. The four nitrogenous bases are **thymine, cytosine, adenine**, and **guanine**. The four nucleotides themselves have more complicated chemical names: deoxythymidine 5'-monophosphate (dTMP), deoxycytidine 5'-monophosphate (dCMP), deoxyadenosine 5'-monophosphate (dAMP), and deoxyguanosine 5'-monophosphate (dGMP). However, for simplicity, each nucleotide in a DNA molecule is usually referred to by the name of its base, and often by the first letter of its base: **T, C, A**, or **G**. C and T, which are called **pyrimidines**, are similar to each other in structure, each having a single six-membered ring in the nitrogenous base. The other two nucleotides, A and G, called **purines**, are also similar to each other in structure; each purine has a nitrogenous base composed of two rings, a six-membered ring fused with a five-membered ring (Figure 2.6a).

The RNA Molecule

RNA is very similar to DNA. As illustrated in Figure 2.6b, RNA is composed of ribonucleotides that differ only

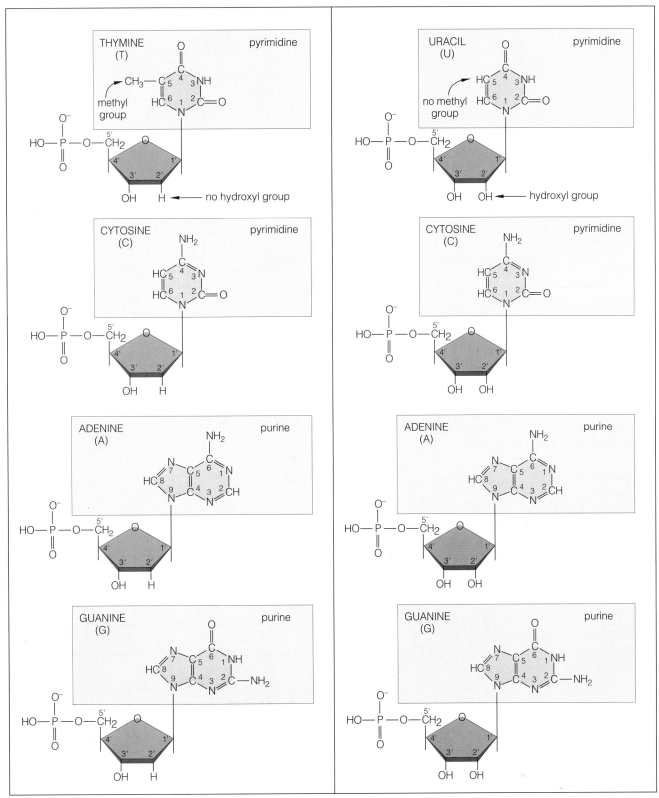

a Deoxyribonucleotides: subunits of DNA

b Ribonucleotides: subunits of RNA

Figure 2.6 The four nucleotide subunits of **(a)** DNA and **(b)** RNA. The deoxyribose or ribose sugar portions are colored red, the phosphates yellow, and the nitrogenous bases blue.

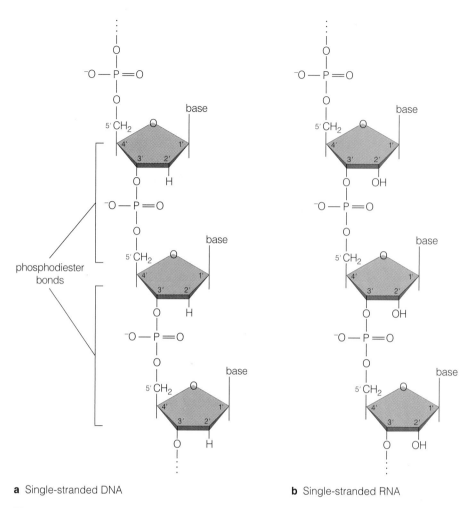

a Single-stranded DNA   **b** Single-stranded RNA

Figure 2.7 Phosphodiester bonds in single strands of **(a)** DNA and **(b)** RNA. All the 5′ carbons point in the same direction (upward in this example).

slightly from their deoxyribonucleotide counterparts in DNA. Each ribonucleotide in RNA consists of a five-carbon ribose sugar that, as in the deoxyribonucleotides, has a nitrogenous base attached to its 1′ carbon and a phosphate group attached to its 5′ carbon. Ribonucleotides, however, have a hydroxyl group attached to the 2′ carbon, whereas deoxyribonucleotides have only a hydrogen in this position. Like DNA, RNA consists of four nucleotides that differ from each other only in the composition of the nitrogenous base. Three of the nitrogenous bases are identical to three of the bases in DNA—cytosine, adenine, and guanine. The fourth nitrogenous base in RNA, **uracil (U)**, is very similar to thymine in DNA. Thymine has a methyl group attached to carbon number 5, but in uracil this methyl group is missing (Figure 2.6b).

DNA and RNA are **polymers** (long chains of repeated subunits). The polymers form as the phosphate group at the 5′ end of one nucleotide attaches to the 3′ carbon of another nucleotide, as illustrated in Figure 2.7. We call the attachments **phosphodiester bonds** because there are two ester linkages, one on either side of the phosphorus

atom. These phosphodiester bonds connect one nucleotide to the next to form a chain of nucleotides called a **single strand** of DNA or RNA. The 5′ carbons of each nucleotide in a single strand all point in the same direction. At one end a strand has a 5′ carbon with a phosphate group attached, and the other end terminates with a 3′ carbon attached to a hydroxyl group. DNA and RNA also exist in double-stranded forms in which two single strands of nucleotides wind around each other to form a double helix. Most DNA molecules are double stranded, whereas most RNA molecules are single stranded.

~

The nucleotides T, C, A, G, and U connect to each other with phosphodiester bonds to form the sugar-phosphate backbones of nucleic acids.

Base Pairing and Models for Double-Stranded DNA

From Phoebus Levene's demonstration in the 1930s that the four bases thymine (T), cytosine (C), adenine (A), and

guanine (G) are present in DNA, most scientists assumed that the four were present in equal amounts. A concept that arose from this conclusion was the *tetranucleotide hypothesis.* According to this now disproven hypothesis, the unit of DNA is a tetramer consisting of one of each of the four nucleotides. This unit is repeated many times in DNA to form a repetitious polymer. Scientists who accepted this hypothesis reasoned that such a molecule could not be sufficiently complex to be the hereditary material; thus little attention was focused on DNA as a candidate for the hereditary material.

Then during the late 1940s and early 50s, Erwin Chargaff showed that in DNA purified from various organisms, A and T are present in essentially equal amounts, and G and C are also present in essentially equal amounts, but the relative amounts of A + T and G + C were not always equal, a conclusion that ran counter to the tetranucleotide hypothesis. This comparison is often called the (A + T)/(G + C) ratio. This ratio was found to be the same for individuals of the same species, but usually different between species. The equality of A and T and of G and C suggested that A and T are somehow chemically related, as are G and C.

In 1951, James Watson arrived in England as a post-doctoral fellow and met Francis Crick, who was a graduate student at the time. They both had their hearts set on studying DNA even though they were assigned to other research topics. They recognized that experiments such as those by Hershey and Chase, and Avery, MacLeod, and McCarty provided good evidence that DNA was the hereditary material, and that discovery of its structure would therefore be of great biological importance. Watson and Crick were fortunate to have access to X-ray diffraction patterns of DNA crystals provided by Rosalind Franklin and Maurice Wilkins (Figure 2.8). Combining the X-ray observations with additional information, including Chargaff's experimental observations, Watson and Crick developed a number of models of the structure of DNA in an attempt to find one that matched the experimental evidence and fit the rules imposed by stereochemistry. Early models included a triple-stranded helix, a model in which the phosphate groups pointed toward the center of the helix, and a model in which individual bases paired with themselves, such as adenine with adenine, guanine with guanine, and so on. Of this last idea, Watson wrote:

> For over two hours I happily lay awake with pairs of adenine residues whirling in front of my closed eyes. Only for brief moments did the fear shoot through me that an idea this good could be wrong.

Each of these models failed in some aspect. The X-ray images taken by Franklin predicted that the DNA molecule is a helix of uniform width that repeats its structure every 34 angstroms and has the bases pointing in-

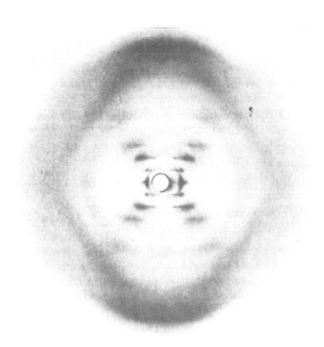

Figure 2.8 An X-ray diffraction image of a DNA molecule, which aided in the discovery of DNA's structure. (Reprinted with permission from Franklin, R. E., and R. G. Gosling. 1953. Molecular configuration in sodium thymonucleate. *Nature* 171: 740–741. Copyright 1953 Macmillan Magazines Limited.)

ward. Chargaff's data predicted that adenine and thymine are present in equal amounts, and that guanine and cytosine are present in equal amounts. Each of the early models failed to meet at least one of these restrictions and had problems with rules of chemistry. However, the final model, in which adenine paired with thymine and guanine paired with cytosine, agreed with the experimental data and the rules of stereochemistry. Once Watson and Crick revealed their model, scientists who had criticized the previous models recognized that this one was correct. Watson and Crick published their results in 1953 in what is now a classic paper in genetics. You can access a link to their paper at the website http://www.brookscole.com./biology. Figure 2.9 shows Watson and Crick standing next to their final model of the DNA molecule.

Let's now examine some of Chargaff's data to see how they correspond with the chemical structure of DNA.

Example 2.2   Chargaff's data on nucleotide composition from eleven different sources.

Erwin Chargaff, in a 1951 paper (Federal Proceedings 10:654−659), reported the following ratios for the four nucleotides in DNA, obtained from eleven different sources, including a virus, several bacteria, yeast, a plant, several animals, and human.

Source	Ratios				
	A:G	T:C	A:T	G:C	Purine: Pyrimidine
Ox	1.29	1.43	1.04	1.00	1.1
Human	1.56	1.75	1.00	1.00	1.0
Hen	1.45	1.29	1.06	0.91	0.99
Salmon	1.43	1.43	1.02	1.02	1.02
Wheat	1.22	1.18	1.00	0.97	0.99
Yeast	1.67	1.92	1.03	1.20	1.0
Haemophilus influenzae, type C	1.74	1.54	1.07	0.91	1.0
E. coli, K-12	1.05	0.95	1.09	0.99	1.0
Avian tubercule bacillus	0.4	0.4	1.09	1.08	1.1
Serratia marcescens	0.7	0.7	0.95	0.86	0.9
Hydrogen organism *Baccillus* Schatz	0.7	0.6	1.12	0.89	1.0

Problem: What do these data suggest about the similarity of DNA among different species and the chemical relationships of the four nucleotides?

Solution: Because the A:G (purine:purine) ratios and T:C (pyrimidine:pyrimidine) ratios vary substantially among species, the DNA of different species differs in nucleotide composition. Because the A:T, G:C, and purine:pyrimidine ratios are close to 1 in all species examined, there must be chemical relationships between A (purine) and T (pyrimidine), and between G (purine) and C (pyrimidine).

Base-Pairing Rules

Suppose that we have a double-stranded DNA molecule, and that a segment of one strand has the sequence AATAGCCA. Can we determine what the nucleotide sequence of the corresponding segment of the other strand is? To answer this question, we need to think about one of the most remarkable aspects of double-stranded DNA. That aspect, which makes exact replication of the molecule possible, is the precise pairing of nucleotides between the two strands, as described by Watson and Crick. As mentioned earlier, a strand of DNA forms as phosphodiester bonds link the 5′ and 3′ carbons of adjacent nucleotides. The bonding produces a long strand of nucleotides, each with a nitrogenous base extending out from the 1′ carbon. A double-stranded DNA molecule is formed when two such strands wind around each other in a double-helical shape, with the nitrogenous bases of each strand facing each other (Figure 2.10).

As indicated by —H— in Figure 2.10, the two strands are held together by hydrogen bonds that connect the ni-

Figure 2.9 James Watson (left) and Francis Crick (right) with their model of the DNA molecule. (A. C. Barrington Brown/ Photo Researchers)

trogenous bases on one strand with those on the other strand. Hydrogen bonds form when slight positive charges ($\delta+$) in the bases align with slight negative charges ($\delta-$). A hydrogen bonded to a nitrogen carries a slight positive charge. An oxygen double-bonded to a carbon, or a nitrogen bonded to carbons carries a slight negative charge. In both A-T and G-C pairs, negative charges align with positive charges and the bases attract each other, forming hydrogen bonds, as illustrated in Figure 2.11.

This results in some very precise nucleotide pairing rules for DNA: T pairs with A, and C pairs with G. We can now answer the question posed at the beginning of this section. The paired nucleotide sequence of the double-stranded DNA segment would be as follows:

```
5′ AATAGCCA 3′
3′ TTATCGGT 5′
```

The two strands must be **antiparallel**, which means that the 5′→3′ orientations of the two strands in the double helix run in opposite directions (Figure 2.10).

A close look at nucleotide structure reveals the chemical basis for the base-pairing rules. The nitrogenous bases fit much like the interlocking pieces of a jigsaw puzzle. A pyrimidine must pair with a purine. Remember that purines have two rings in the nitrogenous base, whereas pyrimidines have only one ring, making purines larger than pyrimidines. If a purine paired with another

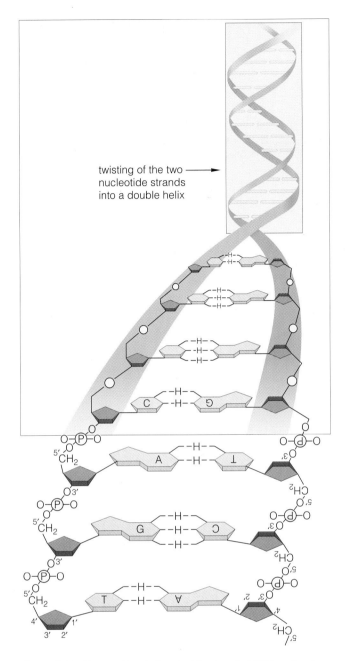

twisting of the two nucleotide strands into a double helix

Figure 2.10 Base pairing in the double-stranded DNA molecule.

purine, there would be four rings across the axis of the DNA molecule, causing it to bulge at that point:

Purine-purine pair is too wide.

Conversely, if a pyrimidine paired with another pyrimidine, there would be only two rings across the axis of the molecule. The DNA molecule would be constricted at that point:

Pyrimidine-pyrimidine pair is too narrow.

However, the DNA molecule has a precisely uniform width throughout its length because purines only pair with pyrimidines.

If a purine must always pair with a pyrimidine, why don't C-A and T-G pairs form? The answer lies in the hydrogen bonding, as illustrated in the example that follows.

Example 2.3   Restrictions on base pairing in DNA.

Problem: Diagram T paired with G and C paired with A and explain why these base pairs do not form.

Solution:

Positive charges align with positive charges, and negative charges align with negative charges. Under these circumstances, the bases repel each other, and hydrogen bonds do not form.

Hydrogen bonds between base pairs hold the antiparallel strands of a DNA molecule together.

The Double Helix

In the Watson-Crick model of double-stranded DNA, the nitrogenous base pairs lie flat against each other, forming

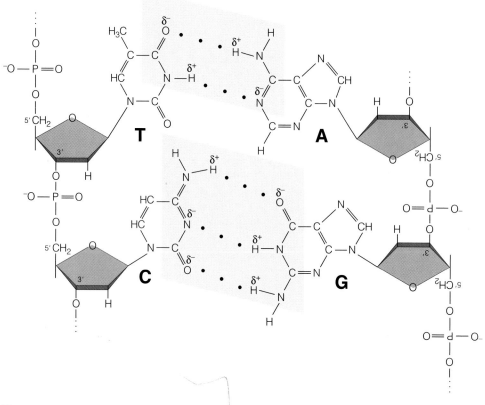

Figure 2.11 A close look at hydrogen bonding between base pairs. In both A-T and G-C pairs, slight positive charges ($\delta+$) align with slight negative charges ($\delta-$), and the bases attract each other.

planes that are almost perpendicular to the axis of the molecule, much like the stairs of a spiral staircase. The deviation from perpendicular is about 6°. The bases stack tightly against each other. In the stack, the axis of each base pair is rotated approximately 36° relative to the previous base, so the molecule makes one complete turn of 360° about every ten base pairs (Figure 2.12).

The turns of a helix may be either right-handed or left-handed; this orientation is consistent regardless of the direction from which the helix is viewed. For instance, the threads on a typical bolt or wood screw are right-handed, meaning that bolts and screws are tightened by turning them in a clockwise direction. The groove in the threads of a screw is a single helix. You can visualize the single helix by placing the point of a felt-tip pen in the groove at the tip of a screw, then rotating the screw clockwise, holding the pen in the groove until it reaches the head of the screw. The pen will deposit ink in the single groove throughout the entire length of the screw.

In the Watson-Crick model, DNA is a right-handed double helix. Because it is a double helix, there are two separate grooves that twist around the molecule between the sugar-phosphate backbone, much like the red and white stripes on a barbershop pole. Looking at the molecule from its side, the two different grooves appear once in each full turn of the helix (Figure 2.12).

The two grooves in DNA are not equal in size. The **major groove** is approximately 22 Å wide, and the **minor groove** 12 Å wide, making the linear distance for one complete turn of the helix about 34 Å, with 3.4 Å between each base pair. This form of DNA, with the bases stacked nearly perpendicular to the axis of the molecule at about a 36° angle of rotation for each base pair and with major and minor grooves in a right-handed double helix, is referred to as the **B form** (or **B-DNA**) and is the most common model for DNA. Three models of B-DNA are illustrated in Figure 2.12.

Alternative Forms of DNA

The conditions in living cells are such that B-DNA predominates. However, scientists have detected alternative forms in purified DNA that has been exposed to different conditions of relative humidity and salt concentration. Most alternative forms are found only under conditions not normally present in cells. However, under certain conditions, an alternative form of DNA may appear in certain regions of DNA within living cells. The alternative forms of DNA

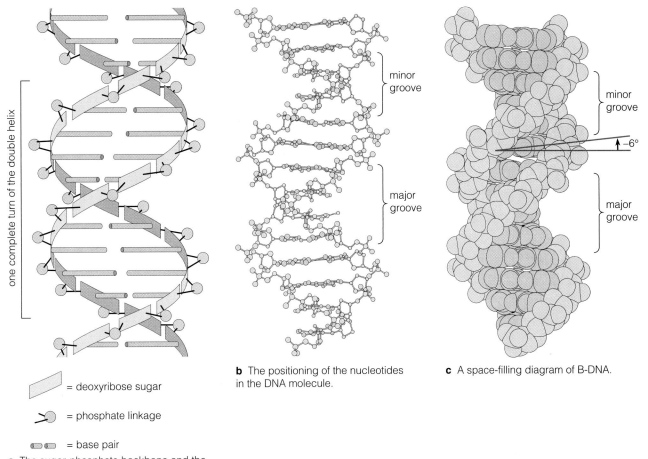

one complete turn of the double helix

= deoxyribose sugar

= phosphate linkage

= base pair

a The sugar-phosphate backbone and the nitrogenous base pairs in the DNA molecule.

b The positioning of the nucleotides in the DNA molecule.

c A space-filling diagram of B-DNA.

minor groove

major groove

minor groove

−6°

major groove

Figure 2.12 The Watson-Crick or B-DNA model. This represents the most common form of DNA under cellular conditions. (Redrawn from Saenger, W. 1984. *Principles of Nucleic Acid Structure.* New York: Springer-Verlag.)

differ from each other primarily in the degree and direction of helical winding and in the stacking of the bases.

In **A-DNA** (Figure 2.13b), the bases are tilted at about 20° from perpendicular to the axis of the molecule. The helix is wound less tightly than in B-DNA, with a 32.7° rotation per base pair. Scientists can induce DNA to assume the A form under laboratory conditions, but A-DNA does not exist in living cells. However, double-stranded **DNA-RNA hybrids** and **double-stranded RNA**, which are found in living cells, assume a conformation like A-DNA.

Z-DNA (Figure 2.13c) is unusual in that its double helix is wound in a left-handed conformation, the molecule is quite narrow, and there are more base pairs per helical turn than in B-DNA. As viewed from outside the space-filling representation, the sugar-phosphate backbones of Z-DNA form a zigzag pattern, which is the reason for the Z designation. Although typical cellular conditions do not favor the formation of Z-DNA, certain situations may cause it to appear within limited regions of the DNA molecule. These may include cases

where specialized proteins cause left-handed twisting of the B-DNA molecule or when the nucleotide sequence favors the Z-DNA form. Antibodies specific for Z-DNA bind to a few regions of DNA in cells, suggesting that limited regions of Z-DNA are present.

H-DNA is a triple-helical form of DNA. A short region of double-helical B-DNA may rearrange to become the H-DNA triple helix, as diagrammed in Figure 2.13d. There is good evidence that short segments of H-DNA are present in living cells and that the H-DNA may help control the expression of genes.

Still, B-DNA appears to be by far the most common form in living cells. Other forms known as C, D, and E-DNA can be induced to form under laboratory conditions but have not been found in living cells.

~

In a typical DNA molecule, two sugar-phosphate backbones wind together in a double helix with their attached nitrogenous bases pointing toward the center of the helix. There are several alternative forms of DNA, although B-DNA is the most common in living cells.

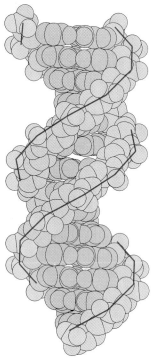

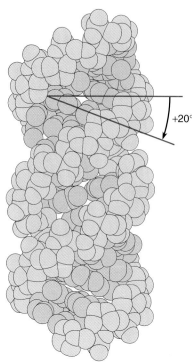

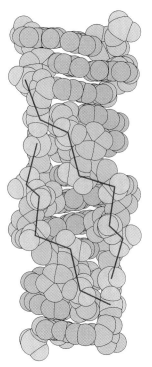

+20°

a B-DNA model with a line superimposed to show the smoothness of the double helix.

b A-DNA model. The bases tilt at a greater angle than in B-DNA, and the right handed double helix is wound less tightly.

c Z-DNA model. The double helix is left-handed, and the molecule is narrower than B-DNA. A superimposed line shows that the sugar-phosphate backbones of Z-DNA form a helix that zigzags, in contrast to the smooth helix of B-DNA.

B-DNA   H-DNA (triple stranded)

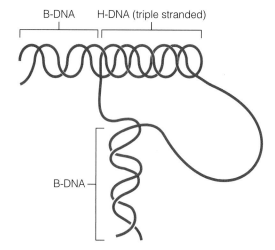

B-DNA

d Rearrangement of B-DNA to form the triple helix of H-DNA.

Figure 2.13 Alternative forms of DNA. (Redrawn from Saenger, W. 1984. *Principles of Nucleic Acid Structure.* New York: Springer-Verlag.)

2.3 SEMICONSERVATIVE DNA REPLICATION

Among the most important aspects of DNA is its ability to replicate. When Watson and Crick first described the structure of DNA, they recognized that the base pairing suggested the mode of replication. In their paper describing the structure of DNA, they pointed this out, stating:

It has not escaped our notice that the specific pairing we have postulated immediately suggests a possible copying mechanism for the genetic material.

The basic model of DNA replication is quite simple. The two strands "unzip" by breaking the hydrogen bonds that hold the paired nitrogenous bases together, creating two single-stranded portions of the molecule and exposing the unpaired bases on these portions. The place

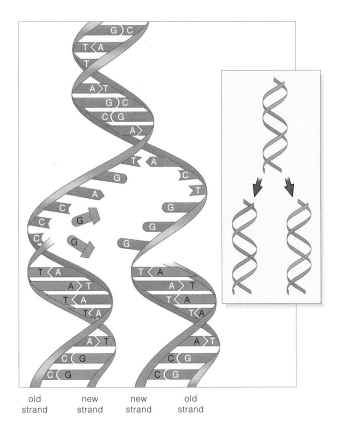

old strand new strand new strand old strand

Figure 2.14 Semiconservative replication of DNA. Each new molecule consists of one strand from the parental molecule and one newly synthesized strand. The parental strands serve as templates for synthesis of the new strands.

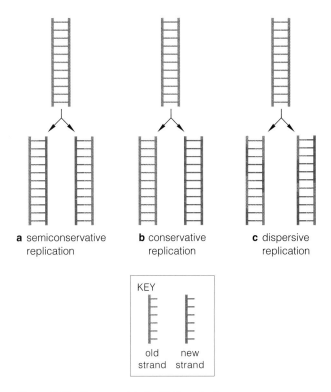

a semiconservative replication **b** conservative replication **c** dispersive replication

KEY

old strand new strand

Figure 2.15 Three models for DNA replication.

at which the two strands separate is called the **replication fork**. Free nucleotides that are present in the cell may then pair with the nucleotides on each of the single strands according to the base-pairing rules, making two identical molecules. This model, illustrated in Figure 2.14, can be used for a general discussion of DNA replication, but it is overly simplistic. We will use this basic model initially to learn about replication, and then see a more detailed model in Figure 2.22.

Picture from the model in Figure 2.14 that, as the molecule unwinds, the replication fork moves farther along until the molecule is fully replicated into two identical double-stranded DNA molecules. This mode of replication is referred to as **semiconservative**, meaning that the two original strands are conserved, each serving as a **template** on which a new strand forms. The two daughter molecules each contain an original parental strand and a newly synthesized strand. Although the structure of DNA led Watson and Crick to devise the model of semiconservative replication, there was no experimental evidence to confirm this hypothesis. The Watson-Crick model allowed for other models, and two were proposed. In the **conservative replication** model, the original double-stranded molecule was conserved, and both strands of the replicated molecule were new. In the **dis-**

persive replication model, each strand on both daughter molecules combined parental and newly synthesized DNA (Figure 2.15).

In 1958, Matthew Meselson and Franklin Stahl published the results of experiments designed to reveal which model is correct. They provided actively dividing *E. coli* cells with a heavy isotope as their source of nitrogen (^{15}N instead of the more common ^{14}N). As the cells grew, they incorporated the ^{15}N into the nitrogenous bases of their DNA. After 17 cell generations, all DNA in the bacteria was labeled with ^{15}N. The labeled bacteria were then transferred to a medium containing only ^{14}N. Some were allowed to divide once (thus replicating their DNA once), and others were allowed to divide twice (thus replicating their DNA twice). The DNA from these groups of bacteria was isolated and centrifuged at high speed in cesium chloride salt. When a concentrated solution of cesium chloride is placed under high centrifugal force, it generates a density gradient with the highest density at the bottom of the centrifuge tube and increasingly less dense matter toward the top. When DNA is included with the salt during centrifugation, it moves to an equilibrium point in the tube, at which the centrifugal forces moving it toward the bottom equal the buoyant forces formed by the density gradient

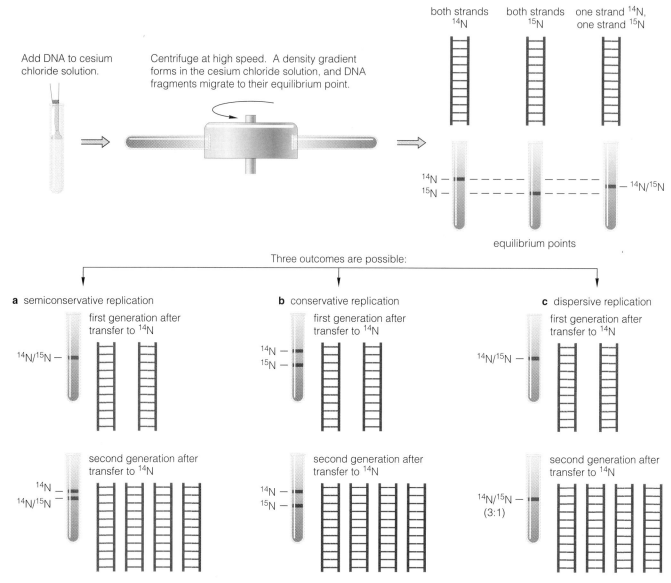

Figure 2.16 The experiments of Meselson and Stahl. The results they obtained correspond with those diagrammed in part a, confirming that DNA replicates semiconservatively.

pushing it toward the top. DNA containing ^{15}N is more dense than DNA containing ^{14}N and reaches equilibrium at a lower point in the tube than the DNA with ^{14}N. Using this method, Meselson and Stahl could distinguish the heavier ^{15}N DNA from the lighter ^{14}N DNA, as well as DNA containing half ^{15}N and half ^{14}N.

As shown in Figure 2.16, the three models for DNA replication predict where the DNA will equilibrate in the gradient. The conservative DNA replication model predicts that after one cell division, there will be two types of DNA that appear as separate fractions in the tube, one heavy (^{15}N) and one light (^{14}N). Both semiconservative and dispersive replication models predict that DNA isolated after one cell division will move to a point in the centrifuge tube that is halfway between the points for ^{15}N

DNA and ^{14}N DNA; the DNA would be at the intermediate point because half the DNA in each molecule would contain ^{14}N and the other half would contain ^{15}N. Meselson and Stahl observed a single fraction at the intermediate point, which ruled out conservative replication as the correct model.

DNA isolated from the cells that had divided twice could then distinguish between dispersive and semiconservative replication. With dispersive replication, all DNA molecules would contain some ^{14}N and some ^{15}N and thus appear between the positions for ^{14}N and ^{15}N DNA. With semiconservative replication, there would be two fractions of DNA, one appearing at the point for ^{14}N DNA and the other at the intermediate point. Meselson and Stahl observed these two fractions in the

DNA from cells that had divided twice, which eliminated the dispersive model and confirmed that DNA replicates semiconservatively.

~

DNA replicates semiconservatively, meaning that the two original strands are conserved, each serving as a template for the formation of a new strand. The two daughter molecules will each contain an original parental strand and a newly synthesized strand.

2.4 THE PROCESS OF DNA REPLICATION

Many aspects of DNA replication are similar in prokaryotes and eukaryotes, but there are some important differences. Because DNA replication has been studied more fully in prokaryotes than in eukaryotes, we will examine the replication process first in prokaryotes, then point out some of the aspects where prokaryotic and eukaryotic replication differ. In our discussion, we will focus on *E. coli* because nearly all aspects of DNA replication were first discovered in this species.

As we mentioned earlier, the model of DNA replication illustrated in Figure 2.14 is overly simplistic. In reality, DNA replication is a complex process requiring several steps. Let's now look at those steps.

Separating and Stabilizing DNA Strands

For DNA replication to proceed, the two strands must separate into single strands. As the strands separate, the molecule must be allowed to unwind or excessive tension will build up in it. To visualize this, take a string and loop it around a pencil so that the two ends of the string are about equal in length. Then rotate the pencil to twist the two strings into a tight double helix, as illustrated in Figure 2.17a. While keeping the pencil stationary, pull the ends of the string apart. Tension will build up in the portions of the string that are still wound, and eventually the wound portions will form **supercoils**, structures that look somewhat like knots although they actually represent the formation of secondary coils (Figure 2.17b). If the pencil is allowed to rotate freely, however, the tension is released as the supercoils and coils unwind (Figure 2.17c).

The same principle applies to DNA. As the molecule unwinds for replication, the part ahead of the replication fork may supercoil. The supercoiling is said to be positive because it turns in the right-handed direction, the same orientation as the original winding of the DNA. The tension of supercoiling must be released during DNA replication. Enzymes called **helicases** unwind the DNA as the replication fork proceeds along the molecule, while enzymes called **topoisomerases** relieve the positive tension created by the unwinding. Although topoisomerase

a The string looped around a pencil and twisted represents a double helix without supercoils.

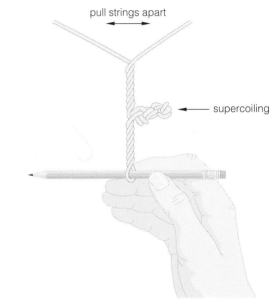

b As the string unwinds, supercoils form ahead of the unwinding if the helix is not allowed to rotate.

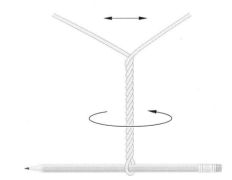

c Rotation relieves the tension from supercoiling.

Figure 2.17 Demonstration of supercoiling.

function is not fully understood, these enzymes apparently nick one strand of the double-stranded DNA ahead of the replication fork. The molecule then rotates at these nicks, relieving the tension of supercoiling. There is some indication that topoisomerases may also create negative supercoils ahead of the replication fork. (Negative supercoils turn in the direction opposite that of the original winding in the DNA, i.e., in a left-handed orientation.) These negative supercoils may assist helicases because as the negative tension is relieved, this may help drive unwinding as the replication fork proceeds.

The two single-stranded segments of DNA that result from the unwinding must be stabilized to prevent them from reattaching to one another. This is accomplished by **single-strand binding proteins (SSBs)**. These proteins rapidly bind to the newly separated single-stranded DNA and maintain it in a single-stranded state until that portion of DNA is replicated. SSBs are referred to as **cooperative tetramers** because four protein subunits form a tetramer that binds to single-stranded DNA, and once an SSB tetramer is bound to single-stranded DNA, it facilitates the binding of an adjacent SSB tetramer. This process rapidly continues from one SSB to the next until all of the single-stranded DNA is bound with SSBs.

Synthesizing New DNA Strands

Once the single strands of DNA are stabilized, they may serve as templates upon which new strands are synthesized by enzymes referred to collectively as **DNA polymerases**. Several DNA polymerases are known, each with the capability of adding new nucleotides in a chain, using a single strand of DNA as a template. In *E. coli*, **DNA polymerase III**, a complex enzyme consisting of seven different protein subunits, synthesizes most of the DNA molecule.

DNA polymerases can add new nucleotides only to the 3'-OH group on a nucleotide already paired with the template strand, which imposes two restrictions, as illustrated in Figure 2.18. First, DNA polymerases can only extend a chain of nucleotides in one direction. Because nucleotides are added only to the 3' end of a DNA strand, DNA synthesis is said to proceed only in the 5'→3' direction. Second, DNA polymerases *cannot initiate* DNA synthesis on a single-stranded DNA template; they can only *extend* a chain of DNA from a nucleotide already paired to the template strand.

What, then, initiates DNA synthesis? Unlike DNA polymerases, **RNA polymerases** are capable of adding ribonucleotides (the subunits of RNA) to a single-stranded DNA template that has no nucleotides paired with it. RNA is very similar to DNA in composition and structure, and its nucleotide subunits may pair with the nucleotides in DNA to form a double-stranded DNA-RNA hybrid molecule. For initiation of DNA replication, an RNA polymerase called **primase** synthesizes a short

segment of RNA (10 to 60 nucleotides) on the DNA template, forming a short RNA-DNA hybrid. The segment of RNA is called a **primer**, and it initiates DNA synthesis. At one end of the primer is a terminal nucleotide with a 3'-OH group onto which DNA polymerases can add a new deoxyribonucleotide (Figure 2.19). Once the first deoxyribonucleotide is added, using the single-stranded DNA as a template, it exposes a 3'-OH group onto which another nucleotide may be added. The DNA polymerase may then continue to add a nucleotide onto the 3'-OH group of each previous nucleotide, eventually creating a double-stranded DNA molecule.

The free nucleotides that are incorporated into the DNA strands are in the triphosphate form—that is, they have three phosphate groups attached to the 5' carbon (Figure 2.18a). As each nucleotide is added to the growing molecule, the two terminal phosphate groups are cleaved from the nucleotide. Because the triphosphate form of a nucleotide contains more energy than the monophosphate form, removal of phosphates releases energy, driving DNA replication.

~

DNA polymerases synthesize new DNA strands using a single strand of DNA as a template. DNA polymerases add nucleotides only in the 5'→3' direction. Because DNA polymerases cannot initiate synthesis, an RNA primer is required for initiation.

Semidiscontinuous DNA Replication

We now face another dilemma. Because DNA strands are antiparallel, only one strand may be synthesized continuously. In a replication fork, only one strand has its 3'-OH group facing the replication fork. On this strand, DNA polymerases may add new nucleotides to each previous one, synthesizing DNA in the direction in which the replication fork is moving. As the replication fork proceeds along the molecule, DNA polymerases add new nucleotides continuously. The new strand of DNA that is synthesized continuously is called the **leading strand**.

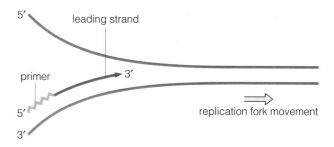

The situation on the other new strand is different. Newly added nucleotides have the 3'-OH group facing away from the replication fork, meaning that DNA is synthesized in the direction *opposite* that in which the replication fork is moving.

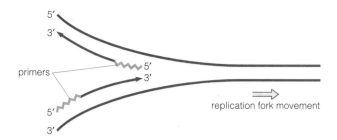

As the DNA molecule continues to unzip and the replication fork moves ahead, the newly exposed single-stranded DNA on the upper template strand cannot be used as a template for replication until a new primer is added, because the previous primer has a 5'-phosphate group facing the replication fork. Thus the strand on the upper template must be synthesized discontinuously (i.e., in fragments) rather than continuously. For this reason it is often called the **lagging strand**.

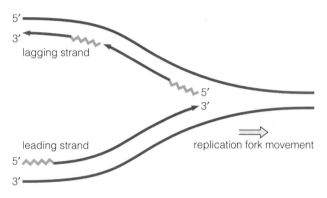

The fragments of DNA formed by discontinuous DNA synthesis of the lagging strand are approximately 1000–2000 nucleotides long in prokaryotes and are much shorter (about 100–200 nucleotides) in eukaryotes. These DNA fragments have been named **Okazaki fragments** after their discoverer, Reiji Okazaki. Because one DNA strand is synthesized continuously and the other discontinuously, DNA replication is said to be **semidiscontinuous**.

Synthesis of each fragment of the lagging strand requires a new initiation of DNA synthesis with a new RNA primer. How can an intact, double-stranded DNA molecule form when there are fragments, each beginning with an RNA primer? Somehow, the RNA primers must be removed and DNA must be synthesized in the space formerly occupied by the primers. Then, adjacent DNA fragments must be linked to one another. In *E. coli*, DNA polymerase III terminates DNA synthesis when it reaches the primer of the previous fragment, leaving a nick between the 3' end of the newly synthesized fragment and the 5' end of the primer (Figure 2.20a). At that point, the previous fragment's RNA primer must be removed and deoxyribonucleotides must be added in its place. An enzyme called **DNA polymerase I** can do this because, in addition to its 5' → 3' polymerase activity, it also has a

5' → 3' exonuclease activity, meaning that it can remove nucleotides ahead of it in the 5' → 3' direction. DNA polymerase I recognizes the nick between the DNA fragment and the primer and binds at this site (Figure 2.20b). It then migrates along the molecule, removing the primer ribonucleotides ahead of it. At the same time, it adds deoxyribonucleotides to the template using its 5' → 3' polymerase activity (Figure 2.20c). After the final deoxyribonucleotide is added, a nick remains between its 3'-OH group and the 5'-phosphate group of the adjacent nucleotide (Figure 2.20d). An enzyme called **DNA ligase** (Figure 2.20e) seals this nick by bonding the 3'-OH group to the 5'-phosphate group, creating a continuous double-stranded segment of DNA.

Ligase works very quickly, making it difficult to recover Okazaki fragments experimentally. In the following example, let's see how researchers used a mutant bacterial strain to isolate Okazaki fragments.

Example 2.4   Accumulation of Okazaki fragments in ligase-deficient *E. coli*.

In 1969, Pauling and Hamm (*Proceedings of the National Academy of Sciences, USA* 64:1195–1202) recovered Okazaki fragments using a temperature-sensitive strain of *E. coli* that lacked ligase activity at 40°C but had ligase activity at 25°C. After raising the temperature to 40°C and transferring the strain for 5 minutes into medium containing radioactive nucleotides, Pauling and Hamm detected, from the radioactivity, newly synthesized Okazaki fragments in the temperature-sensitive bacteria. These fragments did not appear in an *E. coli* strain with normal ligase activity at 40°C when the researchers subjected it to the same procedure. Instead, the newly synthesized DNA in the normal strain was much longer.

Problem: **(a)** Why was it necessary to use a temperature-sensitive, ligase-deficient strain of *E. coli* instead of a strain that lacked ligase activity altogether? **(b)** Why did Okazaki fragments accumulate in the temperature-sensitive strain but not in the strain that had normal ligase activity at 40°C?

Solution: **(a)** A bacterial strain that lacks ligase activity altogether fails to survive because DNA cannot be fully replicated without ligase. However, a temperature-sensitive strain can be grown indefinitely at 25°C because ligase is active at this temperature and the cells can fully replicate their DNA. The effect of ligase deficiency can be studied in this strain by raising the temperature to 40°C, which halts ligase activity. **(b)** After primer removal, ligase joins Okazaki fragments to form a long continuous strand of DNA. With no ligase activity at 40°C in

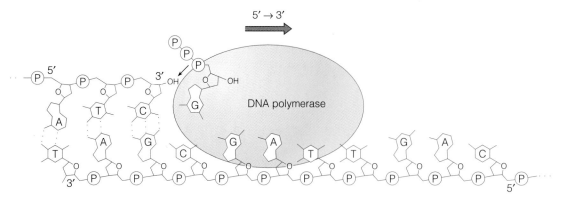

a DNA polymerase can extend the nucleotide chain in the 5′ → 3′ direction.

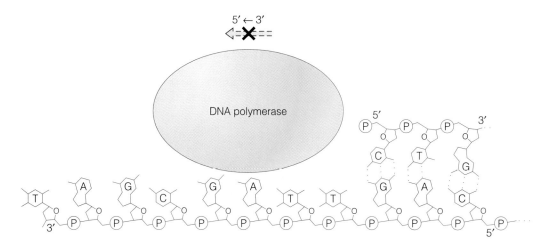

b DNA polymerase cannot extend the nucleotide chain in the 3′ → 5′ direction.

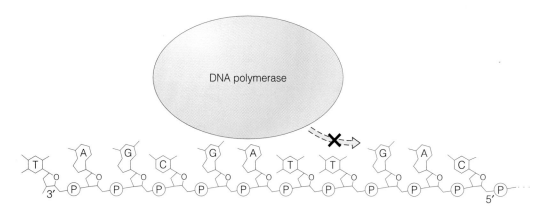

c DNA polymerase cannot initiate synthesis on a template strand without paired nucleotides.

Figure 2.18 Restrictions on DNA polymerases.

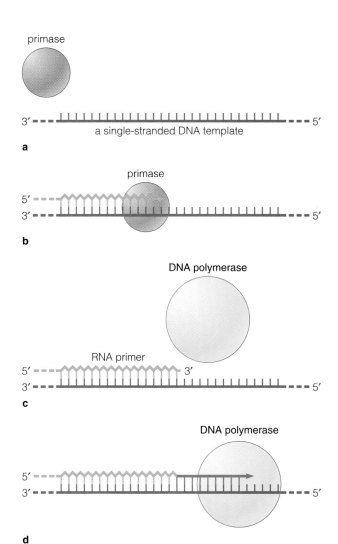

a

a single-stranded DNA template

b

c

d

RNA primer

DNA polymerase

DNA polymerase

Figure 2.19 An RNA primer for initiation of DNA synthesis. The straight line represents a DNA strand, and the zigzag line represents an RNA strand.

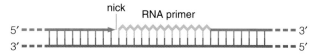

a DNA polymerase III leaves a nick between the 3′ end of the newly sythesized fragment and the 5′ end of the RNA primer.

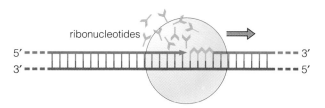

b DNA polymerase I binds at the nicked site.

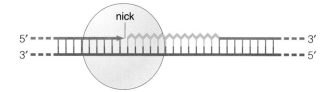

c DNA polymerase I replaces ribonucleotides with deoxyribonucleotides.

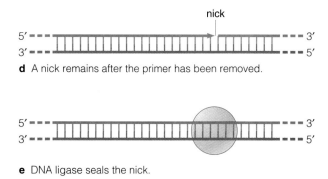

d A nick remains after the primer has been removed.

e DNA ligase seals the nick.

Figure 2.20 Primer removal by DNA polymerase I.

the temperature-sensitive strain, Okazaki fragments cannot be joined and instead accumulate in the bacteria. In the strain with normal ligase activity at the higher temperature, Okazaki fragments do not accumulate because they are joined as usual by ligase.

Simultaneous Synthesis of Leading and Lagging Strands

Most models of DNA replication depict synthesis of the leading and lagging strands by separate DNA polymerase molecules. There is, however, evidence that the two strands are synthesized simultaneously. In *E. coli*, DNA polymerase III is a large enzyme composed of

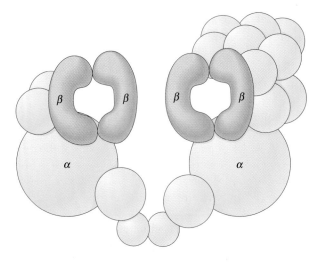

Figure 2.21 The molecular structure of the DNA polymerase III holoenzyme in *E. coli*.

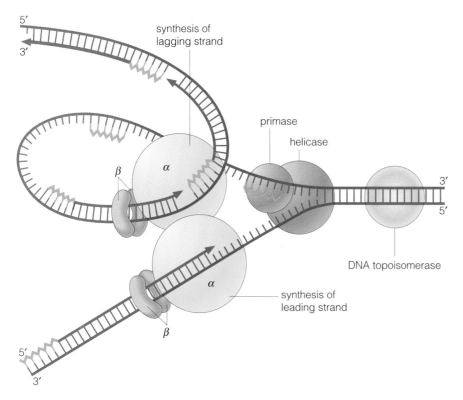

5′
3′
synthesis of
lagging strand
β
α
primase
helicase
3′
5′
DNA topoisomerase
α
synthesis of
leading strand
β
5′
3′

Figure 2.22 Model of DNA replication in prokaryotes with simultaneous synthesis of leading and lagging strands.

many protein subunits (typically 20 or more). The subunits that replicate DNA are designated α. Figure 2.21 shows how the subunits fit together in the whole molecule, called the **DNA polymerase III holoenzyme**. Noting that there are two α subunits, investigators hypothesized that one subunit might synthesize the leading strand while the other synthesizes the lagging strand. The β subunits assemble to form guides through which double-stranded DNA can pass.

So how can a single protein complex simultaneously synthesize two DNA strands that are antiparallel? If the template for the lagging strand forms a loop, then the looped portion points in the same direction as the template for the leading strand. Then, the DNA polymerase III holoenzyme can simultaneously synthesize both strands in the same direction, one continuously and the other discontinuously, as illustrated in Figure 2.22.

An important aspect of this model is the integration of many molecules into a unified complex of DNA polymerases, helicase, and primase that operates at the replication fork. This complex of proteins is called the **replisome**. There is now a large body of evidence to support the DNA replication model shown in Figure 2.22.

~

DNA replication is semidiscontinuous: only the leading strand is synthesized continuously; the lagging strand is synthesized discontinuously, giving a series of fragments that must be joined together. DNA polymerase III holoenzyme synthesizes both the leading and lagging strands simultaneously.

Proofreading Newly Synthesized DNA

If prokaryotic DNA replication relied entirely on chemical interactions between nitrogenous bases for accurate base pairing during the synthesis of a new strand of DNA, bases would be mispaired about once every 1000 nucleotides. However, actual errors during DNA replication are far fewer, about once every billion or more nucleotides. This remarkable fidelity in DNA replication is due mostly to **proofreading** by DNA replication enzymes. In addition to their $5' \rightarrow 3'$ polymerase activity, prokaryotic DNA polymerases also have a $3' \rightarrow 5'$ **exonuclease activity**. This means that these polymerases are capable of moving backward and removing nucleotides from DNA in the direction opposite that of DNA synthesis. When the DNA polymerase adds an incorrect nucleotide during DNA synthesis, the DNA polymerase apparently recognizes the error and uses its $5' \rightarrow 3'$ exonuclease activity to remove the mismatched nucleotide. It then adds the correct nucleotide in its place and continues DNA synthesis in the $5' \rightarrow 3'$ direction (Figure 2.23).

Table 2.1 Prokaryotic DNA Polymerases

Polymerase	Functions
DNA polymerase I	Removal of nucleotides during DNA repair (5′→3′ exonuclease);
	synthesis of DNA during repair;
	synthesis of short gaps in DNA;
	primer removal (5′→3′ exonuclease);
	proofreading (3′→5′ exonuclease)
DNA polymerase II	Synthesis of DNA during repair;
	proofreading (3′→5′ exonuclease)
DNA polymerase III	DNA synthesis; proofreading (5′→3′ exonuclease)

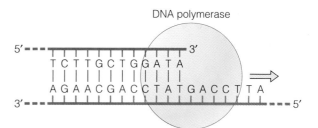

a DNA polymerase adds correctly paired nucleotides in the 5′ → 3′ direction.

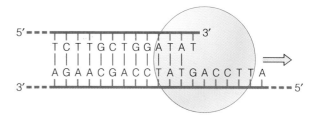

b DNA polymerase adds a mispaired nucleotide.

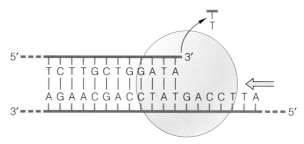

c DNA polymerase reverses direction and acts as a 3′ → 5′ exonuclease to remove the mispaired nucleotide.

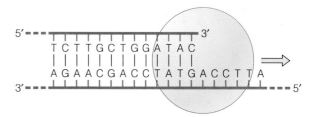

d DNA polymerase resumes DNA synthesis in the 5′ → 3′ direction.

Figure 2.23 DNA proofreading in prokaryotes.

Proofreading is not the only mechanism that cells use for correcting errors in DNA. Nucleotides may be mismatched and DNA may be damaged in a number of ways. If left uncorrected, the errors may become mutations that alter the sequence of nucleotides. The main function of the *E. coli* enzyme **DNA polymerase II** is to synthesize DNA during repair of DNA damage. Table 2.1 lists the DNA polymerases in *E. coli* and their functions.

There are many mechanisms that cells may use to correct nucleotide mismatches or repair DNA damage. Although highly effective, DNA proofreading and repair mechanisms are not perfect, so occasionally DNA mutates and the mutations persist during subsequent DNA replication. Chapter 5 provides a detailed description of mutation and DNA repair mechanisms.

~

Proofreading by DNA polymerases corrects most errors in DNA replication.

Replication of Eukaryotic DNA

The general mechanism for DNA replication in eukaryotes is similar to that in prokaryotes, but the enzymes differ. Most of the enzymes responsible for DNA replication in yeast and in mammals have now been characterized. There are five known DNA polymerases in mammals, named α, β, δ, ε, and γ (Table 2.2). DNA polymerases α and δ are both essential for replication and are probably the DNA polymerases responsible for most of the DNA synthesis. There is strong evidence that α is responsible for synthesis of the lagging strand, and δ is responsible for synthesis of the leading strand.

DNA polymerase α also functions as a primase, synthesizing RNA primers on a DNA template for initiation of replication (Figure 2.24). The two smallest subunits (48 and 58 kD*) in association may act as a primase on their own (Figure 2.24a). They may also act as a primase

*The abbreviation kD stands for kiloDaltons, or 1000 Daltons, which is a unit of measurement often used to designate the molecular mass of proteins. A Dalton is the mass of one hydrogen atom, so a 48 kD protein has a molecular mass of about 48,000.

Table 2.2 Eukaryotic DNA Polymerases

Mammalian Polymerase	Corresponding Polymerase in Yeast	Functions
α	pol I	Synthesis of lagging strand; primer synthesis
β	none	Synthesis of DNA during repair
δ	pol III	Synthesis of leading strand; proofreading ($3' \rightarrow 5'$ exonuclease)
ε	pol II	Synthesis of DNA during repair; proofreading ($3' \rightarrow 5'$ exonuclease)
γ	mitochondrial DNA polymerase	Mitochondrial DNA synthesis; proofreading ($3' \rightarrow 5'$ exonuclease)

when they are part of DNA polymerase α (Figure 2.24b). The 180 and 70 kD subunits of DNA polymerase α function in DNA replication. The combined DNA polymerase and primase activities of DNA polymerase α constitute strong evidence that this enzyme synthesizes the Okazaki fragments that become the lagging strand. DNA polymerase δ has no primase activity. DNA polymerases ε and β help repair damaged DNA and associate with exonucleases to remove damaged DNA. The fifth DNA polymerase in eukaryotes, γ, is found only in mitochondria. This enzyme synthesizes and proofreads the organelle's DNA.

DNA polymerases δ, ε, and γ have $3' \rightarrow 5'$ exonuclease activities, indicating that they are capable of proofreading. None of the eukaryotic DNA polymerases have a $5' \rightarrow 3'$ exonuclease activity like that of DNA polymerase I in *E. coli*. Instead, there are several separate enzymes in eukaryotes that remove nucleotides in the $5' \rightarrow 3'$ direction.

A model for simultaneous synthesis of both DNA strands in eukaryotes has been devised. It is similar to the prokaryotic model shown in Figure 2.22. The eukaryotic model is based on the observation that DNA polymerases α and δ associate with each other in a single complex. In this model (Figure 2.25), α synthesizes the lagging strand, while δ synthesizes the leading strand simultaneously.

~

The model for eukaryotic DNA replication is similar to the prokaryotic model. However, the eukaryotic enzymes differ somewhat in function from their prokaryotic counterparts.

2.5 REPLICATION OF ENTIRE DNA MOLECULES

DNA molecules vary in their sizes and conformations among living organisms. The molecules may be circular or linear, and they range in size from a few thousand nucleotide pairs in viral DNAs to more than a hundred million nucleotide pairs in some eukaryotic chromosomes. Cells and viruses employ several different strategies for replicating their DNA molecules. Which strategy is best depends on the molecule's size and conformation, and on specific nucleotide sequences that determine how the molecule will replicate. In this section, we will examine the various strategies for replication of entire DNA molecules.

Origins of Replication

The points on a DNA molecule where synthesis begins, called **origins of replication**, are fixed sites along the DNA. In other words, replication does not begin at random along the molecule, but rather at predetermined positions that are used each time the DNA replicates. The nucleotide sequences at some origins of replication have been identified, and they tend to be rich in A-T base pairs. Because each A-T pair is held together by only two hydrogen bonds, whereas a G-C pair is held together by

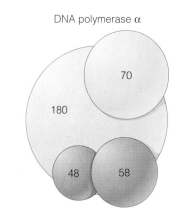

DNA polymerase α

primase

a The 48 and 58 kD subunits of DNA polymerase α may function on their own as a primase.

b They may also function as a primase when part of the entire enzyme.

Figure 2.24 DNA polymerase α in mammals.

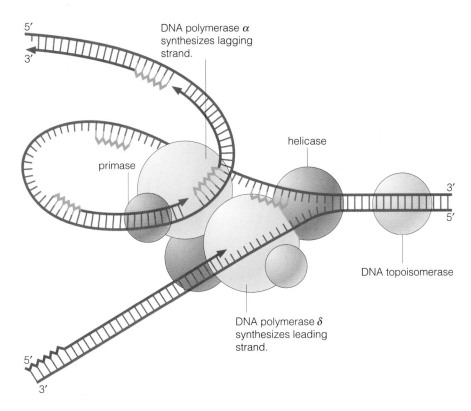

5′
3′

DNA polymerase α
synthesizes lagging
strand.

primase

helicase

3′
5′

DNA topoisomerase

DNA polymerase δ
synthesizes leading
strand.

5′
3′

Figure 2.25 Model of DNA replication in eukaryotes with simultaneous synthesis of leading and lagging strands.

three, the strands in a segment of DNA containing mostly A-T pairs are more easily separated than those in a segment containing many G-C pairs.

E. coli has a single origin of replication called *oriC* in its large, circular DNA molecule. The *oriC* origin of replication consists of 245 nucleotide pairs including three segments having the same sequence of 13 nucleotide pairs, called a 13-mer, followed by four segments having the same sequence of 9 nucleotide pairs (a 9-mer). Notice from Figure 2.26a that 11 of the 13 nucleotides in the 13-mers are A-T pairs. As diagrammed in Figure 2.26b, the DNA unwinds first in the region of the 13-mers when replication is initiated. The 9-mers direct a set of proteins, called DnaA proteins, to bind to the DNA in the region of the 9-mers. The DnaA protein binding causes the DNA to coil into a loop, which, in turn, causes the DNA in the region of the 13-mers to unwind, allowing replication to begin (Figure 2.26b).

~

Origins of replication are fixed sites in DNA that separate easily because they are rich in A-T pairs.

Bidirectional DNA Replication

Presumably, a replication fork could begin at an origin of replication on one end of a linear DNA molecule and proceed in a single direction throughout the molecule until the entire molecule was replicated. Or, in a circular

molecule, a replication fork could begin at a single origin of replication in the circle and proceed around in one direction until the entire circle was replicated. Replication that proceeds in this way is known as **unidirectional replication**. DNA replication is unidirectional in some viruses and bacterial plasmids (small circular DNA molecules that are separate from the larger DNA molecule of bacteria). However, in most cases DNA replication begins at an origin of replication somewhere within the molecule, and then two replication forks begin in opposite directions from the origin, a phenomenon known as **bidirectional replication**. In *E. coli,* for example, two replication forks move at *oriC* and move in opposite directions until they meet, replicating the entire molecule.

Most eukaryotic DNA molecules are linear and typically orders of magnitude longer than bacterial DNA molecules. Each of these long molecules has many origins of replication that can initiate DNA replication simultaneously, enabling the DNA to replicate more quickly than it could from a single origin of replication. As replication forks proceed in both directions from the origins, they form "bubbles" of replicated DNA that eventually merge until the entire molecule is replicated (Figure 2.27).

Let's turn our attention to an experiment that shows how unidirectional and bidirectional replication can be distinguished.

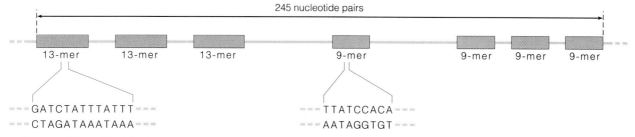

245 nucleotide pairs

13-mer　13-mer　13-mer　　9-mer　　9-mer　9-mer　9-mer

GATCTATTTATTT　　　　　TTATCCACA
CTAGATAAATAAA　　　　　AATAGGTGT

a Structure of the *ori C* origin of replication.

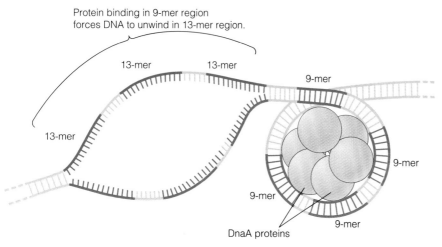

Protein binding in 9-mer region
forces DNA to unwind in 13-mer region.

13-mer　　13-mer

9-mer

13-mer

9-mer

9-mer

9-mer

DnaA proteins

b Function of the DnaA proteins that cause the DNA to unwind.

Figure 2.26　The *oriC* origin of replication in *E. coli*.

Example 2.5　DNA replication in *Bacillus subtilis*.

In 1973, Gyurasits and Wake (*Journal of Molecular Biology* 73:55–63) reported the results of an experiment to determine whether DNA replication was unidirectional or bidirectional in the bacterium *Bacillus subtilis*. They germinated bacterial spores on growth medium containing low-level radioactive (^{3}H) thymidine. As the DNA in the cells replicated, it incorporated labeled thymidine into the newly synthesized DNA. After 120 minutes of growth with low levels of radioactive thymidine, the researchers supplied the culture with high-level radioactive (^{3}H) thymidine. After 20 additional minutes, replication was halted and the DNA was exposed to X-ray film to detect the radioactivity. The DNA labeled with high-level radioactive thymidine appeared more dense on the developed film than the DNA labeled with low-level radioactive thymidine. Figure 2.28a shows one of their autoradiographs (an image on X-ray film that portrays radioactivity).

Problem:　**(a)** What type of replication, unidirectional or bidirectional, is indicated by this autoradiograph? **(b)** Why was it important to use two levels of radioactivity, instead of just labeling with a single level?

Solution:　**(a)** If DNA replication is bidirectional, two regions of higher radioactivity, one at each replication fork, should appear, as diagrammed in Figure 2.28b. If replication is unidirectional, only a single replication fork should be labeled with high-level radioactivity. The autoradiograph in Figure 2.28a shows two replication forks labeled with high-level radioactivity, indicating that DNA replication is bidirectional in this molecule. **(b)** Had the researchers germinated spores on a medium with a single level of radioactivity and not altered the level of radioactivity, it would not have been possible to distinguish bidirectional from unidirectional replication. In both cases, the DNA would have appeared as a uniformly labeled replication bubble.

From an origin of replication, two replication forks move in opposite directions to replicate DNA bidirectionally.

Strategies for Replicating Circular DNA

Circular DNA molecules are found in nearly every cell. The large DNA molecules of bacterial cells are circular, as are the smaller plasmid molecules. In eukaryotic cells, mitochondria contain circular molecules, as do chloro-

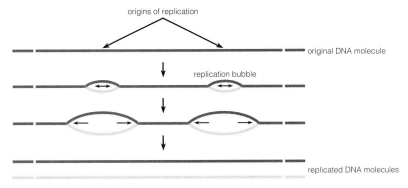

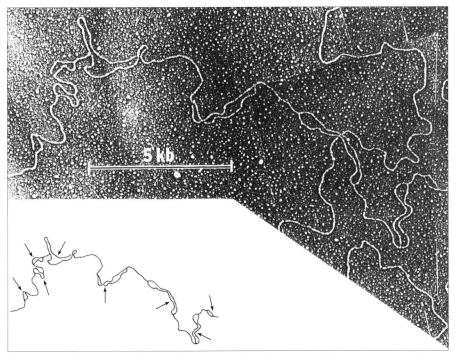

a Replication "bubbles" forming from two origins of replication.

b Replication bubbles (indicated by arrows) in *Drosophila melanogaster* DNA.

Figure 2.27   Bidirectional replication. (Photo courtesy of D. S. Hogness.)

plasts in plant cells. Many viruses also replicate their DNA in a circular conformation. There are three basic strategies for replicating circular molecules.

The first strategy is called **θ-mode replication**. Most large circular DNA molecules in bacteria have a single origin of replication from which replication forks proceed bidirectionally to form a θ (Greek theta)–shaped intermediate when replication is partially completed:

Eventually, the two replication forks meet and the two circular molecules separate. Figure 2.29 shows an autoradiograph of a θ intermediate in *E. coli.*

The second strategy of DNA replication, called **σ-mode** or **rolling circle replication**, and has a σ (Greek sigma)–shaped intermediate. Several bacterial viruses use σ-mode replication to rapidly make multiple copies of the viral DNA. A nick is made at the origin of replication in one strand of the double-stranded DNA (Figure 2.30a), and the nicked strand separates from its complementary strand beginning at the 5′ end of the nicked site (Figure 2.30b). The nicked strand is used as a template for discontinuous DNA synthesis forming Okazaki fragments, while a second strand of DNA is synthesized continuously using the unnicked circular strand as a template.

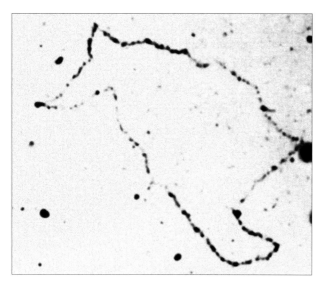

a The bacterial DNA labeled with high-level radioactivity appears more dense on the developed film than the DNA labeled with low-level radioactivity.

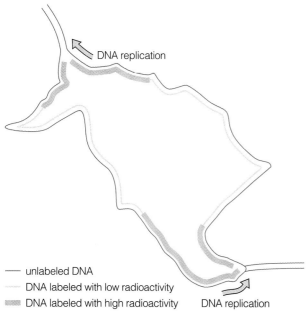

— unlabeled DNA
— DNA labeled with low radioactivity
▨▨▨ DNA labeled with high radioactivity

DNA replication

b Labeling of DNA that exhibits bidirectional replication. Two replication forks are labeled with high radioactivity.

Figure 2.28 DNA replication in *Bacillus subtilis*. (Photo from Gyurasits and Wake. 1973. Bidirectional chromosome replication in *Baccillus* subtilis. *Journal of Molecular Biology.* Reprinted by permission of Academic Press.)

As shown in red in part c of Figure 2.30, this second strand fills in the space that is vacated as the nicked strand extends away from the circle. Replication usually continues for many rounds, generating a linear double-stranded DNA molecule that contains many tandem linear copies of the circular molecule (Figure 2.30d). The tandem copies are cleaved from one another to generate linear copies that can then arrange themselves into circles (Figure 2.30 e and f).

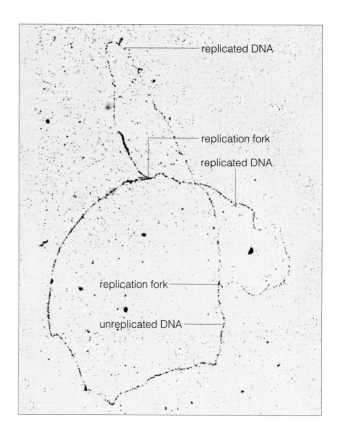

Figure 2.29 Electron micrograph of θ-mode replication in *E. coli*. (Photo from Cairns, J. 1963. The chromosome of *Echerichia coli. Cold Spring Harbor Symposia on Quantitative Biology* 28:43–46. Reprinted by permission of the publisher.)

The circular DNA molecules of mitochondria and chloroplasts utilize the third strategy, called **D-loop replication**. As diagrammed in parts a and b of Figure 2.31, the two DNA strands in a circular molecule separate at an origin of replication, and continuous DNA synthesis begins using one of the strands as a template. As synthesis progresses, the newly synthesized strand displaces one of the old strands, creating a **displacement loop** (abbreviated as **D-loop**). Eventually, DNA synthesis passes a second origin of replication, which becomes available for initiation of DNA synthesis in the single-stranded D-loop (Figure 2.31c). A second newly synthesized strand is initiated from this origin, proceeding in the direction opposite to the synthesis of the first strand. Synthesis is complete when each strand reaches its respective origin (Figure 2.31d).

Some circular DNA molecules replicate with a single D-loop, as depicted in Figure 2.31; however, other molecules may generate more than one D-loop to complete their replication. For example, mammalian mitochondrial DNAs replicate with a single D-loop, whereas most chloroplast DNAs replicate with two D-loops.

~

The three major strategies for replicating circular DNA are θ-mode replication, σ-mode or rolling circle replication, and D-loop replication.

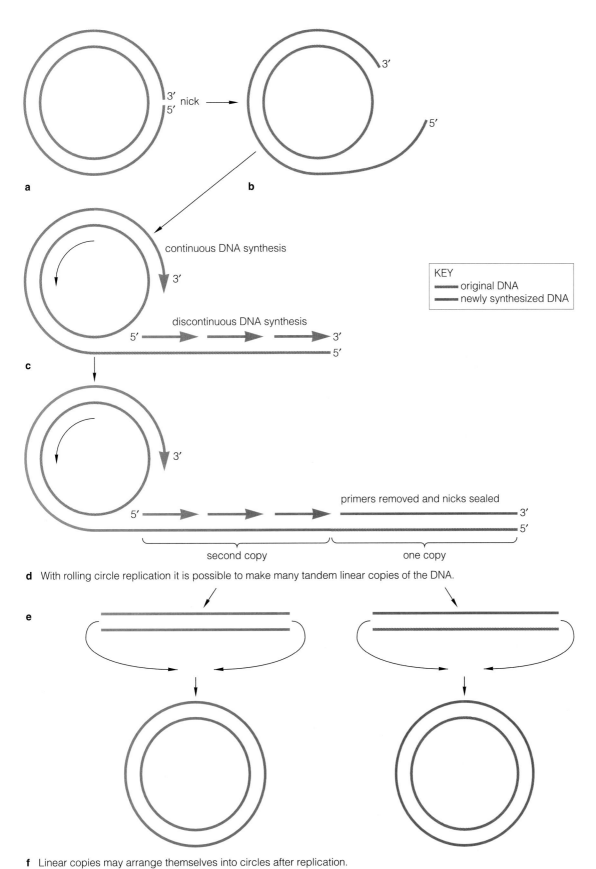

a

b

continuous DNA synthesis

3′

c

discontinuous DNA synthesis

5′ ⟶ ⟶ ⟶ 3′
5′

KEY
original DNA
newly synthesized DNA

3′

primers removed and nicks sealed

5′ ⟶ ⟶ ⟶ 3′
5′

second copy   one copy

d With rolling circle replication it is possible to make many tandem linear copies of the DNA.

e

f Linear copies may arrange themselves into circles after replication.

Figure 2.30 σ-mode (rolling circle) replication.

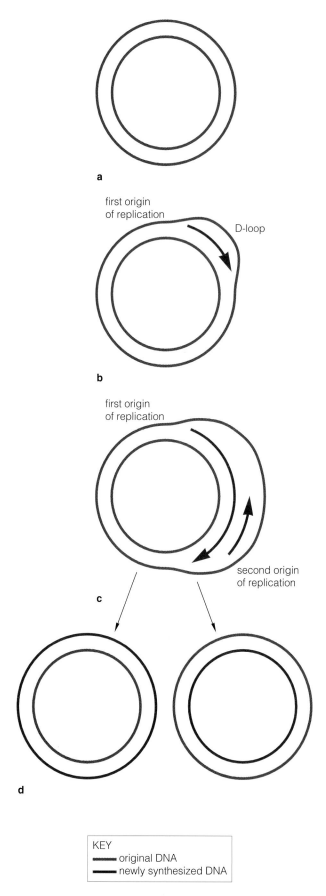

a

first origin
of replication

D-loop

b

first origin
of replication

second origin
of replication

c

d

KEY
━━━ original DNA
━━━ newly synthesized DNA

Figure 2.31 D-loop replication.

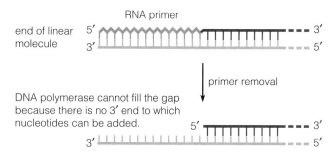

RNA primer

end of linear
molecule

DNA polymerase cannot fill the gap
because there is no 3' end to which
nucleotides can be added.

primer removal

Figure 2.32 The linear DNA replication paradox. How can the gap at the end of the linear molecule be filled?

Strategies for Replicating Linear DNA

The large DNA molecules in eukaryotic chromosomes are linear. What we have learned so far about DNA replication cannot explain how the ends of linear molecules can be fully replicated. To understand why the ends of DNA molecules cannot be fully replicated using semidiscontinuous replication, think about primers. If a primer forms at the end of a linear DNA molecule to initiate replication of the leading strand, as diagrammed in Figure 2.32, a single-stranded gap remains at that end after DNA has been synthesized and the primer has been removed. DNA polymerases cannot fill the gap because there is no 3'-OH group onto which nucleotides can be added. If this gap were left unfilled, a linear DNA molecule would become shorter at each end by the length of the primer with each replication. After replicating many times, the molecule would eventually disappear. This problem is called the **linear DNA replication paradox**. If cells fail to fully replicate the ends of their DNA molecules, eventually the molecules erode at the ends, losing important information, which halts cell growth and division. For cells to grow and divide for more than a few generations, they must have a strategy for overcoming the linear DNA replication paradox.

Linear DNA molecules found in nature employ several different strategies for overcoming the linear DNA replication paradox. In perhaps the simplest strategy, a linear DNA molecule arranges itself into a circle in preparation for replication. The *E. coli* bacterial virus lambda uses this strategy. Its DNA exists in linear form in the virus particle, and the virus injects the linear DNA into a cell during infection. When the viral DNA prepares to replicate, however, it arranges itself into a circle, then generates many linear copies through rolling circle replication. Each of the linear DNA copies then assembles with proteins into a virus particle.

The large DNA molecules in eukaryotic chromosomes use one of the most elaborate strategies. A clue

Figure 2.33 A model for replicating the ends of linear chromosomes in eukaryotes. ▶

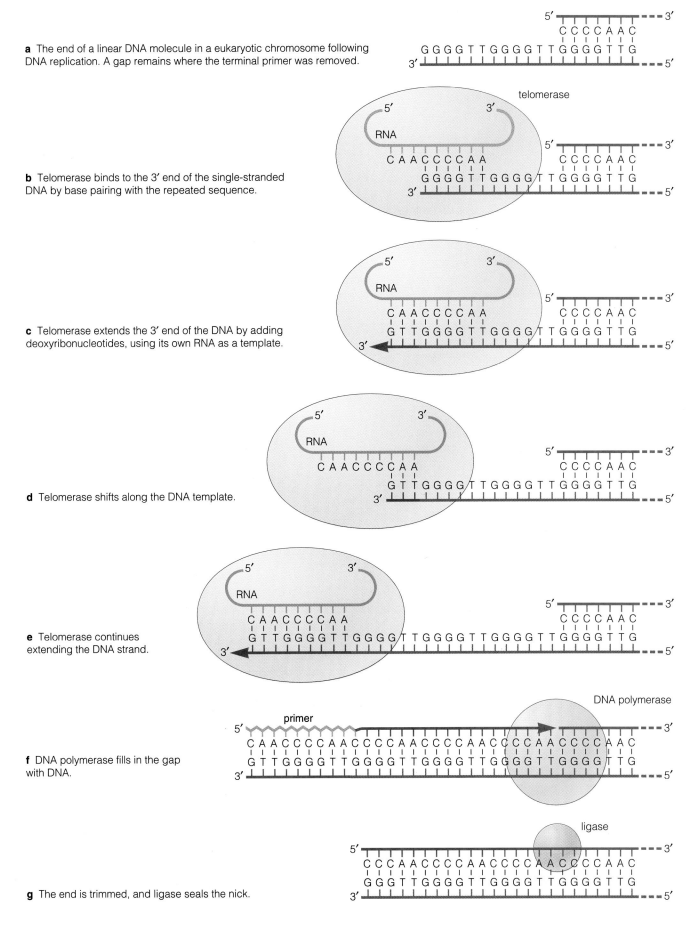

a The end of a linear DNA molecule in a eukaryotic chromosome following DNA replication. A gap remains where the terminal primer was removed.

b Telomerase binds to the 3′ end of the single-stranded DNA by base pairing with the repeated sequence.

c Telomerase extends the 3′ end of the DNA by adding deoxyribonucleotides, using its own RNA as a template.

d Telomerase shifts along the DNA template.

e Telomerase continues extending the DNA strand.

f DNA polymerase fills in the gap with DNA.

g The end is trimmed, and ligase seals the nick.

to how eukaryotic chromosomes overcome the problem of replicating the ends of the molecules comes from the DNA sequences found at the ends of linear chromosomes. At the ends are DNA sequences that are repeated as many as 50–100 times. For example, in *Tetrahymena* (a ciliated protozoan and the first organism in which these repeated sequences were studied) the repeated sequence is

```
5' TTGGGG 3'
3' AACCCC 5'
```

In the intact chromosome, these repeated sequences on the ends of the chromosome form a structure called a *telomere*, which we will discuss in more detail in Chapter 10. These repeated sequences are synthesized by a unique enzyme called *telomerase*. As we discuss telomere synthesis, you may wish to follow the steps illustrated in Figure 2.33.

Once a linear DNA molecule has replicated, a single-stranded 3' overhang remains at each end of the molecule where the primer was removed (Figure 2.33a). *Tetrahymena* telomerase contains a single strand of RNA 159 nucleotides long. Within this RNA sequence is the six-nucleotide repeated sequence, 5' CCCCAA 3', which is complementary to the repeated region. Telomerase binds to the 3' end of the single-stranded DNA and uses its own RNA template to extend the 3' end of the DNA, adding new copies of the TTGGGG repeated region, as diagrammed in parts b, c, and d of Figure 2.33. These repeats further extend the single-stranded gap left after primer removal (Figure 2.33e). However, primers may now be added to the repeated segments of DNA, and a DNA polymerase may now fill in much of the gap with DNA (Figure 2.33f). This extends the double-stranded DNA to a point where it is at least as long as (and perhaps longer than) the original parent molecule. A single-stranded portion will still be left on the end, and may be trimmed off (Figure 2.33g). Because the extension and trimming of telomeric DNA is not precise, the number of telomeric repeats may vary among molecules.

Eukaryotic linear DNA molecules have telomeric sequences on the ends, and they are similar in all species studied. For example, the human telomere repeated sequence

```
5' TTAGGG 3'
3' AATCCC 5'
```

differs by only one nucleotide pair from the *Tetrahymena* sequence.

The first studies on telomerase were conducted by Carol Greider and Elizabeth Blackburn, who used *Tetrahymena,* a single-celled organism that requires telomerase for every cell division. Later, after telomerase had been identified in multicellular organisms, including hu-

mans, an intriguing discovery was made. Telomerase apparently is active as cells in the germ line divide, eventually giving rise to sperm and egg cells. However, in most cells of the body, telomerase is not present when the cells divide, indicating that the DNA molecules in these cells must grow shorter with each cell division. This is of little consequence to most cells in a human being because the cells undergo a limited number of cell divisions and then cease dividing. Cancer cells, on the other hand, may continue dividing until their host dies. When researchers examined cancer cells for the presence of telomerase, the enzyme was found in abundance, indicating that the linear DNA molecules in cancer cells were being fully replicated. This implies that telomerase is essential for cancer cells to continue dividing. Research is now under way to determine if blocking the activity of telomerase may be an effective treatment for cancer.

~

DNA polymerases cannot complete synthesis of the ends of linear DNA molecules. Telomerase extends the ends of linear molecules to complete their synthesis.

SUMMARY

1. DNA is the universal genetic material for all forms of life, with the exception of some viruses, although until the middle of the twentieth century the hereditary material was generally thought to be protein.

2. DNA is composed of four different deoxyribonucleotide subunits, generally referred to by the names of the bases attached to them: thymine (T), cytosine (C), adenine (A), and guanine (G). The nucleotides T and C consist of a single six-membered ring in the nitrogenous base and are both referred to as pyrimidines. The other two nucleotides, A and G, each contain a nitrogenous base composed of two rings and are referred to as purines.

3. The nucleotide subunits of DNA are attached to each other with phosphodiester bonds linking the 3' carbon in the sugar of one nucleotide to the 5' carbon in the sugar of the adjacent nucleotide. This 3'→5' linkage is repeated, forming a sugar-phosphate backbone. The DNA molecule is formed of two strands wound into a double helix, with the nitrogenous bases pointing toward the center of the helix. The sugar-phosphate backbone of one strand is antiparallel to the other, meaning that the 5' carbons of one strand face in the opposite direction of the 5' carbons in the other strand. The two strands are held together by hydrogen bonding between specific base pairs: thymine will pair only with adenine, and cytosine will pair only with guanine.

4. DNA may exist in several forms. The most common form of DNA in living cells is a right-handed helix called the B form (or B-DNA). A-DNA apparently does not exist in living cells, but double-stranded DNA-RNA hybrids

and double-stranded RNA assume a conformation similar to A-DNA in living cells. Z-DNA consists of a helix wound in a left-handed conformation where the sugar-phosphate backbones of Z-DNA form a zigzag pattern instead of a smooth helix. Z-DNA may be present in certain locations within DNA molecules in the cell. H-DNA occurs when a short region of a double-helical molecule rearranges to form a triple helix.

5. DNA replicates in a semiconservative fashion, meaning that the two original strands are conserved, each serving as a template for the formation of a new strand. The two daughter molecules will each contain an original parental strand and one newly synthesized strand.

6. DNA is replicated by a group of enzymes collectively referred to as a replisome. The helix is unwound by enzymes called helicases, and the tension built up by unwinding is relieved by transient single strand breaks in the molecule that allow it to rotate ahead of the replication fork, a process catalyzed by topoisomerases. The resulting single-stranded segments of DNA are stabilized by single-stranded binding proteins (SSBs) that prevent the strands from reattaching. Nucleotides may then be added to the single-stranded DNA by enzymes referred to collectively as DNA polymerases.

7. All known DNA polymerases must add nucleotides to the 3′ end of a nucleotide already present and therefore cannot initiate replication. Initiation of replication requires an RNA primer, a short segment of RNA synthesized on the DNA template strand, from which DNA polymerases can extend the chain with DNA. The primer is later removed and the space it occupied filled with the appropriate deoxyribonucleotides.

8. DNA replication is semidiscontinuous. Since the DNA strands are antiparallel and DNA may be synthesized only in the 5′→3′ direction, only one strand may be synthesized continuously in the direction in which the replication fork is moving. That strand is referred to as the leading strand. The other strand, the lagging strand, must be synthesized discontinuously as a series of fragments synthesized in the direction opposite to that in which the replication fork is moving. These fragments, known as Okazaki fragments, each require a separate initiation event.

9. Errors in DNA replication are corrected by a proof-reading mechanism. About once every 1000 nucleotides, an incorrect nucleotide is added. Prokaryotic DNA polymerases have a 3′→5′ exonuclease activity that removes the incorrect nucleotide; the polymerase then continues DNA synthesis, adding the correct nucleotide where it removed the incorrect one.

10. DNA replication is bidirectional in most cases. This means that replication forks proceed from an initiation site in both directions along the DNA strand. Circular

prokaryotic DNAs generally have a single initiation site from which bidirectional replication proceeds around the circle until the two replication forks meet. Large eukaryotic DNAs are linear and typically have many initiation sites along the molecule from which bidirectional replication proceeds. Initiation sites appear to be fixed sites in the DNA that are rich in A-T base pairs. Small linear molecules can be replicated by strand displacement. Large linear molecules require telomerase to finish synthesis at the ends of the molecule.

QUESTIONS AND PROBLEMS

1. Even though DNA was discovered in the nineteenth century, why was it not recognized as the hereditary material until the 1940s–50s?

2. Miescher's formula for DNA was $C_{29}H_{49}N_9P_3O_{22}$. **(a)** How many nucleotides are represented in Miescher's formula? **(b)** If each of the four nucleotides is represented equally in DNA, what would be the chemical formula for four nucleotides? **(c)** How closely does this answer correspond proportionally with Miescher's formula? **(d)** Why is it not possible to provide an exact chemical formula for DNA?

3. Although Griffith's experiments did not demonstrate that DNA is the transforming principle, they did indicate that the transforming principle is inherited. Using his experimental results as a model, demonstrate why we can conclude that the transforming principle is inherited.

4. Why is the prefix *deoxy* used in the name *deoxyribonucleic acid*?

5. Since base pairing in DNA consists of purine-pyrimidine pairs, why can't A-C and G-T pairs form?

6. At one point Watson proposed a model in which like bases paired with like, that is, C-C, T-T, G-G, and A-A. Describe why this model fails to explain Chargaff's data. Since X-ray diffraction data indicate that the DNA helix is of uniform width, what additional problem does this model have?

7. According to Chargaff (1950, *Experientia* 6:201–209), the proportion of adenine in human DNA is about 30%. What are the proportions of the other four nucleotides?

8. Suppose the nucleotide composition of a DNA virus was found to be 30% G, 20% C, 20% A, and 30% T. What is the most reasonable explanation of these results?

9. Assume that the DNA molecule in a particular chromosome is 50 mm long (about the average length of a DNA molecule in a human chromosome). How may nucleotide pairs are in this DNA molecule?

10. In the diagram of DNA replication below, which points represent 3′ ends of molecules and which

represent 5′ ends? Notice that one DNA strand is being synthesized continuously and the other discontinuously.

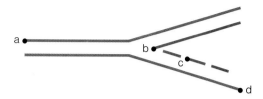

11. Temperature-sensitive strains of *E. coli* that lack topoisomerase activity at higher temperatures but retain it at lower temperatures have been found. What would be the results in terms of DNA replication and survivability if these strains were placed at the temperature that eliminated topoisomerase activity?

12. As described in Example 2.4, temperature-sensitive strains of *E. coli* that lack ligase activity at higher temperatures but retain it at lower temperatures have been found. What would be the results in terms of DNA replication and survivability if these strains were placed at the temperature that eliminated ligase activity?

13. If you had a mixture of single-stranded DNA fragments, all four deoxyribonucleotides (in the triphosphate form), a DNA polymerase, and the appropriate buffers and salts for DNA replication, what additional component would be necessary in a test tube for the DNA polymerase to synthesize new strands using the single-stranded DNA fragments as templates?

14. Diagram how the results of Meselson and Stahl's experiments would have appeared had DNA replication been conservative instead of semiconservative.

15. *E. coli* cells that lack DNA polymerase II activity survive and reproduce but have a higher rate of mutation. Why is this so?

16. (a) Would it be possible for a circular DNA molecule with a single origin of replication to fully replicate unidirectionally? Why or why not? **(b)** Would it be possible for a linear DNA molecule with a single origin of replication located near the center of the molecule to fully replicate unidirectionally? Why or why not?

17. What is the significance of the observation that origins of replication tend to be rich in A-T pairs?

18. Would Okazaki fragments occur during replication of a circular DNA molecule that replicated unidirectionally? Why or why not?

19. Why is it essential for linear eukaryotic DNA molecules to have multiple origins of replication?

20. No known DNA polymerase is capable of synthesizing DNA in the 3′→5′ direction. What factor causes problems with 3′→5′ synthesis?

21. (a) What structural form of double-helical DNA is the most common in living cells? **(b)** What is the direction of the helix (right- or left-handed)? **(c)** At what angle are the bases relative to the vertical axis of the molecule?

22. Which forms of double-helical DNA, other than the one in the previous question, may possibly occur in living cells?

23. Below is a diagram of a DNA molecule where a portion of the molecule has been disassociated for DNA replication.

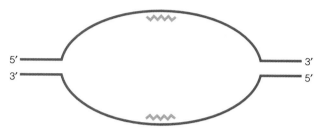

Assume that DNA replication will proceed bidirectionally from the origin of replication. Using zigzag lines to represent RNA and straight lines to represent DNA, draw the newly synthesized strands that would appear if the entire single-stranded templates were paired with newly synthesized DNA. Draw three Okazaki fragments in each strand that replicates discontinuously. Also, in your drawing, assume that none of the primers have been removed and that none of the nicks have been sealed. On the end of each DNA strand, place an arrowhead to show in which direction synthesis has proceeded. For example, a properly drawn DNA strand connected to an RNA strand and paired to a template strand of DNA would look like this:

The first two primers at the origin of replication have already been added for you.

24. In the conclusion of their classic 1953 paper (*Nature* 171:737–738), Watson and Crick stated: "It has not escaped our notice that the specific pairing we have postulated immediately suggests a possible copying mechanism for the genetic material." What aspects of the B form of double-helical DNA suggest a mechanism for replication?

25. Look at the model of rolling circle (σ-mode) replication in Figure 2.30. Is rolling circle replication **(a)** conservative, dispersive, or semiconservative, **(b)** unidirectional or bidirectional, and **(c)** continuous, discontinuous, or semidiscontinuous?

26. Rolling circle replication begins when the 5′ end of one strand dissociates from the circle and continuous replication displaces it. Why must the 5′ end rather than the 3′ end dissociate from the circle?

27. In humans, why is it essential for telomerase to be active in dividing germ-line cells but not other cells?

FOR FURTHER READING

Perhaps the most popular book about DNA is **Watson, J. D. 1968.** *The Double Helix.* **New York: Atheneum**. This book is a readable account from Watson's point of view of the years leading up to the discovery of the structure of DNA. Crick published his own view in **Crick, F. 1988.** *What Mad Pursuit: A Personal View of Scientific Discovery.* **New York: Basic Books**. A biography of Rosalind Franklin tells of her contribution: **Sayre, Anne. 1975.** *Rosalind Franklin and DNA.* **New York: Norton**. Two of the best overall accounts of the history of DNA are **Olby. R. 1974.** *The Path to the Double Helix.* **Seattle: University of Washington Press**; and **Judson, H. F. 1979.** *The Eighth Day of Creation.* **New York: Simon & Schuster**. The most complete review of DNA replication is a full-length book on the subject: **Kornberg, A., and T. A. Baker. 1992.** *DNA Replication, 2nd ed.* **New York: Freeman**. There are also several good review articles on DNA replication, including **Campbell, J. L. 1986. Eukaryotic DNA replication.** *Annual Review of Biochemistry* 55:733–771; **Kornberg, A. 1988. DNA replication.** *J. Biol. Chem.* 263:1–4; **Radman, M., and R. Wagner. 1988. The high fidelity of DNA replication.** *Scientific American* 259 (Aug):40–46; and **Greider, C. W., and E. H. Blackburn. 1996. Telomeres, telomerase, and cancer.** *Scientific American* 267 (Feb):92–97. Griffith's paper on the transformation of R to S bacteria is **Griffith, F. 1928. The significance of pneumococcal types.** *Journal of Hygiene* 27:141–144. Avery, MacLeod, and McCarty's paper identifying DNA as the transforming principle is **Avery, O. T., C. M. MacLeod, and M. McCarty. 1944. Studies on the chemical nature of the substance inducing transformation of the pneumococcal types.** *Journal of Experimental Medicine* 79:137–158. Hershey and Chase's paper describing their experiments that showed DNA to be the inherited material in bacteriophage is **Hershey, A. D., and M. Chase. 1952. Independent functions of viral protein and nucleic acid in growth of bacteriophage.** *Journal of General Physiology* 36:39–56. Chargaff published several papers dealing with the nucleotide composition of DNA. Two of the most important are **Chargaff, E. 1950. Chemical specificity of nucleic acids and mechanism for their enzymatic degradation. Experientia 6:201–209;** and **Chargaff, E. 1951. Structure and function of nucleic acids as cell constituents.** *Federal Proceedings* 10:654–659. Franklin's X-ray diffraction patterns and their interpretation can be found in **Franklin, R. E., and R. G. Gosling. 1953. Molecular configuration in sodium thymonucleate.** *Nature* 171:740–741. Watson and Crick's original paper on the structure of DNA is **Watson, J. D., and F. C. Crick. 1953. A structure for deoxyribose nucleic acid.** *Nature* 171:737–738. Watson and Crick published a second paper on the genetic implications of DNA a month later: **Watson, J. D., and F. C. Crick. 1953. Genetical implications of the structure of deoxyribonucleic acid.** *Nature* 171:964–967. Meselson and Stahl's classic paper demonstrating semiconservative DNA replication is **Meselson, M., and F. W. Stahl. 1958. The replication of DNA in** *Escherichia coli. Proceedings of the National Academy of Sciences, USA* 44:671–682.

For additional reading, go to InfoTrac College Edition, your online research library at: http: www.infotrac-college.com/brookscole

CHAPTER 3

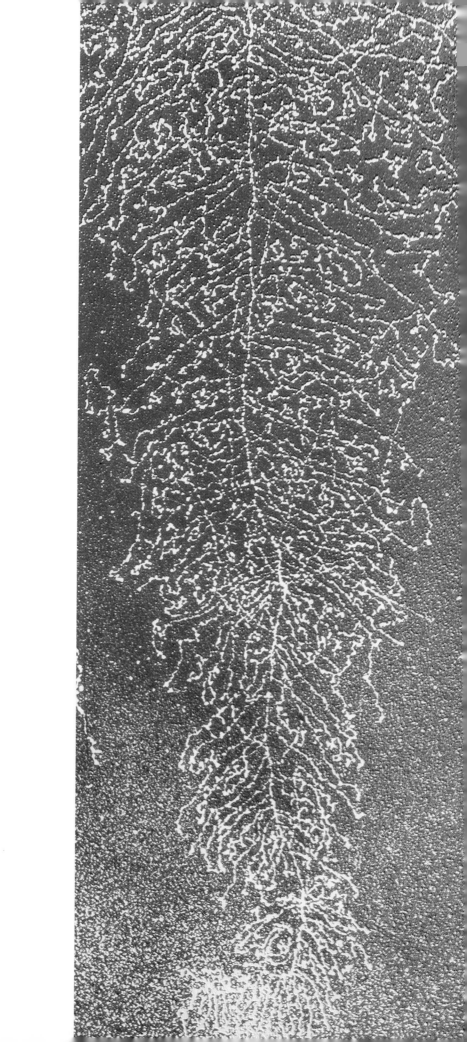

KEY CONCEPTS

A gene is a segment of DNA that is transcribed into an RNA molecule.

~

The central dogma of molecular genetics is that the information in a gene is transcribed into RNA, and the information in RNA is then translated into protein.

~

Transcription proceeds through three steps: initiation, elongation, and termination.

~

To become functional, most RNA molecules must be modified after transcription.

TRANSCRIPTION AND RNA PROCESSING

The concept of the **gene** is central to genetics. The word *gene* was coined in 1909 by Wilhelm Johannsen to describe "the fundamental unit of inheritance," and this remains a useful and valid definition. The establishment of DNA as the genetic material led to much more complete understanding of what a gene is and does, allowing us to define a gene at the molecular level as "a segment of DNA that encodes an RNA molecule." Our current concept of the gene extends from the nucleotide sequence of a gene in DNA to the inheritance of genes in populations (see section 1.5). In this chapter, we look at the gene in detail at the molecular level, discussing how cells transcribe the information in DNA and then process the RNA that is the product of this transcription. Before we do this, however, let's set the stage for transcription by briefly examining gene function at the molecular level.

3.1 THE CENTRAL DOGMA

As we discussed in the previous chapter, proteins were long thought to be the genetic material. Although proteins are not the genetic material, they are very much a part of the expression of genetic information encoded in DNA. The genes in DNA carry information that is expressed in proteins.

Both DNA and proteins are molecules composed of linear chains of subunits. The subunits of DNA are nucleotides, and the subunits of proteins are amino acids. Once it became clear that DNA is the inherited material and that genes consist of DNA, it soon became apparent that the linear sequence of nucleotides in the DNA molecule must be related to the linear sequence of amino acids in proteins. Because DNA resides in the cell nucleus and proteins are synthesized outside of the nucleus, it seemed that there must be an intermediate substance that transferred the information from DNA to protein. Recalling the events during the winter of 1952, James Watson reminisced:

> Virtually all the evidence then available made me believe that DNA was the template upon which RNA chains were made. In turn, RNA chains were the likely candidates for the templates for protein synthesis.... The idea of the genes' being immortal smelled right, and so on the wall above my desk I taped up a paper sheet saying DNA → RNA → protein. The arrows did not signify chemical transformations, but instead expressed the transfer of

genetic information from the sequences of nucleotides in DNA molecules to the sequences of amino acids in proteins.

By the late 1950s, the idea of an RNA intermediate between DNA and protein was well accepted. This understanding led to what is now called the **central dogma** of molecular genetics, diagrammed in Figure 3.1: *A gene composed of DNA is transcribed into RNA, which is then translated into a polypeptide, which is then processed to become a protein.*

The statement of the central dogma introduces two important processes in gene function, **transcription** and **translation**. These two terms can perhaps best be understood by thinking about their definitions outside of biology. When words are *transcribed*, they are copied verbatim. For instance, you transcribe a speech by writing down every word the speaker says in order. The only thing that has changed is the medium through which the words are communicated. In a sense, gene transcription follows the same pattern. The information of a gene contained in a DNA molecule is copied "verbatim" into an RNA molecule, which, like DNA, is composed of nucleotides. The information in the RNA molecule is essentially identical to the information in the DNA; only the medium that carries the information has changed.

When words are *translated*, on the other hand, verbal information is converted from one language to another. Whereas the meaning remains the same, the words are different. Translation of a gene follows this pattern as well. DNA and RNA are composed of nucleotides. In translation, the information that has been transcribed into an RNA molecule is "converted" from the language of nucleotides to the language of amino acids to form a **polypeptide**, a linear chain of amino acids. After translation is complete, the linear polypeptide undergoes further processing and eventually assumes a three-dimensional structure in which the linear chain folds back upon itself in a particular conformation. Once the polypeptide has assumed its three-dimensional structure, it is called a **protein**.

Many genes encode proteins that are enzymes responsible for controlling which chemical reactions take place in a cell and the rate of those reactions. Some genes encode other types of proteins, such as hormones, structural proteins, proteins that carry molecules in the bloodstream, storage proteins, and antibodies. Other genes encode RNAs that are not translated into proteins, such as ribosomal RNAs, which are structural components of the ribosome, and transfer RNAs, which participate in translation. Following a detailed examination of translation in Chapter 4, we discuss how these various gene products operate in the cell. For now, we focus on transcription and RNA processing, the first part of the function of genes as expressed in the central dogma.

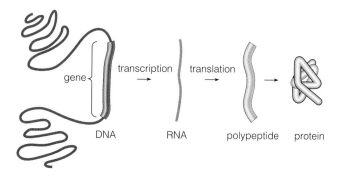

Figure 3.1 The central dogma of molecular genetics.

3.2 OVERVIEW OF TRANSCRIPTION AND RNA PROCESSING IN PROKARYOTES AND EUKARYOTES

Transcription in prokaryotes and eukaryotes is similar in most respects, but there are some important differences. Let's take a brief overview of some of the more important aspects, focusing on similarities and differences between prokaryotes and eukaryotes, before we embark on a detailed discussion of transcription.

Transcription in both prokaryotes and eukaryotes is similar to DNA replication. RNA synthesis is accomplished by an RNA polymerase that synthesizes a complementary RNA molecule using a DNA strand as a template, much as a DNA polymerase synthesizes a complementary DNA strand using a template DNA strand during DNA replication.

Transcription in both prokaryotes and eukaryotes proceeds through the steps of initiation, elongation, and termination. For **initiation**, the RNA polymerase binds to DNA at a specific site near the beginning of a gene. The DNA separates into single strands, and the RNA polymerase begins transcription using one of the DNA strands as a template. During **elongation**, the RNA polymerase extends the RNA chain, which rapidly separates from the DNA, allowing the DNA double helix to re-form once the RNA polymerase has passed. At **termination**, the RNA polymerase reaches the end of the gene, ceases transcribing, and leaves the DNA (Figure 3.2).

There are three major classes of RNA molecules. Transcription produces an RNA molecule that is the initial product of a gene. However, not all RNA molecules will be translated into proteins. There are three major classes of RNA molecules in both prokaryotes and eukaryotes: **messenger RNA (mRNA)**, **ribosomal RNA (rRNA)**, and **transfer RNA (tRNA)**. An mRNA contains information for translation into a protein. The sizes and nucleotide sequences of mRNA molecules are quite varied. An rRNA is a structural component of the **ribosome**, which is the site where the mRNA is translated into a protein. Each ribosome is composed of several different

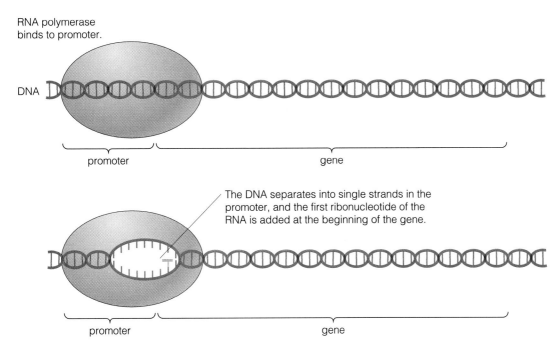

a Initiation

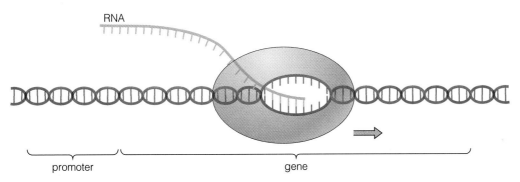

b Elongation

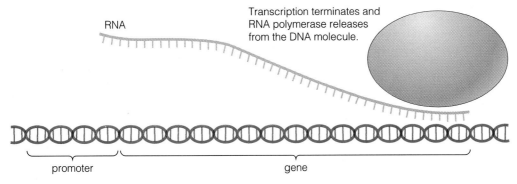

c Termination

Figure 3.2 The three stages of transcription.

rRNAs and proteins. A tRNA is very small, about 70–80 nucleotides long, and carries amino acids to the ribosome for incorporation into the protein.

There is a diverse array of other RNA molecules. Besides the three major classes of RNAs, several other types of RNA are present in cells. These additional RNAs are generally much less abundant than the major classes of RNAs, are rather small (100–300 nucleotides long), and serve a variety of functions. For example, as we saw in Chapter 2, the primers in DNA synthesis are made of RNA, and telomerase RNA is a part of the enzyme telomerase, in which it serves as a template for eukaryotic DNA synthesis. Small nuclear RNAs (snRNAs) are involved in eukaryotic mRNA processing (discussed later in this chapter). Another small RNA assists in directing secretory proteins to receptors in the cell membrane, and a few RNAs called ribozymes act as enzymes in that they catalyze biochemical reactions.

There are three major classes of RNA polymerases in eukaryotes, but only one in prokaryotes. In prokaryotes, a single RNA polymerase transcribes genes encoding mRNA, rRNA, and tRNA. The RNA polymerases of eukaryotes, however, are more diverse and structurally more complicated. They all transcribe DNA in the nucleus of the cell, although transcription of various genes is somewhat compartmentalized within the nucleus. There are three general classes: RNA polymerase I, RNA polymerase II, and RNA polymerase III. As summarized in Table 3.1, each of these polymerases is responsible for transcription of a particular class of RNAs. **RNA polymerase I** transcribes most genes that code for rRNA. These genes encode large rRNAs that in mammals may exceed 13,000 nucleotides in length. **RNA polymerase II** transcribes genes that code for mRNAs. **RNA polymerase III** transcribes very short genes (about 100 nucleotides in length), including those that code for tRNA and very small rRNAs.

Transcription and translation are coupled in prokaryotes. In eukaryotes, transcription and translation are compartmentalized. In prokaryotes, the mRNA is ready to be translated as soon as it is transcribed. In fact, translation begins before transcription has terminated, a process called **cou-**

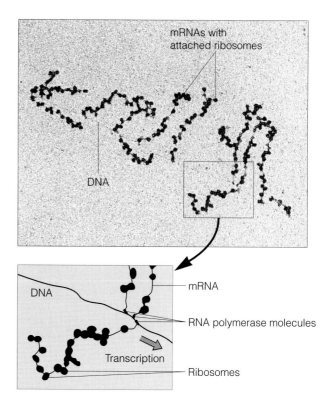

Figure 3.3 Coupled transcription and translation in a prokaryotic cell. Ribosomes begin translating the mRNA while RNA polymerase is still transcribing it. (Photo courtesy of O. L. Miller, Jr., B. A. Hamkalo, and C. A. Thomas, Jr.)

pled transcription and translation. As soon as the growing mRNA chain separates from DNA, ribosomes attach to it and begin translation on the 5′ end of the molecule, following right behind the RNA polymerase while it is still transcribing the mRNA (Figure 3.3).

In eukaryotes, transcription is compartmentalized rather than coupled. RNA molecules are transcribed in the nucleus. The RNA—whether mRNA, rRNA, or tRNA—is then exported from the nucleus into the cytoplasm for translation.

Messenger RNAs are processed in eukaryotes but not in prokaryotes. Prokaryotic mRNAs undergo no processing. With coupled transcription and translation, translation begins immediately after transcription has been initiated, so there is no opportunity to modify the mRNA prior to translation. Eukaryotic mRNAs, on the other hand, are substantially modified in the nucleus before they enter the cytoplasm. Both ends of the RNA are modified, and some internal segments of the RNA are removed to make a mature mRNA that is ready for translation.

rRNAs and tRNAs are processed in both prokaryotes and eukaryotes. Most rRNAs are transcribed as a single large precursor RNA that is then cleaved into its final products. Many of the individual nucleotides in tRNAs are chemically modified to produce the final tRNA.

Table 3.1	Major Eukaryotic RNA Polymerases and the RNAs They Transcribe
RNA Polymerase	**RNA Transcribed**
RNA polymerase I	Large rRNAs
RNA polymerase II	mRNAs
RNA polymerase III	tRNAs, small rRNAs, and other small RNAs

Having reviewed some of the fundamental aspects of transcription and RNA processing, let's now look at some of them in detail, beginning with initiation of transcription.

3.3 INITIATION OF TRANSCRIPTION

Initiation of transcription begins at a point in the double-stranded DNA molecule called the **promoter**, which marks the beginning of a gene (see Figure 3.2). A promoter consists of a specific region of DNA that directs the appropriate RNA polymerase to bind to the DNA and initiate transcription. Along with determining where in the DNA transcription will be initiated, the promoter also determines which of the two DNA strands will serve as the template for transcription. After RNA polymerase has bound to the promoter, the DNA molecule temporarily opens up, exposing a region of two single strands. RNA polymerase then initiates transcription by synthesizing a chain of ribonucleotides paired to one strand of the DNA molecule, following the same base-pairing rules for DNA replication, with one exception: whenever RNA polymerase encounters adenine (A) in the template DNA strand, it adds uracil (U) rather than thymine (T) to the RNA chain. The ribonucleotides pair with the deoxyribonucleotides, and for a short space, there is a portion of the molecule in which one strand consists of DNA and the other of RNA.

Although a gene is composed of a double-stranded DNA molecule, only one of the two strands serves as a template for transcription. The DNA strand that serves as the template has a nucleotide sequence that is complementary to the RNA. The other DNA strand does not serve as a template for transcription and contains essentially the same nucleotide sequence as the RNA (except that there is a U in the RNA at every place where there is a T in this strand of DNA).

The DNA strand that contains the same nucleotide sequence as the RNA is the **sense strand** or **nontemplate strand**. The DNA strand that serves as the template for transcription and contains the nucleotide sequence complementary to that of the RNA, is called the **antisense strand** or **template strand** (Figure 3.4). These designations indicate that the sense strand contains the "same sense" or same information as the RNA (even though the sense strand is not directly involved in transcription), and that the antisense strand carries the complementary information and, therefore, serves as the template for transcription. (You should note that in some textbooks the definitions of the sense and antisense strands are the reverse of those just given. However, thinking of the template strand as the antisense strand will make it easier for you to read research journals because this convention is common in the professional literature.)

The many genes located on the same DNA molecule do not all have the same strand as the sense strand. While one gene has a particular sense strand, an adjacent gene on the same DNA molecule may or may not use this strand as its sense strand (Figure 3.5).

Once the first ribonucleotide is in place, RNA polymerases add nucleotides to the 3' end of a previous

The RNA and the sense strand in the DNA have the same nucleotide sequence in the 5' → 3' direction (substituting U for T in the RNA).

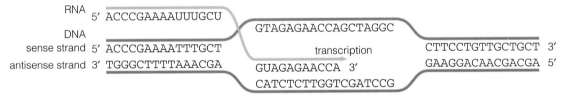

The antisense strand serves as the template for transcription.

Figure 3.4 The two strands of a gene.

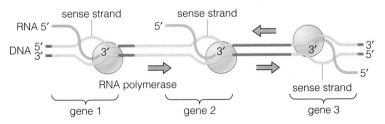

Figure 3.5 Sense and antisense strands and direction of transcription for three adjacent genes. The arrows show the direction of transcription of each gene.

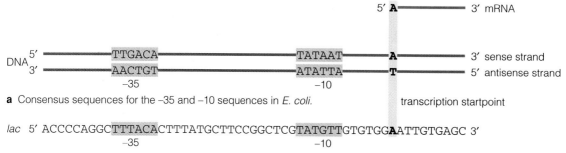

a Consensus sequences for the −35 and −10 sequences in *E. coli*.

lac 5′ ACCCCAGGCTTTACACTTTATGCTTCCGGCTCGTATGTTGTGTGGAATTGTGAGC 3′
　　　　　　　　　　−35　　　　　　　　　　　　　　　−10

b DNA sequence of the *lac* promoter region in *E. coli*. Conserved sequences are boxed.

Figure 3.6 Conserved sequences in prokaryotic promoters.

nucleotide, just as DNA polymerases do. This means that transcription for any given gene proceeds in one direction only, 5′→3′, like a stream that always flows downhill. This leads us to two common terms that are often used when describing transcription. Sites on the DNA may be referred to as **upstream** or **downstream** of each other, downstream being in the direction of transcription and upstream being in the opposite direction. For example, the promoter is generally upstream from the transcribed region, and the point where transcription terminates is downstream from the promoter.

Transcription begins at the **transcription startpoint**, which is the site on the DNA where RNA polymerase adds the first ribonucleotide. This is an important landmark for denoting the positions of other nucleotides. The transcription startpoint is designated zero. Nucleotides upstream of the startpoint are assigned negative numbers, whereas those downstream have positive numbers. For example, a nucleotide numbered −35 is located 35 nucleotides upstream from the startpoint.

Conserved Sequences in Promoters

Promoters are the sites where RNA polymerase binds to the DNA to initiate transcription; they are typically located upstream from the transcription startpoint. However, not all of the DNA sequence information within the promoter region is essential for RNA polymerase binding and initiation of transcription. Researchers have determined which portions of the promoter region are essential by comparing the DNA sequences of a large number of promoters both within and among species and identifying **conserved sequences**, regions of DNA that are identical or similar among the promoters studied. For example, a nucleotide found at a certain position in 90% of the promoters examined is said to be highly conserved. That nucleotide is probably an important part of the promoter sequence. On the other hand, a nucleotide found at a certain position in only 25–30% of the promoters is not highly conserved and might simply be there at random.

When researchers find a conserved sequence in many different genes, the researchers derive a **consensus sequence** to identify the conserved sequence. A consensus sequence is the most representative version of a conserved sequence in many genes. The consensus sequences of several conserved sequences in both prokaryotic and eukaryotic promoters have been identified.

David Pribnow was the first to identify conserved sequences in *E. coli* promoters. When he compared the DNA sequences of promoter regions in six genes, he found a conserved sequence located approximately 10 nucleotides upstream from the transcription startpoint in all six promoters. The high level of nucleotide similarity implied that this region is probably important in promoter function. Later studies confirmed the presence of this conserved sequence in the promoters of many *E. coli* genes.

~

Transcription is initiated at a promoter, which is a segment of DNA containing sequence elements where RNA polymerase binds and begins transcription.

Conserved Sequences in Prokaryotic Promoters

Researchers usually identify conserved promoter sequences by their position relative to the transcription startpoint. The transcription startpoint can be identified by matching the 5′ end of the mRNA sequence with the DNA sequence. The transcription startpoint is an A in most *E. coli* genes. Examination of a number of promoters in *E. coli* has revealed two important conserved sequences, as illustrated in Figure 3.6. The center of a six-nucleotide conserved sequence is located about 10 nucleotide pairs upstream from the transcription startpoint, although the actual distance may vary from 9 to 18 nucleotide pairs. Because of its position, the sequence is referred to as the **−10 sequence**, or sometimes the **Pribnow box** in honor of its discoverer. The consensus sequence is TATAAT.

The other conserved sequence in prokaryotic promoters is located about 35 nucleotides upstream of the

transcription startpoint and is called the **−35 sequence**. Its consensus sequence is TTGACA. A consensus sequence does not mean that every promoter has precisely these nucleotide sequences at the −10 or −35 sites, but that these are the most frequent nucleotides at these sites when all the known promoter sequences in *E. coli* are compared.

Mutations that change nucleotides of conserved sequences typically affect the efficiency of transcription. In general, mutations that cause a conserved sequence to deviate from its consensus sequence reduce or eliminate transcription, whereas mutations that increase the similarity of a conserved sequence to the consensus sequence generally increase transcription. The nucleotides found between the startpoint and the −10 sequence, and those between the −10 sequence and the −35 sequence, are sometimes called **spacer regions**. Spacer regions are important in maintaining the appropriate positions of the −10 and −35 sequences, but apparently it does not matter which nucleotide sequences are actually present in the spacer regions.

The −10 sequence and the −35 sequence in *E. coli* genes each have specific roles in transcription. The −35 sequence is necessary for the initial binding of RNA polymerase to the promoter. This is illustrated by experiments in which the −35 sequence is removed by DNA-degrading enzymes after RNA polymerase has bound to the DNA. In this case, the bound RNA polymerase synthesizes an mRNA, but no additional RNA polymerase molecules can bind to the promoter.

The −10 sequence is the site where the two DNA strands separate in preparation for initiation of tran-

scription. Although this has been demonstrated experimentally, the −10 consensus sequence itself provides a clue as to its function. Its consensus sequence consists entirely of A-T pairs. Recall that A-T pairs are held together by two hydrogen bonds, whereas G-C pairs are held together by three. Therefore, it requires less energy to separate an A-T pair than a G-C pair; and in order to separate the two strands of DNA, separating a continuous sequence of A-T pairs requires the least energy of any double-stranded DNA segment.

The following example shows how DNA sequences from different genes may be compared and analyzed. DNA sequence analysis is currently a very active area in genetics research. It is often complicated, involving thousands of nucleotides. For this reason, DNA sequence analysis is usually conducted with the aid of a computer and specially designed software. In this example, we will examine Pribnow's data to see how he identified the −10 sequence in *E. coli* promoters.

Example 3.1 DNA sequence comparison and analysis.

In 1975, Pribnow (*Proceedings of the National Academy of Sciences, USA* 72:784–788) compared the DNA sequences in the promoter regions of six *E. coli* genes to determine which nucleotides among the promoters were similar. The sense-strand DNA sequences of the six promoter regions compared by Pribnow and the 5′ ends of the mRNA sequences for each of these genes are provided in Table 3.2.

Table 3.2 Information for Example 3.1: DNA Sequences and 5′ mRNA Sequences for Five *E. coli* Genes

Gene	Promoter DNA Sequence
T7	AAGTAAACACGGTACGATGTACCACATGAAACGACAGTGAGTCA
fd	TGCTTCTGACTATAATAGACAGGGTAAAGACCTGATTTTTGA
SV40	TTTATTGCAGCTTATAATGGTTACAAATAAAGCAATAGCA
λ P$_L$	CCACTGGCGGTGATACTGAGCACATCAGCAGGACGCACTGAC
Tyr tRNA	CGTCATTTGATATGATGCGCCCCGCTTCCCGATAAGGGAGCA
Lac w.t.	TTCCGGCTCGTATGTTGTGTGGAATTGTGAGCGGATAACAA

Gene	5′ mRNA Sequence
T7	AUGAAACGACAGUGAGUCA . . .
fd	GUAAAGACCUGAUUUUUGA . . .
SV40	AAAUAAAGCAAUAGCA . . .
λ P$_L$	AUCAGCAGGACGCACUGAC . . .
Tyr tRNA	CGCUUCCCGAUAAGGGAGCA . . .
Lac w.t.	AAUUGUGAGCGGAUAACAA . . .

Table 3.3 Solution to Example 3.1: Aligned Promoter DNA Sequences

Gene	Aligned Promoter DNA Sequences
	−10 sequence
T7	AAGTAAACACGG TACGAT GTACCAC↓ATGAAACGACAGTGAGTCA
fd	TGCTTCTGAC TATAAT AGACAGG↓GTAAAGACCTGATTTTTGA
SV40	TTTATTGCAGCT TATAAT GGTTAC↓AAATAAAGCAATAGCA
λ P_L	CCACTGGCGGT GATACT GAGCAC↓ATCAGCAGGACGCACTGAC
Tyr tRNA	CGTCATTTGA TATGAT GCGCCCC↓GCTTCCCGATAAGGGAGCA
Lac w.t.	CTTCCGGCTCG TATGTT GTGTGG↓AATTGTGAGCGGATAACAA

Arrows point to the transcription startpoints.

Problem: (a) Identify the transcription startpoints and align the DNA sequences so that the −10 sequences coincide in each of the promoters. Identify the −10 sequences by boxing in the sequences, and identify the transcription startpoints with arrows. (b) What evidence is there in the promoter sequences that the −10 sequences represent a conserved sequence?

Solution: (a) By matching the mRNA sequence with the DNA sequence of its corresponding sense strand, it is possible to identify the transcription startpoint (which corresponds to the first nucleotide on the 5′ end of the mRNA) in each of the genes and begin to align the DNA sequences with one another by their transcription startpoints. Notice that four of the six mRNAs begin with an A (as predicted by the consensus sequence) and two do not. The *center* of the −10 sequence in each of the genes should lie about 10 nucleotides upstream from the transcription startpoint. However, the −10 sequence is not exactly 10 nucleotides upstream in every case, so some minor adjustments may be necessary to make the final alignment. Matching the nucleotide sequence of the promoter sequence makes it possible to adjust the alignment so that the −10 sequences are aligned. The best alignment for all six genes are shown in Table 3.3. (b) A consensus sequence is the most representative version of a similar sequence in DNA. The −10 sequences in this example above are similar but not identical. The exact consensus sequence TATAAT is present in two of the promoters, whereas each of the other promoters has a −10 sequence that differs from the consensus sequence by one or two nucleotides. Consensus sequences can be conclusively identified only after a large number of DNA sequences have been examined. In this example, Pribnow proposed the −10 consensus sequence as TATPuATG (where Pu denotes "purine"). This consensus sequence was later refined to TATAAT as additional promoter DNA sequences were examined. Notice that the transcription startpoints do not align exactly in each of the genes, indicating that the center of the −10 sequence is not always exactly 10 nucleotides upstream from the startpoint.

Prokaryotic RNA Polymerase

In prokaryotes, of which *E. coli* is the best studied species, a single type of RNA polymerase transcribes every gene, including genes that code for rRNAs, tRNAs, and mRNAs. This polymerase has several polypeptide subunits, and the whole enzyme is called the **holoenzyme**. There is a **core enzyme** made up of four polypeptides: two copies of the **α subunit** and one copy each of the **β** and **β′ subunits**. The holoenzyme consists of this core enzyme combined with a fifth polypeptide known as the **sigma (σ) factor** (Figure 3.7a).

The RNA polymerase holoenzyme binds to the promoter and initiates transcription. Once bound to the promoter, the polymerase usually makes several abortive attempts at transcription, often synthesizing a very short RNA molecule of less than 10 nucleotides before ceasing transcription and starting over. Once transcription has succeeded beyond about 10 nucleotides, the sigma factor is released, and the core enzyme proceeds through the elongation phase (Figure 3.7b).

Conserved Sequences in Eukaryotic Promoters

Eukaryotic promoters also contain conserved sequences, often called **boxes,** that have important roles in transcription. Three conserved sequences in eukaryotic promoters and two of their corresponding consensus sequences are diagrammed in Figure 3.8.

In eukaryotic genes that encode mRNA, the sense-strand transcription startpoint is typically the central A within a short conserved sequence of three nucleotides,

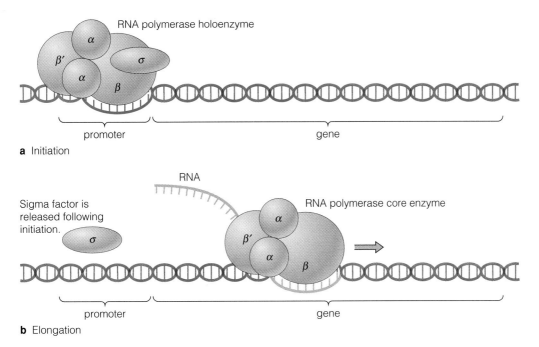

a Initiation

Sigma factor is released following initiation.

RNA polymerase core enzyme

b Elongation

Figure 3.7 Initiation of transcription in prokaryotes.

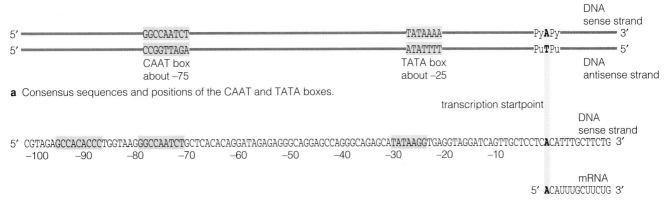

a Consensus sequences and positions of the CAAT and TATA boxes.

5′ CGTAGAGCCACACCCTGGTAAGGGCCAATCTGCTCACACAGGATAGAGAGGGCAGGAGCCAGGGCAGAGCATATAAGGTGAGGTAGGATCAGTTGCTCCTCACATTTGCTTCTG 3′

−100 −90 −80 −70 −60 −50 −40 −30 −20 −10

mRNA
5′ **A**CAUUUGCUUCUG 3′

b DNA sequence of the mouse β-major globin gene promoter region. Conserved sequences are shaded in blue.

Figure 3.8 Conserved sequences in eukaryotic promoters.

PyAPy (where Py refers to either of the two pyrimidines, C or T). Because transcription begins at the A in this sequence, the first nucleotide on the 5′ end of the mRNA is usually an A followed by a pyrimidine.

Most eukaryotic mRNA genes have a conserved DNA sequence at about −25 to −30 called the **TATA box** or **Hogness box**, which has the consensus sequence TATAAAA. The **CAAT box**, with the consensus sequence GGCCAATCT, is present in many promoters at about −75 to −80, although this position and sequence may vary significantly. Upstream of the CAAT box, most eukaryotic promoters have additional conserved sequences. Two of the most common are the **GC box** (GGGCGG) and the **CACCC box** (GCCACACCC). Although these sequences appear to identify the most common features of promoters for eukaryotic genes that encode mRNAs, some mRNA genes lack one or more of these sequences, so we cannot say that they are essential parts of every promoter.

The experimental work described in the following example shows how mutations in conserved sequences can influence transcription.

Example 3.2 The influence of conserved sequences on transcription in a eukaryotic gene.

In 1986, Myers, Tilly, and Maniatis (*Science* 232:613–618) used methods that permitted them to test the effects of many different mutations in the promoter region of the mouse β-major globin gene. β-globin is one of two different protein subunits in

hemoglobin, an abundant substance in blood that carries oxygen and gives blood its red color. The mouse gene had been introduced into the DNA of bacterial cells, where the bacteria replicated the mouse gene as if it were bacterial DNA. This permitted the researchers to obtain as many copies of the gene as they needed simply by growing the bacterial cultures. They isolated the gene from bacterial cells and treated the DNA chemically to induce mutations that randomly altered only one nucleotide in each copy of the DNA. Using this procedure, they recovered 130 different mutations in the promoter region. They introduced each mutant gene, along with a nonmutant version of the gene, into cultured human cells, where both the mutant and nonmutant versions of the gene were simultaneously transcribed. They then compared the transcription activity of the mutant gene and non-mutant genes for each of the 130 mutations. The results of these experiments are presented in Figure 3.9a.

Problem: (a) There are several places in the DNA sequence where mutations cause a relative transcription level (RTL) that is substantially different from 1.0. Identify these places in the DNA sequence. (b) There are three distinct regions where mutations affect transcription. What do these regions represent? (c) Most of the mutations outside of these three regions have a relative transcription level close to 1.0. What does this suggest about the effect of these mutations on transcription? (d) What do these results suggest about the importance of promoter consensus sequences?

Solution: (a) The regions in the DNA at which mutations altered transcription levels are shaded in blue in the DNA sequence in Figure 3.9b. (b) The three blue-shaded regions in Figure 3.9b represent the TATA box at -28, the CAAT box at -75, and a CACCC box at -91. Notice that most mutations in conserved sequences cause a decrease in transcription, but mutations at two positions in the CAAT box cause an increase. Outside of these three conserved sequences, there are only two other sites where mutations significantly alter transcription. One is the transcription startpoint. The other is a single nucleotide at position -37. (c) If the relative transcription level is 1.0, the mutation has no effect on transcription. There are several regions where mutations result in transcription levels very close to 1.0, suggesting that the nucleotide sequences in these regions are not important for transcription (although the *number* of nucleotides in these regions probably *is* important in providing appropriate spacing for conserved sequences). (d) The results of these experiments demonstrate that mutations within a conserved sequence alter the level of transcription, whereas most mutations outside of the conserved

sequences have no effect on transcription. The results of the experiments illustrated here show that conservation of promoter consensus sequences is needed to maintain the appropriate level of transcription.

Transcription Factors and the Basal Eukaryotic Transcription Complex

In eukaryotes, RNA polymerase does not recognize conserved sequences in promoters on its own and bind to them. Instead, proteins called **transcription factors** first bind to the conserved sequences and then assist the RNA polymerase in transcription initiation. Transcription factors are usually designated by the letters TF (for "transcription factor") followed by a roman numeral I, II, or III to indicate whether they bind RNA polymerase I, II, or III, respectively. This designation is then followed by a letter because there may be more than one transcription factor for a particular RNA polymerase. For example, TFIIB and TFIIE are different transcription factors for RNA polymerase II.

Several transcription factors are required for a eukaryotic RNA polymerase to initiate transcription. As an example, let's consider a model for the assembly of transcription factors and RNA polymerase II to form the **basal eukaryotic transcription complex**. According to this model, TFIID is the first transcription factor to bind with the promoter, and it does so at the TATA box. TFIID is a complex that consists of a protein called TBP (for TATA-binding protein) and any of several additional proteins depending on the promoter. TFIIA binds to TFIID either before or after TFIID binds to the DNA. When it binds before, TFIIA appears to accelerate the binding of TFIID to the DNA. A third transcription factor, TFIIB, then binds to the DNA-bound TFIID-TFIIA complex downstream of the TATA box and sets up a conformation that allows RNA polymerase II to bind to the complex and the DNA. Two other transcription factors, TFIIE and TFIIF, then bind to the complex. One of them, TFIIF, acquires energy by hydrolyzing ATP to ADP. TFIIH and TFIIJ also bind to the complex. TFIIH is a helicase that apparently assists in unwinding the DNA into single strands in order for transcription to begin. Once initiation has begun, RNA polymerase II separates from the transcription initiation complex and proceeds on its own through elongation. Figure 3.10 shows the basal eukaryotic transcription complex in place for the initiation of transcription.

~
Transcription factors help RNA polymerases bind to eukaryotic promoters and initiate transcription.

Enhancers

The promoters of most genes are located immediately upstream of the transcription startpoint. However, for many genes there are additional regions of DNA, usually up-

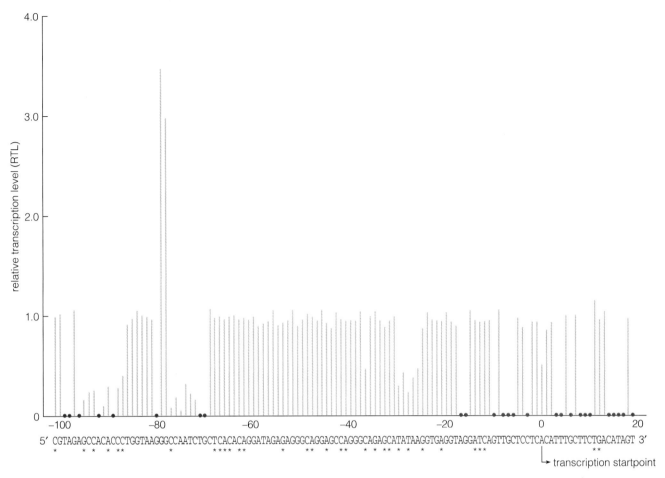

a The DNA sequence of the sense strand is on the *x* axis. The nucleotide positions are numbered to indicate their distance from the transcription startpoint. Solid circles (•) indicate that a mutation was not recovered for the corresponding nucleotide, while asterisks (∗) indicate that more than one mutation was recovered for the corresponding nucleotide. Vertical lines represents the relative transcription level (RTL) for mutations of the corresponding nucleotide. Relative transcription level (the *y* axis) is the level of transcription of the mutant sequence relative to the nonmutant sequence. A value of 1 indicates that transcription for the mutant gene was at the same level as that of the nonmutant gene. A value less than 1 indicates that the mutant gene was transcribed less frequently that the nonmutant gene, whereas a value greater that 1 indicates that the mutant gene was transcribed more frequently than the nonmutant gene.

5′ CGTAGAGCCACACCCTGGTAAGGGCCAATCTGCTCACACAGGATAGAGAGGGCAGGAGCCAGGGCAGAGCATATAAGGTGAGGTAGGATCAGTTGCTCCTCACATTTGCTTCTGACATAGT 3′

       CACCC box   CAAT box   −37   TATA box   transcription startpoint

b Solution to part a of Example 3.2. All nucleotides for which a mutation altered the relative transcription level are shaded in blue.

Figure 3.9 Information for Example 3.2: The effect of mutations in the promoter region on transcription for the mouse B-major globin gene. (Redrawn with permission from Myers, R. M., K. Tilly, and T. Maniatis. 1986. Fine structure genetic aanlysis of a beta-globin promoter. *Science* 232:613–618. Copyright 1986 American Association for the Advancement of Science.)

stream of the promoter, that influence transcription. These DNA regions are known as **enhancers**. Proteins must bind at enhancers to initiate transcription of some genes. Also, enhancers contain conserved sequences similar to those of promoters, and mutations in the conserved sequences usually reduce the rate of transcription. The locations of enhancers relative to their genes differ from those of promoters, however. An enhancer may be located as many as several thousand nucleotides from the transcribed region.

Most enhancers are located upstream from the promoter, but a few are located downstream from the transcribed region or even within it. Enhancers may be thought of as an extension of the promoter sequences dispersed throughout a large region of DNA.

Enhancers typically function as sites for binding proteins called **activators**. Other proteins called **coactivators** bind to the basal transcription complex. Then the activator and coactivator proteins bind to each other, causing

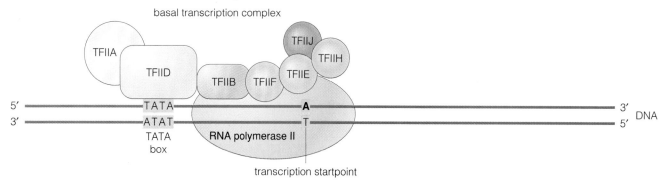

a The basal transcription complex positions RNA polymerase II for initiation of transcription.

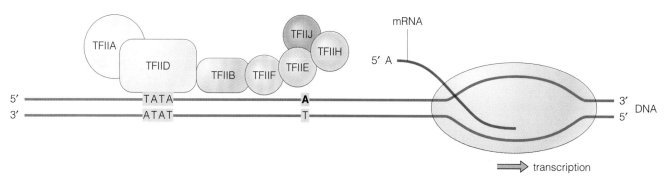

b Once transcription is initiated, RNA polymerase II separates from the basal transcription complex and proceeds to transcribe the gene.

Figure 3.10 The basal eukaryotic transcription complex and initiation of transcription in eukaryotes.

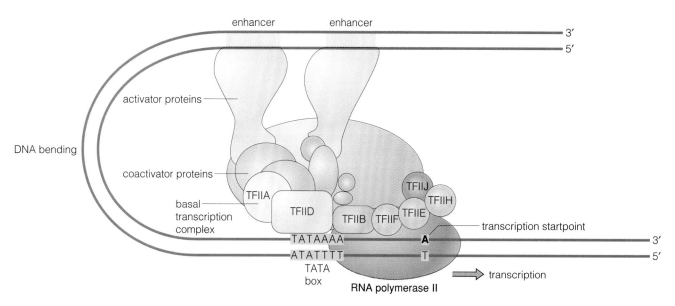

Figure 3.11 Interaction of enhancers, activators, and coactivators with the basal transcription complex for initiation of transcription in eukaryotes. The DNA bends, bringing the activator proteins into contact with coactivator proteins bound to the basal transcription complex. (Adapted from an original drawing by Jared Schneidman Design in Tijan, R. 1995. Molecular machines that control genes. *Scientific American* 272 (Feb 95):54–61. Reprinted by permission.)

the DNA to bend, as illustrated in Figure 3.11. It has been speculated that DNA bending somehow allows RNA polymerase to begin transcribing the gene.

~

Enhancers, located outside promoter regions and either outside or within transcribed regions, bind activators. The activators bend the DNA and bind to coactivators on the basal transcription complex. This, in turn, stimulates the basal transcription complex to initiate transcription.

3.4 ELONGATION

Elongation consists of the synthesis of a single-stranded RNA molecule using the antisense strand of the DNA as a template. As in DNA replication, transcription proceeds in the $5' \rightarrow 3'$ direction. In other words, after the first ribonucleotide pairs with its corresponding deoxyribonucleotide in DNA, each newly added ribonucleotide attaches to the 3'-OH group of the previous ribonucleotide. RNA is synthesized in one direction, using only one strand of DNA as a template, and the synthesis is continuous, like that of the leading strand in DNA synthesis. Because the RNA is single stranded, there is no need for discontinuous synthesis of the complementary strand as in DNA replication.

As in DNA replication, the energy to drive transcription comes from the nucleotides themselves. Free ribonucleotides are in the triphosphate form. As each ribonucleotide joins the growing RNA strand, two of the phosphates are cleaved from it, releasing energy (Figure 3.12).

The double-stranded RNA-DNA hybrid is very transient. During elongation, as RNA polymerase moves down the DNA strand, the newly synthesized RNA rapidly separates from the DNA, increasing the length of the growing single-stranded RNA molecule. At any given time during transcription, the number of nucleotides of RNA that remain paired with the DNA template may be as many as 12 or as few as 2.

Once the RNA polymerase has read the information in the DNA, the portion of the template DNA that is no longer paired with the RNA quickly pairs again with its complementary DNA strand. For this reason, there is no need to nick the DNA to relieve the tension of unwinding during transcription, because only a small length (about 18 nucleotide pairs) of the molecule is unwound at any time.

Elongation was once perceived to proceed at a constant rate and to slow only when approaching termination. Several studies have shown, however, that the rate of elongation is not constant. *E. coli* RNA polymerase apparently elongates the RNA in periodic bursts of RNA synthesis, each followed by a brief pause. Also, certain proteins affect the rate of elongation. Although the core enzyme of *E. coli* RNA polymerase is capable of elongation in the absence of any other proteins in vitro, other proteins present in the cell control the rate of elongation

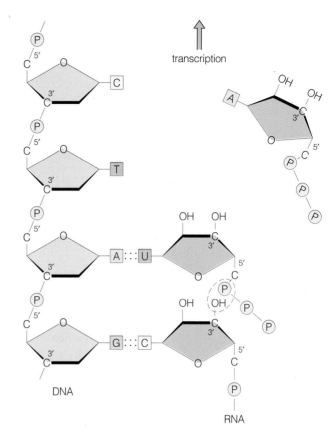

Figure 3.12 Energy for transcription. Cleavage of two phosphate groups from the ribonucleotide triphosphate releases energy.

in vivo. The best studied of these proteins is **NusA**, which binds to the core enzyme in place of the σ factor. NusA slows elongation when RNA polymerase encounters certain nucleotide sequences, ensuring that transcription proceeds at the proper rate. NusA apparently keeps the rate of transcription similar to the rate of translation so that the ribosomes are able to follow on the RNA molecule right behind the RNA polymerase.

Several proteins enhance elongation in eukaryotes. The best studied is a transcription factor called TFIIS that comes into play when RNA polymerase II stalls. Occasionally, RNA polymerase II enters into a configuration or encounters a nucleotide sequence that causes the polymerase to stall on the DNA, halting transcription with the RNA still attached. When this happens, TFIIS causes the RNA polymerase to move backward, and then TFIIS removes the 3' end of the RNA, permitting the RNA polymerase to attempt elongation again over the point where the stall occurred.

3.5 TERMINATION

Transcription terminates at the end of a gene. At this point, the completed RNA molecule is released from the DNA. Although there are some similarities between termination of transcription in prokaryotes and that in

eukaryotes, including the importance of conserved sequences, the two differ in several significant respects. For this reason, we will deal with them separately.

Prokaryotic Termination

In prokaryotes, two modes of termination, called intrinsic termination and rho-dependent termination, can be distinguished using in vitro experiments. Let's look first at intrinsic termination.

Intrinsic termination is a mechanism in which the nucleotide sequence near the end of the transcribed RNA specifies where transcription terminates. The nucleotide sequence at the transcription termination site is called an **intrinsic terminator**. Intrinsic terminators contain a conserved sequence with the consensus sequence UUUU-UUA in the RNA and a **hairpin structure** in the RNA just upstream from (and sometimes including part of) this UUUUUUA conserved sequence. The hairpin forms when complementary nucleotides in the RNA pair with each other to form a double-stranded segment. A prokaryotic RNA polymerase temporarily ceases transcription whenever a hairpin structure forms in the recently transcribed RNA behind it, but a hairpin structure alone does not terminate transcription. A hairpin followed by a string of U's, however, is sufficient for intrinsic termination (Figure 3.13). Possibly, as the polymerase pauses due to the hairpin, the RNA separates from the DNA at the UUUUUUA conserved sequence. Because this sequence is formed entirely of the weaker U-A and A-T nucleotide pairs (recall that these nucleotide pairs share only two hydrogen bonds), separation of the RNA from the DNA at this point requires a minimum of energy.

Let's now take a close look at the nucleotide sequence of an intrinsic terminator in *E. coli* to see how the hairpin structure can form.

Example 3.3 **Identification of an intrinsic terminator in *E. coli.***

In 1976, Squires et al. (*Journal of Molecular Biology* 103:351–381) reported the DNA sequence of the *trpA* intrinsic terminator in *E. coli*. A portion of the sense-strand sequence is

5′...ACCCAGCCCGCCTAATGAGCGGGCTTTTTTTTTTGAAC...3′

Problem: **(a)** Draw the most probable structure of the mRNA that includes this sequence of nucleotides when the hairpin structure has formed.
(b) Notice in the mRNA that the string of U's at the end of the sequence does not include the A of the consensus sequence. Given the function of the string of U's, why should the lack of an A not be problematic?

Solution: **(a)** For intrinsic termination, complementary bases in the RNA form a hairpin structure, and the hairpin is followed by a conserved sequence identical or similar to the UUUUUUA consensus sequence. The sense-strand sequence given may form a hairpin structure in the RNA followed by a string of U's, as illustrated below:

```
          A   A
        U       U
        G — G
        C — G A
        C — C
        C — G
        G — G
        C — G
        C — C
5′···ACCCA — UUUUUUUUU3′
```

Notice that the strands in the stem of the hairpin structure are not exactly complementary; an A must loop out in order for the nucleotide pairs to form.
(b) The absence of the A from the string of U's is of little consequence because this sequence allows the RNA to separate from the DNA with the lowest possible expenditure of energy, and substitution of a U for an A does not increase the energy requirement.

Rho-dependent termination takes its name from a protein known as rho. Rho, which is present at a constant level within the prokaryotic cell, binds to RNA, especially when it encounters nucleotide sequences with a high C and low G content. Rho apparently interacts with the mRNA and the RNA polymerase at a site called a **rho-dependent terminator**, causing the RNA polymerase to cease transcription and separate from the DNA.

A consensus sequence for rho-dependent terminators cannot be defined because the nucleotide sequences differ among the known terminators. In all genes with rho-dependent terminators, however, a high C and low G content upstream of the terminator signals rho to bind to the mRNA. Downstream from the rho binding site is a sequence that causes RNA polymerase to pause. Rho moves along the mRNA and catches up to the RNA polymerase at the point where RNA polymerase has paused. Rho then causes RNA polymerase to separate from the DNA, terminating transcription.

Within a single gene there may be more than one rho binding site followed by a site where RNA polymerase pauses. These sites are called **potential terminators** because they have nucleotide sequences that should cause rho-dependent termination but rarely do. If there are potential terminators in a gene, how does transcription terminate at the correct site? This question can be answered by recalling that transcription and translation are coupled in prokaryotes. As diagrammed in Figure 3.14, ribosomes follow closely behind RNA polymerase,

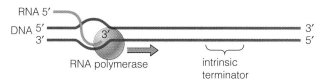

a An RNA molecule grows as RNA polymerase transcribes the DNA.

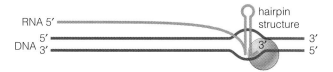

b A sequence of complementary nucleotides folds back on itself to form a hairpin structure, causing the RNA polymerase to pause.

Figure 3.13   Intrinsic termination of transcription in prokaryotes.

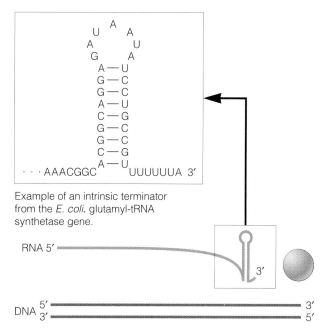

Example of an intrinsic terminator from the *E. coli.* glutamyl-tRNA synthetase gene.

c At a string of U's at the end of the hairpin, the RNA separates from the DNA, and the RNA polymerase detaches from the DNA and RNA.

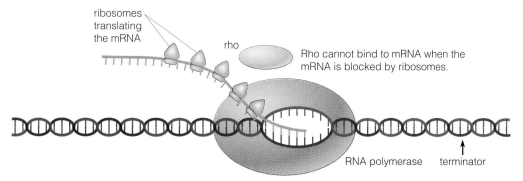

Rho cannot bind to mRNA when the mRNA is blocked by ribosomes.

a Transcription continues as long as ribosomes cover potential binding sites for rho on the mRNA.

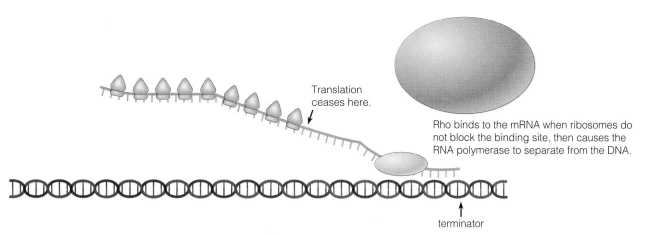

Translation ceases here.

Rho binds to the mRNA when ribosomes do not block the binding site, then causes the RNA polymerase to separate from the DNA.

b Transcription terminates beyond the termination site for translation because ribosomes do not cover a binding site for rho at the end of the mRNA.

Figure 3.14   Rho-dependent termination of transcription in prokaryotes.

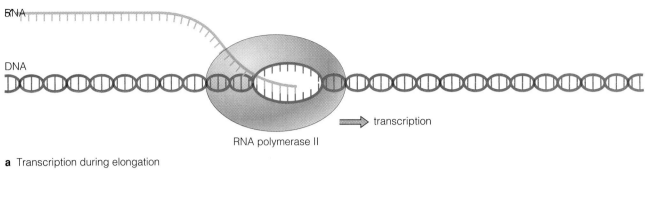

a Transcription during elongation

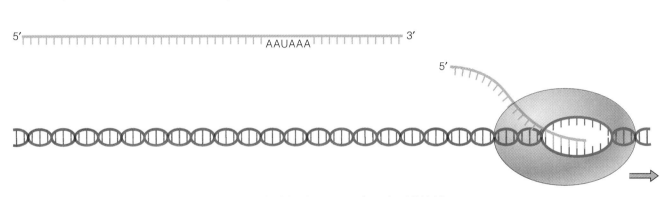

b Transcription of the AAUAAA conserved sequence

c Cleavage of the mRNA: mRNA is cleaved 11–30 nucleotides downstream from the AAUAAA sequence.

Figure 3.15 RNA cleavage that precedes termination of transcription by RNA polymerase II in eukaryotes.

blocking potential sites on the mRNA where rho could bind. Without rho on the RNA, RNA polymerase continues transcribing even though it may pause temporarily when it encounters a potential terminator. Here is where the NusA protein is important in elongation. As we saw earlier, NusA causes RNA polymerase to pause at certain DNA sequences. This allows the ribosomes to stay close to RNA polymerase, preventing rho from binding to the mRNA. Eventually, the ribosomes arrive at a sequence that terminates translation, and they dissociate from the mRNA. Beyond the translation termination site, ribosomes do not bind to the mRNA, so rho is free to bind to the mRNA. When RNA polymerase reaches the terminator, the polymerase pauses. Rho catches up to the paused RNA polymerase and causes it to disassemble from the DNA, terminating transcription.

Eukaryotic Termination

Termination of transcription in eukaryotic genes is more complicated than termination in prokaryotic genes. Cleavage at a specific site in the RNA may precede termination of transcription, so the 3' end of the RNA may not represent the true termination site, making it difficult for researchers to identify the termination site. Also, eukaryotic termination varies depending on the RNA polymerase that transcribes the gene.

RNA polymerase I transcribes identical copies of the same gene (a gene that codes for a large precursor rRNA) and has a specific site for termination. This site is characterized by the 17-nucleotide consensus sequence

AGGTCGACCAG(A/T)(A/T)NTCG

where N represent any one of the four nucleotides. This consensus sequence is recognized by a transcription-termination factor (TTFI), which aids in termination. The large precursor rRNA, however, is cleaved at a site about 18 nucleotides upstream of the termination site before RNA polymerase I reaches it. Consequently, the sequence for termination does not appear in the precursor rRNA molecule.

The RNA molecules transcribed by RNA polymerase II are usually cleaved about 11–30 nucleotides downstream from a highly conserved sequence in the RNA with the consensus sequence AAUAAA (Figure 3.15). After the cleavage, RNA polymerase II continues transcribing the template strand of DNA. Many genes transcribed by this polymerase lack a specific site for termination. Transcription may terminate anywhere from a few hundred to a thousand nucleotides downstream from the cleavage site. The RNA fragment produced between the cleavage site and termination is subsequently degraded. Little is known about the sequences or factors for termination of transcription by RNA polymerase II. As the RNA has already been cleaved at a specific site, the lack of specific termination is of no consequence because the cleavage ensures that all RNAs transcribed from a gene will be identical in length.

Because RNA polymerase III transcribes the same products *in vitro* as *in vivo*, and because mutations near the 3′ end of the transcribed region alter termination, RNA polymerase III apparently recognizes a specific site in the genes it transcribes as a termination site.

∼

Some transcription terminating mechanisms, both prokaryotic and eukaryotic, use conserved sequences. Cleavage of the RNA precedes termination of transcription in some eukaryotic genes.

3.6 TRANSCRIPTION OF rRNAs AND tRNAs

Ribosomal RNAs are structural units of the ribosome. In *E. coli*, there are only three rRNAs, called the 5S, 16S, and 23S rRNAs. These designations refer to the sizes of the rRNAs measured in Svedberg units, which denote how rapidly a molecule sediments under centrifugation. Svedberg units carry the name of the inventor of ultracentrifugation and are written as a number followed by an uppercase S. Larger particles contain more mass and generally sediment more rapidly than smaller particles, giving them a higher Svedberg value. Mass is not the only criterion, however. Density is also a factor, because the greater the density, the faster the sedimentation and therefore the higher the Svedberg value. For RNA molecules, Svedberg units correspond to the size of RNA molecules in nucleotides, but the relationship is not linear. For example, in *E. coli* the 16S rRNA has 1542

nucleotides, whereas the 23S rRNA is larger with 2904 nucleotides. Svedberg values cannot be added. For instance, mammalian ribosomes contain two subunits with individual Svedberg values of 40S and 60S. However, these two rRNAs bind to one another to form an 80S unit.

Large Precursor rRNAs

The three rRNAs in *E. coli*, 5S, 16S, and 23S, are transcribed as a single large precursor rRNA (about 5600 nucleotides long), which is then cleaved to form the individual rRNAs, as we explain later in this chapter. Because each ribosome requires one copy of each of these rRNAs, transcribing them as a unit ensures that there will be no surplus or deficiency of any particular rRNA. There are seven copies of the precursor rRNA gene in *E. coli*, all identical in nucleotide sequence, and transcription of these genes accounts for a large portion, up to 80–90% by mass, of all RNA transcribed in the cell.

In eukaryotic ribosomes, there are typically three to five different rRNAs. All eukaryotic ribosomes contain 5S, 16S, and 28S rRNAs, which actually vary somewhat in size among species. In addition, a 5.8S rRNA is present in mammalian and amphibian ribosomes, and 2S and 7.8S rRNAs have also been found in some species. RNA polymerase I transcribes genes that code for a single large precursor rRNA. This precursor RNA varies substantially in size among species, ranging from 7200 nucleotides in yeast to over 13,000 nucleotides in some mammals. The mammalian precursor rRNA is referred to in Svedberg units as the 45S precursor rRNA. This precursor rRNA is processed after transcription to yield the 5.8S, 16S, and 28S rRNAs.

Many copies of the gene that encodes the precursor rRNA in eukaryotes are located in tandem along a single DNA molecule. Figure 3.16 shows precursor rRNA genes being transcribed by RNA polymerase I, and illustrates their tandem organization. Between each precursor rRNA gene is a nontranscribed spacer sequence. The genes for the precursor rRNA occupy a specific region of the nucleus called the **nucleolus**. In electron micrographs, the nucleolus appears as a dark-staining region within the nucleus (Figure 3.17).

The primary function of the nucleolus is rRNA transcription and processing, and synthesis of the ribosomal subunits. The portion of cellular DNA that contains the tandemly repeated precursor rRNA genes is called the **nucleolus organizer region (NOR)** and is contained within the nucleolus. RNA polymerase I transcribes only in the nucleolus. RNA polymerases II and III transcribe in the part of the nucleus outside of the nucleolus, an area called the **nucleoplasm** (see Figure 3.17).

Because rRNAs are needed in abundance to synthesize the cell's many ribosomes, the transcription activity of RNA polymerase I accounts for up to 80–90% of the total transcription in the nucleus, in terms of the total

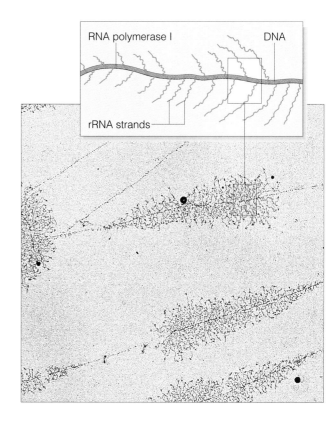

RNA polymerase I   DNA

rRNA strands

Figure 3.16 Electron micrograph of rRNA transcription. Each "Christmas tree" shape represents a single large rRNA precursor gene and its products. The "trunk" is the DNA template, and the "branches" are RNA molecules that have just been transcribed. The shorter RNA molecules are near the beginning of the gene, whereas the longer molecules are near the end of the gene. Notice that there are short nontranscribed spacer regions between the tandemly organized precursor rRNA genes. (Photo courtesy of O. L. Miller, Jr.)

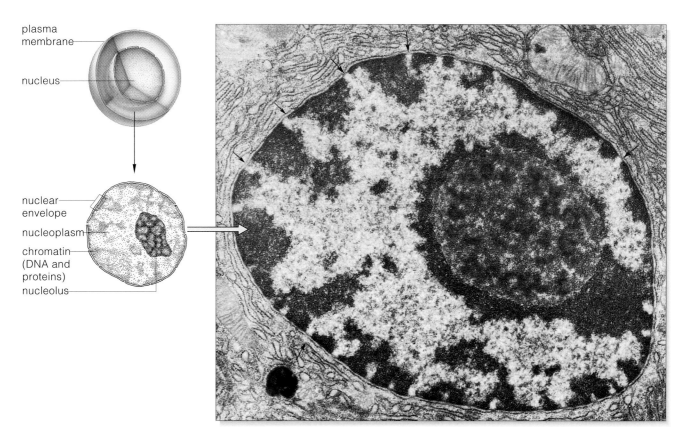

plasma membrane

nucleus

nuclear envelope

nucleoplasm

chromatin (DNA and proteins)

nucleolus

Figure 3.17 Electron micrograph of a cell nucleus showing the nucleolus as a dark-staining region. The black arrows in the micrograph point to pores in the nuclear envelope through which RNA molecules leave the nucleus following transcription. (Photo courtesy of S. L. Wolfe.)

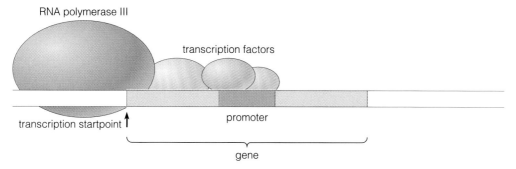

transcription startpoint ↑

promoter

gene

Figure 3.18 Location of the promoter within the transcribed region for eukaryotic 5S rRNA and tRNA genes.

mass of RNA transcribed. Similar to the transcription of a single large precursor rRNA in prokaryotes, transcription of a single eukaryotic precursor rRNA ensures that the 5.8S, 16S, and 28S rRNAs are present in equal quantities in the cell. Each ribosome contains one copy of each rRNA, so this manner of transcription ensures that there will be neither a surplus nor a deficiency of any of these rRNAs. As is the case for genes encoding mRNAs, the promoter for the precursor rRNA gene is located immediately upstream of the transcribed sequence, and is specific for RNA polymerase I.

5S rRNAs and tRNAs

In addition to the rRNAs just discussed, eukaryotes have a small rRNA called the 5S rRNA. It is only about 120 nucleotides long—only slightly larger than tRNAs, which are usually 70–80 nucleotides long. In a few eukaryotic species, RNA polymerase I transcribes the 5S rRNA as part of the large precursor rRNA. In most eukaryotes, however, RNA polymerase III transcribes the 5S rRNA from a separate gene. RNA polymerase III also transcribes all tRNA genes.

RNA polymerase III transcribes very small genes, those about 70–120 nucleotides long. These genes have an unusual, and at first glance puzzling, feature. The promoters for 5S rRNA genes and tRNA genes are located *within* the transcribed region of the gene, about 55 nucleotides downstream from the transcription startpoint. Transcription factors bind to the promoter within the gene, then direct RNA polymerase III to orient itself at the transcription startpoint, where it initiates transcription (Figure 3.18).

3.7 mRNA PROCESSING

Because transcription and translation are coupled in prokaryotes, there is no opportunity to process their mRNAs prior to translation. Eukaryotic mRNAs, however, must undergo substantial posttranscriptional processing before they can be translated. From this point on

in our discussion, mRNA that has been processed is called mRNA, and unprocessed mRNA is called **pre-mRNA** (also called heterogeneous nuclear RNA or hnRNA). We examine next the three basic steps of eukaryotic mRNA processing: (1) addition of a cap to the 5' end, (2) polyadenylation of the 3' end, and (3) intron removal and exon splicing.

5' Capping

As RNA is transcribed in the 5'→3' direction, the first portion of the molecule to be released from the DNA during transcription is the 5' end. The first nucleotide that was added at initiation of transcription retains its triphosphate group. Soon after the 5' end of the pre-mRNA is released from the DNA, and well before transcription is complete, this 5' triphosphate end is capped. The **5' cap** consists of a nucleotide with the base 7-methylguanine, a guanine base with a methyl group attached to nitrogen number 7. The enzyme guananyl transferase attaches the cap to the 5' triphosphate end of the pre-mRNA shortly after transcription is initiated. The attachment is through a 5'-5' triphosphate linkage. In other words, the orientation of the guanine cap is opposite to that of the other nucleotides in the pre-mRNA molecule; the cap is bound to the 5' terminal nucleotide by three phosphate groups (Figure 3.19).

The cap apparently protects the RNA molecule from degradation at the 5' end. It also appears to serve as the site where the ribosome binds to the mRNA for translation.

3' Polyadenylation

After transcription has passed the cleavage site, the pre-mRNA is cleaved at the 3' end. From 100 to 300 adenine ribonucleotides are then added to the 3' end, a process called **3' polyadenylation**. This results in a poly (A) tail on the 3' end of most eukaryotic mRNA molecules (Figure 3.20). The poly (A) tail is composed of RNA, but there are no long poly (T) sequences in the template strand of the DNA from which the poly (A) tail could

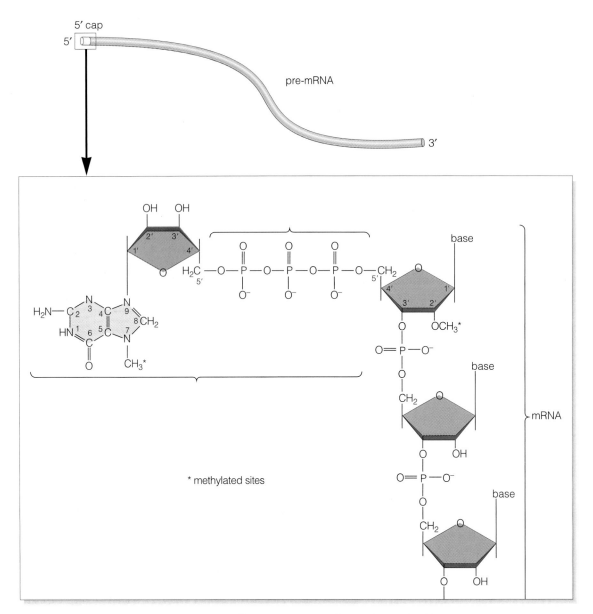

Figure 3.19 The 5′ cap of eukaryotic mRNAs. The 5′ cap consists of a guanine nucleotide attached by a 5′–5′ triphosphate linkage to the 5′ end of the mRNA. The guanine base is methylated on nitrogen number 7. The 2′ hydroxyl group of the nucleotide bound to the cap is also methylated.

be transcribed. Instead, poly (A) is added by an enzyme known as **poly (A) polymerase**, which uses adenosine triphosphate (ATP) as the source of adenine ribonucleotides. The poly (A) tail protects the mRNA from degradation at the 3′ end.

~

Eukaryotic pre-mRNA are capped on the 5′ end and polyadenylated on the 3′ end.

Intron Removal and Exon Splicing

The third step in pre-mRNA processing is intron removal and exon splicing. In general, pre-mRNA molecules are much larger than the mature mRNA, often by several-

fold. Addition of the 5′ cap and the 3′ poly (A) tail lengthens the pre-mRNA, but removal of interior portions shortens it. The removed portions are called **introns**. Once introns are removed, the remaining portions of the pre-mRNA, called **exons**, are spliced together to form the mature mRNA.

Introns might be compared to the unintelligible portions of the following sentence: "The spotted dog bwehfwe njsdf bhuef chased a rubber fgvbhru bvg truiyhth ball down fhdgb retfyt fjwdsvpl the hill." By removing the meaningless portions of the sentence and connecting the remaining portions together, you can form a sentence that makes sense: "The spotted dog chased a rubber ball down the hill."

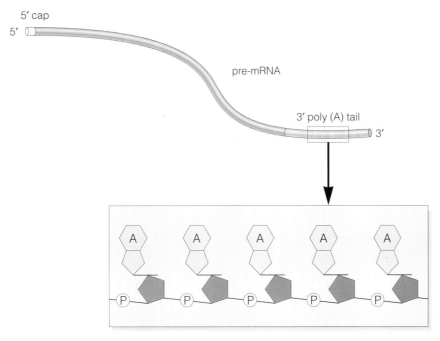

Figure 3.20 The 3′ poly (A) tail of eukaryotic mRNAs.

Introns are not an exclusive feature of pre-mRNAs, however. They have been found in tRNA genes, in rRNA genes, in genes in the DNA of mitochondria and chloroplasts, and in a few genes in bacteriophages (bacterial viruses). The introns of nuclear genes that code for mRNAs have similar structural features and a common splicing mechanism. The introns of other genes (tRNA, rRNA, mitochondrial, and chloroplastic) have their own distinct structural features and splicing mechanisms.

In nuclear genes that encode mRNAs, several conserved sequences define an intron. Conserved sequences are found at both ends of the intron. The 5′ end has the dinucleotide sequence GU, and the 3′ end has the dinucleotide sequence AG. These nucleotides are invariant in all introns from nuclear mRNA genes studied. This observation is often referred to as the **GT-AG rule** (using the sense-strand DNA sequence) and is useful in identifying introns in DNA sequences. Mutations that change these nucleotides prevent splicing at that junction, and may cause splicing at another site.

Adjacent to these terminal nucleotides are conserved sequences that vary slightly from intron to intron. The 5′ consensus sequence is GUAAGU, and the 3′ consensus sequence is 6PyNCAG (where "6Py" refers to a string of six pyrimidines, and N indicates that each of the four nucleotides is equally likely at that position). An additional consensus sequence, PyNPyPyPuAPy (where Py represents a pyrimidine and Pu a purine), is usually found about 18–40 nucleotides upstream from the 3′ end of the intron. The A in this sequence is invariant and the pyrimidine just downstream from the A rarely varies.

To summarize, the consensus sequences of a pre-mRNA intron can be diagrammed as

$$\begin{array}{cc} \downarrow & \downarrow \\ \end{array}$$
5′. . . GUAAGU — PyNPyPyPuAPy — 6PyNCAG . . . 3′

where the arrows represent the splice sites. The nucleotides between the arrows are removed at exactly those sites. This high degree of accuracy is essential. Otherwise, translation could be altered at the splice sites.

The Mechanism of Intron Removal and Exon Splicing

Intron removal from mRNA-encoding genes requires the participation of a particle called a **spliceosome**. It is made up of **small nuclear ribonucleoproteins (snRNPs)**, which are composed of protein and small RNA molecules known as **small nuclear RNAs** or **snRNAs**. Small nuclear RNAs generally range in size from 100 to 250 nucleotides, about the size range of tRNAs and small rRNAs. Some snRNAs are transcribed by RNA polymerase III, others by RNA polymerase II.

Intron removal follows the orderly sequence illustrated in Figure 3.21. First, an snRNP known as U1 binds to the GUAAGU conserved sequence, while an snRNP known as U2 binds to the PyNPyPyPuAPy sequence, as illustrated in Figure 3.21a. U1 then binds to U2, causing the intron to loop around, bringing the 5′ end of the intron in contact with the PyNPyPyPuAPy conserved sequence. Additional snRNPs known as U4, U5, and U6 join

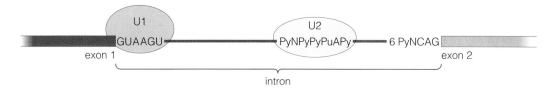

a U1 binds to the GUAAGU sequence at the 5' end of the intron, and U2 binds to the PyNPyPyPuAPy sequence.

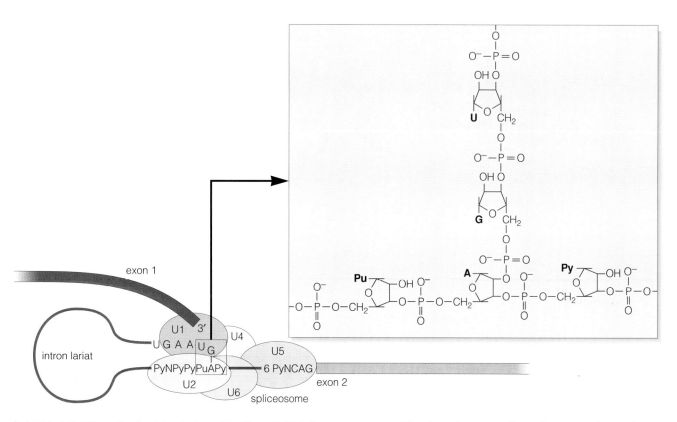

b U1 binds to U2 causing the intron to loop. U4, U5, and U6 join the complex, completing the spliceosome. The spliceosome cleaves the intron from exon 1 and the G on the 5' end of the intron bonds to the A in the PyNPyPyPuAPy sequence. The spliceosome holds the 3' end of exon 1 in place.

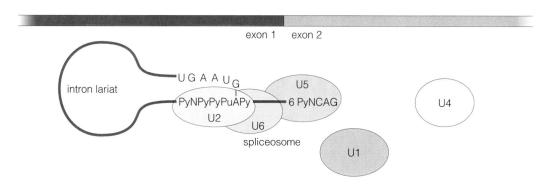

c U1 and U4 disassemble from the spliceosome. The 3' of the intron is cleaved from exon 2, and exons 1 and 2 are joined. The intron is released with U2, U5, and U6 attached. Enzymes then degrade the intron.

Figure 3.21 Intron removal from eukaryotic pre-mRNAs.

the complex to form the complete spliceosome (Figure 3.21b). The spliceosome cuts the RNA at the exon-intron junction, leaving a G on the 5′ end of the intron. This terminal G then forms a 5′-2′ bond with the sixth nucleotide (underlined) in the PyNPyPyPuAPy conserved sequence (see inset in Figure 3.21b). This sixth nucleotide is an A, giving the intron a **lariat structure** because of this unusual bond. At this point, U1 and U4 disassemble from the spliceosome. The intron is cut at the 3′ end of the G in the 6PyNCAG conserved sequence, and the remaining snRNPs join the two exons (Figure 3.21c). The intron is released and degraded.

Intron removal and exon splicing in mitochondrial and chloroplastic RNAs, tRNAs, and rRNAs differs. A single intron is present in some tRNA transcripts. This intron is removed at the splice sites by a splicing enzyme, and the exons are then joined by a splicing ligase. These introns are very small and do not follow the GT-AG rule. The introns in some mitochondrial genes and precursor rRNAs are said to be self-splicing, which means that the introns are removed by autocatalytic action of the RNA itself. In other words, the RNA from which the intron is being cleaved acts as an enzyme (or more appropriately, a ribozyme) to catalyze the cleavage, confirming that some RNAs have catalytic activity.

It is possible to view the effect of intron removal with electron microscopy. The following example shows how this is done.

Example 3.4 Demonstration of intron removal.

When single-stranded RNA is mixed with DNA that has been separated into single strands, the RNA may base-pair with its complementary sequence in the DNA, forming a DNA-RNA hybrid molecule. This process is called DNA-RNA hybridization. In 1978, Tilghman et al. (*Proceedings of the National Academy of Sciences, USA* 75:1309–1313) reported the results of DNA-RNA hybridization experiments with the mouse β-major globin gene (the same gene described in Example 3.2). They recovered two types of RNA encoded by this gene, one approximately 1500 nucleotides long and another about 600 nucleotides long. The smaller RNA was much more abundant than the larger one in cell extracts. They hybridized these two RNAs with DNA from the β-major globin gene and examined the DNA-RNA hybrid molecules using electron microscopy, as shown in parts a and b of Figure 3.22.

Problem: **(a)** Draw diagrams of the DNA-RNA hybridization represented in the electron micrographs, distinguishing between DNA and RNA. **(b)** What do the larger and smaller RNA molecules represent? **(c)** What does the loop in Figure 3.22b

represent? **(d)** Why is there no loop in Figure 3.22a? **(e)** What do these results suggest about the timing of intron splicing?

Solution: **(a)** The diagrams are provided in parts c and d of Figure 3.22. **(b)** The larger RNA is the pre-mRNA encoded by the β-major globin gene prior to intron splicing. The smaller RNA is the mature mRNA after intron splicing. **(c)** The loop represents double-stranded DNA that fails to hybridize with the RNA (Figure 3.22d). This looped DNA is an intron that is present in the DNA but absent in the mature mRNA. **(d)** There is no loop in Figure 3.22a because the intron has not been spliced from the pre-mRNA. Therefore, the pre-mRNA is free to hybridize with DNA along its entire length (Figure 3.22c). **(e)** The presence of a pre-mRNA that includes an intron suggests that this intron is spliced after transcription has been completed.

This study shows evidence of a single intron in the β-globin gene. However, Tilghman et al. in a previous study (*Proceedings of the National Academy of Sciences, USA* 75:725–729) found evidence of a small second intron. Comparison of the DNA sequence with the mRNA sequence has revealed that the mouse β-major globin gene indeed has two introns, one that is 116 nucleotides long and a second that is 653 nucleotides long. The smaller intron failed to appear as a loop in Figure 3.22b, probably because of its small size.

3.8 PROCESSING OF rRNAS AND tRNAS

To become functional, both rRNAs and tRNAs must be processed. Both prokaryotes and eukaryotes process their rRNAs and tRNAs. In this section we will examine rRNA processing first, then see how tRNAs are processed.

rRNA Processing

In both prokaryotes and eukaryotes, the large precursor rRNA molecule must be cleaved into its constituent RNAs. In prokaryotes, the genes encoding the precursor rRNA are similar but not identical to those in eukaryotes, and they are not clustered in a particular region as in eukaryotes. In *E. coli*, there are seven precursor rRNA genes. All are transcribed into a single large 30S precursor unit that has the three rRNAs within it in the same order—16S, 23S, 5S—and also several sequences that are cleaved to form tRNAs (Figure 3.23). The seven precursor rRNA genes differ in the type and number of tRNAs that they encode. The mature products are cleaved from the large precursor by a single enzyme named ribonuclease III (RNase III).

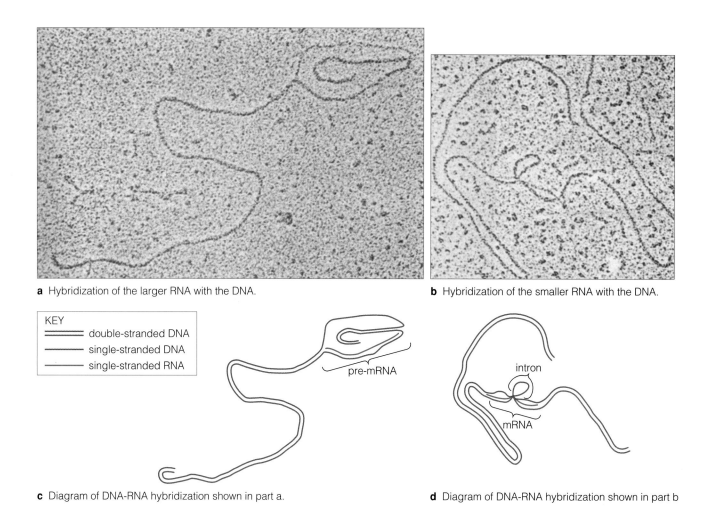

a Hybridization of the larger RNA with the DNA.

b Hybridization of the smaller RNA with the DNA.

KEY
═══ double-stranded DNA
─── single-stranded DNA
─── single-stranded RNA

pre-mRNA

intron

mRNA

c Diagram of DNA-RNA hybridization shown in part a.

d Diagram of DNA-RNA hybridization shown in part b

Figure 3.22 Information for Example 3.4: Electron micrographs of DNA-RNA hybridization. Double-stranded nucleic acids (either double-stranded DNA or double-stranded DNA-RNA hybrids) are thicker than single-stranded DNA or RNA in the micrographs. (From Tilghman, S. M., P. J. Curtis, D. C. Tiemeier, P. Leder, and C. Weissmann. The intervening sequence of mouse beta-globin gene is transcribed within the 15s beta-globin mRNA precurser. 1978. *Proceedings of the National Academy of Sciences, USA* 75:1309–1313. Used by permission of the author.)

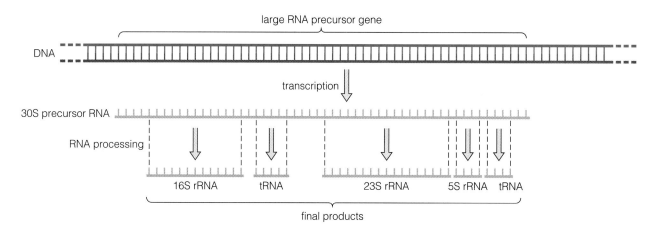

large RNA precursor gene

DNA

transcription

30S precursor RNA

RNA processing

16S rRNA   tRNA   23S rRNA   5S rRNA   tRNA

final products

Figure 3.23 Processing of a large precursor rRNA in *E. coli*.

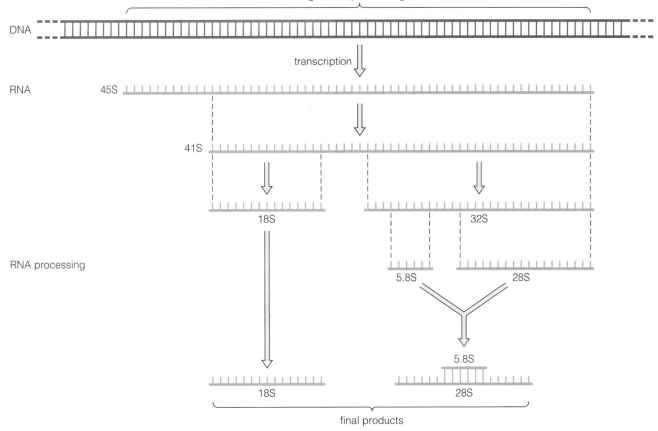

Figure 3.24 One mode of processing the mammalian large precursor rRNA.

The large eukaryotic precursor rRNA is cleaved in a stepwise fashion to release the three mature rRNAs within the transcript. A generalized mode for processing the mammalian 45S precursor is illustrated in Figure 3.24. The 45S precursor is first cleaved to form a smaller 41S precursor that contains all three rRNAs. At this point, there are several minor differences in the pathways used among cells that have been studied. In general, the 18S rRNA is cleaved from the transcript, leaving a 32S precursor that contains the 28S and 5.8S rRNAs. This 32S precursor is then cleaved to release the 28S and 5.8S rRNAs. The 5.8S rRNA then base-pairs with a homologous segment on the 28S rRNA. The mature rRNAs fold back on themselves to assume substantial secondary structure and are incorporated with proteins to form the ribosomal subunits.

Some nucleotides in the rRNAs undergo methylation as part of the RNA processing after transcription. **Methylation** is the enzyme-catalyzed addition of a methyl group ($-CH_3$) to the nitrogenous base on a nucleotide (Figure 3.25). In mammalian rRNAs, about 2% of the nucleotides are methylated after transcription.

tRNA Processing

In prokaryotes, tRNAs are transcribed as large precursor RNAs that are then cleaved to form the final product. As just described, some tRNAs are cleaved from the large precursor rRNA. Others are transcribed as a cluster of tRNAs within a single precursor RNA that is then cleaved to yield the individual tRNAs. In eukaryotes, tRNA genes are clustered, but RNA polymerase III transcribes each gene individually rather than as a large precursor.

tRNAs are substantially processed following transcription. Any introns that are present are removed. All tRNAs have CCA on the 3' end. In some cases, the CCA is transcribed from the DNA. In other cases, the CCA is not found in the DNA and must be added to the 3' end after transcription. Many of the nucleotides in tRNAs are modified. This modification is rather extensive, in some cases affecting as much as two-thirds of the nucleotides in the molecule. Most of the modifications are methylations; others are some rather unusual modifications of the

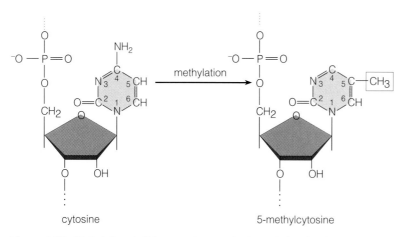

Figure 3.25 Methylation. In this example, cytosine is methylated to form 5-methylcytosine.

bases in the nucleotides. Some of these modified bases are illustrated in Figure 3.26.

In all cases, nucleotides are modified after transcription. During transcription, only the four standard ribonucleotides (C, U, A, and G) are incorporated into the tRNA. After transcription, specific RNA-processing enzymes modify certain nucleotides. Some nucleotide modifications are found at the same position in nearly all tRNAs. For example, pseudouridine is found at the same position in all tRNAs studied. Other modifications may be unique to one or a few specific tRNAs.

The purpose of some nucleotide modifications is known. Modifications in a portion of the tRNA known as the anticodon make the wobble effect possible (a topic we will discuss in Chapter 4). Other modifications assist in forming the secondary structure of the molecule.

Let's now take a look at an example of a tRNA and its gene to see what modifications are made in the tRNA.

Example 3.5   Comparison of a yeast tRNA with its gene.

In 1987, Valenzuela et al. (*Proceedings of the National Academy of Sciences, USA* 75:190–194) compared the nucleotide sequence of a tRNA with the nucleotide sequence of the gene that encodes it. The tRNA sequence is given in Figure 3.27a; the nucleotide sequence of the DNA that encodes it is given in Figure 3.27b.

Problem: Describe the posttranscriptional modifications in the tRNA.

Solution: There are a total of 76 nucleotides in the tRNA. Of these, 14 (18.4%) are modified. Nine are methylated (eight designated with "m" and

one that is a U modified to a T by methylation), two are U modified to D (dihydrouracil), two are U modified to Ψ (pseudouracil), and one is G modified to Y (wyosine). A 19-nucleotide-pair intron has been removed, and CCA has been added on the 3' end after transcription, as illustrated in Figure 3.27c.

Processing of rRNAs and tRNAs includes cleavage into smaller molecules and nucleotide modification.

SUMMARY

1.   A gene may be defined as a segment of a DNA molecule that is transcribed into an RNA molecule.

2.   The genetic information in the DNA of a gene is transcribed into RNA, which is then translated into a polypeptide, which is then processed into a protein. This is the central dogma of molecular genetics.

3.   Transcription and translation are coupled in prokaryotes. In eukaryotes, transcription is compartmentalized within the nucleus. The RNA transcripts are exported from the nucleus to the cytoplasm, where they are translated.

4.   There are three general classes of RNA molecules: messenger RNA (mRNA), ribosomal RNA (rRNA), and transfer RNA (tRNA). Messenger RNA contains the information that will be translated into proteins. Ribosomal RNA is a structural component of ribosomes, making up about one-half of the intact ribosome (the other half is made of protein). Transfer RNAs are small RNA molecules that carry amino acids to the ribosome for translation.

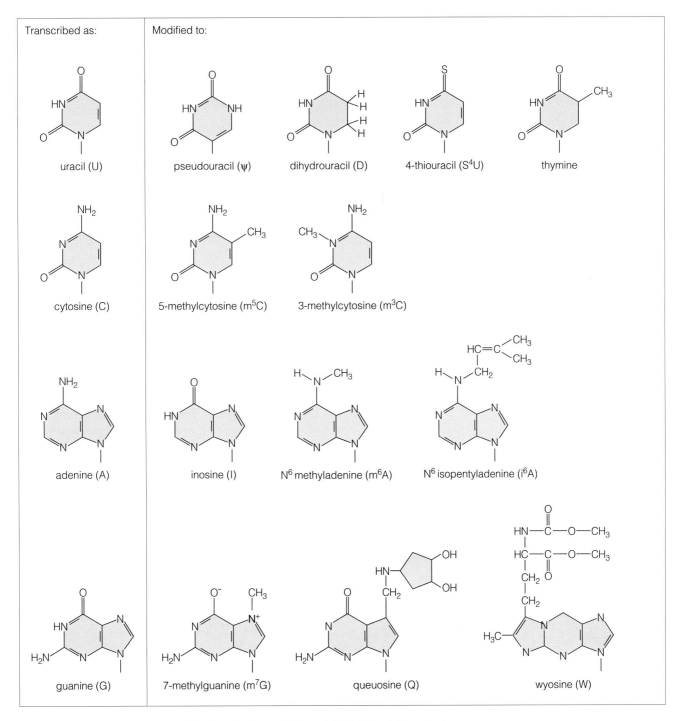

Transcribed as:	Modified to:

uracil (U)

pseudouracil (ψ)

dihydrouracil (D)

4-thiouracil (S⁴U)

thymine

cytosine (C)

5-methylcytosine (m⁵C)

3-methylcytosine (m³C)

adenine (A)

inosine (I)

N⁶ methyladenine (m⁶A)

N⁶ isopentyladenine (i⁶A)

guanine (G)

7-methylguanine (m⁷G)

queuosine (Q)

wyosine (W)

Figure 3.26 Examples of base modifications in the nucleotides of tRNA. In the left-hand column are the bases as they are originally transcribed. In the right-hand column are some common modifications.

5. Transcription of RNA in prokaryotes is accomplished by a single RNA polymerase. However, in eukaryotes there are three RNA polymerases. RNA polymerase I transcribes the larger rRNA genes. RNA polymerase II transcribes genes that code for mRNA. RNA polymerase III transcribes tRNA genes and genes that code for other small RNAs, including the small rRNAs.

6. Transcription is initiated at a promoter, a specific sequence of nucleotides in the DNA that determines where the RNA polymerase should begin transcribing.

7. During transcription, RNA polymerase adds one ribonucleotide to the previous ribonucleotide in the 5′→3′ direction, the same as in DNA synthesis. One

                    m   m   m m   m   m m   m
tRNA 5′ GCGGAUUUAGCUCAGDDGGGAGAGCGCCAGACUGAAYAψCUGGAGGUCCUGUGUψCGAUCCACAGAAUUCGCACCA 3′

a The nucleotide sequence of a tRNA in *E. coli*. Methylated nucleotides are designated with an "m" above the nucleotide. ψ indicates pseudouracil, D indicates dihydrouracil, and Y indicates wyosine.

5′ GCGGATTTAGCTCAGTTGGGAGAGCGCCAGACTGAAGAAAAAGAGACTTCGGTCAAGTTATCTGGAGGTCCTGTGTTCGATCCACAGAATTCGCATAA 3′
DNA (sense strand only)

b The nucleotide sequence of the DNA that encodes the tRNA in part a.

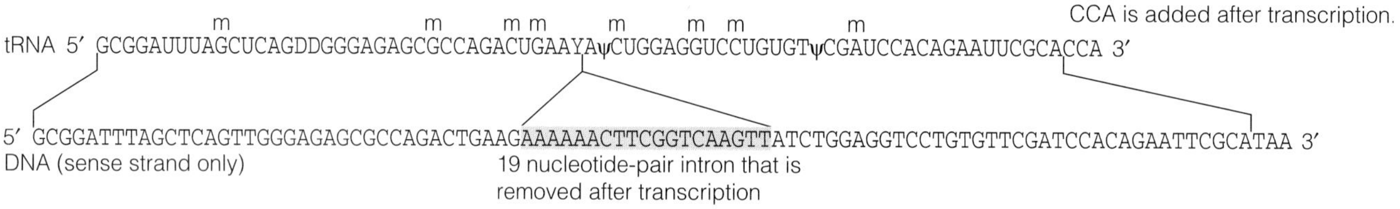

                    m   m   m m   m   m m   m   CCA is added after transcription.
tRNA 5′ GCGGAUUUAGCUCAGDDGGGAGAGCGCCAGACUGAAYAψCUGGAGGUCCUGUGUψCGAUCCACAGAAUUCGCACCA 3′

5′ GCGGATTTAGCTCAGTTGGGAGAGCGCCAGACTGAAGAAAAAACTTCGGTCAAGTTATCTGGAGGTCCTGTGTTCGATCCACAGAATTCGCATAA 3′
DNA (sense strand only)   19 nucleotide-pair intron that is
                                             removed after transcription

c Comparison of that tRNA and DNA sequence from parts a and b.

Figure 3.27 Information for Example 3.5: A comparison of the nucleotide sequence of a tRNA and the gene that encodes it.

DNA strand serves as the template to determine the nucleotide sequence of the RNA using essentially the same base-pairing rules as in DNA replication, with one exception. Whenever RNA polymerase encounters A in the DNA strand, it incorporates U instead of T in the RNA strand.

8. Only one DNA strand is used as a template for transcription of RNA. This strand contains the complementary or antisense nucleotide sequence of the RNA, so it is called the antisense strand. The DNA strand not used as a template has the same nucleotide sequence, or same sense, as the RNA, and is therefore called the sense strand. Coding strand and nontemplate strand are synonymous with sense strand, whereas anticoding strand and template strand are synonymous with antisense strand.

9. Promoters contain conserved sequences that serve as signals for RNA polymerase to initiate transcription. The sequences and positions of conserved sequences differ in prokaryotes and eukaryotes.

10. Two modes of termination of transcription are known in prokaryotes. The first is intrinsic termination, which requires a conserved sequence at the end of the gene that causes a hairpin structure to form in the mRNA. The second is rho-dependent termination, which requires a protein known as rho factor.

11. Termination of transcription in eukaryotes differs depending on the RNA polymerase that transcribes the gene. RNA polymerase I terminates transcription at a conserved sequence that binds a transcription factor. However, the rRNA precursor is cleaved prior to termination. RNA molecules transcribed by RNA polymerase II are also usually cleaved before the RNA polymerase reaches the termination site. Cleavage is determined by conserved sequences.

12. Most prokaryotic and eukaryotic rRNAs are transcribed as a single large precursor molecule that is then processed to yield the final rRNA products. In E. coli cells, there are seven copies of the large precursor rRNA gene. In eukaryotes, there are numerous copies of the precursor rRNA gene organized in tandem along a single DNA molecule in a region of the nucleus called the nucleolus.

13. In eukaryotes, the 5S rRNA, tRNAs, and several other small RNAs are transcribed by RNA polymerase III. The promoter for these small genes is located within the transcribed region.

14. Eukaryotic mRNAs begin as a pre-mRNA, which is processed to become the mature mRNA. A 59 cap, consisting of a methylated guanine nucleotide, is added to the 59 end of the molecule. A poly (A) tail is added to the 39 end of the molecule. Introns are removed from the transcript to yield the mature mRNA.

15. The large precursor RNA in eukaryotes is cleaved in a stepwise fashion to form the final rRNA products. Nucleotides may also be modified, typically by methylation. Many of the nucleotides in tRNAs are also modified.

QUESTIONS AND PROBLEMS

1. Describe the central dogma of molecular genetics.

2. Provide a description of a gene based on the central dogma.

3. List several of the differences between prokaryotes and eukaryotes regarding transcription and RNA processing.

4. Genes that encode rRNA and tRNA do not encode proteins. Why are they called genes if they do not encode proteins?

5. Several of the examples in this chapter highlighted a gene encoding the β subunit of hemoglobin. This gene encodes a protein that has a particular physiological function in the body. (a) What is this function? (b) What is the relationship between the expression of this gene at the molecular level and the physiological function of the protein encoded by the gene?

6. Typically, DNA is double stranded and RNA is single stranded, although there are exceptions to this general rule. What reasons can account for this difference in structure?

7. Describe the characteristics of the three major types of RNA and their functions in the cell.

8. Describe the functions of three types of RNA that do not fall into the three major classes of RNA.

9. Describe the role of σ factor, and distinguish between the holoenzyme and the core enzyme for RNA polymerase in prokaryotes.

10. RNA polymerase II may encounter situations where it stalls, halting transcription before termination should occur. How can this situation be remedied?

11. Provide two reasons that explain why coupled transcription and translation is not possible in eukaryotes.

12. Why is it important for ribosomes to follow close behind prokaryotic RNA polymerase during elongation?

13. The (A + T)/(G + C) ratios of E. coli and Mycobacterium tuberculosis (the bacterium that causes tuberculosis) differ significantly. In which of the following classes of genes would there most likely be a large difference in the (A + T)/(G + C) ratios between these two species? The genes that code for (a) tRNAs, (b) rRNAs, or (c) mRNAs? Please explain your answer.

14. In what way does the position of promoters recognized by RNA polymerase III differ from the position of promoters recognized by RNA polymerases I and II?

15. For each of the following statements, indicate which of the three eukaryotic RNA polymerases is being described: (a) The RNA polymerase that transcribes the greatest diversity of genes in terms of nucleotide sequence. (b) The RNA polymerase that transcribes the smallest genes. (c) The RNA polymerase that transcribes most of the RNA in the cell. (d) The RNA polymerases that use promoters on the 5' end of the gene. (e) The RNA polymerase located primarily in the nucleolus organizer region. (f) The RNA polymerase that transcribes pre-mRNAs. (g) The RNA polymerase that transcribes 5S rRNAs. (h) The RNA polymerase that uses promoters contained within the transcribed region.

16. Compare initiation of transcription in prokaryotic and eukaryotic cells. Be sure to use the following terms in your comparison: −10 and −35 sequences, consensus sequence, TATA box, CAAT box, GC box, promoter, and enhancer.

17. What aspects are similar when the 210 sequence in E. coli and the TATA box in eukaryotic cells are compared? In what way might their functions be similar?

18. Describe the roles of the following proteins: rho, NusA, TFIIA, TFIIB, TFIID, TFIIE, TFIIF, and TFIIS.

19. The promoter region of the Tet gene in E. coli contains the following sense-strand sequence in the DNA (from Rosenberg and Court. 1979. Annual Review of Genetics 13:319–353).

```
5' ATTCTCATGTTTGACAGCTTATCATCGAT-
        AAGCTTTAATGCGGTAGTTTATCACAGT 3'
```

The 5' end of the mRNA has the sequence

```
5' GUUUAUCA ...
```

Identify the transcription startpoint and the −10 and −35 sites, and compare them to the consensus sequences.

20. How is transcription initiated and terminated in E. coli?

21. Elongation does not proceed at a constant rate in E. coli. (a) Describe how it proceeds and what factors may influence the rate of transcription. (b) Why is it essential for cells to control the rate of elongation in prokaryotes?

22. What effect does a hairpin structure in the mRNA have on transcription in prokaryotes?

23. Why is hairpin formation unnecessary in rho-dependent terminators?

24. Compare termination of transcription in prokaryotic and eukaryotic cells. Distinguish between the terms *cleavage site* and *termination site*.

25. In this chapter we discussed how RNA polymerase may bypass a potential rho-dependent terminator and continue transcription. There are known cases where potential intrinsic terminators may be bypassed for essentially the same reason that potential rho-dependent terminators are bypassed. The mechanism for bypassing both types of potential terminators is based on the same principle. Using this information, propose a model that explains how a potential intrinsic terminator may be bypassed.

26. Many prokaryotic genes that encode polypeptides have potential terminators (either rho-dependent or intrinsic) that are bypassed by RNA polymerase. However, bypassed potential terminators are conspicuously absent from genes that encode rRNA in prokaryotes. Explain why this observation is expected.

27. The following sense-strand DNA sequence (from Rosenberg and Court. 1979. *Annual Review of Genetics*

13:319–353) is part of the *his* operon gene of *E. coli*. The sequence contains an intrinsic terminator.

```
5'  AACGCATGAGAAAGCCCCCGGAAGATCACCT-
       TCCGGGGGCTTTATATAATTAGCGCGGTTGAT 3'
```

Diagram the hairpin structure in the mRNA in the terminator region.

28. In the saturation mutagenesis experiments described in Example 3.2, mutation of the A located at the 5' end of the pre-mRNA to another nucleotide resulted in a reduction of transcription. Numerous DNA sequencing experiments have shown that an A is present at the transcription startpoint for most eukaryotic mRNA genes. **(a)** What do these results suggest about the importance of an A at the 5' end of mRNAs? **(b)** What possible role might this A play in the mRNA?

29. Pribnow (*Proceedings of the National Academy of Sciences, USA* 72:784–788) identified the following double-stranded DNA sequence as a region protected by *E. coli* RNA polymerase in footprinting experiments. Sixteen nucleotides of the 5' end of the mRNA encoded by the corresponding gene are also given.

DNA

```
5'  GTAAACACGGTACGATGTACCACATGAAACGACAGTGAGTCA 3'
3'  CATTTGTGCCATGCTACATGGTGTACTTTGCTGTCACTCAGT 5'
```

mRNA

```
5'  AUGAAACGACAGUGAG . . .
```

(a) Which strand of DNA is the sense strand? **(b)** Identify the transcription initiation site and the −10 sequence. **(c)** How closely does the observed −10 sequence match its corresponding consensus sequence?

30. Below is a 167 nucleotide segment from the human β-globin gene (from Lawn et al. 1980. *Cell* 21:647–651). The entire gene is over 1600 nucleotides in length. The sequence below includes a single intron that is 129 nucleotides long.

```
TTGGTGGTGAGGCCCTGGGCAGGTTGGTATCAAGGTTAC-
AAGACAGGTTTAAGGAGACCAATAGAAACTGGCATGTGG-
AGACAGAGAAGACTCTTGGGTTTCTGATAGGCACTGACT-
CTCTCTGCCTATTGGTCTATTTTCCCACCCTTAGGCTGC-
TGGTGGTCTAC
```

Identify the splice sites of the intron, and compare the sequences at the splice sites to the consensus sequences. How much do these sequences differ from their corresponding consensus sequences?

31. Human blood contains a blood clotting factor known as factor VIII that coagulates the blood and strengthens the clot that forms when blood vessels are severed. People who lack functional factor VIII suffer from a genetic disorder known as hemophilia, which is characterized by excessive bleeding. Factor VIII is a protein that consists of two a subunits and two b subunits. Below is a 165-nucleotide segment of the gene that encodes the b subunit (from Bottenus et al. 1990. *Biochemistry* 29:11195–11209). The entire gene is over 28,000 nucleotides in length. The sequence below includes a single intron that is 87 nucleotides long.

```
ATACGTTGTGAAGATGGAAAATGGACAGAACCTCCAAAATGC-
ATTGGTTACTAACACCTTGAAGAAGCTTTGCTAAAATGAAAT-
CTGCATGTGTAGCTAAATGGTCAGTTCTCTTTCAACTGTACC-
TTTTCAGAAGGACAGGAGAAGGTAGCCTGTGAGGACCCA
```

(a) Identify the intron splice sites, and **(b)** provide the nucleotide sequence of the segment of mRNA encoded by this sequence.

32. Proudfoot and Brownlee (*Nature* 263:211–214) compared the following nucleotide sequences from the 3' ends of mRNAs from six different eukaryotic genes:

```
UGGUCUUUGAAUAAAGUCUGAGUGAGUGGCpoly(A)
UGGCUAAUAAAGGAAAUUUAUUUUCAUUGCpoly(A)
UGGUCUUUGAAUAAACUCUGAGUGGGCGGCpoly(A)
UGCCUAAUAAAAAACAUUUAUUUUCAUUGCpoly(A)
AAUAUUCAAUAAAGUGAGUCUUUGCACUUGpoly(A)
CCUUUAAUCAUAAUAAAAACAUGUUUAAGCpoly(A)
```

(a) Identify the consensus sequence that appears in all six of these mRNAs, and align the mRNAs by this consensus sequence. **(b)** How many nucleotides appear between the consensus sequence and the polyadenylation site in each of these mRNAs?

33. In 1984, Wickens and Stephenson (*Science* 226: 1045–1051) reported that the following mutant sequences of the AAUAAA consensus sequences prevented cleavage of the pre-mRNA: AACAAA, AAUUAA, AAUACA, and AAUGAA. Manly (1988. *Biochimica et Biophysica Acta* 950:1–12) reported that 80–90% of all eukaryotic mRNAs sequenced had the AAUAAA sequence conserved exactly. About 10% carried the sequence AUUAAA. **(a)** Is there a contradiction between these two findings? Explain your answer briefly. **(b)** What do these results suggest about the importance of this consensus sequence and its conservation?

FOR FURTHER READING

Detailed information on transcription and RNA processing can be found in Chapters 7, 11, 28, and 30 of **Lewin, B. 1997.** *Genes VI.* **Oxford: Oxford University Press**; Chapters 14 and 15 of **Wolfe, S. L. 1993.** *Molecular and Cellular Biology.* **Belmont, CA: Wadsworth** and Chapters 6 and 8 of **Alberts,**

B., D. Bray, J. Lewis, M. Raff, K. Roberts, and J. D. Watson. 1994. *Molecular Biology of the Cell*, 3rd ed. New York: Garland. A thorough review of the roles of consensus sequences for prokaryotic initiation and termination is Rosenberg, M., and D. Court. Regulatory sequences involved in the promotion and termination of RNA transcription. 1979. *Annual Review of Genetics* 13:319–353. An excellent review of both prokaryotic and eukaryotic elongation and termination is Richardson, J. P. 1993. Transcription termination. *Critical Reviews in Biochemistry and Molecular Biology* 28:1–30. The basal eukaryotic transcription complex and its interaction with enhancers, activators, and coactivators is described in three reviews: Tijan, R. 1995. Molecular machines that control genes. *Scientific American* 272 (Feb):54–61; Zawel, L., and D. Reinberg. 1995. Common themes in assembly and function of eukaryotic transcription complexes. *Annual Review of Biochemistry* 64:533–561; and Tijan, R., and T. Maniatis. 1994. Transcriptional activation: A complex puzzle with few easy pieces. *Cell* 77:5–8. One of the first reviews to describe introns and exons in eucaryotic genes is Breathnach, R., and P. Chambon. 1981. Organization and expression of eucaryotic split genes coding for proteins. *Annual Review of Biochemistry* 50:349–383. Reviews of the intron splicing reactions are Sharp, P. 1987. Splicing of messenger RNA precursors. *Science* 235:766–771; Lührmann, R., B. Kastner, and M. Bach. 1990. Structure of spliceosomal snRNPs and their role in pre-mRNA processing. *Biochimica et Biophysica Acta* 1087:265–292; and Green, M. R. 1991. Biochemical mechanisms of constitutive and regulated pre-mRNA splicing. *Annual Review of Cell Biology* 7:559–599. Polyadenylation of pre-mRNA was reviewed by Manley, J. L. 1988. Polyadenylation of mRNA precursors. *Biochimica et Biophysica Acta* 950:1–12. The presence and nature of mRNA was described by Brenner, S., F. Jacob, and M. Meselson. 1961. An unstable intermediate carrying information from genes to ribosomes for protein synthesis. *Nature* 180:576–581. Consensus sequences in *E. coli* promoters were identified by Pribnow, D. 1975. Nucleotide sequence of an RNA polymerase binding site at an early T7 promoter. *Proceedings of the National Academy of Sciences, USA* 72:784–788. Among the best articles describing eukaryotic consensus sequences and how mutations may alter transcriptional activity is Myers, R. M., K. Tilly, and T. Maniatis. Fine structure genetic analysis of a β-globin promoter. *Science* 232:613–618. Comparison of R-looping in pre-mRNA and mature mRNA when hybridized with DNA is described in Tilghman, S. M., P. J. Curtis, D. C. Tiemeier, P. Leder, and C. Weissmann. 1978. The intervening sequence of a mouse β-globin gene is transcribed within the 15S β-globin mRNA precursor. *Proceedings of the National Academy of Sciences, USA* 75:1309–1313.

For additional reading, go to InfoTrac College Edition, your online research library at: http: www.infotrac-college.com/brookscole

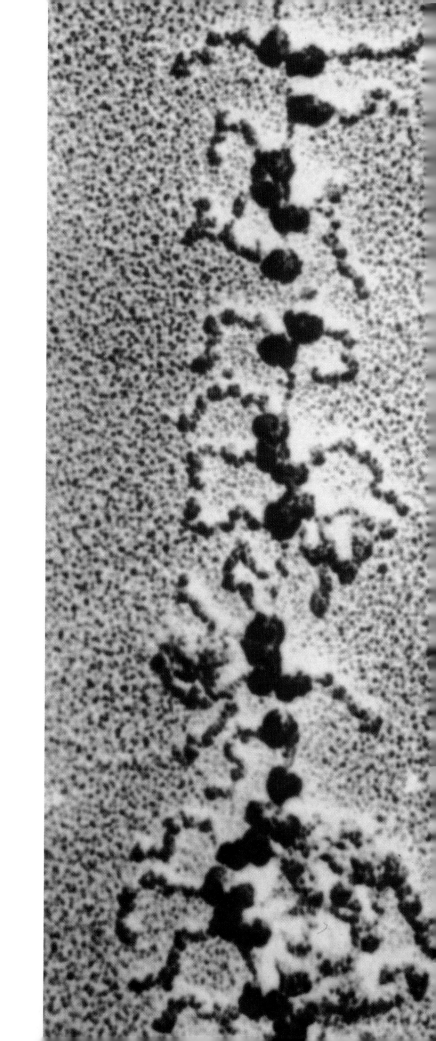

CHAPTER 4

TRANSLATION AND PROTEIN FUNCTION

The first part of the central dogma is transcription, the transfer of genetic information from DNA to RNA. In this chapter, we will examine the rest of the central dogma—translation and protein processing, the transfer of information from RNA to protein. Proteins are the final products of most genes and constitute the most diverse class of biological molecules. Many proteins act as enzymes, and enzymes regulate nearly every biochemical reaction. Some proteins are responsible for the transport of substances across cell membranes. Other proteins may accumulate where they are stored for later use, such as in bird eggs and plant seeds. Certain proteins act as structural components, such as those that compose the bulk of muscle tissue, hair, and fingernails. Still other proteins play some rather unique roles. The primary component of rattlesnake venom is protein (as is that of some other toxins in poisonous animals and plants). The waterproof adhesive that allows mussels to remain attached to rocks amid the constant beating of powerful ocean waves is protein. Whatever their functions, all proteins are encoded by genes and are synthesized through translation.

4.1   AMINO ACIDS AND POLYPEPTIDES

Because translation is the conversion of information from the language of nucleotides to the language of amino acids, we begin by examining amino acids, the building blocks of protein. **Amino acids** have a common structure consisting of a central carbon atom with a carboxyl group, an amino group, and a hydrogen atom bound to it. This leaves one binding site to which a side group may bind. The side group of an amino acid is denoted generically as the **R group** and is the only part of the molecule that differs from one amino acid to the next.

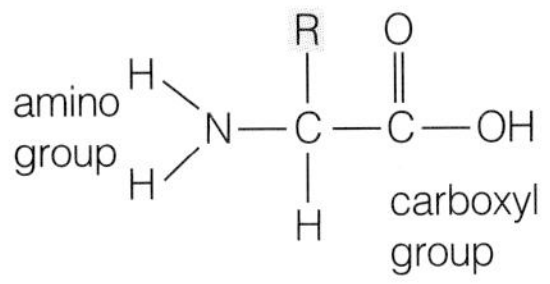

$$
\text{amino group} \quad
\begin{array}{c}
H \\ \diagdown \\ N \\ \diagup \\ H
\end{array}
-
\begin{array}{c}
R \\ | \\ C \\ | \\ H
\end{array}
-
\begin{array}{c}
O \\ \| \\ C \\ \\
\end{array}
-OH \quad \text{carboxyl group}
$$

There are many different amino acids, but only 20 that are utilized in translation. Proteins consist of varying combinations of these 20 amino acids. Each of the amino acids utilized in translation is distinguished by its

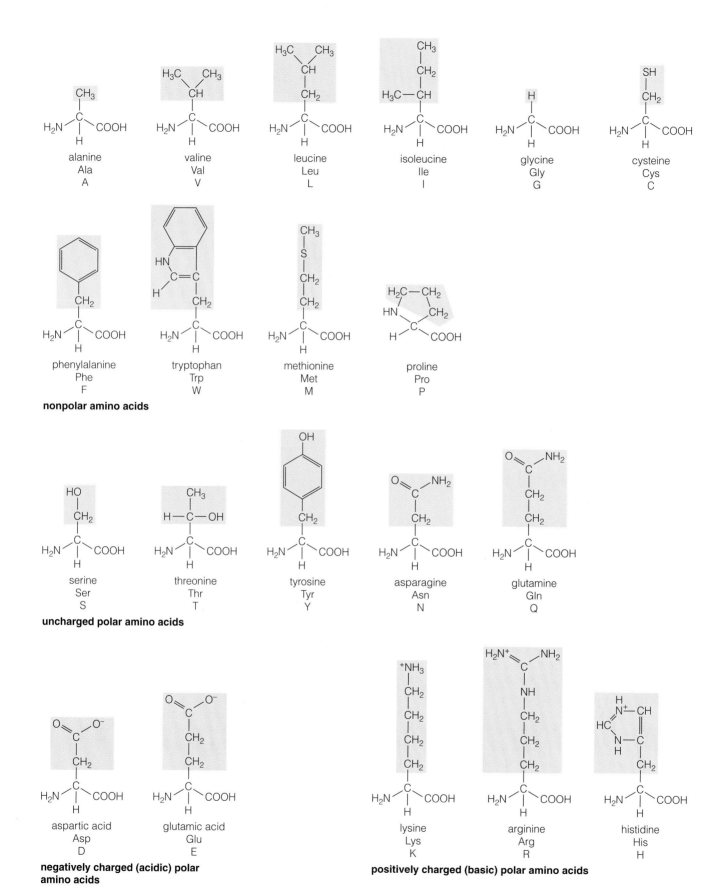

Figure 4.1 The 20 amino acids used in translation, arranged by category. Each amino acid has a three-letter and a one-letter abbreviation. The differing R groups are shaded green.

R group, and the chemical properties of the R groups define categories of amino acids, as shown in Figure 4.1. The proportions and locations of amino acids and the chemical properties of their R groups determine the overall function of the protein.

In translation, amino acids are attached to each other through the formation of **peptide bonds**. A peptide bond forms in an enzyme-catalyzed reaction between the carboxyl group of one amino acid and the amino group of another. This reaction takes place on the ribosome during translation. The reaction removes a water molecule as it forms the peptide bond, which is a covalent bond between the carbon in the carboxyl group of one amino acid and the nitrogen in the amino group of another amino acid (Figure 4.2).

Many amino acids may attach to each other by peptide bonds, forming a long chain called a **polypeptide** (Figure 4.3). Each polypeptide has an amino group on one end and a carboxyl group on the other end. Polypeptides are synthesized in the amino → carboxyl direction, and amino acid sequences are written in this direction. Following translation, a polypeptide may be modified, and in some cases combined with another polypeptide, to become a functional protein.

~

The 20 amino acids that are utilized in translation differ in their chemical properties.

4.2 THE GENETIC CODE

The "message" in a messenger RNA is the segment of nucleotides that translates into a protein. This implies that there is a correspondence between the information contained in the mRNA, which is composed of a specific sequence of nucleotides, and the information contained in a polypeptide, which is composed of a specific sequence of amino acids. In fact, the mRNA and the polypeptide are **colinear**, which means that the $5' \rightarrow 3'$ linear sequence of nucleotides in a segment of the mRNA corresponds directly to the amino → carboxyl linear sequence of amino acids in the polypeptide. This was demonstrated in 1964 by the experimental work highlighted in Example 4.1.

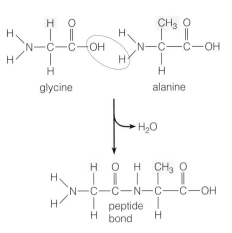

Figure 4.2 Formation of a peptide bond between two amino acids.

From Nucleotides to Amino Acids

Convinced that the nucleotide sequence in mRNA determines the amino acid sequence of a protein, we may ask, How many nucleotides are needed to specify each amino acid? This was the first question facing researchers in the 1950s who wanted to decipher the **genetic code**, the pattern by which nucleotides in an mRNA correspond to the amino acids in the encoded protein. Let's trace through the reasoning they used to answer this question.

Proteins are composed of varying combinations of up to 20 different amino acids, whereas RNA is composed of only four nucleotides. If one nucleotide corresponded to one amino acid, it would be possible to code for only four amino acids. If each amino acid were specified by a combination of two nucleotides, it would be possible to code for 16 amino acids because there are 4^2, or 16, possible combinations of two nucleotides each. We are still four combinations short of specifying all 20 amino acids. If we increase the number of nucleotides per amino acid to three, then there are 4^3, or 64, possibilities—far more than needed.

It turns out that the genetic code indeed consists of triplet nucleotide combinations. This means that the genetic code must have some redundancy in it. In other words, some amino acids may each be encoded by several

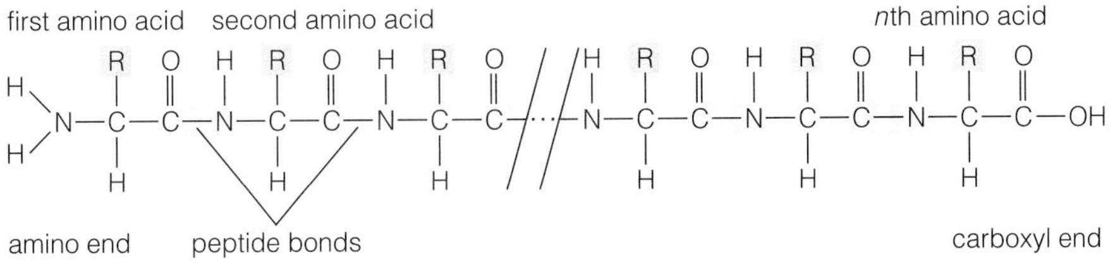

Figure 4.3 A polypeptide, a linear chain of amino acids linked by peptide bonds.

second base of codon

first base of codon	U	C	A	G	third base of codon
U	UUU Phe F / UUC	UCU / UCC Ser S / UCA / UCG	UAU Tyr Y / UAC	UGU Cys C / UGC	U / C
U	UUA Leu L / UUG	UCA / UCG	UAA / UAG Ter	UGA Ter / UGG Trp W	A / G
C	CUU / CUC Leu L / CUA / CUG	CCU / CCC Pro P / CCA / CCG	CAU His H / CAC — CAA Gln Q / CAG	CGU / CGC Arg R / CGA / CGG	U / C / A / G
A	AUU / AUC Ile I / AUA — AUG Met M	ACU / ACC Thr T / ACA / ACG	AAU Asn N / AAC — AAA Lys K / AAG	AGU Ser S / AGC — AGA Arg R / AGG	U / C / A / G
G	GUU / GUC Val V / GUA / GUG	GCU / GCC Ala A / GCA / GCG	GAU Asp D / GAC — GAA Glu E / GAG	GGU / GGC Gly G / GGA / GGG	U / C / A / G

Figure 4.4 The genetic code. The three-letter and one-letter abbreviations of the amino acids are given. "Ter" denotes a termination codon. The initiation codon, AUG, is shaded green, and the three termination codons—UAA, UAG, and UGA—are shaded red.

triplet combinations of nucleotides. Because of this redundancy, the genetic code is said to be a degenerate code.

The questions of which triplets of nucleotides code for which amino acids were all answered in the early 1960s when the genetic code was deciphered. The genetic code has 64 triplet nucleotide combinations, each called a **codon**. All but 3 of the 64 codons specify amino acids. The 3 codons that do not specify amino acids (UAA, UAG, and UGA) are signals for termination of translation and are therefore called **termination codons** (or **stop codons**). Figure 4.4 gives the genetic code. One of the most intriguing aspects of the genetic code is its near universality. The same genetic code is used for translation in every organism from bacteria to mammals, with only a few isolated exceptions.

As the genetic code is preserved in all forms of life, it has strong evolutionary significance. It also has practical significance. Because the genetic code of humans is identical to the genetic code of bacteria, it is possible to synthesize large quantities of valuable proteins from human genes that have been inserted into bacteria. For example, human insulin is a protein hormone needed for the proper metabolism of glucose. Many people who suffer from diabetes can be treated with supplementary insulin. For years, the source of insulin for diabetics was bovine (cattle) or porcine (hog) insulin. Porcine insulin differs from human insulin by one amino acid, and bovine insulin differs by two. These slight differences may cause adverse immune reactions in some people. However, once a synthetic gene for human insulin was constructed based on the genetic code and inserted into bacteria, it became possible to produce human insulin in bacterial cultures. This genetically engineered "human" insulin is now used routinely, with no adverse reaction, by many who suffer from diabetes.

Let's now review the experiments that first demonstrated the colinear relationship between a segment of mRNA and the polypeptide it encodes.

Example 4.1 Colinearity of a gene and its protein.

In 1964, Yanofsky et al. (*Proceedings of the National Academy of Sciences, USA* 51:266–272) reported the results of experiments in which they compared the positions of mutations in the *trpA* (tryptophan synthetase A) gene in *E. coli* with amino acid alterations in the protein encoded by this gene. Figure 4.5a shows the relative locations in the gene of six mutations (designated 23, 46, 58, 78, 169, and 187). At the time, the DNA sequence of the *trpA* gene had not been determined. Yanofsky et al. identified the relative positions of the mutations using genetic analysis. Figure 4.5b provides the amino acid sequences of the segment of protein encoded by this portion of the gene for the nonmutant and mutant types. In 1979, Nichols and Yanofsky (*Proceedings of the National Academy of Sciences, USA* 76:5244–5248) published the DNA sequence of the *trpA* gene. Figure 4.5c shows the DNA sequence of the segment of the *trpA* gene in which these six mutations are found.

Problem: (a) In their 1964 article, Yanofsky et al. concluded that the positions of the mutations in the *trpA* gene correspond in a linear fashion to the positions of amino acid substitutions in the *trpA* proteins of the mutant bacterial strains. Using the data from Figures 4.5a and b, show how the mutations and amino acid sequences are colinear. (b) Each of the mutations represents a substitution of a single nucleotide in the DNA. Using the genetic code, determine the nucleotide substitution for each of the mutations. (c) Align the DNA sequence with the mutations in the gene and the amino acid chain, and show the positions of the mutations in the DNA sequence.

Solution: (a) Figure 4.6a shows that the positions of mutations in the gene are colinear with the positions of amino acid substitutions in the protein. (b) It is possible to identify the nucleotide substitutions in DNA that caused each of the amino acid substitutions by determining which codons encode the relevant amino acids in the original and mutant amino acid sequences. This has been done in Table 4.1. (c) When the DNA sequence is aligned with the positions of mutations and the amino acid sequence, each mutation in the DNA sequence can

Table 4.1 Information for Example 4.1: Nucleotide Sequences of the Nonmutant and Mutant Versions of Each Codon and the Nucleotide Change in the DNA*

Mutation	Original Amino Acid and Codon	Mutant Amino Acid and Codon	Nucleotide Substitution in DNA
23	Gly GGPu	Arg AGPu	G → A
46	Gly GGPu	Glu GAPu	G → A
58	Gly GGPy	Asp GAPy	G → A
78	Gly GGPy	Cys UGPy	G → T
169	Ser UCPu	Leu UUPu	C → T
187	Gly GGN	Val GUN	G → T

*An N refers to any of the four nucleotides, Pu indicates that a purine is present in this position, and Py indicates that a pyrimidine is present in this position.

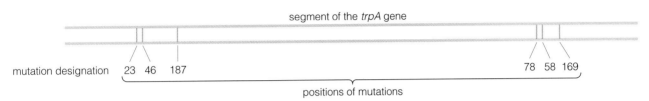

a Positions of six mutations in a segment of the *trpA* gene of *E. coli*.

Nonmutant
ProProLeuGlnGlyPheGlyIleSerAlaProAspGlnValLysAlaAlaIleAspAlaGlyAlaAlaGlyAlaIleSerGlySerAlaIleValLys

Mutant 23
ProProLeuGlnArgPheGlyIleSerAlaProAspGlnValLysAlaAlaIleAspAlaGlyAlaAlaGlyAlaIleSerGlySerAlaIleValLys

Mutant 46
ProProLeuGlnGluPheGlyIleSerAlaProAspGlnValLysAlaAlaIleAspAlaGlyAlaAlaGlyAlaIleSerGlySerAlaIleValLys

Mutant 58
ProProLeuGlnGlyPheGlyIleSerAlaProAspGlnValLysAlaAlaIleAspAlaGlyAlaAlaGlyAlaIleSerAspSerAlaIleValLys

Mutant 78
ProProLeuGlnGlyPheGlyIleSerAlaProAspGlnValLysAlaAlaIleAspAlaGlyAlaAlaGlyAlaIleSerCysSerAlaIleValLys

Mutant 169
ProProLeuGlnGlyPheGlyIleSerAlaProAspGlnValLysAlaAlaIleAspAlaGlyAlaAlaGlyAlaIleSerGlyLeuAlaIleValLys

Mutant 187
ProProLeuGlnGlyPheValIleSerAlaProAspGlnValLysAlaAlaIleAspAlaGlyAlaAlaGlyAlaIleSerGlySerAlaIleValLys

b Amino acid sequences from the nonmutant strain and six mutant strains. Amino acid substitutions are shaded in red.

CCTCCATTGCAGGGATTTGGTATTTCCGCCCCGGATCAGGTAAAAGCAGCGATTGATGCAGGAGCTGCGGGCGCGATTTCTGGTTCGGCCATTGTTAAA

c DNA sequence of the nonmutant gene segment

Figure 4.5 Information for Example 4.1: The effect of mutations on amino acid sequence.

be identified and matched with its corresponding amino acid substitution (Figure 4.6b). The mutations in the DNA sequence are colinear with the positions of the mutations in the gene and the amino acid substitutions.

The universal genetic code has 64 triplet nucleotide codons, of which 61 specify amino acids and 3 signal termination of translation.

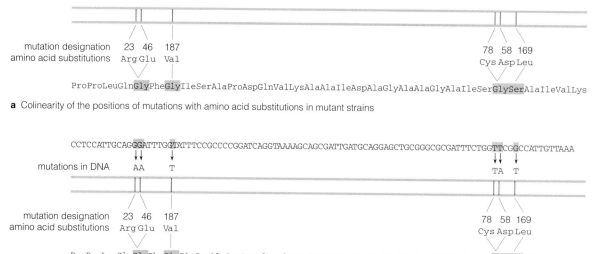

a Colinearity of the positions of mutations with amino acid substitutions in mutant strains

CCTCCATTGCAGGGATTTGGTATTTCCGCCCCGGATCAGGTAAAAGCAGCGATTGATGCAGGAGCTGCGGGCGCGATTTCTGGTTCGGCCATTGTTAAA

mutations in DNA   AA   T   TA   T

mutation designation   23   46   187   78   58   169
amino acid substitutions   Arg Glu   Val   Cys Asp Leu

ProProLeuGlnGlyPheGlyIleSerAlaProAspGlnValLysAlaAlaIleAspAlaGlyAlaAlaGlyAlaIleSerGlySerAlaIleValLys

b Colinearity of the DNA sequence, positions of mutations, and amino acid substitutions in mutant strains

Figure 4.6 Solution to Example 4.1: Colinearity of mutations and alterations of amino acid sequence.

The Reading Frame for Translation

To translate an mRNA, ribosomes bind to the mRNA and read the codons that specify the polypeptide. Translation does not begin immediately at the 5′ end of the mRNA. Instead, somewhere near the 5′ end is an **initiation codon** (or **start codon**) that codes for the first amino acid in the polypeptide. The initiation codon is usually the codon AUG, which codes for methionine in eukaryotes and N-formyl methionine in prokaryotes. At sites beyond the initiation site, the codon AUG codes for methionine.

The initiation codon not only indicates where translation begins, it also establishes the **reading frame** for translation. The reading frame is governed by two facts: (1) codons in the mRNA do not overlap, and (2) there are no intervening nucleotides between codons. In other words, the next three nucleotides after the initiation codon form the second codon, and the following three nucleotides form the third codon, and so on—a pattern that continues uninterrupted throughout the reading frame. For instance, suppose that the following sequence of ribonucleotides, which includes the initiation codon, is part of an mRNA:

5′ . . . GCCACCAUGGCUGGGAGUCAC . . . 3′

The boldface AUG sequence is the initiation codon and codes for methionine. The second codon is composed of the next three nucleotides, GCU, and codes for alanine. The third codon, GGG, codes for glycine. Translation continues moving in the 5′→3′ direction, adding amino acids according to each triplet codon to produce the following:

5′ . . . GCCACC**AUG**GCUGGGAGUCAC . . . 3′

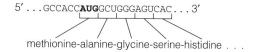

methionine-alanine-glycine-serine-histidine . . .

Looking at the first two codons, AUGGCU, we find the sequence GGC within it (AU<u>GGC</u>U). If these three nucleotides were translated as a codon, they would code for glycine. However, as codons do not overlap, the nucleotide sequence GGC is out of frame and will not be recognized as a codon.

Translation continues from codon to codon uninterrupted until a termination codon is reached. At this point, the ribosomes separate from the mRNA, and transcription ceases. Because termination codons do not code for any amino acid, the amino acid encoded by the codon just before the termination codon is the last amino acid in the polypeptide.

We now have the requisite information for recognizing the reading frame of an mRNA and deciphering the message. In the following example, we'll derive the amino acid sequence of a polypeptide from the nucleotide sequence in the mRNA.

Example 4.2 Determining the amino acid sequence of a polypeptide from the nucleotide sequence.

In 1979, Bell et al. (*Nature* 282:525–527) published the nucleotide sequence of the mature mRNA encoded by the human insulin gene, shown in Figure 4.7a.

Problem: Using this mRNA sequence, identify the initiation codon (in this example, the first AUG codon in the mRNA) and the amino acid sequence of the polypeptide that arises through translation.

Solution: The first AUG codon is 60 nucleotides downstream from the 5′ cap. As the reading frame in a mature mRNA consists of nonoverlapping and

5′ cap AGCCCUCCAGGACAGGCUGCAUCAGAAGAGGCCAUCAAGCAGAUCACUGUCCUUCU-

GCCAUGGCCCUGUGGAUGCGCCUCCUGCCCCUGCUGGCGCUGCUGGCCCUCUGGGG-

ACCUGACCCAGCCGCAGCCUUUGUGAACCAACACCUGUGCGGCUCACACCUGGUGG-

AAGCUCUCUACCUAGUGUGCGGGGAACGAGGCUUCUUCUACACACCCAAGACCCGC-

CGGGAGGCAGAGGACCUGCAGGUGGGGCAGGUGGAGCUGGGCGGGGGCCCUGGUGC-

AGGCAGCCUGCAGCCCUUGGCCCUGGAGGGGUCCCUGCAGAAGCGUGGCAUUGUGG-

AACAAUGCUGUACCAGCAUCUGCUCCCUCUACCAGCUGGAGAACUACUGCAACUAG-

ACGCAGCCUGCAGGCAGCCCCACACCCGCCGCCUCCUGCACCGAGAGAGAUGGAAU-

AAAGCCCUUGAACCAGC 3′ poly(A)

a The nucleotide sequence of the mature mRNA encoded by the human insulin gene

5′ cap AGCCCUCCAGGACAGGCUGCAUCAGAAGAGGCCAUCAAGCAGAUCACUGUCCUUCU-

```
        Met Ala Leu Trp Met Arg Leu Leu Pro Leu Leu Ala Leu Leu
    GCC AUG GCC CUG UGG AUG CGC CUC CUG CCC CUG CUG GCG CUG CUG

    Ala Leu Trp Gly Pro Asp Pro Ala Ala Ala Phe Val Asn Gln His
    GCC CUC UGG GGA CCU GAC CCA GCC GCA GCC UUU GUG AAC CAA CAC

    Leu Cys Gly Ser His Leu Val Glu Ala Leu Tyr Leu Val Cys Gly
    CUG UGC GGC UCA CAC CUG GUG GAA GCU CUC UAC CUA GUG UGC GGG

    Glu Arg Gly Phe Phe Tyr Thr Pro Lys Thr Arg Arg Glu Ala Glu
    GAA GGC CGA UUC UUC UAC ACA CCC AAG ACC CGC CGG GAG GCA GAG

    Asp Leu Gln Val Gly Gln Val Glu Leu Gly Gly Gly Pro Gly Ala
    GAC CAG CUG GUG GGG CAG GUG GAG CUG GGC GGG GGC CCU GGU GCA

    Gly Ser Leu Gln Pro Leu Ala Leu Glu Gly Ser Leu Gln Lys Arg
    GGC AGC CUG CAG CCC UUG GCC CUG GAG GGG UCC CUG CAG AAG CGU

    Gly Ile Val Glu Gln Cys Cys Thr Ser Ile Cys Ser Leu Tyr Gln
    GGC AUU GUG GAA CAA UGC UGU ACC AGC AUC UGC UCC CUC UAC CAG

    Leu Glu Asn Tyr Cys Asn Ter
    CUG GAG AAC UAC UGC AAC UAG ACGCAGCCUGCAGGCAGCCCCACACCCG-

    CCGCCUCCUGCACCGAGAGAGAUGGAAUAAAGCCCUUGAACCAGC 3′ poly(A)
```

b Solution to Example 4.2: Amino acids and their corresponding codons.

Figure 4.7 Information for Example 4.2: The human insulin mRNA and the polypeptide it encodes.

uninterrupted codons, it is a simple matter to determine the amino acids of the polypeptide using the genetic code, as illustrated in Figure 4.7b. The polypeptide encoded by the insulin gene is known as preproinsulin and is processed substantially before becoming functional insulin.

The genetic code is nonoverlapping. Once the reading frame is established from the initiation codon, it proceeds every three nucleotides from that point on, adding amino acids until a termination codon is reached.

4.3 RIBOSOMES, THE SITES OF TRANSLATION

Ribosomes are composed of rRNA molecules and numerous proteins. The actual number of rRNAs and proteins and the final size of the intact ribosome vary among species. For example, mammalian ribosomes are composed of four rRNAs and about 80 proteins, whereas *E. coli* ribosomes are composed of three rRNAs and 52 proteins. Although there are many more proteins than rRNAs in a ribosome, the rRNAs are much larger than

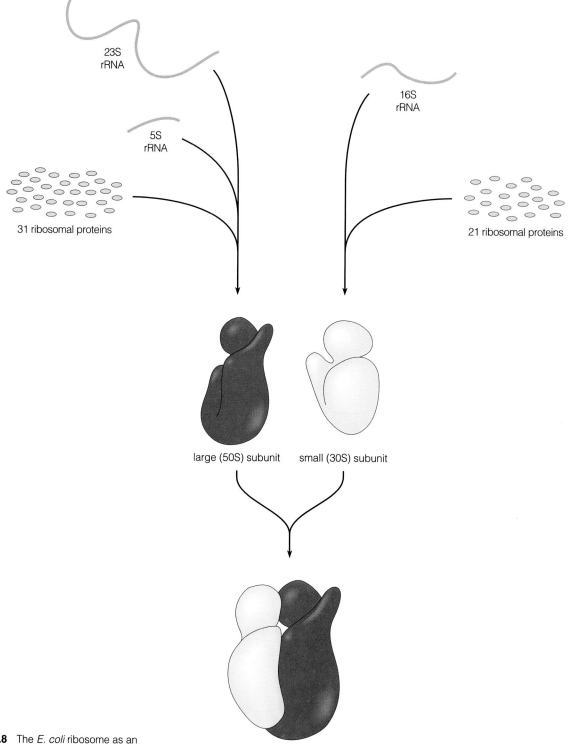

23S
rRNA

5S
rRNA

16S
rRNA

31 ribosomal proteins

21 ribosomal proteins

large (50S) subunit

small (30S) subunit

intact ribosome (70S)

Figure 4.8 The *E. coli* ribosome as an example of prokaryotic ribosomes.

the proteins and generally account for over 60% of the mass of the ribosome.

Prior to translation, the rRNAs and proteins are assembled into two separate subunits, the **large subunit** and the **small subunit**. These subunits form an intact ribosome by attaching to an mRNA during initiation of translation. Both prokaryotic and eukaryotic ribosomes have large and small subunits with similar structures, although ribosomes are somewhat larger in eukaryotes than in prokaryotes.

The *E. coli* ribosome has been well studied and is representative of prokaryotic ribosomes (Figure 4.8). The in-

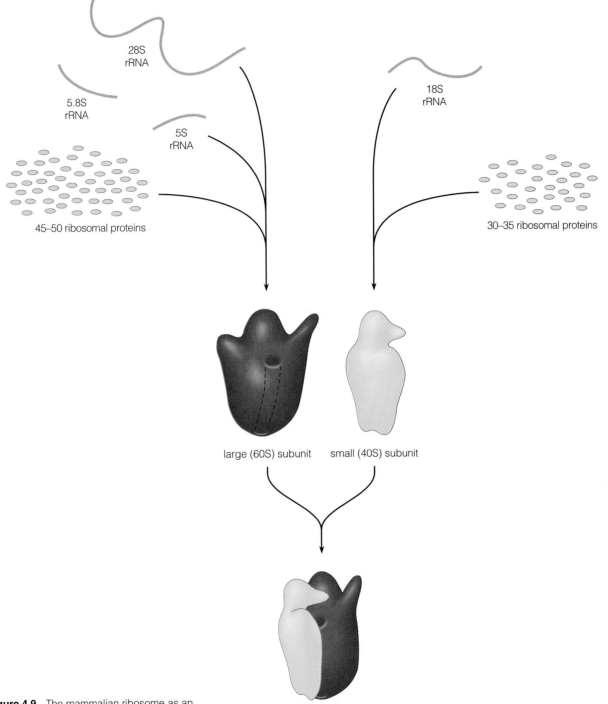

28S
rRNA

5.8S
rRNA

5S
rRNA

18S
rRNA

45–50 ribosomal proteins

30–35 ribosomal proteins

large (60S) subunit   small (40S) subunit

intact ribosome (80S)

Figure 4.9 The mammalian ribosome as an
example of eukaryotic ribosomes.

tact ribosome has a Svedberg value of 70S and consists of
a 50S large subunit and a 30S small subunit (remember
that Svedberg values are not additive). The large subunit
contains two rRNAs, a 23S rRNA (2904 nucleotides) and
a 5S rRNA (120 nucleotides), along with 31 proteins. The
small subunit contains a single 16S rRNA (1541 nu-

cleotides) and 21 proteins. Each subunit has a specific
three-dimensional shape that allows the two subunits to
interlock with each other.

The mammalian ribosome is among the best characterized of eukaryotic ribosomes (Figure 4.9). The intact
ribosome has a Svedberg value of 80S with a large 60S

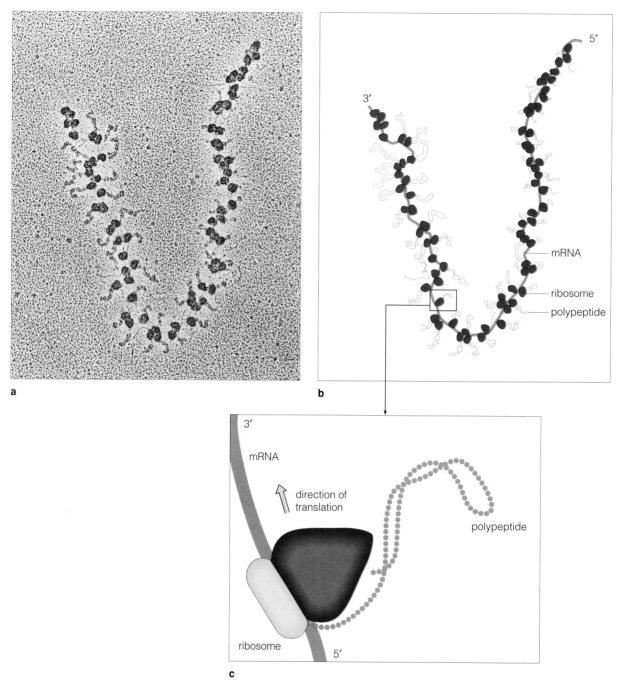

Figure 4.10 A eukaryotic polysome composed of multiple ribosomes translating a single mRNA molecule. Notice how the polypeptide chains grow as the ribosomes move farther along the mRNA. The longer polypeptide chains have folded back on themselves. (Photo from Franke, C., J. E. Edström, and O. L. Miller, Jr., 1992. Electron microscopic visualization of a discrete class of giant translation units in salivary gland cells of *Chironomus tentans. EMBO Journal* 1:59–62.)

subunit and a small 40S subunit. The large subunit contains three rRNAs, a 28S rRNA (4718 nucleotides), a 5.8S rRNA (160 nucleotides), and a 5S rRNA (120 nucleotides). The small subunit contains an 18S rRNA (1874 nucleotides). The number of proteins in eukaryotic ribosomes varies among species, but is about 45–50 in the large subunit and 30–35 in the small subunit.

During translation, several ribosomes may attach one after the other onto an mRNA and proceed through translation as a chain of ribosomes, each ribosome translating one polypeptide. When this happens, the assembled ribosomes and the mRNA together are called a **polysome** (Figure 4.10).

~

Ribosomes attach to an mRNA and translate the codons into a polypeptide.

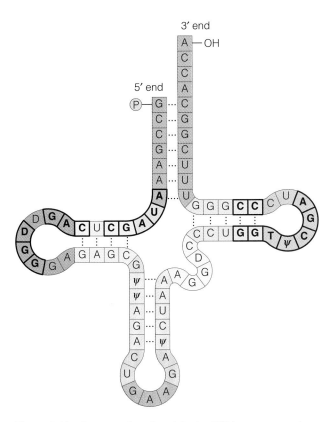

Figure 4.11 A mammalian phenylalanine tRNA as an example of the cloverleaf representation of base pairing in tRNAs. The modified bases are: D = dihydrouracil, T = thymine, and ψ = pseudouracil, all originally transcribed as uracil.

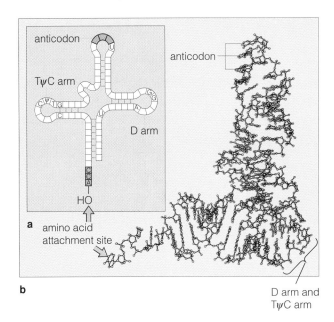

Figure 4.12 The two-dimensional representation **(a)** and three-dimensional structure **(b)** of tRNA. Helical winding in the double-stranded portions of the molecule bends the molecule into an **L** shape. (Adapted from a drawing by Sung Hou Kim.)

4.4 THE ROLE OF tRNA

By itself, a ribosome cannot place the appropriate amino acids in the growing polypeptide chain. This is the function of transfer RNAs (tRNAs). Each amino acid attaches to a tRNA that is responsible for carrying the amino acid to the ribosome and placing it in the proper order according to the nucleotide sequence in the mRNA.

tRNA structure

Transfer RNAs are remarkably similar in all organisms. They are quite small, only 74–95 nucleotides, and they all have the same general structure. Following transcription and processing, over half of the nucleotides base-pair with one another to form a cloverleaf shape when drawn in two dimensions. As shown in Figure 4.11, the cloverleaf consists of four major arms and a small extra arm.

The **acceptor arm** is so named because it forms the part of the molecule that accepts the amino acid. It includes both the 5' and 3' ends of the linear molecule and has four unpaired nucleotides on the 3' end extending beyond the paired nucleotides. The final three nucleotides on the 3' end are CCA. The amino acid attaches

to the terminal A on the 2' or 3' hydroxyl group. This position on the acceptor arm where the amino acid attaches is called the **amino acid attachment site**.

The **anticodon arm** lies opposite the acceptor arm in the cloverleaf and consists of a base-paired stem and a loop of seven unpaired nucleotides, the center three of which form the **anticodon** (Figure 4.11). This site is critical because the three nucleotides in an anticodon pair with the three nucleotides in a codon in the mRNA during translation.

Two other arms, the **D arm** and the **TψC arm**, each have a base-paired stem and a loop of unpaired nucleotides. These arms help form the three-dimensional structure necessary for enzymes to recognize the tRNA and for the tRNA to interact with the ribosome. The Greek letter ψ (psi) refers to the modified base pseudouracil that is between a T and C in the unpaired loop of the TψC arm. The T is thymine that is initially transcribed as U but later methylated to form T. The letter D refers to dihydrouracil, a modified nucleotide in the D arm.

There is also a small **extra arm**, usually consisting entirely of unpaired nucleotides, between the anticodon arm and the TψC arm. Much of the variation among tRNAs in the number of nucleotides and their sequence is in the extra arm.

The cloverleaf shape of tRNAs exists only in two-dimensional representations of their base pairing. In the cell, tRNAs assume a three-dimensional structure that has an L shape with the amino acid attachment site at one end and the anticodon at the other (Figure 4.12). The TψC arm and the D arm are adjacent to one another on the outside corner of the L. The three-dimensional

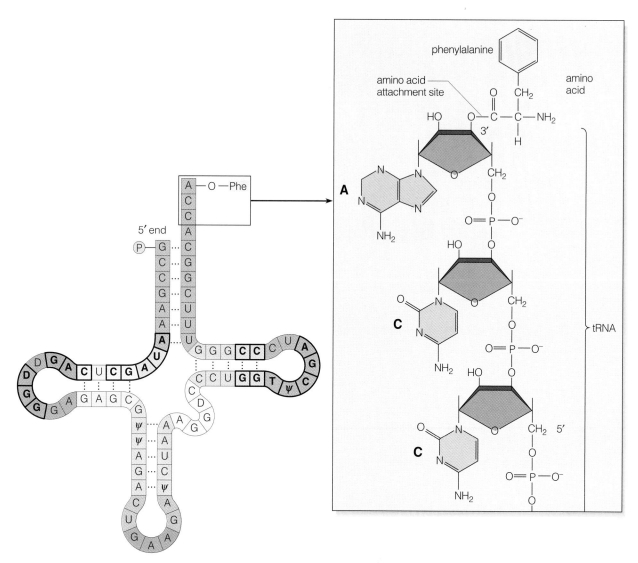

Figure 4.13 A charged tRNA. The inset highlights the bond between the amino acid (in this case, phenylalanine) and the amino acid attachment site on its tRNA. The 3' end of each tRNA ends with the nucleotides CCA. The 2' or 3' OH group of the terminal nucleotide, A, binds to the carboxyl group of the amino acid.

structure of all tRNAs must be essentially the same, because they all must fit into the same sites in the ribosome during translation. However, there are minor structural differences that distinguish one type of tRNA from another.

~

Represented in two dimensions, transfer RNAs have a cloverleaf shape with an amino acid attachment site in the acceptor arm and an anticodon in the anticodon arm. The three nucleotides in an anticodon pair with the three nucleotides in a codon of the mRNA during translation.

Amino Acid Specificity of tRNAs

A particular tRNA accepts only one of the 20 amino acids at the amino acid attachment site—the amino acid that corresponds, according to the genetic code, to the sequence of nucleotides in the anticodon. For example, a tRNA with the anticodon sequence 3' UAC 5' accepts only methionine, because its anticodon pairs with AUG, the codon for methionine.

Note that the anticodon is written in the $3' \rightarrow 5'$ direction. Although nucleotide sequences are conventionally written in the $5' \rightarrow 3'$ direction, anticodons are an exception. We write them in the $3' \rightarrow 5'$ direction to correspond with their complementary codons, which are written in the usual $5' \rightarrow 3'$ direction. Thus, the codon for methionine, AUG, is written $5' \rightarrow 3'$, whereas its corresponding anticodon, 3' UAC 5', is written $3' \rightarrow 5'$. The codon-anticodon pairing can be written as follows:

codon   5' AUG 3'
anticodon   3' UAC 5'

In this book, anticodons are written in the $3' \rightarrow 5'$ direction and are always labeled with 3' and 5' designations to indicate this exception to convention.

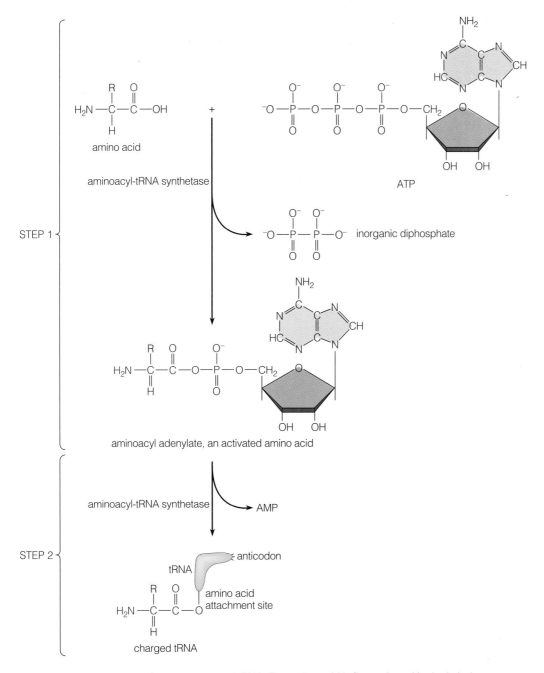

Figure 4.14 The reactions forming a charged tRNA. The amino acid is first activated by hydrolysis of two phosphate groups from ATP to attach AMP (adenosine monophosphate) to the carboxyl group of the amino acid. The activated amino acid is then attached to the amino acid attachment site of the tRNA, releasing AMP.

A tRNA may be in either of two forms. It may be free, with no amino acid attached, or it may have its specific amino acid bound to the amino acid attachment site on its acceptor arm. A **charged (aminoacylated) tRNA** is one that has its amino acid attached. A charged tRNA has its amino acid attached by a bond between the 2' or 3' hydroxyl group on the 3' end of the tRNA and the carboxyl group of the amino acid, as illustrated in Figure 4.13.

Charging is catalyzed by enzymes called **aminoacyl-tRNA synthetases**. These enzymes bind amino acids to the amino acid attachment site through a two-step process (Figure 4.14). First, the aminoacyl-tRNA synthetase causes the amino acid to react with ATP to form aminoacyl adenylate, which is an amino acid with an adenosine nucleotide attached to its carboxyl group by a phosphodiester bond. Energy for the reaction is derived from breaking the phosphate bonds in the ATP molecule that contributes the adenosine. At this point the amino acid is activated, or in other words, prepared for attachment to the tRNA. Second, the synthetase bonds the activated amino acid to the amino acid attachment site on the tRNA, releasing adenosine mono-phosphate (AMP)

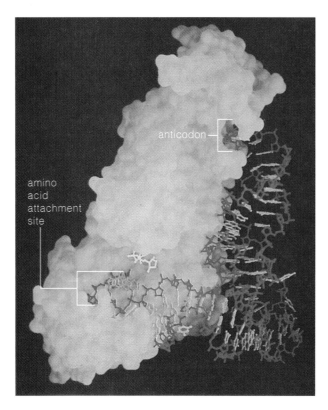

Figure 4.15 An aminoacyl-tRNA synthetase interacting with the tRNA and ATP as the tRNA is charged with its amino acid. The enzyme is colored green, the sugar-phosphate backbone of the tRNA is colored red, and the bases are orange. The yellow molecule is ATP. (Reprinted with permission from Rould, M. A., J. J. Perona, D. Söll, and T. A. Steitz. 1989. Structure of *E. coli* glutaminyl tRNA complexed with tRNA^Gln and ATP at 2.8Å resolution. *Science* 246:1135–1142. Copyright 1989 American Association for the Advancement of Science.)

Example 4.3 Nucleotide sequence and secondary structure of a tRNA.

In 1965, Holley et al. (*Science* 147:1462–1465) published the complete nucleotide sequence of a tRNA from yeast. The sequence is provided in Figure 4.16a.

Problem: **(a)** Construct the base-pairing structure of this tRNA, labeling each of the arms, the amino acid attachment site, and the anticodon. **(b)** This tRNA is specific for which amino acid?

Solution: **(a)** Figure 4.16b provides the structure of the tRNA. **(b)** The anticodon in this tRNA is 3′ CGI 5′. According to the base-pairing rules, the first two nucleotides in the codon recognized by this tRNA must be GC. All four codons that begin with GC encode alanine; therefore, this tRNA is specific for alanine. Inosine, a modified nucleotide, occupies the third position of the anticodon. At this position in the anticodon, inosine may pair with U, C, or A (according to the wobble hypothesis, which we will discuss momentarily). Therefore, this tRNA can pair with the codons GCU, GCC, and GCA, all of which encode alanine.

Each tRNA binds to a specific amino acid. When the amino acid is attached, the tRNA is charged.

Degeneracy and the Wobble Hypothesis

Recall that the genetic code is said to be degenerate. Because there are 61 codons that specify amino acids, there might be 61 anticodons and, therefore, 61 different tRNAs. However, the actual number of anticodons and tRNAs is much lower because codon-anticodon pairing in the ribosome is precise only for the first two nucleotides of the codon. Codon-anticodon pairing at the third position in the codon does not follow the pairing rules as closely as pairing at the other two positions. Within the ribosome, the conformation of the anticodon loop is such that some anticodons can pair with two, or even three, codons. The base-pairing rules that apply only to the third nucleotide in codon-anticodon pairing are provided in Table 4.2.

The set of base-pairing rules in Table 4.2 is called the **wobble hypothesis**. There are several interesting features of the wobble hypothesis. The two purines, A and G, in the codon may both pair with U in the wobble site of the anticodon. The two pyrimidines, U and C, in the codon may both pair with G in the wobble site. Three nucleotides in the codon, U, C, and A, may pair with a modified nucleotide, inosine (I), in the wobble site. Inosine is a modified purine that is similar to guanine but lacks the amino group attached to the number 2 carbon in gua-

in the process. A three-dimensional model of an aminoacyl-tRNA synthetase catalyzing the charging of a tRNA is shown in Figure 4.15.

There is at least one aminoacyl-tRNA synthetase for each amino acid. The aminoacyl-tRNA synthetase specific for an amino acid also recognizes all the different tRNAs that carry that amino acid. Certain amino acids encoded by more than one codon may be carried by tRNAs with different anticodons. For example, the amino acid serine is encoded by six different codons and has three different tRNAs to match these codons (the wobble hypothesis, which is explained in the next section, explains why fewer than six tRNAs suffice). Each of these tRNAs is recognized by the aminoacyl-tRNA synthetase that is specific for serine. Different tRNAs that carry the same amino acid (and are therefore recognized by the same aminoacyl-tRNA synthetase) are called **cognate tRNAs**.

A tRNA from yeast was the first nucleic acid from nature to be fully sequenced. In the following example, we'll use that sequence to identify how the base pairs will form in the tRNA and to determine which amino acid attaches to this tRNA.

5′ GGGCGUGUmGGCGCGUAGDCGGDAGCGCdmGCUCCCUUIGCmI ψGGGAGAGDCUCCGGT ψCGAUUCCGGACUCGUCCACCA 3′

a Nucleotide sequence of a tRNA from yeast. The nucleotides with modified bases in the sequence are: D = dihydrouracil, T = thymine, ψ= pseudouracil, mG = methylguanine, dmG = dimethylguanine, I = inosine, mI = methylinosine. D, T, and ψ were originally transcribed as U, while mG, dmG, I, and mI were originally transcribed as G.

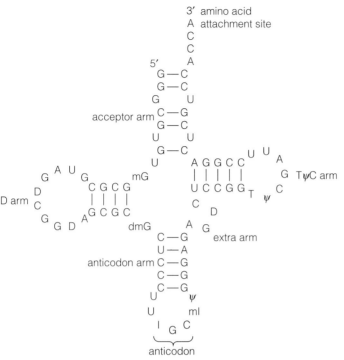

b Solution to Example 4.3, part a

Figure 4.16 Information for Example 4.3: Base pairing in a tRNA. (Redrawn with permission from Holley, R. W., J. Apgar, G. A. Everett, J. T. Madison, M. Marquisee, S. H. Mewrill, J. R. Penswick, and A. Zamir. 1964. Structure of a ribonucleic acid. *Science* 147:1462–1465.Copyright 1964 American Association for the Advancement of Science.)

Table 4.2 Codon-Anticodon Pairing at the Third Position in the Codon According to the Wobble Hypothesis	
5′ Nucleotide in Anticodon	**3′ Nucleotide in Codon**
U	A or G
C	G
A	U
G	U or C
I	U, C, or A

nine. The structure of inosine and some of the details of base pairing according to the wobble hypothesis are illustrated in Figure 4.17.

The four codons that begin with AU provide a good example of how one tRNA can pair with several codons in accordance with the wobble hypothesis (Figure 4.18). As the codon for methionine is AUG, the anticodon on a methionyl tRNA must be 3′ UAC 5′, according to the wobble rules. The three other codons that begin with AU all specify isoleucine, and require just one tRNA, with the anticodon 3′ UAI 5′.

~

The wobble hypothesis, which explains the pattern of degeneracy in the genetic code, provides pairing rules for the third nucleotide in a codon with its corresponding nucleotide in the anticodon.

4.5 DECIPHERING THE GENETIC CODE

The story of how the genetic code was deciphered is a fascinating one. In the mid 1950s, as the central dogma developed, the logical assumption was that the linear sequence of nucleotides in DNA corresponds to the linear sequence of amino acids in the protein through some sort of genetic code. Using mathematical logic (of the type we used at the beginning of section 4.2), geneticists determined that at least three nucleotides were needed to code for one amino acid, at least in some cases.

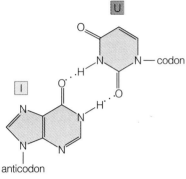

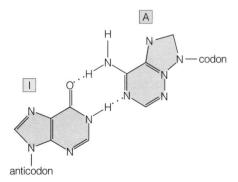

Figure 4.17 Examples of codon-anticodon base pairing at the third position of the codon according to the wobble hypothesis.

Demonstration of Nonoverlapping Codons

Almost nothing was known about tRNA during the 1950s, so it was generally assumed that amino acids were assembled directly on the mRNA molecule. This required not only a linear but also a physical correspondence between amino acids and the nucleotides that

coded for them. The spacing of amino acids in a polypeptide chain is about the same as the spacing of nucleotides in DNA, implying that a one-to-one correspondence between nucleotides and amino acids was likely. As three nucleotides were needed for each amino acid, a genetic code of overlapping triplet codons was the only way to reconcile triplet codons with a one-to-one correspondence of nucleotides and amino acids. For example, the RNA sequence AAGACGCUAC would have eight overlapping triplet codons: AAG, AGA, GAC, ACG, CGC, GCU, CUA, and UAC, moving one nucleotide along the molecule for each codon.

By 1957, partial amino acid sequences for several proteins had been determined. Using these amino acid sequences, Sydney Brenner compared each amino acid with its neighbors and calculated the number of overlapping triplet codons that would be necessary to code for the known sequences. The minimum number of triplet codons required was 70, which is more than the 64 possible combinations—meaning that a genetic code based entirely on overlapping triplet codons was not possible. This conclusion also implied that the assumption of a one-to-one spatial correspondence of nucleotides with amino acids was likewise incorrect.

As partial amino acid sequences of proteins were determined, additional evidence against the overlapping codon hypothesis accumulated. It was logical to assume that a mutation in the DNA would alter the amino acid sequence for a protein. Under the assumptions of the overlapping codon hypothesis, a single-nucleotide substitution mutation should affect three adjacent amino acids, because the altered nucleotide is part of three overlapping triplet codons (Figure 4.19a). However, if codons do not overlap, a single-nucleotide substitution mutation should affect only one amino acid (Figure 4.19b). Studies such as the one highlighted in the example that follows, provided the experimental evidence that researchers used to determine whether or not codons overlapped.

Example 4.4 Testing the overlapping codon hypothesis.

Nitrous acid is a chemical mutagen that causes single-nucleotide substitutions in DNA or RNA. In 1960, Tsugita and Fraenkel-Conrat (*Proceedings of the National Academy of Sciences, USA* 46:636–642) reported that a mutation induced by nitrous acid caused a single–amino acid substitution in the coat protein of tobacco mosaic virus. The amino acid sequences near the C-terminal end of the normal and mutant proteins are provided in Figure 4.20.

Problem: **(a)** What aspect of these results suggests that the genetic code is nonoverlapping? **(b)** What nucleotide substitution in the mRNA caused this amino acid substitution?

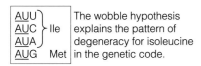

The wobble hypothesis explains the pattern of degeneracy for isoleucine in the genetic code.

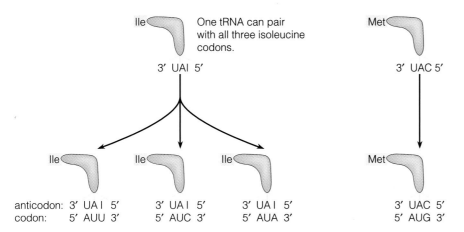

Ile — One tRNA can pair with all three isoleucine codons.

3′ UAI 5′

Met

3′ UAC 5′

Ile   Ile   Ile

Met

anticodon: 3′ UA I 5′   3′ UA I 5′   3′ UA I 5′   3′ UAC 5′
codon: 5′ AUU 3′   5′ AUC 3′   5′ AUA 3′   5′ AUG 3′

Figure 4.18 An example of wobble base pairing. A single isoleucine tRNA with the anticodon 3′ UAI 5′ pairs with three isoleucine codons: AUU, AUC, and AUA. Because G does not pair with I, the codon AUG does not pair with the isoleucine tRNA. AUG pairs with 3′ UAC 5′, which is the anticodon on a methionine tRNA.

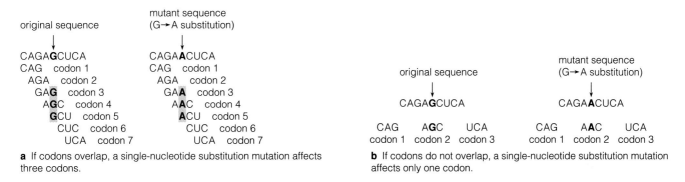

original sequence

mutant sequence
(G→A substitution)

CAGA**G**CUCA
CAG   codon 1
 AGA   codon 2
  GA**G**   codon 3
   A**G**C   codon 4
    GCU   codon 5
     CUC   codon 6
      UCA   codon 7

CAGA**A**CUCA
CAG   codon 1
 AGA   codon 2
  GA**A**   codon 3
   A**A**C   codon 4
    ACU   codon 5
     CUC   codon 6
      UCA   codon 7

a If codons overlap, a single-nucleotide substitution mutation affects three codons.

original sequence

CAGA**G**CUCA

mutant sequence
(G→A substitution)

CAGA**A**CUCA

CAG   A**G**C   UCA
codon 1 codon 2 codon 3

CAG   A**A**C   UCA
codon 1 codon 2 codon 3

b If codons do not overlap, a single-nucleotide substitution mutation affects only one codon.

Figure 4.19 Two competing hypotheses: nucleotides in an mRNA read as **(a)** overlapping and **(b)** nonoverlapping codons.

normal   SerSerPheGluSerSerSerGlyLeuValTyrThrSerGlyProAlaThrCOOH

mutant   SerSerPheGluSerSerSerGlyLeuValTyrThrSerGlyLeuAlaThrCOOH

Figure 4.20 Information for Example 4.4: Amino acid sequence at the carboxyl end of the original and mutant versions of the tobacco mosaic virus coat protein. The amino acids affected by the mutation are shaded red.

Solution: **(a)** A single-nucleotide substitution mutation caused an amino acid substitution of leucine for proline. The two adjacent amino acids were unchanged by the mutation, providing evidence against overlapping codons. In 1961, Crick et al. (1961, *Nature* 192:1227–1232) stated that this experiment provided some of the first empirical evidence that the genetic code is nonoverlapping.

(b) The nucleotide substitution in the mRNA was

C → U. Proline is encoded by four codons with the sequence CCN (where N is any of the four nucleotides), and leucine is encoded by six codons with the sequences CUN and UUPu (where Pu is either of the two purines). A substitution in the mRNA of C → U in the second position of a CCN proline codon yields CUN, which codes for leucine.

Indirect Approaches

The genetic code itself was deciphered during the early 1960s using some rather ingenious methods. Today the nucleotide sequence of a segment of DNA and the amino acid sequence of a polypeptide may be determined routinely. However, in the early 1960s, according to Crick,

> The genetic code could be broken easily if one could determine both the amino acid sequence of a protein and the base sequence of the piece of nucleic acid that codes for it. A simple comparison of the two sequences would yield the code. Unfortunately, the determination of the base sequence of a long nucleic acid molecule is, for a variety of reasons, still extremely difficult. More indirect approaches must be used.

Marshall Nirenberg and Heinrich Matthaei developed the first of these indirect approaches in 1961 by preparing a cell-free protein synthesis system from *E. coli*. They ground the bacterial cells with an abrasive to break them open and release the cellular contents. These cellular contents made a cell-free extract that contained active ribosomes as well as all other necessary components for protein synthesis. They then prepared synthetic RNAs using an enzyme, called polynucleotide phosphorylase, that builds an RNA chain out of nucleotides without a DNA template, so that the nucleotides are incorporated at random. By providing the enzyme with only one type of nucleotide, they could produce a synthetic RNA molecule consisting only of that nucleotide. For example, providing only U resulted in a poly (U) RNA. When this poly (U) RNA was added to the cell-free protein synthesis system, polypeptides consisting entirely of phenylalanine were produced, indicating that UUU must code for phenylalanine. Similarly, poly (C) RNAs produced polypeptides made entirely of proline, meaning that CCC must code for proline.

Several research groups made synthetic RNAs consisting of two or three different nucleotides. The degeneracy of the genetic code soon became evident. Synthetic RNAs containing random combinations of U and A, U and C, or U and G all included leucine in the polypeptides they produced, meaning that several combinations of these nucleotides must code for leucine. A brief look at the genetic code in Figure 4.4 shows that four leucine codons (UUA, UUG, CUU, and CUC) can be made from combinations of U and A, U and C, and U and G.

Later, several tRNAs were isolated and combined with synthetic mRNAs in cell-free systems. Researchers then determined which nucleotides in the mRNA paired with the anticodon in each tRNA. In 1963, Nirenberg put the information from these various studies together and identified the amino acids associated with the nucleotide composition (but not nucleotide order) of 40 codons. When the actual genetic code was deciphered, it turned out that the nucleotide compositions of 35 of the 40 codons proposed by Nirenberg were correct.

A major hurdle to cracking the genetic code was crossed in 1965 when Nirenberg and Philip Leder discovered that ribosomes would read a synthetic "mini mRNA" consisting of three nucleotides (one codon) and would hold the appropriately charged tRNA in the ribosome with its anticodon bound to the synthetic codon (Figure 4.21). Thus each codon could be tested individually to determine which amino acid it retained in the ribosome. Nirenberg and Leder added each of the 64 possible mini mRNAs to a cell-free system and allowed the ribosomes to bind the tRNA charged with the appropriate amino acid. They then separated the ribosome–mini-mRNA–tRNA–amino acid complexes from the remaining components and determined which amino acid was bound in the ribosomes. Using this ribosome-binding test, they were able to identify unambiguously 50 of the 64 possible triplet codons, and to narrow the possibilities for the remaining codons.

The final hurdle was crossed in 1966 when Har Gobind Khorana and his coworkers performed some particularly interesting experiments using synthetic mRNAs. They constructed synthetic polynucleotides consisting of repeating nucleotide combinations. For example, poly (UC), a nucleotide made of repeating UC units (UCUCUCUC... etc.) coded for a repeating polypeptide of alternating serine and leucine (serine-leucine-serine-leucine- ... etc.). No matter where translation starts, UCU and CUC codons alternate within the reading frame. By using repeating dinucleotide, trinucleotide, and tetranucleotide synthetic mRNAs, Khorana and his coworkers were able to confirm the portion of the genetic code determined by the ribosome-binding test and fill in most of the gaps.

Exceptions to the Genetic Code

The genetic code is now one of the axioms of genetics. It has been confirmed in thousands of studies and is nearly universal among all species. Until the mid-1970s, the code was thought to be completely universal. However, comparison of the DNA sequence of several genes with the amino acid sequence of proteins revealed a number of exceptions to the genetic code. In the prokaryote *Mycoplasma capricolum*, the codon UGA is read as tryptophan instead of a termination codon. In the eukaryotic ciliates (protozoans), UAA and UAG are read as glutamine instead of termination codons. The protein synthesis systems contained in mitochondria of various organisms appear to contain one or more differences from the standard genetic code. These differences are not universal in the mitochondria of all species, however, and appear to be restricted to certain phylogenetic groups such as mammals or yeasts.

Comparison of DNA and mRNA sequences have revealed that at least some of the perceived exceptions to the genetic code are actually due to specific nucleotide

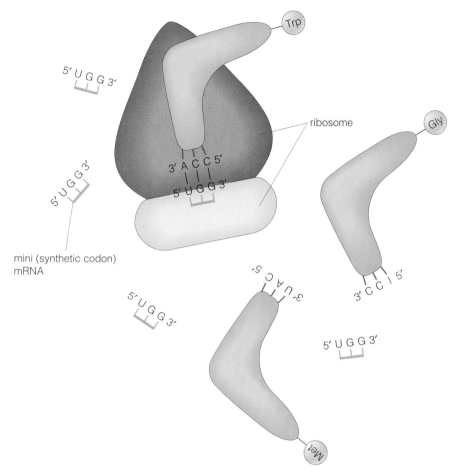

Figure 4.21 Representation of Nirenberg and Leder's experiments with mini mRNAs to decipher the genetic code. In this example, the only mini mRNA in the reaction mixture is UGG, which binds only with a tryptophan tRNA.

modifications in mRNA after it is transcribed, a process called **mRNA editing**. For nearly all genes, the mRNA sequence matches the sequence in the DNA that encoded it. However, for a few genes, mRNA editing causes a nucleotide or two to differ between the sequence in the mRNA and the corresponding sequence in the DNA. Once the mRNA is edited, a codon in the mRNA may be different from the corresponding nucleotide sequence in the DNA. When DNA sequence and protein sequence are compared without comparing the mature mRNA produced in the cell, the difference appears to be an exception in the genetic code. In reality, the mRNA may have been altered, rather than the genetic code. Most known examples of mRNA editing are in mitochondrial genes.

In addition, there are true exceptions in the genetic code that cannot be explained by mRNA editing, indicating that the genetic code is not entirely universal. It should be emphasized, however, that both mRNA editing and true exceptions to the genetic code are rare.

~

With rare exceptions, the genetic code is universal throughout all of nature.

4.6 INITIATION OF TRANSLATION

Having reviewed the genetic code, let's now turn our attention to the process of translation. Translation proceeds through the steps of initiation, elongation, and termination. There are similarities and differences between translation in prokaryotes and that in eukaryotes. Let's examine each step, comparing prokaryotes and eukaryotes.

Prokaryotic Initiation

Before translation can begin in a prokaryote, the small and large ribosomal subunits must separate from each other. This separation is facilitated by a protein initiation factor called IF1. At initiation, the two subunits reassemble on the mRNA with a unique initiator tRNA to form a ribosomal-mRNA initiation complex. The initiation complex forms in stages, as diagrammed in Figure 4.22. First the small (30S) subunit of the ribosome binds to a protein initiation factor called IF3 (initiation factor 3). This complex then binds to an initiation sequence near the 5' end of the mRNA. The initiation sequence in the

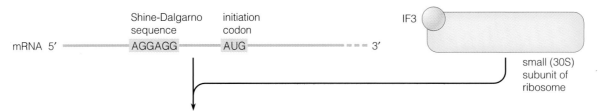

a The small ribosomal subunit binds to the initiation sequence in the mRNA with the assistance of IF3.

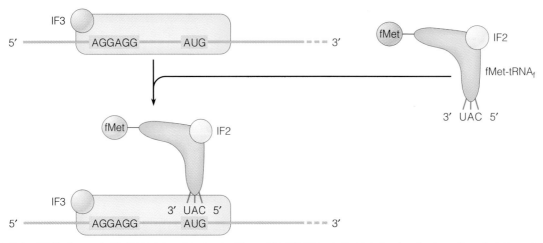

b An initiator tRNA charged with N-formyl methionine and bound to IF2 aligns in the small subunit, where its anticodon pairs with the initiation codon.

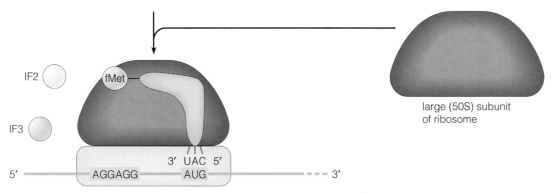

c The large ribosomal subunit binds to this complex to form the complete ribosome. The initiation factors are released.

Figure 4.22 Prokaryotic initiation.

mRNA has two important conserved sequences: The first is the initiation codon (usually AUG, but less frequently GUG or UUG). The second is a six-nucleotide sequence, AGGAGG, about seven nucleotides upstream from the initiation codon, called the **Shine-Dalgarno sequence**, after its discoverers.

The Shine-Dalgarno sequence is complementary to six nucleotides (3' UCCUCC 5') near the 3' end of the 16S rRNA, which is part of the 30S ribosomal subunit. The nucleotides of the Shine-Dalgarno sequence pair with the complementary nucleotides in the 16S rRNA, causing the small ribosomal subunit–IF3 complex to bind to the mRNA.

Initiation requires a unique initiator tRNA called **tRNA$_f^{Met}$** because it is specific for an altered form of methionine, **N-formyl methionine**. tRNA$_f^{Met}$ is used only for initiation of translation. Once charged, the tRNA$_f^{Met}$–amino acid complex is known as **N-formyl-methionyl-tRNA**, or by its abbreviation, **fMet-tRNA$_f$**. Another initiation factor, IF2, binds to fMet-tRNA$_f$. The fMet-tRNA$_f$–IF2 complex then enters the 30S subunit, and the anticodon pairs with the initiation codon. Part b of Figure 4.22 shows the fMet-tRNA$_f$–IF2 complex bound to the AUG initiation codon in the 30S subunit. At this point, the 50S subunit assembles with the 30S subunit to form the complete ribosome (Figure 4.22c).

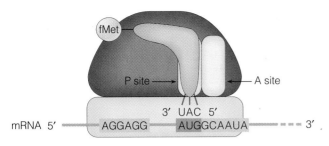

Figure 4.23 P and A sites in the ribosome. There are two sites in the ribosome that can be occupied by tRNAs, the P site and the A site.

Energy for this process is derived from hydrolysis of GTP (guanosine triphosphate). Once the ribosome is complete, IF2 and IF3 are released.

There are two sites within the ribosome that can hold tRNAs: the **A site** (aminoacyl or entry site), and the **P site** (peptidyl or donor site). The ribosome assembles on the mRNA with the A site oriented toward the 3′ end of the mRNA and the P site toward the 5′ end (Figure 4.23). At initiation, the three nucleotides of the initiation codon (AUG) align in the P site, where they pair with the anticodon in the initiator tRNA. This situates the A site over the second codon, which is now ready to receive the appropriate charged tRNA to continue translation.

~

In prokaryotic initiation, the 30S ribosomal subunit binds to the Shine-Dalgarno sequence near the initiation codon and directs fMet-tRNA$_f$ to pair with the initiation codon. The ribosome assembles with fMet-tRNA$_f$ in the P site, situating the A site over the second codon, which receives the next charged tRNA.

Eukaryotic Initiation

Although similar to prokaryotic initiation, eukaryotic initiation differs in several important respects and is more complex, requiring at least nine (and possibly more) initiation factors.

The eukaryotic initiator tRNA, **tRNA$_i$^{Met}**, is charged with methionine but differs from its prokaryotic counterpart in that the methionine is not formylated. tRNA$_i$^{Met} functions only in initiation of translation. Another tRNA, tRNA$_m$^{Met}, is charged with methionine and pairs with internal AUG codons farther downstream. When charged with methionine, the eukaryotic initiator tRNA is usually referred to by its abbreviation, **Met-tRNA$_i$**.

An outline of eukaryotic initiation is illustrated in Figure 4.24. Initiation begins (as shown in part a of Figure 4.24) with the binding of Met-tRNA$_i$ to GTP and a protein initiation factor called eIF2 (for eukaryotic initiation factor 2). This complex then binds to the small ribosomal subunit (part b). All of this happens before mRNA

enters the picture. The small ribosomal subunit–Met-tRNA$_i$–GTP complex then binds to the 5′ cap of the mRNA with the assistance of an initiation factor, eIF4A, that includes a cap binding protein (CBP) that recognizes the 5′ cap (part c).

After binding to the 5′ cap, the complex moves along the mRNA until it encounters the initiation codon through a process called **scanning**. To be an initiation codon, an AUG codon must be embedded in a conserved sequence with the consensus sequence GCCGCCPuCCAUGG, where the underlined AUG is the initiation codon (Figure 4.24d). The most highly conserved nucleotides in this sequence are the AUG codon and the purine three nucleotides upstream from the AUG codon. Mutations that alter this purine usually prevent recognition of the AUG sequence as an initiation codon. The G just beyond the AUG codon is also highly conserved.

In over 90% of genes studied, the first AUG codon downstream from the 5′ cap is contained within the conserved sequence and is the initiation codon. In the remaining 10% of genes, the initiation codon is usually the second or third AUG sequence downstream from the first AUG in the mRNA, as we will see in Example 4.5.

Once the ribosome reaches the initiation codon, the Met-tRNA$_i$ anticodon pairs with the initiation codon, and the large ribosomal subunit attaches to the small ribosomal subunit to form the intact ribosomal-mRNA complex (Figure 4.24e). In the process, the initiation factors are released, and GTP is hydrolyzed to GDP, releasing energy to drive the process. Met-tRNAi, paired to the initiation codon, occupies the P site, positioning the A site over the second codon, where the next charged tRNA enters.

Example 4.5 Identification of the correct initiation codon.

In 1986, Kawakami et al. (*Nucleic Acids Research* 14:2833–2844) reported the RNA sequence of the gene that encodes the enzyme Na, K-ATPase in humans. A portion of the nucleotide sequence near the 5′ end of the mRNA is

```
GAAUUCAUGCUAAAUUGCUGGAAGGCUGCGUCUCUGC-
UGUGGGUGUCAGUUCCGGAUGCCUCAUCGCCAGGGGCG-
CGCCGCAGCCACCCACCCUCCGGACCGCGGCAGCUGC-
UGACCCGCCAUCGCCAUGGCCCGCGGGAAAGCCAAG-
GAGGAGGGC...
```

The first 10 amino acids in the polypeptide are

```
MetAlaArgGlyLysAlaLysGluGluGly...
```

Problem: Identify the initiation codon in the mRNA sequence.

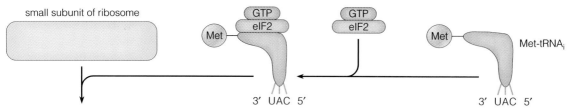

a An initiator tRNA charged with methionine complexes with eIF2 and GTP.

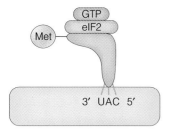

b The tRNA complex enters the small ribosomal subunit.

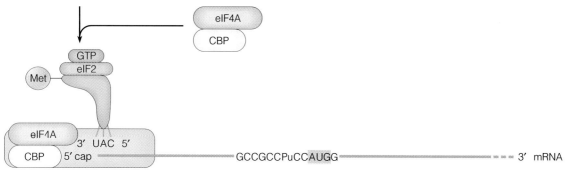

c The small ribosomal subunit–tRNA complex binds to the 5′ cap of the mRNA with the assistance of eIF4A, which contains the cap binding protein (CPB).

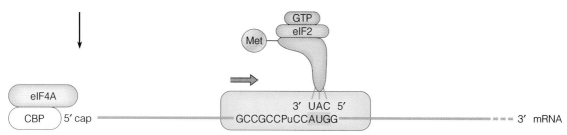

d The complex scans the mRNA until it reaches the conserved initiation sequence, where the UAC anticodon pairs with the AUG initiation codon.

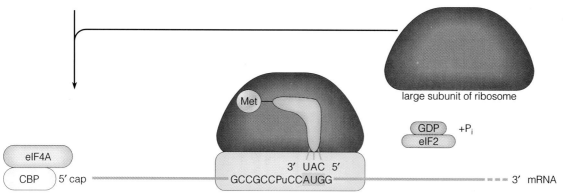

e Once the complex reaches the initiation codon, the large ribosomal subunit binds to it, forming a complete ribosome. The ribosome is positioned to enter the elongation phase.

Figure 4.24 Eukaryotic initiation.

Solution: There are three AUG sites in the mRNA sequence.

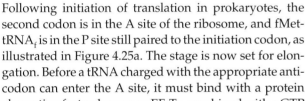

GAAUUC<u>AUG</u>CUAAAUUGCUGGAAGGCUGCGUCUCUGC-
UGUGGUGUCAGUUCCGG<u>AUG</u>CCUCAUCGCCAGGGGCG-
CGCCGCAGCCACCCACCCUCGGACCGCGGCAGCUGC-
UGACCCGCCAUCGCC<u>AUG</u>GC<u>CCG</u>CGGGAAAGCCAAG-
GAGGAGGGC . . .

Only the third AUG codon (GCCAUCGCC<u>AUG</u>G) is embedded in a sequence similar to the consensus sequence; the sequence differs from the consensus sequence by two nucleotides. Comparison of the amino acid sequence of the polypeptide with the codons following this AUG codon confirms that this is the initiation codon. This is one example from the 10% of eukaryotic mRNAs in which the first AUG is *not* the initiation codon.

In eukaryotic initiation, the small ribosomal subunit and tRNA$_i$^{Met} bind to the 5′ cap, then move along the mRNA until they encounter the initiation codon. The ribosome assembles with tRNA$_i$^{Met} paired with the initiation codon in the P site. The second codon is in the A site and accepts the next charged tRNA.

4.7 ELONGATION

Following initiation of translation in prokaryotes, the second codon is in the A site of the ribosome, and fMet-tRNA$_f$ is in the P site still paired to the initiation codon, as illustrated in Figure 4.25a. The stage is now set for elongation. Before a tRNA charged with the appropriate anticodon can enter the A site, it must bind with a protein elongation factor, known as EF-Tu, combined with a GTP molecule. This complex then enters the A site, and its anticodon pairs with the second codon (Figure 4.25b). The GTP complexed with EF-Tu is hydrolyzed to GDP, releasing energy, and the GDP-EF-Tu complex is released from the ribosome.

Charged tRNAs now occupy both the P and A sites, bringing the two amino acids in close proximity for formation of a peptide bond between them. The complex fMet-tRNA$_f$ occupies the P site with the carboxyl group of N-formyl methionine facing the amino group of the amino acid attached to its tRNA in the A site. **Peptidyl transferase** catalyzes the formation of a peptide bond between the carboxyl group of the amino acid in the P site and the amino group of the amino acid in the A site. This enzyme is not free in the cell, but is a part of the large (50S) ribosomal subunit. Formation of the peptide bond breaks the bond holding the carboxyl group of the amino acid in the P site to its tRNA, and simultaneously attaches that carboxyl group to the amino group of the amino acid occupying the A site (Figure 4.25c).

The formation of the peptide bond frees the tRNA in the P site from its amino acid. The ribosome then moves exactly three nucleotides along the mRNA, expelling the uncharged tRNA from the P site as the charged tRNA that occupied the A site moves into the P site, a process known as **translocation**. The A site is now situated over the third codon and is ready to receive the appropriate charged tRNA (Figure 4.25d).

Translocation requires another elongation factor, EF-G, which is complexed with GTP when it interacts with the ribosome. During translocation, the GTP is hydrolyzed to GDP, releasing energy, and the GDP-EF-G complex is then released from the ribosome. At this point, the entire process repeats itself as a charged tRNA enters the A site, a peptide bond forms, and translocation moves the ribosome to the next codon—and so on for each codon in the mRNA.

Elongation in eukaryotes is very similar to elongation in prokaryotes. A eukaryotic elongation factor, eEF1, assists in bringing charged tRNAs into the A site of the ribosome, analogous to the function of EF-Tu in prokaryotes. Like EF-Tu, eEF1 requires the hydrolysis of one GTP to complete its function. Another eukaryotic elongation factor, eEF2, assists in translocation of the ribosome, which also requires hydrolysis of one molecule of GTP. The function is analogous to that of EF-G in prokaryotes.

Elongation begins when a charged initiator tRNA is paired with the initiation codon and occupies the P site of the ribosome. A charged tRNA enters the A site and pairs with the second codon. A peptide bond forms between the two amino acids. The ribosome moves three nucleotides along the mRNA, expelling the initiator tRNA. The process repeats codon by codon, extending the amino acid chain.

4.8 TERMINATION

Elongation continues until the ribosome reaches any one of three stop or termination codons: UAA, UAG, or UGA. None of these codons codes for an amino acid, and there are no tRNAs with anticodons that are complementary to these codons (with the exception of some rare mutant tRNAs). Consequently, when a termination codon appears in the A site of the ribosome, no tRNA enters that site, and the process of termination begins.

Termination in prokaryotes, diagrammed in Figure 4.26, is catalyzed by either of two release factors, RF1 or RF2, when a termination codon is in the A site. RF1 catalyzes the reaction for the codons UAA and UAG, and RF2 catalyzes for UGA. The bond holding the amino acid in the P site with its tRNA is cleaved, the polypeptide chain is released from the ribosome, and the ribosomal subunits separate, releasing the ribosome from the mRNA.

Termination in eukaryotes is accomplished by a single release factor, eRF, that recognizes all three termination codons.

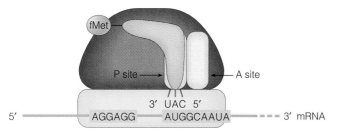

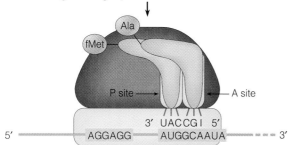

a When elongation begins, the P site of the ribosome is situated over the initiation codon and the A site over the second codon in the mRNA.

b The correct charged tRNA (an alanine tRNA in this example) enters the A site (after bonding with EF-Tu and GTP), and its anticodon pairs with the second codon.

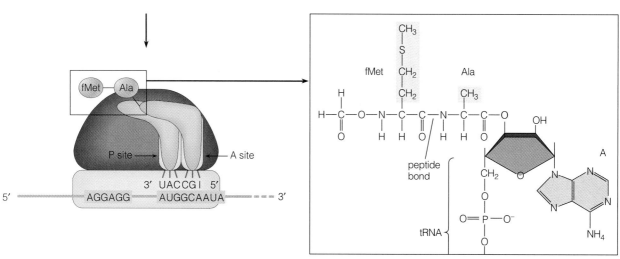

c The bond between N-formyl methionine and its tRNA is broken as the N-formyl methionine is joined to the amino end of the second amino acid via a peptide bond. The carboxyl end of the second amino acid remains bound to its tRNA.

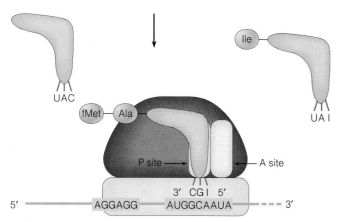

d With the help of EF-G and GTP, the ribosome moves exactly one codon downstream, moving the tRNA paired to the second codon into the P site. This locates the A site over the third codon, where another tRNA (an isoleucine tRNA in this example) may now pair with its codon.

Figure 4.25 Prokaryotic elongation.

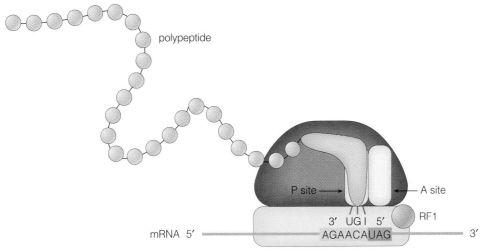

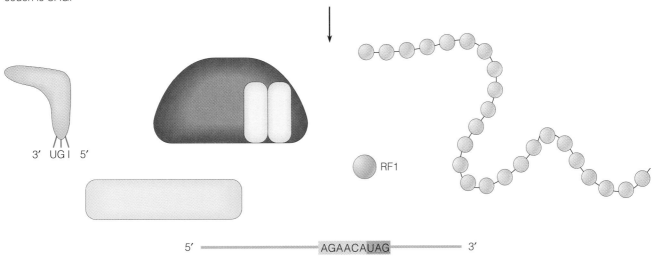

a The A site is positioned over the termination codon. Termination factor RF1 facilitates termination in this example because the termination codon is UAG.

b At termination, all components separate from the mRNA.

Figure 4.26 Prokaryotic termination.

~
When the ribosome encounters a termination codon in the A site, release factors cause the polypeptide and ribosome to separate from one another and the mRNA.

4.9 PROTEIN STRUCTURE AND FUNCTION

A polypeptide is the immediate product of translation, and its amino acid sequence is determined by the nucleotide sequence in the mRNA. However, rarely is the protein ready to function immediately after translation. In nearly all cases, proteins must be modified and processed before they can function in the cell. They must also be transported to the location in the cell where they are needed. In some cases, proteins move out of the cell to function elsewhere. In this section, we will discuss how proteins are processed to assume their final structure, and how their structures determine their functions.

Protein Modification and Processing

Newly synthesized polypeptides undergo **posttranslational modification** during which portions of the polypeptide are removed, amino acids are altered, or additional molecules, such as polysaccharides, are added to the polypeptide. Many polypeptides have a **signal peptide**, consisting of the first 15–30 amino acids at the amino end. According to the signal hypothesis (Figure 4.27), a ribosome that is not attached to the endoplasmic reticulum (ER) binds to the 5′ end of an mRNA and initiates translation. The signal peptide on the amino end is

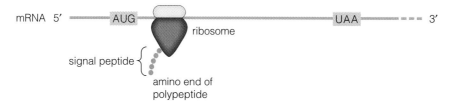

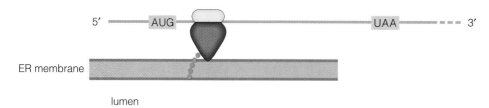

a A ribosome assembles on the mRNA and initiates translation. The amino end of the polypeptide is the first portion to be translated and contains a signal peptide that emerges from the ribosome.

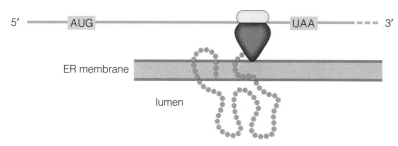

b The signal peptide embeds itself in the endoplasmic reticulum (ER) membrane.

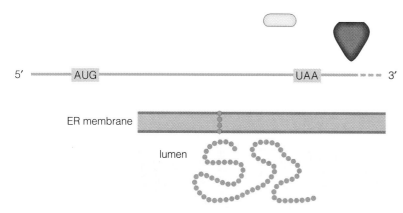

c The ribosome proceeds with translation, depositing the polypeptide into the lumen of the ER.

d Translation terminates at the termination codon, signal peptidase cleaves the polypeptide from the signal peptide, and the polypeptide is free to be transported within the ER.

Figure 4.27 The signal hypothesis.

the first part of the polypeptide to appear (Figure 4.27a). With the assistance of proteins, the signal peptide embeds itself in the ER membrane, attaching the ribosome to the surface of the ER (Figure 4.27b). Translation resumes, and the remainder of the polypeptide is then secreted into the lumen of the ER (Figure 4.27c). The signal peptide is cleaved by an enzyme called signal peptidase, and the polypeptide may then be transported to its destination through the lumen of the ER (Figure 4.27d).

Signal peptides have certain features that are consistent with their function, as we will see in the example that follows.

Figure 4.28 The α helix. Hydrogen bonds that maintain the helix are represented as dotted lines.

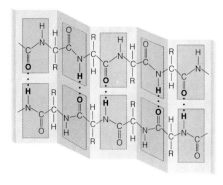

Figure 4.29 The β pleated sheet. The β strands are held together by hydrogen bonds, represented as dotted lines.

proline

Nitrogen cannot participate in hydrogen bonding.

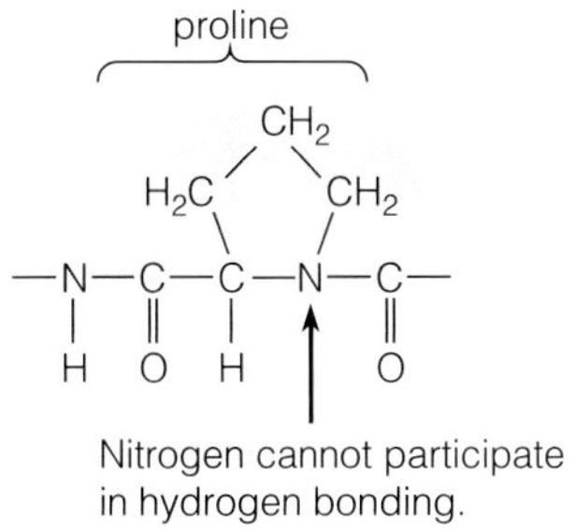

Figure 4.30 A proline in a polypeptide chain.

Example 4.6   An N-terminal signal peptide in the precursor of human insulin.

As reported by Bell et al. (*Nature* 282:525–527), the amino acid sequence encoded by the human insulin gene includes 24 amino acids on the amino end that do not appear in the final protein. The amino acid sequence of these 24 amino acids is

```
MetAlaLeuTrpMetArgLeuLeuProLeuLeuAla-
LeuLeuAlaLeuTrpGlyProAspProAlaAlaAla
```

Problem: What aspects of this amino acid sequence suggest its function?

Solution: The sequence appears at the N-terminus, which is the usual location of a signal peptide. All but two of the amino acid residues (arginine and aspartic acid) are nonpolar and hydrophobic. Bilipid-layer cellular membranes are nonpolar and hydrophobic in their center. The nonpolar nature of this signal peptide allows it to embed itself in cellular membranes.

Once all extra amino acid segments are removed from the polypeptide, it assumes a specific three-dimensional protein conformation. Although some proteins consist of a single polypeptide and are therefore the product of one gene, many proteins consist of two or more polypeptide subunits and may originate from two or more genes. The subunits assemble to form the complete protein.

Protein structural organization is divided into four levels: primary, secondary, tertiary, and quaternary. The **primary structure** is the linear sequence of amino acids attached to each other by peptide bonds. Even before translation is complete, the amino acids in the growing polypeptide chain may begin interacting with one another to form the **secondary structure**. Some of the common secondary structures in proteins are the α helix, the β strand, the β turn, and the random coil.

The **α helix** is a common part of protein secondary structure. In fact, most proteins contain several α helices. The polypeptide chain forms a spiral that is held in place by hydrogen bonds between the double-bonded oxygen in the carboxyl group of one amino acid and the amino group of another amino acid, as illustrated in Figure 4.28.

A **β strand** is a linear strand of amino acids. Although there are no specific hydrogen bonds that hold a β strand in place, the amino acids in a β strand are free to form hydrogen bonds with other amino acids, such as those in an α helix or another β strand. Beta strands often line up side by side and form hydrogen bonds with one another in a **β-pleated sheet**, as illustrated in Figure 4.29.

Random coils and **β turns** are irregular conformations that serve as bridges between two regular secondary structural conformations, such as the bridge between two α helices in the same polypeptide chain. The amino acid proline is often found in random coils or turns. The ring in the R group of proline is bound to the amino group in the peptide backbone, preventing the formation of stable hydrogen bonds with this amino group (Figure 4.30). Proline fails to form hydrogen bonds

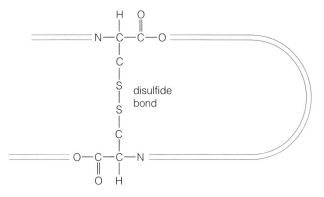

Figure 4.31 Disulfide bond between the R groups of two cysteines in a polypeptide.

that are necessary to maintain some secondary structures. However, random coils and β turns do not require hydrogen bonding.

Interactions between the amino acid residues contained in secondary structures cause the polypeptide to fold back on itself and assume its **tertiary structure**. A common tertiary interaction is a **disulfide bond** between the R groups of two cysteine amino acid residues, as illustrated in Figure 4.31. The tertiary structures of proteins are often illustrated using **ribbon diagrams**, such as those shown in Figure 4.32. Proteins contain many different structural features at the tertiary level. For example, in the enzyme triosephosphate isomerase, there is a

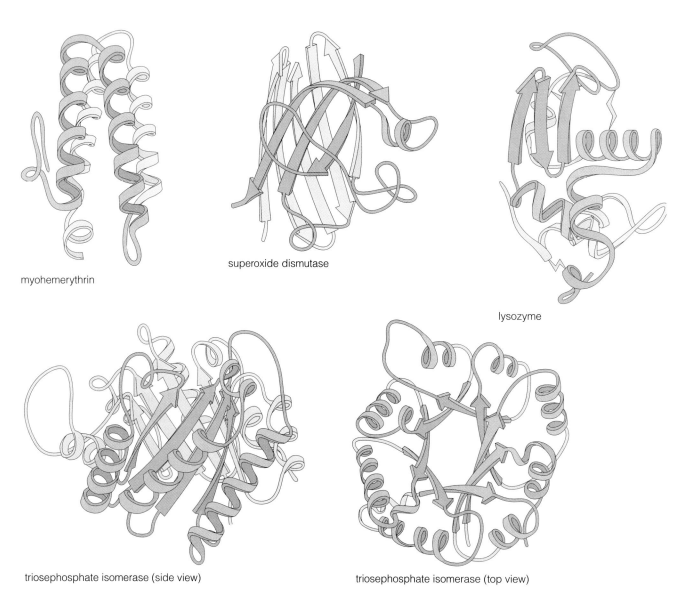

myohemerythrin

superoxide dismutase

lysozyme

triosephosphate isomerase (side view)

triosephosphate isomerase (top view)

Figure 4.32 Ribbon diagrams representing the tertiary structure of four proteins. Alpha helices are represented as coiled ribbons, β strands as arrows, random coils and turns as narrow ribbons, and disulfide bonds as yellow Z-shaped connections. (Redrawn from Richardson, J. S. 1981. *Advanced Protein Chemistry* 34:167.)

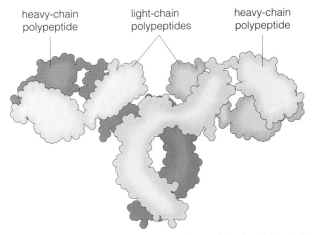

a The antibody consists of four subunits, two identical light-chain polypeptides and two identical heavy-chain polypeptides.

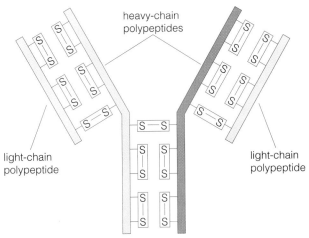

b Disulfide bonds between cysteines holding the four subunits together.

Figure 4.33 The quaternary structure of an antibody molecule.

cylindrical β-pleated sheet, called a **β barrel**, in the center of the protein (Figure 4.32).

Some proteins consist of a single polypeptide, so the tertiary structure is the final level of protein structural organization. However, many proteins contain more than one polypeptide. When two or more protein subunits assemble to form the final protein, the protein assumes its **quaternary structure**. Proteins composed of several subunits often acquire their full function in the quaternary form. Figure 4.33 shows the quaternary structure of an antibody, which is a protein composed of four subunits, two identical heavy-chain polypeptides and two identical light-chain polypeptides.

~

Proteins assume a tertiary or quaternary structure that is dependent on the sequence of amino acids in the protein.

Protein Function

The structure of a protein determines its chemical properties, which, in turn, determine the protein's function. The R groups of amino acids have different chemical properties. As illustrated in Figure 4.1 (page 90), the R groups may be nonpolar, polar, basic, or acidic. When a protein assumes its final structure, the positions of the various amino acids in the protein determine the function of the protein. Some proteins may be water soluble, others water insoluble, depending on which amino acids are located on the protein's surface. Proteins may be acidic in certain regions and neutral or basic in others. The physical conformation of a protein also influences its activities. Some proteins interact with other molecules in such a way that the molecule fits into a site in the protein like one piece of a jigsaw puzzle interlocking with another. The wide variety of possible protein structures and chemical properties leads to enormous diversity in protein function.

Proteins carry out many functions in cells. Table 4.3 lists several classes of proteins and their functions. Let's now look at a membrane transport protein to see how its structure determines its function.

Example 4.7 Relationship of protein structure and function.

In 1995, Alberti et al. (*Infection and Immunity* 63:903–910) reported the tertiary structure of the porin protein OmpK36 from the bacterial species *Klebsiella pneumoniae*. Figure 4.34 shows this protein's tertiary structure.

Problem: This protein is a membrane transport protein that permits small water-soluble molecules to cross the bacterial membrane. What tertiary structural characteristic allows this protein to carry out its function?

Solution: The protein contains a β barrel that serves as a corridor through which molecules can pass across the bacterial membrane.

Enzymes

Enzymes, which function as catalysts, are among the most important classes of proteins in genetics. In this section, we will focus on the relationship between an enzyme's structure and its catalytic function. A **catalyst** is a substance that permits a specific chemical reaction to occur more rapidly than it would under the same conditions in the absence of the catalyst. Although the catalyst participates in the reaction, it is usually not consumed in the reaction. A single enzyme molecule can catalyze the same reaction many times.

Table 4.3 Some Classes of Proteins and Their Functions

Protein Class	Function	Examples
Enzymes	Catalyze biochemical reactions	Phenylalanine hydroxylase: Converts phenylalanine into tyrosine. DNA polymerases: Synthesize DNA.
Immunoglobulins	Provide immunity	Antibodies: Attack foreign substances in the body.
Hormones	Activate and regulate cell, tissue, organ, and body functions	Insulin: Regulates glucose metabolism.
Receptor proteins	Bind specific molecules to cell surface	Insulin receptors: Stimulate glucose uptake when bound by insulin.
Membrane transport proteins	Transport substances across membranes	Maltoporin: Bacterial protein embedded in the cell membrane selectively transports the sugar maltose across the cell membrane.
Carrier proteins	Carry substances in the body	Hemoglobin: Carries oxygen from the lungs through the bloodstream to all tissues of the body.
Structural proteins	Provide tissue structure	Keratin: Is a major structural component of fingernails.
Nucleic acid binding proteins	Stabilize and regulate the functions of DNA and RNA	Histones: Maintain DNA structure and organization. Transcription factors: Bind to DNA and regulate transcription. Single-stranded binding proteins: Stabilize single-stranded DNA
Ribosomal proteins	Form structural and enzymatic functions of ribosomes	*E. coli* S1 ribosomal protein: Is the largest of 21 ribosomal proteins found in the small ribosomal subunit.
Storage proteins	Store amino acids for later use	Seed storage proteins in plants: Serve as a source of amino acids when the seed germinates.
Contractile proteins	Contract and move tissues	Actin and myosin: Are the major protein components of muscle tissue responsible for muscle contraction.

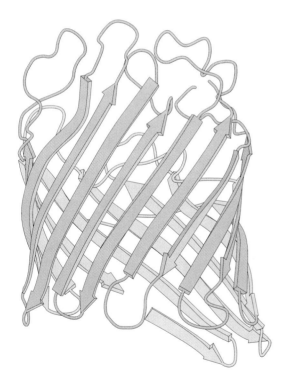

Figure 4.34 Information for Example 4.7: Ribbon diagram of the OmpK36 protein from *Klebsiella pneumoniae*. (Redrawn from Albertí, S., F. Rodrígues-Quiñones, T. Schirmer, G. Rummel, J. M. Tomás, J. P. Rosenbusch , and V. J. Benedí. 1995. A porin from: *Klebsiella pneumoniae* sequence homology, three dimensional model, and complement binding. *Infection and Immunity* 63:903–910.)

Enzymes generally react with one or more substrates that are specific for that enzyme. The **substrate** is the molecule (or molecules) that will be converted into a different molecule (or molecules), called the **product**, during the chemical reaction catalyzed by the enzyme.

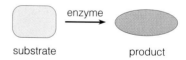

The **active site** is the location in the enzyme where the reaction takes place. The active site often has a three-dimensional structure that fits around the substrate molecule(s) much like interlocking pieces of a jigsaw puzzle. According to the induced-fit model of enzyme-substrate interactions, the substrate molecule(s) enter the active site of the enzyme and alter the enzyme's shape slightly to cause a tight fit between enzyme and substrate, causing the reaction to proceed (Figure 4.35). After conversion of the substrate to the product, the enzyme releases the product, and the enzyme resumes its former shape, ready to bind to another molecule (or molecules) of the substrate.

The concept of enzyme structure and active sites is an important one in genetics. In the next chapter, we will examine mutation in detail. A mutation in DNA may alter a protein's amino acid sequence, which, in turn, may alter the three-dimensional structure of the protein. In the case

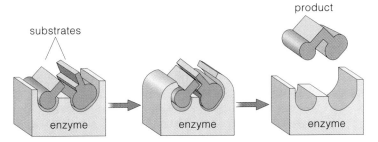

Figure 4.35 The induced-fit model for enzyme-substrate interactions.

of enzymes, a mutant amino acid sequence may alter or eliminate an enzyme's activity, particularly if the mutant amino acid sequence affects the active site of the enzyme. Alteration of enzyme activity affects the substrates and products of the reaction catalyzed by the enzyme. This, in turn, may affect cellular function and may be expressed as a phenotypic difference.

~

Enzymes assume a particular tertiary or quaternary structure that allows them to physically interact with their substrates and convert the substrates into products. The interaction with the substrate takes place in the active site of the enzyme.

SUMMARY

1. A polypeptide is a linear chain of amino acids connected to one another by peptide bonds. There are 20 different amino acids utilized in protein synthesis. The amino acids differ in the R-group portion of the molecule.

2. The nucleic acid sequence of a gene is colinear with the amino acid sequence of the polypeptide encoded by the gene. Each amino acid in a polypeptide is specified by a three-nucleotide codon in the mRNA.

3. The genetic code contains 64 codons, 61 of which code for amino acids. The remaining 3 codons specify termination of translation.

4. Codons are read according to a fixed reading frame that is nonoverlapping and uninterrupted. Translation begins at a specific AUG codon called the initiation codon. The codons are then read in frame from the initiation codon according to the genetic code. Translation continues until the ribosome encounters a termination codon.

5. Ribosomes are composed of a small and a large subunit that assemble on the mRNA for translation. Prokaryotic and eukaryotic ribosomes are similar in structure, although eukaryotic ribosomes are larger.

6. Transfer RNA carries amino acids to the ribosome. The tRNA molecule folds back on itself, forming base pairs in a cloverleaf structure containing an acceptor arm, an anticodon arm, a D arm, a TψC arm, and a small extra arm.

7. The amino acid is attached to the acceptor arm of a tRNA by an aminoacyl-tRNA synthetase. Once the amino acid is attached, the tRNA is said to be charged. The charged tRNA enters the ribosome, where its anticodon pairs with the codon in the mRNA.

8. The genetic code is degenerate, meaning that some amino acids are encoded by more than one codon. Although there are 61 codons, there are fewer than 61 tRNAs, because some anticodons may pair with more than one codon, according to the wobble hypothesis.

9. The genetic code was deciphered through a series of experiments based primarily on the use of synthetic mRNAs in cell-free translation systems. With rare exceptions, the genetic code is universal among all species on Earth.

10. Translation proceeds through the stages of initiation, elongation, and termination.

11. Following translation, polypeptides are processed to form proteins. Processing includes removal of amino acid segments, folding of the polypeptides, and attachment of external groups to the protein.

12. Protein structure is hierarchical. The primary structure is the linear chain of amino acids in a polypeptide. The secondary structure may consist of several structural features, including α helices, β strands, β-pleated sheets, random coils, and β turns. The secondary structures attach to one another through interactions among the R groups, forming the tertiary structure. Two or more protein subunits may associate to form the quaternary structure.

13. A protein's amino acid composition and its structure determine its function. There are many different classes of proteins; enzymes are among the most important in genetics. An enzyme is a catalyst that contains an active site, where the reaction takes place.

QUESTIONS AND PROBLEMS

1. The DNA, mRNA, and amino acid chain diagrammed in Figure 4.36 are from a gene that encodes a subunit of human factor VIII. The factor VIII protein is a blood clotting factor that is missing or nonfunctional in most individuals who have hemophilia. The initiation codon is included in the mRNA sequence given. Several of the nucleotides and one of the amino acids have been deleted from the sequence. **(a)** Which strand is the sense strand? **(b)** Which amino acid should go in the blank space in the amino acid chain? **(c)** Which nucleotide belongs in position 11 (marked with an asterisk) in the mRNA?

2. In 1986, Ow et al. (*Science* 234:856–859) reported that when the gene for luciferase (the enzyme responsible for luminescence in fireflies) was transferred into the DNA of tobacco cells and these cells were regenerated into whole tobacco plants, the plants had the curious ability to glow in the dark. Why would a firefly gene be expressed in a tobacco plant when the two species are so obviously different?

3. Near the 5′ end of the human insulin mRNA is the sequence CCUGUGGAUGCGCCU. This nucleotide sequence codes for the sequence of five amino acids -Leu-Trp-Met-Arg-Leu-. If divided up evenly into triplets, this sequence would be CCU GUG GAU GCG CCU, which should code for ProValAspAlaPro. **(a)** Identify this sequence in the human insulin mRNA given in Example 4.2. **(b)** Explain why the correct sequence of amino acids is LeuTrpMetArgLeu instead of ProValAspAlaPro.

4. Using the sequence of the human insulin mRNA in Example 4.2, explain why the first AUG codon near the 5′ end of the mRNA is the initiation codon.

5. The 16S rRNA in *E. coli* contains the sequence CCUCCU near the 3′ end. Describe the function of this sequence when the small ribosomal subunit is bound to mRNA.

6. Eukaryotic 18S rRNAs are quite similar in nucleotide sequence to the 16S rRNAs of prokaryotes, especially near the 3′ end. However, eukaryotic rRNAs lack the CCUCCU sequence mentioned in the previous question. Why is it possible for eukaryotic rRNAs to function without this sequence, which is important in prokaryotic ribosomes?

7. A single prokaryotic mRNA may have several nonoverlapping reading frames within it and may thus code for several polypeptides. On the other hand, each eukaryotic mRNA encodes only a single polypeptide. Given the differences between prokaryotic and eukaryotic translation, why is it possible for a prokaryotic mRNA to contain more than one reading frame, while eukaryotic mRNAs can contain only a single reading frame?

8. What repeating polypeptide would be produced from the polynucleotide poly (UC) in a cell-free translation system?

9. What repeating polypeptide would be produced from the polynucleotide poly (UCAG) in a cell-free translation system?

10. What polypeptide would be produced from the polynucleotide poly (CUAG) in a cell-free translation system? How might the results from this type of inquiry be used to determine the nucleotide sequence of termination codons?

11. If the portion of the DNA that encoded the 5.8S rRNA in one large rRNA precursor gene were deleted in a mammalian cell, would the cell die due to a lack of protein synthesis?

12. Using the wobble-pairing rules and the genetic code, determine the minimum number of tRNAs that would need to be present in a cell to pair with the 61 codons that specify amino acids.

13. When in translation may a charged tRNA enter the P site without having previously occupied the A site? Which tRNAs may do this?

14. Assume that the following section of an mRNA is translated to produce the polypeptide segment diagrammed below:

```
5′ AAGAUGUUUCUAGUUCAG 3′
   LysMetPheLeuValGln
```

For each of the amino acids, write all possible anticodons that, according to the wobble hypothesis, could be used in the tRNAs that carry these amino acids.

15. Inosine is a common modified nucleotide in the anticodons of tRNA. However, a tRNA with the anticodon 3′ ACI 5′ is not found. Why is this so?

16. Usually there is only a single tRNA that carries tyrosine. What is its anticodon?

17. Why is it essential for tRNAs to bind amino acids at the carboxyl group rather than the amino group?

18. List all the molecular constituents present in a prokaryotic ribosome during translation, and briefly describe their roles.

19. In describing the central dogma, Francis Crick (*Symposium for the Society of Experimental Biology* 12: 138–153) stated that "the transfer of information from nucleic acid to nucleic acid, or from nucleic acid to protein may be possible, but the transfer from protein to protein, or from protein to nucleic acid is impossible." What features of the molecules involved in protein synthesis make it impossible to transfer information from one protein to another or from protein to nucleic acid?

20. The results of Yanofsky et al. described in Example 4.1 demonstrate the colinearity of a gene and its protein.

```
            1       5        10        15        20      25
DNA   5'... C A C _ G A _ G _ T _ A G _ T T G A _ A A A C C _ G A C ...3'
      3'... G T G _ C T T C T A _ T C C _ A C T T T _ T G G A C T G ...5'
                                              *
mRNA  5'... C A C U _ A _ G _ U _ A G G U _ G A A A A A _ C U G _ _ ...3'

polypeptide      methionine-arginine-leucine-lysine-asparagine-_____-threonine-
```

Figure 4.36 Information for problem 1: DNA, mRNA, and amino acid sequences for a subunit of human factor VIII.

Explain how they also confirm that the genetic code is nonoverlapping.

21. Alpha-thalassemias are severe genetic blood disorders in which the α-globin subunit of hemoglobin is either absent or its production is severely reduced. In 1985, Morlé et al. (*EMBO Journal* 4:1245–1250) identified a mutation that was the apparent cause of a particular α-thalassemia. The usual sequence preceding the initiation codon in the α-globin gene is CACCATG. In the mutant version, this sequence is CCCCATG. Under these circumstances, translation of α-globin mRNAs is severely impaired. What aspect of this mutation suggests that it is the cause of the transcription impairment?

22. In 1965, Nishimura, Jones, and Khorana (*Journal of Molecular Biology* 13:302–324) reported that a poly-UC synthetic mRNA incorporated leucine and serine into a polypeptide in a cell-free translation system. They found, however, that when either amino acid is added alone, a polypeptide fails to form. Explain why this result is expected.

23. Chapeville et al. (1962. *Proceedings of the National Academy of Sciences, USA* 48:1086–1092) made synthetic mRNAs consisting of U and G incorporated at random with a U:G ratio of 5:1. These synthetic mRNAs readily incorporated cysteine but not alanine in a cell-free translation system. Charged cysteine tRNAs were treated with Raney nickel, a catalyst that removes sulfur from cysteine, converting the cysteine into alanine. These treated tRNAs, now charged with alanine, were added to a cell-free translation system containing a synthetic mRNA. Under these conditions, alanine was readily incorporated. What do these results suggest about the function of tRNAs?

24. The anticodon in tRNAs that carry tryptophan is typically 3' ACC 5'. However, there are several known cases in which a mutation in the DNA causes the anticodon 3' ACU 5' to appear in the tRNA that carries tryptophan. These mutant tRNAs still carry tryptophan, but are known as suppressor tRNAs. **(a)** What difference (if any) in the polypeptide sequence could occur when these mutant tRNAs are present? **(b)** What amino acid (if any) will be encoded by a UGG codon when the mutant tRNA is present? **(c)** Why are these mutant tRNAs called suppressor tRNAs?

25. In 1962, Weisblum, Benzer, and Holley (*Proceedings of the National Academy of Sciences, USA* 48:1449–1455) isolated two different tRNAs that carried leucine and tested them in cell-free systems. One of the tRNAs recognized a synthetic mRNA composed of U and C (in a ratio of 1:1) incorporated at random, but failed to recognize a synthetic mRNA composed of U and G (in a ratio of 5:1) incorporated at random. The other tRNA recognized a synthetic mRNA composed of U and G (in a ratio of 5:1) incorporated at random, but failed to recognize a synthetic mRNA composed of U and C (in a ratio of 1:1) incorporated at random. **(a)** Using the genetic code and the wobble rules, determine the most probable nucleotide sequences of the anticodons in these two tRNAs. **(b)** Given this information, how many tRNAs would be required to accommodate all six codons for leucine? **(c)** Provide the anticodons of any tRNAs not described in part a. **(d)** Which anticodons, other than those on leucine tRNAs, should recognize these synthetic mRNAs?

26. In 1966, Crick (*Journal of Molecular Biology* 19:548–555) presented the wobble hypothesis. In this article, he used rules of chemistry to show that the base pairings shown in Figure 4.17 were plausible. He also showed that there were sites where hydrogen bonds could form in G-A, U-C, and U-U pairs, but that these pairs should not form during wobble. What general rule about base pairing is met by all of the possible wobble base pairs, but is violated by G-A, U-C, and U-U pairs?

27. Crick (in the same article cited in the previous question) stated that poly-I and poly-A form a double helix. Look at the structure of inosine as shown in Figure 4.17. Why is this double helix expected?

28. Fraenkel-Conrat (1964. *Scientific American* 211 [Oct]: 47–54) summarized the results of several experiments with single-nucleotide substitution mutations in tobacco mosaic virus, a single-stranded RNA virus that infects several plant species. Table 4.4 shows the positions of amino acids in the polypeptide chain and the amino acid substitutions due to mutations. Notice that at positions 20 and 21, two separate mutations affect the same amino acid but substitute different amino acids in the mutant polypeptides. Fill in the blanks for the codons and the nucleotide substitutions in DNA. Be as precise as possible with nucleotide sequences, using Pu, Py, or N when the exact nucleotide cannot be determined.

Table 4.4 Information for Problem 28: Single-Nucleotide Substitution Mutations in Tobacco Mosaic Virus

Position in Polypeptide	Original Amino Acid	Original Codon	Mutant Amino Acid	Mutant Codon	Nucleotide Substitution in DNA
11	Val	_____	Met	_____	_____
20	Pro	_____	Thr	_____	_____
20	Pro	_____	Leu	_____	_____
21	Ile	_____	Thr	_____	_____
21	Ile	_____	Val	_____	_____
25	Asn	_____	Ser	_____	_____
33	Asn	_____	Ser	_____	_____
46	Arg	_____	Lys	_____	_____
61	Arg	_____	Gly	_____	_____
65	Gly	_____	Ser	_____	_____
81	Thr	_____	Ala	_____	_____
97	Glu	_____	Gly	_____	_____
99	Glu	_____	Arg	_____	_____
122	Arg	_____	Gly	_____	_____
126	Asn	_____	Ser	_____	_____
129	Ile	_____	Val	_____	_____
134	Arg	_____	Gly	_____	_____
138	Ser	_____	Phe	_____	_____
148	Ser	_____	Phe	_____	_____
156	Pro	_____	Leu	_____	_____

29. One of the amino acid substitutions in the previous problem is the same one illustrated in Example 4.4. **(a)** Identify which one it is. **(b)** One other substitution is found in the amino acid sequence provided in Example 4.4. Write out the amino acid sequence showing the substitution. **(c)** For each of these amino acid substitutions, adjacent amino acids were unaffected. What aspect of translation is confirmed by this observation?

30. Below is the linear nucleotide sequence of a serine tRNA encoded in the mitochondrial DNA of humans (from Dirheimer et al. 1995. *In* Söll, D., and U. L. RajBhandary, eds. *tRNA: Structure, Biosynthesis, and Function.* Washington, D.C.: ASM Press, pp. 93–126). Mitochondria have their own ribosomes and tRNAs that are separate from their cytosolic counterparts.

```
5' GAGAAAGCUCACAAGAACUGCUAACUCAUGCCCCCAU-
   GUCUAACAACAUGGCUUUCUCACCA 3'
```

(This sequence is derived from the DNA, so posttran-scriptional base modifications are not indicated.) Notice that the number of nucleotides (62) is fewer than the number in nuclear-encoded tRNAs (about 75). **(a)** Determine the base-pairing structure of this tRNA. **(b)** How does this tRNA differ from nuclear and bacterial tRNAs in its structure? **(c)** What is the anticodon sequence?

31. Below is a short segment of the DNA sequence from the coding region of the COXII gene from wheat mito-chondria and the polypeptide segment encoded by this DNA segment (from Covello and Gray. 1989. *Nature* 341:662–666):

```
... GAGATTCGTGGAACTAATCATGCCTTTACGCCTATC ...
    GluIleCysGlyThrAsnHisAlaPheThrProIle
```

The mRNA encoded by this gene has been edited. Write out the edited mRNA segment, encoded by this DNA segment, and identify the nucleotides affected by RNA editing.

FOR FURTHER READING

Detailed information on translation can be found in Chapters 7–10 of **Lewin, B. 1997.** *Genes VI.* Oxford: Oxford University Press; Chapter 17 of **Wolfe, S. L. 1993.** *Molecular and Cellular Biology.* Belmont, Calif.: Wadsworth; and Chapter 6 of **Alberts, B., D. Bray, J. Lewis, M. Raff, K. Roberts, and J. D. Watson. 1994.** *Molecular Biology of the Cell,* 3rd ed. New York: Garland. A good general review of eukaryotic translation is **Merrick, W. C. 1992. Mechanism and regulation of eukaryotic protein synthesis.** *Microbiological Reviews* **56:291–315.** A series of excellent reviews on the genetic code published as the code was being deciphered are **Crick, F. H. C. 1962. The genetic code.** *Scientific American* **207 (Oct):66–74; Nirenberg, M. H. 1963. The genetic code II.** *Scientific American* **208 (Mar):80–94;** and **Crick, F. H. C. 1966. The genetic code III.** *Scientific American* **215 (Oct): 55–62.** Crick proposed and reviewed the wobble hypothesis in **Crick, F. H. C. 1966. Codon-anticodon pairing: The wobble hypothesis.** *Journal of Molecular Biology* **19:548–555.** Exceptions to the genetic code were reviewed in **Parker, J. 1989. Errors and alternatives in reading the universal genetic code.** *Microbiological Reviews* **53:273–298** and **Fox, T. D. 1987. Natural variation in the genetic code.** *Annual Review of Genetics* **21:67–91.** A book with several recent reviews of tRNA is **Söll D., and U. L. RajBhandary, eds.** *tRNA: Structure, Biosynthesis, and Function.* **Washington, D.C.: ASM Press, pp. 93–126.** RNA editing was reviewed by **Weiner, A. M. and N. Maizels. 1990. RNA editing: Guided but not templated?** *Cell* **61:917–920.**

For additional reading, go to InfoTrac College Edition, your online research library at: http: www.infotrac-college.com/brookscole

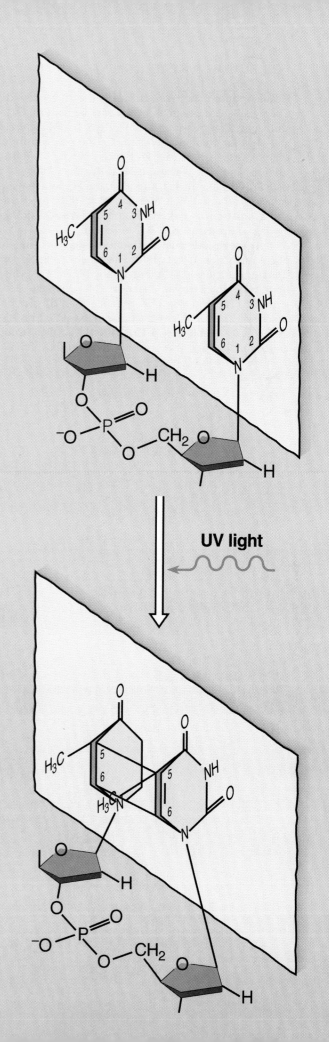

CHAPTER 5

KEY CONCEPTS

Mutation is the only original source of genetic variation.

~

Mutations in the same gene may have different effects depending
on the type and location of the mutation.

~

Errors in DNA replication may mutate genes.

~

Mutagenic agents may damage DNA, causing mutations.

~

Cells have DNA repair mechanisms that restore the original sequence to
damaged DNA.

MUTATION

Mutation is the basis for evolution and the only means by which new genetic variation is created. All other genetic changes simply reorganize variation that is already present. The word *mutation* is derived from the Latin verb *mutare*, which means "to move or change." A genetic mutation is a change in the genetic information in DNA. The word *mutation* was first used in genetics to describe spontaneous heritable changes in the appearance of an organism. With the discovery of DNA as the genetic material, scientists recognized mutations as alterations in the nucleotide sequence of DNA. A mutation in the DNA of a gene may alter the amino acid sequence encoded by that gene, which may alter or eliminate the function of the gene, which, in turn, may affect the phenotype of an individual who inheritsthe mutation.

In this chapter, we will examine the types and causes of mutations, the effects that mutations may have on proteins, cells, and organisms, and how cells counteract the effects of agents that damage DNA.

5.1 TYPES OF MUTATIONS

A **mutation** is a heritable change in the nucleotide sequence of a cell's DNA. The change may be as small as a single-nucleotide substitution, such as a G in place of an A, or as large as the addition or deletion of millions of nucleotides. Because DNA is the substance that mutates, mutations are faithfully replicated—from one cell generation to the next and from one organismal generation to the next. Mutations in a region of DNA that encodes an mRNA are transcribed into RNA and may alter the amino acid sequence of the protein encoded by the gene.

A **point mutation** is a change in a single nucleotide or in a few adjacent nucleotides in the DNA; for this reason, a point mutation usually affects only one gene. There are many different types of point mutations, and many ways in which point mutations arise. In this chapter, we will focus most of our discussion on point mutations.

However, let's take a moment to introduce two types of mutations that are not classified as point mutations. Because of their unique aspects, we will discuss these two types of mutations in detail in later chapters. **Transposable elements** (or **mobile genetic elements**) are large segments of DNA (usually hundreds to thousands of nucleotide pairs in length) that move from one location to another. When a transposable element inserts itself into a gene, it interrupts the gene, causing a mutation and usually causing the gene to lose its function. Excision of the element from the gene may restore the

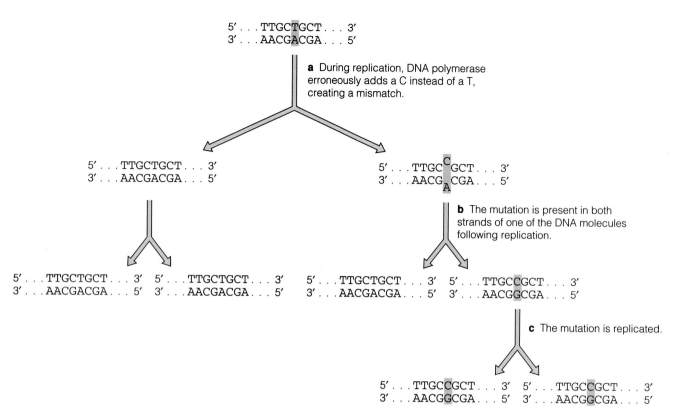

Figure 5.1 Establishment of a mutation in both strands of DNA.

gene's function. We will examine transposable elements in detail in Chapter 22.

Chromosomal changes (or **chromosomal aberrations**) differ significantly from all other types of mutations. Chromosomal changes eliminate, duplicate, or rearrange chromosomal segments that may include hundreds of thousands of nucleotide pairs and several genes. The effects of chromosomal changes are quite different from those of point mutations. We will discuss chromosomal changes in a separate context in Chapter 17.

Substitution, Deletion, and Insertion Mutations

In this section, we will review three types of point mutations, substitutions, deletions, and insertions, drawing examples from the human *PAH* gene, which encodes the enzyme phenylalanine hydroxylase. Mutations in this gene are responsible for the genetic disorder **phenylketonuria (PKU)**, which causes severe mental retardation if left untreated.

The most common type of point mutation is a **substitution mutation**, in which one nucleotide pair is substituted for another. DNA usually mutates during replication, and a nucleotide in only one strand of the DNA mutates. However, once the DNA replicates again, the mutation appears in both strands of one of the progeny molecules and is replicated faithfully in both strands from that point on (Figure 5.1). Geneticists usually study

mutations in DNA that has replicated many times since it mutated, so they observe the mutation as a substitution in both strands when compared to the original double-stranded DNA sequence. Substitution mutations are usually designated by writing the nucleotide change with an arrow, such as G → A, where G is the original nucleotide and A is the nucleotide substituted for it. Only the sense-strand sequence is written although the mutation appears in both strands. A G → A substitution in the sense strand implies a C → T substitution in the antisense strand.

The effect that a substitution mutation has on gene function depends on where in the gene the DNA mutates. For example, Figure 5.2a shows a G → A substitution mutation in the third position of codon 245 of the *PAH* gene. This mutation changes a GUG codon to GUA. Because the original and mutant codons both encode valine, the mutation does not change the amino acid sequence of the polypeptide, leaving protein function unaffected. A mutation that makes no change in the amino acid sequence is called a **same-sense mutation**. Same-sense mutations are also **silent mutations**, mutations that do not alter protein function.

A mutation that causes an alteration in the amino acid sequence of a gene's product is called a **missense mutation**. Figure 5.2b shows a C → T missense substitution mutation in codon 408 of the *PAH* gene. The mutation alters the first nucleotide of the codon, which when

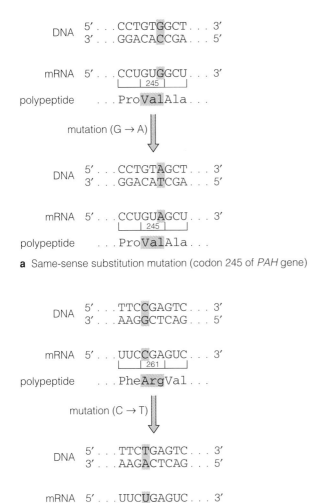

a Same-sense substitution mutation (codon 245 of *PAH* gene)

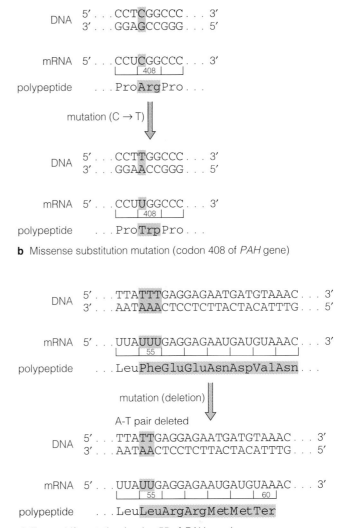

b Missense substitution mutation (codon 408 of *PAH* gene)

c Nonsense substitution mutation (codon 261 of *PAH* gene)

d Frameshift mutation (codon 55 of *PAH* gene)

Figure 5.2 Examples of mutations in the human *PAH* gene.

translated substitutes tryptophan for arginine in the polypeptide. This amino acid substitution eliminates the function of the enzyme encoded by the mutant gene and can cause PKU. Most missense mutations alter the function of a gene. However, some missense mutations cause amino acid substitutions that do not significantly alter protein function and are, therefore, silent mutations.

A missense mutation that creates a new termination codon is called a **nonsense mutation**. Figure 5.2c shows a C → T nonsense substitution mutation in codon 261 of the *PAH* gene. The mutation changes a CGA arginine codon into a UGA termination codon. The product of the mutant gene is a shortened polypeptide that lacks all of the amino acids encoded downstream from the mutant codon.

Substitution mutations are either transitions or transversions. A **transition** is a purine → purine substitution (A → G, G → A), or a pyrimidine → pyrimidine substitution (T → C, C → T). A **transversion** is a purine → pyrimidine substitution (A → T, A → C, G → T, G → C) or a

pyrimidine → purine substitution (T → A, T → G, C → A, C → G). Transitions are more frequent than transversions. For example, over 60% of the known substitution mutations in the human *PAH* gene are transitions. Notice that the substitution mutations illustrated in Figures 5.2a, b, and c are transitions.

Other point mutations include **deletion mutations**, in which one or more nucleotides are deleted from the DNA, and **insertion mutations**, in which one or more nucleotides are inserted into the DNA. Figure 5.2d shows a deletion of a single nucleotide pair in codon 55 of the *PAH* gene, which shifts the reading frame by one nucleotide downstream from the point of the mutation. Notice that all codons downstream from codon 55 are in a different reading frame in the mutant mRNA when compared to the original mRNA. Deletion and insertion mutations that shift the reading frame are called **frameshift mutations**. Any deletion or insertion that is not a multiple of three nucleotides causes a frameshift,

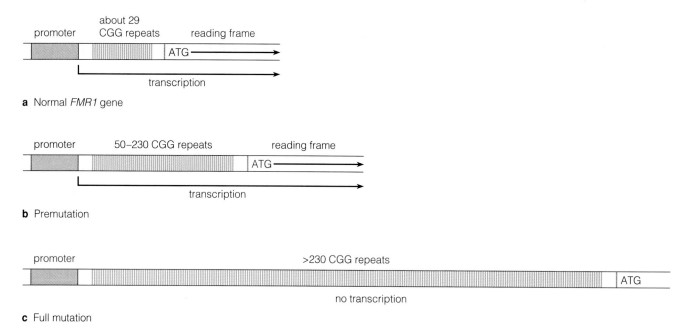

a Normal *FMR1* gene

b Premutation

c Full mutation

Figure 5.3 The expansion of trinucleotide repeats in the human *FMR1* gene.

provided the mutation is located in the reading frame. Deletions or insertions that *are* multiples of three nucleotides cause corresponding deletions or additions of amino acids in the polypeptide but do not shift the reading frame.

Frameshift mutations take the original termination codon out of frame and bring a new termination codon into frame, shortening or lengthening the polypeptide depending on where the first termination codon appears in the mutant reading frame. In Figure 5.2d, the first termination codon in the mutant reading frame appears in codon 60 (5 codons downstream from the mutation), shortening the polypeptide.

Mutations do not have to be in the reading frame to affect the polypeptide. Conserved sequences that lie outside of the reading frame are important for initiation of transcription, intron removal, polyadenylation, and initiation of translation. Mutations in conserved sequences may affect the polypeptide encoded by the gene. For example, the first mutation discovered in the *PAH* gene was a G → A transition, changing the GU at the 5′ end of intron 12 into AU. This mutation disrupts normal intron removal, so the mutant gene fails to encode a functional protein.

~

A same-sense substitution mutation does not alter the amino acid sequence. A missense substitution in the reading frame substitutes a new amino acid or creates a new termination codon. Many deletion and insertion mutations are frameshift mutations, which alter many amino acids and change the position of the termination codon.

Trinucleotide Repeat Expansions

A class of mutation called **trinucleotide repeat expansion**, which denotes an increase in the number of trinucleotide tandem repeats, was discovered in humans in 1991. A genetic disorder called **fragile X syndrome** was found to be due to expansion of a region in DNA where the trinucleotide sequence CGG is repeated several times in tandem. Over 10 genetic disorders in humans are now known to be caused by trinucleotide repeat expansions (Table 5.1). Fragile X syndrome is the most thoroughly studied and offers a good example of some peculiar effects that arise from this type of mutation.

Fragile X syndrome is caused by disruption of a gene on the X chromosome called *FMR1* (for "fragile X mental retardation 1"), which encodes the protein FMRP. This protein binds to mRNAs in brain cells and apparently affects the expression of a number of genes essential for proper brain development. Within the untranslated region at the 5′ end of the gene is a three-nucleotide sequence (CGG) that is repeated anywhere from 6 to 50 times in most people, with 29 copies being the most common number (Figure 5.3a). People with a normal number of repeats show no symptoms of the disorder. The number of repeats can increase or decrease slightly from generation to generation, but rarely exceeds 50. This normal fluctuation has no effect on the size of the polypeptide because the repeated region appears in the mRNA upstream from the reading frame.

Some people, however, may acquire an abnormally large number of repeats. Repeats ranging from 50 to 230 constitute what is known as a **fragile X premutation** (Figure 5.3b). People with the premutation rarely show any symptoms of the disorder, but the descendants of females who carry the premutation have an increased risk of fragile X syndrome. People who have fragile X syndrome often have as many as 700 CGG repeats and sometimes over 1000. When the repeated region expands

Table 5.1 Some Human Genetic Disorders Associated with Trinucleotide Repeat Expansions

Genetic Disorder	Trinucleotide repeat	Symptoms
Fragile X syndrome	CGG	Mental retardation
Fragile X E MR	GCC	Mental retardation
Huntington disease	CAG	Neural degeneration, dementia
Kennedy disease	CAG	Muscle weakness and wasting
Myotonic dystrophy	CTG	Muscle weakness and wasting, cataracts, hypogonadism
Spinocerebeller ataxia	CAG	Ataxic gait, muscular atrophy, diabetes
Dentatorubral-pallidoluysian atrophy	CAG	Neural degeneration, epilepsy, dementia

Adapted from Warren, S. T., and D. L. Nelson. 1994. *Journal of the American Medical Association* 271:538–542.

significantly beyond the premutation size (Figure 5.3c), a **fragile X full mutation** results. The bases in the nucleotides of the repeated region and the promoter become highly methylated, which inhibits transcription of the *FMR1* gene. Ribosomes do not efficiently translate the few mRNAs that are produced. The amount of FMRP produced under these conditions is inadequate, brain development is inhibited, and symptoms of fragile X syndrome appear.

Not all cases of fragile X syndrome are due to trinucleotide repeat expansion. In one case, a missense substitution mutation that substitutes asparagine for isoleucine is responsible. Several deletions of either the entire gene or large portions of it are also known to cause fragile X syndrome.

Let's now examine some data that show a relationship between the repeat number for the *FMR1* gene in parents and repeat expansion in their progeny.

Example 5.1 Trinucleotide repeat expansion and fragile X syndrome.

Fu et al. (1991. *Cell* 67:1047–1058) reported the following DNA sequence that contains the CGG repeat in the *FMR1* gene:

```
GCGGCGGCGGTGACGGAGGCGCCGCTGCCAGGGGGC-
GTGCGGCAGCGCGGCGGCGGCGGCGGCGGCGGCGG-
CGGCGGCGGCGGCGGCGGCGGCGGCTGGGCCTCGAA-
GCGCCCGCAGCCCA
```

They examined unrelated persons who did not have fragile X syndrome and found that the number of CGG repeats ranged from 6 to 54, with 29 repeats being the most common (148 of 492 individuals had 29 repeats). They also examined parents and offspring in which one of the parents had a

premutation (more than 50 repeats). The results of their study are presented in Table 5.2.

Problem: **(a)** Determine the number of tandem CGG repeats in the DNA sequence given above. **(b)** What relationship between the size of the premutation and the probability of expansion is suggested by these data?

Solution: **(a)** There are 16 tandem CGG repeats in this DNA sequence.

```
GCGGCGGCGGTGACGGAGGCGCCGCTGCCAGGGGGC-
GTGCGGCAGCG CGG CGG CGG CGG CGG CGG-
CGG CGG CGG CGG CGG CGG CGG CGG CGG-
CGG CTGGGCCTCGAAGCGCCCGCAGCCCA
```

(b) Every child had a repeat number that differed from that of the parent who had the premutation, and all but two of the differences were increases. These data indicate that the repeat number is unstable for premutations, and that there is a strong tendency toward expansion of a premutation. Moreover, the size of the expansion correlates with the size of the premutation. The probability of expansion to a full mutation is greatest when the premutation has more than 70 repeats.

Several other genetic disorders in humans are due to trinucleotide repeat expansion. **Huntington disease** is characterized by progressive neurological degeneration that usually does not begin until after the age of 30. Once the degeneration begins, the affected person progressively loses motor function, behaves in unusual ways, becomes demented, and eventually dies from the disease, usually within 10–15 years. Huntington disease is usually due to expansion of a CAG repeat in the coding

Table 5.2 Information for Example 5.1: Comparison of Repeat Numbers in the *FMR1* Gene in Parents and Children*

Number of CGG Repeats in Parent's *FMR1* Gene Transmitted to Child	Number of CGG Repeats in Child's *FMR1* Gene Inherited From the Indicated Parent (Multiple numbers on the same line denote siblings)
Fathers	
66	70, 83
86	100
116	163
Mothers	
52	73
54	58, 60, 57, 58, 52
59	54
66	73, 86, F
66	80, 73, 110
70	66, 103
70	F, F
73	F, F, F
73	F, 170
73	113, F
77	F, F, F
80	F
80	F, F
80	100
83	F, F
83	F, F
83	F
83	93, F, F, F
86	F, F, F
86	126/193**
90	F, F
90	F, F
93	F, F
93	F
93	F, F
93	F, F
93	F
100	F, F
100	F
100	F
110	F, F
113	F

*Only results from parents with more than 50 repeats in the *FMR1* gene and their children are shown. Because full mutations were so large that the researchers were unable to determine the exact number of repeats with the methods they were using, full mutations are indicated with an F.
**This individual had different repeat numbers in different cells, 126 and 193 being the most common.

region of the *IT15* gene, which encodes a protein named huntingtin. Each CAG is in frame and is translated as a glutamine. Unaffected persons have 10–35 repeats that are translated as a string of glutamines. Affected persons have 36–121 repeats, and the lengthened string of glutamines alters the protein's function.

~

Trinucleotide repeat expansions increase the number of trinucleotide repeats in a gene, disrupting the function of that gene.

Mutations and Protein Function

The effect of a mutation on protein function depends on several factors: which amino acids in the polypeptide chain are altered, how many amino acids are altered, and what specific changes in amino acid sequence arise from the mutation. As we discussed earlier, a same-sense mutation has no effect on the function of a protein because the amino acid sequence remains unchanged. Missense mutations, however, can have a variety of effects, ranging from no loss of function to complete loss of function or, more rarely, a gain of function. The final effect depends on how and where the mutation alters the protein.

A protein's function depends predominantly on the physical shape of the protein, as determined by its primary, secondary, tertiary, and quaternary structures, the chemical properties of the protein, and the active site of the protein if it is an enzyme. Missense mutations that do not significantly alter any of these will probably have little, if any, effect on the protein's function. For example, valine and isoleucine are both nonpolar amino acids, differing from one another by a single methyl group (see Figure 4.1). Their chemical properties are nearly identical. A missense mutation that substitutes valine for isoleucine should have little, if any, effect on protein function. However, a mutation that alters the protein's active site, or alters the tertiary or quaternary structure, may change the protein's ability to carry out its function or, in many cases, eliminate the protein's function altogether. Most frameshift mutations completely eliminate protein function because they alter nearly every amino acid from the point of the mutation onward and usually change the location of the termination codon. Even if a frameshift mutation does not affect the active site of an enzyme, it usually alters the structure of the enzyme drastically, causing the enzyme to lose its function.

Usually, mutations that have an effect on a protein's activity are **loss-of-function mutations**, which either reduce the protein's ability to function or eliminate its function altogether. Most of the hundreds of known genetic disorders in humans (such as phenylketonuria, sickle-cell anemia, thalassemias, cystic fibrosis, and hemophilias) are caused by loss-of-function mutations.

More rarely, a mutation may cause a gene to *gain* function compared to its previous state, usually by producing a greater-than-normal quantity of protein. These are called **gain-of-function mutations**. Although gain-of-function mutations are more rare than loss-of-function mutations, they are quite important. Many of the mutations that contribute to development of cancer are gain-of-function mutations in genes that regulate cell growth and division, a topic we will discuss in Chapter 24.

Mutations Outside of the Transcribed Region of a Gene

So far, we have discussed mutations in the transcribed regions of genes. What about mutations that reside in the DNA outside of transcribed regions? Can these mutations affect protein activity? Mutations outside of the transcribed region, as a general rule, do not alter the amino acid sequence of a protein. However, depending on where a mutation is located, it may affect the amount of the protein that is produced, thus affecting its level of activity in the cell. For example, a mutation within the promoter region may prevent RNA polymerase from binding and thus prevent transcription, eliminating the protein's synthesis entirely. A promoter mutation may also reduce or, more rarely, enhance the ability of RNA polymerase to transcribe the gene, thus affecting how much protein is produced. In fact, most gain-of-function mutations reside in the promoter or upstream regulatory regions of a gene.

~

The effect of mutations on protein activity depends on what changes, if any, the mutation creates in the physical structure and chemical properties of the protein. Mutations may also affect protein activity by reducing or increasing the level of protein production.

Forward Mutations and Reversions

Most mutations are **forward mutations**. A forward mutation changes an original, usually functional, form of a gene to a mutant form, in which the DNA sequence differs from the most common sequence. Forward mutations are usually detected as a loss of protein function, often due to a change in the protein's amino acid sequence.

Occasionally, a second mutation in the mutant form of a gene may cause the gene to revert back to its functional form, in which it encodes a functional gene product. These second mutations that restore the gene's function are called **reversions**. Sometimes a reversion actually restores the gene's original DNA sequence. For example, suppose that a forward mutation changes a G-C pair to an A-T pair. The A-T pair is then replicated faithfully from one generation to the next, perpetuated indefinitely unless changed by a second mutation. If, however, a second mutation changes the mutant A-T pair *back* to the original G-C pair, then this reversion restores the former function of the gene.

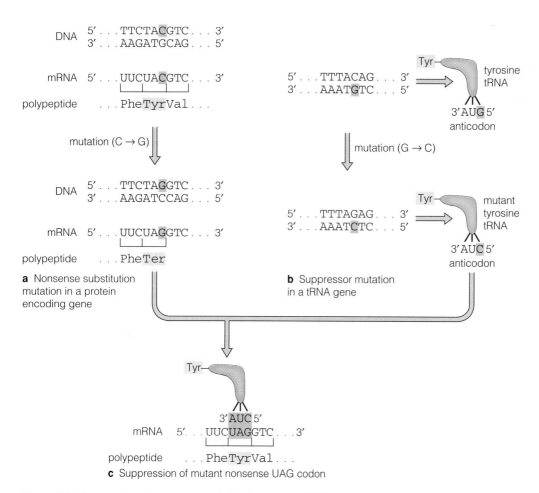

a Nonsense substitution mutation in a protein encoding gene

b Suppressor mutation in a tRNA gene

c Suppression of mutant nonsense UAG codon

Figure 5.4 Suppression of a nonsense mutation by a mutant tRNA.

Reversions, however, do not necessarily require restoration of the gene's original nucleotide sequence. If a second mutation restores the function of a mutant polypeptide by *compensating for* the first mutation while still leaving it in place, the second mutation is also called a reversion. For example, a single-nucleotide deletion in the reading frame of a gene shifts the reading frame by one nucleotide for all codons downstream from the mutation, altering most of the amino acids from that point on. A second mutation that *inserts* a single nucleotide, upstream or downstream from the site of the deletion, restores the normal reading frame after the ribosome has passed both mutations. Only the codons *between* the two mutations remain out of frame. If the altered amino acids are not within a critical portion of the polypeptide, the polypeptide's function may be restored, thus qualifying the second mutation as a reversion.

Forward mutations are more frequent than reversions, usually by an order of magnitude. Because many nucleotides within a gene are essential for the gene to produce a functional product, mutation of any one of these nucleotides may disrupt the function of the gene's product. Reversions, on the other hand, require at least one of a very few specific changes that restore function of the gene's prod-

uct. Because there are many more possibilities for a forward mutation of a functional gene than for reversion of a mutant gene, forward mutations are much more common.

Another type of reversion is caused by **suppressor mutations**. A forward mutation may alter the polypeptide product of a gene, thereby eliminating the function of the polypeptide. However, a mutation within another gene may produce a separate product that compensates for the lost activity of the first. In this case, the second mutation is called a suppressor mutation. Among the best studied suppressor mutations are ones that compensate for nonsense mutations. Suppression of premature UAG termination codons by a mutant tRNA in *E. coli* has been well characterized. A gene that encodes a tyrosine tRNA with the anticodon 3′ AUG′ 5′ (where G′ represents a modified guanine) mutates, changing the anticodon to 3′ AUC 5′. This mutant tRNA is charged with tyrosine but pairs with UAG termination codons. When a nonsense mutation with a premature UAG termination codon, and the mutant tyrosine tRNA with a 3′ AUC 5′ anticodon, are present in the same cell, tyrosine is encoded by the UAG codon, and translation continues beyond the premature termination codon, restoring the functional polypeptide (Figure 5.4).

~
Reversion mutations restore the original function of a gene that has lost its function because of a previous mutation.

5.2 SPONTANEOUS MUTATION

Mutations are either spontaneous or induced. **Spontaneous mutations** are naturally occurring mutations due to errors in DNA replication or natural chemical reactions in DNA. They are not caused by external agents. **Induced mutations** are caused by external agents that chemically alter DNA, usually causing some form of DNA damage. In this section we examine some of the causes of spontaneous mutation. In section 5.3 we will look at some of the agents responsible for induced mutation and see how they affect DNA.

Tautomeric Shifts

Bases in the nucleotides of DNA generally exist in stable forms. However, occasionally a base may shift temporarily to a less stable form that alters its pairing properties. These temporary changes in form are called **tautomeric shifts**. For example, the nucleotides T and G usually exist in the more stable keto forms, but may occasionally shift to the less stable enol forms. When either T or G shifts to the enol form during DNA replication, an unusual G-T pair may appear. Likewise, C and A generally assume the more stable amino forms but may occasionally shift to the less stable imino forms. When either C or A assumes the imino form during DNA replication, an unusual C-A pair may appear. The alternative tautomeric forms of bases and their base pairing are illustrated in Figure 5.5.

Tautomeric shifts may cause transitions. For example, a T in the template strand may undergo a tautomeric shift from the keto to the enol form during DNA replication. If DNA polymerase encounters the T in enol form, the polymerase inserts a G opposite the T, instead of an A, thus substituting G (a purine) for A (also a purine).

~
Spontaneous tautomeric shifts during DNA replication may cause transitions.

Mutations in Repeated DNA Segments

Insertion and deletion mutations are unusually frequent at certain sites in DNA molecules called **mutation hotspots**. DNA sequencing of several mutation hotspots revealed that short, repeated DNA sequences were present at these sites. A repeated DNA sequence may be a mononucleotide repeat (such as AAAAA), a dinucleotide repeat (such as CTCTCTCTCT), a trinucleotide repeat (such as CGGCGGCGGCGGCGG), or longer repeats. During DNA replication, part of the newly synthesized repeated sequence may slip so that some of the nu-

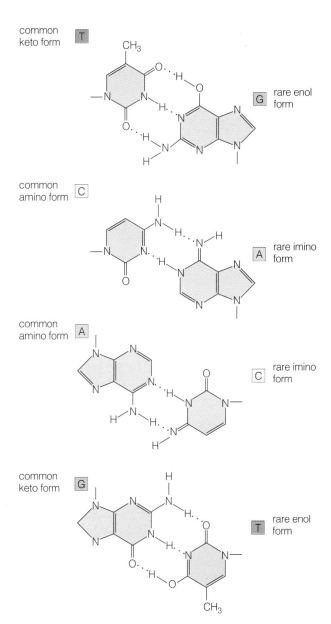

Figure 5.5 Base pairs formed after tautomeric shifts.

cleotides in the repeat no longer pair with the template strand. The unpaired part of the newly synthesized strand then loops out from the DNA double helix. Nucleotides of the repeat beyond the loop in the newly synthesized strand may then pair with nucleotides in the template strand at the point of slippage. These nucleotides are said to be misaligned with the template strand. DNA polymerase continues synthesis, adding as many extra nucleotides as are present in the loop to the newly synthesized strand. The loop is an insertion mutation.

The template strand, rather than the newly synthesized strand, may slip and loop at repeated sequences. In this case, after DNA replication, the newly synthesized strand ends up missing as many nucleotides as are present in the loop of the template strand; in other words, it has a deletion mutation.

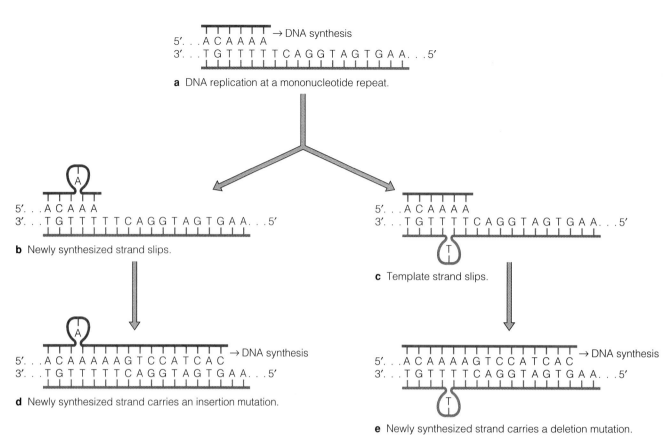

a DNA replication at a mononucleotide repeat.

b Newly synthesized strand slips.

c Template strand slips.

d Newly synthesized strand carries an insertion mutation.

e Newly synthesized strand carries a deletion mutation.

Figure 5.6 Insertion and deletion mutations caused by slippage of either the newly synthesized strand or the template strand at a repeated sequence.

Insertion and deletion mutations caused by strand slippage in repeated sequences are illustrated in Figure 5.6. Slippage has been well documented at mutation hot-spots in bacteria, but was not known in eukaryotes until recently. Several lines of evidence suggest that trinucleotide repeat expansion mutations in humans (such as those responsible for fragile X syndrome and Huntington disease) are due to slippage during DNA replication, especially during synthesis of the lagging strand.

In the following example, we'll learn how mutations at a mutation hotspot are analyzed.

Example 5.2   A mutation hotspot in T4 bacteriophage.

In 1966, Streisinger et al. (*Cold Spring Harbor Symposia on Quantitative Biology* 31:77–84) reported original and mutant amino acid sequences of a segment from the lysozyme gene of bacteriophage T4, a virus that infects *E. coli*:

Original   ... ThrLysSerProSerLeu...

Mutant   ... ThrLysValHisHisLeu...

The nucleotide sequence of the T4 bacteriophage lysozyme gene was published in 1983 by Owen et al. (*Journal of Molecular Biology* 165:229–248). The amino acid sequences above correspond to a hotspot for frameshift mutations with the nucleotide sequence

... ACAAAAAGTCCATCACTT...

Problem: Describe the mutation that encodes the mutant polypeptide, and explain why this is a mutation hotspot.

Solution: The amino acids in the original polypeptide can be matched to their codons in the DNA:

... ThrLysSerProSerLeu...
... ACAAAAAGTCCATCACTT...

Notice that there is a string of five A's in the nucleotide sequence. A deletion of any one of these A's produces a nucleotide sequence that encodes the mutant polypeptide:

... ThrLysValHisHisLeu...
... ACAAAAGTCCATCACTT....

This region is a mutation hotspot for frameshift mutations because the DNA may slip at the mononucleotide AAAAA repeat during replication, causing deletions or insertions.

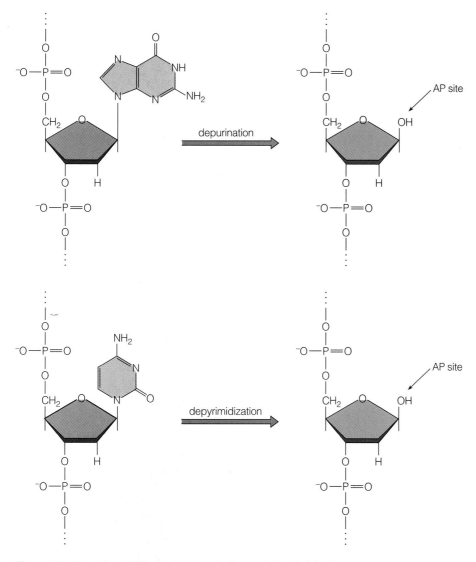

Figure 5.7 Formation of AP sites by depurination and depyrimidination.

Repeated segments of DNA are prone to insertion and deletion mutations due to slippage during DNA replication.

Spontaneous Lesions

Chemical reactions in DNA may modify a nucleotide in such a way that it no longer carries one of the four standard bases (T, C, A, or G). These chemical modifications in DNA are called **lesions**. DNA that contains lesions is called **damaged DNA**. In contrast with lesions, mutations are substitutions, deletions, or insertions of one or more of the four standard nucleotides. A lesion is not a mutation because it cannot be propagated by DNA replication (remember that mutations are *heritable* changes in the nucleotide sequence, which means that they are propagated by DNA replication). However, a lesion may *cause* a mutation when DNA polymerase encounters a lesion in the template DNA strand during DNA replication and incorporates an incorrect nucleotide into the newly synthesized DNA strand.

Spontaneous lesions are natural chemical alterations of the bases in the nucleotides of DNA. **Depurination** is the most frequent cause of spontaneous lesions. In depurination the bond between deoxyribose sugar and a purine base (A or G) is hydrolyzed, releasing the base from the DNA backbone and leaving a hydroxyl group in its place. There is then a sugar but no base at this nucleotide position. **Depyrimidination**, produced by hydrolysis of the bond between deoxyribose sugar and a pyrimidine base (C or T), is less frequent than depurination but produces the same result: a hydroxyl group on the 1′ carbon where a base was previously present. Both of these chemical alterations are illustrated in Figure 5.7. The sites where a base is missing are called **apurinic sites** or **apyrimidinic sites**, depending on whether a purine or a pyrimidine was originally present. As apurinic and apyrimidinic sites are chemically identical, they are called **AP sites** (because both apurinic and apyrimidinic begin with the letters AP).

Table 5.3 Summary of Mutagenic Agents and Their Effects

Agent	Examples	Effect
Radiation		
Ionizing radiation	X-rays; α, β, and γ rays	Indirect damage of DNA through the formation of free radicals, which damage DNA
Nonionizing radiation	UV light	Damage of nucleotides and their bases
Chemical mutagens		
Base analogs	5-bromouracil, 2-amino purine	Transitions
Intercalating agents	Acridine dyes, ethidium bromide, ICR compounds	Insertion and deletion mutations
Deaminating agents	Bisulfite compounds, nitrous acid	Transitions
Hydroxylating agents	Hydroxylamine	Transitions
Aflatoxin B_1		Formation of an AP site by removal of a guanine base from the nucleotide
Alkylating agents	Ethylmethane sulfonate, ethyl ethane sulfonate, mustard gas, nitrogen mustard, polycyclic aromatic hydrocarbons	Transitions (most frequent), transversions, insertion and deletion mutations, chromosome breaks

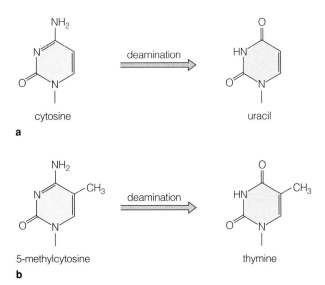

Figure 5.8 Cytosine deamination.

With no base present at an AP site in the template strand, there is nothing to which a newly incorporated nucleotide may pair during DNA replication. Consequently, DNA replication ceases at an AP site. Several repair systems are known that either excise the AP site and replace it—along with some of the adjacent nucleotides—with new nucleotides, or allow DNA polymerase to bypass the AP site during replication. We will discuss these DNA repair mechanisms later in this chapter.

Another spontaneous lesion is **cytosine deamination**, in which the amino group on the number 2 carbon of C is removed and a double-bonded oxygen is left in its place, which converts C to U (Figure 5.8a). Because U is not normally found in DNA, it may be recognized by repair mechanisms as an error and be repaired. If left unrepaired, the U pairs with A instead of G during DNA replication, causing a transition.

Methylation, the addition of a methyl (—CH$_3$) group to the base in a nucleotide, is a common modification of bases in DNA. Cytosines may often be methylated at the number 5 carbon in the base, forming 5-methylcytosine. When 5-methylcytosine is spontaneously deaminated, it becomes a T rather than a U (Figure 5.8b). As just explained, repair systems may recognize U as a lesion in the DNA and correct it. However, because T is one of the four standard nucleotides, repair mechanisms do not recognize it as a lesion. Therefore, deamination of 5-methylcytosine is a C $\rightarrow$ T transition.

Lesions in DNA may cause mutations when the DNA is replicated.

5.3 INDUCED MUTATIONS

The rate of mutation may increase significantly when cells are exposed to certain external agents, either chemical or in the form of radiation. Agents that enter the cell and cause mutations are called **mutagens**. Mutations that are caused by mutagens are induced mutations. All organisms on earth are exposed to radiation from space and to some naturally occurring chemical mutagens. Indeed, naturally occurring radiation and chemical mutagens have played a significant role in creating new

genetic variation by mutation throughout the history of life on earth.

A wide variety of agents may cause DNA to mutate. Some agents, such as ultraviolet (UV) radiation in sunlight and ionizing radiation from space, are naturally present. Other mutagenic agents are the result of human activity. Mutagenic agents can be divided into two general classes: (1) radiation and (2) chemical mutagens. Table 5.3 provides a summary of these agents and their effects. We will consider the two classes in turn.

Radiation

Two types of radiation are mutagenic, ionizing radiation and nonionizing radiation. **Ionizing radiation** is high-energy radiation, including protons, neutrons, X-rays, and alpha, beta, and gamma rays. Ionizing radiation may damage DNA directly when high-energy particles strike DNA molecules. However, indirect damage of DNA is a much more common effect of ionizing radiation. When high-energy rays collide with molecules (water in particular), they often strip atoms of an electron, causing the molecule to be ionized—hence the name *ionizing radiation*. The loss of an electron leaves an unpaired electron within the molecule, which causes the molecule to be highly reactive with other molecules. A highly reactive molecule with an unpaired electron is called a **free radical**. Free radicals may create new free radicals as they collide with other molecules. Free radicals tend to be highly reactive with DNA, attacking and altering the bases. Of even greater consequence, free radicals may attack the phosphodiester bonds in DNA, causing them to break (Figure 5.9). A break in one strand of double-stranded DNA is generally inconsequential because it may be readily repaired, but when both strands break, repair often fails to restore the DNA molecule to what it was before. Therefore, it is generally not the ionizing radiation itself that causes mutation but the free radicals induced by this high-energy radiation.

Ionizing radiation is potentially quite dangerous because it may deeply penetrate living tissue, leaving a trail of free radicals through the tissue. Because of their ability to pass through living tissue, X-rays have proved particularly useful in medical and dental diagnoses; the X-rays pass more easily through some tissues than others, providing an image of internal tissues when the X-rays strike radiographic film. For X-ray photographs, radiation is usually brief and at a low level to minimize the risk of radiation-induced disease.

However, the mutagenic effect of ionizing radiation is cumulative, meaning that repeated exposure to low levels of ionizing radiation can be dangerous over a period of years. This implies that there is no safe level of radiation, particularly for long-term exposure. A short-term exposure to low-level radiation directed at a limited area, such as an X-ray examination for medical purposes, does not

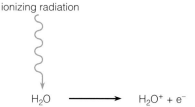

ionizing radiation

$$H_2O \longrightarrow H_2O^+ + e^-$$

a Ionizing radiation strips an electron from a water molecule, ionizing the water molecule with a positive charge.

$$e^- + H_2O \longrightarrow H_2O^-$$

b The electron combines with another water molecule to form a negatively charged water ion.

$$H_2O^+ \longrightarrow H^+ + OH\bullet$$
$$H_2O^- \longrightarrow H\bullet + OH^-$$

c Both of the water ions spontaneously form free radicals (the dots represent unpaired electrons) and hydrogen or hydroxide ions.

$$OH\bullet + RH \longrightarrow R\bullet + H_2O$$

d A hydroxide free radical changes an organic molecule into a free radical. The R represents one of many possible organic groups.

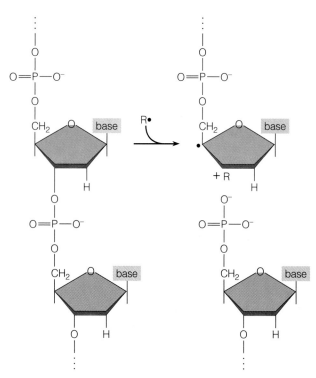

e A free radical breaks a strand of DNA.

Figure 5.9 Formation of free radicals and their interaction with organic molecules.

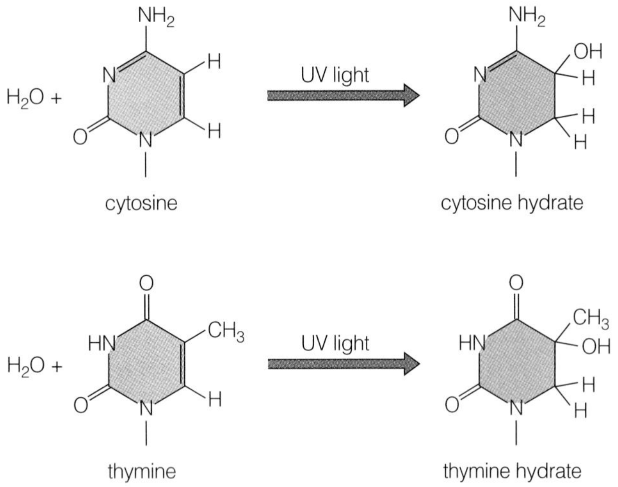

Figure 5.10 Formation of pyrimidine hydrates induced by UV light.

pose a significant danger of harmful mutation, because the exposure is very brief. However, long-term exposure to low levels can be dangerous. For this reason, X-ray technicians use lead aprons and shielding booths to protect themselves from X-ray exposure. Similarly, geneticists who use radioactively labeled nucleotides or amino acids to conduct their research must take adequate precautions to shield themselves from exposure to the radiation.

Nonionizing radiation, unlike ionizing radiation, does not have enough energy to strip electrons from atoms. Instead, it causes electrons to assume temporarily higher energy levels within the atom, a phenomenon called excitation. Ultraviolet (UV) radiation is the most important type of mutagenic nonionizing radiation. With its relatively low energy, nonionizing radiation penetrates only the uppermost layers of cells in living tissue. For this reason, it affects bacteria, plants, and the cells on the surface tissues of animals—mammalian skin cells in particular.

UV light is readily absorbed by the nucleotides of DNA, with maximum absorption—and maximum mutagenesis—at a wavelength of 254 nm. The pyrimidines, C and T, are particularly susceptible to excitation and damage. UV light may convert single pyrimidines into **pyrimidine hydrates** (Figure 5.10), lesions that have altered base-pairing properties. More commonly, UV light causes covalent bonding between the two adjacent pyrimidines forming a lesion called a **pyrimidine dimer** (Figure 5.11). Thymines in particular are highly susceptible to UV-induced dimerization; the most common dimer, and the most common lesion caused by UV light, is the **thymine dimer**. If left unrepaired, a pyrimidine dimer in the template strand can cause mutations in the newly synthesized strand during DNA replication.

~

Radiation is of two types, ionizing and nonionizing. Ionizing radiation penetrates deeply into tissues and forms free radicals, which can damage DNA. Nonionizing radiation does not penetrate deeply into tissues but may damage DNA in exposed cells near the surface of the tissue. Radiation-induced damage in DNA can cause mutations.

Chemical Mutagens

Base analogs are nitrogenous bases that are similar to the four bases of DNA. When present in the form of deoxyribonucleotides, they may be incorporated into DNA during replication in place of the usual nucleotides. Base analogs in a DNA molecule are lesions. The pairing properties of base analogs differ from those of the bases they replace, which mutates nucleotides in the newly synthesized strand during DNA replication when the template strand contains a base analog. Two common base analogs are 5-bromouracil (5-BU) and 2-amino purine (2-AP).

5-BU is an analog of T (Figure 5.12a) and, when present during DNA replication, is usually incorporated into DNA in place of T. Like T, 5-BU pairs with A when in the more stable keto form (Figure 5.13a). However, the keto form of 5-BU is much less stable than the keto form of T, resulting in frequent tautomeric shifts of 5-BU from the keto to the enol form. In the enol form, 5-BU (like T in the enol form) pairs with G (Figure 5.13b). Thus, when 5-BU is in the enol form during DNA replication, it causes a transition (A → G) in the newly synthesized strand. And because 5-BU is much less stable than T, transitions due to tautomeric shifts are more frequent when 5-BU is present in the template strand.

2-AP is an analog of A (Figure 5.12b) and undergoes frequent tautomeric shifts. Depending on its tautomeric form, it may base-pair with either T or C. Thus, when 2-AP is in the template strand, either T or C may be added opposite it, causing a transition when it pairs with C.

Another class of chemical compounds, **intercalating agents**, are similar to the nitrogenous bases in DNA and RNA, which are stacked flat at right angles to the sugar-phosphate backbone. Intercalating agents insert themselves between the stacked bases, a process called **intercalation**, which causes the DNA molecule to stretch at the insertion point (Figure 5.14). If an intercalating agent inserts itself into a single-stranded DNA molecule during replication, it may cause an insertion or deletion mutation, usually of a single nucleotide, causing a frameshift mutation if the mutation is in the reading frame. Intercalating agents include the acridine dyes, such as acridine orange and proflaven, ethidium bromide, and a large class of compounds known as ICR compounds. These agents may intercalate between the bases of both double- and single-stranded DNA and RNA.

Other chemical mutagens express their effects by modifying bases in DNA rather than incorporating themselves into the DNA. **Deaminating agents** include bisulfite compounds and nitrous acid. Bisulfite ions

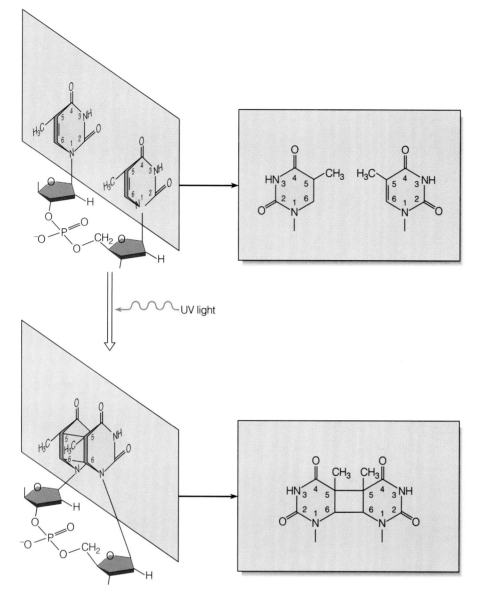

Figure 5.11 Formation of a thymine dimer between two adjacent thymines, induced by UV light. When viewed in three dimensions, the bases are stacked flat within the DNA molecule, so carbons 5 and 6 of both thymines are oriented in the same direction. One bond forms between the number 5 carbons of the adjacent thymines, and the second bond forms between the number 6 carbons.

deaminate C to U, with the same results as spontaneous cytosine deamination, which we described earlier (see Figure 5.8). The mutagenic agent nitrous acid (HNO_2) has the ability to deaminate C to U (which pairs with A instead of G), A to hypoxanthine (which pairs with C instead of T), and G to xanthine (which, like G, pairs with C). Consequently, nitrous acid–induced deamination of C and A can cause transitions during DNA replication (Figure 5.15).

Hydroxylating agents add hydroxyl groups to the bases of nucleotides. The best studied hydroxylating agent is hydroxylamine, which has a very specific mutagenic effect of adding a hydroxyl group to the amino group

attached to the number 4 carbon of C, converting C into hydroxyaminocytosine (Figure 5.16). The addition of the hydroxyl group causes hydroxyaminocytosine to pair with A instead of G. Thus, when present in the template strand, hydroxyaminocytosine causes an A → G transition in the newly synthesized strand during DNA replication.

Aflatoxin B_1 is a powerful mutagen that attaches to guanine and eventually removes the guanine base from the nucleotide, leaving an AP site. In most cases, DNA polymerase inserts an A into the newly synthesized strand across from the AP site, a transversion. Aflatoxin B_1 is a member of a class of large mutagenic molecules that attach to the bases of DNA nucleotides. In some

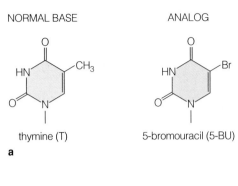

NORMAL BASE

thymine (T)

ANALOG

5-bromouracil (5-BU)

adenine (A)

2-amino purine (2-AP)

a

b

Figure 5.12 Examples of base analogs.

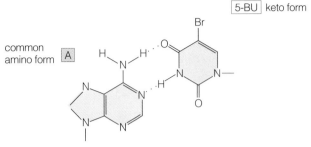

5-BU keto form

common amino form A

a In the keto form, 5-BU pairs with A, resulting in no mutation.

5-BU enol form

common keto form G

b In the enol form, 5-BU pairs with G, resulting in a transition mutation.

Figure 5.13 Alternative base pairing when 5-BU is in the keto and enol forms.

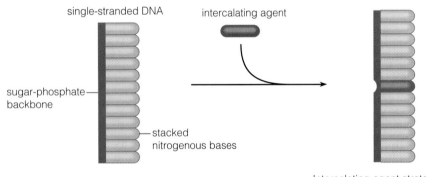

single-stranded DNA

intercalating agent

sugar-phosphate backbone

stacked nitrogenous bases

Intercalating agent stretches the DNA molecule by the thickness of one base.

Figure 5.14 Insertion of an intercalating agent into the stacked bases of a single-stranded DNA molecule.

cases, the modified base is hydrolyzed from the DNA, creating an AP site. In other cases, the large molecule attached to the base prevents pairing with any base.

Alkylating agents add alkyl (usually methyl or ethyl) groups to several positions on all of the nitrogenous bases of DNA, creating lesions. DNA polymeraseses may incorporate incorrect nucleotides in the newly synthesized strand when they encounter alkylated bases in the template strand during DNA replication. Most mutations caused by alkylating agents are transitions, although these agents are also known to cause transversions, insertions, deletions, and chromosome breaks. For example,

polycyclic aromatic hydrocarbons (PAHs) are alkylating agents found in smoke of all kinds. Tobacco smoke contains mixtures of highly mutagenic PAHs that most often attack guanines in the DNA. Adenines are preferentially inserted across from PAH-modified guanines during DNA replication, creating $C \rightarrow A$ transversions in the newly synthesized strand.

Cancer is caused by mutations in certain genes that regulate cell growth and division. Consequently, mutagens are often **carcinogens** (cancer-causing agents). The following example illustrates the effect of mutagenic compounds in tobacco smoke.

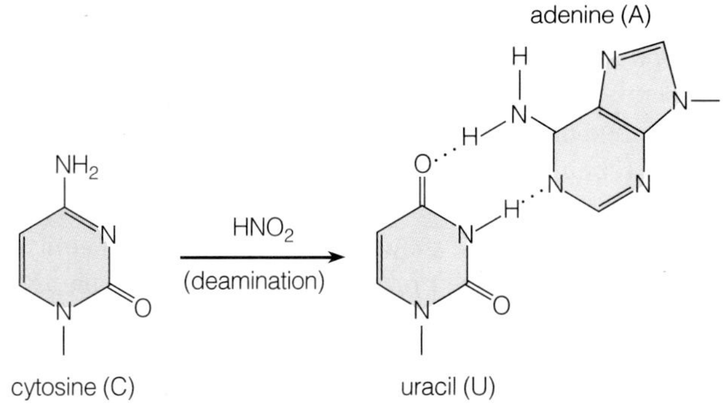

a Pairing causes a transition mutation.

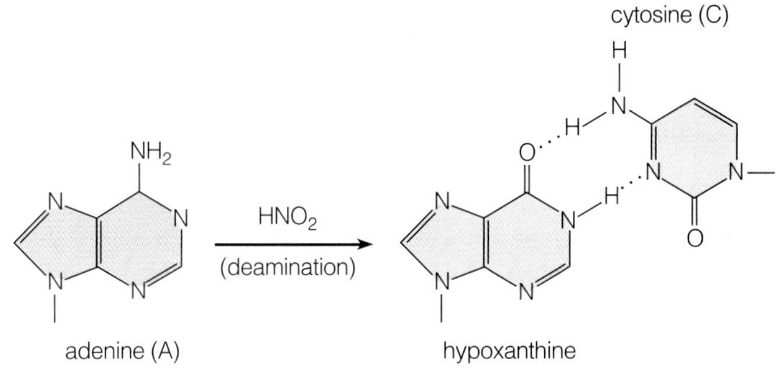

b Pairing causes a transition mutation

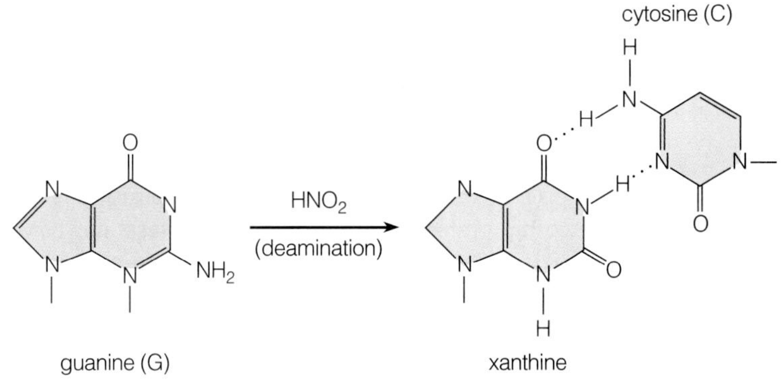

c No mutation

Figure 5.15 Deamination by nitrous acid.

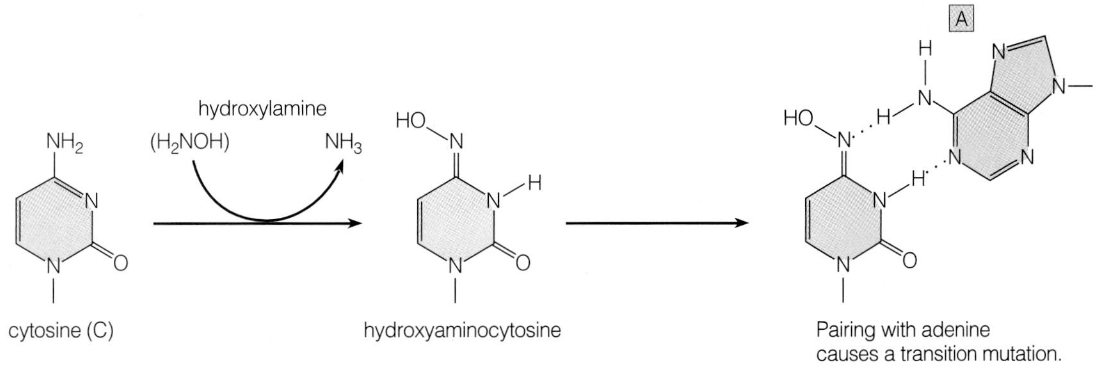

Figure 5.16 Hydroxylation by hydroxylamine.

Example 5.3   **Carcinogens in tobacco smoke and transversions.**

The *TP53* gene in humans encodes a protein, p53, that inhibits cell division. It is one of the most important proteins in preventing cancer. Many cancerous cells have mutations in the *TP53* gene that cause a loss of p53 function in the cells. In 1993, Takeshima et al. (*Lancet* 342:1520–1521) reported a comparison of mutations in the *TP53* gene in lung cancer cells from smokers and persons who had never smoked. All those studied were residents of Hiroshima, Japan, all had lung cancer, and some had been exposed to ionizing radiation fallout from the atomic bomb dropped on Hiroshima in 1945. To determine what mutations were present, the researchers sequenced selected regions of DNA from the *TP53* gene in lung cancer cells from each person. No differences were found between non-smokers exposed to radiation fallout and nonsmokers not exposed to radiation fallout for mutations in the *TP53* gene. However, the researchers did find that transitions were present in the cancer cells of both smokers and nonsmokers, but were significantly more frequent among nonsmokers. They also found no $G \rightarrow T$ or $C \rightarrow A$ transversions among non-smokers, whereas 29% of the smokers had $G \rightarrow A$ or $C \rightarrow A$ transversions. Adjustments for age, sex, and tissue type did not alter these conclusions.

Problem: What do these results suggest about the relationship of smoking to mutations?

Solution: Most substitution mutations—spontaneous and induced—are transitions. For this reason, transitions are expected in nonsmokers with lung cancer. PAHs are known to cause specific transversions. Interaction of PAHs with a guanine in the sense strand of the *TP53* gene should cause a $G \rightarrow T$ transversion in the sense strand, whereas interaction of PAHs with a guanine in the antisense strand should cause a $C \rightarrow A$ transversion in the sense strand. The presence of $G \rightarrow T$ or $C \rightarrow A$ transversions only among the smokers strongly suggests that PAHs in tobacco smoke interacting with guanine bases are responsible for the transversions.

Chemical mutagens cause lesions in DNA, which, in turn, cause mutations when the DNA is replicated.

5.4   DNA REPAIR MECHANISMS

With all the mechanisms for spontaneous and induced mutation, it is surprising that mutations are quite rare. Lesions in DNA, on the other hand, are somewhat common. However, cells are safeguarded from the effects of lesions by several DNA repair mechanisms. The source of most mutations is a template strand with a lesion, such as a modified base, an AP site, or a pyrimidine dimer. Most DNA repair mechanisms recognize the lesion itself, rather than the mutation that would follow during replication, and repair the damaged site before DNA replication. When DNA repair mechanisms fail to recognize and repair a damaged site, this may be because the repair enzymes cannot find all damaged sites before the DNA replicates.

The action of mutagens is virtually identical in prokaryotes and eukaryotes because the chemical nature of DNA is the same in all organisms. Likewise, many prokaryotic and eukaryotic DNA repair mechanisms are similar. Several repair mechanisms in both prokaryotes and eukaryotes have been well characterized. Because of their similarity, we will discuss them together.

Mechanisms that repair lesions in DNA are active at three levels. The first level is preventative: enzymes recognize certain mutagens and detoxify them before they affect the DNA. The second level includes several mechanisms that repair lesions in DNA prior to replication. The third is postreplication repair, in which mutations that occurred *during* replication are recognized and repaired *after* replication. Let's examine these mechanisms, which are summarized in Table 5.4.

Mechanisms That Prevent DNA Damage

At the first level are systems that recognize specific mutagens and detoxify them before they can create lesions in DNA. Among the best studied examples is the two-enzyme system of superoxide dismutase and catalase. As described in section 5.3, ionizing radiation converts water molecules into free radicals, which may then cause mutations. Superoxide dismutase recognizes superoxide free radicals formed from water by ionizing radiation and converts them to hydrogen peroxide. Hydrogen peroxide is also toxic to the cell, but the enzyme catalase immediately converts the hydrogen peroxide to water.

$$2O_2^- + 2H^+ \xrightarrow{\text{superoxide dismutase}} H_2O_2 + O_2$$

$$2H_2O_2 \xrightarrow{\text{catalase}} 2H_2O + O_2$$

Mechanisms That Repair Damaged DNA

Mechanisms at the second level of DNA protection detect lesions in DNA and repair them. An example is a group of enzymes known as **alkyltransferases**, which

Table 5.4 DNA Protection Mechanisms

Level	Examples	Function
Preventative	Superoxide dismutase and catalase	Eliminate superoxide free radicals
Direct DNA repair	Alkyltransferases	Remove alkyl groups from alkylated bases
	Photoreactivation	Restores dimerized pyrimidines to their original form
	Excision repair	Recognizes and excises damaged nucleotides and replaces them with the correct ones
Postreplication repair	Recombination repair	Replaces a lesion with the correct nucleotides by transferring a template DNA strand
	Mismatch repair system	Replaces mismatched nucleotides with the correct ones
	SOS response and mutagenesis	Activate excision, recombination, and error-prone repair mechanisms
	Transcription-repair coupling	Halts transcription at a lesion in the template strand and repairs the lesion

remove alkyl groups from the bases of DNA, countering the mutagenic action of alkylating agents. The best studied of the alkyltransferases is O^6-methylguanine methyl transferase (MGMT). MGMT has been found in all species examined, both prokaryotic and eukaryotic. *E. coli* has two MGMT enzymes, and humans have a single MGMT enzyme. MGMT recognizes very specific damage to DNA: methylation of the oxygen bonded to carbon number 6 of guanine. The enzyme transfers the methyl group from the guanine to a cysteine within the enzyme itself, an action that permanently disables the enzyme. Unlike most enzymes, which serve as catalysts over and over again, each molecule of MGMT catalyzes the transfer of only one methyl group. For this reason, MGMT is called a **suicide enzyme** (Figure 5.17).

Among the best studied examples of DNA repair are the mechanisms that correct UV radiation damage in *E. coli*. One of these is **photoreactivation** by an enzyme called photoreactivating enzyme (PRE). PRE recognizes pyrimidine dimers caused by nonionizing radiation and cleaves the dimerizing covalent bonds, restoring the correct pyrimidines (Figure 5.18). PRE is effective in removing thymine dimers, cytosine dimers, and thymine-cytosine dimers. For PRE to function, light in the blue range of the spectrum must be present. PRE binds to pyrimidine dimers without light but cannot restore the normal bases until light is present. PRE has been found in many different species but is absent from placental mammals, including humans.

Excision Repair Mechanisms

Another group of repair mechanisms, responsible for repair of many types of DNA damage, is called **excision repair**. Excision repair is very versatile in that it recognizes a variety of DNA lesions, including pyrimidine dimers, modified bases, AP sites, and nucleotide mismatches.

There are several excision repair mechanisms but all consist of essentially the same chain of events. An enzyme system recognizes a damaged nucleotide or adjacent damaged nucleotides and makes two cuts in the DNA strand several nucleotides away from the damaged site. The nucleotides between the two cuts, including the lesions, are excised, leaving a gap in the strand that contained the damage. Then a DNA polymerase fills in the gap with the correct nucleotides, using the undamaged strand as a template. Figure 5.19 shows the process of excision repair in *E. coli*.

The uvrABC endonuclease in Figure 5.19 and other enzymes responsible for excision are called **excinucleases**. Excinucleases and their activities have been characterized in both prokaryotes and eukaryotes. Human excinuclease is a complex enzyme requiring 16 polypeptides that function together. Some of the polypeptides are subunits of the transcription initiation factor TFIIH. As we saw in Chapter 3, TFIIH is a helicase that unwinds the DNA strands for initiation of transcription. It functions in the same way in excision repair, unwinding the DNA in preparation for excision.

Another excision repair system present in both prokaryotes and eukaryotes recognizes AP sites and initiates their removal using **AP endonucleases**, which create a nick on the 5' end of an AP site. The AP site may then be removed, along with a few adjacent nucleotides, and the resulting gap filled in by a DNA polymerase. When combined with a **DNA glycosylase**, this mechanism may be used to repair several different types of DNA damage. DNA glycosylases recognize damaged bases and hydrolyze the bond connecting the sugar to the damaged

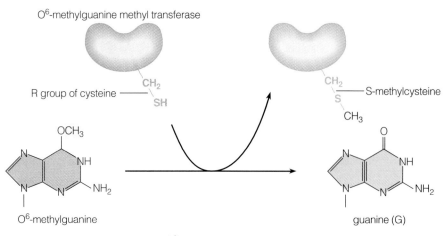

Figure 5.17 Repair that converts O⁶-methylguanine back to guanine. By accepting the methyl group, the suicide enzyme O⁶-methylguanine methyl transferase (MGMT) deactivates itself.

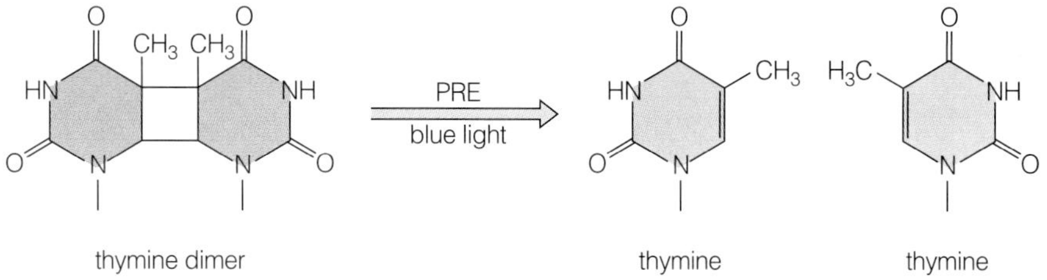

Figure 5.18 Reversal of thymine dimerization by photoreactivating enzyme (PRE).

base, leaving an AP site that can then be recognized by an AP exonuclease, which will initiate the repair.

Some DNA glycosylases are found only in eukaryotes; others have prokaryotic counterparts. These enzymes recognize specific damaged nucleotides and remove them from DNA or alter them so that the correct nucleotide sequence can be restored. For example, among the most important of DNA glycosylases in both prokaryotes and eukaryotes are 8-hydroxyguanine DNA glycosylases. The free radicals produced by ionizing radiation preferentially attack guanines, forming 7,8 dihydro-8-oxyguanine (8oxoG), which is a frequent lesion in damaged DNA. Repair enzymes attack 8oxoG at various levels to protect the DNA. One glycosylase attacks 8oxoGTP in the free nucleotide pool, removing the damaged base from its sugar so that the nucleotide will not be incorporated into DNA. Another enzyme recognizes 8oxoG-A base pairs and corrects them to 8oxoG-C base pairs, preventing transversions. Still another glycosylase removes 8oxoG from DNA strands so the correct nucleotide can be inserted.

Postreplication Repair

In the third level of DNA protection, mechanisms recognize and repair errors during or after DNA replication. One of these mechanisms is **recombination repair**. As

we just discussed, excision repair can only function when DNA is double stranded, because it requires a template to fill in the excised gap. During DNA replication, the parent molecule splits into two single strands, each of which will be the template for synthesis of the new strands. If a lesion is present in one of the single strands, the excision repair mechanism cannot remove the lesion because the DNA is single stranded.

Certain lesions, such as AP sites or pyrimidine dimers, prevent base pairing. A DNA polymerase cannot insert a nucleotide in the newly synthesized strand when it encounters one of these lesions in the template strand. Instead of ceasing replication, the DNA polymerase may simply reinitiate replication at a site beyond the lesion, leaving a gap of unreplicated DNA opposite the lesion. Figure 5.20 shows how recombination repair repairs such a lesion. The damaged strand is red, and the newly synthesized strand with the gap is yellow. The other newly synthesized (green) strand is identical to the red template strand, except that it contains the correct nucleotides at the site of the lesion. RecA, a protein that is part of the recombination repair mechanism, binds to the single-stranded DNA in the red template strand (Figure 5.20b) and causes it to invade the double-stranded DNA in the lower molecule. The invading red strand base-pairs with its blue complement, displacing the new green strand (Figure 5.20c). The part of the blue strand

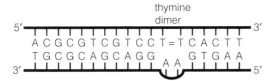

a Double-stranded DNA molecule contains a lesion (a thymine dimer in this example).

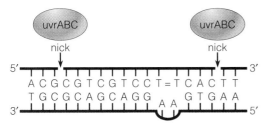

b The enzyme uvrABC endonuclease nicks the DNA strand on both sides of the damaged nucleotides.

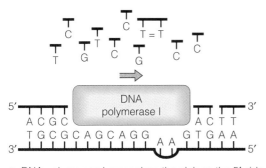

c DNA polymerase I recognizes the nick on the 5' side of the lesion and uses its 5' → 3' exonuclease activity to remove the nucleotides between the nicks including the lesion. Simultaneously, DNA polymerase I uses its 5' → 3' polymerase activity to fill in the region between the nicks.

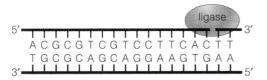

d Ligase seals the nick at the end of the repaired segment.

Figure 5.19 Excision repair in *E. coli*.

that is paired with the red strand is cut on both ends then connected to the yellow strand, where it fills the gap (Figure 5.20d). A DNA polymerase fills in the gap left in the blue strand using the green strand as a template. The excision repair system may now repair the lesion in the red strand, using the transferred part of the blue strand as a template.

Mismatch repair corrects mismatched nucleotides that have escaped correction by proofreading. The mechanism for mismatch repair recognizes the mismatch and replaces the incorrect nucleotide on the newly synthesized strand with the correct one, as identified from the template strand. How does the mismatch repair mechanism recognize which strand is the newly synthesized strand containing the incorrect nucleotide? In *E. coli*, an enzyme called adenine methylase recognizes the nucleotide sequence

```
5' GATC 3'
3' CTAG 5'
```

and adds a methyl group to the A in each of the strands. Because the enzyme methylates the A in the new strand several minutes after it is synthesized, there is a short period when the A in the template strand is methylated but the A in the newly synthesized strand is not. The mismatch repair mechanism uses this difference in methylation to recognize the newly synthesized strand.

SOS Response and Mutagenesis

When *E. coli* cells are exposed to agents that cause numerous lesions in the DNA, they initiate the SOS response. The **SOS response** delays DNA synthesis and cell division, providing time to repair damage in the DNA. It also activates several repair mechanisms, including excision repair and recombination repair. The SOS response also permits error-prone repair mechanisms, which are normally repressed, to operate in the cell. These mechanisms are called *error-prone* because they permit DNA polymerase III to bypass lesions and insert incorrect nucleotides opposite the lesions. Consequently, the SOS response is followed by an increase in the mutation rate, a phenomenon known as **SOS mutagenesis**. The name SOS, derived from the international SOS distress signal, implies that the cell is under distress from DNA damage.

Let's see how exposure to UV radiation induces the SOS response. Before exposure, the cells are in an uninduced state, meaning that the SOS response is not under way. A protein called LexA is produced at a constant rate in uninduced cells. LexA is a repressor protein, which means that it binds to the promoter region of particular genes and prevents them from being transcribed. LexA represses transcription of more than 20 genes that participate in the SOS response. When the cell is exposed to UV radiation, pyrimidine dimers form and prevent completion of DNA replication. The radiation induces production of RecA, a protein that is part of the recombination repair mechanism. RecA stimulates cleavage of LexA, which can then no longer bind to the genes it represses. When LexA is cleaved, the SOS genes are transcribed and translated, initiating the SOS response. The cell is now in an **induced state** (Figure 5.21).

The proteins produced in the SOS response have several effects on an induced cell. Cell division is inhibited until DNA replication can be completed. However, cell

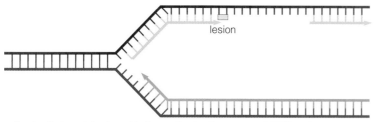

a Lesion in template strand halts synthesis of new strand. DNA polymerase reinitiates synthesis farther downstream, leaving a single-stranded gap.

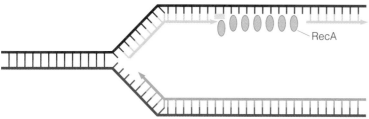

b RecA binds to the single-stranded DNA.

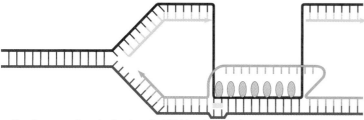

c RecA causes the single-stranded DNA to invade the lower molecule, where it pairs with its complement, displacing the newly synthesized strand.

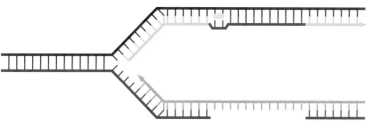

d Part of the template strand from the lower molecule is transferred to the upper molecule, leaving a single-stranded gap in the lower molecule.

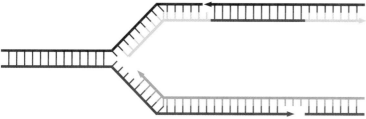

e A template is now present in the upper molecule that can be used to repair the lesion. The gap in the lower molecule is filled in using the newly synthesized strand as a template.

Figure 5.20 Recombination repair.

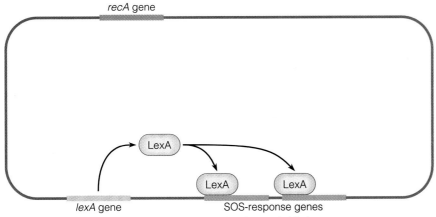

a An uninduced cell. In the uninduced state, the LexA protein binds to SOS-response genes and prevents them from being transcribed.

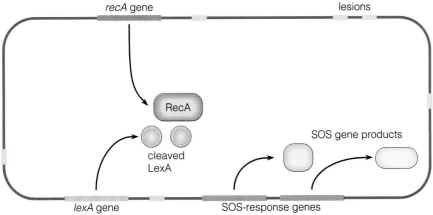

b An induced cell. DNA damage induces SOS-response by activating the *recA* gene. RecA stimulates cleavage of LexA, allowing the SOS-response genes to be transcribed.

Figure 5.21   The SOS response.

growth is not inhibited, so cells elongate as they await completion of DNA replication. In order to complete DNA replication, the SOS response induces **error-prone synthesis**, which increases the mutation rate. Error-prone synthesis is responsible for SOS mutagenesis.

Figure 5.22 shows the process of error-prone synthesis. DNA synthesis ceases when DNA polymerase III encounters a dimer in the template strand (Figure 5.22a). Interacting with RecA, DNA polymerase III inserts an incorrect nucleotide across from the lesion (Figure 5.22b). At this point, DNA synthesis still cannot proceed due to the lesion. Two alternatives can occur. Photoreactivation may repair the dimer, but photoreactivating enzyme uses the misincorporated nucleotide as a template, causing a mutation (Figure 5.22c). Alternatively, two proteins induced by the SOS response, called UmuC and UmuD, allow DNA polymerase III to bypass the lesion, incorporating incorrect nucleotides opposite it (Figure 5.22d). The excision repair mechanism may now repair the lesion, but it does so using the incorrect nucleotides as a template, incorporating a mutation into the DNA (Figure 5.22e).

Transcription-Repair Coupling

Several lines of evidence from studies in both prokaryotes and eukaryotes indicate that DNA is repaired more often and more quickly when it is being transcribed, a phenomenon known as **transcription-repair coupling**. When RNA polymerase encounters a lesion in the DNA during transcription, it stalls at the lesion. In *E. coli*, a protein called transcription-repair coupling factor (TRCF) binds to the stalled RNA polymerase and removes it, along with the unfinished transcript, from the DNA. The unfinished transcript is then degraded. At the same time, the excision repair system recognizes TRCF and initiates excision repair of the lesion. Once the lesion is repaired, transcription may be reinitiated and proceed normally (Figure 5.23).

Transcription-repair coupling has been well studied in humans, and there are some important differences from the prokaryotic system. When RNA polymerase II encounters a lesion in the DNA, it also stalls. Recall from section 3.4 that when RNA polymerase II stalls, TFIIS causes

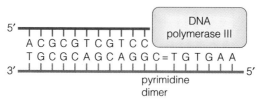

a DNA polymerase III cannot add nucleotides opposite a pyrimidine dimer.

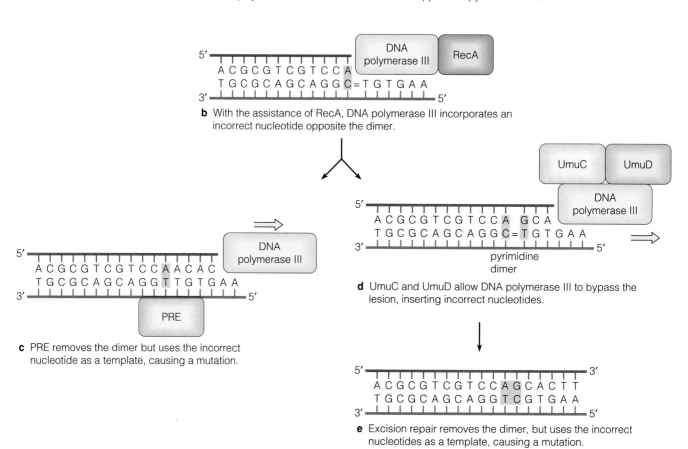

b With the assistance of RecA, DNA polymerase III incorporates an incorrect nucleotide opposite the dimer.

c PRE removes the dimer but uses the incorrect nucleotide as a template, causing a mutation.

d UmuC and UmuD allow DNA polymerase III to bypass the lesion, inserting incorrect nucleotides.

e Excision repair removes the dimer, but uses the incorrect nucleotides as a template, causing a mutation.

Figure 5.22 Error-prone DNA synthesis, which is responsible for SOS mutagenesis.

the polymerase to back up and removes a few of the most recently added nucleotides on the 3′ end of the RNA. Two proteins, CSA and CSB, recognize the stalled complex and bring in proteins of the excision repair system, which repair the DNA while the stalled RNA polymerase II waits. Once repair is complete, the stalled RNA polymerase II continues transcribing the gene (Figure 5.24).

Transcription-repair coupling is specific for the antisense strand, as the experimental evidence in the following example demonstrates.

Example 5.4 Specificity of transcription-repair coupling for the antisense DNA strand.

In 1987, Mellon, Spivak, and Hanawalt (*Cell* 51:241–249) reported that transcription-repair coupling in human and hamster cells selectively re-

paired lesions in the antisense DNA strand used as the template for transcription. They irradiated cultured human and hamster cells with ultraviolet radiation, ruptured the cells at various time intervals following irradiation, isolated the DNA, and measured the number of pyrimidine dimers in the two strands of the *dihydrofolate reductase* (*DHFR*) gene. They found that nearly all pyrimidine dimers were eliminated from the antisense strand within 4 hours after irradiation, but that dimers remained in the sense strand much longer.

Problem: How do these results conform with the model for transcription-repair coupling?

Solution: Transcription-repair coupling is activated when RNA polymerase II stalls at a lesion in the DNA strand it is transcribing, which is the antisense

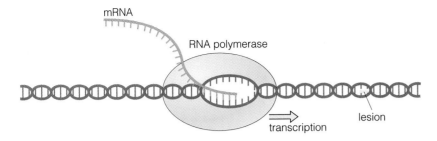

a RNA polymerase approaches a lesion in the template strand of a gene during transcription.

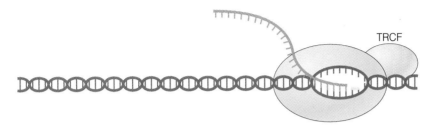

b Upon encountering the lesion, RNA polymerase stalls, and TRCF binds to the stalled polymerase.

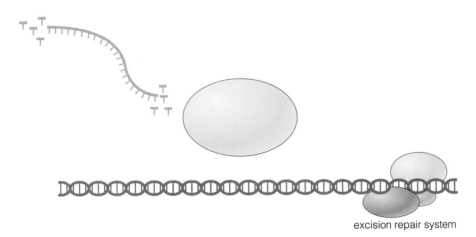

c RNA polymerase and the mRNA separate from the DNA, and the mRNA is degraded while the lesion is repaired.

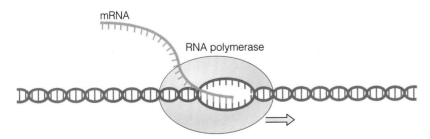

d Transcription is reinitiated after DNA repair.

Figure 5.23 Transcription-repair coupling in prokaryotes.

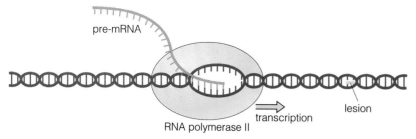

a RNA polymerase II approaches a lesion in the template strand of a gene during transcription.

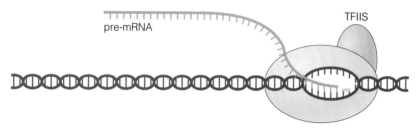

b Upon encountering the lesion, RNA polymerase II stalls; TFIIS recognizes the stalled polymerase.

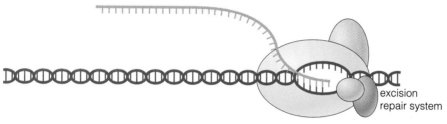

c TFIIS causes RNA polymerase II to back up while the excision repair system repairs the lesion.

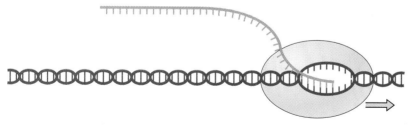

d Once DNA repair is completed, transcription resumes.

Figure 5.24 Transcription-repair coupling in eukaryotes.

strand. Transcription-repair coupling then excises the lesion. RNA polymerase II does not stall at lesions in the sense strand because it does not transcribe the sense strand. The transcription-repair coupling mechanism is thus not deployed for sense-strand lesions, which are repaired only by other mechanisms.

DNA repair mechanisms may eliminate mutagens, repair lesions in DNA before replication, or operate during replication or transcription to repair lesions.

Mutations in Genes That Encode Repair System Enzymes

Mutations that alter repair systems can be serious because the absence of even a single repair system may increase the mutation rate. For example, yeast cells that lack functional uracil DNA glycosylase, an enzyme that removes uracil from DNA, have a mutation rate 20 times that of cells that have the functional glycosylase. In humans, mutations in the genes that encode DNA repair enzymes are often found in cancer cells.

In some cases, a mutation in a gene that encodes a DNA repair enzyme may be responsible for inherited susceptibility to cancer as well as other disorders. Among the best known of these disorders is **xeroderma pigmentosum**, which is characterized by high susceptibility to skin cancer and neurological dysfunction in a subset of patients. Xeroderma pigmentosum may be due to mutations in any one of seven genes that encode seven polypeptides named XPA, XPB, XPC, XPD, XPE, XPF, and XPG (XP for *x*eroderma *p*igmentosum). These polypeptides function as subunits for enzymes in the excision repair system, which fails in patients with xeroderma pigmentosum. An eighth gene, called **XP-variant**, probably encodes a postreplication repair enzyme. Recall that UV radiation in sunlight causes pyrimidine dimers to form in DNA. These dimers are usually removed by the excision repair system. When a mutation disrupts a protein that is essential for excision repair, the dimers are left unrepaired and cause replicating DNA to mutate. Because UV radiation fails to penetrate deeply into tissues, skin is the organ most damaged in patients with xeroderma pigmentosum.

Cockayne syndrome is another genetic disorder caused by failure of a DNA repair mechanism. The syndrome is characterized by premature aging and neuroskeletal abnormalities, including dwarfism and mental retardation. It is due to failure of transcription-repair coupling. Mutations in any one of at least five genes that encode proteins for transcription-repair coupling can cause Cockayne syndrome. Three of these genes encode XPB, XPD, and XPG, which are part of the excision repair system and are also among the proteins whose absence can cause xeroderma pigmentosum. The other two genes encode the CSA and CSB polypeptides that recognize lesions in transcription-repair coupling. The element of premature aging in Cockayne syndrome is one of many links demonstrating the important role of DNA damage in human aging.

~

Mutations in genes that participate in DNA repair may increase overall mutation rates by eliminating repair mechanisms.

SUMMARY

1. Substitution mutations may or may not affect the gene product, depending on where they are located. If a mutation does not change the amino acid sequence of the polypeptide, the mutation is a same-sense mutation, which is also silent. Missense mutations cause a change in the amino acid sequence of the polypeptide. Nonsense mutations create a premature termination codon, causing the polypeptide to be shorter than usual. Insertion and deletion mutations may be frameshift mutations.

2. Trinucleotide repeat expansions are responsible for several genetic disorders in humans, including fragile X syndrome and Huntington disease.

3. The nucleotides of DNA generally exist in stable forms but may occasionally undergo tautomeric shifts to a less stable form. When a nucleotide in the template strand during DNA synthesis happens to be in the less stable form due to a tautomeric shift, its base-pairing properties are altered, and an incorrect nucleotide may be added opposite it during DNA replication.

4. Tautomeric shifts, as well as many other alterations in DNA, cause transitions during DNA replication. A transition is the substitution of the wrong purine for the correct purine or the substitution of the wrong pyrimidine for the correct pyrimidine. Transitions are the most common form of substitution mutations. Less frequent are transversions, in which a purine is substituted for a pyrimidine, or vice versa.

5. Short, repeated nucleotide sequences tend to be hot spots for insertion and deletion mutations. During DNA replication, either the template strand or the newly synthesized strand may slip at a repeated sequence, causing too few or too many nucleotides to be added as DNA replication continues.

6. Depurination and depyrimidization leave a deoxyribose sugar in the DNA backbone with no base attached to it. Such a lesion is called an AP site.

7. Cytosine deamination changes C into U. Because U pairs with A instead of G, a transition takes place during DNA replication.

8. Some mutations are spontaneous, meaning that no known causal agent is involved. Others result from mutagenic agents that induce alterations in DNA; these are called induced mutations.

9. Ionizing radiation (X-rays and alpha, beta, and gamma rays) deeply penetrates biological tissues. The radiation creates free radicals, which, in turn, react chemically with the DNA, altering it. Ionizing radiation frequently causes breaks in the sugar-phosphate backbone of DNA.

10. Nonionizing radiation consists primarily of ultraviolet (UV) light in sunlight. It penetrates only the surface layers of tissues and causes electrons to temporarily assume higher energy levels. Pyrimidines in DNA are particularly sensitive to nonionizing radiation and may form pyrimidine hydrates or pyrimidine dimers.

11. Base analogs are nitrogenous bases that are similar to the bases in DNA and may be incorporated in place of their usual counterparts during DNA replication. Two of the more common base analogs are 5-bromouracil (5-BU) and 2-amino purine (2-AP). These base analogs tend to undergo tautomeric shifts more readily than their

usual counterparts, causing transitions during DNA replication.

12. Intercalating agents, such as ethidium bromide, acridine dyes, and ICR compounds, insert themselves within the stacked bases of the DNA molecule, making a kink in the DNA molecule, which may cause insertion or deletion mutations during DNA replication.

13. Deaminating agents, such as bisulfate compounds and nitrous acid, may remove amino groups from the nitrogenous bases. Alkylating agents add alkyl (usually methyl or ethyl) groups to the nitrogenous bases of DNA. Hydroxylating agents add hydroxyl groups to the nitrogenous bases. All of these lesions may cause mutations during DNA replication.

14. Aflatoxin B_1 removes guanine bases from the DNA molecule, leaving AP sites that halt DNA replication unless the lesion is repaired.

15. There are three levels of DNA protection mechanisms that prevent or eliminate many mutations. The first is preventative, the second is direct DNA repair, and the third is postreplication repair.

QUESTIONS AND PROBLEMS

1. List as many types of DNA damage that result in transitions as you can, and describe why a transition, rather than some other type of mutation, results.

2. Distinguish between loss-of-function and gain-of-function mutations. Why are loss-of-function mutations more frequent than gain-of-function mutations?

3. Distinguish between forward mutations and reversions. Which forward mutation would be more likely to undergo a reversion, a transition or a transversion? Explain.

4. Insertion and deletion mutations often have a greater deleterious effect than single-nucleotide substitutions. Explain why this is so.

5. Although insertion and deletion mutations may have a greater deleterious effect than single-nucleotide substitutions, insertion and deletion mutations are more likely to undergo reversion than substitution mutations. Explain why this is so.

6. Does cytosine deamination result in a transition or a transversion? Explain your answer.

7. In 1978, Farabaugh et al. (*Journal of Molecular Biology* 126:847–857) reported that the original DNA sequence at a mutation hotspot in the *E. coli lacI* gene is

```
TCGGCGCGTCTGCGTCTGGCTGGCTGGCATAAA
```

The amino acid sequence associated with this DNA sequence is

```
SerAlaArgLeuArgLeuAlaGlyTrpHisLys
```

All of the mutations they studied generated premature termination codons. Eighty of the mutations they examined generated the amino acid sequence

```
SerAlaArgLeuArgLeuAlaGlyTrpLeuAla-COOH
```

where –COOH denotes the carboxyl end of the polypeptide. Eighteen of the mutations generated the amino acid sequence

```
SerAlaArgLeuArgLeuAlaGlyIleLys-(9 amino acids)-COOH
```

Identify the DNA sequences of these two groups of mutations, and describe why this is a mutation hotspot.

8. In 1995, Eisensmith and Woo (*Advances in Genetics* 32:199–271) listed several single-nucleotide substitution mutations that resulted in premature termination codons in the human *PAH* gene. **(a)** Fill in the blanks below with all possible alternatives for the original codon, and indicate whether each mutation was a transition or a transversion. The amino acid specified by the original codon is indicated in parentheses.

_____ (Trp) → UGA	_____ (Gly) → UGA
_____ (Gln) → UAG	_____ (Tyr) → UAG
_____ (Arg) → UGA	_____ (Gln) → UAA
_____ (Trp) → UAG	_____ (Tyr) → UAA
_____ (Ser) → UGA	

(b) List all possible codons in the genetic code that could be changed into a termination codon by a single nucleotide substitution. **(c)** Which codons are not represented in part a? **(d)** What characteristic typifies all of the codons listed in part c?

9. Codons that encode glutamine, tryptophan, and arginine are the most likely to be changed into a premature termination codon by a single-nucleotide substitution. In light of the previous question, explain why this is so.

10. Look at the following amino acid alterations. Which could result from single-nucleotide substitution mutations induced by **(a)** nitrous acid, **(b)** 5-BU, **(c)** 2-AP, **(d)** bisulfite compounds, and **(e)** hydroxylamine?

Val → Ile	Thr → Ser	Phe → Leu
Ser → Arg	Pro → Gln	Phe → Met
Ala → Thr	Tyr → Pro	Trp → Ter

11. Suppose that a strain of *E. coli* is treated with an intercalating agent, and a mutant colony is recovered in which an enzyme is no longer functional. Which of the following mutagens is most likely to cause the enzyme to revert to its functional form due to a reversion mutation: **(a)** hydroxylamine, **(b)** 2-AP, **(c)** 5-BU, **(d)** proflavin, **(e)** nitrous acid, or **(f)** aflatoxin B_1?

12. Provide an example of a cellular mechanism that protects DNA from the effects of ionizing radiation before DNA damage occurs.

13. Describe two ways that pyrimidine dimers may be repaired.

14. When DNA polymerases encounter an AP site, they can no longer continue to replicate DNA at that point. Describe three ways that *E. coli* cells overcome the effects of an AP site.

15. Suppose a mutation in *E. coli* altered the gene for adenine methylase so that the enzyme was no longer functional. How would this affect DNA repair?

16. *E. coli* often grows in complete darkness in mammalian intestines, so it does not require light to grow, but it also grows quite well under light. If you grew one group of *E. coli* in cultures under ultraviolet light only, and another group under full-spectrum light with the same intensity of UV radiation, which group would be likely to accumulate more mutations than the other? Explain your answer.

17. Fraenkel-Conrat (1964. *Scientific American* 211 [Oct]: 47–54) summarized the results of several experiments with single-nucleotide substitution mutations induced by nitrous acid in tobacco mosaic virus, a single-stranded RNA virus that infects several plant species. The amino acid substitutions listed in Table 5.5 were observed. **(a)** Use the genetic code to determine which mutations were transitions and which were transversions. **(b)** What proportion of mutations were transitions? **(c)** What do these results suggest about the relative frequency in RNA of transitions as opposed to transversions induced by nitrous acid?

18. Morlé et al. (1985. *EMBO Journal* 4:1245–1250) discovered that a deletion mutation was the cause of an α-thalassemia, a human genetic disorder caused by the absence of the α subunit in hemoglobin. The mutant version of the gene contains a deletion of two nucleotides just upstream from the ATG initiation codon. The normal version has the sequence CCCACCATG, whereas the mutant version had the sequence CCC****C**ATG, where the two asterisks represent the deleted nucleotides and the initiation codon is underlined. **(a)** Many deletion mutations cause a frameshift. Does this particular mutation cause frameshift? Explain why or why not. **(b)** The authors of this study found that the mutant version of this gene produced correctly spliced mRNA. Why did these mRNAs fail to produce the α subunit?

19. In 1993, De Boulle et al. (*Nature Genetics* 3:31–35) reported that a patient with a severe case of mental retardation with symptoms matching those of fragile X syndrome had 25 CGG repeats within the trinucleotide repeat region of the *FMR1* gene. Unlike most fragile X patients, who show reduced transcription of *FMR1*, this patient had normal levels of *FMR1* transcripts. DNA

Table 5.5	Information for Problem 17: Single-Nucleotide Substitution Mutations Induced by Nitrous Acid in Tobacco Mosaic Virus

Amino Acid Substitution	Number of Different Mutations That Caused This Substitution
Val → Met	1
Pro → Thr	1
Pro → Leu	4
Asn → Ser	6
Arg → Lys	1
Arg → Gly	5
Ser → Gly	1
Ser → Phe	6
Thr → Ala	4
Glu → Gly	1
Glu → Arg	1
Ile → Val	1

sequence analysis revealed that a segment of the *FMR1* gene had the sequence AAG CTG AAT CAG GAG, which differs from the normal sequence AAG CTG ATT CAG GAG (the nucleotides are grouped by threes to represent the codons). Analysis of the patient's ancestors and relatives revealed that all had only the normal sequence. **(a)** Which of the boldface terms found in sections 5.1 and 5.2 apply to this mutation? **(b)** Identify the amino acid substitution and the position in the codon affected by this mutation. **(c)** What significance does the number of trinucleotide repeats have in this case? **(d)** What does the DNA sequence analysis in the patient's ancestors and relatives reveal?

20. In 1995, Ashley and Warren (*Annual Review of Genetics* 29:703–728) reviewed trinucleotide expansion mutations in the *FMR1* gene and the association of these mutations with fragile X syndrome. In 1993, De Boulle et al. (*Nature Genetics* 3:31–35) reported that a single T → A nucleotide substitution was responsible for a case of fragile X syndrome. Quan et al. (1995. *American Journal of Human Genetics* 56:1042–1051) summarized results from several cases in which deletions of the entire *FMR1* gene, or deletions of large sections of the gene, also caused fragile X syndrome. Why is it possible for trinucleotide expansion, substitution, and deletion mutations to all cause the same genetic disorder?

21. In 1994, Kunst and Warren (*Cell* 77:853–861) reported on the variation of CGG repeats in the *FMR1* gene (the gene responsible for fragile X syndrome). They found that most people did not have a perfect repeat sequence.

Instead, the 10th and 20th triplets in the repeated segment were AGG rather than CGG triplets. These so-called *cryptic triplets* appeared to stabilize the repeat segment, thereby preventing expansion. Perfect repeats that lacked the AGG triplets were more likely to undergo expansion. The authors used this information, along with additional evidence from this and other genes, to postulate that slippage during DNA replication is the most probable mechanism for trinucleotide repeat expansion. Describe how the presence of cryptic triplets should stabilize a repeat segment against expansion.

22. The genetic code is said to be buffered against substitution mutations, meaning that many of the most likely mutations are silent. Explain this statement, referring to the types of mutations that are most likely, the genetic code, and the chemical properties of amino acids.

23. PKU is caused by mutations in the *PAH* gene. In 1995, Eisensmith and Woo (*Advances in Genetics* 32:199–271) listed the known mutations in the *PAH* gene in humans. The numbers of different substitution mutations they listed are summarized below:

T→C	26	T→A	3	A→T	5
C→T	44	T→G	11	A→C	5
A→G	13	C→A	13	G→T	14
G→A	46	C→G	10	G→C	14

(a) What proportion of the substitution mutations are transitions? **(b)** What do these results suggest about the prevalence of transitions among naturally occurring mutations? **(c)** Why are transitions expected to be more common than transversions? **(d)** Nearly all of these data were derived from people who expressed the symptoms of phenylketonuria. In light of the previous question, why might the actual proportion of transitions be underrepresented in these data?

24. In 1978, Coulondre et al. (*Nature* 274:775–780) reported on their analysis of over 80 nonsense mutations in the *lacI* gene of *E. coli*. They found several mutation hotspots, all of which were transitions and had a methylated cytosine in the original DNA sequence at the site of the mutation. They further found that the mutation hotspots disappeared when the cytosine was not methylated. **(a)** Identify all possible codons in which a nonsense mutation could occur from a transition involving a cytosine residue. **(b)** Given the results above, what is the most probable mechanism for mutation at these hot spots? **(c)** What significance is there in the observation that methylation plays a role in the mutation hotspots?

25. Polycyclic aromatic hydrocarbons (PAHs) typically alkylate guanines in such a way that adenine pairs with the alkylated guanine. Adenines may also be alkylated by PAHs, although this occurs less often. In 1987,

Stezowski et al. (*Journal of Biomolecular Structure & Dynamics* 5:615–637) proposed a model in which an adenine that had been alkylated by a PAH was most likely to pair with adenine. **(a)** What type of mutation (transition or transversion) would this cause? **(b)** What relationship is there between this model and the information presented in Example 5.3?

26. Calos, in 1978 (*Nature* 274:762–765), described a gain-of-function mutation for the *E. coli lacI* gene. The original and mutant DNA sequences are as follows:

Original
GACACCATCGAATGGCGCAAAACCTTTCGCGGTATGGCAT-
GATAGCGCCCGGAAGAGAGTCAA

Mutant
GACACCATCGAATGG**T**GCAAAACCTTTCGCGGTATGGCAT-
GATAGCGCCCGGAAGAGAGTCAA

The single-nucleotide substitution is underlined in the mutant sequence. The 5′ end of the mRNA is

5′ GGAAGAGAGUCAA . . .

(a) Is this a transition or a transversion? **(b)** This mutation does not alter the polypeptide sequence. Describe why this is so. **(c)** Using information from Chapter 3, describe why this change in the DNA sequence of a promoter is expected to be a gain-of-function mutation.

27. The enzyme β-galactosidase is encoded by the *lacZ* gene in *E. coli*. This enzyme breaks down lactose into glucose and galactose. The *lacI* gene encodes a protein that binds to the promoter region of the *lacZ* gene and prevents the *lacZ* gene from being transcribed. In the previous question, we looked at a gain-of-function mutation in the *lacI* gene. The phenotype associated with this mutation is a loss of β-galactosidase activity. Why is it correct to call this mutation a gain-of-function mutation when the result of the mutation is a loss of enzyme activity?

28. Genes with mononucleotide, dinucleotide, or tetranucleotide repeats are prone to frameshift mutations. Genes with trinucleotide repeats, however, are not prone to frameshift mutations. Explain why this is so.

29. In 1983, Owen et al. (*Journal of Molecular Biology* 165:229–248) described the *eJ382* mutation in T4 bacteriophage, which is located at a mutation hotspot. The original and mutant amino acid sequences encoded by the part of the gene with the mutation are as follows:

Original   AlaGluLysLeuPheAsn

Mutant   AlaGluLysThrLeuTer

Reconstruct (as much as possible) the nucleotide sequence that encodes this part of the gene, and describe why this is a mutation hotspot.

30. In 1995, Eisensmith and Woo (*Advances in Genetics* 32:199–271) identified several silent mutations in the *PAH* gene region. Some of these mutations are listed below:

Mutation	Location
T → C	−348 in promoter region
G → A	−224 in promoter region
A → C	−71 in promoter region
T → C	19th nucleotide in intron 2
GTG → GTA	codon 245 (exon 7)
CAG → CAA	codon 304 (exon 8)
GTA → GTT	codon 399 (exon 11)
TAC → TAT	codon 414 (exon 12)

Explain why each of these mutations had no effect on enzyme function.

FOR FURTHER READING

An excellent book that provides a detailed discussion of mutation and DNA repair in both prokaryotes and eukaryotes is **Friedberg, E. C., G. C. Walker, and W. Siede. 1995.** *DNA Repair and Mutagenesis.* **Washington, D.C.: ASM Press**. A thorough review of trinucleotide repeat expansion in humans and the genetic disorders associated with this type of mutation is **Ashley, C. T. and S. T. Warren. 1995. Trinucleotide repeat expansion and human disease.** *Annual Review of Genetics* **29:703–728**. A review article that summarizes the known DNA repair mechanisms in humans is **Sancar, A. 1995. DNA repair in humans.** *Annual Review of Genetics* **29:69–105**. Nucleotide excision repair in eukaryotes was reviewed by **Hoeijmakers, J. H. J. 1993. Nucleotide excision repair II: From yeast to mammals.** *Trends in Genetics* **9:211–217**.

For additional reading, go to InfoTrac College Edition, your online research library at: http: www.infotrac-college.com/brookscole

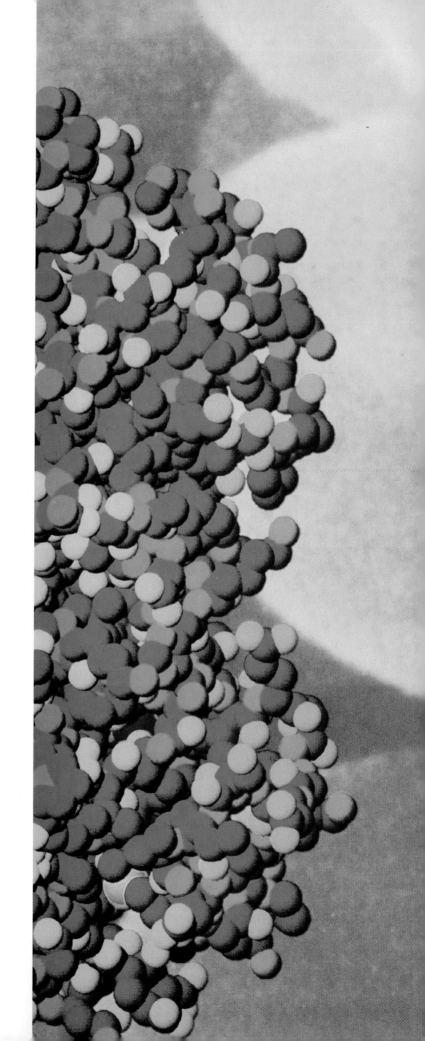

CHAPTER 6

AN INTEGRATED VIEW OF GENE EXPRESSION

No two people (with the exception of identical twins) have the same nucleotide sequences in each of their genes. As we saw in Chapter 5, mutations create variation in the nucleotide sequences of genes and the proteins they encode. Variation in the nucleotide sequences of genes is ultimately expressed as **phenotypic variation**, the differences in observed characteristics among individuals. For example, hair color in humans is a genetic characteristic that varies among people (Figure 6.1). Hair color ranges from dark black to completely white, with an almost infinite variety of brown, red, and blond shades in between. **Melanins** are the pigments that give hair, skin, and eyes their color; and the amounts and types of melanin vary from person to person. However, melanins are not proteins and are not encoded by genes.

How can genes cause variation in a substance that they do not encode? The answer lies in the understanding that proteins, the end products of genes, regulate the synthesis of nearly all compounds in cells. Specific enzymes in cells in the hair follicles regulate melanin synthesis. Differences among people in the nucleotide sequences of the genes that encode these enzymes cause enzyme activities to differ. These differences are ultimately expressed as phenotypic variation for hair color.

This example illustrates the relationship between phenotype and genotype. **Phenotype** is the outward characteristic that can be observed or measured. **Genotype** is the underlying genetic constitution that confers the phenotype. For example, when we say that a person has dark-brown hair, we are describing that person's phenotype. The genotype is the particular nucleotide sequences that person carries in the genes that influence hair color.

This chapter presents an integrated view of **gene expression**, the process through which the nucleotide sequence of a gene is ultimately manifest in the phenotype of an organism. In previous chapters, we studied how genes are transcribed and translated, how proteins undergo posttranslational modification and perform their function, and how mutation can alter that function. In this chapter, we will explore all of these topics together in an integrated way. First, we will look at the expression of a single gene in detail, starting with its nucleotide sequence and proceeding through transcription, RNA processing, translation, and polypeptide processing to the final product of the gene. We will explore how the gene's product functions and how that function confers a phenotype. Then we will see how mutations alter the

Figure 6.1 A group of genetics students exemplify variation for hair color in humans.

function of the gene's product and ultimately alter the phenotype. Finally, we will turn our attention to enzymes, one of the most important classes of proteins, to see how a set of enzymes participate with one another and how mutations in genes that encode enzymes affect the phenotype.

6.1 THE ANATOMY OF A GENE AND ITS EXPRESSION: THE HUMAN β-GLOBIN GENE

The human β-globin gene provides an excellent example of how a eukaryotic gene is organized and expressed, and how mutations in the gene affect the phenotype. This gene encodes the protein **β globin**, which is a subunit of adult hemoglobin. **Hemoglobin** is a protein essential for the cells of the body to carry out aerobic respiration. It is located in erythrocytes (red blood cells) and is responsible for binding oxygen and carrying it from the lungs to the various tissues of the body. In the tissues, erythrocytes enter capillaries, tiny blood vessels that allow erythrocytes to pass through one at a time. In the capillaries, hemoglobin delivers oxygen to the tissues and binds carbon dioxide. Hemoglobin then carries the carbon dioxide to the lungs, where it exchanges the carbon dioxide for oxygen. Because hemoglobin provides an es-

sential function for all tissues in the body, mutations that alter hemoglobin function can cause widespread organ and tissue damage, serious illness, and death.

Figure 6.2 shows the normal nucleotide sequence of the human β-globin gene and its promoter. The locations of conserved sequences and the codons for the polypeptide are identified in the DNA sequence. The entire gene region, including the promoter, is about 1700 nucleotide pairs in length and includes two introns that constitute about 60% of the nucleotides in the gene. This gene is among the smallest protein-encoding genes in humans, making it a simple one to study. Most genes are much larger, many with over 10,000 nucleotides and as many as 20 introns.

The promoter of the β-globin gene is a good example of a eukaryotic promoter. It contains two CACCC boxes, one at −106 and a second at −90. The CCAAT box is located at −75, and the TATA box at −29. The CACCC and CCAAT boxes match their consensus sequences exactly, whereas the TATA box deviates from its consensus sequence by one nucleotide (CATAAAA, compared to the consensus sequence TATAAAA). The transcription startpoint is an A between two pyrimidines. After it has been transcribed, the RNA is cleaved 18 nucleotides beyond a fully conserved AAUAAA site, yielding a pre-mRNA that is 1604 nucleotides in length. The A at the 5′ end of the pre-mRNA is capped, and a poly (A) tail is added to the 3′ end.

CACCC box

ATATCTTAGAGGGAGGGCTGAGGGTTTGAAGTCCAACTCCTAAGCCAGTGCCAGAAGAGCCAAGGACAGGTACGGCTGTCATCACTTAGACCTCA

CACCC box   CCAAT box   TATA box
CCCTGTGGAGCCACACCCTAGGGTTGGCCAATCTACTCCCAGGAGCAGGGAGGGCAGGAGCCAGGGCTGGGCATAAAAGTCAGGGCAGAGCCATC

reading frame begins ↓

↓ transcription startpoint   Met Val His Leu Thr Pro Glu Glu Lys
TATTGCTT ACATTTGCTTCTGACACAACTGTGTTCACTAGCAACCTCAAACAGACACC ATG GTG CAC CTG ACT CCT GAG GAG AAG

Ser Ala Val Thr Ala Leu Trp Gly Lys Val Asn Val Asp Glu Val Gly Gly Glu Ala Leu Gly Ar5' intron splice site
TCT GCC GTT ACT GCC CTG TGG GGC AAG GTG AAC GTG GAT GAA GTT GGT GGT GAG GCC CTG GGC AG GTTGGTA

PyNPyPyPuAPy
TCAAGGTTACAAGACAGGTTTAAGGAGACCAATAGAAACTGGGCATGTGGAGACAGAGAAGACTCTTGGGTTTCTGATAGGCACTGACTCTCTCT

3' intron splice site   g Leu Leu Val Val Tyr Pro Trp Thr Gln Arg Phe Phe Glu Ser Phe Gly
GCCTATTGGTCTATTTTCCCACCCTTAG G CTG CTG GTG GTC TAC CCT TGG ACC CAG AGG TTC TTT GAG TCC TTT GGG

Asp Leu Ser Thr Pro Asp Ala Val Met Gly Asn Pro Lys Val Lys Ala His Gly Lys Lys Val Leu Gly Ala
GAT CTG TCC ACT CCT GAT GCT GTT ATG GGC AAC CCT AAG GTG AAG GCT CAT GGC AAG AAA GTG CTC GGT GCC

Phe Ser Asp Gly Leu Ala His Leu Asp Asn Leu Lys Gly Thr Phe Ala Thr Leu Ser Glu Leu His Cys Asp
TTT AGT GAT GGC CTG GCT CAC CTG GAC AAC CTC AAG GGC ACC TTT GCC ACA CTG AGT GAG CTG CAC TGT GAC

Lys Leu His Val Asp Pro Glu Asn Phe Arg 5' intron splice site
AAG CTG CAC GTG GAT CCT GAG AAC TTC AGG GTGAGTCTATGGGACCCTTGATGTTTTCTTTCCCCTTCTTTTCTATGGTTAAGTT

CATGTCATAGGAAGGGGAGAAGTAACAGGGTACAGTTTAGAATGGGAAACAGACGAATGATTGCATCAGTGTGGAAGTCTCAGGATCGTTTTAGT

TTCTTTTATTTGCTGTTCATAACAATTGTTTTCTTTTGTTTAATTCTTGCTTTCTTTTTTTTTCTTCTCCGCAATTTTTACTATTATACTTAATG

CCTTAACATTGTGTATAACAAAAGGAAATATCTCTGAGATACATTAAGTAACTTAAAAAAAAAACTTTACACAGTCTGCCTAGTACATTACTATTT

GGAATATATGTGTGCTTATTTGCATATTCATAATCTCCCTACTTTATTTTCTTTTATTTTTAATTGATACATAATCATTATACATATTTATGGGT

TAAAGTGTAATGTTTTAATATGTGTACACATATTGACCAAATCAGGGTAATTTTGCATTTGTAATTTTAAAAAATGCTTTCTTCTTTTAATATAC

TTTTTTGTTTATCTTATTTCTAATACTTTCCCTAATCTCTTTCTTTCAGGGCAATAATGATACAATGTATCATGCCTCTTTGCACCATTCTAAAG

AATAACAGTGATAATTTCTGGGTTAAGGCAATAGCAATATTTCTGCATATAAATATTTCTGCATATAAATTGTAACTGATGTAAGAGGTTTCATA

PyNPy
TTGCTAATAGCAGCTACAATCCAGCTACCATTCTGCTTTTATTTTATGGTTGGGATAAGGCTGGATTATTCTGAGTCCAAGCTAGGCCCTTTTGC

PyPuAPy   3' intron splice site Leu Leu Gly Asn Val Leu Val Cys Val Leu Ala His His Phe Gly
TAATCATGTTCATACCTCTTATCTTCCTCCCACAG CTC CTG GGC AAC GTG CTG GTC TGT GTG CTG GCC CAT CAC TTT GGC

AAA GAA TTC ATC CCA CCA GTG CAG GCT GCC TAT CAG AAA GTG GTG GCT GGT GTG GCT AAT GCC CTG GCC CAC
Lys Glu Thr Phe Ile Pro Pro Val Gln Ala Ala Tyr Gln Lys Val Val Ala Gly Val Ala Asn Ala Leu Ala His

↓ reading frame ends
Lys Tyr His Ter
AAG TAT CAC TAA GCTCGCTTTCTTGCTGTCCAATTTCTATTAAAGGTTCCTTTGTTCCCTAAGTCCAACTACTAAACTGGGGGATATTAT

AAUAAA polyadenylation signal   ↓ transcript cleavage site
GAAGGGCCTTGAGCATCTGGATTCTGCCTAATAAAAAACATTTATTTTCATTGC AATGATGTATTTAAATTATTTCTGAATATTTTACTAAAAA

GGGAATGTGGGAGGTCAGTGCATTTAAAAACATAAAGAAATGATGAGCTGTTCAAACCTTGGGAAAATACACTATATCTTAAACTCCATGAAAGAA

Figure 6.2 The nucleotide sequence of the human β-globin gene. Conserved sequences are boxed, and the encoded amino acids are written above their respective codons.

There are two introns in the pre-mRNA. The first intron is 130 nucleotides in length, and the second is 850 nucleotides in length. Both introns are within the reading frame, so precise intron removal is essential to prevent a frameshift. The introns adhere to the GT-AG rule. The 5' splice site of the first intron contains the conserved sequence GTTGGT, which differs by two nucleotides from the GTAAGT consensus sequence. The 3' splice site contains the conserved sequence CCACCCTTAG, which differs by two nucleotides from the 6PyNCAG consensus sequence. The sequence CACTGAC, which corresponds

to the consensus sequence PyNPyPyPuAPy, is 39 nucleotides upstream from the 3' splice site. The second intron has similar conserved sequences. At the 5' splice site is the sequence GTGAGT, which differs by one nucleotide from the consensus sequence. At the 3' splice site is the sequence CCTCCCACAG, which matches its 6PyNCAG consensus sequence perfectly. The sequence TGCTAAT lies 35 nucleotides upstream from the 3' splice site and corresponds to the PyNPyPyPuAPy consensus sequence. Once the introns are removed, the mature mRNA is 624 nucleotides long, not counting the poly (A) tail.

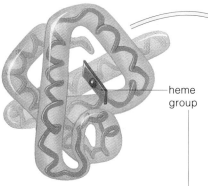

a The tertiary structure with eight α helices. The red structure is a heme group.

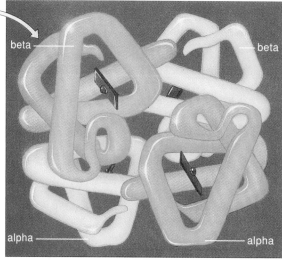

c The tertiary structure of the hemoglobin molecule, with two β-globin subunits and two α-globin subunits.

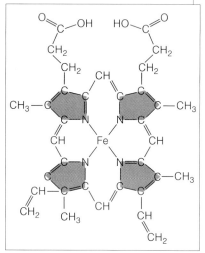

b Detail of a heme group structure.

Figure 6.3 Tertiary and quaternary structures of human β globin. (Source: Parts a and c: Starr and Taggart UDL 7th ed. Fig. 3.19, p. 45. Part b: R. L. Nagel. Disorders of hemoglobin function and stability. In R. I. Handin, S. E. Lux, and T. P. Stossel (eds.). 1995. *Blood: Principles and Practice of Hematology.* Philadelphia: J. B. Lippincott Co.)

The reading frame begins with the first AUG in the mRNA, which is located 50 nucleotides downstream from the transcription startpoint. The AUG initiation codon lies within the sequence ACAGACACCAUGG, which differs from the GCCGCCPuCCAUGG translation initiation consensus sequence by three nucleotides. The purine at −3 from the AUG codon and the G at +1 from the AUG codon, which (except for the AUG) are the two most important nucleotides in the consensus sequence, are fully conserved. The first in-frame stop codon is UAA, which is the 148th codon in the reading frame, yielding a polypeptide that is 147 amino acids in length.

Following translation, the methionine at the amino end of the polypeptide is removed, leaving 146 amino acids in the primary structure. Most of the amino acids coil into eight α helices that fold around a single heme group to form the tertiary structure (Figure 6.3a). Each **heme group** consists of an iron atom surrounded by a protoporphyrin ring (Figure 6.3b). The heme groups

are examples of a nonprotein substance added to each protein subunit to form the tertiary structure during post-translational modification of the protein. Once it has assumed its tertiary structure, a β subunit binds to an α subunit (encoded by an α-globin gene) to make a dimer. Two dimers then attach to each other to form the tetrameric hemoglobin protein (Figure 6.3c).

Hemoglobin carries oxygen from the lungs through the bloodstream to the tissues and releases oxygen when the erythrocyte arrives in a capillary. As it releases oxygen, the hemoglobin molecule picks up carbon dioxide and carries it through the bloodstream to the lungs, where the carbon dioxide is expelled. Thus, hemoglobin must be capable of alternatively binding oxygen or carbon dioxide.

Hemoglobin binds oxygen by drawing an O_2 molecule into a pocket located in each of the four subunits. The pocket is called the **heme pocket** because a heme group resides within it. Heme groups are hydrophobic, which is essential for oxygen binding. The pocket is lined with

hydrophobic amino acids that prevent water from entering the heme pocket. The path into the pocket is just large enough for an O_2 molecule to penetrate and bind to the heme group.

The amino acids on the outer surface of the hemoglobin molecule are mostly hydrophilic, making the hemoglobin molecule soluble in water. This is an effective way to allow a water-insoluble heme group to function in the aqueous interior of an erythrocyte.

When hemoglobin encounters carbon dioxide in the capillaries, arginines located near the carboxyl end of the α-globin subunits bind carbon dioxide. Even though the binding sites for oxygen and carbon dioxide are different, hemoglobin does not bind both molecules simultaneously. Carbon dioxide binding alters the conformation of hemoglobin, causing the hemoglobin molecule to lose its affinity for oxygen. This facilitates the exchange of oxygen and carbon dioxide as hemoglobin delivers oxygen to the tissues. When hemoglobin is in the lungs, it releases carbon dioxide, which increases hemoglobin's affinity for oxygen.

Having discussed the normal function of hemoglobin, we will now examine the effects of mutations in the β-globin gene.

6.2 MUTATIONS IN THE β-GLOBIN GENE

Numerous mutations have been identified in the human β-globin gene. Some of these mutations are especially frequent among people of certain ancestral groups, making hemoglobin disorders among the most important genetic disorders throughout the world. Let's look at two genetic disorders caused by mutations in the β-globin gene, sickle-cell anemia and β-thalassemia.

Sickle-Cell Anemia

Sickle-cell anemia is a serious genetic disorder characterized by weakness, fatigue, heart failure, joint and muscular impairment, abdominal pain and dysfunction, impaired mental function, and eventual death, often during childhood. The mutation that causes sickle-cell anemia in humans was among the first mutations characterized at the molecular level in any organism. The sixth amino acid in the mature β-globin polypeptide is glutamic acid, which is encoded by GAG. An A $\rightarrow$ T substitution in the second nucleotide of the codon changes the codon from GAG to GUG, substituting valine for glutamic acid in the polypeptide. A normal hemoglobin molecule, with glutamic acid at position 6 in the β subunits, is called **Hemoglobin A (Hb A)**, whereas the mutant molecule, with valine at position 6 in the β subunits, is called **Hemoglobin S (Hb S)**. Glutamic acid is acidic and hydrophilic, whereas valine is nonpolar and hydrophobic

(see Figure 4.1). This substitution therefore alters the chemical properties of hemoglobin.

The sixth amino acid is not located within the heme pocket, so it does not directly prevent oxygen binding. However, when all the hemoglobin molecules are HbS (as is the case for individuals with sickle-cell anemia), the HbS molecules attach to one another when they are deoxygenated, forming long chains. Remember that oxygen is bound in the heme pocket when hemoglobin is oxygenated. When oxygen is not bound, the hydrophobic valine at position 6 embeds itself into the hydrophobic heme pocket in a β subunit of another deoxygenated hemoglobin molecule (Figure 6.4a). Interactions between several amino acids within the mutant HbS molecules cause them to attach to one another, forming a long double-stranded polymer (Figure 6.4b). Seven of these double-stranded polymers then wind around each other to form a sickle hemoglobin fiber (Figure 6.5).

The HbS molecules in sickle hemoglobin fibers have reduced oxygen affinity (in part because many of the heme pockets are occupied by valines) and cause erythrocytes to assume a sickle shape instead of the normal disk shape, which is the reason for the name *sickle-cell anemia* (Figure 6.6). Sickled cells do not pass through capillaries as readily as normal disk-shaped cells, so sickled cells often clog capillaries, denying oxygen to the cells served by them. Lack of oxygen and accumulation of carbon dioxide cause widespread tissue damage throughout the body. As the spleen removes sickled cells, it becomes enlarged and suffers damage. The phenotypic symptoms of sickle-cell anemia mentioned at the beginning of this section are a consequence of all these effects.

β-Thalassemia

β-thalassemia is a genetic disorder caused by a reduction of β-globin synthesis (called β^+-thalassemia) or the absence of β globin (called β^0-thalassemia). When a reduction of β-globin synthesis disrupts the 1:1 ratio of α and β subunits, hemoglobin synthesis is impaired and the excess α subunits aggregate and precipitate. This damages erythrocytes and leads to their premature destruction. The phenotypic symptoms of severe β-thalassemia include expansion of bone marrow, which causes skeletal deformities and bone fractures. The reduction of functional hemoglobin also causes iron overloading, which may lead to cirrhosis of the liver, endocrine dysfunction, diabetes, and cardiac failure. Like patients with sickle-cell anemia, many patients with severe β-thalassemia suffer an early death. Because the amount of functional β globin may vary among patients, the severity of symptoms may vary as well.

Over 30 different mutations in the β-globin gene are known to cause β-thalassemia. Nearly half of these mutations disrupt intron removal. Let's now look at one of them in detail.

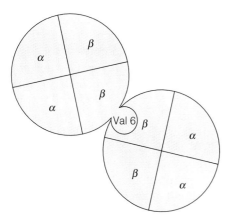

a The valine at position 6 in a β subunit of an HbS molecule embeds itself into the heme pocket in a β subunit of another HbS molecule.

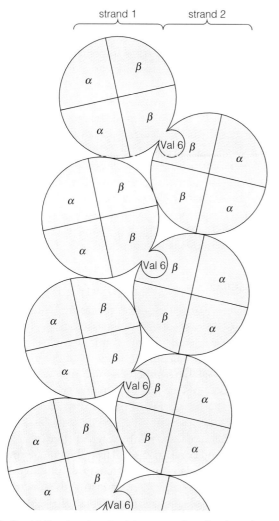

strand 1   strand 2

b The HbS molecules attach to one another, forming a long double-stranded polymer.

Figure 6.4 Polymerization of deoxygenated Hb S molecules. (Reprinted from Wishner, B. C., K. B. Ward, E. E. Lattman, and W. E. Love. 1975. Crystal structure of sickle-cell deoxyhemoglobin. Journal of Molecular Biology 98:178–194. Used by permission of the publisher, Academic Press.)

The DNA sequence surrounding the 5′ splice site of the first intron in the β-globin gene is shown in Figure 6.7a. A G → A transition at the 5′ junction of the intron eliminates cleavage of the intron at this site. The intron, however, is still removed, but it is cleaved incorrectly at alternative sites known as **cryptic splice sites**, shown in Figure 6.7b. The cryptic splice sites are used only when a mutation prevents cleavage of the intron at the correct splice site. Notice that each of these three cryptic splice sites resembles the 5′ splice-site consensus sequence GTAAGT, and that each has GT on the 5′ end.

The first intron is located within the reading frame for translation, so splice site mutations affect the amino acid sequence of the polypeptide. When the intron is cleaved at either of the first two cryptic splice sites, several codons are deleted from the mRNA. When it is cleaved at the third cryptic splice site, several new codons are inserted in the mRNA. All three of these cryptic splice sites are out of frame with the original splice site, so all three cause frameshift mutations. The changes caused by faulty intron removal are so drastic that the mutant gene fails to produce any functional β globin and can cause $β^0$-thalassemia.

The following example shows the effect of another mutation in the human β-globin gene—this time one that causes $β^+$-thalassemia.

Figure 6.5 A sickle hemoglobin fiber.

Example 6.1 A mutation that causes β^+-thalassemia.

In 1984, Collins and Weissman (*Progress in Nucleic Acid Research and Molecular Biology* 31:315–462) discussed the effects of several mutations in the human β-globin gene that caused β-thalassemia. Among the mutations they discussed is a T → C transition that altered the sequence

    CTGCCTAATAAAAAA

to

    CTGCCTAACAAAAAA

(the affected nucleotide is underlined). This mutation causes β^+-thalassemia.

Problem: **(a)** Identify the location of this mutation in the β-globin gene's nucleotide sequence, shown in Figure 6.2. **(b)** Provide a hypothesis that explains why this mutation causes β^+-thalassemia rather than β^0-thalassemia.

Solution: **(a)** The mutation is located near the 3' end of the gene, where it changes the AATAAA polyadenylation signal to AACAAA. **(b)** The mutation does not prevent transcription or correct intron removal, but it disrupts normal polyadenylation of the pre-mRNA. The absence of normal polyadenylation reduces mRNA stability, reducing the amount of β-globin mRNA available for translation. This, in turn, reduces the amount of β globin translated from the mRNAs. The β globin that is produced is fully functional because the mutation does not change the amino acid sequence or the polypeptide. The disorder caused by this mutation is β^+-thalassemia because functional β globin is reduced but not eliminated.

The human β-globin gene offers an excellent example of how the product of a gene is expressed as a functional protein. Mutations in the gene are expressed phenotypically as the genetic disorders sickle-cell anemia and β-thalassemia.

6.3 ENZYMES, BIOCHEMICAL PATHWAYS, AND MUTATION

Our examination of the human β-globin gene showed how mutations in a gene may alter the protein product, which may in turn alter the phenotype. However, it is erroneous to conclude that a particular mutant pheno-

normal cell

sickled cell

Figure 6.6 Normal and sickled erythrocytes. (Source: Bill Longcore/Photo Researchers.)

type is always caused by mutations in the same gene. Mutations in any gene that reduces or eliminates hemoglobin function may cause similar symptoms. For example, several mutations in the α-globin gene cause α-thalassemias, which are similar in their phenotypic effects to the β-thalassemias. Also, mutations in the genes that encode enzymes for heme synthesis may reduce or eliminate functional hemoglobin due to alterations in heme production. Mutations in any of the several genes that participate in hemoglobin synthesis may be expressed as similar phenotypes because all of the mutations have a similar result: the reduction or absence of functional hemoglobin.

Most phenotypic characteristics are influenced by the combined effects of several genes, although a mutation in any one gene may alter the phenotype. In this section, we will focus on enzymes and biochemical pathways to see how several genes collectively influence a single phenotype and how mutations in those genes can change the phenotype.

Biochemical Pathways

Most compounds in a cell arise through a series of steps known as a **biochemical pathway**. Each step is determined and regulated by a specific enzyme, and each enzyme is encoded by one or more genes. The pathway begins with a **pathway substrate**, a compound that the first enzyme in the pathway converts into another compound, which is called an **intermediate**. Each intermediate is the product of one enzyme, and is also the substrate for the next enzyme in the pathway. Enzymes continue step by step, converting one intermediate into another until the last intermediate is converted into the **end product** (Figure 6.8).

Let's return to the synthesis of melanins, the pigments in hair, eyes, and skin, that we discussed at the beginning

AAGGTGAACGTGGATGAAGTTGGTGGTGAGGCCCTGGGCAG**GTTGGT**ATCAAGGGTTACAAAG

a The nucleotide sequence surrounding the 5' splice site of the first intron in the nonmutant human β-globin gene. The conserved sequence at the 5' splice site is shaded red.

AAGGTGAACGTGGATGAAGTTGGTGGTGAGGCCCTGGGCAG**ATTGGT**ATCAAGGGTTACAAAG

b A G→A substitution mutation in the conserved sequence at the 5' splice site for the first intron in the human β-globin gene. The conserved sequence at the mutant 5' splice site is shaded red, and the mutant nucleotide is designated by an arrow. Three cryptic splice sites are shaded blue.

Figure 6.7 A splice site mutation in the human β-globin gene.

enzyme 1   enzyme 2   enzyme 3   enzyme 4

Pathway ⇒⇒⇒A ⇒⇒⇒B ⇒⇒⇒C ⇒⇒⇒ end
substrate   product
             ⎵_____⎵
                  intermediates

Figure 6.8 Components of a biochemical pathway.

of this chapter, as an example of a biochemical pathway. Figure 6.9 summarizes the melanin synthesis pathway. The pathway substrate is the amino acid phenylalanine. People ingest phenylalanine from proteins in the diet. Enzymes in the digestive system break down the proteins in food into the amino acids, one of which is phenylalanine. These amino acids are absorbed into the bloodstream, where the blood transports them to various cells in the body, including melanocytes, the cells that synthesize melanins. An enzyme known as phenylalanine hydroxylase (PAH), which is encoded by the *PAH* gene, converts phenylalanine into tyrosine, the first intermediate in the melanin synthesis pathway. The cellular pool of tyrosine comes from two sources: (1) the conversion of phenylalanine into tyrosine, and (2) tyrosine derived from proteins in the diet.

Melanocytes produce an enzyme called tyrosinase that converts tyrosine into DOPA and then dopaquinone. Dopaquinone may then be converted into several compounds that are eventually converted into melanins through reactions catalyzed by several enzymes. Melanins are the end products of the pathway, which now branches to form the different types of melanins. Different melanins have different colors, so a mixture of melanins determines the pigmentation of hair, eyes, or skin. Consequently, pigmentation of hair, eyes, and skin is controlled by the combined effects of several genes rather than a single one. The nucleotide sequences of the genes that encode the enzymes for melanin synthesis ultimately determine a person's pigmentation phenotype. Variations in the DNA sequences of these genes is responsible for phenotypic variation in pigmentation.

~
Enzymes function in a biochemical pathway to produce an end product. Variation in the nucleotide sequences of genes that encode enzymes in a biochemical pathway is expressed as variation in the end product of the pathway, which, in turn, causes phenotypic variation.

Mutations and Biochemical Pathways

Because all of the genes that encode the enzymes of a biochemical pathway function to produce the same end product of the pathway, mutations in any one of the genes may cause the same or a similar mutant phenotype. Let's now expand our view of gene expression to see how several genes produce the different enzymes in a single biochemical pathway and how mutations affect that pathway.

The idea that genes were expressed as enzymes dates back to just after 1900. Evidence available in the latter part of the nineteenth century indicated that biochemical pathways were controlled by enzymes. In 1902, Archibald Garrod recognized the relationship between the absence of a functional enzyme and an inherited disorder, implying that genes influenced enzymes. Garrod had been studying a disorder in humans called **alkaptonuria**, which he had identified as an inherited disorder on the basis of pedigree analysis. This disorder can be easily diagnosed in affected individuals because their urine turns dark when exposed to the air. Garrod identified a substance, called *alkapton* in his day, that was present in large amounts in the urine of affected individuals but in only trace amounts in that of unaffected individuals. We now call this substance **homogentisic acid**. When exposed to

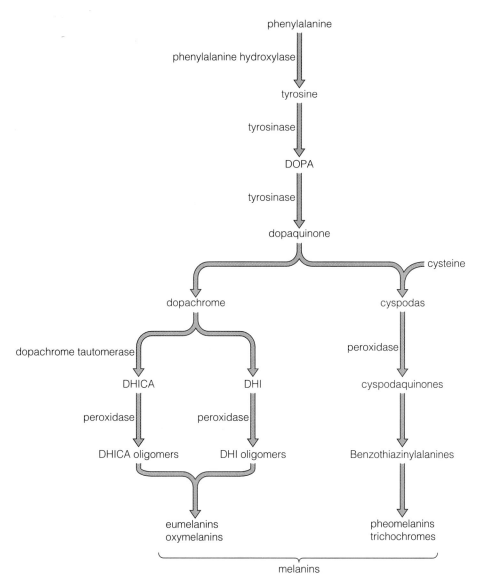

Figure 6.9 The biochemical pathway for synthesis of melanins. Several steps are regulated by enzymes. The known enzymes are indicated at the steps they catalyze. Other steps may be catalyzed by enzymes that are not yet known. (Adapted from G. Prota 1992. *Melanins and melanogenesis*. San Diego: Academic Press. Reprinted by permission.)

oxygen, homogentisic acid turns a dark color, causing the evacuated urine of people with alkaptonuria to darken. Garrod concluded that the excess of homogentisic acid in the urine of people with alkaptonuria was due to the absence of an enzyme that normally degrades homogentisic acid. When this enzyme is absent, homogentisic acid accumulates because there is nothing to degrade it.

The relevant biochemical pathway, including the enzyme that Garrod studied, has been identified and is illustrated in Figure 6.10. A mutation in the gene that encodes the enzyme **homogentisic acid oxidase** eliminates production of the functional enzyme, blocking the pathway and causing accumulation of homogentisic acid. We can

compare a biochemical pathway to a street and the enzymes controlling the flow of the pathway to the stoplights on that street. The absence of a functional enzyme is like a defective stoplight that fails to turn green: traffic passes through all the stoplights that are functioning normally, but backs up behind the defective one.

Geneticists discovered that mutations in different genes sometimes confer the same phenotype. In many cases, these different genes encode different enzymes that participate in the same biochemical pathway. For example, mutations in several different genes in *Drosophila melanogaster* confer a brilliant-red color to the eyes instead of the usual brick-red color. A classic study by George

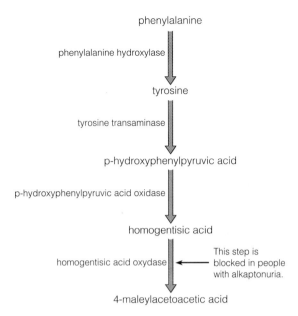

phenylalanine

phenylalanine hydroxylase

tyrosine

tyrosine transaminase

p-hydroxyphenylpyruvic acid

p-hydroxyphenylpyruvic acid oxidase

homogentisic acid

homogentisic acid oxydase ◄—— This step is blocked in people with alkaptonuria.

4-maleylacetoacetic acid

Figure 6.10 The biochemical pathway for degradation of phenylalanine into 4-maleylacetoacetic acid. People with alkaptonuria lack functional homogentisic acid oxidase. Notice that the first step of this pathway is identical to the melanin synthesis pathway shown in Figure 6.9.

Beadle and Boris Ephrussi, summarized in the following example, demonstrated that two of these genes encode different enzymes in the same pathway. These experiments also revealed the order of the enzymes in the pathway, as we shall see.

Example 6.2 Beadle and Ephrussi's experiments with the eye-color mutations *vermilion* and *cinnabar* in *Drosophila melanogaster*.

Drosophila melanogaster develops through a process known as metamorphosis by which each individual progresses through four distinct stages: (1) fertilized egg, (2) larva, (3) pupa, and (4) adult. Early in development, the embryonic cells that are destined to develop into adult organs are partitioned into masses of tissue called imaginal disks, which remain undeveloped in the larva then develop into the organs during pupation. Each disk is specific for a certain organ, such as the eye disk, which will develop into the adult's eyes. When an eye disk is removed from a larva and transplanted into another larva, eye tissue develops in the body cavity at the site of transplantation while the natural eyes also develop as usual. For instance, an eye disk transplanted into the abdomen of a larva develops into eye tissue in the adult fly's abdomen.

Adult flies usually have two types of pigment in their eyes: a brilliant-red pigment and a brown

pigment. A typical fly has brick-red eyes due to the combination of the red and brown pigments. However, mutations in any one of several different genes may prevent the brown pigment from forming. One mutation, in a gene called *vermilion*, blocks synthesis of the brown pigment. In the absence of brown pigment, the bright-red pigment (which is normally dulled by the brown pigment) give the eyes a bright-red color. Another mutation, in a gene called *cinnabar*, has the same phenotypic effect as the *vermilion* mutation: no brown pigment forms, and the eyes appear bright red.

Beadle and Ephrussi (1937. *Genetics* 22:76–86) conducted a series of transplantation experiments with eye disks from mutant *vermilion* and *cinnabar* larvae as well as larvae with no eye-color mutations (called wild type) to determine how the *vermilion* and *cinnabar* genes interacted with one another. The results of these experiments are summarized in Figure 6.11. When they transplanted mutant *vermilion* eye disks into wild-type larvae (experiment 1), brown pigment was synthesized in the transplanted tissue even though the transplanted tissue contained the *vermilion* mutation. They observed the same phenomenon when they transplanted *cinnabar* eye disks into wild-type larvae (experiment 2): brown pigment was synthesized in the transplanted tissue. When they transplanted mutant *vermilion* eye disks into mutant *cinnabar* larvae (experiment 3), brown pigment formed in the transplanted tissue, but no brown pigment formed in the natural eyes of the recipient larvae. When they transplanted mutant *cinnabar* eye disks into mutant *vermilion* larvae (experiment 4), no brown pigment formed in the transplanted tissue, although a small amount of brown pigment formed in the natural eyes of the recipient larvae.

Problem: Using this information, determine the order of these two mutations in the pathway for brown-pigment synthesis.

Solution: The *vermilion* and *cinnabar* genes encode different enzymes in the biochemical pathway for brown-pigment synthesis. Let's call these enzymes 1 and 2, where enzyme 1 precedes enzyme 2 in the pathway:

Enzyme 1   Enzyme 2
A ——————→ B ——————→ C ——————→ Brown pigment

Our task now is to determine which of the enzymes is encoded by the *vermilion* gene and which by the *cinnabar* gene.

A mutation in the gene that encodes enzyme 1 prevents synthesis of the functional enzyme, so the pathway is blocked and no brown pigment forms.

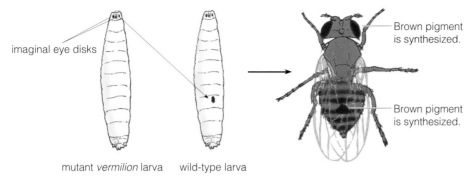

imaginal eye disks

mutant *vermilion* larva   wild-type larva

Brown pigment is synthesized.

Brown pigment is synthesized.

Experiment 1: Imaginal disks are transplanted from mutant *vermilion* larvae into wild-type larvae.

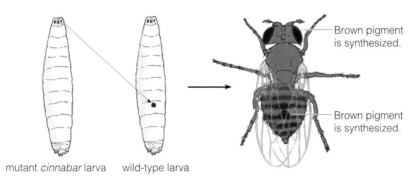

mutant *cinnabar* larva   wild-type larva

Brown pigment is synthesized.

Brown pigment is synthesized.

Experiment 2: Imaginal disks are transplanted from mutant *cinnabar* larvae into wild-type larvae.

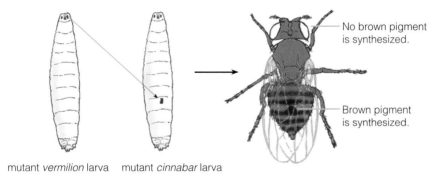

mutant *vermilion* larva   mutant *cinnabar* larva

No brown pigment is synthesized.

Brown pigment is synthesized.

Experiment 3: Imaginal disks are transplanted from mutant *vermilion* larvae into mutant *cinnabar* larvae.

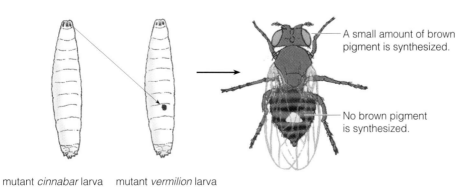

mutant *cinnabar* larva   mutant *vermilion* larva

A small amount of brown pigment is synthesized.

No brown pigment is synthesized.

Experiment 4: Imaginal disks are transplanted from mutant *cinnabar* larvae into mutant *vermilion* larvae.

Figure 6.11   Information for Example 6.2: Summary of transplantation experiments conducted by Beadle and Ephrussi with *Drosophila melanogaster*.

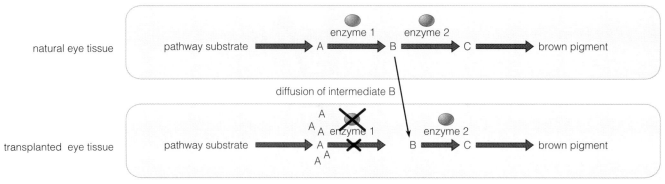

a A mutation in the transplanted tissue eliminates the function of enzyme 1. Intermediate A accumulates due to the blocked pathway. However, intermediate B, produced in the natural eye tissue, diffuses to the transplanted tissue and restores the pathway. Brown pigment is synthesized.

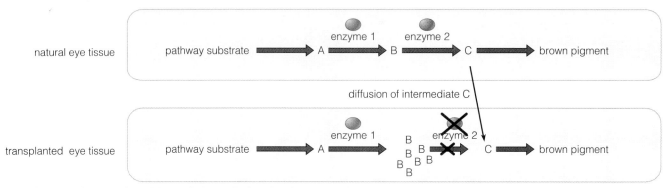

b A mutation in the transplanted tissue eliminates the function of enzyme 2. Intermediated B accumulates due to the blocked pathway. However, intermediate C produced in the natural eye tissue, diffuses to the transplanted tissue and restores the pathway. Brown pigment is synthesized.

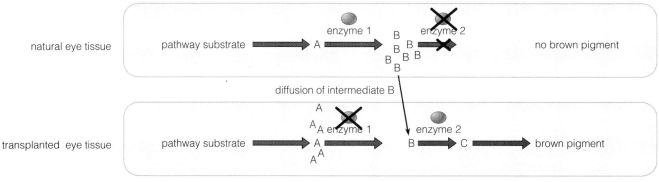

c A mutation in the natural eye tissue eliminates the function of enzyme 2, causing intermediate B to accumulate. A mutation in the transplanted tissue eliminates the function of enzyme 1. Intermediate B, which had accumulated in the natural eye tissue, diffuses to the transplanted tissue and restores the pathway there. Brown pigment is synthesized.

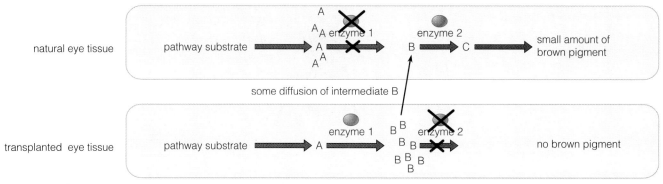

d A mutation in the natural eye tissue eliminates the function of enzyme 1. A mutation in the transplanted tissue eliminates the function of enzyme 2, causing intermediate B to accumulate. A small amount of intermediate B diffuses to the natural eye tissue, causing a small amount of brown pigment to form there.

Figure 6.12 Solution to Example 6.2: Analysis of experimental results.

A mutation in the other gene prevents synthesis of functional enzyme 2, and has the same phenotypic effect: the absence of brown pigment.

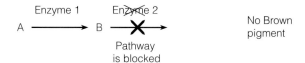

In experiments 1 and 2 as shown in Figure 6.11, mutant *vermilion* eye disks and mutant *cinnabar* eye disks transplanted into wild-type larvae synthesized brown pigment even though they were mutant. This is expected because the wild-type recipient larvae carried nonmutant genes that encode functional enzymes. The enzymes produced in the natural eye tissue of the wild-type larvae do not diffuse to the transplanted eye tissue, but intermediates in the biochemical pathway do. Wild-type flies have complete pathways in their natural eyes, so the diffusible intermediates in the pathway are present in the natural eye tissue and diffuse through the larval body to the transplanted tissue. In experiments 1 and 2, intermediates supplied by the natural eye tissue overcome the effect of the mutations in the transplanted tissue, as illustrated in parts a and b of Figure 6.12.

The results of experiments 3 and 4 (see Figure 6.11) reveal which enzyme precedes the other in the pathway. Figure 6.12c shows what happens when an eye disk that is mutant for enzyme 1 is transplanted into a larva that is mutant for enzyme 2. The functional enzyme 1 in the natural eye tissue produces intermediate B, which then diffuses to the transplanted eye tissue. The functional enzyme 2 in the transplanted eye tissue converts B into C, and brown pigment forms in the transplanted tissue but not in the natural eyes. Beadle and Ephrussi observed this result in experiment 3 when they transplanted a mutant *vermilion* eye disk into a mutant *cinnabar* larva. Therefore, enzyme 1 is the *vermilion* enzyme, and enzyme 2 is the *cinnabar* enzyme, meaning that the *vermilion* enzyme precedes the *cinnabar* enzyme in the biochemical pathway:

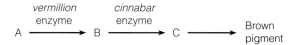

The results of experiment 4 (see Figure 6.11) confirm this conclusion. As illustrated in Figure

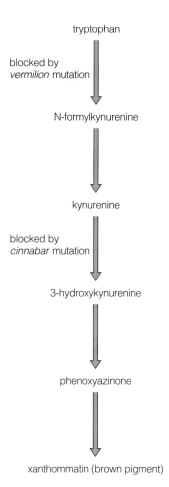

Figure 6.13 Pathway for brown-eye pigment synthesis in *Drosophila melanogaster.*

6.12d, when an eye disk that is mutant for enzyme 2 is transplanted into a larva that is mutant for enzyme 1, intermediate B is produced in the transplanted tissue but not in the natural eyes. A small amount of intermediate B diffuses from the transplanted tissue to the natural eyes, permitting brown pigment to form in the natural eyes. Much larger quantities of intermediates are produced in the natural eyes than in the transplanted eye tissue, so very little intermediate diffuses from the transplanted tissue to the natural eyes. Beadle and Ephrussi observed a small amount of brown pigment in the natural eyes in experiment 4, which is consistent with the conclusion that enzyme 1 is the *vermilion* enzyme and enzyme 2 is the *cinnabar* enzyme.

The actual intermediates in this pathway, including those affected by *vermilion* and *cinnabar*, were determined several years after Beadle and Ephrussi conducted their experiments and are shown in Figure 6.13. The *vermilion* enzyme catalyzes the conversion of tryptophan to N-formylkynurenine. The *cinnabar* enzyme catalyzes

the conversion of kynurenine to 3-hydroxykynurenine. As Beadle and Ephrussi's results demonstrated, the *vermilion* enzyme precedes the *cinnabar* enzyme in the pathway.

The One Gene–One Polypeptide Hypothesis and Biochemical Pathways

After his experiments with *Drosophila melanogaster*, Beadle, together with Edward Tatum, began experiments with fungi of the genus *Neurospora*, a group of bread-mold fungi that can be easily grown in culture plates and studied genetically. *Neurospora* can be grown on a simple medium called **minimal medium**, composed of a sugar, several minerals, and the vitamin biotin. The cells are able to synthesize all other molecules they require from the substances in the minimal medium. Beadle and Tatum treated cultures of *Neurospora* with X-rays to induce mutations. They then recovered three different mutant strains that could *not* grow on minimal medium but could grow when the medium was supplemented with malt extract and yeast extract. They concluded that substances not present in the minimal medium were present in the extracts and that these substances allowed the mutant strains to grow.

This type of mutation, which confers a requirement for supplementation of minimal medium to overcome the effects of the mutation, is called an **auxotrophic mutation**. By supplementing minimal medium with individual vitamins and amino acids, Beadle and Tatum determined which substances overcame the effects of each auxotrophic mutation. At the time, it was known that biochemical reactions were mediated by enzymes. However, many scientists believed that genes acted directly on the phenotype, rather than through enzymes. From their experimental results, Beadle and Tatum concluded that a mutation in one gene eliminated the function of one enzyme, which led them to propose the **one gene–one enzyme hypothesis**, the idea that each gene produced one enzyme.

The one gene–one enzyme hypothesis constituted a major shift in thinking. Beadle and Tatum proposed that, instead of a gene acting directly on the phenotype, genes produce enzymes that catalyze biochemical reactions, and the products of these reactions determine the phenotype. Researchers later discovered that some enzymes consist of two or more protein subunits that together make the functional enzyme, implying that some enzymes are encoded by two or more genes. Also, there are many proteins that are not enzymes. For these reasons, the one gene–one enzyme hypothesis was modified to the **one gene–one polypeptide hypothesis**, which reflects more accurately the relationship of genes and their products. Beadle and Tatum received a Nobel prize in 1958 for their work in discovering the relationship of

genes and enzymes. In his acceptance speech, Beadle appropriately credited Garrod, stating: "In this long, roundabout way, we had discovered what Garrod had seen so clearly so many years before."

Researchers soon discovered many different auxotrophic fungal strains. Among these were strains that required the same substance for growth but had mutations in different genes. Because these different mutant strains required the same substance for growth, they probably carried mutations in genes that encoded the different enzymes in the same biochemical pathway. Among the best examples was a set of auxotrophic mutant strains of *Neurospora* that all required the amino acid arginine in the medium. *Neurospora* cells are normally capable of synthesizing all 20 amino acids when grown on minimal medium. A mutation in a gene that encodes an enzyme in the arginine synthesis pathway prevents synthesis of arginine, so the strain fails to grow unless arginine is supplied in the growth medium.

Whereas all mutant strains that are auxotrophic for arginine grow when supplemented with arginine (the end product of the pathway), they differ in their abilities to grow when the different intermediates in the pathway are added to the medium. By adding the intermediates one at a time to each mutant strain, researchers can determine the relative order of the genes in the biochemical pathway.

The following example describes how geneticists identified the relative orders of intermediates and enzymes in the arginine biosynthesis pathway through analysis of auxotrophic mutations.

Example 6.3 Using auxotrophic mutations to determine the steps in the biochemical pathway for arginine synthesis in *Neurospora*.

From previous work that had been done on arginine synthesis in cells, Srb and Horowitz (1944. *Journal of Biological Chemistry* 154:129–139) concluded that two substances, ornithine and citrulline, were intermediates in the biochemical pathway for arginine synthesis. They tested each of seven auxotrophic mutant strains on minimal medium and minimal medium supplemented with arginine only, ornithine only, and citrulline only. Table 6.1 summarizes their results.

Problem: Determine the relative order of citrulline and ornithine in the biochemical pathway for arginine synthesis and the relative order of the enzymes disrupted by the mutations.

Solution: In order to understand this type of experiment, let's suppose that the biochemical pathway for arginine synthesis has three intermediates, A, B,

Table 6.1 Information for Example 6.3: Growth of Mutant Strains of *Neurospora* on Different Growth Media

Mutant Strain	Growth Medium			
	Minimal	Minimal & Arginine	Minimal & Citrulline	Minimal & Ornithine
arg 1	−	+	−	−
arg 2	−	+	+	−
arg 3	−	+	+	−
arg 4	−	+	+	+
arg 5	−	+	+	+
arg 6	−	+	+	+
arg 7	−	+	+	+

*A "+" indicates that the strain grew on the medium. A "−" indicates that the strain failed to grow on the medium.

substrate ⟹ A ⟹ B ⟹ C ⟹ arginine

a A biochemical pathway for arginine synthesis with three intermediates.

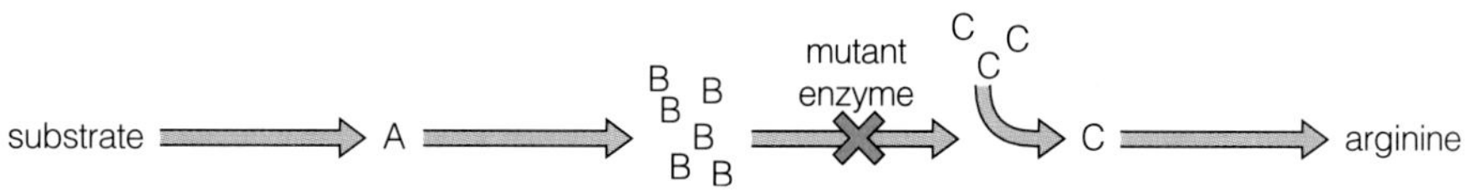

b A mutant enzyme fails to convert intermediate B into intermediate C. However, if intermediate C is added to the medium, the pathway is restored and arginine is synthesized.

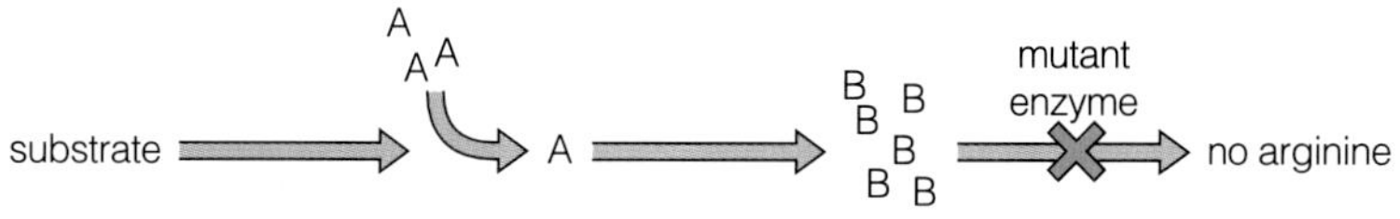

c A mutant enzyme fails to convert intermediate B into intermediate C. If intermediate A is added to the medium, the pathway remains blocked and arginine is not synthesized.

Figure 6.14 Information for Example 6.3: The effect of auxotrophic mutations and supplementation on biochemical pathways.

and C, and that each step is catalyzed by a different enzyme, as illustrated in Figure 6.14a. If a mutant enzyme fails to catalyze a step in the pathway before the point where the added intermediate appears in the pathway, then the added intermediate is converted into the end product, as illustrated in Figure 6.14b. However, if the mutant enzyme catalyzes a step in the pathway beyond the point where the intermediate is added, then the pathway remains blocked, as illustrated in Figure 6.14c.

Because strains with the mutations *arg 4*, *arg 5*, *arg 6*, and *arg 7* grow when either citrulline or ornithine is provided, these mutations must affect

enzymes that precede both ornithine and citrulline in the pathway. Strains with the mutations *arg 2* and *arg 3* can grow when citrulline is provided but cannot grow when ornithine is provided. These mutations must affect enzymes that follow ornithine but precede citrulline in the pathway, meaning that ornithine precedes citrulline in the pathway. The strain with the mutation *arg 1* is incapable of growing when either citrulline or ornithine is provided. This mutation must therefore affect an enzyme that follows both citrulline and ornithine in the pathway, but one that precedes arginine because the strain can grow when arginine is provided.

In summary, the relative order of the intermediates and the mutations in the biochemical pathway can be diagrammed as follows:

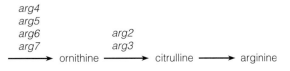

The analysis of auxotrophic mutations has proven to be valuable in the elucidation of biochemical pathways. Such experiments reveal which intermediates are in a biochemical pathway, the relative order of the intermediates, and the relative order of the enzymes.

~

A mutation in any gene that encodes an enzyme in a biochemical pathway may block the pathway, cause an intermediate to accumulate, and prevent synthesis of the end product of the pathway.

SUMMARY

1. Gene expression is the process through which the nucleotide sequence of a gene is manifest in the phenotype of an organism.

2. The human β-globin gene provides an excellent example of how the sequence of nucleotides in a gene is eventually expressed as a functional protein. Mutations in the DNA may alter the protein function, which may cause an altered phenotype.

3. Many compounds in cells are synthesized through biochemical pathways. A biochemical pathway begins with a substrate, which is converted through a series of intermediates into the end product of the pathway.

4. Mutations in genes that encode enzymes may block a biochemical pathway, causing an intermediate to accumulate.

5. The phenotypic effects of mutations that block biochemical pathways can be overcome by supplying either the end product of the pathway or an intermediate of the pathway that appears farther along the pathway than the enzyme affected by the mutation.

6. Mutations in the genes that encode enzymes in the same biochemical pathway can be used to identify the relative orders of intermediates and enzymes in the pathway.

QUESTIONS AND PROBLEMS

1. Define gene expression, and describe, in terms of protein function, how mutations in a gene can affect the phenotype.

2. Describe how the positions of hydrophilic and hydrophobic amino acids in the tertiary and quaternary structures of the hemoglobin molecule affect its ability to carry out its functions.

3. Describe how a Glu $\rightarrow$ Val amino acid substitution in the sixth amino acid of β globin can cause sickle-cell anemia in humans even though this altered amino acid is not part of the active site of the protein.

4. Both α and β subunits of hemoglobin have no signal peptides encoded in their genes. What does this suggest about the cellular location of mature hemoglobin?

5. Mutations in several different genes that encode enzymes in the pathway for heme group synthesis may cause a genetic disorder called methemoglobinemia. The symptoms of this disorder range from mild (such as a blue cast to the skin and some increased fatigue with exercise) to death in the most severe cases. The enzyme cytochrome b_5 reductase (b5R) is one of the enzymes in the pathway for heme group synthesis. This enzyme is responsible for reducing the iron in the heme group to the ferrous state so that it can bind oxygen. Mutations in the *b5R* gene are often responsible for the most severe cases of methemoglobinemia. Kobayashi et al. (1990. *Blood* 75:1408–1413) reviewed several of the mutations in this gene. One of these mutations causes a Ser $\rightarrow$ Pro substitution at the 102nd amino acid residue, located within an α helix of the enzyme. **(a)** Describe the mutation that causes this amino acid substitution, including the position of the mutation in the codon and whether the mutation is a transition or transversion. **(b)** Why does the substitution of proline for serine inactivate the enzyme?

6. Example 6.3, illustrating the experiments of Srb and Horowitz, shows the mutations *arg 4, arg 5, arg 6,* and *arg 7* at the same step in the biochemical pathway. Explain why these four mutations are not necessarily in the same gene even though they appear at the same step in the diagram.

7. Collins and Weissman (1984. *Progress in Nucleic Acid Research and Molecular Biology* 31:315–462) categorized the mutations causing β-thalassemias in humans into three classes: (1) chain-terminator (nonsense and frameshift) mutations, (2) defective promoter mutations, and (3) defective RNA processing mutations. All of the chain-terminator mutations cause β^0-thalassemias, and all of the defective promoter mutations cause β^+-thalassemias, whereas the defective RNA processing mutations cause β^0- and β^+-thalassemias. Explain why these observations are expected.

8. Mutations that disable the enzyme phenylalanine hydroxylase prevent phenylalanine from being converted into tyrosine (Figure 6.9). People who lack functional phenylalanine hydroxylase suffer from phenylketonuria (PKU), a genetic disorder characterized by severe mental retardation if left untreated. People with PKU tend to have much lighter skin and hair than their unaffected siblings due to a reduction in melanins. Using your understanding of how biochemical pathways are related to mutation,

Table 6.2 Information for problem 12: Growth of Mutant Strains of *Neurospora crassa* on Different Growth Media

Mutant Strain	Growth Medium				
	Minimal	Minimal & Methionine	Minimal & Homocysteine	Minimal & Cysteine	Minimal + Cystathionine
me-1	−	+	−	−	−
me-2	−	+	+	−	−
me-3	−	+	+	−	+
me-4	−	+	+	+	+

*A "+" indicates that the strain grew on the medium. A "−" indicates that the strain failed to grow on the medium.

Figure 6.15 Information for problem 13: Pathway for tryptophan synthesis in *Aspergillus nidulans*. (Adapted from from Hütter and de Moss. 1966. Enzyme analysis ot the tryptophan pathway in *Aspergillus nidulans. Genetics* 55:241–247. Reprinted by permission.)

explain why the level of melanins is reduced in people with PKU.

9. Using the information from your answer to the previous question, and assuming that the enzyme phenylalanine hydroxylase is completely disabled in people with PKU, why is the quantity of melanins reduced rather than eliminated entirely?

10. Using the information from the previous two questions and Figures 6.9 and 6.10, should individuals with phenylketonuria also have high levels of homogentisic acid in their urine, like individuals with alkaptonuria? Explain your answer in terms of the effects of the mutations that cause phenylketonuria and alkaptonuria.

11. In 1944, Tatum, Bonner, and Beadle (*Archives of Biochemistry* 3:477–478) reported that a strain of *Neurospora crassa* that was auxotrophic for tryptophan was unable to grow when anthranilic acid was supplied in the medium, but was able to grow when indole was supplied. A second strain that was auxotrophic for tryptophan was able to grow when either anthranilic acid or indole was supplied. Determine the order of anthranilic acid, indole, and tryptophan in the biochemical pathway.

12. In 1947, Horowitz (*Journal of Biological Chemistry* 171:255–264) presented the data summarized in Table 6.2 for mutant strains of *Neurospora crassa* that were auxotrophic for methionine. Determine the order of intermediates in the biochemical pathway, and indicate which step is affected by each mutation.

13. Figure 6.15 shows the pathway for tryptophan synthesis in *Aspergillus nidulans,* as described by Hütter and de Moss (1966. *Genetics* 55:241–247). Hütter and de Moss tested one normal strain (A160) and 14 mutant strains that were auxotrophic for tryptophan for the activity of each of the enzymes in Figure 6.15. Their results are shown in Table 6.3. Notice that strains C473, C485, C612, and C647 are active for only two of the five enzymes. How does this result contradict the one gene–one enzyme hypothesis? Propose a possible model that explains these results. How could your model be tested experimentally?

14. In 1995, Coleman et al. (*Proceedings of the National Academy of Sciences,* USA 92:6828–6831) described the molecular basis of a mutation affecting a seed storage protein in maize. The mutant protein was 21 amino acids longer on its N-terminal end than the normal protein. Analysis of the DNA sequence revealed a GCG→GTG substitution in the 21st codon of the mutant gene. The authors also concluded that the mutant protein remained attached to the endoplasmic reticulum in the cell instead of being transported to protein bodies as the normal protein was. Explain why the mutant protein was larger than the normal protein and why the mutant protein remained attached to the endoplasmic reticulum in the cell.

15. In 1995, Eisensmith and Woo (*Advances in Genetics* 32:199–271) reviewed the molecular genetics of mutations in the *phenylalanine hydroxylase (PAH)* gene that

Table 6.3 Information for problem 13: Enzyme Activity in Different Strains of *Aspergillus nidulans*

Strain	Enzyme				
	Anthranilate Synthetase	PR Transferase	PRA Isomerase	InGP Synthetase	Tryptophan Synthetase
A160	+	+	+	+	+
A21	−	+	+	+	+
A459	−	+	+	+	+
A622	−	+	+	+	+
B26	+	+	+	+	−
B479	+	+	+	+	−
C473	−	+	−	−	+
C485	−	+	−	−	+
C612	−	+	−	−	+
C647	−	+	−	−	+
D462	+	−	+	+	+
D554	+	−	+	+	+
E17	+	+	+	+	+
E542	+	+	+	+	+
E646	+	+	+	+	+

*A "+" indicates that enzyme activity was detected. A "−" indicates that no enzyme activity was detected.

cause PKU. This gene is very large, consisting of over 65,000 nucleotide pairs, most of which are in introns. The gene has 13 exons and 12 introns. The reading frame in the mature mRNA represents only a fraction of the entire gene, with 1356 nucleotides that encode 452 amino acids. The authors of this article compiled a list of all known mutations in the *PAH* gene. All but a few of these mutations were identified in people who had PKU as a result of these mutations. About 20% of all mutations appeared in exon 7 (which represents less than 20% of the reading frame), and 80% of the mutations were found in exons 5–12. Of the mutations located outside of these exons, less than 50% caused single amino acid substitutions. Instead, most of the mutations were frameshift mutations, deletions, mutations that altered intron removal, mutations that altered the initiation codon, or mutations that did not cause PKU but were detected simply as an alteration in DNA sequence. Within exons 5–12, however, most of the mutations were nucleotide substitutions that caused single amino acid substitutions. These authors also noted that the highest similarity in amino acid sequence among different species was for the portion of the protein encoded by exons 7–9. **(a)** Explain why the known mutations are clustered within a particular region of the gene and why many of the mutations outside of this region were not amino acid substitutions. **(b)** What do these results suggest about the active site of this enzyme?

FOR FURTHER READING

A concise textbook that covers genetics and biochemical pathways is **Woods, R. A. 1980. *Biochemical Genetics, 2nd ed.* London: Chapman and Hall.** The genetics and enzymology of melanin synthesis are reviewed in detail in **Prota, G. 1992. *Melanins and Melanogenesis.* San Diego: Academic Press.** The DNA sequence of the human β-globin gene was published by **Lawn, R. M., A. Efstratiadis, C. O'Connell, and T. Maniatis. 1980. The nucleotide sequence of the human β-globin gene. *Cell* 21:647–651.** An excellent review of the molecular biology and evolution of globin genes in humans is **Collins, F., and S. M. Weissman. 1984. The molecular genetics of human hemoglobin. *Progress in Nucleic Acid Research and Molecular Biology* 31:315–462.** Several chapters in the book **Handin, R. I., S. E. Lux, and T. P. Stossel, eds. 1995. *Blood: Principles and Practice of***

Hematology. Philadelphia: Lippincott, provide excellent detailed information on the structure and function of hemoglobin and the effects of mutations responsible for anemias, thalassemias, and faulty heme production. These chapters include **Lux, S. E. Introduction to anemias (pp. 1383–1398), Sassa, S., and A. Kappas. Disorders of heme production and catabolism. (pp. 1473–1524), Forget, B. G., and H. A. Pearson. Hemoglobin synthesis and the thalassemias. (pp. 1525–1590), Nagel, R. L. Disorders of hemoglobin function and stability. (pp. 1591–1644), and Platt, O. S. The sickle syndromes. (pp. 1645–1700)**. A review that summarizes the known functions of the *PAH* gene in humans and the numerous mutations that have been characterized at the DNA level is **Eisensmith, R. C., and S. L. C. Woo. 1995. Molecular genetics of phenylketonuria: From molecular anthropology to gene therapy.** *Advances in Genetics* **32:199–271**. Beadle and Ephrussi's pioneering work with biochemical pathways in *Drosophila melanogaster* was published in **Beadle, G. W., and B. Ephrussi. 1937. Development of eye colors in *Drosophila* Diffusible substances and their interrelations.** *Genetics* **22:76–86**. Beadle and Tatum's work in identifying auxo-trophic mutations and developing of the one gene–one enzyme hypothesis is presented in **Beadle, G. W., and E. L. Tatum. 1941. Genetic control of biochemical reactions in *Neurospora*.** *Proceedings of the National Academy of Sciences, USA* **27:499–506**. Srb and Horowitz's classic paper on *Neurospora* mutations in the arginine synthesis pathway is **Srb, A. M., and N. H. Horowitz. 1944.** *Journal of Biological Chemistry* **154:129–139**.

For additional reading, go to InfoTrac College Edition, your online research library at: http: www.infotrac-college.com/brookscole

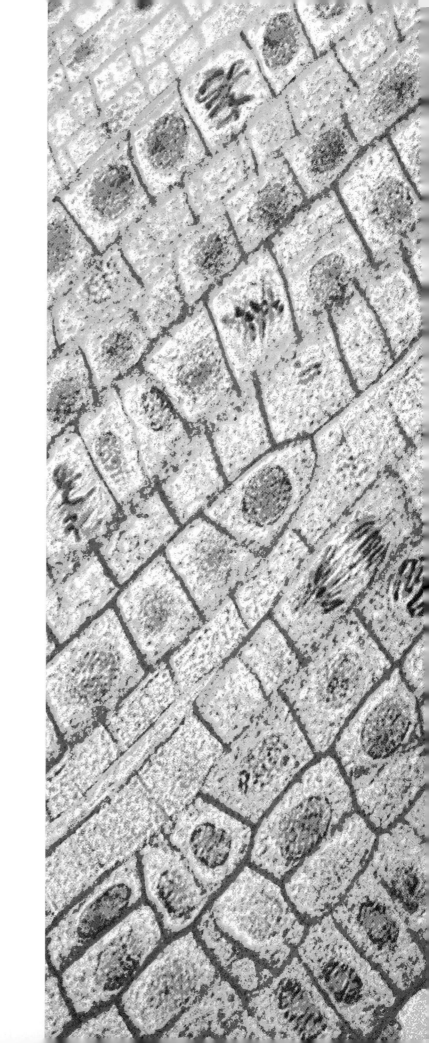

PART II

GENETICS OF CELLS

CHAPTER 7

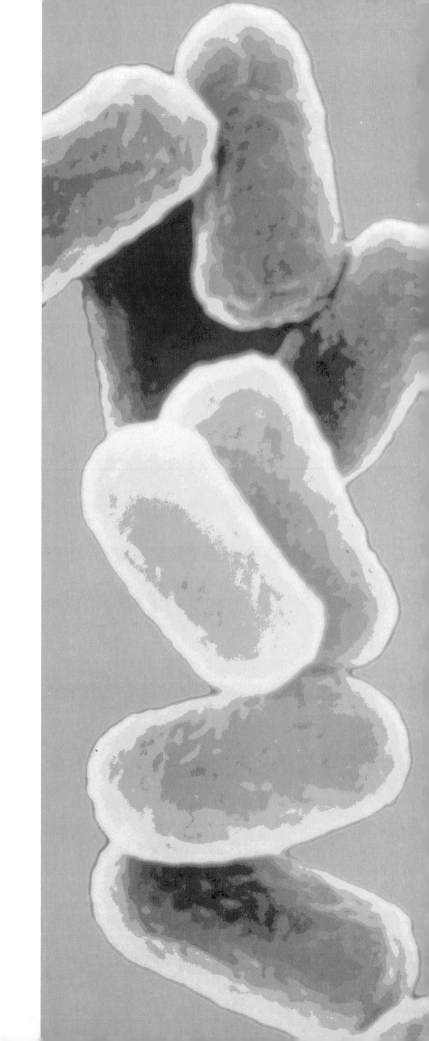

KEY CONCEPTS

Many important genetic concepts were discovered first in bacteria. Bacteria also serve as tools for practical applications of genetics.

~

Most bacterial genes are found on a circular DNA molecule called the bacterial chromosome.

~

DNA may be transferred from one bacterial cell to another, where the transferred DNA then recombines with the chromosomal DNA.

~

DNA transfer and recombination are used to map bacterial genes.

~

Determination of complete DNA sequences in bacterial species is the first step toward understanding the complete biology of those species.

BACTERIAL GENETICS

Bacterial genetics is enormously important—both academically and practically. Much of what we know about DNA structure and replication, gene expression, and mutation was discovered first in bacteria. Geneticists use bacteria as research tools to develop a better understanding of eukaryotic genetics through the application of bacteria-based recombinant DNA techniques. Bacteria are among the most important organisms in medicine and agriculture: they cause many diseases in humans and in the animals and plants we use for food. An understanding of bacterial genetics helps scientists and health care specialists control these diseases. Certain types of bacteria are important tools in the multibillion-dollar biotechnology industry. In this industry, bacteria serve as carriers of genetically engineered genes and provide many of the enzymes used in research. Researchers use bacteria to replicate important genes isolated from humans, plants, and animals. Valuable pharmaceutical products, such as human insulin and interferon, are produced commercially in genetically engineered bacteria. Certain bacteria are also used to transfer new genes into genetically engineered plants and animals.

In order to understand bacterial genetics, it is important to know how bacteria are cultured and manipulated in the laboratory. We will begin our discussion of

bacterial genetics with an introduction to bacterial culture and the methods researchers use to identify genetic differences in bacteria. We will then see how the bacterial DNA is organized in the cell and how bacterial cells divide. Our next topics will be the ways in which bacteria and their viruses transfer DNA from one cell to another and how researchers use DNA transfer to map the positions of genes on bacterial DNA. Finally, we will examine one of the most significant accomplishments in genetics research, the determination of the complete DNA sequence of a cellular organism, which was accomplished first in bacteria.

7.1 BACTERIAL CULTURE

In nature, bacteria thrive wherever they can find sufficient water and nutrients. They are present in the food we eat, the water we drink, and the air we breathe. They live inside of us and on our skin. They are the most ubiquitous organisms on earth.

In the laboratory, researchers usually culture bacteria in a medium (plural, *media*) that contains all the substances needed for the bacterial cells to grow and divide. The medium is either liquid, consisting of water and all

the nutrients needed for growth, or semisolid, consisting of liquid medium solidified with agar, a gel-like substance extracted from sea kelp (Figure 7.1).

Because of the ever present bacteria in our environment, all bacterial media must be sterilized (made free of living cells) to prevent unwanted microorganisms from growing in the medium. Researchers usually sterilize media with heat or filtration. For heat sterilization, researchers heat the medium to about 130°C (which kills all known microorganisms) in an **autoclave**, an instrument that heats the medium in a pressure chamber to prevent the medium from boiling at the sterilization temperature. For agar-solidified medium, the heat in the autoclave sterilizes the medium while at the same time melting the powdered agar into a liquid that dissolves in the liquid medium. The molten agar medium can be poured into sterile petri plates, where the agar solidifies into a gel as it cools. Bacteria can then be cultured on the surface of the gelled medium.

Filtration is used to sterilize media that contain heat-sensitive substances, such as certain antibiotics. Researchers force the liquid medium through a filter with pores that are too small for a bacterium to pass through. The filter traps all bacterial cells, sterilizing the filtered medium.

Because bacteria are present in the air around us, researchers keep containers with bacterial cultures closed, opening the containers under sterile conditions in a laminar flow hood, an instrument that forces sterile air over the cultures and blows any bacterial cells that drop from the researchers' hands or clothing away from the cultures. Researchers often use tools, such as spatulas and wire loops, to manipulate bacterial cultures. They sterilize these tools with a flame just before use to prevent contamination of cultures.

Bacterial Growth Media

Bacterial growth media consist of water, a carbon source (usually a sugar, such as glucose, sucrose, maltose, or lactose), minerals, and any amino acids or vitamins a particular type of bacterium needs to grow. Some media are made with extracts from yeast, animals, or plants that already contain all necessary items for growth. **Minimal media** are media that contain only the minimum number of substances that bacterial cells require for growth: water, a simple carbon source (usually glucose), and inorganic salts. Wild-type bacteria are able to synthesize all the other molecules they need (such as amino acids and nucleotides) from the substances present in the minimal medium.

Minimal media may be supplemented with various substances to identify cells that are auxotrophic for a particular substance. **Auxotrophy** is the inability of cells to grow and divide unless the medium is *supplemented* with

a Agar-solidified medium. Bacterial cells grow and divide on the gelled surface of the medium in petri plates.

b Liquid medium. Bacterial cells are suspended in the medium, where they move freely, grow, and divide.

Figure 7.1 Bacterial culture media.

a particular substance. For example, a bacterial strain that is auxotrophic for the amino acid arginine requires arginine in the growth medium in order for the cells to grow and divide. **Prototrophy** is the ability of cells to grow and divide when a particular substance is *absent* from the medium. For example, bacterial strains that grow and divide on medium without arginine are prototrophic for arginine. Auxotrophic mutant strains are usually designated by an abbreviation of the substance for which they are auxotrophic followed by a superscript minus sign. For example, an Arg⁻ mutant strain is auxotrophic for arginine. An Arg⁺ strain, on the other hand, is prototrophic for arginine.

Detection of Mutations and Preparation of Pure Cultures

We now need to define several important terms. We have defined some of these terms in previous chapters, but here we will discuss them in a bacterial context. An **allele** is a particular version of a gene represented by its nucleotide sequence. A mutation that changes the nucleotide sequence of a gene, even by one nucleotide, creates a different allele. The alleles most commonly found in nature are called **wild-type alleles**. Most wild-type alleles encode a functional product. Mutation of a wild-type allele creates a **mutant allele**. Most mutant alleles fail to encode a functional product.

There are many possible alleles for any one gene, because any of the hundreds to thousands of nucleotides in the gene may mutate. However, different alleles can cause the same observable result. For example, several different alleles in a single gene can cause a bacterial cell to be auxotrophic for arginine. We now need to reacquaint

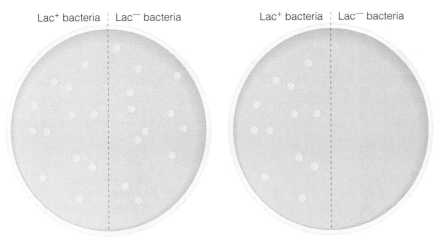

Lac⁺ bacteria | Lac⁻ bacteria   Lac⁺ bacteria | Lac⁻ bacteria

a Nonselective medium: contains glucose. Both Lac⁺ and Lac⁻ bacteria grow.

b Selective medium: lactose is the sole carbon Only Lac⁺ bacteria grow.

Figure 7.2 Selective and nonselective media.

ourselves with the difference between phenotype and genotype. The **phenotype** is the outward characteristic that can be observed or measured. Bacterial phenotypes are usually typeset in roman letters with the first letter capitalized. For example, the phenotype of a cell that is auxotrophic for methionine is Met⁻, whereas the phenotype of a cell that is prototrophic for methionine is Met⁺.

The **genotype** is the underlying genetic constitution that causes a particular phenotype. In bacteria, a gene is designated with an italicized abbreviation of the gene's name with a superscript number, letter, or other symbol to designate the allele. For example, one of the several genes that cause auxotrophy for methionine in *E. coli* is called *metB*. Mutant alleles in that gene that confer auxotrophy for methionine are designated as *metB⁻* alleles, whereas wild-type alleles that confer prototrophy for methionine are designated *metB⁺*. By italicizing genotypic designations, but not phenotypic designations, the distinction between genotype and phenotype is apparent.

In addition to **auxotrophic mutations** (mutations that cause auxotrophy), three other classes of mutant alleles are often studied in bacterial genetics. **Carbon source mutations** confer dependency on certain carbon sources for growth. A carbon source mutant strain is unable to grow on a particular type of carbohydrate on which wild-type strains can grow. Carbon source mutant phenotypes are designated by an abbreviation that denotes the carbon source on which the mutant strain cannot grow. For example, a Lac⁻ mutant strain of *E. coli* cannot grow when lactose is the only carbon source, whereas wild-type Lac⁺ bacteria are able to grow when lactose is the sole carbon source.

Antibiotic-resistant strains can grow on medium supplemented with an **antibiotic**, a substance that nor-

mally prevents or inhibits bacterial growth. Penicillin, streptomycin, erythromycin, and tetracycline are examples of common antibiotics. Phenotypes of antibiotic-resistant strains are designated by an abbreviation of the antibiotic followed by a superscript "r" to indicate resistance. The corresponding antibiotic-sensitive types are designated by the superscript "s" to indicate sensitivity. For example, a Tetʳ mutant strain can grow on medium supplemented with tetracycline, whereas a Tetˢ strain cannot.

Temperature-sensitive mutations permit cells to grow and divide at the usual culture temperature of 37°C but prevent growth and division when the temperature is raised. In this case, no supplements are added to the medium. The cultures are simply placed in an incubator with a higher temperature to identify the mutant phenotype by the failure of cultures to grow at the higher temperature.

With respect to a particular mutant phenotype, a bacterial medium may be considered selective or nonselective. A **selective medium** is one on which the mutant bacterial strain can be distinguished from the wild-type strain. Wild-type and mutant strains cannot be distinguished on a **nonselective medium**. For example, Lac⁻ mutant strains cannot be distinguished from the normal Lac⁺ strains when both strains are grown on medium with glucose as a carbon source, because both strains are able to utilize glucose. However, if lactose is the sole carbon source in the medium, Lac⁺ cells grow and divide, whereas Lac⁻ cells do not. The glucose medium is nonselective, whereas lactose medium is selective for Lac⁻ mutant strains (Figure 7.2). In the case of an antibiotic-resistant strain, the medium supplemented with the antibiotic is selective, whereas the antibiotic-free medium is nonselective.

Single-celled derived colonies

Figure 7.3 Photograph of a bacterial culture plate showing colonies derived from single cells. Each colony consists of bacterial cells that are the progeny of a single cell placed at that spot on the medium surface.

Often it is necessary to grow a bacterial culture from a single cell to ensure that all the progeny cells are genetically identical. In a liquid medium, all cells are mixed together and divide into many more cells. While it is very difficult to distinguish the phenotypes of individual cells in mixed culture, it is simple to identify phenotypes in the progeny of a single cell. Liquid bacterial cultures can be diluted and spread on the surface of an agar medium so that individual cells are separated from one another. If a single cell is able to grow and divide on the medium, it produces a clump of genetically identical cells called a **colony**. Figure 7.3 shows several bacterial colonies, each derived from a single cell.

Many types of bacteria are identified by their inability to grow on a selective medium. If they cannot grow on the selective medium that identifies them, how can pure cultures of the mutant types be recovered and grown on a nonselective medium? The procedure for recovering such mutant colonies is called **replica plating**, illustrated in Figure 7.4. The colonies are grown first on nonselective agar-solidified medium in a master plate. A researcher then places a sterile piece of velvet on the surface of the medium, picking up bacteria from each colony in the place where they were located on the master plate. The researcher then presses the velvet onto the surface of a plate with selective medium (called a replica plate), leaving bacteria from each colony on the replica plate in the same positions they occupied on the master plate. Phenotypic differences can be identified in colonies that fail to grow on the selective medium. For example, a Lac⁻ mutant colony can be identified as one that grows on a medium with glucose (a nonselective medium) but fails to grow on a medium with lactose as the sole carbon source (a selective medium).

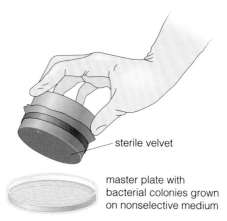

sterile velvet

master plate with bacterial colonies grown on nonselective medium

a A sterile piece of velvet is pressed on a master plate with bacterial colonies growing on nonselective medium. The velvet picks up cells from the colonies.

replica plate with selective medium

b The velvet is pressed onto a replica plate with selective medium so that it places cells on the plate in the same positions as the colonies on the master plate.

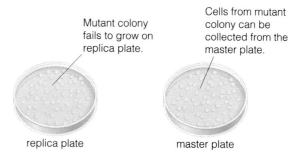

Mutant colony fails to grow on replica plate.

Cells from mutant colony can be collected from the master plate.

replica plate   master plate

c The cells on the replica plate are allowed to grow into colonies. Any colony present on the master plate but missing from the replica plate has a mutation that prevents its growth on the selective medium. Cells from the mutant colony can then be collected from the master plate.

Figure 7.4 Replica plating.

Colonies that grow on a selective medium can be recovered and propagated simply by taking them directly from the selective medium and culturing them. But what about auxotrophic or carbon source mutant colonies that fail to grow on a selective medium—how can they be recovered? Let's use a Lac⁻ mutant colony as an example.

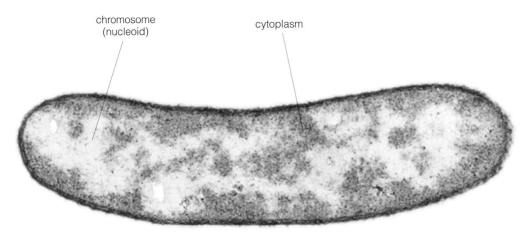

Figure 7.5 Electron micrograph of a cross section of a bacterial cell. The chromosome occupies about one-third of the cell. (Photo courtesy of G. Cohen-Bazire.)

A Lac⁻ mutant colony is identified as a colony that is present on the nonselective glucose plate but is absent on a replica plate of selective lactose medium. Once the mutant colony has been identified by its absence on the selective medium, the researcher can collect and culture cells from the corresponding colony on the original nonselective master plate, as illustrated in Figure 7.4c.

~

Mutant types of bacteria are typically identified by differences in their ability to grow and divide on selective media.

7.2 STRUCTURE OF BACTERIAL CHROMOSOMES

Each bacterial cell contains a large circular DNA molecule called the **bacterial chromosome**. The bacterial chromosome is also called the **nucleoid** because it occupies a central region in the cell, much as the nucleus of a eukaryotic cell does, but is not surrounded by a nuclear membrane as are eukaryotic nuclei (Figure 7.5). The bacterial chromosome contains most of the cell's genes. Bacterial cells may also carry one or more **plasmids**, small pieces of DNA that contain a few genes. Plasmids are usually circular and are much smaller than the chromosomal DNA (Figure 7.6). Typically, plasmids are separate from the chromosome, but in some cases a plasmid may integrate into the chromosomal DNA and be replicated as part of it. Plasmids that are integrated into the chromosome are called **episomes**.

Genes in plasmids are usually not essential for survival of the cell, so most bacterial cells can live without a plasmid. However, the genes on a plasmid may confer a significant advantage to a bacterial cell. For example, some genes that confer resistance to antibiotics are located in plasmids. Bacterial cells with a plasmid that contains an antibiotic-resistance gene may survive and

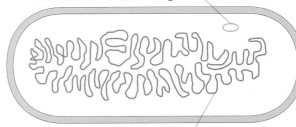

a Diagram of the bacterial chromosome and a plasmid in a cell

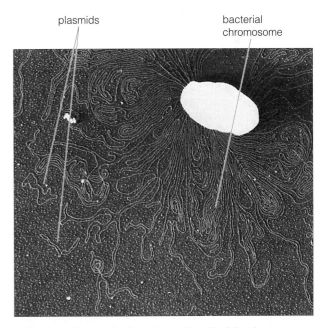

b Electron micrograph of a ruptured *E. coli* cell that has released its DNA.

Figure 7.6 DNA molecules in bacterial cells. (Photo courtesy of H. Potter and D. Dressler.)

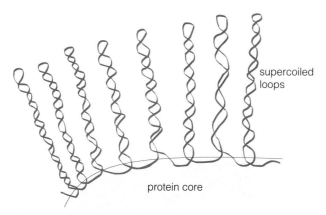

supercoiled loops

protein core

The bacterial chromosome consists of a central protein core from which loops of supercoiled DNA extend.

Figure 7.7 Bacterial chromosome structure.

reproduce in the presence of an antibiotic, whereas the cells lacking the plasmid fail to grow and divide.

The circular chromosomal DNA in *E. coli* is 1530 micrometers (μm) in circumference and must be squeezed into a cell that is about 2 μm in diameter. Without some sort of organization, a molecule of DNA this large would become hopelessly tangled. A central core of proteins stabilizes the chromosomal DNA and holds it in an organized form. Supercoiled loops of DNA extend out from the protein core, as illustrated in Figure 7.7. Even though the DNA in a bacterial chromosome is condensed and supercoiled, it is in constant contact with the cytoplasm, a feature that makes simultaneous transcription and translation of genes possible (see Figure 3.3).

Bacterial cells divide through a process called **binary fission**. The chromosomal DNA is attached at one point to the inner surface of the cell membrane (Figure 7.8a). When the chromosomal DNA replicates, each of the two daughter molecules remains attached to the bacterial membrane (Figures 7.8b and c). As the cell grows in size in preparation for division, the growing membrane pulls the two DNA molecules apart (Figures 7.8d and e). The cell wall grows inward at the cell's midpoint, forming two compartments, each containing one of the DNA molecules (Figure 7.8f). The two compartments separate from one another, forming two cells, each with chromosomes that are identical in DNA sequence (Figure 7.8g). Figure 7.9 shows an electron micrograph of a bacterial cell in the process of binary fission.

~

The DNA molecule of the prokaryotic chromosome is circular and is stabilized by proteins that hold it in an organized conformation. Prokaryotic cells divide by binary fission, duplicating the chromosomal DNA once for each cell division.

7.3 DNA TRANSFER AND RECOMBINATION

Most people think of bacteria as asexual organisms. Bacterial cells usually reproduce asexually through binary fission, in which the DNA in a single cell is duplicated, then the cell divides to produce two genetically identical cells. Bacterial cells, however, have ways of acquiring DNA from other cells. They can take up DNA directly from their environment, or they can receive DNA from bacterial viruses that infect the cells. In some species, two bacterial cells may connect to one another with a tube called a conjugation bridge, allowing one cell to transfer some of its DNA to the other. Technically, these processes do not constitute true sexual reproduction, but they do accomplish the same purpose: recombination of DNA and the genes contained in that DNA. **Recombination** is the process through which cells acquire DNA from another cell then exchange the introduced DNA with a similar sequence in their own DNA.

Frederick Griffith was one of the first scientists to observe the effects of recombination in bacteria. His classic experiments with R and S cells, which we discussed in section 2.1, showed that a substance was transferred from S cells to R cells that transformed the R cells into S cells. Avery, MacLeod, and McCarty's experiments, also discussed in section 2.1, demonstrated that the substance was DNA. Some of the DNA from S cells had entered the R cells, where it recombined with the DNA in the R cells. Two years after Avery, MacLeod, and McCarty published their results, Joshua Lederberg and Edward Tatum reported some important experiments demonstrating that recombination, rather than mutation, explained the appearance of new bacterial strains when two strains were mixed. Let's examine their experiments to see how they arrived at this conclusion.

Example 7.1 Experiments demonstrating recombination of bacterial genes.

In 1946, Lederberg and Tatum (*Cold Spring Harbor Symposium on Quantitative Biology* 11:113–114, and *Nature* 158:558–559) published the results of experiments designed to determine whether mutation or recombination explained the appearance of new bacterial strains when cultures were mixed. Lederberg and Tatum used two mutant *E. coli* strains, each with three different auxotrophic mutations. One strain, called Y-10, was auxotrophic for threonine, leucine, and thiamin. The other, called Y-24, was auxotrophic for biotin, phenylalanine, and cysteine. They mixed the two strains together and grew the mixed culture in nonselective liquid medium. They also grew pure (unmixed) cultures

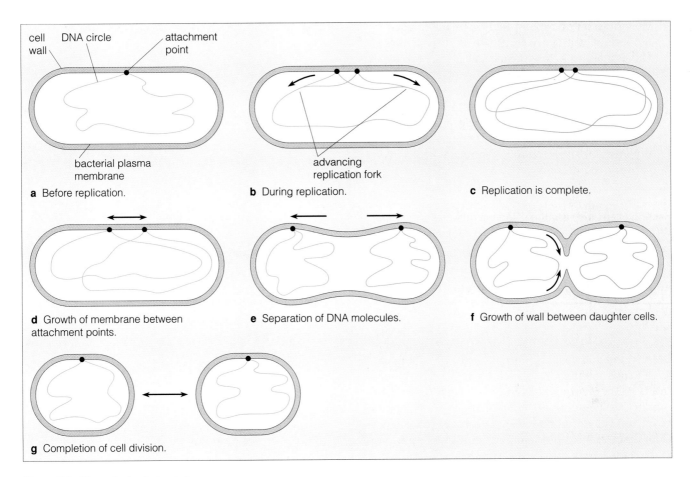

cell wall   DNA circle   attachment point

bacterial plasma membrane

a Before replication.

advancing replication fork

b During replication.

c Replication is complete.

d Growth of membrane between attachment points.

e Separation of DNA molecules.

f Growth of wall between daughter cells.

g Completion of cell division.

Figure 7.8 Binary fission in bacteria.

of both strains at the same time in nonselective liquid medium as control experiments. Following culture on nonselective media, they transferred cells from the mixed and pure cultures to various types of selective media, including minimal medium and media with each of the six substances in all possible combinations. As they stated in one of their papers (*Nature* 158:558–559).

> The only new types found in "pure" cultures of the individual mutants were occasional forms which had reverted for a single factor, giving strains which required only two of the original three substances. In mixed cultures, however, a variety of types has been found. These include wild-type strains with no growth factor deficiencies and single mutant types requiring only thiamin or phenylalanine. In addition, double requirement types have been obtained, including strains deficient in the synthesis of biotin and leucine, biotin and threonine, and biotin and thiamin, respectively.

Problem: Lederberg and Tatum concluded from these results that recombination, rather than mutation, was responsible for the new types recovered from mixed culture. What aspects of their results favor this conclusion?

Solution: Lederberg and Tatum identified several strains from the mixed culture that were capable of growing on minimal medium, which required changes in three genes. Also, they identified strains from the mixed culture that were auxotrophic for only one or two substances. Among those auxotrophic for two substances were strains auxotrophic for biotin and leucine, biotin and threonine, and biotin and thiamin. These three strains also required changes in three genes. While mutation might explain such changes, the probability of separate mutations in each of three genes was quite low. Nonetheless, Lederberg and Tatum tested the possibility that mutation might explain their results by

including control experiments where the original triple-mutant strains were grown as pure cultures. Had mutations in three genes caused the new strains to arise in the mixed culture, then similar strains should have arisen in the pure cultures. However, the only alterations they found in the pure cultures were changes in one gene, probably by mutation. Recombination is therefore the best explanation for the multiple gene changes observed in the mixed cultures.

Bacterial cells may acquire DNA from other cells, and that DNA may recombine with the bacterial chromosome in the recipient cell.

Recombination by Single-Strand Displacement

Looking back on Griffith's experiments, Avery, MacLeod, and McCarty's experiments, and Lederberg and Tatum's experiments, we clearly see that bacterial cells can acquire DNA from other cells. The simple acquisition of DNA, however, is insufficient to create a new recombinant strain. Pieces of acquired DNA do not replicate on their own in a recipient cell. To replicate, the new DNA must be incorporated into the recipient cell's chromosome. Once incorporated, the acquired DNA then replicates as if it were part of the original chromosome, creating a recombinant strain of bacterial cells.

Bacterial DNA may enter the cell as a single strand or as a double strand. Let's look first at how a single strand of DNA recombines with DNA in the chromosome. When DNA enters the cells of certain bacterial species, the cell degrades one of the strands, leaving the introduced DNA as a single-stranded fragment. The single-stranded fragment then invades the chromosomal DNA, where it encounters **homology**, or similarity between two DNA sequences. For two DNA sequences to be **homologous**, their nucleotide sequences must be highly similar but not necessarily identical.

The single-stranded DNA invades the chromosomal DNA and displaces a homologous strand of chromosomal DNA. RecA, a protein that has affinity for single-stranded DNA, binds to the 3' end of the single-stranded DNA fragment and facilitates pairing of the free 3' end with the complementary sequence in the chromosomal DNA (Figure 7.10a). Single-strand displacement creates **heteroduplex DNA**, double-stranded DNA in which a strand from one DNA molecule pairs with a strand from a different DNA molecule. It also creates a **D-loop**, a displaced chromosomal strand that loops out as single-stranded DNA (Figure 7.10b). Once the pairing has been initiated, the single-stranded DNA fragment continues to form base pairs with the chromosomal DNA along its entire length. While the invading DNA is homologous,

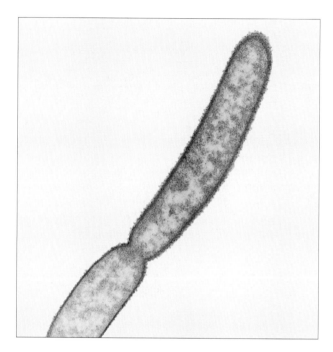

Figure 7.9 Electron micrograph of a bacterium. (Photo courtesy of W. M. Hess.)

it is not necessarily identical to the DNA it displaces. At any location where there is a difference between the nucleotide sequence of the invading DNA and the DNA it displaced, a nucleotide mismatch will occur in the heteroduplex DNA (Figure 7.10c). Enzymes degrade the D-loop, and the invading DNA is connected to the chromosomal DNA, forming a continuous DNA molecule (Figure 7.10d). If nucleotide pair mismatches are left uncorrected, the sequence of the introduced strand will be faithfully replicated when this strand is used as a template during DNA replication. One of the daughter cells will carry the sequence of the introduced DNA, the other cell the original sequence. The introduced DNA has now recombined with the chromosomal DNA (Figure 7.10e).

Let's now examine some experimental evidence of single-strand displacement in bacteria of the genus *Pneumococcus*.

Example 7.2 Single-strand displacement in bacteria.

In 1962, Lacks (*Journal of Molecular Biology* 5:119–131) published the results of experiments that revealed the stage at which introduced DNA was made single stranded in *Pneumococcus*. Lacks cultured cells for many cell generations in a medium with radioactive phosphorus (^{32}P) so that all of the DNA in these cells contained radioactive phosphorus. He then extracted the radioactive DNA from the cells and added it to cultures of cells whose DNA was

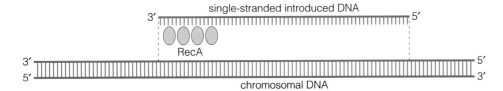

single-stranded introduced DNA

RecA

chromosomal DNA

a RecA binds to the single-stranded DNA, and this DNA finds its homologous sequence in the chromosome.

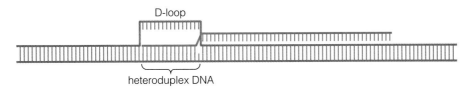

D-loop

heteroduplex DNA

b The 3' end of the single-stranded DNA molecule displaces the corresponding strand of chromosomal DNA, forming heteroduplex DNA and a D-loop.

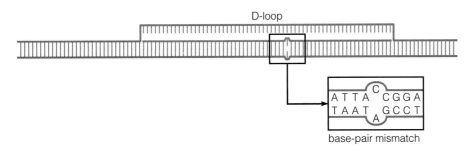

D-loop

```
      C
ATTA   CGGA
TAAT   GCCT
      A
```

base-pair mismatch

c The invading strand s nucleotides pair with those in the chromosome except where the sequences differ.

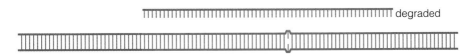

degraded

d The strands are broken and resealed to form intact double-stranded DNA.

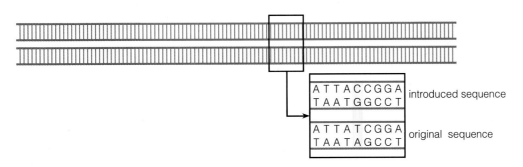

```
ATTACCGGA
TAATGGCCT   introduced sequence
```

```
ATTATCGGA
TAATAGCCT   original sequence
```

e Replication of the DNA produces molecules that differ in DNA sequence.

Figure 7.10 Single-strand recombination.

not radioactive. Immediately after adding the radioactive DNA, he isolated some of the cells and analyzed the DNA that they contained. He found that half of the radioactivity in these cells was in a long single-stranded DNA molecule; the remaining half of the radioactivity was in oligonucleotides (short single-stranded DNA molecules) and inorganic phosphates. When he examined the cells that remained in the culture, he found that the radioactive DNA had entered the chromosomes. In 1964, Fox and Allen (*Proceedings of the National Academy of Sciences, USA* 52:412–419) reported the results of experiments in which they added radioactive DNA to bacterial cell cultures, then extracted the DNA from the cells before they had time to replicate their DNA and divide. They found that part of the chromosomal DNA contained one strand that was radioactive and one strand that was not. They also found that the radioactive DNA segment was part of a continuous DNA strand with nonradioactive DNA on either side of the radioactive segment.

Problem: What do these results suggest about the mechanism of recombination in *Pneumococcus*?

Solution: Lacks's results show that one strand of DNA is degraded upon entry into the cell, leaving a single strand of introduced DNA in the cell. Lacks also found that the single strand of DNA incorporates into the chromosomal DNA. Fox and Allen showed that the introduced single strand of DNA incorporates into the chromosomal DNA, where it becomes part of a heteroduplex segment. This observation suggests that the introduced strand had displaced a strand in the chromosomal DNA. Fox and Allen's observation that the radioactive DNA was part of a continuous DNA strand in the chromosome shows that the single strand is fully incorporated as part of the chromosome.

Recombination of Double-Stranded DNA

In some bacterial species, introduced DNA remains double stranded following entry into the cell. As shown in Figure 7.11a, recombination is facilitated by RecA and a complex of proteins called RecBCD. RecBCD has the ability to unwind double-stranded DNA at the end of a linear molecule, such as an introduced double-stranded DNA fragment. RecA binds to the 3' end of the single-stranded DNA, which then invades double-stranded chromosomal DNA, where it encounters homology. The 3' end of the invading molecule enters the chromosomal DNA, forming a D-loop. A region of single-stranded DNA in the introduced molecule may now form base pairs with the single-stranded DNA from the D-loop,

making two double-stranded heteroduplex molecules connected where the two DNA strands are exchanged (Figure 7.11b). This point of exchange is called a **Holliday junction**, named after Robin Holliday, who first described the structure. The Holliday junction migrates along the two molecules, forming heteroduplex DNA in both molecules as they exchange strands (Figure 7.11c). Any differences in nucleotide sequence between the introduced DNA and the chromosomal DNA appear as base-pair mismatches. When the invading strand differs from the original strand, recombination takes place as in single-strand displacement.

~

A single strand of DNA that has been introduced into a bacterial cell may recombine with the chromosomal DNA by displacing a homologous strand and incorporating itself into the chromosomal DNA.

Transfer Mechanisms

In order to recombine with chromosomal DNA, a DNA fragment must enter a bacterial cell. Although bacterial cells are generally resistant to the introduction of foreign DNA, there are three mechanisms that permit bacterial cells to transfer DNA from one cell to another:

1. *Transformation*: the direct uptake of DNA from a cell's environment.

2. *Conjugation*: the direct transfer of DNA from one bacterial cell to another.

3. *Transduction*: the indirect transfer of DNA from one bacterial cell to another via a virus.

Transformation permits the entry of DNA from any species, but in most cases, only DNA from the same bacterial species can recombine with chromosomal DNA. Most other types of DNA are destroyed by the cell. Conjugation and transduction transfer DNA between cells of the same species. In these ways, bacterial cells of the same species can recombine genes while protecting themselves from DNA of other species.

In the following sections, we will discuss transformation, conjugation, and transduction in detail.

7.4 TRANSFORMATION

Bacterial cells can take up DNA from their surroundings through **transformation**, the transfer of DNA into a bacterial cell from the cell's environment. Griffith's experiments and Avery, MacLeod, and McCarty's experiments described in section 2.1 are examples of transformation. In Griffith's experiments, the heat-killed virulent S cells ruptured, releasing their DNA into their surroundings. Some of the live avirulent R cells took up this S-cell DNA,

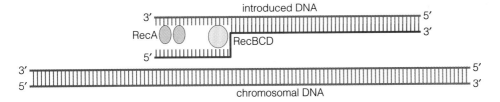

a An introduced double-stranded DNA molecule is unwound by RecBCD. RecA binds to the single-stranded DNA, where it will assist in finding the homologous sequence in the chromosomal DNA.

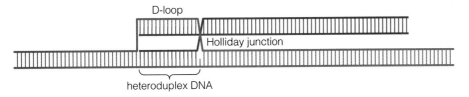

b One strand from the introduced molecule displaces its corresponding strand in the chromosome, forming heteroduplex DNA. The other strand of the introduced DNA pairs with the D-loop. A Holliday junction forms at the site of strand exchange.

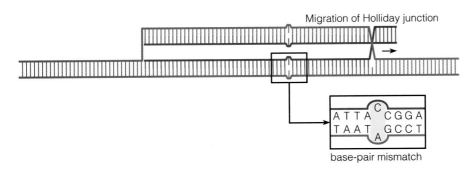

c The Holliday junction migrates, lengthening the heteroduplex DNA, which may contain mismatches.

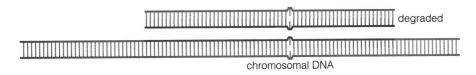

d The strands are broken and resealed to form intact double-stranded DNA in the chromosome.

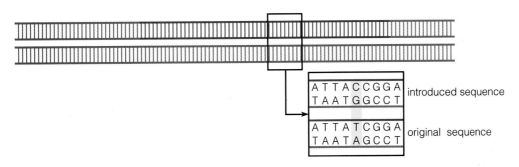

e Replication of the DNA produces daughter cells differing in DNA sequence where there were mismatches.

Figure 7.11 Double-strand recombination.

and the S-type DNA recombined with the chromosomal DNA in the R cells, converting them into S cells. The same thing happened in Avery, MacLeod, and McCarty's experiments: They added DNA extracted from S cells to live R cells. Some of the R cells took up the DNA, and recombination changed the R cells into S cells.

In nature, transformation can take place when DNA from a ruptured bacterial cell is taken up by another bacterial cell. If the introduced DNA is homologous with a portion of the chromosomal DNA in the recipient cell, the introduced DNA can recombine with chromosomal DNA. Transformation is a rare event both in nature and in the laboratory. Some species of bacteria do not take up DNA from their environment and cannot be transformed. In other species, the rate of transformation can be increased by treatments such as exposure to high concentrations of calcium ions, which increases the permeability of the bacterial cell to DNA. Cells that have been treated to increase their ability to take up DNA are called **competent cells**.

Plasmid introduction represents a type of transformation that does not require recombination. Plasmids are closed circular molecules of DNA that must be introduced as a complete and intact molecule in order to function in the cell. They enter the cell as double-stranded DNA and do not invade the chromosomal DNA. Because a plasmid contains its own origin of replication, it can replicate independently as a separate entity from the chromosomal DNA. Some plasmids replicate so profusely that a single cell may contain up to hundreds of copies of the plasmid. Other plasmids do not replicate as readily and may be present in only one or two copies per cell. Because plasmid DNA does not invade the chromosomal DNA, there is no need for homology between the plasmid and the chromosomal DNA. Typically, cells that are capable of plasmid transformation must be treated to make them competent. *E. coli* is an example of a species that is capable of plasmid transformation when its cells are competent.

Bacterial species that are capable of natural transformation can be used for transformation-based gene mapping. **Gene mapping** in bacteria is the process of determining the locations of genes in the chromosome. In the case of transformation, cells of one strain containing certain genes can be ruptured and the DNA isolated. Competent cells of a second strain that carries different alleles of these genes are then incubated in the presence of excess DNA from the first strain. The DNA from the first strain is usually fragmented at random, so random pieces of DNA enter the competent cells. The rate of transformation for any particular gene can be determined by counting the number of cells transformed for that gene and dividing by the total number of cells. Transformation of a cell by two or more genes is called **cotransformation**. The probability that a cell will be cotransformed by two

genes on two different fragments is the product of their individual frequencies of transformation. However, if two genes are located close to one another, some DNA fragments should contain both genes, and the observed frequency of cotransformation should therefore be higher than the predicted frequency. The closer the two genes are, the more likely they are to be found on the same fragment and to be cotransformed. Consequently, the rate of cotransformation is inversely related to the distance between two genes. The following example shows how genes can be mapped by measuring cotransformation rates.

Example 7.3 Transformation-based gene mapping in *Bacillus subtilis*.

In 1963, Yoshikawa and Sueoka (*Proceedings of the National Academy of Sciences, USA* 49:559–566) published the results of transformation-based gene mapping experiments in *Bacillus subtilis*. In one set of experiments, they added DNA from a wild-type Ile⁺ Met⁺ strain to cultures of an auxotrophic Ile⁻ Met⁻ strain. Their results are summarized in Table 7.1.

Problem: What aspects of these data suggest that the genes for methionine and isoleucine auxotrophy are close to one another on the DNA?

Solution: The frequency of Ile⁺ & Met⁺ cotransformation should be the product of the individual transformation frequencies if the genes are not located close to each other on the DNA. The expected frequencies are calculated in Table 7.2 for each of the DNA concentrations. In each case, the expected frequency of Ile⁺ & Met⁺ cotransformation is less than one per ten million (10^{-7}). However, the observed frequencies were more than 1000 times higher than the expected frequencies indicating that the two genes have close proximity to one another.

Transformation is the uptake of DNA into a bacterial cell from the cell's surroundings. Once inside the cell, the introduced DNA can recombine with the chromosomal DNA. Recombination can be used to map genes on the chromosome.

7.5 CONJUGATION

The results of Lederberg and Tatum's experiments, summarized in Example 7.1, could be explained by transformation. But, as mentioned earlier, transformation is a rare

Table 7.1 Information for Example 7.3: Frequencies of Transformed Cells in Different *Bacillus subtilis* Strains

Concentration of DNA (μg/ml)	Strain		
	Ile⁺ Only	Met⁺ Only	Ile⁺ & Met⁺
0.04	158×10^{-6}	162×10^{-6}	27×10^{-6}
0.008	106×10^{-6}	100×10^{-6}	18×10^{-6}
0.004	48×10^{-6}	52×10^{-6}	9×10^{-6}
0.000	0	0	0

Table 7.2 Solution to Example 7.3: Expected Frequencies of Ile⁺ & Met⁺ Cotransformation

Concentration of DNA (μg/ml)	Expected Frequency if Genes are not on the Same Fragment
0.04	$158 \times 10^{-6} \times 162 \times 10^{-6} = 2.56 \times 10^{-8}$
0.008	$106 \times 10^{-6} \times 100 \times 10^{-6} = 1.06 \times 10^{-8}$
0.004	$48 \times 10^{-6} \times 52 \times 10^{-6} = 0.25 \times 10^{-8}$

event and requires cell rupture to release DNA. The following example describes experiments in which two bacterial cultures were grown in the same liquid medium but kept out of contact by a filter, thereby allowing DNA to pass across the filter but preventing the bacterial cells on either side of it from actually contacting each other. These experiments permitted researchers to distinguish between transformation and other methods of DNA transfer.

Example 7.4 U-tube experiments.

In 1950, Davis (*Journal of Bacteriology* 60:507–508) reported the results of experiments designed to test whether or not cell contact was required for recombination of the type described by Lederberg and Tatum (see Example 7.1). Davis used two strains of *E. coli*: Y-10, which is auxotrophic for threonine, leucine, and thiamin, and 58-161, which is auxotrophic for methionine. Davis constructed a U-tube apparatus, diagrammed in Figure 7.12, that had a glass-disk filter separating each half of the U tube. The pores in the filter were too small for bacterial cells to pass through but large enough to allow DNA molecules to pass freely. Davis placed yeast-extract liquid medium on both sides of the U tube and inoculated one side with the Y-10 strain and the other with the 58-161 strain. He then applied alternating suction and pressure on one side of the tube so that about half of the medium crossed the filter with each suction/pressure cycle at three cycles per hour. For control experiments, he inoculated the same liquid medium in test tubes with Y-10 only, 58-161 only, and a mixture of Y-10 and 58-161. After 4 hours, he plated samples of the cultures from each side of the U tube on minimal medium. He also plated the pure cultures and the mixed culture from the control experiments on minimal medium. The mixed culture from a test tube yielded 58 colonies

on minimal medium, whereas all other cultures yielded no colonies.

Problem: **(a)** Explain why Davis used control experiments in this example. **(b)** What conclusions can be drawn from the results of his experiments?

Solution: **(a)** The experiments conducted with pure and mixed strains in test tubes are the control experiments and are essentially the same experiments conducted by Lederberg and Tatum. The results from these control experiments reinforce the conclusions drawn from the U-tube experiment—namely, that cell contact is required for DNA transfer. For example, the presence of prototrophs in the mixed culture demonstrated that at least one strain transferred DNA to the other when the strains were in contact. Had no prototrophs arisen in the mixed culture, the conclusion that cell contact was required for recombination could not be made. **(b)** These experiments demonstrated that bacterial cell contact was required for DNA transfer, and that transformation could not explain recombination in *E. coli*. When the cells were not in contact, there was no DNA transfer, even when the liquid medium in which the strains were cultured was moved back and forth across the filter.

The F Factor

When viewed under a microscope, *E. coli*, as well as several other bacterial species, forms **conjugation bridges**, microscopic tubes that connect two cells, as shown in Figures 7.13 and 7.14. Through **conjugation**, one cell transfers DNA directly to the other cell. In *E. coli*, the transferred DNA is usually, though not always, located on a circular plasmid called the **F factor**, for "fertility" factor. The F factor is a circular molecule that usually remains separate from the chromosome as a plasmid but is also able to insert itself into the chromosomal DNA as an episome.

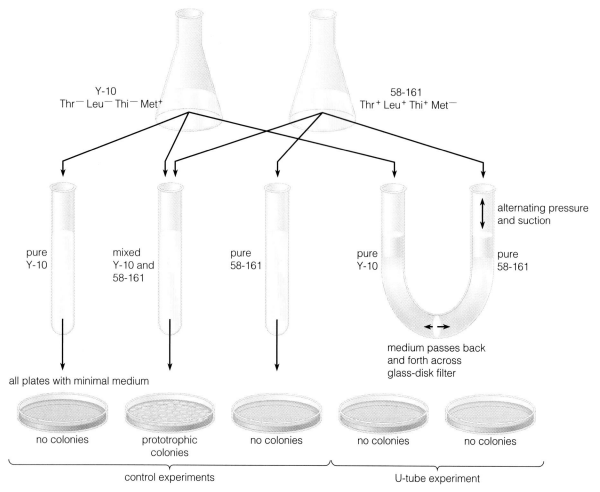

Figure 7.12 Davis's U-tube experiments. Only the cells from the mixed culture produced prototrophic colonies.

The F factor contains genes that regulate conjugation and transfer of DNA from one cell to another. A cell that has an F factor is designated F$^+$, whereas cells without an F factor are designated F$^-$. Cells transfer DNA in one direction only: An F$^+$ cell conjugates only with an F$^-$ cell and transfers a copy of the F factor to it. F$^+$ cells have protrusions on the surface of the cell, called **pili** (singular, *pilus*). When a pilus comes in contact with an F$^-$ cell, the pilus develops into a conjugation bridge between the two cells. The pili on the F$^+$ cell and the conjugation bridge between two E. coli cells can easily be seen in the micrograph in Figure 7.13.

The F$^+$ cell is called the donor cell, and the F$^-$ cell the recipient cell (Figure 7.14a). Once the conjugation bridge is established between the two cells, the F factor initiates σ-mode (rolling circle) replication and the nicked strand enters the conjugation bridge while being replicated (Figure 7.14b). The F factor undergoes a single round of replication, producing a linear copy of the F factor that moves through the conjugation bridge, and leaving a circular copy in the F$^+$ donor cell (Figure 7.14c). The linear

F factor arranges itself into a circle once it has entered the recipient cell, making the recipient cell F$^+$ (Figure 7.14d). The entire process requires only a few minutes. Both cells are now F$^+$ cells and can act as donor cells. When F$^+$ cells divide through binary fission, the F factor replicates by θ-mode replication, and each cell receives one copy of the F factor.

The F factor is not very prevalent among *E. coli* cells in nature, probably because concentrated populations, where many cells of a pure *E. coli* culture are in close contact, are rare in nature. Thus, while conjugation can and does occur naturally, it is a rare event. On the other hand, laboratory cultures of *E. coli* are usually pure and concentrated. The cells are in frequent contact with one another, permitting frequent conjugation. If a small number of F$^+$ cells are added to an F$^-$ culture, the F$^+$ cells rapidly convert all F$^-$ cells into F$^+$ cells by conjugation.

~

An F$^+$ cell may conjugate with an F$^-$ cell and transfer the F factor to the F$^-$ cell.

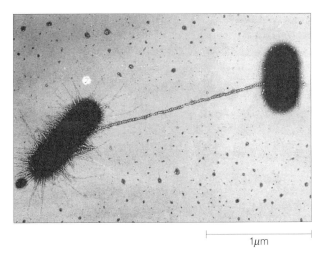

Figure 7.13 Conjugation in *E. coli*. (Photo courtesy of C. C. Brinton, Jr. and J. Carnahan.)

1μm

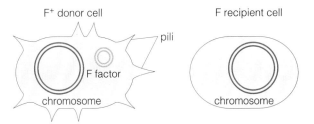

a F⁺ cells contain an F factor and have pili that radiate out from the cell. F⁻ cells lack the F factor and pili.

b A pilus from an F⁺ cell contacts an F⁻ cell and becomes a conjugation bridge. The F factor replicates by σ-mode (rolling circle) replication, sending a copy of the F factor into the conjugation bridge.

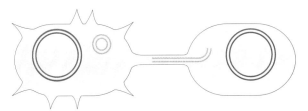

c A linear copy of the F factor enters the F⁻ cell.

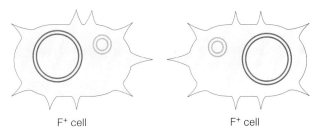

d After transfer is complete, the F factor arranges itself into a circle, and the conjugation bridge is broken. The F⁻ cell has been converted to an F⁺ cell.

Figure 7.14 Bacterial conjugation and transfer of the F factor.

Hfr Cells

By itself, the F factor provides little opportunity for recombination because its genes are different from those in the bacterial chromosome. Consequently, recombination is a very rare event when an F factor is transferred from one cell to another. Under certain circumstances, however, the F factor can facilitate recombination of genes normally located on the chromosome. As mentioned earlier, the F factor is an episome, which means it is able to integrate into the chromosome. Once the F factor has integrated, recombination of chromosomal genes increases significantly. Cells with an integrated F factor are called **Hfr cells** for "high-frequency recombination" (Figure 7.15).

The high frequency of recombination associated with Hfr cells is due to the ability of the F factor to initiate σ-mode replication and pass through a conjugation bridge. Because the F factor is part of the chromosomal DNA in Hfr cells, the chromosomal DNA is like a large F factor where the integrated F factor's origin of replication can initiate σ-mode replication of the entire chromosome, causing chromosomal genes to pass through the conjugation bridge (Figure 7.16). Complete transfer of the entire chromosomal DNA from an Hfr cell requires 100 minutes in *E. coli*. Rarely does a conjugation bridge remain intact long enough for the entire chromosome from an Hfr cell to be transferred. Usually, the conjugation bridge breaks after partial transfer has occurred, leaving a fragment with only a part of the F factor attached to a piece of chromosomal DNA in the recipient cell (Figure 7.16c). The recipient cell remains F⁻ because it does not contain a complete F factor (Figure 7.16d).

The introduced fragment cannot replicate because it usually lacks an origin of replication. The only way for the genes in the introduced DNA to be salvaged is for them to recombine with the chromosomal DNA in the recipient cell. The introduced fragment, called the **exogenote**, contains chromosomal DNA that is homologous to a segment of chromosomal DNA, called the **endogenote**, in the recipient cell. The homology between the exogenote and the endogenote allows them to recombine, resulting in transfer of a portion of the DNA

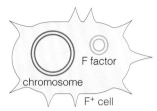

a The F factor is a plasmid that is separate from the chromosome in most F⁺ cells.

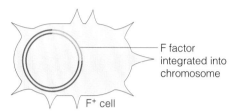

b The F factor is integrated as an episome into the chromosome in an Hfr cell. Because an Hfr cell contains the F factor, it is also an F⁺ cell.

Figure 7.15 The F factor as **(a)** a plasmid or **(b)** an episome.

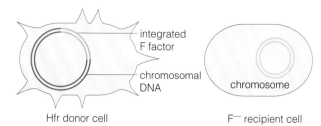

a An Hfr cell contains an F factor that has integrated as an episome into the chromosome.

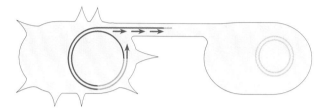

b During conjugation, the integrated F factor initiates σ-mode (rolling circle) replication of the Hfr chromosome and begins to transfer the chromosome to the F⁻ recipient cell.

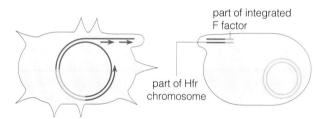

c The conjugation bridge breaks and conjugation is interrupted before transfer of the Hfr chromosome is complete.

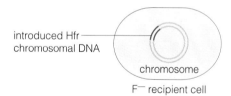

d The introduced segment of the Hfr chromosome recombines with the F⁻ cell's chromosome. The cell remains F⁻ because it did not receive a complete copy of the F factor.

Figure 7.16 Conjugation between an Hfr cell and an F⁻ cell.

sequence from the Hfr donor cell's chromosome to the F⁻ recipient cell's chromosome.

When an Hfr cell conjugates with an F⁻ cell, the genes most likely to be transferred are those located on the portion of chromosomal DNA that first enters the recipient cell. The first gene to enter is called the **leading gene** (Figure 7.17). The more distant a gene is from the leading gene, the lower its probability of being transferred.

The likelihood of transfer can be controlled with some precision through a technique called **interrupted mating**, in which researchers mix F⁻ cells with an excess of Hfr cells so that all the F⁻ cells quickly come into contact with Hfr cells and rapidly establish conjugation at the same time. The researcher allows conjugation and DNA transfer to proceed for a predetermined period of time. At the end of the allotted period of time, the researcher physically disrupts conjugation by agitating the cells, such as whirring the liquid culture in a blender. The cells are then immediately plated on selective media to determine the rate of recombination for certain genes.

Those genes close to the leading gene are transferred first and recombine in a short period of time. Genes that are farther from the leading gene require more time to enter the cells and recombine. Those genes that are quite far from the leading gene fail to enter the recipient cell. In this way, genes can be mapped by determining when they enter the cell. The more time required to enter the cell, the farther the gene is from the leading gene. This process is called **time-of-entry gene mapping**; the following example shows how it is done.

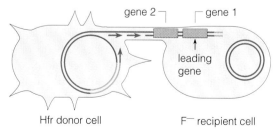

Figure 7.17 The leading gene.

Table 7.3 Information for Example 7.5: Genotypes of Two *E. coli* Strains Used to Map Four Genes

F⁻ Strain (P678)	Hfr Strain (HfrH)
*str*ʳ (streptomycin resistant)	*str*ˢ (streptomycin susceptible)
thr⁻ (auxotrophic for threonine)	*thr*⁺ (prototrophic for threonine)
leu⁻ (auxotrophic for leucine)	*leu*⁺ (prototrophic for leucine)
*azi*ʳ (resistant to sodium azide)	*azi*ˢ (susceptible to sodium azide)
*tonA*ʳ (resistant to phage T1 infection)	*tonA*ˢ (susceptible to phage T1 infection)
lac⁻ (unable to utilize lactose)	*lac*⁺ (able to utilize lactose)
galB⁻ (unable to utilize galactose)	*galB*⁺ (able to utilize galactose)

Example 7.5 Mapping four genes in *E. coli* with time-of-entry mapping.

In 1956, Wollman, Jacob, and Hayes (*Cold Spring Harbor Symposia of Quantitative Biology* 21:141–162) presented data from a time-of-entry mapping experiment in *E. coli*. They used F⁻ and Hfr strains with the genotypes listed in Table 7.3. Their objective was to map the four genes *azi, tonA, lac,* and *galB*. They knew beforehand that *thr* and *leu* were very close to the origin of replication and that *str* was quite distant from the origin of replication in the Hfr chromosomal DNA. They also knew that *thr* and *leu* were closer to the origin of replication than the four genes they were mapping and that *str* was farther from the origin than these four genes.

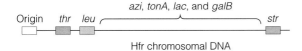

In the mixed culture where mating occurs, the two parent strains and recombinant cells that arose through conjugation should all be present in the culture. However, only the recombinant cells were of value for mapping. The researchers needed a way to eliminate the parental cells, leaving only recombinant cells. The *thr, leu,* and *str* genes allowed them to do this. The Hfr donor cells were Thr⁺, Leu⁺, and Str⁵. The F⁻ recipient cells were Thr⁻, Leu⁻, and Strʳ. Any cell with the phenotype Thr⁺, Leu⁺, and Strʳ must therefore be a recombinant cell. Recombinant colonies could easily be selected by culturing cells on medium that contained streptomycin and lacked threonine and leucine. This procedure eliminated the parental cells and selected only recombinant cells for mapping.

Conjugation was carried out in liquid medium. At intervals of 10, 15, 20, 25, 30, 40, 50, and 60 minutes, samples of the mixed culture were collected and agitated in a blender to interrupt conjugation. The percentages of recombinant cells that were Aziˢ, T1ˢ, Lac⁺, or Gal⁺ for each of the samples are illustrated in Figure 7.18.

Problem: Using the data in Figure 7.18, determine the relative distances in minutes among the *azi, tonA, lac,* and *galB* genes.

Solution: To read the graph in Figure 7.18, look straight up from the 20-minute mark. Among the Thr⁺−Leu⁺−Strʳ recombinant cells sampled at 20 minutes, 10% were Lac+, 65% were T1ˢ, and 85% were Aziˢ. Following this same procedure for each

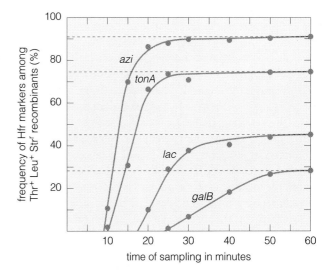

Figure 7.18 Information for Example 7.5: Times of entry for four genes in an Hfr × F⁻ mating in *E. coli*. (Adapted from Wollman, E. L., F. Jacob, and W. Hayes 1956. *Cold Spring Harbor Symposia on Quantitative Biology* 21:141–162. Reprinted by permission.)

time interval, the researchers placed points along the graph and drew curves connecting the points for each gene. The point at which each curve intersects the *x* axis represents the time at which the gene entered the cell.

In this example, *tonA* entered 2 minutes after *azi*, *lac* entered about 7 minutes after *tonA*, and *galB* entered 6 minutes after *lac*. The map of these four genes is therefore as follows:

Hfr chromosomal DNA

Notice in Figure 7.18 that the frequency of recombinant cells carrying a particular gene levels off at a certain point. For example, at 30 minutes, 75% of the recombinant cells are susceptible to T1 infection, and the frequency remains about 75% from that point on. The farther a gene is from the origin of replication, the lower the frequency at which stability is reached. This is due to natural interruption of mating. For example, mating interrupts naturally in about 25% of the recombinant cells before the *tonA* gene enters the cell. Thus the maximum frequency for *tonA* recombinants is 75% regardless of how much time elapses.

Time-of-entry mapping is most effective for genes located near the leading gene. The chromosomal location at which the F factor integrates determines which gene is the leading gene. The F factor does not have a predetermined site on the chromosome for integration, meaning that different Hfr strains may have different leading genes. Also, the F factor may insert itself in either of two possible orientations, one called a clockwise orientation and the other a counterclockwise orientation. When the F factor inserts itself in a clockwise orientation, the genes enter the cell in a particular order (Figure 7.19a and b). When the F factor inserts itself in a counterclockwise orientation at another site, the same genes may enter the cell in the reverse order (Figure 7.19c and d). Figure 7.20 shows the locations and orientations of several integrated F factors in the *E. coli* genome.

Each Hfr strain permits mapping of several genes in relation to one another. When the results of time-of-entry maps from mating experiments with different Hfr strains are combined, some of the maps overlap, eventually allowing researchers to construct a complete circular map covering the entire *E. coli* chromosome. The following example illustrates how the results from different time-of-entry mapping experiments can be combined into a single map.

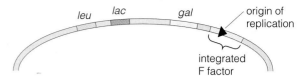

a F factor oriented clockwise.

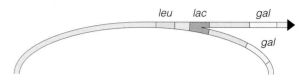

b The order of entry into the recipient cell is *gal-lac-leu*.

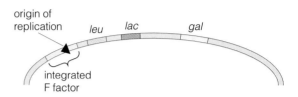

c F factor oriented counterclockwise.

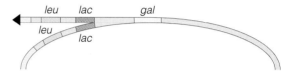

d The order of entry into the recipient cell is *leu-lac-gal*.

Figure 7.19 Clockwise and counterclockwise orientations of the F factor in Hfr strains.

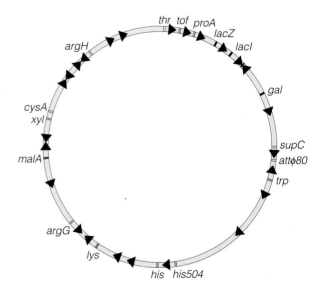

Figure 7.20 F-factor orientation and map location of 25 Hfr strains of *E. coli*. Each arrowhead denotes the position of the F factor in an Hfr strain and the direction in which it enters a recipient cell during conjugation. The gene following each arrowhead is the leading gene for that Hfr strain. (Adapted from Ayala, F. J. and J. A. Kiger. 1980. *Modern Genetics.* Menlo Park, CA: The Benjamin Cummings Publishing Co.)

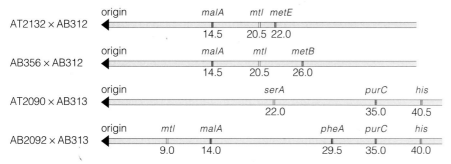

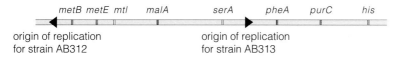

a Time-of-entry maps for four matings in *E. coli*.

origin of replication
for strain AB312

origin of replication
for strain AB313

b Consolidation of the maps in part a.

Figure 7.21 Solution to Example 7.6: Time-of-entry mapping.

Example 7.6 Overlapping maps from two Hfr strains.

In 1964, Taylor and Thoman (*Genetics* 50:659–677) reported data from several Hfr × F⁻ matings to construct a complete circular map of the *E. coli* chromosomal DNA from time-of-entry mapping. Table 7.4 provides data from several matings with two Hfr strains.

Problem: Construct a time-of-entry map of the genes in this example showing the distances between genes in minutes.

Solution: To solve this problem, let's make a time-of-entry map for each cross. Figure 7.21a shows the four maps derived from the four crosses. The arrows denote the origins of replication and the direction that the Hfr DNA enters the F⁻ cell. Notice that two genes, *mtl* and *malA*, are transferred by both AB312 and AB313 and that they are in opposite

orientations as they enter the F⁻ cells from the two strains. This means that the F factor in strain AB312 integrated in the opposite orientation when compared to the F factor in strain AB313. Thus, the genes are transmitted in opposite orders from the two strains. This observation permits consolidation of the four maps into a single map, shown in Figure 7.21b. This map covers 40 minutes, which equals 40% of the *E. coli* chromosome.

Detailed *E. coli* maps have been constructed using time-of-entry mapping, such as the one shown in Figure 7.22. Although it is represented as linear segments of 5 minutes each, the map is actually circular, representing the entire *E. coli* chromosome. Notice that the map is measured in minutes. The minutes depicted on the map correspond to the number of minutes that separate entry of genes during time-of-entry mapping. The entire map covers 100 minutes, which is the time required for the

---

Table 7.4 Information for Example 7.6: Number of Minutes Required for Particular Genes to Enter the F⁻ Cell in Four Matings									
Mating		**Genes**							
F⁻ Strain ×	**Hfr Strain**	*malA*	*mtl*	*metE*	*metB*	*serA*	*pheA*	*purC*	*his*
AT2132	AB312	14.5	20.5	22.0					
AB356	AB312	14.5	20.5		26.0				
AT2090	AB313					22.0		35.0	40.5
AT2092	AB313	14.0	19.0				29.5	35.0	40.0

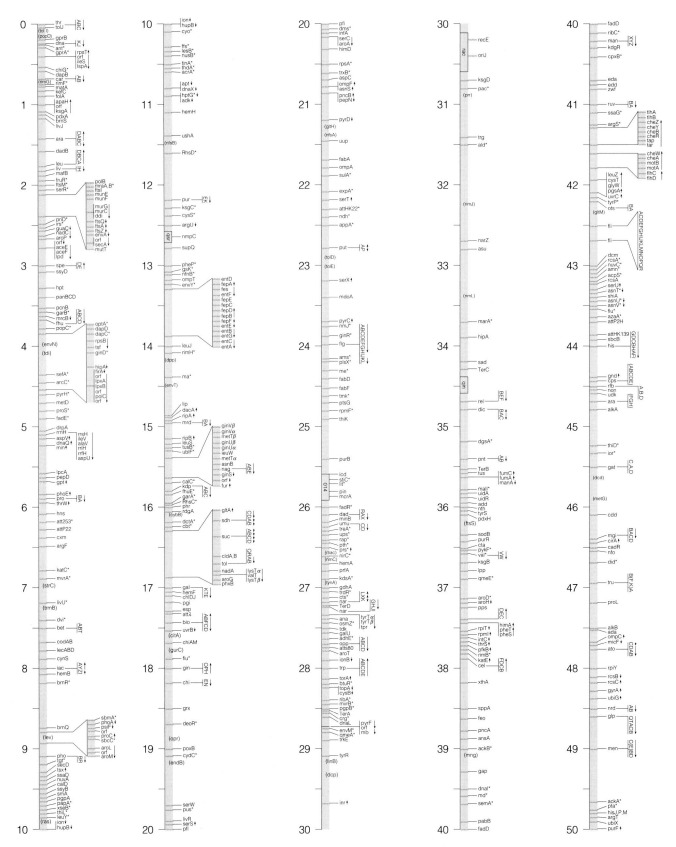

Figure 7.22 A consolidated time-of-entry map of the circular *E. coli* chromosome with 1403 mapped genes. The circular map is broken into segments of 5 minutes each in order to represent the enormous number of mapped genes. The distances are measured in minutes that are compiled from time-of-entry mapping experiments. Notice that the entire chromosome is 100 minutes long which represents the time required for the entire chromosome to pass through a conjugation bridge. (From Bachmann, B. J. 1990. Linkage map of *Echerichia coli* K-12, edition 8. *Microbiological Reviews* 54:130–197. Reprinted by permission.)

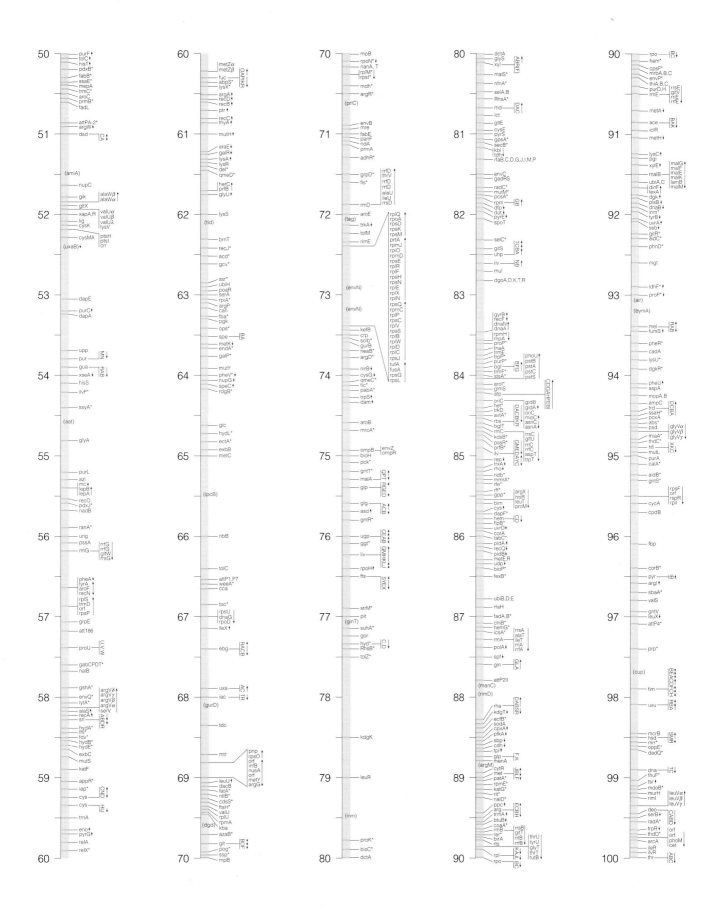

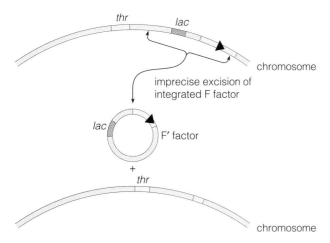

Figure 7.23 Formation of an F' factor. Imprecise excision of an integrated F factor from the chromosome forms an F' factor, which contains some chromosomal DNA, often with one or more genes.

rare cases in which the entire chromosome is replicated and transferred during conjugation. Geneticists who work with *E. coli* agreed on an arbitrary starting point of 0 minutes at *thr*, so each gene on the map is measured in minutes from there. Notice that some of the genes we have discussed in previous chapters can be found in Figure 7.22. For example, the gene that encodes rho factor (section 3.5) is located at 84.75 minutes. The genes *polC* (4.3 minutes), *dnaQ* (5.25 minutes), *dnaZ* (10.9 minutes), and *dnaX* (11.0 minutes) encode subunits of DNA polymerase III (section 2.4).

~

Genes can be mapped in Hfr × F⁻ matings in *E. coli* by measuring the time required for a gene to enter the recipient cell following initiation of conjugation.

F' Factors

An integrated F factor in an Hfr cell can excise itself from the chromosome to once again become an F factor plasmid. Occasionally, the excision is not precise and the F factor takes some chromosomal DNA along with it, forming a circular F factor with a piece of integrated chromosomal DNA that may carry one or more chromosomal genes (Figure 7.23). An F factor with a segment of chromosomal DNA is called an **F' factor**. F' factors can carry anywhere from a small amount of chromosomal DNA up to nearly half the chromosome in an enormous F' factor. Most F' factors carry only a few genes, however. Conjugation with an F' factor is a very efficient way to transfer chromosomal genes from one strain to another in the laboratory. Under laboratory conditions, an F' factor spreads to every cell of a culture in a short period of time, just as a normal F factor does.

Once an F' factor enters a recipient cell, the recipient cell has two copies of the piece of chromosomal DNA introduced on the F' factor. Cells that have received a second copy of genes are somewhat similar genetically to eukaryotic cells that have two copies of every gene. For that reason, geneticists have borrowed some terms normally used to describe eukaryotic cells. The transfer of an F' factor is called **sexduction** (or sometimes **F-duction**) because it creates a situation similar to that caused by sexual reproduction in eukaryotes. Once the recipient cell has received the F' factor, it is called a **merozygote** because it is somewhat like a eukaryotic zygote with two copies of DNA. It is also called a **partial diploid**, *diploid* indicating the presence of two copies of a segment of DNA. (We will redefine the terms *zygote* and *diploid* in their true eukaryotic context in later chapters.)

Complementation Analysis

Merozygotes are advantageous for genetic analysis because it is possible to place two mutant alleles that confer the same phenotype into the same cell and investigate whether or not the alleles belong to the same gene. This procedure is called **complementation analysis**. The genes that encode the enzymes for lactose metabolism in *E. coli* provide a good example of how this analysis is done. In the late 1950s and early 1960s, François Jacob and Jacques Monod conducted a series of landmark experiments using F' factors and lactose-metabolizing genes in *E. coli* to demonstrate how genes are regulated. We will discuss their experiments in detail in the next chapter. For now, let's look at the part of their experiments that illustrates complementation analysis.

Lactose is a disaccharide composed of a glucose molecule bonded to a galactose molecule. *E. coli* cannot use lactose directly. Instead, it must first convert lactose into glucose and galactose. Eventually, *E. coli* converts galactose to glucose as well. Two of the genes that encode lactose-metabolizing enzymes in *E. coli* are *lacY* and *lacZ*. The *lacY* gene encodes galactoside permease, which causes lactose to enter the cell. The *lacZ* gene encodes β-galactosidase, which converts lactose into glucose and galactose (Figure 7.24). *E. coli* cells that lack either galactoside permease or β-galactosidase are unable to use lactose, so mutations in either gene cause a Lac⁻ phenotype (unable to grow when lactose is the sole carbon source).

The *lacY* and *lacZ* genes are located adjacent to one another in the chromosomal DNA, so an F' factor that contains a *lacY* gene usually carries the *lacZ* gene as well. Suppose that an F' factor carries the alleles *lacY⁻ lacZ⁺* and that this F' factor enters an *F⁻* cell with the genotype *lacY⁺ lacZ⁻*. As illustrated in Figure 7.25a, the *lacY⁻* allele in the F' factor fails to produce functional galactosidase permease, but this is overcome by the *lacY⁺* gene in the chromosomal DNA, which does produce functional

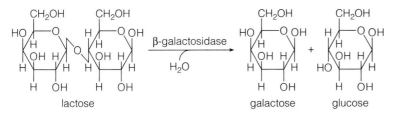

Figure 7.24 Conversion of lactose into glucose and galactose.

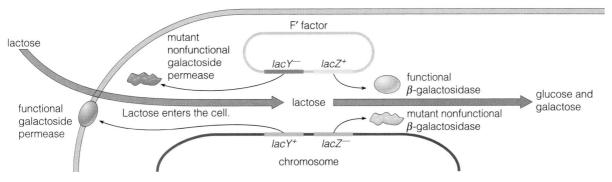

a Genes in the F′ factor and the chromosome complement one another.

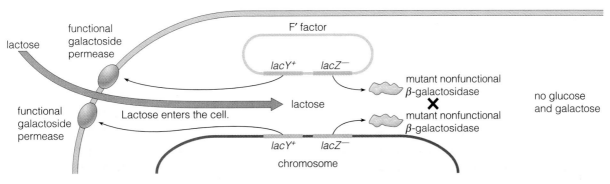

b Genes in the F′ factor and the chromosome fail to complement one another.

Figure 7.25 Complementation and lack of complementation in merozygotes.

galactosidase permease. The functional galactosidase permease facilitates entry of lactose into the cell. The *lacZ*⁻ allele in the chromosomal DNA cannot produce functional β-galactosidase, but this shortcoming is overcome by the *lacZ*⁺ allele in the F′ factor, which encodes functional β-galactosidase and converts lactose to glucose and galactose. The merozygote acquires a Lac⁺ phenotype. The genes in the chromosomal DNA and the genes in the F′ factor complement one another.

On the other hand, suppose that an F′ factor carries the alleles *lacY*⁺ *lacZ*⁻ and that it enters an F⁻ cell with the genotype *lacY*⁺ *lacZ*⁻. As illustrated in Figure 7.25b, the *lacY*⁺ alleles on both the chromosomal DNA and the F′ factor produce functional galactoside permease, allowing lactose to enter the cell. However, neither

the *lacZ*⁻ allele in the chromosomal DNA nor the *lacZ*⁻ allele in the F′ factor produces functional β-galactosidase. In the absence of functional β-galactosidase, lactose is *not* converted into glucose and galactose, and the cell retains its Lac⁻ phenotype. In this case, the genes in the chromosomal DNA and the genes in the F′ factor fail to complement one another.

In complementation, the F′ factor and the chromosomal DNA each supply an essential function that the other lacks. In other words, alleles that complement each other restore the wild-type phenotype even though each allele causes the same mutant phenotype individually. Alleles that complement each other are usually in different genes (such as alleles in the *lacZ* and *lacY* genes in this example). Alleles that fail to complement each

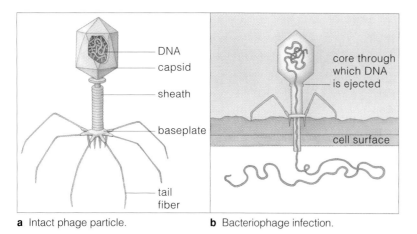

a Intact phage particle.   **b** Bacteriophage infection.

Figure 7.26 Structure and function of a bacteriophage. The virus particle injects its DNA into the bacterial cell, leaving the capsid on the cell's surface.

other cause a mutant phenotype when they are together in the same merozygotic cell, usually because they are mutant alleles of the same gene.

Complementation tests are valuable in determining whether or not two mutations belong to the same gene. Keep in mind, however, that occasionally the results of complementation tests may be misleading. Sometimes mutations in different genes fail to complement one another, leading to the false conclusion that the mutations are in the same gene. Also, occasionally two mutations in the same gene may complement each other, leading to the false conclusion that they belong to different genes.

The *lacZ* gene is a good example of how some mutant alleles of the same gene may complement one another. β-galactosidase is a tetrameric protein consisting of four identical polypeptide subunits encoded by the *lacZ* gene. Certain alleles encode mutant subunits that fail to form a functional enzyme when all four mutant subunits are identical. But in rare cases, two different mutant alleles of the *lacZ* gene may encode different types of subunits that form a functional enzyme when the different mutant subunits combine to form the tetrameric enzyme. Complementation by different mutant alleles of the same gene is called **intragenic complementation**, which is rare. In most cases, the results of complementation tests accurately reveal whether or not two mutant alleles belong to the same gene. We will discuss additional aspects of complementation tests in eukaryotes in Chapter 16.

~

Merozygotes contain F′ factors that allow researchers to conduct complementation analysis.

7.6 PHAGE LIFE CYCLES

We encountered bacterial viruses in the experiments of Hershey and Chase in section 2.1. Bacterial viruses are also called **bacteriophages** or simply **phages**. When they are outside of a bacterial cell, phages consist of a DNA molecule encased in a protein coat called a **capsid**. The intact phage, with its DNA and capsid intact, is called a **phage particle**. Phage particles outside of a bacterial cell do not grow or reproduce but remain in a biologically inactive state, awaiting the chance to infect a bacterial cell. When a phage particle comes in contact with a bacterial cell, it adsorbs to the cell's outer surface and injects the DNA molecule into the cell, leaving the capsid outside of the cell (Figure 7.26). Some phages have a polyhedral head, where the DNA resides, and a tail through which the DNA passes during cell infection. In some cases, the tail has fibers that act like tiny legs, assisting the phage as it adsorbs to the cell's surface.

The phage DNA molecule contains genes that synthesize the capsid proteins and regulate phage reproduction in the cell. Phage genes have the same structure and organization as bacterial genes, which is not surprising because the phage genes are expressed in bacterial cells. Some phages rely on the bacterium's RNA polymerase to transcribe all of their genes. Other phages use the bacterial RNA polymerase for transcription of the first genes, one of which is a phage RNA polymerase that transcribes the remaining phage genes. All phage mRNAs, however, are translated on the bacterial cell's ribosomes. In many cases, the phage genes completely take over gene expression in the bacterium at the expense of the cell's own genes. In fact, shortly after infection, many phages produce enzymes that degrade the bacterial DNA into nucleotides that the phage uses to synthesize its own DNA, leaving the bacterial cell devoid of its own genes.

When the time arrives for the phage to reproduce, it synthesizes many copies of its DNA, each destined to become a phage particle. At a certain point, phage genes that encode capsid proteins and their assembly are expressed. The capsids assemble around the phage DNA molecules, forming intact phage particles. The cell, already devoid of its own DNA, is now simply a host

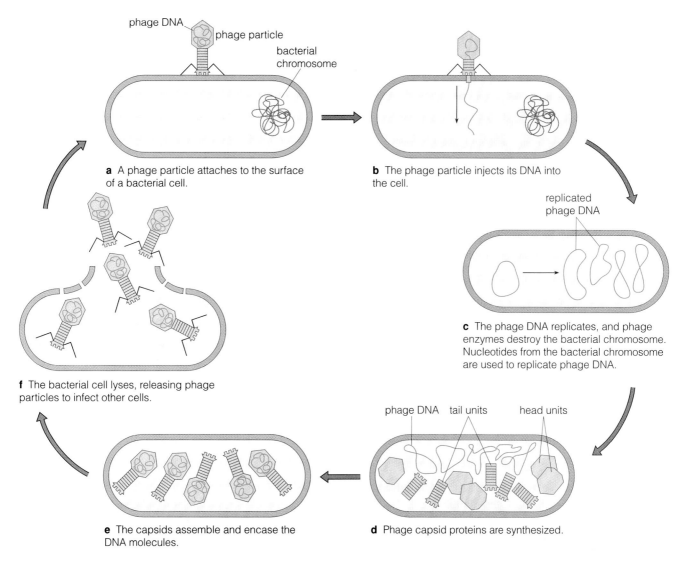

a A phage particle attaches to the surface of a bacterial cell.

b The phage particle injects its DNA into the cell.

c The phage DNA replicates, and phage enzymes destroy the bacterial chromosome. Nucleotides from the bacterial chromosome are used to replicate phage DNA.

f The bacterial cell lyses, releasing phage particles to infect other cells.

e The capsids assemble and encase the DNA molecules.

d Phage capsid proteins are synthesized.

Figure 7.27 The lytic cycle of a virulent phage.

where phage particles can develop. Once the phage particles have fully assembled, they burst the cell (a process called **lysis**) and release the phage particles into the cell's surroundings to infect other cells. This entire process is called the **lytic cycle** (Figure 7.27).

Phage infection and lysis can easily be detected in bacterial cultures. Typically, bacterial cells are cultured at a high concentration in the uppermost layer of an agar plate so that they form a uniform layer of bacterial cells in the agar medium. This bacterial layer in a petri plate is called a bacterial **lawn**. Phage infection and lysis can be detected as a clear spot where cells die in the bacterial lawn (Figure 7.28). A single phage particle on the medium adsorbs to a growing bacterial cell and proceeds though the lytic cycle in that cell, eventually releasing the progeny phage particles following lysis of the cell. These particles then infect adjacent cells and proceed through the lytic cycle in them. Several rounds of infection and

lysis produce a transparent spot where the cells have been lysed in the otherwise opaque lawn. This transparent spot is called a **plaque**.

Presumably, the phage particles could spread throughout the lawn, killing all the cells and leaving the entire plate transparent. However, the lytic cycle can only proceed in cells that are actively dividing. In a bacterial lawn, cells reach a point where they have used all the available nutrients, so they cease growing and dividing. Once they have stopped growing, the lytic cycle is arrested and the plaque remains fixed in size, appearing as a circular transparent spot in the lawn (Figure 7.28). Each plaque represents one phage particle that was originally present in the center of the plaque on the growth medium.

Some phages, however, do not immediately enter the lytic cycle after infecting the cell. Instead, the phage DNA may insert itself into the chromosomal DNA,

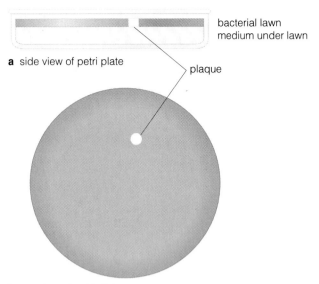

a side view of petri plate

bacterial lawn
medium under lawn

plaque

b top view of petri plate

Figure 7.28 A transparent plaque (clear spot) in a bacterial lawn. The center of the plaque is the original location of a single phage particle in the bacterial medium.

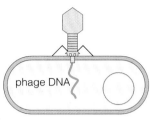

phage DNA

a A phage that is capable of lysogeny injects its DNA into a bacterial cell.

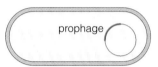

prophage

b The phage DNA inserts itself into the bacterial chromosome, where it becomes a prophage.

c The prophage replicates as part of the chromosomal DNA when the cell divides.

Figure 7.29 Lysogeny.

becoming a part of the chromosomal DNA and replicating with it during cell division. A phage DNA molecule that is integrated into the chromosomal DNA is called a **prophage**. The prophage prevents transcription of genes that initiate the lytic cycle, so it can replicate as part of the bacterial DNA and be passed on to each daughter cell during cell divisions. In this way, many cells receive the prophage through inheritance rather than infection, an effective way for a phage to increase its numbers. Phage insertion and replication in the chromosomal DNA is called **lysogeny** (Figure 7.29).

Lysogenic bacterial cells (cells that carry a prophage) grow and divide normally even though they contain a deadly pathogen. It is as if the prophage is hiding within the cell, utilizing the cell's resources for its own reproduction, waiting to emerge from hiding and destroy the cell in which it resides. When the time comes, the prophage excises itself from the chromosome and enters the lytic cycle, producing many progeny phages and lysing the cell. The phage particles, now outside of the cell, can infect other cells.

Prophage excision and entrance into the lytic cycle is called **induction** because it is usually a response to some external stimulus. When lysogenic *E. coli* cells are exposed to ultraviolet light or other agents that damage the cells, the prophages undergo induction and enter the lytic cycle. This is an excellent survival strategy for the

phage. As long as lysogenic cells are growing and dividing, the prophages replicate as part of the cells' chromosomes. However, as soon as the cells' survival is in danger, the prophages excise themselves, replicate, and destroy the cells in which they reside. The phage particles then remain in a chemically inert state outside of the cells until the danger to the cells has passed. Once the phages encounter cells that are growing and dividing, they infect the cells, where they are once again able to replicate.

Not all phages are capable of lysogeny. Phages that rely entirely on the lytic cycle for reproduction are called **virulent phages**. Phages that are capable of lysogeny are called **temperate phages**. Lysogeny has some advantages in that many copies of phage DNA can be produced without infection, allowing the prophages to persist through many cell generations and move wherever the cells are carried.

When temperate phages infect cells in a bacterial lawn, some of the phages immediately start the lytic cycle while others enter lysogeny. This means that some infected cells contain phages that rapidly lyse the cells, while other cells contain prophages in the chromosomal DNA. These lysogenic cells live, grow, and divide, replicating the phage DNA in the process. Plaques caused by temperate phages are **turbid plaques**, meaning that they have a cloudy, rather than clear, appearance. The

lysed cells cause the plaque to have some transparency, whereas the lysogenic cells grow and divide within the plaque, giving it its cloudy appearance.

~

Phages may be virulent or temperate. A virulent phage enters the lytic cycle soon after infection. A temperate phage may enter the lytic cycle soon after infection, or it may enter the lysogenic cycle, where it integrates into the host cell's chromosomal DNA and replicates as part of the chromosomal DNA.

7.7 TRANSDUCTION

Having reviewed phage life cycles, we are now ready to discuss transduction, the third major mechanism for DNA transfer in bacteria. Typically, phage particles carry their own DNA, which includes only genes devoted to carrying out the phage life cycle. However, on rare occasions, a capsid may package bacterial DNA instead of phage DNA, and transmit the bacterial DNA from one cell to another through phage infection. This process is called **transduction**.

The discovery of transduction was a milestone in bacterial genetics. In the following example, we shall see how Norton Zinder and Joshua Lederberg discovered transduction in *Salmonella typhimurium*.

Example 7.7   Discovery of transduction.

In 1952, Zinder and Lederberg (*Journal of Bacteriology* 64:679–699) published the results of experiments demonstrating transduction in *Salmonella typhimurium*. They mixed several strains of bacteria that were auxotrophic for different substances in pairwise combinations and recovered fully prototrophic wild-type colonies from the mixed cultures. These results were essentially the same as those that Lederberg and Tatum observed in *E. coli* (see Example 7.1). However, when Zinder and Lederberg subjected these strains to a U-tube analysis (like the experiments described in Example 7.4), they consistently recovered prototrophic bacteria in one experiment. Presumably, a substance passed across the filter and transmitted DNA to the culture on the other side of the filter. They called the substance FA for "filterable agent." They purified FA and found that when they inoculated auxotrophic cells with it, prototrophic colonies arose. Among many experiments designed to determine the composition of FA was one in which they treated it with deoxyribonuclease, an enzyme that degrades DNA. Following this treatment, FA still

caused prototrophy to arise in the treated bacterial cells. The researchers knew that one of the cultures contained a lysogenic phage known as PLT-22. The filter they used in the U-tube experiment had a pore size large enough to allow PLT-22 to cross but too small to allow bacterial cells to cross. They conducted experiments with filters of varying pore sizes and found that FA passed only through those filters with pores that were the size of the PLT-22 phage or larger. Also, the size of FA was too large for it to be a simple DNA molecule, as in Avery, MacLeod, and McCarty's experiments. A series of additional experiments collectively showed a consistent association between PLT-22 and the observation of prototrophy, leading Zinder and Lederberg to conclude that the FA was the PLT-22 phage, and that this phage carried bacterial DNA. They called this new process of gene transfer *transduction*.

Problem: What aspects of these experiments demonstrated that transduction could not be accounted for by transformation or conjugation?

Solution: The U-tube experiments ruled out conjugation. The pores of the filter were too small to allow contact between bacterial cells in the two cultures. The U-tube experiments, however, did not rule out transformation. Avery, MacLeod, and McCarty (see section 2.1) found that treatment of the transforming principle with deoxyribonuclease eliminated the transforming activity, one of several observations that led them to conclude that the transforming principle was DNA. Zinder and Lederberg found that FA was unaffected by deoxyribonuclease, meaning that if FA were composed of DNA, the DNA must be protected from deoxyribonuclease degradation. Their discovery that transduction was associated with the virus provided an explanation for the failure of deoxyribonuclease to destroy FA's activity. The DNA within the virus is protected by a protein capsid that protects the DNA from degradation.

There are two types of transduction, specialized transduction and generalized transduction. In **specialized transduction**, phage particles contain mostly phage DNA but carry a specific fragment of chromosomal DNA attached to the phage DNA. They infect bacterial cells, transferring the specific segment of bacterial DNA along with the phage DNA into the cell. In **generalized transduction**, phage particles contain DNA from any part of the bacterial chromosome. Usually, generalized transducing phage particles contain no phage DNA, only chromosomal DNA in its place.

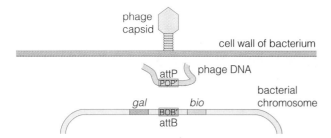

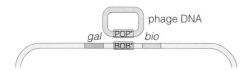

a Lambda (λ) phage injects its DNA as a linear molecule into a bacterial cell.

b The phage DNA arranges itself into a circle.

c The phage DNA recombines with the chromosomal DNA where the POP′ and BOB′ sites have homology.

integrated prophage

d The prophage is integrated into the chromosome.

Figure 7.30 Integration of lambda (λ) into the *E. coli* chromosome during initiation of lysogeny.

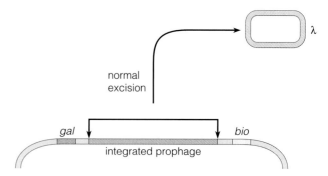

a Normal excision of a lambda (λ) prophage produces a circular λ DNA molecule.

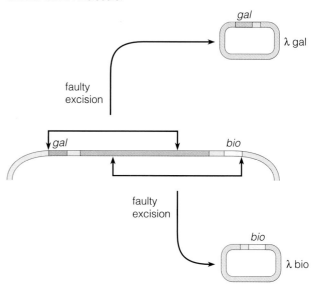

b Faulty excision of a λ prophage may produce a circular λ DNA molecule with either the *gal* or the *bio* gene.

Figure 7.31 Formation of λbio and λgal transducing phages.

Specialized Transduction

The temperate *E. coli* phage lambda (λ) is an example of a phage capable of specialized transduction. When it infects a cell, the λ phage particle inserts the DNA as a linear molecule (Figure 7.30a). Once inside the cell, the linear DNA arranges itself into a circle (Figure 7.30b). When the λ DNA inserts itself into the chromosomal DNA for lysogeny, it does so at a specific site. There is a region within the circular λ DNA molecule that is homologous to a region in the chromosomal DNA. This region in λ is called attP for "phage attachment site." The homologous region in the chromosome is called attB for "bacterial attachment site." The attB site is located between two genes, *gal* and *bio*, so λ inserts itself between these two genes. The attP site is sometimes called POP′, and the attB site BOB′ (P for phage, B for bacterium), designations that are intended to illustrate how the phage inserts itself into the bacterial DNA. When the circular phage prepares to insert itself into the chromosome, the

homologous POP′ and BOB′ sites align with one another. An enzyme called integrase causes the two sites to recombine with one another, inserting the phage DNA into the chromosome (Figure 7.30c). At one end of the prophage is the recombined segment, now BOP′, and at the other end, the segment POB′ (Figure 7.30d).

Upon induction, the prophage excises itself from the chromosomal DNA following steps b–d in Figure 7.30 in reverse. Usually this exchange is so precise that the excised circular λ phage DNA is identical to the circular λ phage DNA before insertion. However, sometimes recombination is imprecise, creating a λ phage that carries a portion of the chromosomal DNA and lacks a portion of its own DNA, which is left behind in the chromosome. Imprecise recombination may take place on either side of the prophage, so in some cases the excised phage carries the *bio* gene and in other cases it carries the *gal* gene. These recombinant phages are called λbio and λgal (Figure 7.31). In the cell, these phages replicate and form phage particles just like any other phage, even

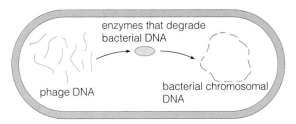

a In preparation for lysis, phage genes produce proteins that degrade the bacterial DNA.

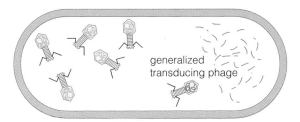

b A generalized transducing phage is formed when a capsid mistakenly packages bacterial DNA.

Figure 7.32 Formation of a generalized transducing phage.

though they lack some of the phage DNA. When a λbio or λgal phage enters a new cell through infection, the *bio* or *gal* gene in the phage may recombine with its homologous counterpart in the chromosome. These phages are called **specialized transducing phages** because they are specialized for transfer of specific genes, *gal* or *bio* in this example.

Generalized Transduction

The *E. coli* phage P1 provides a good example of generalized transduction. As part of its lytic pathway, this phage produces an enzyme that cuts the bacterial DNA into fragments. This fragmentation prevents bacterial genes from being expressed, and the nucleotides from the bacterial DNA can be used to make phage DNA. Occasionally, some of these bacterial DNA fragments are about the same size as phage DNA and are mistakenly incorporated into phage capsids in place of phage DNA (Figure 7.32). A phage with chromosomal DNA in place of phage DNA is called a **generalized transducing phage**, *generalized* because the DNA may be any fragment of chromosomal DNA that is about the same size as phage DNA. A generalized transducing phage may, therefore, carry any chromosomal gene, rather than a specific gene.

Generalized transducing phage particles are released at lysis along with the normal phage particles. No genes are expressed within any phage particle because a phage particle has no enzymes or ribosomes. This means that infection is purely a function of the capsid, and that the DNA in it is biologically inactive at the moment of infection. So any DNA contained within a capsid is equally

infective regardless of what genes it carries. A generalized transducing phage injects bacterial DNA into cells upon infection. Once inside the cell, the bacterial DNA recombines with its homologous sequences in the chromosome. The transduced DNA segment may carry several genes, all of which may recombine with the chromosome. The end result is the same as for transformation: a segment of bacterial DNA enters a cell where it can recombine. For this reason, transducing phages can be used to map genes in the same way that genes are mapped in transformation.

~

Transduction is the transfer of DNA from one cell to another through a phage intermediate. Specialized transducing phages transfer bacterial DNA that is attached to phage DNA. Generalized transducing phages package and transmit bacterial DNA in the place of phage DNA.

7.8   GENOME SEQUENCING

Most bacterial and phage genetic maps show genes whose locations have been determined with gene recombination methods. Some of these maps are quite extensive, showing the relative locations of many genes, such as the *E. coli* map shown in Figure 7.22. The ultimate map, however, is the entire nucleotide sequence of a **genome**, the complete genetic complement of the organism. Entire genome sequencing is the first step toward developing an understanding of the complete biology of an organism.

Frederick Sanger and his colleagues were the first to sequence an entire genome. In 1977, they published the nucleotide sequence of the bacteriophage φX174, which contains 5386 nucleotide pairs. Since that time, many phage genomes have been fully sequenced. However, the genome of a bacterium is much larger than a phage genome. Even with automated DNA sequencing, substantial effort and expense are required to sequence an entire bacterial genome. It was not until 1995 that Robert Fleischmann and his colleagues published the first complete genomic sequence of a bacterium. They sequenced the genome of *Haemophilus influenzae*, a bacterium that causes respiratory infections and meningitis in humans. In 1996, entire genome sequences of four other bacterial species were published. Because of *E. coli's* importance as a model organism, determination of its entire genome sequence, which was much larger than the genomes of the previously sequenced species, represented an important milestone in genetics. Frederick Blattner and his colleagues accomplished this feat in 1997. Table 7.5 provides genome characteristics revealed by DNA sequencing in four bacterial species.

Determination of the complete nucleotide sequence of an organism represents a significant leap that stands to benefit all researchers who work with the organism or

Table 7.5   Genome Characteristics Revealed by Complete Genome Sequencing in Four Bacterial Species

Species	Genome Size in Nucleotide Pairs	Number of Potential Polypeptide-Encoding Genes
Escherichia coli	4,639,221	4288
Haemophilus influenzae	1,830,137	1743
Methanococcus jannaschii	1,664,976 (chromosome)	1682
	58,407 (large plasmid)	44
	16,550 (small plasmid)	12
Mycoplasma genitalium	580,070	470

with related species. Every polypeptide, every enzyme, every tRNA or rRNA that the organism produces is encoded in the genome's nucleotide sequence. Potential genes can be identified by searching the entire nucleotide sequence for conserved sequences that are arranged in the form expected in a gene. The amino acid sequences of the polypeptides encoded by these potential genes can be determined directly from the DNA sequence. Additional features of the genome, such as regulatory regions, origins of replication, and repeated DNA sequences, can also be identified in the DNA sequence.

Let's look at the genome of *Mycoplasma genitalium*, which has the smallest genome of any free-living organism, as an example. Analysis of the genomic DNA sequence revealed the locations and nucleotide sequences of the origin of replication and all rRNA and tRNA genes. A total of 470 potential polypeptide-encoding genes were found on the basis of their DNA sequence organization (Figure 7.33 and Table 7.6). Of these, 374 were identified as genes with polypeptides of known function, leaving 96 potential genes with no known function and no known homology to proteins in any other organism. In the case of *E. coli*, about 40% of the 4288 polypeptide-encoding sequences could not be associated with a function. These potential genes provide an exciting challenge because they represent enzymes or proteins whose biological function has yet to be discovered.

Information from recombination-based genetic maps and genomic DNA sequences can be combined to identify the exact location and DNA sequence of a gene. For decades, many genes have been placed on genetic maps (like the one shown in Figure 7.22) based on the phenotypes associated with mutations in each of these genes. While a gene's position on a genetic map may be known, the gene's DNA sequence, the amino acid sequence of the polypeptide, and the function of the protein encoded by the gene may not be known. A gene's location in the genetic map can be matched with the nucleotide sequence at that same position in the genomic DNA sequence. Also, any additional information about the gene

or its product can be used to match the identity of the gene with its DNA sequence. The following example shows how this is done.

Example 7.8   Assignment of a DNA sequence to a mapped gene in *E. coli*.

In 1994, Sofia et al. (*Nucleic Acids Research* 22:2576–2586) published their DNA sequence analysis of the region in the *E. coli* genome spanning 76.0 to 81.5 minutes as a part of the overall effort to sequence the entire *E. coli* genome. Among the genes in this region is *dctA*, which had been mapped at minute 80 in the *E. coli* genetic map. The *dctA* gene was known to encode a membrane protein that binds dicarboxylates. A potential gene in the sequenced region was identified by its conserved sequence elements and was named *f428*. Analysis of the polypeptide encoded by the reading frame of *f428* revealed an amino acid sequence that was similar to the amino acid sequences of dicarboxylate permease proteins from two other bacterial species. The authors determined that *f428* was *dctA*.

Problem: What evidence led these authors to match the DNA sequence of *f428* with the *dctA* gene?

Solution: The time-of-entry map position of *dctA* was the first clue. Knowing that *dctA* was located at 80 minutes allowed the researchers to narrow their search to a limited region of DNA. They could then compare potential genes, identified by DNA analysis, in this region with the predicted function of the protein known to be encoded by *dctA*. Comparison of the predicted amino acid sequence from *f428* with proteins from other species revealed that *f428* encoded a protein with the predicted function of the *dctA* product. This led them to conclude that *f428* was the *dctA* gene.

KEY:

16S	ribosomal RNA
MgPar	MgPa Repeat
T	transfer RNA

— 1 kb

- ☐ amino acid biosynthesis
- ☐ biosynthesis of cofactors, prosthetic groups, carriers
- ☐ cell envelope
- ☐ cellular processes
- ☐ central intermediary metabolism

- ☐ transport/binding proteins
- ☐ translation
- ☐ transcription
- ☐ other categories
- ☐ hypothetical
- ☐ unknown

- ☐ energy metabolism
- ☐ fatty acid and phospholipid metabolism
- ☐ purines, pyrimidines, nucleosides, and nucleotides
- ☐ regulatory functions
- ☐ replication

Figure 7.33 Map of all genes identified by DNA sequence analysis in the *Mycoplasma genitalium* chromosome. Arrows indicate the direction of transcription. The genes are color-coded by their function. The numbered genes are identified by name in Table 7.6. (Reprinted with permission from Fraser, C. M., et al. 1995. The minimal gene complement of *Mycoplasma genitalium*. *Science* 270:397–403. Copyright 1995 American Association for the Advancement of Science.)

Once the gene's DNA sequence is known, the amino acid sequence and most probable secondary and tertiary structures of the polypeptide encoded by the gene can be identified. This information helps researchers determine the biological function of the gene's product and purify the product so it can be studied.

Complete genome sequencing is under way in other bacterial species, some of which have relatively large genomes. Also, genomic sequencing of several eukaryotes, including *Drosophila melanogaster*, *Caenorhabditis elegans* (a microscopic nematode worm used as a model

organism in developmental genetics), *Arabidopsis thaliana* (a plant used as a model organism), and *Homo sapiens*, is under way. As a result of international collaboration among several laboratories, the first complete DNA sequence of a eukaryote, *Saccharomyces cerevisiae* (brewer's yeast) was published in 1996. We will discuss these eukaryotic genomes in later chapters.

~

Entire genome sequencing has been accomplished in several bacterial species. This is the first step toward understanding the complete biology of an organism.

Table 7.6 List of genes in *Mycoplasma genitalium* Grouped According to the Functions of the Gene Products*

MG #	Identification	% ID
	Amino acid biosynthesis	
	Serine family	
394	serine hydroxymethyltransferase (glyA)	55
	Biosynthesis of cofactors, prosthetic groups, and carriers	
	Folic acid	
013	5, 10-methylene-tetrahydrofolate DHase (folD)	33
228	dihydrofolate RDase (dhfr)	33
	Heme and porphyrin	
259	protoporphyrinogen oxidase (hemK)	31
	Thioredoxin, glutaredoxin, and glutathione	
124	thioredoxin (trx)	36
102	thioredoxin RDase (trxB)	39
	Cell envelope	
	Membranes, lipoproteins, and porins	
318	fibronectin-BP (fnbA)	25
040	membrane lipoprotein (tmpC)	31
086	prolipoprotein diacylglyceryl Tase (lgt)	29
	Surface polysaccharides, lipopolysaccharides and antigens	
137	dTDP-4-dehydrorhamnose RDase (rfbD)	32
356	lic-1 operon prt (licA) motif	28
060	LPS biosyn prt (rfbV) motif	36
269	surface prt antigen precursor (pag) motif	26
025	TrsB	28
	Surface structures	
192	114 kDa prt, MgPa operon (mgp)	100
191	attachment prt, MgPa operon (mgp)	100
315	cytadherence-accessory prt (hmw1)	42
312	cytadherence-accessory prt (hmw1)	39
386	cytadherence-accessory prt (hmw1)	34
313	cytadherence-accessory prt (hmw1)	53
317	cytadherence-accessory prt (hmw3)	41
459	surface exclusion prt (prgA) (Plasmid pCF10)	28
	Cellular processes	
	Cell division	
457	cell division prt (ftsH)	50
297	cell division prt (ftsY)	36
224	cell division prt (ftsZ)	31
434	mukB suppressor prt (smbA)	41
	Cell killing	
146	hemolysin (tlyC)	26
220	pre-procytotoxin (vacA)	36
	Chaperones	
019	heat shock prt (dnaJ)	34
002	heat shock prt (dnaJ) motif	40
200	heat shock prt (dnaJ) motif	34
392	heat shock prt (groEL)	52
201	heat shock prt (grpE)	32
393	heat shock prt 60-like prt (PggroES)	40
305	heat shock prt 70 (hsp70)	57
	Detoxification	
008	thiophene and furan oxidizer (tdhF)	32
	Protein and peptide secretion	
138	GTP-binding membrane prt (lepA)	48
179	haemolysin secretion ATP-BP (hlyB) motif	35
072	preprotein translocase (secA)	44
170	preprotein translocase secY sub (secY)	39
210	prolipoprotein signal peptidase (lsp)	32
048	signal recognition particle prt (ffh)	43
	Transformation	
316	competence locus E (comE3) motif	30
	Central intermediary metabolism	
	Degradation of polysaccharides	
217	bifunctional endo-1, 4-beta-xylanase xyla precursor (xynA) motif	38
	Other	
357	acetate kinase (ackA)	43
038	glycerol kinase (glpK)	47
293	glycerophosphoryl diester phosphodiesterase (glpQ)	30
299	phosphotransacetylase (pta)	45
	Phosphorus compounds	
351	inorganic pyrophosphatase (ppa)	39
	Energy metabolism	
	Aerobic	
039	glycerol-3-phospate DHase (GUT2)	43
460	L-lactate DHase (ldh)	50
275	NADH oxidase (nox)	39
	ATP-proton motive force interconversion	
405	adenosinetriphosphatase (atpB)	36
401	ATP Sase alpha chain (atpA)	63
403	ATP Sase B chain (atpF)	37
399	ATP Sase beta chain (atpD)	81
404	ATP Sase C chain (atpE)	50
402	ATP Sase delta chain (atpH)	34
398	ATP Sase epsilon chain (atpC)	37
400	ATP Sase gamma chain (atpG)	38
	Glycolysis	
063	1-phosphofructokinase (fruK)	26
215	6-phosphofructokinase (pfk)	39
407	enolase (eno)	54
023	fructose-bisphosphate aldolase (tsr)	46
301	G3PD (gap)	56
111	phosphoglucose isomerase B (pgiB)	35
300	phosphoglycerate kinase (pgk)	51
430	phosphoglycerate mutase (pgm)	45
216	pyruvate kinase (pyk)	35
431	triosephosphate isomerase (tim)	40
	Pentose phosphate pathway	
264	6-phosphogluconate DHase (gnd)	30
066	transketolase 1 (TK 1) (tktA)	33
	Pyruvate DHase	
272	dihydrolipoamide acetyltransferase (pdhC)	45
271	dihydrolipoamide DHase (pdhD)	38
274	pyruvate DHase E1-alpha sub (pdhA)	43
273	pyruvate DHase E1-beta sub (pdhB)	55
	Sugars	
112	D-ribulose-5-phosphate 3 epimerase (cfxEc)	33
050	deoxyribose-phosphate aldolase (deoC)	83
396	galactosidase acetyltransferase (lacA)	40
053	phosphomannomutase (cpsG)	39
	Fatty acid and phospholipid metabolism	
212	1-acyl-sn-glycerol-3-phosphate acetyltransferase (plsC)	32
437	CDP-diglyceride Sase (cdsA)	38
368	fatty acid/phospholipid synthesis prt (plsX)	29
085	hydroxymethylglutaryl-CoA RDase (NADPH)	23
344	lipase-esterase (lip1)	27
114	phosphatidylglycerophosphate Sase (pgsA)	29
	Purines, pyrimidines, nucleosides, and nucleotides	
	2'-Deoxyribonucleotide metabolism	
231	ribonucleoside-diphosphate RDase (nrdE)	54
229	ribonucleotide RDase 2 (nrdF)	50
227	thymidylate Sase (thyA)	57
	Nucleotide and nucleoside interconversions	
382	uridine kinase (udk)	34
	Purine ribonucleotide biosynthesis	
107	5'-guanylate kinase (gmk)	43
171	adenylate kinase (adk)	32
058	phosphoribosylpyrophosphate Sase (prs)	44
	Salvage of nucleosides and nucleotides	
276	adenine PRTase (apt)	34
052	cytidine deaminase (cdd)	38
330	cytidylate kinase (cmk)	40
268	deoxyguanosine-deoxyadenosine kinase(I) sub 2	30
458	hypoxanthine-guanine PRTase (hpt)	38
049	purine-nucleoside phosphorylase (deoD)	44
034	thymidine kinase (tdk)	48
051	thymidine phosphorylase (deoA)	53
006	thymidylate kinase (CDC8)	28
030	uracil PRTase (upp)	45
	Sugar-nucleotide biosynthesis and conversions	
118	UDP-glucose 4-epimerase (galE)	34
453	UDP-glucose pyrophosphorylase (gtaB)	48
	Regulatory functions	
024	GTP-BP (gtp1)	47
384	GTP-BP (obg)	40
387	GTP-BP (era)	27
248	major sigma factor (rpoD)	28
448	pilin repressor (pilB)	53
408	pilin repressor (pilB) motif	49
104	virulence-associated prt homolog (vacB)	29
	Replication	
	Degradation of DNA	
032	ATP-dependent nuclease (addA)	27
	DNA replication, restriction, modification, recombination, and repair	
469	chromosomal replication initiator prt (dnaA)	31
004	DNA gyrase sub A (gyrA)	100
003	DNA gyrase sub B (gyrB)	99
244	DNA helicase II (mutB1)	36
254	DNA ligase (lig)	38
262	DNA polymerase I (polI) motif	30
031	DNA polymerase III (polC)	38
261	DNA polymerase III alpha sub (dnaE)	32
001	DNA polymerase III beta sub (dnaN)	100
420	DNA polymerase III sub (dnaH)	49
007	DNA polymerase III sub (dnaH) motif	23
250	DNA primase (dnaE)	27
010	DNA primase (dnaE) motif	26
122	DNA topoisomerase I (topA)	39
235	endonuclease IV (nfo)	29
421	excinuclease ABC sub A (uvrA)	48
073	excinuclease ABC sub B (uvrB)	48
206	excinuclease ABC sub C (uvrC)	28
379	glucose-inhibited division prt (gidA)	40
380	glucose-inhibited division prt (gidB)	25
358	Holliday junction DNA helicase (ruvA)	26
359	Holliday junction DNA helicase (ruvB)	35
184	MTase (ssoIM)	43
339	recombination prt (recA)	47
094	replicative DNA helicase (dnaB)	33
438	restriction-modification enzyme EcoD specificity sub (hsdS)	25
047	S-adenosylmethionine Sase 2 (metX)	44
091	single-stranded DNA BP (ssb)	22
204	DNA topoisomerase IV sub A (parC)	100
203	DNA topoisomerase IV sub B (parE)	100
097	uracil DNA glycosylase (ung)	33
	Transcription	
	Degradation of RNA	
367	ribonuclease III (rnc)	30
465	RNase P C5 sub (rnpA)	40
	RNA synthesis, modification, and DNA transcription	
308	ATP-dependent RNA helicase (deaD)	23
425	ATP-dependent RNA helicase (deaD)	32
018	helicase (mot1) motif	44
141	N-utilization substance prt A (nusA)	36
177	RNA polymerase alpha core sub (rpoA)	31
341	RNA polymerase beta sub (rpoB)	39
340	RNA polymerase beta' chain (rpoC)	47
022	RNA polymerase delta sub (rpoE)	29
249	RNA polymerase sigma A factor (sigA)	44
054	transcription antitermination factor (nusG)	31
	Translation	
	Amino acyl tRNA synthetases and tRNA modification	
292	Ala-tRNA Sase (alaS)	34
378	Arg-tRNA Sase (argS)	34
113	Asn-tRNA Sase (asnS)	41
036	Asp-tRNA Sase (aspS)	41
253	Cys-tRNA Sase (cysS)	34
462	Glu-tRNA Sase (gltX)	43
251	Gly-tRNA Sase	36
035	His-tRNA Sase (hisS)	31
345	Ile-tRNA Sase (ileS)	33
266	Leu-tRNA Sase (leuS)	43
136	Lys-tRNA Sase (lysS)	46
365	Met-tRNA formyltransferase (fmt)	24
021	Met-tRNA Sase (metS)	38
083	peptidyl-tRNA hydrolase homolog (pth)	38
195	Phe-tRNA Sase alpha chain (pheT)	26
194	Phe-tRNA Sase beta chain (pheS)	35
283	Pro-tRNA Sase (proS)	23
182	pseudouridylate Sase I (hisT)	27
005	Ser-tRNA Sase (serS)	43
375	Thr-tRNA Sase (thrSv)	39
445	tRNA (guanine-N1)-MTase (trmD)	41
126	Trp-tRNA Sase (trpS)	41
455	Tyr tRNA Sase (tyrS)	39
334	Val-tRNA Sase (valS)	39
	Degradation of proteins, peptides, and glycopeptides	
391	aminopeptidase	45
324	aminopeptidase P (pepP)	31
239	ATP-dependent protease (lon)	44
355	ATP-dependent protease binding sub (clpB)	48
067	glutamic acid specific protease (SPase)	29
219	IgA1 protease	32
183	oligoendopeptidase F (pepF)	30
020	proline iminopeptidase (pip)	38
310	proline iminopeptidase (pip)	29
046	sialoglycoprotease (gcp)	36
238	trigger factor (tig)	25
	Protein modification and translation factors	
089	elongation factor G (fus)	59
026	elongation factor P (efp)	26
433	elongation factor Ts (tsf)	39
451	elongation factor TU (tuf)	100
106	formylmethionine deformylase (def) motif	37
173	initiation factor 1 (infA)	49
172	methionine amino peptidase (map)	36
258	peptide chain release factor 1 (RF-1)	43
108	prt phosphatase 2C homolog (ptc1) motif	28
109	prt serine-threonine kinase motif	34
142	prt synthesis initiation factor 2 (infB)	46
435	ribosome releasing factor (frr)	35
282	transcription elongation factor (greA)	40
196	translation initiation factor IF3 (infC)	31
	Ribosomal proteins: synthesis and modification	
082	ribosomal prt L1	48
361	ribosomal prt L10	30
081	ribosomal prt L11	52
418	ribosomal prt L13	40
161	ribosomal prt L14	63
169	ribosomal prt L15	42
158	ribosomal prt L16	64
178	ribosomal prt L17	35
167	ribosomal prt L18	43
444	ribosomal prt L19	49
154	ribosomal prt L2	58
198	ribosomal prt L20	58
232	ribosomal prt L21	38
233	ribosomal prt L21 homolog	100
156	ribosomal prt L22	49
153	ribosomal prt L23	39
162	ribosomal prt L24	45
234	ribosomal prt L27	64
426	ribosomal prt L28	36
159	ribosomal prt L29	42
151	ribosomal prt L3	43
257	ribosomal prt L31	37
363	ribosomal prt L32	48
325	ribosomal prt L33	58
466	ribosomal prt L34	67
197	ribosomal prt L35	60
174	ribosomal prt L36	78
152	ribosomal prt L4	39
163	ribosomal prt L5	58
166	ribosomal prt L6	46
362	ribosomal prt L7/L12 "A" type)	48
093	ribosomal prt L9	33
150	ribosomal prt S10	49
176	ribosomal prt S11	48
087	ribosomal prt S12	75
175	ribosomal prt S13	63
164	ribosomal prt S14	70
424	ribosomal prt S15	48
446	ribosomal prt S16	49
160	ribosomal prt S17	51
092	ribosomal prt S18	45
155	ribosomal prt S19	59
070	ribosomal prt S2	35
157	ribosomal prt S3	47
311	ribosomal prt S4	43
168	ribosomal prt S5	56
090	ribosomal prt S6	24
012	ribosomal prt S6 modification prt (rimK) motif	31
088	ribosomal prt S7	65
165	ribosomal prt S8	47
417	ribosomal prt S9	52
252	rRNA methylase	39
	Transport and binding proteins	
	Amino acids, peptides, and amines	
226	aromatic amino acid transport prt (aroP)	25
180	membrane transport prt (gl Q)	37
303	membrane transport prt (gl Q)	32
079	oligopeptide transport ATP-BP (amiE)	48
080	oligopeptide transport ATP-BP (amiF)	47
078	oligopeptide transport permease prt (dciAC)	33
077	oligopeptide transport permease prt (oppB)	28
042	spermidine-putrescine transport ATP-BP (potA)	42
043	spermidine-putrescine transport permease prt (potB)	27
044	spermidine-putrescine transport permease prt (potC)	29
	Anions	
410	peripheral membrane prt B (pstB)	51
409	peripheral membrane prt U (phoU)	27
411	periplasmic phosphate permease homolog (AG88)	31
	Carbohydrates, organic alcohols, and acids	
187	ATP-BP (msmK)	41
062	fructose-permease IIBC component (fruA)	43
033	glycerol uptake facilitator (glpF)	36
061	hexosephosphate transport prt (uhpT)	31
188	membrane prt (msmF)	22
189	membrane prt (msmG)	27
119	methylgalactoside permease ATP-BP (mglA)	33
429	PEP-dependent HPr prt kinase phosphoryltransferase (ptsI)	46
041	phosphohistidinoprotein-hexose phosphotransferase (ptsH)	49
069	phosphotransferase enzyme II, ABC component (ptsG)	43
129	PTS glucose-specific permease	25
120	ribose transport permease prt (rbsC)	27
	Cations	
071	cation-transporting ATPase (pacL)	34
	Other	
290	ATP-BP P29	32
289	high affinity transport prt P37 (P37)	36
390	lactococcin transport ATP-BP (lcnDR3)	22
322	Na⁺ ATPase sub J (ntpJ)	31
014	transport ATP-BP (msbA)	28
015	transport ATP-BP (msbA)	32
291	transport permease prt P69 (P69)	28
406	transport permease prt P69 (P69) motif	40
	Other categories	
	Adaptations and atypical conditions	
454	osmotically inducible prt (osmC)	28
	phosphate limitation prt (sphX)	31
470	SpoOJ regulator motif	27
277	spore germination apparatus prt (gerBB) motif	31
383	sporulation prt (outB) motif	36
	Drug and analog sensitivity	
463	high-level kasgamycin resistance (ksgA)	36
	Other	
298	115 kD prt (p115)	33
190	29 kDa prt, MgPa operon (mgp)	62
065	heterocyst maturation prt (devA)	35
467	heterocyst maturation prt (devA)	40
099	hydrolase (aux2)	32
131	hypothetical prt (GB:M31161_3)	22
218	macrogolgin	25
327	magnesium chelatase 30 kD sub (bchO)	27
304	membrane-associated ATPase (cbiO)	30
364	mobilization prt (mob13) motif	31
336	nitrogen fixation prt (nifS)	26
287	nodulation prt F (nodF)	35
098	p48 eggshell prt	23
100	PET112 prt	31
037	pre-B cell enhancing factor (PBEF)	34
288	prt L	31
328	prt V (fcrV)	28
145	prt X	29
280	sensory rhodopsin II transducer (htrII) motif	16
059	small prt (smpB)	33
360	UV protection prt (mucB)	22

*MG # corresponds to the gene's number in Figure 7.33. % ID denotes the percentage of similarity between the amino acid sequence of the gene in *M. genitalium* and the most similar polypeptide from known amino acid sequences of another species.

Reprinted with permission from Fraser, C. M., et al. 1995. The minimal gene compliment of *Mycoplasma genitalium*. *Science* 270:397–403. Copyright 1995 American Association for the Advancement of Science.

SUMMARY

1. Bacteria are usually studied in laboratory cultures. They are grown on a liquid or agar-solidified medium that contains all the nutrients needed for the bacterial cells to grow. Minimal medium contains only those substances needed for growth.

2. Bacteria that are prototrophic for a substance are capable of growing on medium that does not contain that substance. Bacteria that are auxotrophic for a substance are unable to grow unless the medium is supplemented with that substance.

3. Genetic variation used in bacterial genetics research includes auxotrophic mutations, carbon-source mutations, antibiotic-sensitive or resistant genes, and temperature-sensitive mutations.

4. Each unique mutation within a single gene creates a new allele of that gene. Wild-type alleles are the versions of a gene found most often in nature. In most cases, wild-type alleles produce a functional protein product. Mutant alleles arise by mutation of a wild-type allele and usually fail to encode functional products.

5. The observed characteristics of a bacterial culture are called the phenotype. The underlying genetic constitution that confers the phenotype is called the genotype.

6. Medium can be selective or nonselective. A mutant bacterial strain can be distinguished from a nonmutant strain on selective medium. Both mutant and nonmutant strains grow equally well on nonselective medium.

7. There are three types of DNA transfer in bacteria: transformation, conjugation, and transduction.

8. Transformation is the uptake of DNA into a bacterial cell from the cell's surroundings. The introduced DNA recombines with its homologous counterpart on the chromosome. Recombination can be used to map genes relative to one another on the chromosome.

9. Conjugation is the transfer of bacterial DNA from one cell to another through a conjugation bridge. The transferred DNA is contained within an F factor, a plasmid that becomes an episome when it inserts itself into the chromosomal DNA. A bacterial cell with an F factor is an F$^+$ cell, whereas a cell without the F factor is an F$^-$ cell.

10. In conjugation, DNA is transferred in one direction only: from an F$^+$ cell to an F$^-$ cell. The F factor replicates in the F$^+$ cell using σ-mode replication, and one copy of the F factor passes through the conjugation bridge to the F$^-$ cell, converting it into an F$^+$ cell.

11. The F factor can remain separate from the chromosome and replicate autonomously using θ-mode replication, or it can insert itself into the chromosome, where it is replicated along with the chromosomal DNA. A bacterial cell with an integrated F factor is called an Hfr cell.

12. The integrated F factor in an Hfr cell may initiate DNA transfer to an F$^-$ cell during conjugation. The F factor causes σ-mode replication of the chromosome and carries the chromosomal DNA along with it into the F$^-$ cell. Usually mating ceases before transfer of the entire chromosome, meaning that only a part of the F factor and a piece of the chromosome is transferred.

13. The transferred chromosomal DNA may recombine with its homologous counterpart in the chromosome. The time at which genes enter the cell and recombine during conjugation between an Hfr cell and an F$^-$ cell allows the genes to be mapped on the chromosome. Numerous genes on the *E. coli* chromosome have been mapped in this way, providing detailed maps of the *E. coli* chromosome.

14. An integrated F factor may excise itself from the chromosome. Sometimes the excision is faulty and the F factor takes some chromosomal DNA along with it. An F factor with chromosomal DNA integrated into it is called an F′ factor. The F′ factor may be transferred during conjugation, resulting in two copies of the integrated DNA in the cell: one in the F′ factor, the other in the chromosome. A cell with two copies of part of its chromosomal DNA is called a merozygote or partial diploid.

15. Complementation analysis can be done in merozygotes to determine whether or not two mutations belong to the same gene. If the two mutations fail to complement each other, they are probably alleles of the same gene. If they do complement each other, they are probably alleles of different genes.

16. Bacteriophages are bacterial viruses. They may be either virulent or temperate. A virulent phage multiplies into numerous progeny phage particles soon after infection, then lyses the cell, releasing the progeny phage particles to infect other cells. Temperate phages may also enter the lytic cycle soon after infection, or they may enter lysogeny, where the phage integrates into the chromosomal DNA and is replicated along with the chromosomal DNA as the bacterium divides. An integrated lysogenic phage is called a prophage. At some point, the prophage may excise itself from the chromosome and enter the lytic cycle.

17. DNA may be transferred from one cell to another through a phage intermediate, a process called transduction.

18. There are two types of transduction. In specialized transduction, phage particles contain a specific segment of chromosomal DNA connected to the phage DNA. In generalized transduction, a phage capsid mistakenly packages chromosomal DNA instead of phage DNA. The phage particle that contains chromosomal DNA injects the chromosomal DNA into another cell when the phage infects the cell.

19. The complete nucleotide sequences of several bacterial and phage genomes have been determined. These sequences are the first step toward understanding the complete biology of an organism.

QUESTIONS AND PROBLEMS

1. **(a)** Why are bacterial growth media autoclaved before use? **(b)** Why is filtration sometimes used to add substances after autoclaving?

2. For each genotype and phenotype listed in Table 7.7, indicate what type of allele the strain carries (prototrophic, auxotrophic, carbon source mutant, temperature-sensitive mutant, antibiotic resistant, or antibiotic susceptible).

3. Table 7.8 shows results reported by Taylor and Thoman (1964. *Genetics* 50:659–677) for one Hfr strain used for time-of-entry mapping in *E. coli.* **(a)** Given this information, construct a time-of-entry map for these genes. **(b)** How many minutes are represented in this map?

4. Which of the following bacterial strains would require replica plating to identify and recover a colony: **(a)** arg^-, **(b)** met^-, **(c)** amp^r, **(d)** amp^s, **(e)** lac^-, **(f)** gal^-, and **(g)** dies at 40°C?

5. Look at Example 7.1 illustrating the experiments of Lederberg and Tatum. In their control experiments, some of the colonies gained prototrophy for one gene. What was the reason for these changes?

6. How were Lederberg and Tatum able to distinguish mutation from recombination?

7. Lederberg and Tatum concluded that their results could be explained if two cells fused with each other. Describe how their experiments could also be explained by transformation or conjugation.

8. What subsequent experiments suggested that Lederberg and Tatum's results were best explained by conjugation?

9. Outline the steps from the moment that a portion of an Hfr donor chromosome enters an F⁻ recipient cell to the appearance of a recombinant cell.

10. Design an experiment to select for new λgal and λbio transducing phages. Be sure to include the genotypes of the donor and recipient strains, the types of selective media used, and the method for recovering the transducing phage.

11. Why is it advantageous to choose an Hfr strain that is susceptible to an antibiotic and an F⁻ recipient strain that is resistant to the same antibiotic for time-of-entry mapping experiments?

12. Explain why the gene selected for antibiotic resistance in time-of-entry mapping experiments should be distant from the region being mapped.

13. Look at Example 7.5 illustrating the experiments described by Wollman, Jacob, and Hayes. What were the purposes of the *thr*, *leu*, and *str* genes?

14. Assume that you wish to distinguish lac^-, leu^-, or str^r strains from a wild-type strain that is $lac^+ leu^+ str^s$. In Table 7.9, write an S for "selective" if the medium in each row can distinguish the strain indicated in each column from the wild-type ($lac^+ leu^+ str^s$) strain for at least one of the three phenotypes. If the medium cannot be used to distinguish the strain from the wild-type strain, write N for "nonselective."

15. Table 7.10 shows results reported by Taylor and Thoman (1964. *Genetics* 50:659-677) for two Hfr strains used for time-of-entry mapping in *E. coli.* The F-factor origin of replication in strain AB261 is in reverse orientation when compared to the origin of replication for strain AB259. **(a)** Given this information, construct a time-of-entry map for these genes. **(b)** How many minutes are represented in this map?

Table 7.7 Information for Question 2: Determining the Underlying Allele Type

Genotype	Phenotype
arg^-	Cannot grow in the absence of arginine
gal^-	Cannot grow when galactose is the only sugar present
kan^s	Cannot grow in the presence of kanamycin
bio^-	Cannot grow in the absence of biotin
lac^+	Can grow when lactose is the only sugar present
amp^r	Can grow in the presence of ampicillin
thr^+	Can grow in the absence of threonine

Table 7.8 Information for Question 3: Time-of-Entry Mapping Data*

		Genes		
F⁻ Strain	Hfr Strain	*his*	*aroD*	*trpB*
AT2121	AB311	7.0		22.0
AT1372	AB311	6.0		21.0
AT1359	AB311	6.0	12.0	
AT1360	AB311	5.5	11.5	

*The numbers denote the number of minutes elapsed before a gene enters the F⁻ cell.

Table 7.9 Information for Question 14: Determining the Selectivity of Different Media for Different Bacterial Strains

Medium	Genotype				
	lac⁺ leu⁻ str^s	*lac⁻ leu⁻ str^s*	*lac⁻ leu⁺ str^s*	*lac⁺ leu⁺ str^r*	*lac⁻ leu⁻ str^r*
Minimal					
Minimal + leucine					
Minimal − glucose + lactose					
Minimal + streptomycin					
Minimal + leucine + streptomycin					
Minimal + lactose − glucose + streptomycin					
Minimal + leucine + lactose − glucose + streptomycin					

Table 7.10 Information for Question 15: Time-of-Entry Mapping Data*

F⁻ Strain	Hfr Strain	Genes							
		proA	thr	purD	purE	gal	purB	proB	leu
AB1133	AB261	5.5	12.0						
AT1380	AB261	5.5		23.0					
AT2213	AB259		7.25						7.75
AT2270	AB259	15.0			19.5	24.0			
AB1325	AB259	15.0				24.0	32.0		
AB2217	AB259							16.0	8.0

*The numbers denote the number of minutes elapsed before a gene enters the F⁻ cell.

16. The orientation of the F-factor origin of replication in *E. coli* strain AB311 is the same as the F-factor origin of replication in strain AB313. The *trpB* and *gal* genes are separated by 8 minutes. The *metB* and *purD* genes are separated by 1 minute. From this information, and from the data in Example 7.6 and the previous problem, construct a complete circular map of the *E. coli* chromosomal DNA.

17. The discovery of transduction by Zinder and Lederberg, reported in 1952, was summarized in Example 7.7. Three years later, in 1955, Morse, Lederberg, and Lederberg (*Genetics* 41:142–156) published additional data on transduction of genes responsible for galactose metabolism in *E. coli*. In one experiment, they used Gal$^+$ and Gal$^-$ strains of *E. coli*. One-half of a petri plate that contained medium with galactose as the sole carbon source was inoculated with phages that came from lysed cells of the Gal$^+$ strain. The other half of the plate was left uninoculated. A Gal$^-$ strain that did not contain λ was then placed on the entire plate. Numerous Gal$^+$ colonies appeared on the half of the plate inoculated with phage, whereas only a few appeared on the half that was not inoculated with phage. **(a)** Describe how carbon source mutations were used in this experiment. **(b)** Why does this experiment demonstrate that transduction, instead of transformation or conjugation, was responsible for the Gal$^+$ colonies?

18. Morse, Lederberg, and Lederberg, in the same article discussed in the previous question (1955. *Genetics* 41:142–156), compared their results of transduction in *E. coli* with the results of Zinder and Lederberg (1952. *Journal of Bacteriology* 64:679–699), who studied transduction in *Salmonella typhimurium*. Zinder and Lederberg were able to transduce to prototrophy strains of *S. typhimurium* that were auxotrophic for different substances. Morse, Lederberg, and Lederberg, on the other hand, attempted to transduce *E. coli* strains that were auxotrophic for several different substances, including histidine, leucine, methionine, proline, glycine, serine, and tryptophan. Transduction to prototrophy was not observed for any of these amino acids. The authors only observed transduction for *gal* genes. What could account for this difference between transduction in *S. typhimurium* and transduction in *E. coli*?

19. In this same article (1955. *Genetics* 41:142–156), Morse, Lederberg, and Lederberg described experiments in which they used a high concentration of λ phages from lysed Gal$^+$ cells to transduce Gal$^-$ cells. They selected for transduced cells by growing them on galactose medium. The transduced cells were all lysogenic and Gal$^+$. When the lysogenic transduced cells were induced to undergo lysis by treatment with ultraviolet light, an unusually high percentage of the phage particles (in some cases, nearly all the phage particles) released by lysis were *gal$^+$*

transducing phages. They called these HFT lysates (for "high-frequency transduction"). Arber, Kellenberger, and Weigel (1957. *Schweizerische Zeitschrift fur Allgemeine Pathologie und Bakteriologie* 20:659–665) found that the lysates of HFT cells had high numbers of transducing phages with *gal$^+$* genes as well as normal λ phages (with no *gal* genes). Devise a model of phage infection, lysogeny, and induction that would explain these observations.

20. Look at the "Replication" section of Table 7.6 for enzymes encoded by known genes in the *Mycoplasma genitalium* genome. Describe the functions of as many of the gene products in this section as you can.

21. Look at Figure 7.33. What evidence is there that genes of similar function may be clustered together on the genome?

22. In the same article highlighted in Example 7.8 (Sofia et al. 1994. *Nucleic Acids Research* 22:2576–2586), the authors were faced with the task of identifying the sequence of the gene *pscA*. Mutations in *pcsA* cause cold sensitivity. The *pcsA* gene had been mapped at 82 minutes. All potential genes in this region had already been assigned to known genes, leaving no unassigned potential genes as candidates for *pcsA*. Certain mutant alleles of the *rfaG* gene, which also mapped to 82 minutes, caused a phenotype similar to *pscA* mutants. From this information, the authors concluded that *rfaG* and *pcsA* were probably the same gene. **(a)** What evidence led the authors to arrive at this conclusion? **(b)** What type of experiments discussed in this chapter might be used to test this conclusion?

FOR FURTHER READING

An excellent textbook that covers in detail all aspects of bacterial and phage genetics is **Maloy, S. R., J. E. Cronan, Jr., and D. Freifelder. 1994. *Microbial Genetics*, 2nd ed. Boston: Jones and Bartlett.** An informative presentation describing key research in bacterial genetics is presented at the beginning of the book **Adelberg, E. A. 1966. *Papers on Bacterial Genetics*, 2nd ed. Boston: Little, Brown.** This book also contains copies of some of the most significant articles on bacterial genetics. Lederberg and Tatum's original papers on recombination in *E. coli* are **Lederberg, J., and E. L. Tatum. 1946. Novel genotypes in mixed cultures of biochemical mutants of bacteria. *Cold Spring Harbor Symposium on Quantitative Biology* 11:113–114; and Lederberg, J., and E. L. Tatum. 1946. Gene recombination in *Escherichia coli*. *Nature* 158:558–559.** A discussion of the molecular mechanisms of recombination in bacteria can be found in **Lewin, B. 1994. *Genes V*. Oxford: Oxford University Press, pp. 978–983.** Early work on transformation was published by **Griffith, F. 1928. The significance of pneumococcal types. *Journal of Hygiene* 27:141–144; and Avery, O. T., C. M.**

MacLeod, and M. McCarty. 1944. Studies on the chemical nature of the substance inducing transformation of the pneumococcal types. *Journal of Experimental Medicine* 79:137–158. Discoveries on the mechanism of transformation are found in Lacks, S. 1962. Molecular fate of DNA in genetic transformation of *Pneumococcus*. *Journal of Molecular Biology* 5:119–131; and Fox, M. S. and M. K. Allen. 1964. On the mechanism of deoxyribonucleate integration in pneumococcal transformation. *Proceedings of the National Academy of Sciences, USA* 52:412–419. The U-tube experiments demonstrating the need for bacterial cell contact for gene recombination in *E. coli* were published by Davis, B. D. 1950. Nonfiltrability of the agents of recombination. *Journal of Bacteriology* 60:507–508. Transduction was first described in a classic paper by Zinder, N. D. and J. Lederberg. 1952. Genetic exchange in *Salmonella*. *Journal of Bacteriology* 64:679–699. An extensive genetic map of *E. coli* constructed from a compilation of Hfr conjugation experiments was published by Bachmann, B. J. 1990. Linkage map of *Escherichia coli* K-12, Edition 8. *Microbiological Reviews* 54:130–197. Articles describing sequencing of entire bacterial genomes are Fleischmann, R. D., et al. 1995. Whole genome random sequencing and assembly of *Haemophilus influenzae* Rd. *Science* 269:496–512; Fraser, C. M., et al. 1995. The minimal gene complement of *Mycoplasma genitalium*. *Science* 270:397–403; Bult, C. J. 1996. Complete genome sequence of the methanogenic archaeon, *Methanococcus jannaschii*. *Science* 273:1058–1073; and Blattner, F. R., et al. 1997. The complete genome sequence of *Escherichia coli* K-12. *Science* 277:1453–1462.

For additional reading, go to InfoTrac College Edition, your online research library at: http: www.infotrac-college.com/brookscole

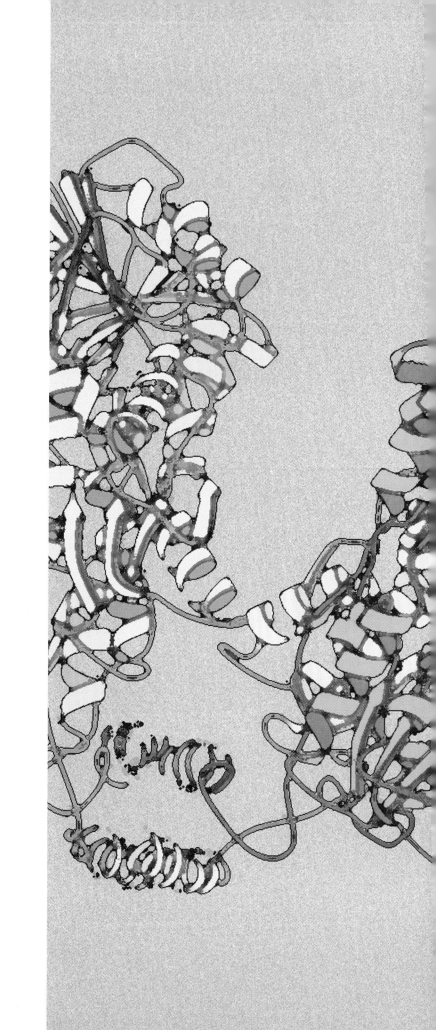

CHAPTER 8

REGULATION OF GENE EXPRESSION

A single human cell contains over 100,000 different genes. Yet only a small fraction of those genes are expressed in any one cell. The cells of all multicellular organisms are specialized to perform the specific functions of the tissues in which they reside. Most of the genes within a cell do not encode products that are needed for the cell to function, so those genes are not transcribed. For example, a cell within the retina of your eye expresses a particular set of genes that it requires to carry out its function. A cell in the lining of your intestinal tract expresses a different set of genes to carry out its function.

Prokaryotes are unicellular, so cells do not differentiate into tissues and organs. Every cellular function must be carried out by the single cell itself. Even so, prokaryotic cells regulate the expression of their genes, often in response to different environments. For example, *E. coli* may live in a mammalian intestine, thriving on the food that its host eats, or it may live outside of its host in a much colder environment and with a different source of nourishment. A bacterial cell has no tissues to protect it from changes in its environment, and no multicellular network to regulate its temperature in response to environmental changes. Instead, it must rely on the ability of its genes to respond rapidly to changes in the environment.

In this chapter, we will examine **gene regulation**, the ability of cells to determine which genes should be expressed, when those genes should be expressed, and the extent to which each gene should be expressed. There are four levels of gene regulation that apply to both prokaryotes and eukaryotes:

1. *Transcriptional regulation*: Determination of whether or not, and to what extent, a gene will be transcribed.

2. *Posttranscriptional regulation*: Regulation of the processing, transport, and longevity of mRNAs. In eukaryotes this level includes regulation of the rate and extent of processing pre-mRNA into mature mRNA, the transport of mRNA into the cytoplasm, and the rate of RNA degradation.

3. *Translational regulation*: Determination of whether or not and to what extent an mRNA will be translated.

4. *Posttranslational regulation*: Regulation of protein modification and activity.

Most genes and their products are regulated at all four levels, although transcriptional regulation is usually the most important. We'll begin our discussion with bacterial gene regulation, and then move on to eukaryotes in the latter part of the chapter. Most of our discussion will

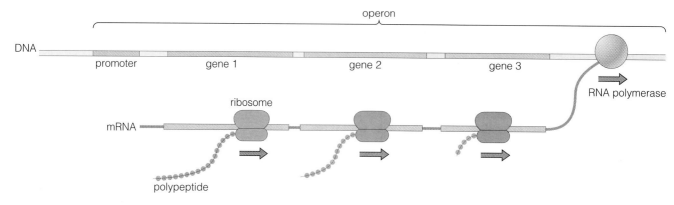

DNA

promoter   gene 1   gene 2   gene 3

RNA polymerase

ribosome

mRNA

polypeptide

Figure 8.1 A prokaryotic operon containing three genes. RNA polymerase transcribes an operon as a single mRNA. Ribosomes translate each gene in the mRNA separately.

focus on transcriptional regulation because it is the most important and the best understood of the four levels of gene regulation.

8.1 THE *lac* OPERON

A common mechanism for transcriptional gene regulation in prokaryotes is the operon. An **operon** is a cluster of different genes that are transcribed together as a single mRNA, so transcriptional regulation affects all genes in the operon equally. In most cases, the genes in an operon encode the different enzymes required for a single biochemical pathway, providing an effective means to jointly regulate the production of those enzymes. The genes within an operon share a single promoter and are transcribed jointly as a single large mRNA, which is then translated into the several different proteins (Figure 8.1).

How can a single mRNA code for several different polypeptides? Recall from section 4.6 that translation in prokaryotes does not begin at the 5′ end of the mRNA but at a conserved sequence in the mRNA, called the Shine-Dalgarno sequence, followed by an initiation codon. The ribosomes assemble at the initiation site, then translate the mRNA until they reach a termination codon, where they separate from the mRNA. An operon containing three genes has three initiation sites, each followed by a reading frame and a termination codon. Ribosomes may assemble simultaneously at the three initiation sites and translate all three genes simultaneously (Figure 8.1).

When an operon is actively transcribed, the products of each of the genes are produced at the same time in approximately equal amounts. When an operon's gene products are not needed, transcription of the operon is blocked, and production of all the gene products encoded by the operon ceases.

The first operon discovered is the *lac* operon in *E. coli.* The *lac* operon not only illustrates the structure and function of an operon, but also provides a good example of how genes are regulated in response to environmental stimuli. Before we look at the structure of the *lac* operon, let's take a few moments to discuss its function.

Function of the *lac* Operon

E. coli is a bacterium that lives in the intestinal tracts of mammals, including humans, and in certain places outside of the mammalian intestine, such as waterways and sewers. Only under limited circumstances does *E. coli* come in contact with lactose, the principal sugar found in mammalian milk. From birth until weaning, mammalian offspring rely mostly on milk for their nutrition, making lactose the most abundant carbohydrate in their intestines. *E. coli* cells living in the intestine of an unweaned mammal must rely on lactose as their source of energy. However, an *E. coli* cell cannot use lactose directly for energy. Instead, it produces enzymes that transport the lactose into the cell and break it down into two sugars, glucose and galactose. The *lac* operon contains the genes that encode these enzymes (Figure 8.2).

With the exception of humans and their mammalian pets, most mammals cease using milk when they are weaned during infancy, meaning that lactose is no longer present in their intestines after weaning. It would be wasteful of energy and resources for *E. coli* cells in a weaned mammal to produce the enzymes for lactose metabolism when lactose is no longer present. The *lac* operon ensures that these enzymes are produced only when lactose is the principal source of carbohydrates in the environment. Lactose is, therefore, the environmental stimulus that activates transcription of the *lac* operon.

Structure and Regulation of the *lac* Operon

The *lac* operon contains three adjacent genes within a transcribed region, as diagrammed in Figure 8.2. The transcribed region of the *lac* operon encodes a single mRNA with three genes, each of which encodes an enzyme for lactose metabolism. The first gene, *lacZ*, encodes the enzyme **β-galactosidase**, which converts lactose into glucose and galactose. Enzymes encoded by another

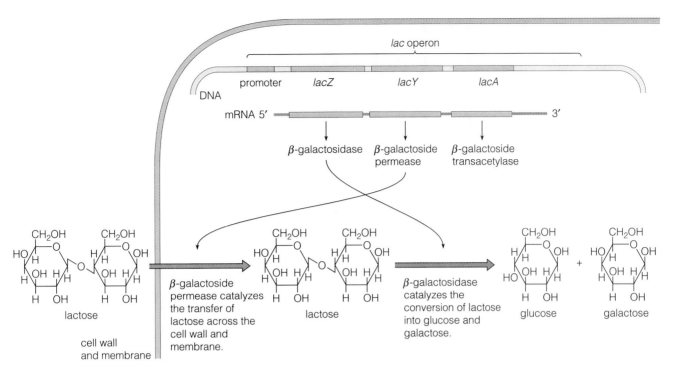

Figure 8.2 The *lac* operon and the functions of two of the three enzymes it produces.

operon, called the *gal* operon, convert galactose to glucose-1-phosphate, a derivative of glucose. Thus, lactose is eventually converted to glucose or one of its derivatives, sugars that the bacterium can metabolize to acquire energy. *E. coli* requires β-galactosidase in order to utilize lactose as an energy source.

The second gene, *lacY*, encodes **β-galactoside permease,** an enzyme that causes lactose to permeate the cell wall and membrane and enter the cell. This enzyme is also required for *E. coli* to use lactose. The third gene, *lacA*, encodes the enzyme **β-galactoside transacetylase**, which acetylates (adds a —CH₂CH₃ group to) lactose or molecules that resemble lactose. This enzyme is not essential for the cell to utilize lactose. Its primary function appears to be detoxification of certain molecules that resemble lactose and are toxic to the cell if left unacetylated.

Upstream from the transcribed region is a regulatory region that contains a promoter, where RNA polymerase binds to initiate transcription. The *lac* operon also contains three **operators**, sites in the DNA where regulatory proteins bind. The first operator, called the **principal operator (O_1)**, occupies a position downstream from the promoter. The other two operators are called **auxiliary operators**, because they interact with a protein that binds first to the principal operator. The auxiliary operators are called O_2, located within the *lacZ* gene, and O_3, located upstream from O_1. Figure 8.3 shows the DNA sequence of the regulatory region in the *lac* operon, the promoter, and the three operators.

An additional gene that resides outside of the operon, *lacI*, is called the **regulator gene** and encodes the *lac* **repressor protein**, a protein that recognizes and binds to

the operators in the *lac* operon and blocks transcription. The *lac* operon is classified as an **inducible operon**, which means that the pathway's substrate (lactose in this example) induces transcription of the operon. When lactose is present, the operon can be transcribed, producing all three enzymes. When lactose is absent, however, the operon is not transcribed, and the three enzymes are not produced.

This process offers a good illustration of the regulatory system employed by operons. The regulator gene for the *lac* operon, *lacI*, is located just a short distance from the operon itself, although for some other operons the regulator gene may be quite distant from the operon. The *lacI* gene encodes the identical subunits of the *lac* repressor protein. Shortly after being translated, the four identical protein subunits encoded by the *lacI* gene bind to each other to form the functional tetrameric repressor protein (Figure 8.4). The *lacI* gene has a **constitutive promoter**, a promoter that causes RNA polymerase to transcribe the gene at a constant rate under all circumstances. RNA polymerase does not have a high affinity for the *lacI* promoter, so although the gene is transcribed at a constant rate, that rate is fairly low relative to that of other genes. The rates of transcription, translation, and protein degradation are such that there are about 10 repressor protein molecules in each cell at any one time.

Each subunit of the repressor protein has two active sites. The first site, called the **DNA-binding site**, is where the repressor molecule binds to DNA. The DNA-binding site has a very high affinity for the operator sequences in the *lac* operon. When lactose is absent, the repressor protein binds tightly to the operators and prevents RNA

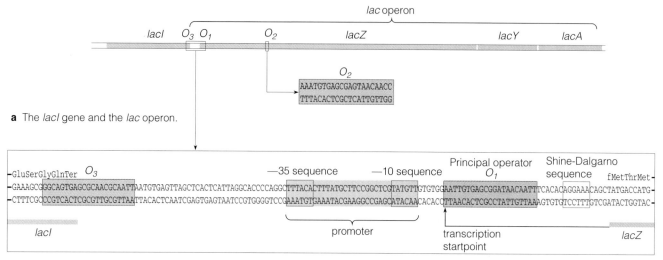

a The *lacI* gene and the *lac* operon.

O_2

```
AAATGTGAGCGAGTAACAACC
TTTACACTCGCTCATTGTTGG
```

-GluSerGlyGlnTer O_3 —35 sequence   —10 sequence   Principal operator   Shine-Dalgarno
                                                                                              O_1 sequence   fMetThrMet-
-GAAAGCGGGCAGTGAGCGCAACGCAATTAATGTGAGTTAGCTCACTCATTAGGCACCCCAGGCTTTACACTTTATGCTTCCGGCTCGTATGTTGTGTGGAATTGTGAGCGGATAACAATTTCACACAGGAAACAGCTATGACCATG-
-CTTTCGCCCGTCACTCGCGTTGCGTTAATTACACTCAATCGAGTGAGTAATCCGTGGGGTCCGAAATGTGAAATACGAAGGCCGAGCATACAACACACACCTTAACACTCGCCTATTGTTAAAGTGTGTCCTTTGTCGATACTGGTAC-

       lacI   promoter   transcription   *lacZ*
                                                                                          startpoint

b Regulatory region of the *lac* operon.

Figure 8.3 The *lac* operon and its regulatory region. The −35 and −10 sequences are promoter elements (see section 3.3). The Shine Dalgarno sequence is a signal for initiation of translation.

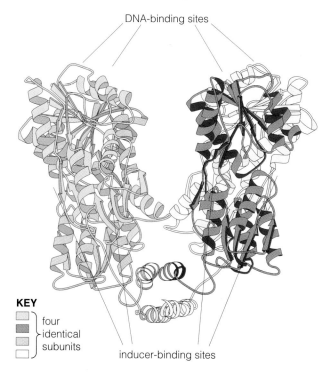

DNA-binding sites

KEY

four identical subunits

inducer-binding sites

Figure 8.4 Structure of the *lac* repressor. The protein is a tetramer formed of four identical subunits. Each subunit has a DNA-binding site and an inducer-binding site. The light blue circle is a plutonium ion and the dark blue circles are mercury ions that function as cofactors. (Reprinted from Lewis, M., G. Chang, N. C. Horton, M. A. Kercher, H. C. Pace, M. A. Schumacher, R. G. Brennan, and P. Lu. 1996. Crystal structure of the lactose operon repressor and its complexes with DNA and inducer. *Science* 271:1247–1254. Copyright 1996 American Association for the Advancement of Science. Reprinted by permission.)

polymerase from binding to the promotor and initiating transcription of the operon.

The repressor protein has the greatest affinity for binding to O_1 and a lesser affinity for binding to O_2 and O_3. The repressor protein also has an affinity for binding to DNA in general, so the repressor proteins not bound to operators probably bind to random sites along the DNA. Two of the four subunits in the repressor protein bind to O_1, and the DNA forms a loop, causing either O_2 or O_3 to bind to the other two subunits of the repressor protein. The repressor protein usually binds to O_1 and O_3, as illustrated in Figure 8.5a. When the repressor protein is bound to operators, the *lac* operon cannot be transcribed, and none of the three enzymes is produced (Figure 8.5a).

The interaction of the repressor protein with the operators is another aspect of the *lac* operon. The DNA in each operator is a sequence of nucleotides that repeats itself in reverse orientation, forming a symmetrical pattern for 28 of the 35 nucleotides in the operator. This phenomenon, where a DNA sequence is repeated symmetrically in reverse orientation, is called a **palindrome** (Figure 8.6). Each side of the palindrome binds to the DNA-binding site in one of the repressor subunits, as illustrated in Figure 8.7.

When lactose enters the cell, β-galactosidase converts some of the lactose molecules to **allolactose,** a derivative of lactose, diagrammed in Figure 8.8. The ability of β-galactosidase to convert lactose to allolactose is secondary to its primary activity of converting lactose to glucose and galactose. Allolactose is the **inducer,** a molecule that binds to a repressor protein causing it to leave the DNA. Each subunit of the *lac* repressor protein contains an **inducer-binding site,** where the inducer binds to the repressor protein.

The allolactose rapidly binds to the inducer-binding sites on all of the repressor protein molecules, including the one bound to the operators. When allolactose enters the inducer-binding sites, the repressor protein undergoes a structural change that alters its DNA-binding sites,

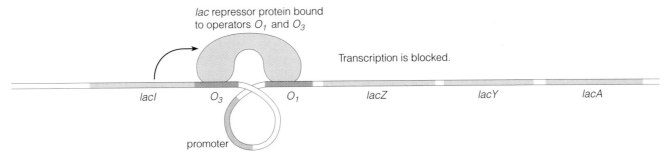

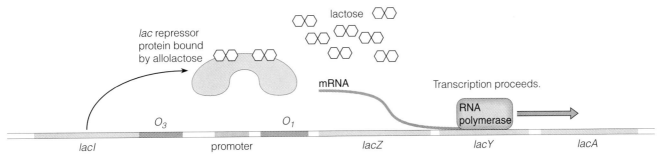

a When lactose is absent, the repressor binds to the operators and blocks transcription of the *lac* operon

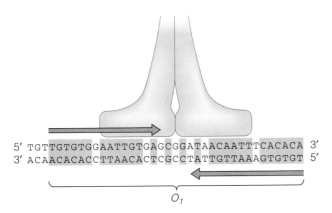

b When lactose is present, some of the lactose is converted to allolactose, which binds to the repressor. When bound to allolactose, the repressor cannot bind to the operators, and transcription may proceed.

Figure 8.5 Regulation of the *lac* operon.

```
5' TGTTGTGTGGAATTGTGAGCGGATAACAATTTCACACA 3'
3' ACAACACACCTTAACACTCGCCTATTGTTAAAGTGTGT 5'
```

O_1

Figure 8.6 Palindromic symmetry of the DNA sequence in the *lac* principal operator, O_1. When read in the 5'→3' direction, the two DNA strands are the same in the boxed regions. Two subunits of the *lac* repressor protein bind to the symmetrical sequences in the operator.

causing the protein to separate from the DNA. So, when allolactose binds to the inducer-binding sites, the repressor protein immediately leaves the operator and no longer prevents RNA polymerase from initiating transcription. The operon can now be transcribed, and all three enzymes can be produced. The enzymes then permit lactose to enter the cell and convert the lactose molecules into glucose and galactose (Figure 8.5b).

When lactose is no longer available, all the remaining lactose and allolactose molecules, including those in the inducer-binding sites, are broken down into glucose and

galactose. With no allolactose in the inducer-binding site, the repressor proteins resume their original conformation, and one of them binds to the operators and prevents further transcription. The mRNA transcribed from the *lac* operon is very unstable, as are the enzymes themselves. Within minutes after transcription of the *lac* operon ceases, the lactose-metabolizing enzymes are rapidly degraded. The enzymes are no longer needed because lactose is no longer available. The repressor protein bound to the operator ensures that the enzymes are no longer produced.

In spite of the elegant nature of this system, you may have noticed a paradox. When lactose first appears in the cell's environment, the lactose must first enter the cell and then be converted to allolactose before the repressor protein can be removed from the operator. Entry of lactose into the cell requires β-galactoside permease, and conversion of lactose to allolactose requires β-galactosidase. Both of these enzymes are encoded by the *lac* operon, which will not be transcribed until allolactose enters its binding site on the repressor protein. How can lactose enter the cell and be converted to allolactose when the genes for these functions are shut down?

The answer to this paradox is that in the absence of lactose, prevention of transcription is not completely effective. The *lac* operon is occasionally transcribed; the operon produces a small number of each of the lactose-metabolizing enzymes. As soon as lactose is present in the cell's environment, some lactose molecules enter the cell due to the low level of β-galactoside permease. A few

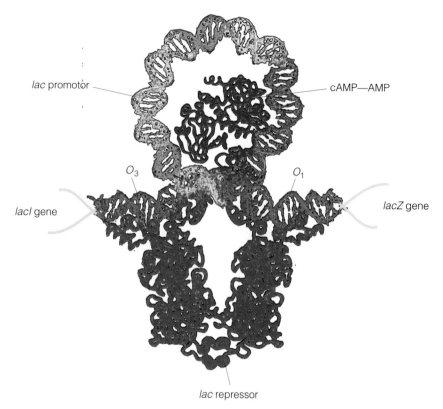

lac promotor

cAMP—AMP

O_3

O_1

lacI gene

lacZ gene

lac repressor

Figure 8.7 The tetrameric *lac* repressor protein binding to O_1 and O_3 in the DNA. The DNA forms a loop to allow the repressor to bind to both operators. cAMP-CAP is a protein complex we will encounter later in this chapter. (Reprinted with permission from Lewis, M., G. Chang, N. C. Horton, M. A. Kercher, H. C. Pace, M. A. Schumacher, R. G. Brennan, and P. Lu. 1996. Crystal structure of the lactose operon repressor and its complexes with DNA and inducer. *Science* 271:1247–1254. Copyright 1996 American Association for the Advancement of Science. Reprinted by permission.)

of these are converted to allolactose by the small number of β-galactosidase molecules in the cell. These allolactose molecules then enter the inducer-binding site on the repressor protein, and the repressor protein leaves the operators. RNA polymerase may now freely transcribe the operon, making large quantities of the lactose-metabolizing enzymes, causing abundant quantities of lactose to enter the cell. This chain of events sets off a flood of enzyme production. When lactose is present in a cell's environment, as much as 10% of the total protein in the cell may be β-galactosidase.

~

The *lac* operon encodes a single mRNA that contains three genes, each encoding an enzyme for lactose metabolism. The organization of the *lac* operon is such that transcription is repressed in the absence of lactose and permitted in its presence.

8.2 MUTATIONS IN THE *lac* OPERON

The components of the *lac* operon function in delicate balance among themselves. When each component functions properly, the enzymes encoded by the operon are produced only when they are needed, and in the proper

amounts. However, mutations in any of the components can disrupt the balance, causing a variety of effects depending on where the mutation occurs. Let's now look at how different mutations affect the function of the *lac* operon.

Repressor Gene Mutations

Mutations in the *lacI* gene, which encodes the repressor protein, have a significant effect on lactose metabolism even though they cause no change in the amino acid sequences of the three operon enzymes. Instead, these mutations affect how and when the three lactose-metabolizing enzymes are produced. Because each subunit of the repressor protein has two active sites, the DNA-binding site and the inducer-binding site, mutations in the *lacI* gene may affect either or both of these sites. A mutation that eliminates the function of the DNA-binding sites prevents the repressor protein subunits from binding to the operators regardless of whether lactose is present or absent (Figure 8.9a). Even though allolactose may still enter the inducer-binding sites, this has no effect on operon transcription because the repressor protein cannot bind to the operator under any conditions. RNA polymerase transcribes the *lac* operon at the same

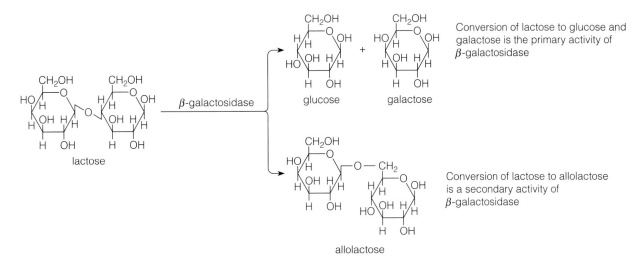

Figure 8.8 Two activities of β-galactosidase.

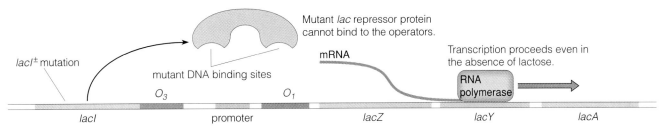

a A *lacI*[±] mutation eliminates the function of the DNA-binding sites of the repressor, so it cannot bind to the operators and block transcription.

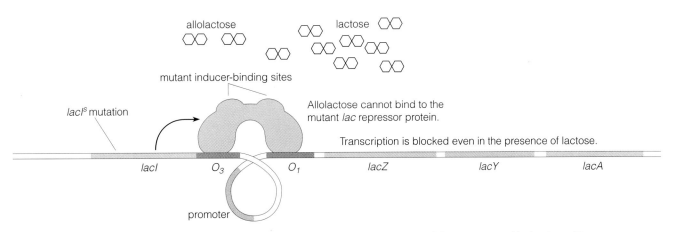

b A *lacI*^s mutation eliminates the function of the inducer-binding sites of the repressor, so allolactose cannot bind to them. The repressor remains permanently bound to the operators and constantly blocks transcription, even in the presence of lactose.

Figure 8.9 The effect of mutations in the *lacI* gene.

rate in the presence and absence of lactose, because the mutant repressor protein is unable to prevent transcription. Mutations that eliminate the function of the DNA-binding sites in the *lac* repressor protein create *lacI*⁻ mutant alleles, the superscript minus sign indicating that functional *lacI* gene product is absent.

A mutation that eliminates the function of the inducer-binding sites on the repressor protein has the

opposite effect of a mutation that eliminates the function of the DNA-binding sites (Figure 8.9b). Because the DNA-binding sites are still intact, the repressor protein binds to the operators and prevents transcription of the operon. However, because there are no functional inducer-binding sites, allolactose cannot bind to the repressor protein. In this case, the repressor protein remains permanently bound to the operator and prevents

transcription of the *lac* operon. Because transcription cannot occur, the lactose-metabolizing enzymes are not produced and the cell is incapable of using lactose. A mutation that eliminates the function of the inducer-binding sites creates a *lacI^s* allele; the *s* stands for "super-repressed," meaning that transcription is permanently repressed.

Operator Mutations

Mutations within the operators affect repressor protein binding. In most cases, mutations in operators either prevent repressor protein binding or reduce the binding strength. Most of the mutations that have been studied are in O_1 because this is the most important of the three operators. A mutation in O_1 that prevents the repressor protein from binding allows RNA polymerase to transcribe the *lac* operon constantly. A mutation that permits constant transcription of a gene under all conditions is called a **constitutive mutation**. Constitutive mutations in O_1 are called *lacO_1^c* mutations, *O* standing for "operator" and *c* for "constitutive." A mutation in O_1 that lessens the binding strength of the repressor protein allows reduced transcription in the absence of lactose, and normal transcription in the presence of lactose.

A few operator mutations *increase* the binding strength between the repressor protein and O_1. Under these circumstances, allolactose is not as successful in removing the repressor protein from the operator. Transcription of the operon is reduced below normal levels in the presence of lactose due to the increased binding ability of the repressor protein. An interesting pattern characterizes some of these mutations. As Figure 8.6 shows, O_1 consists of 35 base pairs in a nearly symmetrical palindrome. On the right side are 3 nucleotide pairs that do not match their counterparts on the left. Mutations that alter the nucleotide pairs on the right side so that they match their counterparts on the left side increase the binding strength of the repressor protein.

Mutations in Enzyme Genes

If the *lacZ* gene mutates and alters the amino acid sequence of β-galactosidase, the enzyme may lose its activity. Such a mutation creates a *lacZ^-* mutant allele. Because the mRNA is formed normally, one might expect the other two enzymes, β-galactoside permease and β-galactoside transacetylase, to be translated normally and carry out their functions. This, however, is not the case. A *lacZ^-* mutation may eliminate production of β-galactoside permease and β-galactoside transacetylase, as well as β-galactosidase. This is because the inducer is not lactose but allolactose, which is formed when β-galactosidase converts lactose to allolactose. If lactose cannot be converted to allolactose because of the absence of functional β-galactosidase, the inducer is not produced, the repressor remains bound to the operator, and

the *lac* operon is not transcribed, even in the presence of lactose. This is an example of a **polar mutation**, a mutation that affects the genes downstream from it as well as the gene in which it occurs.

A mutation within the *lacY* gene may eliminate functional β-galactoside permease in the cell, creating a *lacY^-* mutant allele. With no functional β-galactoside permease, lactose cannot enter the cell, and the cell cannot use lactose as an energy source. Like *lac Z^-* mutant alleles, *lacY^-* mutant alleles cause a Lac^- phenotype because they eliminate the cell's ability to use lactose.

Mutations in the *lacA* gene may eliminate functional β-galactoside transacetylase. However, because this enzyme is not essential for lactose metabolism, cells with mutations in this gene are fully capable of utilizing lactose. Therefore, *lacA^-* mutations do not cause a Lac^- phenotype.

Another example of a polar mutation is a frameshift mutation in the *lacZ* gene that brings the termination codon out of frame and causes the ribosome to translate the mRNA past the normal termination codon. The ribosome then enters the *lacY* gene carrying a polypeptide from the mutant *lacZ* gene. The *lacY* gene cannot be translated normally because ribosomes block the *lacY* gene's translation initiation site. No functional β-galactosidase or β-galactoside permease is produced.

Trans and Cis Effects of Mutations in Merozygotes

François Jacob and Jacques Monod developed the operon model for gene regulation in bacteria in the 1950s, for which they received a Nobel prize in 1965. Some of the best evidence that they used to devise the operon model came from experiments with merozygotes, bacterial cells with an extra copy of part of their genome contained within the F factor (see section 7.5). Let's look at the results of some of their experiments with merozygotes that have extra copies of the *lac* operon and the *lacI* gene in an F' factor.

In order to describe the results of these experiments, we first need to define some terms. In designating alleles in a *lac* operon and the *lacI* gene, the "*lac*" designation is often left out for simplicity, so *I*, *O*, *Z*, *Y*, and *A* stand for *lacI*, O_1, *lacZ*, *lacY*, and *lacA*, respectively. As before, a superscript plus sign indicates that a gene or operator is capable of functioning as it does in wild-type cells, whereas a superscript minus sign indicates that the gene or operator fails to function. A superscript *c* designates a constitutive mutation. Because a merozygote has two copies of the entire *lac* operon, the designations for genes or operators on the F factor are separated from the designations for genes or operators on the chromosome with a slash. For example, a merozygote with a wild-type *lac* operon in its F' factor and a *lac* operon with a mutation in the *lacZ* gene in the chromosome is designated F' *I^+ O^+ Z^+ Y^+ A^+* / *I^+ O^+ Z^- Y^+ A^+* and is diagrammed in Figure 8.10.

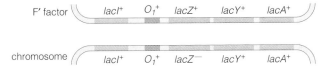

F′ factor: *lacI*⁺ *O₁*⁺ *lacZ*⁺ *lacY*⁺ *lacA*⁺

chromosome: *lacI*⁺ *O₁*⁺ *lacZ*⁻ *lacY*⁺ *lacA*⁺

Figure 8.10 Diagram of a merozygote with the genotype F′ I^+ O^+ Z^+ Y^+ A^+ / I^+ O^+ Z^- Y^+ A^+.

Any gene or DNA sequence that affects the expression of a gene located on its same DNA molecule is called a **cis-acting element**:

Cis-acting element

A DNA sequence that affects genes on a *different* DNA molecule than the one on which it is located is called a **trans-acting element**:

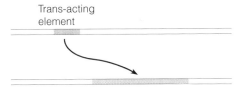

Trans-acting element

Most cis-acting elements are not genes but are DNA sequences to which regulatory proteins bind. Most trans-acting elements are protein-encoding genes that regulate the expression of other genes. It is possible to identify certain cis- and trans-acting elements in merozygotes for the *lac* operon, and their behavior can be readily explained using the operon model.

A cell with the chromosomal genotype I^- O^+ Z^+ Y^+ A^+ transcribes the *lac* operon constitutively because there is no functional repressor protein to block transcription. If an F′ factor with the genotype F′ I^+ O^+ Z^- Y^- A^- is introduced to make the merozygote F′ I^+ O^+ Z^- Y^- A^- / I^- O^+ Z^+ Y^+ A^+, the *lac* operon in the chromosome functions normally, as if the chromosome contained a functional *lacI* gene. In this case, the functional *lacI* gene on the F′ factor is trans acting. The *lacI*⁺ allele in the F′ factor encodes functional repressor protein, which in the absence of lactose, binds to the operator in the chromosome, nullifying the effect of the mutant *lacI*⁻ allele (Figure 8.11a). In fact, the *lacI* gene is both cis and trans acting because the repressor protein it encodes may bind to operators on its same DNA molecule as well as to operators on a different DNA molecule in the same cell.

On the other hand, constitutive operator mutations are cis acting. A *lacO₁*ᶜ mutation affects only the expression of genes on its same molecule. For example, the merozygote F′ I^+ O^c Z^- Y^- A^- / I^+ O^+ Z^+ Y^+ A^+ does not produce the enzymes encoded by the chromosomal *lac* operon constitutively (Figure 8.11b). Instead, the chromosomal *lac* operon functions as an inducible operon, in-

dicating that the O^c constitutive mutation in the F′ factor has no effect on the *lac* operon in the chromosome. However, the merozygote F′ I^+ O^+ Z^- Y^- A^- / I^- O^c Z^+ Y^+ A^+ does produce functional enzymes constitutively because the O^c mutation is in the same DNA molecule as the functional enzyme genes (Figure 8.11c). This indicates that the operator does not encode a product but probably acts in some other way, in this case as a binding site for a product produced elsewhere.

Let's now look at some of the experiments with merozygotes that Jacob and Monod used to develop the operon model.

Example 8.1 **Interaction of the *lacI* and *lacZ* genes in E. coli.**

Some of the most important experimental evidence that led to development of the operon model was published in 1959 by Pardee, Jacob, and Monod (*Journal of Molecular Biology* 1:165–178). They cultured cells that were merozygotes for the *lac* operon and the *lacI* gene in the presence and absence of inducer to investigate the effects of mutations in the *lacZ* and *lacI* genes. The researchers used isopropyl-thio-β-galactoside (IPTG) instead of lactose as the inducer in these experiments. IPTG is a derivative of lactose that binds to the inducer-binding sites in the repressor protein like allolactose. However, unlike allolactose, IPTG is not affected by β-galactosidase. Also, IPTG does not require β-galactoside permease to enter the cell.

Let's examine the results of two experiments. In the first experiment, the researchers introduced cells with the F′ genotype F′ I^+ Z^+ into a culture of F^- cells with the chromosomal genotype I^- Z^-. Conjugation between these cells formed merozygotes with the genotype F′ I^+ Z^+ / I^- Z^-. In the absence of IPTG, these merozygotes produced β-galactosidase at a higher level than induced wild-type cells, indicating that expression of β-galactosidase was high and constitutive. However, the researchers found that F′ I^+ Z^+ / I^- Z^- merozygotes (the type used in the first experiment) lost constitutive expression of β-galactosidase after 90 minutes in the absence of inducer. From that point on, β-galactosidase expression was inducible in these merozygotes, and was no longer constitutive.

In the second experiment, the researchers introduced cells with the F′ genotype F′ I^- Z^- into a culture of F^- cells with the chromosomal genotype I^+ Z^+. Conjugation between these cells formed merozygotes with the genotype F′ I^- Z^- / I^+ Z^+. These merozygotes failed to produce β-galactosidase in the absence of inducer, but did produce β-galactosidase in the presence of inducer.

The merozygotes in these two experiments had the same alleles in them, differing only in

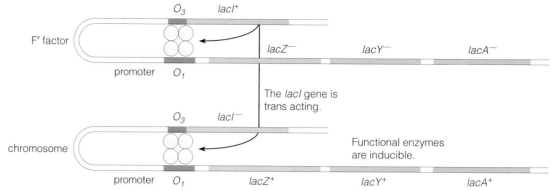

a F' I^+ O_1^+ Z^- Y^- A^- / I^- O_1^+ Z^+ Y^+ A^+ merozygote.

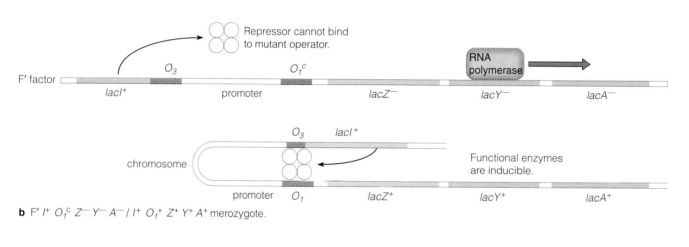

b F' I^+ O_1^c Z^- Y^- A^- / I^+ O_1^+ Z^+ Y^+ A^+ merozygote.

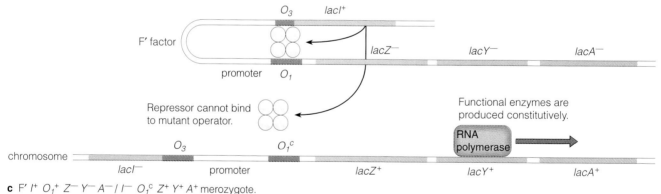

c F' I^+ O_1^+ Z^- Y^- A^- / I^- O_1^c Z^+ Y^+ A^+ merozygote.

Figure 8.11 The effect of I^- and O^c mutations in *lac* operon merozygotes.

which alleles were in the F' factor and which were in the chromosome. Yet in the first experiment, β-galactosidase expression was constitutive at first then inducible after 90 minutes, whereas in the second experiment, β-galactosidase expression was inducible throughout.

Problem: **(a)** Why was IPTG a better choice than lactose as the inducer? **(b)** At the time these experiments were conducted, the researchers had not yet discovered the operators in the *lac* operon. Could a

constitutive operator mutation explain the high levels of constitutive expression of β-galactosidase in the merozygotes from the first experiment? **(c)** Explain the observation of constitutive expression during the first 90 minutes of the first experiment, and the absence of constitutive expression in the second experiment.

Solution: **(a)** β-galactoside permease (the product of the *lacY* gene) is required for lactose to enter the cell, and a basal level of β-galactosidase is required

to convert lactose to allolactose, the natural inducer of the *lac* operon. Lactose is broken down by β-galactosidase, so repression increases as lactose is metabolized. Therefore, lactose is not a good choice as an inducer in these experiments because the gene products affect it. IPTG functions directly as an inducer but not as a substrate of β-galactosidase. Because IPTG is not metabolized, its concentration in the cell remains constant, unlike that of lactose. Also, IPTG permeates the cell membrane with or without β-galactoside permease. Many *lacZ* mutations are polar, disrupting the function of the *lacY* gene, thus preventing lactose from entering the cell. When IPTG is used as the inducer, mutations in the *lacZ* gene that affect induction can be investigated without confounding effects due to possible loss of β-galactoside permease function. **(b)** These observations do not suggest a constitutive operator mutation. Had a constitutive operator mutation been present, the F′ I^+ Z^+ / I^- Z^- merozygotes should have retained constitutive expression rather than losing it after 90 minutes. **(c)** In the first experiment, the F′ factor with the Z^+ and I^+ alleles entered an F^- cell that had neither β-galactosidase nor functional *lac* repressor protein in it (because the F^- cell had Z^- and I^- alleles in the chromosome). In the absence of functional *lac* repressor protein, the Z^+ allele on the F′ factor is expressed at high levels immediately upon entry into the cell. The I^+ allele on the F′ factor is expressed at a low but constant level. Thus, once the I^+ gene entered the F^- cell, it was transcribed and translated, but because its expression level was low, it took some time (90 minutes in this case) for the repressor protein to accumulate. During the first 90 minutes, the absence of sufficient repressor protein allowed the introduced Z^+ allele to be expressed constitutively at high levels until enough repressor had accumulated to repress expression. In the second experiment, the F^- cell already had both Z^+ and I^+ alleles in it prior to conjugation. Therefore, in both the presence and absence of the inducer, repressor protein was already present in these cells, making the merozygotes from the second cross inducible, but not constitutive, for β-galactosidase.

The *lac* repressor protein is trans acting, and the operators are cis acting. These observations contributed to the development of the operon model.

Negative and Positive Regulation of the *lac* Operon

The gene regulation we have discussed so far for the *lac* operon is called **negative regulation**, which is the inhi-

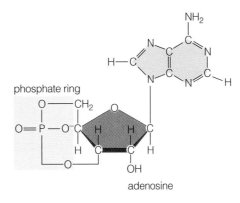

Figure 8.12 Cyclic adenosine monophosphate (cAMP), which has a phosphate ring attached to the 5′ and 3′ carbons of adenosine.

bition of transcription when a regulatory protein binds to DNA. A repressor protein exerts negative regulation because it inhibits transcription in the presence of an inducer. On the other hand, a regulatory protein that stimulates transcription when it binds to DNA exerts **positive regulation**.

The *lac* operon is subject to both negative and positive regulation, for very good reasons. Glucose is the preferred energy source for *E. coli*, and the purpose of the enzymes encoded by the *lac* operon is to derive glucose from lactose. If ample glucose is available to the cell, it would be inefficient to spend energy expressing genes for lactose metabolism because the cell can use glucose directly, even if lactose is also available.

Two criteria must be met in order for RNA polymerase to transcribe the *lac* operon. First, the repressor protein must not be bound to the operators. This criterion is met when lactose is present and allolactose binds to the repressor protein. Second, a protein called **catabolite activator protein (CAP)** must bind to the DNA at a site just upstream from the *lac* operon promoter. CAP is a positive regulator of the *lac* operon. In the presence of glucose, CAP fails to bind to the DNA and the *lac* operon is not transcribed, even if lactose is present. In the absence of glucose, CAP binds to the DNA, stimulating transcription, provided the repressor protein is not bound to the operators.

The concentration of glucose does not directly affect CAP. Instead, when glucose concentrations in the cell decrease, the concentration of a molecule called **cyclic adenosine monophosphate (cAMP)** increases. When glucose levels increase, cAMP levels decrease. You may have noticed a similarity in the name of cAMP and AMP, a nucleotide in RNA. Cyclic AMP is like AMP, but in cAMP the single phosphate group is attached to both the 5′ and the 3′ carbons of the ribose sugar, forming a phosphate ring (Figure 8.12).

Cyclic AMP interacts directly with CAP. When the concentration of glucose is low, the concentration of

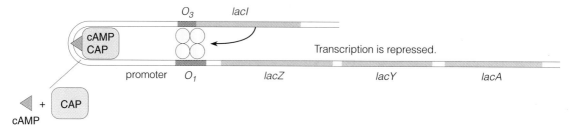

a When lactose is absent, the repressor protein binds to the operators and blocks transcription of the *lac* operon.

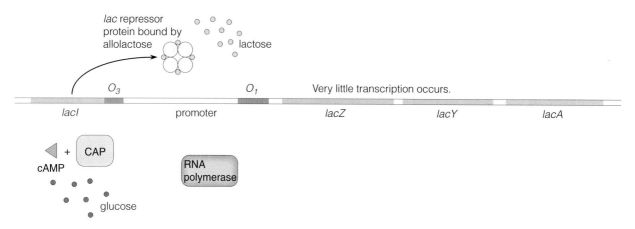

b When lactose and glucose are both present, neither the repressor protein nor CAP binds to the DNA. The level of transcription is very low because RNA polymerase does not bind readily to the promoter when CAP is not bound to the DNA.

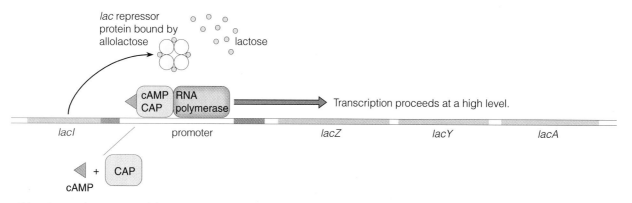

c When lactose is present and the concentration of glucose is low, CAP binds to the DNA and the repressor protein does not bind to it. RNA polymerase transcribes the *lac* operon at a high level.

Figure 8.13 Summary of combined positive and negative regulation of the *lac* operon.

cAMP is high; cAMP then binds to a site on CAP and converts the protein into a form that binds to the DNA just upstream from the *lac* operon promoter at a site called the **CAP-binding site**. CAP bends the DNA into a 90° angle, unwinding the DNA for transcription. When CAP is bound to the CAP-binding site, and the repressor protein is not bound to the operators, RNA polymerase transcribes the *lac* operon. When the concentration of glucose is high, the concentration of cAMP is low, and it separates from CAP. With no cAMP bound to it, CAP separates from the DNA. Without CAP bound to the DNA, tran-

scription of the *lac* operon is substantially reduced, regardless of the presence or absence of lactose.

With both positive and negative regulation operating on it, the *lac* operon is not transcribed in the absence of lactose regardless of glucose concentration, because the repressor protein is bound to the operator. The *lac* operon is also not transcribed when the concentration of glucose is high, even if lactose is present, because CAP does not bind to DNA in the presence of glucose. In order for the *lac* operon to be transcribed, lactose must be present, and the concentration of glucose must be low. Figure 8.13

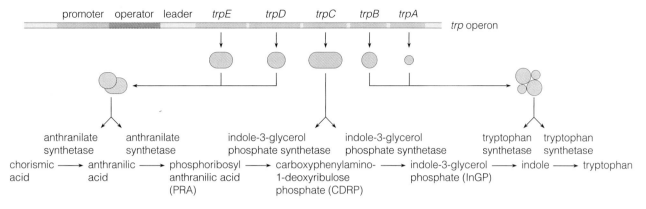

Figure 8.14 Expression of the genes in the *trp* operon. (Adapted with permission from Tanemura, S. and R. H. Bauerle. 1980. Conditionally expressed misseuse mutations: the basis for the unusual phenotype of an apparent *trpD* nonsense mutant of *Salmonella typhimurium*. *Genetics* 95:545–559.)

summarizes both positive and negative regulation of the *lac* operon. Looking back at Figure 8.7, we can now identify the blue-colored molecule as cAMP-CAP attached to its binding site on the DNA.

The *lac* operon is not the only operon affected by cAMP levels. The *gal* operon, mentioned earlier, encodes enzymes that convert galactose into glucose-1-phosphate. The cAMP-CAP complex exerts positive regulation over the *gal* operon, just as it does over the *lac* operon.

~

The *lac* operon is subject to both positive and negative regulation.

8.3 THE *trp* OPERON

The *trp* operon provides another example of negative regulation, but with an intriguing twist called attenuation. **Attenuation** is a form of regulation that permits initiation of transcription in an operon but terminates transcription before the operon's genes can be transcribed. Attenuation is a regulatory mechanism for several operons that encode enzymes for amino acid synthesis. Before describing regulation of the *trp* operon and attenuation, let's take a brief look at the tryptophan synthesis pathway and the functions of the enzymes encoded by the *trp* operon.

Function and Regulation of the trp Operon

As mentioned at the beginning of this chapter, *E. coli* uses gene regulation to cope with changing environmental conditions. The *E. coli* cells living in a mammalian intestine are in a fortunate situation compared to *E. coli* cells that do not happen to be in an intestine. Food consumed by a bacterial cell's mammalian host contains proteins that are broken down into their constituent amino acids by enzymes and acids in the digestive tract. *E. coli* simply absorbs these amino acids to make its own proteins.

However, when *E. coli* cells are outside of an intestine, certain amino acids may no longer be available. In order to continue protein synthesis, the cell must synthesize the needed amino acids from substances that are available. The *trp* operon contains five genes (*trpA* through *trpE*) transcribed into a single mRNA that encodes five polypeptides. These five polypeptides make three enzymes needed to synthesize the amino acid tryptophan, as illustrated in Figure 8.14.

The enzymes that synthesize tryptophan are produced only when tryptophan is not readily available in the cell's environment. Tryptophan is the regulatory factor and is also the product of the pathway. Thus, unlike lactose, which *induces* transcription of the *lac* operon, tryptophan must *repress* transcription of the *trp* operon.

This brings up a general feature of operons. For **inducible operons**, such as the *lac* operon, high concentrations of the *substrate* of the biochemical pathway *induce* transcription. For **repressible operons**, on the other hand, high concentrations of the *product* of the pathway *repress* transcription. The *trp* operon is, therefore, a repressible operon. In a sense, the *lac* operon is both inducible and repressible, because high concentrations of the substrate (lactose) induce transcription, and high concentrations of the product (glucose) repress transcription. However, the *trp* operon is unaffected by the level of chorismic acid, the pathway's substrate, making the *trp* operon solely a repressible operon.

As with the *lac* operon, the *trp* operon is regulated by a repressor protein that binds to the operator. The *trpR* gene, which encodes the trp repressor protein, is located far from the *trp* operon on the chromosome. The trp repressor protein binds to DNA and represses transcription when tryptophan concentrations rise. When tryptophan concentrations fall, the repressor protein separates from the DNA and permits transcription of the *trp* operon. Figure 8.15 illustrates how this process works. The repressor protein has a domain that is specific for

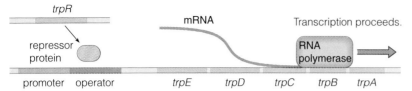

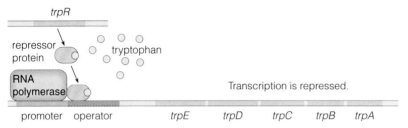

a When the availability of tryptophan is low, the repressor protein does not bind to DNA, permitting transcription to proceed.

b When the concentration of tryptophan is high, tryptophan binds to the repressor protein, causing it to bind to the DNA and repress transcription of the *trp* operon.

Figure 8.15 Function of the *trp* repressor protein.

tryptophan binding. When tryptophan is present in the cell, it binds to the repressor protein, causing a conformational change in the repressor protein that increases its affinity for the *trp* operator. Thus, an increase in tryptophan concentration represses transcription. Tryptophan is called a **corepressor** because it acts in conjunction with the repressor protein to repress transcription.

When the repressor protein–tryptophan complex is bound to the operator, transcription of the operon is reduced about 70-fold. When the concentration of tryptophan decreases, tryptophan separates from the repressor protein, and the repressor protein separates from the operator. With the repressor protein no longer bound to the operator, transcription of the *trp* operon may proceed.

The trp repressor protein acts on genes other than the *trp* operon. First of all, it regulates its own synthesis by regulating transcription of *trpR*, the gene that encodes the repressor protein. When the concentration of the trp repressor protein increases, the protein binds to an operator adjacent to the promoter of the *trpR* gene and represses transcription. When the concentration of the trp repressor protein falls, the protein leaves the operator and allows transcription. This form of regulation, where the product of a gene regulates the gene itself, is called **autogenous regulation**.

Attenuation of the *trp* Operon

Transcription of the *trp* operon is reduced significantly when the repressor protein binds to the operator. However, this reduction is still not adequate to reduce wasteful transcription in the presence of tryptophan. A second mechanism called attenuation reduces transcription another 10-fold beyond the reduction caused by the repressor protein. The 70-fold reduction in transcription

caused by repression, and the 10-fold reduction caused by attenuation, combine to reduce transcription by 700-fold when tryptophan is present.

When attenuation is active, transcription ends prematurely at a terminator sequence called the **attenuator**, which is located slightly downstream from the transcription initiation site and upstream from all enzyme genes. The product of attenuation is a short RNA that contains none of the genes that encode enzymes. When tryptophan is abundant, transcription terminates at the attenuator. When tryptophan is in short supply, RNA polymerase transcribes through the attenuator and on to the genes in the operon.

In order to understand how this process works, we need to review some concepts from previous chapters. As we discussed in section 3.5, there are two ways for transcription to terminate in *E. coli*: rho-dependent termination and intrinsic termination. In the RNA, the attenuator contains an intrinsic terminator consisting of a GC-rich sequence that can form a hairpin structure, followed by a string of eight uridines (Figure 8.16).

So how can RNA polymerase recognize a terminator when tryptophan is present, but fail to recognize a terminator when tryptophan is absent? There are several features of the DNA sequence near the 5′ end of the mRNA that determine whether or not transcription will terminate at the attenuator. Figure 8.17a shows the nucleotide sequence of the first 180 nucleotides on the 5′ end of the mRNA and the features that are important in attenuation. Looking at Figure 8.17b, we see that a hairpin structure forms between nucleotides 54 and 91. We'll call it hairpin #1, so we can identify it easily in our discussion. This hairpin is not followed by a string of uridines, so it does not cause termination of transcription. Then another hairpin forms between nucleotides 114 and

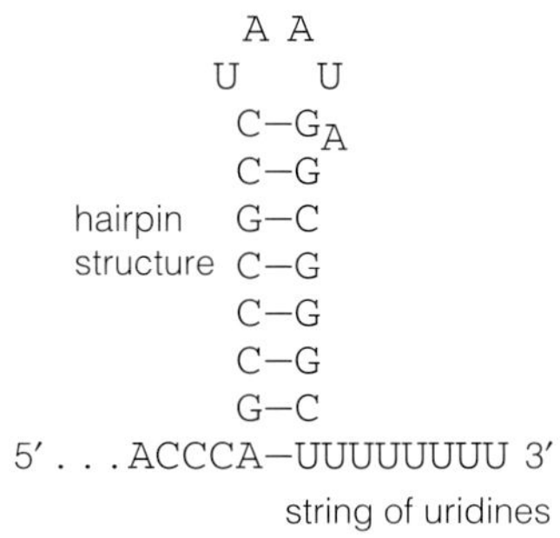

```
           A  A
        U       U
          C—G
          C—G  A
hairpin   G—C
structure C—G
          C—G
          C—G
          G—C
5' . . . ACCCA—UUUUUUUU 3'
          string of uridines
```

Figure 8.16 The *trp* operon attenuator in hairpin form, an intrinsic terminator of transcription.

134, and this one *is* followed by a string of uridines. We'll call it hairpin #3 (we'll get to hairpin #2 in a moment). Hairpin #3 is part of the attenuator. Hairpins #1 and #3 may form simultaneously, and when they do, transcription terminates at the attenuator. An alternative single hairpin, which we'll call hairpin #2, may form between nucleotides 74 and 119 (Figure 8.17c). Hairpin #2 is not immediately followed by a string of uridines and therefore is not a terminator. It also prevents formation of hairpins #1 and #3 because it ties up some of the nucleotides that are needed to form them. When hairpin #2 forms, transcription does not terminate at the attenuator because hairpin #3 fails to form.

When the concentration of tryptophan is high, hairpin #3 forms (with or without the formation of hairpin #1), and transcription terminates at the attenuator, preventing transcription of the genes in the operon. When the concentration of tryptophan is low, on the other hand, hairpin #2 forms, preventing termination at the attenuator and allowing transcription to continue into the enzyme-encoding genes.

How does the presence or absence of tryptophan influence which hairpins will form? Looking again at the nucleotide sequence at the 5' end of the mRNA in Figure 8.17a, we see that at nucleotide 27 there is an AUG initiation codon. Fourteen codons after the initiation codon is a termination codon. Thus, a small peptide is translated before any of the five genes on the mRNA. This small peptide is called the **leader peptide**. The 10th and 11th codons both encode tryptophan, and they determine which hairpins form and whether or not transcription continues.

Let's look first at the case where tryptophan is in short supply, as illustrated in Figure 8.18a. Recall from section 3.2 that transcription and translation are coupled in bacteria. As soon as RNA polymerase has transcribed the AUG initiation codon for the leader peptide, a ribosome assembles on the RNA and begins translating the leader peptide. When the ribosome reaches the two tryptophan codons, it stalls because there is too little tryptophan available for the ribosome to continue building the amino acid chain. The two tryptophan codons (UGGUGG) are located at nucleotides 54–59, which are within the se-

quence of nucleotides that forms hairpin #1 (nucleotides 54–91).

As the ribosome translates the mRNA, it disrupts any base pairing, so as it reaches hairpin #1, it straightens the hairpin structure as it translates the nucleotides that form it. When the ribosome stalls at the tryptophan codons (nucleotides 54–59), hairpin #1 remains disrupted and cannot re-form because the ribosome blocks its nucleotides. In the meantime, RNA polymerase continues transcription past the sequence for hairpin #2 (nucleotides 74–119). Because the stalled ribosome blocks none of these nucleotides, hairpin #2 forms after it has been transcribed. RNA polymerase continues transcription through the attenuator. Hairpin #3 of the attenuator sequence (nucleotides 114–134) cannot form because some of the nucleotides needed to form it are already base-paired in hairpin #2. Consequently, RNA polymerase transcribes through the attenuator (hairpin #3) and continues on to transcribe the rest of the operon.

Now, let's see what happens when tryptophan is abundant (Figure 8.18b). After RNA polymerase has transcribed the AUG initiation codon for the leader peptide, a ribosome assembles and translates the leader peptide. Because there is sufficient tryptophan, the ribosome does not stall at the tryptophan codons but proceeds to the termination codon. The ribosome disrupts hairpin #1 as it translates the nucleotides within it. In the meantime, RNA polymerase finishes transcribing the nucleotides for hairpin #2 (nucleotides 74–119). Hairpin #2 cannot completely form, however, because when the ribosome reaches the termination codon for the leader peptide, it is so large that it blocks formation of hairpin #2. RNA polymerase continues transcription through the sequence for hairpin #3 (nucleotides 114–134), and hairpin #3 forms because there is no hairpin #2 to block its formation. The attenuator now acts as an intrinsic terminator, and transcription terminates before any of the genes in the operon are transcribed. As long as the ribosome is free to transcribe past the adjacent tryptophan codons, attenuation terminates transcription.

If the leader peptide sequence is not translated by a ribosome (Figure 8.18c), hairpin #1 is the first to form, thereby preventing formation of hairpin #2. RNA polymerase continues transcription past hairpin #3, and hairpin #3 forms because hairpin #2 has not formed. Again, transcription terminates at the attenuator. Thus, RNA polymerase transcribes the entire *trp* operon only when ribosomes stall at the tryptophan codons in the leader peptide sequence due to a shortage of tryptophan.

Small peptides, including the leader peptide, are unstable. Shortly after its formation, the leader peptide breaks down into its constituent amino acids. The sole function of the leader peptide sequence is to regulate attenuation.

In *E. coli*, seven operons that encode enzymes for amino acid synthesis use attenuation. The basic model is the same for all of these operons: The ribosome stalls at

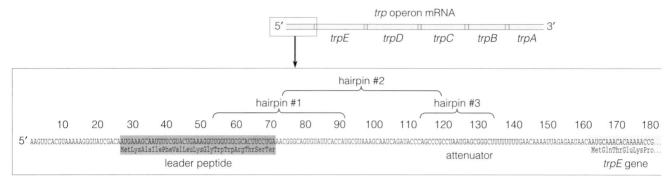

a The 5′ end of the *trp* operon mRNA carries the sequences that regulate attenuation.

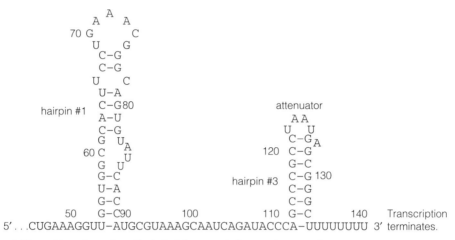

b When hairpin #3 forms (with or without formation of hairpin #1), transcription terminates at the attenuator.

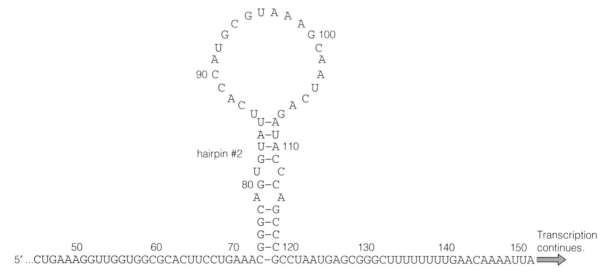

c When hairpin #2 forms, hairpin #3 cannot form, and transcription continues through the operon.

Figure 8.17 Features at the 5′ end of the *trp* operon mRNA. **(a)** The nucleotide sequence of the 5′ end of the mRNA. Included in this diagram are the leader peptide sequence, the three potential hairpins, one of which is the attenuator, and the beginning of the *trpE* gene. **(b)** Hairpins #1 and #3. **(c)** Hairpin #2.

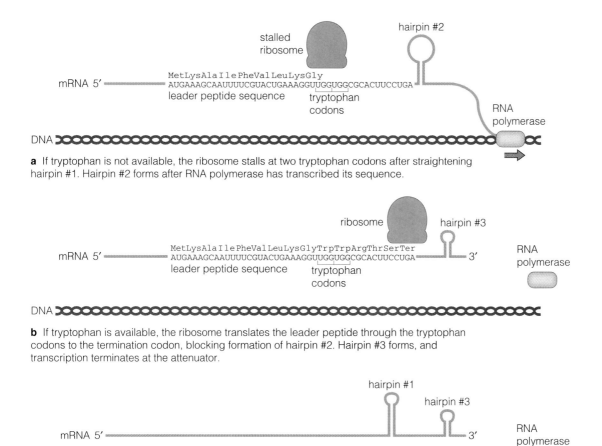

a If tryptophan is not available, the ribosome stalls at two tryptophan codons after straightening hairpin #1. Hairpin #2 forms after RNA polymerase has transcribed its sequence.

b If tryptophan is available, the ribosome translates the leader peptide through the tryptophan codons to the termination codon, blocking formation of hairpin #2. Hairpin #3 forms, and transcription terminates at the attenuator.

c If the leader peptide is not translated, hairpin #1 forms first, preventing formation of hairpin #2. Then hairpin #3 forms, and transcription terminates at the attenuator.

Figure 8.18 Function of the leader peptide sequence in attenuation.

adjacent codons that encode the amino acid synthesized by the enzymes encoded by the operon. This prevents formation of a terminator hairpin, allowing RNA polymerase to transcribe DNA through the attenuator. The number of adjacent codons may vary among operons. For example, the coding sequence for the leader peptide in the *his* operon contains seven adjacent histidine codons.

~

The *trp* operon is regulated by a combination of repression and attenuation.

8.4 REGULATION OF LYSIS AND LYSOGENY

Phage genes are subject to the same types of regulation as the genes in the bacterial chromosome. Among the best studied cases of phage gene regulation is the competition between lysis and lysogeny in *E. coli* cells that are infected with the temperate phage lambda (λ).

After wild-type λ phage particles infect cells in an *E. coli* lawn on a culture plate, turbid (instead of clear) plaques appear. A turbid plaque has a cloudy appearance because some of the cells have lysed while other cells in the plaque are lysogenic and fail to lyse. Because both lysis and lysogeny take place in the same plaque, what determines a particular cell's fate? Will it be lysed soon after infection, or spared by lysogeny? The answer lies in the expression of the *cI* gene in the phage DNA, which encodes a repressor protein called pcI.

When the *cI* gene is expressed, it produces pcI, which represses transcription of the genes for the lytic pathway, resulting in lysogeny. When the *cI* gene is not expressed, no repressor is produced, and the genes of the lytic pathway are expressed freely, resulting in lysis. The presence of pcI is, therefore, the determining factor for lysis versus lysogeny: pcI binds to operators and exerts negative regulation of the operons that initiate the lytic pathway.

The name *cI* stands for "*clear I*," because mutations in the *cI* gene cause clear, rather than turbid, plaques. A

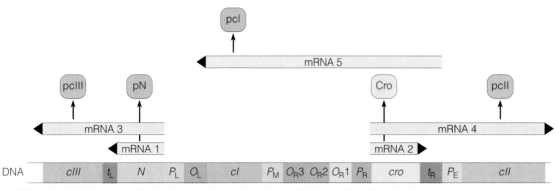

a When lambda infects a cell, the first proteins produced are pN, Cro, pcIII, pcII, and pcI.

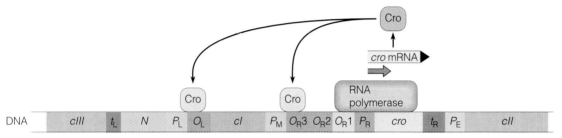

b If pcI succeeds in binding to O_R1 and O_R2, it establishes and maintains lysogeny by permitting transcription of *cI* and repressing transcription of the remaining genes.

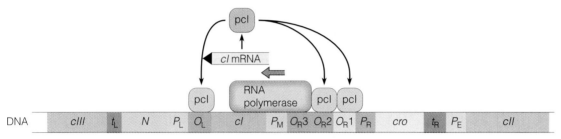

c If Cro succeeds in binding to O_R3, it initiates lysis and prevents lysogeny by repressing transcription of *cI* and other genes that are required for lysogeny.

Figure 8.19 Competition between lysogeny and lysis in lambda. The functions of the genes and DNA sequence elements are described in Table 8.1.

clear plaque indicates that all the cells within the plaque have lysed and none are lysogenic. Most mutations in the *cI* gene eliminate functional pcI, allowing expression of the genes for the lytic pathway. Lambda phages that are mutant for *cI* cannot enter lysogeny because the lytic pathway is not repressed.

When lambda first enters a cell, the *cI* gene is not immediately expressed. If it were, then lysogeny would be immediately established in all infected cells. Instead, several other genes are expressed first, initiating what is, in a sense, a race between establishment of the lytic versus the lysogenic pathway. In some cells, lysis wins and new phage particles are produced to infect other cells. In other cells, lysogeny wins and the cells do not die. When some cells lyse while others do not because of lysogeny, a turbid plaque appears.

The *cI* gene is flanked on either side by promoter-operator complexes (Figure 8.19). On the left side, there is a single operator, O_L, and a single promoter, P_L. On the right side is the *cI* gene's own promoter, P_M, and a series of three operators, O_R3, O_R2, and O_R1, followed by another promoter, P_R. The L, R, and M subscripts stand for "left," "right," and "maintenance." Both P_L and P_R are promoters for genes that initiate the lytic pathway as well as genes that stimulate transcription of the *cI* gene to initiate lysogeny. The genes that succeed in being expressed first determine whether lysis or lysogeny is established in the cell.

This process is both fascinating and complicated. It may help to refer to Table 8.1 and follow along in Figure 8.19 as we discuss what happens. When lambda first enters the cell, it arranges itself into a circle, and the bacte-

Table 8.1 Genes and DNA Sequence Elements in Lambda That Affect Lysis and Lysogeny

Gene or DNA Sequence Element	Product	Function
cI	pcI (repressor protein)	Binds preferentially to O_R1, O_R2, and O_L, repressing promoters P_R and P_L and activating promoter P_M
cro	Cro	Binds preferentially to O_R3 and O_L, repressing promoters P_M and P_L
t_R	none	Terminates transcription
t_L	none	Terminates transcription
N	pN	Antiterminator protein for t_L and t_R terminators
cII	pcII	Assists RNA polymerase in binding to P_E
cIII	pcIII	Protects pcII from degradation by HflA (a protein encoded by a gene on the chromosome)
O_R1	none	Operator with high binding affinity for pcI, low binding affinity for Cro
O_R2	none	Operator with high binding affinity for pcI, low binding affinity for Cro
O_R3	none	Operator with low binding affinity for pcI, high binding affinity for Cro
O_L	none	Operator with binding affinities for both pcI and Cro
P_R	none	Promoter for *cro* and *cII*
P_L	none	Promoter for *N* and *cIII*
P_M	none	Promoter for *cI*
P_E	none	Promoter for backward transcription through *cro* into *pcI* for initiation of lysogeny

rial RNA polymerase binds to P_L and P_R, initiating transcription at these sites. The *N* gene is transcribed on the left side, while the *cro* gene is transcribed on the right side. Both of these genes have transcription terminators at their ends, so initially transcription proceeds no farther than the ends of these genes, producing mRNAs 1 and 2 in Figure 8.19. However, the *N* gene product, pN, is an antiterminator protein, a protein that allows RNA polymerase to transcribe past the t_L and t_R terminators. Once the pN protein is produced, transcription can proceed beyond both *N* and *cro*, producing mRNAs 3 and 4 in Figure 8.19. (We'll look at the *cro* gene product a little later.)

Once transcription proceeds beyond both *N* and *cro*, the next transcribed genes are *cII* and *cIII*, so named because mutations in them cause clear plaques. This implicates them as genes necessary for establishment of lysogeny because their mutant forms cause lysis. The *cII* protein, pcII, is the essential protein for lysogeny, but it is very unstable and can be readily degraded by a protein called HflA, which is encoded by a gene on the bacterial chromosome. The *cIII* gene product, pcIII, has the

sole function of protecting pcII from degradation by HflA. Mutations in *cII* usually eliminate functional pcII, so pcII cannot function in establishing lysogeny. Without establishment of lysogeny, lysis proceeds and the plaques are clear. Mutations in *cIII* typically eliminate functional pcIII, so HflA degrades pcII and lysogeny cannot be established. Lysis proceeds and the plaques are clear. Both pcII and pcIII are, therefore, necessary for establishment of lysogeny.

The protein pcII assists RNA polymerase in binding to still another promoter, P_E, (E for "establishment" of lysogeny). This promoter is only active after pcII is produced. Once bound to P_E, RNA polymerase transcribes backward through the *cro* gene, backward through P_R, O_R1, O_R2, and O_R3, forward through P_M, and forward through the *cI* gene, making a large mRNA (mRNA 5 in Figure 8.19) that contains only one gene transcribed in the proper direction, *cI*. So only the *cI* gene product, which is pcI, can be translated from this mRNA.

Once pcI is present, it can bind to O_L and O_R1, O_R2, and O_R3, blocking RNA polymerase from P_R and P_L. When RNA polymerase can no longer access P_R and P_L,

all genes that rely on these promoters, including *cro*, *N*, *cII*, and *cIII*, are no longer transcribed. Another gene, *int*, codes for an enzyme called integrase. Once repression of P_R and P_L is established, integrase is produced and causes the phage DNA to integrate into the chromosome. Lysogeny is established.

In order for lysogeny to be maintained, the pcI must be continuously produced. If its production ceases, the pcI proteins bound to P_R and P_L are degraded, these promoters become active once again, and the cell enters the lytic pathway. Something must regulate production of the *cI* gene. The regulator happens to be pcI itself, which provides both positive and negative regulation of the *cI* gene in a remarkable way. The gene product pcI has a greater affinity for O_R1 and O_R2 than it does for O_R3. When there is a limited amount of pcI available, it binds preferentially to O_R1 and O_R2, leaving O_R3 vacant. Under these circumstances, the pcI proteins bound to O_R1 and O_R2 serve a dual function: First, they prevent RNA polymerase from binding to P_R, which prevents transcription of *cro* and additional genes downstream. Second, they activate RNA polymerase bound to P_M, stimulating transcription of the *cI* gene and continued production of pcI.

When the amount of pcI becomes too high, the excess pcI binds to O_R3. This inhibits RNA polymerase from binding to P_M, halting transcription of *cI*. As the amount of pcI falls, it leaves O_R3 first, reinitiating transcription of the *cI* gene. This is another example of autogenous regulation, where the product of a gene regulates its own production. This dynamic system ensures that the correct amount of pcI is constantly available for maintenance of lysogeny.

Now let's consider *cro*. The *cro* gene is one of the first to be expressed. It encodes a protein called Cro that binds to the same operators as pcI. However, Cro's binding affinities are different. Its greatest affinity is for O_R3, where it inhibits transcription of the *cI* gene. This prevents synthesis of pcI, favoring the lytic pathway. As the level of Cro increases, it binds to O_R1 and O_R2 as well as O_L. This reduces transcription of the genes under the regulation of these operators, including *cII*, so lysogeny cannot be initiated or reinitiated. Cro also regulates its own production by autogenous regulation. When Cro binds to O_R1 and O_R2, lysis is the cell's fate. If Cro beats pcI to the operators, lysis is the result. If pcI beats Cro to the operators, lysogeny is the result (Figure 8.19).

There is an interesting side effect of λ lysogeny. Think about the lysogenic cells in a turbid plaque. They are surrounded by λ phage particles from their lysed neighbors. Why don't these λ particles simply infect the lysogenic cells and cause them to enter the lytic pathway? The answer is that lysogenic cells are immune to lysis by further infection. This immunity is due to the presence of pcI in lysogenic cells. λ DNA that enters a lysogenic cell is immediately inactivated because the pcI produced by the prophage already present in the cell binds to O_R1, O_R2,

and O_L on the infecting DNA. The infecting DNA never has a chance to express any gene but *cI*, so lysis is immediately prevented. In other words, pcI is trans acting because protein produced from the genes of a prophage can bind to the operators on an infecting phage.

We described induction briefly in section 7.6. Induction takes place when an integrated λ prophage excises itself from the chromosome and enters the lytic pathway, usually when the cell is exposed to a dangerous environment, such as ultraviolet radiation. Induction can be readily explained in terms of the competition between pcI and Cro. When *E. coli* is exposed to ultraviolet radiation, the RecA protein is activated and acquires a proteolytic activity, causing it to cleave certain proteins (see section 5.4). The protein pcI is among those that are highly susceptible to cleavage by activated RecA. Activated RecA cleaves pcI bound to the operators and any pcI produced from the *cI* gene. The cleaved pcI is inactive and cannot bind to the operators. Once pcI is no longer bound to O_R1 and O_R2, Cro is free to bind to O_R3, initiating the lytic pathway.

Let's now see how researchers have used recombinant DNA methods to assist them in discovering the functions of pcI and Cro.

Example 8.2 Mutations that turn Cro into a transcriptional activator.

When bound to O_R1 and O_R2, pcI exerts both positive and negative regulation. It represses transcription of genes under the regulation of P_R and activates transcription at P_M, the promoter for its own gene. Mutations that alter the amino acid composition of an acidic patch on the surface of pcI eliminate its transcriptional activation function but not its repressor function. Evidence suggests that pcI carries out its activation function by bringing the acidic patch into contact with the RNA polymerase already bound to P_M. Cro, on the other hand, is purely a repressor protein, with no transcriptional activation function. Bushman and Ptashne (1988. *Cell* 54:191–197) hypothesized that Cro could mimic pcI as a transcription activator if it had pcI's acidic patch in the correct position. Using recombinant DNA methods, Bushman and Ptashne replaced a segment of the *cro* gene with DNA that encoded the four amino acids of the acidic patch on pcI. They named the product of this altered gene Cro67. They introduced four substitution mutations into O_R3 and found that under these conditions, Cro67, bound preferentially to O_R1. They then introduced four mutations into O_R2, making it similar to O_R1, so that Cro67 bound to O_R1 and O_R2 equally. The altered λ phage, with specifically designed mutations in O_R2, O_R3, and *cro*, produced Cro67,

which bound equally to O_R1 and O_R2 and activated transcription at P_M, just as pcI does.

Problem: **(a)** What do these results demonstrate about the function of the acidic patch on λ repressor? **(b)** Why was it necessary to alter the DNA sequences of O_R2 and O_R3?

Solution: **(a)** These results confirmed the conclusions from mutational studies that the acidic patch on pcI functions to activate transcription. They also demonstrated that, with a few mutational changes, Cro could mimic the functions of pcI. **(b)** Cro preferentially binds to O_R3. Because the DNA binding domain of Cro67 had not been altered, it would bind preferentially to O_R3, which would prevent RNA polymerase from binding to P_M. Mutations in O_R3 prevented Cro67 from binding to O_R3, causing Cro67 to bind preferentially to O_R1. The gene product pcI binds equally to O_R1 and O_R2. In order to have Cro67 truly mimic pcI, it was necessary to have it bind to O_R1 and O_R2 equally. This was possible once O_R2 was mutated to make it resemble O_R1. This research illustrates how recombinant DNA techniques (the subject of the next chapter) can be used to address fundamental questions about gene function.

Phage genes are subject to regulation. Among the best studied examples is the regulation that determines whether a cell infected by λ undergoes lysis or lysogeny.

8.5 TRANSCRIPTIONAL GENE REGULATION IN EUKARYOTES

There are many similarities between prokaryotic and eukaryotic gene regulation. For example, many eukaryotic genes are regulated in response to environmental stimuli by proteins that bind to DNA in the promoter regions. However, not surprisingly, eukaryotic gene regulation systems are much more complex than their prokaryotic counterparts. Even single-celled eukaryotes, such as yeasts that live and grow much like bacteria, have highly complex gene regulation mechanisms when compared to bacteria.

Most eukaryotic species are multicellular, often with highly specialized organs and tissues designed to carry out specific functions within the organism. Cell differentiation during development is closely tied to cascades of gene regulation and expression, where expression of one gene activates a set of different genes, which activates yet another set, and so forth. Ultimately, each differentiated cell expresses only those genes that are necessary for the cell to carry out its function, even though the cell contains all genes necessary for every function in the entire organism.

Gene regulation in eukaryotes is far too extensive to discuss in a single chapter. We will look at several aspects of eukaryotic gene regulation in detail in later chapters. For example, gene regulation in eukaryotes is related to packaging of DNA in chromosomes, a topic we will discuss in Chapter 10. The study of the relationships between gene regulation and cell differentiation during development is called **developmental genetics** and is currently one of the most active areas of genetics research. Chapter 23 is devoted entirely to this topic. Cell differentiation is closely tied to the cell cycle, the stages through which a cell passes in preparation for and during cell division. The cell cycle is tightly regulated to prevent haphazard cell growth, division, and proliferation. Cancer often results when cell cycle regulation goes awry. Chapter 24 discusses regulation of the cell cycle and its relationship to cancer. The immune systems of many eukaryotic species depend on specific types of gene regulation, many of which rearrange DNA sequences of genes that participate in immunity. This is the topic of Chapter 25.

For now, let's look at some of the basic mechanisms that apply widely to gene regulation in eukaryotes, particularly at the transcriptional level. This will serve as a foundation for the more specialized aspects of eukaryotic gene regulation that we will encounter later.

Transcription Factors and Eukaryotic Gene Regulation

To regulate prokaryotic operons, repressor proteins and other protein complexes bind to regulatory DNA sequence elements, usually in the promoter regions of genes. Similar mechanisms of gene regulation exist in eukaryotes. Any time a particular protein binds to DNA to activate or inhibit transcription, regulation of that factor may be used for gene regulation. We discussed some of the eukaryotic transcription factors and DNA sequence elements in section 3.3, including promoter consensus sequences, transcription factors and the basal eukaryotic transcription complex, and enhancers. Certain transcription factors participate in the regulation of all eukaryotic genes. Others are specific to one gene or a certain group of genes.

In order to understand how transcription is regulated in eukaryotic genes, let's briefly review some information about transcription factors and DNA sequence elements that we discussed in Chapter 3. In the context of gene regulation, transcription factors are the eukaryotic counterparts of proteins that exert positive and negative regulation in prokaryotes. Some transcription factors bind directly to DNA at conserved sequences. Other transcription factors bind to proteins that are part of protein complexes bound to DNA. The DNA sequences to which transcription factors bind are called **sequence elements**. For example, the transcription factor TFIID, which is part of the basal eukaryotic transcription complex, binds to the TATA box, a promoter sequence element.

Certain transcription factors called **activators** bind to sequence elements called **enhancers**. When an activator

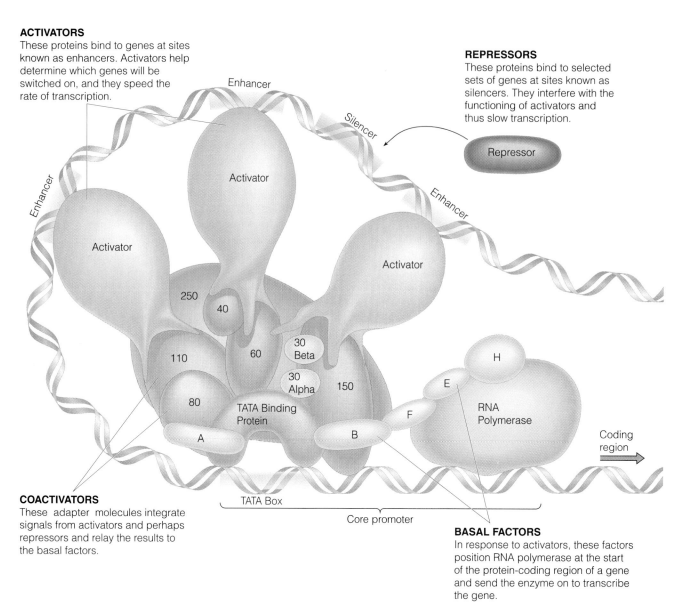

ACTIVATORS
These proteins bind to genes at sites known as enhancers. Activators help determine which genes will be switched on, and they speed the rate of transcription.

REPRESSORS
These proteins bind to selected sets of genes at sites known as silencers. They interfere with the functioning of activators and thus slow transcription.

COACTIVATORS
These adapter molecules integrate signals from activators and perhaps repressors and relay the results to the basal factors.

BASAL FACTORS
In response to activators, these factors position RNA polymerase at the start of the protein-coding region of a gene and send the enzyme on to transcribe the gene.

Figure 8.20 Interaction of transcription factors and DNA sequence elements to initiate transcription of a eukaryotic gene. (Adapted from an original drawing by Jared Schneidman Design published in Tijan, R. 1995. Molecular machines that control genes. *Scientific American* 272 (Feb):55–61.)

is bound to its enhancer, it stimulates transcription of the gene regulated by the enhancer. Other transcription factors called **repressors** bind to sequence elements called **silencers**. A repressor bound to a silencer reduces or prevents transcription of the gene regulated by the silencer. Coactivators are transcription factors that facilitate binding of activators with the basal eukaryotic transcription complex to stimulate initiation of transcription. Figure 8.20 shows how transcription factors and sequence elements may interact with one another for initiation of transcription.

Some transcription factors, such as those in the basal eukaryotic transcription complex, regulate transcription generally for nearly all genes. Other transcription factors may be specific for certain genes, and regulation of the transcription factor may regulate the expression of all the genes

affected by it. For example, a group of transcription factors are classified as **steroid receptors**, proteins that bind steroid molecules. Once specific steroid molecules bind to these transcription factors, the transcription factors undergo conformational changes and bind to regulatory DNA sequences called response elements near the promoters of certain genes, activating transcription of those genes. **Response elements** are DNA sequence elements that regulate transcription of a particular gene or group of genes. Once a steroid molecule binds to its steroid receptor, the steroid-steroid receptor complex binds to its response element and stimulates transcription (Figure 8.21). Steroid receptor response elements contain a conserved sequence with the consensus sequence TGGTACAAATGTTCT.

Steroid hormones are usually produced at sites in the body that are not target sites for the hormone. The hor-

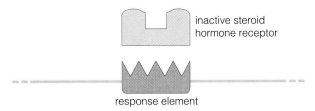

a When not bound to a steroid hormone, a steroid hormone receptor cannot bind to its response element in DNA.

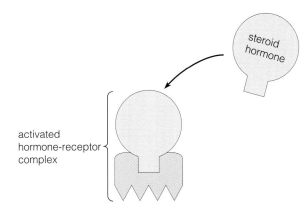

b When a steroid hormone enters the cell, it binds to its receptor, activating the receptor.

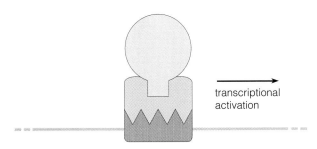

c The activated hormone-receptor complex binds to its response element in DNA, activating transcription of the gene under the influence of the response element.

Figure 8.21 Transcriptional activation by a steroid hormone.

mones are secreted into the circulatory system, where they are transported to target cells. The hormones enter the target cells and bind to the steroid receptors, which then bind to the glucocorticoid response elements in the DNA, inducing transcription. In this way, different genes in different cells may all be activated by a single hormone, provided those genes all recognize the same steroid receptor. In the absence of the steroid molecule, the transcription factors do not bind to DNA, and the genes they affect are not transcribed. Several hormones, including testosterone, estrogen, corticosteroids, and vitamin D, are steroid molecules that bind to steroid receptors.

Certain transcription factors bind to sequence elements in the DNA. Most of these DNA-bound transcription factors must also bind to proteins to exert their action, as illustrated in Figure 8.20. In order to do this, the transcription factors have a DNA-binding domain in one part of the molecule and a protein-binding domain in another part of the molecule. Several structural features of the DNA-binding domains present in transcription factors have been well characterized, including zinc fingers, helix-turn-helix domains, and leucine zippers.

Zinc fingers consist of two β strands and an α helix held together in four places by a zinc ion bound to cysteine and histidine residues in the protein molecule, as diagrammed in Figure 8.22a. Molecules with this structure are called zinc fingers because they protrude like fingers into the major groove of the DNA molecule (Figure 8.22b). Some transcription factors contain zinc fingers that cause the transcription factors to bind to DNA at conserved sequences. Transcription factors that affect RNA polymerases II and III may contain zinc-finger domains.

A **helix-turn-helix domain** consists of two α helices within the protein, one that fits into either the major or minor groove of the DNA molecule, and another that lies at an angle across the DNA, with a sharp turn in the amino acid chain between the two helices (Figure 8.23). The α helix that fits into the DNA groove is called the **recognition helix** because it contains amino acids that bind to a specific nucleotide sequence in the DNA. The other α helix is called the **stabilization helix** because it binds to the DNA at any sequence to stabilize binding of the recognition helix to its target sequence.

A large number of regulatory proteins in both prokaryotes and eukaryotes have helix-turn-helix domains. Some of these we have already discussed. For instance, the *trp* repressor protein and the catabolite activator protein (CAP) in *E. coli* have helix-turn-helix motifs in their DNA-binding domains.

A **leucine zipper** is a segment of amino acids where every seventh amino acid is a leucine. These leucine residues act like the teeth in a zipper. The leucine zippers of two protein subunits may bind to each other like a zipper that is zipped up, holding the two subunits of the protein together. Extending from the two polypeptides are recognition helices that bind to the major groove in the DNA at a conserved nucleotide sequence. Figure 8.24 illustrates the characteristics of leucine zippers and how they bind to DNA.

These three structural features of regulatory proteins are not mutually exclusive in the same protein. For instance, several proteins with zinc fingers and leucine zippers have been well characterized.

The human transcription factor **Sp1 (specificity protein 1)** provides a good example of how the DNA-binding and protein-binding domains of a transcription factor function. Sp1 is an activator that binds to GC boxes generally found at about 90 nucleotides or more upstream from the transcription startpoint for many genes. The upstream regulatory region of a gene may contain more than one GC box. The DNA-binding domain of Sp1 contains zinc fingers that bind to GC boxes in the DNA,

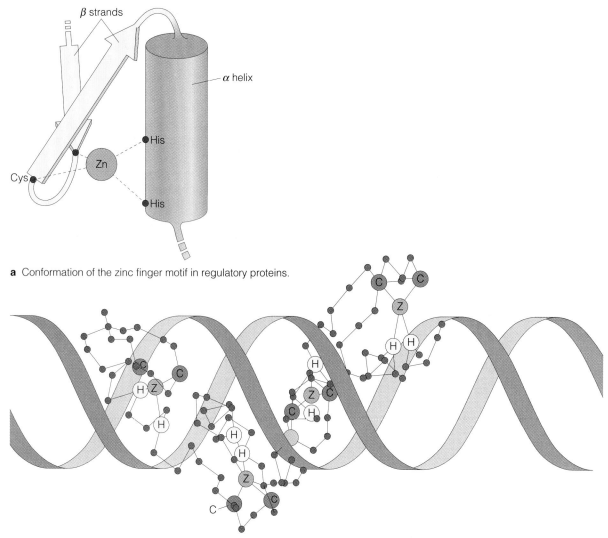

a Conformation of the zinc finger motif in regulatory proteins.

b A series of zinc fingers in the major groove of the DNA molecule.

Figure 8.22 Zinc fingers. (Redrawn from original art courtesy of J. M. Berg, reproduced with permission from *Annual Review of Biophysical Chemistry* 19:405. Copyright © 1990 by Annual Reviews, Inc.)

as illustrated in Figure 8.26. In another part of the Sp1 protein is a glutamine-rich domain that binds to a coactivator protein that binds to the basal eukaryotic transcription complex. When all proteins are bound to their DNA sequence elements or to one another, transcription initiation is enhanced.

Environmental Stimuli That Activate Gene Expression

Bacterial operons are often transcribed in response to environmental stimuli. Similar systems, in which environmental stimuli affect gene expression, exist in eukaryotes. Perhaps the best studied example is the GAL4/80 system in yeast. When yeast cells encounter the sugar galactose, they must produce several enzymes to metabolize it. Unlike the bacterial system, there is no operon for

galactose-metabolizing genes. Instead, the genes that encode these enzymes are located at different locations, but their expression is coordinated.

The genes encoding galactose-metabolizing enzymes respond to two specific transcription factors, GAL4 and GAL80. GAL4 has a DNA-binding domain that binds to response elements in the regulatory regions of genes that encode the galactose-metabolizing enzymes (Figure 8.27a). GAL80, on the other hand, does not bind to DNA, but instead binds to a specific domain in the GAL4 protein. In the absence of galactose, GAL80 remains bound to GAL4, and transcription is inhibited (Figure 8.27b). Galactose (or one of its derivatives) binds to GAL80, causing GAL80 to change its conformation so it no longer can bind to GAL4. Without GAL80 bound to GAL4, a loop forms in the DNA, and GAL4 binds to a transcription factor bound to the TATA box, stimulat-

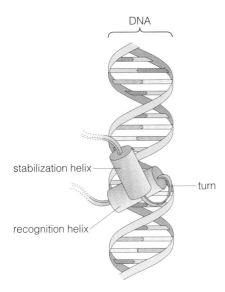

Figure 8.23 The helix-turn-helix domain in regulatory proteins. A recognition helix fits into a groove in the DNA molecule, where it is held by a stabilization helix. The two protein helices are connected by a turn in the amino acid chain.

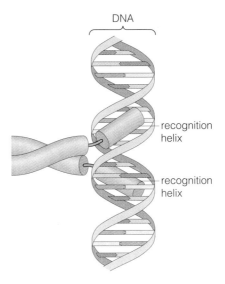

Figure 8.24 The leucine zipper domain in regulatory proteins. The recognition helices fit into the grooves of the DNA molecule.

ing initiation of transcription, as illustrated in Figure 8.27c.

Notice that this system is similar to that of the *lac* operon in *E. coli*, with GAL80 filling the role of the repressor protein. However, instead of binding to an operator, GAL80 binds to another protein.

Let's now look at the structure of GAL4 and its response element in DNA to see how the DNA sequence corresponds to the protein structure.

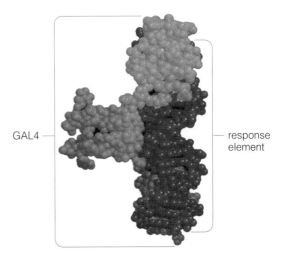

Figure 8.25 Information for Example 8.3: GAL4 binding to its response element in DNA. (From Marmorstein R., M. Carey, M. Ptashne, and S. C. Harrison. 1992. DNA recognition by GAL4: structure of a protein-DNA complex. *Nature* 356:408–414. Copyright 1992 Macmillan Magazines Limited. Reprinted with permission.)

Example 8.3   GAL4 and its response element.

In 1992, Marmostein, et al. (*Nature* 356:408–414) described the structure of GAL4 and its interaction with DNA. The molecule is a dimer, composed of two identical subunits, that binds to DNA, as diagrammed in Figure 8.25. The consensus sequence for the response element found in genes that are activated by GAL4 is

    5′ CGGAGGACTGTCCTCCG 3′
    3′ GCCTCCTGACAGGAGGC 5′

where the three nucleotides on each end of the sequence are the most highly conserved.

Problem: What characteristic of the DNA sequence is consistent with the dimeric nature of GAL4 and its binding, as illustrated in Figure 8.25?

Solution: The DNA sequence is a palindrome with symmetry centered around the A-T pair that is nine nucleotides from either end of the consensus

sequence. Each subunit of the dimer binds to the same sequence of nucleotides on each end of the conserved sequence. The high conservation of the CGG sequences on each end suggests that these nucleotides are the contact points for the GAL4 dimer. The contact points illustrated in Figure 8.26 are indeed the CGG sequences on either end of the consensus sequence.

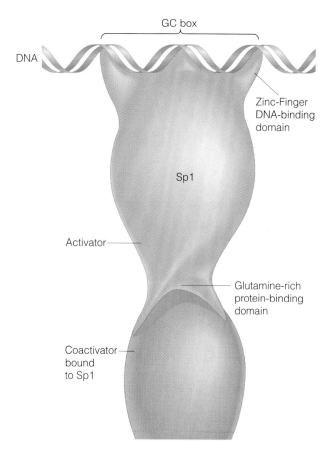

Figure 8.26 The transcription factor Sp1 in place as a transcriptional activator. (Adapted from an original drawing by Jared Schneidman Design published in Tijan, R. 1995. Molecular machines that control genes. *Scientific American* 272 (Feb):55–61.)

DNA Methylation

In many eukaryotes, some of the nucleotides in the DNA of each cell are methylated. Most of the methylation takes place on cytosines that are converted to 5-methylcytosine (Figure 8.28). The proportion of methylated cytosines varies considerably from one species to another. *Drosophila melanogaster* has essentially no methylation, whereas about 2–7% of the cytosines in mammals are methylated.

Cytosines are methylated most frequently in a CG doublet in the DNA. In most cases, both cytosines are methylated:

fully methylated
CG doublet

```
        m
5′ A C T T C G A A C A 3′
3′ T G A A G C T T G T 5′
            m
```

m = methyl group

Sometimes only one of the cytosines is methylated, in which case the doublet is said to be hemimethylated:

hemimethylated
CG doublet

```
        m
5′ A C T T C G A A C A 3′
3′ T G A A G C T T G T 5′
```

m = methyl group

Evidence that CG doublet methylation is important in gene regulation comes from the observation that many promoter regions contain CG doublet–rich regions within them. There is a strong correlation between the absence of methylation of these CG doublet–rich regions and transcriptional activity of the gene. In transcriptionally active genes, the CG doublet–rich regions are usually not methylated, whereas they tend to be methylated in cells where the gene is not transcribed.

When a methylated CG doublet is replicated, the cytosines added in the newly synthesized strand are not methylated, resulting in a hemimethylated doublet. In most cases, the unmethylated cytosines in a hemimethylated doublet are methylated shortly following DNA replication. However, if the cytosines in a CG doublet are not methylated in the original DNA, the cytosines in newly synthesized strands are also not methylated following replication. An enzyme called methyltransferase recognizes hemimethylated CG doublets and methylates the methyl-free cytosines, creating fully methylated CG doublets. If neither of the cytosines in a CG doublet is methylated, methyltransferase bypasses the site, leaving it unmethylated. Thus, methylated sites and unmethylated sites tend to perpetuate themselves through DNA replication.

This situation is important in cell differentiation. When the regulatory regions of a gene are highly methylated, the gene is not transcribed, whereas transcriptionally active genes are undermethylated, meaning that there are no methylated cytosines in the promoter region and fewer methylated cytosines within the gene when compared to transcriptionally inactive genes. The genes that are transcriptionally active in a differentiated cell are undermethylated and remain undermethylated when DNA replicates and the cell divides.

When the DNA replicates, the newly synthesized DNA is methylated in the same places as the original DNA, ensuring that methylated genes remain methylated after DNA replication and cell division. The daughter cells usually have the same active and inactive genes as their parent cell, eventually resulting in a multicelled tissue of similar specialized cells.

Although cytosine methylation is perpetuated through DNA replication and cell division, it is not necessarily permanent. An inactive gene can be demethylated in order to activate it, or an active gene methylated to inactivate it.

Methylation may directly inactivate transcription, or it may exert its effect indirectly through protein intermediates. Let's now examine some experiments that were designed to determine whether the effect of methylation on transcription is direct or indirect.

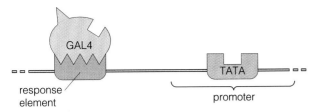

a GAL4 binds to DNA upstream from the TATA box.

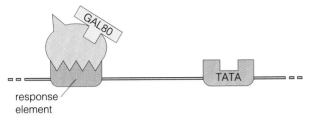

b When galactose is absent, GAL80 binds to GAL4, preventing transcription.

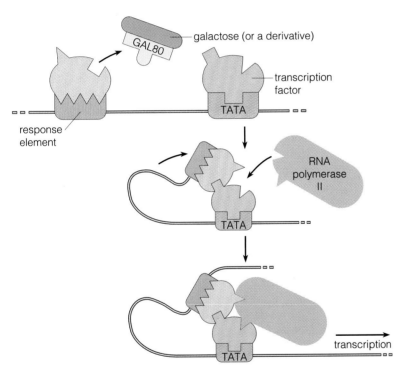

c When galactose is present, GAL80 cannot bind to GAL4, which is then free to interact with the transcription factor and RNA polymerase II at the promoter to initiate transcription.

Figure 8.27 The model for regulation of galactose-metabolizing genes by GAL4 and GAL80 in yeast.

Example 8.4   The role of methylation in gene inactivation.

Two models have been proposed for the role of methylation in gene inactivation. In the first, methylation acts directly by preventing transcription factors from binding to methylated DNA. In the second model, the effect of methylation is indirect. Methylated genes can be transcribed, but proteins that are specific for methylated DNA may inhibit transcription. In 1991, Boyes and Bird (*Cell* 64:1123–1134) published the results of experiments designed to distinguish between these two models. They chose to study promoters for four different mouse genes that were known to be methylated

when inactive. They conducted in vitro studies using nuclear extracts and in vivo studies using cultured cells. In one of the in vitro studies, they found that when the concentration of DNA in the nuclear extracts was low, initiation of transcription at methylated promoters was inhibited. However, when the researchers increased the concentration of methylated promoters, transcription increased. Also, when they added methylated DNA that did not include promoters to a nuclear extract that contained DNA with methylated promoters, they found that transcription increased to the level of unmethylated promoters. They postulated that a protein, MeCP-1, that binds to methylated DNA inhibited transcription of methylated promoters. To test this, they introduced a gene with a methylated promoter into cultured cells that lacked MeCP-1. Transcription from methylated promoters was higher in these cells when compared to cells with normal levels of MeCP-1. From these results, Boyes and Bird concluded that inhibition of transcription by methylation is indirect and exerts its effect when inhibitory proteins bind to methylated sequences.

Problem: (a) Why did adding methylated DNA to the nuclear extracts increase transcription? **(b)** How does the in vivo experiment suggest that the indirect model is correct?

Solution: (a) If methylation acts directly to inhibit transcription, then adding methylated DNA should not increase transcription. On the other hand, if proteins inhibit transcription, then the nuclear extract should have a limited amount of inhibitory proteins within it. When methylated DNA is added to the extract, this added DNA should bind most of the inhibitory proteins, leaving the methylated genes free to be transcribed. The observation that increasing the concentration of methylated DNA increased transcription supports the indirect model. **(b)** If the direct model is correct, then cells that are deficient in a protein that binds to methylated DNA should not have increased levels of transcription of methylated genes. If the indirect model is correct, then a methylated DNA-binding protein, such as MeCP-1, may inhibit transcription by binding to methylated DNA. The observation that cells deficient in MeCP-1 show increased transcription of methylated DNA suggests that MeCP-1 inhibits transcription when bound to methylated DNA in the promoter region.

Eukaryotic genes are regulated by the interaction of transcription factors and DNA sequence elements. DNA methylation also regulates eukaryotic gene expression.

8.6 GENE REGULATION BEYOND TRANSCRIPTION

Transcriptional regulation is the first and most important level of gene regulation in both prokaryotes and eukaryotes. For the genes in a cell that are transcribed, the RNA and protein products are then subject to additional regulation. In a sense, this additional regulation is a way of fine-tuning the activity of a gene product.

Transcript Processing

Because transcription and translation in prokaryotes are coupled, there is no transcript processing and, therefore, no regulation at this level. However, in eukaryotes, there is abundant opportunity for gene regulation during transcript processing. Approximately 75% of pre-mRNAs are degraded within the nucleus and never become mRNAs, providing the opportunity for selective degradation as a means of gene regulation. Even so, selective degradation of mRNA does not represent a common method of gene regulation.

In some cases, there may be more than one alternative for pre-mRNA processing into mRNA. Alternative processing permits the synthesis of different versions of a protein. Among the best studied examples is alternative splicing of introns in the pre-mRNAs encoded by genes that are transcribed in muscle cells. There are different types of muscle tissue, each requiring its own set of proteins. For example, smooth muscle tissue, found in the involuntary muscles of internal organs, is different than the striated tissue of skeletal muscles. Some of the variation in proteins between smooth and striated muscle tissues is due to alternative splicing of the same pre-mRNA.

For example, the α-tropomyosin pre-mRNA in rats includes 13 exons separated by 12 introns. However, the final mRNAs do not contain all 13 exons, as shown in Figure 8.29. The predominant form of mRNA in striated muscle cells contains 10 exons, whereas the form found most often in smooth muscle cells contains 9 exons, 7 of which are common between the two mRNAs. These two mRNAs encode proteins that are similar in some respects but different in others, even though both proteins are encoded by the same gene.

Beyond the nucleus, there are additional opportunities for regulation. Some mRNAs are unstable and have a very short life in the cytoplasm, whereas others may be much more stable, permitting them to be translated many times. The production of a gene product may be regulated by altering the stability of a specific mRNA. This appears to be common in mammals for genes regulated by hormones. A good example is the production of casein, an abundant protein in milk. Prolactin, a hormone that stimulates milk production in the mammary gland, sub-

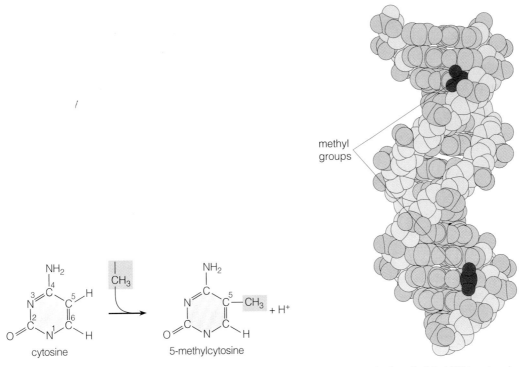

methyl
groups

a Addition of a methyl group to cytosine in DNA.

cytosine

5-methylcytosine

CH_3

$+ H^+$

b A methylated DNA molecule.

Figure 8.28 Cytosine methylation in DNA. (Source: Part b adapted from W. Saenger. 1983. *Principles of Nucleic Acid Structure*. New York. Springer-Verlag.)

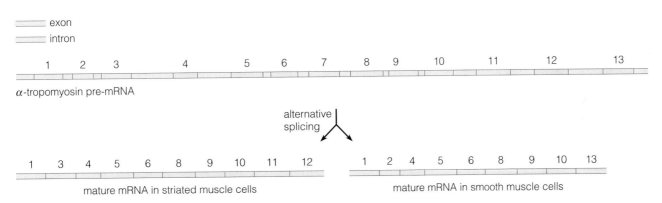

exon
intron

1 2 3 4 5 6 7 8 9 10 11 12 13

α-tropomyosin pre-mRNA

alternative
splicing

1 3 4 5 6 8 9 10 11 12

mature mRNA in striated muscle cells

1 2 4 5 6 8 9 10 13

mature mRNA in smooth muscle cells

Figure 8.29 Alternative processing of pre-mRNAs encoded by the α-tropomyosin gene in striated and smooth muscle cells. (Adapted from Lees-Miller, J., and Helfman, D. 1991. The molecular basis for tropomyosin isoform diversity. *BioEssays* 13:429–437. Copyright © 1991. Reprinted by permission of John Wiley & Sons, Inc.)

stantially increases the stability of mRNAs that encode casein, causing increased translation of casein when prolactin is present.

Translational Regulation

There are several known examples of regulation at the level of translation. One well-studied case is the storage of maternal mRNAs in the eggs of some animals. These mRNAs are stored in the egg cytoplasm as the egg awaits fertiliza-tion. Upon fertilization, the maternal mRNAs are freed to be translated profusely, providing the new embryo with an abundance of proteins without the need for transcription.

The human immunodeficiency virus (HIV), which is responsible for AIDS, has an intriguing method of regulating translation of its genes. The virus encodes a protein that binds to the 5′ end of the viral mRNAs, greatly increasing their rate of translation—thus favoring translation of the viral mRNAs over the native mRNAs in an infected cell. This phenomenon may offer an opportunity

to treat HIV infection. If a drug could be found that interferes with the function of this protein, then gene expression of HIV genes could be greatly reduced without hindering expression of the genes normally present in the cell. While such a treatment would not eliminate the virus altogether, it could slow or stop the progression of the disease, or prevent disease symptoms in individuals known to be infected who have not yet exhibited any symptoms.

Posttranslational Regulation

Of all types of regulation beyond transcription, posttranslational regulation is probably the most common. Many enzymes are regulated after they have been translated and are functioning. Among the most common mechanisms of posttranslational regulation in both prokaryotes and eukaryotes is **allosteric feedback inhibition**, the inhibition of an enzyme's activity by the end product of the biochemical pathway. Typically, when a large amount of the end product of a pathway accumulates, the end product binds to a site, called the **allosteric site**, on one of the enzymes in the pathway, inhibiting that enzyme's function. This binding blocks the pathway at the step catalyzed by the inhibited enzyme, which is usually the first enzyme in the pathway.

A prokaryotic example of allosteric feedback inhibition is the isoleucine biosynthetic pathway. In *E. coli*, isoleucine is synthesized from threonine through a pathway consisting of five steps (Figure 8.30). The first step in the pathway is conversion of threonine to α-ketobutyrate, which is catalyzed by the enzyme threonine dehydratase. Threonine dehydratase has an allosteric site to which isoleucine can bind. Once bound, isoleucine causes a conformational change in threonine dehydratase, causing it to lose its enzymatic activity and blocking the pathway at that point. When the isoleucine concentration is high, the excess isoleucine inactivates threonine dehydratase. Once the isoleucine concentration drops, the isoleucine molecules bound to threonine dehydratase separate from the enzyme, restoring its activity and allowing the pathway to synthesize more isoleucine.

Feedback inhibition is common for amino acid synthesis and is a form of autogenous regulation at the posttranslation level. The tryptophan synthesis pathway in *E. coli*, which we discussed in section 8.3, is regulated by allosteric feedback inhibition. The enzyme inhibited by tryptophan is the first enzyme in the pathway, anthranilate synthetase. This enzyme is the product of the *trpE* and *trpD* genes in the *trp* operon. Consequently, there are at least three mechanisms that regulate tryptophan production: repression, attenuation, and allosteric feedback inhibition. These three mechanisms operate jointly to ensure that the proper amount of tryptophan is available to the cell.

Among the most interesting eukaryotic examples of feedback inhibition is the pathway for synthesis of lysine, threonine, isoleucine, and methionine in plants. Lysine, threonine, isoleucine, and methionine are synthesized in developing plant seeds through the pathway diagrammed in Figure 8.31. The first enzyme in the pathway, aspartate kinase (AK), catalyzes the conversion of aspartate to aspartyl-phosphate. Lysine and threonine inhibit this enzyme, thus regulating the synthesis of all four amino acids. When lysine and threonine concentrations are high, AK activity is inhibited, and none of the four amino acids are synthesized. When lysine and threonine concentrations are low, AK is active, and production of the four amino acids increases. Two other enzymes in the pathway, homoserine dehydrogenase (HSDH) and dihydrodipicolinate synthase (DHPS), are also feedback-inhibited. HSDH is feedback-inhibited by threonine, and DHPS by lysine.

Mutations in the genes that encode these enzymes are of commercial importance in agriculture and human nutrition. Humans and many other animals cannot synthesize certain amino acids, as plants and bacteria can, but must consume these amino acids in the diet. The amino acids that must be included in the diet are called **essential amino acids**. One of the essential amino acids is lysine. Plant seed proteins from seeds in the grass family are a major source of protein for humans and grain-fed animals worldwide. The grass family includes the cereal grains, such as wheat, rice, corn, oats, barley, sorghum, and millets. The protein in these seeds is inherently low in lysine. For this reason, many people in less developed countries of the world suffer from lysine deficiency because their diets are based primarily on rice, corn, or other cereal grains.

Scientists have developed an ingenious way to select for mutations that overproduce lysine in the seeds of these grain crops. Plant cells are grown as undifferentiated cell masses in culture and are supplied with high amounts of lysine and threonine but no methionine in their growth medium. The lysine and threonine inhibit the enzymes in the amino acid synthesis pathway, and the cells die of methionine starvation, because no methionine is present in the growth medium and the pathway for its synthesis is blocked. However, mutations that eliminate the function of the allosteric site for feedback inhibition make the enzymes insensitive to lysine and threonine concentrations. Cells that are resistant to lysine and threonine concentrations produce lysine, threonine, isoleucine, and methionine even when lysine and threonine levels are high. Plant cells with this mutation survive in cultures supplemented with high concentrations of lysine and threonine and can eventually be regenerated into whole plants. In several instances, mutations recovered from cultured cells increase the lysine content in seeds.

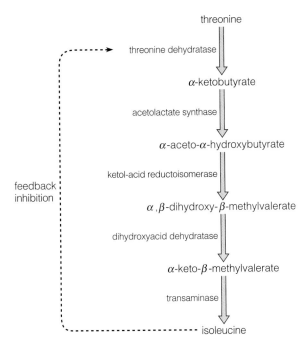

Figure 8.30 The biochemical pathway for synthesis of isoleucine from threonine in *E. coli*. Accumulation of the end product, isoleucine, inhibits the first enzyme, threonine dehydratase.

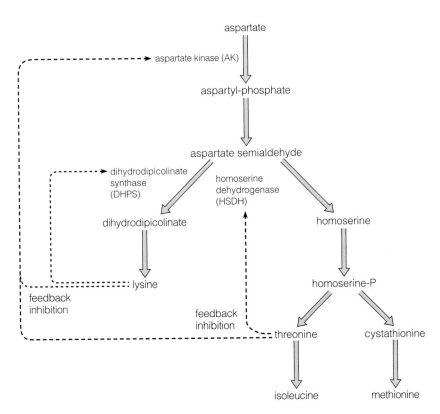

Figure 8.31 The aspartate-derived amino acid synthesis pathway in plant seeds. Lysine, methionine, isoleucine, and threonine are all synthesized from aspartate. Several enzymes in the pathway are regulated by feedback inhibition. (Adapted from Muehlbauer, G. J., D. A. Somers, B. F. Matthews, and B. G. Gengenbach. 1994. Molecular genetics of the maize (*Zea mays* L.) aspartate kinase-homoserine dehydrogenase gene family. *Plant Physiology* 106:1303–1312.)

The following example shows how an understanding of feedback inhibition in a biochemical pathway can be used to achieve the goal of increased lysine synthesis.

Example 8.5 Increased lysine synthesis through feedback inhibition.

In 1994, Muehlbauer, Gengenbach, Somers, and Donovan (*Theoretical and Applied Genetics* 89:767–774) reported that two mutations in the AK gene in maize caused overproduction of threonine and methionine but only slight increases in lysine production.

Problem: **(a)** Using the biochemical pathway for aspartate-derived amino acids shown in Figure 8.31, explain how mutations in the AK gene could cause overproduction of threonine and methionine but not lysine. **(b)** What additional mutations should cause lysine overproduction? **(c)** How might a selection procedure that is specific for lysine overproduction be designed?

Solution: **(a)** The pathway branches after aspartate semialdehyde is formed. The enzyme HSDH converts aspartate semialdehyde into homoserine, which eventually is converted to methionine, threonine, and isoleucine. The enzyme DHPS converts aspartate semialdehyde into dihydrodipicolinate, which is then converted into lysine. These two enzymes and AK are inhibited by lysine or threonine or both lysine and threonine. The selection procedure for cultured cells relies on methionine starvation to kill nonmutant cells. Only mutant feedback-insensitive enzymes in the methionine biosynthesis pathway are selected. Notice that methionine is synthesized in the pathway with AK and HSDH but not DHPS. [Thus, selection in tissue culture for lysine and threonine resistance selects for feedback-insensitive AK and HSDS mutations but not for feedback-insensitive DHPS mutations.] If DHPS remains inhibited by lysine, and AK and HSDS are not, threonine and methionine may be overproduced but lysine may not be overproduced. **(b)** Mutations that render DHPS insensitive to feedback inhibition, combined with AK insensitivity, should cause lysine overproduction. **(c)** In order to select specifically for lysine overproduction, cells must be starved for lysine. This cannot be done by adding lysine to the medium. However, a chemical analog of lysine that inhibits DHPS but cannot be used by the cell in place of lysine, should starve the cells for lysine while selecting for mutations that render DHPS insensitive to feedback inhibition. An example of such a lysine analog that has been used for selection of lysine-overproducing cells is S-2-aminoethly-L-cysteine (AEC).

Examples of gene regulation beyond the transcriptional level include alternative splicing of pre-mRNAs, regulation of mRNA longevity and the rate of translation, and allosteric feedback inhibition.

SUMMARY

1. Many bacterial genes are organized into operons. Each operon consists of a set of genes encoding enzymes for the same biochemical pathway. The genes in an operon are all transcribed as a single mRNA, which is then translated into the individual polypeptides. The operon model allows a set of enzymes to be coordinately regulated at the level of transcription.

2. The *lac* operon is one of the most fully characterized operons and provides a good example of operon structure and function. It contains three genes that encode three enzymes for lactose metabolism.

3. One characteristic of an operon is the presence of at least one operator, a short segment of DNA located near the promoter, to which a regulatory protein may bind.

4. The *lac* operon is subject to negative regulation, which means that the operon is repressed (not transcribed) unless an inducer molecule is present. The inducer molecule is allolactose, which is a derivative of lactose. When lactose is present, some of the lactose is converted to allolactose, and the operon is induced. In the presence of lactose, the lactose operon may be transcribed. In the absence of lactose, the operon remains repressed and is not transcribed.

5. Repressor proteins bind to operators and block transcription. When allolactose is present, it binds to the repressor protein, changing its conformation so that it will no longer bind to the operator. When the repressor protein is no longer bound to the operator, transcription is no longer blocked.

6. The *lac* operon is subject to positive as well as negative regulation. Glucose is the preferred source of carbohydrate, and when it is present it is preferentially utilized instead of lactose. Glucose exerts positive regulation through a protein called catabolite activator protein (CAP), which activates transcription when it binds to the *lac* operon promoter. CAP is induced by cyclic AMP (cAMP), which is regulated by the concentration of glucose. When glucose levels are low, cAMP is high, activating CAP and activating transcription of the *lac* operon. When glucose levels are high, cAMP is low, reducing transcription of the *lac* operon.

7. With positive and negative regulation acting together, transcription is repressed when lactose is absent regardless of glucose levels. When glucose concentration is high, the *lac* operon is not transcribed, even in the presence of lactose, because glucose is the preferred carbon source. The *lac* operon is transcribed only when lactose is present and when glucose concentrations are low.

8. The *trp* operon encodes five polypeptides that form three enzymes for tryptophan synthesis. The *trp* operon provides a good example of attenuation, an additional type of gene regulation in operons. The attenuator is an intrinsic terminator of transcription in the mRNA upstream from the enzyme-encoding genes. A short leader peptide sequence precedes the attenuator. When tryptophan is in short supply, the ribosome stalls at two tryptophan codons and prevents the attenuator from assuming its terminator conformation. Transcription then continues on into the enzyme genes. When tryptophan is abundant, the ribosome translates past the tryptophan codons and allows the attenuator to assume its terminator conformation. Transcription terminates at the attenuator before the enzyme genes are transcribed.

9. Gene regulation determines whether the temperate phage lambda (λ) will enter the lytic or lysogenic pathway after it infects a cell. There is a competition between the proteins pcI and Cro. If pcI succeeds first in binding to operators, lysogeny is established. If Cro succeeds first in binding to operators, lysis proceeds.

10. Eukaryotic gene regulation has some similarities with prokaryotic gene regulation, but is much more complex.

11. Eukaryotic gene regulation at the transcriptional level includes interactions between transcription factors and DNA response elements. When bound to a DNA response element, transcription factors may activate or repress transcription.

12. Transcription factors may contain DNA-binding and protein-binding domains. Common DNA-binding domains include zinc fingers, leucine zippers, and helix-turn-helix domains.

13. Response elements are DNA sequence elements that are common to certain genes that have similar functions. These elements bind transcription factors that are specific to the gene or group of genes that they regulate. In this way, a set of genes can be coordinately regulated by a single signal, such as a hormone.

14. Gene regulation is associated with DNA methylation in many species. Actively transcribed genes are undermethylated, whereas genes that are transcriptionally inactive tend to be methylated.

15. Most genes are regulated at the transcriptional level and at levels beyond transcription.

16. Expression of some genes is regulated through pre-mRNA processing. Alternative splicing of the pre-mRNAs from the same gene may be used to produce different, but related, proteins.

17. In translational regulation, cells regulate mRNA longevity and the rate of translation.

18. Allosteric feedback inhibition is a form of posttranslational regulation of enzymes after they have been translated and have assumed their function. Typically, the first enzyme in a pathway is inhibited by the end product of the pathway. Thus, when the end product is abundant, the first enzyme is inhibited, blocking synthesis of the end product at the step catalyzed by that enzyme. When the end product is deficient, the enzyme remains active, allowing synthesis of the end product to proceed.

QUESTIONS AND PROBLEMS

1. Most genes are regulated predominantly at the level of transcription, which is understandable because there is little reason to waste energy transcribing a gene when its product will not be used. Nonetheless, there are many examples of gene regulation beyond transcription. Explain the relationship of gene expression beyond transcription with gene regulation at the transcriptional level.

2. Explain three ways in which tryptophan acts as a regulator of its own synthesis. Include examples of both transcriptional and posttranslational regulation.

3. Transcription of the *lacI* gene is unaffected by the presence or absence of lactose. **(a)** What would happen to regulation of the *lac* operon if the *lacI* gene were transcribed only in the presence of lactose? **(b)** What would happen to regulation of the *lac* operon if the *lacI* gene were transcribed only in the absence of lactose?

4. Frameshift mutations in the first gene of an operon often cause the ribosome to translate through the gene's termination codon, reaching a new termination codon somewhere in the second gene, resulting in a longer-than-normal polypeptide. These mutations are called polar mutations because a mutation in one gene affects a second gene downstream from it. A frameshift mutation in the *lacZ* gene is known that places a *premature* termination codon in frame, resulting in a shorter-than-normal polypeptide. This mutation is also a polar mutation even though it does not interfere with translation of the *lacY* gene. Explain why this mutation is polar in the *lac* operon and why a similar mutation in another operon would not necessarily be polar.

5. Table 8.2 lists various situations for lactose metabolism genes in *E. coli*. Column 1 lists 11 mutations that are possible. Columns 2–5 list possible carbohydrate sources for the cell. In the blank spaces in columns 2–5 for each mutation, put a plus sign if the *lac* operon genes would be actively transcribed under the conditions given and a minus sign if transcription would be reduced or eliminated.

6. Would lactose be metabolized under conditions of high lactose and low glucose in the merozygote F′ I^+ O^+ Z^- Y^+ A^- / I^- O^+ Z^+ Y^- A^+? Explain your answer.

7. Would lactose be metabolized under conditions of high lactose and low glucose in the merozygote F′ I^- O^+ Z^- Y^- A^- / I^- O^c Z^+ Y^+ A^+? Explain your answer.

Mutation	Carbohydrate Concentrations			
	Lactose High, Glucose High	Lactose High, Glucose Low	Lactose Absent, Glucose High	Lactose Absent, Glucose Low
A mutation in *lacI* prevents the DNA-binding site in the protein from interacting with DNA.				
A mutation in *lacI* alters the inducer-binding site, preventing the inducer from binding to the repressor protein.				
A mutation in the CAP gene prevents cAMP from binding to CAP.				
A mutation in the CAP gene prevents CAP from binding to the DNA.				
A frameshift mutation in *lacA* results in a longer-than-normal polypeptide.				
A same-sense mutation in *lacZ*.				
A mutation in the operator prevents the repressor protein from binding.				
A substitution mutation in *lacZ* results in a complete loss of β-galactosidase activity.				
The merozygote F′ I^+ O^+ Z^- Y^- A^- / I^- O^+ Z^+ Y^+ A^+.				
The merozygote F′ I^+ O^+ Z^- Y^- A^- / I^- O^c Z^+ Y^+ A^+.				
The merozygote F′ I^- O^+ Z^- Y^- A^- / I^- O^c Z^+ Y^+ A^+.				

8. Would lactose be metabolized under conditions of high lactose and low glucose in the merozygote F′ I^- O^+ Z^- Y^+ A^- / I^- O^c Z^+ Y^- A^+? Explain your answer.

9. Why is attenuation a common regulation mechanism for amino acid synthesis operons, but is not found in other types of operons?

10. Suppose that a mutation in the *trpR* gene prevented the *trp* repressor from binding to the DNA. Would transcription of the *trp* operon be reduced when tryptophan levels are high? Explain how you arrived at your answer.

11. Suppose that an insertion mutation of an adenine is found at position 50 in the mRNA of the *trp* operon (see Figure 8.17 for the nucleotide sequence of the leader portion of the mRNA and the numbered positions of nucleotides). **(a)** What would be the amino acid sequence of the leader peptide? **(b)** Using what you have learned about attenuation in the *trp* operon, predict what hairpins would form and whether or not attenuation would take place when tryptophan levels are high and when they are low.

12. Suppose that a substitution mutation in the DNA of the *trp* operon changed the G at position 29 to an A in the mRNA (see Figure 8.17 for the nucleotide sequence of the leader portion of the mRNA and the numbered positions of nucleotides). **(a)** What would be the amino acid sequence of the leader peptide? **(b)** Using what you have learned about attenuation in the *trp* operon, predict what hairpins would form and whether or not attenuation would take place when tryptophan levels are high and when they are low.

13. When a cell divides, the portions of its DNA that are methylated in the parent cell also tend to be methylated in the progeny cells of subsequent generations. Explain why patterns of methylation are retained from one cell generation to the next.

14. How might a mutation in a gene that encodes a steroid receptor protein affect the expression of other genes?

15. Using the GAL4/80 model shown in Figure 8.27, predict what effect a mutation in the *GAL80* gene that prevents GAL80 from binding to GAL4 would have on transcription **(a)** in the presence of galactose and **(b)** in the absence of galactose.

16. Promoters in genes that encode mRNAs are usually located at the 5′ end of the gene. Enhancers, on the other hand, may be located some distance from the gene on either side of it. Explain how distantly located enhancers may affect transcription of a gene and why they can be located on either side of the gene.

17. In 1951, before the operon model of gene expression had been discovered, Joshua Lederberg found that wild-type *E. coli* could not utilize the sugar neolactose, a derivative of lactose. He discovered a mutant type that

Table 8.3 Information for Problem 18: Enzyme Activities for β-Galactosidase and β-Galactoside Permease in the Absence and Presence of Inducer for Six Genotypes in *E. coli

Genotype	Inducer Absent		Inducer Present	
	β-Galactosidase	β-Galactoside Permease	β-Galactosidase	β-Galactoside Permease
$I^+ \, O^+ \, Z^+ \, Y^+$	<1	0	100	100
$F' \, I^+ \, O^+ \, Z^- \, Y^+ \, / \, I^- \, O^+ \, Z^- \, Y^+$	<1	0	320	100
$F' \, I^+ \, O^c \, Z^+ \, Y^+ \, / \, I^- \, O^+ \, Z^- \, Y^+$	36	33	270	100
$F' \, I^+ \, O^c \, Z^+ \, Y^+ \, / \, I^+ \, O^+ \, Z^- \, Y^+$	110	50	330	100
$F' \, I^+ \, O^c \, Z^- \, Y^+ \, / \, I^+ \, O^+ \, Z^+ \, Y^-$	<1	—	100	—
$F' \, I^+ \, O^c \, Z^+ \, Y^- \, / \, I^+ \, O^+ \, Z^- \, Y^+$	60	0	300	100

*Enzyme activity is expressed as a percentage of wild-type enzyme expression.

Table 8.4 Information for Problem 19: Activities of Seven Enzymes, Each Encoded by a Single Gene in the *his* Operon of *E. coli

Enzyme

Strain	Isomerase	Phosphatase	Transaminase	Dehydrogenase	Cyclase	Pyrophos-phorylase	Amido transferase
Wild-type	1.0	1.0	1.0	1.0	1.0	1.0	1.0
Mutation 1	0.0	0.8	0.8	1.1	0.0	1.1	—
Mutation 2	0.2	0.0	0.8	0.9	0.1	0.9	0.1
Mutation 3	0.3	0.1	0.0	0.7	0.3	0.8	0.1
Mutation 4	0.3	0.2	0.2	0.0	0.3	1.0	0.2
Mutation 5	0.3	0.4	0.2	0.3	0.3	0.0	0.2
Mutation 6	0.2	0.3	0.2	0.3	0.4	0.0	0.2
Mutation 7	0.4	0.4	0.4	0.3	0.6	0.0	0.5

*The activity of each enzyme is expressed as a portion of the enzyme activity in the wild-type strain.

could utilize this sugar and found that the mutant type produced β-galactosidase constitutively. Provide a plausible explanation for these observations.

18. In 1960, Jacob, Perrin, Sanchez, and Monod, in what has become a classic paper in bacterial genetics (*Comptes Rendus des Séances de l'Academie des Sciences* 250: 1727–1729), described the basic aspects of the *lac* operon. Among the data they presented was a comparison of merozygotes with different genotypes for their production of β-galactosidase and β-galactoside permease, included in Table 8.3. What evidence is there in these data that O^c mutations are cis acting and that *lacI*⁻ mutations are trans acting?

19. In 1963, Ames and Hartman (*Cold Spring Harbor Symposia on Quantitative Biology* 28:349–356) described the structure of the histidine operon in *E. coli*. Part of their analysis was based on polar mutations in the genes that encode enzymes. Table 8.4 shows their data on how polar mutations affect enzyme expression. To the extent possible, identify the order of the genes in the operon and the gene in which each of the seven mutations is present. Remember that polar mutations typically affect genes downstream from the mutated gene.

20. In 1977, Humayun, Jeffrey, and Ptashne (*Journal of Molecular Biology* 112:265–277) published the complete DNA sequences of the λ phage operators, illustrated in Figure 8.32. All six operators bind pcI and share similarities in sequence. **(a)** Write a consensus sequence for all six operators. **(b)** Which nucleotides are fully conserved in all six operators? **(c)** Describe the symmetry in the consensus sequence. **(d)** Both λ repressor and Cro bind to each operator as a dimer. What does this observation suggest about the sequence symmetry? **(e)** Cro preferentially binds to $O_R 3$, whereas λ repressor preferentially binds to $O_R 1$ and $O_R 2$. What similarities are there between $O_R 1$ and $O_R 2$ that are not found in $O_R 3$?

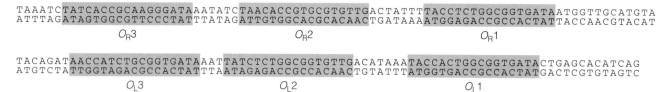

```
TAAATCTATCACCGCAAGGGATAAATATCTAACACCGTGCGTGTTGACTATTTTACCTCTGGCGGTGATAATGGTTGCATGTA
ATTTAGATAGTGGCGTTCCCTATTTATAGATTGTGGCACGCACAACTGATAAAATGGAGACCGCCACTATTACCAACGTACAT
        O_R3                        O_R2                        O_R1
```

```
TACAGATAACCATCTGCGGTGATAAATTATCTCTGGCGGTGTTGACATAAATACCACTGGCGGTGATACTGAGCACATCAG
ATGTCTATTGGTAGACGCCACTATTTAATAGAGACCGCCACAACTGTATTTATGGTGACCGCCACTATGACTCGTGTAGTC
        O_L3                        O_L2                        O_L1
```

Figure 8.32 Information for Problem 20: The DNA sequences of the operators in λ. (Reprinted by permission of Academic Press.)

21. Look at the DNA sequence of the *lac* operator shown in Figure 8.3. Find the DNA sequence symmetry in the operator. Given the information from the previous problem, what is the most probable purpose of the DNA sequence symmetry?

22. Most *lacI*⁻ mutations are clustered near the 5′ end of the gene in codons 1–60. What does this imply about the function of the amino acids encoded in this region?

23. The *lacI*⁻ mutations clustered in codons 1–60 are called *lacI*⁻ᵈ mutations because they have a dominant effect when combined with a *lacI*⁺ allele in a merozygote. In other words, if an F′ *lacI*⁺ factor is introduced into an F⁻ *lacI*⁻ᵈ cell, the repressor protein still fails to bind to the operator. The *lacI*⁻ mutations in codons downstream from codon 60 do not have this dominant effect and are called *lacI*⁻ mutations. If an F′ *lacI*⁺ factor is introduced into an F⁻ *lacI*⁻ cell (where the *lacI*⁻ mutation is downstream from codon 60), functional repressor protein may form and bind to the operator. What do these observations imply about the DNA-binding domains of the *lac* repressor protein?

24. The *lacI*ˢ mutations are only in codons downstream from codon 62. There are nine clusters of *lacI*ˢ mutations, six of which are in regular spacings of 78 nucleotide pairs from the adjacent clusters. What do these observations imply about the inducer-binding sites?

25. Sauer, Ross, and Ptashne (1982. *Journal of Biological Chemistry* 257:4458–4462) found that RecA cleaves λ repressor at a particular site between an alanine and a glycine. They also found that the repressor for the temperate phage P22 is cleaved by RecA at a similar site between alanine and glycine. The cleavage separates the DNA-binding domain from the domains that interact to form dimers. Once cleaved, the repressors no longer bind to the DNA. What does this result suggest about the necessity of dimer formation and DNA binding?

26. Sauer, Ross, and Ptashne, in the same article cited in the previous question found that mutant λ repressor that lacked the Ala-Gly cleavage site was no longer cleaved by RecA. What is the expected consequence of such a mutation in terms of lysogeny and induction of lysis?

27. Li, Bestor, and Jaenisch (1992. *Cell* 69:915–926) cultured mutant mouse cells that produced substantially reduced levels of DNA methyltransferase. The mutant cells appeared the same as normal cells in culture, but cytosine methylation in these cells was only one-third that of wild-type cells. Mutant mouse embryos with reduced levels of DNA methyltransferase, however, were stunted and poorly developed. All of the mutant embryos died during gestation. Some of the most important model organisms, such as *Drosophila melanogaster* and yeast, lack methylation altogether, calling into question the role of methylation in gene regulation. What do these research results with mouse embryos suggest about the necessity of DNA methylation? Formulate a hypothesis that would explain these results in light of the observation that in many eukaryotic species inactive genes tend to be methylated whereas active genes tend to be undermethylated.

28. Methylation typically is found on the cytosines in CG doublets in DNA. Under these circumstances, the methylated cytosine is prone to undergo mutation to thymine through deamination (see section 5.2). As Bird (1992. *Cell* 70:5–8) pointed out, one-third of all point mutations responsible for human genetic disorders are due to mC → T mutations in CG doublets, even though CG doublets are fairly rare in the human genome. Thus, cytosine methylation is apparently a cause of increased mutation. In light of the previous question, why might methylation be present in some species but not others?

29. In 1994, Muehlbauer, Somers, Matthews, and Gengenbach (*Plant Physiology* 106:1303–1312) reviewed the effects of lysine and threonine inhibition of aspartate kinase in maize. Three forms of the enzyme had been characterized, one that was inhibited by lysine only, one that was inhibited by lysine and S-adenosyl methionine, and one that was inhibited by threonine only. These authors identified an enzyme with dual aspartate kinase (AK) and homoserine dehydrogenase (HSDH) activity that was inhibited only by threonine. Given these observations, why is it essential to include both lysine and threonine in plant cell culture medium to recover feedback-insensitive mutant cells?

FOR FURTHER READING

Chapter 7 of **Maloy, S. R., J. E. Cronan, Jr., and D. Freifelder. 1994.** *Microbial Genetics*, **2nd ed. Boston: Jones and Bartlett**, provides an excellent summary of gene regulation

in prokaryotes. Chapter 17 of **Wolfe, S. L. 1993. Molecular and Cellular Biology. Belmont, Calif.: Wadsworth**; and Chapter 9 of **Alberts, B., D. Bray, J. Lewis, M. Raff, K. Roberts, and J. D. Watson. 1994.** *Molecular Biology of the Cell*, **3rd ed. New York: Garland**, cover gene regulation in both prokaryotes and eukaryotes. **Lewin, B. 1997.** *Genes VI.* **Oxford: Oxford University Press**, provides detailed discussions of prokaryotic gene regulation in Chapters 11–13, and of eukaryotic gene regulation at the transcriptional level in Chapter 29. Mark Ptashne, one of the leading researchers in gene regulation, wrote a succinct and informative book that focuses entirely on gene regulation, covering the basic concepts and experimental evidence underlying those concepts: **Ptashne, M. 1992.** *A Genetic Switch: Phage λ and Higher Organisms*, **2nd ed. Cambridge, Mass.: Cell Press, Blackwell Scientific**. The development of the operon model of gene regulation was first fully formulated by **Jacob, F., D. Perrin, C. Sanchez, and J. Monod. 1960. L'opéron: groupe de gènes à expression coordonnée par un opérateur.** *Comptes Rendus des Séances de l'Academie des Sciences* **250:1727–1729**; and was summarized by **Jacob, F., and J. Monod. 1961. Genetic regulatory mechanisms in the synthesis of proteins.** *Journal of Molecular Biology* **3:318–356.** The structure and function of the *lac* repressor and its interaction with operators, the inducer, and CAP are described in detail in **Lewis, M., G. Chang, N. C. Horton, M. A. Kercher, H. C. Pace, M. A. Schumacher, R. G. Brennan, and P. Lu. 1996. Crystal structure of the lactose operon repressor and its complexes with DNA and inducer.** *Science* **271:1247–1254**. The competition between Cro and the λ repressor for establishment of lysis or lysogeny is reviewed in detail in **Johnson, A. D., A. R. Poteete, G. Lauer, R. T. Sauer, G. K. Ackers, and M. Ptashne. 1981. l repressor and cro—components of an efficient molecular switch.** *Nature* **294:217–223**; and in **Ptashne, M., A. Jeffrey, A. D. Johnson, R. Maurer, B. J. Meyer, C. O. Pabo, T. M. Roberts, and R. T. Sauer. 1980. How the λ repressor and cro work.** *Cell* **19:1–11.** Eukaryotic gene regulation and some of the history behind the development of the current model are discussed in **Tijan, R. 1995. Molecular machines that control genes.** *Scientific American* **272 (***Feb***):55–61**. The possible role of methylation in eukaryotic gene regulation is discussed in a brief but highly informative review by **Bird, A. 1992. The essentials of DNA methylation.** *Cell* **70:5–8.**

For additional reading, go to InfoTrac College Edition, your online research library at: http: www.infotrac-college.com/brookscole

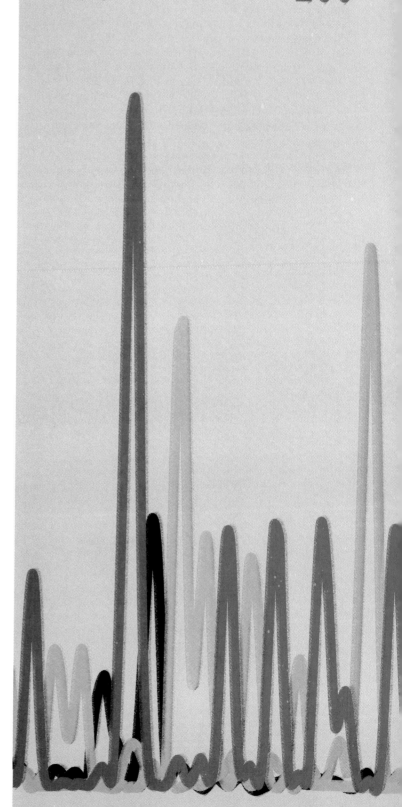

CHAPTER 9

RECOMBINANT DNA AND MOLECULAR ANALYSIS

The discovery that DNA is universal as the genetic material and that the genetic code is essentially universal among cellular organisms has had some astounding implications. About 15 years after the structure of DNA was discovered, one of these implications began to remodel both the study and the practical applications of genetics in medicine and agriculture. Because DNA is the genetic material of all cellular organisms and the genetic code is universal, it should be possible to replicate and express genes from any species in any other species. However, nature has barriers that inhibit the transfer of DNA from one species to another. The challenge to scientists was to overcome these barriers by developing methods through which foreign DNA could be introduced into the cells of a recipient species and integrated into the cell's own DNA.

Scientists have met the challenge, and the potential is now a reality, especially in bacteria. A **DNA clone** is DNA from any species that researchers introduce into bacterial cells, where the introduced DNA replicates as the bacterial cells grow and divide. When modified to function in bacterial systems, eukaryotic genes that have been cloned in bacteria can be expressed in bacteria, often producing the same proteins that those genes produced in their original host. Human insulin and human growth hormone are two examples of human proteins now produced commercially in bacteria.

Practical applications are not the only area where such DNA manipulation has been of value. Researchers in every field of genetics routinely use methods from molecular biology to address fundamental questions about how genes are organized, expressed, and inherited. Much of the information we have already discussed and much of what we will discuss in the remainder of this book was obtained using these methods as research tools.

The purpose of this chapter is to describe the fundamental methods of DNA and protein manipulation and how they are used in genetic analysis. We include this chapter here because these methods are closely tied to bacterial genetics and gene regulation, the subjects of the two preceding chapters.

Applications of the methods of molecular biology to genetics research are now so numerous that it would be impossible to describe even a few of them in a single chapter. Instead, we cover in this chapter the fundamental methods that provide a foundation for the numerous applications we will encounter in every subsequent chapter of this book.

We begin by defining and discussing recombinant DNA. We then examine the purpose and design of cloning

vectors, segments of DNA that researchers have designed to carry and replicate cloned DNA fragments in bacterial cells. Next, we will see how researchers find and utilize specific fragments of cloned DNA as research tools and how researchers can design a eukaryotic gene so that it can be expressed in bacteria. We will discuss the polymerase chain reaction, a widely used method that permits researchers to amplify abundant copies of a DNA fragment in a test tube. Finally, we will discuss how DNA fragments are identified, characterized, and sequenced, and how DNA sequence information is managed in computerized databases.

9.1 RECOMBINANT DNA

As we saw in Chapter 7, it is possible for foreign DNA to enter a bacterial cell. This is not always advantageous for the bacterium. In fact, most bacterial species maintain their genetic integrity with protective mechanisms that prevent foreign DNA from becoming a part of the cell's genome. Among the most important natural protections that bacteria have against foreign DNA invasion are systems that methylate and cleave DNA. By manipulating these natural systems, researchers have developed methods to cut and splice fragments of DNA then introduce these DNA fragments into a bacterial cell, where the DNA can replicate. Let's look first at the natural system, then see how researchers use it for manipulation of DNA.

Restriction Endonucleases and Methylases

As we have seen in previous chapters, DNA is often methylated. Bacterial cells methylate their DNA in specific ways to distinguish their own DNA from foreign DNA that has entered the cell. In some species of bacteria, each protection system consists of a restriction endonuclease and a methylase. A **restriction endonuclease** (also called a **restriction enzyme**) is an enzyme that cleaves DNA at a specific nucleotide sequence. A **methylase** is an enzyme that methylates DNA. Each restriction endonuclease carries out its function cooperatively with a methylase, where both enzymes recognize a particular DNA sequence called a **restriction site**. Let's take a close look at the *Eco*RI system in *E. coli* to illustrate how these systems function.

The enzymes in a restriction endonuclease–methylase system have an italicized three-letter name that represents the bacterial species, followed by a letter and Roman numeral, or sometimes just a Roman numeral, to designate the system. In the *Eco*RI example, *Eco* stands for "*E. coli*," and RI designates the system. Other systems in the same species carry the same species designation but a different system designation. For example, the *Eco*RV system has a restriction endonuclease and methylase that differ from those in the *Eco*RI system.

The restriction site for the *Eco*RI restriction endonuclease and methylase is the six-nucleotide-pair sequence

```
5'...G A A T T C...3'
3'...C T T A A G...5'
```

The methylase adds methyl groups to the two adenine residues near the center:

```
         m
5'...G A A T T C...3'
3'...C T T A A G...5'
            m
```

m = methyl group

Notice that the sequence of nucleotides is **palindromic**, meaning that the nucleotide sequence is the same on both strands when read in a 5' → 3' direction. This ensures that both DNA strands will be marked in the same way at the same place. When the DNA replicates, the original strand remains methylated but the newly synthesized strand is not methylated initially. Shortly after replication, *Eco*RI methylase recognizes the methyl group on the original strand and methylates the corresponding adenine on the newly synthesized strand.

The *Eco*RI restriction endonuclease recognizes the same restriction site as *Eco*RI methylase and cleaves the DNA within the site, creating a staggered (as opposed to a blunt) cut:

```
5'...G A A T T C...3'
3'...C T T A A G...5'
            ↓
5'...G            A A T T C...3'
3'...C T T A A            G...5'
```

However, *Eco*RI restriction endonuclease does not cleave this sequence if the two adenines at the center are methylated. Because an *E. coli* cell methylates these sequences using *Eco*RI methylase, the cell's own *Eco*RI restriction endonuclease does not cleave the cell's DNA at the restriction site. Foreign DNA that enters the cell as a phage, a plasmid, or a piece of transformed DNA often lacks the methylation pattern, and *Eco*RI restriction endonuclease cleaves the DNA wherever the enzyme encounters its restriction site in the DNA.

This sort of system, in which there are separate methylases and restriction endonucleases, is called a type II system. There are other systems, called type I and type III systems, that methylate and cleave DNA. In these systems, a single enzyme methylates and cleaves DNA. Type I and type III enzymes do not cleave the DNA at the restriction site. Instead, they bind to the DNA at the restriction site and cleave the DNA elsewhere. Also, type I and III enzymes require ATP as an energy source for cleavage, but type II restriction endonucleases do not. For these reasons, type I and type III restriction endonucleases are of limited value for DNA manipulation, and we will not discuss them further.

Table 9.1 Some Type II Restriction Endonucleases with Their Restriction Sites, Shown After Cleavage

Restriction Endonuclease	Source	Restriction Site		Type of Cut
AluI	Arthobacter luteus	5′ AG	CT 3′	Blunt
		3′ TC	GA 5′	
BalI	Brevibacterium albidum	5′ TGG	CCA 3′	Blunt
		3′ ACC	GGT 5′	
BamHI	Bacillus amyloliquifaciens	5′ G	GATCC 3′	Staggered
		3′ CCTAG	G 5′	
BglII	Bacillus globigii	5′ A	GATCT 3′	Staggered
		3′ TCTAG	A 5′	
CfoI	Clostridium formoaceticum	5′ GCG	C 3′	Staggered
		3′ C	GCG 5′	
DraI	Deinococcus radiophilus	5′ TTT	AAA 3′	Blunt
		3′ AAA	TTT 5′	
EcoRI	Escherichia coli	5′ G	AATTC 3′	Staggered
		3′ CTTAA	G 5′	
EcoRV	Escherichia coli	5′ GAT	ATC 3′	Blunt
		3′ CTA	TAG 5′	
HaeIII	Haemophilus aegyptius	5′ GG	CC 3′	Blunt
		3′ CC	GG 5′	
HindIII	Haemophilus influenzae	5′ A	AGCTT 3′	Staggered
		3′ TTCGA	A 5′	
PstI	Providencia stuarti	5′ CTGCA	G 3′	Staggered
		3′ G	ACGTC 5′	
SacI	Streptomyces achromogenes	5′ GAGCT	C 3′	Staggered
		3′ C	TCGAG 5′	
SalI	Streptomyces alba	5′ G	TCGAC 3′	Staggered
		3′ CAGCT	G 5′	
TaqI	Thermus aquaticus	5′ T	CGA 3′	Staggered
		3′ AGC	T 5′	
XhoI	Xanthomonas holica	5′ C	TCGAG 3′	Staggered
		3′ GAGCT	C 5′	

There are many type II restriction endonucleases, each with a specific restriction site in the DNA. Table 9.1 provides a partial list of these enzymes along with their restriction sites. Notice that all of the restriction sites in Table 9.1 are palindromic.

Type II restriction endonucleases are used extensively in DNA manipulation. Let's now look briefly at the procedure used to construct a **restriction map**, a diagram of the restriction sites in a DNA molecule. Restriction mapping is one of many applications for genetic research using type II restriction endonucleases.

Example 9.1 Restriction mapping of DNA molecules.

Each mitochondrion in a human cell contains several identical copies of a DNA molecule that is 16,569 nucleotide pairs long. DNA fragments of the lengths listed in Table 9.2 are obtained when mitochondrial DNA is digested (cut into fragments) with the restriction endonucleases EcoRI only, EcoRV only, and both EcoRI and EcoRV together.

Problem: (a) From the information in Table 9.2, determine whether the human mitochondrial DNA molecule is linear or circular. (b) Draw a restriction map of the molecule with the positions of the EcoRI and EcoRV restriction sites labeled.

Solution: (a) Digestion of a circular molecule generates the same number of fragments as there are restriction sites in the molecule, whereas digestion of a linear molecule generates one more fragment than there are restriction sites in the molecule. In this example, three fragments were generated by the single digestion with EcoRI. This corresponds to three EcoRI sites if the molecule is circular and two EcoRI sites if the molecule is linear, as illustrated in Figure 9.1.

Table 9.2 Information for Example 9.1: Sizes of DNA Fragments (in Nucleotide Pairs) That Are Generated by Digestion of Human Mitochondrial DNA with *Eco*RI only, *Eco*RV only, or *Eco*RI and *Eco*RV

	*Eco*RI Only	*Eco*RV Only	*Eco*RI and *Eco*RV
	8050	6877	6877
	7366	6137	5906
	1153	3555	1460
			1153
			942
			231
Totals	16569	16569	16569

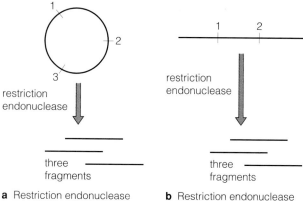

a Restriction endonuclease digestion of a circular molecule yields the same number of fragments as there are restriction sites.

b Restriction endonuclease digestion of a linear molecule yields one more fragment than there are restriction sites.

Figure 9.1 Restriction endonuclease digestion of **(a)** circular and **(b)** linear DNA molecules.

If the mitochondrial DNA is circular prior to digestion, there must be three *Eco*RI sites and three *Eco*RV sites, for a total of six restriction sites. Double digestion with *Eco*RI and *Eco*RV would generate six fragments from such a circular molecule. If the molecule is linear, it must have two *Eco*RI sites and two *Eco*RV sites, for a total of four restriction sites. Double digestion of such a molecule would yield five fragments. As six fragments are produced by double digestion, the molecule must be circular.

(b) Generating a restriction map often requires substantial application of logic, sometimes through a lengthy process of trial and error. Let's work through this example step by step. Because the original molecule is circular, the ends of each fragment must have been cut by a restriction enzyme. A fragment that was cut by *Eco*RI on both ends is called an *Eco*RI fragment. A fragment that was cut by *Eco*RV on both ends is called an *Eco*RV fragment. A fragment that was cut by *Eco*RI on one end and *Eco*RV on the other end is called an *Eco*RI/*Eco*RV fragment. All fragments from the single *Eco*RI digestion are *Eco*RI fragments, and all fragments from the single *Eco*RV digestion are *Eco*RV fragments. Fragments from the double digestion may be *Eco*RI fragments, *Eco*RV fragments, or *Eco*RI/*Eco*RV fragments. Any *Eco*RI fragment in the double digestion would also appear in the *Eco*RI single digestion. There is one such fragment, 1153. Using the same logic, a single *Eco*RV fragment, 6877, can be identified in the double digestion. The remaining four fragments in the double digestion (5906, 1460, 942, and 231) do not appear in the single digestions and, therefore, must be *Eco*RI/*Eco*RV fragments.

Let's add these four *Eco*RI/*Eco*RV fragments in all possible combinations of two to see if any of the sums match an *Eco*RI or *Eco*RV fragment from the single digestions. If any do, then those two fragments must be contiguous to one another. We find two such combinations, 5906 + 1460 = 7366 and 5906 + 231 = 6137, and we can draw maps of these contiguous fragments with the restriction sites labeled:

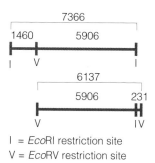

I = *Eco*RI restriction site
V = *Eco*RV restriction site

Because these two pairs of contiguous fragments share the 5906 fragment, we can combine the two into a single map:

The one remaining *Eco*RI/*Eco*RV fragment is 942. It must not be attached to either end of the segment we just mapped because the analysis we just conducted would have revealed such an attachment. Rather, there must be an *Eco*RV fragment attached to the right end of the mapped segment and an *Eco*RI fragment attached to the left end. Let's try the 6877 *Eco*RV fragment first. If it is attached to the 231 fragment, then the sum of these two is 6877 + 231 = 7108. Now, let's try attaching

the 942 *Eco*RI/*Eco*RV fragment to 6877. If this is a correct attachment, the sum of all three fragments should equal one of the *Eco*RI fragments from the single digestion. Indeed, 6877 + 231 + 942 = 8050, which represents the 8050 *Eco*RI fragment from the single digestion, allowing us to expand the map:

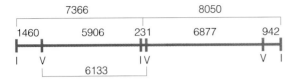

Five of the six fragments have been mapped, leaving only the 1153 *Eco*RI fragment to be placed between the 942 and 1460 fragments to complete the circular map:

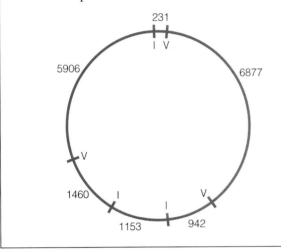

Recombination of DNA Fragments

The ability of many restriction enzymes to make staggered cuts in double-stranded DNA at specific restriction sites is fortuitous for DNA manipulation. The staggered cuts leave short sequences of single-stranded DNA:

The short, single-stranded sequences are called **cohesive ends** (or **sticky ends**) because one cohesive end can form base pairs with the complementary nucleotides in the cohesive end of another fragment of DNA that was cleaved by the same enzyme. For example, suppose separate samples of human DNA and *Drosophila* DNA are both treated with *Eco*RI. In each sample, *Eco*RI cuts the DNA wherever it encounters an *Eco*RI restriction site, creating DNA fragments with cohesive ends. The human DNA fragments and the *Drosophila* DNA fragments have the same cohesive ends because they were cut by the same restriction endonuclease. If the human DNA frag-

ments are mixed with the *Drosophila* DNA fragments, DNA fragments join to one another at random through nucleotide pairing at the cohesive ends:

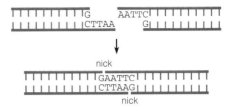

In the resulting mixture of double-stranded DNA molecules, some consist of a human DNA fragment joined to a *Drosophila* DNA fragment. At each joining, there are nicks in the two sugar-phosphate backbones. Addition of the enzyme ligase seals the nicks, making continuous segments of double-stranded DNA:

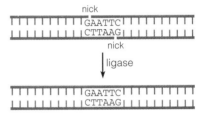

DNA that has been cut and joined in this manner is called **recombinant DNA**, because it consists of DNA that is not normally attached in nature but has been recombined artificially. DNA from the same individual, two or more individuals of the same species, two or more individuals of different species, or any number of individuals from any number of different species can be recombined artificially in this way.

~

After a restriction endonuclease cuts DNA at a restriction site, DNA fragments with cohesive ends can be joined, forming recombinant DNA.

9.2 PLASMID CLONING VECTORS

The recombinant DNA described in the previous section is of little value because it consists of random combinations of DNA that cannot be replicated. It is difficult and impractical for researchers to isolate and work with a single molecule of DNA. In order to manipulate a DNA fragment, researchers must produce many identical copies of the fragment. One of the best ways to amplify a single molecule of DNA into many copies is to insert the DNA into bacterial cells and allow the cells to replicate the DNA as they grow and divide.

In order to replicate in a bacterial cell, an introduced DNA fragment must be inserted into a **cloning vector**, a DNA molecule that has an origin of replication and is capable of replicating in a bacterial cell. Most cloning

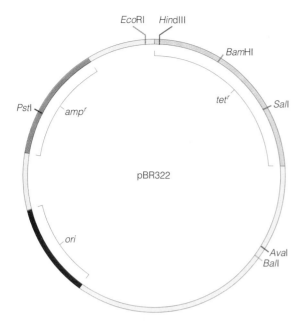

Figure 9.2 Plasmid pBR322. Seven of the unique restriction sites are shown, as well as the two selectable marker genes, *amp^r* and *tet^r*, the ampicillin and tetracycline resistance genes; *ori* represents the origin of replication. (Adapted from Sambrook, J., E. F. Fritsch, and T. Maniatis. 1989. *Molecular Cloning: A Laboratory Manual*, 2nd ed. Cold Spring Harbor, N.Y.: Cold Spring Harbor Laboratory Press. Used by permission)

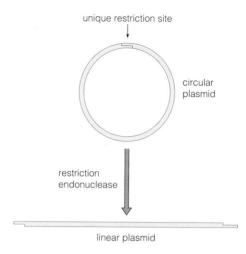

Figure 9.3 Cleavage of a circular plasmid at a unique restriction site.

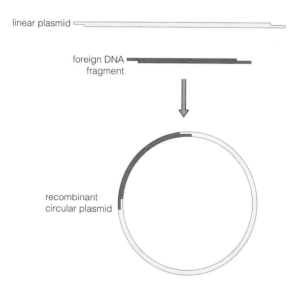

Figure 9.4 Recombination of a DNA fragment and a linear plasmid to form a circular plasmid.

vectors are genetically engineered plasmids or phages. In this section, we will discuss plasmid cloning vectors, leaving phage cloning vectxors for the next section.

If a researcher can insert a fragment of DNA into a plasmid and introduce the plasmid into a bacterial cell, then the plasmid, and the fragment it contains, can replicate to make many copies. Researchers have constructed artificial plasmids, called **plasmid cloning vectors**, into which a fragment of DNA can be readily inserted. The plasmid cloning vector, with its inserted DNA fragment, can then be introduced into a bacterial cell, where it replicates to form many copies within the cell. When the cell divides, the plasmids continue to replicate, eventually synthesizing many copies of the inserted DNA fragment in a culture of cells. In most cases, both the cloning vector and the host cell strain that carries the vector have been genetically designed to facilitate the cloning process.

Let's use the plasmid cloning vector pBR322 as an example to illustrate how DNA cloning is accomplished with plasmids.

An Example of a Plasmid Cloning Vector: pBR322

The plasmid pBR322 is an artificial plasmid that researchers constructed by recombining DNA from several sources. After its introduction in 1977, pBR322 became the standard vector for DNA cloning for more than a decade. It has since been replaced by newer plasmid vectors, but it remains an excellent example of how plasmids are designed and used for DNA cloning. Genetic engineers constructed pBR322 as a cloning vector with certain features, including an origin of replication, two antibiotic resistance genes, and several unique restriction sites, as illustrated in Figure 9.2.

The origin of replication allows the plasmid to be replicated once it is introduced into a bacterial cell. The antibiotic resistance genes are called **selectable markers**, because they confer a means to eliminate all cells that do not contain the plasmid—that is, by growing the cells on medium containing an antibiotic. A **unique restriction site** is one that appears only once in the plasmid. The importance of unique restriction sites is that the circular plasmid is cut in only one place when treated with a restriction endonuclease, making it a linear molecule with cohesive ends, as illustrated in Figure 9.3. When a plasmid that has been cut at a unique restriction site is mixed with a DNA fragment that has been cut with the same restriction enzyme, the fragment may recombine with the plasmid (Figure 9.4).

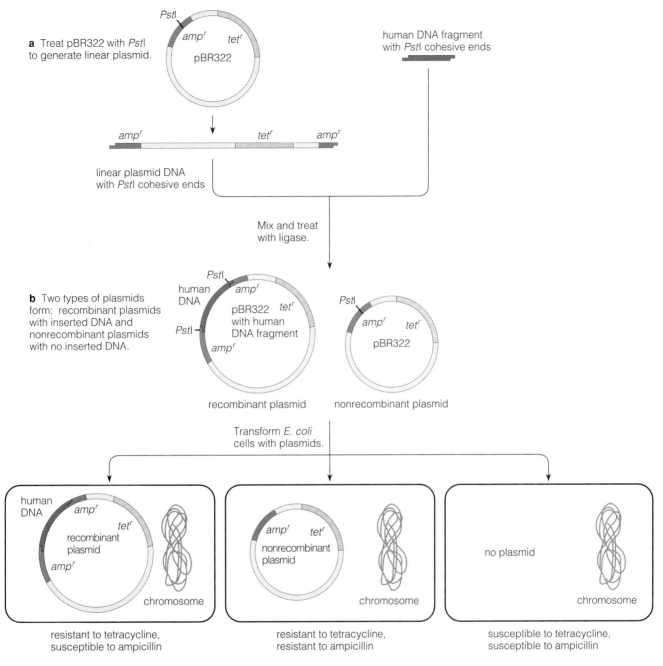

a Treat pBR322 with *Pst*I to generate linear plasmid.

human DNA fragment with *Pst*I cohesive ends

linear plasmid DNA with *Pst*I cohesive ends

Mix and treat with ligase.

b Two types of plasmids form: recombinant plasmids with inserted DNA and nonrecombinant plasmids with no inserted DNA.

recombinant plasmid

nonrecombinant plasmid

Transform *E. coli* cells with plasmids.

resistant to tetracycline, susceptible to ampicillin

resistant to tetracycline, resistant to ampicillin

susceptible to tetracycline, susceptible to ampicillin

Figure 9.5 Construction and identification of recombinant plasmids.

Several of the unique restriction sites in pBR322 lie within one or the other of the antibiotic resistance genes. A fragment of DNA inserted at one of these sites creates an insertion mutation that causes the gene to lose its function. Fragments inserted at the *Pst*I site cause ampicillin susceptibility, and fragments inserted at the *Bam*HI, *Hin*dIII, or *Sal*I sites cause tetracycline susceptibility. This provides a way to select for cells that contain a plasmid with an inserted fragment of DNA.

Let's use an example to illustrate the procedure; as we discuss it, you may wish to follow along in Figure 9.5. Suppose that we have many identical copies of a human DNA fragment generated by digestion with the restric-

tion endonuclease *Pst*I. The fragments have *Pst*I cohesive ends. We first digest pBR322 with *Pst*I, which gives it a linear form with *Pst*I cohesive ends (Figure 9.5a). We then mix the linear pBR322 with the DNA fragments. Some of the plasmids rejoin as circular plasmids without an inserted fragment. These are called **nonrecombinant plasmids**. The remaining plasmids recombine with the human DNA fragments. These are called **recombinant plasmids** (Figure 9.5b). We treat the mixture with ligase to seal the nicks in all the plasmids.

We then add this mixture of plasmids to *E. coli* cells that have been treated to make them competent for transformation with the plasmids. We obtain a mixture of three

types of cells from this procedure: cells that contain recombinant plasmids, cells that contain nonrecombinant plasmids, and cells that have escaped transformation and contain no plasmid (Figure 9.5c). Each cell contains only one type of plasmid; only one copy of pBR322 enters a cell during transformation. Soon after transformation, however, the plasmid replicates several times so that each cell contains about 10–15 copies of a single type of plasmid.

We can use the antibiotic resistance genes to identify and isolate the cells that contain recombinant plasmids. As illustrated in Figure 9.6, the first step is to plate the cells on agar-solidified medium containing tetracycline. Because recombinant and nonrecombinant plasmids contain an active tetracycline resistance gene but untransformed cells do not, plating the cells on tetracycline eliminates all untransformed cells. The next step is to make a replica plate from the tetracycline plate onto a plate containing both tetracycline and ampicillin. Colonies grow on this plate only from cells that have been transformed with nonrecombinant plasmids, which have an uninterrupted ampicillin resistance gene. We can now identify those colonies that grow on the tetracycline plate but not on the replicate tetracycline-ampicillin plate as containing recombinant plasmids.

We can select cells with recombinant plasmids from the original tetracycline plate and grow abundant quantities of them in liquid medium with tetracycline to eliminate any bacteria that happen to lose their plasmids. As the cells grow and divide, the fragment of DNA cloned into the pBR322 vector is replicated as part of the plasmid vector.

After culture growth, recovery of the cloned fragment of DNA is simple. Researchers treat the bacterial cells with a solution that lyses the cells, releasing the cells' DNA into the solution. The plasmid DNA is much smaller than the chromosomal DNA and can be easily isolated and purified. Once the plasmid DNA is purified, the plasmid may be digested with *Pst*I to release the cloned fragment.

Abundant quantities of a DNA fragment may be generated simply by growing bacterial cultures that contain a recombinant plasmid. Also, there is no limit on how long a DNA fragment may be maintained and propagated. Bacteria containing a recombinant plasmid can be mixed with glycerol, placed in a −70°C freezer, and kept there indefinitely. The bacteria do not divide and grow at this cold temperature, but when protected from freezing by glycerol, they remain alive indefinitely, and resume growth and division once they are warmed.

Blunt-End Ligation

The cloning procedure just described is a standard for cloning DNA fragments in a plasmid. However, some DNA fragments have blunt ends (no single-stranded overhangs on the end) instead of cohesive ends. Under these circumstances, a method called **blunt-end ligation** can be used to insert a blunt-ended DNA fragment into

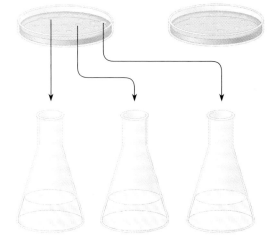

a Colonies growing on a plate containing tetracycline only.

b Colonies growing on a replica plate containing ampicillin and tetracycline. Three colonies are missing when compared to the tetracycline plate.

c The three colonies that are missing from the replica plate must contain recombinant plasmids. Cells from these colonies can be isolated from the tetracycline plate and cultured.

Figure 9.6 Identifying and selecting colonies with recombinant plasmids by replica plating.

a blunt-ended vector. An enzyme called T4 DNA ligase, isolated from *E. coli* cells infected with phage T4, has the ability to join blunt ends of any two DNA molecules to form an intact double-stranded molecule. When both the DNA fragment and the linear plasmid have blunt ends, these blunt ends can be joined using T4 DNA ligase. The recombinant plasmids can then be selected as before using antibiotic resistance.

Homopolymer Tailing

Cohesive ends are usually better than blunt ends for inserting a DNA fragment into a vector. A method called **homopolymer tailing** can be used to synthesize artificial cohesive ends on both the linear plasmid and the DNA fragments using the enzyme terminal transferase. This enzyme adds a single strand of nucleotides onto the 3' end of a DNA molecule. If only one type of nucleotide is supplied, such as dATP, then terminal transferase synthesizes a short, single-stranded poly (A) segment on both ends of the linear molecule. For example, if a linear plasmid is treated with terminal transferase and dATP, it acquires cohesive ends consisting of short poly (A) segments. The DNA fragments to be inserted into this plasmid are then treated with terminal transferase and dTTP, the nucleotide complementary to adenine. This provides the DNA fragments with cohesive ends made of short poly (T) segments. When the fragments and the plasmids are mixed, the poly (A) segments pair with the poly (T) segments. The number of nucleotides may not

always match, and there may be single-stranded gaps in some places. However, these gaps can be easily filled with the proper nucleotides by adding the recombinant fragments to a mixture con-taining a DNA polymerase and a pool of nucleotides. The nicks are then sealed with ligase. Recombinant plasmids can then be selected as before using antibiotic resistance.

Linker DNA

Sometimes researchers need to add a new restriction site into a vector. This can be done by inserting a small fragment of DNA that contains a particular restriction site, called **linker DNA**, at a different restriction site in the vector. It is possible to synthesize any short nucleotide sequence to serve as linker DNA using a machine called a nucleic acid synthesizer. For example, a linker that has a *Bam*HI site for insertion into a *Pst*I site in a plasmid can be constructed by synthesizing a short fragment of linker DNA with a *Bam*HI site flanked by two *Pst*I sites. This linker can then be inserted into a plasmid's *Pst*I site. The new *Bam*HI site can be used for cloning fragments generated with *Bam*HI.

Plasmid Vectors with Polylinkers and Marker Genes for Blue-White Screening

The original workhorse plasmids, such as pBR322, have been replaced by newer plasmids that eliminate the need for replica plating. Colonies with a recombinant plasmid are white, and colonies with a nonrecombinant plasmid are blue, so researchers simply culture cells from the white colonies to select recombinant plasmids, a process called **blue-white screening**. Most of these modern plasmids also have a DNA insertion site called a **polylinker**, a segment of DNA that contains unique restriction sites for several enzymes. A polylinker provides researchers with a wide choice of unique restriction sites, reducing the need for blunt-end ligation, homopolymer tailing, or construction of DNA linkers.

Hundreds of plasmids with polylinkers and genes for blue-white screening are now available from biotechnology companies. Let's take a close look at pUC19, one of the first such plasmids to be developed, to see how plasmids with polylinkers and genes for blue-white screening are constructed.

Blue-white screening with pUC19 is based on an ingenious modification of the *lac* operon. (You may wish to review the material on the *lac* operon in section 8.2 before reading the rest of this section.) As illustrated in Figure 9.7, pUC19 has a single antibiotic resistance gene, *amp^r*, which confers resistance to ampicillin. The plasmid also contains a portion of the *lac* operon region, including the *lacI* gene and a modified version of the *lacZ* gene called *lacZ'*. The *lacZ'* gene encodes the first 146 amino acids of β-galactosidase (about 14% of the polypeptide). By

itself, this fragment of the enzyme does not function. However, a mutant form of β-galactosidase, encoded by a mutant *lacZ* allele in the *lac* operon of the host cell's chromosome, can combine with the *lacZ'* gene's product to form functional β-galactosidase. This ensures that functional β-galactosidase is produced in the mutant host cells only when the plasmid is present.

Expression of both the plasmid and host cell *lacZ* genes can be induced with IPTG, an analog of lactose that induces transcription of the *lac* operon. A substance called X-gal (5-bromo-4-chloro-3-indole-β-D-galactoside) is added to the medium. X-gal is a substrate of β-galactosidase and turns blue in the presence of functional β-galactosidase. Thus, when host cells that contain a nonrecombinant pUC19 plasmid are grown on medium with IPTG, X-gal, and ampicillin, the colonies that arise from these cells are blue.

The DNA insertion site in pUC19 is a polylinker with 13 unique restriction sites, as illustrated in Figure 9.7. The DNA fragment to be cloned can be inserted at any of these restriction sites, so researchers have several options when choosing a restriction enzyme. The polylinker is located within the *lacZ'* gene, so insertion of foreign DNA into the polylinker disrupts the *lacZ'* gene, eliminating functional β-galactosidase. Thus, colonies with recombinant plasmids fail to turn blue and appear instead as white colonies.

Cells with a recombinant pUC19 plasmid can be isolated from other types of cells in a single step with blue-white screening. After cells have been transformed with plasmids, they are plated on medium with ampicillin, IPTG, and X-gal. Cells without plasmids are eliminated by their ampicillin susceptibility, but cells containing either a recombinant plasmid or a nonrecombinant plasmid form colonies. Blue colonies contain nonrecombinant plasmids, and white colonies contain recombinant plasmids. Researchers isolate the white colonies directly from this plate, eliminating the need for replica plating (Figure 9.8).

An additional advantage that many newer plasmids have over older plasmids is the ability to provide high yields of cloned recombinant DNA because the plasmids replicate at a very high rate. For example, pBR322 usually replicates to form 15–20 copies per cell, whereas pUC19 replicates to form 500–700 copies per cell.

Let's now take a close look at the pUC19 polylinker to see how it is designed.

| Example 9.2 Organization of the pUC19 polylinker. |

In 1985, Yanisch-Perron, Vieira, and Messing (*Gene* 33:103–119) published the methods they used to construct the polylinker in the pUC19 plasmid vector and the complete DNA sequence of pUC19.

Problem: **(a)** The polylinker is within the reading frame of the *lacZ'* gene (see Figure 9.7). How can this gene produce a functional product when it contains the polylinker? **(b)** Why is *lacZ'* function

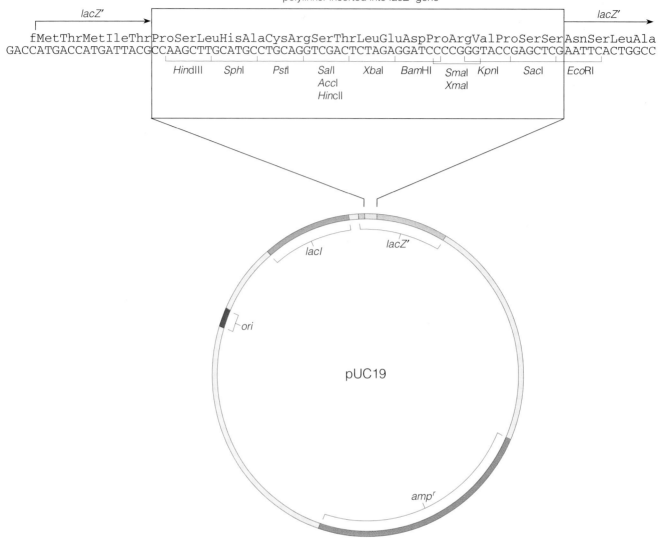

polylinker inserted into *lacZ'* gene

lacZ' → | polylinker inserted into *lacZ'* gene | ← *lacZ'*

fMetThrMetIleThr|ProSerLeuHisAlaCysArgSerThrLeuGluAspProArgValProSerSer|AsnSerLeuAla
GACCATGACCATGATTACG|CCAAGCTTGCATGCCTGCAGGTCGACTCTAGAGGATCCCCGGGTACCGAGCTCG|AATTCACTGGCC

HindIII SphI PstI SalI XbaI BamHI SmaI KpnI SacI EcoRI
                      AccI   XmaI
                      HincII

pUC19

lacI *lacZ'* ori amp^r

Figure 9.7 The pUC19 plasmid vector. (Adapted from Sambrook, J., E. F. Fritsch, and T. Maniatis. 1989. *Molecular Cloning: A Laboratory Manual*, 2nd ed. Cold Spring Harbor, N.Y.: Cold Spring Harbor Laboratory Press. Used by permission.)

lost when foreign DNA is inserted at one of the restriction sites in the polylinker?

Solution: **(a)** The polylinker is 54 nucleotide pairs long, which is evenly divisible by three (54 ÷ 3 = 18). Thus, it does not shift the reading frame. The polylinker is located very near the 3' end of the coding region, only five codons downstream from the initiation codon. The active site of the polypeptide encoded by the *lacZ'* gene resides in amino acids encoded downstream from the polylinker, so the product's function is not eliminated by the 18 additional amino acids encoded by the polylinker. **(b)** DNA insertions in the polylinker interrupt the reading frame of the *lacZ'* gene. For the *lacZ'* gene product to retain its function, the insertion would have to be so small that it did not substantially alter the protein, and the insertion would have to be a

multiple of three nucleotides to avoid a frameshift. However, most DNA fragments inserted into the polylinker are hundreds to thousands of nucleotide pairs long and therefore cause a loss of *lacZ'* function.

DNA cloning is the replication of a DNA fragment contained within a vector. The most frequently used vectors are bacterial plasmids that have been specifically designed for DNA cloning.

9.3   VECTORS FOR CLONING LARGE DNA INSERTS

It would be nice to be able to insert any DNA fragment into any vector. But how large a DNA insert can be depends on the vector used. Although plasmids are prob-

Figure 9.8 Blue-white screening on medium with ampicillian, X-gal, and IPTG. Blue colonies contain nonrecombinant plasmids. White colonies contain recombinant plasmids and can be isolated directly from this plate.

Table 9.3 Size Limits of Inserted DNA Fragments in Cloning Vectors

Vector	Approximate Size Limit of Inserted DNA (in kb)
Plasmids	10
Lambda and its derivatives	23
Cosmids	46
Bacterial artificial chromosomes (BACs)	300
Yeast artificial chromosomes (YACs)	500

ably the most commonly used vectors because of ease in manipulation, they are not the best vectors for cloning large fragments of DNA.

The sizes of DNA fragment are often expressed in **kb**, which stands for **kilobases**, which might be interpreted as 1000 bases. In common usage, however, one kb corresponds to 1000 nucleotide pairs in double-stranded DNA. Some of the most common vectors and the approximate sizes of DNA inserts they can accommodate are listed in Table 9.3.

Many genes are less than 10 kb, so they can be readily inserted into plasmids. However, other types of vectors are better choices than plasmids for cloning large fragments of DNA. Fragments up to about 23 kb may be accommodated in a phage vector, or up to 46 kb in a cosmid vector. Exceptionally large fragments (> 100 kb) can be cloned in a bacterial artificial chromosome (BAC) or a yeast artificial chromosome (YAC). We'll now take a look at phage cloning vectors, cosmid cloning vectors, and BACs. We'll postpone a discussion of YACs until the next chapter, where we can address them in the context of eukaryotic chromosome structure.

Phage Cloning Vectors

Genetic engineers have redesigned several natural bacteriophages as cloning vectors. Phages replicate rapidly during the lytic cycle, and any foreign DNA added to the phage is replicated along with it. Lambda (λ) is the most common phage cloning vector, but it requires substantial

modification from its native form to be used as a cloning vector. With the addition of foreign DNA, a native λ DNA molecule may become too large to fit into its capsid. Also, native λ is relatively large, so it has more than one restriction site for most restriction endonucleases. However, genetically engineered λ cloning vectors do not have these limitations.

Only about 60% of the λ genome is needed for the lytic pathway to proceed, and the genes required for the lytic pathway are clustered together on the ends of the linear DNA molecule, as diagrammed in Figure 9.9. If the region between the *J* and *N* genes is replaced with foreign DNA, λ still goes through the lytic cycle, and replacing this region is a basic element of the design of λ cloning vectors. However, because some of the genes essential for lysogeny are within the replaced region, λ cloning vectors are unable to enter the lysogenic pathway.

Many λ vectors are constructed with a **stuffer fragment**, a segment of DNA that resides in the region between the *J* and *N* genes. The stuffer fragment is cut out and replaced by foreign DNA during the cloning procedure (Figure 9.10). The stuffer fragment serves two purposes. First, because it keeps the λ vector at the correct size for packaging as a phage particle, the vector can be propagated indefinitely as a lytic phage. Second, the stuffer fragment often carries selectable marker genes that are removed when foreign DNA is inserted into the vector. These selectable markers allow nonrecombinant phages to be distinguished from ones that have incorporated foreign DNA.

Let's look at the construction of the Charon 4A λ vector, diagrammed in Figure 9.11, to see how a phage vector is designed. The Charon 4A λ vector contains a 15 kb stuffer fragment that includes the *lac5* and *bio256* selectable marker genes. The DNA insertion site is bordered by two *Eco*RI sites, one near the left end of the *lac5* gene and the other near the right end of the *bio256* gene. When a foreign DNA fragment that was cut with *Eco*RI replaces the stuffer fragment between these sites, both the *lac5* and *bio256* genes are lost.

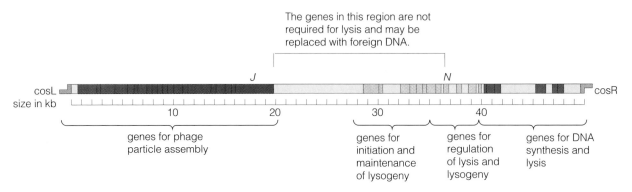

Figure 9.9 Clusters of genes for different functions in the wild-type λ DNA molecule, and the region that may be replaced in a vector.

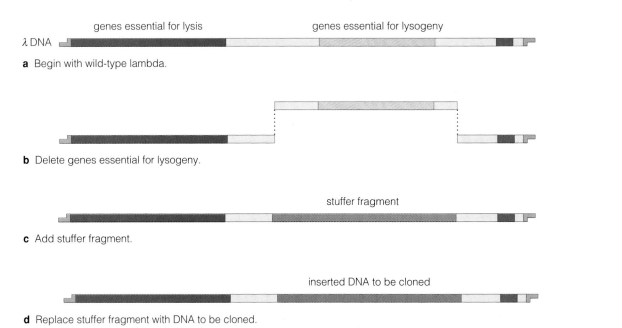

Figure 9.10 General pattern for construction of a λ cloning vector.

The *lac5* gene encodes β-galactosidase. When a Charon 4A λ vector with an intact *lac5* gene infects *lacZ⁻* cells on medium containing X-gal, the resulting plaques are dark blue because of the reaction catalyzed by β-galactosidase. When a foreign DNA segment has replaced the stuffer region, the plaques are clear because there is no *lac5* gene to encode the enzyme required for the reaction. Thus, blue-white screening can distinguish recombinant from nonrecombinant phages.

Lambda has strict size limits for packaging into its capsid. DNA molecules smaller than 38 kb or greater than 51 kb are not packaged. Many λ vectors have been designed with stuffer fragments of various sizes to accommodate a wide size range of inserted DNA fragments. For example, the Charon 4A λ vector is 45.4 kb long with a 15 kb stuffer fragment and can accommodate inserted fragments in the range of 7.6–20.6 kb. The Charon 3A λ vector, on the other hand, has a smaller stuffer fragment and accommodates inserted fragments only up to 9.5 kb in size.

Many different λ vectors are available together with host bacterial types that have been designed to allow for efficient selection of recombinant phages. Which λ vector is best depends on the size of the DNA fragment to be cloned, the restriction enzyme or enzymes to be used, and the methods to be used for selecting recombinant types.

Let's take a few moments to compare the Charon 4A λ plasmid vector with the pUC19 plasmid vector to see how genetic engineers design phage and plasmid vectors.

Example 9.3   The *lacZ* gene as a marker in plasmid and phage vectors.

The wild-type *lacZ* gene of *E. coli* encodes a polypeptide that is 1021 amino acid residues in length. The *lacZ'* gene in the pUC19 vector (see Figure 9.7) encodes only the first 146 amino acids of

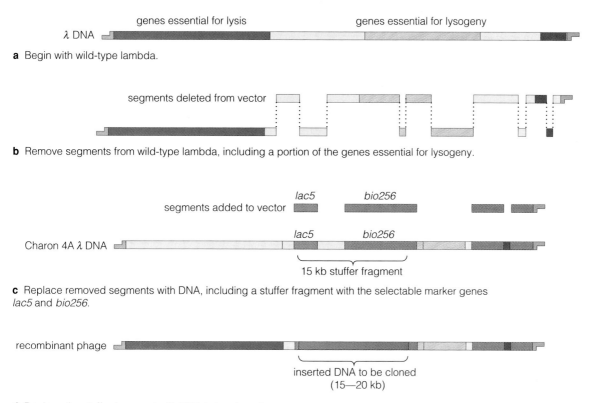

a Begin with wild-type lambda.

b Remove segments from wild-type lambda, including a portion of the genes essential for lysogeny.

c Replace removed segments with DNA, including a stuffer fragment with the selectable marker genes *lac5* and *bio256*.

d Replace the stuffer fragment with DNA to be cloned.

Figure 9.11 Construction of the Charon 4A λ cloning vector. (Adapted from Sambrook, J., E. F. Fritsch, and T. Maniatis. 1989. *Molecular Cloning: A Laboratory Manual,* 2nd ed. Cold Spring Harbor, N.Y.: Cold Spring Harbor Laboratory Press. Used by permission.)

the wild-type polypeptide. This truncated polypeptide interacts with a mutant polypeptide encoded in the chromosome of the host cell to form functional β-galactosidase. This situation permits blue-white screening for bacterial colonies carrying recombinant plasmids. The Charon 4A λ vector (see Figure 9.11) includes a complete *lacZ* gene instead of a truncated version. This gene is also used for blue-white screening.

Problem: (a) Offhand it seems that it would be simpler to include a complete *lacZ* gene in plasmid vectors, then simply introduce the plasmids into *lacZ*⁻ cells for blue-white screening. What advantage does the shorter but more complicated *lacZ'* gene system have over a complete *lacZ* gene in a plasmid vector? (b) Why is the complete *lacZ* gene used in phage vectors (such as Charon 4A λ) instead of the truncated *lacZ'* gene?

Solution: (a) Plasmid vectors are most useful when they are small, leaving more room to accommodate foreign DNA. Also, a smaller vector can be replicated in higher numbers than a larger vector. The *lacZ* gene is fairly large for a bacterial gene. In fact, the *lacZ* gene is 1.3 times the size of the entire pUC19 vector—much too large to include as part of a plasmid vector. Utilizing a *lacZ'* gene in the plas-

mid vector permits blue-white screening without substantially increasing the size of the vector.

(b) A phage vector, such as Charon 4A λ, has no trouble accommodating the entire *lacZ* gene within its stuffer fragment. The Charon 4A λ stuffer fragment is 15 kb long, which easily includes the 3.5 kb *lac5* gene with plenty of room to spare for the *bio256* gene (which is even larger than the *lac5* gene).

It is not possible to do blue-white screening with a phage that contains a *lacZ'* gene. To encode functional β-galactosidase, both the *lacZ'* gene and a mutant chromosomal *lacZ* gene are required. When a virulent phage infects a cell, it rapidly fragments the bacterial chromosome, which destroys the chromosomal *lacZ* gene. This is not a problem when an entire *lacZ* gene (such as *lac5*) is a part of the phage, because the intact gene may be expressed directly from the phage DNA.

Cosmid Vectors, a Combination of λ and Plasmid Vectors

DNA fragments in the range of 30–46 kb cannot be cloned in λ vectors because they are too large, but they can be cloned in cosmid vectors, which combine the essential elements of the plasmid and λ systems. A **cosmid vector**

is a short, circular DNA molecule (about 4–6 kb) that contains the ends of linear λ DNA, a plasmid origin of replication, an antibiotic resistance gene, and several unique restriction sites. Recall from section 7.6 that when λ DNA is encapsulated in its capsid, it is in a linear form. The **cos sites** are short single-stranded segments at the ends of the linear λ DNA. When the phage DNA enters the cell upon infection, the cos sites base-pair with one another to circularize the λ DNA molecule (Figure 9.12). Cosmid is a hybrid name, *cos* for the λ cos sites, and *mid* for plasmid.

As we discuss cosmid vectors, you may wish to follow along in Figure 9.13. Cosmid vectors replicate in bacterial cells as plasmids before foreign DNA is inserted into them. To insert foreign DNA, we extract cosmids from the cells and mix them with a restriction endonuclease to cleave the cosmids at a unique restriction site. We then mix the cleaved cosmids with foreign DNA that is 30–46 kb long and has been cleaved by the same endonuclease. The linear cosmids attach to both sides of the foreign DNA fragments (Figure 9.13b). We then cleave the DNA at the cos sites, which leaves cos sites on both ends of the recombinant DNA molecules (Figure 9.13c). The recombinant cosmids are then packaged into λ capsids to form phage particles (Figure 9.13d). When these phage particles infect bacterial cells, they inject the recombinant cosmid into the bacterial cell, where the recombinant cosmid arranges itself into a circle and replicates as a plasmid (Figure 9.13e).

Cells with recombinant cosmids can be selected by virtue of their antibiotic resistance. There is no need for a second antibiotic resistance gene to determine whether or not the cosmid is recombinant because nonrecombinant cosmids are not packaged effectively into phage particles. Once in a host cell, recombinant DNA fragments in a cosmid can be maintained and recovered in the same way as fragments contained in plasmid cloning vectors.

Bacterial Artificial Chromosomes (BACs)

The cosmid vectors we have just discussed are unable to accommodate fragments that exceed 46 kb in size. However, the F factor of *E. coli* is naturally capable of accommodating exceptionally large segments of DNA. The chromosome of an Hfr cell can be thought of as an enormous F factor that includes the entire *E. coli* chromosome. The F factor is equally capable of accommodating large segments of foreign DNA.

Bacterial artificial chromosomes (BACs) are genetically engineered F factors that carry segments of foreign DNA as long as 300 kb. Figure 9.14 shows the structural organization of pBAC108L, a BAC that has three general features: (1) a set of regulatory genes that control replication and copy number, (2) a chloramphenicol resistance gene, and (3) a cloning segment. The regulatory region includes the genes *oriS* and *repE*, which control F-factor replication, and *parA* and *parB*, which limit the

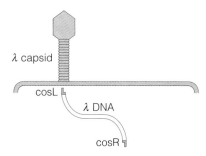

a Lambda injects its DNA as a linear molecule bordered by short single-stranded segments called cos sites.

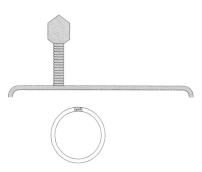

b Shortly after injection, the DNA joins at the cos sites to form a circle.

Figure 9.12 Circularization of λ DNA at the cos sites. CosL is the left cos site, and cosR is the right cos site.

number of copies to one or two per cell. The chloramphenicol resistance gene, CM^r, serves as a selectable marker. The cloning segment includes unique restriction sites for several common restriction enzymes, such as *Hind*III and *Bam*HI, and sites for the rare cutting restriction enzymes *Not*I and *Sfi*I, which cut at eight-nucleotide recognition sites.

Once a large fragment of DNA is inserted into the BACs, the recombinant BACs are introduced into *E. coli* host cells by **electroporation**, a procedure whereby cells and recombinant BACs are mixed and subjected to a brief high-voltage electric current. The current facilitates transfer of large DNA molecules into the cell. Once in a cell, a recombinant BAC replicates like an F factor.

~

Specialized vectors, including phage vectors, cosmid vectors, and bacterial artificial chromosomes, have been designed to carry large fragments of foreign DNA.

9.4 CLONING STRATEGIES

Researchers who clone DNA fragments are often faced with many different situations. Sometimes they have many copies of a purified DNA fragment to clone. Or they may need to extract a specific DNA fragment from a mixture of many different fragments, such as those

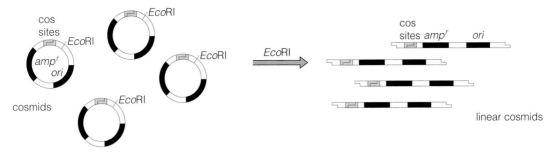

a Cut cosmids with *Eco*RI to obtain linear molecules.

b Mix linear cosmids with large (30—46 kb) DNA fragments that have also been cleaved with *Eco*RI and allow the molecules to recombine. Linear cosmids attach to both ends of the DNA fragments.

c Cut recombinant DNA molecules at cos sites.

d Package recombinant cosmids into λ capsids.

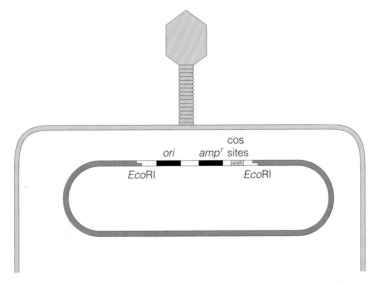

e Infect *E. coli* cells and grow them on medium containing ampicillin. Each recombinant cosmid arranges itself into a circle in the cell and replicates as a plasmid.

Figure 9.13 Inserting DNA into cosmid vectors and packaging the recombinant cosmids for infection of *E. coli* cells.

obtained when the entire cellular DNA of an organism is digested with a restriction endonuclease. **Cloning strategies** are approaches that researchers devise to isolate and replicate a particular fragment of DNA. The strategy of choice depends on the source of DNA and the type of fragment the researcher wishes to clone.

Let's look first at shotgun cloning, one of the simplest but least efficient strategies for cloning a DNA fragment.

Shotgun Cloning

Often, a single fragment of DNA is not readily available for cloning. More typically, a researcher has available the entire genomic DNA isolated from cells. In **shotgun cloning**, a researcher uses a restriction enzyme to digest the entire genomic DNA and clones the fragments in vectors, in the hope that one of the many cloned fragments is the one he or she needs. Then the desired DNA fragment must be isolated from all other cloned fragments present.

The collection of cloned fragments from the entire cellular genome of an organism is called a **genomic DNA library.** An *Eco*RI digest of the human genome, which is about 3 billion nucleotide pairs, yields over 700,000 different DNA fragments, so an *Eco*RI human genomic library may contain more than 700,000 different clones. Finding the single desired fragment in a genomic DNA library is like finding the proverbial needle in a haystack.

To find a specific fragment of DNA, a researcher needs something to identify the bacterial colony that carries the DNA fragment. This identification is usually done with a **DNA probe**, a segment of DNA that is homologous to at least a portion of the DNA fragment of interest. Figure 9.15 illustrates how a probe is used to find a cloned DNA fragment of interest. Thousands of single-cell-derived colonies from the library are grown on an agar-solidified medium (Figure 9.15a). A piece of nylon or nitrocellulose filter is pressed against the plate's surface. The bacterial cells stick to the filter (Figure 9.15c). The cells are lysed, usually by treatment with a detergent solution, to release their DNA. The released DNA binds tightly to the filter, so it is immobilized on the filter at the place where it was released. The filter is then treated with some type of abundant, fragmented DNA, such as fragmented salmon sperm DNA, to bind DNA to all the sites on the filter that could hold DNA. This blocks the entire filter so no more DNA can bind to it. The DNA bound to the filter is then denatured into single strands, usually by treatment with sodium hydroxide (Figure 9.15d). A probe that is labeled (with either a radioactive or nonradioactive label) is also denatured into single strands, then added to the filter. Because all sites for binding to the filter are filled with DNA, the probe cannot bind to the filter itself. However, when the probe encounters its complementary sequence among the many molecules of

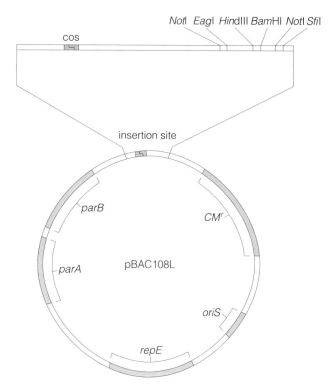

Figure 9.14 pBAC108L, an example of a BAC vector. (Adapted from Shizuya, H., B. Birren, U. Kim, V. Mancino, T. Slepak, Y. Tachiiri, and M. Simon. 1992. Cloning and stable maintenance of 300-kilobase-pair fragments of human DNA in *Escherichia coli* using an F-factor-based vector. *Proceedings of the National Academy of Sciences, USA* 89:8794–8797. Used by permission.)

DNA on the filter, it base-pairs with that sequence and that sequence only. The filter is washed to remove any unpaired probe, leaving the base-paired probe attached to its complementary sequence (Figure 9.15d).

The label on the probe can then be identified on the filter. If the label is radioactive, this is done by exposing X-ray film to the filter (Figure 9.15e). A dark spot appears where the labeled probe is located. Some nonradioactive labels emit small amounts of light and can be detected by exposure to X-ray film, whereas others are detected as a colored spot that appears on the filter when it is treated with an antibody (similar to the color reaction used in home pregnancy tests). The probe identifies which colony carries the DNA fragment of interest. The researcher can then return to the original plate and isolate the colony that contains the cloned fragment of interest and grow that colony in pure culture (Figure 9.15f).

cDNA Cloning

Shotgun cloning is not a very efficient way to identify a clone of interest because a single clone must be selected from as many as several hundred thousand unwanted clones. Scientists have developed alternative methods that reduce the number of unwanted clones. One of these

a Colonies derived from single cells are grown on a culture plate.

b A nylon or nitrocellulose filter is placed on the plate to pick up the bacterial colonies, then removed from the plate.

c The filter is placed in a solution to lyse the cells. The DNA from each colony sticks to the filter. All sites on the filter are then blocked with fragmented DNA. The DNA is denatured into single strands.

d A radioactively labeled DNA probe is added to the filter where binds to DNA fragments that have its complementary sequence.

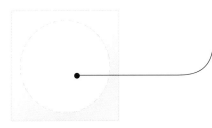

e X-ray film is exposed to the filter to detect the radioactive probe. The image on the filter identifies the colony with the cloned DNA of interest. The colony with the clone of interest can be taken from the original plate and cultured.

Figure 9.15 Use of a DNA probe to identify a bacterial colony containing a specific clone.

is **cDNA cloning**. Suppose we want to clone a gene that is expressed in a particular cell type. For example, the genes for the subunits of hemoglobin (a blood protein) are transcribed in bone marrow cells. Because mRNA has poly (A) tails, it can be isolated from bone marrow cells as follows: Total RNA isolated from the cells is passed through a column filled with cellulose that has poly (dT) (polydeoxythymidine) segments of DNA attached to it. The poly (A) tails base-pair with the poly (dT) segment, retaining the mRNAs in the column while the rest of the

RNA passes through it (Figure 9.16). The trapped mRNA is then treated to break the base pairing, releasing purified mRNAs.

An enzyme called **reverse transcriptase** transcribes a single-stranded DNA molecule from a single-stranded RNA template (the reverse of normal transcription). The DNA transcribed by reverse transcriptase is called **cDNA** (for "complementary DNA"). Once the single-stranded cDNA is made, the RNA template is removed. Double-stranded DNA can then be made using the cDNA as a template.

After synthesizing double-stranded DNA from the purified mRNAs, we have a collection of DNA molecules that represent only the mRNAs in the bone marrow cells. These DNA fragments can be inserted into vectors and cloned in bacterial cells to make a **cDNA library** of bone marrow cells. The cDNAs that correspond to the hemoglobin genes should be in our bone-marrow-cell cDNA library in a much higher proportion than they would be in a genomic library. Selection of the clone having the genes we want should be efficient.

There is another advantage to using cDNA. A cDNA clone has no introns because it is made from an RNA template whose introns have already been removed. For this reason, cDNA clones can be more easily used in the bacterial expression systems we will discuss shortly. A bacterium can transcribe a eukaryotic gene if the gene has a bacterial promoter attached to it. But the bacterium cannot remove introns from the mRNA. For a eukaryotic protein to be produced in a bacterium, the introns must not be present in the inserted gene. A cDNA clone provides the nucleotide sequence of a gene with no introns.

Probe Identification and Synthesis

You may have noticed a paradox in our discussion of probes. To identify a cloned fragment of DNA, we need a probe that is homologous to that DNA. Often, the probe itself is a cloned fragment of DNA. So how can an appropriate probe be found in the first place? There are many methods, and the best method for any particular probe depends on the circumstances. Suppose we want to identify a DNA fragment with the human insulin gene and that we already have a cloned mouse insulin gene. The gene for human insulin should be quite similar to the one for mouse insulin, so we can use the mouse clone as a homologous probe to identify the human clone.

Using a probe from one species to identify clones from another species is a common procedure. However, suppose there is no probe available from any species for a gene of interest. If the purified protein product of the gene is available, part of the amino acid sequence from the protein can be determined, and this information can be used to synthesize a degenerate set of probes using the genetic code as a guide. As few as 5–8 adjacent amino

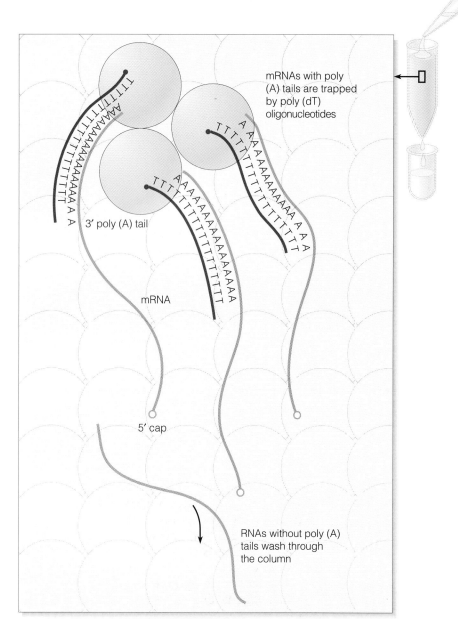

mRNAs with poly (A) tails are trapped by poly (dT) oligonucleotides

3′ poly (A) tail

mRNA

5′ cap

RNAs without poly (A) tails wash through the column

Figure 9.16 Isolation of mRNAs for cDNA cloning.

acids provide enough information for this procedure. A degenerate set of probes includes all possible nucleotide sequences that could encode a particular amino acid sequence. One of the nucleotide sequences in the degenerate set will match the DNA sequence of the clone perfectly.

Synthesis of degenerate sets of probes can be expensive and time-consuming. An alternative strategy is to synthesize a single longer probe (usually 30 or more nucleotides), or a few similar longer probes, that have sufficient homology to permit the probe to bind to the target DNA with only a few nucleotide-pair mismatches. Several steps can be taken to maximize the probability

that a probe matches the target sequence. The first step is selecting a region in the amino acid sequence where the probability of mismatching nucleotides is relatively low. A region high in methionine and tryptophan is a good choice because each of these amino acids is encoded by a single codon. The selected region should be low in leucine, serine, and arginine because each of these amino acids is encoded by six codons. The region of amino acids represented by the smallest number of alternative codons is usually the best choice for probe synthesis.

Additional steps can be taken to reduce the probability of nucleotide mismatches. Some species tend to util-

ize a certain codon more often than the cognate codons. For example, there are four cognate codons for valine (GUU, GUC, GUG, and GUA), but about 50% of all valine codons in human genes are GUG, so G should usually be used in the third position of a valine codon when synthesizing a probe.

There are many other methods to acquire probes when the DNA sequence of a gene is unknown. Which method is best often depends on highly specialized circumstances. The methods are too many and the circumstances too specific to discuss them all here. However, we will discuss several of these methods (such as transposon tagging and chromosome walking) in later chapters when we confront circumstances where these methods are needed.

Let's now look at an example of how researchers select amino acid sequences for degenerate probe synthesis.

Example 9.4   Degenerate probes.

In 1981, Suggs et al. (*Proceedings of the National Academy of Sciences, USA* 78:6613–6617) described a set of degenerate probes that they synthesized to identify clones containing the human β_2-microglobulin gene from a cDNA library. One of the sequences of amino acids they selected for degenerate probe synthesis was TrpAspArgAspMet.

Problem: **(a)** List the entire set of degenerate probes that correspond to this amino acid sequence. **(b)** Why is this sequence of amino acids a good choice for synthesis of degenerate probes?

Solution: **(a)** There are 24 probes that match this sequence of amino acids:

TGGGATCGTGATATG	TGGGACCGTGATATG
TGGGATCGCGATATG	TGGGACCGCGATATG
TGGGATCGAGATATG	TGGGACCGAGATATG
TGGGATCGGGATATG	TGGGACCGGGATATG
TGGGATCGTGACATG	TGGGACCGTGACATG
TGGGATCGCGACATG	TGGGACCGCGACATG
TGGGATCGAGACATG	TGGGACCGAGACATG
TGGGATCGGGACATG	TGGGACCGGGACATG
TGGGATAGAGATATG	TGGGACAGAGATATG
TGGGATAGGGATATG	TGGGACAGGGATATG
TGGGATAGAGACATG	TGGGACAGAGACATG
TGGGATAGGGACATG	TGGGACAGGGACATG

(b) The amino acids in the sequence are tryptophan, methionine, aspartic acid, and arginine. Tryptophan

and methionine are each encoded by one codon, and aspartic acid is encoded by two codons. Arginine is encoded by six codons. Identification of a sequence that includes methionine, tryptophan, and aspartic acid ensured that the number of degenerate probes would be kept small. Unfortunately, arginine was also a part of this sequence, so the number of probes is greater than it would have been had an amino acid encoded by only one, two, or four codons been in the position of arginine.

Cloning strategies are approaches that researchers devise to clone a specific fragment of DNA then identify and isolate the desired clone from all the other clones that may be present.

9.5   EXPRESSION VECTORS

So far, we have discussed how to clone genes, or portions of them, for study or for use as a probe. Genes from any source may not only be replicated but may also be expressed in a bacterial host. Under the right circumstances, a bacterium can replicate, transcribe, and translate a gene from another species; if a bacterium translates an mRNA from an introduced gene, it makes the same polypeptide as is produced in the original species.

When researchers attempt to clone a gene using a restriction endonuclease, there is always a chance that a restriction site for the endonuclease is present within the gene. If that is the case, the clone produced may contain only part of the gene and therefore be of no value for expressing the gene. This possibility illustrates an advantage of using cDNA clones. Because restriction enzymes are not used in the isolation of cDNAs, there is no risk of this type of loss of DNA.

Recall from the previous section that cDNA clones contain no introns. This is highly important, as bacterial cells cannot remove introns from a eukaryotic gene. Any eukaryotic gene that is to be expressed in a bacterial vector must have a coding sequence with no interruptions between the initiation and termination codons.

Once a gene with its uninterrupted coding sequence is available, it must have certain features to be expressed properly in a bacterial host. First, it must have a bacterial promoter so the bacterial RNA polymerase can transcribe it. Second, it needs a Shine-Dalgarno sequence so the bacterial ribosome can bind effectively to the mRNA and translate it. Third, although this is not always necessary, it is helpful to have a rho-dependent terminator to end transcription after the gene has been transcribed.

Promoters are often included as part of the expression vectors. One of the most commonly used promoters is the

λP_L promoter. Because it can be regulated by the λ repressor protein (pcI), researchers can induce cells to turn the introduced gene on and off. Let's look at the pPLa2311 plasmid as an example of how an expression vector with λP_L is used.

Figure 9.17 diagrams pPLa2311 with its essential features. This plasmid is 3.8 kb long and has the P_L promoter, O_L operator, genes for ampicillin and kanamycin resistance, and several unique restriction sites. The two restriction sites of most value for expression are the $EcoRI$ site and the $PstI$ site. The P_L promoter initiates transcription of genes inserted into these sites. Because the $PstI$ site is in the ampicillin resistance gene, genes inserted at this site render the host cell ampicillin sensitive while it remains kanamycin resistant, allowing for easy selection of recombinants.

Recall from Chapter 7 that a prophage is a phage DNA molecule integrated into a bacterial host. The proper bacterial host for pPLa2311 is permanently lysogenic for a mutant λ prophage that cannot excise itself and become lytic. The prophage has a temperature-sensitive cI gene that researchers use to turn genes inserted in pPLa2311 on and off. After a culture of the proper bacterial cells has been infected with pPLa2311, the cells that have picked up the plasmids can be selected because of their kanamycin resistance and ampicillin sensitivity. The cells are then grown at 32°C, a temperature at which the cI gene is active and produces the λ repressor. This repressor binds to O_L on pPPLa2311, preventing transcription of the inserted gene and permitting rapid proliferation of the cells.

Once many cells are available, expression of the inserted gene is activated by incubating the cells at 42°C. This temperature inactivates the temperature-sensitive cI gene, so the repressor is no longer produced. This frees P_L to be active as a promoter, so the inserted gene can be transcribed and translated. This example illustrates the elegant logic often employed in genetic engineering. With a matched host and vector that have been carefully engineered, scientists can turn the expression of a cloned gene on and off by simply altering the temperature.

The challenges to genetic engineers do not end with the transcription of the gene inserted into an expression vector. For the mRNA to be translated effectively, it must have a Shine-Dalgarno sequence. Presumably, this could be added in the vector, but the Shine-Dalgarno sequence needs to be about 7–10 nucleotides upstream from the initiation codon. Many eukaryotic mRNAs have an initiation codon that is some distance from the 5' end, so their cDNAs must be altered before cloning them into an expression vector. In addition, many eukaryotic polypeptides are processed following translation, and the processing controls are not present in the bacterium. This means that scientists must modify most genes before they place them in an expression vector to ensure that the genes are translated into the right product.

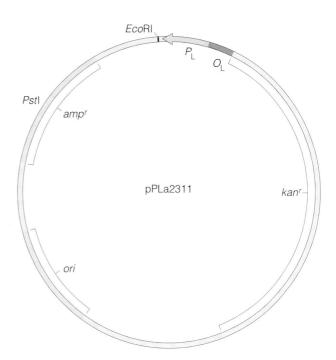

Figure 9.17 The pPLa2311 expression vector. The $P_L O_L$ promoter-operator complex from λ regulates expression of genes inserted at either the $EcoRI$ site or the $PstI$ site. The ampicillin and kanamycin resistance genes are selectable markers. The arrow at the P_L promoter indicates the direction of transcription. (Adapted from Sambrook, J., E. F. Fritsch, and T. Maniatis. 1989. *Molecular Cloning: A Laboratory Manual*, 2nd ed. Cold Spring Harbor, N.Y.: Cold Spring Harbor Laboratory Press. Used by permission.)

In some cases, scientists cut and splice DNA to form a **fusion gene**, a gene that contains parts of two genes. Fusion genes inserted into expression vectors are usually part bacterial and part eukaryotic. The portion of a bacterial gene containing the promoter, operator, Shine-Dalgarno sequence, and initiation codon is attached to a eukaryotic gene in such a way that the bacterial ribosome will translate the correct reading frame of the eukaryotic portion of the gene (Figure 9.18). The fusion gene is regulated by the bacterial promoter-operator region, but it encodes a eukaryotic protein. The regulatory region of the *lacZ* gene in the *lac* operon is often used to construct fusion genes. A eukaryotic gene derived from cDNA is attached a few codons downstream from the initiation codon for the *lacZ* gene so that the eukaryotic reading frame is in the same frame as the initiation codon. This fusion gene is then placed into a vector and replicated in *E. coli* cells.

There is a distinct advantage to using a regulated fusion gene, such as the *lacZ* gene. Scientists are able to regulate expression of the fusion gene simply by altering the medium on which the bacterial cells are grown. When the cells are grown in the absence of a *lac* operon inducer, the introduced gene is not transcribed because transcription is blocked at the *lac* operators upstream from the fusion gene. To express the fusion gene, scientists add an inducer (such as IPTG) to the culture medium. This activates transcription of the fusion gene.

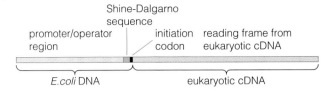

Figure 9.18 Structure of a fusion gene.

Several pharmaceutical companies currently market genetically engineered eukaryotic proteins that are produced in bacteria. Genetically engineered human insulin is probably the best-known example. Development of recombinant human insulin required some elaborate genetic engineering, which we will describe in detail in section 26.4. Among the best examples of how genetic engineers construct fusion genes for expression in bacteria is the genetically engineered fusion gene that produces human growth hormone in bacteria.

Example 9.5 Producing human growth hormone in bacteria using a fusion gene.

Human growth hormone (HGH) is a protein secreted from the pituitary gland that regulates growth rate throughout the body. Children with HGH deficiency suffer from pituitary dwarfism unless they are treated with periodic injections of HGH. Growth hormones from animal species do not function in humans, so treatment requires human growth hormone. Before genetically engineered HGH became available, affected children were treated with HGH from the pituitary glands of persons who had died and bequeathed the use of their body parts in medicine. This meant that the supply of HGH was unreliable and tentative. The problem was exacerbated by virus contamination in some HGH preparations, which caused fatal infections in some treated children. When the techniques of genetic engineering were being developed, scientists recognized that recombinant HGH produced in bacteria would potentially be a reliable and safe source of HGH.

In 1979, Goeddel et al. (*Nature* 281:544–548) reported the development of genetically engineered recombinant HGH. This was the first fully processed human protein produced in bacteria. The work provides an excellent example of the procedure required to genetically engineer a eukaryotic gene for expression in a bacterial cell. To understand the procedure, you may wish to follow along in Figure 9.19.

The natural *HGH* gene encodes a precursor polypeptide that is 217 amino acids in length. The first 26 amino acids constitute a signal peptide that is removed during processing to leave the mature protein of 191 amino acids. There are two *Hae*III

restriction sites in the cDNA clone, as illustrated in Figure 9.19a. *Hae*III cleaves the cDNA on the left side within the 24th codon for the mature protein (which is the 50th codon for the polypeptide before signal peptide cleavage). *Hae*III also cleaves the cDNA on the right side in the untranslated 3′ region downstream from the termination codon.

After digesting the cDNA with *Hae*III, the researchers recovered a 551-nucleotide-pair *Hae*III fragment that included the codons for amino acids 24–191 of the mature protein. They inserted this fragment into pBR322 to replicate the fragment. They named this recombinant plasmid pHGH31 (Figure 9.19a).

The researchers now needed to place the first 24 codons onto the gene. They artificially synthesized a double-stranded DNA fragment with an *Eco*RI site on one end, a *Hin*dIII site on the other end, and an ATG codon near the *Eco*RI site followed by the first 24 codons in mature HGH. They inserted this fragment between the *Eco*RI and *Hin*dIII sites of pBR322 to replicate the fragment, and named this recombinant plasmid pHGH3 (Figure 9.19b).

After replicating the two recombinant plasmids in bacterial cells to generate many copies of the inserted DNA fragments, the researchers cut the inserted fragment from pHGH3 with *Eco*RI and *Hae*III. This gave the fragment a blunt end in codon 24 on the right end of the fragment because *Hae*III cleavage creates blunt ends (see Table 9.1). They cut the inserted fragment from pHGH31 with *Hae*III to give the fragment a blunt end in codon 24 on the left end. They cut this fragment further with *Xma*I to trim off some of the extra DNA downstream from the termination codon on the right end. They then attached the two fragments at the blunt ends in codon 24 using blunt-end ligation to generate a complete reading frame for the mature HGH polypeptide. They inserted the engineered *HGH* fragment into an expression vector with a *lac* promoter and named the recombinant expression vector pHGH107-1 (Figure 9.19c).

The genetically engineered *HGH* gene is a fusion gene with the *lac* promoter region from *E. coli*, 24 codons of artificially synthesized DNA, and the rest of the gene from a human cDNA clone. When bacteria with pHGH107-1 are grown in medium supplemented with IPTG, the bacteria produce genetically engineered HGH that is identical to natural HGH.

Problem: **(a)** Why did the researchers cleave the cDNA with *Hae*III, then artificially synthesize the first 24 codons of the genetically engineered gene? **(b)** Why was IPTG used in the medium to express the genetically engineered *HGH* gene?

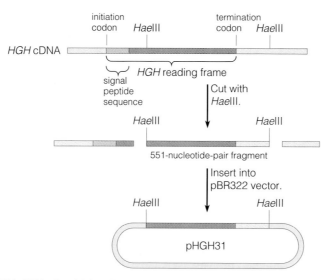

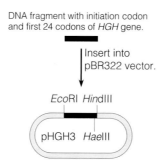

a Starting with cloned HGH cDNA, Goeddel et al. cut the cDNA with the restriction endonuclease *Hae*III and recovered a 551-nucleotide-pair fragment with most of the HGH reading frame. They inserted this fragment into pBR322 and named the recombinant plasmid pHGH31.

b The researchers artificially synthesized a DNA fragment with an initiation codon and the first 24 codons of the *HGH* gene. They inserted the fragment into pBR322 and named the recombinant plasmid pHGH3. A *Hae*III restriction site was present in codon 24.

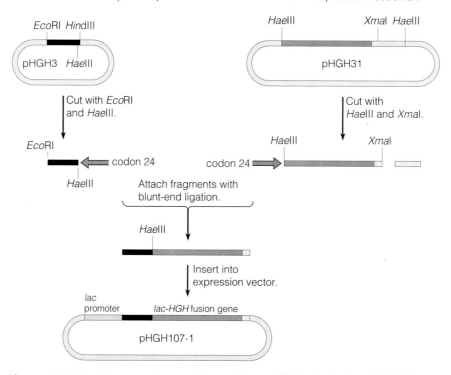

c After replicating both cloned fragments in *E. coli*, the researchers cut the insert from pHGH3 with *Eco*RI and *Hae*III, leaving a blunt end in codon 24 on the right end. They cut pHGH31 with *Hae*III and *Xma*I leaving a blunt end in codon 24 on the left end. They then attached the two fragments at codon 24 with blunt-end ligation and inserted the fragment into an expression vector.

Solution: **(a)** The gene from the cDNA included 26 codons for a signal peptide. These codons had to be removed because bacterial cells do not remove eukaryotic signal peptides. The *Hae*III site provided a convenient cleavage site that removed the signal peptide while leaving most of the reading frame intact. The researchers then artificially synthesized the 24 codons that were removed by *Hae*III cleavage and attached this synthetic fragment to the gene.

(b) IPTG is an inducer of the *lac* operon that is not consumed by the bacterial cells. In the presence of IPTG, the *lac* promoter for the engineered HGH is transcriptionally active. This provides a convenient way for researchers to turn expression of the cloned gene on or off by choosing whether or not to add IPTG to the growth medium. In the absence of IPTG, the gene is repressed, and no HGH is produced. In the presence of IPTG, the gene is transcribed and translated, and HGH is produced.

Expression vectors are plasmids designed for expression of a foreign gene. Typically, eukaryotic genes must be modified through genetic engineering to express the correct products in a bacterial host.

9.6 THE POLYMERASE CHAIN REACTION (PCR)

In 1987, the **polymerase chain reaction (PCR)** was introduced as a rapid and efficient way to replicate specific fragments of DNA. Like cloning, PCR amplifies DNA, but it requires substantially less time and effort. If we think of the genome of an individual as being similar to the information in a book, PCR is like selecting a specific page from the book and photocopying it millions of times. Because it is much easier than cloning, PCR has swept the world of genetics by storm since its inception. Most geneticists use PCR routinely.

The purpose of PCR is to amplify a specific region of DNA in a genome, called the target region. To do so, researchers must know beforehand the sequences of 15–30 nucleotides on both ends of the target region. They use this sequence information to synthesize single-stranded segments of DNA, called **PCR primers**, that are complementary to the nucleotide sequences at the ends of the target region. The primers are mixed in a test tube with a DNA polymerase, deoxyribonucleotides in their triphos-

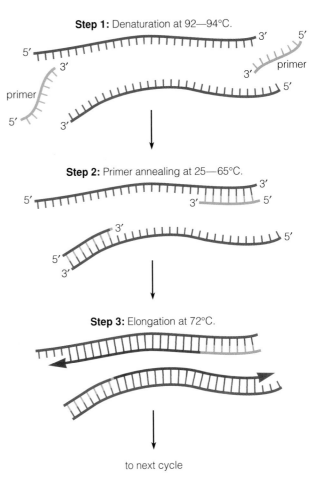

Figure 9.20 The three steps in one cycle of the polymerase chain reaction. The template DNA is blue, the primers are green, and the newly synthesized DNA is red.

phate form (dNTPs), an appropriate buffer to maintain pH, magnesium (a cofactor for DNA polymerases), and cellular DNA to serve as the template for replication.

Figure 9.20 illustrates the three steps in a PCR cycle. First, the mixture is heated to 92–94°C, nearly the boiling temperature of water, to separate the strands of the double-stranded cellular DNA. Second, the temperature is then cooled to a preselected temperature between 25 and 65°C (usually about 50°C). The primers form base pairs with their complementary sequences in the single-stranded cellular DNA at the boundaries of the target region. Third, the temperature is raised to 72°C. The DNA polymerase used in PCR is called *Taq* DNA polymerase, a DNA polymerase that is not destroyed by the hot temperatures used in PCR. *Taq* DNA polymerase is from a bacterium called *Thermus aquaticus* that lives in hot springs, and the polymerase functions best at 72°C, about the temperature of the water coming out of a hot water tap. At this temperature, the polymerase synthesizes DNA strands using the 3' ends of the primers as start points and the cellular DNA strands as templates. This completes the first cycle, which generates two double-stranded copies of the target region.

◄ **Figure 9.19** Information for Example 9.5: Construction of an expression vector that produces human growth hormone in *E. coli*. (Adapted from Goeddel, D. V., H. L. Heynecker, T. Hozumi, R. Arentzen, K. Itakura, D. G. Yansura, M. J. Ross, G. Miozzari, R. Crea, and P. H. Seeburg. 1979. Direct expression in *Escherichia coli* of a DNA sequence coding for human growth hormone. *Nature* 281:544–548. Copyright 1979 Macmillan Magazines Limited. Used by permission.)

The second cycle repeats the first. The mixture is heated to 92–94°C to separate the double-stranded DNA of the target region. Then the temperature is cooled to between 25 and 65°C for the primers to bind. Again, the temperature is raised to 72°C to allow the polymerase to synthesize new strands of DNA. The second cycle generates four double-stranded copies of the target region.

Each subsequent cycle doubles the number of double-stranded copies of the target region. As illustrated in cycle 1 of Figure 9.21, each DNA strand synthesized directly from the cellular DNA template extends beyond the boundaries of the target region. These are called **variable-length strands** because their lengths depend on how far the polymerase synthesizes DNA from the template. On the other hand, all DNA strands synthesized from a template that was replicated in a previous cycle include only the target region and are all the same length. These are called **constant-length strands**. As illustrated in Figure 9.21, the first cycle produces two variable-length strands and no constant-length strands. The first constant-length strands appear in the second cycle, and their number increases exponentially with each subsequent cycle. The number of variable-length strands, however, increases only linearly. After a few cycles, nearly all DNA molecules in the mixture contain only constant-length strands.

PCR is usually continued for about 25–45 cycles. In later cycles, the efficiency of amplification decreases as primers and nucleotides are consumed and the polymerase starts to lose its activity. Beyond 45 cycles, DNA synthesis is reduced considerably, so there is little reason to continue. By that time, the target DNA sequence has been amplified several million times. The entire process usually requires less than 3 hours.

PCR is valuable because the cellular template DNA does not need to be purified, nor is a large quantity of template needed. In fact, the amount of DNA in a single cell may be sufficient for PCR to amplify a target region. DNA that has been degraded into fragments may also be amplified with PCR. As long as there is at least one copy of the complete target region, the DNA can be amplified. Small amounts of DNA extracted from blood stains, single hair follicles, or small amounts of tissue have been amplified successfully with PCR. The PCR cycles are automated in a machine called a **thermal cycler** that raises and lowers the temperatures of the samples according to a preset program (Figure 9.22).

The principal consideration in PCR is the choice of primers. The primer sequences determine which fragment of DNA in the entire genome is amplified. To amplify a gene, or a piece of a gene, it is necessary to know enough about the gene's DNA sequence to synthesize primers that will form base pairs with both ends of the target region. In the case of PCR, primers are made of single-stranded DNA, instead of RNA as natural primers are. Because of this, the primers do not need to be replaced with DNA as they do in living cells. Instead, the primers become part of the newly synthesized DNA strands.

The polymerase chain reaction (PCR) is a rapid and effective method for replicating specific fragments of DNA.

9.7 GEL ELECTROPHORESIS

The various applications of gel electrophoresis are among the most important methods in genetic analysis. Gel electrophoresis permits researchers to separate DNA fragments or proteins from one another and visualize differences in the DNA or proteins that represent genetic differences. Gel electrophoresis is used to identify the molecular weights of DNA fragments and proteins, and the enzymatic activities of proteins. It is also an important component of DNA sequencing. Let's look first at electrophoresis of DNA.

Gel Electrophoresis of DNA

DNA or RNA fragments of different sizes can be readily separated from one another and purified using gel electrophoresis. Gel electrophoresis is often used to separate DNA fragments that have been created by restriction enzyme digestion. For example, the restriction endonuclease *Hin*dIII cleaves λ DNA into eight DNA fragments, each a different length. Electrophoresis separates these fragments in a gel slab, allowing the researcher to identify the fragments by size.

The gel slabs for DNA analysis are either agarose or polyacrylamide. When the DNA fragments to be analyzed are longer than about 200 nucleotide pairs, agarose is typically used, except that when all fragments are shorter than about 500 nucleotide pairs, polyacrylamide is usually a better choice. The ends of the gel are immersed in a buffer solution in reservoirs at each end of a tray. The orientation of the slab may be vertical or horizontal. In horizontal electrophoresis, the entire gel may be immersed in the buffer. Each buffer reservoir has an electrode of platinum wire that runs along the width of the reservoir. Once the gel has been immersed in the buffer, each DNA sample is mixed with a concentrated solution of glycerol or sugar to make it denser than the buffer. Then the DNA samples are added to depressions in the gel slab called wells. After all the DNA samples have been added, an electric current is passed through the gel between the electrodes, with the positive charge at the opposite end of the gel from the wells. Because DNA is negatively charged at each of its phosphate groups, it is attracted to the positive electrode. To get there, the DNA must migrate through the gel. Figure 9.23 shows an apparatus for agarose gel electrophoresis of DNA.

DNA fragments migrate through a gel at a rate that is inversely proportional to their size. In other words, smaller fragments migrate faster than larger fragments. The researcher allows the electric current to run for a

Before amplification: The target region is a segment of DNA from cellular DNA that will be amplified many times. Each primer is complementary to one end of the target region.

Cycle 1: DNA synthesis makes two double-stranded copies of the target region. The newly synthesized strands are called variable-length strands because they extend beyond the target region on one end.

Cycle 2: DNA synthesis makes four double-stranded copies of the target region. DNA strands synthesized from an amplified template are called constant-length strands because both ends terminate at the ends of the target region.

Cycle 3: DNA synthesis makes eight double-stranded copies of the target region. In two copies, both strands are constant-length strands.

Cycle 4: The number of copies with constant-length strands increases exponentially while the number of copies with variable-length strands increases linearly. This pattern continues in subsequent cycles.

KEY

primer DNA
cellular DNA
newly synthesized DNA
DNA synthesized in a previous cycle

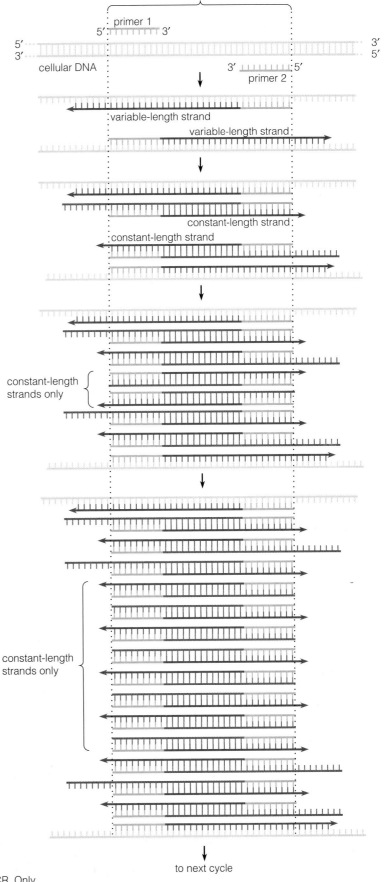

Figure 9.21 Four cycles of DNA amplification using PCR. Only the elongation steps of each cycle are shown.

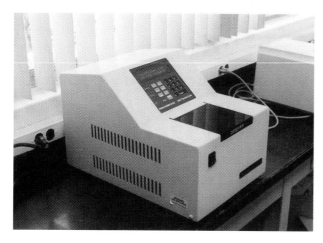

Figure 9.22 A DNA thermal cycler, which automates the PCR reactions.

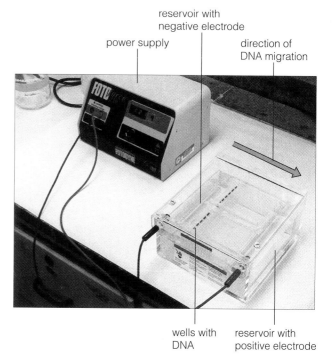

Figure 9.23 An agarose gel electrophoresis apparatus for separation of DNA fragments. DNA that has been mixed with a blue dye has been added to the wells. The DNA migrates through the gel toward the positive electrode (attached to the red wire) and away from the negative electrode (attached to the black wire).

set period of time to separate the DNA fragments into an array ranging from the largest to the smallest fragments. The current is turned off before the fragments reach the end of the gel. The fragments remain separated in the gel by size, with the larger fragments closer to the wells and the smaller fragments farther from them.

The DNA can then be visualized in a number of ways. One of the most common is staining with ethidium bromide, a mutagenic dye that intercalates between the bases and fluoresces a pinkish-orange color under ultraviolet light of about 260 Å in wavelength. Thus, the DNA fragments can be viewed and photographed under UV light transmitted through the gel. The researcher must wear protective clothing and gloves to avoid exposure to ethidium bromide, and use UV-protective eyewear to prevent damage to the eyes from the intense UV light. Figure 9.24 shows a black-and-white photograph of a gel stained with ethidium bromide after electrophoresis. The white bands in the gel are the fragments from a *Hin*dIII digestion of λ DNA that have been separated by electrophoresis. The fragment at the top of the gel is the largest, and the fragments are progressively smaller moving downward through the gel.

Gel electrophoresis is used for a variety of analytical purposes, including estimating DNA fragment size, analyzing DNA fragments amplified by PCR, identifying genetic similarities and differences among individuals, and estimating cloned fragment sizes in recombinant plasmid DNA. Electrophoresis may also be used for preparative purification of DNA fragments. When a mixture of DNA fragments exists, the fragments may be separated by electrophoresis and a piece of the gel containing the fragment of interest cut from the gel. The DNA can then be extracted from the gel and purified.

The following example shows how researchers can estimate the sizes of DNA fragments that have been electrophoresed in a gel.

Example 9.6 Estimating the size of a DNA fragment with electrophoresis.

The size of a DNA fragment that has been electrophoresed in a gel can be estimated by measuring the distance that the fragment migrates relative to the distances migrated by fragments of known sizes. Scientists routinely include a lane of fragments of known sizes, called marker fragments, in each gel. Figure 9.25a shows a gel where lane 1 contains a PCR fragment of unknown size, and lane 2 contains marker fragments of known sizes.

Problem: Estimate the size of the PCR fragment.

Solution: The size of the fragment can be roughly estimated as between 400 and 500 nucleotide pairs because the fragment migrated to a position between the 400 and 500 nucleotide-pair markers. However, a more accurate estimate can be obtained mathematically. The first step is to analyze the relationship between size and distance migrated in the marker fragments. Figure 9.25b shows a graph on which the distance in millimeters that each marker fragment migrated is plotted on the *x* axis and its size in nucleotide pairs is plotted on the *y* axis. The relationship is logarithmic rather than linear, which is the rule for migration of DNA fragments in elec-

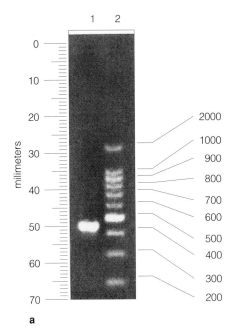

Figure 9.24 Electrophoretic separation of DNA fragments generated by digestion of λ DNA with *Hind*III. The gel was stained with ethidium bromide and photographed under UV light. The numbers denote the sizes of the fragments in nucleotide pairs.

trophoretic gels. A logarithmic transformation of the fragment sizes plots as a straight line, as illustrated in Figure 9.25c.

A least-squares fit of a line to the points in Figure 9.25c can be expressed as a linear equation in the form $y = mx + b$, where m is the slope and b the y intercept:

$$y = -0.026x + 3.945$$

With this linear equation, we can compute the size of the PCR fragment. The distance that the PCR fragment migrates is substituted for x in the equation, and the corresponding value of y is determined. The antilogarithm of the value for y is then an estimate of the fragment's size. In this example, the fragment in lane 1 migrated 49 mm, which when substituted into the equation gives $y = 2.671$. As the antilogarithm of 2.671 is 468.8, we estimate the fragment to be 469 nucleotide pairs long. Computer software is now available that does this type of calculation automatically from digitized images of gels, providing instant estimations of fragment sizes.

DNA fragments can be separated from one another and their lengths estimated using gel electrophoresis.

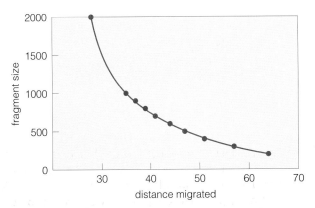

b The relationship between a fragment s size and the distance it migrates is logarithmic.

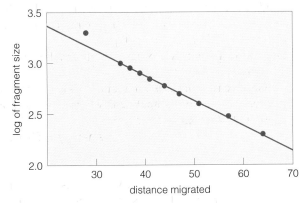

c After conversion of fragment sizes to their logarithms, the relationship is linear.

Figure 9.25 Information for Example 9.6: Identifying the size of a DNA fragment using markers of known sizes. (Photo by P. F. Randall and D. J. Fairbanks.)

Gel Electrophoresis of Proteins

DNA and RNA are not the only substances that can be analyzed by gel electrophoresis. Proteins are also routinely analyzed in this way. Because they are smaller than most DNA fragments, proteins are usually analyzed in polyacrylamide gels, which have smaller pores than most agarose gels. There are electrophoretic methods that separate proteins by molecular weight alone, by the net charge of the protein alone, or by a combination of net charge and molecular weight.

To separate proteins by molecular weight alone, the usual method of choice is **SDS polyacrylamide gel electrophoresis (SDS-PAGE)**. In preparation for electrophoresis, the proteins are heated in a solution of sodium dodecyl sulfate (SDS), which denatures the proteins, causing them to become linear chains of amino acids with no secondary, tertiary, or quaternary structure. The SDS molecules, which are negatively charged, surround each polypeptide chain, neutralizing any positive charges in the polypeptide.

The denatured polypeptides are then added to a polyacrylamide gel that contains SDS. When an electric field is applied to the gel, the proteins migrate through the gel toward the positive pole, separating according to their molecular weight. Because the rate of migration is inversely proportional to the molecular weight of the polypeptide, polypeptides of lower molecular weight migrate more quickly than heavier ones. After the polypeptides have been separated sufficiently, the gel is stained to make the polypeptides become visible, usually as dark-colored bands against a clear background in the gel. Figure 9.26 shows several polypeptides of different molecular weights that have been separated by SDS-PAGE.

Each protein carries a net charge. The acidic amino acids glutamic acid and aspartic acid carry negative charges in aqueous solution. The basic amino acids lysine, arginine, and histidine carry positive charges in aqueous solution. The net charge of a protein is determined by the ratio of the acidic and basic amino acids of which it is composed. Proteins can be separated electrophoretically on the basis of net charge alone using a method called **isoelectric focusing (IEF)**.

IEF gels are usually composed of a low concentration of polyacrylamide, which allows proteins of all molecular weights to pass through at the same rate, so molecular weight has no influence on their migration. A pH gradient is set up in the gel, then native (undenatured) proteins are added to the gel. An electric current is applied, and proteins migrate through the gel according to their net charge. At some point in the gel, each protein reaches its **isoelectric point**, where the electric field pulling the protein is countered by the pH gradient in the gel. At this point, the protein stops migrating and remains focused at a particular spot in the gel. Figure 9.27 shows proteins separated using IEF.

IEF and SDS-PAGE are sometimes combined in a method called **two-dimensional polyacrylamide gel electrophoresis (2D-PAGE)**. Proteins are first separated by charge using IEF. The IEF gel with the focused proteins is then immersed in an SDS solution to denature the proteins in the gel. This gel is then placed on one end of an SDS gel, and the polypeptides are separated in the

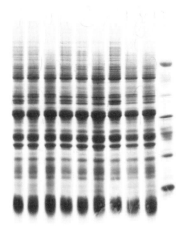

Figure 9.26 Separation of denatured polypeptides by molecular weight alone using SDS-polyacrylamide gel electrophoresis. The first nine lanes contain polypeptides extracted from plant seeds. The far-right lane contains polypeptides of known molecular weights to permit the researchers to estimate the molecular weights of the seed polypeptides. Differences in the mobilities of polypeptides represent genetic differences for the genes that encode those polypeptides. (From Fairbanks, D. J., K. W. Burgener, L. R. Robison, W. R. Andersen, and E. Ballon. 1990. Electrophoretic characterization of quinoa seed proteins. *Plant Breeding* 104:190–195.)

Figure 9.27 Separation of proteins by net charge alone using isoelectric focusing. Each lane represents proteins from a different individual. Differences in the patterns among individuals represent genetic differences. (Photo by Y. J. Lu and D. J. Fairbanks.)

second dimension by molecular weight using SDS-PAGE. 2D-PAGE allows highly detailed analysis of mixtures that contain many proteins. Figure 9.28 shows a two-dimensional gel of proteins from a wheat leaf.

When enzymes are separated using SDS-PAGE, protein denaturation eliminates all enzyme activity. IEF may also eliminate or reduce an enzyme's activity because the enzyme may focus at a pH at which it cannot function. **Native protein gel electrophoresis** is a method that permits most enzymes to retain their full enzymatic activities. The method is essentially the same as SDS-PAGE, but because SDS is not added to the protein sample or the gel, the proteins retain their tertiary and quaternary structures during electrophoresis. Under these conditions, the rate of migration for any particular protein is determined by both the net charge and the molecular weight. The enzymes can then be identified in the gel by their enzymatic activities. Figure 9.29 shows how genetic variation for a particular enzyme can be identified in a native gel.

~

Electrophoresis can separate proteins from one another by molecular weight alone, by net charge alone, or by a combination of molecular weight and net charge.

9.8 DNA, RNA, AND PROTEIN BLOTTING

As we have seen, after a small DNA molecule, such as λ DNA, is digested with a restriction endonuclease, the DNA fragments may be separated by electrophoresis, stained, and visualized directly in the gel. DNA fragments generated by PCR can easily be visualized the same way. Most species, however, have large genomes, and restriction endonuclease digestion generates so many different fragments that they cannot be distinguished from one another in a gel. For instance, an *Eco*RI digestion of cellular human DNA yields over 700,000 different fragments. When so many fragments are separated by gel electrophoresis, they appear as a smear in a gel lane and cannot be resolved from one another (Figure 9.30).

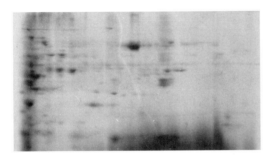

Figure 9.28 Two-dimensional electrophoresis of proteins from a wheat leaf. The proteins are separated first by charge on the horizontal axis. Then they are denatured into polypeptides and separated by molecular weight on the vertical axis. When there is a mixture of many polypeptides, it is often not possible to separate all of them from one another in a single dimension. Two-dimensional electrophoresis separates individual polypeptides in a complex mixture. (Photo by Y. J. Lu and D. J. Fairbanks.)

Figure 9.29 Native protein gel electrophoresis. The researchers who made this gel wanted to detect expression of an enzyme called iron reductase. After native proteins were separated electrophoretically, they detected the iron reductase by its ability to convert Fe^{3+} to Fe^{2+}. Fe^{2+} turns a red color in the presence of certain chemicals that were added to the gel. The dark-red bands in the gel show where the enzyme is active. (From Cook, K. A., V. D. Jolley, D. J. Fairbanks, and L. R. Robison. 1996. Identification of iron reductase isozymes in soybeans. *Journal of Plant Nutrition* 19:457–467.)

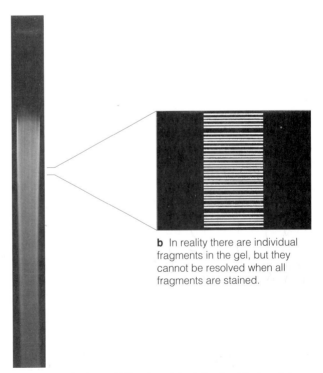

b In reality there are individual fragments in the gel, but they cannot be resolved when all fragments are stained.

a Photograph of an ethidium-bromide-stained gel that contains total cellular sugarcane DNA that has been digested with *Eco*RI. The smear fades out near the bottom of the gel because there is less mass in the smaller DNA fragments.

Figure 9.30 A DNA smear from genomic DNA digested with a restriction enzyme and separated electrophoretically. (Photo by R. E. Zook and D. J. Fairbanks.)

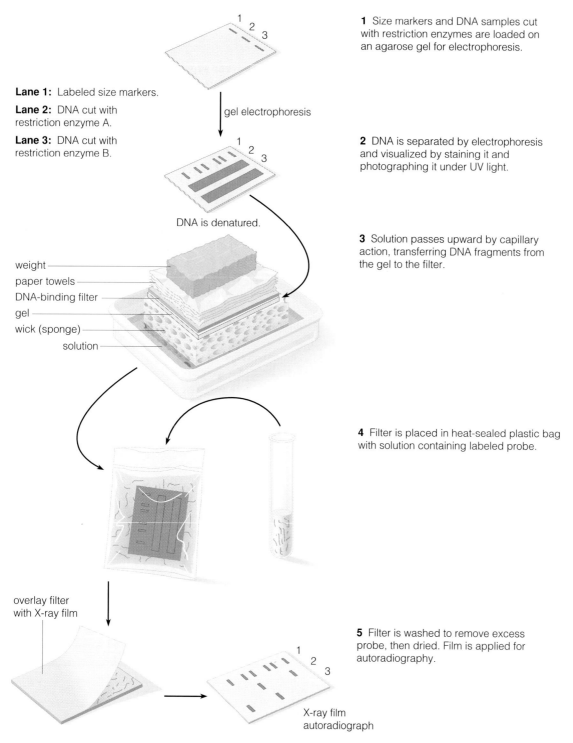

1 Size markers and DNA samples cut with restriction enzymes are loaded on an agarose gel for electrophoresis.

Lane 1: Labeled size markers.

Lane 2: DNA cut with restriction enzyme A.

Lane 3: DNA cut with restriction enzyme B.

gel electrophoresis

2 DNA is separated by electrophoresis and visualized by staining it and photographing it under UV light.

DNA is denatured.

3 Solution passes upward by capillary action, transferring DNA fragments from the gel to the filter.

weight
paper towels
DNA-binding filter
gel
wick (sponge)
solution

4 Filter is placed in heat-sealed plastic bag with solution containing labeled probe.

overlay filter with X-ray film

5 Filter is washed to remove excess probe, then dried. Film is applied for autoradiography.

X-ray film autoradiograph

Figure 9.31 Southern blotting for detection of DNA fragments using a probe. (Adapted from Cummings, M. R. 1997. *Human Heredity: Principles and Issues*. 4th ed. Belmont, Calif.: Wadsworth Publishing Co.)

A DNA fragment from a particular gene is present in the gel, but it cannot be distinguished at this stage because other DNA fragments obscure it.

The same problem with visualizing individual proteins or RNAs may exist following electrophoresis of complex mixtures of these molecules. Techniques have been developed for researchers to visualize only the molecules of interest, rather than all fragments. In each method, DNA fragments, RNA molecules, or proteins are transferred from a gel onto a filter, where individual molecules can be detected. Southern blotting is the transfer method for DNA, Northern blotting for RNA, and Western blotting for protein. Let's examine each of them.

Southern Blotting and Analysis

To detect a specific fragment of DNA among many other fragments, researchers hybridize a labeled probe to the DNA. A probe hybridizes to a DNA fragment that is similar to it in nucleotide sequence, allowing the fragment to be detected without visualizing any of the other thousands of DNA fragments. Unfortunately, probes cannot be used in gels directly. Instead, the DNA must be transferred from the gel onto a filter. The transfer procedure is called **Southern blotting**, named after E. M. Southern, who developed it. The steps are diagrammed in Figure 9.31.

First, DNA fragments (often those generated by restriction endonuclease digestion) are loaded onto an agarose gel and are separated using electrophoresis (steps 1 and 2 in Figure 9.31). Then the gel is placed on a nylon or nitrocellulose filter (which has the appearance and feel of smooth white paper) and a liquid capillary flow through the gel and the filter is set up as shown in step 3 of Figure 9.31. The DNA moves with the fluid out of the gel and onto the filter, but is trapped by the filter as the fluid flows through it. Once on the filter, the DNA fragments bind tightly to the filter and remain immobilized. The DNA fragments are deposited on the filter in the same positions they occupied in the gel.

After the DNA is immobilized on the filter, the filter is treated and hybridized with a probe in the same way as described in Figure 9.15. The probe hybridizes only to those fragments that share sequence similarity to it. Figure 9.32 shows how specific DNA fragments appear on the x-ray image of a Southern blot.

Northern and Western Blotting

Northern blotting is a modification of Southern blotting for detecting RNA instead of DNA. Researchers often isolate various types of RNA and wish to determine which molecule of RNA corresponds to a cloned fragment of DNA. RNA fragments can be separated electrophoretically, blotted onto a nylon or nitrocellulose filter, then hybridized to a DNA probe. The procedures are like those for DNA. The appearance of a Northern blot following hybridization is essentially the same as a Southern blot.

Western blotting is for detecting protein instead of DNA or RNA. As we saw earlier, proteins may be electrophoresed in a number of different ways that separate the proteins by molecular weight, net charge, or a combination of the two. Once separated electrophoretically, the proteins may be blotted onto a nylon or nitrocellulose filter using procedures similar to those used in Southern blotting. Like DNA and RNA, proteins remain immobilized on a filter in the same positions they occupied in the gel. An individual protein may then be detected if an antibody is used as a probe.

An **antibody** is a protein produced by an animal's immune system. Each antibody binds tightly to a specific foreign substance called its **antigen**. Whenever a foreign substance, such as a foreign protein, enters the bloodstream of a mammal, the immune system produces antibodies that recognize that single foreign substance and bind tightly to it. In this way, antibodies mark their antigens for destruction by the immune system.

Scientists may inject a purified protein into a laboratory animal, such as a mouse or rabbit, causing the animal to raise specific antibodies against the protein. Antibody-producing cells from the animal may be extracted and cultured to produce usable quantities of the antibody. The antibody can then be used to probe a Western blot. The antibody binds to its specific antigen, usually the same protein that was injected into the antibody-producing animal. The antibody can then be detected on the blot. In this way, a single protein may be identified in a Western blot while all the other proteins on the blot remain invisible. Figure 9.33 shows a Western blot treated with an antibody probe.

~

After DNA or RNA fragments are separated by electrophoresis, they may be transferred from the gel to a filter on which the DNA or RNA can be hybridized to a labeled probe. Proteins may also be transferred from gels to filters for identification of specific proteins with antibodies.

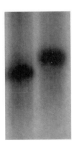

Figure 9.32 Detection of specific DNA fragments from a Southern blot. The differences in electrophoretic mobility for the two fragments represent genetic differences. (From Ruas, C. F., D. J. Fairbanks, W. R. Andersen, R. P. Evans, H. C. Stutz, and P. M. Ruas. 1998. Male-specific DNA in the dioecious species *Atriplex garrettii* (Chenopodiaceae). *American Journal of Botany* 85:162–167.)

Figure 9.33 A Western blot of a protein gel. An antibody probe identifies only a specific protein. In this case, the antibody identifies the large subunit (upper, very dark band) and small subunit (lower, lighter band) of the enzyme ribulose 1,5-bisphosphate carboxylase/oxygenase. (Photo by Y. J. Lu and D. J. Fairbanks)

a Many copies of a DNA fragment cloned in a vector are mixed with a radioactively labeled primer, a DNA polymerase, and all four dNTPs for DNA synthesis. The mixture is divided into four equal fractions, and each fraction receives one of the four ddNTPs.

b The DNA in each sample is denatured, and DNA synthesis is initiated. In each tube, DNA synthesis terminates whenever a ddNTP (indicated with an asterisk) is incorporated into the newly synthesized strand. When the DNA synthesis reactions are complete, there are fragments ending with every nucleotide in the region to be sequenced.

c The DNA is denatured into single strands, and the single-stranded DNA fragments in each sample are separated by length using gel electrophoresis. X-ray film is exposed to the gel to reveal the positions of the labeled fragments. The sequence is read from bottom to top. In this example, it is 5' TCCATGGACCAGAGA 3'.

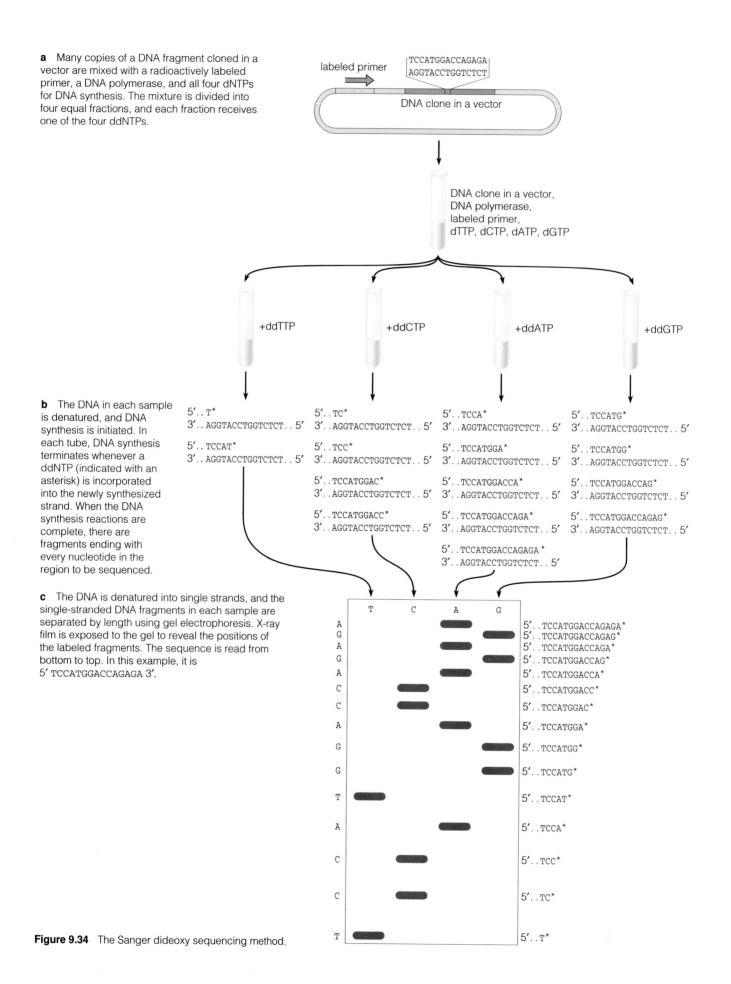

Figure 9.34 The Sanger dideoxy sequencing method.

9.9 DNA SEQUENCING

One of the goals of genetic analysis is to determine the nucleotide sequence of important genes. Ultimately, the nucleotide sequence of the entire genome of an organism can be obtained. The entire DNA sequences of several viruses were determined in the late 1970s and early 1980s. But such an accomplishment in a bacterial or eukaryotic organism was little more than a dream just a decade ago. Now, the genomes of several bacterial species and one eukaryotic species (*Saccharomyces cerevisiae*, brewer's yeast) have been fully sequenced. Among the most ambitious scientific projects ever undertaken is the combined effort of many laboratories to determine all 3 billion nucleotides of the human genome. This is a daunting and extremely expensive task, but it has already yielded academically and medically valuable information. We will discuss the Human Genome Project in more detail in later chapters.

Sanger Dideoxy Sequencing

There are several methods for sequencing DNA, but most current protocols are derived from the **Sanger dideoxy sequencing** method developed by Frederick Sanger in 1977. The steps in this method are illustrated in Figure 9.34. A DNA fragment to be sequenced is inserted into a vector that has a primer binding site so that a new DNA strand can be synthesized using one strand of the inserted DNA as a template. Many copies of the cloned fragment in the vector are mixed with a DNA primer that has been radioactively labeled, a DNA polymerase, and all four deoxynucleotide triphosphates (dNTPs). The N in dNTP stands for "nucleotide" and represents any of the four bases, A, T, C, or G.

The mixture is divided into four equal fractions, and one of the four dideoxynucleotide triphosphates is added to each fraction. A **dideoxynucleotide triphosphate (ddNTP)** lacks hydroxyl groups on both the 2′ and the 3′ carbons (Figure 9.35). When a DNA polymerase adds nucleotides to the growing chain, it can place a ddNTP in the chain just as easily as a normal dNTP. However, once a dideoxynucleotide is in place, another nucleotide cannot be added to the chain. To attach the next nucleotide, DNA polymerase must use the 3′ hydroxyl group. Because there is no hydroxyl group on the 3′ carbon of the dideoxynucleotide, chain elongation terminates.

Each of the four fractions has a different ddNTP added to it. One has ddTTP, another ddCTP, a third ddATP, and the fourth ddGTP (see Figure 9.34a). This means that the newly synthesized nucleotide chains in each fraction terminate at a different nucleotide. The nucleotide chains in the fraction with ddCTP all terminate with ddCMP, for example. Each reaction mixture also has all four normal dNTPs in it, so the chains in a mixture do not all terminate at the same place (see Figure 9.34b).

Once the reactions are complete, the DNA strands within each mixture are separated in a polyacrylamide gel that separates DNA strands that differ in length by only one nucleotide. The four mixtures are placed in adjacent wells in a gel, and after electrophoresis, an X-ray film is exposed to the gel. The radioactivity in the primers makes bands on the film where there are fragments in the gel. The DNA sequence can then be read from bottom to top across the four gel lanes, reading each band in order, as illustrated in Figure 9.34c. The X-ray image from a DNA sequencing gel is shown in Figure 9.36.

Automated DNA sequencing uses some modifications of the Sanger dideoxy sequencing procedure and an automated DNA sequencing machine. Instead of using a radioactively labeled primer, each ddNTP is labeled with a different fluorescent dye. Because these dyes cause the labeled nucleotides to fluoresce differently, each nucleotide can be distinguished from the others. Incorporation of the four labeled ddNTPs is done in the same

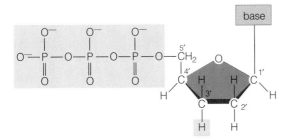

Figure 9.35 Structure of a dideoxynucleotide triphosphate. A hydrogen (blue), instead of a hydroxyl group, is attached to the 3′ carbon.

Figure 9.36 The image on X-ray film from a sequencing gel.

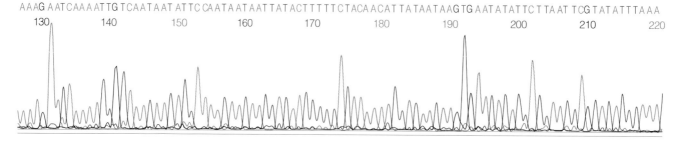

AAAG AATCAAA ATTG TCAATA AT ATTC CAATA ATAAT TATACT TTTTTCTACA ACAT TA TAATAAGTG AAT ATA TATTCT TTA AT TCG TATATTTAAA

130   140   150   160   170   180   190   200   210   220

Figure 9.37 Printout of a DNA sequence from an automated DNA sequencer. The peaks in the diagram represent absorbance by labeled dideoxynucleotides as they come off the gel in an automated DNA sequencing machine.

reaction tube instead of four separate tubes. The sample is then electrophoresed through a gel to separate the fragments by length. Each fragment reaches the end of the gel in order of its size, starting with the smallest fragment and ending with the largest. As the fragments come off the gel, they pass through a laser beam, and the DNA sequencing machine measures their fluorescence. The machine can then identify both the terminal nucleotide and its position in the chain, and from this information compile the DNA sequence. Figure 9.37 shows a DNA sequence read by an automated DNA sequencer.

Automated DNA sequencers are now used for nearly all DNA sequencing applications. In fact, automated DNA sequencing in centralized facilities has become so inexpensive and readily available that many scientists no longer conduct DNA sequencing in their own laboratories. Instead, they send cloned DNA to a commercial DNA sequencing laboratory or to a centralized DNA sequencing facility at their institution and pay a predetermined fee for the sequencing. This frees laboratory workers to conduct research without having to spend large amounts of time on DNA sequencing.

DNA Sequence Information on the Internet

Scientists often submit DNA sequences to a centralized computerized database, where the information is catalogued and made accessible to scientists worldwide. The advent of the Internet has facilitated the dissemination of many types of scientific information, including rapid cataloguing and sharing of DNA sequence information. Several databases are maintained worldwide, but all major public databases exchange submitted sequences daily and update their databases accordingly. The largest DNA sequence database in the United States, named GenBank, is maintained by the National Center for Biotechnology Information under the auspices of the National Institutes of Health and the National Library of Medicine. Figure 9.38 shows an entry for a DNA sequence in GenBank.

Scientists who determine a new DNA sequence may submit the sequence along with explanatory information

to GenBank, where the submission is reviewed, assigned an accession number, and entered into the database. Scientists may also use features associated with the database to compare their DNA sequences with all sequences in the database for nucleotide sequence or amino acid sequence similarity. The entire process usually requires only a few minutes for each DNA sequence. In this way, it is possible to quickly compare DNA sequences across many species for similarities and differences.

The following example shows how the DNA sequence in Figure 9.38 was compared to other DNA sequences in GenBank to detect similarities with other recorded sequences.

Example 9.7 Comparison of DNA sequences using GenBank, BLASTN, and BLASTX.

DNA and amino acid sequence analyses have become routine for scientists worldwide with the availability of databases and analysis software through the Internet. GenBank can be accessed by all interested scientists worldwide, free of charge. The software programs BLASTN and BLASTX, described by Altschul et al. (1990. *Journal of Molecular Biology* 215:403–410), are also freely available for analysis of sequences in GenBank. BLASTN compares both strands of a submitted nucleotide sequence to all sequences in GenBank, or to a subset of sequences defined by the submitter, and identifies those GenBank sequences with nucleotide sequence similarity. BLASTX takes the submitted nucleotide sequence and reads it in all possible reading frames, then compares the resulting amino acid sequences to amino acid sequences of proteins catalogued in GenBank and identifies those proteins with amino acid similarity. In this example, we'll look at a male-specific DNA sequence from a plant species that was compared to GenBank sequences using BLASTN and BLASTX.

While most animal species have separate male and female individuals, most plants are hermaphroditic, with both male and female reproductive

```
LOCUS        AGU79997      2055 bp    DNA              PLN        06-MAR-1998
DEFINITION   Atriplex garrettii male-specific DNA sequence.
ACCESSION    U79997
NID          g2160708
KEYWORDS     .
SOURCE       Atriplex garrettii.
  ORGANISM   Atriplex garrettii
             Eukaryotae; mitochondrial eukaryotes; Viridiplantae;
             Charophyta/Embryophyta group; Embryophyta; Magnoliophyta;
             Magnoliopsida; Caryophyllales; Chenopodiaceae; Atriplex.
REFERENCE    1  (bases 1 to 2055)
  AUTHORS    Ruas,C.F., Fairbanks,D.J., Evans,R.P., Stutz,H.C., Andersen,W.R.
             and Ruas,P.M.
  TITLE      Male-specific DNA in Atriplex garrettii
  JOURNAL    Am. J. Bot. 85, 162-167 (1998)
REFERENCE    2  (bases 1 to 2055)
  AUTHORS    Ruas,C.F., Fairbanks,D.J., Evans,R.P., Stutz,H.C., Andersen,W.R.
             and Ruas,P.M.
  TITLE      Direct Submission
  JOURNAL    Submitted (26-NOV-1996) Botany and Range Science, Brigham Young
             University, 401 WIDB, Provo, UT 84602, USA
FEATURES             Location/Qualifiers
     source          1..2055
                     /organism="Atriplex garrettii"
                     /db_xref="taxon:55308"
     misc_feature    1..2055
                     /note="This sequence was found in all male plants examined
                     and was absent in all female plants examined."
     misc_feature    555..648
                     /note="similar to the Antirrhinum majus (snapdragon) CEN
                     gene product encoded by GenBank Accession Number S81193 and
                     Arabidopsis thaliana TFL1 gene product encoded by GenBank
                     Accession Number U77674; possible pseudogene; The reading
                     frame is interrupted by repetitive DNA 31 codons upstream
                     from the stop codon at the point where similarity to the
                     CEN and TFL1 genes is lost"
BASE COUNT      708 a    311 c    338 g    698 t
ORIGIN
        1 tccatggacc agagacactg ggacaagata tatagttcta taaagataaa tagaaataaa
       61 agaatcaaaa ttgtcaataa tattccaata ataattatac tttttctaca acattataat
      121 aagtgaatat attcttaatt cgtatattta aatacttgcc atgagtcctt gaagagacac
      181 gtattgaaga atttttctac aacattataa taagtgaata tattcttaat tcgtataatt
      241 atacctttc tatcagcatt attaagttag aaagagggg cccccctct tggtattcaa
      301 cataatacag agctttatg tacttctagt cctaagtcta tgactcatcg ctctatagaa
      361 tttgtacgta ataaatgagt ttttgttcag cctcaactat ttcacaggct ttcatgggat
      421 ttaactcact gtaagtgtgc ttacaactat caccattatt aagttagaaa gaggggggc
      481 cctcttggta ttcaacatgc ttacaactat cagcattatt aagttagaaa gaggggcttc
      541 ggtacggttc agctgtctag cgccttcgag cagtggtttc cctttggcaa ttgaagtaga
      601 cagcagcaac aggaaggcct agctggttct ctacagcaaa ttttcgggta atgtaagttt
      661 tcatccgtac ataaagttaa gtcgaattta attgtcgttt tcatccgtac cattcattaa
      721 atgcactctg ctgtttttctt aatattcgtt gggcacaatt aaaatgtcgt gggtaatgta
      781 agttttatcc aaggcatggg aaaggaaact atactgggta atgacccctt cataatctat
      841 tgattatcat gaattatgca tgaaagaggg ggcccctccc tctttgaatt aatgattctg
      901 atataaatca ttattcaaaa gtaacaagtt tacaaacgga gtagtattac tagtttacta
      961 ctcgtacata aatgaaagag acaacccaac aaactaaaaa gggtgaatag ctaattcgta
     1021 ctgtagttta ggtgaaactt caaagcttaa aactttaacg aaatgaatgc aaagaatgag
     1081 aagggaaaaa actttgggca attgtgtgta acaataatgt caccatgtta tgtaagaatc
     1141 atctgcttta ttgacgacaa ttgtatcaaa atgaatggag gagcctggat ttgaagagaa
     1201 aaggtgaaac atgagcatca gatgcaagac tagcttataa agatagcaaa taaataaaaa
     1261 ggacaacagc aaaggtagaa tgtaggctga gatgcaagaa aattagactt gaataaaaag
     1321 taccatatat attgacttag caactttagt actattgatg agtcatattg atgtatccta
     1381 catcaaggtt ttattgtaca tgttgatgta caatttgcga agatatatag accattgagc
     1441 caaatgcaaa gaaacaacgt gacttttgtc tttactttca tcaatgcaaa taattttata
     1501 attacatcta atacaaatta aattcgatct attaatgctc ttctaaattt gattaatgcc
     1561 tattcattcg tatatttgca cttaggtttt tttttttttt ttttttttt ttttttttt
     1621 gaaaggatat ttgcacttag gttaacacct agatatatgg taatattgag aatatatgta
     1681 ttattagaac tagacaacta gttatatgac gactgacgag tgttgatcat attttggttg
     1741 ttctttttata tatgtatgga atgttcaata atttatcaa gatttttacca aaaatataatg
     1801 gggccaatta taataacact attactattt acaacatgct ttcgtccatg gtttcatcaa
     1861 ctagatcaac ttgtatgtaa ttctctgtga agattccatt atgaaggatg tttcatttt
     1921 aattaaataa aaatactgca ctgcaattaa ttacatgaga taatgtccgg aaatatcat
     1981 caatttttaaa actatatatg acagtattcc tcaatttata tacttcacaa ttccataatt
     2041 agcacataat gtcct
```

Figure 9.38 Example of GenBank accession for a submitted DNA sequence. (From the GenBank database at the National Center for Biotechnology Information website, http://www.ncbi.nlm.nih.gov.)

Atriplex garretti

```
ThrArgLysPheAlaValGluAsnGlnLeuGlyLeuProValAlaAlaValTyrPheAsnCysGlnArgGluThrThrAlaArgArgArgTer
ACCCGAAAATTTGCTGTAGAGAACCAGCTAGGCCTTCCTGTTGCTGCTGTCTACTTCAATTGCCAAAGGGAAACCACTGCTCGAAGGCGCTAGAC
```

```
ACGAGAAAATTCACACAGGAAAATGAATTGGGCCTCCCTGTTGCCGCTGTCTTCTTCAATTGCCAGCGCGAAACCGCTGCCAGAAGGCGTTGAAC
ThrArgLysPheThrGlnGluAsnGluLeuGlyLeuProValAlaAlaValPhePheAsnCysGlnArgGluThrAlaAlaArgArgArgTer
```

Antirrhinum majus

Figure 9.39 Information for Example 9.7: Comparison of the nucleotide and amino acid sequences for two DNA sequences matched by BLASTN and BLASTX. Identical amino acids are indicated in red, and identical nucleotides are indicated in blue.

organs present in a single individual, often within the same flower. However, some plant species have separate male and female individuals. In 1998, Ruas et al. (*American Journal of Botany* 85:162–167) reported the discovery of a DNA sequence that appeared only in male plants of the species *Atriplex garrettii*, a desert shrub called Garrett saltbush that grows along the banks of the Colorado River in western North America. When Ruas et al. compared their DNA sequence to sequences in GenBank using BLASTN and BLASTX, both programs identified a portion of the DNA sequence from the *CEN* gene of *Antirrhinum majus* (snapdragon) that was homologous to the male-specific *Atriplex garrettii* sequence.

The two sequences are aligned in Figure 9.39 to show the homology for nucleotide and amino acid sequences. The sequence in *Atriplex* corresponds to the 3′ end of a coding region for the *CEN* gene. GenBank identified the publication that described the *CEN* gene (Bradley et al. 1996. *Nature* 379: 791–797). Bradley et al. demonstrated that the *CEN* gene regulates the timing of flowering and flower position in snapdragon.

Problem: (a) What proportion of the nucleotides and amino acids in the two sequences are identical? (b) Assuming that these genes are of common evolutionary origin, what could explain the differences in similarity between nucleotide and amino acid sequences? (c) Snapdragon does not have separate male and female plants, so the *CEN* gene is not responsible for sex determination in snapdragon. What possible significance might the information about this gene's function in snapdragon have regarding the possible role of the male-specific DNA sequence in Garrett saltbush?

Solution: (a) As illustrated in Figure 9.39, 69 of the 95 nucleotides are identical in the two sequences, for 72.6% similarity, and 26 of 31 codons encode the same amino acid (or termination point), for 83.9% similarity. (b) Amino acid sequence similarity is higher than nucleotide sequence similarity, which is not unexpected. Most of the differences in the nu-

cleotide sequence are due to same-sense mutations, and thus have no effect on protein function or phenotype. As natural selection acts on the phenotype, it does not remove same-sense mutations, so as evolution proceeds, amino acid similarity is maintained at a higher level than nucleotide sequence similarity. (c) Snapdragon and Garrett saltbush are not closely related. The same gene may now be carrying out different roles in regulation of flowering in the two species. In snapdragon, this gene regulates the timing of flowering and flower position. In Garrett saltbush, the same gene may determine whether male or female flowers develop.

DNA sequencing is based on chain termination by dideoxynucleotides. Automated DNA sequencing has made possible rapid and inexpensive acquisition of DNA sequences. Scientists may submit DNA sequence information for inclusion in centralized databases that are freely available to scientists worldwide.

SUMMARY

1. DNA manipulation is possible because of the universality of DNA replication and the genetic code. It includes the transfer of DNA from one species to another, the in vitro modification and replication of DNA, and determination of the nucleotide sequence of DNA fragments.

2. Restriction endonucleases cleave DNA at specific sites. Researchers may recombine DNA from different sources at the sites cut by the restriction endonucleases to make recombinant DNA.

3. Geneticists use recombinant DNA technology to insert DNA fragments into bacterial vectors, where the DNA replicates easily in bacterial cells. Bacterial vectors include specially designed plasmids, phages, cosmids, and bacterial artificial chromosomes (BACs). The choice of a vector depends on the size of the DNA fragment.

4. Geneticists choose a cloning strategy to identify and isolate a desired clone from other clones that may be present.

5. Expression vectors are plasmids that have been modified for expression of foreign genes in bacteria. They often contain a promoter and a transcription terminator that can be recognized by the bacterial host. They are also constructed so that the inserted gene can be easily turned on or off by altering culture conditions. Eukaryotic genes must be modified before insertion into an expression vector in order to produce the desired product in bacterial cells.

6. The polymerase chain reaction (PCR) replicates a specific fragment of DNA, just as cloning does, but more rapidly and efficiently.

7. Cloned fragments of DNA can be sequenced with chain termination methods. Automated DNA sequencing permits rapid and inexpensive DNA sequencing.

8. DNA and amino acid sequences are catalogued in computer databases that are freely available to scientists worldwide.

QUESTIONS AND PROBLEMS

1. Define the term *recombinant DNA* as it is used in this chapter.

2. **(a)** What is the purpose of restriction endonucleases in nature? **(b)** How does the use of restriction endonucleases in recombinant DNA research differ from their function in nature?

3. Why are type II restriction endonucleases used more often in recombinant DNA research than type I or type III enzymes?

4. In the polymerase chain reaction, the amplified DNA fragments recovered from the reaction consist almost entirely of fragments bordered on both ends by the primer binding sites. Why is this so?

5. In the polymerase chain reaction, why do constant-length fragments amplify in an exponential fashion while variable-length fragments do not?

6. Below are the sequences of two PCR primers that amplify a fragment of DNA from the sequence in Figure 9.38.

        5′ CTAGCGCCTTCGAGCAGTGG 3′
        5′ ATCAACACTCGTCAGTCGTC 3′

Determine the length and the DNA sequence of the fragment amplified by these primers. When solving this problem, remember that the primers become part of the amplified sequence and that the two primers must bind to opposite strands.

7. In a procedure known as asymmetric PCR, many copies of a single-stranded DNA fragment can be produced by reducing the concentration of one of the two primers 10-fold so that the ratio of the two primers is 10:1 (instead of the usual 1:1 ratio for standard PCR). **(a)** Why are single-stranded fragments produced with asymmetric PCR, whereas double-stranded fragments are produced using standard PCR? **(b)** Why are both primers, rather than just one, included in asymmetric PCR, given that the objective is to amplify a single DNA strand?

8. Distinguish between cloning vectors and expression vectors. What are the essential differences in their structures and functions?

9. What is the practical function of each of the following features in the pBR322 plasmid cloning vector (see Figure 9.2): **(a)** the *amp*^r gene when the *Pst*I site is used for cloning, **(b)** the *tet*^r gene when the *Pst*I site is used for cloning, **(c)** the *ori* segment, and **(d)** the *Bam*HI site?

10. **(a)** What is the natural function of reverse transcriptase? **(b)** How is this enzyme used for the construction of cDNA libraries?

11. Cosmid cloning vectors are similar to plasmid cloning vectors in that they are circular pieces of DNA. However, up to 46 kb (kilobases) of DNA can be cloned in a cosmid but only up to about 10 kb of DNA can be cloned in pBR322. Explain why.

12. A 12-residue amino acid sequence in the human growth hormone gene is ThrProSerAsnArgGlu (from Goeddel et al., 1979. *Nature* 281:544–548). Provide a list of degenerate DNA probes that you could use to identify this gene.

13. Why do the ends of different restriction fragments carry the same nucleotide sequences when cut by the same restriction endonuclease?

14. Why does insertion of a piece of DNA into the *Bam*HI site of pBR322 inactivate the gene for tetracycline resistance?

15. Why wouldn't a human insulin gene cloned from a cDNA library into pUC19 be expressed?

16. What is the practical function of each of the following features in the pPLa2311 expression vector (see Figure 9.17): **(a)** the *amp*^r gene, **(b)** the *kan*^r gene, **(c)** the *ori* segment, **(d)** the *Eco*RI site, **(e)** the *Pst*I site, and **(f)** the P_L segment?

17. **(a)** What features of DNA molecules and gels make it possible to separate DNA fragments by length using gel electrophoresis? **(b)** Why does DNA carry a net negative charge?

18. Why do ddNTPs terminate DNA synthesis?

19. Why can a single sequencing reaction be used in an automated DNA sequencer, whereas four reactions are required for Sanger dideoxy sequencing?

Table 9.4	Information for Problem 20: Recovered Fragment Sizes (in Nucleotide Pairs) When Lambda DNA Is Digested	
EcoRI	**HindIII**	**EcoRI and HindIII**
21226	23130	21226
7421	9416	5148
5804	6557	4973
5643	4361	4268
4878	2322	3530
3530	2027	2027
	564	1904
	125	1584
		1375
		947
		831
		564
		125

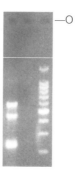

Figure 9.40 Information for problem 21: The left lane contains fragments of unknown size generated by PCR. The right lane contains fragments of known sizes. From top to bottom, the sizes are 2000, 1000, 900, 800, 700, 600, 500, 400, 300, and 200 nucleotide pairs. The "O" denotes the origin in the gel. (Photo by P. F. Randall and D. J. Fairbanks.)

20. The λ DNA molecule is 48,502 nucleotide pairs in length and may exist in linear or circular form. Table 9.4 lists the fragment sizes that are recovered when λ DNA is digested with *Eco*RI only, *Hin*dIII only, and *Eco*RI and *Hin*dIII together. **(a)** Was the λ DNA molecule used in this example linear or circular prior to digestion? **(b)** Construct a map of the λ DNA molecule showing the positions of the *Eco*RI and *Hin*dIII restriction sites.

21. Estimate the size of the DNA fragments in the left lane of Figure 9.40.

22. In 1951, Sanger and Tuppy (1951. *Biochemical Journal* 49:463–490) reported the following amino acid sequence of the B chain of bovine insulin:

```
PheValAsnGlnHisLeuCysGlySerHisLeuValGluAlaLeu-
TyrLeuValCysGlyGluArgGlyPhePheTyrThrProLysAla
```

Determine which five-amino acid sequence would yield the smallest number of degenerate oligonucleotide probes, and indicate how many probes would be needed for a complete degenerate set.

23. According to Lathe (1985. *Journal of Molecular Biology* 183:1–12), CG doublets should be avoided when synthesizing probes from amino acid sequences. What is the reason for this recommendation?

24. Example 9.5 presented the procedure that Goeddel et al. (1979. *Nature* 281:544–548) used to construct a genetically engineered human growth hormone gene that is expressed in *E. coli*. When the authors inserted the engineered gene into the expression vector, the ATG initiation

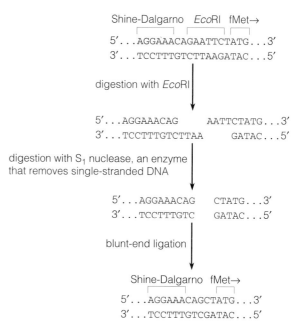

Figure 9.41 Information for problem 25: Procedure for moving the initiation codon closer to the *lac* Shine-Dalgarno sequence. (Adapted from Goeddel, D. V., H. L. Heynecker, T. Hozumi, R. Arentzen, K. Itakura, D. G. Yansura, M. J. Ross, G. Miozzari, R. Crea, and P. H. Seeburg. 1979. Direct expression in *Escherichia coli* of a DNA sequence coding for human growth hormone. *Nature* 281:544–548. Copyright 1979 Macmillan Magazines Limited. Used by permission.)

codon was 11 nucleotides downstream from the *lac* Shine-Dalgarno sequence. As illustrated in Figure 9.41, the researchers cleaved the plasmid with *Eco*RI and treated it with S₁ nuclease, which digests single-stranded DNA but not double-stranded DNA. This removed four nucleotides of the *Eco*RI cohesive ends, leaving blunt ends. They then reattached the blunt ends through blunt-end ligation. Return to Chapter 8 and look at Figure 8.3, then describe why the researchers moved the initiation codon closer to the lac Shine-Dalgarno sequence.

25. Determine the nucleotide sequence of the DNA molecule sequenced in Figure 9.36. Be sure to write the sequence in the $5' \rightarrow 3'$ direction.

FOR FURTHER READING

Detailed information relating to the concepts discussed in this chapter can be found in Chapters 5, 6, and 7 of **Watson, J. D., M. Gilman, J. Witkowski, and M. Zoller. 1992.** *Recombinant DNA, 2nd ed.* **New York: Scientific American Books**; and in Chapter 20 of **Lewin B. 1997.** *Genes VI.* **Oxford: Oxford University Press**. Details of vectors used for molecular biology as well as step-by-step descriptions of laboratory methods in molecular biology can be found in **Sambrook, J., E. F. Fritsch, and T. Maniatis. 1989.** *Molecular Cloning: A Laboratory Manual, 2nd ed.* **Cold Spring Harbor, N.Y.: Cold Spring Harbor Laboratory Press**. An entertaining article that describes how the idea for the polymerase chain reaction was discovered during a late-night drive in the mountains is **Mullis, K. B. 1990. The unusual origin of the polymerase chain reaction.** *Scientific American* **262 (Apr):56–65.** Frederick Sanger discussed the development of DNA sequencing in the review article **Sanger, F. 1988. Sequences, sequences, and sequences.** *Annual Review of Biochemistry* **57:1–28.**

For additional reading, go to InfoTrac College Edition, your online research library at: http: www.infotrac-college.com/brookscole

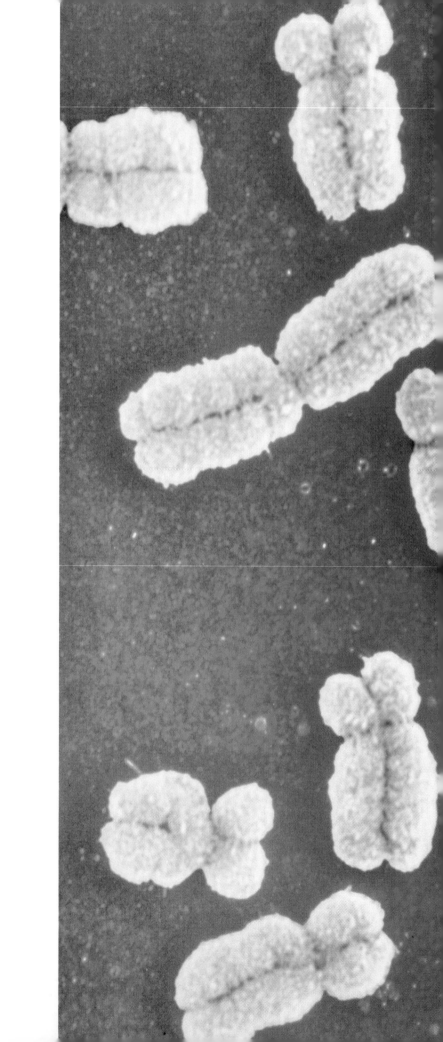

EUKARYOTIC GENOME ORGANIZATION

Eukaryotes are much more complex than prokaryotes. Even unicellular eukaryotes, such as yeasts, are more complex in their cellular structure, their reproductive and life cycles, and their interactions with the environment than prokaryotes. Because of this higher complexity, most eukaryotes require many more genes than prokaryotes, and larger genomes to accommodate those genes.

Let's compare the sizes of genomes among several species. With the exception of some unicellular algae, eukaryotic genomes are larger than prokaryotic genomes. The smallest known genome (other than viral genomes) belongs to the bacterium *Mycoplasma genitalium*, with just a little over 580,000 nucleotide pairs. Some unicellular eukaryotes have exceptionally small genomes, such as the unicellular alga *Pyrenomas salina*, with about 660,000 nucleotide pairs. The genomes of other single-celled eukaryotes are about two to three times the size of most prokaryotic genomes. For example, the *E. coli* genome (a prokaryote) is about 4.2 million nucleotide pairs. The genome of brewer's yeast (*Saccharomyces cerevisiae*), a unicellular eukaryote, contains slightly more than 13 million nucleotide pairs, about three times the size of the *E. coli* genome.

Although genome sizes among prokaryotes and unicellular eukaryotes do not differ substantially, the genomes of multicellular eukaryotes are larger than prokaryotic genomes by at least an order of magnitude. For example, the microscopic nematode worm *Caenorhabditis elegans* has a genome of about 80 million nucleotide pairs. The *Drosophila melanogaster* genome has about 140 million nucleotide pairs, and the human genome contains over 3 billion nucleotide pairs. The genomes of some amphibians and plants may exceed 75 billion nucleotide pairs. These numbers represent the size of one genome. Because most eukaryotic cells are diploid (two genomes per cell), these numbers must be doubled to obtain the amount of DNA in the nucleus of just one cell.

In this chapter, we will examine the organization of eukaryotic genomes. We begin by comparing repetitive and nonrepetitive DNA. Then we examine the organization of chromosomes and their structural features. We will then look at how eukaryotic genes are organized. Finally, we will examine the organization of entire genomes.

10.1 REPETITIVE AND NONREPETITIVE DNA

When we say something is *complex*, we usually mean that it is made up of many different components related to each other in a complicated way. Although this is an apt

description of eukaryotic genomes, the term *complexity* means something much more specific in terms of a genome. The **complexity** of a genome is defined as the length (usually expressed in nucleotide pairs) of different DNA sequences. For example, a DNA sequence of 300 nucleotide pairs that is repeated 100 times in the genome has a total length of 30,000 nucleotide pairs but contributes only 300 nucleotide pairs to the genome's complexity. Whereas a DNA sequence that is 30,000 nucleotide pairs long and is not repeated, either within itself or anywhere else in the genome, contributes 30,000 nucleotide pairs to the genome's complexity. DNA sequences that are repeated many times in the genome are called **repetitive DNA sequences**, whereas those that appear only once in the genome are called **unique** (or **nonrepetitive**) **DNA sequences**.

The genomes of most eukaryotes are made up of both repetitive and unique DNA sequences. In some eukaryotic species, such as brewer's yeast, only a small proportion of the genome consists of repetitive DNA sequences. In most multicellular species, however, a major fraction of the genomic DNA contains repetitive sequences. And in some plants and amphibians, repetitive DNA may account for more than 70% of the genome.

It is possible to determine the proportions of unique and repetitive DNA sequences in a genome through DNA renaturation experiments like those described in Example 10.1. The repetitive DNA sequences are often divided into two classes: **moderately repetitive DNA sequences** and **highly repetitive DNA sequences**. Thus, there are three general classes of DNA sequences based on the degree of repetition:

1. *Unique (or single-copy) DNA sequences*, which appear only once within the genome.

2. *Moderately repetitive DNA sequences*, which are repeated a moderate number (usually tens to thousands) of times.

3. *Highly repetitive DNA sequences*, which are repeated many (typically, more than 100,000) times.

Repetitive DNA

It is not possible to state an exact number of repetitions that distinguishes moderately from highly repetitive DNA sequences. In fact, most genomes have a continuum of repetitive sequences, from those repeated twice up to those repeated millions of times. The distinction between moderately and highly repetitive sequences is usually based on function as well as number. Most highly repetitive sequences are short sets of nucleotides that are repeated in tandem hundreds of thousands to millions of times. For example, in *Drosophila virilis*, most of the highly repetitive DNA contains repeats of a six-nucleotide sequence that varies only in the second and fifth nucleotides:

```
            5' ACAACT 3'
or
            5' ATAACT 3'
or
            5' ACAATT 3'
```

These sequences are repeated millions of times throughout the genome, often in tandem repeats.

Transposable Elements

Among repetitive DNA sequences is a particular class of DNA segments called **transposable elements**, which can insert themselves at sites throughout the genome. Some transposable elements transpose by excising themselves from their original sites and inserting themselves elsewhere in the genome. Other transposable elements transpose by replicating and inserting the replicated copy elsewhere in the genome, leaving the original element at its site. Sometimes a transposable element inserts itself within a gene where the insertion causes a mutation.

Transposable elements were first found in maize, *Drosophila*, and *E. coli*. They have since been found in all species examined. In some species, transposable elements may account for a significant part of the highly or moderately repetitive fractions of DNA. Chapter 22 examines the various types of transposable elements and the mechanisms they use to transpose. For now, we will look at how some types of transposable elements contribute to repetitive DNA.

Two types of transposable elements, SINEs and LINEs, have been well characterized in mammals and a few other species. **LINEs (long interspersed nuclear elements)** are sequences of DNA that are several thousand nucleotide pairs long, are repeated as many as 100,000 times, and are interspersed throughout the genome. **SINEs (short interspersed nuclear elements)** are usually less than 500 nucleotide pairs long, are repeated hundreds of thousands of times, and are also interspersed throughout the genome. Both LINEs and SINEs are a particular type of transposable element called a **retroposon**, which transposes through an RNA intermediate. The DNA of a retroposon has a promoter and may be transcribed into an RNA molecule that is reverse-transcribed into DNA. The reverse-transcribed DNA inserts itself at a new site in the genome, an event called **retrotransposition**. The original retroposon remains in its location, while the new copy inserts itself elsewhere. Because retrotransposition leaves the original retroposon in place, the number of retroposons in the genome increases with each retrotransposition. Over many generations, retroposons may increase in number as they spread throughout the genome, eventually becoming highly repetitive DNA sequences.

The best known LINE is L1, which is found in mammals. L1 differs somewhat in each species, so names such

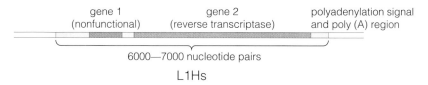

gene 1
(nonfunctional)

gene 2
(reverse transcriptase)

polyadenylation signal
and poly (A) region

6000—7000 nucleotide pairs

L1Hs

Figure 10.1 The structure of L1Hs.

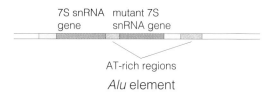

7S snRNA
gene

mutant 7S
snRNA gene

AT-rich regions

Alu element

Figure 10.2 The consensus structure of *Alu* elements in humans.

as L1Hs (L1 *Homo sapiens*) for humans and L1Md (L1 *Mus musculus domesticus*) for mouse have been given to the LINEs isolated from these species. There are about 100,000 copies of L1 in the human genome.

The structure of L1Hs is diagrammed in Figure 10.1. L1Hs is about 6000 nucleotide pairs in length and contains a polyadenylation signal (AATAAA) followed by a poly (A) sequence. These structural features imply that L1Hs originated from reverse transcription of an mRNA. L1Hs also contains two genes. The gene on the left in Figure 10.1 contains many mutations and is nonfunctional. The gene on the right is similar to reverse transcriptase genes from viruses.

SINEs are about 75–500 nucleotide pairs in length, much shorter than LINEs. The best-characterized human SINEs are a group of similar elements called the *Alu* family. The *Alu* family of elements is named after the *Alu*I restriction endonuclease, which cleaves DNA at its restriction site (AGCT) found within *Alu* elements. The members of the *Alu* family average 300 nucleotide pairs, and there are about 500,000 copies in the human genome, constituting about 5% of the genome.

Alu elements are distributed at random, and thus are often found when DNA sequences are analyzed. A sufficient number of *Alu* sequences have been characterized to suggest the original consensus element. As diagrammed in Figure 10.2, a complete *Alu* element contains two copies of a gene that encodes a 7S snRNA. One of the two genes (the left one in Figure 10.2) is complete and contains an RNA polymerase III promoter. The other gene (on the right in Figure 10.2) is nonfunctional. Its promoter has mutated so that it no longer functions, and the gene contains a 31-nucleotide-pair insertion mutation. The sequences of the two genes are about 68% similar, suggesting that they arose by duplication of a single gene. Between the two genes is a short segment of AT-rich DNA. A slightly longer AT-rich region is located on the right end of the element. Because the right promoter can-

not function, the left promoter initiates transcription of the entire element.

Alu sequences in the human genome vary by an average of about 14%. Much of the variation is due to mutations that inactivate the elements, so only a small fraction of the *Alu* sequences are still capable of retrotransposition. RNA transcripts that arise from active *Alu* elements apparently arise from elements that insert themselves at a site that stimulates transcription. *Alu* elements lack the necessary regulatory sequences for efficient transcription, so even most nonmutant *Alu* elements are not transcribed. However, there are some examples of recent retrotransposition of *Alu* sequences in humans, such as cancer patients who have tumors that arose following retrotransposition of *Alu* elements.

Functions of Repetitive and Unique DNA

For many years, it was thought that highly repeated DNA sequences had no essential function in the genome. Some scientists referred to highly repetitive DNA as "junk DNA" or "selfish DNA," implying that the DNA simply replicated without accomplishing anything beneficial for the organism. Although there still is no known function for many types of repetitive DNA, the functions of some types have become evident. Most highly repeated DNA sequences are located in well-defined regions of chromosomes, mostly the centromeres and telomeres, where these sequences perform important cellular functions. Telomeric repeats are important in replication of linear DNA molecules, as discussed in section 2.4. We will discuss the role of repetitive DNA in the centromere shortly.

Most moderately repetitive and unique DNA sequences are of great importance. Recall from Chapters 3 and 4 that rRNA and tRNA genes are repeated, often many times, within eukaryotic genomes. These genes fall within the moderately repetitive DNA class. Genes that encode mRNAs are usually unique DNA sequences.

The C-Value Paradox

The **C value** is the total amount of DNA in a single genome of a species. Although the exact number of genes in a genome has yet to be determined for all but the simplest of eukaryotes, it is possible to come up with a rough estimate for some species. In prokaryotes, most of the

DNA in the cell is contained within genes. This is evident in the genome map of the bacterium *Mycoplasma genitalium* shown in Figure 7.33. Thus, the C value for a prokaryote is nearly equal to the number of nucleotides within genes.

In many eukaryotes, however, the C value is often an order of magnitude greater than the number of nucleotides thought to be within genes, meaning that much of the DNA must not be included within the sequences for genes. Also, some species have much greater C values than closely related species, whereas the similarity of related species implies that the number of genes should be similar. This discrepancy between C value and the number of genes in an organism is called the **C-value paradox**.

It is now possible to account for much of the DNA not found in genes. A large proportion is highly repetitive. Also, some segments of both repetitive and unique DNA serve as spacer regions between genes, where some of the DNA functions in regulating gene expression. Introns also account for a portion of the DNA that is not included within the coding sequences of genes. Nonetheless, it is remarkable how much of the DNA in eukaryotes is outside of genes. In humans, only about 5% of the DNA in the genome encodes amino acid sequences.

DNA Renaturation Experiments

DNA denaturation is the separation of double-stranded DNA molecules into single-stranded molecules. The most common way to denature DNA in the laboratory is to heat it. When a solution of DNA is heated, the energy breaks the hydrogen bonds. If the temperature is elevated sufficiently, all hydrogen bonds are broken and the DNA separates completely into single strands. The temperature at which denaturation is complete depends on the relative proportions of A-T and G-C pairs. Because A-T pairs have only two hydrogen bonds, they separate more easily than G-C pairs. Thus, a DNA sample with a high proportion of A-T pairs denatures at a lower temperature than a sample with a high proportion of G-C pairs. When plotted against time, denaturation of a DNA sample follows a denaturation curve like the one illustrated in Figure 10.3. In this example, denaturation begins at 70°C and is complete at 95°C. The temperature at which half the nucleotide pairs in the sample have separated is called the **melting temperature** and is designated T_m, about 85°C in Figure 10.3. Usually, a temperature of 92–95°C ensures that all DNA in a solution is completely denatured. This is advantageous in research, because it is possible to heat DNA until it completely denatures without boiling the water in which it is dissolved.

Once DNA is denatured, it may renature with complementary strands to re-form double-stranded helices if the temperature is cooled. Renaturation is accomplished by reducing the temperature to about 25° below T_m, which allows the hydrogen bonds to re-form. At the molecular level, the single-stranded DNA molecules collide with one another in the solution and form hydrogen bonds wherever an A encounters a T or a C encounters a G. If nucleotides adjacent to the A-T or G-C pair do not match properly for pairing, the paired bases separate because a single base pair cannot hold the two strands together. Thus, noncomplementary strands do not pair effectively with each other. However, if complementary strands collide and a few nucleotides form base pairs, the remaining nucleotides pair with one another rapidly, and the double-stranded molecule re-forms.

Renaturation is dependent on DNA concentration. When the concentration of DNA is high, molecular collisions are more frequent and renaturation is therefore more rapid than when DNA concentrations are low. When the rate at which DNA renatures is measured, it follows an S-shaped renaturation curve. The time at which half of the DNA is renatured is designated as $t_{1/2}$. Figure 10.4 shows how DNA at a higher concentration renatures faster (with a lower $t_{1/2}$) than DNA at a lower concentration.

In DNA renaturation experiments, DNA that has been extracted from cells is randomly sheared to form double-stranded fragments that are about 100–300 kb long. These are heated to completely denature the strands, then the temperature is cooled to about 65°C to permit renaturation of the fragments. The unique and repetitive DNA sequences renature at different rates. Highly repeated DNA sequences renature more rapidly than moderately repeated sequences, which renature more rapidly than unique DNA sequences. Highly repeated sequences start renaturing immediately, but unique DNA sequences may require weeks to months to fully renature. Renaturation rates differ because a single-stranded DNA fragment that is highly repeated is much more likely to encounter a complementary fragment than a unique DNA sequence is.

The different rates of renaturation for highly repetitive, moderately repetitive, and unique DNA fractions follow a renaturation curve like the one illustrated in Figure 10.5. Three fractions—highly repetitive, moderately repetitive, and unique DNA—can be identified and their relative proportions in the genome determined. In Figure 10.5, 10% of the DNA is highly repetitive, 15% is moderately repetitive, and 75% is unique.

The renaturation curve in Figure 10.5 is often called a **Cot curve**, derived from the equation

$$C/C_0 = 1/(1 + kC_0t)$$

where C is the concentration of single-stranded DNA at time t, C_0 is the concentration of single-stranded DNA at the beginning of the experiment, and k is a constant determined by the conditions of the reaction (such as the renaturation temperature) and the complexity of the DNA. The vertical axis on the graph in Figure 10.5 represents

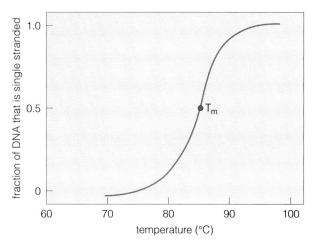

Figure 10.3 Denaturation curve for DNA.

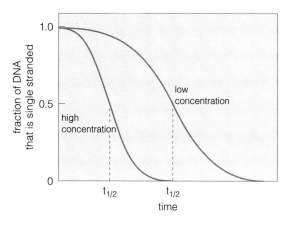

Figure 10.4 The effect of DNA concentration on DNA renaturation. When the DNA concentration is high, the DNA renatures more quickly than when the concentration is low.

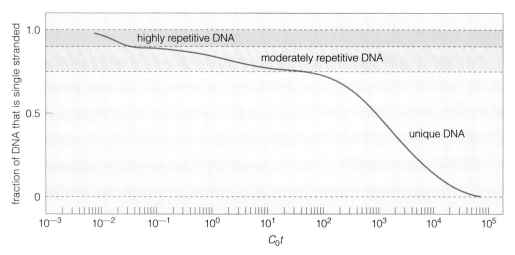

Figure 10.5 DNA renaturation curve for mouse genomic DNA. There are three fractions of DNA, distinguished by their rate of renaturation: highly repetitive DNA, which is the first to renature; moderately repetitive DNA, which is the next to renature; and unique, or single-copy, DNA, which is the last to renature. (Adapted from McConaughy, B. L., and B. J. McCarthy. 1970. Related base sequences in DNA of simple and complex organisms. VI. The extent of base sequence divergence among the DNAs of various rodents. *Biochemical Genetics* 4:425–446.)

the values of C/C_0, which is equivalent to the proportion of DNA that is double stranded. These are plotted against the corresponding values of C_0t on a logarithmic horizontal axis. Each of the individual fractions has its own value for k, because the complexity of each fraction differs. For each fraction it is possible to determine a $C_0t_{1/2}$ at which half of the DNA of that fraction has renatured. For each fraction:

$$C_0t_{1/2} = 1/k$$

which defines the rate of renaturation for each individual fraction, and makes it possible to calculate k.

Let's now discuss the results of some DNA renaturation experiments to see how DNA renaturation is related to DNA complexity.

Example 10.1 DNA renaturation experiments and genome complexity.

In 1968, Britten and Kohne (*Science* 161:529–540) used DNA renaturation experiments to compare genome complexity of several organisms. The results of some of their experiments are illustrated in Figure 10.6.

Problem: What do these results demonstrate about DNA complexity?

Solution: The $C_0t_{1/2}$ values of DNA samples are correlated with genome complexity. The sample with the lowest $C_0t_{1/2}$ is the Poly (U) + Poly (A) sample, which has a complexity of 1 nucleotide

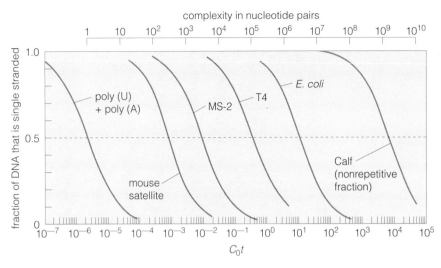

Figure 10.6 Renaturation curves for several different samples of DNA. The $C_0t_{1/2}$ values are positively correlated with the complexities of the DNA samples. (Adapted from Britten, R. J., and D. E. Kohne. 1968. Repeated sequences in DNA. *Science* 161:529–540. Copyright 1968 American Association for the Advancement of Science. Used by permission.)

pair, because the same nucleotide pair is repeated at every position in the molecule. Mouse satellite DNA, which is the highly repetitive fraction of mouse DNA, renatures quickly because of its low complexity. The MS-2 and T4 viral genomes are more complex than mouse satellite DNA but less complex than the *E. coli* genome. The most complex DNA in the experiment is the unique fraction of calf DNA. It is the only sample that contains unique DNA from a multicellular eukaryote, so we expect it to have the highest complexity of all the DNA samples tested.

The DNA in a eukaryotic genome can be divided into three general classes: (1) unique (or single-copy) DNA sequences, (2) moderately repetitive DNA sequences, and (3) highly repetitive DNA sequences. Most protein-encoding genes are within the unique sequences.

10.2 EUKARYOTIC DNA PACKAGING

Unlike the circular DNA molecules of prokaryotes, nuclear DNA in eukaryotes consists of linear double-stranded DNA molecules that are packaged with proteins into individual chromosomes. As large as eukaryotic genomes are, it is essential that the DNA be packaged in a highly organized way to prevent it from becoming hopelessly tangled. DNA packaging must also be highly organized in order for the DNA to replicate and its genes to be expressed.

When packaged DNA is at its highest level of condensation, a tremendous amount of DNA is held tightly in a small space. This is well illustrated in the micrograph in Figure 10.7, where a condensed chromosome has been treated with detergents to disrupt the proteins that hold the DNA in condensed form. Once the proteins are removed, the tightly packed DNA spills out of the chromosome into its surroundings. All of the fiberlike material surrounding the residual proteins in Figure 10.7 is double-stranded DNA that was packaged into a chromosome.

DNA packaging follows a hierarchy with several stages of coiling. At each stage of coiling, specific proteins hold the DNA in its coiled state. The DNA and its associated proteins are collectively called **chromatin**.

The hierarchical packaging follows a set pattern, diagrammed in Figure 10.8. First, DNA winds around individual cores composed of proteins called **histones**. The histone cores act as spools that hold the coiled DNA around them. DNA winds around a histone core to form a beadlike structure called a **nucleosome**, illustrated in Figure 10.8a. When visualized with electron microscopy, the nucleosomes look like a set of beads along the DNA to make a string (Figure 10.8b). In the cell the nucleosomes form at regular intervals along the DNA to make a string that is 10 nm in diameter and is called the **10 nm fiber**. This level of packaging compresses the DNA to around $1/6$th to $1/7$th its original length.

In the next level of packaging, the 10 nm fiber coils into a helix called a **solenoid**, as illustrated in Figure 10.8c. Each coil of the solenoid contains six nucleosomes. The solenoid is 30 nm in diameter and is called the **30 nm fiber**. This level of packaging compresses the DNA to about $1/40$th of its original length. Chromatin may be condensed even more, particularly when chromosomes condense in preparation for cell division, as illustrated in Figure 10.8d.

The level of DNA packaging is not constant; it changes over time during different stages of the cell's life. Compare the relaxed chromosomes in the nucleus in

DNA

residual proteins

Figure 10.7 A condensed chromosome has been treated with detergents that disrupt some of the proteins holding the DNA in a condensed state. The DNA then spills out of the chromosome into its surroundings, making it possible to visualize the tremendous amount of DNA contained within the chromosome. (From Paulson, J. R. and U. K. Laemmli. 1977. The structure of histone-depleted metaphase chromosomes. *Cell* 12:817–828. © 1977 Cell Press. Used by permission.)

Figure 10.9a with the condensed chromosomes of a cell preparing to divide in Figure 10.9b. When the cell is not preparing to divide, the chromosomes are decondensed and cannot be distinguished from one another microscopically, even at magnifications possible with an electron microscope. Even so, much of the DNA is still packaged as 30 nm fiber loops connected to a nuclear protein matrix. However, after DNA replicates and the cell prepares for division, the chromosomes condense into compact, easily distinguished masses within the cell.

Not only does the level of DNA packaging vary over time, it also varies within each chromosome. Nearly all eukaryotic DNA in the cell nucleus is condensed to at least the level of 10 nm fibers, and in most parts of the chromosome the DNA is condensed to 30 nm fibers, which may be coiled even further.

Nucleosomes

Let's now take a more detailed look at the structure of nucleosomes. A nucleosome forms when 140–180 nucleotide pairs of DNA wind two times around the histone core. The DNA within the nucleosome is called **core DNA**. Between each nucleosome is a segment of DNA called **linker DNA** that connects each nucleosome to the next, giving the DNA the beads-on-a-string appearance. The number of nucleotide pairs in each segment of linker DNA averages about 50 but may vary from as few as 8 to over 100.

Histones contain an unusually large proportion of basic amino acids, which confer a net positive charge to histones. The phosphate groups in DNA are negatively charged and bind to the positively charged histones.

There are eight histone molecules in a nucleosome core: two molecules each of histones H2A, H2B, H3, and H4 (Table 10.1). These subunits are arranged as illustrated

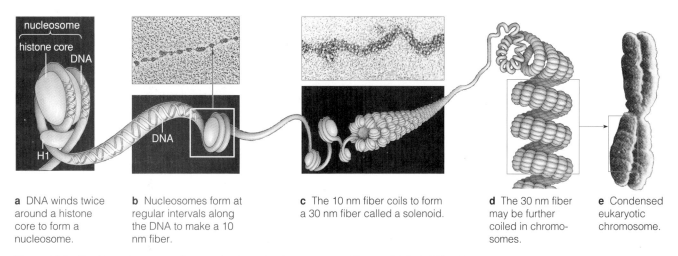

a DNA winds twice around a histone core to form a nucleosome.

b Nucleosomes form at regular intervals along the DNA to make a 10 nm fiber.

c The 10 nm fiber coils to form a 30 nm fiber called a solenoid.

d The 30 nm fiber may be further coiled in chromosomes.

e Condensed eukaryotic chromosome.

Figure 10.8 The hierarchical organization of eukaryotic chromosomes. (Photos: **(b)** O. L. Miller, Jr., S. L. McKnight, **(c)** B. Hamkalo, **(d)** Harrison, C. J., T. D. Allen, and R. Harris. 1983. Scanning electron microscopy of variations in human metaphase chromosome structure revealed by Giesma banding. *Cytogenetics and Cell Genetics* 35:21–27 © 1983 S. Karger A.G., Basel.)

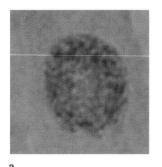 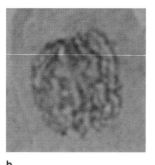

a b

Figure 10.9 Comparison of a cell that is not dividing **(a)** with one that is in the process of dividing **(b)**. Notice how the chromosomes are in a much more condensed state in the dividing cell. (Photos courtesy of A.S. Bajer, University of Oregon.)

Table 10.1 Characteristics of Some Histone Proteins

| Histone | Size | |
	Molecular Weight	Number of Amino Acids
H1	17,000–28,000	200–265
H2A	13,900	129–156
H2B	13,800	121–148
H3	15,300	135
H4	11,300	102

in Figure 10.10, with DNA wound around them. Most nucleosomes contain an additional histone called a **linker histone**. Linker histones have two binding sites for DNA, where they link the DNA coil in the nucleosome. Linker histones include H1, H5, and H°. Most models of nucleosome structure predict that a single linker histone resides on the outer surface of the nucleosome, where it links the DNA coil in the nucleosome, as diagrammed in Figure 10.11a. However, the results of some experiments have challenged this model, and suggest that the linker histone may reside within the nucleosome, as diagrammed in Figure 10.11b. In fact, the two models need not be mutually exclusive. In some instances the linker histone may reside outside of the nucleosome, and in others it may reside inside of the nucleosome.

Linker histones play an important role in gene regulation. Experiments in which a mutant *H1* gene fails to encode functional H1 show that expression of many genes is then altered. Some genes are transcribed more often in the absence of H1, while others are transcribed less. These results suggest that H1 may have a positive or a negative effect on transcription, depending on the interaction of H1 with other transcription factors.

Some histones are among the most highly conserved proteins in nature, meaning that the histones differ very little in amino acid sequence from one species to another. For example, there is no difference in the amino acid sequence of histones H3 and H4 from several plants and animals, suggesting that these proteins play a vital and identical role among a wide variety of organisms. Linker histones, on the other hand, are much more varied among species, and are even absent in certain species.

DNA Packaging and Regulation of Transcription

In a specialized cell, DNA packaging prevents transcription of most genes. RNA polymerase cannot physically access the genes in highly condensed DNA. In order for a eukaryotic gene to be transcribed, its DNA must be de-

condensed. How far decondensation proceeds depends on how intensively the gene is transcribed. For instance, rRNA genes are often saturated with RNA polymerase I molecules transcribing the genes (Figure 10.12). These intensively transcribed rRNA genes, as well as the spacer DNA between them, are free of nucleosomes.

Most genes that encode mRNAs are not transcribed intensively, and the histones remain attached to DNA during transcription, although the 30 nm fiber must be unraveled. RNA polymerase molecules are about the same size as nucleosomes, making RNA polymerases too large to transcribe DNA when it is bound tightly to histones. Nucleosome structure is therefore temporarily altered during transcription making the DNA more accessible to RNA polymerase. Once RNA polymerase has passed, the nucleosome resumes its normal conformation.

When an mRNA-encoding gene is actively transcribed, the DNA in the gene's promoter must be completely free of nucleosomes for the basal transcription complex to initiate transcription. Transcription factors and other proteins stabilize the nucleosome-free portion of DNA and prevent nucleosomes from forming in it. A gene with a promoter wound into nucleosomes is not likely to be transcribed.

Nucleosomes and DNA Replication

Because DNA strands must separate in order for the DNA to replicate, nucleosomes must be at least partially disrupted during replication. As illustrated in Figure 10.13, which shows a replication fork in *Drosophila* DNA, nucleosomes are in the DNA behind and ahead of the replication fork, suggesting that new nucleosomes are synthesized shortly after DNA is replicated. Experiments, such as those highlighted in the following example, have demonstrated that nucleosomes do not fully disassemble during DNA replication and that histone molecules present in the original DNA molecule are retained in both of the newly replicated DNA molecules.

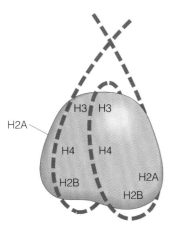

Figure 10.10 The positions of histones H2A, H2B, H3, and H4 in the nucleosome.

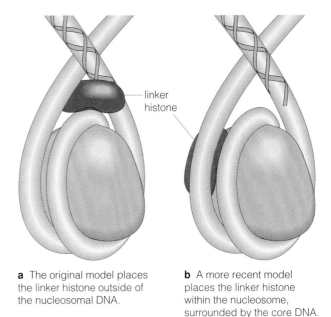

a The original model places the linker histone outside of the nucleosomal DNA.

b A more recent model places the linker histone within the nucleosome, surrounded by the core DNA.

Figure 10.11 Two models of linker histone placement in the nucleosome.

Example 10.2 Nucleosome conservation during DNA replication.

During DNA replication, the two strands separate and serve as templates for synthesis of two new strands. What is the fate of the histones that are bound to a DNA molecule when the DNA replicates? To address this question, Bonne-Andrea, Wong, and Alberts (1990. *Nature* 343:719–726) transferred a limited number of histones to a plasmid that could be replicated in a cell-free mixture using the replication enzymes of T4 bacteriophage. The histones assembled into normal nucleosomes in the plasmid DNA. The researchers found that the plasmid DNA replicated when nucleosomes were in it and that intact nucleosomes were present on both daughter molecules. They also found that when a 10-fold excess of nucleosome-free, nonreplicating DNA was added to the nucleosome-bound plasmid DNA before the plasmid DNA replicated, the histones were associated with the plasmid DNA only, both before and after it replicated. The researchers also found that the histone proteins bound to one another normally in the nucleosomes after replication, even though no new histones were added to the mixture.

Problem: **(a)** What was the purpose of the experiment in which nonreplicating, nucleosome-free DNA was added to the preparation, and what conclusion can be derived from this experiment? **(b)** In nature, histones are found only in eukaryotic cells. This experiment was conducted with bacterial DNA and bacterial DNA replication enzymes. In light of this, what conclusions can be drawn about the affinity of histones for DNA?

Solution: **(a)** If histones leave the DNA during replication then reassemble on the DNA after repli-

cation, some of the histones should have assembled on the nonreplicating DNA. The absence of histones on the nonreplicating DNA indicates that the histones remained bound to the plasmid DNA during replication. **(b)** Nucleosomes can form in the absence of eukaryotic proteins (other than the histones themselves) and can form on bacterial DNA.

DNA packaging follows a set hierarchical pattern. The DNA winds around histones to form the 10 nm fiber, which coils to form a solenoid of histones called the 30 nm fiber. DNA may further condense into loops held in place by nonhistone proteins.

10.3 THE EUKARYOTIC CHROMOSOME

In eukaryotic cells, each chromosome contains one long linear molecule of DNA that extends from one end of the chromosome to the other. The DNA is packaged as chromatin at different levels of condensation throughout the chromosome. Chromosomes are usually diagrammed in their most condensed state, during metaphase of mitosis, because mitosis is the only point at which individual chromosomes can be observed microscopically. When you view diagrams of chromosomes, you must keep in mind that throughout most of a cell's life, the chromosomes are not going through mitosis and are not condensed as highly as the diagrams tend to suggest.

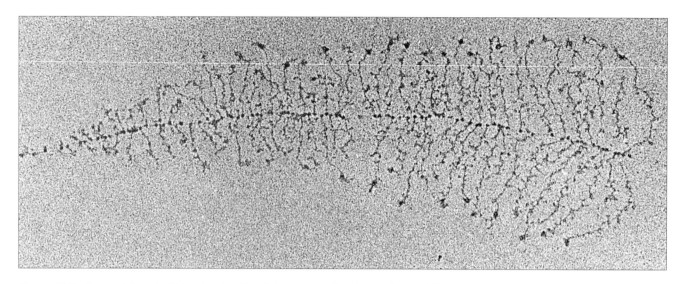

Figure 10.12 Transcription of rRNA molecules. The Christmas tree–like structure is an rRNA gene that is being transcribed. The trunk is DNA, and the branches are rRNA transcripts. The RNA polymerase I molecules are visible as dark bodies on the DNA, each with an rRNA transcript attached. There are no nucleosomes in the DNA of this gene. (Photo courtesy of O. L. Miller, Jr. and B. R. Beatty.)

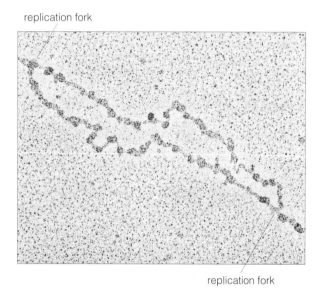

Figure 10.13 Bidirectional DNA replication in *Drosophila*. Notice that nucleosomes are present on the unreplicated DNA in front of the replication forks and on the newly replicated DNA behind the replication forks. (McKnight, S.L. and O. L. Miller, Jr. 1977. Electron microscopic analysis of chromatin replication in the cellular blastoderm *Drosophila melanogaster* embryo. *Cell* 12:795–804. © 1977 Cell Press. Used by permission.)

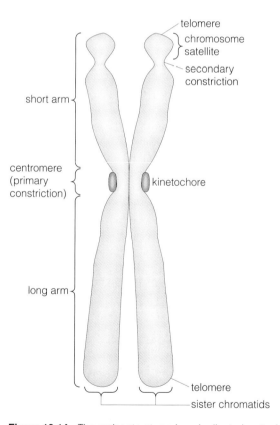

Figure 10.14 The major structures in a duplicated metaphase chromosome.

Structural Features of Eukaryotic Chromosomes

Metaphase chromosomes have certain structural characteristics that define their size and shape. Figure 10.14 provides a diagram of a condensed metaphase chromosome showing the structural features that are common. Metaphase chromosomes have an X-shaped appearance because they have replicated and contain two DNA molecules.

The two DNA molecules are attached at a constricted site called the **centromere**. On the outside surface on either side of the centromere are protein bodies called **kinetochores**, which participate in separation of the replicated chromosome during mitosis. Once the replicated portions

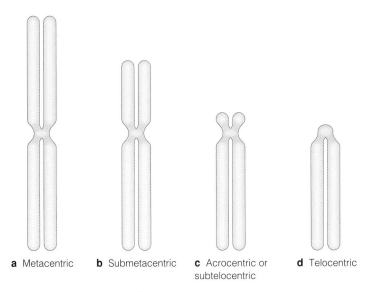

a Metacentric   **b** Submetacentric   **c** Acrocentric or subtelocentric   **d** Telocentric

Figure 10.15 Classification of chromosomes by structure.

of the chromosome separate, two complete chromosomes form, each with a single DNA molecule.

The centromere is a constricted region that is often located near the middle of the chromosome. A chromosome with a centromere at or near the middle of the chromosome is called a **metacentric** chromosome (see Figure 10.15a). The centromere may also be located somewhat off center, in which case the chromosome is **submetacentric** (Figure 10.15b). When the centromere is close to one end of the chromosome, the chromosome is **subtelocentric** or **acrocentric** (Figure 10.15c). The term *acrocentric* is usually used when describing human and animal chromosomes, whereas *subtelocentric* is the more common term for describing plant chromosomes. When the centromere is located at one end of the chromosome, the chromosome is said to be **telocentric** (Figure 10.15d). There are no telocentric human chromosomes; all can be classified as metacentric, submetacentric, or acrocentric.

In all but telocentric chromosomes, the centromere divides the chromosomes into two segments called **chromosome arms**. Unless the centromere falls exactly in the center of the chromosome, it divides the chromosome into a **long arm** and a **short arm** (Figure 10.14). In human chromosomes, the short arm is called the **p arm**, and the long arm the **q arm**. Alternatively, in some species chromosome arms are designated arbitrarily as the **right arm** and the **left arm**. This is the convention for *Drosophila* and yeast chromosomes. It is not unusual to describe the location of a gene by its position on a particular arm of a chromosome. For example, the gene in humans that encodes phenylalanine hydroxylase is located on the q arm (long arm) of chromosome 12.

The centromere is also called the **primary constriction** because it is a constricted site found on every chromosome. Occasionally, another constriction, called a **secondary constriction**, is found near one end of the chromosome. The small portion of the chromosome that extends beyond the secondary constriction is called a **chromosome satellite** (Figure 10.14). Often, a secondary restriction and satellite are located at the nucleolus organizer region (NOR), where the rRNA genes are clustered.

Chromatids

When the DNA in the chromosome is duplicated, as in metaphase chromosomes, there are two short arms and two long arms attached to a single centromere. The duplicated chromosome is still considered to be one chromosome, so the term **chromatid** denotes the identical duplicated portions of the chromosome. The long arm and the short arm on one side of a duplicated chromosome are called a chromatid, which contains one linear DNA molecule. The other long and short arm, which contain the other DNA molecule, are also called a chromatid. The two chromatids of a single duplicated chromosome are called **sister chromatids** (Figure 10.14). Once the sister chromatids separate from one another, they become individual chromosomes and are no longer called chromatids.

Centromeres

The constricted region of a centromere is readily apparent in the electron micrograph of a mammalian chromosome shown in Figure 10.16. The centromeric regions of metaphase chromosomes are condensed to the level of the 30 nm fiber, while the remainder of the chromosome is condensed further into loops, causing a constriction to appear throughout the centromeric region.

The DNA sequence of the entire genome of *Saccharomyces cerevisiae* has been determined, and much research has focused on comparison of centromeric DNA

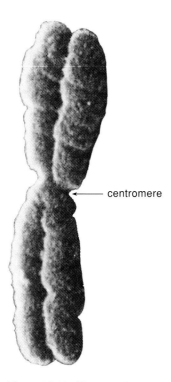

Figure 10.16 Electron micrograph of a metaphase chromosome. Notice the constriction at the centromere, where the two sister chromatids are attached. (From Harrison, C. J., T. D. Allen, and R. Harris. 1983. Scanning electron microscopy of variations in human metaphase chromosome structure revealed by Giesma banding. *Cytogenetics and Cell Genetics* 35:21–27. © 1983 S. Karger A. G., Basel.)

sequences among the chromosomes. Each of the chromosomes contains a conserved centromeric region (called CEN) that is similar among all 16 chromosomes. The CEN region of each chromosome consists of about 160–220 nucleotide pairs that can be divided into three elements, called CDEI, CDEII, and CDEIII.

CDEI is an 8-nucleotide-pair sequence that has the consensus sequence

$$\substack{A\\C}TCAC\substack{A\\C}TG$$

This sequence is important for proper centromere function but not essential. When this sequence is deleted, mitosis is not eliminated, but chromosome stability during mitosis declines. CDEII is a sequence of 78–86 nucleotide pairs that contains about 90% A-T pairs. It likewise is important for centromere function because deletion of CDEII causes a loss of some chromosomes during cell division. CDEIII is a 26-nucleotide-pair element with the consensus sequence

TGTTTTTGNTTTC**C**GAAANNNAAAAA

where the underlined sequence has palindromic symmetry on both sides of a cytosine residue (in boldface). CDEIII is essential for centromere function. If a substi-

tution mutation alters the cytosine in the center of the palindrome, the centromere no longer functions. Figure 10.17 shows the complete DNA sequence of the CEN region in a yeast centromere.

The sequence elements of *S. cerevisiae* centromeres are not found in other eukaryotes, not even in other yeast species. DNA sequence elements in centromeres are highly diverse among eukaryotes. Efforts to identify centromere consensus sequences among eukaryotes have, however, revealed some similarities. The centromeres of multicellular eukaryotes are surrounded by highly repetitive DNA, and the centromeres themselves often contain tandem repeats. For example, human chromosomes contain a repeat sequence called alphoid (also called α-satellite DNA) that is about 170 nucleotides in length and is repeated many times in a tandem array in the centromere.

Alphoid sequences contain a conserved sequence called the CENP-B box with the consensus sequence

PyTTCGTTGGAAPuCGGGA

which serves as a binding site for proteins associated with the centromere. The CENP-B box is found in other mammalian species and may be present in *Drosophila*. The CENP-B box may be a common sequence amid the highly diverse centromeric sequences of multicellular eukaryotes.

Telomeres

The ends of eukaryotic chromosomes are specialized structures called **telomeres** (Figure 10.14), which stabilize the ends of the chromosomes. When a piece of chromosome is broken off, the broken end tends to be sticky in that other chromosome pieces readily fuse to it. A telomere stabilizes the end of the chromosome and prevents other chromosome fragments from fusing to it. Also, the telomere protects the end of the DNA molecule in the chromosome from enzyme degradation.

Telomeres consist of short DNA sequences that are repeated many times in tandem. The repeated sequences are similar among all species that have been studied, and play an important role in replication of linear DNA molecules. These repeated sequences and their relationship to telomerase, an enzyme that synthesizes the repeats, were discussed in detail in section 2.4.

~

Eukaryotic chromosomes have centromeres, which function in chromosome separation during mitosis, and telomeres, which protect the ends of the linear chromosomes.

Chromosome Banding

When metaphase chromosomes are prepared for observation under a microscope, they must be stained to be visualized. Several stains reveal specific patterns called

ATCACGTGCTATAAAAATAATTATAATTTAAATTTTTTAATATAAAATATATAAATTAAAAATAGAAAGTAAAAAAAGAAAAAATAGTTTTTGTTTTCCGAAGATGTAAAA

Figure 10.17 DNA sequence of the CEN6 region from a yeast centromere. (Adapted from Sears, D. D., J. H. Hegemann, J. H. Shero, and P. Hieter. 1995. Cis-acting determinants affecting centromere function, sister chromatid cohesion, and reciprocal recombination during meiosis in *Saccharomyces cerevisiae*. *Genetics* 139:1159–1173.)

chromosome bands, alternating light-staining and dark-staining regions that give each chromosome a particular pattern. Among the most frequently used chromosome stains, particularly for mammalian chromosomes, is one called Giesma, which reveals G-bands (G for Giesma). Figure 10.18 shows G-banded human metaphase chromosomes and an example of how G-bands are represented in chromosome diagrams.

Notice that each chromosome has a particular pattern of alternating light and dark G-bands. The patterns allow researchers to discriminate different chromosomes that are the same size and have the same centromeric position. For example, human chromosomes 6–12 are highly similar in size and centromere position, so they are difficult to distinguish from one another. However, when stained with Giesma, each chromosome has a particular pattern of G-bands, which allows it to be identified.

The dark G-bands are mostly AT-rich chromosomal material that is more highly condensed than the light bands. The light bands tend to be GC-rich and are less condensed. There is compelling evidence that G-banding represents an aspect of chromosome organization. Genes are not randomly scattered along the chromosome but are grouped, usually within the light-staining G-bands.

Several other stains reveal chromosome banding patterns. The choice of a stain often depends on the species under study. For example, most mammalian chromosomes can be best characterized with G-banding, but the chromosomes of many other species cannot. Also, the choice of stain may depend on the chromosomal features the researcher wishes to study. For example, NOR-banding permits clear identification of the nucleolus organizer regions in chromosomes.

The chromosomes of most species can only be distinguished microscopically when the cell is preparing to divide. Metaphase chromosomes are an example of such chromosomes. During the major part of a cell's life, the chromosomes are dispersed and intermingled with one another in the nucleus and cannot be identified microscopically. However, there are some important exceptions. Chromosomes in the larvae of *Drosophila melanogaster* and several other insect species are very large and can be observed microscopically. These large chromosomes are called **polytene chromosomes**.

Drosophila larvae are small, white, wormlike creatures. When they first emerge from their eggs they are about 1–2 mm in length, but rapidly grow to about 7–8 mm before pupating. During the rapid larval growth stage, the

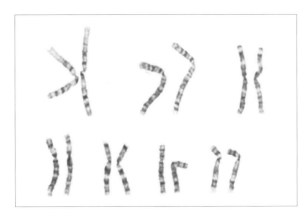

a G-banded human chromosomes. The chromosomes have been stained with Giesma, which reveals alternating light and dark regions called G-bands. The G-band patterns can be used to distinguish one chromosome from another. Most genes are located within the light G-bands.

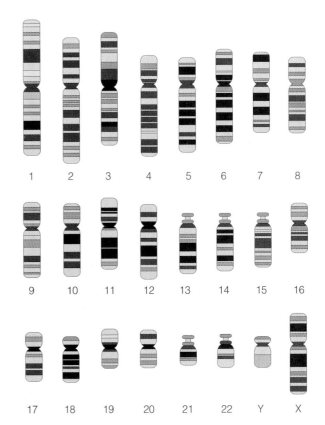

b Diagram of human chromosomes showing the positions of G-bands.

Figure 10.18 G-bands. (From Schnedl, W. 1974. Banding patterns in chromosomes. *International Review of Cytology* suppl. 4:237–272.)

larva's cells need large quantities of proteins. Polytene chromosomes contain many copies of DNA molecules that have replicated several times side by side to provide additional copies of DNA as templates for transcription. The extra copies of DNA cause the chromosomes to expand to an unusually large size, so they can easily be visualized under a microscope. Each chromosome has a detailed banding pattern that permits identification of specific chromosomal regions. The banding pattern can be seen in the segment of a polytene chromosome shown in Figure 10.19. Each band in the *Drosophila* polytene chromosomes has been identified and numbered, and many *Drosophila* genes have been localized to particular bands.

Transcription of individual genes can also be detected in polytene chromosomes. Transcribed genes appear as **chromosome puffs**, regions where decondensed DNA and RNA transcripts make a puffed shape on the chromosome (Figure 10.20).

In Situ Hybridization

As we saw in Chapter 9, labeled DNA probes hybridize to their complementary DNA sequence and make it possible to detect specific fragments of DNA. Researchers can hybridize a labeled probe to a chromosome and detect the site of a gene in the chromosome using a procedure called **in situ hybridization** (*in situ* is Latin for "on site"). Detecting genes through in situ hybridization has been possible for many years with *Drosophila* polytene chromosomes. Because these chromosomes have many side-by-side copies of DNA, there are many DNA copies to which a probe can hybridize. The DNA in polytene chromosomes is denatured into single strands but remains within the chromosome, allowing a labeled probe to hybridize with the DNA. The probe's label shows the chromosomal location of the DNA that is complementary to the probe. For example, Figure 10.21 shows a probe containing DNA from a *β*-tubulin gene hybridized to a polytene chromosome at the position of that gene.

At first glance, this appears to be a good method to apply to other species to find the locations of genes on chromosomes. However, most species do not have polytene chromosomes like *Drosophila*, and when probes are hybridized to metaphase chromosomes that have only a single copy of DNA in each chromatid, only one molecule of the probe hybridizes to each chromatid. The signal emitted by radioactive probes is too weak for in situ hybridization. However, the development of fluorescent labels for probes has made possible **fluorescence in situ hybridization**, often abbreviated as **FISH**. FISH probes carry a nonradioactive fluorescent label that can be detected even in very small quantities using fluorescence microscopy. Figure 10.22 shows the positions of several genes identified on a human chromosome using FISH.

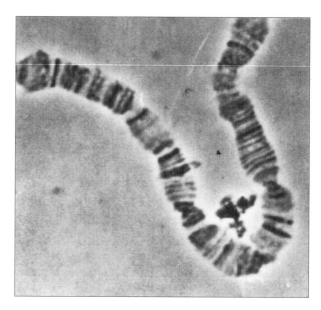

Figure 10.19 A segment of a polytene chromosome from *Drosophila virilis*. (Photo courtesy of D. E. Jeffery.)

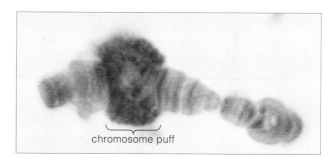

chromosome puff

Figure 10.20 Chromosome puffing in a polytene chromosome from a midge larva. The puff is a site where gene transcription is actively under way. (Photo courtesy of W. Beerman.)

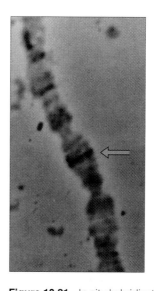

Figure 10.21 In situ hybridization of a probe to a *β*-tubulin gene in a *Drosophila virilis* polytene chromosome. The blue-colored band (indicated with an arrow) is the position where a DNA probe with a nonradioactive label hybridized to its DNA sequence in the chromosome. (Photo courtesy of D. E. Jeffery.)

Euchromatin and Heterochromatin

When cell division is not imminent, the chromosomes are decondensed within the cell nucleus and certain genes are transcribed. In this state, the chromosomes are more condensed in some regions and less condensed in others. Regions of the chromosome that are less condensed are classified as **euchromatin**, and those that are highly condensed are classified as **heterochromatin**. The DNA in euchromatin may be condensed up to the 30 nm fiber level, but it is not as highly condensed as heterochromatin. When nuclei are properly stained for electron or light microscopy, heterochromatin appears as dark regions, whereas euchromatin appears as light regions. This is well illustrated in Figure 10.23, where the dark regions are condensed heterochromatin and the light regions are euchromatin.

Most transcribed genes are in euchromatin. Notice in Figure 10.23 that each of the nuclear pores, where the mRNAs leave the nucleus and enter the cytoplasm (indicated by arrows), is connected with a euchromatic region. Heterochromatin, on the other hand, contains mostly repetitive DNA that is not transcribed.

Facultative heterochromatin is a type of chromatin found in chromosomes that are highly condensed. The term *facultative* implies that the heterochromatin is not invariably in a heterochromatic state. An excellent example of facultative heterochromatin is the condensed X chromosome found in the cells of mammalian females. Mammalian females have two X chromosomes in each cell, whereas males have only one. In order to equalize transcription of genes on the X chromosome in males and females, one of the X chromosomes in each cell of a female is condensed into facultative heterochromatin, and no genes on this condensed chromosome are expressed.

Constitutive heterochromatin consists of highly repetitive DNA that is condensed into heterochromatin in all chromosomes where it is found. At one time, constitutive heterochromatin was thought to be devoid of genes. However, there are several genes embedded within the repetitive sequences of constitutive heterochromatin that surround the centromere in *Drosophila*. In fact, as illustrated in the following example, these genes are transcribed only when they are embedded in heterochromatin. This, however, is an exception rather than the rule. Most genes are located in euchromatin.

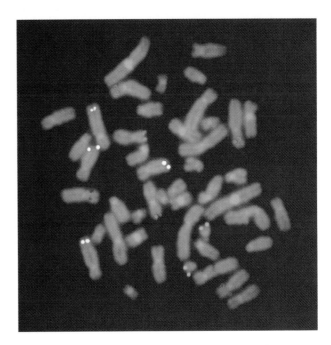

Figure 10.22 Fluorescence in situ hybridization (FISH). The colored spots on the chromosome are the positions where probes with a fluorescent label hybridized to their DNA sequences in the chromosome. In this example, several different probes have been hybridized to their genes. Each probe fluoresces a different color.

tive heterochromatin on expression of the *rolled* gene. The researchers used gamma rays to induce chromosome breakage and rearrangement of chromosome segments, several of which moved the *rolled* gene to another chromosomal position. When the *rolled* gene with a small portion of heterochromatin was moved into a block of euchromatin, gene expression was substantially reduced. However, when a large block of constitutive heterochromatin was moved close to the relocated *rolled* gene, expression was restored.

Problem: What do these results indicate about the effect of constitutive heterochromatin on expression of the *rolled* gene?

Solution: These results indicate that the *rolled* gene requires heterochromatin for normal transcription. This is the opposite of most genes, which must be located within euchromatin for expression.

Example 10.3 Regulation of a gene in centromeric heterochromatin.

The *rolled* gene of *Drosophila melanogaster* is embedded within repetitive DNA in the constitutive heterochromatin that surrounds the centromere of chromosome II. In 1993, Eberl, Duyf, and Milliker (*Genetics* 134:277–292) reported the results of experiments in which they examined the role of constitu-

Constitutive heterochromatin is usually found in the centromeric and telomeric regions and is found only rarely at other chromosomal locations, as illustrated in Figure 10.24. The centromeric and telomeric heterochromatic regions contain highly repetitive DNA that is organized in large blocks. The fact that most constitutive heterochromatin is found in the centromeric and telomeric regions does not mean that repetitive DNA is

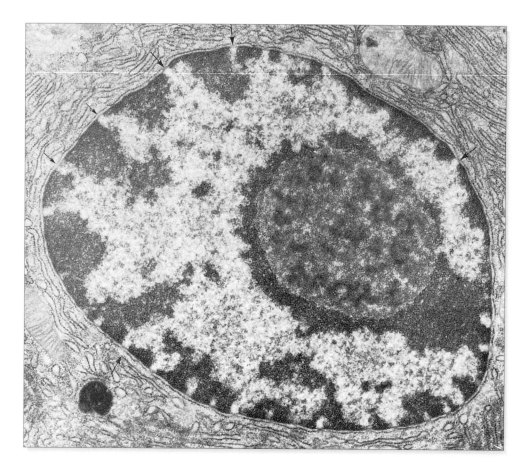

Figure 10.23 Euchromatin (lighter regions) and heterochromatin (darker regions) in the nucleus of a eukaryotic cell. Notice how the nuclear pores (marked by arrows) are connected to euchromatin. (Photo courtesy of S. L. Wolfe.)

found nowhere else. In fact, repetitive DNA is scattered throughout eukaryotic genomes, although usually not in large heterochromatic blocks.

Euchromatic repetitive DNA is often found in **short tandem repeats** (also called **microsatellites**), short repeated sequences, such as di-, tri-, or tetranucleotide sequences, repeated several times in tandem. Short tandem repeats are often found within genes, such as the CGG trinucleotide repeat found in the *FMR1* gene, which is responsible for fragile-X syndrome in humans (see section 5.1). The number of tandem repeats in any particular repeat sequence often varies among individuals.

~

Chromatin may be classified as euchromatin or heterochromatin. Euchromatin is less condensed than heterochromatin and contains most of the genes. Heterochromatin is highly condensed and contains repetitive DNA.

Yeast Artificial Chromosomes (YACs)

Geneticists usually clone DNA in plasmid or phage cloning vectors. A eukaryotic chromosome is many times larger than a plasmid or phage and should be able to carry a large segment of foreign DNA. Genetic engineers have constructed a remarkable cloning vector called a **yeast artificial chromosome (YAC)**, a segment of DNA that replicates as a yeast chromosome when it contains foreign DNA. YACs are valuable because they can carry as much as 500 kb of foreign DNA. YACs are designed to replicate as a plasmid in bacteria when there is no foreign DNA inserted into the vector. Once a fragment of DNA is inserted, YACs are transferred into yeast cells, where they replicate as eukaryotic chromosomes.

YACs contain a yeast centromere and two yeast telomeres. They also have a bacterial origin of replication and bacterial selectable markers derived from bacterial plasmids to allow nonrecombinant YACs to replicate as plasmids in *E. coli* cells (Figure 10.25a). To convert a YAC plasmid to a yeast chromosome, a researcher inserts a large segment of foreign DNA into a unique restriction site (the *Eco*RI site in Figure 10.25) and cleaves the plasmid with another restriction endonuclease (*Bam*HI in Figure 10.25) that removes a fragment of DNA and causes the YAC to become linear. The telomeres are now on the two ends of the linear molecule. Insertion of a large fragment of DNA into the YAC makes the YAC similar in size and structure to yeast chromosomes. Once it is in a yeast cell, the recombinant YAC replicates as a chromosome, replicating the foreign DNA in the process (Figure 10.25b).

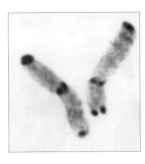

Figure 10.24 Chromosomes stained with a chemical that makes constitutive heterochromatin appear dark. Notice that the dark-stained regions are at the centromeres and telomeres. (Photo courtesy of E. N. Jellen.)

The following example shows how YACs have been used as a research tool to address questions about centromere function.

Example 10.4 Alphoid sequences in centromere formation.

Alphoid sequences are found in the centromeric DNA of human chromosomes and play an important role in centromere structure and function. In 1994, Larin, Fricker, and Tyler-Smith (*Human Molecular Genetics* 3:689–695) reported the results of experiments to determine how introduced alphoid sequences behaved in mammalian cells. A YAC that contained about 120 kb of human alphoid DNA from the Y chromosome was introduced into cultured hamster cells and cultured human cells. A second YAC that contained about 420 kb of noncentromeric human DNA, also from the Y chromosome, was also introduced into hamster and human cell cultures. In each case, the YAC DNA (including the inserted human DNA) integrated into the chromosomes of the cultured hamster and human cells. The integrated alphoid DNA formed a secondary constriction in the chromosomes at the site of integration and interacted with antibodies that were specific for proteins normally found at the centromere. Noncentromeric human DNA sequences introduced into the cells also integrated into the chromosomes but failed to form secondary constrictions and did not interact with the antibodies.

Problem: **(a)** What do these results demonstrate about the function of alphoid DNA? **(b)** What was the purpose of the experiments with YACs that contained noncentromeric human DNA? **(c)** Why were YACs used as the cloning vectors in these experiments? **(d)** Each YAC contains a yeast centromere that allows the YAC to function as a chromosome in yeast cells. Why didn't the YAC centromeres form constrictions in the hamster and human cells?

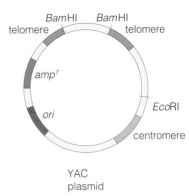

YAC
plasmid

a A nonrecombinant YAC is a circular plasmid that replicates in *E. coli*. It contains a plasmid origin of replication (*ori*), an ampicillin resistance gene (*amp*^{r}), a yeast centromere sequence, and two yeast telomere sequences.

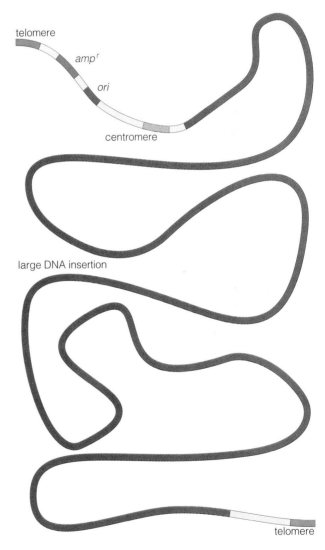

b A large segment of DNA (up to about 500 kb) can be inserted into the *Eco*RI site, and the plasmid can be cleaved with *Bam*HI at the ends of the two telomere sequences. This generates a long linear DNA molecule with telomeres on the ends, a centromere, and a large DNA insertion. The recombinant molecule replicates as a chromosome in yeast cells.

Figure 10.25 Organization of a yeast artificial chromosome (YAC).

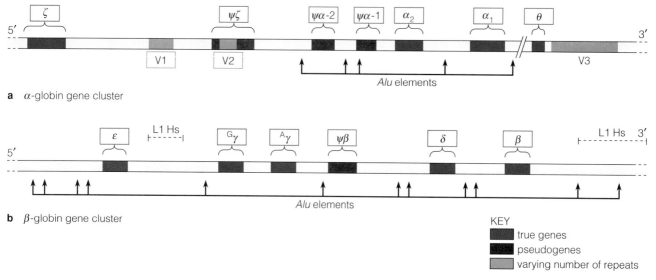

a α-globin gene cluster

b β-globin gene cluster

KEY
- ■ true genes
- ■ pseudogenes
- ■ varying number of repeats

Figure 10.26 Organization of the human globin gene families. Pseudogenes are marked with the symbol ψ. The regions marked with a V are regions where the number of repeated copies of DNA varies among individuals. The arrows point to *Alu* elements that are present in the gene clusters. The locations of L1 elements are also shown.

Solution: **(a)** The introduced alphoid DNA formed a constriction and had centromeric proteins bound to it, both of which are characteristics of a mammalian centromere. **(b)** The YACs that contained noncentromeric human DNA were used as controls. These DNA sequences failed to form constrictions and did not interact with antibodies. These results confirmed that the observations of secondary constrictions and antibody associations in alphoid sequences were true functions of alphoid sequences and not artifacts of the experimental procedure. **(c)** YACs were used because they could accommodate the large fragments of DNA used in these experiments. Alphoid sequences are repeated many times in mammalian centromeres, so the centromeric segment that contained the alphoid sequences was large. **(d)** YACs contain a yeast centromere. The DNA sequences of yeast centromeres differ substantially from the DNA sequences of mammalian centromeres. Therefore, yeast centromeres do not function as centromeres in mammalian cells.

Yeast artificial chromosomes are cloning vectors that carry large fragments of inserted DNA and replicate as chromosomes in yeast cells.

10.4 GENE ORGANIZATION IN CHROMOSOMES

As we saw in Chapter 7, prokaryotic genes that encode enzymes for the same biochemical pathway are often clustered into operons to coordinate gene regulation.

There are no known operons in eukaryotic cells. Instead, a eukaryotic promoter controls a single gene that encodes a single polypeptide. Genes that encode enzymes for the same biochemical pathway may be scattered among several different chromosomes, but the genes may be coordinately regulated by the same regulatory proteins.

Many eukaryotic genes, however, are organized into clusters within a chromosomal region. The genes that encode the large rRNA precursor in eukaryotes are clustered into one or more chromosomal regions known as **nucleolus organizer regions**, or **NORs**. This organization facilitates the transcription of rRNA genes. In humans, there are five NORs, located at secondary constrictions in chromosomes 13, 14, 15, 21, and 22. Other species may have a single NOR, such as maize, where the NOR is located on chromosome 6. Some tRNA genes are also clustered in specific chromosomal regions.

Many mRNA-encoding genes are clustered in groups of similar genes called **gene families**. Gene families are common for genes that specify proteins needed in large amounts. For instance, large amounts of histone proteins are required during DNA replication for nucleosome synthesis. Histone genes are repeated from ten to several hundred times depending on the species. In mammals there are about 20 copies of the histone genes, whereas in *D. melanogaster* there are about 100 copies. In some species, particularly those with many copies, the histone genes are clustered in tandem repeat families into one or more chromosomal regions. In other species, the repeated histone genes may be dispersed among the chromosomes.

The genes for seed storage proteins in plants are also organized into gene families. Most of the proteins in plant seeds are seed storage proteins. When a seed germinates, the growing embryo utilizes the abundant seed storage

proteins as a source of amino acids until the seedling is large enough to begin synthesis of its own amino acids. For this reason, the seed storage proteins must be available in large quantities.

Perhaps the best studied gene families are the globin gene families in vertebrates, which encode the protein subunits of hemoglobin. The genes are clustered into two gene families, called the **α cluster** and the **β cluster**. In humans, the α cluster contains four true genes and three pseudogenes, and the β cluster contains five true genes and one pseudogene (Figure 10.26a).

A **pseudogene** is a sequence of DNA that has the structural elements of a gene but is not expressed as a gene. Pseudogenes serve no function but are still faithfully replicated along with the functional genes each time DNA is replicated. The α-globin and β-globin pseudogenes in humans are similar in DNA sequence to the functional genes, and in some cases include the same introns. However, because of mutations, the pseudogenes are no longer functional. Most of the mutations in pseudogenes are deletions or frameshift mutations that would render the gene product inactive if the gene were transcribed and translated. Pseudogenes are probably relics of evolution. Presumably, they were once active genes but have become inactive because of debilitating mutations, and their functions are now carried out by the functional copies of the genes.

The four transcribed α-globin genes (Figure 10.26a) are expressed at different times during development. The first gene to be expressed is the ζ gene, which is transcribed early during fetal development. Shortly thereafter, still during fetal development, transcription of the ζ gene ceases, and the two α genes are then transcribed to produce α globin. These two genes remain active throughout the adult life of the individual. The θ gene is also located in the α cluster. It has no mutations that should render it inactive, and it might be transcribed, but so far its product has not been found.

The human β cluster (Figure 10.26b) is located on chromosome 11 in humans and includes five transcribed genes and one pseudogene. The ε and γ genes are expressed during embryonic development. During later development of the fetus, transcription of the ε gene ceases, and the two γ genes (Gγ and Aγ) continue to be expressed. At birth, transcription of the two γ genes drops dramatically, and the δ and β genes take over. The β gene supplies about 97% of the total β globin in adult humans, and the δ gene supplies about 2%. Low-level transcription of the γ genes supplies the remaining 1%. The developmental regulation of all the globin genes is summarized in Figure 10.27.

Another interesting feature of the globin gene clusters, illustrated in Figure 10.26, is the large number of repetitive DNA sequences found in the DNA between the genes. Several *Alu* elements reside in both the α and β clusters. Some repetitive sequences vary in number from

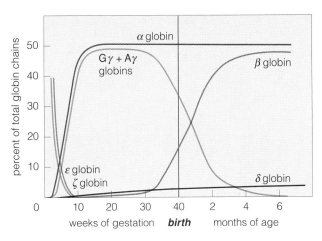

Figure 10.27 Expression of the globin genes during early human development. (Adapted from Cummings, M. *Human Heredity*, 4th ed. Belmont, Calif.: West/Wadsworth Publishing Co.)

one person to another. These repeated sequences are called **variable regions**. The V1 variable region in the α cluster contains 36 nucleotide pairs that are repeated 58 times in some chromosomes but only 32 times in others. The large number of repeated sequences and introns means that only about 8% of the DNA in the globin gene clusters actually encodes mature mRNA.

Gene families are a common feature of many structural genes, particularly when pseudogenes are considered as members of a gene family. Pseudogenes fall into two classes, unprocessed pseudogenes and processed pseudogenes. **Unprocessed pseudogenes** arise through duplication of a gene followed by a mutation that converts a functional gene into a nonfunctional one. Unprocessed pseudogenes typically have all the features of active genes, including promoter sequences and introns. **Processed pseudogenes**, on the other hand, lack introns and in many cases have a long sequence of adenines on the 3' end. Because introns are removed from RNA rather than DNA, and poly (A) tails are found in mRNA but not in the gene's DNA, processed pseudogenes must originate from processed mRNA. They probably arise by reverse transcription, in which reverse transcriptase synthesizes a DNA strand from a mature mRNA template. The reverse-transcribed DNA then inserts itself into the chromosomal DNA and replicates as a pseudogene from that point on.

Let's now compare the DNA sequence of a pseudogene with the DNA sequence of its functional copy.

Example 10.5 Comparison of a pseudogene with a functional gene.

In 1982, Proudfoot and Gill (*Cell* 31:553−563) compared the DNA sequences of the human ζ-globin gene with the ψζ-globin pseudogene, both in the

α-globin gene family. The two genes are highly similar in sequence, with only three differences in the coding region. The pseudogene, however, has significantly larger introns. Figure 10.28 shows the DNA sequences of the two genes in the region of the initiation codon.

Problem: **(a)** Is the ψζ-globin pseudogene a processed or unprocessed pseudogene? **(b)** How many mutations are in the DNA segment of the pseudogene shown in Figure 10.28? **(c)** Which, if any, of these mutations prevents the pseudogene from encoding a functional product?

Solution: **(a)** Processed pseudogenes have no introns. The ψζ-globin pseudogene has introns, so it is an unprocessed pseudogene. **(b)** There are three mutations in the ψζ-globin pseudogene. One is in the untranslated leader sequence, and the other two are within the reading frame. **(c).** The T → C transition in the untranslated leader sequence probably has no effect. At the seventh codon in the reading frame is a G → T transversion that changes a GAG codon into a UAG termination codon. This mutation terminates the reading frame at the seventh codon and prevents the gene from encoding a functional product. The third mutation, an A → G transition, changes an AGG arginine codon into a GGG glycine codon. However, this change is of no consequence because the reading frame has already terminated in the previous codon.

Many eukaryotic genes are organized into gene families, some of which contain pseudogenes.

10.5 WHOLE GENOME ORGANIZATION

Nineteen ninety-six was a landmark year for genetics. This year marked the completion of a collaborative international project to sequence the entire genome of *Saccharomyces cerevisiae* (brewer's yeast), which was the first eukaryote to have its entire DNA sequence determined. Complete genome sequencing efforts are under way in several eukaryotes, including *Drosophila melanogaster, Arabidopsis thaliana* (a plant species), and humans. The projected completion date for determining the entire human genome sequence is 2005.

Complete genome sequencing answers many questions about genome organization. It reveals the total number of genes, how they are organized within the DNA, and how the noncoding DNA is organized relative to the genes. The genome of *S. cerevisiae* has a little more than 13 million nucleotide pairs of DNA in 16 chromosomes. The genome contains 5885 sequences that represent

Figure 10.28 Comparison of DNA sequences in the ζ-globin gene and the ψζ-globin pseudogene. Nucleotides that match one another are in blue; those that do not match are in red. The initiation codons are boxed.

known or potential mRNA genes, 140 rRNA genes, 275 tRNA genes, and 40 snRNA genes. The relative simplicity of the *S. cerevisiae* genome is evident when compared to the human genome, which has 3 billion nucleotide pairs in 23 chromosomes with an estimated 50,000–100,000 mRNA genes.

The genomic DNA sequence of *S. cerevisiae* reveals several important aspects of genome organization. Among the most important is the level of redundancy within a simple eukaryotic genome. The regions near the telomeres (called subtelomeric regions) of several chromosomes share the same sequences and, in some cases, copies of the same genes. For example, several copies of the sugar fermentation genes (some of which are pseudogenes) and a set of 23 similar genes of unknown function called the seripauperines (because they are conspicuously low in serine codons) are found in the subtelomeric regions. An example of the repetition in the subtelomeric regions is evident in the diagram of chromosome I, the smallest of the yeast chromosomes, shown in Figure 10.29.

There are many **cluster homology regions (CHRs)** scattered throughout the yeast genome. A CHR is a region of DNA, with several genes in a particular arrangement, that is repeated at least once somewhere else in the genome. For example, CHRs on chromosomes II and IV contain 13 mRNA genes and 5 tRNA genes that are organized in a similar fashion. These two regions probably arose by duplication of a large segment of DNA. The two sequences have diverged somewhat. One is 120 kb in size, whereas the other is 170 kb. The presence and positions of introns differ in the genes, and different pseudogenes have arisen. Also, transposable elements are found at different sites within the two sequences.

At least some of the redundancy in the yeast genome may be explained by duplication followed by divergence of gene function over evolutionary time. For example, the gene that encodes citrate synthase appears more than once in the genome. One copy of this gene encodes citrate synthase that is transported to mitochondria, where it is utilized. Another copy encodes a slightly different version of the enzyme that is transported to peroxisomes. Gene duplication followed by divergence appears to be a common evolutionary theme in eukaryotic genome organization, as exemplified by the duplication and

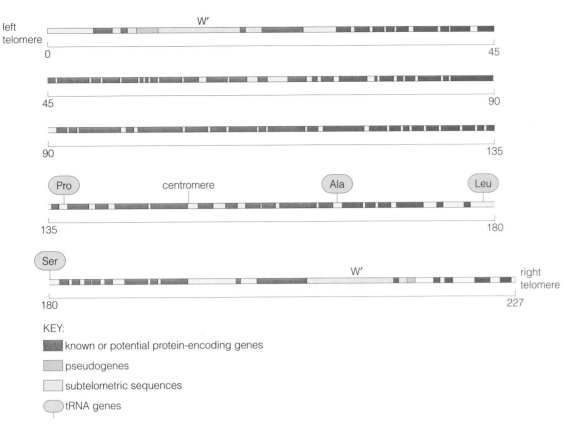

KEY:

■ known or potential protein-encoding genes

▨ pseudogenes

☐ subtelometric sequences

⬭ tRNA genes

Figure 10.29 Organization of chromosome I from *Saccharomyces cerevisiae* as determined from the complete DNA sequence. The scale is in kb. (Adapted from Bussey, et al. 1995. The nucleotide sequence of chromosome I from *Saccharomyces cerevisiae. Proceedings of the National Academy of Sciences, USA* 92:3809–3813. Used by permission.)

divergence of the mammalian globin genes, discussed earlier.

Let's now examine the function of a duplicated segment in the yeast genome.

Example 10.6   Function of a repeated sequence in chromosome I of *Saccharomyces cerevisiae*.

Bussey et al. (1995. *Proceedings of the National Academy of Sciences, USA* 92:3809–3813) summarized several of the organizational features of DNA sequences in chromosome I of *Saccharomyces cerevisiae*. Some of those features are illustrated in Figure 10.29. Chromosome I is the smallest of the yeast chromosomes. The W' regions in the subtelomeric regions of this chromosome contain no genes, yet they account for about 9% of the entire chromosome, a fairly significant portion.

Problem: Researchers have observed that recombinant YACs (which are generally smaller than chromosome I) tend to be unstable. Given this information, what function might the W' sequences in chromosome I have?

Solution: The instability of YACs is probably due to their small size relative to the other yeast chromosomes. The W' regions of chromosome I are probably DNA sequences that extend the size of chromosome I to improve its stability.

Determination of an entire genomic DNA sequence is the first step toward deciphering the complete biology of an organism. Now that the nearly 6000 genes of the yeast genome have been identified, the challenge is to identify the functions of all those genes, a daunting but exciting challenge. Once the genomes of higher eukaryotes have been fully sequenced, the task of identifying the functions of all genes will be even greater. Human genome sequencing has already yielded the complete sequences of genes associated with many genetic disorders. As the functions of these genes are determined, development of more effective treatments for the genetic disorders can begin. As the complete human DNA sequence becomes available over the next several years, the opportunities to discover essential functions in cells, tissues, and the entire body will be greater than at any other time in the history of science.

Entire genome sequencing has been completed in one eukaryotic species (*Saccharomyces cerevisiae*) and will soon be completed in several others. Whole genome sequencing reveals many important aspects about eukaryotic genome organization.

SUMMARY

1. Most eukaryotic genomes are significantly larger than prokaryotic genomes.

2. Eukaryotic genomes may be described in terms of complexity. Repeated sequences contribute only once to a genome's complexity.

3. Eukaryotic genomes have three general classes of DNA sequences: unique (or single-copy) DNA sequences, moderately repetitive DNA sequences, and highly repetitive DNA sequences.

4. A major fraction of eukaryotic genomes contains repetitive DNA.

5. Linear DNA molecules in the nucleus are packaged into chromosomes. Chromosomes are composed of chromatin, an orderly winding of DNA with proteins that hold the DNA in a condensed state.

6. DNA packaging follows a set hierarchical organization. First, the linear double-stranded DNA molecule is wound around cores of histone proteins to form a string of nucleosomes. Second, the nucleosomes coil to form a solenoid. Third, the solenoids loop out from a scaffold of nonhistone proteins. DNA packaging is dynamic: it varies over time and along the length of a chromosome.

7. Metaphase chromosomes have certain structural characteristics that define their size, shape, and function. These include the primary constriction or centromere, chromosome arms, telomeres, secondary constrictions, knobs, and chromosome satellites.

8. Chromosomes have two major types of chromatin: euchromatin and heterochromatin. Euchromatin is generally less condensed and contains nearly all of the active genes. Heterochromatin is more condensed and contains few active genes.

9. Many eukaryotic genes are organized into gene families. A gene family is a series of identical or similar genes that are clustered. The genes within a gene family are often expressed differently at predetermined stages during the organism's development.

10. Whole genome DNA sequencing is under way in several eukaryotic species and has been completed in *Saccharomyces cerevisiae*. Determining the DNA sequence of an entire genome reveals much about genome organization and is the first step toward understanding the complete biology of an organism.

QUESTIONS AND PROBLEMS

1. Determine the complexity in nucleotide pairs of the following DNA sequences: (a) six copies of a 1-million-nucleotide-pair sequence (total = 6 million nucleotide pairs); (b) one copy of a unique 2-million-nucleotide-pair sequence, 200,000 copies of a 5-nucleotide-pair sequence, and 20 copies of a 100,000-nucleotide-pair-sequence (total = 5 million nucleotide pairs); (c) one copy of a unique 2-million-nucleotide-pair sequence.

2. Describe the C-value paradox and how it is related to DNA complexity.

3. The graph in Figure 10.3 represents a genome with A-T and G-C pairs in an equal ratio. Using this graph as a starting point, what would happen to T_m if the DNA were (a) predominantly A-T pairs, and (b) predominantly G-C pairs?

4. In 1970, McConaghy and McCarthy (*Biochemical Genetics* 4:425–446) presented a DNA renaturation curve for the rat genome, which is illustrated in Figure 10.30. What percentage of the rat genome is unique DNA?

5. There are many different repetitive DNA sequences in eukaryotes. Some of these repetitive sequences perform important cellular functions, whereas others have no known function. List as many types of repetitive DNA as you can and describe the cellular functions associated with the repetitive DNA when a function is known.

6. Assuming that an average segment of linker DNA is 50 nucleotide pairs and that the number of nucleotides in nucleosomal DNA is 180, (a) how many nucleosomes should be present in a linear DNA molecule of 1 million base pairs, and (b) approximately how long (in Å) would the DNA molecule be as a linear DNA molecule with no nucleosomes, a 10 nm fiber, and a 30 nm fiber? (Use information from Chapter 2 to determine the length of a nucleosome-free DNA molecule.)

7. Devise an explanation based on the function of linker histones to describe why they vary among species more than other types of histones.

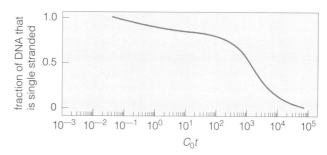

Figure 10.30 Information for Problem 4: DNA renaturation curve for rat genomic DNA. (Adapted from McConaughy, B. L. and B. J. McCarthy. 1970. Related base sequences in DNA of simple and complex organisms. VI. The extent of base sequence divergence among the DNAs of various rodents. *Biochemical Genetics* 4:425–446.)

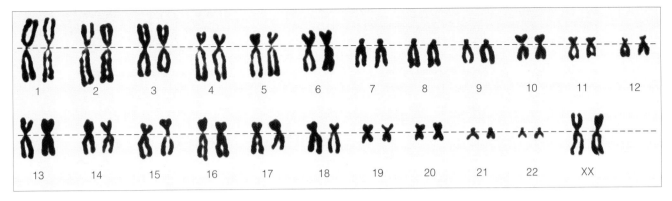

Figure 10.31 Information for Problems 8 and 11: human metaphase chromosomes arranged by size.

8. Figure 10.31 shows a set of human metaphase chromosomes lined up by size. Classify each of the chromosomes as metacentric, submetacentric, or acrocentric.

9. In 1964, Levan, Fredga, and Sandberg (*Hereditas* 52:201–220) proposed a mathematical system for chromosome classification in plants. In this system, chromosomes are classified as metacentric, submetacentric, subtelocentric, or telocentric using the arm ratio, which is the length of the long arm divided by the length of the short arm. If the arm ratio is between 1.00 and 1.70, the chromosome is classified as metacentric. From 1.71 to 3.00, the chromosome is submetacentric. From 3.01 to 7.00, the chromosome is subtelocentric; and greater than 7.00, the chromosome is considered telocentric. In what ways is this system similar to or different from the system used for classifying human chromosomes?

10. How many **(a)** chromosomes, **(b)** chromatids, and **(c)** DNA molecules are in the photograph of human chromosomes shown in Figure 10.31?

11. Are highly repetitive sequences more likely to be found in euchromatin or heterochromatin?

12. Constitutive heterochromatin is usually found in distinct regions of nearly every eukaryotic chromosome. **(a)** What structural features of the chromosome do these regions represent? **(b)** Why do these regions contain constitutive heterochromatin?

13. Describe the most likely origins of **(a)** a pseudogene that contains introns and is located within an active gene cluster, and **(b)** a pseudogene with no introns that is located on a different chromosome than its gene cluster.

14. Pseudogenes often contain many mutations when compared to functional genes. Why do pseudogenes have many mutations rather than just one?

15. Thalassemias are inherited diseases caused by hemoglobin deficiencies; they are due to mutations in the globin genes. Which of the following mutations could cause thalassemia, and at what point during development should the thalassemia arise: **(a)** mutation in the ζ gene, **(b)** mutation in the α_2 gene, **(c)** mutation in the θ gene, **(d)** mutation in the $\gamma\beta$ gene, **(e)** mutation in the δ gene, and **(f)** mutation in the ϵ gene?

16. In most genes, the nucleosomes remain in place during transcription. However, rRNA genes are often devoid of nucleosomes in every cell. Why?

17. Which portions of chromatin are DNase hypersensitive? Why?

18. DNA sequencing studies have revealed many rRNA and tRNA unprocessed pseudogenes. From an evolutionary perspective, why might rRNA and tRNA genes be especially susceptible to conversion into unprocessed pseudogenes?

19. In 1983, Sawada et al (*Nucleic Acids Research* 11:8087–8101) noted that the chimpanzee $\psi\alpha$-globin pseudogene had the same 20-nucleotide-pair deletion as the human $\psi\alpha$-globin pseudogene. There is an *Alu* element in the same position near the $\psi\alpha$-globin pseudogene in both humans and chimpanzee. What do these observations suggest about the origin of the $\psi\alpha$-globin pseudogene?

20. Unprocessed pseudogenes are typically found adjacent to or very near their functional counterparts. Processed pseudogenes, on the other hand, are usually found in another part of the genome than the functional gene. Explain this pattern.

21. A pseudogene is present in the complete *Alu* element. What are the characteristics that make it a pseudogene?

22. In 1996, Goytisolo et al. (*EMBO Journal* 15:3421–3429) described two positively charged sites on linker histones. The researchers found that if mutations altered either of the two sites, the linker histone was no longer able to properly assemble in the nucleosome. They also found that a region of DNA in the nucleosome that is typically protected from DNase digestion was no longer protected when mutant linker histone with one altered site was present. **(a)** What is the most probable function of these positively charged sites on the linker histones, and why are

Table 10.2	Information for Problem 27: Comparison of Four Human Genes Associated with Genetic Diseases and the Functions of Similar Yeast Genes		
Human Disease Associated with Gene	**Human Gene Name**	**Yeast Gene Name**	**Yeast Gene Function**
Hereditary nonpolyopsis colon cancer	MSH2	MSH2	DNA repair protein
Hereditary nonpolyopsis colon cancer	MLH1	MLH1	DNA repair protein
Bloom's syndrome	BLM	SGS1	Helicase
Amyotrophic lateral sclerosis	SOD1	SOD1	Superoxide dismutase

the sites positively charged? **(b)** What does the presence of two of these sites suggest about the role of linker histones in the nucleosome?

23. In 1995, Bernardi (*Annual Review of Genetics* 29: 445–476) reviewed research on the organization of the human genome. He noted that the genome could be divided into AT-rich regions and GC-rich regions and that most genes were located in the GC-rich regions. He also stated that the nucleotides that occupy the third position of codons are preferentially G or C in humans and in many other species. **(a)** Why is a high GC content in the third position of codons expected? **(b)** Is constitutive heterochromatin expected to be AT-rich or GC-rich?

24. Goffeau et al. (1996. *Science* 274:546, 563–567), in reviewing the entire DNA sequence of *Saccharomyces cerevisiae*, noted that the genome could be divided into GC-rich regions and AT-rich regions. At which positions in the chromosome are the AT-rich regions most likely to be found?

25. Compare the three centromeric elements of CEN6 in Figure 10.17 with the consensus sequences. **(a)** What percentages of the CDEI and CDEIII sequences match the consensus sequences? **(b)** What proportion of CDEII is composed of A-T pairs?

26. According to Basset et al. (1996. *Nature* 379:589–590) and Goffeau et al. (1996. *Science* 274:546, 563–567), over half of all human disease genes that have been sequenced are similar to genes found in *Saccharomyces cerevisiae*. What does this observation imply about the importance of whole genome sequencing of model organisms?

27. Table 10.2 (derived from Basset et al. 1996. *Nature* 379:589–590) shows four pairs of *Saccharomyces cerevisiae* and human genes with high amino acid sequence similarity. Given the functions of each of these yeast genes, why is amino acid sequence similarity between the human and yeast genes expected in organisms that are so different? (You may wish to refer to information from Chapter 5 to help you answer this question.)

28. Look at Figure 10.29. The subtelomeric regions of yeast chromosome I contain a repeated sequence called W'. Look at the genes surrounding both copies of W' on chromosome I. What similarities and differences are there in the organization of these two regions?

29. According to Bussey et al. (1995. *Proceedings of the National Academy of Sciences, USA* 92:3809–3813), W' is present in the subtelomeric regions of yeast chromosome I and is also present in the subtelomeric region of the right arm of yeast chromosome VIII. Identify and describe a general feature of the yeast genome that is illustrated by this situation.

30. Even in the smallest eukaryotic genomes, the C value is smaller than the genome size. In most higher eukaryotes, the C value is substantially smaller than the genome size. What can account for the discrepancy between C value and genome size?

31. Genome sequencing projects have revealed that duplication of large blocks of genetic material is common in even the smallest eukaryotic genomes. Thus, significant duplication appears to be a characteristic of all eukaryotic genomes. **(a)** How do these large blocks of duplicated DNA differ from the highly repetitive DNA of heterochromatin? **(b)** What are the various fates of genes contained within duplicated blocks of DNA? **(c)** In what ways is gene duplication an advantage or disadvantage to the organism?

FOR FURTHER READING

A brief but informative view of the functions of DNA sequences that do not encode proteins was published by **Nowak, R. 1994. Mining treasures from "junk DNA".** *Science* 263:608–610. An extensive description of the detection of repetitive DNA in DNA renaturation experiments is **Britten, R. J., and D. E. Kohne 1968. Repeated sequences in DNA.** *Science* 161:529–540. The current understanding of nucleosome structure and the function of histones was

briefly summarized by **Pennisi, E. 1996. Linker histones, DNA's protein custodians, gain new respect.** *Science* 274:503–504. Centromeres and their functions were reviewed by **Sunkel, C. E. and P. A. Coelho. 1995. The elusive centromere: Sequence divergence and functional conservation.** *Current Opinion in Genetics and Development* 5:756–767. Structural features of mammalian centromeres were reviewed by **Rattner, J. B. 1991. The structure of the mammalian centromere.** *BioEssays* 13:51–56. The resurgence of interest in heterochromatin and its function was reviewed by **Lohe, A. R. and A. J. Hilliker. 1995. Return of the H-word (heterochromatin).** *Current Opinion in Genetics and Development* 5:746–755. A description of information gleaned from complete sequencing of the *Saccharomyces cere-* *visiae* genome was published by **Goffeau, A., et al. 1996. Life with 6000 genes.** *Science* 274:546, 563–567. Recent reviews of human genome organization are **Schuler, G. D. et al. 1996. A gene map of the human genome.** *Science* 274:540–546; and **Bernardi, G. 1995. The human genome: Organization and evolutionary history.** *Annual Review of Genetics* 29:445–476.

For additional reading, go to InfoTrac College Edition, your online research library at: http: www.infotrac-college.com/brookscole

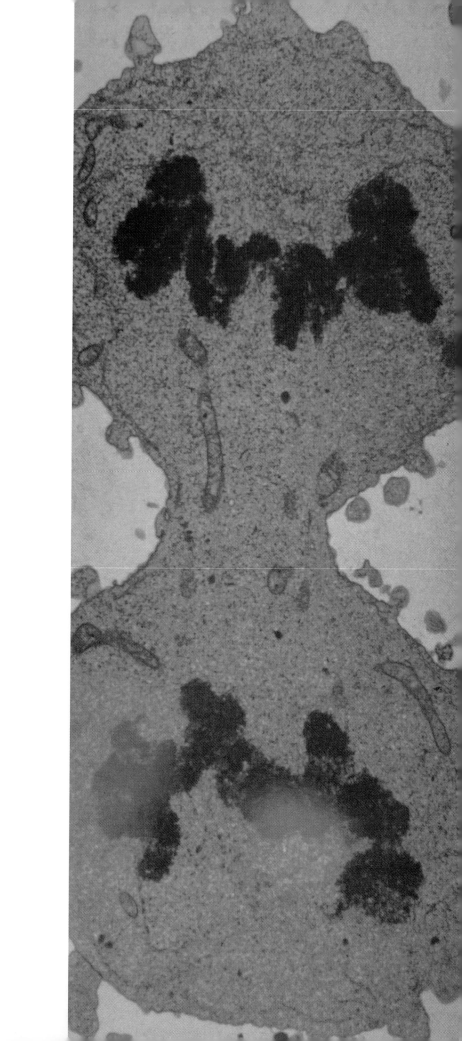

CHAPTER 11

KEY CONCEPTS

Life continues from one generation to the next because cells
and organisms reproduce.

~

A particular species reproduces sexually, asexually, or in both ways.

~

Cells reproduce through mitosis (division of the nucleus) and
cytokinesis (division of the cell).

~

Each cell from a mitotic division receives a complement of chromosomes
that are genetically identical to those in the parent cell.

~

Meiosis consists of two cell divisions that form cells that develop
into sperm or egg cells.

~

In the life cycles of most eukaryotes, a haploid stage alternates
with a diploid stage.

MITOSIS, MEIOSIS, AND LIFE CYCLES

Life perpetuates itself through replication. When your father's sperm united with your mother's egg, you were for a brief time a single-celled organism. Then that single cell prepared to divide into two cells. The DNA replicated, making two identical copies of all the chromosomes in the cell. The cell separated the two identical copies of each chromosome to either side of the cell. With the chromosomes grouped on opposite sides, the cell divided into two genetically identical cells. This process repeated itself over and over to form more cells as you became an embryo, a fetus, and finally an infant ready to be born. Even as you read this, cells within your body are still dividing. The lining of your small intestine is replaced about every 5 days by cells that arise from ongoing divisions. When you cut your finger, cells divide and grow to heal the wound. However, at this point in your life, most cells within your body will never divide again. Instead, they are devoted to carrying out specific functions in the tissues where they reside.

The eukaryotic process of chromosome replication and partitioning during cell division is called **mitosis**. Mitosis followed by the division of the cytoplasm produces two new somatic cells. Every cell in the body, except the germ cells, is a **somatic cell**. For single-celled eukaryotes, such as yeasts and algae, each mitotic division represents a new organismal generation as well as a new cell generation. Some multicelled eukaryotes also produce new organismal generations through mitosis. Potatoes, for example, reproduce vegetatively by mitotic cell divisions. As a potato develops, certain cells divide mitotically and differentiate into tiny embryonic plants. These embryos are located within depressions on the surface of the potato called eyes. For planting, a potato is cut into pieces, each piece containing at least one eye. Then the potato grower plants the pieces in the ground, and the embryo in the eye sprouts into a new potato plant. Because the embryo arose entirely through mitosis, the cells of the new potato plant are genetically identical to the cells of the parent plant.

Reproduction without the union of sperm and egg, such as that of the potato, provides little opportunity for genetic recombination. A species that reproduces entirely by mitosis acquires new genetic combinations primarily through mutations. Sexual reproduction, on the other hand, produces new genetic combinations each organismal generation, because genes from two parents join in unique combinations when gametes (sperm and egg) unite at fertilization. Gametes form in **meiosis**, a process of specialized cell divisions beginning with a somatic cell produced by mitosis. The stages of meiosis ensure that

each gamete receives half the number of chromosomes present in the original somatic cell. Thus, when two gametes unite, their chromosomes added together restore the original chromosome number.

This chapter focuses on mitosis and meiosis, emphasizing how these processes transmit genetic material from one cell generation and from one organismal generation to the next. We also explore the alternation of mitosis and meiosis in the **life cycle**, the span of an organism's life from fertilization to the time it reproduces sexually.

11.1 THE CELL CYCLE AND MITOSIS

As an organism's tissues grow and develop, somatic cells pass through a cycle of events in which chromosomes replicate and divide. To begin the discussion of these events, we need to review some basic aspects of chromosome number and function within the cell.

Chromosome Number and Homology

A typical somatic cell in most animal and many plant species has two similar but usually nonidentical sets of chromosomes. One set of chromosomes came from the paternal parent, the other set from the maternal parent. The somatic cells are said to be **diploid**, a term indicating that there are two complete sets of chromosomes in each cell. For example, the diploid somatic cells of *Drosophila melanogaster* have eight chromosomes per cell. Four came from the paternal parent and four from the maternal parent, so there are four chromosomes in a complete set, eight in two complete sets. The four chromosomes in one complete set of *Drosophila melanogaster* chromosomes have been assigned numbers to distinguish them from one another. Chromosome I includes both the X and Y chromosomes and is usually designated X or Y instead of I. The largest chromosome is chromosome III. The next largest is chromosome II, which is slightly smaller in length than chromosome III. Chromosome IV is very small compared to the others (Figure 11.1).

A female fly has two of each of these chromosomes in her somatic cells. She has, for example, a chromosome II from her paternal parent and one from her maternal parent. These two copies of chromosome II are **homologous chromosomes**, meaning that they are identical in size and structure and are *similar* in nucleotide sequence. But they are probably not *identical* in nucleotide sequence. Typically, each copy of chromosome II carries the same genes at the same locations on the chromosome, but due to mutation throughout many generations, the nucleotide sequences may differ somewhat between the two chromosomes. The same holds true for the other homologous chromosomes: they are similar in size, structure, and nucleotide sequence, but are not necessarily identical. Chromosomes with dif-

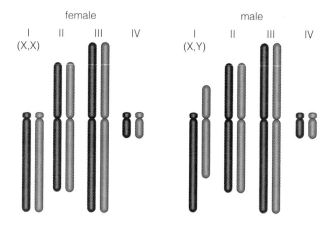

Figure 11.1 The chromosomes of *Drosophila melanogaster*. Chromosomes of maternal origin are purple, and those of paternal origin are blue.

ferent numbers are substantially different from each other and are called **nonhomologous chromosomes**. The two chromosomes of a homologous pair are called **homologues** (sometimes spelled **homologs**).

In *Drosophila melanogaster*, as in many other animals, an individual's sex is determined by chromosome constitution (a topic discussed in detail in Chapter 14). Males have one X chromosome, one Y chromosome, and two of every other chromosome in their somatic cells. The Y chromosome differs in size from the X chromosome and contains different genes, but behaves as if it were the homologue of the X chromosome. In fact, a small portion of the nucleotide sequence in the Y chromosome matches a part of the sequence of the X chromosome.

The number of chromosomes in a gamete (a male or female germ cell) is designated by the term *n* and is called **the haploid chromosome number**. In *Drosophila*, $n = 4$ because there are four chromosomes in a gamete. Each diploid somatic cell has two haploid sets of chromosomes, one set that originally came from the maternal parent and another set from the paternal parent, so it has a chromosome number of $2n$. In *Drosophila*, $2n = 8$. In humans, the haploid set of chromosomes is $n = 23$, so diploid somatic cells have $2n = 46$ chromosomes.

Conventions for chromosome numbering differ slightly from one species to another. Often, chromosomes are numbered by size, starting with the largest chromosome and proceeding down to the smallest. The X chromosome of *D. melanogaster* is designated chromosome I, but the sex chromosomes in many species are not given a number. Human chromosomes are numbered 1 through 22, with no numbering for the X and Y chromosomes. Figure 11.2 shows the human male **karyotype**, a photographic arrangement in which the chromosomes have been ordered by number.

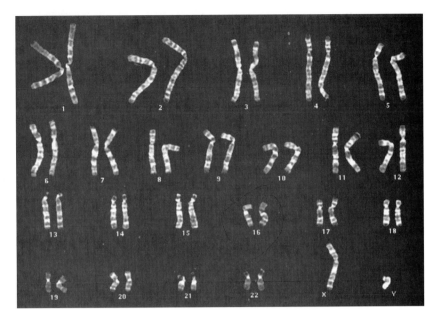

Figure 11.2 The karyotype of a human male. Each of these chromosomes is duplicated at the metaphase stage of mitosis. The homologous chromosomes have been placed next to each other, showing their similar size, shape, and patterns. Notice that, with the exception of the X and Y chromosomes, the chromosomes are numbered in order of decreasing size, although chromosome 21 is actually slightly smaller than chromosome 22 (because chromosome-preparation techniques and microscopy were not sufficiently well developed to accurately distinguish their sizes when numbers were assigned to the chromosomes).(Keryotype courtesy of A. P. Brathman and B. Issa.)

~

A diploid cell contains two sets of chromosomes, one inherited from the maternal parent, the other from the paternal parent. Each chromosome from the maternal parent is similar (but not identical) to a homologous chromosome from the paternal parent.

The Cell Cycle

As chromosomes undergo mitosis and cells divide, a predictable cycle of events called the **cell cycle** occurs. As a somatic cell prepares to divide, it goes through a period of growth called G_1. The G in G_1 is an abbreviation for the word *gap*, meaning that the G_1 phase is the first "gap" in the cell cycle; it comes between mitosis and DNA synthesis. At the end of G_1, DNA replication begins, initiating the **S phase** (S for "synthesis"). The DNA is replicated during the S phase, so each chromosome duplicates into sister chromatids that remain attached at the centromere. The S phase is followed by a second gap called G_2, which comes between DNA synthesis and mitosis. This entire period consisting of the G_1, S, and G_2 phases is called **interphase**, the period that intervenes between one mitosis and the next (Figure 11.3). Most of your cells are arrested in the G_1 phase of interphase, where they carry out their specialized functions. For example, the nerve cells in your brain are held permanently at G_1 and will never divide again, whereas fibroblast cells in your skin

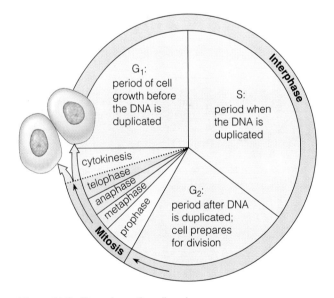

Figure 11.3 The eukaryotic cell cycle.

are held at G_1 but will divide if necessary to repair a wound. Once a cell enters the S phase, it is irreversibly destined to complete the cell cycle back to the G_1 phase in a relatively short period of time.

The times for each phase vary from one species to another and from one cell type to another. With the exception of G_1, which may range from a period of a few

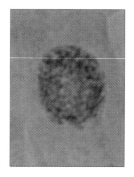
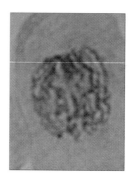

MITOSIS

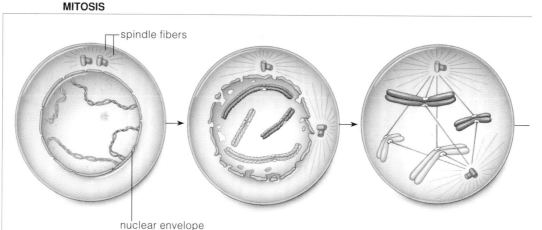

nucleus   pair of centrioles

spindle fibers

plasma membrane   DNA (not condensed)

nuclear envelope

INTERPHASE

As the cell prepares for division, the DNA in the nucleus replicates to make two identical copies of every DNA molecule.

EARLY PROPHASE

The chromatin has started to condense into visible chromosomes. The two larger chromosomes are homologous to each other, as are the two smaller chromosomes. The purple chromosomes are of maternal origin, and the blue chromosomes are of paternal origin.

LATE PROPHASE

The chromosomes continue to condense. Spindle fibers composed of microtubules assemble at the centrioles. The two centriole pairs migrate to opposite poles. The nuclear envelope disintegrates.

TRANSITION TO METAPHASE

Spindle fibers attach to the kinetochores at the centromere of each chromosome and begin to orient the chromosomes at the spindle equator.

Figure 11.4 The stages of mitosis. Light micrographs of the four main stages are included above the drawings.

minutes up to an indefinite period, each stage is usually measured in minutes to hours. For example, cultured mammalian cells typically spend about 5–10 hours in G_1, 7–9 hours in the S phase, 3–4 hours in G_2, and 1 hour in mitosis.

Mitosis

The stages of mitosis are diagrammed in Figure 11.4, which also includes light micrographs of actual cells at each stage. During interphase, the chromosomes are dispersed within the nucleus and usually cannot be distinguished microscopically. The first stage of mitosis is **prophase**. Near the beginning of prophase, the chromosomes (which duplicated during the S phase of interphase) condense sufficiently within the nucleus to be distinguished from each other microscopically. The chro-

mosomes continue to shorten and condense to the point that the sister chromatids can often be seen attached at a single centromere. The nucleolus, where genes encoding rRNA are transcribed, disappears, and transcription of nearly all RNA molecules ceases.

Also during prophase, **spindle fibers**, a collection of microtubules that will later direct the chromosomes to either side of the cell, begin to form outside of the nucleus in the cytoplasm. Eukaryotic cells, except those of vascular plants, have organelles called **centrioles** to which the spindle fibers attach. The centrioles duplicate during interphase to form two pairs. As the spindle fibers form, the centriole pairs separate: one pair migrates toward one side of the cell while the other migrates to the opposite side. The centrioles themselves are composed of microtubules, and microtubules radiate out from the centrioles, forming **asters** (Figure 11.5). In the late stages of pro-

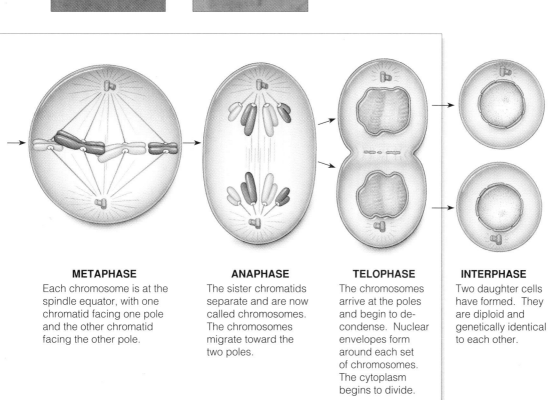

METAPHASE
Each chromosome is at the spindle equator, with one chromatid facing one pole and the other chromatid facing the other pole.

ANAPHASE
The sister chromatids separate and are now called chromosomes. The chromosomes migrate toward the two poles.

TELOPHASE
The chromosomes arrive at the poles and begin to de-condense. Nuclear envelopes form around each set of chromosomes. The cytoplasm begins to divide.

INTERPHASE
Two daughter cells have formed. They are diploid and genetically identical to each other.

phase, the nuclear envelope dissipates so that the chromosomes are no longer contained within the nucleus and may now come into contact with the spindle fibers.

During the transition to metaphase, the cell takes on a distinct form with two poles from which the spindle fibers radiate toward a central axis called the **spindle equator**. The spindle fibers move at random, and some come into contact with the **kinetochores**, proteinaceous regions located on either side of the centromere on each of the duplicated chromosomes (Figure 11.6). The spindle fibers attach to the kinetochores and move the chromosomes toward the spindle equator.

At **metaphase** the chromosomes orient at the spindle equator. In this position, the centromeres are held by the spindle fibers at the equator, and the chromosome arms dangle out from the centromeres. Until late metaphase, the sister chromatids remained attached by specialized protein connections that cause the sister chromatids to adhere to one another throughout their lengths until they separate. This phenomenon, known as **sister chromatid cohesion**, persists through prophase and metaphase. During the transition from metaphase to the next stage, the sister chromatids separate from one another, and each one becomes an independent chromosome. The spindle fibers direct the chromosomes to either side of the cell.

The kinetochores play several roles. As mentioned earlier, they serve as an attachment site for the spindle fibers. Because there is a kinetochore on either side of the centromere, the chromosomes align properly on the equatorial plane during metaphase, with the two kinetochores facing opposite poles. As the chromatids separate from one another, one kinetochore moves toward the pole it faces while the other moves toward its pole, each pulling its attached chromosome with it.

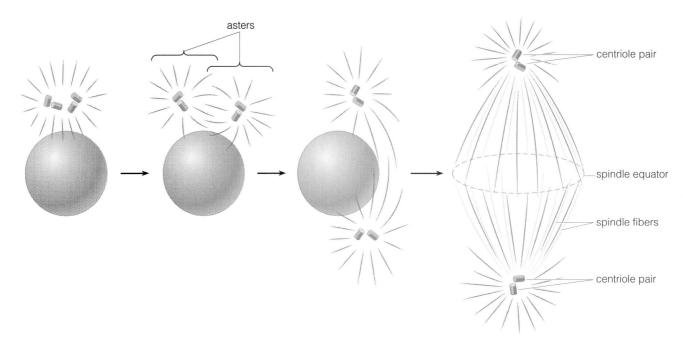

Figure 11.5 Migration of centrioles and asters in formation of the spindle apparatus during mitosis.

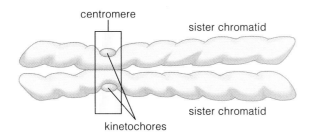

Figure 11.6 Structure of a eukaryotic metaphase chromosome.

Separation of the chromatids marks the beginning of **anaphase**. During anaphase, the spindle fibers, still attached to the kinetochores, pull the centromeres toward the poles. The arms of each chromosome trail behind as they move toward the pole, giving the chromosome mass a fan-shaped appearance in some cells (see the anaphase photograph in Figure 11.4).

When the centromeres reach the poles, the cell is in **telophase**. During telophase, the chromosomes begin to decondense back into their interphase state. Two nuclear envelopes form, one around each of the chromosome masses at either pole of the cell.

Cytokinesis, the division of the cytoplasm, has often started by the time the cell reaches telophase. In plant cells, new membranes and cell walls form between the two new cells. In animal cells, a furrow forms on the surface of the cell and new membranes form, separating the cytoplasm into two complete cells. When the nuclear envelopes have formed and cytokinesis is complete, the cell cycle has gone full circle, and the two new daughter cells are in the G_1 phase of interphase.

During mitosis, homologous chromosomes do not interact with one another. Each is replicated and partitioned as an independent unit. In this way, mitosis ensures that each cell receives one copy of each maternally derived chromosome and one copy of its paternally derived homologue. This noninteraction is one of several important distinctions between mitosis and meiosis. One of the most important events of meiosis, which we will discuss momentarily, is the interaction between homologous chromosomes.

Let's first look at an example of research that delineated one detail in the sequence of events during the cell cycle.

Example 11.1   Replication of centromeric DNA in yeast.

For many years, it was thought that heterochromatin located at or near the centromere was not replicated until late metaphase, and that the unreplicated DNA was responsible for holding sister chromatids together. It was assumed that after this centromeric DNA was replicated, the sister chromatids could separate at anaphase. In 1988, McCarroll and Fangman (*Cell* 54:505–513) reported the results of experiments designed to determine

when yeast centromeric DNA is replicated. They grew yeast cells for many generations in medium containing ^{13}C as the sole carbon source and ^{15}N as the sole nitrogen source, thereby labeling the DNA in the cells with these heavy isotopes of carbon and nitrogen. After the DNA was labeled, the researchers cultured the cells under conditions that arrested the cells between the G_1 and S phases of the cell cycle. The cells were then transferred to medium containing the usual light isotopes of carbon (^{12}C) and nitrogen (^{14}N) and the cells were induced to enter the S phase at the same time, a process called synchronization. The researchers then isolated DNA from cells sampled at various times during the S phase, digested the DNA with restriction enzymes, and separated the fully replicated DNA fragments from the unreplicated DNA fragments using cesium chloride density gradient centrifugation. The DNA fractions were blotted onto nitrocellulose filters and hybridized with centromeric DNA probes to determine which fraction (replicated or unreplicated) contained centromeric DNA. The researchers studied nine different yeast centromeres and discovered that the centromeric DNA from all nine centromeres was fully replicated in the S phase.

Problem: **(a)** Why was it necessary to synchronize the cells in these experiments? **(b)** How could replicated DNA be distinguished from unreplicated DNA using cesium chloride density gradient centrifugation? **(c)** What do the results of these experiments suggest about the hypothesis that unreplicated centromeric DNA holds sister chromatids together during prophase and metaphase?

Solution: **(a)** A single cell provides too little DNA to use for these studies, so DNA had to be derived from samples containing many cells. Cell synchronization ensured that all cells sampled at any particular time during the S phase were at exactly the same stage of DNA replication. **(b)** The unreplicated DNA was heavier than the replicated DNA because the unreplicated DNA contained two strands with heavy isotopes, whereas the replicated DNA consisted of one strand with heavy isotopes and one strand with light isotopes. The unreplicated and replicated DNA could be separated from one another using cesium chloride density gradient centrifugation, similar to the Meselson-Stahl experiments discussed in section 2.3. **(c)** Because centromeric DNA is replicated early in the S phase, something other than unreplicated centromeric DNA must hold the sister chromatids together during prophase and metaphase.

The cell cycle proceeds through the G_1, S, and G_2 stages of interphase and through the prophase, metaphase, anaphase, and telophase stages of mitosis.

11.2 MEIOSIS

Mitosis produces two daughter somatic cells that are genetically identical to their parent cell. Meiosis, on the other hand, produces gametes that have half the chromosome number of their parental somatic cell. While mitosis ensures genetic identity among cells, meiosis contributes to genetic variability by forming new combinations of chromosomes in gametes and new combinations of chromosome segments within a single chromosome.

Meiosis halves the chromosome number in cells by going through two cell divisions with only one round of DNA replication. Each of these two divisions has the same stages as in mitosis: prophase, metaphase, anaphase, and telophase. To distinguish between the two divisions, each stage of the first division is designated by the Roman numeral I, and each stage of the second division by the Roman numeral II. Thus, prophase I is prophase during the first division, and anaphase II is anaphase in the second division. Although the phases in each meiotic division are similar to those of mitosis, the chromosomes behave differently during meiosis.

During development of a multicellular organism, cells divide mitotically. At a certain point, groups of cells become destined to eventually go through meiosis and produce gametes. These cells are called the **germ line**. In all but the simplest plants and animals, the germ-line cells are located in the reproductive organs.

The stages of meiosis are diagrammed in Figure 11.7. As a cell in the germ line prepares to go through meiosis, the DNA replicates essentially as it does during the S phase that precedes mitosis. The cell then proceeds into prophase I of meiosis, where the duplicated chromosomes condense and the nuclear membrane begins to dissipate. Here the differences between mitosis and meiosis become apparent. In mitosis, homologous chromosomes remain independent of one another throughout the entire process. In meiosis, the duplicated homologous chromosomes pair with each other early during prophase I and eventually undergo a close association called **synapsis**, also during prophase I. During synapsis, pairing of homologous chromosomes is very precise. Corresponding nucleotide sequences line up parallel with one another, so that the same genes are aligned.

This precise pairing permits an exchange of DNA molecules between homologous chromosomes called **crossing-over** or **nonsister chromatid exchange**. As illustrated in Figure 11.8, a **crossover** results when two nonsister chromatids exchange segments with each other. The exchange is remarkably precise, down to individual nucleotide pairs. When chromosomes pair precisely,

spindle equator
(midway between
the two poles)

one pair of
homologous
chromosomes

PROPHASE I
Duplicated chromosomes
condense and pair with their
homologues. Nonsister
chromatids in the homologous
pairs exchange segments, a
process called crossing over.

METAPHASE I
Chromosome pairs align at
the spindle equator so that
one homologue faces one
pole and the other
homologue faces the
opposite pole.

ANAPHASE I
Homologous chromosomes
separate from one another
and migrate to opposite
poles.

TELOPHASE I
The chromosomes arrive
at opposite poles, and
the cytoplasm divides to
form two cells.

Figure 11.7 The stages of meiosis.

there is no loss or gain of DNA in either chromosome, simply a direct exchange of homologous chromosome segments. In Chapters 15 and 16, we will take a closer look at the genetic consequences of crossing-over and discuss models that describe at the DNA level how the precision is achieved.

During prophase I, the nuclear envelope dissipates completely, and spindle fibers attach to kinetochores and move the paired chromosomes toward the spindle equator. The chromosomes remain paired in a formation called a tetrad (Figure 11.9) as they line up on the spindle equator during **metaphase I**. The sister chromatids of one homologue orient toward one pole, and those of the other homologue orient toward the other pole.

The sister chromatids do not separate at **anaphase I**, as they did during anaphase of mitosis, but remain attached at the centromere for the remainder of meiosis I. The two homologues of each pair segregate from each other during anaphase I; one homologue goes to one pole while the other homologue goes to the other pole. Thus,

corresponding genes on homologous chromosomes segregate from each other during meiosis I. This **segregation** of homologous chromosomes and their genes during meiosis I is one of the most fundamental aspects of inheritance.

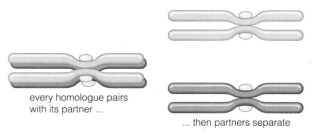

every homologue pairs
with its partner ...

... then partners separate

How one homologous chromosome pair orients at the spindle equator during metaphase I has no influence on the orientation of any other chromosome pair. This means that nonhomologous chromosomes assort independently into the two daughter cells during meiosis I, as illustrated in Figure 11.10. **Independent assortment** of nonhomol-

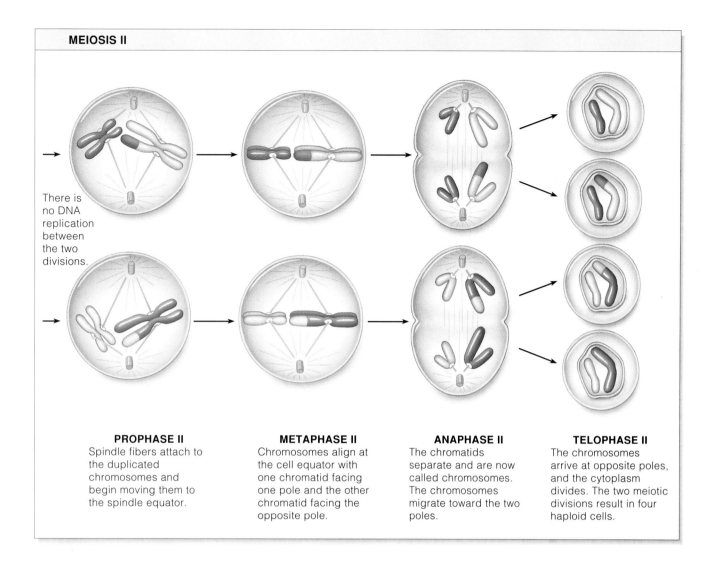

There is no DNA replication between the two divisions.

PROPHASE II
Spindle fibers attach to the duplicated chromosomes and begin moving them to the spindle equator.

METAPHASE II
Chromosomes align at the cell equator with one chromatid facing one pole and the other chromatid facing the opposite pole.

ANAPHASE II
The chromatids separate and are now called chromosomes. The chromosomes migrate toward the two poles.

TELOPHASE II
The chromosomes arrive at opposite poles, and the cytoplasm divides. The two meiotic divisions result in four haploid cells.

ogous chromosomes, like segregation of genes, is an important aspect of meiosis, with consequences for inheritance. It allows for a tremendous number of chromosome combinations in different gametes, 2^n possibilities where n equals the haploid chromosome number. In humans there are $2^{23} = 8,388,608$ possible combinations of chromosomes in the cells that form after the first meiotic division due to independent assortment alone. Now factor in the additional recombinations from crossing-over, and it is highly unlikely that any two gametes from the same individual will be genetically identical.

Following telophase I, the two daughter cells may enter a brief interphase in which the chromosomes unfold and lengthen, although this interphase is virtually nonexistent in some species. No DNA replication occurs during this interphase. As the cells enter **prophase II**, the chromosomes condense again, and the nuclear membrane dissipates. The sister chromatids of each chromosome remain attached at the centromere, meaning that the chromosomes are still duplicated. As illustrated in Figure 11.7, the centriole pair in each daughter cell separates for the assembly of a spindle. The chromosomes line up along the spindle equator during **metaphase II**. At the transition from metaphase II to anaphase II, the sister chromatids separate and become independent chromosomes that are drawn to opposite poles during **anaphase II**. During **telophase II**, nuclear envelopes form around the haploid sets of chromosomes, and the cytoplasm divides to form new cells with a haploid chromosome number.

Meiosis I is sometimes called the **segregational division** because the paired homologous chromosomes segregate from one another during anaphase I. Meiosis II is called the **equational division** because sister chromatids separate from one another. In this sense, meiosis II is similar to mitosis, in which joined sister chromatids also align along the spindle equator and then separate.

Let's now see how the positioning of kinetochores on duplicated chromosomes differs in meiosis I and mitosis.

a Nonsister chromatids in paired homologous chromosomes come into contact with one another.

b The nonsister chromatids exchange segments.

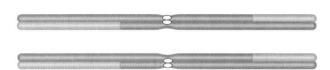

c The resulting chromosomes are a hybrid patchwork of maternal and paternal segments.

Figure 11.8 Crossing-over during prophase I of meiosis. Nonsister chromatids in paired homologous chromosomes exchange segments.

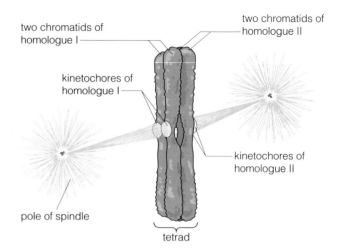

Figure 11.9 The structure of a tetrad during metaphase I of meiosis.

Example 11.2 Positioning of kinetochores during meiosis I.

Miyazaki and Orr-Weaver (1993. *Annual Review of Genetics* 28:167–187) pointed out that the two kinetochores on a mitotic metaphase chromosome are located opposite one another on the centromere. On the other hand, the two kinetochores of meiotic metaphase I chromosomes are both located on the same side of the centromere.

Problem: What is the reason for this difference in kinetochore location?

Solution: In mitosis, the two sister chromatids are pulled apart from one another, each moving to one of the two poles. Throughout meiosis I, however, the two sister chromatids of each chromosome remain attached as a single duplicated chromosome. Chromosome movement during anaphase I of meiosis separates the paired homologous chromosomes, not the sister chromatids. The location of the kinetochores on the same side of the centromere ensures that the two homologous chromosomes will segregate from one another, as illustrated in Figure 11.11.

In cells that will give rise to gametes, meiosis reduces the number of chromosomes to one set. Meiosis proceeds through two cell divisions, each divided into prophase, metaphase, anaphase, and telophase. Nonsister chromatids of paired homologous chromosomes exchange segments during prophase I.

11.3 A CLOSER LOOK AT PROPHASE I

Some of the most critical events in meiosis occur during prophase I, notably chromosome pairing and synapsis. Chromosome pairing is highly important. It provides a way for the cell to segregate homologues from one another. It also permits crossing-over between nonsister chromatids. Because pairing, synapsis, and crossing-over are complicated events, prophase I generally takes longer than any of the other stages of meiosis. In some species, it may last for a very long time. In human females, germ cells may remain in prophase I for many years.

Stages of Prophase I

Chromosomes undergo a series of changes during prophase I that can be represented as stages, as shown in Figure 11.12. The stages in order are leptonema, zygonema, pachynema, diplonema, and diakinesis. The first four stages have corresponding adjectives: leptotene, zygotene, pachytene, and diplotene (for example, a "pachytene chromosome," or the "zygotene stage").

The first stage is **leptonema** (Figure 11.12a), in which the chromosomes begin to condense. The sister chromatids of each chromosome associate tightly with each other, making the chromosomes appear like single threads. The chromatids attach by their telomeres to the nuclear envelope. Homologous chromosomes are not yet synapsed, but they seem to search for homologous regions where they can begin to pair.

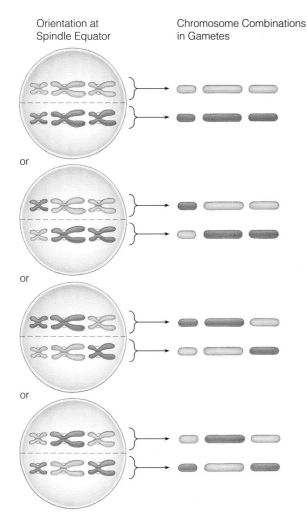

Orientation at Spindle Equator   Chromosome Combinations in Gametes

or

or

or

Figure 11.10 Independent assortment of nonhomologous chromosomes during meiosis I. Because the orientation toward the poles of one chromosome pair has no influence on the orientation of the other pairs, the nonhomologous pairs assort independently. Each of the four combinations is equally likely.

The close association of sister chromatids during leptonema may be due to regions where the sister chromatids are held together because the DNA has not yet replicated. Some of this unreplicated DNA is called **zygDNA**, which stands for "zygotene DNA," because this DNA is replicated during zygonema, the next stage of prophase I. Most of the remaining unreplicated DNA is called **P-DNA**, for "pachytene DNA," because it is replicated during pachynema, the stage that follows zygonema. Portions of zygDNA and P-DNA are transcribed during zygonema and pachy-nema and may produce products that regulate the progress of the cell through prophase I.

During **zygonema** (Figure 11.12b) synapsis begins and chromosomes condense further. Chromosome pairing and synapsis are separate events. Chromosome pairing begins during leptonema and is a physical association of homologous chromosomes. Synapsis follows pairing and may begin at several points where homologous chromo-

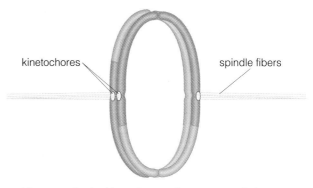

a Alignment of paired homologous chromosomes during metaphase I.

b Segregation of paired homologous chromosomes during anaphase I.

Figure 11.11 Functional positioning of kinetochores on metaphase I and anaphase I duplicated chromosomes.

somes are paired, although in many cases synapsis begins at the telomeres, where the chromosomes are attached to the nuclear envelope. Synapsis is a very tight alignment of homologous chromosome segments in a parallel fashion with a uniform width of about $2\mu m$ between the synapsed chromatids. Within this space is a highly organized structure called the **synaptonemal complex**, which maintains tight association between the nonsister chromatids of homologous chromosomes (Figure 11.13).

By **pachynema** (Figure 11.12c), synapsis extends from one telomere to the other throughout the entire chromosome. Crossing-over takes place during pachynema. Pachytene chromosomes appear as thick threadlike structures, representing the tight association of chromatids in the synaptonemal complex.

Crossing-over creates a structure called a **chiasma** (plural, *chiasmata*), an X-shaped structure formed where two nonsister chromatids of the synapsed chromosomes bind tightly to each other along a short segment. The chiasmata in pachynema can usually be visualized only with an electron microscope. However, during **diplonema** (Figure 11.12d), the chromosomes decondense somewhat, the paired chromosomes begin to separate, except where a chiasma is present, and sister chromatids can be distinguished under a light microscope. The *diplo-* prefix means "double" and refers to the double-thread appearance of the adjacent sister chromatids.

Chromosomes decondense during diplonema to permit transcription, and decondensation varies with cell

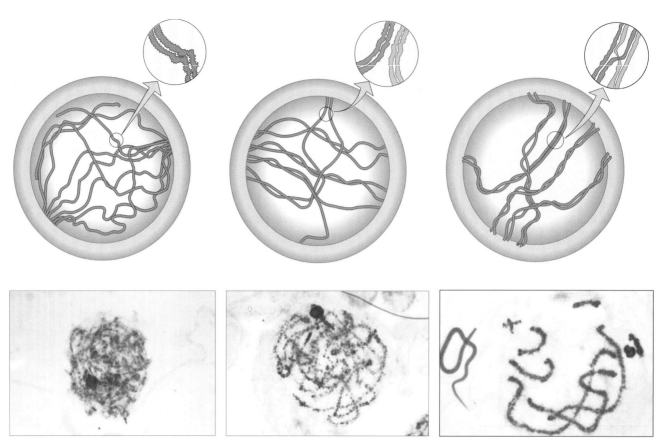

a Leptonema. Chromosomes condense and begin to pair, but do not synapse.

b Zygonema. Chromosome pairs begin to synapse.

c Pachynema. Synapsed chromosome pairs initiate crossing over.

Figure 11.12 The stages of prophase I of meiosis. (Photos courtesy of J. L. Walters.)

type. In cells destined to become eggs, transcription may be extremely intensive and require substantial decondensation, whereas in cells destined to become sperm, decondensation may not be as extensive.

During diplonema, genes are transcribed to provide products needed by the developing gamete. For example, in birds and reptiles, the egg yolk develops during an extended diplotene stage during which there is intensive transcription of rRNA, ribosomal protein genes, and other genes that synthesize the carbohydrates, lipids, and proteins required for nourishment of the embryo. The oocytes of amphibians produce intensively transcribed lampbrush chromosomes, which have loops of DNA, visible in Figure 11.14, that are transcribed at very high levels. (They are called lampbrush chromosomes because, under the microscope, their appearance resembles a lampbrush.) Cells may remain in diplonema for a period of weeks to years. For example, in human females, reproductive cells reach diplonema during fetal development and remain at this stage for years until the cell resumes prophase I just prior to ovulation in a sexually mature female.

After diplonema is complete, the cell passes into **diakinesis** (Figure 11.12e), during which the chromosomes condense once again. The condensation pushes the chiasmata toward the telomeres, a process called **terminalization**, so the chiasmata that can be viewed in diakinesis cells are no longer at the positions where crossing-over took place (Figure 11.15).

Terminalized chiasmata play an important role in chromosome movement during metaphase I in preparation for segregation of paired homologous chromosomes during anaphase I. Paired homologous chromosomes are not attached to one another at the centromeres by the time the chromosomes have reached diakinesis. The homologous chromosomes instead remain associated by chiasma attachments at or near the telomeres. In this state, the spindle fibers move the chromosomes toward the equatorial plane and cause one chromosome to face one pole and its homologue to face the other pole. In experiments in which chiasmata formation is prevented or disrupted, homologous chromosomes fail to properly segregate.

The following example investigates the role of the synaptonemal complex in meiosis and crossing-over.

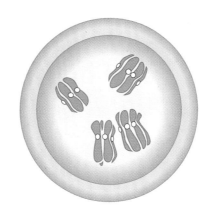

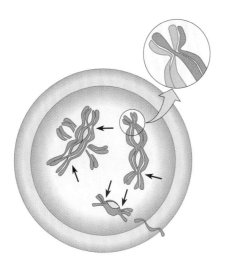

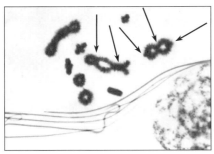

d Diplonema. Sister chromatids and chiasmata become visible with light microscopy. The arrows point to chiasmata.

e Diakinesis. Paired chromosomes condense, and chiasmata terminalize.

The synaptonemal complex consists of two lateral elements and a central element that hold homologous nonsister chromatids in close association beginning in zygonema and persisting through pachynema (see Figure 11.13). In 1993, Sym, Engebrecht, and Roeder (*Cell* 72:365–378) reported the discovery and characterization of a protein called ZIP1 that is part of the central element in the synaptonemal complex of yeast. These researchers discovered that cells carrying the *zip1* mutation fail to produce functional ZIP1. The mutant cells form the lateral elements but not the central element. The homologous chromosomes in these cells pair with one another but fail to synapse during pachynema. Meiosis then ceases during diplonema or diakinesis. The researchers discovered that crossing-over was initiated in the mutant cells before the lateral elements formed, but was not completed. However, if after initiating but failing to complete meiosis, the cells reverted to vegetative growth, the crossovers were completed in the vegetative cells.

Problem: (a) What do these results indicate about the necessity of the synaptonemal complex in meiosis? **(b)** Is the synaptonemal complex required for crossing-over?

Solution: (a) The observation that meiosis ceases late in prophase I in the mutant cells suggests that the synaptonemal complex is essential for completion of prophase I and the remaining meiotic stages. **(b)** The observation that crossing-over was initiated in the absence of a functional synaptonemal complex and could be completed in the vegetative cells suggests that the synaptonemal complex is not essential for crossing-over. However, the synaptonemal complex is probably necessary for the cross-overs to become chiasmata, which enable paired homologous chromosomes to segregate properly in vegetative cells.

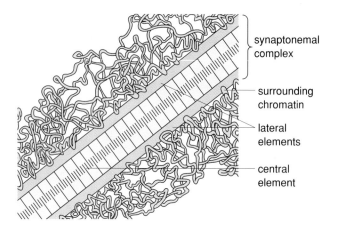

Figure 11.13 Diagrammatic representation of the synaptonemal complex.

synaptonemal complex

surrounding chromatin

lateral elements

central element

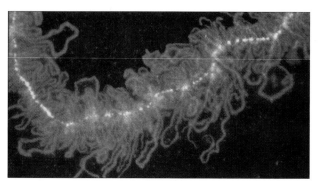

Figure 11.14 A lampbrush chromosome from an amphibian oocyte. The loops represent regions of intensively transcribed DNA. (Photo courtesy of M. Roth and J. Gall.)

Crossing-over

Crossing-over has some very important consequences from a genetic standpoint. As chromosomes are passed from parent to offspring, DNA sequences at different places on the same chromosome are not inextricably linked. Crossing-over permits recombination between homologous chromosomes of maternal and paternal origin. Often there are several crossovers in a single chromosome pair, making the resulting chromosomes hybrid patchworks of paternal and maternal segments. In your somatic cells, each chromosome is of maternal or paternal origin, with one maternal homologue and one paternal homologue for each chromosome pair. However, because of crossing-over during meiosis in your parents, each of your maternally derived chromosomes contains segments inherited from both of your maternal grandparents, and each of your paternally derived chromosomes consists of segments from both of your paternal grandparents. Similarly, the chromosomes you pass on to your children will be composed of segments from both your father and your mother.

Another consequence of crossing-over is the formation of truly unique chromosomes. Chromosomes rarely exchange segments at the same places in two different meiotic events, so every chromosome differs from the corresponding chromosome in other gametes from the same individual. The combined effects of crossing-over and independent assortment make each gamete genetically unique.

~

Prophase I is the stage at which synapsis and crossing-over take place. It is the longest and most complex of the phases in meiosis I and can be divided into five stages: leptonema, zygonema, pachynema, diplonema, and diakinesis.

a Chiasmata form at the points of crossing over.

b The chiasmata migrate away from the centromere.

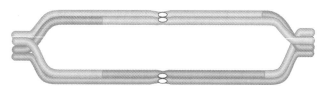

c The chiasmata terminalize at the telomeres and keep the homologous chromosomes paired.

Figure 11.15 Terminalization of chiasmata during diakinesis.

11.4 LIFE CYCLES

In the life cycles of sexually reproducing eukaryotes, there are two major stages: a diploid stage and a haploid stage. The haploid stage follows meiosis, and the diploid stage follows fertilization. In animals and vascular plants, the haploid stage is typically quite short. However, for many unicellular eukaryotes and multicellular fungi, the haploid stage may be the longer of the two stages. In some fungi, meiosis shortly follows fertilization, and the organisms spend the major part of their life cycles in the haploid stage. The haploid cells may divide mitotically for many generations before

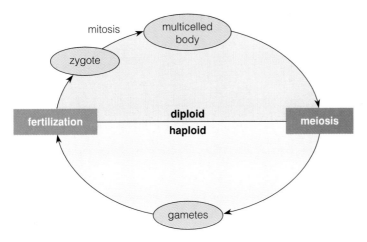

a Generalized life cycle for animals. The haploid stage is short when compared to the diploid stage. There are no mitotic cell divisions in the haploid stage.

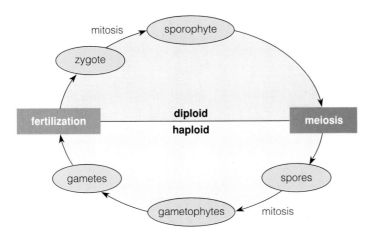

b Generalized life cycle for flowering plants. The haploid stage is short when compared to the diploid stage, but the haploid cells divide a few times mitotically.

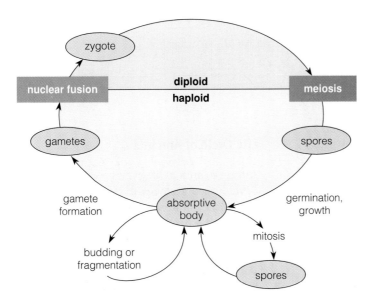

c Generalized life cycle for fungi. The haploid stage is longer than the diploid stage. Haploid cells divide many times mitotically.

Figure 11.16 Generalized life cycles for animals, plants, and fungi.

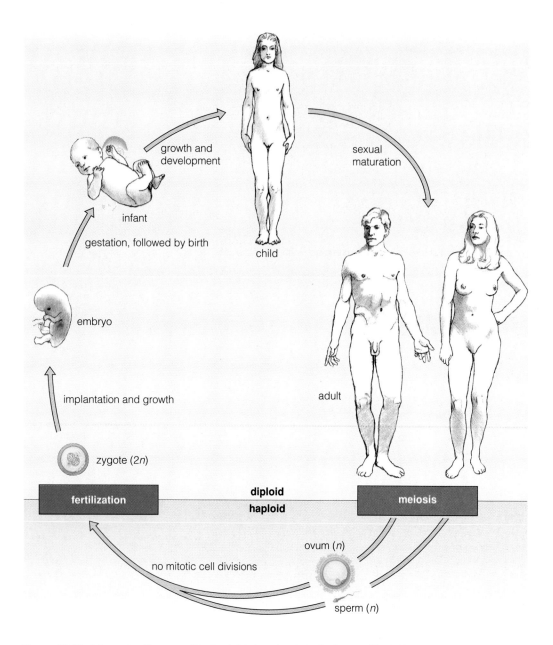

Figure 11.17 Life cycle of humans. The haploid stage is relatively short and includes no mitotic cell divisions.

undergoing fertilization. Because there is no pairing of homologous chromosomes and each chromosome duplicates and separates independently during mitosis, haploid cells can go through mitosis quite easily.

When studying the genetics of a species, researchers must understand the life cycle of that species. The eukaryotic species most used as model organisms in genetic research can be classified as animals, flowering plants, or fungi. Figure 11.16 shows generalized life cycles for these three groups of eukaryotes. To see the variety of life cycles in which mitosis alternates with meiosis, let's review a few examples, starting with animals, which have the shortest haploid stage.

Life Cycle of Animals

Although animal life cycles comprise a great diversity of developmental stages, the underlying alternation of haploid and diploid stages is quite similar among species. Animal life cycles are characterized by a brief haploid stage in which the products of meiosis develop into gametes that unite at fertilization to form diploid zygotes. There are no mitotic divisions of haploid cells. After fertilization, mitotic cell divisions form a multicelled organism during the diploid stage. The diploid stage is usually quite long compared to the haploid stage. The human life cycle, illustrated in Figure 11.17, shows the

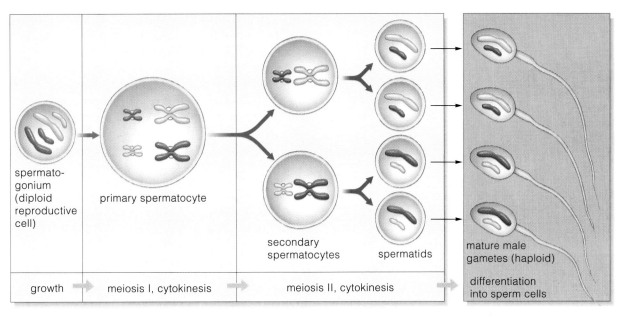

Figure 11.18 Spermatogenesis in animals.

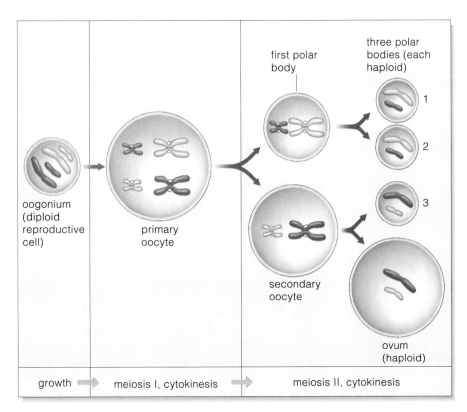

Figure 11.19 Oogenesis in animals.

brief single-celled haploid stage and the long multicelled diploid stage that are typical of animals.

Gamete formation in animals differs in males and females. Sperm cells are formed through **spermatogenesis** (Figure 11.18). A diploid spermatogonium becomes a primary spermatocyte, in which the chromosomes have duplicated.

The primary spermatocyte goes through meiosis I and divides into two secondary spermatocytes. These cells divide at the end of meiosis II to form the haploid spermatids. All four spermatids develop into mature sperm cells.

In **oogenesis** (egg formation), illustrated in Figure 11.19, a diploid oogonium becomes a primary oocyte

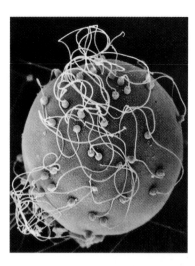

Figure 11.20 Fertilization of a clam ovum by sperm cells. Notice the great difference in size between the ovum and the sperm cells.

with duplicated chromosomes. This cell divides at the end of meiosis I into one secondary oocyte and one polar body. Meiosis I divides the chromosomes equally, but the cytoplasm is unequally divided in the cytokinesis that follows. Most of the cytoplasm goes into the secondary oocyte, leaving only a small amount for the first polar body. Both the secondary oocyte and the first polar body go through meiosis II. When the secondary oocyte divides, most of the cytoplasm goes to one daughter cell, the ovum. The three haploid polar bodies each contain very little cytoplasm. The ovum is the egg, and the three polar bodies are eventually degraded. The ovum is much larger than the sperm cells, as is apparent in the electron micrograph in Figure 11.20, which shows sperm fertilizing a clam egg.

Life Cycles of Flowering Plants

Like animals, flowering plants have a relatively short haploid stage, but the haploid nuclei undergo a few mitotic divisions before the gametes are produced. Figure 11.21 diagrams the life cycle of a flowering plant. Starting with the mature sporophyte (upper left in Figure 11.21), the male and female reproductive organs form within the flower. In the anthers (male reproductive organs), a diploid microspore mother cell goes through meiosis to form four haploid microspores, each of which divides by mitosis to form two nuclei within each pollen grain.

In the meantime, a megaspore mother cell in each ovule goes through meiosis to form four haploid megaspores. Three of the megaspores disintegrate, and the fourth divides through three rounds of mitosis without

cytokinesis to form eight haploid nuclei. The cytoplasm then divides to form seven cells, six with one haploid nucleus each, and one with two haploid nuclei in the mature embryo sac.

When a pollen grain lands on the stigma of a flower (lower left of Figure 11.21), it germinates, forming a pollen tube. One of the two nuclei enters the pollen tube and divides mitotically to produce two haploid sperm nuclei. The pollen tube grows down the style and enters the ovule. One sperm nucleus fertilizes the egg to make a diploid zygote. The other sperm nucleus fertilizes the endosperm mother cell to make a triploid (three copies of each chromosome) endosperm cell. The diploid zygote divides mitotically many times to become the sporophyte, the plant with leaves and flowers. The triploid endosperm remains in the seed.

In the seeds of some species, endosperm cells divide several times by mitosis to form the major portion of the seed. For example, most of a wheat or corn seed is triploid endosperm tissue that contains most of the starch and protein. The tiny embryo within the seed is diploid and grows into the mature plant after the seed germinates (Figure 11.22). Notice that there are two fertilizations, one that forms the zygote and one that forms the endosperm. For this reason, fertilization in flowering plants is called **double fertilization**.

Life Cycles with a Predominant Haploid Phase

Because most of us are familiar with the life cycles of animals and plants, in which the diploid phase dominates, it is often difficult to conceive of organisms with a dominant haploid stage. There are many organisms whose diploid stage is relatively minor, however. The life cycle of the black bread mold (*Rhizopus stolonifer*), diagrammed in Figure 11.23, is one in which the haploid stage makes up most of the life cycle. A diploid cell forms from the fusion of two haploid nuclei. This diploid cell, which is called a zygosporangium, has a protective coating that allows it to remain dormant. When the zygosporangium comes in contact with moisture under the right conditions (such as on a loaf of bread), it immediately undergoes meiosis to form haploid spores. These haploid spores go through rounds of mitosis to generate many haploid cells, which grow as thin filaments called hyphae. The hyphae grow into a network called a mycelium, which looks like a patch of dirty cotton on moldy bread.

Haploid cells have no trouble going through mitosis because each chromosome is independent of all other chromosomes during mitosis. Haploid fruiting bodies may develop by mitosis and release additional haploid spores, which can be carried by the wind over long distances. When they find the right conditions, these spores also germinate and grow mitotically into a haploid mycelium. Black bread mold is ubiquitous on Earth. Its spores are carried by the wind to the most remote parts

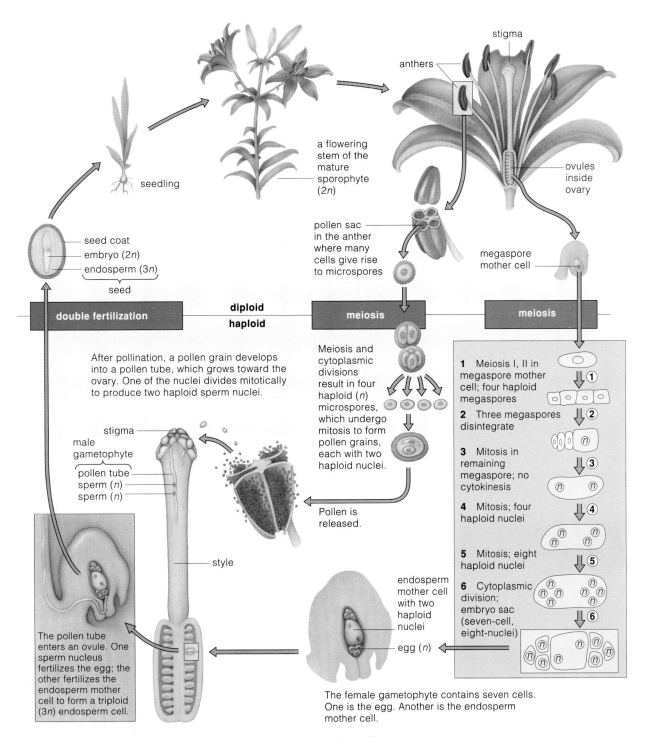

Figure 11.21 Life cycle of lily, a higher plant. Notice that the haploid stage is more predominant here than in animals.

of the planet, including the North and South Poles. When haploid hyphae from mycelia of different mating types come into contact, they fuse to make gametangia between them, which contain several haploid nuclei. The gametangia fuse to form a zygosporangium. Within the zygosporangium, two haploid nuclei fuse to form the diploid zygosporangium, completing the cycle.

Life cycles consist of a haploid stage and a diploid stage. Animals have a short haploid stage in which the haploid products of meiosis develop into gametes with no mitotic divisions. The haploid cells of flowering plants undergo a few mitotic divisions before the gametes develop. Many fungi spend most of their life cycles as haploid cells that divide mitotically many times.

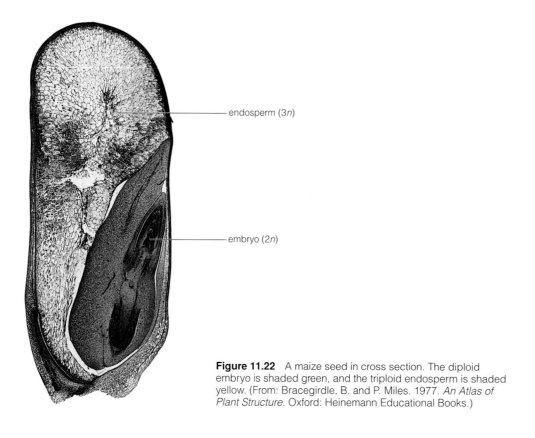

Figure 11.22 A maize seed in cross section. The diploid embryo is shaded green, and the triploid endosperm is shaded yellow. (From: Bracegirdle, B. and P. Miles. 1977. *An Atlas of Plant Structure.* Oxford: Heinemann Educational Books.)

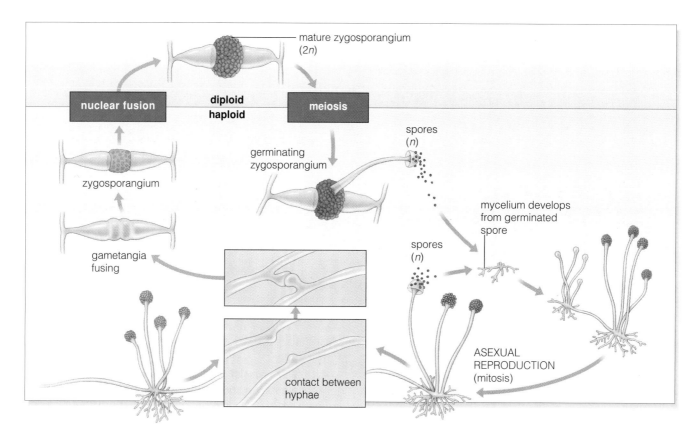

Figure 11.23 Life cycle of the black bread mold (*Rhizopus stolonifer*). Most of the organism's life takes place in the haploid stage.

SUMMARY

1. Organismal reproduction may be sexual or asexual. Sexual reproduction allows for recombination of genetic material at each generation, whereas asexual reproduction does not.

2. The somatic cells of most plants and animals are diploid, which means there are two copies of each chromosome in the nucleus, one copy inherited from the maternal parent, the other from the paternal parent. These two chromosomes are said to be homologous, because they are similar in structure, location of genes, and nucleotide sequence, although there are often some differences in nucleotide sequence.

3. The number of chromosomes in a haploid genome is designated n. A diploid cell has $2n$ chromosomes in the nucleus. Diploid human cells, for example, have $2n = 46$ chromosomes.

4. Conventions for chromosome numbering vary among species, but chromosomes are often numbered consecutively from the largest to the smallest. Human chromosomes are numbered in this manner, except for the X and Y chromosomes, which are not numbered.

5. Somatic cells pass through a series of events called the cell cycle. The cell cycle consists of two main stages: mitosis and interphase, which are divided into substages.

6. Interphase begins with G_1, during which most cells carry out their function. Cells that are not in the process of dividing are typically in the G_1 phase. When a cell prepares to divide, it enters the S phase, during which the DNA is replicated. Following the S phase is a relatively short G_2 phase before the cell enters mitosis.

7. Mitosis begins with prophase, during which the nuclear envelope dissipates and the duplicated chromosomes condense. At metaphase, the duplicated chromosomes align at the equatorial plate of the cell. During anaphase, the duplicated chromosomes separate, with one copy of each chromosome moving to each of the two cell poles. In telophase, the chromosomes reach the poles, and nuclear envelopes re-form, surrounding each set of chromosomes. Cytokinesis (cell division) typically begins during telophase. Mitosis ensures that each daughter cell receives an identical complement of chromosomes.

8. Haploid gametes arise from meiosis, a series of two cell divisions with one round of DNA replication. The two divisions of chromosomes are called meiosis I and meiosis II.

9. Meiosis I begins with prophase I, during which the chromosomes condense, duplicated homologous chromosomes pair with each other to form tetrads, and the nuclear membrane dissipates. At this stage, nonsister chromatids exchange segments. During metaphase I, the tetrads align at the cell equator; and at anaphase, the tetrads separate to each pole. Nonhomologous chromosomes assort independently. The cell is in telophase I when the chromosomes arrive at the poles.

10. Meiosis II begins shortly after telophase I, with an abbreviated (or sometimes nonexistent) interphase. During prophase II, the chromosomes condense, and the nuclear envelope dissipates. The duplicated chromosomes align at the cell equator during metaphase II; and chromatids separate from each other toward opposite poles during anaphase II, becoming true chromosomes. During telophase II, the chromosomes arrive at the poles, and the nuclear envelopes re-form. The cells are now haploid and may develop further into mature gametes.

11. Prophase I is one of the most lengthy and important phases of meiosis. It is subdivided into several stages. During leptonema, the chromosomes condense and begin to pair, but do not synapse. During zygonema, the chromosomes condense further and begin to synapse. By pachynema, chromosome synapsis is complete, and chiasmata form at the sites of crossing-over. During diplonema, chromosomes decondense in order to be transcribed. Diakinesis is the final stage of prophase I, when the chromosomes condense once again and prepare to complete meiosis.

12. Crossing-over recombines genes on paired homologous chromosomes.

13. Life cycles have diploid and haploid stages. Animals typically have a very abbreviated haploid stage and a long diploid stage, whereas the haploid stage is somewhat increased in flowering plants. In fungi, the haploid stage may dominate the life cycle.

QUESTIONS AND PROBLEMS

1. Aphids reproduce by both sexual and asexual means. During the summer season, every aphid is a female that gives birth to live females produced asexually. As the fall approaches, some males are born and reproduce sexually with the females, who then lay eggs to overwinter (survive the winter). Given that aphids are preyed upon very heavily, how might this combination of asexual and sexual reproduction be advantageous?

2. Some of the most troublesome weed plants, such as quackgrass, johnsongrass, and bindweed, reproduce asexually from underground stems and roots. They also reproduce sexually through seeds that may remain dormant (ungerminated but still alive) for years. Given the way humans manage farmland and home gardens, why are these weeds so successful?

3. Describe the differences between (a) meiosis I and mitosis, and (b) meiosis II and mitosis.

4. Cells in the haploid phase of the black bread mold's life cycle undergo repeated normal mitotic divisions (see Figure 11.23). Likewise, after meiosis in higher plants, haploid cells divide by normal mitosis. In the endosperm tissue of higher plants, the triploid cells divide several times by mitosis to form the endosperm tissue. Although they can undergo normal mitosis, haploid and triploid cells cannot undergo normal meiosis, but diploid cells can. What differences in mitosis and meiosis make it possible for haploid and triploid cells to undergo normal mitosis but not normal meiosis?

5. Watermelons are seedless when the sporophyte (the plant itself) is triploid. Using what you learned answering the previous question, describe why this is so.

6. On rare occasions, all paired homologous chromosomes fail to segregate from one another during anaphase I of meiosis, and there is no cytokinesis following meiosis I. Instead, the cell with all chromosomes still in it passes into meiosis II. How many sets of chromosomes would each gamete have when this rare event occurs?

7. What proportion of the chromosomes are of maternal origin in the endosperm tissue of a higher plant?

8. In which of the human cells listed here is the diplotene stage most likely to be extended for the longest period of time: **(a)** primary spermatocyte, **(b)** primary oocyte, **(c)** secondary spermatocyte, **(d)** secondary oocyte, **(e)** spermatid, or **(f)** ovum? Explain your answer.

9. Describe the events that characterize each stage of prophase I.

10. The names of the stages in prophase I are derived from Greek terms. Leptonema is derived from *leptos*, which means "thin"; zygonema from *zygon*, which means "yoke"; pachynema from *pachus*, which means "thick"; diplonema from *diplos*, which means "double"; and diakinesis from *dia*, which means "through," and *kinein*, which means "to move." How do the meanings of the Greek terms correspond with the chromosomal appearance at each of these stages?

11. In somatic cells produced by mitosis, each homologue is either of maternal or paternal origin. Can the same be said of chromosomes found in gametes? Explain why or why not.

12. In meiosis I in the maize plant, 10 tetrads assort independently during anaphase I. If crossing-over is not taken into account, how many different combinations of maternal and paternal chromosomes are possible due to independent assortment?

13. In 1950, Rhoades (*Journal of Heredity* 41:58–67) provided one of the first descriptions of the stages of prophase I using maize meiotic chromosomes as examples. He noted that chiasmata in paired diplotene chromosomes may be located at many different positions along the chromosome arms, but that chromosomes at diakinesis and metaphase I are connected by chiasmata at or near the telomeres. What is the function of the terminalized chiasmata?

14. As chromosomes approach pachynema, occasionally an arm of one chromosome may by chance be trapped between the paired nonsister chromatids of another chromosome that are in the process of synapsing, a phenomenon called chromosome tangling. Chromosome tangling is sometimes evident in zygonema, but is resolved by the time the chromosomes are fully synapsed in pachynema. The experiments of Sym, Engebrecht, and Roeder (1993. *Cell* 72:365–378) that we described in Example 11.3 revealed that chromosome tangling disappeared in both wild-type and *zip*1 mutant cells. What does this observation suggest about the relationship between the synaptonemal complex and chromosome untangling?

15. Rockmill and Roeder (1990. *Genetics* 126:563–574) studied the effect of a mutation called *red*1 in yeast that causes failure of synaptonemal complex formation. They found that the mutant cells had a reduction in crossing-over, but that crossing-over was not eliminated. They also found that paired homologous chromosomes in the mutant cells often failed to segregate properly during anaphase I. This usually resulted in migration of both homologous chromosomes to a single pole. What do these results suggest about the relationship of synaptonemal complex formation, crossing-over, and segregation of paired homologous chromosomes?

16. In another study, Rockmill and Roeder (1994. *Genetics* 136:65–74) described results from another yeast mutation, called *med*1, which caused meiosis to fail. Mutant cells had normal synaptonemal complexes and about 50% as many crossovers as nonmutant cells. However, paired homologous chromosomes often failed to segregate properly during meiosis I. In some cases, sister chromatids segregated to opposite poles during meiosis I. The authors concluded that this mutation causes failure of sister chromatid adhesion. **(a)** What would be the consequences of segregation of sister chromatids during meiosis I? **(b)** What do these results suggest about the role of sister chromatid adhesion?

17. Henderson et al. (1996. *The Journal of Cell Biology* 134:1–12) studied the telomeres of human interphase chromosomes using fluorescence in situ hybridization (see section 10.3) and found that telomeres were heterogeneous in size among chromosomes and among cells. They further found that the telomeres of older cells that had undergone more mitotic divisions were shorter than the telomeres of younger cells that had undergone fewer mitotic divisions. Why are telomeres expected to shorten with each mitotic division? (You may wish to use information from Chapter 2 to help you answer this question.)

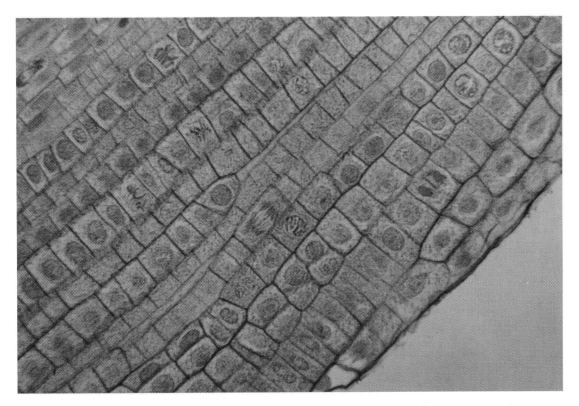

Figure 11.24 Information for question 19: Cells in the tip of an onion root undergoing mitosis.

18. Rhoades (1950. *Journal of Heredity* 41:58–67) noted that a large globular structure is attached to chromosome 6 at prophase I in maize. This structure is the nucleolus. What is the nucleolus, and why is it attached to chromosome 6?

19. Figure 11.24 shows cells in the growing tip of an onion root. Several of the cells are at various stages of mitosis. Identify all those cells that are undergoing mitosis and indicate the stage of mitosis for each of the cells.

FOR FURTHER READING

Detailed discussions of the cellular events in mitosis and meiosis can be found in Chapters 24 and 25 of **Wolfe, S. L. 1993. Molecular and Cellular Biology. Belmont, Calif.: Wadsworth**. Several informative minireviews of current literature on mitosis and meiosis have been published in *Cell*, one of the leading journals in cell biology and molecular biology. Some of the minireviews that are pertinent to this chapter include **Koshland, D. 1994. Mitosis: Back to the basics. Cell 77:951–954; Holm, C. 1994. Coming undone: How to untangle a chromosome. Cell 77:955–957; Carpenter, A. T. C. 1994. Chiasma function. Cell 77:959–962;** and **Hawley, R. S., and T. Arbel. 1993. Yeast genetics and the fall of the classical view of meiosis. Cell 72:301–303.** Sister chromatid adhesion and its role in mitosis and meiosis were reviewed by **Miyazaki, W. T. and T. L. Orr-Weaver. 1994. Annual Review of Genetics 28:167–187.** A classic description of the events in prophase I that remains one of the best articles on the subject is **Rhoades, M. M. 1950. Meiosis in maize. *Journal of Heredity* 41:58–67.**

For additional reading, go to InfoTrac College Edition, your online research library at: http: www.infotrac-college.com/brookscole

PART III

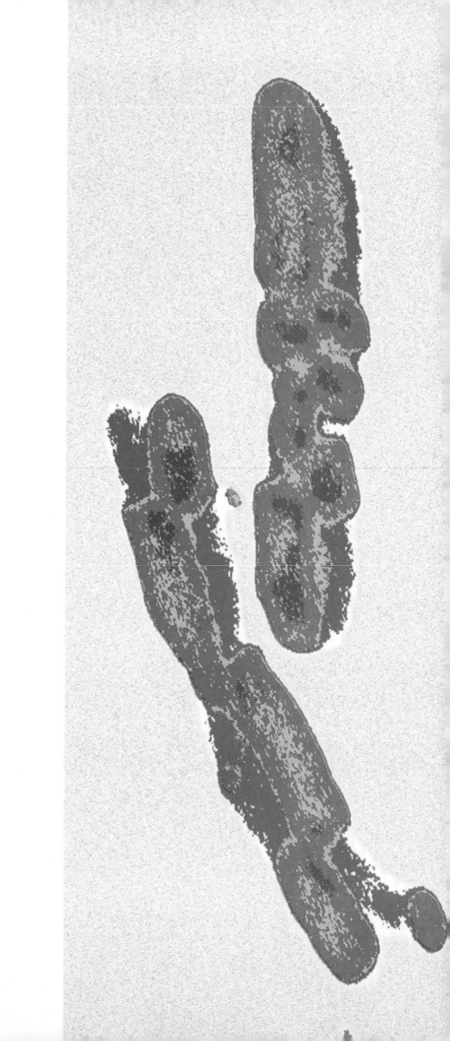

GENETICS OF ORGANISMS

CHAPTER 12

Mendelian principles of inheritance reflect the fundamental patterns
of chromosome partitioning during meiosis and can be explained
at the molecular level.

~

The principle of segregation describes how alleles located on homologous
chromosomes segregate from each other during meiosis.

~

The principle of independent assortment describes how genes on
nonhomologous chromosomes assort during meiosis.

~

Mendelian principles function in accordance with the laws of probability.

MENDELIAN GENETICS

Mendelian genetics, named after Gregor Johann Mendel (see Figure 1.5), who first described the basic patterns of inheritance, is central to genetic analysis. Mendel was an Augustinian monk who spent his scientifically productive years in the St. Thomas Monastery (Figure 12.1) in the city of Brünn (now Brno in the Czech Republic), about 120 km north of Vienna. Mendel is often portrayed as a secluded monk who lived in an out-of-the-way place, had little contact with others, and grew peas simply as a hobby. However, such an image is far from the truth. Brünn was an educational and cultural center with close ties to Vienna and other parts of Europe, and the St. Thomas Monastery played an important role in the intellectual life of the city. The monks at the monastery included scientists, musicians, composers, agriculturists, linguists, theologians, historians, and writers. They served as professors in the local schools and were members of professional societies and on the boards of businesses and financial institutions.

Mendel was among the most active in these pursuits. He had studied physics, mathematics, chemistry, biology, geology, and natural sciences at the University of Vienna and served as a science teacher in Brünn. He conducted experiments with several plant species; cultivated gardens, vineyards, and orchards; kept bees; maintained detailed meteorological records; served on the board of directors of the regional bank; and was a member of several professional societies. He eventually became abbot of the monastery, a position with much prestige and influence.

Mendel's training in physics and mathematics proved invaluable. Among the most important of his contributions to science was his application of

Figure 12.1 The St. Thomas Monastery in Brno, where Mendel conducted his experiments. Part of the monastery is now a museum devoted to Mendel.

mathematics and statistics to the study of plant hybrids and their offspring, an application unique in his day. Although many scientists had written books and articles on plant hybrids prior to Mendel's work, these writings consisted almost entirely of descriptive results with little in the way of interpretation or derivation of theory. Yet Mendel recognized that his studies went beyond the experimental and (in his words) "entered the rational domain."

In the mid-1860s, he wrote what is now considered a classic paper on his studies with peas. After more than 130 years, Mendel's paper remains one of the finest examples in all of science of careful experimental design and interpretation. From his results, he developed a theory of how traits are transmitted and recombined in the progeny of hybrids. His results and his theoretical interpretations of those results provide the basis for several of the most fundamental and important principles of inheritance.

12.1   MENDEL'S EXPERIMENTS

Mendel found predictable patterns among the offspring of pea hybrids. On the basis of his studies, he hypothesized that the determinants of inherited traits remained discrete during hybridization, rather than being blended as was the general opinion among scientists of his day. He also concluded that the offspring of hybrids (in his words) "are variable and follow a definite law in their variations." Two of the most important principles that can be derived from this "definite law" are (1) the segregation of dominant and recessive characters and (2) the independent assortment of characters. The hereditary determinants that Mendel described are now known as genes located on chromosomes. The principle of segregation describes the inheritance of alleles on homologous chromosomes. The principle of independent assortment describes the inheritance of genes on nonhomologous chromosomes.

In this section, we'll review some of Mendel's key experiments, including as much as possible his own terminology, symbols, and interpretations. As we discuss these experiments, try to view the results and interpretations as Mendel might have. Such an exercise reveals much about the process of scientific discovery and inference. In the next section, we'll examine Mendel's principles in a modern context, relating them to our current understanding of molecular genetics and cell biology.

Mendel intentionally chose the garden pea (*Pisum sativum*) as his experimental organism because this species lent itself well to the type of experiments he intended to conduct. Left to themselves, pea plants are self-pollinating, meaning that the pollen produced in the male portion of one flower pollinates the female portion of the same flower. Only with human intervention does pollen from one plant pollinate a flower on a different plant. A plant that produces offspring that all have the same trait as the parent plant is said to be **true breeding** for that trait. Self-pollinating plants are usually true breeding because different alleles from other plants are seldom introduced during the reproduction of a self-pollinating plant. For example, white-flowered pea plants typically produce all white-flowered offspring when self-pollinated, and purple-flowered plants produce all purple-flowered offspring. Mendel chose 22 different pea varieties, each with a different set of traits. To confirm that the varieties were true breeding, he grew them for 2 years and allowed them to self-pollinate, before initiating his hybridization experiments.

From published reports of pea hybridization and his own observations, Mendel selected for study seven traits with contrasting forms. These seven traits and their contrasting forms are (1) seed shape (round or wrinkled), (2) seed color (yellow or green), (3) flower color (purple or white), (4) pod shape (inflated or constricted), (5) pod color (green or yellow), (6) flower position (axial or terminal), and (7) plant stature (tall or dwarf), as summarized in Figure 12.2. The form that each plant exhibits is called its phenotype. For example, there are two phenotypes for the trait of seed color: yellow and green.

Mendel's Monohybrid Experiments and the Principle of Segregation

Although pea plants naturally self-pollinate, a person can cross-pollinate them to obtain hybrid offspring. Mendel selectively hybridized different varieties of plants using the procedure diagrammed in Figure 12.3. We'll use Mendel's experiment with tall and dwarf pea plants as an example. Mendel removed the anthers (the male, pollen-producing organs) from the flower buds of a dwarf plant when the anthers were still immature and had not yet released their pollen, a process called emasculation. Without the pollen-producing anthers, the flowers could not self-pollinate. Mendel then took flowers with ripe anthers from a tall plant and brushed their pollen onto the stigmata (the sticky surfaces that receive the pollen) of the emasculated flowers on the dwarf plant. The pollen from the tall plant's flowers fertilized the egg cells in the emasculated flowers on the dwarf plant.

Mendel's initial experiments were **monohybrid experiments**, in which the two true-breeding parents differ in their phenotypes for only one trait. In the plant stature experiment, he selected as parents varieties that differed in height (tall and dwarf) but had identical phenotypes for the six other traits. For each of the seven traits, he selected varieties that had identical phenotypes for six traits and differed only in one. The varieties that serve as parents for a cross are called the

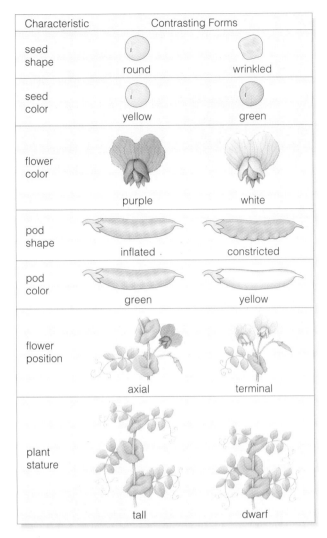

Characteristic	Contrasting Forms	
seed shape	round	wrinkled
seed color	yellow	green
flower color	purple	white
pod shape	inflated	constricted
pod color	green	yellow
flower position	axial	terminal
plant stature	tall	dwarf

Figure 12.2 The seven contrasting traits of pea plants that Mendel studied.

parental generation. The first-generation progeny of a cross between two true-breeding parents is called the **F$_1$ generation** (F$_1$ for "first-filial"). From each of the seven crosses of true-breeding varieties differing in only a single trait, Mendel found that all the F$_1$ progeny had the same phenotype as one of the parents but not the other. For example, from the cross between tall and dwarf plants, he found that all the F$_1$ progeny were tall, just like their tall parent. No dwarf plants appeared among the F$_1$ progeny even though one of their parents was dwarf.

Each F$_1$ individual was a **monohybrid**—a hybrid for one trait. Mendel allowed the monohybrids to self-fertilize. The progeny of F$_1$ individuals that have been self-fertilized or crossed with one another is the **F$_2$ generation.** In the F$_2$ plants produced by the self-fertilization of Mendel's F$_1$ plants, both of the contrasting parental phenotypes appeared in the progeny, and in each case the phenotypes appeared in a similar ratio.

In the plant height experiment, Mendel observed 787 tall F$_2$ plants and 277 dwarf F$_2$ plants, a ratio of 2.84:1.

Table 12.1 summarizes the data that Mendel recorded for the F$_1$ and F$_2$ generations in each of the seven monohybrid experiments. In each experiment, one of the two contrasting phenotypes appeared in all the F$_1$ progeny. The same phenotype then appeared in about ¾ of the F$_2$ progeny, while the phenotype of the other parent appeared in the remaining ¼ of the progeny. The phenotype that appeared in the F$_1$ and in ¾ of the F$_2$ progeny is the **dominant phenotype**. The other phenotype, which was absent from the F$_1$ generation but reappeared in ¼ of the F$_2$ progeny, is the **recessive phenotype**. In the plant stature example, tall is dominant and dwarf is recessive.

Mendel formulated a hypothesis to explain these observations. He hypothesized that each parent carried hereditary factors that determined its phenotype. In each of the monohybrid experiments, one form of the hereditary factors conferred the dominant phenotype, and the other conferred the recessive phenotype. These hereditary factors are what we now call *genes,* and the differing forms of a single gene are *alleles* of that gene. In Mendel's experiments, each gene had two alleles, and one of these alleles conferred the dominant phenotype whereas the other conferred the recessive phenotype. Thus, we can refer to the alleles, as well as the phenotypes, as dominant and recessive. Mendel designated the dominant allele with an uppercase letter and the recessive allele with a lowercase letter. In the plant stature example, the dominant allele for tall plants is *A,* and the recessive allele for dwarf plants is *a.* Mendel represented the F$_1$ hybrid plants as *Aa* because they had inherited an *A* allele from their tall parent and an *a* allele from their dwarf parent. Because the *A* allele is dominant over *a*, all the F$_1$ plants were tall.

Mendel then further developed his hypothesis using information from the cell theory of his day. He assumed that a single sperm cell carried by a pollen grain fertilized a single egg cell to form the zygote. He further assumed that as the individual developed, its somatic cells received the two alleles of each gene, but that during the formation of the next generation of pollen and egg cells, the alleles segregated from one another so that each sperm cell and each egg cell carried only a single allele of each gene. Because an F$_1$ plant carries both *A* and *a* alleles, and these alleles segregate from one another, half of the sperm cells produced by the plant will carry the *A* allele, and half will carry the *a* allele. The same is true of the egg cells. This is **Mendel's principle of segregation**. In Mendel's own words,

> the differing elements succeed in escaping from the enforced association only at the stage at which the reproductive cells develop. In the formation of these cells . . . those [elements] that differ separate from one another.

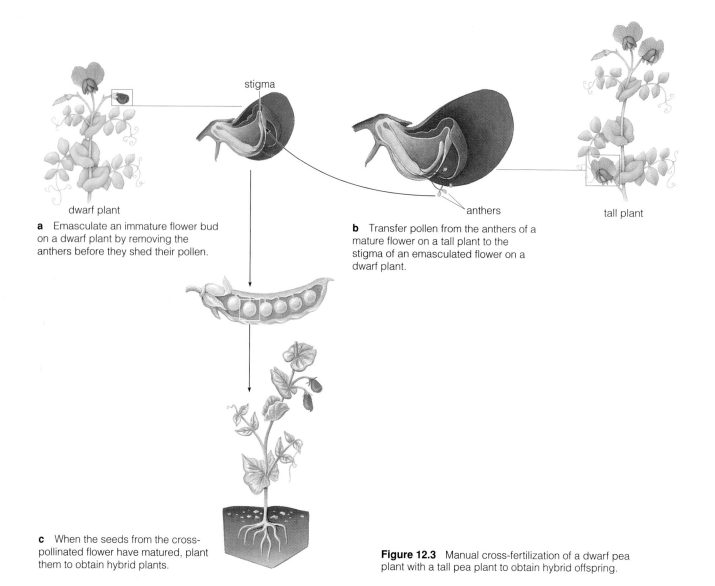

stigma

dwarf plant

a Emasculate an immature flower bud on a dwarf plant by removing the anthers before they shed their pollen.

anthers

tall plant

b Transfer pollen from the anthers of a mature flower on a tall plant to the stigma of an emasculated flower on a dwarf plant.

c When the seeds from the cross-pollinated flower have matured, plant them to obtain hybrid plants.

Figure 12.3 Manual cross-fertilization of a dwarf pea plant with a tall pea plant to obtain hybrid offspring.

After segregation of the alleles into sperm and egg cells, random union of the sperm cells with the egg cells during self-fertilization of the F_1 plants results in a mathematical series of all possible combinations of alleles in the F_2 generation. Mendel represented this series as

$$\frac{A}{A} + \frac{A}{a} + \frac{a}{A} + \frac{a}{a}$$

where the symbols in the numerator of each fraction represent the alleles from the sperm cells and the symbols in the denominator of each fraction represent the alleles from the egg cells. Each of these four possible combinations is equally probable in any single plant of the F_2 generation.

Three of the four combinations have at least one dominant A allele, which causes plants to be tall. One of

the four combinations has only a, which causes the plants to be dwarf. Thus, about three-fourths of the progeny are expected to be tall, and the other one-fourth is expected to be dwarf. The principle of segregation and the random unions of sperm cells with egg cells explain Mendel's observations for all seven traits.

Mendel then conducted a series of experiments to test his hypothesis that alleles segregate during formation of sperm and egg cells. He allowed the F_2 plants to self-fertilize, and he observed the progeny in the F_3 and subsequent generations. He found that all F_2 plants with the recessive phenotype bred true for that phenotype. For example, dwarf F_2 plants produced only dwarf F_3 progeny. He further found that about two-thirds of the F_2 plants with the dominant phenotype produced F_3 progeny with both the dominant and recessive phenotypes, and one-third bred true for the dominant pheno-

Phenotypes of True-Breeding Parents		F₁ Phenotypes	F₂ Phenotypes		F₂ Ratio
Round seeds	Wrinkled seeds	All round seeds	5474 round seeds	1850 wrinkled seeds	2.96:1
Yellow seeds	Green seeds	All yellow seeds	6022 yellow seeds	2001 green seeds	3.01:1
Purple flowers	White flowers	All plants with purple flowers	705 plants with purple flowers	224 plants with white flowers	3.15:1
Inflated pods	Constricted pods	All plants with inflated pods	882 plants with inflated pods	299 plants with constricted pods	2.95:1
Green pods	Yellow pods	All plants with green pods	428 plants with green pods	152 plants with yellow pods	2.82:1
Axial flowers	Terminal flowers	All plants with axial flowers	651 plants with axial flowers	207 plants with terminal flowers	3.14:1
Tall plants	Dwarf plants	All tall plants	787 tall plants	277 dwarf plants	2.84:1

type. For example, in one of his experiments he grew 100 purple-flowered F₂ plants and allowed them to self-fertilize, then planted the seeds produced on these plants. He found that 64 of the purple-flowered F₂ plants produced both purple- and white-flowered progeny, while 36 of them bred true, producing only purple-flowered progeny, close to a 2:1 ratio.

In the series

$$\frac{A}{A} + \frac{A}{a} + \frac{a}{A} + \frac{a}{a}$$

one of the four combinations has only the recessive *a* allele and when self-pollinated, breeds true for the recessive phenotype. Three of the combinations have at least one dominant *A* allele. However, one of these three has only the *A* allele. Therefore, one-third of the F₂ plants with the dominant phenotype breed true. The other two-thirds carry both the dominant *A* allele and the recessive *a* allele. When these plants self-fertilize, both dominant and recessive phenotypes appear among their progeny. Mendel allowed F₂ plants with the dominant phenotype from all seven monohybrid crosses to self-fertilize and found the same pattern in all of these experiments, further confirming the principle of segregation (Table 12.2).

Mendel's Dihybrid Experiments and the Principle of Independent Assortment

The next question Mendel addressed was whether the inheritance of one trait (such as seed shape) influences the inheritance of another trait (such as seed color). To test this question, he conducted a series of **dihybrid experiments**, in which the original parents differed in two traits but were the same for the other five. Mendel conducted dihybrid experiments for all possible combinations of the seven traits, but he reported the results of only one of these experiments. In this experiment, he took plants that were true breeding for yellow, round seeds and crossed them with plants that were true breeding for green, wrinkled seeds. All the F₁ progeny had yellow, round seeds, which he expected because these phenotypes are both dominant. His results for the F₂ progeny were as follows:

Yellow, round	315
Yellow, wrinkled	101
Green, round	108
Green, wrinkled	32
Total	556

Each of the traits alone shows the expected 3:1 ratio. Looking at seed color alone, we see that the number of yellow seeds is 315 + 101 = 416 and the number of green seeds is 108 + 32 = 140, a 2.97:1 ratio. Looking at seed shape alone, we see that the number of round seeds is 315 + 108 = 423 and the number of wrinkled seeds is 101 + 32 = 133, a 3.18:1 ratio. These 3:1 ratios indicate that both traits, now in the same cross, adhere to the principle of segregation.

Now, if seed shape is inherited independently of seed color, then among the yellow seeds there should be a 3:1 ratio of round and wrinkled seeds, and among the green seeds there should also be a 3:1 ratio of round and wrinkled seeds. Among the round seeds, there should be a 3:1 ratio of yellow and green seeds, and among the wrinkled seeds there should also be a 3:1

Trait	F$_2$ Plants Whose F$_3$ Progeny Had Both Dominant and Recessive Phenotypes	F$_2$ Plants Whose F$_3$ Progeny All Had The Dominant Phenotype	Observed Ratio	Expected Ratio
Seed shape	372	193	1.93:1	2:1
Seed color	353	166	2.13:1	2:1
Flower color	64	36	1.78:1	2:1
Pod shape	71	29	2.45:1	2:1
Pod color	60	40	1.50:1	2:1
Flower position	67	33	2.03:1	2:1
Plant stature	72	28	2.57:1	2:1
Pod color (repeat)	65	35	1.86:1	2:1
Totals	1124	560	2.01:1	2:1

ratio of yellow and green seeds. In Mendel's experiment, all of these ratios were indeed close to 3:1 , as shown in Figure 12.4.

This exercise, in which partitioning the results consistently reveals 3:1 ratios, indicates that seed color and seed shape are inherited independently of one another. In other words, the inheritance of seed color has no influence on the inheritance of seed shape. This is known as **Mendel's principle of independent assortment**.

Mendel developed a model that describes how independent assortment operates, illustrated by the results of this experiment. He designated the dominant allele responsible for round seeds A, and the recessive allele responsible for wrinkled seeds a. He designated the dominant allele responsible for yellow seeds B, and the recessive allele responsible for green seeds b. If these alleles assort independently during formation of the

gametes in the F$_1$ plants, then four types of gametes,

$$AB \quad Ab \quad aB \quad ab$$

should be present in equal frequencies in both the sperm and egg cells. Mendel assumed that these four types of gametes unite at random during formation of the F$_2$ generation. This should result in a series of all possible combinations of gametes, which Mendel represented as

$$\frac{AB}{AB} + \frac{AB}{Ab} + \frac{AB}{aB} + \frac{AB}{ab} + \frac{Ab}{AB} + \frac{Ab}{Ab} + \frac{Ab}{aB} + \frac{Ab}{ab} +$$
$$\frac{aB}{AB} + \frac{aB}{Ab} + \frac{aB}{aB} + \frac{aB}{ab} + \frac{ab}{AB} + \frac{ab}{Ab} + \frac{ab}{aB} + \frac{ab}{ab}$$

where the numerators represent the gametes contributed by the sperm cells and the denominators the gametes contributed by the egg cells. There are 16 possible combinations. Nine of them confer yellow, round seeds:

$$\frac{AB}{AB} + \frac{AB}{Ab} + \frac{AB}{aB} + \frac{AB}{ab} + \frac{Ab}{AB} + \frac{Ab}{aB} + \frac{aB}{AB} + \frac{aB}{Ab} + \frac{ab}{AB}$$

Three confer yellow, wrinkled seeds:

$$\frac{Ab}{Ab} + \frac{Ab}{ab} + \frac{ab}{Ab}$$

Three confer green, round seeds:

$$\frac{aB}{aB} + \frac{aB}{ab} + \frac{ab}{aB}$$

And one confers green, wrinkled seeds:

$$\frac{ab}{ab}$$

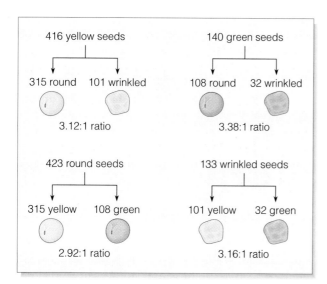

Figure 12.4 Evidence of independent assortment in the data from Mendel's dihybrid experiment.

In other words, the four phenotypes should appear in a 9:3:3:1 ratio.

Let's see how close Mendel's actual results were to this predicted ratio.

Example 12.1 **Matching observed and expected ratios.**

Problem: Mendel's dihybrid cross for the traits seed shape and seed color produced 556 F_2 progeny. What would a perfect 9:3:3:1 ratio be for 556 progeny, and how close are Mendel's observed results to a perfect 9:3:3:1 ratio?

Solution: To determine the numbers of individuals in a perfect 9:3:3:1 ratio, take the total number of individuals and multiply this by each fraction in the ratio. In this case, $556 \times \frac{9}{16} = 312.75$, $556 \times \frac{3}{16} = 104.25$, and $556 \times \frac{1}{16} = 34.75$. When we compare these values with Mendel's results, we see that Mendel's results are very close to a 9:3:3:1 ratio.

Phenotype	Observed	Expected
Yellow, round	315	312.75
Yellow, wrinkled	101	104.25
Green, round	108	104.25
Green, wrinkled	32	34.75

Mendel's experimental results led him to develop mathematical theories to explain the basic principles of inheritance, including the principle of segregation and the principle of independent assortment. Mendel confirmed his theories through additional experimentation.

12.2 THE MOLECULAR BASIS OF DOMINANCE

When Mendel conducted his experiments, nothing was known about the molecular basis of inheritance, and little was known about its cellular basis. Mitosis and meiosis were yet to be discovered, and the connection between DNA and inheritance would not be made for nearly a century. Given the lack of background information available to him, it is a tribute to Mendel's genius that he was able to develop theories of inheritance that remain essentially unchanged today. It is now possible to explain each of Mendel's principles in light of our current understanding of molecular and cellular biology. As we do this, we need to define some important terms that are essential in a modern discussion of Mendelian inheritance. Each term is presented in boldface type. Although most of these terms were used long before DNA was known to be the hereditary material,

they can now be defined using our understanding of DNA and gene expression. We have encountered some of these terms before in this book but will redefine them now in a Mendelian context.

Basic Genetic Terminology

Several of the genes that Mendel studied have now been well characterized. Let's use the gene responsible for flower color to illustrate the molecular basis for dominance. The gene that Mendel studied that governs flower color in peas has been designated *A* and is located on the short arm of chromosome I. When we refer to a gene in terms of its location on a chromosome or within a DNA molecule, we are referring to the gene **locus** (plural, *loci*). Although often used synonymously with *gene*, the word *locus* is more specific because it refers to the gene in terms of its chromosomal location.

A DNA sequence at the *A* locus that has undergone a mutation may no longer encode a functional enzyme. Following a mutation, the locus remains the same, but the DNA sequence at the mutant locus is different. Different DNA sequences at the same locus are called different **alleles** of the same gene. Any change in the DNA sequence of a gene technically creates a new allele, although different alleles are usually identified by the phenotype they cause, such as purple or white flowers. The purple-flower phenotype is due to one allele, the allele that encodes a functional enzyme, and the white-flower phenotype is due to another allele at the same locus, a mutant allele that fails to encode a functional enzyme.

Recall that there are two homologous chromosomes in each somatic cell. In pea plant somatic cells, there are two homologous copies of chromosome I, each with the *A* locus on the short arm. Because each somatic cell has two chromosomes of each type, each diploid individual carries two alleles at each locus. The two alleles can be identical (in having the same DNA sequence at the locus on both homologous chromosomes), or they can be different. In the flower color example, if there is one allele (called *A*) that encodes the functional enzyme and a mutant allele (called *a*) that fails to encode a functional enzyme, then there are three possible combinations of the two alleles:

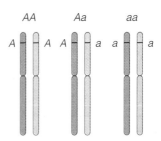

In two cases, *AA* and *aa*, the two alleles on the homologous chromosomes are identical. In both of these

cases, the cell is said to be **homozygous**. The *AA* plants are homozygous for the *A* allele, and the *aa* plants are homozygous for the *a* allele. When the alleles differ at the same locus on homologous chromosomes, as in *Aa*, the cell is **heterozygous**.

True-breeding individuals, such as the plants Mendel used as his original parents, are typically homozygous for the alleles in question. For example, the plants that bred true for purple flowers were homozygous *AA*, whereas those that bred true for white flowers were homozygous *aa*. Mendel's parental-generation cross between purple- and white-flowered plants can be diagrammed as *AA* × *aa*. All the F₁ plants from this cross are heterozygous *Aa* and have purple flowers because *A* is dominant over *a*.

AA and *Aa* both cause flowers to be purple, while *aa* causes flowers to be white. We need a way to distinguish between the genetic composition (*AA*, *Aa*, or *aa*) and what we actually see (purple or white flowers). The **genotype** of an individual is its genetic composition, and its **phenotype** is what we actually see. *AA* and *Aa* are different genotypes because their genetic composition is different, but they cause the same phenotype (purple flowers).

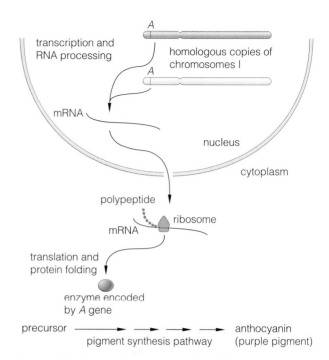

Figure 12.5 Transcription and translation of the *A* gene on chromosome I to produce an enzyme that catalyzes one of the first steps in the synthesis of anthocyanin.

Genotype, Phenotype, and Dominance

Let's now look at the relationship of genotype with phenotype at the molecular level, again using the flower color example. The purple flowers contain a purple pigment called anthocyanin. Anthocyanin is not a protein, meaning that it is not the direct product of the *A* gene. Instead, anthocyanin is synthesized through a series of steps in a biochemical pathway with several genes that encode several enzymes. As illustrated in Figure 12.5, the *A* gene is on the short arm of chromosome I and encodes one of the first enzymes in the pigment synthesis pathway. As long as functional enzymes are present at each step in the pathway, the purple pigment is produced and all the flowers on that plant are purple. However, if an enzyme in the pathway is nonfunctional or absent, the pathway is blocked, and the purple pigment is not produced. When no pigment is present to color the flowers, the flowers are white. The DNA sequence at the *A* locus encodes an enzyme necessary for synthesizing one of the early intermediates in the pigment synthesis pathway. The *A* allele encodes functional enzyme, whereas the *a* allele is a mutant allele that fails to encode functional enzyme.

Let's now see what happens in the pathway for pigment synthesis in each of the three genotypes. If a plant is homozygous for the *A* allele (*AA*), then both homologues of chromosome I contain the *A* allele, which encodes functional enzyme. Because the functional

enzyme is present, the pathway is complete, the pigment forms, and all the flowers on the plant are purple (Figure 12.6a). If a plant is homozygous for the *a* allele (*aa*), then both homologues of chromosome I contain the *a* allele, which fails to encode functional enzyme. Because no functional enzyme is present, the pathway is blocked at the step catalyzed by the enzyme, no pigment is produced, and the flowers on the plant are white (Figure 12.6b). In the heterozygous genotype *Aa*, one homologue of chromosome I contains the *A* allele and the other contains the *a* allele. Because the *A* allele on one of the homologues encodes functional enzyme, the pathway is complete, the full amount of pigment is produced, and the flowers on the plant are purple. The functional enzyme encoded by the *A* allele compensates for the failure of the *a* allele to produce functional enzyme. For this reason, the *A* allele is said to be dominant over the *a* allele, and the *a* allele is said to be recessive to the *A* allele (Figure 12.6c).

This method of naming alleles by upper- and lowercase letters is common and was introduced by Mendel. The letter itself refers to the locus. Dominant alleles are usually written as uppercase letters and recessive alleles as lowercase letters, just as we have done here.

In the following example, we'll look at one of the genes that Mendel studied in detail at the molecular level.

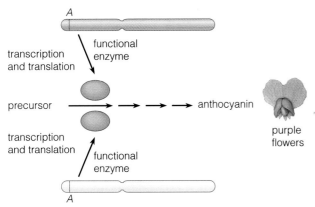

a Homozygous *AA*. The *A* alleles on both chomosomes encode functional enzyme, and anthocyanin is produced. Anthocyanin colors the flowers purple.

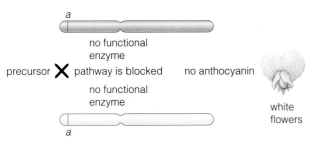

b Homozygous *aa*. The mutant *a* alleles on both chromosomes fail to encode functional enzyme, and no anthocyanin is produced. In the absence of anthocyanin, the flowers are white.

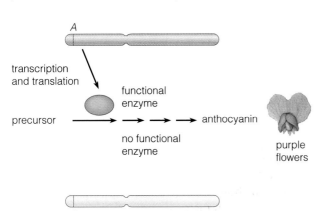

c Heterozygous *Aa*. The *A* allele on one chomosome encodes functional enzyme. The *a* allele on the other chromosome fails to encode functional enzyme. Because functional enzyme is present, anthocyanin is produced, and it colors the flowers purple. The *A* allele is therefore dominant over *a*.

Figure 12.6 The molecular basis for dominance, illustrated by comparison of anthocyanin synthesis in the *AA*, *aa*, and *Aa* genotypes.

Of the seven pea genes that Mendel studied, the best characterized at the molecular level is the *R* locus, which affects seed shape. In 1990, Battacharyya et al. (*Cell* 60:115–122) reported that the *R* locus is the gene that encodes SBEI (starch branching enzyme I), an enzyme that is responsible for synthesis of branched starch molecules during early development of pea seeds. The authors also reported that the mutation that created the recessive allele *r* is an insertion of approximately 800 nucleotide pairs. Comparison of the gene's DNA sequence (Burton et al. 1994. *GeneBank* Accession X80009) with the DNA sequence flanking the mutation reveals that the insertion is in the reading frame, 62 codons upstream from the termination codon. In a 1993 paper, Battacharyya et al. (*Plant Molecular Biology* 22:525–531) reported that the wrinkled-seed phenotype was due to a reduction in branched starch molecules and an increase in the sugar content of the seed. The round seeds contain more branched starch and less sugar than the wrinkled seeds.

Problem: Explain at the molecular level why the *r* allele is recessive.

Solution: The insertion mutation in the *r* allele disrupts the gene within the reading frame and thus prevents production of functional SBEI enzyme. The dominant *R* allele, however, encodes functional SBEI enzyme, so functional enzyme is present in *Rr* heterozygotes. The functional enzyme in the heterozygotes permits the starch branching pathway to proceed, so *Rr* heterozygotes have the full amount of branched starch, which confers the dominant round-seed phenotype. The *r* allele is recessive because *R* compensates for it.

Most dominant alleles encode a functional product, and most recessive alleles are mutant and fail to encode a functional product. When an individual is heterozygous for dominant and recessive alleles, the dominant phenotype appears because the functional product encoded by the dominant alleles compensates for the recessive allele's failure to encode a functional product.

12.3 THE CELLULAR BASIS FOR SEGREGATION

Mendel's principle of segregation is the consequence of chromosome segregation during meiosis. Let's use the flower color example to see how meiosis results in

segregation. Realizing that the true-breeding purple-flowered parent was homozygous *AA* and that the true-breeding white-flowered parent was homozygous *aa*, we can see what happens to these alleles during meiosis and fertilization in a cross between the two in which the purple-flowered plant is the male parent and the white-flowered plant the female parent. In the purple-flowered plant, all sperm nuclei in each pollen grain in every flower on the plant are haploid and carry one copy of the *A* allele. In the white-flowered plant, all eggs in the ovules carry one copy of the *a* allele. (Chromosomes from the male parent are shown as violet, and chromosomes from the female parent are shown as blue.)

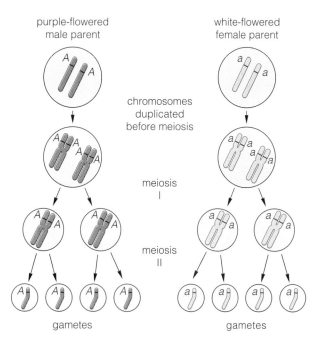

When Mendel pollinated emasculated white flowers with pollen from true-breeding purple-flowered plants, each sperm nucleus had the haploid genotype *A* and each egg had the haploid genotype *a*. With the union of two gametes, the zygote was heterozygous *Aa*. One chromosome in the zygote carried *A* from the paternal parent, and the homologous chromosome carried *a* from the maternal parent. Thus, every F₁ plant from this cross had the genotype *Aa* and purple flowers.

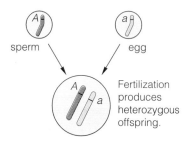

Mendel then allowed the F₁ plants to self-pollinate to produce F₂ progeny. Because each F₁ plant was heterozygous, he was essentially taking a heterozygous male parent and allowing it to pollinate a heterozygous female parent (itself). Had he cross-pollinated two separate heterozygous F₁ progeny plants, the results would have been the same.

Let's see what happens in an F₂ generation. Both male and female F₁ parents are heterozygous *Aa*. In prophase I the chromosome carrying the *A* allele pairs with its homologue carrying the *a* allele. These two chromosomes, now duplicated, segregate from each other during meiosis I, so one of the cells produced after the first meiotic division carries only the *A* allele, and the other cell carries only the *a* allele.

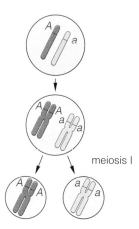

These two cells then go through meiosis II, producing a total of four haploid cells, two with the *A* allele and two with the *a* allele.

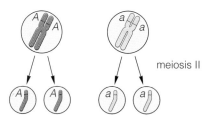

In other words, meiosis ensures that when an individual is heterozygous at any locus, half of its gametes will carry one allele, and half will carry the other.

The union of sperm and egg at fertilization is a random event. Whether an egg has *A* or *a* has no influence on which type of sperm unites with it. This means that all three genotypes (*AA*, *Aa*, and *aa*) are possible in the F₂ zygotes. What proportions of the three genotypes and two phenotypes do we expect in the F₂ generation? One way to predict the outcome of self-fertilization of F₁ plants is to use a **Punnett square**, named after the geneticist Reginald C. Punnett, who first used it. Knowing that each F₁ plant produces *A* and *a* gametes in equal frequencies, and that the union of sperm and egg is a

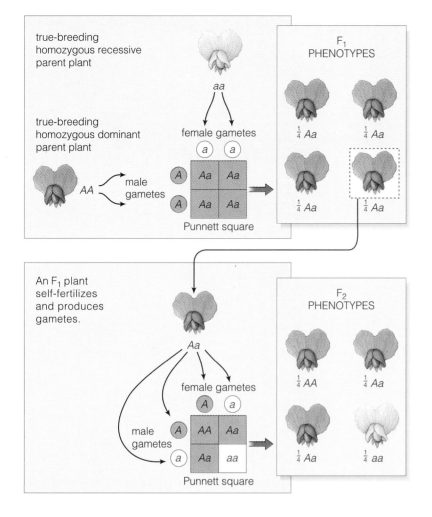

Figure 12.7 The use of Punnett squares to predict the outcomes of genetic experiments when the parental genotypes are known. The genotypes of the offspring are obtained by combining the gamete from each column with the gamete from each row, as shown in the cells of the square.

random event, we can predict the four possible ways in which zygotes may arise from fertilization using the Punnett square shown in Figure 12.7, and we can then determine the genotypic and phenotypic proportions.

According to this Punnett square, there is a 25% chance (one of four equally likely possibilities) that a random union of gametes will produce a homozygous AA zygote, and a 25% chance that it will produce a homozygous aa zygote. Because there are two ways for the heterozygous genotype to arise, there is a 50% chance that a zygote will be heterozygous Aa. Zygotes are heterozygous when the female donates the A allele and the male donates the a allele, and also when the male donates the A allele and the female donates the a allele.

Because AA and Aa plants have purple flowers, we add their probabilities to obtain a 75% probability that an F_2 plant will have the purple-flowered phenotype. Only aa confers white flowers, so there is a 25% chance that an F_2 plant will have the white-flowered phenotype. The Punnett square thus predicts an expected phenotypic ratio of 3:1 (Figure 12.7). Mendel's actual results for this experiment were 705 purple-flowered plants and 224 white-flowered plants in the F_2 generation, a ratio of 3.15:1, very close to the expected 3:1 ratio.

Monohybrid Testcross Experiments

We can use Punnett squares to determine the probable outcomes of other types of crosses as well. Take, for instance, a cross between a purple-flowered, heterozygous Aa pea plant and a white-flowered, homozygous aa plant. As illustrated in Figure 12.8a, each individual in the first-generation offspring from this cross has a 50% chance of being Aa (purple-flowered), and a 50% chance of being aa (white-flowered). This type of cross, called a **monohybrid testcross**, can be used to determine whether a purple-flowered plant is homozygous AA or heterozygous Aa. Only a heterozygous Aa parent can have white-flowered offspring when testcrossed with a homozygous aa plant. When a large number of progeny are produced, about half are purple-flowered Aa and half are white-flowered aa. If the purple-flowered parent is homozygous AA, then all offspring are purple-flowered Aa (Figure 12.8b).

We can now state Mendel's principle of segregation in modern terms:

~

The Principle of Segregation. **Because homologous chromosomes segregate from each other during meiosis,**

alleles at the same locus on homologous chromosomes also segregate from each other, so half the gametes receive one allele and half receive the other.

Let's now look at a mouse testcross experiment to see how geneticists can draw conclusions about an individual's genotype from its testcross progeny.

Example 12.3   Testcrossing.

In mice, the *B* allele causes a black coat color when homozygous *BB* or heterozygous *Bb*. The *b* allele causes a brown coat color when homozygous *bb*. In 1918, Detlefsen (*Genetics* 3:599–607) reported the results of numerous genetic experiments with mice. In one experiment, a black female mouse was mated with a brown male mouse and produced a litter of three black pups and one brown pup.

Problem: (a) What are the genotypes of the parents? **(b)** Why is the ratio 3:1 in the offspring of this cross?

Solution: (a) Because the male is brown, his genotype must be *bb*. The female must be *Bb* because she had a brown pup. If she were *BB*, then all her offspring would be black. **(b)** A black:brown ratio close to 1:1 is expected in the offspring of the cross *Bb* × *bb*, provided that a large number of progeny are produced. The fact that the actual ratio was 3:1 is not especially unusual because there were only four progeny. Had the number of progeny been much larger, such as 40, then an observed ratio of 3:1 would be much less likely.

Parental Equivalence

In each experiment, Mendel used plants of each genotype as both male and female parents. For example, in crossing true-breeding purple-flowered plants with true-breeding white-flowered plants, he took pollen from a purple-flowered plant and used it to pollinate a white-flowered plant. He then performed the **reciprocal cross**, in which he took pollen from a white-flowered plant and used it to pollinate a purple-flowered plant. For each of the seven traits he studied, the same patterns appeared in the progeny of reciprocal crosses, indicating that the segregation of alleles was the same in both male and female parents. This allows us to derive an additional principle from Mendel's results and interpretations that is an extension of the principle of segregation. Stated in modern terms:

~

The Principle of Parental Equivalence. **In the formation of both male and female gametes, segregation of alleles is the same.**

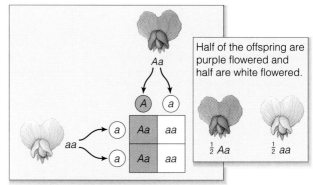

a A monohybrid testcross between a heterozygous *Aa* purple-flowered plant and a homozygous *aa* white-flowered plant.

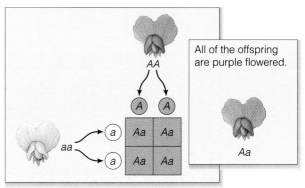

b A monohybrid testcross between a homozygous *AA* purple-flowered plant and a homozygous *aa* white-flowered plant.

Figure 12.8 Punnett squares of monohybrid testcrosses.

The principle of parental equivalence does not apply to all genes. For genes located on sex chromosomes (such as the human X and Y chromosomes), the pattern of inheritance may differ depending on whether the allele in question is contributed by the female or the male parent. We will discuss the inheritance of genes on sex chromosomes in Chapter 14. Also, in mammals and possibly other species, the expression of certain genes is influenced by which parent contributes the gene, a phenomenon known as imprinting. We will discuss imprinting and its implications in Chapter 17. In addition, genes located in the DNA of mitochondria and chloroplasts are not subject to meiosis, so the traits influenced by these genes may not conform to the principle of parental equivalence. Genes inherited on mitochondrial or chloroplastic DNA fall under the category of extranuclear inheritance, which is the topic of Chapter 18.

12.4   THE CELLULAR BASIS OF INDEPENDENT ASSORTMENT

Mendel's principle of independent assortment of genes can be explained by the independent assortment of nonhomologous chromosomes during meiosis. Let's use Mendel's dihybrid experiment for the traits of seed color and seed shape to illustrate this idea. The locus responsible for seed color is called *I* and is located on chromosome I. The dominant allele *I* causes seeds to

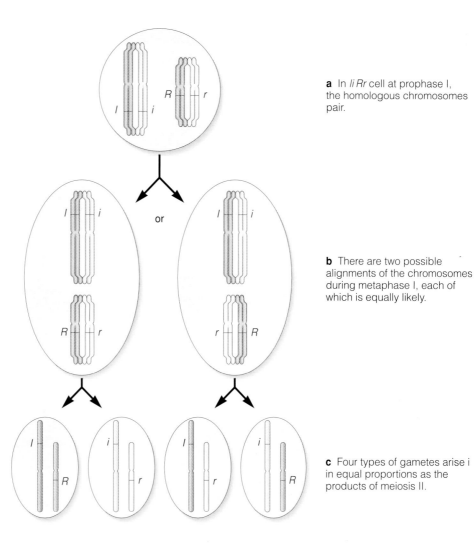

a In *Ii Rr* cell at prophase I, the homologous chromosomes pair.

b There are two possible alignments of the chromosomes during metaphase I, each of which is equally likely.

or

c Four types of gametes arise i in equal proportions as the products of meiosis II.

Figure 12.9 Independent assortment of nonhomologous chromosomes during meiosis I.

be yellow when the seed's genotype is homozygous *II* or heterozygous *Ii*. The recessive allele *i* causes seeds to be green when the seed's genotype is homozygous *ii*. The locus responsible for seed shape is called *R* and is located on chromosome VII. The dominant allele *R* causes seeds to be round when the seed's genotype is homozygous *RR* or heterozygous *Rr*. The recessive allele *r* causes seeds to be wrinkled when the seed's genotype is homozygous *rr*.

In Mendel's dihybrid experiment, one parent was homozygous *II RR* and the other was homozygous *ii rr*. All F₁ progeny were *Ii Rr*. During prophase I of meiosis in the F₁ plants, homologous chromosomes pair with one another, as diagrammed in Figure 12.9a. During metaphase I, the chromosome pairs with the *I* and *R* loci align at the cell's equatorial plane in one of two possible orientations, as illustrated in Figure 12.9b. When meiosis II is completed, these two possible alignments result in four possible types of gametes. Because the alignment of one pair of chromosomes has no influence on the alignment of any other pair of chromosomes, the two possible alignments are equally likely, resulting in equal frequencies of the four possible gametes (Figure 12.9c).

When the F₁ plants self-fertilize, there are 16 possible ways in which the gametes can unite, as represented in the Punnett square shown in Figure 12.10. Because the union of sperm and egg cells is a random event, each of the 16 unions is equally likely. Notice that nine of the unions give the genotypes *II RR, II Rr, Ii RR,* or *Ii Rr,* all of which result in yellow, round seeds. Three of the unions give the genotypes *ii RR* or *ii Rr,* both of which result in green, round seeds. Three of the unions give the genotypes *II rr* or *Ii rr,* both of which result in yellow, wrinkled seeds. Only one union gives the genotype *ii rr,* which confers green, wrinkled seeds. This explains the 9:3:3:1 ratio observed by Mendel and also conforms to his mathematical model.

We can now state Mendel's principle of independent assortment in modern terms:

~

***The Principle of Independent Assortment.* Because genes located on nonhomologous chromosomes assort independently during meiosis, the inheritance of alleles at one locus does not influence the inheritance of alleles at another locus.**

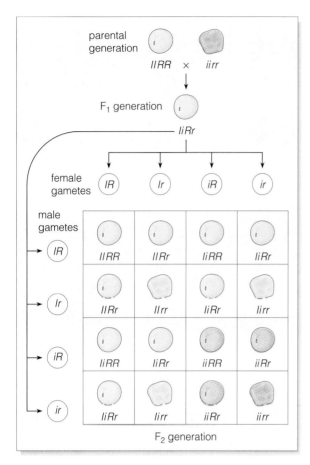

Figure 12.10 Punnett square illustrating the genotypes and phenotypes of the F₂ generation from Mendel's dihybrid experiment.

Dihybrid Testcross Experiments

Earlier we discussed how testcrosses could be used to identify the genotype of an individual with the dominant phenotype for a single trait. The same logic used for a single trait can be extended to determine the genotype of an individual with the dominant phenotypes for two or more traits. In any testcross, an individual with one or more dominant traits is crossed with an individual that is homozygous recessive for all traits in question. This homozygous recessive "tester" parent—the key to the test—produces a single type of gamete, with recessive alleles only. Because this parent has no dominant alleles to pass to its offspring, none of its offspring can be homozygous for a dominant allele. Thus all testcross progeny with the dominant phenotype must be heterozygous for the dominant allele. This ensures that in testcross progeny, each possible phenotype represents only one genotype.

For example, suppose we did not know whether a purple-flowered, tall plant was homozygous or heterozygous at the two loci. We write its genotype as $A\_B\_$. The blank spaces indicate that we do not yet know what its other two alleles are. The purple-

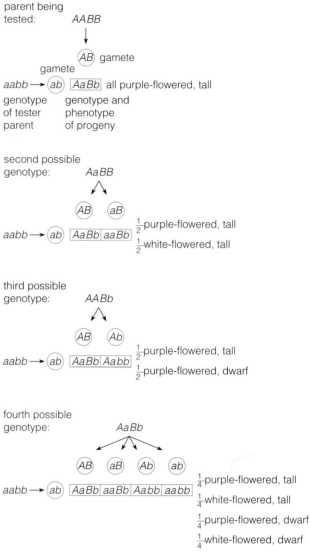

Figure 12.11 The four possible genotypes of a purple-flowered, tall plant and the progeny genotypes and phenotypes produced from testcrosses of each of these four genotypes with a white-flowered, dwarf plant.

flowered, tall plant is crossed with a white-flowered, dwarf plant, and the phenotypes observed in the first-generation progeny let us know what the genotype of the purple-flowered, tall parent is. Figure 12.11 shows the four possible genotypes of the purple-flowered, tall parent and the phenotypes of testcross progeny from each of these four parental genotypes.

Trihybrid Experiments

Mendel demonstrated that genes also assort independently in **trihybrid experiments**, in which the original true-breeding parents differ in three traits. Let's use the symbols A, B, and C to designate the three loci (as Mendel did). The F₁ plants, which have the geno-

type *AaBbCc*, produce eight types of gametes in equal proportions:

This means that there are 64 possible ways for male and female gametes to unite to form the F_2 generation. A Punnett square describing such a cross must be an 8×8 lattice with 64 spaces in it. At this point, Punnett squares become cumbersome to use because they are so large. Another approach that provides the same information is the **forked-line method**. Figure 12.12 shows how the forked-line method can be used to determine genotypic ratios of F_2 progeny in a trihybrid experiment. At the *A* locus, the expected proportions of F_2 progeny are ¼ *AA*, ½ *Aa*, and ¼ *aa*, so we write these three genotypes with their expected proportions in a column. Within each of these three classes, we expect the *B* locus to segregate ¼ *BB*, ½ *Bb*, and ¼ *bb*, so we write these three genotypes and their expected proportions in the second column and connect them to the genotypes for the *A* locus with forked lines. We then repeat the procedure for the *C* locus. In the end, we can determine the expected proportion of any genotype by multiplying the expected proportions for each locus. For example, let's determine the expected proportion of F_2 individuals with the genotype *AaBbcc*. The expected proportion of all individuals with the genotype *Aa* is ½. Among them, the expected proportion of individuals with the genotype *Bb* is ½. And among them, the expected proportion of individuals with the genotype *cc* is ¼. We multiply the three expected proportions to obtain the result: ½ × ½ × ¼ = ¹⁄₁₆.

At this point it should be obvious that genetic variation between parents at just a few loci can lead to a large amount of genetic variation among offspring. When two parents are heterozygous at just three loci, there are 27 possible genotypes and 8 possible phenotypes in their progeny. We can easily determine the number of possible gametes, genotypes, and phenotypes for any number of independently segregating loci with some simple formulas that Mendel derived. Where *n* equals the number of heterozygous loci, each with one dominant and one recessive allele, there are 2^n possible types of gametes, 3^n possible genotypes in the progeny, and 2^n possible phenotypes in the progeny. If we apply the formulas to Mendel's experiments in which the variation was at just seven loci, we see that the variation consists of 128 possible types of gametes, 2187 possible genotypes, and 128 possible phenotypes.

Linked Genes

What about genes that are located on the same chromosome? Do they also assort independently? If the two loci are close to each other on the chromosome, they tend to

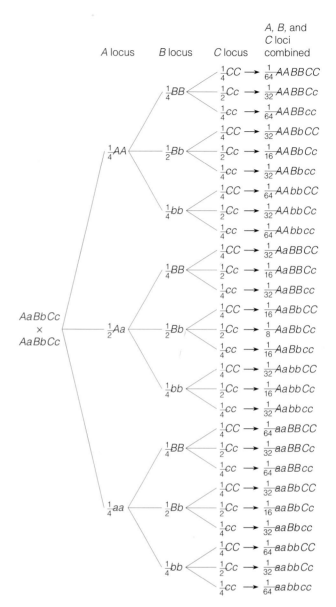

Figure 12.12 The forked-line method for determining genotypic and phenotypic frequencies in the F_2 generation of a trihybrid cross.

remain associated during meiosis rather than assorting independently. When genes do not assort independently because they are close together on the same chromosome, they are said to be **linked genes**. However, alleles of linked genes can be separated by crossovers. The farther apart two genes are on the chromosome, the more likely their alleles are to be separated by a crossover. In fact, some genes on the same chromosome are so far apart that crossovers between them are frequent enough that the alleles assort independently in genetic experiments.

Mendel analyzed the inheritance of seven traits in peas, and peas happen to have seven chromosomes. Was he lucky enough to find each locus on a different chromosome? Even though Mendel reported that all seven traits assorted independently of one another, some of the

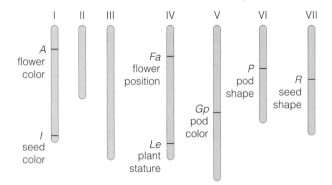

Figure 12.13 The chromosomal locations of the loci that Mendel probably studied. The loci on chromosomes I and IV are so far apart that alleles at these loci assort independently.

genes he studied were in fact on the same chromosome. Two of the loci he studied are on chromosome I and two are on chromosome IV; the other three are all on different chromosomes (Figure 12.13). However, the loci on the same chromosomes are far enough apart that the high frequency of crossovers accounts for the independent assortment of alleles. In Chapter 15, we will discuss in more detail how linked genes are inherited. For now, the following example shows how linked genes do not assort independently.

Example 12.4 Linkage as a deviation from independent assortment.

In 1908, Bateson and Punnett (*Reports to the Evolution Committee of the Royal Society, Report IV*) provided data for the F$_2$ progeny from a dihybrid experiment in sweet pea (*Lathyrus odoratus*, which is related to the garden pea, *Pisum sativum*, which Mendel studied). The two traits they studied were flower color and pollen grain shape. They crossed a true-breeding purple-flowered plant that had long pollen grains with a true-breeding red-flowered plant that had round pollen grains, and tabulated the following results for the F$_2$ progeny:

Purple flowers, long pollen grains	296
Purple flowers, round pollen grains	19
Red flowers, long pollen grains	27
Red flowers, round pollen grains	85
Total	427

Problem: **(a)** Which traits are dominant and which are recessive? **(b)** Do these experimental results conform to Mendel's principle of segregation? **(c)** Do these experimental results conform to Mendel's

principle of independent assortment? **(d)** Propose an explanation for any lack of conformity to these principles.

Solution: **(a)** Purple flowers and long pollen grains are the dominant phenotypes. Red flowers and round pollen grains are the recessive phenotypes. **(b)** When flower color is considered alone, the number of purple-flowered plants is 296 + 19 = 315. The number of red-flowered plants is 27 + 85 = 112. This corresponds to a ratio of 2.81:1, which is close to 3:1. When pollen grain shape is considered alone, the number of plants with long pollen grains is 296 + 27 = 323, and the number of plants with round pollen grains is 19 + 85 = 104, a ratio of 3.11:1, which is also close to 3:1. The inheritance of both flower color and pollen grain shape conforms to Mendel's principle of segregation. **(c)** If these traits assort independently, then the data should not deviate substantially from a 9:3:3:1 ratio. The total number of progeny is 427. The expected numbers of progeny in a perfect 9:3:3:1 ratio are $9/16 \times 427 = 240.1875$, $3/16 \times 427 = 80.0625$, and $1/16 \times 427 = 26.6875$. Let's compare these expected numbers with the observed numbers:

Phenotype	Observed	Expected
Purple flowers, long pollen grains	296	240.1875
Purple flowers, round pollen grains	19	80.0625
Red flowers, long pollen grains	27	80.0625
Red flowers, round pollen grains	85	26.6875

The observed deviations from expected values are so large that these data do not conform to Mendel's principle of independent assortment. **(d)** Notice that there is a pattern in the data. In the two parents, long pollen grains are associated with purple flowers, and round pollen grains with red flowers. These associations persist in the progeny, as we see when we notice that the two phenotypic classes with the highest numbers of progeny are the same as the parental phenotypic classes. This pattern of the parental associations of genes persisting in the progeny is typical of linked genes. We will return to this example when we discuss linkage further in Chapter 15.

Genes that are closely linked to each other on the same chromosome do not assort independently during meiosis.

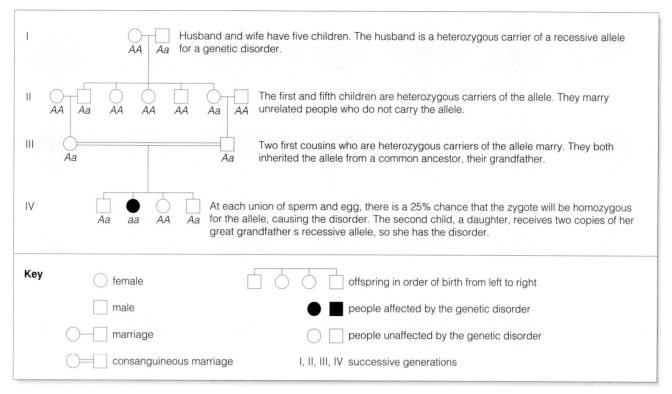

Figure 12.14 A human pedigree illustrating the appearance of a recessive phenotype in the fourth generation of a family. In the third generation, two first cousins who marry have a common ancestor (their grandfather) who carried a rare recessive allele that causes a genetic disorder when in the homozygous condition. They both inherited the recessive allele, so there is a 25% chance that each child they have will be homozygous for the allele and will be affected by the disorder. One of their four children is affected.

12.5 MENDELIAN INHERITANCE IN HUMANS

The three scientists who rediscovered Mendel's principles in 1900 demonstrated Mendelian inheritance in plant species. By 1902, Mendelian inheritance had been described in animals. With these discoveries, one of the most obvious and pressing questions was whether or not the principles of Mendelian inheritance applied to humans. For a variety of reasons—such as small family size, a relatively long generation time, and the ethical inappropriateness of conducting experimental matings—humans do not lend themselves well to the type of genetic experimentation and analysis that made it possible to demonstrate Mendelian inheritance in many plant and animal species.

Instead, demonstration of Mendelian inheritance had to depend on **pedigree analysis** of the type diagrammed and described in Figure 12.14. Within the first few years after Mendelian inheritance was rediscovered, Mendelian inheritance of several traits was identified in human pedigrees. The first was brachydactyly, a genetic disorder that causes malformed hands with fingers that are short and thick (Figure 12.15a). The first

pedigree of brachydactyly, published in 1905, suggested that a dominant allele caused the disorder, as every individual with the disorder had an affected parent (Figure 12.15b). By 1913, numerous pedigrees of individuals with ocular-cutaneous albinism (the absence of pigmentation in skin, hair, and eyes; Figure 12.16), had been analyzed. The pedigrees clearly demonstrated that the allele for ocular-cutaneous albinism was recessive and inherited in a simple Mendelian fashion. Now, thousands of dominant and recessive traits that are inherited in simple Mendelian patterns in humans have been described. Many of these traits are genetic disorders that cause noticeable abnormalities. Table 12.3 lists a few of these disorders.

With the advent of tools such as chromosome analysis, biochemical genetics, and recombinant DNA, many of the limitations associated with traditional genetic analysis in humans have been circumvented. Now our species is among the best understood and most researched in genetics. Because of these tools, many of the genetic disorders listed in Table 12.3 have been localized to a site on a chromosome, and the DNA sequence of mutant alleles, as well as the gene product and its function, are now known. Most of the DNA sequences and

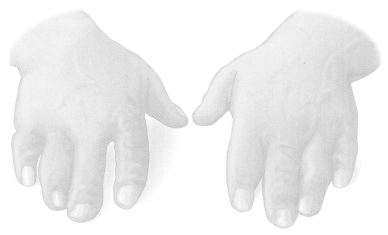

a Brachydactyly is characterized by malformed hands with fingers that are short and thick.

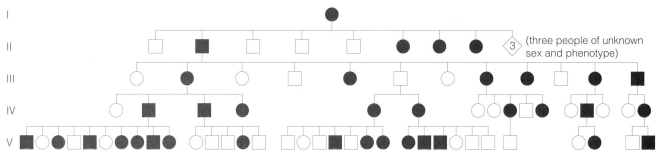

b The first pedigree of brachydactyly, published in 1905 by Farabee. All offspring with the trait have an affected parent, indicating that the responsible allele is dominant. This pedigree does not show the spouses of the family members whose progenies are shown.

Figure 12.15 Brachydactyly, the first human trait shown to exhibit a Mendelian pattern of inheritance. (Source: Adapted from Bateson, W. 1913. *Mendel's Principles of Heredity.* Cambridge: Cambridge University Press, and Farabee, W. C. 1905. *Papers of Peabody Museum of American Archaeology and Ethnnology.* Harvard University III. 3, p. 69.)

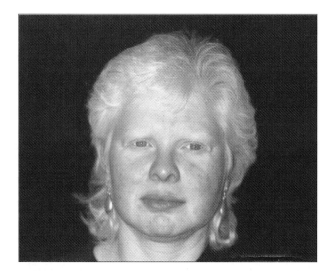

Figure 12.16 Ocular-cutaneous albinism. The absence of melanins (pigments) in the skin, hair, and eyes, due to homozygosity for a recessive allele, causes this disorder. Dr. P. Marazzi/Science Photo Library.)

gene products have been identified in the past decade. Summary information on the inheritance and molecular characteristics of all known human genes is available on the Internet at the Online Mendelian Inheritance in Man (OMIM) website. A link to OMIM is available at the website for this book (www.wadsworth.com).

12.6 MENDELIAN GENETICS AND PROBABILITY

Mendel was able to discern patterns in his experiments because his observed data were close to theoretical ratios. Observed data from genetic experiments often approximate theoretical ratios, but rarely do the data match the theoretical ratios exactly, as evidenced in Mendel's experiments (see Table 12.1 and Example 12.1). Deviations from theoretical ratios are often observed because chromosome assortment during meiosis and union of gametes at fertilization are random events.

Table 12.3 Examples of Genetic Disorders in Humans That Are Inherited in a Simple Mendelian Fashion

Disorder	Phenotype	Gene Dominant or Recessive
Achondroplasia	Dwarfism	Dominant
Achromatopsia	Complete color blindness	Recessive
Alkaptonuria	Failure to metabolize homogentisic acid; dark urine	Recessive
Brachydactyly	Malformed hands with thick, short fingers	Dominant
Campodactyly	Bent little finger	Dominant
Cystic fibrosis	Fibrous cyst formation in pancreas, excessive mucus production, excessive salt in perspiration	Recessive
Familial hypercholesterolemia	Abnormally high serum cholesterol	Dominant
Hereditary deafness	Deafness from birth	Recessive
Huntington disease	Degeneration of nervous system during ages 20–40	Dominant
Marfan syndrome	Tall, thin individual with long arms and legs, weak aorta	Dominant
Nail-patella syndrome	Absence of kneecaps, fingernails, and toenails	Dominant
Neurofibromatosis	Abnormal skin pigmentation and neurofibromas	Dominant
Ocular-cutaneous albinism	Lack of pigmentation in skin, eyes, and hair	Recessive
Phenylketonuria	Failure to metabolize phenylalanine; mental retardation	Recessive
Polycystic kidney disease	Cyst formation in kidneys	Dominant
Polydactyly	More than five digits on hands and/or feet	Dominant
Sickle-cell anemia	Production of defective hemoglobin; red blood cells sickle when starved for oxygen; capillary blockage	Recessive
α-thalassemias	Defective α subunit of hemoglobin; symptoms vary	Recessive
β-thalassemias	Defective β subunit of hemoglobin; symptoms vary	Recessive
Xeroderma pigmentosum	Defective DNA repair, high susceptibility to skin cancer	Recessive
Tay-Sachs disease	Nerve degeneration and early death	Recessive

Under these conditions, two rules of probability, the product rule and the sum rule, can be applied to predict and analyze the outcomes of genetic experiments.

The Rules of Probability

The first rule of probability that we will examine is the **product rule**, which can be stated as follows:

~

The Product Rule. **When two events are independent of one another, the probability that they will occur together is the product of their individual probabilities.**

This rule can easily be illustrated by flipping two fair coins, such as a penny and a nickel, one time. The outcome of one flip has no influence on the outcome of the other, meaning that the two coin flips are inde-pendent—as events must be for the product rule to apply. The probability that the penny will turn up heads is 50%. The probability that the nickel will turn up heads is also 50%. The probability that *both* will turn up heads is the product of the individual probabilities, which is 0.5 × 0.5 = 0.25, or 25%. (A probability may be expressed as either a fraction or a percentage.)

We have already seen the results of the product rule in genetics. For instance, we multiply probabilities to determine the likely outcomes of independent assortment of two or more traits. Let's use the example of seed color and seed shape, which we described in Figure 12.10. The probability of any seed in the F_2 generation being green is $\frac{1}{4}$. The probability of any seed in the F_2 generation being wrinkled is also $\frac{1}{4}$. Because seed color is independent of seed shape, the probability of any seed in the F_2 generation being *both* green and wrinkled is $\frac{1}{4} \times \frac{1}{4} = \frac{1}{16}$, just as the Punnett square in Figure 12.10

predicts. The probability of any seed being yellow is $\frac{3}{4}$, so the probability of any seed being both yellow and wrinkled is $\frac{3}{4} \times \frac{1}{4} = \frac{3}{16}$, again exactly as the Punnett square predicts. The probabilities associated with the other two possible phenotypes are calculated in the same way, resulting in a 9:3:3:1 ratio.

The other law of probability that often applies in genetics is the **sum rule**:

~

The Sum Rule. **If two events are mutually exclusive, the probability that one of the two events will occur is the sum of their individual probabilities.**

This rule can also be illustrated quite simply by flipping a fair coin. When a coin is flipped, it must come up either heads or tails. It cannot come up both heads and tails in the same flip, so heads and tails are mutually exclusive of one another, and we may therefore apply the sum rule. The probability that a coin will come up *either* heads or tails on a single flip is the sum of the respective probabilities: probability of heads = 0.5, probability of tails = 0.5, and 0.5 + 0.5 = 1.0, so the probability of a single coin flip coming up either heads or tails is 100%, which is obvious because there are no possibilities other than heads and tails. We can also apply the sum rule to rolling a fair die. When a die is rolled, there are six mutually exclusive possibilities: 1, 2, 3, 4, 5, and 6. What is the probability that either a 1 or a 4 will come up in a single roll? Using the sum rule, we determine that the probability of rolling a 1 is $\frac{1}{6}$ and the probability of rolling a 4 is also $\frac{1}{6}$, so the probability of rolling either a 1 or a 4 is $\frac{1}{6} + \frac{1}{6} = \frac{1}{3}$, or 33%.

We have already seen the sum rule applied to genetics. In the F_2 generation of a monohybrid experiment with alleles A and a, there are three mutually exclusive possibilities for genotypes: AA, Aa, and aa. If A is dominant, the probability that any one F_2 individual will have the dominant phenotype is the sum of the probabilities that the individual will have the genotype AA or Aa, which is $\frac{1}{4} + \frac{1}{2} = \frac{3}{4}$, or 75%.

Binomial Distributions

Recall that homozygosity for a recessive allele causes ocular-cutaneous albinism in humans. If a man and a woman who are both heterozygous for this allele have four children, we cannot say with certainty that one child will have albinism and three will not, even though the expected ratio is 3:1. However, we can say that the probability of any one child having ocular-cutaneous albinism is 25%, and the probability of any one child *not* having ocular-cutaneous albinism is 75%, using the rules of probability just discussed.

The rules of probability can be used to determine the probabilities of phenotypes in the next generation when the genotypes of the parents are known—as we did in the previous paragraph. A convenient way to determine certain probabilities is to use the **binomial distribution**. The binomial distribution is simply $(p + q)^n$, which is the expansion of $(p + q)$ raised to the *n*th power, where p represents the probability of one outcome, q the probability of another mutually exclusive outcome, and n is the total number of events, or individuals to be classified, in a finite sample. The distribution is called *binomial* because it applies to situations in which there are only two mutually exclusive outcomes, or classes, into which an event or individual could fall, such as heads or tails, purple or white flowers, male or female.

The binomial distribution requires two assumptions. First, the two possible outcomes must be mutually exclusive: an event can have one or the other outcome but not both. Second, each event must be independent of every other event in the series: for example, one coin toss cannot influence the outcome of a second coin toss. These assumptions are usually met in genetic situations.

One of the most common uses of the binomial distribution is to determine the probability of a particular outcome in a human family. For example, there are six possible combinations of males and females in a human biological family of five children: the children could be all females, four females and one male, three females and two males, two females and three males, one female and four males, or all males. Together, these six combinations represent the binomial distribution for a biological family of five children.

The probability (P) of any particular outcome in a binomial distribution can be calculated as

$$P = \frac{n!}{x!y!}p^x q^y \qquad [12.1]$$

where n is the total number of individuals, x is the number of individuals in one class (such as females), y is the number in the other class (such as males), p is the probability of falling into the class with x individuals, and q is the probability of falling into the class with y individuals. The symbol "!" stands for "factorial," which means that the number is multiplied by all descending whole numbers down to unity, such as $5! = 5 \times 4 \times 3 \times 2 \times 1 = 120$. Also, $0! = 1$, which can be important in the binomial distribution because x or y may equal zero. The quantity $x + y$ always equals n, and $p + q$ always equals unity in a binomial distribution. Figure 12.17 shows the six possible outcomes for a biological family of five children and their respective probabilities as calculated using equation 12.1. For example, given that five children are born, let's calculate the probability of there being three females and two males. We'll assume that

there is a 1:1 probability of any individual being male or female (the probability at birth is actually 1.05 males:1 female, but we'll use a 1:1 approximation for simplicity). Then, $n = 5$, $x = 3$, $y = 2$, $p = 0.5$, and $q = 0.5$, so $P = \dfrac{5!}{3!2!} 0.5^3 0.5^2 = 0.3125$, or 31.25%, as illustrated in Figure 12.17.

The $\dfrac{n!}{x!y!}$ portion of equation 12.1 is called the **binomial coefficient**. Rather than calculating it using factorials, you can obtain it from a table called **Pascal's triangle** (Table 12.4), in which each value in the triangle is the sum of the two values immediately above it. Each row in Pascal's triangle corresponds to a value of n. Within the triangle, the value at the far left in each row is the binomial coefficient when $x = n$ and $y = 0$. Moving left to right along the row, the binomial coefficients correspond to descending values of x and ascending values of y. For instance, to find the binomial coefficient for $n = 7$, $x = 5$, and $y = 2$, go to the row for $n = 7$ and count across three values ($x = 7$, $x = 6$, $x = 5$), and you arrive at a binomial coefficient of 21. Indeed, $\dfrac{7!}{5!2!} = 21$.

The following example shows how the binomial distribution can be used to predict the probability of a recessive genetic disorder in a human biological family of a particular size. Applications of the binomial distribution like the one described in this example are often used in genetic counseling.

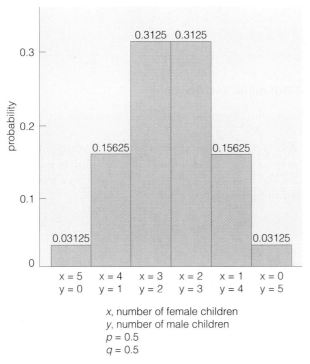

Figure 12.17 Binomial distribution for $n = 5$. What is the probability that a biological family of five children will have three females and two males?

Table 12.4 Pascal's Triangle

n	Coefficients
0	1
1	1 1
2	1 2 1
3	1 3 3 1
4	1 4 6 4 1
5	1 5 10 10 5 1
6	1 6 15 20 15 6 1
7	1 7 21 35 35 21 7 1
8	1 8 28 56 70 56 28 8 1

Example 12.5 The binomial distribution: Recessive alleles in human families.

Problem: Phenylketonuria (PKU) is a human genetic disorder characterized by severe mental retardation if left untreated. It is caused by homozygosity for a recessive mutant allele. If two parents are heterozygous for the allele, and they plan on having four children, what is the probability that at least one child will have PKU?

Solution: Let x be the number of children without PKU and y the number of children with PKU. The probability of a child without PKU is $p = 0.75$, and the probability of a child with PKU is $q = 0.25$. With four children, there are four possible outcomes in which at least one child has PKU: $y = 1$, $y = 2$, $y = 3$, and $y = 4$. Because the four outcomes are mutually exclusive, we can determine the probability of each outcome, then add the probabilities according to the sum rule:

$4(0.75)^3(0.25)^1 = 0.421875$

$6(0.75)^2(0.25)^2 = 0.2109375$

$4(0.75)^1(0.25)^3 = 0.046875$

$\underline{(0.75)^0(0.25)^4 = 0.00390625}$

Total   0.68359375

The probability of at least one child with PKU is about 68%.

There is an easier way to calculate this value. In the family of four children, there are five possible

outcomes, only one of which ($y = 0$) does not include a child with PKU. The probability of this outcome is $P = \frac{4!}{4!0!}0.75^4 0.25^0$, which reduces to $0.75^4 = 0.31640625$. The sum of the probabilities of all possible outcomes in a binomial distribution equals unity. Therefore, the probability of at least one child with PKU is $1 - 0.31640625 = 0.68359375$.

Having reviewed the binomial distribution, we can now look at an example from Mendel's experiments that Sir Ronald A. Fisher analyzed in a famous article published in 1936. Fisher was a brilliant scientist who was among the most influential people in establishing the science of statistics. This example is now famous among geneticists and statisticians as a demonstration of how probability can affect the conclusions derived from experimental results.

Example 12.6 The probability of genotypic misclassification in Mendel's experiments.

As shown in Table 12.1, Mendel found near 3:1 ratios of dominant:recessive phenotypes in the F_2 progeny in all seven monohybrid experiments. Mendel proposed that in each of these experiments two-thirds of the F_2 individuals with the dominant phenotype were heterozygous and the remaining one-third were homozygous, giving a genotypic ratio of 2:1. To test this hypothesis, he allowed F_2 plants with the dominant phenotype to self-fertilize, and then he recorded the phenotypic traits of the F_3 progeny. Two of the traits, seed shape and seed color, were manifest in the F_3 seeds rather than the plants, so they did not require any space for growing F_3 plants. He studied the F_3 seeds collected from 565 F_2 plants for seed shape and 519 F_2 plants for seed color. For the five other traits, he had to grow the F_3 plants to observe their phenotypes. Because of space restrictions, he limited the number of F_3 plants that he grew. In each experiment for the five traits, he chose 100 F_2 plants with the dominant phenotype, allowed them to self-fertilize, and planted 10 F_3 seeds from each F_2 plant. If any one of the 10 F_3 plants had the recessive phenotype, then Mendel classified its F_2 parent as heterozygous. If all 10 F_3 plants had the dominant phenotype, then Mendel classified their F_2 parent as homozygous for the dominant allele. The experiment with pod color yielded results that Mendel felt were too far from the predicted ratio of 2:1, so he repeated that experiment, obtaining the second time results that were more acceptable to him (see Table 12.2).

Problem: In 1936, Fisher (*Annals of Science* 1: 115–137) pointed out that, given the numbers of F_3 plants observed in the last six experiments listed in Table 12.2, Mendel should have misclassified some heterozygous F_2 plants as homozygous. What is the probability of misclassifying a heterozygous F_2 plant as homozygous on the basis of 10 F_3 progeny?

Solution: This question can be resolved by applying the binomial distribution. Mendel grew 10 F_3 progeny from each F_2 plant to classify that plant as homozygous or heterozygous. Mendel expected that if the F_2 plant was heterozygous, at least one of the 10 F_3 progeny from that plant would have the recessive phenotype. However, the binomial distribution reveals that there are 11 possible outcomes among the 10 progeny when the F_2 parent is heterozygous. One of these outcomes results in misclassification. In the binomial distribution, $n = 10$, $p = 0.75$, $q = 0.25$, x is the number of plants with the dominant phenotype, and y is the number of plants with the recessive phenotype. The outcome that results in misclassification is the case in which all 10 progeny of a heterozygous plant have the dominant phenotype, which is written in mathematical terms as $x = 10$ and $y = 0$. The probability of this outcome is

$$P = \frac{10!}{10!0!}0.75^{10}0.25^0 = 0.75^{10} = 0.0563.$$

Mendel probably misclassified about 5.6% of the heterozygous F_2 plants as homozygous. We will discuss some of the implications of this conclusion in section 12.7.

Multinomial Distributions

Sometimes, there are more than two classes into which individuals can fall, in which case a multinomial distribution applies. The equations for calculating multinomial probabilities are simply an extension of equation 12.1. For example, a particular outcome in a trinomial distribution is calculated as

$$P = \frac{n!}{x!y!z!}p^x q^y r^z \qquad [12.2]$$

where n, x, y, p, and q have the same meanings as before, and r represents the probability of an individual's falling into the class consisting of z individuals. More terms can be added in the same way to construct higher multinomial equations.

~

Binomial and multinomial distributions allow researchers to determine the probability of any possible outcome in genetic experiments.

Sampling Error

The binomial distribution shows us why large deviations from expectation are common when sample sizes are very small, and why large samples tend to approximate expected ratios more closely. Suppose you crossed a true-breeding purple-flowered pea plant with a true-breeding white-flowered one and allowed the F_1 progeny to self-pollinate. If you then grew just four F_2 plants from this cross, the probability that all four plants would be purple-flowered is about 0.31, a fairly high probability. However, the probability of all purple-flowered plants in a comparable sample of 40 F_2 plants is 0.00001, and with 400 plants it is 10.6×10^{-51}. When the number of F_2 individuals is large, we expect the segregation to deviate less from a 3:1 ratio than in small samples. In Mendel's monohybrid experiments for flower color, he observed 929 F_2 plants. Thus we expect about 697 plants (929×0.75) to be purple flowered and about 232 plants (929×0.25) to be white flowered. The actual results were 705 purple-flowered plants and 224 white-flowered plants, which are quite close to the expected values.

The deviation from expected ratios due to random variation is called **sampling error**. Sampling error is always a possibility with samples of finite size, and the magnitude of sampling error tends to be greater with smaller sample sizes. The probability of an all-girl family of two children is much greater than the probability of an all-girl family of 10 children. In genetics, the smaller the number of progeny, the greater the probability that observed ratios will deviate from predicted ratios due to sampling error.

Chi-Square Analysis

It is fairly obvious that the observed numbers of progeny having a particular genotype or phenotype do not always match exactly the expected numbers. The binomial distribution shows us that any possible combination of genotypes or phenotypes is theoretically possible, but the closer a combination is to expectation, the more likely it is to be observed, *if* the deviation is due solely to sampling error. Alternatively, the observed deviation may *not* be due to sampling error alone, but may be in part because the hypothesis used to determine expected values is incorrect. Statistical tests are available to determine whether the probability of an observed outcome is high enough to accept sampling error as an explanation for the deviation, or whether the probability of an observed outcome is so low that it calls into question the validity of the hypothesis. A statistical test that is commonly used to make this determination is called chi-square analysis.

Chi-square analysis compares observed values with hypothesized expected values and determines the probability of observing that outcome under the assumption that the hypothesis is correct. The equation for chi-square analysis is

$$\chi^2 = \sum \frac{(O - E)^2}{E} \qquad [12.3]$$

where χ^2 is chi-square, O represents the number of individuals observed in a particular phenotypic (or genotypic) class, and E represents the number of individuals expected in that class from hypothesis. When there is more than one expected class (e.g., white flowers and purple flowers), the chi-square values for the classes are summed (the reason for the symbol Σ). Once the chi-square value has been obtained, that value can be compared to theoretical values to see how closely the observed data fit the expectations. The theoretical values can be obtained from a chi-square table. The portion of a chi-square table most useful for geneticists is provided in Table 12.5. For a more detailed chi-square table, see Appendix I.

The degrees of freedom for most chi-square calculations are one less than the number of classes being analyzed. For instance, there are two phenotypic classes in the example of white flowers and purple flowers, so there is one degree of freedom. There are four phenotypic classes in the F_2 generation in a dihybrid experiment—for example, yellow, round seeds; yellow, wrinkled seeds; green, round seeds; and green, wrinkled seeds. With four classes, there are three degrees of freedom.

If the chi-square value calculated from observed data exceeds the theoretical chi-square value listed in the table, then the observed ratio is said to differ significantly from expected, because the probability that the observation would deviate that much due to sampling error alone is quite low. The theoretical value tells us how frequently we should expect to see that deviation due to sampling error if our hypothesis is correct. For example, at the 0.05 probability level, we expect to see

Table 12.5 Theoretical Chi-Square Values

Degrees of Freedom	Probability Level	
	0.05	0.01
1	3.84	6.64
2	5.99	9.21
3	7.82	11.34
4	9.49	13.28
5	11.07	15.09

an observed chi-square value of 3.84 or greater due to sampling error in only 5% of the cases. If the chi-square value does not exceed the theoretical value, then there is no statistical evidence to reject the hypothesized pattern, because the deviations may be reasonably explained by random sampling error.

Generally, there are two levels of significance used in statistical analysis, the 0.05 level, which is said to be **significant**, and the 0.01 level, which is said to be **highly significant**. At the 0.05 level, we expect to erroneously reject a correct hypothesis about 5% of the time. At the 0.01 level, we expect to erroneously reject a correct hypothesis only about 1% of the time. The erroneous rejection of a correct hypothesis is called a **type I error**. The value chosen reflects the level of type I error that the experimenter is willing to tolerate. The 5% level for significance and the 1% level for high significance are standards accepted for most experiments. Most computer analyses of genetic data now provide an exact probability value associated with the chi-square value, telling the researcher the probability of a type I error for any particular experiment.

Let's start with a simple example of the chi-square test to illustrate its use.

Example 12.7 Chi-square analysis: Significance testing.

Problem: Suppose you flip a fair coin 20 times. The expected outcomes are 10 heads and 10 tails under the hypothesis that the probability of heads is 50% and that of tails is 50%. However, there is a high likelihood that the observed number will deviate from the expected. Suppose your sample came up 15 heads and 5 tails. Is this a statistically significant deviation from the expected 1:1 ratio?

Solution: Using equation 12.3, take the observed number of heads and subtract the expected number of heads, then square the result and divide this number by the expected number: $(15 - 10)^2 \div 10 = 2.5$. Then do the same for tails: $(5 - 10)^2 \div 10 = 2.5$. Add the two values together to obtain the chi-square value: $2.5 + 2.5 = 5$. As there are two classes (heads and tails), there is one degree of freedom. The calculated chi-square value of 5 is greater than the theoretical value of 3.84 for the 0.05 probability level, but less than the theoretical value of 6.64 for the 0.01 probability level. This means that if you flip a fair coin 20 times, the probability of a deviation as great or greater than 15 heads and 5 tails is less than 5% but greater than 1%. We conclude that the deviation from expected is statistically significant, but not highly significant. This is often considered sufficient evidence to reject a hypothesis. Rejection of

the hypothesis that the probability of heads is 50% and the probability of tails is 50% on the basis of this particular chi-square value is an example of a type I error. When using the 5% level for rejecting hypotheses, an experimenter expects to make a type I error about 5% of the time.

Now let's apply the chi-square test to some actual results of genetic experimentation that we have already seen, to illustrate how chi-square analysis is used in genetics.

Example 12.8 Chi-square analysis applied to genetics.

Problem: Perform chi-square analyses of **(a)** Mendel's dihybrid experiment highlighted in Example 12.1 and **(b)** Bateson and Punnett's dihybrid experiment highlighted in Example 12.4 to test the hypothesis that the results of these two experiments conform to Mendel's principle of independent assortment. Draw appropriate conclusions for these experiments based on hypothesis testing.

Solution: **(a)** As listed in Example 12.1, the observed and expected numbers for Mendel's dihybrid cross, under the assumption of independent assortment, are as follows:

Phenotype	Observed	Expected
Yellow, round	315	312.75
Yellow, wrinkled	101	104.25
Green, round	108	104.25
Green, wrinkled	32	34.75

To perform a chi-square test, take the observed value for each case, subtract the expected value, square the result, then divide that number by the expected value. Sum the results of all four cases to calculate the chi-square value. In this example, $\chi^2 = [(315 - 312.75)^2 \div 312.75] + [(101 - 104.25)^2 \div 104.25] + [(108 - 104.25)^2 \div 104.25] + [(32 - 34.75)^2 \div 34.75] = 0.47$ with 3 degrees of freedom. The chi-square value of 0.47 is less than the critical value of 7.82, so the probability of a deviation this great or greater due to sampling error under the hypothesis of independent assortment is greater than 5%. The deviation is not statistically significant, and we do not reject the hypothesis that these results conform to Mendel's principle of independent assortment.

(b) As listed in Example 12.4, the observed and expected numbers for Bateson and Punnett's dihy-

brid cross, under the assumption of independent assortment, are as follows:

Phenotype	Observed	Expected
Purple flowers, long pollen grains	296	240.1875
Purple flowers, round pollen grains	19	80.0625
Red flowers, long pollen grains	27	80.0625
Red flowers, round pollen grains	85	26.6875

In this example, $\chi^2 = [(296 - 240.1875)^2 \div 240.1875] + [(19 - 80.0625)^2 \div 80.0625] + [(27 - 80.0625)^2 \div 80.0625] + [(85 - 26.6875)^2 \div 26.6875] = 222.12$ with 3 degrees of freedom. The chi-square value of 222.12 is substantially greater than the 0.01 critical value of 11.34, so the deviation is highly significant. The probability of a deviation this great or greater due to sampling error under the hypothesis of independent assortment is much less than 1%. This probability is so low that we reject the hypothesis that these results conform to Mendel's principle of independent assortment. Something other than sampling error should explain the deviation from independent assortment. In this case, linkage explains the deviation.

Chi-square analysis is used to determine whether or not sampling error can reasonably explain the deviation of observed data from hypothesized values.

12.7   ABOUT MENDEL

The discovery and rediscovery of Mendelian genetics is one of the best stories in the history of science and is worth telling here because it had such a dramatic impact on the beginning of modern genetics. Johann Mendel was born to a peasant family in 1822 in Heinzendorf bei Odrau (now Hynčice), a small Silesian village in what is now part of the Czech Republic (Figure 12.18). He later took on the name Gregor, which means "in religion" when he chose to become a priest. As a child, Johann worked on his family's farm, where he learned much about cultivation of plants. Through considerable sacrifice, his family sent him away to school at age 11. At one point, the family resources ran out and his younger sister, Theresia, gave a portion of her dowry so that he could remain in school. He later repaid her manyfold, supporting her three sons in their education.

Figure 12.18 Mendel's birthplace in Hynčice, Czech Republic.

In 1843, Mendel entered the St. Thomas Monastery in Brünn (now Brno), which is now in the Czech Republic but was under Austrian control at the time. In 1850, he failed an examination for teacher certification in the sciences. Because of this, he enrolled in a course of study that included mathematics, physics, chemistry, and various subjects in biology at the University of Vienna. Returning to his monastery in 1853, he began studying peas.

Mendel initiated his genetic experiments in 1856 and worked for 8 years collecting data. He presented the results and interpretations of his experiments with peas and a few experiments with beans to the Brünn Society for Natural History in 1865 and published them as a scientific paper entitled "Versuche über Pflanzen-Hybriden" ("Experiments on Plant Hybrids") in the *Proceedings of the Brünn Natural Historical Society* in 1866 (Figure 12.19). Links to the original German version and an English translation of Mendel's paper can be found at the website for this book, www.wadsworth.com.

Mendel's paper is one of the finest scientific papers ever written. He organized it in the same way that most scientific papers are organized today, beginning with an introduction, where he stated the rationale and objectives of his experiments and set the work in the context of previous research. He then described the plant materials and the methods he used. The remaining major portion of his paper includes his results and his interpretations of those results. Mendel interpreted his results mathematically, which was unusual for biologists of his day. In doing so, he developed an explanation of inheritance that remains essentially unchanged.

Mendel's appropriate use of hypothesis development and testing is obvious in his paper. In several cases, he formulated a hypothesis based on the results of some crosses, then conducted additional experiments to test the hypothesis. The principles attributed to Mendel, particularly segregation and independent

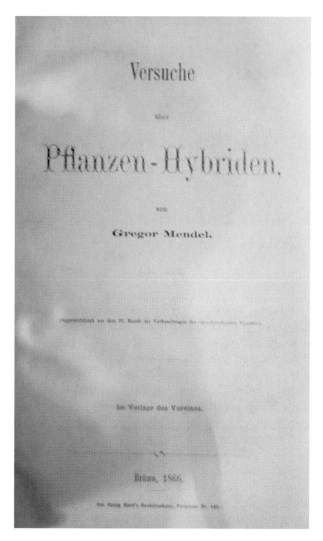

Figure 12.19 The title page of Mendel's classic paper, "Versuche über Pflanzen-Hybriden" ("Experiments on Plant Hybrids"), published in 1866.

assortment, were not given those titles by Mendel, but they are evident in the context of the whole paper.

Mendel presented his paper in two sessions and distributed reprints to several scientists. The volume in which it was published was held in over 120 libraries in Europe and in nine libraries in the United States. Nonetheless, his work received little attention. It was not completely ignored, being cited some 16 times prior to 1900, but never as an important discovery. According to William Bateson,

> The total neglect of [Mendel's] work is known to have been a serious disappointment to him, as well it might. He is reported to have had confidence that sooner or later it would be noticed, and to have been in the habit of saying *"Meine Zeit wird schon kommen!"* ["My time will surely come!"] This episode in the history of science is not a very pleas-

ant one to contemplate. There are of course many similar examples, but there must be few in which the discovery so long neglected was at once so significant, so simple, and withal so easy to verify.

Some writers paint a fanciful picture of Mendel as a frustrated genius trying to convey the importance of his work to a world of apathetic scientists whose educational backgrounds were insufficient for them to appreciate what he had done. Mendel recognized that his work was important. However, another passage from his paper reveals his reluctance to draw general conclusions about inheritance from experiments on only one or two species:

> A final decision can be arrived at only when we have before us the results of detailed experiments made on plants belonging to the most diverse orders.

Mendel intended to resolve this question by carrying out the very experiments he suggested. Near the end of his paper, he stated:

> It must be the object of further experiments to ascertain whether the law of development discovered for *Pisum* applies to the hybrids of other plants. To this end several experiments were recently commenced.

He initiated studies on several plant species, including maize, beans, stocks, hawkweed, four-o'clock, columbine, toadflax, morning glory, wallflower, and nasturtium, among others. He presented some preliminary results with beans in his classic paper on peas. Apparently, he saw the same patterns of inheritance in beans, maize, stocks, and four-o'clock as he did in peas.

Much of the information that we have regarding Mendel's later studies comes from letters he wrote to a prominent botanist named Karl von Nägeli. Nägeli was especially interested in hawkweed (*Hieracium*) and encouraged Mendel to work with this plant. Mendel published a short paper on hybridization studies with several species of hawkweed in 1870. While the garden pea was a superb choice as a model organism for genetic studies, hawkweed was among the worst possible choices. Neither Mendel nor Nägeli knew it, but hawkweed often produces seed by apomixis, a mechanism in which seed is produced from unfertilized ova that do not go through meiosis. In other words, the seeds inherit their genes only from the female parent, and therefore do not show any of the segregation patterns Mendel had documented in peas. In his hawkweed paper, Mendel concluded that

> In *Pisum* the hybrids, obtained from the immediate crossing of two forms, have in all cases the same type, but their posterity, on the contrary, are variable and follow a definite law in their variations. In *Hieracium* according to the present experiments the exactly opposite phenomenon seems to be exhibited.

Thus, only a few years after reporting his experiments with peas, Mendel concluded that the principles he discovered were not universal.

While his studies with hawkweed must have confused Mendel, other duties apparently took him away from the long hours of careful study that were required to conduct and analyze experiments like those reported in his classic paper. In 1868, just two years after his paper on peas was published, Mendel was elected prelate (abbot) of his monastery, bringing increased administrative duties. His letters indicate that he continued genetic experiments for 3 more years, but probably had to give up most of this work after 1871.

Unfortunately, most of the information Mendel collected in his later genetic studies was never published. He confided in a letter to Nägeli that he hoped his administrative duties would lessen and allow him more time to conduct his experiments. Sadly, this was not to be. Shortly after his death, when no one could foresee the significance his research would later have, the pages of notes containing his original data were burned.

Aside from plant genetics, Mendel had other scientific interests. He kept bees and attempted some genetic experiments with them. He raised mice and may have conducted genetic experiments with them as well. He observed and recorded the movement of sunspots and hypothesized that they had an effect on weather patterns. Over several years, he measured ozone concentrations in the air. He recorded daily meteorological observations and groundwater levels in the monastery's well until his death and wrote a scientific paper about his observations of a tornado.

By 1875, Mendel became embroiled in a time-consuming dispute over the taxation of church property. In spite of overwhelming pressure from high government officials, colleagues, and acquaintances, including offers of honorary awards if he would cease his opposition, he remained firm throughout the rest of his life in his resolve that the tax was unjust. He died in 1884, unaware that 16 years later his work in genetics would resurface from obscurity, and that he would come to be considered the founder of modern genetics.

The Rebirth of Mendelian Genetics

After Mendel's paper had lain unrecognized for 35 years, three botanists, Hugo De Vries, Carl Correns, and Erich von Tschermak, published in 1900 results from independent experiments that included some of the same ratios Mendel had earlier reported. As each of these scientists became aware of Mendel's paper, they gave Mendel credit for discovering the principles of inheritance 35 years earlier than they had. All three botanists observed Mendelian patterns of inheritance in plant species, but the confirmation in animals was soon to follow. Two other scientists, William Bateson and Lucien

Cuénot, independently confirmed Mendel's principles in fowl and mice, respectively, in 1902. With this surge of independent experiments confirming Mendel's principles of inheritance in a variety of species, Mendelism was accepted—rapidly by some, reluctantly by others—and the science of genetics was born. Perhaps the most active proponent of Mendelism was Bateson, who coined the term *genetics* in 1905, as well as many of the other terms we use in the discipline today. Bateson also wrote the first genetics textbook, which was published in 1909 under the title *Mendel's Principles of Heredity*. From that time onward, Mendel has been universally recognized as the founder of modern genetics.

Criticisms of Mendel's Work

In spite of the universal recognition of Mendel's contributions, he has not been immune from criticism. His most extreme critics claim that he intentionally fabricated his results. This conclusion is based primarily on a very well written paper by Sir Ronald A. Fisher (cited in Example 12.6). Fisher used chi-square analysis to demonstrate that Mendel's results were remarkably close to expected ratios. In fact, from a probability standpoint, the results were so close to the expected ratios that the truthfulness of Mendel's data has been called into question. Fisher's statistical analysis of Mendel's results is correct, implying that taken as a whole Mendel's data are "too good to be true." This observation has convinced several modern writers that Mendel was a fraud, although Fisher concluded from his extensive study of Mendel's experiments that Mendel was truthful.

There are several credible explanations of Mendel's results that do not imply fraud. Mendel stated in his paper that he had conducted many more experiments than those he reported, and it is clear from his letters to Nägeli that he planned to publish a more detailed account of all his experiments later. In his classic paper, he probably reported the results of those experiments that had ratios close to those he expected and, therefore, best demonstrated his principles. Some scientists have suggested that Mendel unconsciously misclassified questionable phenotypes so the results of his experiments were close to his expectations. Unfortunately, as Mendel's original notes were burned, all explanations (including fraud) of the "too good to be true" data are speculative.

Several aspects of Mendel's paper itself lend weight to the argument that he was truthful. As we saw in Example 12.6, Mendel reported the results of an experiment that deviated from his expectations more than he thought it should. As it turns out, the deviation was not statistically significant according to a chi-square analysis, but in Mendel's mind it was too far from what he expected. So he repeated the experiment and the

second time obtained results that fit the expected ratio more closely. Had Mendel fabricated or manipulated his data, it seems unlikely that he would have reported the results of this first experiment.

Not all of the results Mendel reported in his paper closely fit the patterns he described using the seven traits in peas. Near the end of his paper, Mendel described the results of flower color variations in beans. In the F_2 generation from a cross between purple- and white-flowered plants, he obtained 31 plants that varied in color "from purple to pale-violet and white." Only one of the 31 plants was white, instead of one in four as might be expected based on the experiments with peas. Mendel interpreted this experiment as one in which two or more genes governed flower color. He hypothesized (correctly) that if two genes governed flower color, one plant in 16 should be white flowered, and if three genes governed it, one plant in 64 should be white flowered. As his result (one plant in 31) fell in between these two hypothesized ratios, he could not be sure which was correct. He recognized the limitations of such a small sample size, stating

> It must, nevertheless, not be forgotten that the explanation here attempted is based on a mere hypothesis, only supported by the very imperfect result of the experiment just described.

Had Mendel manipulated data to fit expected ratios, it is unlikely that he would have reported the results of an inconclusive experiment.

Statistical tests, such as chi-square analysis, had not been developed at the time Mendel conducted his experiments. Instead, he relied on his own intuition to interpret his data. Let's see if his interpretation of the results from the experiment with beans is statistically valid.

Example 12.9 Statistical justification of Mendel's interpretation of his experiment with flower color in beans.

In his experiment with flower color in beans, Mendel counted 31 plants with colored flowers and 1 plant with white flowers among a total of 32 F_2 plants. He concluded that these results did not represent a 3:1 ratio. Instead, he concluded that they represented either a 15:1 or a 63:1 ratio and that the number of plants was too small to distinguish between these two ratios.

Problem: Perform chi-square analysis to determine whether or not Mendel's conclusions are justified statistically.

Solution: Let's first test the hypothesis of a 3:1 ratio. The expected values are $\frac{3}{4} \times 32 = 24$ plants with colored flowers and $\frac{1}{4} \times 32 = 8$ plants with white flowers. $\chi^2 = [(31 - 24)^2 \div 24] + [(1 - 8)^2 \div$

8$] = 8.17$ with 1 degree of freedom. The chi-square value exceeds the 0.01 critical value of 6.64, which is highly significant. Mendel's rejection of a 3:1 ratio is justified statistically. Now let's test the hypothesis of a 15:1 ratio. The expected values are $\frac{15}{16} \times 32 = 30$ plants with colored flowers and $\frac{1}{16} \times 32 = 2$ plants with white flowers. $\chi^2 = [(31 - 30)^2 \div 30] + [(1 - 2)^2 \div 2] = 0.533$ with 1 degree of freedom, which is not significant. Now let's test the hypothesis of a 63:1 ratio. The expected values are $\frac{63}{64} \times 32 = 31.5$ plants with colored flowers and $\frac{1}{64} \times 32 = 0.5$ plants with white flowers. $\chi^2 = [(31 - 31.5)^2 \div 31.5] + [(1 - 0.5)^2 \div 0.5] = 0.508$ with 1 degree of freedom, which is also not significant. Because the chi-square values for the 15:1 and 63:1 ratios are both below the 0.05 critical value of 3.84 for 1 degree of freedom, there is no evidence to reject either hypothesis. Even though the chi-square value for the 63:1 hypothesis was slightly lower, neither the 15:1 nor the 63:1 hypothesis can be rejected. Mendel's interpretation of these data is fully justified statistically.

Fisher's observations had a firm statistical foundation and raised some important scientific questions that are still being addressed. However, a number of other unfortunate criticisms with no scientific foundation have been leveled at Mendel that continue to reappear all too often. Most of these criticisms can be easily dismissed simply by reading Mendel's paper. For example, Mendel has been criticized for presenting too simplistic an approach to inheritance. The most common criticism is that he presented data on only seven traits that all obeyed the laws he proposed, but he must have intentionally ignored obvious traits whose inheritance cannot be explained as easily.

This criticism has no foundation. In his paper, Mendel described several traits that could not be as easily studied as the seven traits he examined. These traits included the length of the pea flower peduncle (the flower's "stem"), time of flowering in peas, and (as we mentioned earlier) flower color in beans. The descriptions Mendel gives of these traits provide even further insight into his commitment to careful experimentation as a scientist. In discussing time of flowering, he stressed the need for the researcher to carefully account for differences that may be caused by planting methods and temperature. He also explained his reasoning for not including traits that fail to show discrete differences, and correctly concluded that

> The uniformity of behaviour shown by the whole of the characters submitted to experiment permits, and fully justifies, the acceptance of the principle that a similar relation exists in the other characters which appear less sharply defined in plants, and therefore could not be included in the separate experiments.

Unfortunately, Mendel is usually not given credit for additional principles of inheritance that he described. For instance, the genetic basis for inbreeding (which we will discuss in Chapter 20) is rarely attributed to Mendel, although he described it at some length in his paper. He correctly described pleiotropy, continuous variation, epistasis, and the influence of the environment on gene expression, all topics we will discuss in the next chapter. He also briefly described a case of heterosis, also a topic for Chapter 20. The overall picture that emerges of Mendel is one of a careful and meticulous scientist who was reluctant to draw conclusions until he had substantial convincing evidence.

Another side of Mendel is often lost in our emphasis on his scientific research. We must remember that genetics was not Mendel's profession. Although he taught science, conducted experiments, presented and published his results, corresponded with some of the leading scientists of his day, and had an obvious curiosity about the world that surrounded him, his chosen profession was that of a teacher and priest. Those who knew Mendel best were his students and people whom he served. Hugo Iltis, in writing Mendel's biography, interviewed many of the less influential people who knew Mendel well. The person they remembered was a cheerful teacher who devoted himself to his duties and who used his position and resources to assist those in need. This was perhaps Mendel's greatest legacy, and it merits recognition alongside his contribution to science.

SUMMARY

1. Mendelian genetics is a central theme of genetics. It is named after Gregor Mendel, who formulated the basic principles of inheritance in the nineteenth century based on hybridization experiments with pea plants.

2. Several key terms in Mendelian genetics can be defined in the context of molecular biology. A locus is the position of a gene in the DNA of a chromosome. Different DNA sequences at the same locus are different alleles. Because there are two homologues for each chromosome, there are two copies of each locus, one on each of the homologues. If both homologues have the same allele at a locus, the individual is homozygous for that allele. If the alleles differ between homologues at a locus, the individual is heterozygous for those alleles. If one allele masks the effect of another in a heterozygous individual, then the masking allele is dominant over the other allele. Conversely, the masked allele is recessive to the dominant allele. Genotype describes the genetic composition of alleles at a locus, whereas the phenotype is what we actually see in the individual.

3. Using modern terms, we may state the principle of segregation as follows: Because homologous chromosomes segregate from each other during meiosis, alleles at the same locus on homologous chromosomes also

segregate from each other, so half the gametes receive one allele and half receive the other.

4. Using modern terms, we may state the principle of independent assortment as follows: Because genes located on nonhomologous chromosomes assort independently during meiosis, the inheritance of alleles at one locus does not influence the inheritance of alleles at another locus.

5. Genes that are closely linked to each other on the same chromosome do not assort independently during meiosis.

6. The laws of probability apply to Mendelian genetics and can be used to predict the outcomes of genetic experiments.

7. Researchers can use statistical tests to assist them in drawing conclusions from the results of genetic experiments.

QUESTIONS AND PROBLEMS

1. What are the possible genotypes and their expected frequencies in the first-generation offspring of the crosses (a) *AA BB* × *Aa Bb*, (b) *Aa Bb* × *aa Bb*, (c) *Aa bb* × *aa Bb*, and (d) *Aa Bb Cc* × *aa Bb cc*?

2. Suppose that an individual plant is known to be heterozygous at 15 different loci. (a) How many possible genotypic combinations of gametes can this individual produce? (b) How many possible genotypic combinations can appear in the progeny derived from self-fertilization? (c) How many possible phenotypic combinations can appear if one allele at each locus is dominant over the other allele?

3. Take two coins and flip them simultaneously 48 times. Keep track of the number of times they both turn up heads ("homozygous" heads), both turn up tails ("homozygous" tails), and turn up heads and tails ("heterozygous"). When you are finished, perform a chi-square test to see if the results conform to the hypothesis of a 1:2:1 ratio.

4. In the chi-square analysis you performed for the previous question, how many degrees of freedom are there?

5. Fisher (1936. *Annals of Science* 1:115–137) concluded that Mendel's results were exceptionally close to expected ratios. Do a chi-square analysis on each of Mendel's experiments described in Table 12.1. Are the results of these experiments exceptionally close to 3:1 ratios? Answer this question in terms of the chi-square values and their associated probabilities from the chi-square table in Appendix I.

6. Given that an individual's genotype is *AA Bb Cc dd*, determine the genotypes of all possible gametes that this individual could produce.

7. Most missense mutations create a recessive allele.

Using the biochemical pathway model and the function of enzymes described in this chapter and in Chapter 6, describe why this is the case.

8. In maize, wrinkled kernels are caused by homozygosity for a recessive allele called *su*, for "sugary." The kernels are wrinkled because they have abundant sugar and very little starch. Full kernels have abundant starch and very little sugar and carry at least one dominant allele (*Su*) at the *sugary* locus. Sweet corn (the type we eat as corn on the cob) is often homozygous *su su*, whereas field corn (the type used to make corn chips and tortillas) carries the dominant *Su* allele. Starch is composed of long chains of sugar (glucose) subunits. Propose an explanation of why *Su* is dominant over *su*.

9. Less commonly, missense mutations can create a dominant allele. Using the information you have learned in this and previous chapters, devise one or more models that describe how such a dominant allele might arise by mutation.

10. Technically, a same-sense mutation creates a new allele because it alters the DNA sequence. However, same-sense mutations are not likely to be recognized as new alleles. Why?

11. Among all individuals of a species, (a) can there be more than one dominant allele for a particular locus? (b) Can there be more than one recessive allele for a particular locus? Justify your answers to these questions using mutations in DNA sequence as a basis for your explanation.

12. Using the examples described in this chapter, state whether the observation of a dwarf, white-flowered pea plant indicates its phenotype, its genotype, or both. Explain your answer.

13. Among people of European descent, about 4% have an allele for the genetic disorder cystic fibrosis. Yet only about 1 in 2500 people actually have the disorder. What is the most likely reason for this?

14. Rabbits with inherited albinism have white hair and pink eyes. This phenotype is due to the absence of melanins in the skin, hair, and eyes. (The pink color in the eyes is due to blood vessels in the iris being visible in the absence of pigment.) Why is the albinism allele typically recessive?

15. The recessive disorder phenylketonuria is caused by the absence of functional phenylalanine hydroxylase. Albinism is caused by the absence of functional tyrosinase. Both of these enzymes appear in the pathway for melanin synthesis. Suppose an individual is heterozygous for both the recessive albinism allele and the recessive phenylketonuria allele. Which, if any, of the two disorders will this individual have? Why?

16. Suppose a woman with the genotype described in the previous question marries a man with her same genotype for these alleles. What is the probability that any child of theirs will have (a) albinism, (b) phenylketonuria, (c) albinism and phenylketonuria?

17. Suppose that hybridization of a purple-flowered, tall pea plant with a purple-flowered, dwarf pea plant produced the following progeny:

Purple flowers, tall	58
Purple flowers, dwarf	64
White flowers, tall	15
White flowers, dwarf	21

Use the symbols *A* and *a* to represent the alleles that govern in flower color and *B* and *b* to represent those that govern in plant height. (a) What are the genotypes of the parent plants? (b) Test the hypothesis that these genotypes are correct using chi-square analysis.

18. Using the forked-line approach, determine the expected genotypic and phenotypic ratios of the cross *Aa Bb cc* × *aa Bb Cc*. Assume that each uppercase letter represents a dominant allele, that each lowercase letter represents a recessive allele, and that all loci assort independently.

19. A couple, neither of whom has phenylketonuria, have a child with phenylketonuria as their first child. They plan to have three more children. What is the probability that at least one of these additional three children will have phenylketonuria?

20. In rabbits, black hair is dominant over white. A black male rabbit whose mother was white is mated with a white female rabbit. If these two rabbits have eight offspring, what is the probability that three of the offspring will be black and the other five white?

21. Let's add a slight twist to the previous problem. Suppose that the two rabbits mated in the previous problem have eight offspring. What is the probability that *at least* five will be white?

22. The white-flowered pea plants that Mendel used for his studies were homozygous for a mutant allele at the *A* locus. In 1986, Hrazdina and Weeden (*Biochemical Genetics* 24:309–317) attempted to identify the enzyme encoded by the *A* locus by studying enzyme activity in pea plants with the genotype *aa*. They postulated that the enzyme chalcone synthase, which appears early in the anthocyanin synthesis pathway, may be the enzyme encoded by the *A* locus. They found that chalcone synthase activity was the same in plants with the genotype *AA* and the genotype *aa*. What do these results suggest about the relationship between the *A* locus and chalcone synthase?

23. One of the intermediates in the anthocyanin synthesis pathway in pea flowers is *p*-coumaric acid. Hrazdina and Weeden (1986. *Biochemical Genetics* 24:309–317) found *p*-coumaric acid in plants with the genotype *aa*.

What do these results suggest about the location of *p*-coumaric acid in the anthocyanin synthesis pathway relative to the enzyme encoded by the *A* locus in peas?

24. In 1974, Stratham and Crowden (*Phytochemistry* 13: 1835–1840) reported that at least six genes, *A, Am, Ar, B, Ce,* and *Cr,* govern flower color in peas. How is it possible for six or more genes to affect a single trait? State your answer in molecular and cellular terms.

25. Mendel hybridized plants that were true breeding for yellow, round seeds and purple flowers with plants that were true breeding for green, wrinkled seeds and white flowers. Mendel designated the alleles as follows:

A	round seeds	*a*	wrinkled seeds
B	yellow seeds	*b*	green seeds
C	purple flowers	*c*	white flowers

Mendel carried this experiment through the F_3 generation so that he could classify the F_2 plants by genotype. The numbers of plants by genotype are provided in Table 12.6. **(a)** Organize the results of this experiment in a forked-line diagram. Include the expected and observed numbers of individuals at each branchpoint in the diagram. **(b)** Consolidate these data into three experiments in which each locus is considered irrespective of the other two loci. Use a chi-square test to examine the observed ratios for each of these consolidated experiments, comparing them to the hypothesized ratio of 1:2:1. **(c)** Conduct a chi-square test of the hypothesis that these data conform to independent assortment using the entire data set with 26 degrees of freedom. (Note that you will need to use the chi-square table in Appendix I to answer this part of the question.) **(d)** What conclusions can be drawn from these chi-square tests?

26. In one experiment, Mendel crossed a purple-flowered, tall plant that had come from previous hybridization (i.e., it was not true breeding) with a white-flowered, dwarf plant. The results were as follows:

Purple flowers, tall	47
White flowers, tall	40
Purple flowers, dwarf	38
White flowers, dwarf	41

Using *A* to represent the locus for flower color and *B* to represent the locus for height, write the genotypes of the two parents and the genotypes of all four classes of progeny.

27. In 1913, Bateson (*Mendel's Principles of Heredity.* Cambridge: Cambridge University Press, p. 153) reported data from a dihybrid experiment with sweet peas.

Table 12.6 Information for Problem 25: Number of F_2 Plants with Each Genotype in Mendel's Trihybrid Experiment

AA BB CC	8	AA BB Cc	22	AA Bb Cc	45
AA BB cc	14	AA bb Cc	17	aa Bb Cc	36
AA bb CC	9	aa BB Cc	25	Aa BB Cc	38
AA bb cc	11	aa bb Cc	20	Aa bb Cc	40
aa BB cc	8	AA Bb CC	15	Aa Bb CC	49
aa BB CC	10	AA Bb cc	18	Aa Bb cc	48
aa bb CC	10	aa Bb CC	19		
aa bb cc	7	aa Bb cc	24	Aa Bb Cc	78
		Aa BB CC	14		
		Aa BB cc	18		
		Aa bb CC	20		
		Aa bb cc	16		

One parent had dark axils and was male-fertile. The other parent had light axils and was male-sterile. All the F_1 plants had dark axils and were male-fertile. The F_2 progeny phenotypes were as follows:

Dark axils, male-fertile	627
Dark axils, male-sterile	27
Light axils, male-fertile	17
Light axils, male-sterile	214

(a) Which traits are dominant and which are recessive? **(b)** Conduct a chi-square test of the hypothesis that the inheritance of axil color conforms to Mendel's principle of segregation. Phrase your conclusion in terms of hypothesis testing. **(c)** Conduct a chi-square test of the hypothesis that the inheritance of male fertility conforms to Mendel's principle of segregation. Phrase your conclusion in terms of hypothesis testing.

28. In another cross with sweet pea, Bateson (1913. *Mendel's Principles of Heredity.* Cambridge: Cambridge University Press, p. 154) reported data from a dihybrid experiment for flower color and flower structure. One parent had colored, erect flowers, and the other parent had white, hooded flowers. All the F_1 plants had colored, erect flowers. The F_2 progeny phenotypes were as follows:

Colored, erect	108
Colored, hooded	36
White, erect	84
White, hooded	28

(a) Which traits are dominant and which recessive? (b) Conduct a chi-square test of the hypothesis that the inheritance of flower color in this cross conforms to Mendel's principle of segregation. Phrase your conclusion in terms of hypothesis testing. (c) Conduct a chi-square test of the hypothesis that the inheritance of flower shape in this cross conforms to Mendel's principle of segregation. Phrase your conclusion in terms of hypothesis testing.

29. Bateson (1913. *Mendel's Principles of Heredity*. Cambridge: Cambridge University Press, p. 226) reported data from several human pedigrees that included individuals with albinism. In each family, neither parent had albinism. Among the children in these families, 115 had albinism and 174 did not. Albinism is due to a recessive allele. Bateson concluded that "the affected are far in excess of expectation (on the hypothesis that they are ordinary recessives"). (a) Is Bateson's conclusion justified statistically? Show your statistical calculations. (b) Bateson did not observe these families directly. Instead, he relied on pedigrees given to him by several people. With this in mind, provide several possible explanations for the discrepancy between the observed and expected values.

30. Among the families that Bateson used in his summary of human albinism was a family of eight children, seven of whom had albinism. What is the probability of this particular outcome assuming that albinism is due to a recessive allele and that neither parent had albinism?

31. Alkaptonuria is a human genetic disorder due to a recessive allele. Garrod (1902. *Lancet* 2:1616–1620) studied alkaptonuria in 17 families. In eight of those families, the parents were first cousins. Why might a genetic disorder due to a recessive allele be more likely in the offspring of close relatives than in the offspring of unrelated parents?

32. Carl Correns published his classic paper in which he reported his rediscovery of Mendel's principles in 1900 (*Berichte der deutchen botanischen Gesellschaft* 18:158–168; English translation in *Genetics* 35, no. 5, pt. 2:33–51). In one experiment with pea plants, Correns crossed a green-seeded variety with a yellow-seeded variety and found that all 51 F_1 seeds were yellow. He grew 19 F_1 plants from these seeds and allowed them to self-fertilize. Among the F_2 seeds, 619 were yellow and 206 were green. Use chi-square analysis to test the hypothesis that these results conform to Mendel's principle of segregation.

33. In his discussion of independent assortment, Correns stated (in translation) "In the case of *two* pairs of traits, *nine* different classes of individuals may occur. However, only *four* groups may be distinguished *externally*; the numbers of individuals in the classes must occur in a ratio of 9:3:3:1." (a) Restate the concept that Correns explained in this passage in terms of genotype and phenotype. (b) A Punnett square for the F_2 progeny of a dihybrid cross has 16 spaces in it. Why did Correns state that there were nine different classes instead of 16?

34. Mendel crossed a true-breeding variety of peas that had yellow pods with one that had green pods. He then collected the seed from 100 self-pollinated F_2 plants with green pods and grew the seed to obtain F_3 plants. In Mendel's own words (in translation), "the offspring of 40 [F_2] plants had only green pods; of the offspring of 60 plants, some had green, some yellow ones." Mendel felt that the observed ratio of 1.5:1 was too far from the expected 2:1 ratio, so he repeated the experiment the next year, the second time obtaining results that were closer to the expected 2:1 ratio. The results of this original experiment and the repeated experiment are listed in Table 12.2. (a) Conduct a chi-square test of the hypothesis that these results represent a 2:1 ratio. (b) Is Mendel's conclusion that these results were too far from the expected ratio justified statistically?

35. Read Mendel's classic paper on peas and make a note of all passages that deal with the principles of segregation, parental equivalence, and independent assortment. (a) Is it appropriate to attribute these principles to Mendel? Justify your answer by quoting and discussing the pertinent passages from his paper. (b) Is it appropriate to say that Mendel perceived alleles to be paired, as they are in prophase I of meiosis?

36. Find all the experimental results in Mendel's paper that demonstrate independent assortment. Of the 21 possible dihybrid experiments, which did Mendel present in his paper?

37. Mendel had a copy of Darwin's *Origin of Species*. On the first page is a note in Mendel's handwriting that reads "pag 302." On page 302, Mendel marked the following passage with double lines:

> The slight degree of variability in hybrids from the first cross or in the first generation, in contrast with their extreme variability in the succeeding generations, is a curious fact and deserves attention.

Identify the generations in this passage that we now call F_1 and F_2. What in this passage makes it obvious that Mendel would be interested in it?

38. Immediately following the marked passage quoted above, Darwin gave his explanation for this phenomenon:

> For it bears on and corroborates the view which I have taken on the cause of ordinary variability; namely, that it is due to the reproductive system being eminently sensitive to any change in the conditions of life, being thus rendered either impotent or at least incapable of its proper function of producing offspring identical with the parent form.

Now hybrids in the first generation are descended from species (excluding those long cultivated) which have not had their reproductive systems in any way affected, and they are not variable; but hybrids themselves have their reproductive systems seriously affected, and their descendants are highly variable.

In what ways does Darwin's explanation differ from the one that Mendel most likely would have given?

39. Shortly after the rediscovery of Mendel's paper in 1900, the applicability of Mendelian genetics to species other than domesticated plants and animals was disputed by some scientists. They believed that 3:1 ratios should be observed in natural populations if Mendelian principles applied to these populations. In dispute was the proposal that brachydactyly in humans was due to a dominant allele. Some scientists argued that, if this were true, 75% of humans should have brachydactyly according to a 3:1 Mendelian ratio. In 1908, Hardy wrote a letter to the editor of *Science* (28:49–50) explaining how natural populations should rarely exhibit Mendelian traits in 3:1 ratios even when Mendel's principles apply fully to the inheritance of those traits. Using what you have learned about Mendelian genetics, explain why natural populations usually do not have 3:1 ratios for Mendelian traits.

40. It is possible to determine how many individuals we need to have to meet a certain probability of obtaining at least one individual of the desired type, through some substitution and rearrangement of the binomial distribution equation. This type of calculation is often used in designing genetic experiments. If we let a equal the probability of obtaining at least one individual of the desired type, and b the probability of obtaining no individuals of the desired type, then $a = (p + q)^n - q^n$. $p + q = 1$, so $a = 1 - q^n$, which can be rearranged to give $q^n = 1 - a$. Taking the logarithm of both sides of the equation, we obtain $n(\log q) = \log(1 - a)$. Solving for n, we obtain

$$n = \frac{\log(1 - a)}{\log q} = \frac{\log b}{\log q}$$

Suppose we had an F_2 population of pea plants segregating for flower color and plant height. How many F_2 plants would we need to grow to be 95% certain of obtaining at least one white-flowered, dwarf plant?

FOR FURTHER READING

Mendel's classic paper was originally published in **Mendel, G. 1866. Versuche über Pflanzen-Hybriden.** *Verhandl naturf Verein Brünn* **4:3–47**. English translations of Mendel's paper are available from several sources. A link to the MendelWeb, with the original German version and an English translation, is available at the website for this book (www.wadsworth.com). Several books include a translation of the paper along with commentary. Among the most recent is **Corcos, A., and F. Monaghan. 1993.** *Gregor Mendel's Experiments on Plant Hybrids: A Guided Study.* **New Brunswick, N.J.: Rutgers University Press**. A book that contains Mendel's paper, Mendel's letters to Nägeli, papers by Mendel's discoverers, and Fisher's review of Mendel's work is **Stern, C. and E. Sherwood, eds. 1966.** *The Origin of Genetics: A Mendel Source Book.* **San Francisco: Freeman**. There are two excellent biographies of Mendel. The most recent was written by the former director of the Museum Mendelianum in Brno: **Orel, V. 1996.** *Gregor Mendel: The First Geneticist.* **Oxford: Oxford University Press**. The other is the first book-length biography of Mendel, which was written by Hugo Iltis and published in German in 1924. An English translation is **Iltis, H. 1966.** *Life of Mendel.* **Trans. Paul, E. and C. Paul. New York: Hafner Publishing Co**. A well-researched account of the development of Mendelian genetics is **Olby, R. C. 1985.** *Origins of Mendelism,* **2nd ed. Chicago: University of Chicago Press**. Fisher's review of Mendel's work is **Fishers R. A. 1936. Has Mendel's work been rediscovered?** *Annals of Science* **1:115–137**. An excellent review of experiments confirming Mendel's work shortly following the rediscovery is **Bateson, W. 1913.** *Mendel's Principles of Heredity.* **Cambridge: Cambridge University Press**. Bateson's book also contains an English translation of Mendel's classic paper and his paper on hawkweed, as well as a brief biography. An excellent demonstration of the binomial distribution, chi-square analysis, and their applicability to genetic experiments is **Detlefsen, J. A. 1918. Fluctuations of sampling in a Mendelian population.** *Genetics* **3:599–607.**

For additional reading, go to InfoTrac College Edition, your online research library at: http: www.infotrac-college.com/brookscole

CHAPTER 13

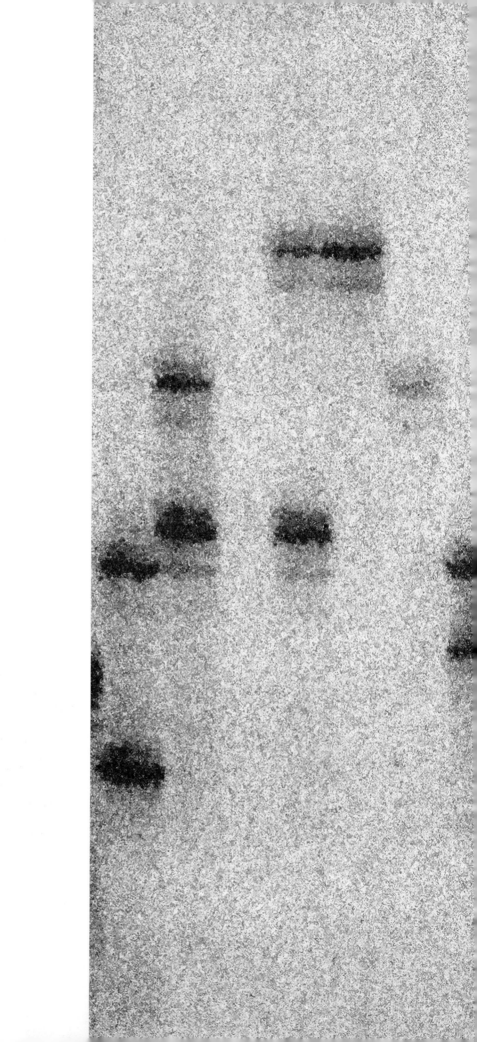

KEY CONCEPTS

Alleles are inherited according to Mendelian principles, but the traits
encoded by some alleles do not appear in typical Mendelian ratios because
of variations in the ways in which genes are expressed phenotypically.

~

Different types of gene expression and gene interaction can be
detected phenotypically.

~

Nucleotides pair with each other according to strict pairing rules
determined by the chemical properties of the nucleotides.

~

DNA markers are inherited like alleles that affect physical characteristics,
but do not have many of the limitations for genetic analysis that
physical traits may have.

VARIATIONS ON MENDEL'S THEME

Some traits do not pass from parent to offspring in accordance with the simple patterns of inheritance that Mendel described for peas. In fact, few traits are always inherited in a simple Mendelian fashion. Take eye color in humans, for instance. Eye color ranges over brown, green, hazel, gray, and blue. Some people have blue eyes because they are homozygous for a recessive allele, and inheritance of this genotype follows simple Mendelian patterns. For example, a family in which one parent is blue-eyed and the other brown-eyed may have children who are either blue-eyed or brown-eyed, just like the parents, with no children having eye colors that are intermediate between those of the parents. In this case, the blue-eyed parent is homozygous for a recessive allele that reduces synthesis of melanins in the iris of the eye. The brown-eyed parent is heterozygous, having a dominant allele that confers brown eyes. There is a 50% chance that any child of these parents will be brown-eyed and a 50% chance that he or she will be blue-eyed.

However, another couple in which one parent has brown eyes and the other blue eyes may have children with brown, green, hazel, gray, or blue eyes. It is even possible for two blue-eyed parents to have a brown-eyed child. The inheritance of eye color varies because eye color pigments are synthesized in a biochemical pathway

of several steps, each controlled by a different enzyme, with each enzyme encoded by a different gene. DNA sequence differences in various alleles of these genes allow for a wide variety of eye colors.

In each of the pairs of alleles Mendel studied, one was fully dominant and the other was completely recessive. But we will see in this chapter that dominance is not always complete. We will also see that in a population there may be many alleles at a single locus, that one gene may affect more than one trait, that several different genes may affect the same trait, and that the effect a gene has on a trait may be dependent on which alleles of other genes the individual carries and on nongenetic factors as well. All of these situations combine to produce the final phenotype of an individual, and phenotypic patterns of inheritance may appear to be quite complex.

This does not mean that Mendelian genetics rarely applies to life in the real world. Because of the way chromosomes segregate and assort during meiosis, each individual gene follows a simple Mendelian pattern of inheritance. However, phenotypes among progeny for many traits depart from typical Mendelian ratios, owing to the genetic complexities discussed in this chapter together with the effect of environmental factors. Mendel himself recognized this when he stated:

The uniformity of behaviour shown by the whole of the characters submitted to experiment permits, and fully justifies, the acceptance of the principle that a similar relation exists in the other characters which appear less sharply defined.

Let's now take a look at several variations on Mendel's general theme that help explain the variety of inheritance patterns we see in nature. During our discussion, we will encounter several types of genetic notation that may not be familiar. The accompanying box explains some of the most common types of genetic notation; you may wish to refer to it whenever you encounter a new type of notation.

13.1 MULTIPLE ALLELES AND DOMINANCE RELATIONS

Mendel studied only two alleles at each locus in his experiments. Although a single diploid individual can have only two alleles for any locus (because an individual has only two copies of each chromosome), there may be many alleles for a single locus within a population. Any alteration in the DNA sequence at a locus creates a new allele, meaning that any one locus may have an enormous number of potential alleles. When more than two alleles for a single locus are present in a population, they are called **multiple alleles**. In this section, we consider multiple alleles, as well as alleles that are not fully dominant, and genes that affect more than one trait.

Incomplete Dominance

With complete dominance, a heterozygous F_1 hybrid appears indistinguishable from its homozygous dominant parent, such as the tall offspring from a cross between tall and dwarf pea plants. However, dominance is not always complete. Long before Mendel conducted his experiments, hybrid plants and animals with phenotypes intermediate between those of their parents had been described. An intermediate phenotype is often due to **incomplete dominance**. A good example of incomplete dominance is inheritance of flower color in snapdragons (Figure 13.1). When a red-flowered plant is hybridized with a white-flowered plant, the F_1 offspring are pink-flowered, as if the two alleles blended with each other. However, there is no blending; the alleles themselves remain distinct. This is evidenced by plants in the F_2 generation, which segregate 25% red-flowered, 50% pink-flowered, and 25% white-flowered. In this case, flower color is governed by two alleles, R^1 and R^2. A homozygous R^1R^1 plant has red flowers; a homozygous R^2R^2 plant has white flowers; and a heterozygous R^1R^2 plant has pink flowers. As neither allele is dominant over the other, both are designated by uppercase letters that are numbered to distinguish them from one another.

homozygous parent × homozygous parent
R^1R^1 R^2R^2

F_1 heterozygotes
R^1R^2

Figure 13.1 Incomplete dominance for flower color in snapdragons. The phenotype of the heterozygote is intermediate between the phenotypes of the two homozygotes. (Photo courtesy of W. E. Ferguson.)

At the biochemical level, incomplete dominance is usually caused by an enzyme the quantity of which influences the *amount* of product that is produced at the end of the pathway. The R^1 allele in snapdragons encodes an enzyme in the pathway for synthesis of the red pigment, whereas the R^2 allele fails to encode a functional enzyme. In the heterozygote, only half the usual amount of enzyme is present. In this case, the amount of enzyme influences the amount of pigment that is produced, so heterozygotes produce less pigment than R^1R^1 homozygotes, and the flowers of heterozygotes are pink instead of red. When an allele confers complete dominance, the enzyme encoded by the allele may not be the limiting factor. A single allele may produce enough enzyme to maintain the biochemical pathway at full production. Alternatively, in some cases of complete dominance, gene regulation may compensate for an allele that fails to produce a functional enzyme by doubling the enzyme production from the functional allele in a heterozygous individual.

The amount of red pigment in the flowers of a heterozygous R^1R^2 snapdragon is about half the amount in the flowers of a homozygous R^1R^1 plant. Incomplete dominance does not require that a heterozygote's phe-

notype be exactly halfway between the two homozygotes. Sometimes the heterozygote may closely resemble one of the homozygotes but still be distinguished from it. For example, in some cases of incomplete dominance for flower color, a heterozygote may have only slightly less pigment in the flowers than the homozygote.

Dominant Mutant Alleles

Alleles are often classified as wild-type alleles or mutant alleles. **Wild-type alleles** are the alleles most often found in nature, and **mutant alleles** are deviations from the wild type. In most cases, a wild-type allele encodes a functional product and is dominant, and a mutant allele fails to encode a functional product and is recessive. However, there are exceptions.

Cancer in humans is often caused by dominant mutant alleles of proto-oncogenes and tumor suppressor genes, the genes responsible for promoting or inhibiting cancer. We will discuss these genes at length in Chapter 24, but let's use them briefly here as examples of how dominant mutations may arise. A proto-oncogene is a gene that encodes a product that stimulates cell division. Once a tissue has developed, its cells rarely divide any longer. Instead, the cells devote their energies (and their genes) to the functions required in the tissues of which they are a part. Regulatory mechanisms shut down the genes that stimulate cell division. Such regulation is usually at the level of transcription. Occasionally, a proto-oncogene mutates in the regulatory region, eliminating the inhibition of transcription. Once regulation over the gene's expression is lost, its product is produced at abnormally high levels. This is called a **gain-of-function mutation** because the mutation increases the amount of the gene's product. Most mutations, however, are **loss-of-function mutations**, which result in either an altered product or no product at all. Loss-of-function mutations generally create recessive alleles, and gain-of-function mutations generally create dominant alleles.

Gain-of-function mutations are often **constitutive**, meaning that the product is produced constantly. In cancer cells, a gain-of-function mutation may induce uncontrolled cell division. Under these circumstances, a single mutant allele causes the product to be produced in abnormally high amounts. If a proto-oncogene on one chromosome mutates, the gene is expressed. The mutation is dominant because a single allele causes the mutant phenotype when the other chromosome carries a normal allele (Figure 13.2).

The same gene may have some loss-of-function mutations that are dominant and others that are recessive. Let's look at the *TP53* gene in humans (OMIM 191170) as an example. This gene is an important tumor suppressor gene, a gene that suppresses cell division and is important in preventing cancer. Over half of all human cancers are associated with loss-of-function mutations in the *TP53*

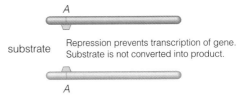

a Wild-type allele of gene *A* is repressed.

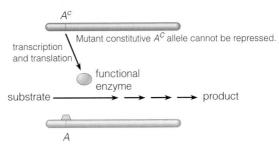

b Dominant gain-of-function mutation eliminates repression.

Figure 13.2 A dominant gain-of-function mutation. A mutation in one chromosome is dominant in causing constitutive expression of a gene that was repressed.

gene. We will discuss this gene in detail in Chapter 24, but for now it provides good examples of dominant and recessive loss-of-function mutations.

Whether particular loss-of-function mutations in the *TP53* gene are dominant or recessive has to do with the quaternary structure of the p53 protein. This protein is composed of four identical subunits, encoded by the *TP53* gene, that form a functional tetramer (Figure 13.3a). A missense point mutation may alter the functionality of a protein subunit, but this nonfunctional subunit may still associate with other subunits to form a tetramer. A cell that is heterozygous for a normal *TP53* allele and a mutant *TP53* allele produces both functional and nonfunctional subunits, which associate with one another to form tetramers. A tetramer may be composed of any possible combination of functional and nonfunctional subunits. However, if any of the subunits in the tetramer is nonfunctional, the tetrameric protein is nonfunctional. Functional and nonfunctional subunits in a heterozygous cell are expected to combine at random to give a binomial distribution in which $15/16$ of the tetramers have at least one nonfunctional subunit and only $1/16$ of the tetramers are composed entirely of functional subunits (Figure 13.3b). In reality, the nonfunctional subunits are less subject to degradation than the functional subunits, increasing the proportion of nonfunctional p53 proteins to greater than $15/16$. The proportion of functional tetramers in heterozygous cells is so low that the net effect is that of a dominant mutant allele—the phenotype of the heterozygote is indistinguishable from that of the mutant homozygote.

Dominant loss-of-function mutant alleles, such as the *TP53* mutant allele we just discussed, illustrate the concept of **threshold effects** in the expression of certain genes.

Mendel introduced the first genetic notation. He used an uppercase letter to symbolize a dominant allele (such as *A*) and the same lowercase letter to symbolize the recessive allele of the same gene (such as *a*). He symbolized different genes by different letters (*A* and *B*, for instance). We used this notation throughout Chapter 12. It is still used today and serves as the basis for some of the more detailed forms of notation that have since been developed.

Genetic notation used for different species is not uniform. Historically, some geneticists developed their own systems of notation, which caused a lack of uniformity for notation, even for a single species. Eventually, geneticists who worked with certain model organisms developed standardized notations for some species. What works well for one species, however, may not work particularly well for another. Differences in notation systems from one species to another reflect differences in the genetic attributes of species and the different conventions accepted by geneticists. In this book we use the type of notation that is generally accepted for the species being discussed. The rules of genetic notation are detailed, so we'll highlight just a few of the more common ones here.

Genetic notation in most species

Genetic notation for most species is derived from Mendel's notation and has been expanded to accommodate such complexities as incomplete dominance, codominance, and multiple alleles. Each locus is designated by a letter or letters. Dominant alleles are designated by the uppercase letter, and recessive alleles by the lowercase letter. When the name of the locus and the letters representing alleles refer to genotypes, they are italicized. However, when the name and the letters refer to the phenotype, they are not italicized. For example, recessive alleles at a locus in maize cause kernels to have a bronze color. The locus is called the *bronze* locus, but the phenotype is bronze (no italics). The locus is designated *bz*. Dominant alleles are designated *Bz*, and recessive alleles *bz*.

Multiple alleles, codominant alleles, and alleles that exhibit incomplete dominance are usually identified by some sort of designation that follows the symbol representing the locus. This designation is often a superscript number or letter. For example, R^1 and R^2 in Figure 13.1 symbolize alleles with incomplete dominance for flower color.

Genetic notation in *Drosophila melanogaster*

Genetic notation for *Drosophila melanogaster* is based on Mendel's original notation, but many modifications have been made. The complete rules are explained on pages 1–4 of Lindsley, D. L., and G. G. Zimm. 1992. *The Genome of Drosophila melanogaster*. San Diego, Calif.: Academic Press.

As for most species, a locus name for *D. melanogaster* has an abbreviation consisting of one or more letters. For example, the *rosy* locus is designated *ry*, and the *brown* locus *bw*. Most wild *D. melanogaster* flies have similar phenotypic characteristics, so the phenotypic traits most often encountered in nature are called *wild type phenotypes*. Alleles that confer a wild-type phenotype are designated with a plus sign (+), whereas mutant alleles at the locus are designated by the locus abbreviation. For example, the wild-type allele at the *brown* locus is designated "+" (or sometimes bw^+ to avoid ambiguity). Most mutant alleles at the *brown* locus are recessive, so they are designated *bw*, usually with a superscript letter or number (such as bw^6) to distinguish among various mutant alleles at the locus. The few mutant alleles that are dominant are designated with an uppercase superscript (such as bw^D for *brown-Dominant*). Loci with mutant alleles that are typically dominant are designated by uppercase letters. For example, the mutant phenotype Curly (wings curled upward) is caused by a dominant mutant allele called *Cy* at the *Curly* locus. A rare recessive mutant allele of a locus that typically has dominant mutant alleles is usually designated by a lowercase superscript. For example, most mutant alleles at the *Henna* locus (a locus affecting eye color) are dominant, and a recessive allele is designated Hn^r for *Henna-recessive*.

Flies that are homozygous for an allele are designated by just one symbol for that allele. When only one symbol is present, it is understood that the flies are homozygous for that allele. For example, a fly designated

```
1.224|7|26|95|1p13.2-p12|SLC16A1, MCT1|P|Solut
1.225|11|20|95|1p13.1|CD2|C|CD2 antigen (p50),
1.226|4|30|93|1p13.1|HSD3B1|C|Hydroxy-delta-5-
1.229|11|20|95|1p13.1|NGFB|C|Nerve growth fact
1.230|6|30|98|1p13.1-q21.3|PTGFRN, FPRP|P|Pros
1.231|10|12|90|1p13|CD58, LFA3|C|CD58 antigen
1.232|5|18|92|1p13|GNAI3|C|Guanine nucleotide-
```

bw^6 is homozygous for the bw^6 allele. Flies that are heterozygous have the two alleles at the locus separated by a slash. For example, a fly heterozygous at the *brown* locus for the wild-type allele and bw^6 is designated $+/bw^6$. A fly heterozygous for the bw^5 and bw^6 alleles is designated bw^5/bw^6.

Flies in research laboratories often have mutant alleles at several loci. This has led to a standardized type of notation for designating the entire mutant genotype of a fly. In the genotype for two loci on the same chromosome, the allele abbreviations are separated by a space. For example, the *brown* and *scute* loci are both located on chromosome II. A fly that is homozygous for the *bw* and *sc* alleles is designated *bw sc*. A fly heterozygous for the wild-type and mutant alleles at both loci is designated *+/bw +/sc*. For two loci on different chromosomes, the designations are separated by a semicolon. For example, a fly that is homozygous for *sc* (located on chromosome II) and heterozygous for wild-type and mutant alleles at the *rosy* (*ry*) locus (located on chromosome III) is designated *sc; +/ry*.

Genetic notation in humans

For many years, genetic notation in humans was not standardized but generally followed Mendel's notation style, with upper- and lowercase letters representing dominant and recessive alleles, respectively. Now, genetic notation has been standardized by the Human Gene Nomenclature Committee and reflects the impact that molecular biology has made on human genetics. The rules are listed in Shows et al. 1987. Guidelines for human gene nomenclature. *Cytogenetics and Cell Genetics* 46:11–28.

The current names and symbols for each human gene are available from the Online Mendelian Inheritance in Man (OMIM), a website that provides frequently updated information on all human genes that have been identified. A link to the OMIM database can be found at the website http://brookscole.com/biology. From this point onward, we will provide the OMIM entry number for each human gene we discuss.

Ideally, human loci are named for the enzyme or gene product they encode, although they may be named for a phenotype when the gene product is not known. If the gene product is discovered, the name of a locus can be changed. For example, a locus responsible for synthesis of a blood antigen called the H substance was originally named the *H* locus, with two alleles, *H* and *h*. This locus has since been renamed the *fucosyltransferase-1* locus after the enzyme it encodes. The symbol of a locus consists of uppercase letters, sometimes combined with numbers, that reflect the name of the locus. For example, the *fucosyltransferase-1* locus is designated *FUT1* (OMIM 211100). Alleles are designated by letters or numbers and are separated from their locus designation by an asterisk. For instance, the two alleles formerly called *H* and *h* are now named *FUT1*H* and *FUT1*O*, respectively. Heterozygous genotypes are written with a slash separating the two alleles, as in *Drosophila*. For example, an individual heterozygous for the two *FUT1* alleles has the genotype *FUT1*H/*O*. Notice that it is not necessary to repeat the locus name for the second allele in heterozygotes.

Once the nucleotide and amino acid sequences of a locus are known, mutant alleles are designated in a way that denotes the alteration in the nucleotide or amino acid sequence. For example, phenylketonuria (PKU) is a recessive genetic disorder in humans caused by mutations in the gene that encodes the enzyme phenylalanine hydroxylase (PAH). The locus is named *PAH* (OMIM 261600) after the enzyme it encodes, and many mutant alleles have been characterized. One mutant allele is called *PAH*R408W*. The designation *R408W* indicates that a substitution mutation in codon 408 caused arginine (R) to be replaced by tryptophan (W). Alternatively, substitution mutations may be indicated by the three-letter designations of the amino acids, such as *PAH*Arg408Trp*, which is the same as *PAH*R408W*. Mutations other than substitutions can also be indicated in the allele designation. For example, *PAH*ΔtF55fs* represents a mutant allele that arose from a deletion mutation (Δ) of a thymine (t) in codon 55, which encodes phenylalanine (F). The *fs* designation indicates that this mutation causes a frameshift. The allele designation *PAH*IVS2nt5* represents a mutation at the fifth nucleotide (*nt5*) in the second intron (*IVS2* = second intervening sequence).

family 16 (monocarboxylic acid transporters), m
blood cell receptor||186990|REa, A, RE|||||3(
hydrogenase, 3 beta- and steroid|delta-isomeras
162030|REa, H, A, Fd, RE|same 310kb fragment as
F2 receptor negative regulator||601204|Psh, REc
e function-associated antigen 3)||153420|S, REa
otein (G-protein), alpha-inhibiting|activity po

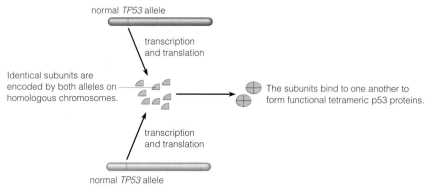

a Normal situation for *TP53* gene.

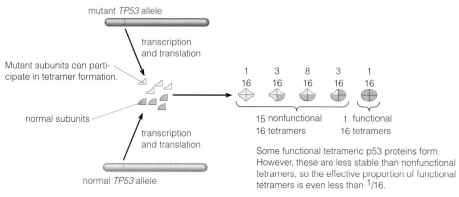

15 nonfunctional   1 functional
16 tetramers   16 tetramers

Some functional tetrameric p53 proteins form. However, these are less stable than nonfunctional tetramers, so the effective proportion of functional tetramers is even less than ¹/16.

b Dominant loss-of-function mutation in *TP53* gene.

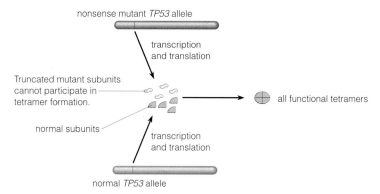

all functional tetramers

c Recessive loss-of-function mutation in *TP53* gene.

Figure 13.3 Loss-of-function mutations in the *TP53* gene.

For some genes, the normal phenotype is only expressed when a particular threshold level of gene-product function is attained. When function drops below the threshold level, the mutant phenotype appears because there is insufficient product to confer the normal phenotype, even though there is *some* functional product present. Heterozygosity for a dominant loss-of-function allele in the *TP53* gene permits some functional product to be produced, but the amount of functional product is below the required threshold, so the mutant phenotype appears.

Some loss-of-function mutant alleles of the *TP53* gene are recessive, however. These are usually due to frameshift or nonsense mutations that cause the mutant subunits to be substantially different from the functional subunits. These highly altered subunits cannot participate in forming a tetramer. Under these conditions, only fully functional tetramers are formed in heterozygous cells, all from the functional subunits encoded by the normal allele (Figure 13.3c).

Codominance

The locus that governs the ABO histo-blood group in humans provides one of the best examples of a phenome-

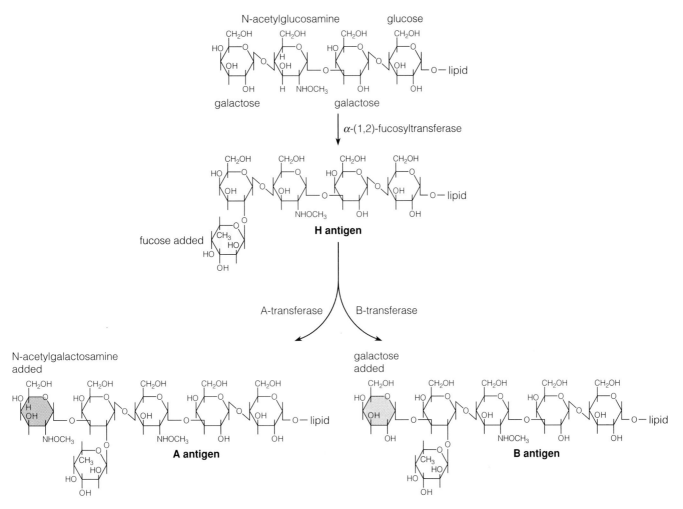

Figure 13.4 The two final steps in synthesis of ABO blood group antigens.

non called codominance, and also an example of multiple alleles. On the surface of human epithelial cells and red blood cells, there is a glycolipid, a lipid molecule with sugars attached. The lipid has an attached oligosaccharide that is usually composed of five or six sugars (in rare cases only four). Differences in the sugar composition of this oligosaccharide cause the four major ABO blood types: A, B, AB, and O. If blood types are not matched in particular combinations for a blood transfusion, there may be a fatal immune reaction to the transfusion. Because they can elicit an immune reaction, the different oligosaccharides are called **antigens**, substances that interact with antibodies produced by the immune system during an immune reaction.

The oligosaccharides are produced in a biochemical pathway that has several steps. We are concerned with the last two steps (Figure 13.4). In the next-to-last step, an enzyme called α-(1,2)-fucosyltransferase, encoded by the *FUT1* locus, converts a lipid-bound four-sugar oligosaccharide into a five-sugar oligosaccharide called the H antigen. Nearly every person has functional α-(1,2)-fucosyltransferase that produces the H antigen. In the last

step, another enzyme adds a sixth sugar to the H antigen. This enzyme is encoded by the *ABO* locus (OMIM 110300). There are three major alleles at the *ABO* locus: *ABO*A*, *ABO*B*, and *ABO*O*. (See the Genetic Notation box for an explanation of this notation.) The *ABO*A* allele encodes an enzyme called A-transferase (an abbreviation for α-3-N-acetyl-D-galactosaminyltransferase), which adds N-acetylgalactosamine to the H antigen, converting it into the A antigen. The *ABO*B* allele encodes a slightly different version of the enzyme called B-transferase (an abbreviation for α-3-D-galactosyltransferase), which adds galactose to the H antigen, converting it into the B antigen. The third allele, *ABO*O*, fails to encode a functional enzyme, so in the cells of a homozygote for this allele, no sugar is added to the H antigen.

Because a person has two alleles at a locus, there are six possible genotypes: *ABO*A/*A*, *ABO*A/*O*, *ABO*B/*B*, *ABO*B/*O*, *ABO*A/*B*, and *ABO*O/*O*. Figure 13.5 illustrates how these six genotypes give rise to the four phenotypes: type A, type B, type AB, and type O blood. Persons with type A blood display the A antigen on the surface of their red blood cells, those with type B blood

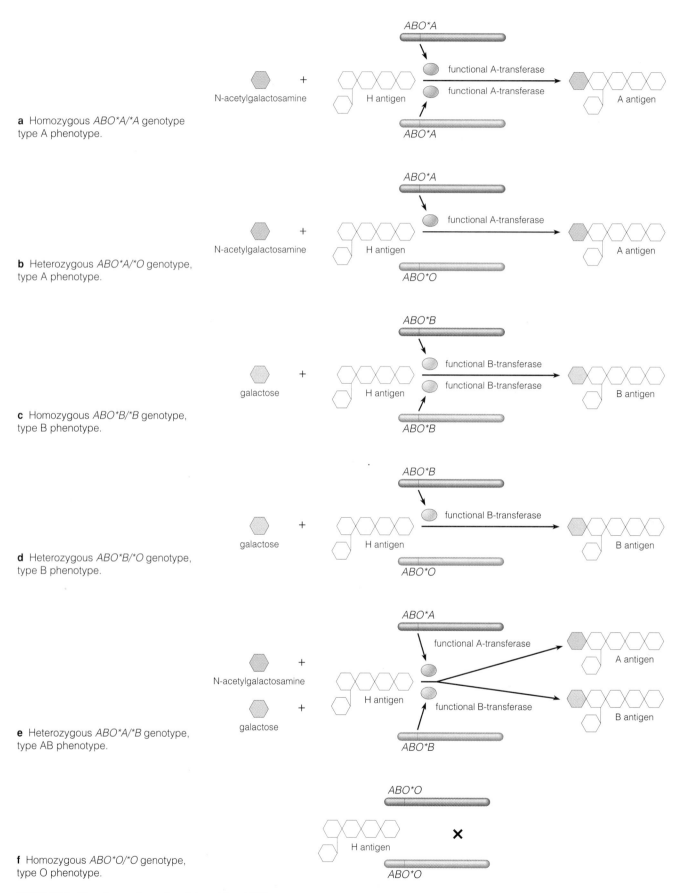

a Homozygous *ABO*A/*A* genotype type A phenotype.

b Heterozygous *ABO*A/*O* genotype, type A phenotype.

c Homozygous *ABO*B/*B* genotype, type B phenotype.

d Heterozygous *ABO*B/*O* genotype, type B phenotype.

e Heterozygous *ABO*A/*B* genotype, type AB phenotype.

f Homozygous *ABO*O/*O* genotype, type O phenotype.

Figure 13.5 The six genotypes for alleles at the *ABO* locus.

display the B antigen, those with type AB blood display both A and B antigens, and those with type O blood display the H antigen.

The *ABO*A* and *ABO*B* alleles are both completely dominant to the *ABO*O* allele, as diagrammed in Figures 13.5b and d. However, the *ABO*A* and *ABO*B* alleles are codominant to each other, as illustrated in Figure 13.5e. Unlike incomplete dominance, in which heterozygotes have a phenotype intermediate between the phenotypes of the two homozygotes, **codominance** is the distinct expression of both phenotypes in heterozygotes. In *ABO*A/*B* heterozygotes, both the *ABO*A* and the *ABO*B* alleles encode their respective functional forms of the enzyme, so some of the H antigen is converted into A antigen, and the remaining H antigen is converted into B antigen. Therefore, *ABO*A/*B* heterozygotes display both A and B antigens on the surface of their red blood cells, and for this reason have type AB blood. Both phenotypes (type A and type B) are distinctly present in the heterozygote.

Using this information, we can see why certain blood transfusions are successful and others are not (Table 13.1). The immune system produces antibodies that attack and destroy substances perceived as foreign to the body. Someone with type O blood cannot receive blood from anyone with type A, B, or AB blood because people with type O blood do not have either the A or B antigens and their immune systems perceive the A and B antigens as foreign. For this reason, a person with type O blood can only receive type O blood. People with type O, A, B, or AB blood can receive type O blood because they all have the H antigen as part of the biochemical pathway. People with type A blood can receive type A or O blood, but not B or AB because their immune systems recognize the B antigen as foreign. Similarly, someone with type B blood can receive type B or O blood, but not A or AB because the A antigen is perceived as foreign. People with type AB blood can receive type O, A, B, or AB blood because the H, A, and B antigens are all present in their cells.

Blood antigens encoded at other loci, such as the Rh factor, must also be taken into account for blood transfusions to succeed. To ensure maximum compatibility, most hospitals administer donated blood that exactly matches the blood type of the recipient, even though other transfusions may be successful.

A rare recessive allele at the *FUT1* locus (*FUT1*O*) causes an unusual phenotype when homozygous. This allele fails to encode functional α-(1,2)-fucosyltransferase, so when a person is homozygous *FUT1*O/*O*, the four-sugar oligosaccharide is *not* converted into the H antigen, so neither the A nor the B antigens can be produced regardless of the alleles the person carries at the *ABO* locus (Figure 13.6). This rare condition is called the **Bombay phenotype** because it was discovered in a woman from Bombay, India. People with the Bombay phenotype are usually diagnosed as having type O blood because they

Table 13.1 Compatible Blood Transfusions for the ABO Blood Groups

Blood Type of Recipient	Blood Type of Compatible Donor
Type A	Type A or type O
Type B	Type B or type O
Type AB	Type A, type B, type AB, or type O
Type O	Type O

have neither the A nor the B antigens. However, in terms of both the antigen structure (four instead of five sugars in the oligosaccharide) and the genetic behavior of the *FUT1*O* allele, the Bombay phenotype is distinctly different from type O. The blood type is designated O_h to indicate that the person lacks A, B, and H antigens.

It should be obvious at this point that the type of dominance a particular allele exhibits is dependent on the allele at the corresponding locus on the homologous chromosome. For example, *ABO*A* is completely dominant over *ABO*O* but codominant with *ABO*B*. The *ABO* locus offers a rather simple illustrative example of how multiple alleles may show different types of dominance. In this example, we dealt with only three alleles at the *ABO* locus. In reality, there are more than three alleles at this locus, most of them subgroups of *ABO*A*. There are often many alleles at a single locus in populations. For example, hundreds of different alleles have been identified at both the *white* and the *rosy* loci in *Drosophila melanogaster*.

Let's now take a close look at the alleles of the *ABO* locus at the molecular level.

Example 13.1 Molecular characterization of the *ABO* alleles.

In 1990, Yamamoto et al. (*Nature* 345:229–233) described the differences in DNA sequences for the major *ABO* alleles in humans and the amino acid sequences of the polypeptides encoded by these alleles. The researchers first isolated a cDNA clone of the *ABO*A* allele. This clone was then used as a probe to identify additional cDNA clones of the *ABO*A*, *ABO*B*, and *ABO*O* alleles. The DNA sequences of these clones were determined and used to derive amino acid sequences. The *ABO*A* and *ABO*B* alleles differ in four single-nucleotide substitutions, causing four amino acid differences in the two enzymes. The *ABO*O* allele differs from the *ABO*A* allele by only a single nucleotide deletion near the 5' end of the coding region. This deletion

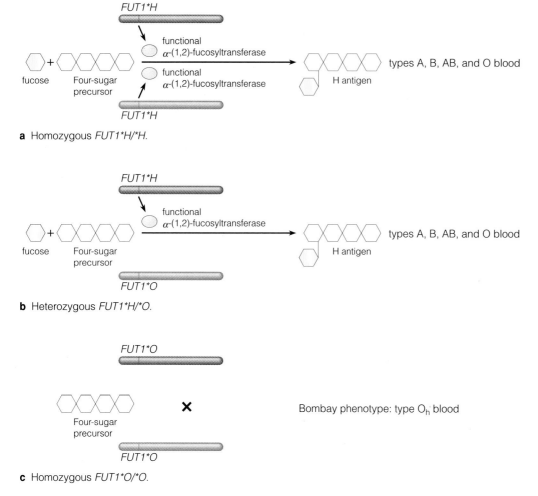

a Homozygous *FUT1*H/*H.*

b Heterozygous *FUT1*H/*O.*

Bombay phenotype: type O_h blood

c Homozygous *FUT1*O/*O.*

Figure 13.6 The Bombay phenotype.

is a frameshift mutation that brings a termination codon into frame and causes premature termination of translation.

Problem: Explain why the *ABO*A* and *ABO*B* alleles are codominant and the *ABO*O* allele recessive.

Solution: The products produced by the *ABO*A* and *ABO*B* alleles differ from each other in four amino acid residues. The two proteins encoded by these alleles are nearly identical but differ in their enzymatic functions. When present simultaneously in an *ABO*A/*B* heterozygote, both enzymes function to produce the A and B antigens. Because both antigens are produced, the two alleles are codominant. The *ABO*O* allele contains a frameshift mutation that causes the allele to encode a shortened polypeptide, which fails to function as an enzyme. The *ABO*O* allele is recessive because the *ABO*A* and *ABO*B* alleles compensate for its lack of function in *ABO*A/*O* and *ABO*B/*O* heterozygotes.

Dominance is not always complete. Alleles may show incomplete dominance or codominance. Although a diploid individual carries only two alleles at a locus, a population may have more than two alleles at a locus.

Leaky Recessive Alleles

A recessive mutant allele may encode a mutant enzyme that has some function, but only a reduced amount. Such alleles are called **leaky recessive alleles** because they do not block a biochemical pathway altogether when homozygous, but allow some of the final product to "leak" through. Several of the more than 100 known alleles of the *white* locus in *Drosophila melanogaster* are leaky alleles. The first allele discovered at this locus was a recessive allele that caused the fly to have white rather than brick-red eyes, which gave the locus its name. This allele (now called w^1) completely blocks synthesis of the two eye color pigments: a bright-red pigment and a brown pigment. Other alleles at the *white* locus reduce, rather than completely block, synthesis of one or both pigments.

a Wild-type fly with brick-red eye color.

b Fly homozygous for the w^1 (*white*) allele. All pigment synthesis is blocked.

c Fly homozygous for the leaky recessive w^a (*white-apricot*) allele. This genotype produces a small amount of eye pigment.

Figure 13.7 A leaky recessive allele for eye color in *Drosophila melanogaster*.

For instance, the w^a allele allows synthesis of a small amount of pigment, which gives the eyes a pale orange color (Figure 13.7). The designation w^a stands for *white-apricot*, because the color of the eyes is similar to that of an apricot. Another allele, w^{co}, blocks synthesis of most of the red pigment but allows brown pigment to be produced, giving the eyes a brown color. Both of these are examples of leaky recessive alleles.

~

When homozygous, leaky recessive alleles allow some of the dominant phenotype to appear, but at reduced levels.

Compound Heterozygotes and Dominance Series

Because many different alleles of the same gene may be present in a population, it is possible for an individual to be heterozygous at a single locus for two different mutant alleles. An individual heterozygous for two different recessive alleles has a recessive phenotype. For example, there are over 190 known mutant alleles of the *PAH* gene for the recessive genetic disorder phenylketonuria (PKU) in humans. Most people who have PKU inherited different mutant alleles from their two parents and are, therefore, heterozygous for two different mutant alleles even though they have a recessive phenotype. Individuals who are heterozygous for different alleles that confer the same phenotype are said to be **compound heterozygotes**.

Not all mutant alleles for the same gene confer the same phenotype when homozygous. Leaky recessive alleles, for example, may allow some of the wild-type phenotype to appear in individuals who are homozygous for the allele, whereas homozygosity for a different mutant allele at the same locus may completely eliminate the wild-type phenotype.

When more than two alleles are present in a population, there is the potential for a **dominance series** among the alleles. The classic example of this is the *c* locus in rabbits, which affects coat color. There are four common alleles at this locus: C, c^{ch}, c^h, and c. The C allele is a fully dominant, wild-type allele, and a rabbit with one or two copies of C has the agouti phenotype—the mottled brown-gray color of most wild rabbits. The appearance of the fur comes from hair shafts that are dark-colored

Figure 13.8 Hair shafts from an animal with the agouti phenotype.

(black, brown, or dark gray) at the tip and base with a yellow band in between (Figure 13.8). The c^{ch} (chinchilla) allele makes the coat light gray when homozygous or when heterozygous with c^h or c. The c^h (Himalayan) allele gives a white coat with black hair on the extremities when homozygous or when heterozygous with c. The c (albino) allele prevents pigment formation altogether when homozygous, giving a white coat. The dominance series can be expressed as

$$C > c^{ch} > c^h > c$$

where each allele is dominant over all alleles to its right and recessive to all alleles to its left. Figure 13.9 shows all possible genotypes for these four alleles and the phenotypes associated with each genotype.

~

The dominance relations of multiple alleles at a locus may form a series.

Lethal Alleles

A mutant allele that eliminates a function essential for survival may kill the organism. Recessive lethal alleles are not uncommon. Because they are expressed only in the homozygous condition, they may persist in a population for many generations in unaffected heterozygotes. Dominant lethal alleles rarely remain in a population beyond the first generation because they usually prevent the affected individual from reproducing and passing the allele to subsequent generations. The exception to this is a dominant lethal allele whose effect does not appear until later in life. In some cases, a dominant allele may exert its lethal effect after an individual has reproduced and possibly passed the allele to offspring. An example

agouti	chinchilla	himalayan	albino
CC, Cc^h, or Cc	$c^{ch}c^{ch}$, $c^{ch}c^h$ or $c^{ch}c$	c^hc^h or c^hc	c

Figure 13.9 Four coat color phenotypes in rabbits and their associated genotypes.

of this is the dominant allele for Huntington disease in humans, whose effects usually appear only after age 40. Many heterozygous carriers of the allele pass the allele to some of their children.

Lethal alleles are usually detected as deviations from Mendelian ratios. Often, lethal alleles affect the zygote or the embryo long before it develops beyond microscopic size. Such lethal alleles can be detected from a conspicuously absent class of progeny. An example of this is an allele causing yellow coat color in mice. The normal coat color in mice is agouti. The wild-type allele causing the agouti color is designated A^W, and the allele A^y, when heterozygous with A^W, causes the coat to be yellow. In determining coat color, A^y is dominant over A^W. However, when homozygous, the A^y allele is lethal, meaning that in determining life or death it is recessive to A^W. When yellow (A^yA^W) mice are crossed with each other, their offspring appear in a ratio of $1/3$ agouti to $2/3$ yellow. The underlying ratio is in fact 1:2:1, but only two of the three genotypic classes complete gestation, giving a 1:2 ratio at birth. The third class consists of the homozygous A^yA^y embryos, which die during embryonic development and are never born (Figure 13.10).

The molecular basis for the phenotypes associated with the A^y allele has been determined and is highlighted in the following example.

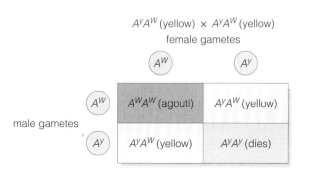

phenotypic ratio: 2 yellow:1 agouti

Figure 13.10 Punnett square of the expected genotypes and phenotypes of the offspring from the cross $A^yA^W \times A^yA^W$ in mice.

Example 13.2   Molecular structure of the A^y recessive lethal allele.

The inheritance of the mutant A^y allele in mice was described in 1905 by Cuénot (*Archives de Zoologie Experimentale et Generale*, quatrième série 3: cxxiii–cxxxii) and has long been the classic example of a recessive lethal allele. In 1994, Duhl et al. (*Development* 120:1695–1708) determined the molecular structure of the A^y allele. The structure revealed why the mutant allele was dominant for the yellow coat color but recessive in determining life or death. The A locus encodes a protein that serves as a molecular switch for coat color. Mice homozygous for the wild-type A^W allele have a yellow band on each dark hair shaft, which gives

the agouti phenotype (see Figure 13.8). Most loss-of-function recessive alleles at this locus reduce or eliminate the yellow band and thus cause the coat to be dark without mottling. The A^y allele is a gain-of-function, dominant mutant allele that eliminates the dark color, leaving each hair shaft completely yellow in heterozygous mice.

About 120,000 nucleotide pairs of the wild-type DNA sequence of the A^W allele are missing from the mutant A^y allele. The deletion eliminates all of the 5′ regulatory region of the A locus but leaves the polypeptide-coding region intact. Upstream, the deletion also eliminates all of the polypeptide-coding region of a second gene called *Merc*, which is a constitutively expressed gene important in embryonic development. The deletion in the A^y allele brings the coding region of the A locus under the control of the *Merc* promoter (Figure 13.11). Thus, the A^y allele is expressed constitutively, overproducing the A gene product, while expression of *Merc* has been eliminated.

Problem: Given this information, explain why the A^y allele is dominant with respect to coat color but recessive in determining life or death.

Solution: Mice that are heterozygous A^yA^W overexpress the protein encoded by the A locus and thus have a yellow coat color. For the A gene product,

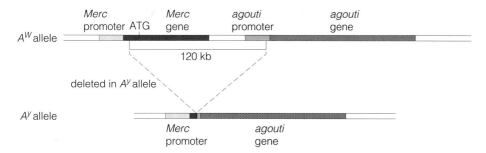

Figure 13.11 Deletion of a 120 kb segment that characterizes the A^y allele of the locus governing coat color in mice. (Adapted from Duhl, D. M. J., M. E. Stevens, H. Vrieling, P. J. Saxon, M. W. Miller, C. J. Epstein, and G. S. Barsh. 1994. Pleiotropic effects of the mouse lethal yellow (A^y) mutation explained by deletion of a maternally expressed gene and the simultaneous production of agouti fusion RNAs. *Development* 120:1695–1708. © 1994 Company of Biologists Ltd.)

the A^y allele is a gain-of-function mutation. On the homologous chromosome, heterozygous mice have a wild-type allele for the *Merc* gene, which produces the *Merc* gene product and permits embryonic development. Because homozygous $A^y A^y$ embryos cannot express the *Merc* gene, they lack the *Merc* gene product, which is essential for normal development, and die as embryos. Thus, the large deletion affects two adjacent genes, in turn affecting two phenotypes, coat color and embryonic development. For the *A* gene, the deletion is a dominant gain-of-function mutation. For the *Merc* gene, the deletion is a recessive loss-of-function mutation.

Lethal alleles cause death and are often detected as a class of progeny that is absent.

Pleiotropy

Pleiotropy is the situation in which a single gene influences more than one phenotypic trait. Mendel observed pleiotropy in noticing that some genes affect more than one part of the pea plant. For example, he saw that the gene responsible for flower color also affected the axils of the leaves and the color of the seed coats. Purple-flowered pea plants have purple rings at the leaf axils and produce colored seed coats. White-flowered plants do not have purple rings at the leaf axils and produce transparent seed coats (Figure 13.12). Anthocyanin (the same pigment that makes flowers purple) causes the purple rings at the leaf axils and colors the seed coats. White-flowered plants lack purple rings and have transparent seed coats because anthocyanin is absent throughout the plant. The enzyme encoded by the *A* gene is expressed in the flower petals, leaf axils, and seed coats, so a mutation that affects this gene affects the phenotypes at these three locations in the plant.

Many genes have pleiotropic effects. An example is the sickle-cell anemia gene in humans. Figure 13.13 dia-

Purple-flowered pea plants have purple rings at the leaf axils, and their seed coats are colored.

White-flowered pea plants do not have purple rings at the leaf axils, and their seed coats are transparent. The yellow color of the seeds is the inner part of the seed showing through the transparent seed coat.

Figure 13.12 Pleiotropy in pea plants.

grams the multiple effects of the recessive sickle-cell allele when it is present in the homozygous condition.

~

A pleiotropic gene affects the phenotype of more than one trait.

13.2 EPISTASIS

As there are several enzymes in the typical biochemical pathway, several enzyme-encoding genes help form the pathway's final product. For this reason, one gene in the pathway may influence how another is expressed, a phenomenon called **epistasis**. Epistasis brings about

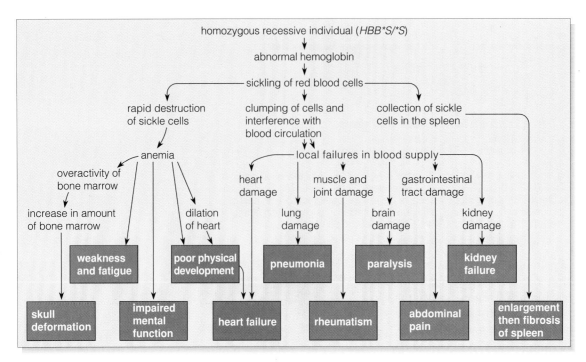

Figure 13.13 Pleiotropic effects of the *HBB*S* allele that, when homozygous, causes sickle-cell anemia.

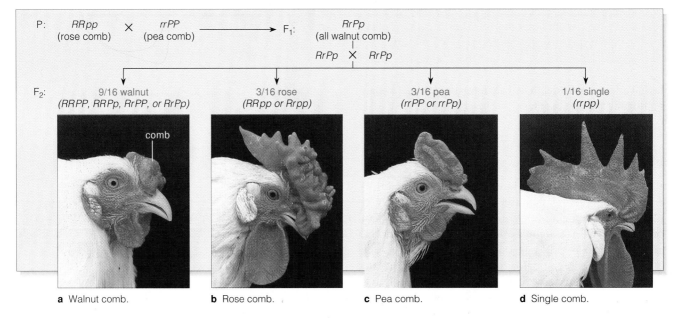

a Walnut comb. b Rose comb. c Pea comb. d Single comb.

Figure 13.14 The 9:3:3:1 F₂ segregation of comb type in chickens. (Photos courtesy of Ted Somes.)

inheritance patterns that may, on the surface, appear to deviate from Mendel's principles, but on closer examination we find that in most cases they do not. Epistatic patterns of inheritance arise from the interaction of at least two and sometimes more loci that influence the phenotypes for a single trait.

In the simplest case of epistasis, two loci, each with dominant and recessive alleles, interact to produce four distinct phenotypes of the same trait in a 9:3:3:1 ratio in F₂ progeny in a dihybrid experiment. One of the earliest cases was described by Bateson shortly after the rediscovery of Mendel's principles at the beginning of this century. In this example, two loci interact to determine four phenotypes for comb shape in chickens, as illustrated in Figure 13.14.

There are many forms of epistasis, each with a predictable pattern of inheritance. Figure 13.15 summarizes

several of the more common types of epistasis in which two genes interact. Let's see how each of the progeny ratios shown is determined.

Complementary Gene Action

Sweet pea is an ornamental pea species (*Lathyrus odoratus*) that differs from the edible garden pea (*Pisum sativum*) that Mendel used. In sweet pea, crosses between certain varieties of true-breeding, white-flowered plants yield F_1 progeny that are all purple-flowered, a dominant phenotype. If the F_1 plants are self-pollinated, the progeny segregate in a 9 purple-flowered:7 white-flowered ratio. These seemingly unusual results can be explained if we assume that two loci interact to determine flower color in this cross. This type of epistatic interaction is called **complementary gene action** (Figure 13.16).

Two loci, *C* and *P*, are located on nonhomologous chromosomes and each encodes an enzyme in the pathway that synthesizes purple pigment in sweet pea flowers, as illustrated in Figure 13.17a. If the first parent plant is homozygous for a recessive allele that prevents pigment synthesis at the *C* locus (*cc*), and is also homozygous for the dominant allele at the *P* locus (*PP*), its flowers are white because pigment synthesis is blocked at the first step in the pathway (Figure 13.17b). If the second parent is homozygous for the dominant allele at the *C* locus (*CC*), synthesis may proceed beyond the first step. However, if this plant is homozygous for a recessive allele that blocks pigment synthesis at the *P* locus (*pp*), then pigment synthesis is blocked at the second step in the pathway, and its flowers are white (Figure 13.17c).

When these two plants are crossed, all F_1 progeny are doubly heterozygous:

$$cc\,PP \times CC\,pp$$
$$\downarrow$$
$$Cc\,Pp$$

Because the double heterozygote has one dominant allele at each of the two loci, it produces functional enzymes at both steps, the pathway is not blocked, pigment is produced, and the flowers are colored (Figure 13.17d).

In other words, for flowers to be colored, dominant alleles must be present at both the *C* and the *P* loci. The genotype for colored flowers is *C_ P_*, and the genotypes for white flowers are *C_ pp*, *aa P_*, and *cc pp*. When the F_1 plants are self-fertilized, the alleles at the two loci assort independently, as diagrammed in the Punnett square shown in Figure 13.16. A look at this Punnett square reveals that the progeny genotypes for purple flowers (*C_ P_*) and white flowers (*C_ pp*, *cc P_*, and *cc pp*) segregate in a 9:7 ratio.

An alternative biochemical explanation of complementary gene action is that the two loci, *C* and *P*, produce different subunits of a single enzyme. The functional enzyme forms when the two subunits come together. If at least one dominant allele is present at each of the two

complementary gene action	9	7	
duplicate gene action	15		1
dominant suppression	13		3
dominant epistasis	12	3	1
recessive epistasis	9	3	4

Figure 13.15 Several types of epistasis and their associated phenotypic ratios in F_2 progenies.

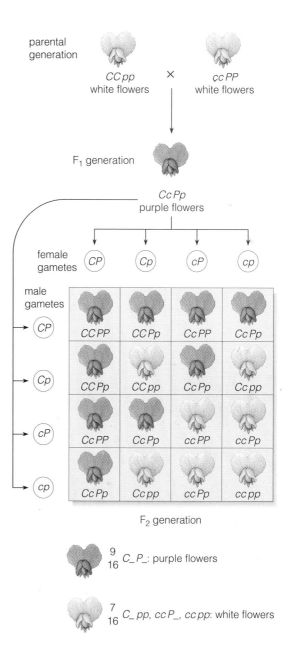

9/16 *C_ P_*: purple flowers

7/16 *C_ pp*, *cc P_*, *cc pp*: white flowers

Figure 13.16 Complementary gene action in the determination of flower color in sweet peas.

loci, both functional subunits are formed and, in the sweet pea example, the pigment is synthesized (Figure 13.18a). Under this alternative explanation, when the recessive allele at either locus is homozygous (*cc P_*, or *C_ pp*), one of the two subunits is not produced. As

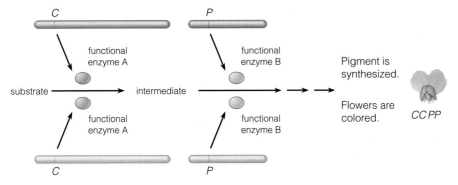

a Loci *C* and *P* encode enzymes in the pigment synthesis pathway.

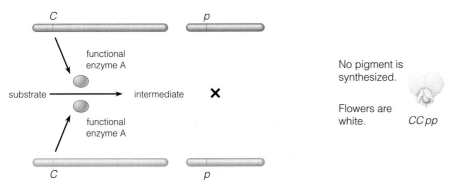

b One parental variety has a genotype that does not encode a functional version of the first enzyme.

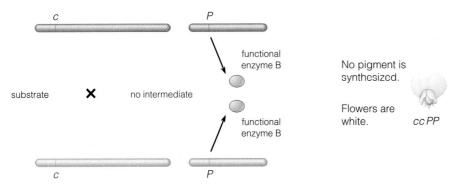

c The other parental variety has a genotype that does not encode a functional version of the second enzyme.

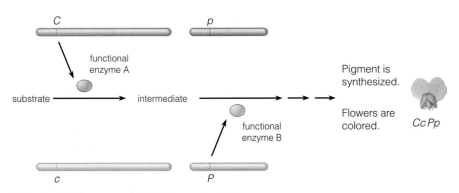

d The doubly heterozygous F₁ plants produce both enzymes.

Figure 13.17 A biochemical explanation for complementary gene action.

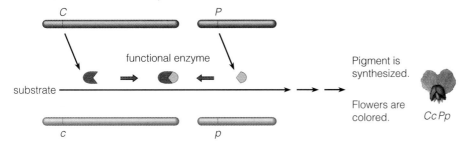

a Loci *C* and *P* encode subunits of an enzyme in the pigment synthesis pathway.

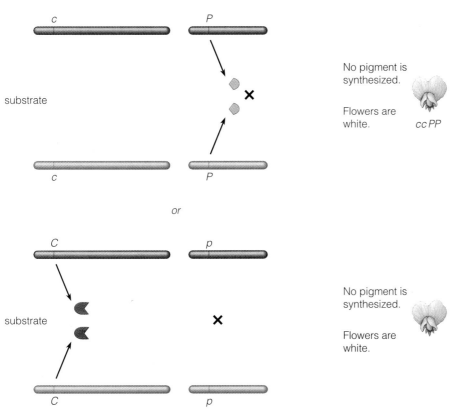

b Absence of either subunit blocks the same step in the pathway.

Figure 13.18 An alternative biochemical explanation for complementary gene action.

both subunits must be present for the enzyme to form properly, no functional enzyme forms, and the pathway is blocked (Figure 13.18b). If the recessive alleles are homozygous at both loci (*cc pp*), neither subunit is produced, and the pathway is again blocked.

Both of these biochemical explanations predict a 9:7 segregation ratio in the F_2 generation. Let's now review some experiments that demonstrated this predicted ratio.

Example 13.3   Complementary gene action.

Bateson and Punnett (as cited in Bateson, W. 1913. *Mendel's Principles of Heredity*. Oxford: Oxford University Press, pp. 88–92) first noticed the epi-

static interaction that determines sweet pea flower color in an experiment conducted for the study of another trait. They wanted to determine the inheritance of pollen shape in sweet pea. They crossed two white-flowered varieties, one with long pollen grains, the other with round pollen grains. They expected the F_1 progeny to have white flowers like their parents, but to their surprise, all the F_1 progeny had colored flowers. Among the F_2 progeny, 36 had colored flowers and 28 had white flowers. They concluded that this was an example of epistasis.

Problem: Formulate a hypothesis for an appropriate epistatic ratio, and test the hypothesis that these results conform to the ratio.

Solution: The pattern of inheritance suggests that complementary gene action determines flower color. If two genes interact, the expected ratio in the F_2 progeny is 9 colored:7 white. The expected numbers of progeny for a 9:7 ratio among 64 progeny are 36:28. As the observed values equal the expected values exactly, the chi-square value is zero. There is no evidence to reject the hypothesis that these results conform to the 9:7 ratio typical of complementary gene action at two loci.

In complementary gene action, the dominant phenotype is encoded by the presence of at least one dominant allele at each locus.

Duplicate Gene Action

Another common epistatic pattern of inheritance gives a 15:1 ratio in the F_2 generation following an intercrossing or self-fertilization of doubly heterozygous F_1 individuals. In the F_2 generation, only the doubly homozygous recessive genotype (*aa bb*) expresses the recessive phenotype. The epistatic interaction determining the 15:1 ratio is called **duplicate gene action**; the inheritance of growth patterns in wheat offers a good example. Wheat varieties may be divided into two types: spring wheat and winter wheat. Spring wheat is planted in the spring and harvested in late summer. Winter wheat is planted in the fall and grows into a short grassy plant before the first frost, after which it remains dormant through the winter. In the spring, the short grassy plant grows into wheat that matures in early summer. If spring wheat is planted in the fall, it cannot survive the cold winter. If winter wheat is planted in the spring, it remains a short grassy plant throughout the summer and fall and fails to produce any grain. Two loci (we'll call them *A* and *B*) determine whether a plant is the spring or winter type. The doubly homozygous recessive *aa bb* is the winter type. All other genotypes are the spring type. If F_1 doubly heterozygous spring type plants (*Aa Bb*) are self-fertilized, the F_2 plants segregate 15 spring:1 winter, as diagrammed in Figure 13.19.

Let's consider how this works at the biochemical level. This sort of segregation is expected if the two loci are duplicate genes that encode the same product. In our example, the product is a protein that causes the spring growth pattern (Figure 13.20a). A single dominant allele at either locus produces the functional product (Figure 13.20b). Only the doubly homozygous recessive genotype (*aa bb*) fails to produce the functional product, thereby causing the winter type (Figure 13.20c).

Mendel's experiment with flower color in beans (see Example 12.9) provides another example of duplicate gene action. Mendel found 1 white-flowered plant out of 31. It is clear from his interpretation of these results that he understood the theoretical basis of duplicate gene

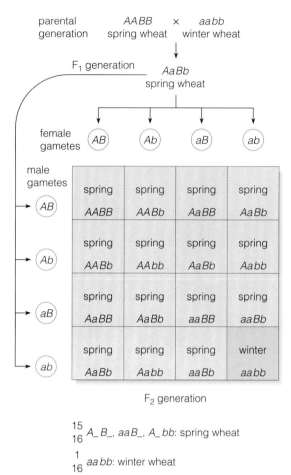

Figure 13.19 Duplicate gene action in the determination of growth pattern in wheat.

action. He concluded that the number of plants was too small to determine with certainty whether two or three genes were interacting.

In duplicate gene action, the dominant phenotype is encoded by the presence of at least one dominant allele at either locus.

Dominant Suppression

In another form of epistasis called **dominant suppression**, a dominant allele at one locus suppresses the effect of a dominant allele at another locus. An example of this is feather color in chickens. A locus called *C* (for colored) produces an enzyme for the synthesis of feather pigments. The dominant allele produces pigment in both the homozygous and heterozygous conditions; *CC* and *Cc* chickens are colored. The recessive allele blocks pigment synthesis when homozygous; *cc* chickens are white. A second locus called *I* (for inhibitor) suppresses pigment production when the dominant allele is present in either the homozygous or heterozygous condition (*II* or *Ii*), regardless of the genotype at the *C* locus; a chicken with

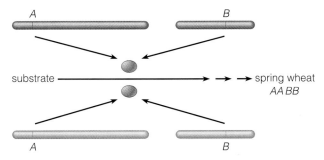

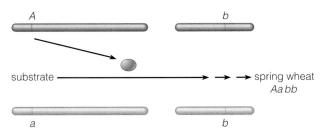

a Dominant alleles at loci *A* and *B* encode the same product, which causes wheat to follow the spring pattern of growth.

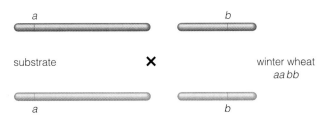

b A single dominant allele at either locus produces enough product for the dominant phenotype.

c Plants that are doubly homozygous recessive do not produce the product and instead follow the winter pattern of growth.

Figure 13.20 A biochemical explanation of duplicate gene action.

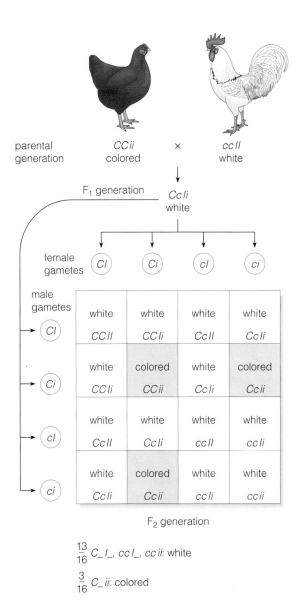

$\frac{13}{16}$ *C_ I_, cc I_, cc ii*: white

$\frac{3}{16}$ *C_ ii*: colored

Figure 13.21 Dominant suppression in the determination of feather color in chickens.

at least one *I* allele is white. Thus, pigment can be produced only when a chicken is homozygous for the recessive allele at the *I* locus (*ii*), and homozygous or heterozygous for the dominant allele at the *C* locus (*C_*). In summary, chickens with the genotypes *C_ ii* are colored, and chickens with the genotype *C_ I_, cc I_,* or *cc ii* are white. The ratio in the F₂ generation from doubly heterozygous F₁ parents is 13 white:3 colored, as diagrammed in Figure 13.21.

Dominant Epistasis

Leaky recessive alleles may cause **dominant epistasis**. A good example comes from onions. At the supermarket, you typically find three types of onions: red, yellow, and white. These colors are determined by two loci, designated *R* and *I*. The *R* locus is responsible for pigment production. The dominant *R* allele causes formation of red

pigment and gives the onion a dark-red color. The *r* allele is a leaky recessive allele that causes a slight amount of pigment to form, giving the onion a yellow color when homozygous *rr*. The *I* locus has a dominant inhibitor allele (*I*) that prevents pigment from forming. When the dominant *I* allele is present, the onion is white (has no pigment) regardless of the genotype at the *R* locus. When homozygous *ii*, the onion is red or yellow depending on the genotype at the *R* locus. In summary, *R_ ii* is red, *rr ii* is yellow, and the *R_ I_* and *rr I_* genotypes are white. The segregation in the progeny of doubly heterozygous parents is 12 white:3 red:1 yellow, as diagrammed in Figure 13.22. This is called dominant epistasis because the dominant *I* allele masks the effect of both alleles at the *R* locus. Dominant epistasis is quite similar to dominant suppression, the only difference being the effect of a leaky recessive allele.

Recessive Epistasis

Recessive epistasis is illustrated by coat color in Labrador retrievers, which is determined by two loci, designated *B* and *E*. The *B* locus codes for an enzyme in the pathway for production of the pigment melanin. The dominant allele *B* causes the coat to be black when homozygous *BB* or heterozygous *Bb*. The leaky recessive allele *b* reduces pigment production when homozygous *bb*, making the coat chocolate colored. The *E* locus determines how much pigment is deposited in the hair. The dominant allele *E* allows normal deposition of pigment when homozygous *EE* or heterozygous *Ee*, so *B_ E_* is black, and *bb E_* is chocolate. The leaky recessive allele *e* when homozygous allows only a slight amount of pigment to be deposited in the hair regardless of the genotype at the *B* locus, giving a yellow coat. Thus, *B_ ee* and *bb ee* are yellow. The segregation in the progeny of two doubly heterozygous parents is 9 black:3 chocolate:4 yellow, as diagrammed in Figure 13.23. This example is called recessive epistasis because only the homozygous recessive genotype *ee* masks the effect of alleles at the *B* locus.

While epistatic patterns of inheritance may seem quite complicated and diverse, they are unified in that they all occur in accordance with the principles of segregation and independent assortment. In all of the cases we just discussed, two loci are involved, and the resulting segregation ratios of phenotypes in the F_2 progeny are always expressed in sixteenths. More complicated epistatic ratios may occur when interactions involve three loci, where the phenotypes of the F_2 generation are expressed in sixty-fourths.

~

Epistasis is the interaction of two or more genes and is detected by observing certain phenotypic patterns in the progeny of heterozygous parents.

13.3 PENETRANCE AND EXPRESSIVITY

Up to this point, we have discussed phenotypes that differ because their underlying genotypes differ. In this section we learn about two ways in which a phenotype may vary among individuals sharing the same underlying genotype.

Penetrance

Sometimes an individual with a certain genotype fails to have the phenotype that is usually associated with that genotype. When this happens, the allele causing the expected phenotype is said to be **nonpenetrant** in that individual. We saw an example of nonpenetrance when we discussed the Bombay phenotype for ABO blood groups earlier in this chapter. Suppose a person is homozygous recessive *FUT1*O/*O* and, therefore, has the Bombay phenotype. Suppose further that this person is homozygous

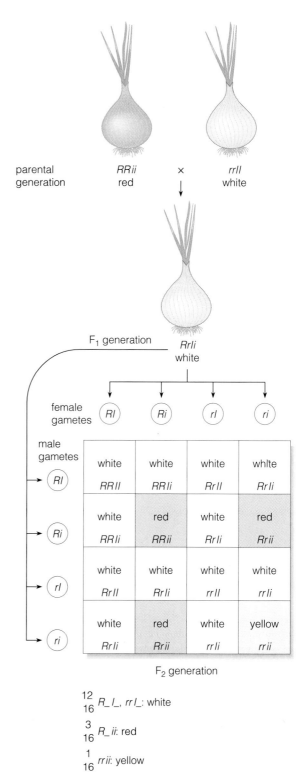

Figure 13.22 Dominant epistasis in the determination of color in onion bulbs.

for the *ABO*A* allele (*ABO*A/*A*). Under most circumstances, we would expect a person with the genotype *ABO*A/*A* to have type A blood, but in this case, no A antigen is produced; the *ABO*A* allele is not expressed even though it is present. For this reason, the *ABO*A* allele is not penetrant in this person.

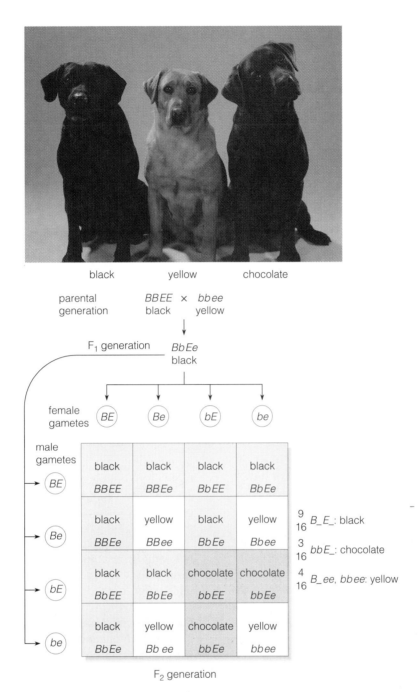

black yellow chocolate

parental generation *BBEE* × *bbee*
      black yellow

F₁ generation *BbEe*
      black

female gametes BE Be bE be

male gametes

BE black *BBEE*	black *BBEe*	black *BbEE*	black *BbEe*
Be black *BBEe*	yellow *BBee*	black *BbEe*	yellow *Bbee*
bE black *BbEE*	black *BbEe*	chocolate *bbEE*	chocolate *bbEe*
be black *BbEe*	yellow *Bbee*	chocolate *bbEe*	yellow *bbee*

$\frac{9}{16}$ *B_E_*: black

$\frac{3}{16}$ *bbE_*: chocolate

$\frac{4}{16}$ *B_ee, bbee*: yellow

F₂ generation

Figure 13.23 Recessive epistasis in the determination of coat color in Labrador retrievers.

Often a lack of penetrance can be attributed to epistasis. The interaction of the *FUT1* locus with the *ABO* locus in causing the Bombay phenotype is an example of complementary gene action. The examples of dominant suppression, dominant epistasis, and recessive epistasis shown in Figures 13.21–13.23 are also examples of incomplete penetrance, because in each case the phenotypic expression of a particular genotype is lacking under certain conditions.

Penetrance does not refer to the degree of expression but only to whether the genotype is expressed in the phenotype. If the phenotype is expressed to any degree, the genotype is penetrant. If the phenotype is not expressed, the genotype is nonpenetrant. Penetrance can be expressed as a percent value in a population. Suppose that 1000 individuals in a population of 10,000 carry a dominant allele *A*. As the allele is dominant, we expect every individual who carries it to express the phenotype associated with it. However, suppose that of the 1000 individuals who carry the allele, 25 do not express the phenotype, whereas the remaining 975 do. The allele is expressed in 975 of the 1000 individuals who have it, so the allele is 97.5% penetrant in the population. For example, the *ABO*A* and *ABO*B* alleles are less than 100% penetrant in human populations because of the Bombay phenotype.

Figure 13.24 Variable expressivity for purple spotting in pea seeds.

Figure 13.25 Temperature effect on gene expression. The c^h allele in Himalayan rabbits is expressed only at temperatures cooler than body temperature, so it is expressed predominantly in the body extremities.

Expressivity

When the genotype is expressed, **expressivity** defines the degree of expression. Mendel recognized variable expressivity in the coat color of pea seeds. As mentioned earlier, coat color in pea seeds is pleiotropic to flower color, so purple-flowered plants have colored seed coats and white-flowered plants have transparent seed coats. Some of the coloration in seed coats may be purple spotting. Figure 13.24 shows variable expressivity for purple spotting on seeds from different varieties of purple-flowered peas. The patterns range from few spots to complete purple coloration on the seed. All of these varieties carry the dominant *A* allele, which permits purple spotting, but they vary in their genotypes for **modifier genes**, genes that have a minor effect on a phenotype determined by a major gene. Over 20 modifier genes are known to influence the coloration of seed coat in peas.

Causes of Nonpenetrance and Variable Expressivity

Often we do not know the exact causes of nonpenetrance and variable expressivity. Frequently they are related to the **genetic background**, which refers to all genes in an individual except the one under study. The influence of the genetic background on the gene being studied may determine whether the gene is expressed (penetrance) and the degree to which it is expressed (expressivity). Nonpenetrance and variable expressivity may also be caused by **nongenetic factors** that influence gene expression. These may include environmental influences and random physiological changes during an organism's development. The phenotypic effects of nongenetic factors may be significant, but these effects are not inherited. There are countless examples of the influence of non-genetic factors on gene expression, such as the influence of diet on height and body weight in humans.

One of the most clear-cut examples of environmental influence on expressivity is the c^h allele in Himalayan rabbits, which controls pigment synthesis in the rabbit's hair. This allele is temperature sensitive. When the temperature of the rabbit's skin remains close to the body temperature of the rabbit, the gene is not expressed and the fur is white. But where the temperature of the skin remains significantly below the body temperature, the gene is expressed and the fur is black. The colder extremities of the rabbit's body, which often lose body heat, such as the feet, ears, and nose, are all colored with black fur. The warmer portions of the body are covered with white fur (Figure 13.25).

The cause of nonpenetrance of polydactyly in humans (additional digits on the hands or feet, as seen in the child pictured in Figure 13.26a) is unknown. Occasionally, the dominant allele responsible for polydactyly is not expressed in a person whose pedigree indicates that he or she must carry the allele, as illustrated in Figure 13.26b. The man labeled with an asterisk must have carried the allele because he had a parent and a child with the trait, but he did not express it. This is a case of nonpenetrance for which the cause is unknown and may be genetic, nongenetic, or both. Variable expressivity is also illustrated in this pedigree in the variation for number of digits on the hands and feet of affected people. Again, the cause of these variations is unknown and may be either genetic, nongenetic, or both.

The following example shows how conclusions about dominance, nonpenetrance, and variable expressivity can be drawn from information in a pedigree.

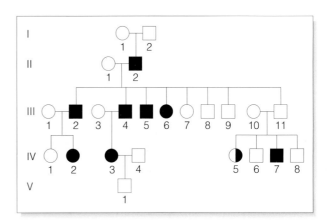

Figure 13.27 Information for Example 13.4: Pedigree of type-D brachydactyly. (Adapted from Gray, E. and V. K. Hurt. 1984. Inheritance of brachydactyly type D. *Journal of Heredity* 75:297–299.)

a An affected child with six digits on each hand.

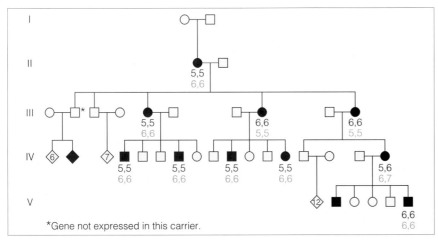

*Gene not expressed in this carrier.

b Pedigree showing the inheritance of the dominant allele that causes polydactyly.

Figure 13.26 Polydactyly in humans. Black numbers in the pedigree indicate the number of digits on each hand, and blue numbers indicate the number of digits on each foot. For an explanation of the symbols used in human pedigrees, see Figure 12.14. (Photo courtesy of V. A. McKusick.)

Example 13.4 Type-D brachydactyly in humans.

Brachydactyly was the first phenotype to be associated with a dominant mutant allele at the *BDD* locus in humans (OMIM 113200). It is characterized by abnormal shortening of the fingers (see Figure 12.15). Type-D brachydactyly is a less severe type in which only the thumbs and sometimes the big toes are affected. In 1984, Gray and Hurt (*Journal of Heredity* 75:297–299) published the pedigree of type-D brachydactyly shown in Figure 13.27. Full shading indicates that both thumbs were affected; half shading indicates that only one thumb was affected.

Problem: **(a)** Assuming that type-D brachydactyly is caused by a single mutant allele, what type of dominance relationship is illustrated by type-D brachydactyly? **(b)** Is there any evidence in the pedigree of nonpenetrance? **(c)** Is there any evidence in the pedigree of variable expressivity?

Solution: **(a)** As the type-D brachydactyly phenotype appears in each generation in approximately a 1:1 ratio, we may assume that it is caused by an allele with complete dominance. **(b)** There is evidence of nonpenetrance. As individual III-11 has two children with type-D brachydactyly, he must be a heterozygous carrier of the allele even though he does not

have the phenotype. **(c)** There is also evidence of variable expressivity. Individual IV-5 has a shortened thumb on only one hand, whereas all other affected individuals have shortened thumbs on both hands.

Penetrance is the expression of the phenotype associated with a particular genotype. Expressivity is the degree to which a penetrant genotype is expressed in the phenotype.

13.4 DNA MARKERS IN GENETIC ANALYSIS

As we have seen in this and past chapters, mutations create different alleles. Geneticists traditionally identify different alleles by the contrasting phenotypes they cause. If differences in DNA sequences could be detected directly, rather than indirectly by observing the phenotype associated with an allele, genetic analysis could be much more efficient. Direct detection of differences in nucleotide sequence can be accomplished with **DNA markers**, fragments of DNA that can be distinguished from one another because of differences in their nucleotide sequences. Researchers use recombinant DNA methods to observe DNA markers, then analyze the information genetically as if they were examining phenotypic traits. Like many traits, DNA markers are inherited in typical Mendelian patterns. Unlike most traits, however, DNA markers are not subject to environmental influences, pleiotropy, epistasis, nonpenetrance, and variable expressivity, so genetic analysis is usually more straightforward and precise with DNA markers than with traits.

As we embark on a discussion of DNA markers, you may wish to review sections 9.6–9.9, which explain restriction enzyme digestion, electrophoresis, Southern blotting, hybridization with probes, PCR, and DNA sequencing, because a basic understanding of these methods is essential for discussing DNA marker analysis.

Restriction Fragment Length Polymorphism (RFLP) Analysis

Among the most effective and widely used DNA markers are **restriction fragment length polymorphisms (RFLPs)**. RFLP analysis relies on differences in DNA sequence that affect the position of restriction enzyme recognition sites in the DNA. Let's use detection of the mutant allele responsible for sickle-cell anemia in humans as an example of how RFLPs are used. As described in sections 6.1 and 6.2, the *HBB* locus (OMIM 141900) encodes the β subunit of hemoglobin. Sickle-cell anemia is a genetic disorder of persons who are homozygous for the mutant allele *HBB*S* (also called *HBB*Q6V*). The mutation is a substitution of valine for glutamic acid in the sixth

amino acid residue of the protein. The DNA sequence surrounding the sixth codon in the normal *HBB*A* allele is

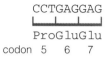

and the mutant *HBB*S* allele contains an A → T transversion in the second nucleotide of the sixth codon:

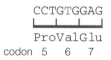

The restriction endonuclease *Dde*I cleaves DNA at the sequence CTNAG, where N represents any of the four nucleotides. The *HBB*A* allele has a *Dde*I cleavage site spanning the fifth and sixth codons:

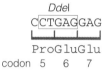

There is a second *Dde*I site 175 nucleotide pairs upstream from the *Dde*I site at the sixth codon, and a third *Dde*I site 201 nucleotide pairs downstream from it. Digestion of the DNA of the *HBB*A* allele with *Dde*I yields a fragment that is 175 nucleotide pairs long and a second fragment that is 201 nucleotide pairs long (Figure 13.28a).

The mutation in the *HBB*S* allele alters the CTGAG sequence to CTGTG, which is not a *Dde*I cleavage site. However, the two *Dde*I sites on either side of this site remain unchanged. Thus, digestion of DNA containing the *HBB*S* allele with *Dde*I yields a single fragment that is 376 nucleotide pairs long (Figure 13.28b).

Homologous DNA fragments from different individuals cut to different lengths by a restriction endonuclease constitute RFLPs. Because the fragments are of different lengths, they migrate to different positions when separated by electrophoresis. When human genomic DNA is digested by *Dde*I (or any other restriction endonuclease), hundreds of thousands of fragments are produced because the genome is so large. To detect an RFLP, a geneticist first separates the numerous fragments by electrophoresis, then blots the fragments onto a membrane, then hybridizes the fragments with a probe that is specific for the RFLP in question.

In the sickle-cell anemia example, the probe is homologous to a region that does not extend beyond the two outer *Dde*I sites, so it hybridizes only to DNA fragments between these two sites. The RFLPs appear as distinct bands at predictable sites. The *HBB*A* allele appears as two fragments, 175 and 201 nucleotide pairs long, while the *HBB*S* allele appears as a single fragment, 376 nucleotide pairs long (Figure 13.28c).

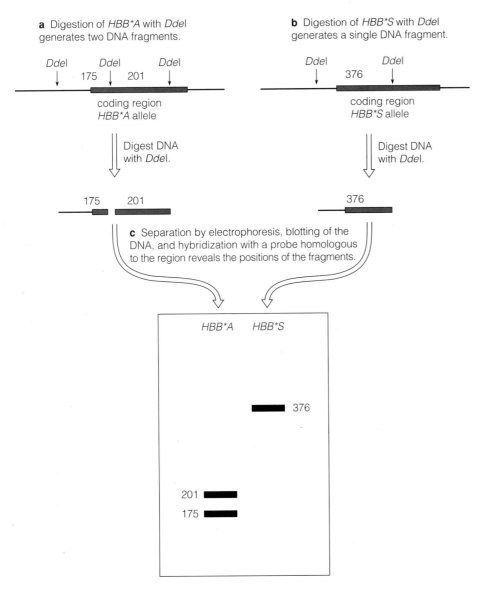

a Digestion of *HBB*A* with *Dde*I generates two DNA fragments.

b Digestion of *HBB*S* with *Dde*I generates a single DNA fragment.

*Dde*I   *Dde*I   *Dde*I
   175   201

coding region
HBB*A allele

Digest DNA
with *Dde*I.

175   201

*Dde*I   *Dde*I
       376

coding region
HBB*S allele

Digest DNA
with *Dde*I.

376

c Separation by electrophoresis, blotting of the DNA, and hybridization with a probe homologous to the region reveals the positions of the fragments.

HBB*A   *HBB*S*

376

201
175

Figure 13.28 Detection of the *HBB*S* mutant allele responsible for sickle cell anemia using RFLPs.

Moreover, homozygotes can be readily distinguished from heterozygotes. For example, as illustrated in Figure 13.29a, the RFLP pattern of a person who is homozygous for the *HBB*A* allele is two fragments, 175 and 201 nucleotide pairs long, because the two homologous chromosomes are both cut by *Dde*I at the same three sites. The RFLP pattern of a person who is homozygous for the *HBB*S* allele is a single fragment, 376 nucleotide pairs long, because the same fragment is produced from digestion of both homologous chromosomes with *Dde*I (Figure 13.29c). The RFLP pattern of a heterozygote (*HBB*A/*S*) is three fragments because the 175 and 201 nucleotide-pair fragments are cut from the chromosome carrying the *HBB*A* allele, and the 376 nucleotide-pair fragment is cut from the chromosome carrying the *HBB*S* allele (Figure 13.29b). Notice that the RFLP pattern of the heterozygote contains the superimposed patterns of the two homozygotes. For this reason, RFLPs are inherited as codominant alleles.

RFLPs do not necessarily have to be within the transcribed region of a gene. Because the DNA is detected directly, rather than the product of a gene or the outward phenotype associated with a gene, RFLPs can be anywhere in the DNA where a mutation has altered the length of a restriction fragment. RFLPs can be used for genetic analysis just like standard alleles manifested as visual differences. In fact, because RFLPs are inherited in codominant patterns, the terms *locus* and *allele* are used for them even though RFLPs do not necessarily represent the DNA sequences of genes.

RFLPs can be anywhere in the DNA, and the number of RFLPs detectable with different probes numbers in the millions. These two factors make RFLP analysis a powerful research and diagnostic tool. RFLPs have been used

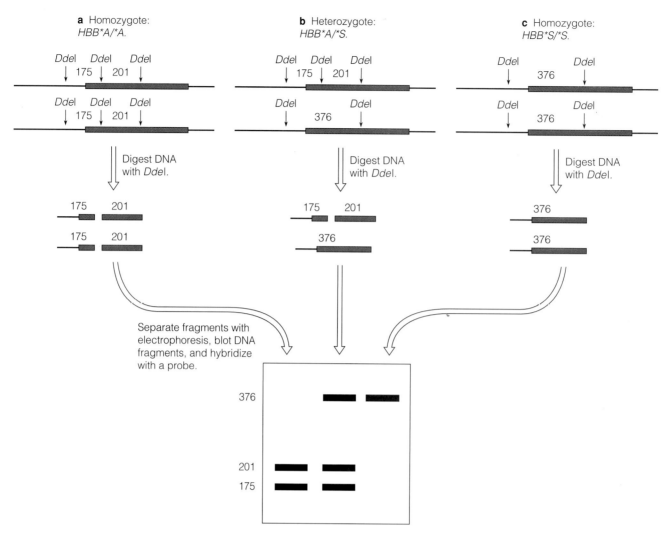

a Homozygote: *HBB*A/*A.*

b Heterozygote: *HBB*A/*S.*

c Homozygote: *HBB*S/*S.*

Figure 13.29 The codominance of RFLPs. The phenotype of the heterozygote shows the phenotypes of both homozygotes.

extensively in many types of genetic analyses, including genetic mapping of chromosomes, paternity testing, evolutionary studies, forensic analysis with DNA fingerprinting, and a host of other applications. We will discuss all of these applications in later chapters.

Tandem Nucleotide Repeat Markers

A special class of RFLPs, based on DNA sequences that are repeated in tandem, has become very useful for genetic analysis. Among the most informative markers are **tandem nucleotide repeat markers**; two types are minisatellites and microsatellites.

As discussed in section 5.1, slippage mutations during DNA replication may cause tandem nucleotide repeats to vary in length from individual to individual. The rate of slippage mutations is usually high enough to produce significant variation over many generations in a species, but not so high that researchers often encounter new mutations when analyzing parents and offspring.

For genetic analysis with DNA markers, this is an ideal situation. Polymorphism for minisatellite and microsatellite markers in humans is so high that only identical twins share all the same patterns.

Minisatellites, which are also called **variable number tandem repeats (VNTRs)**, consist of DNA segments of about 10–100 nucleotide pairs that are repeated several times in tandem. Most minisatellites are bordered by unique DNA sequences. When genomic DNA is digested with a restriction enzyme that does not have a recognition site within a tandem repeat sequence, the restriction enzyme cuts the DNA outside of the tandem repeat region and leaves the tandem repeat intact (Figure 13.30).

Minisatellites have been used extensively for human DNA fingerprinting, which is valuable in forensic analysis and paternity testing. Often a particular tandem repeat may be present at several locations in the genome, so each tandem repeat segment represents a different minisatellite locus. This means that a single probe may sample several minisatellite loci simultaneously.

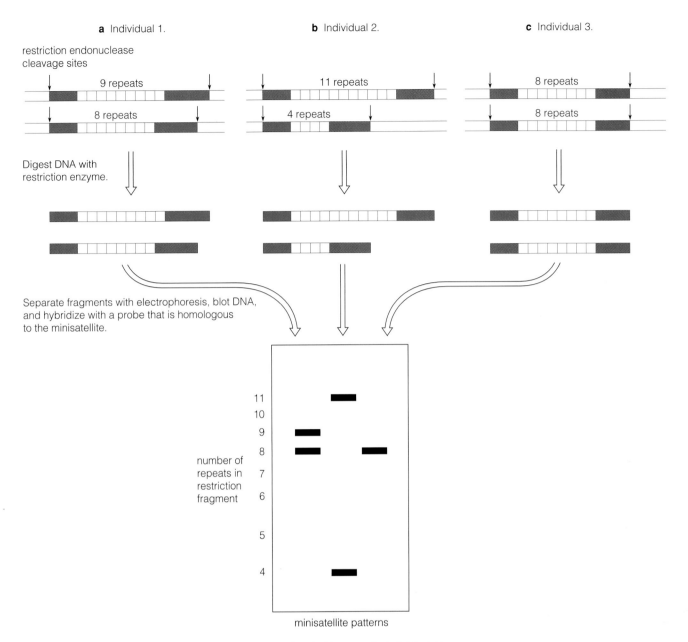

Figure 13.30 Model of minisatellite generation and detection. In this example, individual 1 is heterozygous for minisatellite alleles with 9 and 8 repeats; individual 2 is heterozygous for alleles with 11 and 4 repeats; and individual 3 is homozygous for an allele with 8 repeats. The DNA from each individual is digested with a restriction enzyme that does not have a cleavage site within the repeated region. This generates restriction fragments that contain the entire repeat region. The fragments are separated by electrophoresis, blotted, and hybridized with a probe that is homologous to the repeat region. The separation puts the fragments in order by number of repeats. Two fragments are detected from heterozygotes, while a single fragment is detected from a homozygote.

Microsatellites, which are also called **short tandem repeat polymorphisms (STRPs)**, are similar to minisatellites, but the repeated segment is shorter, usually consisting of di-, tri-, or tetranucleotide repeats. Examples of microsatellite repeats include the following:

dinucleotide repeat (AC)$_n$: ACACACACACACACAC

trinucleotide repeat (CCG)$_n$: CCGCCGCCGCCGCCG

tetranucleotide repeat (CATT)$_n$: CATTCATTCATTCATT

Microsatellite markers are detected in the same way as minisatellite markers. A restriction enzyme that does not cleave the DNA within the tandem repeat is used to digest genomic DNA, and then the restriction fragments are separated electrophoretically, transferred by Southern blotting, and detected by hybridization with a probe that is complementary to the tandem repeat. Like other RFLPs, minisatellites and microsatellites are typically codominant markers, so heterozygotes can be readily distinguished from homozygotes.

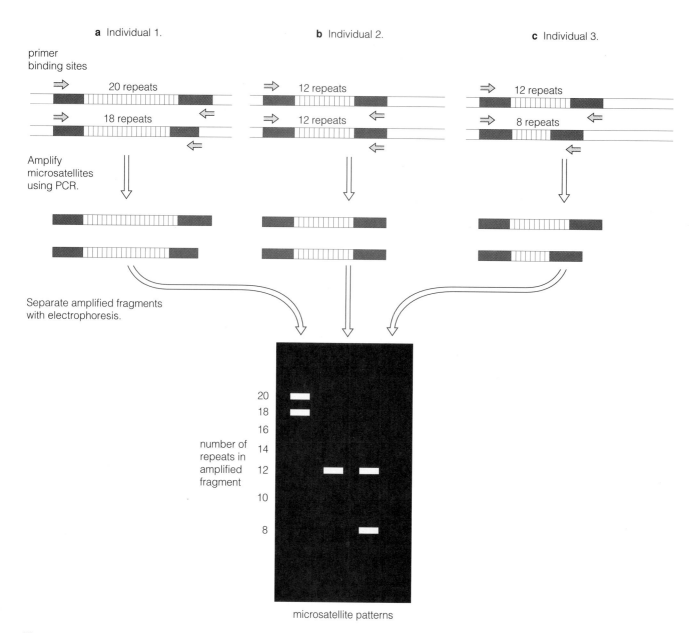

Figure 13.31 PCR amplification of microsatellite markers. DNA from each individual is amplified by PCR with primers that bind on either side of the repeat segment. The amplified DNA is then separated electrophoretically. Microsatellites that differ in length migrate to different positions on the gel, permitting simple and rapid detection. Notice the heterozygosity and homozygosity for the microsatellites.

PCR-Based Markers

The polymerase chain reaction (PCR) can be used to generate DNA markers for genetic analysis. PCR-marker analysis has several advantages over RFLP analysis, mostly in taking less time and costing less. Short tandem repeats, such as di-, tri-, and tetranucleotide repeats, lend themselves especially well to PCR-marker analysis and are a type of microsatellite marker that may be detected directly in a gel, avoiding the need for Southern blotting and hybridization with a probe. An experimenter must know the nucleotide sequences of the unique DNA regions that flank the tandem repeat segment to synthe-size primers complementary to the sequences. The primers, rather than restriction enzymes, select the microsatellites and amplify them by PCR. Variation in the number of tandem repeats can be detected from electrophoresis, as diagrammed in Figure 13.31. PCR-based microsatellite markers have been used extensively in human genetic analysis in recent years because there are thousands of different microsatellite loci available for PCR-marker analysis throughout the human genome.

Let's now look at an example of how PCR microsatellite markers are used to identify carriers of the mutant allele that causes Huntington disease.

Table 13.2 Information for Example 13.5: Analysis of PCR Microsatellites in the Family of a Person with Huntington Disease

Person	Genotype (number of CAG repeats in both alleles at *HD* locus)
Male with Huntington disease	45, 17
Mother	23, 17
Father	30, 19
Paternal uncle	30, 19
Control 1 (unrelated person with Huntington disease)	47, 17
Control 2 (unrelated person with Huntington disease)	44, 18

Example 13.5 Use of microsatellite markers to identify the Huntington disease allele.

Huntington disease is a debilitating, fatal genetic disorder in which the nervous system degenerates. A dominant mutant allele causes the disorder. The phenotype shows delayed onset with symptoms appearing most often in midlife. People who are carriers of the allele can be identified by DNA markers that detect the mutation. The gene associated with Huntington disease is called the *HD* gene (OMIM 143100) and encodes a protein called huntingtin. The gene is 180–200 kb long and contains 67 exons. The mutation responsible for Huntington disease is a CAG trinucleotide repeat expansion in exon 1 that creates a dominant gain-of-function allele. The normal allele contains 10–35 CAG repeats. Afflicted people carry one allele with 36–121 CAG repeats. In 1996, Alford et al. (*American Journal of Medical Genetics* 66:281–286) reported their analysis of PCR microsatellites in the family of a person with Huntington disease. The DNA came from a male with Huntington disease, his parents, and a paternal uncle, as indicated in Table 13.2. DNA from two unrelated individuals with Huntington disease were also included as controls.

Problem: (a) As Huntington disease is caused by a dominant allele, we initially expect that at least one parent carries the allele, but neither parent has Huntington disease. Do the microsatellite markers suggest nonpenetrance of the allele? If not, what do they suggest? **(b)** From which parent did this individual inherit the expanded allele?

Solution: (a) The microsatellite markers indicate that a new mutation, rather than nonpenetrance, explains the Huntington disease of the male whose parents do not have the disease. This male is heterozygous for an expanded mutant allele with 45

CAG repeats, similar in size to the expanded alleles in the control individuals. His other allele contains 17 repeats. His mother is heterozygous for 23 and 17 repeats, and his father is heterozygous for 30 and 19 repeats. Because neither parent has the 45-repeat allele, it must have been created by a new expansion mutation, probably in a parental germ line. **(b)** The affected male inherited a 17-repeat allele from his mother. The trinucleotide repeat probably expanded in his father's germ line and was then transmitted to the son. In fact, additional experiments conducted by the authors of this study confirmed that the expanded allele was inherited from the father. The 30-repeat allele is a premutation that is prone to expansion into a full mutant allele, much like the fragile-X premutations discussed in section 5.1.

Random Amplified Polymorphic DNA (RAPD)

As we have seen, primers for PCR must be synthesized from known nucleotide sequences. Preliminary DNA sequencing is often required to identify DNA sequences that can be used to synthesize primers. However, there are several PCR-based DNA marker methods that do not require prior knowledge of DNA sequences. Among the most utilized of these methods is **random amplified polymorphic DNA** analysis, also known by the acronym **RAPD**, which implies the rapidity of this method compared with other methods. RAPD analysis begins with a single short primer consisting of 10 nucleotides chosen arbitrarily. A primer this short should bind to many sites throughout the genome because the probability of matching a particular random sequence of 10 nucleotides is one in 4^{10}, or about once every 1 million nucleotides. In a human genome of 3 billion nucleotides, a random 10-nucleotide sequence should be represented about 3000 times. Moreover, the number of sites in the genome to which a 10-nucleotide primer binds can be increased by lowering the annealing temperature, because a lower

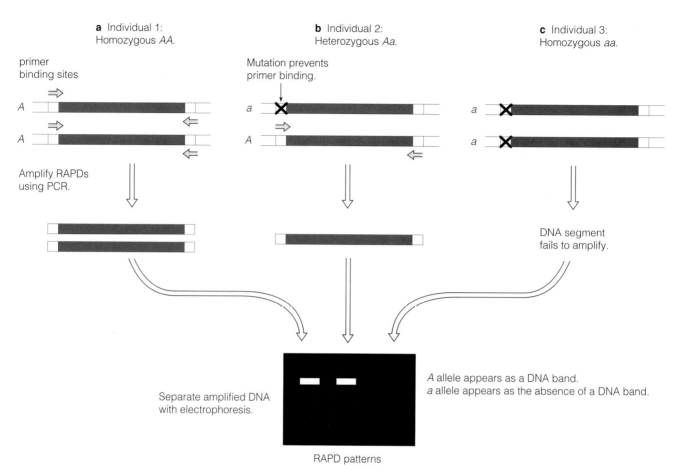

a Individual 1:
Homozygous *AA*.

b Individual 2:
Heterozygous *Aa*.

c Individual 3:
Homozygous *aa*.

primer binding sites

Mutation prevents primer binding.

Amplify RAPDs using PCR.

DNA segment fails to amplify.

Separate amplified DNA with electrophoresis.

A allele appears as a DNA band.
a allele appears as the absence of a DNA band.

RAPD patterns

Figure 13.32 The dominant and recessive phenotypes of an RAPD marker. A DNA band that is present in a gel represents a dominant allele. Both homozygous dominant (*AA*) and heterozygous (*Aa*) genotypes have the same electrophoretic phenotype. The homozygous recessive genotype (*aa*) does not produce a DNA band.

temperature permits some nucleotide pair mismatches between the primer and the DNA. In fact, the temperature used for primer annealing is so low that there are probably many thousands of binding sites in the genome. However, only a few of these sites are sufficiently close to one another for the bound primers to successfully amplify the DNA between them. In a RAPD experiment, each primer typically amplifies about 4–8 segments of DNA when total cellular DNA is used as the template.

If an individual carries a mutation in a RAPD primer binding site that prevents the primer from binding, the DNA segment does not amplify. Because of such mutations, populations are polymorphic for RAPD markers, and the differences among individuals are made visible by gel electrophoresis. Because polymorphic RAPD markers represent differences in DNA sequences, they can be used for genetic analysis in essentially the same way as RFLPs.

There is one major difference, however: RAPD markers are inherited as completely dominant and recessive alleles. The polymorphism has only two phenotypes, presence and absence of the marker on a gel. As a fragment produced from the DNA of either one or both homologous

chromosomes appears on the gel, the individual homozygous for the production of the fragment and the heterozygote have the same phenotype. The fragment is not produced from either chromosome of the recessive homozygote, whose phenotype is absence of the marker, as explained in Figure 13.32. Like other loci with dominant and recessive alleles, RAPD markers segregate in a 3:1 fashion in the progeny of two heterozygous parents, and in a 1:1 ratio in testcross progeny when one parent is heterozygous and the other is homozygous recessive. Figure 13.33 shows a RAPD marker that segregates in testcross progeny.

~

DNA markers detect differences in DNA sequence directly. The inheritance of DNA markers is analyzed in the same way as that of alleles governing physical traits.

SUMMARY

1. Most recessive alleles are mutant alleles in that they differ from the dominant wild-type alleles most often found in nature. However, there are many known cases of dominant mutant alleles.

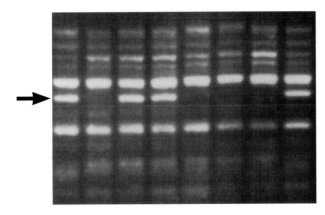

Figure 13.33 RAPD markers used in genetic analysis. Each lane in this gel contains amplified DNA from one individual. All individuals are testcross progeny from the same cross. The arrow points to the position of a polymorphic marker that segregates in a 1:1 ratio in these progeny.

2. Different alleles at a locus do not always show a simple dominant-recessive relationship. Incomplete dominance is a situation in which a heterozygote has a phenotype intermediate between the phenotypes of the two homozygotes. Codominance is a situation in which a heterozygote has the phenotypes of both homozygotes.

3. Although a diploid individual carries only two alleles at a locus, there may be more than two alleles for a single locus in a population of individuals. Multiple alleles may show various dominance relationships with one another, in some cases forming a dominance series.

4. Some recessive alleles do not completely eliminate the dominant phenotype but cause reduced expression of it. These alleles are often called leaky recessive alleles.

5. The phenotype associated with lethal alleles is death, often in zygotic or embryonic stages. Lethal alleles are usually recessive and are only expressed in homozygotes. They are usually detected as the absence of an expected class of progeny. Dominant lethal alleles are much more rare because their carriers usually die before reproducing. However, dominant lethal alleles with delayed onset may be transmitted from one generation to the next.

6. Pleiotropy is a situation in which a single gene affects more than one trait in the phenotype.

7. Epistasis is the interaction of alleles at two or more loci that affect a single trait. It is often detected by observing ratios such as 9:7, 15:1, 13:3, 12:3:1, or 9:3:4 in the offspring of doubly heterozygous parents.

8. Penetrance is the proportion of individuals with a particular genotype that have the phenotype typically associated with that genotype.

9. Expressivity is the degree to which a phenotype is expressed in individuals with a given phenotype.

10. DNA markers detect differences in nucleotide sequence and do not rely on analysis of phenotypes associated with genes.

11. Restriction fragment length polymorphisms (RFLPs) and PCR-based markers are among the DNA markers most often used. DNA markers are usually inherited as codominant markers, although some are inherited as fully dominant and recessive markers.

QUESTIONS AND PROBLEMS

1. The *TP53* locus in humans encodes a protein that protects cells against mutations at other loci that cause cancer. Many cancerous cells have mutations at the *TP53* locus. Some of the mutations in *TP53* are dominant loss-of-function mutations, and others are recessive loss-of-function mutations. At the biochemical level, what distinguishes these two types of mutant alleles?

2. Under what circumstances is there nonpenetrance for ABO blood types?

3. In 1916, Safir (*Genetics* 1:584–590) reported the discovery of a new recessive mutant allele at the *white* locus in *Drosophila melanogaster*, called w^{bf} for *white-buff*. Flies that are homozygous w^{bf}/w^{bf} have buff-colored eyes. Flies that are heterozygous $+/w^{bf}$ have wild-type brick-red eyes. The w^1 allele, reported by Morgan in 1910 (*Science* 32:120–122) causes white eyes when homozygous. Flies that are heterozygous w^1/w^{bf} have buff-colored eyes that are lighter in color than w^{bf}/w^{bf} homozygotes. **(a)** What are the dominance relationships between the w^+, w^1, and w^{bf} alleles? **(b)** Propose a biochemical model that explains these results.

4. Phenylketonuria (PKU) is characterized by severe mental retardation and abnormally high plasma phenylalanine. Over 190 recessive alleles at the *PAH* locus have been characterized. Some people have somewhat elevated plasma phenylalanine but do not suffer from PKU. This situation is called nonphenylketonuria hyperphenylalanemia (non-PKU HPA). Avigad et al. (1991. *American Journal of Human Genetics* 49:393–399) examined 27 families with non-PKU HPA and discovered that this phenotype was the result of compound heterozygosity for mutant alleles at the *PAH* locus in which one of the alleles is a PKU mutant allele and the second is also mutant but causes much milder effects when heterozygous with a PKU mutant allele. These less severe alleles apparently do not increase phenylalanine levels when homozygous. Provide a model based on enzyme function that explains this situation.

5. In 1991, Okano et al. (*New England Journal of Medicine* 324:1232–1238) reported the results of studies on the relationship between genotype and phenotype for PKU in humans. They studied eight different mutant alleles. Some of the alleles caused complete loss of PAH activity, whereas others reduced PAH activity. Table 13.3 lists the PAH activity associated with each of these alleles. Two clinical PKU phenotypes are identified on the basis of serum phenylalanine levels: classic PKU, with highly elevated levels of phenylalanine, and mild PKU, in which

Table 13.3	Information for Problem 5: PAH Activity Associated with Eight Mutant Alleles
Mutant Allele	**PAH Activity (% of normal PAH activity)**
PAH*R243X	< 1
PAH*P281L	< 1
PAH*R408W	< 1
PAH*IVS-12	< 1
PAH*E280K	< 3
PAH*R158Q	10
PAH*R261Q	30
PAH*Y414C	50

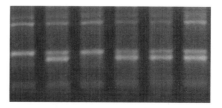

Figure 13.34 Information for Problem 8: RAPD markers in the progeny of a cross between two parents.

phenylalanine levels were not as highly elevated as in classic PKU. Both types of PKU result in mental retardation if left untreated, although mild PKU requires less stringent treatment. All individuals homozygous or heterozygous for the first five alleles listed in Table 13.3 had classic PKU. Compound heterozygotes between PAH*Y414C and any of the first four alleles had mild PKU. PAH*R158Q/*R158Q homozygotes had classic PKU, whereas PAH*R261Q/*R261Q homozygotes and PAH*R261Q/*Y414C heterozygotes had mild PKU. **(a)** Using the allele designations, describe the specific mutations (in terms of codon number and amino acid substitution for the seven substitution mutations) that created each of the alleles. **(b)** Predict the phenotypic PAH activity for every possible homozygote and compound heterozygote for these eight mutant alleles. **(c)** What is the relationship between the predicted PAH activity and the clinical phenotype?

6. In 1900, Carl Correns, who was one of Mendel's rediscoverers, reported the results of crosses between two types of stocks (an ornamental flowering plant), one with purple flowers and hoary leaves, the other with white flowers and smooth leaves (*Botanisches Centralblatt* 84:97–113, p. 19). The F_1 plants had purple flowers and hoary leaves, indicating that these two traits were dominant. In the F_2 progeny, all purple-flowered progeny had hoary leaves, and all white-flowered progeny had smooth leaves, just like the two original parents, and the two phenotypes appeared in a 3:1 ratio. However, stocks with purple flowers and smooth leaves as well as stocks with white flowers and hoary leaves were also known. **(a)** Describe why pleiotropy is or is not a reasonable explanation for these results. **(b)** What phenomena other than pleiotropy can explain these results?

7. What types of dominance (complete dominance, incomplete dominance, or codominance) are illustrated by the DNA markers in **(a)** Figure 13.30, **(b)** Figure 13.31, and **(c)** Figure 13.32?

8. Figure 13.34 shows a photograph of RAPD markers in the progeny of a cross between two parents. One of the markers is polymorphic. **(a)** Identify the polymorphic marker. **(b)** Identify those individuals that display the dominant phenotype for this marker and those that display the recessive phenotype for it.

9. In the garden pea, homozygosity for a recessive allele at the *A* locus causes white flowers, transparent seed coats, and absence of a purple ring at the leaf axils. Homozygosity for a recessive allele at the *Am* locus causes white flowers but does not prevent coloration of the seed coats and leaf axils. Devise a biochemical model that explains why one gene may show pleiotropy and another that affects the same trait does not. In your explanation, be sure to indicate which gene product appears first in the biochemical pathway.

10. Table 13.4 lists data from three RAPD loci in the progeny of a cross between two parents (extracted from data used for Mudge et al. 1996. *Crop Science* 36:1362–1366). **(a)** Are these testcross or F_2 progeny? **(b)** Using chi-square analysis, test the hypothesis that these three RAPD loci assort independently.

11. Hustad et al. (1995. *Genetics* 104:255–265) studied six different recessive alleles at the *agouti* (*A*) locus in mice. All of these alleles when homozygous caused a darker coat color than the wild-type allele. How is it possible to study six different alleles when a diploid cell can have only two alleles at a single locus?

12. None of the alleles studied by Hustad et al. (1995. *Genetics* 104:255–265) were lethal when homozygous. Given the mechanism of lethality for the mutant A^y allele (see Example 13.2), why is homozygous lethality not expected for most other mutant alleles at the *A* locus?

13. The A^y allele has two phenotypes associated with it, yellow coat color and lethality. **(a)** At the level of phenotypic inheritance, is this an example of pleiotropy? **(b)** At the molecular level, is this an example of pleiotropy? Explain your answers in terms of the definition of pleiotropy.

14. The results in Table 13.5 are from several experiments for coat color in mice reported by Bateson in 1913 (*Mendel's Principles of Heredity*. Oxford: Oxford University Press, pp. 81–83). Assume that the original parents were homozygous at all loci governing coat color. **(a)** How

Table 13.4 Information for Problem 10: RAPD Marker Analysis Data

| | RAPD Marker* | | |
Progeny	729s435	E10s469	362s244
1	−	+	−
2	+	−	+
3	−	+	+
4	−	−	−
5	−	−	+
6	+	+	−
7	+	−	−
8	+	−	+
9	+	+	+
10	+	−	+
11	−	+	+
12	−	+	+
13	−	−	+
14	+	−	−
15	−	−	+
16	+	−	+
17	−	+	−
18	+	+	+
19	+	−	+
20	−	−	−
21	+	+	+
22	−	−	+
23	−	−	+
24	+	+	−
25	+	−	+
26	−	+	+
27	−	−	−
28	+	+	+
29	+	−	−
30	−	+	+
31	−	+	+
32	−	−	+
33	−	−	+
34	−	+	+
35	−	−	+
36	−	+	+
37	−	+	−
38	+	−	+
39	+	+	+
40	−	+	−

| | RAPD Marker* | | |
Progeny	729s435	E10s469	362s244
41	+	+	−
42	−	+	−
43	+	−	−
44	+	+	+
45	−	−	−
46	−	−	−
47	+	+	−
48	−	+	−
49	−	−	+
50	−	−	+
51	+	−	−
52	−	+	−
53	+	+	−
54	−	+	−
55	+	+	−
56	+	−	+
57	−	+	−
58	−	−	+
59	+	+	+
60	+	+	+
61	+	+	+
62	−	+	−
63	+	−	−
64	−	−	+
65	+	−	−
66	+	−	−
67	+	+	−

* A "+" indicates that the RAPD marker amplified and a "−" indicates that the marker failed to amplify.

many loci interact in these three crosses? **(b)** Propose an epistatic model that explains the inheritance of coat color in these crosses. **(c)** Use your model to derive expected ratios in each of the crosses, then test the hypothesis that the observed ratios conform to the expected ratios in each cross using chi-square analysis.

15. In the same publication cited in the previous question, Bateson reported the data in Table 13.6 from additional experiments for coat color in mice. Assume that the original parents were homozygous at all loci governing coat color. **(a)** How many loci interact in these three crosses? **(b)** Propose an epistatic model, based on your

Table 13.5 Information for Problem 14: Data from Mouse Coat Color Experiments

Cross	F₁ Generation	F₂ Generation
Blue × chocolate	All black	44 black, 17 blue, 17 chocolate, 17 silver-faun
Black × silver-faun	All black	67 black, 21 blue, 20 chocolate, 5 silver-faun
Blue × silver-faun	All blue	46 blue, 17 silver-faun

Table 13.6 Information for Problem 15: Additional Data from Mouse Coat Color Experiments

Cross	F₁ Generation	F₂ Generation
Silver-faun × albino	All chocolate	19 chocolate, 4 silver-faun, 6 albino
Black × albino	All black	76 black, 24 chocolate, 27 albino
Blue × albino	All black	33 black, 10 blue, 8 chocolate, 2 silver-faun, 12 albino

Table 13.7 Information for Problem 19: Results of Sweet Pea Crosses

Cross	F₁ Phenotypes	F₂ Phenotypes and Ratios
Line 1 × line 2	All colored	9 colored:7 white
Line 3 × line 4	All colored	9 colored:7 white
Line 2 × line 4	All colored	3 colored:1 white
Line 1 × line 4	All colored	3 colored:1 white
Line 3 × (line 3 × line 4)	————————————————→ 1 colored:3 white	

model from the previous question, that explains the inheritance of coat color in these crosses. **(c)** Use your model to derive expected ratios in each of the crosses, then test the hypothesis that the observed ratios conform to the expected ratios in each cross using chi-square analysis.

16. Provide plausible biochemical models to account for each of the following genetic phenomena: incomplete dominance, codominance, the Bombay phenotype, leaky recessive alleles, recessive lethal genes, pleiotropy, epistasis (F₂ ratios: 9:3:3:1, 9:3:4, 9:7, 12:3:1, and 15:1), variable expressivity, and incomplete penetrance.

17. In sweet peas, the C and P loci govern flower color in the manner diagrammed in Figure 13.16. Give the expected phenotypic ratios for colored and white flowers in the first-generation offspring in each of the crosses **(a)** $Cc\ pp \times cc\ Pp$, **(b)** $Cc\ Pp \times Cc\ pp$, **(c)** $ccpp \times Cc\ Pp$, and **(d)** $Cc\ Pp \times Cc\ Pp$.

18. With respect to the C and c alleles and the P and p alleles in sweet pea diagrammed in Figure 13.16, what

proportion of the F₂ progeny are true breeding for their respective flower color when self-fertilized?

19. True-breeding (fully homozygous) lines of sweet pea numbered line 1, line 2, line 3, and line 4 have different genotypes with respect to flower color. The phenotypes are as follows: line 1 = white flowers; line 2 = white flowers; line 3 = white flowers; line 4 = colored flowers. Based on the crosses listed in Table 13.7, determine the genotypes at the C and P loci for flower color for each pure line if line 1 is known to be $cc\ PP$.

20. Pattern baldness is due to an allele that is dominant in men and recessive in women. Yet among men and women with pattern baldness, the degree to which the phenotype is expressed is variable. What could account for this variability if the phenotype is due to an allele at a single locus?

21. Often our perception as to phenotypic dominance varies depending on the aspect of phenotype we are looking at. For instance, an enzyme in rabbits found in the liver (coded for by gene Y) converts xanthophils, a yellow

chloroplast pigment found in plants that the rabbits consume, to a colorless product. Rabbits homozygous for an allele (*y*) that fails to produce a functional enzyme cannot covert xanthophils to the colorless compound. The yellow pigment is fat soluble and accumulates in body fat. This is the basis for the yellow- and white-fat phenotypes of different strains of rabbits. A true-breeding rabbit strain with white fat is mated with a yellow-fat strain. The 40 hybrid rabbits all have white fat, but biochemical analysis reveals that the enzyme activity per liver cell is exactly half of that found in the homozygous white-fat parent. One hundred F_2 rabbits segregate 74 white fat:26 yellow fat based on phenotypic appearance. Enzyme analysis of the F_2 rabbits reveals that 52 of the white-fat rabbits have half the normal activity for the liver enzyme, and 22 of the white-fat rabbits have full normal activity. **(a)** What are the dominance relationships (complete dominance, incomplete dominance, or codominance) for inheritance of fat color and inheritance of enzyme activity? **(b)** Assuming that the *y* allele is mutant, is it caused by a gain-of-function mutation or a loss-of-function mutation? **(c)** Could the *y* allele be considered an example of a dominant mutant allele? Why or why not?

22. In chickens, comb shape is controlled by the interaction of two independent loci called the *R* and *P* gene loci by Bateson and Punnett (see Figure 13.14). Four comb types were seen to be inherited: walnut, rose, pea, and single comb. Bateson and Punnett observed the following when chickens of different comb shape were crossed:

True-breeding rose × true-breeding pea → F_1 all walnut

F_1 walnut × F_1 walnut → F_2 $^9/_{16}$ walnut, $^3/_{16}$ rose, $^3/_{16}$ pea, $^1/_{16}$ single

F_1 walnut × true-breeding single → $^1/_4$ walnut, $^1/_4$ rose, $^1/_4$ pea, $^1/_4$ single

True-breeding rose × true-breeding single → F_1 all rose

F_1 rose × F_1 rose → F_2 $^3/_4$ rose, $^1/_4$ single

True-breeding pea × true-breeding single → F_1 all pea

F_1 pea × F_1 pea → F_2 $^3/_4$ pea, $^1/_4$ single

Determine which phenotypes should appear and in what proportions in the offspring of the following crosses: **(a)** true-breeding rose × true-breeding pea, **(b)** F_1 pea (from pea × single) × F_1 rose (from rose × single), **(c)** F_1 walnut (from walnut × single) × F_1 rose (from rose × single), **(d)** F_1 walnut (from walnut × single) × F_1 pea (from pea × single).

23. In maize several genes govern production of anthocyanin, a purple pigment in seedling leaves. In the absence of anthocyanin, the seedling leaves show only the green chlorophyll pigmentation. In the presence of anthocyanin, the seedlings have a purple cast on top of the green coloration. The gene locus called *colorless 1* (C^1) appears to function as a trans-acting inducer locus for at least two other loci (*ChsA* and *ChsJ*) that encode two

enzymes in the pathway for anthocyanin synthesis. Assume that independent assortment applies to these three loci. Homozygosity for the recessive alleles at any of the three loci cause green seedlings. Determine the phenotypic progeny ratios expected from the crosses **(a)** *ChsA ChsA ChsJ chsJ* C^1c^1 × *ChsA ChsA ChsJ chsJ* C^1c^1, **(b)** *ChsA chsA ChsJ chsJ* C^1C^1 × *ChsA chsA ChsJ chsJ* C^1C^1, and **(c)** *ChsA chsA ChsJ chsJ* C^1c^1 × *chsA chsA chsJ chsJ* c^1c^1.

24. Draw diagrams that explain the biochemical basis for the genetic interaction of the C^1, *ChsA*, and *ChsJ* loci, showing how the pathway for anthocyanin is blocked in homozygous recessive genotypes.

25. In tomatoes presence of anthocyanin in seedlings gives a purple cast to stems and undersides of leaves. Several loci control presence/absence of anthocyanin in the leaves: A^1 (*anthocyaninless*) and A^m (*anthocyanin minus*) produce green stems and green undersides of leaves when either locus is homozygous for the recessive alleles (a^1a^1 or a^ma^m). A cross is made as follows: $a^1a^1 A^mA^m$ × $A^1A^1 a^ma^m$. **(a)** What would be the F_2 dihybrid ratios if the A^1 locus encoded a small protein necessary for induction of the A^m locus and the A^m locus encoded an enzyme in the biosynthetic pathway for anthocyanin synthesis? What would be the genotype and phenotype of the F_1? **(b)** What would be the F_2 dihybrid ratios if the A^1 locus and the A^m locus each encoded an enzyme in the biosynthetic pathway for anthocyanin synthesis? What would be the genotype and phenotype of the F_1? **(c)** What would be the F_2 dihybrid ratios if the A^1 locus and the A^m locus each encoded polypeptide subunits present in equal amounts in a dimeric enzyme in the biosynthetic pathway of anthocyanin pigment? What would be the genotype and phenotype of the F_1?

26. In chickens, the *C* and *I* loci control feather color, as indicated in Figure 13.21. The White Leghorn breed has the genotype *CC II*. The White Wyandotte breed has the genotype *cc ii*. Give the F_2 phenotypes and expected proportions from the cross White Leghorn × White Wyandotte.

27. In cattle, the allele *P* for hornless (polled) is dominant to the allele *p* for horned. Red and white coat colors (*RR* and *R'R'*, respectively) are controlled by codominant alleles, with the heterozygote (*RR'*) being roan. Diagram a cross between a polled red bull (*PP RR*) and a horned white cow (*pp R'R'*) showing expected genotype and phenotype frequencies of the F_1 and F_2.

28. Two loci produce a black pigment on the adult body in *Drosophila melanogaster*. The *ebony* locus is located on chromosome III, and the *black* locus is located on chromosome II. Homozygosity for recessive alleles at either locus (*ee* or *bb*) cause a black body color. Assume that the black and ebony phenotypes are morphologically indistinguishable (actually, they are distinguishable). True-breeding ebony flies (*ee*; b^+b^+) are crossed with true-breeding black flies (e^+e^+; *bb*). **(a)** Determine the

genotype and phenotype of the F_1 and the expected genotypic and phenotypic ratios of the F_2. **(b)** Determine the phenotypic ratios if the F_1 were crossed with a true-breeding ebony strain and a true-breeding black strain.

29. In *Drosophila melanogaster*, the hairless phenotype, discovered by Calvin Bridges, causes missing bristles on the abdomen and is caused by a mutant allele *H* that is dominant to the wild-type allele (H^+). The *H* locus is on chromosome III. A dominant suppressor allele *Su H*, at a locus on chromosome II, was also discovered by Calvin Bridges. In the presence of *Su H*, the effect of the *H* allele is completely suppressed, and the phenotype is indistinguishable from the wild-type phenotype. A true-breeding hairless strain is crossed with a true-breeding wild-type strain that may carry the *Su H* allele. The F_1 is wild type. The F_1 flies are intermated to produce a large F_2 population. The bristle phenotypes of individuals in the F_2 generation have a ratio of 13 wild type:3 hairless. **(a)** What evidence indicates that the suppressor allele is present in this cross? **(b)** Explain these ratios in terms of genotypes and phenotypes and expected ratios (where appropriate) of the parental, F_1, and F_2 generations. **(c)** Postulate a plausible biochemical model for this kind of epistatic gene interaction. **(d)** Is there any evidence of nonpenetrance or variable expressivity in this example? If so, what is that evidence?

30. Devise a set of experiments to determine the genotypes of the F_2 flies from the previous question using testcrosses, and give the expected results of each cross.

31. Purple eye color, due to homozygosity for a recessive allele (*pr*) encoded at a locus on chromosome 2 in *Drosophila melanogaster*, was discovered by Calvin Bridges. Curt Stern later discovered a recessive suppressor locus on chromosome 3 (*su pr*) that, when present in the homozygous condition, completely suppresses *pr*, causing a wild-type phenotype. A true-breeding, purple-eyed strain is crossed with a true-breeding, wild-type strain. The wild-type F_1 is intercrossed to produce a large F_2 population. The eye color phenotype of the F_2 population is classified as 1306 with wild-type eyes:297 with purple eyes. **(a)** Explain these ratios in terms of genotypes and phenotypes and expected ratios (where appropriate) of the parental, F_1, and F_2 generations. **(b)** Postulate a plausible biochemical basis for this kind of epistatic gene interaction. **(c)** Explain why the F_2 dihybrid segregation pattern in this problem is exactly the same as that expected for problem 29, in which we examined a dominant mutant trait and a dominant suppressor locus, whereas in this case we examine a recessive mutant trait and a recessive suppressor locus.

32. In rats, two loci, *A* and *R*, affect coat color in the following manner:

> A_ R_   gray coat
> A_ rr   yellow coat

> aa R_   black coat
> aa rr   cream coat

A pair of gray rats are mated and produce 8 offspring. Six pups are gray coated, 1 pup is yellow coated, and 1 pup has a black coat. What are the genotypes of the gray parents?

33. The coat color phenotypes described in the previous question are expressed only in the presence of a dominant allele at the *C* locus. Homozygosity for the recessive allele (*cc*) causes the albino phenotype (no hair pigment develops, and the rats are white). Two true-breeding albino lines (line 1 and line 2) are mated with a cream line (line 3) with the following results:

Mating	F_1 Phenotype
Line 1 × line 3	Black
Line 2 × line 3	Gray

Write the genotypes of lines 1, 2, and 3.

34. Consider two loci in horses affecting coat color. A dominant coat color allele (*W*) produces a white coat. Homozygosity for the recessive allele (*w*) allows a colored coat to appear. Crosses between two white horses usually yield 2 white:1 colored. A dominant allele *R* at a second locus produces a roan-colored horse. Homozygosity for the recessive allele *r* produces nonroan. Roan × roan matings tend to produce 2 roan:1 nonroan progeny. **(a)** What is the most likely explanation for the 2:1 progeny ratios? **(b)** Give the expected genotypic and phenotypic offspring ratios when horses with the following genotypes mate: *Ww Rr* × *Ww Rr*.

35. Many loci interact with one another to govern coat color in horses. In this question, we will examine four of them, *B, E, C,* and *A*. At the *B* locus, the dominant *B* allele causes a black coat in the genotypes *BB* and *Bb*. The recessive allele, *b*, causes a brown coat in *bb* horses. However, the alleles at the *B* locus cause these phenotypes only in the presence of the dominant *E* allele at the *E* locus. Horses that are homozygous *ee* are chestnut colored regardless of the genotype at the *B* locus. The *C* locus affects pigment development. *CC* permits full development of the colors governed by the *B* and *E* loci. The allele *C* is incompletely dominant over *c*. When *c* is present in the heterozygous condition, it causes a dilution of coat color. Intermediate coat colors, such as buckskin and palomino, are found in horses that are heterozygous *Cc*. Homozygous *cc* horses have a cream-colored coat called cremello. The *A* locus complicates matters but contributes to the delightful variations in coat color in horses. In the *aa* genotype, the phenotypes governed by the other three loci are fully expressed over the body. In the *AA* and *Aa* genotypes, dark pigment is limited to the mane, tail, and lower legs. A bay stallion (*AA BB CC EE* with a light-colored body and black legs, mane, and tail) is mated with a cremello mare (*aa bb cc ee*). Assume that all loci assort independently. **(a)** What is the

expected phenotype of the F$_1$ horses? **(b)** What are the possible phenotypes in the F$_2$ generation, and what are the expected frequencies for each?

36. In 1981, Geever et al. (*Proceedings of the National Academy of Sciences, USA* 78:5081–5085) described the use of *Dde*I to detect the *HBB*S allele in humans using RFLP analysis. The first cloned DNA they used as a probe covered a large region including the 5′ end of the *HBB* locus that contained five *Dde*I sites. They subcloned a portion of this cloned DNA that included the single *Dde*I site spanning codons 5 and 6 of the *HBB* locus. Why is the subcloned DNA a better choice as a probe than the larger cloned DNA?

FOR FURTHER READING

Mendel, in his classic paper **Mendel, G. 1866. Versuche über Pflanzen-Hybriden.** *Verhandl naturf Verein Brünn* 4:3–47, observed, and in some cases correctly interpreted, several of the concepts discussed in this chapter, including epistasis, pleiotropy, variable expressivity, and delayed onset. Bateson, in his genetics textbook, **Bateson, W. 1913.** *Mendel's Principles of Heredity.* **Cambridge: Cambridge University Press**, further identified and developed many of the principles in this chapter. Several recent articles provide detailed information on the molecular bases underlying multiple alleles, codominance, epistasis, lethal alleles, and pleiotropy. The effects of mutations in the *TP53* gene were reviewed by **Vogelstein, B., and K. W. Kinzler. 1992. p53 function and dysfunction.** *Cell* 70:523–526. The molecular basis for ABO blood groups was described by **Yamamoto, F., H. Clausen, T. White, J. Marken, and S. Hakomori. 1990. Molecular genetic basis of the histo-blood group ABO system.** *Nature* 345:229–233. The molecular basis for the lethal yellow effect in mouse was published by **Duhl, D. M. J., M. E. Stevens, H. Vrieling, P. J. Saxon, M. W. Miller, C. J. Epstein, and G. S. Barsh. 1994. Pleiotropic effects of the mouse** *lethal yellow* **(A^y) mutation explained by detection of a maternally expressed gene and the simultaneous production of** *agouti* **fusion RNAs.** *Development* 120:1695–1708. The molecular basis for the interaction of coat color loci and alleles in mouse was reviewed by **Jackson, I. J. 1994. Molecular and developmental genetics of mouse coat color.** *Annual Review of Genetics* 28:189–217. The rules for human genetic notation were summarized by **Shows et al. 1987. Guidelines for human gene nomenclature.** *Cytogenetics and Cell Genetics* 46:11–28. The rules for *Drosophila* genetic notation were summarized by **Lindsley, D. L., and G. G. Zimm. 1992.** *The Genome of Drosophila melanogaster.* **San Diego, Calif.: Academic Press**. A description of how minisatellites are used in human genetic analysis is **Nakamura, Y., et al. 1987. Variable number of tandem repeat (VNTR) markers for human genetic mapping.** *Science* 235:1616–1622. A succinct review of microsatellite markers is **Hearne, C. M. 1992. Microsatellites for linkage analysis of genetic traits.** *Trends in Genetics* 8:288–294. RAPD analysis was described by **Williams, J. G. K., A. R. Kubelik, K. J. Livak, J. A. Rafalski, and S. V. Tingey. 1990. DNA polymorphisms amplified by arbitrary primers are useful as genetic tools.** *Nucleic Acids Research* 18:6531–6535.

For additional reading, go to InfoTrac College Edition, your online research library at: http: www.infotrac-college.com/brookscole

CHAPTER 14

KEY CONCEPTS

Sex determination may be genetic or nongenetic.

~

Genetic sex determination is usually governed by genes on
sex chromosomes.

~

The inheritance of genes on sex chromosomes differs in males and females.

~

Mechanisms known as dosage compensation equalize the effect of genes
located on the X chromosome in males and females.

~

Sex-influenced and sex-limited traits are expressed differently in males and
females, but are governed by genes that are not on sex chromosomes.

SEX DETERMINATION AND SEX-RELATED INHERITANCE

Most species reproduce sexually at some point in their life cycles. Because of sexual reproduction, novel combinations of genes can be produced with each mating, thereby maintaining genetic diversity. Without sexual reproduction, mutation is the only source of genetic variation, and because mutation rates tend to be relatively slow, the lack of genetic diversity may limit a species' ability to adapt to changing environments.

Cross-fertilization encourages heterozygosity, which facilitates exchange and recombination of genes. Although many plant species and some animal species are capable of self-fertilization, there is a wide array of natural mechanisms—genetic and nongenetic—that inhibit or exclude self-fertilization.

In the first section of this chapter, we briefly explore the variety of natural processes that facilitate cross-fertilization. We then examine genetic and nongenetic sex determination, focusing on the genetic systems in mammals and *Drosophila* that have been extensively studied. Finally, we see how genes located on sex-determining chromosomes are inherited and regulated.

14.1 PROMOTION OF CROSS-FERTILIZATION

Nearly all animal and plant species reproduce sexually, and most reproduce through cross-fertilization. There are a few species that reproduce exclusively by asexual means, such as some whiptail lizard species that consist exclusively of females. The females produce asexual offspring through parthenogenesis, the development of offspring from unfertilized diploid egg cells. Some plants also reproduce exclusively by asexual means. These, however, are exceptions to the general rule of sexual reproduction.

The exchange and recombination of genetic material through sexual reproduction is facilitated when something inhibits or prevents self-fertilization. The most effective means is a situation in which each individual is either male or female. Members of such species cannot self-fertilize, so recombination of genetic material at each sexual generation is obligatory. Most animal species have separate males and females, but this situation in animals is not universal. Some animal species consist entirely of **hermaphrodites**, meaning that each individual has both male and female reproductive organs. A few hermaphroditic animals, such as some species of snails, are capable of either self- or cross-fertilization. Others, such as earthworms, are hermaphroditic but are incapable of self-fertilization: when two individuals mate, they each deposit sperm into the other, and both carry fertilized eggs.

Many plants are hermaphroditic, having both male and female reproductive organs in the same flower. Some plant species, including many of those we use for food, self-fertilize and may have mechanisms that prevent cross-fertilization. However, most hermaphroditic plant

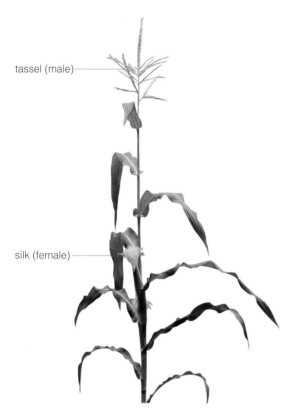

tassel (male)

silk (female)

Figure 14.1 Maize, an example of a monoecious plant. The male and female reproductive structures are physically separated on the same plant.

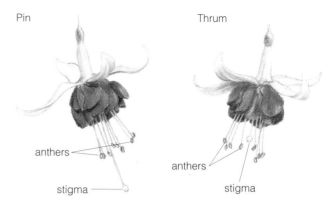

Pin

Thrum

anthers

anthers

stigma

stigma

Figure 14.2 The two types of floral structure in fuchsia: **(a)** pin and **(b)** thrum. (Drawing by Marcus Alan Vincent.)

species have mechanisms that inhibit or exclude self-fertilization. The alfalfa plant, for example, has a flower that cannot release its pollen until a bee trips the flower, causing the reproductive parts of the flower to snap forward, hitting the bee and scattering pollen onto it. When this happens, some pollen grains may self-fertilize the flower, but usually pollen from another plant that is already on the bee cross-fertilizes the flower.

Some hermaphroditic plants have a genetic form of self-incompatibility that prevents self-fertilization. Self-incompatibility alleles prevent plants of the same genotype from successfully fertilizing themselves or each other. To successfully fertilize, pollen must land on the flowers of a plant with a different self-incompatibility genotype. Apples are a good example. A Jonathan apple tree requires another apple tree of a different variety nearby as a source of pollen for fruits to develop. An apple orchard of all Jonathan trees fails to produce apples unless there is a compatible pollen source nearby so that bees can carry the pollen from one type of tree to the other.

Monoecy, in which a plant has male and female flowers that are physically separated on the same plant, is a mechanism that encourages cross-fertilization in some plant species. Maize is a good example of a monoecious plant (Figure 14.1). The tassel at the top of the plant contains the male flowers, and the silks that protrude from the tops of the ear contain the female flowers. Pollen may fall from a maize plant's tassel onto the silks and self-

fertilize some of the eggs, but the eggs from which most kernels develop are cross-fertilized by pollen carried by wind from another plant.

Some plant species are **dioecious,** having separate male and female plants. Asparagus is an example. The male plants have pollen-producing flowers, and the female plants have female flowers and develop small red berries containing the seeds after fertilization. Most dioecious species of plants produce rare hermaphrodites. Some species of plants are **trioecious** in that they consistently produce male, female, and hermaphroditic individuals.

The following example shows one way in which self-incompatibility is determined genetically.

Example 14.1 Self-incompatibility in plants: Pin and thrum flowers in fuchsia.

Fuchsia, an ornamental flowering plant, was one of Mendel's favorite plants. He is seen holding a fuchsia flower in a photograph and was recognized in his obituary for the prize-winning fuchsias he grew. Fuchsia produces plants with two types of flowers, pin and thrum, as illustrated in Figure 14.2. Because of genetic self-incompatibility, a pin plant can only be fertilized by a thrum plant, and a thrum plant can only be fertilized by a pin plant. This excludes self-pollination as well as cross-pollination between like phenotypes. The pin and thrum phenotypes are governed by two self-incompatibility alleles at a single locus, S and s, that are inherited in a Mendelian fashion. The S allele is dominant and causes the thrum phenotype. The s allele causes the pin phenotype when homozygous ss.

Problem: **(a)** As pin and thrum represent two mating types, why can't they be called male and female? **(b)** What proportions of pin and thrum progeny are expected from seed harvested from a thrum plant? **(c)** What proportions of pin and thrum progeny are expected from seed harvested from a pin plant?

Solution: **(a)** On the surface it might appear that pin and thrum plants are different sexes because one can only be fertilized by the other. However, pin and thrum plants produce both male and female gametes, so the plants are hermaphroditic. **(b)** The information in the problem does not tell you directly whether the thrum parent is homozygous *SS* or heterozygous *Ss*. However, because each thrum plant must have a pin (*ss*) parent, all thrum plants must be heterozygous *Ss*. With normal gamete segregation and fertilization, plants with the *SS* genotype cannot arise. Every mating must be *Ss* × *ss*. Half of the progeny harvested from a thrum plant should be *Ss* and half *ss*, a 1:1 ratio for pin and thrum. **(c)** The same is true for the seed harvested from a pin plant. The pin parent on which the seeds are borne is *ss* and contributes the *s* allele to each seed. The fertilizing pollen comes from a thrum plant that is *Ss*, so half the pollen grains carry *S* and half carry *s*. The seeds on a pin plant should be half *Ss* and half *ss*, so the expected ratio is again 1:1 for pin and thrum progeny.

There are many mechanisms in nature that promote cross-fertilization, which makes exchange and recombination of genes possible.

14.2 SEX DETERMINATION

Having separate male and female individuals is the most effective way to ensure cross-fertilization. Most animal and many plant species have separate sexes, but the mechanisms for sex determination among them are quite diverse. Let's take a brief look at several types of sex determination, using some well-studied species as examples.

Environmental Sex Determination

In some species, sex is determined entirely by environmental factors. In certain species of sea turtles and geckos, the temperature at which the eggs hatch determines sex. When the temperature is cool (below about 25°C), the eggs all hatch into females. When the temperature is warm (above about 32°C), they all hatch into males. When the temperature is moderate (between 25° and 32°C), both males and females may hatch from a cluster of eggs.

The slipper limpet (*Crepidula fornicata*) provides a remarkable example of environmental sex determination. The slipper limpet is a molluscan marine organism that lives much of its life as a free-swimming larva that is neither male nor female. When the time comes for a larva to mature and reproduce, it settles on a substrate and remains attached to that substrate for the rest of its life. If

Figure 14.3 A pile of slipper limpets. The individual at the bottom of the pile is female.

the substrate is the seafloor, the larva matures into a female. However, larvae usually seek other slipper limpets that are already attached to a substrate, and settle on top of them to mature, creating piles of mature slipper limpets (Figure 14.3). Larvae that land on top of other slipper limpets mature into males. In the pile, the individual on the bottom is typically female, whereas most of the others are males that bathe the female in sperm for reproduction. If several males are on the pile, one of them may also become a female. After an individual has become a female, it remains female for the rest of its life. When a female dies, the next male in line becomes a female. The factor that determines whether an individual becomes male or female is proximity to a female. When a slipper limpet lands on or near a female, it becomes a male. When it is distant from a female, it becomes a female.

Genetic Sex Determination by One Gene

In many dioecious plant species, sex is determined by two alleles at a single locus. The chromosomes that carry the sex-determining alleles are homologous and pair with each other along their full lengths. Typically, one sex is heterozygous for the alleles and the other sex is homozygous. Asparagus is a good example. A dominant allele, *M*, confers the male phenotype, so females are homozygous *mm*. Males are usually heterozygous *Mm* because they inherit the *m* allele from their female parent. Consequently, most matings are *mm* (female) × *Mm* (male), so males and females appear in a 1:1 ratio. Rarely, an *Mm* plant develops into a hermaphrodite. When the female flowers on a hermaphroditic *Mm* plant are pollinated by an *Mm* male, 25% of the progeny are homozygous *MM* males that are fully viable and fertile, and produce all male *Mm* offspring when mated with an *mm* female.

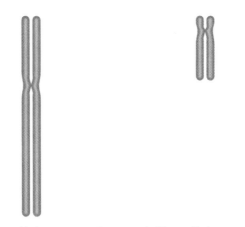

a Human X chromosome at metaphase. There are 164 million nucleotide pairs of DNA in each chromatid.

b Human Y chromosome at metaphase. There are 28 million nucleotide pairs of DNA in each chromatid.

Figure 14.4 The heteromorphic human sex chromosomes.

Ploidy and Sex Determination

Among certain insects, particularly those of the order hymenoptera (bees, wasps, and ants), sex is related to **ploidy**, the number of chromosome sets in each cell. In honeybees, the queen is a fertile diploid female. The male drones are usually haploid and produce haploid sperm cells. During the mating flight, the male drones deposit haploid sperm in the female queen. Most of these sperm fertilize haploid eggs to make diploid zygotes, but a few eggs escape fertilization. The queen lays an egg in each cell of the honeycomb. Most of the diploid eggs become sterile female worker bees. The few haploid eggs develop into male drones, who attempt to mate with queens from other hives during the mating season.

Because females are diploid and males haploid, ploidy appears to be the cause of sex determination, but in reality the relationship between sex determination and ploidy is indirect. Multiple alleles at a single locus determine sex. Individuals who are heterozygous for two different alleles are female, and individuals who carry only one allele, either as haploids or homozygous diploids, are male. There are so many different alleles at the sex-determination locus that only rarely do homozygous drones arise—nearly all diploid individuals are heterozygous females. Homozygous diploid drones are sometimes found, but only when drones and queens with the same allele mate and produce homozygous offspring.

Chromosomes and Sex Determination

In most animals and many dioecious plants, sex determination is associated with a pair of **sex chromosomes**, chromosomes that differ between males and females.

Many species have sex chromosomes that are **heteromorphic**, which means that they differ in size and in the genes that they carry, as exemplified by the human X and Y chromosomes shown in Figure 14.4. The heteromorphic sex chromosomes of most species are designated X and Y, although a few species have heteromorphic sex chromosomes called Z and W. All chromosomes other than sex chromosomes are called **autosomes**. There are three major chromosomal sex-determination systems: the XX-XO system, which is not heteromorphic, the XX-XY system, and the ZZ-ZW system.

Grasshoppers and some other insects have an **XX-XO system** of sex determination in which females have two copies of the X chromosome in their diploid cells and males have only one. There is only one type of sex chromosome, and sex is determined by how many copies of it an individual possesses. Meiosis in the female produces gametes with one copy of each chromosome, and meiosis in the male produces two types of gametes: half with one copy of each chromosome, including the X chromosome, and half with one copy of each autosome but no X chromosome. When sperm and egg cells unite, half of the zygotes receive two X chromosomes and develop as females, and half receive only one X chromosome and develop as males. The chromosome constitution of males is written as XO; the O indicates the absence of a second X chromosome (Figure 14.5a).

All mammalian species, as well as many other animal species and some plant species, have an **XX-XY system** of sex determination in which the X and Y chromosomes are heteromorphic. The X chromosome of many species is larger than the Y chromosome and carries many genes, only some of which govern sex determination. Females carry two X chromosomes, and males carry one X chromosome and one Y chromosome. The Y chromosome usually contains one or more genes that stimulate development of male reproductive organs and other phenotypic sex characteristics. Segregation of X and Y chromosomes during meiosis in males equalizes the numbers of XX and XY zygotes (Figure 14.5b).

In species with an XX-XY sex-determination system, females are the **homogametic sex**, which means that they produce only one type of gamete with regard to sex-determining factors. All gametes from an XX female carry an X chromosome and no Y chromosome. Males are the **heterogametic sex** because their gametes differ for sex-determining factors. Half of the gametes of an XY male carry an X chromosome and half carry a Y chromosome. In species with an XX-XO sex-determination system, males are the heterogametic sex because they produce two types of gametes (X and O), and females are the homogametic sex because they produce only one type of gamete (X).

Galinaceous bird species (such as chickens and turkeys), moths, butterflies, and some fishes have a

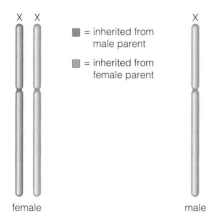

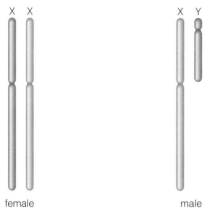

a In the XX-XO system, the female has two X chromosomes and the male has one.

b In the XX-XY system, the female has two X chromosomes and the male has one X chromosome and one Y chromosome.

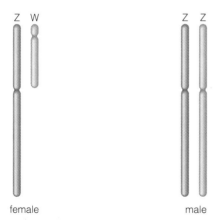

c In the ZZ-ZW system, the female has one Z chromosome and one W chromosome and the male has two Z chromosomes.

Figure 14.5 The three chromosomal systems of sex determination.

ZZ-ZW system of sex determination in which the heterogametic sex is female and the homogametic sex is male, the reverse of the XX-XY system. Males have two Z chromosomes, whereas females have one Z chromosome and one W chromosome. The Z chromosome is usually the larger of the two. As in the XX-XY sex-

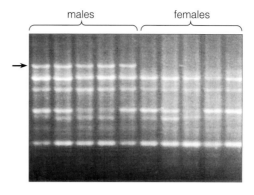

Figure 14.6 Information for Example 14.2: RAPD patterns in *A. garrettii*. (From Ruas, C. F., D. J. Fairbanks, W. R. Andersen, R. P. Evans, H. C. Stutz, and P. M. Ruas. 1998. Male-specific DNA in *Atriplex garrettii. American Journal of Botany* 85:162–167.)

determination system, meiosis in the heterogametic sex equalizes the numbers of males and females among the progeny (Figure 14.5c).

With such a wide variety of sex-determination systems in nature, scientists must examine each species experimentally to identify its sex-determination system. The following example shows how DNA markers can be used to do this.

Example 14.2 **Sex determination in a dioecious plant.**

Atriplex garrettii is a dioecious plant that grows along the Colorado River in the western United States. In 1998, Ruas et al. (*American Journal of Botany* 85:162–167) reported identification of a DNA marker associated with sex determination in *A. garrettii*. Figure 14.6 shows the RAPD patterns amplified with a single primer from five male and five female plants. Notice that one DNA marker (labeled with an arrow) is part of the genotype of the five males but is absent from the genotype of the five females. The authors used the same primer to amplify DNA from 230 randomly chosen plants that could not be identified as male or female because they had not yet flowered. The marker appeared in the genotypes of 114 plants and failed to appear in those of 116 plants. When the plants flowered, all 114 whose genotypes included the marker were male, and all 116 lacking the marker were female.

Problem: **(a)** Conduct a chi-square test of the hypothesis that males and females appear in a 1:1 ratio using the plants selected at random before flowering as the data set. **(b)** What aspects of these results suggest that sex determination in *Atriplex garrettii* is genetic and not environmental? **(c)** Which sex is the heterogametic sex?

Solution: **(a)** The total number of plants is 230, so the expected number of males is 115 and the expected number of females is also 115, for a 1:1 ratio. The chi-square value is [$(116 - 115)^2 + (114 - 115)^2$] $\div 115 = 0.017$ with 1 degree of freedom. This value is substantially below the critical value of 3.84, indicating that there is no evidence to reject the hypothesis that these results conform to a 1:1 ratio. **(b)** Offhand, the 1:1 ratio may seem to be good evidence that sex determination is genetic. However, it is possible for a 1:1 ratio to arise from environmental sex determination. The best evidence that sex is determined genetically is the association of the DNA marker with sex phenotype. If sex determination were purely environmental, then there would be no significant association of a DNA marker with sex phenotype. **(c)** The DNA fragment is amplified in males but not in females, so there must be a segment of DNA that is present in males but absent in females. Because RAPD markers are inherited as dominant alleles, the males are heterozygous and the females are homozygous for the sex-determining factor. Thus, a 1:1 ratio of males and females is expected, which matches the observed ratio. Males are, therefore, the heterogametic sex and females the homogametic sex.

Sex determination may be genetic or nongenetic. Genetic sex determination is often associated with heteromorphic sex chromosomes.

14.3 HETEROMORPHIC CHROMOSOMES AND SEX DETERMINATION

Among those species that have heteromorphic sex chromosomes, there is substantial diversity in the mechanisms that determine sex. Let's compare two of the best studied systems: XX-XY sex determination in *Drosophila melanogaster* and XX-XY sex determination in mammals. Because *Drosophila* and mammals have XX-XY systems, on the surface it may appear that sex determination in *Drosophila* is the same as that in mammals. However, the mechanisms of sex determination in *Drosophila* and mammals are in fact very different.

Sex Determination in *Drosophila melanogaster*

Most male flies have one X chromosome and one Y chromosome, and most females have two X chromosomes, the same as in humans. In the early 1920s, Calvin Bridges studied some flies whose chromosome numbers and sex phenotypes were abnormal. From his observations, Bridges developed a model of how sex is determined in *Drosophila*.

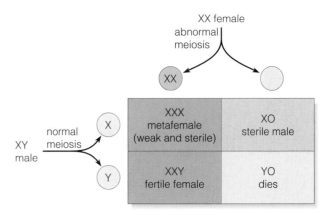

a Abnormal meiosis in the female produces gametes with two X chromosomes and gametes with no X chromosome. When these unite with normal sperm, four possible genotypes can result.

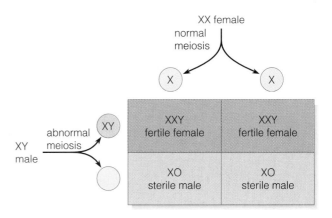

b Abnormal meiosis in the male produces gametes with both X and Y chromosomes and gametes with neither chromosome. When these unite with normal ova, two possible genotypes can result.

Figure 14.7 The consequences of abnormal meiosis for the X and Y chromosomes in *Drosophila melanogaster*.

Very rarely, sex chromosomes do not segregate properly during meiosis in a female fly, causing a gamete to carry two X chromosomes or none at all. Figure 14.7a shows the possible genotypes that can result from union of these abnormal eggs with normal sperm. Figure 14.7b shows that abnormal segregation of chromosomes during meiosis in the male can produce some of the same genotypes. When autosomes are in the normal diploid condition, an XXY fly is a fertile female, an XO fly is a sterile male, and an XXX fly is called a metafemale, a weak, infertile female that is likely to die before emerging from pupation. YO individuals die soon after fertilization because several genes on the X chromosome are essential for embryonic development.

In *Drosophila*, the Y chromosome has little to do with sex determination because XXY flies are fully fertile females. Although the Y chromosome is not essential for a male phenotype to form, it is necessary for a male to be fertile. Sex is determined by the ratio of X chromo-

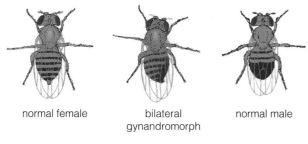

normal female   bilateral   normal male
gynandromorph

Figure 14.8 Sex phenotypes in *Drosophila melanogaster*. The female is larger than the male and differs in banding on the abdomen. The male has small black sex combs on his front legs. A bilateral gynandromorph has the female phenotype on one side of the body and the male phenotype on the other side.

somes to autosomal sets, often called the **X/A ratio**. In normal males, there is one X chromosome and two sets of autosomes, a ratio of $^{1}/_{2}$, or 0.5. In normal females, there are two X chromosomes and two sets of autosomes, a ratio of $^{2}/_{2}$, which equals 1. Bridges found that a ratio of 0.5 causes the male phenotype and a ratio of 1.0 causes the female phenotype. If the ratio is above 1, the individual is a metafemale. When the ratio is between 0.5 and 1.0, such as a fly with three sets of autosomes and two X chromosomes (X/A = $^{2}/_{3}$), the individual is an intersex, having characteristics of both male and female phenotypes.

Several genes on the X chromosome and the autosomes interact to determine the sex phenotype. Mutations in these genes alter sexual phenotype associated with the X/A ratio. Several sex-determination genes on the X chromosome have been identified; they are called **numerator elements** because they influence the numerator of the X/A ratio. Sex-determination genes on the autosomes are called **denominator elements** because they influence the denominator of the X/A ratio.

Some unusual flies provide us with further insight into sex determination in *Drosophila*. Sex is determined by the genotype of individual cells and not by hormones that circulate throughout the body. Occasionally, flies may be **mosaics**, meaning that some somatic cells have one genotype and other somatic cells in the same individual have another genotype. Flies that are a mosaic of XX and XO cells are male in the XO portion of the body and female in the XX portion. Such flies are called **gynandromorphs** (*gyn-* = female; *-andro-* = male; *-morph* = form). Among the most interesting are bilateral gynandromorphs, which are female on one side and male on the other (Figure 14.8). These flies develop when an X chromosome is lost from one of the daughter cells of the first mitotic division of an XX zygote, leaving one cell XX and the other XO. The side of the fly that develops from the XX cell has a female phenotype, while the side that develops from the XO cell has a male phenotype. Gynandromorphs have been observed in many insect species.

Sex Determination in Mammals

Although mammals have an XX-XY system of sex determination, the mechanism is very different from that in *Drosophila*. In mammals, the sex-determining factor is the Y chromosome, or more specifically, a gene on the Y chromosome that initiates development of the male phenotype. Unlike *Drosophila*, in which an XXY individual is female and an XO individual is a sterile male, mammals with an XXY genotype are male and those with an XO genotype are female.

Sex determination in mammals is a complex process in which the products of genes on the X and Y chromosomes and the autosomes interact. As testes develop in a male, hormones produced in the testes influence development of the male phenotype. Similarly, hormones produced in the ovaries of a female influence development of the female phenotype. The factor responsible for the initial stimulus that determines sex phenotype has long been known to reside on the Y chromosome and is called the **testis-determining factor**, or **TDF**, but the gene itself was identified only after a long hunt for it.

The first suggestions that a single gene governs development of the male phenotype came from tissue transplantation experiments in mice. Certain strains of laboratory mice are highly inbred, meaning that the mice are homozygous at nearly every locus, much like Mendel's true-breeding peas. Every individual in an inbred strain is nearly genetically identical to every other. The only major genetic difference among the individuals of an inbred strain is the XY chromosomal constitution of males and the XX chromosomal constitution of females. The genes located on the Y chromosome are found only in males, and the X chromosome is identical in males and females, although females have two copies of it and males only one. Males and females in an inbred strain are said to be **isogenic** except for the Y chromosome, meaning that they are genetically identical to each other in all chromosomes except the Y chromosome.

Mice in an inbred strain are so similar that they can receive skin grafts from one another without the body's immune system rejecting the graft. The one exception is a skin graft from male to female mice, which stimulates an immune reaction in the female recipient. (However, a graft from a female to a male stimulates no such reaction.) The female's immune system recognizes a foreign substance introduced from the male, a substance called the **H-Y antigen**, which is encoded by a gene on the Y chromosome and is present only in males. The H-Y antigen was later found in males of other mammals, including humans (*HYA*, OMIM 426000), and at first was thought to be the testis-determining factor. The locus for this gene resides on the long arm of the Y chromosome.

However, some hints that the H-Y antigen might *not* be the testis-determining factor arose when sex-reversed humans were studied. A sex-reversed person is one who has an XY genotype but a female phenotype, or an XX

genotype with a male phenotype. Presumably, the XY females lack a small portion of the Y chromosome that contains the testis-determining factor, whereas XX males have a small portion of the Y chromosome that contains the testis-determining factor. When XY females were studied, each was found to lack at least a small segment near the telomere of the short arm of the Y chromosome, and all XX males were found to have this small segment of the Y chromosome. As the short arm of the Y chromosome contains the gene (or genes) of interest, the gene that encodes the H-Y antigen cannot be the testis-determining factor because it is located on the long arm of the Y chromosome. A piece of DNA was eventually identified that contains the gene *zinc finger Y* (*ZFY*, OMIM 490000), so named because it is located on the Y chromosome and encodes a protein that has a zinc finger domain (see section 8.5 for an explanation of zinc fingers).

However, several XX males lack the segment of the Y chromosome that contains *ZFY* but have a small segment of the Y chromosome that is located close to *ZFY*. This segment carries a single gene, named *sex-determining region Y* (*SRY*, OMIM 480000), which is present in all XX human males studied. Genes similar to human *SRY* are found in nearly all mammals, indicating that *SRY* and its counterparts in other species are the mammalian testis-determining factor. Confirmation of this conclusion comes from experiments in which a segment of DNA that contains the mouse *Sry* gene was inserted into the cells of XX mouse embryos. Some of these transgenic XX embryos developed into males, an observation that indicates that the *Sry* gene stimulates development of the male phenotype. Let's now examine those experiments in some detail.

Example 14.3   Introduction of the *Sry* gene into XX mice.

In 1991, Koopman et al. (*Nature* 351:117–121) reported the results of experiments in which they injected a cloned 14 kb fragment of DNA that contains the entire mouse *Sry* gene into fertilized mouse eggs. The only gene within the fragment is *Sry*. They transplanted the injected eggs into surrogate mothers, where the eggs developed into embryos, then removed 158 embryos for examination during embryonic development. Most of the embryos did not incorporate the injected DNA into the chromosomes and developed as normal XY male embryos or XX female embryos in approximately a 1:1 ratio. Eight XX embryos contained the inserted *Sry* DNA in their cells. Six of the embryos were phenotypic females and two were phenotypic males. The two XX male embryos were anatomically identical both internally and externally to their normal XY male counterparts. From the injected eggs that were allowed to develop to full term, 93 mice were born. Of these, five had the introduced

Sry gene, one of which was an XX male. This male developed phenotypically as a normal male with no signs of hermaphroditism. His copulation behavior with females was no different from that of normal males. However, he was sterile, which was expected because male mice that naturally have two X chromosomes (such as XXY males) are sterile. Two of the mice that had the injected *Sry* gene were XX females, and the other two were XY males. Neither of the males transmitted the inserted *Sry* gene to progeny. One of the transgenic XX females transmitted the inserted *Sry* gene to XX progeny, which developed into females.

Problem: (a) What evidence from these experiments indicates that the *Sry* gene is responsible for sex determination? **(b)** What could account for the observation that some of the XX embryos with the *Sry* gene developed as females?

Solution: (a) The observation of XX males who carried *Sry* indicates that the *Sry* gene alone is sufficient to induce development of the male phenotype. **(b)** From these observations, it is evident that insertion of an *Sry* gene into the chromosomes of a developing embryo is insufficient to always cause development of the male phenotype. There are several possible reasons for this. First, not all cells in a mouse embryo may contain the gene because its incorporation into the chromosomes may take place after the first division of the zygote. Mice that develop from such an embryo carry the gene in some somatic cells but do not necessarily carry the gene in the cells that develop into reproductive organs, an example of mosaicism. The observation that some transgenic mice did not transmit the gene to offspring is evidence of mosaicism. Second, the activity of a gene is often dependent on where the gene is located in the genome. Genes that have been injected into fertilized eggs may integrate at many different places in the chromosomes. Some of these locations may permit expression of the gene, and others may not. The observation that a female transmitted the *Sry* gene to female progeny is evidence that the gene is in a chromosomal location that prevents expression during development.

The human *SRY* gene contains a single exon that encodes a DNA-binding protein of 223 amino acids that functions as a transcription factor. When bound to DNA, the SRY protein causes DNA bending, as diagrammed in Figure 14.9. The SRY protein binds to the regulatory region of other genes and regulates their expression during embryonic development.

Although *SRY* initiates development of the male phenotype, full development of the sexual organs and fer-

tility is dependent on the interaction of several genes located on the X chromosome and the autosomes. This is well illustrated by a condition in humans known as **testicular feminization**, in which an XY embryo forms testes that produce levels of anti-Müllerian hormone and testosterone that are equivalent to the levels found in the testes of normal XY male embryos. These two hormones are produced in the testes and stimulate development of the male phenotype. However, people with testicular feminization have a recessive allele on the X chromosome that prevents cells from producing receptors that bind to male-determining hormones, so the hormones cannot perform their function. As testes have already formed, no ovaries form in the individual. However, the insensitivity of cells to male-determining hormones prevents the testes from descending, and newborns with this condition have a female phenotype in all outward respects. They develop typical female characteristics throughout the remainder of their lives and are female in every respect, except for chromosomal constitution and the absence of ovaries and uterus, for which reason they are sterile. Females with testicular feminization are usually not diagnosed until they reach puberty and fail to menstruate. The vagina is present but is not attached to a uterus. People with this condition have levels of testosterone equivalent to males, have no secondary male characteristics, and except for bearing children, may lead normal lives as females.

There are several genetic disorders that affect sexual phenotype in humans. These disorders arise because of mutations in one or more of the genes that govern phenotypic sexual differentiation. Recessive X-linked alleles, similar to the allele that causes testicular feminization, can cause partial insensitivity to male hormones and **pseudohermaphroditism**, the presence of partially developed male and female reproductive organs, typically in people with an XY chromosomal constitution. While hermaphroditism is found in many plant and animal species, **true hermaphroditism** in humans, defined as the presence of testicular tissue with well-differentiated tubules, and ovarian tissue with follicles, is extremely rare. The external genitalia of true human hermaphrodites vary in appearance and may be predominantly male, predominantly female, or ambiguous. True hermaphrodites may have XX or XY chromosome constitutions. Some cases are caused by mosaicism, in which some cells are XY and others are XX. Most genetic disorders that affect the sexual phenotype can be treated with surgery and hormone administration so that each affected person can be distinctly male or female.

~

Sex determination mechanisms differ among species with an XX-XY system. In *Drosophila melanogaster*, sex is determined by the ratio of X chromosomes to autosomes. In mammals, sex is determined by the presence of a gene located on the Y chromosome.

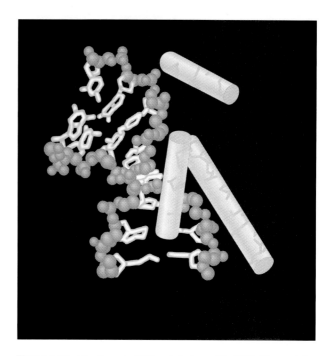

Figure 14.9 The human SRY protein bound to DNA. The protein is colored green and blue, and the DNA to which it is bound is colored red and white. (Source: www.ncbi.nlm.nih.gov/cgi-bin/SCIENCE96/gene?TDF.)

14.4 SEX-LINKED INHERITANCE

In species with heteromorphic sex chromosomes, males and females differ in their chromosomal constitution. Genes located on heteromorphic sex chromosomes are not inherited according to the principle of parental equivalence; phenotypes associated with recessive alleles on sex chromosomes may be expressed more frequently in one sex than in the other. In this section we examine the inheritance of genes located on heteromorphic sex chromosomes.

X- and Y-Linked Inheritance

The X chromosome in many species is larger than the Y chromosome and carries more genes. Any gene located on the Y chromosome is transmitted from father to son and appears only in males. Such genes are said to be **Y-linked** or **holandric genes**. In mammals, very few Y-linked genes are known. They include the *SRY*, *ZFY*, and *HYA* genes mentioned earlier.

No genes that are required for survival can be located on the Y chromosome because females do not have a Y chromosome. Both males and females have X chromosomes, so genes that are essential for survival can be and are located on the X chromosome. In fact, many genes on the X chromosome have nothing to do with sex determination and are expressed in both sexes. Any gene located on the X chromosome is said to be an **X-linked gene**. Although the phenotype governed by an X-linked

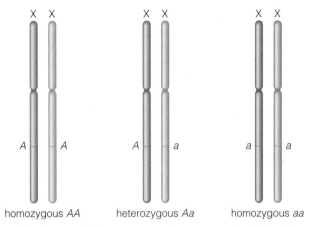

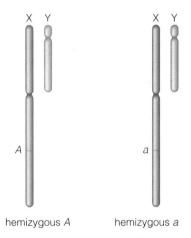

a Females may be homozygous or heterozygous for alleles at a locus on the X chromosome.

b Males are hemizygous for alleles on the X chromosome.

Figure 14.10 Homozygosity, heterozygosity, and hemizygosity for alleles on the X chromosome.

allele may be the same in males and females, the genotypes of males and females are different. Females have two X chromosomes and may be homozygous or heterozygous for X-linked alleles. Males, on the other hand, have only one X chromosome and can be neither homozygous nor heterozygous for X-linked alleles because they carry only one allele at each locus on the X chromosome. Instead, males are **hemizygous** for X-linked alleles, which means that each of their cells has only one copy of the gene (Figure 14.10).

Because males transmit their X chromosome only to their female offspring, and females transmit X chromosomes to both male and female offspring, Mendel's principle of parental equivalence does not apply to X-linked alleles. Instead, the progeny from reciprocal crosses differ in the distribution of alleles in males and females. In *Drosophila melanogaster*, this is illustrated by reciprocal crosses between red-eyed and white-eyed flies, reported in 1910 by Thomas Hunt Morgan. His experiments were the first to demonstrate the inheritance of an X-linked allele. The allele, called w^1, is recessive and causes white eyes when homozygous in females or hemizygous in males. When a white-eyed male is crossed with a female that is homozygous for the dominant wild-type allele, all of the F_1 offspring have the dominant red eyes (Figure 14.11a). The F_1 females are heterozygous and the F_1 males are hemizygous for the dominant wild-type allele. When the F_1 flies are intermated, red-eyed and white-eyed flies appear in the F_2 generation in a 3:1 ratio. However, all of the white-eyed flies are males. Half of the males are hemizygous for the white-eye allele, and the other half are hemizygous for the red-eye allele. Half of the females are homozygous for the red-eye allele, and the other half are heterozygous for it. Thus, all F_2 females from this cross have red eyes.

In the reciprocal cross, a white-eyed female is mated with a wild-type male (Figure 14.11b). The female parent is homozygous for the w^1 allele and transmits that allele to all of her male offspring, so all of the F_1 males are hemizygous for w^1 and have white eyes. The F_1 females are all heterozygous and have red eyes. When these F_1 flies are intermated, both red- and white-eyed flies appear in a 1:1 ratio in both males and females in the F_2 progeny of this cross.

The inheritance pattern shown in Figure 14.11a is sometimes called a crisscross pattern of inheritance. When an X-linked phenotype is governed by a recessive allele and is present in the male parent, it skips the first generation, then reappears in the second generation in the male offspring of the first-generation females. This pattern of inheritance is called a crisscross pattern because the allele is passed from an affected male to heterozygous carrier females, who then pass it to their male offspring, so in a sense, the allele crisscrosses sexes in each generation. From a genetic standpoint, the crisscross pattern is not wholly true because heterozygous females pass the allele to both male and female offspring, but the phenotype appears only in the male offspring. The observation of a crisscross pattern of phenotypic inheritance in human and other natural populations in which mating is not controlled experimentally indicates that the phenotype may be governed by a recessive X-linked allele.

Examples of X-Linked Genetic Disorders in Humans

Several genetic disorders in humans are associated with mutant X-linked alleles. Some of the X-linked genes and the disorders associated with mutant alleles of these genes are listed in Table 14.1. We will discuss three relatively common disorders, hemophilia A, red-green color blindness, and fragile-X syndrome, as examples of X-linked inheritance in humans.

Table 14.1 Examples of X-Linked Genetic Disorders in Humans

Disorder	Gene Responsible	Phenotype Associated with Mutant Alleles
Deuteranopia	GCP	Green color blindness; deficiency of functional pigment involved in green perception
Protanopia	RCP	Red color blindness; deficiency of functional pigment involved in red perception
Hemophilia A	F8C	Blood fails to clot properly; lack of functional clotting factor VIII
Hemophilia B	F9	Blood fails to clot properly; lack of functional clotting factor IX
Fragile-X syndrome	FMR1	Developmental delay, mental retardation; deficiency of functional FMR1 protein
G6PD deficiency	G6PD	Anemia that is expressed only upon exposure to certain foods or chemicals, due to deficiency of functional enzyme glucose-6-phosphate dehydrogenase (G6PD)
Duchenne muscular dystrophy	DMD	Progressive muscle deterioration; eventually fatal
Ocular albinism	OA1	Deficiency of eye pigments; decreased visual acuity
Congenital stationary night blindness	CSNB1	Decreased visual acuity in reduced light
X-linked deafness	DFN1, DFN2, DFN3	Different types of deafness
Severe combined immunodeficiency	SCIDX1	Immune system failure

Hemophilia A is a genetic disorder that prevents normal blood clotting when blood vessels are ruptured. It appears in about 1 in 7000 males. In the most severe cases, an affected person can bleed to death from a bruise or cut. Hemophilia A is caused by the absence of a protein called factor VIII that is essential for blood clotting. Factor VIII is encoded at the *factor VIII c* locus (*F8C*, OMIM 306700). Proteins encoded by several other loci are also required for blood clotting. Mutant recessive alleles at any of these loci can cause hemophilia. Because mutant *F8C* alleles are X-linked, hemophilia A is much more common among males than in females. In the rare cases in which a female has the disorder, her father typically has hemophilia A and her mother is a heterozygous carrier. At one time, hemophilia was nearly always lethal, although affected men sometimes lived into their reproductive years. Now, people with hemophilia can be treated with transfusions of blood products with normal factor VIII to prevent the excessive bleeding. Recombinant human factor VIII produced in genetically engineered bacteria is now available.

About 85% of males with hemophilia have hemophilia A. Most others have **hemophilia B**, caused by a recessive allele at a different X-linked locus, called *F9* (OMIM 306900), that encodes factor IX, another essential blood-clotting protein. The DNA segment containing the *F9* gene has also been cloned and sequenced, and scientists have developed genetically engineered factor IX for treatment of people with hemophilia B.

The probability of a female being born with X-linked hemophilia is very small, theoretically about 1 in 100 million females, based on the probability of a mother who is heterozygous and a father who is hemizygous for the recessive allele. Historically, males with the disorder often died from bleeding-related problems before fathering children, and others elected to not have children, reducing the actual frequency of females born with hemophilia. Matings between close relatives, however, can increase this frequency considerably because the probability that a male with hemophilia will marry a heterozygous female is increased in consanguineous marriages. Some cases of females with hemophilia have been studied in the offspring of related parents. Also, some cases of hemophilia that are due to a new mutation have appeared in females. Historically, females who were born with hemophilia usually did not live beyond puberty because it was not possible to prevent excessive bleeding during menstruation. However, some alleles that cause mild hemophilia allow hemophiliac females to live and in some cases bear children.

Hemophilia was for many years known as "the royal disease" because it affected three generations of males in European royal families during the nineteenth and twentieth centuries. Each affected male carried an allele inherited from Queen Victoria, who was a heterozygous carrier. The allele persisted through three generations of Victoria's descendants, although it is now probably

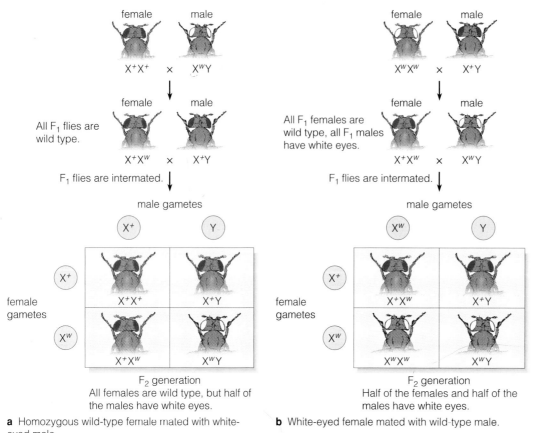

a Homozygous wild-type female mated with white-eyed male.

b White-eyed female mated with wild-type male.

Figure 14.11 X-linked inheritance of the white-eye trait in *Drosophila melanogaster*.

---

lost. The pedigree of European royal families in Figure 14.12 shows affected males and carrier females.

Let's now examine a pedigree in which both hemophilias A and B are found.

Example 14.4   X-linked inheritance of the *F8C* and *F9* genes in humans.

In 1964, Robertson and Truman (*Blood* 24:281–288) described a family in which both hemophilia A and hemophilia B were found. Figure 14.13 shows a pedigree adapted from their article.

Problem: **(a)** Which females must be heterozygous for a mutant allele of the *F8C* gene? **(b)** Which females must be heterozygous for a mutant allele of the *F9* gene? **(c)** Individual 12 carries mutant alleles in both the *F8C* and *F9* genes. Why doesn't she have hemophilia? **(d)** Notice that individual 28 has both hemophilia A and B. Using the pedigree, and assuming that no new mutations arose in this individual or his parents, explain how this situation could arise.

Solution: **(a)** Individuals 3, 5, 8, and 12 must have been heterozygous for a mutant allele at the *F8C*

locus. **(b)** Individuals 5, 7, and 12 must have been heterozygous for a mutant allele at the *F9* locus. Individuals 9, 10, 15, 16, 20, 24, 30, and 31 might also be heterozygous for either mutant allele, but it is not possible to determine their genotypes from the information in the pedigree. **(c)** The two mutant alleles are at different loci. Therefore, individual 12 carries one mutant recessive allele at the *F8C* locus and one mutant recessive allele at the *F9* locus. Because she is heterozygous at both loci, dominant alleles mask the effects of the recessive alleles and she does not have either hemophilia A or B. **(d)** As individual 28 is male, he has only one X chromosome, and this chromosome must carry the mutant alleles at the *F8C* and the *F9* loci. Because he is hemizygous for both mutant alleles, he is deficient in both factor XIII and factor IX. He must have inherited this chromosome from his mother (individual 12), who must have inherited it from her mother (individual 5). However, two of the male offspring of individual 5 have hemophilia B but not hemophilia A (individuals 11 and 14). This situation can best be explained if individual 1 is hemizygous for one of the mutant alleles and individual 2 is heterozygous for the other mutant allele. Individual 5 inherited one mutant allele on the X chromosome

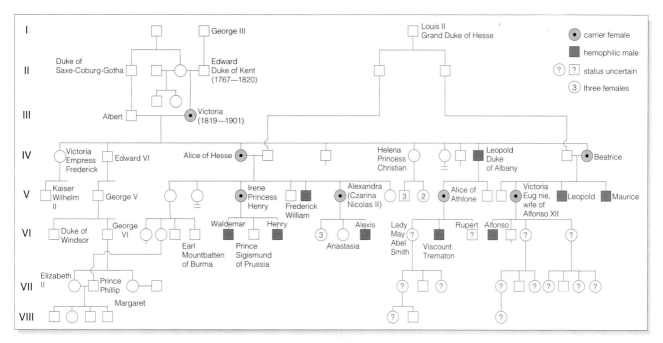

Figure 14.12 A partial pedigree of European royal families. Queen Victoria was a heterozygous carrier of a recessive mutant allele at the *F8C* locus that caused hemophilia A in some of her male descendants. (McKusick, V. A. 1969. *Human Genetics.* Englewood Cliffs, N.J.: Prentice-Hall.)

contributed by her father and the second mutant allele on the X chromosome contributed by her mother. She, therefore, was doubly heterozygous for the mutant alleles, with one mutant allele on each X chromosome. Individuals 11 and 14 inherited from their mother the X chromosome with the mutant allele at the *F9* locus. Individual 12, however, inherited an X chromosome with two mutant alleles, one at the *F8C* locus and the other at the *F9* locus. This chromosome probably arose from a crossover between the *F8C* and the *F9* loci during oogenesis in individual 5. Individual 27 probably inherited from his mother the other X chromosome which carried neither mutant allele.

Red-green color blindness is the most common X-linked genetic disorder in humans. Color perception in humans is governed by an interaction of X-linked genes and autosomal genes. Thus, inherited color blindness can be X-linked or autosomal, although the most common types of red-green color blindness are caused by recessive X-linked alleles. Recessive alleles at either of two loci on the X chromosome can cause red-green color blindness. One locus encodes a pigment that allows perception of the color green, and the other locus encodes a pigment that allows perception of the color red. **Deuteranopia** is the more common form of red-green color blindness and is caused by the absence of the pigment that permits green perception. **Protanopia** is caused by the absence of

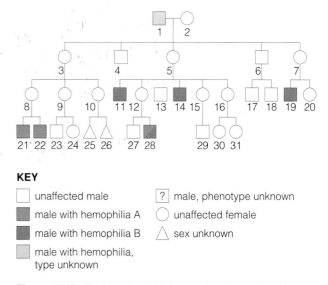

KEY

☐ unaffected male   ⍰ male, phenotype unknown

▩ male with hemophilia A   ◯ unaffected female

▩ male with hemophilia B   △ sex unknown

▨ male with hemophilia, type unknown

Figure 14.13 Pedigree in which hemophilias A and B are found. (Adapted from Robertson, J. H., and R. G. Truman. 1964. *Combined hemophilia and Christmas disease. Blood* 24:281–288.)

functional pigment for red perception. People with deuteranopia or protanopia cannot distinguish green from red, or in less severe cases, have difficulty distinguishing the two colors. The locus for deuteranopia (called *GCP* for "green cone pigment," OMIM 303800) and the locus for protanopia (called *RCP* for "red cone pigment," OMIM 303900) are very close to each other on the long arm of the X chromosome. Because of their

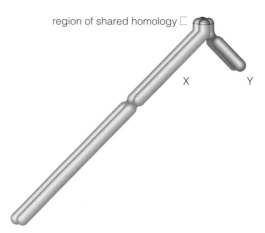

Figure 14.15 Pairing of human X and Y chromosomes during prophase I at the region of shared homology.

Figure 14.14 A fragile-X chromosome. The arrow points to a secondary constriction where a piece of the chromosome is sometimes lost in cultured cells. (Photo courtesy of C. J. Harrison.)

proximity and similar effects, the two loci were at one time thought to be a single locus. The DNA segments containing these two loci have been isolated and studied. The two loci encode light-absorbing proteins that are nearly identical. Each X chromosome carries a single copy of *RCP* but may carry one or more copies of *GCP*. When additional copies are present, they are organized into a tandem array, but apparently the copy of the *GCP* gene that is nearest the *RCP* gene is the only one that is transcribed.

Although the proteins encoded by *RCP* and *GCP* are similar, both are required for normal color perception. Several alleles are known at both loci. Those alleles that fail to encode color perception pigments cause the most severe forms of color blindness. Alleles that reduce the function of the pigments they encode cause less severe forms of color blindness. Red-green color blindness is relatively common; it affects about 8% of males of western European descent to some degree.

Fragile-X syndrome (Martin-Bell syndrome) afflicts about 1 in 1500 human males and is the most common form of inherited mental retardation in humans. It is characterized by slow development in infancy and childhood, followed by mental retardation. Its X-linked inheritance was first described by J. P. Martin and J. Bell in 1943, and for many years it was called Martin-Bell syndrome. It is now called fragile-X syndrome because the X chromosome in cells cultured from afflicted males and carrier fe-

males has a constriction near the tip of the long arm, as pictured in Figure 14.14. This feature is called a **fragile site** because the end of the chromosome tends to be broken off when cultured cells are starved for thymine or cytosine. However, the term *fragile-X syndrome* is a misnomer because chromosome breakage is an artifact of cell culture and is not the cause of the syndrome.

The inheritance of fragile-X syndrome puzzled human geneticists for many years. The chromosome constriction in cultured cells was clearly associated with the disorder, but the patterns of inheritance were not entirely typical of an X-linked allele. Most males whose cells had an X chromosome with the fragile site have fragile-X syndrome, but a few are phenotypically normal, which suggests that the disorder is not completely penetrant in males. The most puzzling pattern of inheritance, however, was a phenomenon called **genetic anticipation**, in which the frequency or severity of an inherited disorder increases with each generation of transmission. In the case of fragile-X syndrome, the frequency of people in a family with the disorder tends to increase with each generation, as if the chromosomal defect somehow becomes more penetrant each time it is transmitted.

The reason for genetic anticipation of fragile-X syndrome became clear when the molecular structure of the mutation was determined. As discussed in section 5.1, fragile-X syndrome is caused by a trinucleotide $(CCG)_n$ repeat expansion mutation in the *FMR1* gene (OMIM 309550). The normal number of repeats varies from 6 to 52 with 30 being the most common. Repeats ranging from 50 to 230 constitute a fragile-X premutation, which is prone to expansion to a full mutation, consisting of over 230 and often as many as 700–1000 repeats. Females and males with the premutation are usually phenotypically normal, but the descendants of heterozygous female carriers of a premutation have an increased risk of expansion to a full mutation. As illustrated in Example 5.1, the larger the premutation, the higher the risk of expansion to a full mutation.

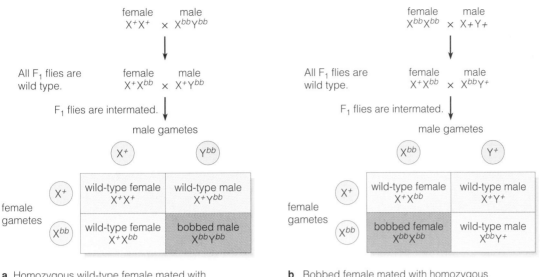

a Homozygous wild-type female mated with bobbed male.

b Bobbed female mated with homozygous wild-type male.

Figure 14.16 Pseudoautosomal inheritance of the bobbed trait in *Drosophila melanogaster*.

Both a full mutation and a premutation in the *FMR1* gene can cause the secondary constriction in fragile-X chromosomes. The constriction is caused by the abnormally high number of repeats. The fact that individuals with a premutation may have the chromosome constriction but may be phenotypically normal, accounts for the incomplete penetrance observed in individuals whose cultured cells have a fragile-X chromosome. The expansion of the premutation into the full mutation accounts for genetic anticipation: A male premutation carrier usually does not have the symptoms of fragile-X syndrome, but he transmits premutation to all of his daughters. His daughters' children are then at risk for expansion of the premutation into the full mutation.

~

Alleles on the X chromosome are inherited in a predictable pattern that does not conform to the principle of parental equivalence. Females may be homozygous or heterozygous for X-linked alleles, and males are hemizygous for X-linked alleles.

Homology Between X and Y Chromosomes

During prophase I of meiosis in males, X and Y chromosomes pair with each other, and they segregate as homologues during anaphase I. The pairing takes place at a small, terminal segment of the short arm of the X chromosome that is homologous to a similar segment of the Y chromosome (Figure 14.15). The other portions of the X and Y chromosomes are not homologous. Alleles located at any locus within the homologous segment can be homozygous or heterozygous in males. Such loci are called **pseudoautosomal** because the inheritance pat-

terns are similar but not identical to those of alleles at autosomal loci. One such locus is known in *Drosophila melanogaster*, the *bobbed* (*bb*) locus. A recessive allele at the *bb* locus causes the bristles on the fly's body to be unusually short when the fly is homozygous for the allele. When reciprocal crosses are made between wild-type and bobbed homozygotes, the inheritance pattern is similar to that of autosomal traits in that all F_1 progeny have the dominant (wild-type) phenotype, and phenotypic segregation in the F_2 generation is 3:1. However, if the male parent is homozygous *bb/bb*, all of the bobbed individuals in the F_2 progeny are male, and if the female parent is homozygous *bb/bb*, all of the bobbed individuals in the F_2 progeny are female, as diagrammed in Figure 14.16.

In human males, one crossover always occurs within the homologous region between the X and Y chromosomes during meiosis. This separates any genes located within this region from genes on the nonhomologous portion of the Y chromosome each generation. So far in humans, only one gene is known in this region, a gene that encodes a cell-surface antigen. In *Drosophila*, there is no crossing over in males, so the genes in the homologous portion of the Y chromosome remain inextricably linked to genes on the nonhomologous portion.

Crossing over between the homologous regions of the X and Y chromosomes in humans explains the rare appearance of XX males and XY females. Recall that studies to determine the location of the testis-determining factor were conducted with people who had these genotypes. *SRY* is located on the short arm of the Y chromosome very close to, but not within, the homologous region. Rarely during meiosis, an errant crossover may transfer the portion of the Y chromosome that carries *SRY*

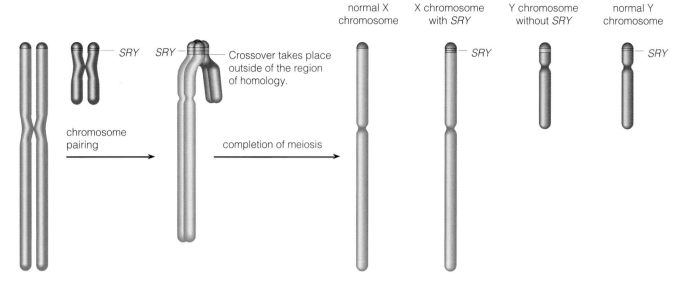

Figure 14.17 An errant crossover transfers a small piece of the Y chromosome containing the *SRY* gene to the X chromosome.

to the X chromosome (Figure 14.17). When the sperm cell that carries an X chromosome with *SRY* unites with an egg with a normal X chromosome, an XX male results. Alternatively, the Y chromosome that arises from this rare crossover event lacks the segment that contains *SRY*. When a sperm cell that carries such a Y chromosome unites with an egg cell that carries a normal X chromosome, an XY female results. The frequency of XX males and XY females in human populations is about 1 in 20,000.

~

X and Y chromosomes pair during meiosis at a region of shared homology. Genes in this homologous region are inherited in a pseudoautosomal fashion.

Attached X Chromosomes in *Drosophila melanogaster*

Lilian Vaughan Morgan, the wife of Thomas Hunt Morgan, discovered an attachment of two X chromosomes at one of their ends in a single fly. The story of this rare fly is both interesting and humorous. Lilian Morgan had anesthetized some flies with ether and was examining them under the microscope. She noticed an unusual female with a mosaic appearance. While she was examining the fly, it recovered from the ether and hopped onto the floor. Because flies regain their ability to walk before they regain their ability to fly, Morgan thoroughly searched the floor for the tiny fly but was unable to find it. After some time had passed, she was certain that the fly was hopelessly lost. In desperation, remembering that flies are attracted to light, she went to the window, where she found the fly sitting on the window pane and captured it. When she examined its progeny, she discovered attached X chromosomes and realized that the female

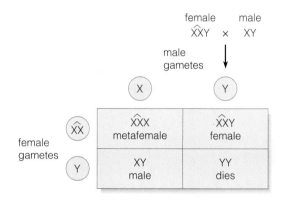

Figure 14.18 Inheritance of attached X chromosomes in *Drosophila melanogaster*.

who had almost escaped produced gametes containing attached X chromosomes.

Because the X chromosomes are attached, they behave as a single chromosome during meiosis. (The attached X chromosomes are often designated X̂X.) Thus, an X̂XY fly is female because the X/A ratio is 1.0. The X̂X chromosome and the Y chromosome segregate from each other during meiosis. Under these circumstances, when an X̂XY female is mated with an XY male, progeny arise according to the Punnett square shown in Figure 14.18. The X̂XX progeny are metafemales and usually die before pupation is complete. The YY flies die early in embryonic development because they lack essential genetic information carried by the X chromosome. The XY males inherit their X chromosome from their male parent and their Y chromosome from their female parent, the opposite of the usual situation. The remaining flies are X̂XY females that inherit the X̂X chromosome from their female parent and the Y chromosome from their male

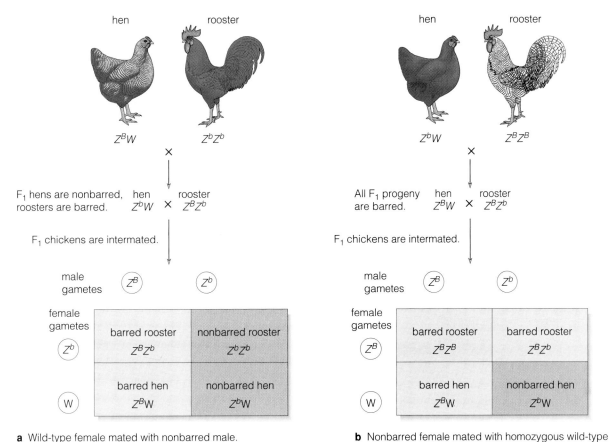

a Wild-type female mated with nonbarred male.

b Nonbarred female mated with homozygous wild-type male.

Figure 14.19 Z-linked inheritance of the barred trait in chickens.

parent. Because the metafemales rarely survive pupation, the free X chromosome is always passed from a male parent to male offspring each generation. Thus, it is possible to maintain an X-linked allele entirely within the male lineage by mating males that are hemizygous for the allele with $\widehat{XX}Y$ females.

Z-Linked Inheritance

In species with a ZZ-ZW system of sex determination, Z-linked traits are inherited in the same manner as X-linked traits, but males are the homogametic sex and females heterogametic. For example, in chickens a dominant allele on the Z chromosome (designated B) causes a barred pattern in feather coloration. The recessive allele (b) prevents the barred pattern from forming in an individual homozygous or hemizygous for the allele. If a barred hen (Z^BW) is mated with a nonbarred rooster (Z^bZ^b), all the female offspring are nonbarred (Z^bW) and all the male offspring are barred (Z^BZ^b), as diagrammed in Figure 14.19. Notice that reciprocal crosses produce different ratios of the two phenotypes in F_2 hens and roosters.

The first case of sex-linked inheritance was discovered in moths, and additional cases were found shortly thereafter in canaries and chickens, all species with ZZ-ZW systems of sex determination. At the time, microscopic examination of cells in several insects, including beetles, flies, and true bugs, revealed that males had an XY chromosomal constitution and females an XX chromosomal constitution. Although both sex-linked traits and sex chromosomes had been discovered, no one was able to determine the relationship between the two. Because sex-linked inheritance was first discovered in species with ZZ-ZW systems of sex determination, the patterns of sex-linked inheritance were the reverse of those expected for species with an XX-XY system of sex determination, which was the heteromorphic system that had been revealed by microscopy. After X-linked inheritance was discovered, and Z and W chromosomes were identified microscopically, the confusion was resolved.

~

Z-linked inheritance is similar to X-linked inheritance, but the inheritance patterns associated with sex are reversed.

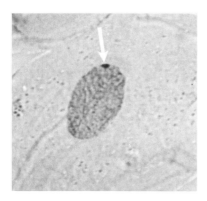

Figure 14.20 A Barr body (indicated by the arrow) in the interphase nucleus of a somatic cell of a human female.

14.5 DOSAGE COMPENSATION

In species with XX-XY, ZZ-ZW, or XX-XO systems of sex determination, it would seem that chromosomes are out of balance between the sexes. Female mammals and fruit flies, for example, have two copies of the X chromosome whereas males have only one. Given that the X chromosome typically contains a large number of genes, many of which are unrelated to sex determination, does this mean that females have twice the level of gene products for X-linked genes than males? For most genes located on the X chromosome, the answer to this question is no. Enzymes encoded by X-linked genes are generally present at equal levels in males and females. This means there must be some mechanism to compensate for the lack of a homologous X chromosome in males. Indeed, species with XX-XY systems of sex determination have several mechanisms to compensate for the differences in gene number due to the different chromosomal constitution of males and females. These mechanisms are collectively referred to as **dosage compensation**.

Dosage Compensation in Mammals

The same single mechanism, called **X-chromosome inactivation**, provides dosage compensation in all mammalian species. In 1949 Murray Barr reported the presence of a distinct body in microscopic preparations of neural-cell nuclei taken from female cats. No such bodies were present in the nuclei of neural cells from male cats. This body is now called a **Barr body** in honor of its discoverer. Barr bodies have since been found in the cells of every mammalian species examined. Barr bodies can be observed microscopically in cells from human females as a body located on the inside wall of the nuclear envelope, as shown in Figure 14.20. No such body is present in cells from males.

The Barr body is an inactivated X chromosome that is condensed into facultative heterochromatin so that all of its genes are inactive. When union of sperm and egg cre-

ates an XX human zygote, genes on both X chromosomes are initially transcribed. However, in the blastocyst stage, about 16 days after conception, one of the X chromosomes in some of the cells condenses into a Barr body, and the genes on the Barr body are no longer transcribed. The genes on the other X chromosome, however, are free to be transcribed. Which X chromosome is inactivated, the one of paternal origin or the one of maternal origin, is purely a matter of chance. On average, about 50% of the cells inactivate the paternal X chromosome and 50% inactivate the maternal X chromosome. After an X chromosome is inactivated in a cell, the same chromosome (maternal or paternal) is inactivated in the daughter cells that arise by mitosis. As a female mammal develops, certain cell lineages have an inactive maternal X chromosome, and other cell lineages have an inactive paternal X chromosome. This means that the somatic cells of a female display mosaicism for paternal and maternal Barr bodies.

The idea of mosaicism in mammalian females was developed by Mary Lyon and is often called the **Lyon hypothesis**, although there is now so much evidence to support it that it hardly can be called a hypothesis any longer. According to the Lyon hypothesis, females who are heterozygous for X-linked alleles should have a mosaic expression of those alleles. In other words, in each organ of the body, some cells express one allele while other cells express the other allele. One of the best-known examples of this phenomenon is tortoiseshell and calico cats. Tortoiseshell cats have a mosaic pattern of orange and black fur, and calico cats have a mosaic pattern of orange, black, and white fur (Figure 14.21).

The orange and black patterns are governed by alleles at an X-linked locus called the *O* locus. (The white patches in calico cats are caused by an allele at an autosomal locus.) The *O* locus has two common alleles, called *O* and *o*. The *O* allele blocks expression of other colors and causes orange-colored hair, and the *o* allele permits other colors (usually black) to be expressed. Female cats who are heterozygous *Oo* have a random tortoiseshell or calico patchwork pattern of orange and black coloration because the *O* allele is active in some cells and the *o* allele is active in others. Nearly all tortoiseshell and calico cats are female. Males are usually either hemizygous for the *O* allele and have all orange fur, or hemizygous for the *o* allele and have all black fur. Only very rarely does a male tortoiseshell or calico cat arise, and in most cases they are sterile, as explained in the following example.

Example 14.5 Rare cases of male tortoiseshell and calico cats.

Male tortoiseshell or calico cats are rare but do appear on occasion. In 1973, Centerwell and Benirschke (*Journal of Heredity* 64:272–278) reviewed all published reports of male tortoiseshell and calico cats in which the chromosomes had been exam-

Figure 14.21 A female calico cat. (Photo courtesy of Jack Carey.)

ined. A total of 25 male tortoiseshell or calico cats were identified. Four of them had an XXY chromosome constitution. The remaining 21 cats were mosaics for XX/XY, XY/XY (with different X chromosomes), XY/XXY, XX/XXY, XY/XXY/XXYY, or XX/XY/XXY/XXYY.

Problem: What aspects of these results explain the rare appearance of male tortoiseshell and calico cats?

Solution: All of these unusual male cats had at least two different X chromosomes. Those cats with an XXY chromosome constitution are male because they have a Y chromosome, but their cells inactivate one of the two X chromosomes. The pattern of inactivation is random (as in XX females), so the coat color phenotype of the XXY male cat is the same as that of a tortoiseshell or calico XX female. The cats with mosaicism also have two different X chromosomes among their somatic cells. Whether or not an X chromosome is inactivated depends on the genotype of the cell. Those cells with one X chromosome express the allele present on that chromosome. Cells with two X chromosomes inactivate one X chromosome and express the allele on the other X chromosome.

A rare genetic condition in humans known as anhidrotic ectodermal dysplasia is caused by a recessive X-linked allele at the *EDA* locus (OMIM 305100) and ap-

pears in a patchwork pattern in females. This allele prevents the normal formation of teeth and of sweat glands in the skin. Males who are hemizygous for the allele have no teeth and no sweat glands. Heterozygous females have random portions of skin without sweat glands and may lack some or all of their teeth, depending on the X-chromosome inactivation pattern in the jaw. This condition is most common in Pakistan.

Some of the best evidence supporting the Lyon hypothesis came from studies of the enzyme glucose-6-phosphate dehydrogenase (G6PD). Two different forms of this enzyme can be detected, each produced by a different allele at the *G6PD* locus (OMIM 305900). Heterozygous females have both forms of the enzyme (an example of codominance). However, when single cells from a heterozygous female are induced to divide mitotically in culture, a process called cell cloning, each cloned colony has either one form of the enzyme or the other, but never both. This observation indicates that only one allele is active in each original cell and that the mitotic progeny of each cell retain the same activity as their parent cell.

What about X-linked conditions such as hemophilia and red-green color blindness? Do they show a patchwork pattern in heterozygous females as well? Human females who are heterozygous for the hemophilia A allele produce normal factor VIII in those cells that produce blood primordia whose active X chromosome carries the dominant allele. The cells that inactivate the dominant allele cannot produce normal factor VIII. Because most heterozygous females have cells of both types, they have normal factor VIII in their blood, but at reduced levels compared to females homozygous for the functional allele. The levels of factor VIII in their blood range from 20% to 100% of the level of homozygous females and hemizygous males. The variation is caused by random chromosome inactivation. Because factor VIII circulates in the blood throughout the body, there is no mosaic pattern for hemophilia, although there is a mosaic pattern for the cells that produce factor VIII.

The circulation of blood throughout the body precludes a patchwork pattern for hemophilia in females, but we might expect protanopia and deuteranopia to have a patchwork pattern. Presumably, some heterozygous females should have normal vision, others should have color blindness in both eyes, and still others should have one color-blind eye and normal color perception in the other eye. However, with rare exceptions, heterozygous females have normal color perception in both eyes. X chromosome inactivation does not take place until the later stages of development of retinal cells. Thus, heterozygous females have a patchwork of retinal cells with normal color perception and faulty color perception in both eyes. Thus, about 50% of the cells in each eye have normal color perception, and these cells compensate for the color-blind cells, so heterozygous females have normal color perception in both eyes.

Dosage Compensation in *Drosophila*

In the cells of *Drosophila* females, both X chromosomes are active, but males and females have equal levels of enzymes produced by X-linked genes. The genes on each X chromosome in the cells of females are transcribed at half the level of genes on the single X chromosome in male cells, making the level of transcription for X-linked genes equal in male and female cells. Females with three X chromosomes and two of each autosome also have normal levels of enzymes encoded by X-linked genes, meaning that each X-linked gene is expressed at one-third the level of that in a normal XY male. Some mechanism very different from X chromosome inactivation in mammals provides dosage compensation in *Drosophila*.

In rare cases, a portion of the X chromosome may be moved to an autosome, or a portion of an autosome may be moved to an X chromosome. These situations provide clues about the mechanism of dosage compensation in *Drosophila*. An X-linked gene translocated to an autosome is subject to dosage compensation even though it is no longer on the X chromosome. On the other hand, an autosomal gene translocated to the X chromosome is not subject to dosage compensation even though it is located on the X chromosome. These results suggest that dosage compensation functions through regulation of individual genes normally located on the X chromosome, rather than regulation of the entire X chromosome.

~

Dosage compensation ensures that expression of X-linked genes is equivalent in males and females. The mechanisms for dosage compensation are not always the same. In mammals, one X chromosome is inactivated in the somatic cells of females. In *Drosophila*, X-linked genes are regulated individually.

14.6   SEX-INFLUENCED AND SEX-LIMITED TRAITS

One of the first questions that usually arises in a discussion of sex-linked inheritance is whether or not pattern baldness in humans (Figure 14.22) is an X-linked trait. It is commonly believed that a man inherits pattern baldness not from his father but from his maternal grandfather. This belief is based on the observation that although women may have pattern baldness, it is much less common in women than in men, a situation that is similar to patterns associated with inheritance of a recessive X-linked allele. However, the common belief is false. Pedigree studies conducted several decades ago indicate that pattern baldness is an inherited trait, but that it is caused by alleles at an autosomal locus that act as dominant alleles in men and recessive alleles in women. If a man's maternal grandfather has pattern baldness, then

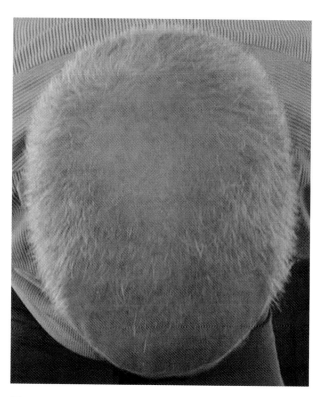

Figure 14.22   Pattern baldness in humans. Hair loss begins at the temples and the crown and may progress throughout the upper part of the head.

there is some likelihood that the man will inherit the allele from his mother. However, if the man's father has pattern baldness, the man may also inherit the allele from his father. Women must be homozygous for pattern baldness alleles to have pattern baldness, so it is rarer in females than in males. Also, females who are homozygous for the pattern baldness allele usually express the phenotype later in life and to a lesser extent than do males who are heterozygous or homozygous for such alleles. The molecular basis for pattern baldness remains unknown.

Sex-linked traits are governed by genes on sex chromosomes. Pattern baldness is not a sex-linked trait but is a **sex-influenced trait**, a trait that appears in both sexes but more often in one sex, and is governed by genes on autosomal chromosomes. **Sex-limited traits** are traits that appear in only one sex and are governed by autosomal genes. Secondary sex characteristics, such as antlers on male deer, brilliant coloration of male birds, breast development in human females, and beard growth in human males, are examples of sex-limited traits. The genes that govern these traits are usually autosomal and are carried by individuals of both sexes, but with rare exceptions they are expressed exclusively in one sex.

Let's now see how genetic experimentation can distinguish among sex-linked, sex-influenced, and sex-limited traits.

Example 14.6 Inheritance of spotting patterns in the pea weevil.

Breitenbecher (1925. *Genetics* 10:261–277) had been conducting genetic experiments with pea weevils of the species *Bruchus quadrimaculatus* when he discovered a culture in which some of the females had an unusual asymmetrical pattern of spotting with black spots on one side and red spots on the other side. Males of this species do not show any spotting pattern. He named this pattern of asymmetrical spotting *piebald*. Table 14.2 summarizes Breitenbecher's experiments with this trait and the results he reported.

Problem: Piebald spotting is caused by a mutant allele at a single locus. **(a)** Is this allele dominant or recessive? **(b)** Determine which pattern of inheritance, sex-linked, sex-limited or sex-influenced, applies to piebald spotting. **(c)** Explain the inheritance patterns in Table 14.2 using appropriate symbols.

Solution: **(a)** All females in the F_1 progeny are red-spotted, and all males are nonspotted. None of the F_1 progeny are piebald even though the female parent was. This suggests that the piebald allele is recessive. This conclusion is confirmed in the F_2, in which females show a 3:1 ratio for red-spotted:piebald. The results of the remaining crosses also support this conclusion. **(b)** The locus for this allele is autosomal and is never expressed in males although they may carry it. This conclusion is derived from the F_2 data. Had the allele been X-linked (presuming an XX-XY or XX-XO system of sex determination), the piebald phenotype should have appeared in half of the F_2 females. Instead, it was present in $\frac{1}{4}$ of the F_2 females, typical of inheritance for an autosomal locus. Because males have no spotting pattern in any instance, this characteristic is sex-limited. **(c)** Let p represent the allele responsible for piebald and P the allele for red-spotted. The genotypes and expected proportions of progeny are summarized in Table 14.3. The observed ratios correspond closely with the expected ratios in each cross.

Table 14.2 Information for Example 14.6: The Results of Breitenbecher's Experiments

| | Numbers of Progeny | | |
| | Females | | Males |
Cross	Red-Spotted	Piebald	Nonspotted
1. P_1 piebald females × P_2 nonspotted males	1426	0	1508
2. F_2 of cross 1	897	290	1192
3. P_2 red-spotted females × F_1 males from cross 1	291	0	286
4. F_1 females from cross 1 × P_2 non-spotted males	325	0	307
5. P_1 piebald females × F_1 males from cross 1	276	273	557

P_1 represents piebald female parents, and P_2 represents male or female parents from a culture that was true breeding for fully red-spotted females and nonspotted males—i.e., there were no piebald individuals in this culture.

Table 14.3 Solution to Part c of Example 14.6: Explanation of Inheritance Patterns

| | Numbers of Progeny | | |
| | Females | | Males |
Cross	Red-Spotted	Piebald	Nonspotted
1. P_1 piebald females (*pp*) × P_2 nonspotted males (*PP*)	$\frac{1}{2}$ *Pp*	0	$\frac{1}{2}$ *Pp*
2. F_2 of cross 1 (*Pp* × *Pp*)	$\frac{3}{8}$ *PP*, *Pp*	$\frac{1}{8}$ *pp*	$\frac{1}{2}$ *PP*, *Pp*, *pp*
3. P_2 red-spotted females (*PP*) × F_1 males from cross 1 (*Pp*)	$\frac{1}{2}$ *PP*, *Pp*	0	$\frac{1}{2}$ *PP*, *Pp*
4. F_1 females from cross 1 (*Pp*) × P_2 nonspotted males (*PP*)	$\frac{1}{2}$ *PP*, *Pp*	0	$\frac{1}{2}$ *PP*, *Pp*
5. P_1 piebald females (*pp*) × F_1 males from cross 1 (*Pp*)	$\frac{1}{4}$ *Pp*	$\frac{1}{4}$ *pp*	$\frac{1}{2}$ *Pp*, *pp*

Sex-influenced and sex-limited traits are governed by genes on autosomes. Sex-influenced traits appear more often in one sex than the other. Sex-limited traits appear exclusively in one sex.

SUMMARY

1. Sex determination may be genetic or nongenetic. Nongenetic sex determination may be influenced by such factors as temperature or proximity to females.

2. Genetic sex determination may be controlled by two alleles of a single gene, by multiple alleles of a single gene, or by genes on heteromorphic sex chromosomes.

3. There are three general types of sex determination caused by heteromorphic chromosomes: (a) the XX-XO system, in which females have two X chromosomes and males have one, (b) the XX-XY system, in which females have two X chromosomes and males have one X and one Y chromosome, and (c) the ZZ-ZW system, in which males have two Z chromosomes and females have one Z chromosome and one W chromosome.

4. Heterogametic sex determination usually ensures that the numbers of males and females are essentially equal.

5. In species with an XX-XO or XX-XY system, females are the homogametic sex and males the heterogametic sex. In the ZZ-ZW system, males are the homogametic sex (ZZ) and females the heterogametic sex (ZW).

6. The mechanism of XX-XY sex determination may differ among species. In *Drosophila melanogaster*, sex is determined by the ratio of X chromosomes to autosomes. When the ratio is 0.5, the individual is male. When the ratio is 1.0, the individual is female. When the ratio is between 0.5 and 1.0, the individual is an intersex. In mammals, however, sex is determined by the presence of a gene on the Y chromosome called the sex-determining region Y (*SRY* in humans). When *SRY* is present, the individual has a male phenotype. When *SRY* is absent, the individual has a female phenotype.

7. Genes on the Y chromosome are transmitted from father to son and are expressed only in males.

8. Genes on the X chromosome may be expressed in both males and females. Because females have two X chromosomes, they may be homozygous or heterozygous for X-linked alleles. Males have only one X chromosome, so they can be neither homozygous nor heterozygous for X-linked alleles; instead, they are hemizygous for X-linked alleles.

9. Recessive X-linked alleles are typically expressed more often in males than in females because females may carry a recessive allele in the heterozygous condition. Hemophilia and red-green color blindness are examples of human genetic disorders caused by recessive X-linked alleles.

10. X and Y chromosomes pair during meiosis at a small region of homology. Genes located within this small region of homology show a pseudoautosomal pattern of inheritance.

11. An attached X chromosome in *Drosophila melanogaster* is designated X^X. Flies with the genotype X^XY are females. In matings of these females with XY males, the males transmit their Y chromosome to X^XY female progeny and their X chromosome to male progeny.

12. In species with a ZZ-ZW system of sex determination, Z-linked traits are inherited in the same manner as X-linked traits, but males are the homogametic sex and females heterogametic.

13. In species with an XX-XY system of sex determination, females have two copies of every X-linked gene, whereas males have only one copy. The expression of X-linked genes, however, is equal in males and females because of dosage compensation.

14. In mammals, dosage compensation is accomplished by inactivation of one X chromosome in each somatic cell. The inactivated chromosome is called a Barr body. X chromosomes are usually inactivated early in development, so a mosaic pattern of expression may appear in females who are heterozygous for X-linked alleles.

15. In *Drosophila melanogaster*, X chromosomes are not inactivated in females. Instead, dosage compensation is accomplished by regulation of individual X-linked genes.

16. Sex-influenced and sex-limited traits are expressed differently in males and females, but are controlled by genes on autosomes rather than sex chromosomes. Sex-influenced traits are expressed in both sexes but preferentially in one sex. Sex-limited traits are expressed exclusively in one sex.

QUESTIONS AND PROBLEMS

1. What is the difference between dioecy and self-incompatibility in plants?

2. Look at Example 14.2, which discusses evidence of genetic sex determination in *Atriplex garrettii*. The authors of this study (Ruas et al. 1998. *American Journal of Botany* 85:162–167) found no differences between male and female karyotypes for chromosome size or structure. All nine pairs of chromosomes were fairly similar to one another in size and structure, with no evidence of a heteromorphic pair. The authors also attempted to find additional male-specific RAPD markers by amplifying DNA from male and female plants with an additional 158 primers. No additional male-specific (or female-specific) markers were found. The authors concluded that sex was not determined by heteromorphic chromosomes. Why do these results suggest that heteromorphic chromosomes are not responsible for sex determination?

3. When considering all dioecious species, we can hypothesize that dioecy may have arisen once or only a few times in evolutionary history, or that it may have arisen

independently many times. Using the information in this chapter, determine which of these two possibilities is most likely to be correct and explain why.

4. Among mammals, is dioecy likely to have arisen once or many times independently? Provide evidence to support your conclusion.

5. Why is an individual with an XXY genotype female in *Drosophila* but male in mammals? Be sure to explain in your answer the differences between *Drosophila* and mammals for mechanisms of sex determination.

6. (a) Can we truly say that sex determination in *Drosophila melanogaster* is due to heteromorphic sex chromosomes? **(b)** Can we truly say that sex determination in humans is due to heteromorphic sex chromosomes? Explain your answers to both questions.

7. Gynandromorphs are found in many species of insects but never in mammals. Based on the mechanisms of sex determination for *Drosophila* and humans, describe why this is the case.

8. (a) What evidence led researchers to exclude *ZFY* and the *H-Y antigen* gene as candidates for the testis-determining factor in humans? **(b)** What evidence suggests that *SRY* is the testis-determining factor in humans?

9. In 1990, Berta et al. (*Nature* 348:448–450) reported that an XY human female carried a mutation in the *SRY* gene on the Y chromosome. The mutation was a G → A transition that substituted isoleucine for methionine in the DNA-binding site of the protein. Neither the father nor the brother of this individual had this mutation in their Y chromosomes. What evidence that *SRY* is the testis-determining factor in humans do these observations provide?

10. In 1990, Jäger et al. (*Nature* 348:452–454) identified a four-nucleotide deletion mutation in the *SRY* gene on the Y chromosome of an XY human female. The mutation was present near the 3′ end of the region that encodes the DNA-binding site of the protein and caused a frameshift. This individual's father did not carry this mutation. What evidence that *SRY* is the testis-determining factor in humans do these observations provide?

11. The previous two problems describe newly discovered mutations in the *SRY* gene in XY human females. In neither case did the father carry the mutation on the Y chromosome in his somatic cells, indicating that the mutations had occurred recently. Explain why most mutations in the *SRY* gene in XY females are expected to be new mutations not present in the Y chromosome from the father's somatic cells.

12. Some of the most common genetic disorders in humans are X-linked. Describe why this is expected.

13. If a trait appears only in males, is this good evidence that the trait is due exclusively to a Y-linked allele? Explain your answer.

14. Traits caused by a dominant X-linked allele will probably be (a) more frequent in males, (b) more frequent in females, or (c) equally frequent in males and females? Explain your answer.

15. A woman's maternal grandfather had hemophilia A, and she wishes to know if she is a heterozygous carrier. **(a)** Based on the information in this chapter, what medical test (that does not require DNA testing) could determine whether or not she is a carrier? **(b)** What is the underlying genetic principle that allows this test to be effective in identifying heterozygous carriers?

16. Members of the current British royal family are descendants of Queen Victoria, yet none of them are at risk for hemophilia. Using the pedigree in Figure 14.12, describe why this is the case. (The current British royal family can be found in the lower left-hand corner of the pedigree.)

17. A woman has a father with X-linked hemophilia and a maternal grandfather with deuteranopia. She and her husband have neither of these traits. One of their sons has deuteranopia, inherited from his great-grandfather through his mother and grandmother. Barring new mutations, would it be possible for this woman to have a child with both hemophilia and deuteranopia? Explain why this is or is not possible.

18. A boy is born with protanopia, although both of his parents have normal color perception. On his maternal side, he has two aunts and six uncles. Two of his maternal uncles but none of his maternal aunts have protanopia. **(a)** Which of his maternal grandparents carried the protanopia allele? **(b)** Did either of his maternal grandparents have protanopia?

19. Winderickx et al. (1992. *Proceedings of the National Academy of Sciences, USA* 89:9710–9714) studied 13 human males with normal color perception for the number of *GCP* (*green cone pigment*) genes and expression of both *RCP* (*red cone pigment*) and *GCP* genes. Of the 13 individuals studied, 2 had one copy of the *GCP* gene and 11 had more than one copy. Ten of the 11 individuals with more than one *GCP* gene carried different alleles of that gene. The two alleles they carried were named *GCP*A* and *GCP*C*, which differed from one another by a single same-sense mutation in exon 5. Regardless of the number of *GCP* alleles an individual carried, only one type of mRNA was detected. What does this observation suggest about the regulation of multiple copies of the *GCP* gene?

20. Ocular albinism is a recessive X-linked trait in humans that causes a lack of pigment in the eyes. Two first cousins marry, a man with ocular albinism and a woman whose father (her husband's uncle) also had ocular albinism. **(a)** What is the probability that any one of their children will have ocular albinism? **(b)** Does the probability differ if the child is known beforehand to be male or female?

Table 14.4 Information for Problem 30: The Results of Bamber and Herdman's Experiments

Parents (female listed first)	Number of Offspring					
	Females			Males		
	Orange	Black	Tortoiseshell	Orange	Black	Tortoiseshell
1. Black × black	0	8	0	0	5	0
2. Orange × black	0	0	3	3	0	0
3. Tortoiseshell × black	0	9	8	8	11	0
4. Orange × orange	3	0	0	3	0	0
5. Black × orange	0	0	2	0	7	0
6. Tortoiseshell × orange	5	0	1	5	1	0

21. In humans, all the male children of mothers who have G6PD deficiency also have the trait. (G6PD deficiency is a benign condition that can cause anemia upon exposure to certain substances.) Rarely, a father with G6PD deficiency also has a son with the trait. **(a)** What is the most likely genetic basis for G6PD deficiency? **(b)** How can the rare occurrence of sons with G6PD deficiency whose father also had the trait best be explained?

22. In humans, what is the expected frequency of phenotypically male offspring in families where at least one of the offspring has testicular feminization?

23. In chickens, a silver hen mated with a gold rooster produces all silver male offspring and all gold female offspring. In the F_2 generation, half the progeny are gold and half are silver regardless of sex. Propose the simplest possible model of how gold versus silver feather color is determined in chickens. Be sure to identify which alleles are dominant and which recessive.

24. Bobbed is a pseudoautosomal recessive trait in *Drosophila melanogaster*. If a bobbed male is mated with an attached X ($\widehat{X}XY$) wild-type female, what are the expected frequencies of bobbed males and females in the F_1 and F_2 generations?

25. If a bobbed female is mated with a wild-type male in *Drosophila*, and all the F_1 flies are wild-type, then the F_1 flies are crossed with a bobbed male, what is the expected proportion of bobbed males and bobbed females in the offspring?

26. Define genetic anticipation, and describe why fragile-X syndrome in humans shows genetic anticipation.

27. Rarely, a male human child has three X chromosomes and a Y chromosome (XXXY). How many Barr bodies should be found in his somatic cells?

28. In *Drosophila*, should a Barr body be found in the cells of an XXY female with an attached X chromosome? Why or why not?

29. Table 14.4 summarizes the results of experiments conducted by Bamber and Herdman (1926. *Journal of Genetics* 18:87–97) for coat color in cats. **(a)** Rewrite the table showing the genotypes of the parents and the offspring. **(b)** Are there any cases of unexpected phenotypes or genotypes in the offspring?

30. Doncaster (1913. *Journal of Genetics* 3:11–23) reported that from 34 crosses between tortoiseshell female cats and orange male cats, the offspring consisted of 47 orange females, 43 tortoiseshell females, 5 black females, 54 orange males, 40 black males, 1 tortoiseshell male, and 5 cats whose sex was not determined. **(a)** Among the cats whose sex could be determined, there are a few individuals with unexpected phenotypes. What are the sexes and phenotypes of these unexpected progeny? **(b)** What could explain the appearance of these unexpected progeny?

31. Incontinentia pigmenti is an inherited disorder caused by a dominant mutant allele at the *IP* locus on the X chromosome in humans. The disorder is observed almost exclusively in heterozygous females because the allele is lethal when homozygous or hemizygous. Some females who are heterozygous for the allele have mild cases of the disorder because of apparent nonrandom inactivation of the X chromosome that bears the mutant allele. In 1993, Coleman et al. (*Journal of Medical Genetics* 30:497–500) reported a case of a female with hemophilia A. Her mother had incontinentia pigmenti and her father had hemophilia A. DNA marker analysis confirmed that this individual was heterozygous for the mutant *IP* allele inherited from her mother and also heterozygous for the mutant *F8C* allele inherited from her father. Explain this occurrence of hemophilia A in a heterozygous female.

32. A man develops pattern baldness. Neither his mother nor his father had pattern baldness to any degree, but both his maternal and paternal grandfathers had pattern baldness. **(a)** Is the man homozygous, heterozygous, or hemizygous for pattern baldness? **(b)** To the extent pos-

Table 14.5 Information for Problem 34: F_2 Phenotypes		
Phenotype	**Females**	**Males**
Brick-red (wild-type) eyes	9/16	9/32
White eyes	1/16	17/32
Bright-red eyes	3/16	3/32
Brown eyes	3/16	3/32

hen-feathered hen   hen-feathered rooster   cock-feathered rooster

Figure 14.23   Feather type phenotypes in chickens.

sible from the information given, draw a pedigree showing the transmission of pattern baldness alleles in this family.

33. In cattle, a red-spotted male is crossed with a mahogany-spotted female. The F_1 females are red-spotted, and the F_1 males are mahogany-spotted. In the F_2 males, ¾ are mahogany-spotted, and ¼ are red-spotted. This situation is the reverse in the F_2 females: ¾ are red-spotted, and ¼ are mahogany-spotted. **(a)** Is red-spotting a sex-linked, sex-influenced, or sex-limited trait? **(b)** Propose a model of inheritance assuming one locus with two alleles.

34. In *Drosophila melanogaster*, a white-eyed male fly is crossed with a white-eyed female fly, and all F_1 offspring are wild-type. In the F_2 generation, the proportions listed in Table 14.5 appear. Using a Punnett square, provide the simplest possible explanation for these results. Be sure to include the number of loci, whether alleles at these loci are dominant or recessive, and whether each locus is X-linked or on an autosome. (Recall that eye color in *Drosophila melanogaster* is caused by two types of pigments, one that is bright red, the other that is brown. When the two pigments are present, the wild-type brick-red eye color results. When both pigments are absent, white eyes are the result.)

35. As illustrated in Figure 14.23, chickens have two different types of feather patterns. Hens always have hen feathering, whereas roosters may have either cock feathering or hen feathering. A hen-feathered rooster from a flock that is true breeding for hen feathering is mated with a hen from a flock where all the roosters are cock-feathered. The F_1 chickens are all hen-feathered regardless of sex. In the F_2 generation, all the hens are hen feathered, ¾ of the roosters are hen-feathered, and ¼ of the roosters are cock-feathered. Assume that feather pattern is influenced by two alleles at a single locus. **(a)** Is cock feathering a sex-linked, sex-influenced, or sex-limited trait? **(b)** Propose an explanation for the inheritance of feather pattern.

36. In *Drosophila*, a female fly with a yellow body is mated with a male fly with vermilion eyes. They produce over 100 F_1 offspring. All the males but none of the females have vermilion eyes, and all the females but none of the

males have yellow bodies. What is the most probable explanation for these results?

37. A special stock of *Drosophila melanogaster* was developed for schoolchildren to analyze inheritance and distinguish a fly's sex without the aid of a microscope. All of the females have wild-type brick-red eyes, whereas all of the males have white eyes in every generation. When white-eyed males with vestigial wings (short, stubby wings) are mated with females from this stock, all the F_1 offspring have wild-type wings, but all the males have white eyes and all the females have wild-type eyes. In the F_2 generation, vestigial wings appear in about 25% of the flies regardless of sex, but again all the males are white-eyed and all the females have wild-type eyes. Explain why the white eye color and vestigial wing traits are inherited in their respective patterns.

FOR FURTHER READING

Morgan's classic paper on X-linked inheritance of white eyes in *Drosophila melanogaster* is **Morgan, T. H. 1910. Sex-limited inheritance in *Drosophila*. Science 32:120–122.** The chromosomal balance theory for sex determination in *Drosophila* was derived by **Bridges, C. 1925. Sex in relation to chromosomes and genes. American Naturalist 59:127–137.** The history of how the sex-determination mechanism in humans was discovered was reviewed by **Schafer, A. J., and P. N. Goodfellow. 1996. Sex determination in humans. BioEssays 18:955–963.** Extensive reviews of the inheritance, function, molecular biology, and phenotypes associated with the X-linked alleles responsible for red-green color blindness and hemophilia in humans, as well as numerous other X-linked and Y-linked genes, are available electronically from **McKusick, V. A. Online Mendelian Inheritance in Man (OMIM) at http://www.ncbi.nlm.nih.gov/Omim.** The broad range of sex-determining mechanisms in plants was reviewed by **Dellaporta, S. L., and A. Calderon-Urrea. 1993. Sex determination in flowering plants. The Plant Cell 5:1241–1251.**

For additional reading, go to InfoTrac College Edition, your online research library at: http: www.infotrac-college.com/brookscole

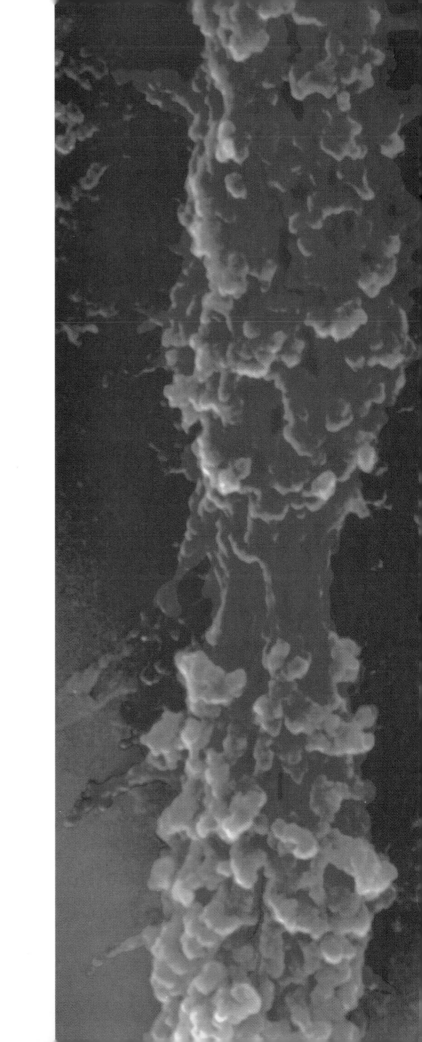

CHAPTER 15

Genes that are located close to each other on the same chromosome
are linked and do not assort independently.

~

The genetic distance between linked genes is measured as the
frequency with which the genes recombine.

~

Genetic maps are constructed by consolidating map distances
among genes and genetic markers.

~

Physical chromosome maps show the positions of genes and genetic
markers on a chromosome viewed microscopically.

~

Genetic mapping has several practical applications.

CHROMOSOME MAPPING

Mendel showed that genes undergo **genetic recombination** during inheritance—that is, some progeny contain different combinations of genes than their parents. Mendel's principle of independent assortment explained recombination of genes in his experiments. Shortly after the rediscovery of Mendel's principles, geneticists were faced with a dilemma. Genes were known to be located on chromosomes, and independent assortment of chromosomes explained independent assortment of genes. However, there were many more genes than chromosomes, so some genes had to be located on the same chromosome. These genes should remain inextricably associated with one another, failing to recombine through independent assortment. Yet sometimes they do recombine. Hugo de Vries, a rediscoverer of Mendel's principles, attempted to resolve this dilemma by proposing that homologous chromosomes freely exchange genes with one another when the chromosomes pair during meiosis. If this is true, all genes on the same chromosome should assort independently, just like genes on nonhomologous chromosomes.

Although de Vries's proposal was incorrect, he was not far from the truth. Through crossing-over, paired homologous chromosomes exchange segments, but the exchanges are not frequent enough for all genes to assort

independently. Genes located on the same chromosome are called **syntenic genes**. Some syntenic genes are so distant that they assort independently during meiosis because crossing-over between them is frequent. On the other hand, some syntenic genes are **linked genes**: they are so close to one another that crossovers may recombine them, but the crossovers are not frequent enough to permit independent assortment. All linked genes are syntenic, but syntenic genes are not necessarily linked (Figure 15.1).

The frequency of recombination between two genes depends on the distance between them on the chromosome. The closer two genes are, the lower the frequency of crossovers between them. The farther apart they are, the greater the frequency of crossovers. At one point geneticists concluded that crossovers are random events along the length of the chromosome, which explains why distantly linked genes are more likely to have a crossover between them than closely linked genes. Actually, crossing-over is not entirely random; it is less frequent in certain regions of the chromosome, such as the region surrounding the centromere. But the general rule that crossing-over is more frequent between distant genes than closely linked genes still holds true. Thus, scientists can measure the frequency of recombination between linked genes to

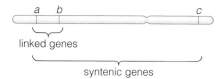

Figure 15.1 Comparison of linked and syntenic genes. Syntenic genes are located on the same chromosome. Genes that are sufficiently close to one another on the same chromosome do not assort independently and are linked. Syntenic genes are not necessarily linked genes.

determine the relative positions of genes on a chromosome, a process called **chromosome mapping**.

In this chapter we will learn how to determine the relative positions of linked genes from recombination frequencies and how to construct genetic maps. We will also see how scientists map genes to their physical locations on chromosomes. We conclude with some of the practical applications of chromosome mapping.

15.1 THE DISCOVERY OF LINKAGE AND CROSSING-OVER

Linkage was first reported in 1900 by Carl Correns, a rediscoverer of Mendel's principles. Correns reported a case of complete linkage, a situation in which two genes remain completely associated with each other when transmitted from parents to offspring. Correns hybridized two types of stocks (an ornamental flowering plant), one with purple flowers and hoary leaves, the other with white flowers and smooth leaves. The F_1 plants had purple flowers and hoary leaves, indicating that these two traits were dominant. In the F_2 progeny, Correns expected the two traits to assort independently. However, all purple-flowered F_2 progeny had hoary leaves, and all white-flowered F_2 progeny had smooth leaves, just like the two original parents. These two phenotypes appeared in a 3:1 ratio. Such an observation could be attributed to pleiotropy. Recall Mendel's experiments in which flower color and seed coat color were associated with one another because the two traits were governed by the same gene, an example of pleiotropy. In Correns's experiments, the purple flowers in stocks were always associated with colored seeds, which Correns correctly interpreted as pleiotropy. However, Correns knew of stocks with purple flowers and smooth leaves as well as plants with white flowers and hoary leaves, an indication that the two traits are not fully dependent on one another. Correns correctly concluded that flower color and leaf texture were governed by different genes that (usually) fail to assort independently.

The first case of incomplete linkage was reported by William Bateson and Reginald Punnett in 1908 in sweet pea. They observed that in the F_2 progeny of a cross be-

tween a purple-flowered plant with long pollen grains and a red-flowered plant with round pollen grains, these traits failed to assort independently, although all four possible phenotypes were present (see Example 12.4). However, linkage analysis in F_2 progeny can be a complicated situation, as we will soon see, and Bateson and Punnett initially developed an incorrect theory to explain the deviation from independent assortment.

Thomas Hunt Morgan was the first geneticist to correctly identify linkage. He and his colleagues hybridized flies with several mutant alleles at different X-linked loci and discovered that the genes did not assort independently among the offspring. Morgan concluded that the deviations from independent assortment were caused by linkage. Other scientists had reported the appearance of chiasmata in prophase I chromosomes. Morgan assumed that they represented exchanges of chromosomal segments, which he called *crossovers*, and that crossovers were responsible for recombination of syntenic genes. The term crossover has been used ever since.

One of Morgan's students, Alfred Sturtevant, later compiled data that he and others had collected and used these data to determine the relative positions of five genes on the X chromosome of *Drosophila*. Although the title of Sturtevant's paper ("The linear arrangement of six sex-linked factors in *Drosophila*, as shown by their mode of association") suggests that he mapped six genes, he actually mapped only five loci. Two of the genes he studied, *white* and *eosin*, were both eye color alleles and showed complete linkage. It was later discovered that they were different alleles of the *white* (*w*) locus. Sturtevant also studied mutant alleles at the *y* (yellow body), *v* (vermilion eyes), *r* (rudimentary wings), and *m* (miniature wings) loci, all on the X chromosome. After analyzing data from 31 different crosses, Sturtevant created the chromosome map shown in Figure 15.2. The numbers show the frequencies of recombination (expressed as percentage values) between alleles of the *w*, *v*, *r*, and *m* loci with alleles of the *y* locus, to which Sturtevant assigned a position numbered zero. This was the first chromosome map ever made. At the time, Sturtevant was an undergraduate student at Columbia University in New York, where he worked in Morgan's laboratory. One day, during a conversation with Morgan, it occurred to Sturtevant that genes on the same chromosome might not assort independently and that the frequencies of crossing-over among genes might be related to the relative distances that separated the genes on the chromosome. He took the data they had on inheritance of X-linked genes home with him and, in his own words, "spent most of the night (to the neglect of my undergraduate homework) in producing the first chromosome map."

Sturtevant, together with Calvin Bridges, later discovered linkage on autosomes in *Drosophila*. Eventually,

y   w   v   r   m
───

0.0   1.0   30.7   33.7   57.6

Figure 15.2 The first genetic map. Sturtevant mapped five genes on the X chromosome in *Drosophila melanogaster*. (Adapted from Sturtevant, A. H. 1913. The linear arrangement of six sex-linked factors in *Drosophila*, as shown by their mode of association. *Journal of Experimental Zoology* 14:43–59.)

detailed genetic maps for several model organisms, such as the *Drosophila melanogaster* map depicted in Figure 15.3, were made from the results of experiments conducted over a period of several decades.

The units used for mapping genes on chromosomes represent the frequency of crossovers as a percentage value and are often called **centimorgans (cM)** in honor of Morgan's pioneering work. Mapping is one of the most important areas of current genetics research. Human genome mapping is under way and, as we will see later in this chapter, mapping genes and DNA markers in a large number of species, including humans, is one of the most active areas in genetic research.

~

Genetic map distances between linked genes are measured in centimorgans (cM), which are defined as the percentage of crossing-over between linked genes. The genetic map distances between many linked genes can be combined to construct a genetic map.

15.2   TWO-FACTOR LINKAGE ANALYSIS

Analysis of two linked genes is called **two-factor linkage analysis**. Typically, two-factor linkage is diagrammed using **linkage notation**, which consists of a line representing the chromosome, and the alleles of two linked genes designated by letters:

A   B
─────────────────────

In linkage notation, alleles have their usual designations. The homologous chromosome is represented by a second line with the alleles on it. For instance, when loci *A* and *B* are linked, the cross *AA BB* × *aa bb* produces the F$_1$ heterozygote *Aa Bb*. In linkage notation, this cross is shown as follows:

A   B   a   b
────────────   ×   ────────────
A   B   a   b

A   B
────────────
a   b

Crossovers between two genes can be detected in the progeny of an individual that is heterozygous for two linked genes. The chromosomes that arise from a crossover between two genes are called **recombinant chromosomes** (or **crossover-type chromosomes**):

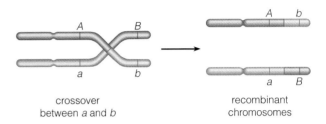

crossover
between *a* and *b*

recombinant
chromosomes

In recombinant chromosomes, the alleles at two linked loci have recombined when compared to the original chromosomes. When there is no crossover between two linked genes, the chromosomes that arise are called **nonrecombinant chromosomes** (or **noncrossover-type chromosomes**):

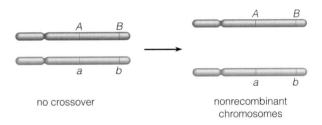

no crossover

nonrecombinant
chromosomes

The **recombination frequency** between two genes is calculated by dividing the number of recombinant chromosomes by the total number of chromosomes (recombinant + nonrecombinant chromosomes). For example, if 15 out of 100 chromosomes in the progeny of a doubly heterozygous individual are recombinant, then the recombination frequency is 0.15. The value in centimorgans is found by multiplying the recombination frequency by 100, so 0.15 × 100 = 15 cM.

Two-Factor Linkage in Testcross Progeny

Linkage analysis is usually done with testcross progeny (rather than F$_2$ progeny) because each of the four phenotypes in testcross progeny represents one of the four possible genotypes. In testcross progeny, researchers can readily distinguish recombinant from nonrecombinant

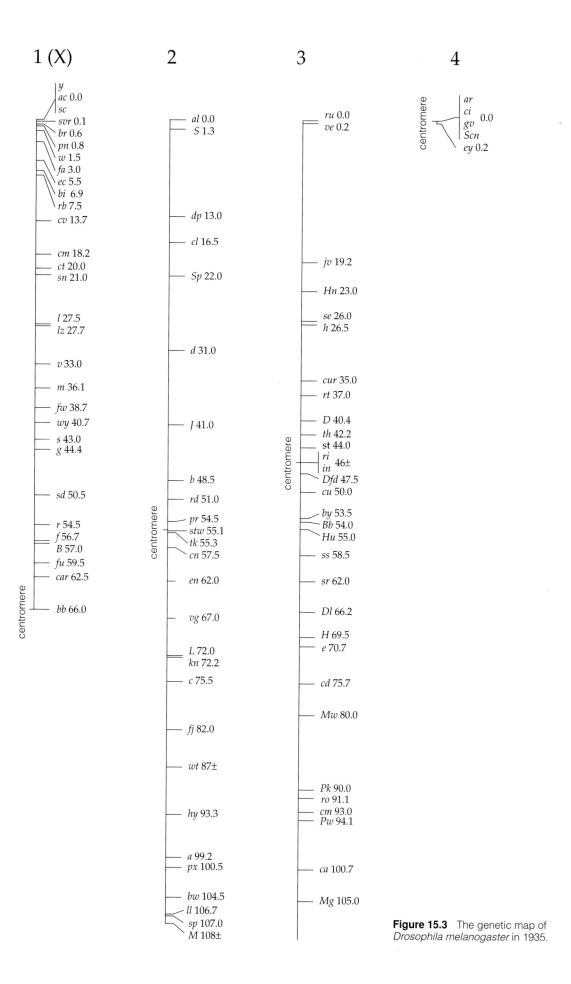

1 (X)

y
ac 0.0
sc
svr 0.1
br 0.6
pn 0.8
w 1.5
fa 3.0
ec 5.5
bi 6.9
rb 7.5
cv 13.7
cm 18.2
ct 20.0
sn 21.0
l 27.5
lz 27.7
v 33.0
m 36.1
fw 38.7
wy 40.7
s 43.0
g 44.4
sd 50.5
r 54.5
f 56.7
B 57.0
fu 59.5
car 62.5
bb 66.0
centromere

2

al 0.0
S 1.3
dp 13.0
cl 16.5
Sp 22.0
d 31.0
J 41.0
b 48.5
rd 51.0
pr 54.5
stw 55.1
tk 55.3
cn 57.5
centromere
en 62.0
vg 67.0
L 72.0
kn 72.2
c 75.5
fj 82.0
wt 87±
hy 93.3
a 99.2
px 100.5
bw 104.5
ll 106.7
sp 107.0
M 108±

3

ru 0.0
ve 0.2
jv 19.2
Hn 23.0
se 26.0
h 26.5
cur 35.0
rt 37.0
D 40.4
th 42.2
st 44.0
ri
in 46±
Dfd 47.5
cu 50.0
centromere
by 53.5
Bb 54.0
Hu 55.0
ss 58.5
sr 62.0
Dl 66.2
H 69.5
e 70.7
cd 75.7
Mw 80.0
Pk 90.0
ro 91.1
cm 93.0
Pw 94.1
ca 100.7
Mg 105.0

4

centromere
ar
ci
gv 0.0
Scn
ey 0.2

Figure 15.3 The genetic map of *Drosophila melanogaster* in 1935.

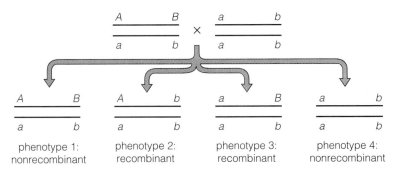

Figure 15.4 The four genotypes and phenotypes that arise from a two-factor testcross. Two are nonrecombinant and two are recombinant.

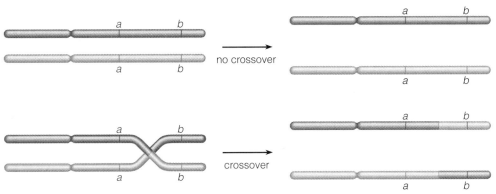

Figure 15.5 A crossover between two linked genes in a double homozygote. The combination of genes is the same in the chromosomes regardless of whether or not the chromosomes exchanged segments.

chromosomes, as illustrated in Figure 15.4. Notice that we count only the crossovers in the doubly heterozygous parent (*Aa Bb*). We cannot detect crossovers in the homozygous *aa bb* parent because the recombinant and nonrecombinant chromosomes it produces are identical (Figure 15.5).

When analyzing data from a dihybrid testcross, we must first ask if there is evidence to indicate that the two genes are linked. Because linkage is manifest as a deviation from independent assortment, we first determine whether or not the deviation is significant using a chi-square test. If the deviation is not significant, there is insufficient evidence to conclude that linkage is present. If the deviation is significant, then linkage may be the most probable explanation for the deviation, as illustrated in the following example.

Example 15.1 Two-factor linkage in maize.

Problem: In maize a recessive allele *c1* confers a colorless kernel when homozygous, and the recessive allele *wx* confers a kernel with a waxy texture when homozygous. The dominant allele *C1* gives a colored kernel, and the dominant allele *Wx* gives a starchy texture to the kernel. In 1918, Bregger

(*American Naturalist* 52:57–61) reported data from a cross between a plant grown from a homozygous colored, starchy kernel (*C1 C1 Wx Wx*) and a plant grown from a homozygous colorless, waxy kernel (*c1 c1 wx wx*). The F_1 kernels were all doubly heterozygous (*C1 c1 Wx wx*) and were colored and starchy. When Bregger testcrossed an F_1 plant with a plant grown from a homozygous colorless, waxy kernel (*c1 c1 wx wx*), he observed the following phenotypes in the testcross progeny:

Colored, starchy (*C1 c1 Wx wx*)	858
Colored, waxy (*C1 c1 wx wx*)	310
Colorless, starchy (*c1 c1 Wx wx*)	311
Colorless, waxy (*c1 c1 wx wx*)	781
Total	2260

(a) Are the *c1* and *wx* loci linked? **(b)** If so, what is the genetic map distance between them in centimorgans?

Solution: **(a)** With independent assortment, a 1:1:1:1 ratio is expected in testcross progeny. It is obvious that Bregger's data do not conform to a

1:1:1:1 ratio, but let's test them with chi-square analysis to illustrate its use. The total number of progeny is 2260, so the expected number of progeny in each class under the assumption of independent assortment is $0.25 \times 2260 = 565$. The chi-square calculation is as follows: $\chi^2 = [(858 - 565)^2 + (310 - 565)^2 + (311 - 565)^2 + (781 - 565)^2] \div 565 = 463.8$ with 3 degrees of freedom, which is substantially greater than the theoretical value of 11.34 at the 0.01 probability level (from Table 12.5). We therefore reject the hypothesis that these alleles assort independently. Linkage is the most probable explanation for the deviation from independent assortment. **(b)** We calculate the genetic map distance by summing the number of recombinant progeny (progeny with a recombinant chromosome) and dividing this value by the total number of progeny. We can reconstruct the cross using linkage notation:

$$\frac{C1 \qquad Wx}{c1 \qquad wx} \quad \times \quad \frac{c1 \qquad wx}{c1 \qquad wx}$$

So,

Colored, starchy (*C1 c1 Wx wx*) 858 Nonrecombinant

Colored, waxy (*C1 c1 wx wx*) 310 Recombinant

Colorless, starchy (*c1 c1 Wx wx*) 311 Recombinant

Colorless, waxy (*c1 c1 wx wx*) 781 Nonrecombinant

The recombination frequency is

$$(310 + 311) \div (858 + 310 + 311 + 781)$$
$$= 621 \div 2260 = 0.275 \text{ or } 27.5 \text{ cM}$$

which can be diagrammed in linkage notation as

$$\underline{c1 \qquad\qquad 27.5 \qquad\qquad wx}$$

In Example 15.1, both dominant alleles (*C1* and *Wx*) were on the same chromosome, and both recessive alleles (*c1* and *wx*) were on the homologous chromosome in the F_1 double heterozygote:

$$\frac{C1 \qquad Wx}{c1 \qquad wx}$$

In this situation, the dominant alleles are said to be in **coupling conformation** (or **cis conformation**) because the two dominant alleles are coupled to one another on the same chromosome. The two recessive alleles are also in coupling conformation.

Suppose the cross for generating the doubly heterozygous genotype had been between a plant grown from a homozygous colored, waxy kernel (*C1 C1 wx wx*) and a plant grown from a homozygous colorless, starchy kernel (*c1 c1 Wx Wx*). The chromosomes contributed by these parents are

$$\underline{C1 \qquad wx} \quad \text{and} \quad \underline{c1 \qquad Wx}$$

giving an F_1 double heterozygote with the genotype

$$\frac{C1 \qquad wx}{c1 \qquad Wx}$$

In this genotype, the dominant alleles are in **repulsion conformation** (or **trans conformation**), which means that the two dominant alleles are not coupled to one another; one dominant allele is on each of the homologous chromosomes. Likewise, the recessive alleles are in repulsion to one another. Linkage analysis with alleles in repulsion conformation is illustrated in the following example.

Example 15.2 Two-factor linkage in repulsion conformation.

Bregger (1918. *American Naturalist* 52:57–61) examined linkage in repulsion conformation. An F_1 plant from the cross *C1 C1 wx wx* × *c1 c1 Wx Wx* was crossed with a plant with the genotype *c1 c1 wx wx*, and the progeny were as follows:

Colored, starchy (*C1 c1 Wx wx*) 115

Colored, waxy (*C1 c1 wx wx*) 340

Colorless, starchy (*c1 c1 Wx wx*) 298

Colorless, waxy (*c1 c1 wx wx*) 92

Total 845

Problem: **(a)** Diagram the genotype of the F_1 plant in linkage notation. **(b)** Determine the genetic map distance between the two loci.

Solution: **(a)** The original homozygous parents of the doubly heterozygous F_1 plant each carried a dominant allele at one locus and a recessive allele at the other locus:

$$\frac{C1 \qquad wx}{C1 \qquad wx} \quad \times \quad \frac{c1 \qquad Wx}{c1 \qquad Wx}$$

So the F_1 plant had the genotype

$$\frac{C1 \qquad wx}{c1 \qquad Wx}$$

The alleles are in repulsion conformation in the F_1 generation.

(b) The recombinant and nonrecombinant progeny are the reverse of those in the coupling conformation analyzed in Example 15.1:

Colored, starchy (*C1 c1 Wx wx*)   115   Recombinant

Colored, waxy (*C1 c1 wx wx*)   340   Nonrecombinant

Colorless, starchy (*c1 c1 Wx wx*) 298   Nonrecombinant

Colorless, waxy (*c1 c1 wx wx*)   92   Recombinant

The genetic map distance is calculated by adding the number of recombinant progeny and dividing this sum by the total number of progeny:

$$(115 + 92) \div (115 + 340 + 298 + 92)$$
$$= 207 \div 845 = 0.245 \text{ or } 24.5 \text{ cM}$$

The calculations in Examples 15.1 and 15.2 estimate the genetic map distance between *c1* and *wx*. Because of sampling error, we cannot assume that *c1* and *wx* will always map at exactly the same distance. For instance, in Example 15.1, *c1* and *wx* mapped at 27.5 centimorgans, and in Example 15.2 they mapped at 24.5 centimorgans. This difference has nothing to do with coupling or repulsion conformation but is the result of sampling error. The data from the two experiments can be combined to give a map distance of 26.7 centimorgans. More accurate map distances are usually obtained by combining the results of several experiments that together provide a very large number of progeny.

~

Genetic map distances between two genes are typically calculated by conducting a two-factor testcross and dividing the number of recombinant progeny by the total number of progeny. Two linked genes may be in either coupling or repulsion conformation.

Genetic Recombination and Physical Detection of Crossovers

For many years, researchers assumed that a physical exchange of chromosome segments caused genetic recombination. Although such an assumption was supported by all the data obtained in linkage analysis experiments, no one had demonstrated experimentally that chromosomes exchange segments when genes recombine. Harriet Creighton and Barbara McClintock addressed this topic with the experiments highlighted in the following example.

Example 15.3   Association of genetic recombination with a physical exchange of chromosome segments.

In 1931, Creighton and McClintock (*Proceedings of the National Academy of Sciences, USA* 17:492–497) reported the results of what is now considered a classic experiment to determine whether genetic recombination of linked genes is associated with a physical exchange of chromosomal segments. They used the two genes we just discussed, the *c1* locus and the *wx* locus on chromosome 9 in maize, from plants that had **cytological markers**, differences in the chromosomes that can be observed microscopically. Creighton and McClintock found a chromosome 9 with two cytological markers, a knob on one end and a piece from another chromosome on the other end. Thus, both ends of this chromosome could be distinguished from normal copies of chromosome 9 using microscopy. This chromosome carried the dominant *C1* allele and the recessive *wx* allele. From the cross

they observed the following progeny:

c1 c1 Wx wx	knobless	no translocation	5
c1 c1 Wx Wx	knobless	no translocation	5
c1 c1 Wx __	knobless	no translocation	1
C1 c1 wx wx	knob	translocation	3
c1 c1 Wx Wx	knobless	translocation	2
c1 c1 Wx wx	knobless	translocation	2
c1 c1 wx wx	knobless	translocation	2
C1 c1 Wx __	knob	no translocation	8
Total			28

Problem: What do these results reveal about the association between genetic recombination and a physical exchange of chromosome segments?

Solution: As diagrammed in Figure 15.6, the last three classes in the list above contain 12 individuals that arose from genetic recombination between *c1* and *wx*, and all 12 had an exchange of cytological markers. Of the 16 individuals that did not arise from a crossover between *c1* and *wx*, 2 had an exchange of cytological markers because of crossovers between *wx* and the translocation. From this

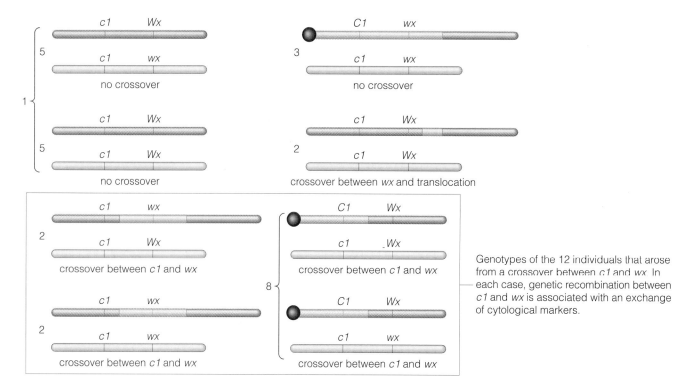

Figure 15.6 The results of Creighton and McClintock's experiment: Genotypes and chromosomal constitutions of the testcross progeny. Twelve individuals (those within the box) arose from recombination between *c1* and *wx*. In each case, genetic recombination between *c1* and *wx* is associated with an exchange of cytological markers.

evidence, Creighton and McClintock concluded that a physical exchange of chromosome segments causes genetic recombination.

Crossing-Over in Duplicated Chromosomes

Up until now, we have illustrated crossing-over using diagrams of unduplicated chromosomes. This simplifies the illustration of crossing-over but is a little misleading. Chromosomes initiate crossing-over during prophase I of meiosis after they have duplicated, so a homologous chromosome pair contains four chromatids: two sister chromatids in one chromosome, and two sister chromatids in its paired homologue. Crossing-over between two sister chromatids, also known as meiotic sister chromatid exchange, is probably rare in meiosis and has no genetic consequence. Sister chromatids are identical to one another, so a crossover between two sister chromatids generates identical chromatids and has no effect on inheritance (Figure 15.7).

For crossovers to be detected, nonsister chromatids on paired homologous chromosomes of a doubly heterozygous individual must exchange segments. As illustrated in Figure 15.8, it does not matter which nonsister

chromatids exchange segments; the result is the same. For this reason, the simple diagrams of unduplicated chromosomes give the same results as diagrams with duplicated chromosomes.

~

Recombination of linked genes is associated with cytological observation of crossing-over. Crossing-over is initiated between nonsister chromatids after chromosomes have been duplicated.

15.3 MAP DISTANCE CORRECTION

As genetic mapping is dependent on crossover frequency, any factors that alter crossover frequencies need to be taken into account in order to calculate accurate genetic map distances. The chromosomal location of a gene, the age, sex, or genotype of an individual, and environmental influences may all affect crossover frequency. In some cases, these effects may be negligible, but in other cases they may be significant. For example, in *Drosophila melanogaster*, there are no crossovers in male flies, and as female flies grow older, the frequency of crossovers decreases. Temperatures either warmer or cooler than 22°C tend to increase the frequency of crossing over in

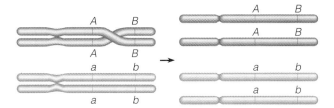

Figure 15.7 Meiotic sister chromatid exchange. The result is the same as if there had been no crossover.

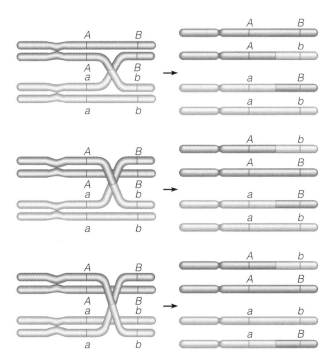

Figure 15.8 Meiotic nonsister chromatid exchange. A crossover between any two nonsister chromatids produces the same result: two recombinant chromosomes and two nonrecombinant chromosomes.

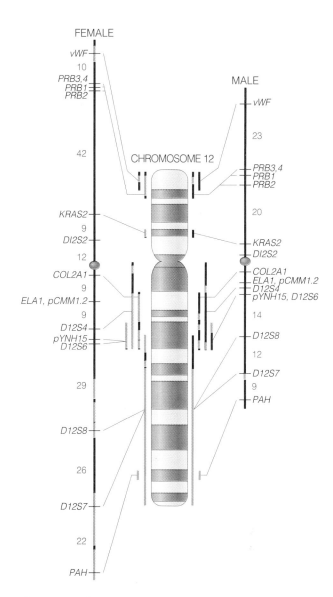

Figure 15.9 Genetic maps of chromosome 12 in humans from females (left) and males (right). The frequency of crossing-over tends to be less in human males than in human females. (Redrawn from an illustration by Joan Starwood in White, R. and J. M. Lalouel. 1988. Chromosome mapping with DNA markers. *Scientific American* 258 (Feb):40–48.)

Drosophila females, and the levels of calcium and magnesium in the diet also influence crossover frequencies. Researchers can compensate for these factors by designing crosses so that the female is always the heterozygous parent in a testcross, by using young female parents in mapping crosses, and by controlling the environment in which the flies are raised.

Factors that alter crossover frequency tend to differ among species, so researchers need to be aware of the peculiarities for the species in question. Generally, experiments can be designed to overcome these alterations, or corrections can be made if the nature of the alteration is known beforehand. For example, in humans crossover frequencies differ in males and females. As illustrated in Figure 15.9, map distances are lower when derived from male parents than from female parents, an observation that implies a lower crossover frequency in males. For this reason, human geneticists must adjust map distances to compensate for the differences in crossover frequency between males and females.

Mapping Function

Two-factor mapping experiments tend to underestimate the actual map distance when the true genetic map distance exceeds 20 cM. For example, suppose that genes

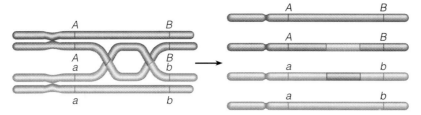

a Two-strand double crossover produces all nonrecombinant chromosomes.

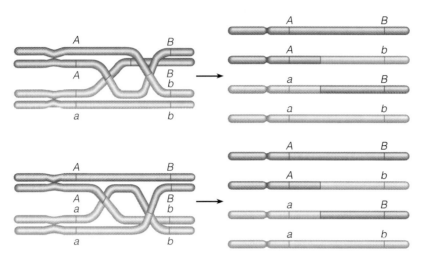

b Two types of three-strand double crossovers produce two recombinant and two nonrecombinant chromosomes.

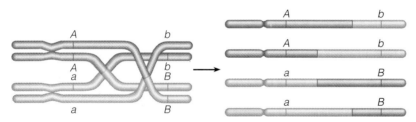

c Four-strand double crossover produces all recombinant chromosomes.

Figure 15.10 The effect of double crossovers on the frequency of recombinant progeny.

A and *B* map at 13 cM from each other, and that genes *B* and *C* map at 18 cM from each other in separate experiments. *A* and *C* should map at 31 cM from each other based on the additive distances of *A–B* and *B–C*. However, the experimental results from the progeny of a two-factor cross for *A* and *C* consistently yield a map distance less than 31 cM.

Underestimation of map distances is the result of undetected double crossovers. A **double crossover** is the simultaneous occurrence of two crossovers between linked genes. Chromosomes exchange segments during prophase I when paired chromosomes are already duplicated (see Figure 15.8). This means that any two non-

sister chromatids can participate in the crossover. As diagrammed in Figure 15.10, there are four possible types of double crossovers, some of which are detected phenotypically, but some of which are not.

A single crossover produces two recombinant and two nonrecombinant chromosomes (see Figure 15.8). Thus, for every single crossover, two of four chromosomes are recombinant after meiosis is completed. Notice in Figure 15.10a that a two-strand double crossover produces four nonrecombinant chromosomes, even though there are two crossovers. Phenotypically, this appears as though there were no crossovers. Three-strand double crossovers produce two recombinant chromosomes and

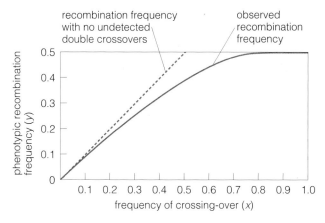

Figure 15.11 Relationship between genetic map distance and phenotypic recombination frequency. As the true map distance exceeds 20 cM, there is a tendency to underestimate the map distance because of undetected double crossovers; in other words, the observed recombination frequency is less than the actual recombination frequency.

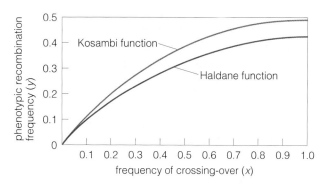

Figure 15.12 Comparison of Haldane and Kosambi functions.

two nonrecombinant chromosomes (Figure 15.10b). Phenotypically, these appear as a single crossover. A four-strand double crossover produces four recombinant chromosomes (Figure 15.10c). Phenotypically, this type of crossover produces the same number of recombinant chromosomes per crossover as single crossovers do. In a two-factor cross, two-strand and three-strand double crossovers result in undetected crossovers. The frequencies of the four types of double crossovers are equal. Thus, half of the crossovers from double crossovers are not detected phenotypically.

The expected frequency of double crossovers between two genes is the square of the genetic map distance between the two genes. This means that the probability of double crossovers should increase exponentially as the distance between genes increases. When genetic map distances are very small, the frequency of double crossovers (and other multiple crossovers) is so small that the probability of undetected crossovers may be negligible. In addition, the presence of one crossover interferes with initiation of a second crossover in its immediate vicinity, a phenomenon known as **interference**. When the distances between genes are small (less than about 5 cM), the probability of a double crossover between them is close to zero. At increasing distances between genes, the probability of double crossovers increases exponentially, meaning that at some point the number of undetected crossovers begins to significantly affect the calculation of genetic map distances. As Figure 15.11 illustrates, this point is reached in most species when the frequency of crossing-over is greater than 0.2 (a map distance of 20 cM).

Figure 15.11 also shows that the frequency of crossing-over has a curvilinear relationship to the phenotypic recombination frequency. Phenotypic recombination

frequencies can be corrected to approximate crossover frequencies with equations called **mapping functions**. The first mapping function, often called the **Haldane function**, was derived by J. B. S. Haldane in 1919, and is expressed as

$$x = -0.5\ln(1 - 2y) \qquad \text{[15.1]}$$

in which x is an estimate of the actual frequency of crossing-over, "ln" stands for "natural logarithm," and y is the phenotypic recombination frequency. This is a theoretical equation and does not take interference into account, so it departs from the linear mapping function at zero. In reality, interference causes the mapping function to remain almost linear between 0 and 20 cM. Haldane recognized this in his original paper and developed a second function based on empirical data instead of theory.

Since Haldane's attempt to derive a mapping function to match empirical data, several researchers have derived mapping functions from empirical data for particular species. Perhaps the most widely used is the **Kosambi function**. This function, derived in 1944 by D. D. Kosambi, can be used to correct phenotypic recombination frequencies measured in *Drosophila* and many other species. As in equation 15.1, x estimates the frequency of crossing-over and y equals the phenotypic recombination frequency:

$$x = 0.25\ln\frac{1 + 2y}{1 - 2y} \qquad \text{[15.2]}$$

The theoretical Haldane function tends to overestimate the true genetic map distance when interference is present; functions derived from empirical data are usually better because they take interference into account. Figure 15.12 compares the Haldane and Kosambi functions.

Let's now apply mapping functions to some data we have already examined to see how a mapping function corrects the map distance.

Example 15.4 Mapping functions.

Problem: **(a)** Combine the genetic map distances between *c1* and *wx* in maize from Bregger's data in Examples 15.1 and 15.2. **(b)** Correct the genetic map distance with the Haldane function and the Kosambi function. **(c)** Which function more accurately estimates the true map distance of 30 cM?

Solution: **(a)** The combined genetic map distance from these two experiments is calculated by summing the recombinant progeny from both experiments, then dividing this sum by the total number of progeny from both experiments:

$$(621 + 207) \div (2260 + 845)$$
$$= 828 \div 3105 = 0.267 \text{ or } 26.7 \text{ cM}$$

Notice that it is not appropriate to average the map distances calculated individually from the two experiments. A smaller number of progeny were used in one experiment, so averaging the map distances biases the results. **(b)** Using the Haldane function (equation 15.1), we get

$$x = -0.5\ln[1 - 2(0.267)] = 0.382 \text{ or } 38.2 \text{ cM}$$

Using the Kosambi function (equation 15.2), we get

$$x = 0.25\ln \frac{1 + 2(0.267)}{1 - 2(0.267)} = 0.298 \text{ or } 29.8 \text{ cM}$$

(c) The Kosambi function more accurately estimates the true genetic map distance of 30 cM.

There is a tendency to underestimate the map distance when distances exceed about 20 cM because of undetected double crossovers. Mapping function equations compensate for undetected crossovers and make it possible to estimate map distances more accurately.

Two-Factor Linkage Analysis in F_2 Progeny

Calculation of genetic map distances with F_2 progeny is complicated because several genotypes may be combined into a single phenotype if any of the alleles show complete dominance. Some genotypes carry two recombinant chromosomes, some carry one recombinant and one nonrecombinant chromosome, and others carry two nonrecombinant chromosomes, all of which may be found in a single phenotypic class. For this reason, calculation of

map distances from phenotypic frequencies is difficult and not highly reliable statistically, so whenever possible, it is best to do genetic mapping in testcross progeny.

Phenotypically, linkage in F_2 progeny is manifest as a deviation from the 9:3:3:1 ratio predicted for independent assortment. Linkage estimates from F_2 progeny data are best made by finding in a table the theoretical frequencies that most closely match the observed frequencies from an experiment. Table 15.1a contains the theoretical values for the four phenotypes in F_2 progeny for coupling conformation, and Table 15.1b contains the theoretical values for repulsion conformation. These values do not take undetected double crossovers into account and may need to be corrected using a mapping function.

It is important to point out that these tables assume equal probabilities of crossing-over in the male and female parents, which is a reasonable assumption for many species. There are, however, some very important exceptions, such as *Drosophila* and humans. Before using F_2 progeny for mapping, or designing testcrosses for mapping, it is important to account for any peculiarities in the species being studied.

The following example shows how to use a table of theoretical frequencies to determine map distances from F_2 progeny.

Example 15.5 Using F_2 data for two-factor linkage analysis.

The first published data with incomplete linkage were reported in 1908 by Bateson and Punnett (*Reports to the Evolution Committee of the Royal Society, Report IV*). They observed the results of incomplete linkage in the F_2 generation of sweet pea, but failed to develop a correct explanation of their results, in part because of the difficulties associated with F_2 linkage analysis. Let's use their results to estimate the map distance between the linked genes. The dominant allele *R* causes purple flowers, and the recessive allele *r* causes red flowers when homozygous. The dominant allele *Ro* causes long pollen grains, and the recessive allele *ro* causes round pollen grains when homozygous. In the F_2 generation from a cross between a double heterozygous purple-flowered plant with long pollen grains and a red-flowered plant with round pollen grains, they tabulated the following results:

Purple flowers, long pollen grains	296
Purple flowers, round pollen grains	19
Red flowers, long pollen grains	27
Red flowers, round pollen grains	85
Total	427

Table 15.1a Expected Proportions of F₂ Progeny for Two Linked Genes with Complete Dominance in Coupling Conformation

y (frequency of recombination)	Genotype			
	A_ B_	A_ bb	aa B_	aa bb
0.01	0.7450	0.0050	0.0050	0.2450
0.02	0.7401	0.0099	0.0099	0.2401
0.03	0.7352	0.0148	0.0148	0.2352
0.04	0.7304	0.0196	0.0196	0.2304
0.05	0.7256	0.0244	0.0244	0.2256
0.06	0.7209	0.0291	0.0291	0.2209
0.07	0.7162	0.0338	0.0338	0.2162
0.08	0.7116	0.0384	0.0384	0.2116
0.09	0.7070	0.0430	0.0430	0.2070
0.10	0.7025	0.0475	0.0475	0.2025
0.11	0.6980	0.0520	0.0520	0.1980
0.12	0.6936	0.0564	0.0564	0.1936
0.13	0.6892	0.0608	0.0608	0.1892
0.14	0.6849	0.0651	0.0651	0.1849
0.15	0.6806	0.0694	0.0694	0.1806
0.16	0.6764	0.0736	0.0736	0.1764
0.17	0.6722	0.0778	0.0778	0.1722
0.18	0.6681	0.0819	0.0819	0.1681
0.19	0.6640	0.0860	0.0860	0.1640
0.20	0.6600	0.0900	0.0900	0.1600
0.21	0.6560	0.0940	0.0940	0.1560
0.22	0.6521	0.0979	0.0979	0.1521
0.23	0.6482	0.1018	0.1018	0.1482
0.24	0.6444	0.1056	0.1056	0.1444
0.25	0.6406	0.1094	0.1094	0.1406
0.50*	0.5625	0.1875	0.1875	0.0625

*Expected values for independent assortment (9:3:3:1).

Problem: Determine the genetic map distance between the two genes.

Solution: The frequencies of each phenotypic class can be determined by dividing the number of progeny in each class by the total number of progeny:

Purple flowers, long pollen grains	0.6932
Purple flowers, round pollen grains	0.0445
Red flowers, long pollen grains	0.0632
Red flowers, round pollen grains	0.1991
Total	1.0000

One of the original parents (in the P generation) had both dominant phenotypes, and the other had both recessive phenotypes. This indicates that the dominant alleles are in coupling conformation in the F_1 progeny. The F_2 data confirm this conclusion. The two most frequent phenotypic classes are the class with both dominant phenotypes (purple flowers, long pollen grains) and the class with both recessive phenotypes (red flowers, round pollen grains), the result expected with coupling. Using Table 15.1a, we find that a recombination frequency of $y = 0.12$ corresponds most closely with the data.

Table 15.1b Expected Proportions of F$_2$ Progeny for Two Linked Genes with Complete Dominance in Repulsion Conformation

y (frequency of recombination)	Genotype			
	A_ B_	A_ bb	aa B_	aa bb
0.01	0.5000	0.2500	0.2500	0.0000
0.02	0.5001	0.2499	0.2499	0.0001
0.03	0.5002	0.2498	0.2498	0.0002
0.04	0.5004	0.2496	0.2496	0.0004
0.05	0.5006	0.2494	0.2494	0.0006
0.06	0.5009	0.2491	0.2491	0.0009
0.07	0.5012	0.2488	0.2488	0.0012
0.08	0.5016	0.2484	0.2484	0.0016
0.09	0.5020	0.2480	0.2480	0.0020
0.10	0.5025	0.2475	0.2475	0.0025
0.11	0.5030	0.2470	0.2470	0.0030
0.12	0.5036	0.2464	0.2464	0.0036
0.13	0.5042	0.2458	0.2458	0.0042
0.14	0.5049	0.2451	0.2451	0.0049
0.15	0.5056	0.2444	0.2444	0.0056
0.16	0.5064	0.2436	0.2436	0.0064
0.17	0.5072	0.2428	0.2428	0.0072
0.18	0.5081	0.2419	0.2419	0.0081
0.19	0.5090	0.2410	0.2410	0.0090
0.20	0.5100	0.2400	0.2400	0.0100
0.21	0.5110	0.2390	0.2390	0.0110
0.22	0.5121	0.2379	0.2379	0.0121
0.23	0.5132	0.2368	0.2368	0.0132
0.24	0.5144	0.2356	0.2356	0.0144
0.25	0.5156	0.2344	0.2344	0.0156
0.50*	0.5625	0.1875	0.1875	0.0625

*Expected values for independent assortment (9:3:3:1).

15.4 THREE-FACTOR LINKAGE ANALYSIS

Genetic map distances for three or more linked genes can be determined simply by adding the distances calculated in two-factor linkage experiments. However, it is possible to map three or more linked genes in the progeny of a single cross. In a testcross situation the map distance between two genes is calculated by counting the number of progeny that arose from a crossover between the two genes and dividing that number by the total number of progeny. In a three-factor testcross, this procedure is done twice. With three linked genes

$$A \qquad B \qquad C$$

the genetic map distance between A and B and the genetic map distance between B and C are calculated separately using progeny from the same cross.

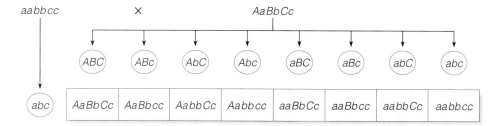

Figure 15.13 The eight genotypes that are possible in the progeny from a three-factor testcross.

However, before calculating the map distances, we must determine the order of the three genes. With three genes, *A*, *B*, and *C*, there are three possible orders:

$$\underline{A \qquad B \qquad C}$$

$$\underline{B \qquad A \qquad C}$$

$$\underline{A \qquad C \qquad B}$$

You may be asking, Aren't there six possible orders? What about *C B A*, *C A B*, and *B C A*? These latter three possibilities are actually the same as the former three. We usually write the order of genes on a chromosome from left to right, because that is the direction we read. But in a three-factor testcross, we determine the order of three genes in relation to one another. Under these circumstances, the left-right orientation does not matter; *A B C* is the same as *C B A*.

We determine the correct order from the proportions of individuals in each class of testcross progeny. The three-factor testcross *Aa Bb Cc* × *aa bb cc* produces eight genotypic (and phenotypic) classes, as diagrammed in Figure 15.13. If the alleles at the three loci assort independently, the eight genotypes should appear in a 1:1:1:1:1:1:1:1 ratio. Linkage causes a significant deviation from this ratio. Let's suppose for the time being that the genes are linked and that their order is *A B C*, so the heterozygous parent has the genotype $\frac{A\,B\,C}{a\,b\,c}$. Of the eight possible gametes that could arise from the heterozygous parent, two gametes arise from no crossovers in the region of interest (*A B C* and *a b c*), two from a single crossover between *A* and *B* (*A b c* and *a B C*), two from a single crossover between *B* and *C* (*A B c* and *a b C*), and two from a double crossover (one crossover between *A* and *B* and a second crossover between *B* and *C*: *A b C* and *a B c*), as diagrammed in Figure 15.14.

Double-crossover progeny are the least frequent because double-crossover chromosomes are the least likely to arise. On the other hand, the most frequent chromo-

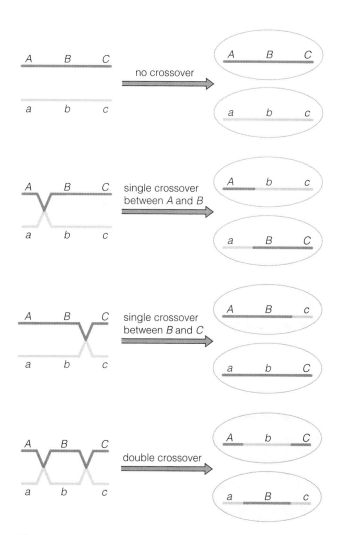

Figure 15.14 The eight types of gametes that can arise in an individual who is heterozygous for three linked genes.

somes are the noncrossover types. Using this information, we can assign each class of progeny to its crossover type. The double-crossover classes are the least frequent, and the noncrossover types are the most frequent. To determine the gene order on the chromosome, we diagram the three possible orders and determine which produces

the observed double-crossover class. If the gene order on the chromosome is <u>A B C</u>, then the double-crossover classes are

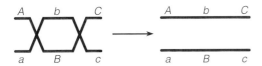

If the gene order on the chromosome is <u>B A C</u>, then the double-crossover classes are

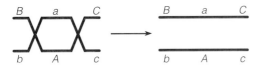

And if the gene order on the chromosome is <u>A C B</u>, then the double-crossover classes are

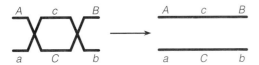

After we have determined the correct gene order, we calculate the genetic map distance. Let's assume that the gene order is <u>A B C</u>. As with two-factor testcross progeny, we calculate the genetic map distance between A and B by determining the number of progeny that represent a crossover between A and B and dividing this number by the total number of progeny. This means that we add the totals for progeny that represent a single crossover between A and B and all double-crossover progeny, then divide this sum by the total number of progeny. (We include double-crossover progeny in the numerator because all progeny that arose from a crossover between A and B must be included. The double-crossover classes arose from a crossover between A and B as well as a crossover between B and C.) To calculate the map distance between B and C, we add the totals for progeny that represent a single crossover between B and C and all double-crossover progeny, then divide this sum by the total number of progeny.

Let's now look at a real example to see how three-factor gene mapping is done.

Example 15.6 Three-factor linkage analysis for alleles with complete coupling.

In maize, genotypes for alleles at three linked loci cause the following phenotypes:

$V\_\_$: normal seedlings vv: virescent seedlings

$Gl\_\_$: normal leaves $gl\ gl$: glossy leaves

$Va\_\_$: normal fertility $va\ va$: variable sterility

Beadle (as reported in Emerson, Beadle, and Fraser. 1935. *Cornell University Agricultural Experiment Station Memoir* 180:1–83) testcrossed a plant that was heterozygous at these three loci with a fully homozygous recessive plant and obtained the following results:

Normal	235
Glossy, variably sterile	62
Variably sterile	40
Variably sterile, virescent	4
Glossy	7
Glossy, virescent	48
Virescent	60
Virescent, glossy, variably sterile	270
Total	726

Problem: Determine the genetic map for these three genes.

Solution: Our first task is to determine the gene order. From the original parents of the triply heterozygous F_1 plant, we know that all three alleles are in coupling conformation in this plant because one original parent was fully homozygous dominant and the other was fully homozygous recessive. There are three possible orders of genes in the triply heterozygous F_1 plant:

$$\frac{(Gl \qquad V \qquad Va)}{(gl \qquad v \qquad va)}$$

$$\frac{(V \qquad Gl \qquad Va)}{(v \qquad gl \qquad va)}$$

$$\frac{(Gl \qquad Va \qquad V)}{(gl \qquad va \qquad v)}$$

The parentheses indicate that we do not yet know the correct gene order. To determine which of these three is correct, we need to determine which one conforms to the observed double-crossover class of progeny. The class that is variably sterile, virescent, and the class that is glossy are the two double-crossover types because they include the smallest numbers of progeny.

A double crossover with the <u>Gl V Va</u> gene order produces the following:

Gl   V   Va   Gl   v   Va (virescent)
gl   v   va   gl   V   va (glossy, variably sterile)

The phenotypes do not match those that arise from double crossovers, so *Gl V Va* is not the correct gene order.

A double crossover with the *V Gl Va* gene order produces the following:

This is the correct gene order because it produces the observed double-crossover phenotypes.

The third gene order, *Gl Va V*, is incorrect because a double crossover produces the following:

Now that we know the correct gene order, we can determine the genetic map distances. We can identify the phenotypic classes by crossover type as follows:

Normal	235	No crossover
Glossy, variably sterile	62	Single crossover between *V* and *Gl*
Variably sterile	40	Single crossover between *Gl* and *Va*
Variably sterile, virescent	4	Double crossover
Glossy	7	Double crossover
Glossy, virescent	48	Single crossover between *Gl* and *Va*
Virescent	60	Single crossover between *V* and *Gl*
Virescent, glossy, variably sterile	270	No crossover
Total	726	

Let's first determine the genetic map distance between *v* and *gl*. A single crossover between *v* and *gl* causes the glossy, variably sterile phenotype and the virescent phenotype. The numbers of progeny for these two phenotypes are 62 and 60, respectively. To these, we add the number of double-crossover-type progeny and divide the sum by the total number of progeny: (62 + 60 + 4 + 7) ÷ 726 = 0.183 or 18.3 cM.

Now we need to determine the genetic map distance between *gl* and *va*. A single crossover between *gl* and *va* causes the variably sterile phenotype and the glossy, virescent phenotype. The number of progeny for these two phenotypes are 40 and 48, respectively. To these, we add the num-

ber of double-crossover-type progeny and divide the sum by the total number of progeny: (40 + 48 + 4 + 7) ÷ 726 = 0.136 or 13.6 cM. We can now draw the map:

So far, we have examined three-factor linkage analysis for alleles that are in complete coupling conformations. Genes may also be in repulsion in three-factor linkages. For example, in the chromosomes

A and *B* are in coupling conformation, but *C* is in repulsion to *A* and *B*.

When we know the genotypes of the original parents of a triple heterozygote, it is easy to tell which genes are in repulsion and which are in coupling. However, we can also tell from the testcross progeny alone which genes are in coupling and which are in repulsion, without ever knowing the genotypes of the original parents. Let's now use an example to illustrate how this is done.

Example 15.7   Three-factor linkage analysis for alleles with repulsion.

In 1922, Hutchison (*Cornell University Agricultural Experiment Station Memoir* 60:1419–1473) reported the results of a testcross for alleles at three linked loci (*c1*, *wx*, and *sh*) on chromosome 9 in maize. One parent was colored, starchy, and full, and heterozygous at all three loci, and the other parent was colorless, waxy, and shrunken. Colorless, waxy, and shrunken are the recessive phenotypes. The results of the testcross are as follows:

Colored, starchy, full	4
Colored, starchy, shrunken	2538
Colored, waxy, full	113
Colored, waxy, shrunken	601
Colorless, starchy, full	626
Colorless, starchy, shrunken	116
Colorless, waxy, full	2708
Colorless, waxy, shrunken	2
Total	6708

Problem: (a) Diagram the genotype of the colored, starchy, full parent in linkage notation. **(b)** Determine the genetic map distances.

Solution: **(a)** The first step is to find which genes are in repulsion and which are in coupling in the triply heterozygous parent. We do this by identifying the noncrossover-type classes, which are the two with the largest numbers of progeny:

Colored, starchy, shrunken (*C1 c1 Wx wx sh sh*)   2538

Colorless, waxy, full (*c1 c1 wx wx Sh sh*)   2708

From this information, we conclude that *c1* and *wx* are in coupling conformation and *sh* is in repulsion to the other two. In other words, the noncrossover chromosomes are (*C1 Wx sh*) and (*c1 wx Sh*). (Again, the parentheses indicate that we have not yet determined the correct gene order.) Knowing what the coupling-repulsion situation is, we can now determine gene order. The double-crossover types are the two classes with the fewest numbers of progeny: colored, starchy, full; and colorless, waxy, shrunken. The genotype

produces the correct double-crossover phenotypes, so it is the genotype of the heterozygous parent:

(b) Now we can determine the genetic map distances. A single crossover between *c1* and *sh* causes the phenotype colored, full, waxy, and the phenotype colorless, shrunken, starchy. The number of progeny for these two phenotypes are 113 and 116, respectively. To these, we add the number of double-crossover-type progeny and divide the sum by the total number of progeny: (113 + 116 + 4 + 2) ÷ 6708 = 0.035 or 3.5 cM.

A single crossover between *sh* and *wx* causes the phenotype colored, shrunken, waxy, and the phenotype colorless, full, starchy. The number of progeny for these two phenotypes are 601 and 626, respectively. To these, we add the number of double-crossover-type progeny and divide the sum by the total number of progeny: (601 + 626 + 4 + 2) ÷ 6708 = 0.184 or 18.4 cM. We can now draw the map:

Three-factor linkage analysis is used to determine the order and genetic map distances among three linked genes in the progeny of a single testcross.

Interference

After map distances have been calculated in three-factor linkage analysis, we can predict the frequency of double crossovers if there is no interference. Using Example 15.7, we determine that the expected frequency of crossovers between *c1* and *sh* is 0.035, and between *sh* and *wx* it is 0.184. According to the product rule, the expected frequency of double crossovers (one between *c1* and *sh* and one between *sh* and *wx*) is 0.035 × 0.184 = 0.00644, assuming that there is no interference. The total number of progeny is 6708, so the expected number of double-crossover progeny is 0.00644 × 6708 = 43.2. The observed number of double-crossover progeny is 6. In nearly every three-factor linkage experiment, the number of double-crossover progeny is less than expected because of interference, the tendency of a chiasma to interfere with the formation of another chiasma in the same vicinity. Interference is quantified as the **coefficient of coincidence (C)**, which is calculated as follows:

$$C = \frac{\text{observed number of double crossovers}}{\text{expected number of double crossovers}}$$

$$= \frac{\text{observed frequency of double crossovers}}{\text{expected frequency of double crossovers}} \quad [15.3]$$

Using the data from Example 15.7, we determine that the coefficient of coincidence is $C = 6 \div 43.2 = 0.139$, which means that interference reduced the frequency of double crossovers to 13.9% of the frequency we would expect assuming no interference. Interference (*I*) is defined as

$$I = 1 - C \quad [15.4]$$

so in this example, $I = 1 - 0.139 = 0.861$.

Positive interference is defined as $I > 0$, which is the usual situation in three-factor linkage analysis. Complete interference is defined as $I = 1$ (no observed double crossovers). When three genes are very closely linked, complete interference is not uncommon. Negative interference is defined as $I < 0$, and is hardly ever observed.

~

The coefficient of coincidence (C) is calculated as the observed frequency of double crossovers divided by the expected frequency of double crossovers. Interference (I) is calculated as 1 − C.

Mapping Genes on Heteromorphic Chromosomes

Recall from Chapter 14 that sex-determining chromosomes such as X and Y chromosomes or Z and W chromosomes are heteromorphic. Mapping genes on the X chromosome in a species with an XX-XY system of sex determination requires essentially the same procedure as mapping genes on autosomes, with one exception: heterozygous females must be testcrossed with hemizygous recessive males. Crossovers in the X chromosome take

place only in females; because males have only one X chromosome, there is no corresponding homologus X chromosome for crossing-over.

The same procedure applies to species with an XX-XO system of sex determination. For mapping testcrosses, the XX females must be heterozygous for the loci being mapped, and they are testcrossed with recessive XO males. In *Drosophila*, there is no crossing-over in the male, so regardless of the chromosome being mapped, all linkage analysis experiments in *Drosophila* must be designed with the female as the heterozygous parent. In species with a ZZ-ZW system of sex determination, the situation is reversed. To map the Z chromosome, the male must be the heterozygous parent that is testcrossed with a hemizygous recessive female.

Let's now look at an example of linkage analysis of genes on the X chromosome in *Drosophila melanogaster*.

Example 15.8 **Three-factor linkage analysis of alleles on the X chromosome in *Drosophila melanogaster*.**

Three loci, *vermilion* (*v*), *cut* (*ct*), and *garnet* (*g*) are linked on the X chromosome in *Drosophila melanogaster*. In 1926, Bridges and Olbrycht (*Genetics* 11:41–56) reported the following results from a cross between a female heterozygous for all three recessive alleles and a male hemizygous for all three recessive alleles:

Wild type	1370
Vermilion, cut, garnet	1015
Cut	249
Vermilion, garnet	254
Garnet	185
Vermilion, cut	159
Vermilion	9
Cut, garnet	8
Total	3249

Problem: Estimate the map distances between these three genes and the interference.

Solution: First of all, we know that all three genes are in coupling conformation, as indicated by the distribution of phenotypes in the progeny. The next step is to determine the correct gene order, which is

$$\begin{array}{ccc} + & + & + \\ \hline \hline ct & v & g \end{array}$$

The distance between *ct* and *v* is (249 + 254 + 9 + 8) ÷ 3249 = 0.160 or 16.0 cM. The distance between

v and *g* is (185 + 159 + 9 + 8) ÷ 3249 = 0.111 or 11.1 cM. The expected frequency of double crossovers is 0.160 × 0.111 = 0.01776. The observed frequency is 17 ÷ 3249 = 0.00523. The coefficient of coincidence is $C = 0.00523 \div 0.01776 = 0.2945$. The interference is $I = 1 - 0.2945 = 0.7055$.

15.5 MULTIFACTOR CHROMOSOME MAPPING

It is possible to calculate map distances with four or more linked genes by extending the procedures used for three-factor mapping. With each added factor, the number of possible genotypes (and phenotypes) in testcross progeny doubles. This tends to make mapping beyond a three-factor cross rather cumbersome when done by hand. For example, there are 16 possible genotypes with four-factor linkage. In four-factor linkage, the coupling-repulsion relationships of four genes must be determined and noncrossover-, single-, double-, and triple-crossover-type progeny must be identified.

Fortunately, there are computer programs that can take information for any number of segregating genetic markers in a cross, determine which are linked and which assort independently, and construct a genetic map from the data. Genetic markers may include alleles that confer outward phenotypic differences or DNA markers (such as RFLPs, minisatellites, microsatellites, and RAPDs).

Genetic markers are initially mapped to linkage groups. A **linkage group** is a chromosomal segment that contains two or more linked genetic markers. As large numbers of genetic markers are mapped in a species, linkage groups increase in size and may be coalesced as markers from one linkage group are found to be linked to markers in another linkage group (Figure 15.15). Eventually, the number of linkage groups equals the haploid number of chromosomes, and each linkage group represents one chromosome.

Before computer programs for mapping were available, information from many different crosses was combined to form linkage groups, which were eventually coalesced into chromosome maps like the one shown in Figure 15.3. These chromosome maps took decades to construct. Now it is possible using DNA markers and computer programs to construct chromosome maps in a few months.

DNA markers are now used routinely for genetic map construction. For traditional gene mapping, at least two polymorphic markers must be present at a locus. **Polymorphic markers** are genetic markers that are distinguishable as different alleles of a single locus. The word *polymorphic* means literally "of many forms." Polymorphic markers may be alleles that differ in some identifiable respect from each other. For example, the alleles that caused purple and white flowers in Mendel's

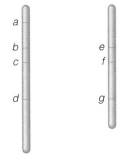

linkage group 1   linkage group 2

a Two linkage groups contain different sets of linked genes.

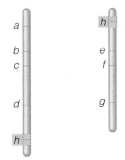

linkage group 1   linkage group 2

b Gene *h* is linked to *d* and also to *e*. Genes *d* and *e* are not linked, but are syntenic.

c Linkage groups 1 and 2 are coalesced into a single linkage group.

Figure 15.15 Coalescence of two linkage groups into a single linkage group.

peas are polymorphic to each other since they cause an identifiable difference. DNA markers, such as RFLPs or RAPDs, are also polymorphic. Polymorphic DNA markers are homologous segments of DNA located at the same position on homologous chromosomes that can be distinguished from one another because of differences in DNA sequence. As the chromosomes on which these segments of DNA reside are inherited in a Mendelian fash-

ion, polymorphic markers are likewise inherited in a Mendelian fashion, and linked DNA markers behave genetically just like linked genes.

Linkage analysis with DNA markers is no different than analysis with traditional genetic markers. There are some distinct advantages to DNA markers, however, that allow rapid chromosome mapping. The main advantage is the large number of potential markers available. In many species, a genetic map of all chromosomes may be made in the progeny of a single cross using DNA markers; many different crosses are typically required for traditional chromosome mapping. When DNA markers are used, a map can be constructed quickly by identifying DNA markers that are polymorphic between the two parents and mapping those markers in the progeny. With traditional morphological markers (markers that cause a difference in the appearance of the organism), this is not possible in a single cross. With maize, for example, one cannot expect two single plants to differ in kernel shape, color, and texture; leaf shape, pattern, texture, and color; plant height and growth habit; male fertility; disease resistance; etc., without going through years of effort to breed differences for these traits into single plants. However, the number of DNA markers that can be tested for polymorphism numbers in the thousands, so it is often possible to identify large numbers of polymorphic DNA markers in two parents and map all of them in the progeny of a single cross. Figure 15.16 shows a map of DNA markers and morphological markers in chromosome 9 of maize, as generated by a computer program called Mapmaker. Notice that the *c1*, *sh*, and *wx* loci we studied in Example 15.7 are mapped relative to DNA markers.

~

Multifactor chromosome mapping is typically done using computer programs that construct genetic maps from data sets. With DNA marker analysis, it is possible to construct a genetic map of all chromosomes from the progeny of a single cross.

15.6 GENETIC MAPPING IN HUMANS

Humans are not well suited to genetic mapping with traditional methods. Numbers of offspring tend to be small compared to other species used for genetic analysis, and human crosses cannot be designed to meet experimental needs. Even with these limitations, however, statistical methods for use in human genetic mapping were developed in the 1950s. Unfortunately, only limited mapping could be done because of an insufficient number of polymorphic genetic markers. With extensive efforts, some genes were mapped on the X chromosome, and a few autosomal linkages were found. Several methods for physical mapping of human chromosomes emerged and were used to establish the physical (but not the genetic) locations of some genes on their respective chromosomes. We will discuss physical chromosome mapping in the

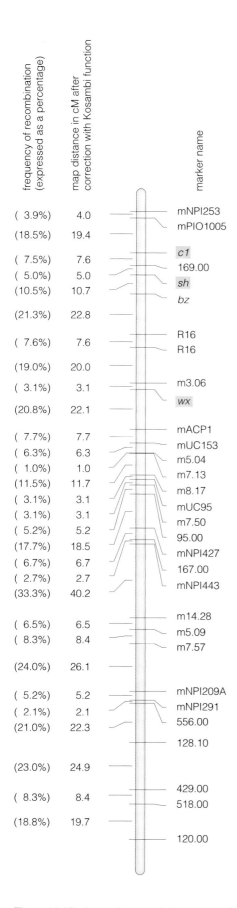

frequency of recombination (expressed as a percentage)	map distance in cM after correction with Kosambi function		marker name
(3.9%)	4.0		mNPI253
(18.5%)	19.4		mPIO1005
(7.5%)	7.6		*c1*
(5.0%)	5.0		169.00
(10.5%)	10.7		*sh*
			bz
(21.3%)	22.8		
(7.6%)	7.6		R16
			R16
(19.0%)	20.0		
(3.1%)	3.1		m3.06
(20.8%)	22.1		*wx*
(7.7%)	7.7		mACP1
(6.3%)	6.3		mUC153
(1.0%)	1.0		m5.04
(11.5%)	11.7		m7.13
(3.1%)	3.1		m8.17
(3.1%)	3.1		mUC95
(5.2%)	5.2		m7.50
(17.7%)	18.5		95.00
(6.7%)	6.7		mNPI427
(2.7%)	2.7		167.00
(33.3%)	40.2		mNPI443
(6.5%)	6.5		m14.28
(8.3%)	8.4		m5.09
			m7.57
(24.0%)	26.1		
(5.2%)	5.2		mNPI209A
(2.1%)	2.1		mNPI291
(21.0%)	22.3		556.00
			128.10
(23.0%)	24.9		
			429.00
(8.3%)	8.4		518.00
(18.8%)	19.7		
			120.00

Figure 15.16 A genetic map of chromosome 9 in maize. The map contains DNA markers and morphological markers. The *c1*, *sh*, and *wx* loci are highlighted.

next section. While extensive genetic maps of model organisms, such as *Drosophila*, mouse, and maize, were being developed over a period of decades, human genetic mapping remained almost stagnant.

All of this changed when DNA marker analysis became available. Tens of thousands of polymorphic tandem repeat markers, such as minisatellites and microsatellites, can be detected in humans. Most tandem repeat markers are codominant, so homozygotes and heterozygotes can be readily distinguished from one another. In order for a pedigree to provide information that can be used in mapping, at least one parent must be doubly heterozygous for linked markers. Because large numbers of tandem repeat markers are available, researchers simply need to look for many tandem repeat markers that are heterozygous in at least one parent.

Statistical analysis can be done in markers derived from individuals in multigenerational pedigrees to determine whether any two polymorphic markers assort independently or are linked. If they are linked, map distances can be estimated by identifying the numbers of recombinant and nonrecombinant chromosomes wherever possible in the pedigrees examined. When this sort of analysis is done with large numbers of markers and many individuals from known pedigrees, genetic maps of human chromosomes can be constructed. Although the statistical methods used for constructing human genetic maps are more complicated than those used in designed testcrosses, the fundamental calculations for determining map distances based on the frequency of recombination between linked markers are essentially the same.

DNA markers have been used to construct detailed human genetic maps. Figure 15.17 is an example of one chromosome from a map of microsatellite markers in humans. These maps are made from DNA sampled from individuals in a set of reference families where pedigree information showing the relationships among individuals in the reference families has been determined. The pedigrees consist of multiple families and multiple generations. Centers for human genetic mapping use the same reference families and share data so that information collected in one center can be combined with information collected in other centers to construct a single genetic map. Detailed information derived from human genetic maps can now be accessed freely over the Internet by any interested person. Links to human chromosome maps, as well as several other important sites providing genetic information, can be found at the website for this book, http://www.brookscole.com/biology.

15.7 PHYSICAL CHROMOSOME MAPPING

Genetic chromosome maps show the relative locations of genes on a chromosome as determined by the frequencies of recombination between linked genes. The

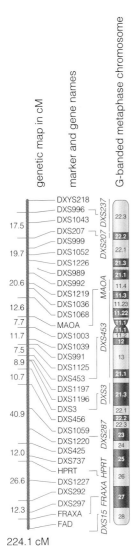

224.1 cM

Figure 15.17 A genetic map of the human X chromosome. Most of the mapped markers are microsatellites. Four genes (*MAOA*, *HPRT*, *FRAXA*, and *FAD*) are also included. Nine of the markers have been mapped to their physical locations on the G-banded metaphase chromosome. (Adapted from Buetow, K. H. et al. 1994. Human genetic map: Genome maps V. *Science* 265:2055–2070. Copyright 1994 American Association for the Advancement of Science.)

genetic map distances in such maps, however, do not necessarily represent physical distances between genes on a chromosome. Several methods permit geneticists to identify the physical locations of genes on chromosomes observed microscopically. In this section, we will explore the most widely used methods for physical chromosome mapping.

Deletion Mapping

In some species, chromosomes have been well characterized microscopically. As discussed in section 10.3, certain stains used in microscopy reveal banding patterns

on chromosomes. Using banding patterns, researchers can identify not only chromosomes but chromosomal regions. A deletion is a mutation in which a segment of a chromosome is missing. Small chromosomal deletions usually include large amounts of DNA that may contain several genes. Many deletions are lethal, particularly if they are large and include several genes. However, some deletions are small enough that they are not lethal when heterozygous, because the homologous chromosome carries copies of the genes in the deleted region. Most deletions are lethal when homozygous.

Many deletions can be identified microscopically in chromosomes as missing regions in the banding pattern. **Deletion mapping** is the identification of genes located in the deleted region of a deletion heterozygote. An individual that is homozygous for a recessive point-mutation allele in the gene of interest is crossed with an individual that is heterozygous for a deletion. If the gene is within the deleted region, half of the F_1 progeny display the recessive phenotype, as explained in Figure 15.18.

Appearance of the recessive phenotype in half of the progeny is a good indication that the gene lies within the deletion and can be assigned to that chromosomal region. An example of this is the *white* locus in *Drosophila melanogaster*. Several deletions that are homozygous lethal in females and hemizygous lethal in males have been found in the *white* region on the X chromosome and can be readily identified in the polytene chromosomes as missing bands. When female flies that are heterozygous for a deletion are crossed with flies with a recessive allele at the *white* locus, half of the female progeny display the mutant phenotype. Such a result leads researchers to conclude that at least part of the *white* locus is within the deleted DNA.

Somatic Cell Hybridization Mapping

A special type of deletion mapping has been used to physically map genes on human chromosomes. **Somatic cell hybridization** is the fusion of human cells with rodent cells (such as mouse or Chinese hamster cells) growing in culture. A fused cell initially contains all the rodent and human chromosomes. However, as the cell prepares to divide, the rodent chromosomes replicate and enter mitosis more quickly than the human chromosomes. As the cells divide, some of the unreplicated human chromosomes are lost. After several cell divisions, a cell line can be grown from a single cell. Chromosomes are lost at random, so a cell line may contain a particular combination of human chromosomes, and another cell line may contain a different combination. The genes on human and rodent chromosomes are expressed in the cell lines. In some cases, enzymes differ enough in rodent and human cells that they can be distinguished from each other. When a cell line lacks a particular human enzyme, the gene for that enzyme must be located on a chromosome that is missing from the cell

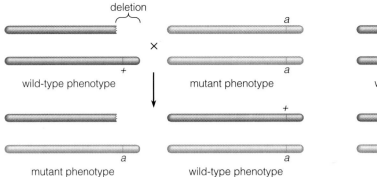

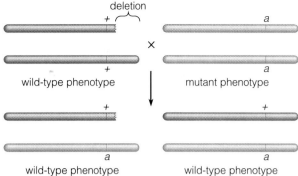

a If gene *a* is in the deleted region, half of the progeny have the deletion and display the mutant phenotype.

b If gene *a* is ouside of the deleted region, half of the progeny have the deletion, but all display the wild-type phenotype.

Figure 15.18 Deletion mapping. Cross the deletion heterozygote that carries the dominant wild-type allele (+) with an individual that is homozygous for a recessive point-mutation allele of gene *a*. Two outcomes are possible.

line. When a cell has a particular human enzyme activity, the gene for that enzyme must be located on a chromosome that is present in the cell line. By determining which chromosomes are present and absent in cell lines with and without a human enzyme, researchers can eliminate certain chromosomes and eventually narrow the possibilities to one chromosome for the gene specifying the enzyme under study.

In somatic hybrid cell culture, deletions are more frequent in human chromosomes. Also, deletions can be induced in hybrid cell lines by treating them with ionizing radiation. When a human chromosome is known to carry a certain enzyme gene, and a portion of the chromosome is deleted, the enzyme activity is lost if the gene is located within the deletion. If the enzyme is not lost after the deletion, then the gene that encodes the enzyme must lie outside the deleted region. Geneticists have used somatic cell hybridization and deletion mapping to determine the physical locations of over 2000 genes on human chromosomes. Figure 15.19 shows the physical locations of several genes on the X chromosome, many of which were located with somatic cell hybridization and deletion mapping.

Somatic cell hybridization mapping has played a very important role in the human genome mapping effort. For many years, somatic cell hybridization maps were the only detailed human genetic maps available, and they provided much information about the structure and organization of human chromosomes. DNA marker mapping (described in the previous section) and in situ hybridization mapping (described later in this section) have now moved to the forefront of human genetic mapping because researchers can map more genes in shorter periods of time with these methods than with somatic cell hybridization.

Let's now see how somatic cell hybridization analysis is used to identify which chromosome contains a human gene.

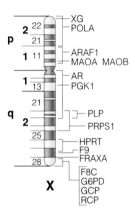

Figure 15.19 Physical map of the human X chromosome. (Adapted from McKusick, V. A. 1992. *Mendelian Inheritance in Man*, 10th ed. Baltimore, Md.: The Johns Hopkins University Press. Copyright 1992.)

Example 15.9 Identification of the human chromosome that carries a particular gene using somatic cell hybridization.

In 1985, Carson (*Science* 229:991–993) used human-mouse cell hybrids to identify the chromosome that carries the human gene *F3* (OMIM 134390), which encodes blood coagulation factor III. Table 15.2 shows data obtained from several cell lines.

Problem: On which human chromosome is the gene that encodes factor III located?

Solution: A chromosome that is present in a hybrid cell line that lacks human factor III is not the chromosome that carries the *F3* gene. Conversely, a chromosome that is absent in a cell line that has factor III is not the chromosome that carries the *F3* gene. Hybrid cell lines WIL1, WIL6, WIL7, TSL1,

Table 15.2 Information for Example 15.9: Somatic Cell Hybridization Analysis		
Hybrid Cell Line	Human Factor III Detection in Cell Line ("+" = factor III present, "−" = factor III absent)	Human Chromosomes Present in Hybrid Cell Line
WIL1	−	8, 14, 17, 21, X
WIL6	−	2, 4, 5, 6, 7, 8, 10, 11, 14, 17, 19, 20, 21, X
WIL7	−	2, 3, 5, 6, 8, 10, 11, 13, 14, 17, 18, 21, X
WIL14	+	1, 3, 7, 8, 10, 12, 14, 15, 17, X
SIR3	+	1, 2, 3, 4, 5, 6, 7, 9, 10, 11, 12, 13, 16, 17, 18, 19, 20, 21, 22, X
TSL1	−	3, 4, 10, 11, 13, 14, 16, 17, 18, 20
TSL2	−	2, 3, 5, 6, 10, 12, 17, 18, 21, 22, X

and TSL2 do not produce factor III. Chromosomes 1, 9, and 15 are missing from all of these cell lines, so the *F3* gene must be on one of these three chromosomes. Hybrid cell lines WIL14 and SIR3 produce factor III and both contain chromosome 1. Line WIL14 lacks chromosomes 9, and line SIR3 lacks chromosome 15. Therefore, the *F3* gene must be located on chromosome 1.

Physical Chromosome Mapping with in situ Hybridization

In section 10.3 we discussed the use of in situ hybridization to identify the locations of genes in chromosomes. A labeled probe that contains some of the DNA sequence from a gene can be hybridized with its complementary sequence in the DNA of a chromosome. The probe can then be detected at its position in the chromosome to reveal the physical location of the gene on the chromosome. In situ hybridization to polytene chromosomes in certain insect species (such as *Drosophila melanogaster*) has been routine for many years (see Figure 10.21). The development of fluorescent in situ hybridization (FISH) has made it possible to identify the physical locations of genes in human chromosomes (see Figure 10.22).

Correspondence of Genetic and Physical Maps

Physical chromosome maps correspond with genetic chromosome maps for the order of genes on the chromosome, but genetic map distances do not correspond closely with physical map distances. Varied levels of condensation within chromosomes observed microscopically and variation in crossover frequencies throughout a chromosome are two factors that contribute to the lack of correspondence between physical and genetic map

distances. Genetic map distances may also vary significantly among species. In yeast, each cM unit averages about 3000 nucleotide pairs, whereas in humans each cM unit averages about 1,000,000 nucleotide pairs. Therefore, we cannot assume that genetic map units correspond to a constant physical distance.

~

Genes can be mapped to their physical locations on the chromosome using deletion mapping or in situ hybridization. Although the order of genes in genetic maps corresponds with the order in physical maps, genetic map distances do not necessarily correspond with physical map distances.

15.8 PRACTICAL APPLICATIONS OF CHROMOSOME MAPPING

Genetic maps, particularly maps that include DNA markers, are used routinely in medicine and agriculture. DNA markers that are coupled to an allele of an important gene can be used to identify the allele and track its inheritance. DNA markers can also be used as a starting point to find and clone the DNA of a gene that is closely linked to the marker. In this section, we look at how researchers have used such applications in medicine and agriculture.

Tracking the Inheritance of an Allele with Coupled DNA Markers

Sometimes it is not easy to determine whether or not an individual carries a certain allele. Ideally, the DNA sequence of the gene can be used to design a molecular marker that can detect the mutant allele directly. For example, the *HBB*S* mutant allele that causes sickle-cell anemia can be detected directly with RFLPs, as described

in section 13.4. However, until at least a portion of the gene has been cloned, it is not always possible to detect the gene directly using DNA markers. In these cases, a polymorphic DNA marker that is closely coupled to a mutant allele can be used to track the inheritance of the allele. The mutant allele responsible for Huntington disease is a good example. In 1983, before the gene responsible for Huntington disease had been cloned and sequenced, an RFLP marker was identified that was closely coupled to the mutant allele. Clinical geneticists used this marker to determine whether people who had a parent with Huntington disease were heterozygous carriers of the allele. The gene responsible for Huntington disease has been cloned and sequenced, so a DNA test that detects mutant alleles directly, instead of a coupled marker, is now available, as described in Example 13.5.

Molecular characterization of human genes has progressed at a rapid pace. The genes for most major genetic disorders have been cloned and sequenced, and DNA tests that detect mutant alleles directly are now available. However, until a gene has been cloned and sequenced, a coupled DNA marker is usually the best choice for tracking the inheritance of alleles of that gene.

Plant and animal breeders routinely track important alleles with coupled DNA markers. After a genetic map consisting of DNA markers has been established in an important agricultural species, plant and animal breeders can quickly map genes of economic importance and find DNA markers that are closely coupled to alleles of those genes. Once closely coupled DNA markers have been found, it is possible to select for the marker and, in so doing, indirectly select for the allele.

For example, many agricultural plants are susceptible to certain diseases. In most cases, the best method for controlling the disease is to use plants that are genetically resistant to it. When disease-resistant plants are used, there is no need to fight the disease with chemicals that may be expensive and potentially harmful to people or the environment. Plant breeders must screen plants for resistance to common diseases and eliminate all that are susceptible in early generations. To determine whether or not a plant is susceptible to the disease, breeders usually cultivate the disease organism and inoculate the plants with it. Sometimes this must be done in the field, and field testing may pose unwanted risks because the disease-causing organism may be released into the environment, where it may infect other plants. However, if polymorphic DNA markers that are closely coupled to the alleles that confer resistance can be identified, DNA can be extracted from a small amount of leaf tissue in plant seedlings, and susceptible and resistant plants identified by testing for the DNA marker. The plants do not need to be inoculated with the disease-causing organism. Large numbers of plants may be screened in this way, thereby avoiding the risks and expenses associated with large-scale inoculation.

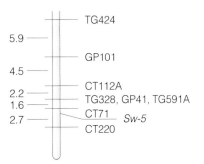

Figure 15.20 Genetic map of the telomeric region of the long arm of chromosome 9 in tomato showing the *Sw-5* locus and linked DNA marker loci. (Adapted from Stevens, M. R., E. M. Lamb, and D. D. Rhoads. 1995. Mapping the *Sw-5* locus for tomato spotted wilt virus resistance in tomatoes using RAPD and RFLP analyses. *Theoretical and Applied Genetics* 90:451–456.)

A DNA marker that is closely coupled to an allele can be used to identify nearly all individuals that carry the allele. A few individuals, however, may be misclassified by detection of the DNA marker because a crossover separated the DNA marker from the allele. Misclassification can be prevented by the use of two flanking DNA markers, markers that are located on either side of the allele. When flanking markers are used, only double crossovers separate both markers from the allele. Often flanking markers are so close to the allele that interference prevents double crossovers, and the allele can be identified with complete accuracy. This form of selection, in which genetic markers are used to indirectly select for a coupled allele, is called **marker-assisted selection**.

The following example shows how marker-assisted selection is used to select for an allele that confers disease resistance in tomato.

Example 15.10 Identification of DNA markers coupled to the *Sw-5* allele in tomato.

The *Sw-5* allele is a dominant allele that confers resistance to tomato spotted wilt virus, a disease that causes serious losses in commercial tomato production. In 1995, Stevens, Lamb, and Rhoads (*Theoretical and Applied Genetics* 90:451–456) identified several DNA markers that were closely coupled to the *Sw-5* allele. Figure 15.20 shows a map of the *Sw-5* locus and linked DNA marker loci near the telomere of chromosome 9 in tomato.

Problem: How can the DNA markers be used for marker-assisted selection of the *Sw-5* resistance allele?

Solution: The CT71 marker locus is closely linked to the *Sw-5* locus, so polymorphic CT71 markers that are coupled to the resistance allele can be used

in a breeding program to identify resistant plants. The CT71 and CT220 marker loci flank the *Sw-5* locus, so markers at the CT71 and CT220 loci can be used to identify crossovers that separate the resistance allele from its coupled marker loci.

Chromosome Walking and Jumping

When tracking the inheritance of genes with DNA markers, it is better to have a DNA clone of the gene itself, rather than a closely linked marker. Might it be possible to use the linked DNA marker as a starting point to work along the DNA and eventually arrive at the gene? Indeed, some important genes have been cloned in this way using procedures called chromosome walking and chromosome jumping. Let's use the gene responsible for cystic fibrosis in humans as an example.

Cystic fibrosis is the most common genetic disorder in humans of European ancestry. The disorder disrupts normal functioning of the exocrine system, including the pancreas, sweat glands, and glands that produce mucus. Its name derives from the fibrous cysts that form in the pancreas as pancreatic ducts become clogged. Sweat glands release abnormal amounts of salt, and mucus glands secrete excessively thick mucus. The thick mucus accumulates in the lungs, where it often facilitates bacterial infection. The most common cause of death in patients with cystic fibrosis is bacterial infection because of lung congestion. The disorder cannot be cured, but it can be treated with several types of therapy, including antibiotic treatments to prevent infection. Children who are treated may live into their teens. However, most people with cystic fibrosis die before reaching adulthood.

For decades researchers have known that the disorder was due to homozygosity for a recessive autosomal allele. In 1985, the gene was mapped to the long arm of chromosome 7. Little was known about the cellular mechanism of the disorder. The DNA sequence, mRNA, and protein product of the gene were unknown at the time, so there was no information to develop a probe for cloning.

Several linked RFLP loci had been identified, however. One of the RFLP loci was used as a starting point to work toward the gene through chromosome walking and chromosome jumping. **Chromosome walking** is the reconstruction of a long DNA sequence from many cloned segments of DNA by identifying overlapping clones. One clone, whose position on the chromosome is known, may hybridize with a second clone whose position is not known. The second clone has a portion of its sequence in common with the first clone, so their DNA sequences overlap. Part of the sequence in the second clone may extend beyond the limits of the first clone. Once the correct position of the second clone is established relative to the first clone, the second clone can be used to identify a third clone with an overlapping sequence. By repeating the process clone by clone, it is possible to "walk" along a chromosome from a closely linked DNA marker locus to the target gene locus, without knowing the gene's sequence (Figure 15.21).

Chromosome jumping makes it possible to cover large regions of DNA from a clone with a known position to a second clone whose position is unknown. It can help speed up the process of chromosome walking. The procedure is illustrated in Figure 15.22. Unlike chromosome walking, chromosome jumping can detect two linked segments of DNA that do not overlap and are separated by up to tens of thousands of nucleotide pairs. Chromosome jumping is especially valuable when the distances between a linked marker and the target gene are large (several hundred thousand nucleotide pairs). For chromosome jumping, DNA is digested with a **rare cutting restriction enzyme**, an enzyme that cuts DNA at an eight-nucleotide target sequence. On average, an enzyme with an eight-nucleotide target sequence cuts every 65,536 nucleotide pairs. The DNA fragments generated with a rare cutting restriction enzyme are usually tens of thousands of nucleotide pairs long (Figure 15.22a). The large fragments are mixed with a gene for a suppressor tRNA (see the final paragraph of section 5.1 for an explanation of suppressor tRNAs), and the fragments are treated to encourage their arrangement into circles. Some of the circular DNAs contain large DNA fragments connected to the suppressor tRNA gene (Figure 15.22b). The mixture is treated with the restriction enzyme *Eco*RI, which does not have a cleavage site

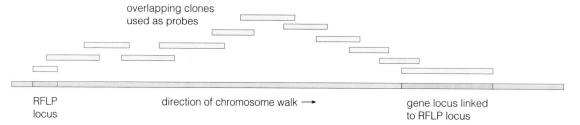

overlapping clones
used as probes

RFLP
locus

direction of chromosome walk →

gene locus linked
to RFLP locus

Figure 15.21 Identifying a gene for cloning by chromosome walking. The cloned probe for an RFLP locus is used to identify cloned DNA in a genomic library that overlaps the RFLP locus. The overlapping probe is then used to identify another fragment of cloned DNA that overlaps it. The process is repeated so that researchers can "walk" along the chromosome with overlapping clones until they reach the gene.

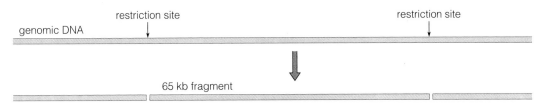

a Digest genomic DNA with rare cutting restriction enzyme. Fragments are large, on average about 65 kb.

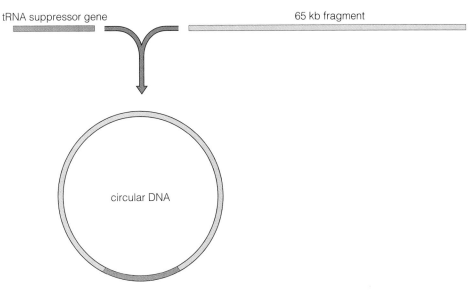

b Add tRNA suppressor gene to digested fragments to circularize them.

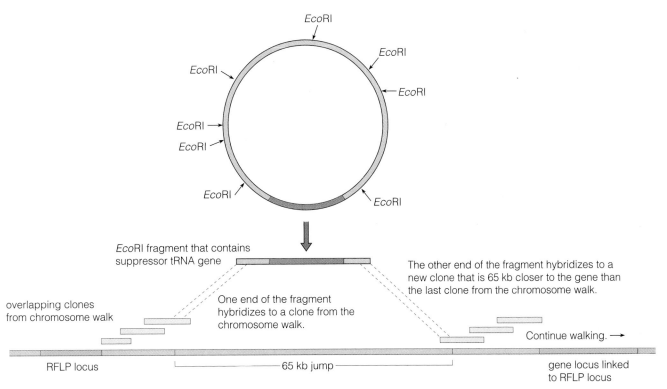

c Digest circular DNA with *Eco*RI and identify a fragment that hybridizes to a clone from the chromosome walk. The other end of the fragment hybridizes to a clone that is linked by 65 kb to the clone from the chromosome walk.

Figure 15.22 Chromosome jumping.

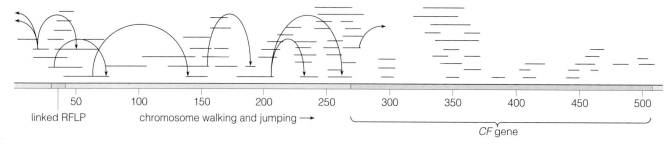

| 50 | 100 | 150 | 200 | 250 | 300 | 350 | 400 | 450 | 500 |

linked RFLP   chromosome walking and jumping ⟶

CF gene

Figure 15.23 Chromosome walking and jumping to arrive at the cystic fibrosis gene in humans from a linked RFLP locus. The horizontal lines show the positions of overlapping clones used for chromosome walking, and the arcs show the chromosome jumps. The numbers show the distances on the DNA molecule in kb. (Adapted from a drawing by Lap-Chee Tsui published in Marx, J. L. 1989. The cystic fibrosis gene is found. *Science* 245:923–925.)

in the tRNA suppressor gene. Because the circular DNA fragments are so large, they usually have several *Eco*RI cleavage sites. The enzyme trims the DNA on either side of the tRNA gene, leaving the tRNA gene flanked on either side by the ends of the large fragment (Figure 15.22c). This fragment is cloned into a phage vector with a nonsense mutation that is suppressed by the tRNA suppressor gene. The tRNA suppressor gene acts as a selectable marker and allows only those phages that contain the tRNA suppressor gene to replicate. If one of the ends of the cloned DNA hybridizes to a clone that has been positioned in the chromosome walk, then the DNA on the other side of the tRNA gene should hybridize to another clone that is linked to the first clone, but at a large distance (tens of thousands of nucleotide pairs). The DNA on the other end can then be used for continued walking or for additional jumps.

Figure 15.23 shows the overlapping clones and the chromosome jumps that Lap-Chee Tsui, Francis Collins, and their collaborators used to find and clone the human *CF* (cystic fibrosis) gene (OMIM 219700).

~

Mapped DNA markers can be used for marker-assisted selection, genetic testing, and chromosome walking and jumping.

SUMMARY

1. Linked genes are located on the same chromosome and do not assort independently. Genes that are located on the same chromosome are syntenic but are not necessarily linked because genes that are distant on the same chromosome may assort independently.

2. Alleles of linked genes may be in coupling or repulsion conformation. Alleles that are coupled are on a single chromosome of a homologous pair of chromosomes. Alleles in repulsion to one another are on different chromosomes of a homologous chromosome pair.

3. Analysis of two linked genes is called two-factor linkage analysis. Recombinant chromosomes are those that arise from a crossover between the two genes under consideration. Nonrecombinant chromosomes are those that do not arise from a crossover between the two genes under consideration and are, therefore, the same as the original parental chromosomes.

4. Linkage is best analyzed in a testcross situation in which recombinant and nonrecombinant chromosomes are represented by recombinant and nonrecombinant phenotypes in the progeny. Under these conditions, the recombination frequency can be determined by totaling the number of recombinant progeny and dividing by the total number of progeny.

5. Linkage may also be measured in F_2 progeny, although the procedure is not as accurate as in testcross progeny. Testcross situations should be used whenever possible for linkage analysis.

6. Genetic map distances may be underestimated when recombination frequencies exceed 0.2 because of undetected double crossovers. Underestimations can be corrected with mapping functions.

7. Three-factor and higher linkages can be calculated in progeny of the same cross. It is also possible to estimate the degree of interference in three-factor crosses.

8. Genetic map distances can be combined to construct a genetic map. Linkage groups are initially identified and can be coalesced into maps of individual chromosomes as additional linkages are identified.

9. Human chromosome mapping progressed very slowly until DNA marker analysis was available. With DNA marker analysis, chromosome mapping in humans has progressed at a rapid pace, and many genes have now been mapped.

10. Chromosome mapping has several practical applications, including marker-assisted selection, DNA testing, chromosome walking, and chromosome jumping.

QUESTIONS AND PROBLEMS

1. A North American wild strawberry, *Fragaria bracteata*, has 28 chromosomes in its somatic cells. How many linkage groups should there be once all linkage groups have been coalesced as far as possible?

2. In chickens, the allele *F* confers normal feathers and is dominant to the allele *f*, which causes frizzled feathers when homozygous. At a linked locus, the dominant allele *I* causes white feathers, and its recessive counterpart *i* causes black feathers when homozygous. A third locus that is linked to the other two has two alleles: *Cr*, which is dominant and causes crested plumage, and *cr*, which is recessive and causes noncrested plumage when homozygous. In 1935, Warren and Hutt (*American Naturalist* 70:379–394) reported the results listed in Table 15.3 from three separate two-factor testcrosses. **(a)** For each of the testcrosses, determine whether the alleles were in repulsion or coupling conformation in the heterozygous parent. **(b)** Which of the three alleles is between the other two in the linkage map? **(c)** Write the recombination frequencies for the genes in each testcross with no correction, and after correction with the Haldane and Kosambi functions.

Recombination Frequencies

Genes	No Correction	Haldane Function	Kosambi Function
Testcross 1	————	————	————
Testcross 2	————	————	————
Testcross 3	————	————	————

(d) Indicate which of the three calculations (no correction, correction with the Haldane function, or correction with the Kosambi function) provides the most consistent map (best additivity of map distances among genes) when data from the three testcrosses are combined.

3. How do undetected crossovers affect recombination frequencies between linked genes?

4. On the X chromosome of *Drosophila melanogaster*, loci are assigned positions that depend on their map distance in centimorgans from the *y* (*yellow*) locus, which is located near the telomere on the long arm. The *y* locus is located at position 0.0, the *v* (*vermilion*) locus is at position 33.0, and the *r* (*rudimentary*) locus is at position 54.5. Assuming that the Kosambi function applies to mapping the X chromosome in *Drosophila*, predict the expected frequencies of phenotypes in the progeny from a cross between a yellow, vermilion, rudimentary male and a wild-type female that is heterozygous for the recessive alleles at all three loci.

5. Three loci, *C*, *B*, and *Y*, are linked in rabbits. The *C* allele causes a rabbit to be fully colored, whereas *c* causes the rabbit to be Himalayan when homozygous. The *B* allele causes black fur color, whereas *b* causes brown fur color when homozygous. The *Y* allele causes white body fat, whereas *y* causes yellow body fat when homozygous. In 1933, Castle (*Genetics* 19:947–950) reported the following data from a testcross:

Phenotypes	Genotype(s) of Gametes from Heterozygous Parent	Number of Progeny
Black, full-colored, white fat	————————	33
Black, full-colored, yellow fat	————————	48
Black, Himalayan, white fat	————————	151
Black, Himalayan, yellow fat	————————	2
Brown, full-colored, white fat	————————	11
Brown, full colored, yellow fat	————————	142
Brown, Himalayan, white fat	————————	67
Brown, Himalayan, yellow fat	————————	23

(a) Write the genotype of the heterozygous parent in linkage notation. **(b)** Fill in the blanks above with the

---

Table 15.3 Information for Problem 2: Three Separate Two-Factor Testcrosses

Testcross 1	Testcross 2	Testcross 3
168 frizzled feathers, white	129 frizzled feathers, noncrested	321 white, noncrested
172 normal feathers, black	128 normal feathers, crested	330 black, crested
43 frizzled feathers, black	40 frizzled feathers, crested	36 black, noncrested
35 normal feathers, white	47 normal feathers, noncrested	45 white, crested
418 total	344 total	732 total

genotypes for each phenotypic class. **(c)** Calculate the genetic map distances for the three linked loci. **(d)** Calculate the interference.

6. In tomato, the recessive allele *bk* causes a beaked fruit when homozygous. The allele *wv* is at another locus and causes yellow-white growing points at the stem tip when homozygous. According to the Report of the Tomato Genetics Cooperative, No. 21 (1971), the loci at which these two alleles are found are linked at 6 cM on chromosome 2. What gametes are expected and in what proportions from a doubly heterozygous parent when the alleles are in **(a)** coupling and **(b)** repulsion conformation?

7. In tomato, there is a recessive allele that reduces the frequency of crossing-over at all locations when homozygous. **(a)** How should homozygosity for this allele affect the gametic frequencies calculated in the previous question? **(b)** What effect should homozygosity for this allele have on estimating map distances between linked loci?

8. Suppose two tomato plants are heterozygous *Bk bk* and *Wv wv*. Plant 1 has the dominant alleles in coupling conformation, whereas plant 2 has them in repulsion conformation. Both plants are self-pollinated. What are the expected frequencies of genotypes and phenotypes in the progeny of both plants?

9. In 1996, Mudge et al. (*Crop Science* 36:1362–1366) published a genetic map of sugarcane based on RAPD markers. Linkage group 12 consists of four linked markers (169s313, S15s407, A12a1213, and pA8s309). From the following distances in cM, establish a genetic map of the linkage group:

Linkage Interval	Distance (in cM)
169s313–S15s407	12.0
169s313–A12a1213	2.9
169s313–pA8s309	5.7
S15s407–A12a1213	9.1
S15s407–pA8s309	17.7
A12a1213–pA8s309	8.6

10. In tomato, the dominant allele *O* confers round fruits, and the recessive allele *o* confers elongate fruits when homozygous. At another locus, the dominant allele *S* confers simple inflorescence, and the recessive allele *s* confers compound inflorescence when homozygous. In 1928, MacArthur (*Genetics* 13:410–420) reported the following data from F$_2$ progeny:

Round fruit, simple inflorescence	126
Elongate fruit, simple inflorescence	66
Round fruit, compound inflorescence	63
Elongate fruit, compound inflorescence	4

(a) Using chi-square analysis, test the hypothesis of independent assortment. **(b)** Determine the recombination frequency for these alleles.

11. In tomato, the dominant allele *P* confers smooth fruits, and its recessive counterpart *p* confers peach fruits when homozygous. This locus is linked to the *o* locus described in the previous question. In 1928, MacArthur (*Genetics* 13:410–420) reported the following data from testcross progeny:

Smooth, round	12
Smooth, elongate	123
Peach, round	133
Peach, elongate	12

(a) Are the alleles in coupling or repulsion conformation in the smooth, round parent? **(b)** What is the frequency of recombination?

12. In *Drosophila melanogaster*, scute flies lack bristles on the scutellum, and crossveinless flies lack crossveins on the wings. Both traits are due to recessive alleles (*sc* and *cv*, respectively) on the X chromosome. Below are the combined results of several crosses between females that were heterozygous for both alleles and males that were hemizygous for both recessive alleles, from Bridges and Olbrycht (1926. *Genetics* 11:41–56):

Wild type	8577
Scute	1678
Crossveinless	1718
Scute, crossveinless	8812
Total	20,785

(a) What is the recombination frequency between these two loci? **(b)** Are the alleles in coupling or repulsion conformation in the heterozygous female parents? **(c)** Correct the map distances using the Haldane and Kosambi functions. **(d)** Which of the three values you calculated in parts (a) and (c) most closely matches the map distance in the *Drosophila melanogaster* map shown in Figure 15.3?

13. The diagram in Figure 15.24 represents paired homologus chromosomes in prophase I of meiosis. Linked on these chromosomes are three loci, *A*, *B*, and *C*. The individual is heterozygous at all three loci. Three crossovers are indicated in the diagram. What are the genotypes of the four gametes produced from this meiosis?

14. In 1923, Bridges and Morgan (1923. *Carnegie Institution Washington* publ. 327) reported the following data from a three-factor linkage testcross involving chromosome 3 in *Drosophila melanogaster*. The three loci are *ca* (*claret*), *e* (*ebony*), and *ro* (*rough*). All three mutant phenotypes are due to recessive alleles.

Wild type	49
Ebony	1
Claret	395
Rough	119
Ebony, rough	370
Ebony, claret	89
Rough, claret	1
Ebony, rough, claret	66
Total	1090

(a) Are all three alleles in coupling conformation in the heterozygous parent? If not, which allele is in repulsion to the other two? **(b)** Construct a genetic map for these three loci showing the distances between loci in cM. **(c)** Calculate the interference.

15. In the same 1923 publication as cited in the previous question, Bridges and Morgan reported the following data from a three-factor linkage testcross involving chromosome 3 in *Drosophila melanogaster*. The three loci involved are *se* (*sepia*), *ss* (*spineless*), and *ro* (*rough*). All three mutant phenotypes are due to recessive alleles.

Wild type	338
Sepia	96
Spineless	46
Rough	156
Sepia, spineless	173
Sepia, rough	43
Spineless, rough	114
Sepia, spineless, rough	370
Total	1336

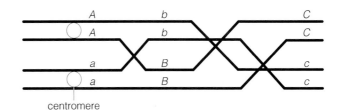

Figure 15.24 Information for Question 13: A triple crossover.

(a) Are all three alleles in coupling conformation in the heterozygous parent? If not, which allele is in repulsion to the other two? **(b)** Construct a genetic map for these three loci showing the distances between loci in cM. **(c)** Calculate the interference. **(d)** Correct the map distances using the Kosambi function, and draw a new linkage map. **(e)** Which values (corrected with the Kosambi function or uncorrected) correspond most closely to the map distances in the *Drosophila melanogaster* map shown in Figure 15.3?

16. A third experiment reported in Bridges and Morgan's 1923 publication includes the following data from a three-factor linkage testcross for alleles on chromosome 3 of *Drosophila melanogaster*. The three loci are *sp* (*spineless*), *sr* (*stripe*), and *D* (*delta*). The spineless and stripe phenotypes are due to recessive mutant alleles, and the delta phenotype is due to a dominant mutant allele.

Delta	658
Spineless, stripe	606
Stripe	25
Spineless, delta	22
Wild type	41
Spineless, stripe, delta	35
Total	1387

(a) Why are there only six instead of eight classes of progeny? **(b)** Are all three alleles in coupling conformation in the heterozygous parent? If not, which allele is in repulsion to the other two? **(c)** Construct a genetic map for these three loci showing the distances between loci in cM. **(d)** Calculate the interference.

17. In 1935, de Winton and Haldane (*Journal of Genetics* 31:67–100) reported the following results from three three-factor testcrosses for four linked loci (*S*, *B*, *L*, and *G*) in evening primrose (*Primula sinensis*). The genotypes of parents and progeny are given for each cross.

Cross #1: *Ss Bb Ll* × *ss bb ll*
Progeny:

Ss Bb Ll	457
ss bb ll	469
Ss bb ll	38
ss Bb Ll	45
Ss Bb ll	256
ss bb Ll	284
Ss bb Ll	11
ss Bb ll	20
Total	1580

Cross #2: *Bb Gg Ll* × *bb gg ll*
Progeny:

Bb Gg Ll	582
bb gg ll	551
Bb gg ll	272
bb Gg Ll	326
Bb Gg ll	28
bb gg Ll	25
Bb gg Ll	5
bb Gg ll	6
Total	1795

Cross #3: *Ss Bb Gg* × *ss bb gg*
Progeny:

Ss Bb Gg	1673
ss bb gg	1654
Ss bb gg	127
ss Bb Gg	125
Ss Bb gg	710
ss bb Gg	783
Ss bb Gg	42
ss Bb gg	45
Total	5159

(a) From the data, construct a three-factor map for each cross. **(b)** Using the data, construct a single four-factor map by combining results of the three crosses. **(c)** Correct each three-factor map with the Kosambi function.

(d) How does data correction using the Kosambi function alter the maps? **(e)** Does use of the Kosambi function help overcome problems with additivity of map distances?

18. Using the information from the previous question, assume that a plant with the genotype *SsBbGg* is self-pollinated. **(a)** What are the possible genotypes, and their expected frequencies, of the gametes in the parent plant? **(b)** Using these gametic frequencies, determine the expected phenotype frequencies among the progeny.

19. (a) Construct a linkage map for the genetic markers *a*, *b*, *c*, *d*, and *e* with distances indicated in cM using the following two-factor recombination frequencies:

a–b	0.10
b–c	0.15
c–d	0.20
d–e	0.05
a–c	0.22
b–d	0.31
c–e	0.22

(b) Why are the distances not additive in all cases? **(c)** Correct the distances using Haldane and Kosambi functions. **(d)** Does correction with a mapping function improve the additivity?

20. The following diagram summarizes the RFLP marker recombination frequencies calculated from two experiments. The first was a two-factor testcross with RFLP A and RFLP B. The second was a three-factor testcross with RFLP A, RFLP B, and RFLP C.

Experiment 1:

RFLP A	0.31	RFLP B

Experiment 2:

RFLP A	0.15	RFLP C	0.20	RFLP B

What percentage of double crossovers were undetected in experiment 1 based on the map distances calculated in experiment 2?

21. The loci *sh* (shrunken kernels), *wx* (waxy endosperm), and *gl* (glossy leaves) are located on chromosome 3 in maize. The approximate map distances in cM are *sh* 30 *wx* 10 *gl*. What are the expected frequencies of gametes from a plant with the genotype *Sh sh Wx wx Gl gl* if **(a)** there is no interference, and **(b)** interference is 0.6?

22. As we saw in Example 15.7, the loci *c1*, *sh*, and *wx* are linked on chromosome 9 in maize. One of the largest

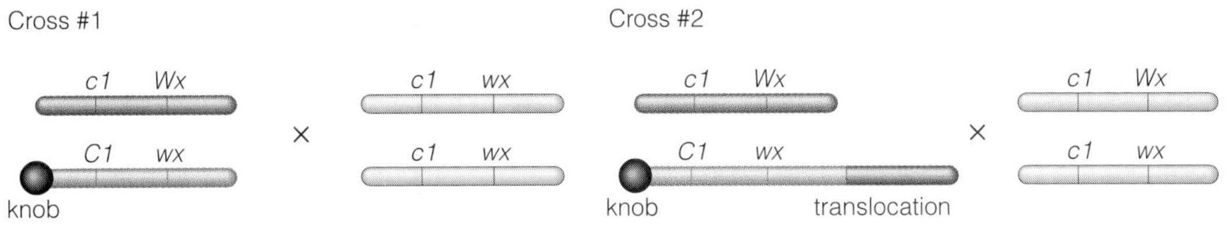

Cross #1

c1 Wx / C1 wx (knob) × c1 wx / c1 wx

Progeny:

Genotype	Knob		Count
C1c1 wxwx	knob		12
C1c1 wxwx	knobless		5
c1c1 Wxwx	knob		5
c1c1 Wxwx	knobless		34
C1c1 Wxwx	knob		4
C1c1 Wxwx	knobless		0
c1c1 wxwx	knob		0
c1c1 wxwx	knobless		3
Total			63

Cross #2

c1 Wx / C1 wx (knob, translocation) × c1 Wx / c1 wx

Progeny:

Genotype	Knob	Translocation	Count
C1c1 wxwx	knob	translocation	3
c1c1 wxwx	knobless	translocation	2
C1c1 Wx__	knob	no translocation	8
c1c1 Wxwx	knobless	translocation	2
c1c1 Wxwx	knobless	no translocation	5
c1c1 WxWx	knobless	translocation	2
c1c1 WxWx	knobless	no translocation	5
c1c1 Wx__	knobless	no translocation	1
Total			28

Figure 15.25 Results from the experiments of Creighton and McClintock.

mapping experiments for these loci was conducted by Stadler (as cited in Strickberger, M. W. 1985. *Genetics*, 3rd ed. New York: Macmillan Publishing Co., p. 312) with the following results:

Colored, full, starchy	17,959
Colorless, shrunken, waxy	17,699
Colored, shrunken, waxy	509
Colorless, full, starchy	524
Colored, full, waxy	4455
Colorless, shrunken, starchy	4654
Colored, shrunken, starchy	20
Colorless, full, waxy	12
Total	45,832

(a) Are all three alleles in coupling conformation in the heterozygous parent? If not, which allele is in repulsion to the other two? **(b)** Construct a genetic map for these three loci showing the distances between loci in cM. **(c)** Calculate the interference. **(d)** How do these map distances compare to those we determined in Example 15.7? **(e)** What factors might explain the discrepancies in map distances?

23. Creighton and McClintock's classic paper (1931. *Proceedings of the National Academy of Sciences, USA* 17:492–497) is used in nearly every genetics textbook (including this one) as evidence that genetic recombination between linked genes is associated with recombination of chromosome markers. Most textbooks (including this one) provide only part of their results in order to simplify the interpretation. The actual results are more complicated than most textbooks imply, but still provide firm evidence of a correlation between cytological markers and genetic recombination. Figure 15.25 shows their results from two crosses. **(a)** In the progeny of cross #1, which classes represent a crossover between the knob and the *c1* locus? **(b)** Although the data in cross #1 are limited, use them to calculate recombination frequencies between the knob and *c1*, and between *c1* and *wx*. **(c)** Which classes are missing from the progeny of cross #2? What types of crossover progeny do these missing classes represent? **(d)** Take each of the classes in the progeny of cross #2 and identify where the crossover (if any) occurred.

24. In *Drosophila melanogaster*, yellow body is due to a recessive X-linked allele, and white eyes are due to a recessive X-linked allele at another locus. T. H. Morgan crossed a yellow-bodied female with white eyes with a wild-type male, then allowed the F_1 flies to intermate,

forming an F$_2$ generation, and scored the following phenotypes in the F$_2$:

Phenotype	Females	Males	Total
Wild type	647	512	1159
White eyes	6	11	17
Yellow body	7	5	12
Yellow body, white eyes	543	474	1017
Totals	1203	1002	2205

(a) Calculate the genetic map distance between these two loci. **(b)** Crossing-over does not happen in male *Drosophila*, so why is it possible to use F$_2$ progeny for mapping in this experiment instead of performing a testcross?

25. The pedigree in Figure 15.26 shows a human family in which X-linked deuteranopia (red-green color blindness) and hemophilia A are present. A crossover coupled the two mutant alleles during meiosis at some point in the pedigree. Is it possible to identify with certainty the individual in which the crossover took place? If so, indicate which individual this is. If not, indicate in which individuals it might have taken place.

26. Below are summaries of two-factor linkage data used by Sturtevant (1913. *Journal of Experimental Zoology* 14:43–59) to derive a map of the X chromosome in *Drosophila melanogaster*:

Loci	Recombinant Progeny	Total Progeny
y–w	214	21,736
y–v	1464	4551
y–r	115	324
y–m	260	693
w–v	471	1584
w–r	2062	6116
w–m	406	898
v–r	17	573
v–m	109	405

(a) Using these data, determine how Sturtevant obtained the map distances on his map (Figure 15.2). **(b)** Why are the map distances not additive?

27. Among the experiments listed in Sturtevant's paper (1913. *Journal of Experimental Zoology* 14:43–59) was a cross between a white-eyed, miniature-winged female and a wild-type male. All of the F$_1$ females were wild type, and all of the males had white eyes and minature

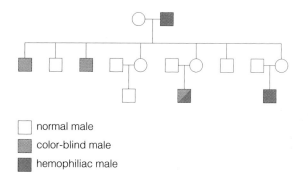

Figure 15.26 Information for Problem 25: Pedigree analysis of crossing-over.

Table 15.4	Information for Problem 27: Phenotypes of Flies in the F$_2$ Generation	
Phenotype	**Females**	**Males**
Wild type	193	202
White eyes	109	114
Miniature wings	124	123
White eyes and miniature wings	208	174
Totals	634	613

wings. The phenotypes in the F$_2$ generation of this experiment are listed in Table 15.4. **(a)** Calculate the genetic map distance between *w* and *m* separately for males and females. **(b)** The genetic map distances calculated from males and females do not differ appreciably. Why is this result expected in this experiment? **(c)** Why is it possible in this cross to calculate map distances from F$_2$ progeny without using Tables 15.1a or 15.1b?

FOR FURTHER READING

The early history of research related to genetic mapping can be found in Chapters 6 and 7 of **Sturtevant, A. H. 1965. *A History of Genetics*. New York: Harper and Row.** The first chromosome map was published by **Sturtevant, A. H. 1913. The linear arrangement of six sex-linked factors in *Drosophila*, as shown by their mode of association. *Journal of Experimental Zoology* 14:43–59.** Creighton and McClintock's classic paper demonstrating the association of cytological crossing-over with genetic recombination of linked genes is **Creighton, H. B., and B. McClintock. 1931. A correlation of cytological and genetical crossing over in *Zea mays*. *Proceedings of the National Academy of Sciences, USA* 17:492–497.** The development of mapping functions was reviewed by **Crow, J. F. 1990. Mapping functions. *Genetics* 125:669–671.** Extensive chromosome maps of many different

species and bibliographic references for those maps are available in the most recent edition of *Genetic Maps*. **Bethesda, Md.: Laboratory of Viral Carcinogenesis, National Cancer Institute, National Institutes of Health**. A review of mapping human chromosome using human-rodent cell hybrids is **Ruddle, F. H., and R. S. Kucherlapati. 1974. Hybrid cells and human genes.** *Scientific American* **(Jul):36–44**. In 1935, Calvin Bridges compared the physical locations of genes on polytene chromosomes with their map distances in what have become classic drawings of the polytene chromosomes in **Bridges, C. 1935. Salivary chromosome maps.** *Journal of Heredity* **26:60–64**. These drawings have been reproduced along with photographs, a comparison of physical and genetic maps for *Drosophila*, genetic and molecular fine structure maps, and information and references for nearly all mutants described prior to 1992 in **Lindsley, D., and G. Zimm. 1992. The genome of** *Drosophila melanogaster*. **San Diego, Calif.: Academic Press**. The use of chromosome walking and jumping to identify and clone the human cystic fibrosis gene was described in **Rommens, J. M., M. C. Iannuzzi, B. Kerem, M. L. Drumm, G. Melmer, M. Dean, R. Rozmahel, J. L. Cole, D. Kennedy, N. Hidaka, M. Zsiga, M. Buchwald, J. R. Riordan, L.-C. Tsui, and F. S. Collins. 1989. Identification of the cystic fibrosis gene: Chromosome walking and jumping.** *Science* **245:1059–1065.**

For additional reading, go to InfoTrac College Edition, your online research library at: http: www.infotrac-college.com/brookscole

CHAPTER 16

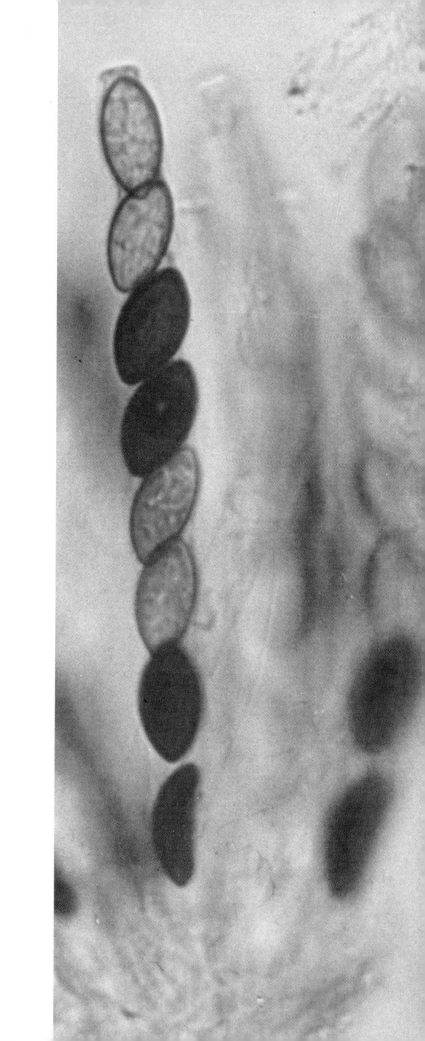

KEY CONCEPTS

Genetic analysis in fungi has provided much information about the mechanism of crossing-over.

~

The gene is a linear segment of DNA that is divisible by crossing-over.

~

The units of recombination and mutation are the individual nucleotides of DNA.

~

Alleles can be identified as belonging to the same gene, or to different genes, through a process called complementation analysis.

GENETIC FINE STRUCTURE

The concept of crossing-over as the mechanism for recombination of linked genes arose soon after the rediscovery of Mendelian principles and was confirmed by numerous experiments. However, without an understanding of the chromosome as a linear molecule of DNA and genes as linear segments of that DNA, the mechanism of crossing-over could not be determined, nor was there any reason to believe that a crossover *within* a gene was possible. Genes were visualized as points along the chromosome, and crossovers as exchanges between the points.

Our understanding of the nature of the gene and crossing-over has advanced considerably. We now view genes as linear segments of DNA and crossovers as events that recombine DNA molecules. The units of recombination and mutation are the individual nucleotides of DNA. We begin this chapter with a discussion of genetic analysis in fungi. The results of genetic analysis in fungi led researchers to develop models of recombination at the DNA level, which is our next topic. We also discuss the genetic consequences of crossing-over within a single gene and see how scientists identify alleles of the same gene through a method known as complementation analysis.

16.1 TETRAD ANALYSIS

Most of the early experiments on chromosome mapping were conducted with *Drosophila*, tomato, maize, and mice, and from these species a large body of information was obtained. In each case, however, researchers discovered how crossing-over behaves in meiosis by observing multicelled individuals that were several somatic cell generations removed from the original meiotic events. Some intriguing questions about recombination were answered using fortuitous atypical situations, such as the unusual chromosome 9 in maize with a knob and an attached segment of another chromosome that demonstrated the association between chromosomal crossing-over and genetic recombination of linked genes (see Figure 15.6). Such experiments illustrate the brilliance and logic employed to determine the underlying mechanisms of genetic recombination from progeny whose cells are far removed from the mechanisms themselves.

Spore Formation in *Neurospora crassa*

However successful such experiments were, much more could be learned if researchers could study the haploid products of meiosis itself, rather than just diploid

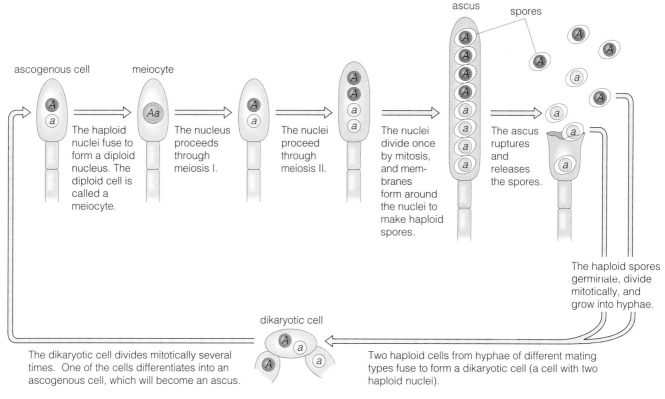

The haploid nuclei fuse to form a diploid nucleus. The diploid cell is called a meiocyte.

The nucleus proceeds through meiosis I.

The nuclei proceed through meiosis II.

The nuclei divide once by mitosis, and membranes form around the nuclei to make haploid spores.

The ascus ruptures and releases the spores.

The haploid spores germinate, divide mitotically, and grow into hyphae.

The dikaryotic cell divides mitotically several times. One of the cells differentiates into an ascogenous cell, which will become an ascus.

Two haploid cells from hyphae of different mating types fuse to form a dikaryotic cell (a cell with two haploid nuclei).

Figure 16.1 Life cycle of *N. crassa* with a focus on meiosis and spore development.

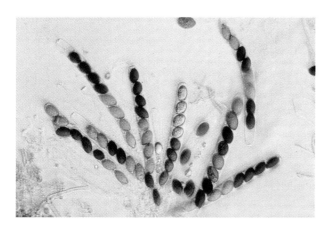

Figure 16.2 A photomicrograph of *N. crassa* asci with two genetically different types of spores, a dark-colored type and a light-colored type from the ascomycete *Sordaria brevicollis*.

progeny derived from those products. As we discussed in section 11.4, many fungal species are haploid during most of their life cycles. Some belong to a group of fungi known as ascomycetes, or sac fungi, and include yeasts, molds, cup fungi, and truffles. Ascomycetes carry mature **spores**, the haploid cells that are the products of meiosis, in a membranous sac called an **ascus** (plural, **asci**). Eventually, the ascus ruptures, releasing the haploid spores into the environment, where they then germinate and multiply by mitosis. Figure 16.1 shows the life cycle

of the black bread mold *Neurospora crassa* as an example of meiosis and spore formation in ascomycetes.

The ascus contains the products of meiosis, and these cells may be observed directly in the ascus. In some species (such as *Saccharomyces cerevisiae*), the four products of meiosis mature into four spores, in which case the ascus is called a **tetrad**. In other species (such as *N. crassa*, shown in Figure 16.1), each of the four nuclei that arise from meiosis divides once mitotically before mature spores form. The result is an ascus with eight spores, called an **octad**. However, the four pairs of spores in an octad are often analyzed as if the octad were a tetrad because the two members of each pair of spores that arose mitotically are genetically identical.

Some species produce asci with **unordered spores**, which do not correspond spatially to the meiotic divisions. In other species the asci contain **ordered spores**, whose spatial arrangement in the ascus is determined by the meiotic divisions. *S. cerevisiae* is an example of an ascomycete with unordered spores. *N. crassa* is an example of a species with ordered spores. The ascus sac of *N. crassa* is a tight, tubelike structure with a diameter about the same size as the spores that fit inside it; and the spores are stacked on top of one another in a linear fashion (Figure 16.1). There is no room for the spores to move around or switch places with one another. When a nucleus within an ascus divides, the two daughter nuclei remain adjacent to one another. A diploid cell undergoes meiosis within the

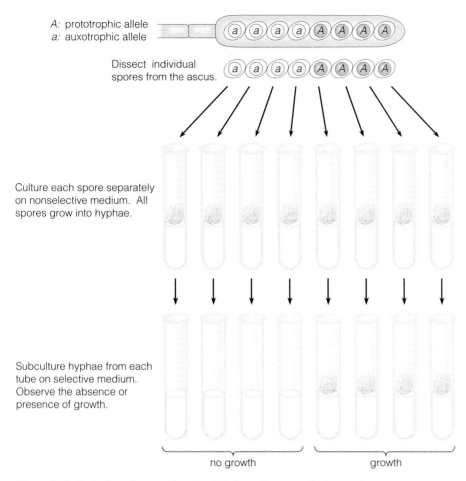

A: prototrophic allele
a: auxotrophic allele

Dissect individual spores from the ascus.

Culture each spore separately on nonselective medium. All spores grow into hyphae.

Subculture hyphae from each tube on selective medium. Observe the absence or presence of growth.

no growth   growth

Figure 16.3 Detection of an auxotrophic allele by culture on selective medium.

ascus, so the products of meiosis remain ordered within the ascus according to the two meiotic divisions.

At its earliest stage, the developing ascus of *N. crassa* contains a single diploid nucleus and is called a **meiocyte** (Figure 16.1). The nucleus within the meiocyte proceeds through meiosis I and meiosis II to form four haploid nuclei, the final products of meiosis. The nuclei then divide mitotically into eight haploid nuclei within the ascus. Membranes and cell walls eventually partition the eight nuclei into eight spores. The linear array of the eight spores represents the sequence of meiotic divisions and allows researchers to observe the products of meiosis ordered according to the meiotic divisions. The results of segregation and independent assortment can be readily observed in the ordered spores.

Some alleles cause visible phenotypes, such as the white spore phenotype shown in Figure 16.2. Asci with a visible phenotype can easily be counted under the microscope. However, many of the mutant alleles in ascomycetes produce auxotrophic mutant phenotypes, which cannot be identified as a single spore within an ascus. It is still possible to determine which spores within the ascus carry auxotrophic alleles while tracking the po-

sition of the spore in the ascus. After segregation of any visual alleles has been recorded, the ascus is dissected under the microscope, and each spore is numbered so its original position in the ascus can be recalled. Then each spore is germinated individually on nonselective medium and allowed to grow into hyphae of haploid cells. Portions of the hyphae can then be subcultured on selective media to determine which auxotrophic alleles they carry. In this way, the combination of alleles in each spore can be determined for genetic analysis (Figure 16.3).

~

Ascomycetes are used as model organisms for studying the direct products of meiosis, which are present within the ascus.

Centromere Mapping

Tetrad analysis is the method used to determine which genes are linked and to calculate the genetic map distances between them. Tetrad analysis may also be used to calculate the genetic map distance from a gene to the centromere of its chromosome. Let's look first at mapping a single gene to its centromere. Suppose a meiocyte is

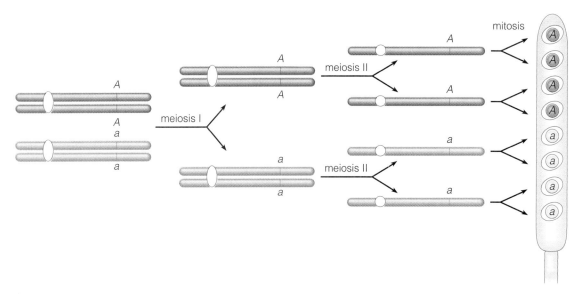

a In first-division segregation, one orientation of chromosomes at metaphase I produces an ascus in which the top four spores receive the *A* allele and the lower four spores receive the *a* allele.

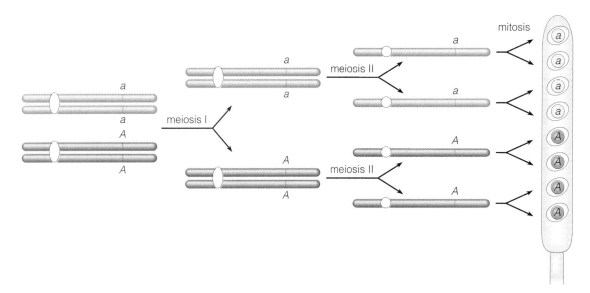

b The other orientation produces an ascus in which the top four spores receive the *a* allele and the lower four spores receive the *A* allele.

Figure 16.4 First-division segregation of a heterozygote in *Neurospora crassa*.

heterozygous *Aa*. If there is no crossover between the *a* locus and the centromere, the two alleles segregate from each other at the first meiotic division. This produces a pattern called **first-division segregation**, in which four spores of one type are grouped at the top half of the ascus and four spores of the other type are grouped at the bottom half. Which allele is in the top half and which is in the bottom half depends on the orientation of the paired chromosomes in the meiocyte. The orientation is determined at random, so in the heterozygote *Aa*, about half of the first-division-segregation asci have *A* in the top half and *a* in the bottom half (Figure 16.4a), and the

remaining first-division-segregation asci have the opposite orientation (Figure 16.4b).

If a single crossover takes place between a gene and the centromere in a heterozygote, the alleles do not segregate completely from each other as in first-division segregation. As illustrated in Figure 16.5a, the alleles segregate completely only in the second meiotic division, a pattern called **second-division segregation**. There are four types of second-division-segregation patterns that can arise from a crossover between a gene and the centromere (Figure 16.5b). Which pattern arises depends on the orientation of chromatids during meiosis. Because the

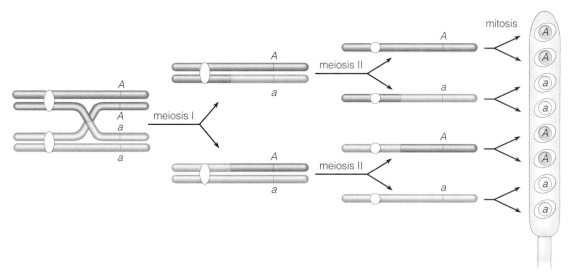

a Segregation of recombinant and nonrecombinant chromosomes to produce one of the four possible patterns that can arise from a single crossover between a gene and the centromere in a heterozygote with the genotype *Aa*.

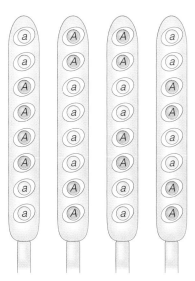

b The four patterns that may arise from a single crossover between a gene and the centromere in a heterozygote with the genotype *Aa*.

Figure 16.5 Second-division segregation of a heterozygote in *Neurospora crassa*.

chromatid orientation is random during crossing-over and segregation, the four patterns are equally likely in second-division-segregation asci. By counting the relative numbers of first-division-segregation asci and second-division-segregation asci, we can determine the genetic map distance between the locus of interest and the centromere.

Genetic map distance is defined as the number of recombinant progeny (in this case, spores, not asci) divided by the total number of progeny. Each second-division-segregation ascus has four recombinant spores and four nonrecombinant spores, so only half the spores arise

from a crossover (see Figure 16.5a). To calculate the map distance between a gene and the centromere, we count the number of recombinant spores and divide by the total number of spores. An easier method in *N. crassa* is to count the number of second-division-segregation asci and divide by 2 (because only half of the spores in a second-division-segregation ascus are crossover types), then divide this number by the total number of asci:

$$\text{map distance in cM} = \frac{\left(\begin{array}{c}\text{number of second-division-} \\ \text{segregation asci}\end{array}\right) \div 2}{\text{total number of asci}} \times 100 \quad \textbf{[16.1]}$$

The following example shows how to use equation 16.1 to calculate the genetic map distance between a gene and the centromere of its chromosome.

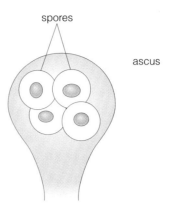

spores

ascus

Figure 16.6 An unordered tetrad. The four spores are not spatially arranged in a particular order.

Example 16.1 Mapping a gene to the centromere in *Neurospora crassa*.

In *Neurospora crassa*, haploid mycelia grown from spores that carry mutant alleles of the *leu-1* locus cannot grow on minimal medium without leucine. Below are the results of several mapping experiments for mutant alleles at the *leu-1* locus:

Experiment	Number of First-Division- Segregation Asci	Number of Second-Division- Segregation Asci
1	65	8
2	107	22
3	22	6
4	33	23
5	55	12
Totals	282	71

(Data from Barrat et al. 1954. *Advances in Genetics* 6:1–93.)

Problem: Estimate the genetic map distance between the *leu-1* locus and the centromere in each experiment, then in all five experiments combined.

Solution: Let's use equation 16.1 to estimate the genetic map distance in each experiment, then total the results of all five experiments for a composite map distance.

Experiment 1: $[(8 \div 2) \div (65 + 8)] \times 100 = 5.48$ cM

Experiment 2: $[(22 \div 2) \div (107 + 22)] \times 100 = 8.53$ cM

Experiment 3: $[(6 \div 2) \div (22 + 6)] \times 100 = 10.71$ cM

Experiment 4: $[(23 \div 2) \div (33 + 23)] \times 100 = 20.54$ cM

Experiment 5: $[(12 \div 2) \div (55 + 12)] \times 100 = 8.96$ cM

Composite: $[(71 \div 2) \div (282 + 71)] \times 100 = 10.06$ cM

The outcomes of the individual experiments differ because of sampling error. The composite result is more likely to approximate the true map distance.

The genetic map distance between a gene and the centromere of its chromosome can be estimated as half of the second-division-segregation asci divided by the total number of asci.

16.2 UNORDERED TETRAD ANALYSIS

Some fungal species have unordered spores in the ascus. The major advantage of studying ordered spores is that they allow us to map the centromere, because the order of the spores indicates which asci contain recombinant spores. When centromere mapping is not needed, however, unordered tetrad analysis is often used even with species like *Neurospora crassa* that have ordered tetrads. With *N. crassa*, it is more convenient to allow the asci to rupture and eject the spores onto an agar plate than to dissect individual asci to remove the ordered spores. The ejected spores from each ascus remain clumped in groups of eight and can be analyzed as unordered tetrads. Even though there are eight spores in each ascus, they can be analyzed as a tetrad because each spore has an identical mitotic counterpart. The spores are counted and classified, then the numbers of each type are divided by 2. Other species, such as *Saccharomyces cerevisiae*, have asci with four unordered spores (Figure 16.6) and can be analyzed directly as unordered tetrads.

There are three types of unordered tetrads:

1. A **parental ditype** has two types of spores, both of which are identical to the two parental types. For example, in the progeny of the cross *AB* × *ab*, the parental ditype has four spores with the same genotypes as the haploid parents: *AB*, *AB*, *ab*, and *ab*.

2. A **tetratype** has four different types of spores. For example, in the progeny of the cross *AB* × *ab*, there are two nonrecombinant (parental) types: *AB* and *ab*, and two recombinant (nonparental) types: *Ab* and *aB*.

3. A **nonparental ditype** has two types of spores, neither of which matches the parental types; they are all recombinant spores. In the progeny of the

cross *AB* × *ab*, the nonparental ditype has spores with the genotypes *Ab, Ab, aB,* and *aB.*

Two linked genes can be mapped in relation to one another in any type of tetrad, ordered or unordered. A dihybrid cross may be written *AB* × *ab* (remember that the parents in an ascomycete cross are haploid, so the cross is not written *AA BB* × *aa bb* as with diploid organisms). This cross produces a double heterozygote in the meiocyte with the alleles in coupling conformation if the genes are linked. If the genes are unlinked, independent assortment should be observed as a 1:1:1:1 ratio of the four spore types, *AB, Ab, aB,* and *ab.* A significant deviation from independent assortment indicates that linkage is likely.

One of the advantages of tetrad analysis is that it allows us to detect the reciprocal products of crossing-over in a single nucleus and to estimate the number of undetected double crossovers from that information. The spore genotypes for noncrossover, single-crossover, and double-crossover types are diagrammed in Figure 16.7. For single crossovers (Figure 16.7b), paired homologous chromosomes include four chromatids, only two of which exchange segments. Two of the spores are recombinant, and the other two are nonrecombinant. As all four products of the same meiosis are present within a tetrad, the two recombinant chromatids are represented as spores with reciprocal recombinant genotypes. The other two spores have the two parental genotypes because they arise from nonrecombinant chromatids.

As we saw in Figure 15.10, there are four types of double crossovers between two linked genes: two-strand, two types of three-strand, and four-strand double crossovers. Two-strand and three-strand double crossovers cannot be identified as double crossovers in unordered tetrad analysis. Tetrads that arise from a two-strand double crossover are indistinguishable from noncrossover tetrads, whereas tetrads that arise from three-strand double crossovers are indistinguishable from single-crossover-type tetrads. However, four-strand double crossovers can be readily identified (Figure 16.7c). A four-strand double crossover produces four recombinant spores within the tetrad, a nonparental ditype. Four-strand double crossovers represent one-fourth of all double crossovers, so the number of nonparental ditype tetrads is multiplied by 4 to estimate the total number of tetrads that arose from double crossovers. If we let *x* be the number of nonparental ditypes, then the number of parental ditypes that arose by a two-strand double crossover is *x*, and the number of tetratypes that arose from a three-strand double crossover is 2*x* (giving a total of 4*x*).

We can use this information to compensate for undetected crossovers instead of using a mapping function. The number of undetected crossovers in the parental ditype class is 2*x* because each parental ditype that arises

from a two-strand double crossover represents two undetected crossovers. The number of undetected crossovers in the tetratype class is also 2*x* because each tetratype that arises from a three-strand double crossover represents only one undetected crossover, but there are twice as many three-strand double crossovers as two-strand double crossovers. If we add these together, we get 2*x* + 2*x* = 4*x*. Thus, we must include 4 times the number of nonparental ditypes to account for *undetected* crossovers. To these we must then add the number of *detected* crossovers, which we identify in two ways. (1) Each tetratype represents one crossover; and (2) each nonparental ditype represents two crossovers, so the number of crossovers from this class is 2*x*. Thus the formula for adding up *all* crossovers is

$$4x + \text{number of tetratypes} + 2x$$

or simply

$$\text{number of tetratypes} + 6x$$

To calculate the map distance, we can use the following equation:

$$\text{map distance in cM} = \frac{\left[\left(\substack{\text{number of}\\\text{tetratypes}}\right) + 6\left(\substack{\text{number of}\\\text{nonparental ditypes}}\right)\right] \div 2}{\text{total number of asci}} \times 100$$

[16.2]

The following example shows how equation 16.2 can be used for two-factor linkage analysis.

Example 16.2 Two-factor mapping using unordered tetrad analysis.

In 1939, Lindegren and Lindegren (*Genetics* 24:1–7) reported 143 parental ditypes, 131 tetratypes, and 4 nonparental ditypes in the progeny of a cross between fluffy and peach strains of *Neurospora crassa.*

Problem: What is the genetic map distance between the *fluffy* and *peach* loci based on these data?

Solution: Using equation 16.2, we calculate the map distance as follows:

{[131 + (6 × 4)] ÷ 2} ÷ (143 + 131 + 4) × 100 = 27.88 cM

In unordered tetrad analysis, the effect of double crossovers on the frequency of recombination can be estimated directly from the progeny to calculate a map distance between two genes.

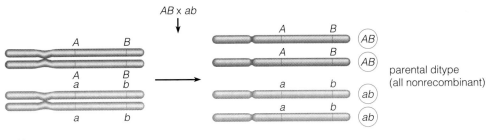

a No crossover.

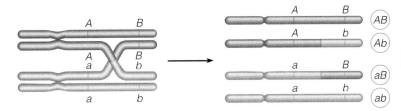

b Single crossover.

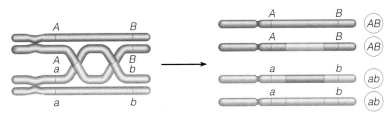

two-strand double crossover

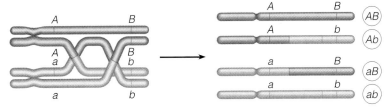

three-strand double crossover

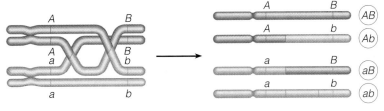

three-strand double crossover

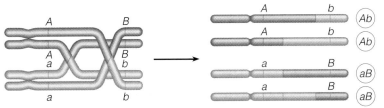

four-strand double crossover

c Four different types of double crossovers.

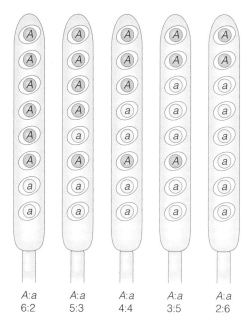

| A:a | A:a | A:a | A:a | A:a |
| 6:2 | 5:3 | 4:4 | 3:5 | 2:6 |

Figure 16.8 Aberrant spore segregation patterns in asci of *Neurospora crassa*.

16.3 GENE CONVERSION AND THE MECHANISM OF CROSSING-OVER

The expected segregation patterns for spores that arise from a heterozygous meiocyte are shown in Figures 16.4 and 16.5. In each case, the ratio of *A:a* alleles in the spores is 4:4, and each spore is always paired with an identical spore. However, on occasion aberrant patterns such as those shown in Figure 16.8 appear. In the 5:3, aberrant 4:4, and 3:5 asci, some of the paired spores are not identical. This is unexpected because paired spores arise by mitosis, which should result in identical products. In all but the aberrant 4:4 asci, the number of spores with *A* is not equal to the number of spores with *a*. This is also unexpected because, according to the patterns of meiosis and mitosis, spore segregation should appear in a 4:4 ratio whether or not there is a crossover.

In 6:2 and 5:3 segregation patterns, an *a* allele has been converted to *A*. In 3:5 and 2:6 segregation patterns, an *A* allele has been converted to *a*. This phenomenon is called **gene conversion**, meaning that one allele in a heterozygote is converted into the allele on the homologous chromosome. Gene conversion could be explained as a mutation that changes *a* to *A* or *A* to *a*, but even though gene conversion is rare, it is much more frequent than mutation. Also, when flanking markers (genetic markers

◄ **Figure 16.7** Possible outcomes of spore segregation for linked genes with **(a)** no crossover, **(b)** a single crossover, and **(c)** two-strand, three-strand, and four-strand double crossovers.

located on both sides of the locus of interest) are present, so crossovers can be detected, about 50% of all gene conversions are associated with a crossover. This observation suggests that gene conversion is somehow related to crossing-over.

The following example shows how gene conversion is detected and how it can be distinguished from mutation.

Example 16.3 Gene conversion in *Sordaria brevicollis*.

Sordaria brevicollis is an ascomycete that has several spore color mutants that permit easy visual scoring of spores within the ascus. In 1967, Fields and Olive (*Genetics* 57:483–493) reported the data listed in Table 16.1.

Problem: **(a)** Determine the frequency of aberrant asci in each cross. **(b)** Why is gene conversion the best explanation for the appearance of aberrant asci instead of mutation or crossing-over? **(c)** Gene conversion apparently uses the DNA on one chromosome to convert the allele on the homologous chromosome. Assuming that this is the case, devise an appropriate experiment to determine whether mutation could explain the results.

Solution: **(a)** Table 16.2 shows the frequencies of aberrant asci. **(b)** Mutation might be viewed as a possible cause of the observed patterns, but the frequencies are too high for mutations. The lowest frequency is 0.5×10^{-3}, which is 1 in 2000. Mutations are typically one or two orders of magnitude less frequent. **(c)** Although the frequencies of aberrant asci suggest that mutation is an unlikely explanation, it is appropriate to include control experiments to exclude mutation as a cause, which the authors of this paper did. Gene conversion requires a wild-type DNA sequence opposite the mutant site on the homologous chromosome as a pattern for conversion. A mutation in a gene, on the other hand, is not influenced by the DNA sequence on the homologous chromosome. Crosses between strains carrying the same mutation (such as C28 × C28) produce a meiocyte that is homozygous and should generate asci with 8 identical mutant spores among the progeny, unless a mutant allele reverts to the wild-type allele by mutation. There should be no gene conversions because there is no wild-type DNA at the mutant site to serve as a pattern for conversion of the mutant allele to the wild type. A reversion mutation in a mutant homozygote should appear as a 2:6 (wild type:mutant) ascus. The authors of this paper stated that "in crosses designed to detect back mutations, no such reversions were

Table 16.1 Information for Example 16.3: Numbers of aberrant asci in five crosses

| Cross* | Number of Asci Examined | Aberrant Asci | | | | |
		2:6	3:5	4:4	5:3	6:2
C17 × WT	13,000	8	4	0	2	0
C22 × WT	13,600	6	4	0	2	0
C27 × WT	11,928	3	4	0	2	2
C28 × WT	23,980	9	1	0	0	2
C29 × WT	16,857	5	0	0	5	1

*The numbers preceded by a C are designations of different mutant strains. WT refers to a wild-type strain.

Table 16.2 Solution to Example 16.3: Frequencies of aberrant asci

Mutant Strain	Number of Asci Examined	Total Number of Aberrant Asci	Frequency
C17	13,000	14	1.08×10^{-3}
C22	13,600	12	0.88×10^{-3}
C27	11,928	11	0.92×10^{-3}
C28	23,980	12	0.50×10^{-3}
C29	16,857	11	0.65×10^{-3}
Totals	79,365	60	0.76×10^{-3}

found among several million asci surveyed." Therefore, the rate of reversion mutations was less than one per million, which is at least two orders of magnitude lower than the observed rates of gene conversion.

Gene conversion, the conversion of one allele into the other in a heterozygote, is associated with crossing-over.

The Holliday Model of Crossing-Over

With the discovery of DNA's structure, scientists were able to devise models that explain crossovers and account for gene conversion. Robin Holliday described the first such model, called the **Holliday model**, in 1964. Several other models have since been devised. Most of the later models are similar to the Holliday model with refinements to account for observations that it could not explain. It is not certain which model is correct. The Holliday model as well as several other models were derived from data obtained from experiments with ascomycetes. There is now substantial evidence to suggest that the Holliday model may not be entirely correct and that the mecha-

nism for crossing-over may not be the same among all species. Nonetheless, the Holliday model does explain a number of observations, including gene conversion, and is therefore a good place to begin our discussion of the molecular mechanism for crossing-over. After discussing the Holliday model, we will look at the most important modifications that have been proposed.

In the simplest model of crossing-over, both strands of the DNA molecules in two paired nonsister chromatids break at the same place and exchange segments. Such a model explains the effects of crossing-over but does not account for gene conversion. The Holliday model proposes that single strands of DNA break at the same place in nonsister chromatids when homologous chromosomes are paired. The DNA strands unwind over a short segment of the DNA beginning at the breaks (Figure 16.9a). The broken, unwound strands exchange places and pair with the corresponding nucleotides on the homologous chromatids to form **heteroduplex DNA**, double-stranded DNA with strands that originated in different molecules. The nicks are sealed, and the two DNA molecules are connected by crossed DNA strands called a **Holliday junction** (Figure 16.9b). The junction may migrate along the molecules, extending the segments of heteroduplex DNA (Figure

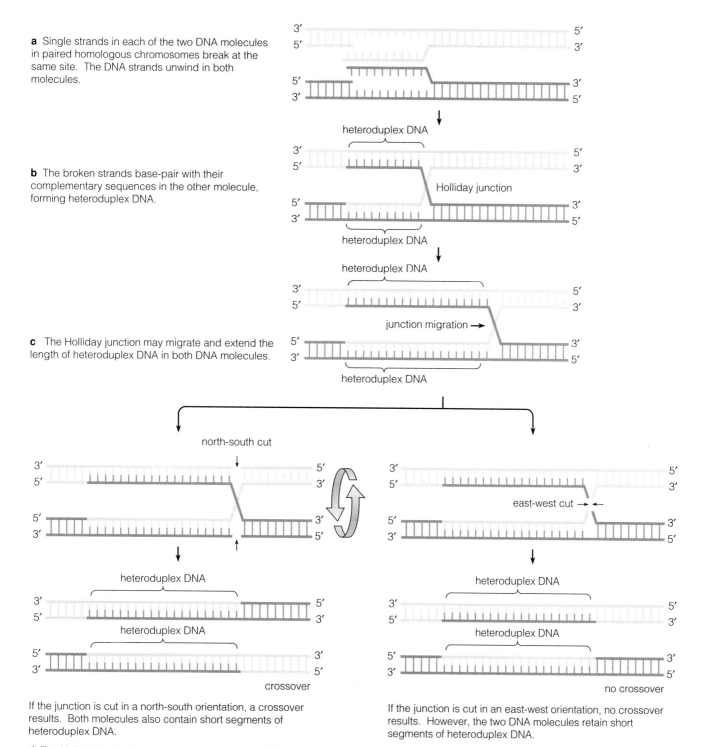

a Single strands in each of the two DNA molecules in paired homologous chromosomes break at the same site. The DNA strands unwind in both molecules.

b The broken strands base-pair with their complementary sequences in the other molecule, forming heteroduplex DNA.

c The Holliday junction may migrate and extend the length of heteroduplex DNA in both DNA molecules.

heteroduplex DNA

Holliday junction

heteroduplex DNA

heteroduplex DNA

junction migration →

heteroduplex DNA

north-south cut

heteroduplex DNA

heteroduplex DNA

crossover

If the junction is cut in a north-south orientation, a crossover results. Both molecules also contain short segments of heteroduplex DNA.

east-west cut → ←

heteroduplex DNA

heteroduplex DNA

no crossover

If the junction is cut in an east-west orientation, no crossover results. However, the two DNA molecules retain short segments of heteroduplex DNA.

d The Holliday junction is removed by cutting the DNA in one of two orientations.

Figure 16.9 The Holliday model of recombination at the DNA level.

16.9c). Because strand exchange is reciprocal, segments of heteroduplex DNA form in both DNA molecules.

At some point, the Holliday junction needs to be cut in one of two ways to free the two chromosomes. As depicted in Figure 16.9d, a north-south cut, in which the two unexchanged strands of DNA are cut, results in a crossover. An east-west cut, in which the two exchanged strands of DNA are cut, restores the original molecules and does not result in a crossover. Both events are equally frequent, so only half of the exchanges become crossovers. In both cases, short segments of heteroduplex DNA remain near the site of the Holliday junction

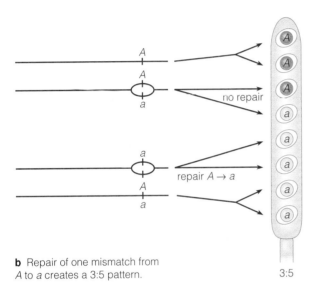

a No repair of mismatches in heteroduplex DNA creates an aberrant 4:4 pattern.

aberrant 4:4

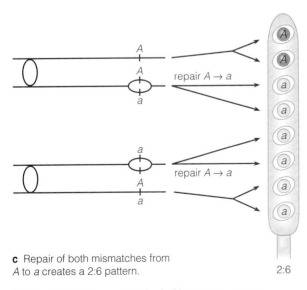

b Repair of one mismatch from *A* to *a* creates a 3:5 pattern.

3:5

c Repair of both mismatches from *A* to *a* creates a 2:6 pattern.

2:6

Figure 16.10 Gene conversion in *Neurospora crassa*.

(Figure 16.9d). This heteroduplex DNA is required for gene conversion.

Let's suppose that the DNA strands exchange places at a site where a single nucleotide differs between the two chromosomes. The heteroduplex DNA in both chromosomes contains a single nucleotide mismatch:

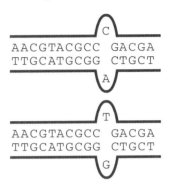

A mismatch may or may not be repaired, so one of three gene conversion patterns can appear. (1) As illustrated in Figure 16.10a, if both strands are left unrepaired, an aberrant 4:4 pattern appears. (2) Mismatch repair mechanisms may recognize a mismatch and repair one of the nucleotides, using the other strand as the template. This converts *A* to *a*, or *a* to *A*, depending on which of the two mismatched nucleotides is replaced to create a correct match. If one of the two DNA molecules is repaired, a 3:5 pattern results (Figure 16.10b). (3) If both DNA molecules are repaired in the same direction (such as *A* to *a*), then a 2:6 pattern results (Figure 16.10c).

The Meselson-Radding (Asymmetric Strand-Transfer) Model of Recombination

An important modification of the Holliday model was proposed in 1975 by Matthew Meselson and Charles Radding. It is called the **Meselson-Radding Model**, or the **asymmetric strand-transfer model**, and differs from the Holliday model in that it proposes a transfer of only one DNA strand (instead of two). The Meselson-Radding model was proposed to account for observations in *Saccharomyces cerevisiae* and *Ascobolus immersus* that could not be explained by the Holliday model.

In the asymmetric strand-transfer model, a single strand in one DNA molecule breaks and unravels for a short distance (Figure 16.11a). The broken strand invades the DNA of the paired homologous chromosome and displaces one of the DNA strands as it forms base pairs with its complementary DNA sequence. The displaced strand becomes a D-loop and is digested by enzymes, and the space left vacant where the broken strand unraveled is filled in with nucleotides by DNA synthesis (Figure 16.11b and c). The DNA molecules then isomerize (rearrange chemically) so that the two strands that remain intact are crossed (Figure 16.11d). At this point, the crossed branch may or may not migrate along the DNA

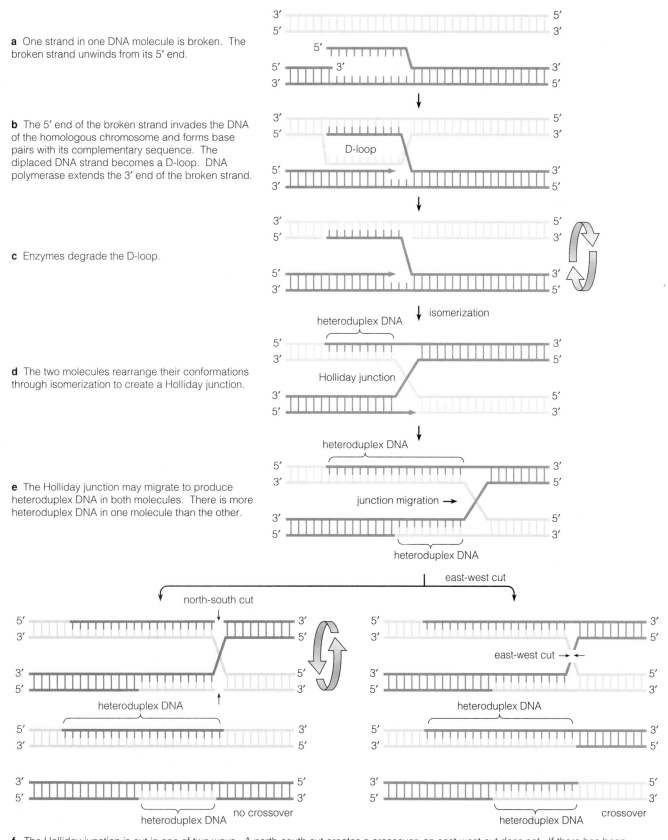

a One strand in one DNA molecule is broken. The broken strand unwinds from its 5′ end.

b The 5′ end of the broken strand invades the DNA of the homologous chromosome and forms base pairs with its complementary sequence. The diplaced DNA strand becomes a D-loop. DNA polymerase extends the 3′ end of the broken strand.

c Enzymes degrade the D-loop.

d The two molecules rearrange their conformations through isomerization to create a Holliday junction.

e The Holliday junction may migrate to produce heteroduplex DNA in both molecules. There is more heteroduplex DNA in one molecule than the other.

f The Holliday junction is cut in one of two ways. A north-south cut creates a crossover; an east-west cut does not. If there has been branch migration, then all molecules retain heteroduplex DNA.

Figure 16.11 The Meselson-Radding or asymmetric strand-transfer model of recombination at the DNA level.

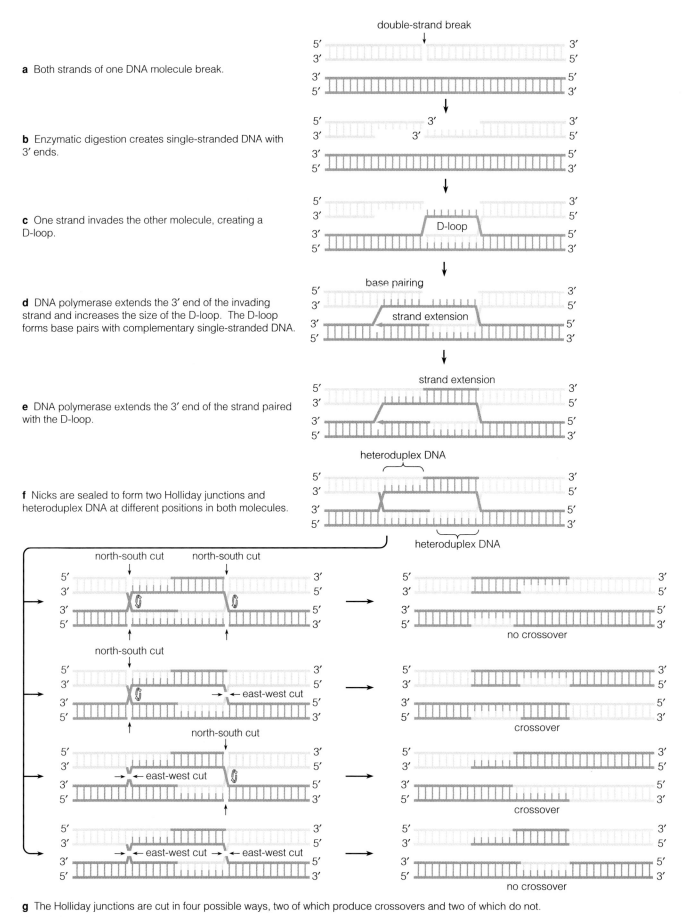

a Both strands of one DNA molecule break.

b Enzymatic digestion creates single-stranded DNA with 3′ ends.

c One strand invades the other molecule, creating a D-loop.

d DNA polymerase extends the 3′ end of the invading strand and increases the size of the D-loop. The D-loop forms base pairs with complementary single-stranded DNA.

e DNA polymerase extends the 3′ end of the strand paired with the D-loop.

f Nicks are sealed to form two Holliday junctions and heteroduplex DNA at different positions in both molecules.

g The Holliday junctions are cut in four possible ways, two of which produce crossovers and two of which do not.

Figure 16.12 The double-strand-break repair model of recombination at the DNA level.

molecule. If there is no migration, heteroduplex DNA is present in only one DNA molecule. If the branch migrates, heteroduplex DNA is present unequally in both DNA molecules (Figure 16.11e). At some point, the DNA strands are cut in either a north-south or east-west manner as in the Holliday model. Depending on which way they are cut, there may or may not be a crossover (Figure 16.11f).

In the Holliday model, heteroduplex strands are equally present in both molecules for every gene conversion. In the asymmetric strand-transfer model, a heteroduplex portion can be present on only one molecule if there is no branch migration. If the branch migrates, heteroduplex portions are present in both molecules, although not equally. The Holliday model predicts that gene conversions should be equally frequent in both DNA molecules. In some cases, gene conversion is more frequent in one molecule than the other, a phenomenon that is explained by the unequal heteroduplex DNA in the Meselson-Radding model.

The Double-Strand-Break Repair Model of Recombination

A model called the **double-strand-break repair model** was devised to explain several observations that cannot be accounted for by either the Holliday model or the Meselson-Radding model. We'll discuss these observations in a moment. For now let's look at the model. As explained in Figure 16.12, a double-strand break in one molecule creates two Holliday junctions and four possible outcomes, two that cause a crossover and two that do not. However, all outcomes may result in gene conversion. Mismatch repair in a heteroduplex region causes gene conversion. Heteroduplex regions are present in both DNA molecules, but unlike in the Holliday model, the heteroduplex regions are located at different positions in the two DNA molecules.

Several lines of evidence support the double-strand-break repair model. Among them is the observation that predetermined sites, called **recombination hotspots**, are apparently prone to double-strand breaks. This conclusion is based on two general observations of genetic data. First, crossing-over is rarely observed between two alleles of the same gene in fungi (which, as we will see in the next section, is different from the situation in *Drosophila* and other higher eukaryotes). Second, gene conversions in fungi are **polar**—meaning that they are more frequent at one end of a gene than the other. Polarity is expected if crossing-over is initiated at a predetermined hotspot. Mutations that are closer to the site of recombination tend to convert more frequently than more distant mutations. These genetic observations in ascomycetes have been recently confirmed by DNA analysis. For example, the recombination hotspot for the *ARG4* gene in yeast has been localized to the promoter region of the gene, upstream from the transcribed region (Figure 16.13).

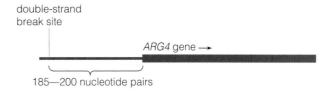

Figure 16.13 Location of the double-stranded break site that is a recombination hotspot for the *ARG4* gene in yeast. (Adapted from Sun, H., D. Treco, and J. W. Szostak. 1991. Extensive 3'-overhanging, single-stranded DNA associated with the meiosis-specific double-strand breaks at the *ARG4* recombination initiation site. *Cell* 64:1155–1161. © 1991 Cell Press. Used by permission.)

There is much evidence to support the double-strand-break repair model in yeast and other ascomycetes. There are also indications that some form of double-strand-break repair may be the mechanism for recombination in other eukaryotes.

Gene Conversion in *Drosophila*

Gene conversion was readily demonstrated in *Neurospora*, yeast, and other ascomycetes. Unfortunately, it was not easy to identify gene conversion in multicellular eukaryotes. Gene conversion can be identified in ascomycetes because all products of a single meiosis can be observed. The aberrant patterns typical of gene conversion are **nonreciprocal**, which means that for each converted gene there is not necessarily a corresponding gene converted in the other direction on the homologous chromosome. Crossing-over is a reciprocal exchange, but gene conversion is not reciprocal, which allows the two to be distinguished when all the products from a single meiosis are available for analysis. In multicellular eukaryotes, it is difficult to distinguish gene conversion from a single or double crossover because the products of a single meiosis cannot be directly observed in most cases. In females, only one mature gamete forms from a single meiosis; the remaining three products are lost as polar bodies. In males, all four products of a single meiosis become gametes, but they are mixed with the gametes from millions of other meioses and cannot be separately distinguished. In order to identify gene conversion in a multicellular eukaryote, a situation or procedure that allowed the observation of products from the same meiosis needed to be found.

Arthur Chovnick and his colleagues were able to induce such a situation in *Drosophila melanogaster* using X-ray treatment of chromosomes. The experimental design that they used to recover reciprocal products of meiosis is complicated and brilliant—but also well beyond the scope of this chapter. Their paper is cited in the list of references at the end of this chapter and is well worth reading. They discovered strong evidence of gene conversion at the *rosy* locus in *D. melanogaster*, indicating that gene conversion was not limited to fungi.

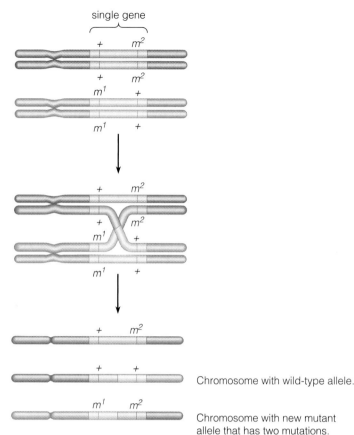

single gene

a Paired chromosomes in a compound heterozygote with two different mutant alleles (m^1 and m^2).

b A crossover recombines the two mutant alleles.

c Recombination between the two mutant alleles creates a wild-type allele.

Chromosome with wild-type allele.

Chromosome with new mutant allele that has two mutations.

Figure 16.14 Creation of a functional, wild-type allele from a crossover between two different mutant alleles (m^1 and m^2) within the same gene.

~
Models of recombination at the DNA level explain crossing-over and gene conversion. All models propose the formation of heteroduplex DNA at the site of recombination.

16.4 INTRAGENIC RECOMBINATION

The structure of the DNA molecule and the models of crossing-over suggest an important point. Because each gene is a linear segment along a DNA molecule, rather than a point on the DNA molecule, crossovers may occur within genes as well as between them. Even before DNA was known to be the genetic material, evidence had accumulated to show that genes could be rearranged by crossovers within them. Different alleles of a gene represent different mutations within the DNA sequence of a gene. Mutations that create new alleles do not necessarily alter nucleotides at the same place. Recall from Chapter 13 that a compound heterozygote is an individual with a mutant phenotype who carries two different mutant alleles in the same gene. In compound heterozygotes, it is possible for a crossover between the mutant

sites of two different recessive mutant alleles to create a functional wild-type allele (Figure 16.14). This phenomenon is called **intragenic recombination**.

The first case of intragenic recombination was reported in 1940 by Clarence P. Oliver for the *lozenge* (*lz*) locus of *Drosophila melanogaster*. Most mutations in the *lz* locus create recessive alleles that alter the structure of eye facets and reduce pigment synthesis. The site of the mutation that created the *lz*^{g} allele is located in a different part of the gene than the mutation that created the *lz*^{s} allele (Figure 16.15a). A crossover between the mutant sites of the two alleles in a compound heterozygote produces a functional wild-type allele, which confers a wild-type phenotype, and a double mutant allele with both the *lz*^{g} and *lz*^{s} mutations in it (Figure 16.15b). A reversion mutation, rather than crossing-over, could be the reason for the appearance of wild-type progeny. However, in Oliver's crosses, flanking markers showed that each wild-type fly that arose in the progeny had a crossover between the flanking markers as well (Figure 16.15c), thus ruling out reversion mutation.

This was powerful evidence that a crossover between two mutations creates a functional wild-type allele. The

lozenge locus

a The lz^g and lz^s alleles are caused by mutations in different parts of the gene.

b A crossover between lz^g and lz^s produces a wild-type allele and an allele with two mutations.

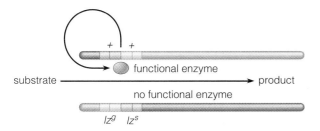

c Recombination of flanking markers reveals that the wild-type allele arises from a crossover and not from a reversion mutation.

Figure 16.15 Recombination in a compound heterozygote at the *lz* locus in *Drosophila melanogaster*.

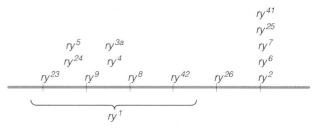

Figure 16.16 Function of a wild-type allele that arose from a crossover between two mutations in a compound heterozygote. The crossover combines the functional portions of the mutant alleles to restore the wild-type DNA sequence, which encodes a functional enzyme.

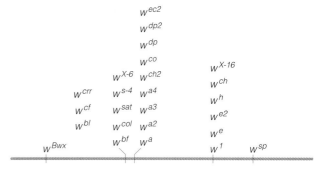

a *white* locus.

b *rosy* locus.

Figure 16.17 Fine-structure genetic maps of alleles at the **(a)** *white* locus and **(b)** *rosy* locus in *Drosophila melanogaster*. The ry^1 allele is a deletion that spans the region indicated. (Adapted from Judd, 1964. The structure of intralocus duplication and deficiency chromosomes produced by recombination in *Drosophila melanogaster* with evidence for polarized pairing. *Genetics* 49:253–265; and Chovnick, A., A. Schalet, R. P. Kernaghan, and M. Krauss. 1964. The *rosy* cistron in *Drosophila melanogaster*. Genetic fine-structure analysis. *Genetics* 50:1245–1259 (*rosy*). Used by permission.)

crowning evidence would be the recovery of the double-mutant allele that is the reciprocal product of a crossover between the two mutant sites. Unfortunately, Oliver was not able to identify this allele because he could not distinguish it phenotypically from single-mutant alleles. In 1949, however, Melvin M. Green and K. C. Green reported the discovery of the double-mutant, as well as wild-type, alleles that arose from crossovers within the *lozenge* locus, confirming Oliver's conclusions.

The wild-type allele that arises from the crossover is dominant because the crossover takes the functional portions of two mutant alleles and puts them together, restoring the DNA sequence that encodes a functional enzyme, as diagrammed in Figure 16.16.

Researchers can map mutant alleles in the same locus by tabulating recombination frequencies between the mutant alleles in compound heterozygotes. Mapping within genes, however, requires very large numbers of progeny to recover the rare wild-type recombinants. Intragenic map distances are usually very small, typically on the order of 0.01 to 0.05 cM. Figure 16.17 shows fine-structure maps of the *white* and *rosy* loci in *Drosophila melanogaster*.

Although crossing-over within genes has been demonstrated in *Drosophila* and several other multicellular eukaryotes, it is conspicuously absent or reduced in fungi. Instead, gene conversion accounts for most intragenic recombination events in fungi. A double-strand break is induced at a recombination hotspot, and genes convert according to the double-strand-break repair model. Most, if not all, crossover and gene conversion events in fungi are at the hotspots. Hotspots are rarely found within a gene or between the mutant sites in different alleles of a gene, so intragenic recombination is either absent or rare in fungi. The presence of intragenic crossing-over in *Drosophila* and other higher eukaryotes compared to the absence or rarity of intragenic recombination in fungi, indicates that recombination in higher eukaryotes is not tied to specific hotspots as it is in fungi.

~

Intragenic recombination in compound heterozygotes may create a functional wild-type allele.

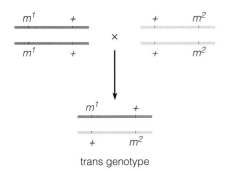

trans genotype

Figure 16.18 Formation of the trans genotype from a cross between two homozygotes.

16.5 COMPLEMENTATION

Different mutant alleles of the same gene can cause the same mutant phenotype because they affect the same protein. Mutant alleles in different genes that encode enzymes that function in the same pathway can also cause the same mutant phenotype. How do we know whether two mutant alleles that cause the same phenotype are at the same locus or are at different loci? A cis-trans test is considered the appropriate way to answer this question and is used to determine whether two recessive mutations are **allelic** (located at the same locus) or **nonallelic** (located at different loci). A cis-trans test is similar to complementation analysis in bacteria, which was described in section 7.5.

If alleles are in the same gene, a cis-trans test analyzes the phenotypes when the two recessive mutations are in cis (coupling) and trans (repulsion) conformation. The trans genotype is the easiest to construct and gives the most information. It is obtained in the F_1 offspring of a cross between an individual that is homozygous for one of the mutant alleles (m^1) and an individual that is homozygous for the other mutant allele (m^2), as diagrammed in Figure 16.18. If the two alleles are at the same locus, the F_1 progeny have the mutant phenotype because the two alleles fail to complement one another: neither encodes the gene's functional product (Figure 16.19a). However, if two recessive mutant alleles are at different loci, the F_1 progeny have the wild-type phenotype because the alleles complement one another: the wild-type alleles at the two loci produce functional products (Figure 16.19b).

This analysis completes the trans portion of a cis-trans test. In most cases, the results of the trans test correctly indicate whether or not two mutant alleles are allelic, and often researchers do not proceed any further. There are instances, however, when the results of the trans test may be misleading. The cis portion of the test is a control experiment to indicate whether or not the trans test results

are valid. To conduct the cis portion of the test, linked mutant alleles must be in cis (coupling) conformation:

Under these conditions, the phenotype from the cis test must be wild type to validate the results of the trans test. If the cis phenotype is not wild type, the results of a trans test may be called into question. Alleles that fail to produce a wild-type phenotype in a cis test include dominant, codominant, or incompletely dominant mutant alleles, polar mutations (alleles that affect adjacent genes), and alleles that reside outside the transcribed region that affect gene expression rather than the amino acid sequence of the polypeptide. The cis genotype is often difficult to obtain because it requires a crossover within a restricted region. Unless there is compelling evidence to indicate that the information from the trans test may be misleading, researchers usually forgo the cis test.

Alleles that produce a wild-type phenotype in the trans test complement each other, and for this reason the trans portion of the test is called a **complementation test**. Through a series of trans tests, alleles can be placed in complementation groups. Alleles that fail to complement one another are placed in the same **complementation group**. Generally, as complementation groups are made, any two alleles in different groups complement one another, and any two alleles in the same group fail to complement one another. For example, suppose there are 10 mutations that all cause the same phenotype. If we let "+" indicate complementation and "−" indicate no complementation, the results of all pairwise complementation tests can be summarized in a table such as Table 16.3.

To identify the complementation groups, we place all mutations that fail to complement each other in the same group. To do this, look at the column under mutation 1 in Table 16.3 first. Mutations 3, 4, 6, and 10 fail to complement mutation 1, and thus mutations 1, 3, 4, 6, and 10 fall tentatively into a single group. The group can be confirmed by looking at columns 3, 4, 6, and 10. Notice that in each case, mutations 1, 3, 4, 6, and 10 fail to complement each other in any pairwise comparison of two alleles in the group. A similar analysis of each column reveals that there are three complementation groups:

Group 1: mutations 1, 3, 4, 6, 10

Group 2: mutations 5, 8

Group 3: mutations 2, 7, 9

These data can best be explained if there are at least three loci that govern the phenotype and each complementation group includes alleles of the same locus. We cannot conclude that only three loci govern the phenotype because we may not have found mutations at other relevant loci.

Let's now analyze the results of a trans complementation test in *Neurospora crassa*.

Table 16.3 The Results of All Pairwise Complementation Tests for 10 Mutations*

Mutation Number	Mutation Number									
	1	2	3	4	5	6	7	8	9	10
1	−									
2	+	−								
3	−	+	−							
4	−	+	−	−						
5	+	+	+	+	−					
6	−	+	−	−	+	−				
7	+	−	+	+	+	+	−			
8	+	+	+	+	−	+	+	−		
9	+	−	+	+	+	+	−	+	−	
10	−	+	−	−	+	−	+	+	+	−

*The symbol "+" indicates complementation, as shown by a wild-type phenotype, and "−" indicates no complementation, as shown by a mutant phenotype.

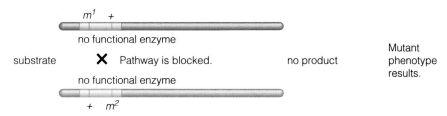

a If two mutant alleles are at the same locus in the trans heterozygote, functional enzyme is not produced, and the recessive phenotype results.

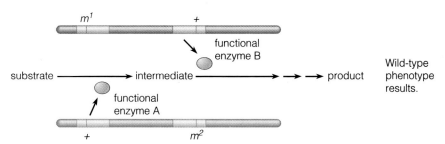

b If two mutant alleles are at different loci in the trans heterozygote, functional enzyme is produced, and the dominant phenotype results.

Figure 16.19 Biochemical basis of complementation.

Example 16.4 Complementation of osmotic mutants in *Neurospora crassa*.

In many ascomycetes, hyphae may be composed of either haploid cells or dikaryotic cells (see Figure 16.1). A dikaryotic cell contains two haploid nuclei in the same cytoplasm. Dikaryotic cells are often heterokaryotic, which means that the two nuclei differ for alleles at a particular locus, a situation analogous to heterozygosity. Heterokaryotic cells are similar to diploid heterozygous cells because two homologous copies of each chromosome are present in the same cell, even though they are in different nuclei. Genes may complement each other in a heterokaryotic cell because their mRNAs are translated in the same cytoplasm and the

Table 16.4 Information for Example 16.4: The Results of All Pairwise Complementation Tests for 10 Mutations*

Mutant Strains	Mutant Strains									
	A11	A52	A70	A71	B135	*flm-2*	L0	L5	M1	S1
A11	−									
A52	+	−								
A70	+	+	−							
A71	+	+	−	−						
B135	−	+	+	+	−					
flm-2	+	+	+	+	+	−				
L0	−	+	+	+	−	+	−			
L5	+	+	−	−	+	+	+	−		
M1	+	−	+	+	+	+	+	+	−	
S1	+	+	+	+	+	+	+	+	+	−

* The symbol "+" indicates that the heterokaryotic hyphae grew on high salt medium, and "−" indicates that they failed to grow on the medium.

products of both alleles are present in the same cell. Complementation studies in *Neurospora* and other ascomycetes are often done with heterokaryotic cells.

Certain mutants in *N. crassa* called osmotic mutants prevent cells from growing on media with high levels (> 4%) of salt. In 1969, Mays (*Genetics* 63:781–794) reported the results of complementation analysis of heterokaryotic cells for 23 osmotic mutants. The complementation results of 10 of those mutants are summarized in Table 16.4.

Problem: Assign the mutant strains to complementation groups.

Solution: Starting with the column under the first mutant strain, we determine that A11, B135, and L0 fall into one complementation group. From the column under the second mutant strain, we determine that A52 and M1 fall into a second group. From the column under the third mutant strain, we determine that A70, A71, and L5 fall into a third group. The columns under all other mutant strains except *flm*-2 and S1 confirm the conclusions derived from the columns under these first three mutant strains. The mutant strains *flm*-2 and S1 complement all other mutations, so each is alone in its own complementation group. In summary, the mutants can be assigned to five complementation groups:

Group 1: A11, B135, and L0

Group 2: A52 and M1

Group 3: A70, A71, and L5

Group 4: *flm*-2

Group 5: S1

Seymour Benzer coined the term **cistron** to describe a complementation group identified by a series of cis-trans tests. The term was derived from "cis-trans" and is often used as a synonym for *gene* to define a segment of DNA that encodes one polypeptide. Complementation usually suffices to define a gene on the basis of its mutant alleles, although there are exceptions. We discussed intragenic complementation in bacteria in the final two paragraphs of section 7.5. In those paragraphs we showed how the combination of two mutant subunits can sometimes create a functional enzyme in bacteria. The same is true for intragenic complementation in eukaryotes.

~

Recessive mutant alleles that produce a wild-type phenotype in a trans test complement one another. In most cases, complementation means that the mutant alleles belong to different loci. When recessive mutant alleles fail to complement one another, they usually belong to the same locus.

16.6 FINE-STRUCTURE MAPPING

Fine-structure mapping is identification of the relative positions of mutant alleles within the same gene. We saw examples of fine-structure maps of the *white* and *rosy* loci in *Drosophila melanogaster* in Figure 16.17. In this section, we will see how fine-structure maps are constructed.

One of the most extensive fine-structure maps was reported by Seymour Benzer in 1955 in what is now a classic paper in genetics. His experiments were important in a historical sense because they were among the first to demonstrate that genes consisted of linear sequences of

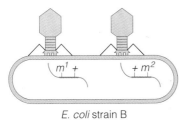

a Infect *E. coli* strain B cells with two different types of mutant phages. Some cells receive DNA from both phage types.

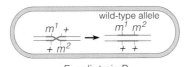

b Recombination between the two mutations can create a wild-type allele.

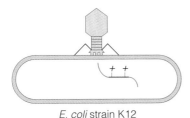

c Allow recombinant phages to develop in infected strain B cells. Infect *E. coli* strain K12 cells with recombinant phages. Only wild-type phages lyse strain K12 cells.

Figure 16.20 Benzer's procedure for identifying recombinant phages.

DNA and that DNA could mutate at many different points along the molecule. His results also demonstrated that crossing-over was possible anywhere within a viral gene and that mutations were not random along the DNA molecule.

He studied the *r*II region in *E. coli* bacteriophage T4. Mutant alleles in this region cause rapid lysis of host cells, which causes plaques that are significantly larger than usual. Among the unusual properties of *r*II mutant strains is their inability to reproduce in certain strains of *E. coli*, such as strain K12. Wild-type phages reproduce in K12, but the mutant phages do not. This provides a powerful system for selecting wild-type recombinant phages. On the other hand, both wild-type and *r*II mutant phages reproduce in *E. coli* strain B. The advantage of using a phage system is the ability to select for rare events among billions of progeny phages in a single day within a few petri plates. At the end of his experiment, Benzer had analyzed and mapped over 2400 mutations.

Benzer infected *E. coli* strain B with two different mutant phage types so that both phage particles were present in the same cell (Figure 16.20a). This created a trans test for the phage mutations. Both mutant phage types

reproduced in strain B. When both phages were in the same cell, recombination between two different mutations in the same gene produces a wild-type allele and an allele with two mutations (Figure 16.20b). The lysed cells from the strain B culture could then be added to *E. coli* strain K12, in which only the wild-type recombinant phages replicated, lysed cells, and formed plaques (Figure 16.20c). Benzer controlled the number of phage particles added to K12 by dilution and determined the frequency of recombination by counting the number of plaques and dividing by the number of phage particles added. With this procedure, he could identify recombination frequencies as low as one per billion.

Benzer isolated different rapid-lysis mutant strains and added them in pairwise combinations to *E. coli* strain B. He found two complementation groups within the *r*II region and concluded that the two complementation groups represented two genes. He then mapped mutations in both *r*II genes and determined the frequency of recombination between any two *r*II mutations within the same gene. Although his procedure allowed him to detect recombination frequencies as low as 0.0001%, the lowest frequency he found was 0.01% (1 per 10,000). Thus the minimal unit of recombination (which Benzer called a *recon*) was about one-ten-thousandth of the gene. The gene, therefore, consisted of a linear array of recons. Although Benzer did not have the tools to demonstrate it at the time, his recon corresponded to one nucleotide pair in DNA.

Benzer found some *r*II mutations that failed to recombine with several alleles mapped to a particular region. These unusual mutations, however, did recombine with alleles mapped to more distant regions. He correctly concluded that these unusual mutations were deletions. Deletion mutations fail to recombine with any point mutations that fall within the deleted region because one DNA molecule cannot recombine in a region that is missing in the other molecule (Figure 16.21). Alleles that fail to recombine with a deletion mutation fall within the deleted region, whereas those that do recombine fall outside the deletion. Benzer identified several deletion mutations and mapped them on this basis.

After Benzer had mapped a large number of point mutations, he determined the rate of reversion for each point mutation by infecting strain K12 with each single-mutant strain individually. Wild-type plaques could only appear when a mutation caused a mutant allele to revert to wild type. His results are illustrated in Figure 16.22, in which each square represents a reversion. He concluded that reversions are more common for certain mutant alleles than for others. Two in particular showed very high rates of reversion. Benzer called these **mutational hotspots**. He also coined the term *mutons* to describe the individual mutational units, which we now recognize as individual nucleotides in the DNA molecule. The terms *cistron, muton,* and *recon* were used in publications for many years before DNA analysis was

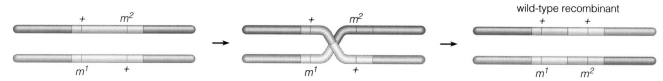

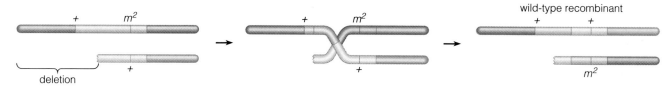

a Recombination between two point mutations in the same gene restores the wild-type allele.

b Recombination between a deletion mutation and a point mutation in the same gene restores the wild-type allele if the point mutation lies outside of the deleted region.

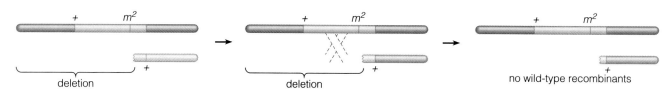

c Recombination between a deletion mutation and a point mutation in the same gene is not possible if the point mutation lies within the deleted region.

Figure 16.21 Deletion mapping in phages.

incorporated into genetic research. These terms are now used less often because geneticists speak in terms of *nucleotides* instead of *mutons* and *recons*, and use *gene* instead of *cistron*.

Benzer was not the first to show that the gene was divisible by crossing-over. This had already been done in *Drosophila*, first with the *lozenge* locus and later with other loci. The importance of Benzer's research is that it demonstrated through exhaustive analysis that the gene is composed of a linear array of mutational and recombinational units and that these units are indivisible. The structure of DNA had been described only 3 years before Benzer reported his experiments. His results were explained when scientists recognized that the units of mutation and the units of recombination are the individual nucleotides in DNA.

Benzer was able to accomplish extensive fine-structure mapping because he analyzed millions of progeny phages in a short period of time and in a small area. Fine-structure mapping in multicellular eukaryotes is much more time-consuming and difficult. Nonetheless, several fine-structure mapping experiments have been completed in *Drosophila melanogaster* and a few other species. The following example describes fine-structure mapping experiments for the *rosy* locus in *D. melanogaster*.

Example 16.5 Fine-structure mapping of the *rosy* locus in *Drosophila melanogaster*.

In a paper describing recombination among 15 alleles at the *rosy* locus in *D. melanogaster*, Chovnick et al. (*Genetics* 50:1245–1259) reported the recombination data listed in Table 16.5 for the alleles *ry¹*, *ry⁸*, *ry⁹*, *ry²³*, and *ry²⁶*. The *ry²⁶* allele is located to the right of the other four alleles on the genetic map.

Problem: (a) Show relative positions of these five alleles in the *rosy* locus. (b) Explain why *ry¹* failed to recombine with three of the other alleles. (c) Construct a fine-structure map showing the relative distances between these genes.

Solution: (a) The recombination frequency for each pair of alleles is calculated by dividing the number of wild-type progeny by the total number of progeny, then multiplying the result by 2. The recombination frequency is multiplied by 2 because wild-type recombinant progeny represent only half of the recombinant progeny. The other half are the double-mutant recombinant types, which are not

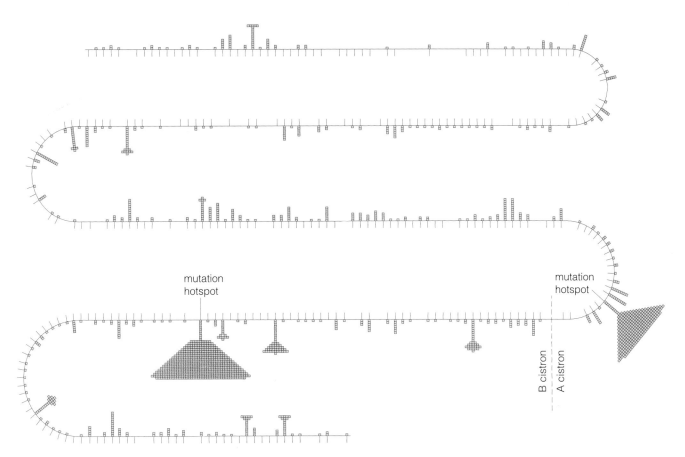

Figure 16.22 Benzer's map of reversion mutations in the A and B cistrons of the *r*II region. Each box represents one reversion mutation. Notice that reversion mutations are much more likely at certain places, which are called mutational hotspots. (Adapted from Benzer, S. 1961. On the topography of the genetic fine structure. *Proceedings of the National Academy of Sciences, USA* 47:403–415. Used by permission of the author.)

Table 16.5 Information for Example 16.5: Recombination Data for Five Alleles at the *rosy* Locus in *D. melanogaster*		
Trans Test	**Number of Wild-Type (Recombinant) Progency**	**Total Number of Progeny***
ry^1–ry^8	0	515,970
ry^1–ry^9	0	749,610
ry^1–ry^{23}	0	660,210
ry^1–ry^{26}	3	776,480
ry^8–ry^9	2	1,129,950
ry^8–ry^{23}	2	468,210
ry^8–ry^{26}	4	285,320
ry^9–ry^{23}	3	436,380
ry^9–ry^{26}	2	155,000
ry^{23}–ry^{26}	14	620,490

*Counting the numbers of progeny indicated here is a daunting task. Chovnick et al. designed their crosses to include lethal alleles in flanking genes that eliminated 95% of the nonrecombinant progeny. This made it possible to count 95% fewer progeny than they would have counted had they not designed their crosses in this way. The total number of progeny in each cross was estimated by multiplying the surviving number of progeny by 20. For this reason, all total progeny estimates end with zero.

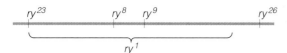

Figure 16.23 A fine-structure map of mutant alleles of the *ry* locus in *Drosophila melanogaster*, as determined from the data in Table 16.6.

detected. To obtain the map distance in cM, the recombination frequency is multiplied by 100. Table 16.6 provides the map distances in cM for each trans test.

The *ry²⁶* allele is located to the right of all other alleles *on the genetic map*, so we can create an initial map by placing *ry²⁶* on the right and all other alleles to the left of it based on their map distances from *ry²⁶*:

$$ry^{23}\text{—}ry^{8}\text{—}ry^{9}\text{—}ry^{1}\text{—}ry^{26}$$

(b) The map distances are not additive, but they are consistent with each other for the order of the alleles on the map, with the exception of the *ry¹* allele. It failed to recombine with *ry²³*, *ry⁸*, or *ry⁹*, yet based on its short distance from *ry²⁶* it should have recombined with the other alleles. These data suggest that *ry¹* is a deletion that spans the region where *ry²³*, *ry⁸*, and *ry⁹* are found. DNA analysis has shown that *ry¹* is indeed a deletion of about 1000 nucleotide pairs, that *ry²³*, *ry⁸*, and *ry⁹* are contained within the region deleted in *ry¹*, but that *ry²⁶* is outside of that region, confirming the genetic data. **(c)** The map derived from the data in Table 16.6 is shown in Figure 16.23.

The map Chovnick et al. published (see Figure 16.17b) was constructed using more alleles than we have analyzed in this example. Notice that in Figure 16.17b the positions of *ry⁸* and *ry⁹* are reversed from the order shown in Figure 16.23. Chovnick et al. used flanking markers in their crosses, and the markers showed that *ry⁹* is located to the left of *ry⁸* rather than to the right as the recombination data suggest. This discrepancy is not especially surprising. The two alleles fall very close to each other (map distance of only 0.35×10^{-3} cM), so identification of their relative positions may be influenced by sampling error. As recombination frequencies are exceptionally low in fine-structure mapping, flanking markers are more reliable indicators for positioning alleles on the map than the very small calculated map distances.

Analysis of intragenic recombination in large numbers of progeny permits fine-structure mapping of genes.

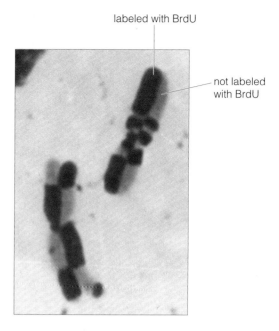

Figure 16.24 Mitotic sister chromatid exchange revealed by labeling chromosomes with BrdU. Notice that each exchange is reciprocal. (Photo courtesy of Berenice Quinzani Jordão.)

16.7 MITOTIC CROSSING-OVER

So far, we have dealt only with meiotic crossovers. Are crossovers ever detected during mitosis? The answer is yes, but in most cases there is no genetic or phenotypic consequence. During mitosis, sister chromatids exchange segments readily. This is well illustrated by the chromosomes shown in Figure 16.24. Chromosomes labeled with 5-bromodeoxyuridine (BrdU) are allowed to go through two rounds of replication and are then observed microscopically at metaphase in the second mitosis after labeling. The BrdU-labeled segments stain a different color and reveal that sister chromatids have exchanged segments, providing striking evidence of reciprocal sister chromatid exchange. However, sister chromatid exchange has no genetic consequence. Sister chromatids are identical, so an exchange between them creates chromatids that are identical to those before the exchange.

Homologous chromosomes do not pair during mitosis, so there is little opportunity for a mitotic crossover between nonsister chromatids of homologous chromosomes. However, in rare cases a chromosome happens by chance to be next to its homologue during mitosis, and crossovers between nonsister chromatids are possible. Curt Stern recognized mitotic crossing-over in *Drosophila melanogaster* and published his observations in 1936. He studied two X-linked loci, the *yellow* locus and the *singed* locus. The recessive allele *y* causes a yellow body color when homozygous in females, and the recessive allele *sn* produces bristles that appear singed. Stern studied female flies that were heterozygous for the *y* and *sn* alleles in repulsion

Table 16.6 Solution to Example 16.5: Recombination Frequencies for Five Alleles at the *rosy* Locus in *D. melanogaster*

Trans Test	Number of Wild-Type (Recombinant) Progeny	Total Number of Progeny	Map Distance (in cM)
ry^1–ry^8	0 × 2 ÷	515,970 × 100 =	0
ry^1–ry^9	0 × 2 ÷	749,610 × 100 =	0
ry^1–ry^{23}	0 × 2 ÷	660,210 × 100 =	0
ry^1–ry^{26}	3 × 2 ÷	776,480 × 100 =	0.77×10^{-3}
ry^8–ry^9	2 × 2 ÷	1,129,950 × 100 =	0.35×10^{-3}
ry^8–ry^{23}	2 × 2 ÷	468,210 × 100 =	0.85×10^{-3}
ry^8–ry^{26}	4 × 2 ÷	285,320 × 100 =	2.80×10^{-3}
ry^9–ry^{23}	3 × 2 ÷	436,380 × 100 =	1.37×10^{-3}
ry^9–ry^{26}	2 × 2 ÷	155,000 × 100 =	2.58×10^{-3}
ry^{23}–ry^{26}	14 × 2 ÷	620,490 × 100 =	4.51×10^{-3}

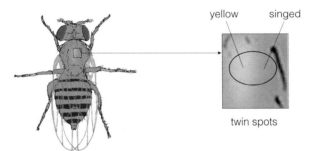

yellow   singed

twin spots

Figure 16.25 Yellow-singed twin spots in *Drosophila melanogaster*.

conformation. Phenotypically, the female flies were wild-type, but occasionally he found wild-type females with yellow-singed twin spots, a spot on the body that was yellow with wild-type bristles and an adjacent spot that had singed bristles and wild-type color (Figure 16.25). The twin spots arose from a mitotic crossover between the *singed* locus and the centromere. One of the daughter cells that arises after the crossover is homozygous *y/y*, and the other daughter cell is homozygous *sn/sn* (Figure 16.26). The two cells divide to produce daughter cells like themselves, which, in turn, produce cell lines that are homozygous for the recessive alleles. As the two daughter cells that arose from the crossover were adjacent to each other after division, they produce adjacent spots, one with yellow color and the other with singed bristles.

~

Crossing-over between sister chromatids during mitosis is frequent but has no genetic consequence because sister chromatids are identical. Mitotic crossing-over between homologous chromosomes is very rare but is possible when homologous chromosomes are close enough for partial pairing during mitosis.

SUMMARY

1. Ascomycete fungi are important in genetics for a number of reasons. Among the most important is that they allow us to observe all four haploid products of a single meiosis within the ascus.

2. Ordered tetrad analysis in ascomycetes makes it possible to map a gene to its centromere.

3. Unordered tetrad analysis can be used to map genes in relation to one another. The frequency of undetected double crossovers can be estimated directly from the data.

4. Most intragenic recombination events in fungi are gene conversions in which one allele is converted to another. Unlike intragenic crossing-over, gene conversion is nonreciprocal.

5. About half of all gene conversions are associated with crossing-over between flanking markers. In fungi, gene conversions are usually polar because of crossing-over at recombination hotspots.

6. Gene conversion data and molecular analysis have led to several models for recombination at the DNA level. The most widely accepted model for fungi is the double-strand-break repair model.

7. Gene conversion has also been demonstrated in multicellular eukaryotes, although it is very difficult to study in these species.

8. Although intragenic crossing-over is rare in fungi, it is common in most other eukaryotes.

9. Intragenic crossing-over between two mutant alleles restores the wild-type allele.

10. Two mutant alleles of the same gene usually fail to complement one another in a trans test. Mutant alleles of

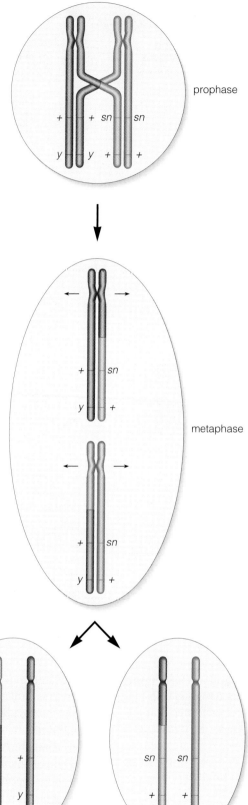

a During prophase of mitosis, homologous chromosomes happen to be adjacent to one another and a crossover takes place between the *sn* locus and the centromere.

prophase

b During metaphase, the chromosomes align in such a way that both chromatids with *y* face the same pole and both chromatids with *sn* face the other pole.

metaphase

c After the cells divide, one is homozygous *y/y*, and the other is homozygous *sn/sn*. Further cell divisions produce adjacent cell lines that are homozygous for these two alleles.

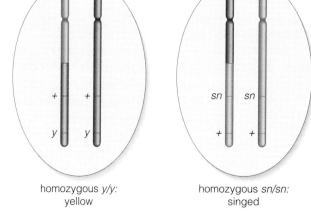

homozygous *y/y*:
yellow

homozygous *sn/sn*:
singed

Figure 16.26 Mitotic crossing-over that produces yellow-singed spots in *Drosophila melanogaster*.

different genes that cause the same phenotype complement one another.

11. A series of trans tests with all pairwise combinations of several mutant alleles allows researchers to assign alleles to complementation groups. A complementation group represents a single gene.

12. Fine-structure mapping reveals the locations of mutant alleles within a single gene.

13. Mitotic crossing-over has been observed between homologous chromosomes, but it is very rare.

QUESTIONS AND PROBLEMS

1. Why is it necessary to divide the number of second-division-segregation asci by 2 when calculating map distances between a gene and its centromere in *Neurospora*?

2. Diagram the results of two-strand, three-strand, and four-strand double crossovers between a gene and its centromere in *Neurospora*. What are the consequences of double crossing-over in gene-to-centromere mapping?

3. Devise an appropriate experimental approach to compensate for undetected double crossovers in gene-to-centromere mapping in *Neurospora*.

4. Asci with a 5:3 segregation are referred to as postmeiotic segregants. Why is the term *postmeiotic* used, and under what conditions might a 5:3 segregation arise? Use DNA models of recombination to illustrate your answer.

5. What evidence suggests that crossing-over takes place primarily (or perhaps solely) at recombination hotspots in fungi, but that recombination is not associated with hotspots in multicellular eukaryotes?

6. What evidence suggests that double-strand breaks are common at recombination hotspots in fungi?

7. What aspects of multicellular eukaryotes make it more difficult to detect gene conversion when compared to fungi?

8. **(a)** Why can nonparental ditype asci be used to estimate the number of undetected double crossovers in two-factor linkage experiments in ascomycetes? **(b)** Why must the number of nonparental ditype asci be multiplied by 6? **(c)** On what assumptions is the multiplication of nonparental ditype asci by 6 based?

9. In experiments with fungi to detect gene conversion, researchers often include a control experiment in which a mutant strain is crossed with a strain that carries the same mutation to produce homozygous meiocytes. Explain why wild-type spores may arise through gene conversion from a meiocyte that is heterozygous for different alleles at the same locus, and why wild-type spores may arise only by mutation from a meiocyte that is homozygous for a single mutant allele.

10. Summarizing the results of several experiments, Mortimer and Schild (*Microbiological Reviews* 49:181–212) reported totals of 635 first-division-segregation asci and 57 second-division-segregtion asci for mapping the *ade1* locus to its centromere in yeast. What is the recombination frequency between the gene and its centromere?

11. In 1967, Fields and Olive (*Genetics* 57:483–493) reported the data listed in Table 16.7 for mapping mutant alleles to their centromeres in the ascomycete *Sordaria brevicollis*. In this fungus, partial spindle overlap results in asymmetrical second-division-segregation asci that do not necessarily represent a crossover type. For this reason, the estimate of true second-division-segregation asci is calculated as twice the number of symmetrical second-division-segregation asci. Calculate the recombination frequency between the mutant alleles and their centromeres.

12. In the same paper used in the previous question (1967. *Genetics* 57:483–493), Fields and Olive reported the data listed in Table 16.8 for asci from a single three-factor cross. **(a)** Construct a genetic map for these three mutations. **(b)** Why is the absence of nonparental ditype asci an expected result in this cross?

13. Suppose that a new X-linked recessive allele is discovered in *Drosophila* that causes white eyes when homozygous in females and hemizygous in males. **(a)** Design experiments that would determine whether or not the allele is a new allele of the *white* locus. **(b)** What

Table 16.7 Information for Problem 11: Numbers of Asci with Different Segregation Types				
		Number of Second-Division-Segregation Asci		
Cross*	**Number of First-Division-Segregation Asci**	**Asymmetrical**	**Symmetrical**	**Total Number of Asci**
C17 × WT	1200	735	159	2094
C22 × WT	1664	1217	278	3159
C29 × WT	619	400	65	1084

*The numbers preceded by a C or an R are designations of different mutant strains. WT refers to a wild-type strain.

Table 16.8 Information for Problem 12: Numbers of Asci with Different Segregation Types

Gene Interval*	Number of Parental Ditype Asci	Number of Nonparental Ditype Asci	Number of Tetratype Asci	Total Number of Asci
R83–C28	207	0	6	213
R155–C28	212	0	1	213
R83–R155	208	0	5	213

*The numbers preceded by a C or an R are designations of different mutant strains.

results would you expect if the allele is of the *white* locus? **(c)** What results would you expect if the allele is of another locus? **(d)** If the new allele belongs to another locus that is linked to the *white* locus, how could you determine the map distance between the two loci? **(e)** If the new allele is at another locus that is syntenic but not linked (it assorts independently from *white*), how could you determine the map distance between the two loci?

14. An eye color locus called *brown* is on chromosome II in *Drosophila melanogaster*. Suppose a new autosomal recessive mutation is discovered that causes a phenotype identical to the phenotype caused by alleles of the *brown* locus. **(a)** Design experiments that would determine whether or not the new allele is at the *brown* locus or some other locus. **(b)** If it is at another locus, design an experiment that determines whether or not the new allele is at a locus that is linked to the *brown* locus. **(c)** The allele bw^5 is a deletion that includes at least part of the *brown* locus. It is homozygous lethal. Diagram a cross in which bw^5 could be used to identify whether or not the new allele belongs to the *brown* locus. **(d)** What proportion of the progeny would have the mutant brown phenotype from the cross in part c if the new allele is indeed at the *brown* locus?

15. The gene conversion experiments used in Example 16.3 had no aberrant 4:4 asci, although 2:6, 3:5, 5:3, and 6:2 asci were observed among the progeny. According to which model(s) of DNA recombination and under what circumstances would the absence of aberrant 4:4 asci among the progeny be expected? Use diagrams of DNA recombination models to justify your answer.

16. In 1963, Stadler and Towe (*Genetics* 48:1323–1344) published a paper describing experiments designed to recover wild-type recombinants between alleles of a single gene that cause auxotrophy for cysteine in *Neurospora crassa*. Their crosses were designed so that crossovers between flanking markers could be detected. The flanking markers were *lys*, located to the left on the genetic map of the *cys* locus, and *ylo*, located to the right of the *cys* locus. The general design of the crosses was

$$lys\ cysA + \times + cysB\ ylo$$

in which *cysA* and *cysB* represent any two different mutant alleles at the *cys* locus. Table 16.9 shows a portion

of their results. **(a)** Describe why these results must be explained by gene conversion and cannot be explained by intragenic crossing-over between *cys* mutants or by reversion mutations. **(b)** Determine the frequency of gene conversion for each cross listed in the table. **(c)** What proportion of the gene conversions are associated with a crossover between the *lys* and *ylo* loci when the data listed in the table are totaled? **(d)** What evidence is there of polarity with regard to gene conversion?

17. Green and Green (1949. *Proceedings of the National Academy of Sciences, USA* 35:586–591) made the crosses in *Drosophila melanogaster* listed in Table 16.10. The phenotypes associated with the flanking markers are *ct* = cut, *sn* = singed, *ras* = raspberry, and *v* = vermilion. The phenotype "like lozenge-spectacled" is the phenotype of males with the double-mutant allele. This double-mutant allele could only be detected in males. **(a)** Why can't mutation be invoked as a cause of all the "wild-type for lozenge" progeny? **(b)** Why are the "wild-type for lozenge" and "like lozenge-spectacled" progeny more likely the result of intragenic crossing-over rather than gene conversion? **(c)** The number of "like lozenge-spectacled" progeny is half the number of "wild-type for lozenge" progeny when the results from the three crosses are totaled. Why is this expected? **(d)** Draw a genetic map of the three mutant *lozenge* alleles in relation to the flanking markers. Use flanking markers rather than map distances to position the alleles on the map.

18. In Green and Green's paper on intragenic recombination at the *lozenge* locus in *Drosophila* (1949. *Proceedings of the National Academy of Sciences, USA* 35:586–591), the authors concluded that the *lozenge* locus must be three separate loci that are closely linked because it was possible to separate three *lozenge* alleles by crossing-over. Nine years earlier, Oliver (1940. *Proceedings of the National Academy of Sciences, USA* 26:452–454) speculated that there might be two duplicate *lozenge* loci based on crossing-over between two *lozenge* alleles. Given that DNA was not recognized as the genetic material at the time, why might these authors have arrived at these conclusions when the mutant *lozenge* alleles actually belong to a single locus?

19. In reference to the previous two questions, **(a)** diagram the crosses needed for a complete cis-trans test for

Cross	Number of Spores Screened	Wild-Type *cys* Recombinants			
		Parental		Recombinant	
		+ + *ylo*	*lys* + +	+ + +	*lys* + *ylo*
+ *cys⁹ ylo* × *lys cys⁴* +	170,700	16	72	59	19
+ *cys⁴ ylo* × *lys cys⁹* +	96,000	36	9	9	46
+ *cys⁷ ylo* × *lys cys⁶⁴* +	81,300	45	13	12	13
+ *cys¹⁷ ylo* × *lys cys⁶⁴* +	42,000	12	10	8	15
+ *cys⁶⁴ ylo* × *lys cys¹⁷* +	60,900	12	23	20	10
+ *cys⁹ ylo* × *lys cys¹⁷* +	267,700	44	109	100	44
+ *cys⁷ ylo* × *lys cys⁹* +	80,000	60	13	8	9
+ *cys⁴ ylo* × *lys cys¹⁵* +	163,900	66	42	37	51
+ *cys¹⁵ ylo* × *lys cys⁴* +	221,100	97	103	47	58
+ *cys⁷ ylo* × *lys cys¹⁵* +	106,200	55	27	27	26

Cross	Genotype of Female Parent	Genotype of Male Parent	Total Number of Offspring	Phenotype	
				Wild-Type for Lozenge (males and females)	Like Lozenge-Spectacled (males only)
1	+ + *lzᴮˢ* + + + *sn lz⁴⁶ ras v*	+ + *lzᵍ* + *v*	20,554	9, all singed	5, all raspberry, vermilion
2	+ + *lzᴮˢ* + + *ct* + *lzᵍ* + *v*	+ + *lzᵍ* + *v*	16,255	13, all cut	5, all vermilion
3	+ + *lz⁴⁶* + + *ct* + *lzᵍ* + *v*	+ + *lzᵍ* + *v*	16,098	4, all cut	3, all vermilion

the *lzᴮˢ* allele and the *lz⁴⁶* allele. **(b)** Did Green and Green recover the flies needed for the parents in a complete cis-trans test? **(c)** What results would be expected in a cistrans test?

20. Suppose that *gene A* and *gene B* in the diagram in Figure 16.27 are closely linked and that they produce different enzymes in the same biochemical pathway, so the mutant phenotypes for both genes are phenotypically indistinguishable. Also make the following assumptions: (1) These are autosomal genes in *Drosophila melanogaster*. (2) The arrows point to the locations of point mutations. (3) If there is no arrow or deletion within a gene, it produces the functional enzyme. (4) All mutations created recessive mutant alleles that cause the same mutant phenotype when homozygous. (5) Deletions are homozygous lethal but viable as heterozygotes. **(a)** Construct a complementation table showing all possible trans tests. Indicate a wild-type phenotype with a "+," a mutant

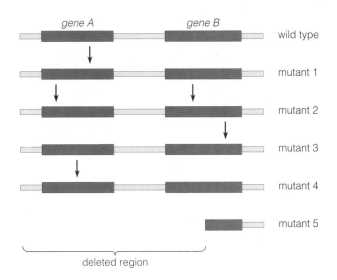

Figure 16.27 Information for problem 20: Locations of mutations in 2 genes.

Table 16.11   Information for Problem 22: Numbers of Asci with Different Segregation Types

Cross	Number of Parental Ditype Asci	Number of Nonparental Ditype Asci	Number of Tetratype Asci
NHS1 × hom3	610	1	9
NHS1 × his1	562	2	41
NHS1 × trp2	158	12	234

Table 16.12   Information for Problem 23: Ascus Phenotypes for All Possible Crosses Among 6 Strains

Cross	8 white spores	6:2 (white:black)	4:4 (white:black)
188 × 188	> 35,000	0	0
188 × 63	5,264	4	0
188 × 46	4,008	5	0
188 × W	8,047	40	0
188 × 137	8,595	81	1
188 × B	25,210	0	0
63 × 63	> 280,000	0	0
63 × 46	11,278	6	0
63 × W	5,974	17	0
63 × 137	2,100	27	0
63 × B	18,027	4	0
46 × 46	>190,000	0	0
46 × W	12,021	9	0
46 × 137	8,600	99	0
46 × B	22,345	0	0
W × W	>186,000	0	0
W × 137	18,072	150	1
W × B	23,214	0	0
137 × 137	>222,000	0	0
137 × B	22,490	0	0
B × B	>150,000	0	0

21. Using Figure 16.27, suppose the following crosses were made:

Cross 1: heterozygous mutant 1/mutant 2 × homozygous mutant 2

Cross 2: heterozygous mutant 1/mutant 3 × homozygous mutant 2

Cross 3: heterozygous mutant 1/mutant 4 × homozygous mutant 2

Cross 4: heterozygous mutant 1/mutant 5 × homozygous mutant 2

Cross 5: heterozygous mutant 2/mutant 3 × homozygous mutant 2

Cross 6: heterozygous mutant 2/mutant 4 × homozygous mutant 2

Cross 7: heterozygous mutant 2/mutant 5 × homozygous mutant 2

Cross 8: heterozygous mutant 3/mutant 4 × homozygous mutant 2

Cross 9: heterozygous mutant 3/mutant 5 × homozygous mutant 2

Cross 10: heterozygous mutant 4/mutant 5 × homozygous mutant 2

Rank the crosses for expected frequency of wild-type progeny. Assume that single crossovers are possible but that double crossovers are not because of complete interference between and within the genes.

22. In 1983, Takahashi and Sakai (*Bulletin of Brewing Science* 28:1–5) reported the data listed in Table 16.11 for mapping the *NHS1* locus to other linked loci in *Saccharomyces cerevisiae*. **(a)** Calculate the genetic map distance in each of the two-factor crosses listed in the table. **(b)** Given that the *hom3, his1,* and *trp2* loci are all located on the genetic map to the right of *NHS1*, draw a genetic map of these loci showing the distances in centimorgans between each adjacent pair.

23. In 1960, Lissouba and Rizet (*Academie des Science Comptes Rendus* 250:3408–3410) reported the recovery of black (wild-type) spores in several crosses in which white spore mutant strains were crossed in all possible

phenotype with a "−," and a lethal phenotype with an L. **(b)** Suppose an individual that is heterozygous for the mutant 5 and wild-type alleles is crossed with an individual that is homozygous for the mutant 1 allele. What proportion of the progeny should have a mutant phenotype?

combinations in *Ascobolus immersus*, an ascomycete with octad asci. Their results are summarized in Table 16.12. **(a)** What evidence in the table excludes reversion mutation as a cause of all the black spores? **(b)** What is the most reasonable explanation for the appearance of black spores? **(c)** All crosses with two different mutant alleles resulted in a few black spores, except for all but one of the crosses with mutation *B*. What does this suggest about the nature of mutation *B*? **(d)** Four black spores arose in the cross 63 × *B*, whereas no black spores arose in any other cross with *B*. What is the most likely reason for the appearance of black spores in this cross?

24. Significant discoveries in science are often made when the results of experiments do not turn out as expected. Stern's report in 1936 (*Genetics* 21:625–730) of somatic crossing-over is an example. He was conducting experiments to detect the loss of an X chromosome in the somatic cells of female flies. The female flies were heterozygous *y* + / + *sn*[3]. Stern expected to find a few female flies with single singed spots, or single yellow spots, due to loss of an X chromosome. Among 381 females, Stern found 2 with a single singed spot, 2 with a single yellow spot, and 11 with twin yellow-singed spots. Stern did not expect to see twin spots but invoked somatic crossing-over as an explanation for their appearance. **(a)** Why must somatic crossing-over, rather than chromosome loss, be the cause of the twin spots? **(b)** Single singed spots or single yellow spots may be the result of chromosome loss, or they may arise from mitotic crossovers. Explain how chromosome loss could cause single spots, and how mitotic crossing-over could cause single spots.

25. In ascomycetes, the relationship of interference to genetic map distance between two markers can be estimated directly using the frequency of nonparental ditype asci in two-factor crosses. As predicted by mapping functions, interference increases with decreasing map distances. This relationship holds true until markers are very close to each other, such as different alleles within the same gene. For these very closely linked markers, the situation is reversed: high negative interference is observed. Explain the reasons for high positive interference for closely linked genes and high negative interference for different alleles of the same gene.

FOR FURTHER READING

Summaries of original data and references for gene-to-centromere and two-factor mapping experiments using tetrad analysis in fungi are in **Barrat, R. W., D. Newmeyer, D. D. Perkins, and L. Garnjobst. 1954. Map construction in *Neurospora crassa*. *Advances in Genetics* 6:1–93**; and in **Mortimer, R. K., and D. Schild. 1985. Genetic map of *Saccharomyces cerevisiae*, edition 9. *Microbiological Reviews* 49:181–212**. Nonreciprocal recombination of alleles in fungi was described by **Mitchell, M. B. 1955. Aberrant re-**combination of pyroxidine mutants of *Neurospora*. *Proceedings of the National Academy of Sciences, USA* 41:215–220. This research was followed up by many researchers who demonstrated gene conversion in several fungal species. A paper that shows most of the unique characteristics of fungal gene conversion is **Stadler, D. R., and A. M. Towe. 1963. Recombination of allelic cysteine mutants in *Neurospora*. *Genetics* 48:1323–1344**. Gene conversion in *Drosophila* was described in a brilliant manner by **Chovnick, A., G. H. Ballantyne, D. L. Ballie, and D. G. Holm. 1970. Gene conversion in higher organisms: Half-tetrad analysis of recombination within the *rosy* cistron of *Drosophila melanogaster*. *Genetics* 66:315–329**. The Holliday model of recombination was described by **Holliday, R. 1964. A mechanism for gene conversion in fungi. *Genetical Research* 5:282–304**. The Meselson-Radding model was described by **Meselson, M. S., and C. M. Radding. 1975. A general model for genetic recombination. *Proceedings of the National Academy of Sciences, USA* 72:358–361**. Several double-strand-break repair models were described by **Szostak, J. W., T. L. Orr-Weaver, R. J. Rothstein, and F. W. Stahl. 1983. The double-strand-break repair model for recombination. *Cell* 33:25–35**. This model was modified to account for observations using DNA analysis in yeast by **Sun, H., D. Treco, and J. W. Szostak. 1991. Extensive 3'-overhanging, single-stranded DNA associated with the meiosis-specific double-strand breaks at the *ARG4* recombination initiation site. *Cell* 64:1155–1161**. A review of eukaryotic recombination at the DNA level was written by **Stahl, F. W. 1994. The Holliday junction on its 30th anniversary. *Genetics* 138:241–246**. The first case of intragenic recombination was reported by **Oliver, C. P. 1940. A reversion to wild-type associated with crossing-over in *Drosophila melanogaster*. *Proceedings of the National Academy of Sciences, USA* 26:452–454**; and was followed by **Green, M. M., and K. C. Green. 1949. Crossing-over between alleles at the lozenge locus in *Drosophila melanogaster*. *Proceedings of the National Academy of Sciences, USA* 35:586–591**. An excellent paper on fine-structure mapping of alleles at the *rosy* locus that illustrates some innovative fine-structure mapping techniques was published by **Chovnick, A., A. Schalet, R. P. Kernaghan, and M. Krauss. 1964. The *rosy* cistron in *Drosophila melanogaster*: Genetic fine structure analysis. *Genetics* 50:1245–1259**. Seymour Benzer reviewed his own work in an excellent and very readable article, **Benzer, S. 1962. The fine structure of the gene. *Scientific American* 206 (Jan):70–84**. His classic original paper on intragenic recombination and deletion mapping was published in **Benzer, S. 1961. On the topography of the genetic fine structure. *Proceedings of the National Academy of Sciences, USA* 47:403–415**. Curt Stern's extensive and classic paper on mitotic crossing-over is **Stern, C. 1936. Somatic crossing over and segregation in *Drosophila melanogaster*. *Genetics* 21:625–730**.

For additional reading, go to InfoTrac College Edition, your online research library at: http:www.infotrac-college.com/brookscole

CHAPTER 17

KEY CONCEPTS

Alterations in chromosome number and structure influence phenotypic expression of traits and speciation.

~

Aneuploidy is an aberration in which chromosomes are missing or added so that chromosome sets are unbalanced.

~

Polyploidy, in which there are three or more sets of chromosomes in a cell, is common in plants and is important in plant evolution, but is rare in animals.

~

Alterations in chromosome structure include deletions, duplications, inversions, translocations, fissions, and fusions. These alterations have important genetic consequences.

~

Most chromosome alterations in humans do not permit survival of the fetus. Those chromosome alterations that permit survival usually have medical syndromes associated with them.

ALTERATIONS IN CHROMOSOME NUMBER AND STRUCTURE

Most people know or have seen a child who has Down syndrome (Figure 17.1a). Down syndrome is a collection of symptoms that include mental retardation, short stature, hands that are reduced in length, and an epicanthic fold over the eye (a prolongation of a fold of the skin of the upper eyelid). Children with Down syndrome require significant medical care and special education. Most people with Down syndrome lead happy lives, although their life spans are usually shortened by a variety of health problems, such as congenital heart defects. Down syndrome affects between 1 in 700 and 1 in 1000 children born. An extra copy of chromosome 21, the smallest of human chromosomes, is the usual cause of Down syndrome (Figure 17.1b). Down syndrome is perhaps the best-known example of aneuploidy, an alteration in chromosome number.

Alterations in chromosome number and structure are all around us. Many of the plants we eat are polyploids, meaning that they have three or more sets of chromosomes. Some individual plants and animals are aneuploids, in which either a chromosome is missing or there is an extra chromosome. In addition to Down syndrome, Klinefelter and Turner syndromes (mentioned briefly in Chapter 14) are also examples of aneuploidy in humans. We will discuss all of these syndromes in detail in this chapter. Chromosome structure may be altered in significant ways. Chromosome segments may be deleted, duplicated, or shuffled around by inversion and translocation. These alterations usually have major genetic and phenotypic effects. Alterations in chromosome number and structure are important for genetic change on an evolutionary scale, more important than the types of simple mutations we discussed in Chapter 5. We will discuss some of the evolutionary implications in Chapter 21. For now, let's take a close look at what these chromosome changes are, how they arise, how they are inherited, and what effects they have.

17.1 ANEUPLOIDY

As we begin our discussion of aneuploidy, we need to define a few important terms. **Ploidy** describes the number of chromosome sets in a cell. **Euploidy** is the situation in which each cell contains complete sets of chromosomes, with no missing or extra chromosomes. The genetic material in a euploid cell is **balanced**; it has the correct number of chromosomes and the correct amount of genetic material in those chromosomes. On the other hand, **aneuploidy** is a situation in which each cell has

a a child with Down syndrome.

b The karyotype of a male with Down syndrome.

Figure 17.1 Down syndrome in humans. The arrows in part b point to the three copies of chromosome 21. (Karyotype courtesy of A. R. Brathman and B. Issa.)

at least one extra chromosome or is missing at least one chromosome. Aneuploid cells are **unbalanced** because there is extra or missing genetic material. For example, most people with Down syndrome have aneuploid cells with 47 chromosomes; each cell has three copies of chromosome 21 and two copies of every other chromosome (Figure 17.1b).

Chromosome numbers in somatic cells and gametes are often designated with the symbol n, which represents the haploid number of chromosomes for the species. The haploid number of chromosomes is the number of chromosomes in a balanced gamete. For example, $n = 23$ in humans because there are 23 chromosomes in a typical sperm or egg cell. Most somatic cells are diploid, which is represented as $2n$. Human somatic cells have $2n = 46$ chromosomes.

The most common form of aneuploidy is **trisomy**, in which there is one extra chromosome. The next most common form of aneuploidy is **monosomy**, in which one chromosome is missing. Trisomy is described numerically as $2n + 1$, which means that the individual has the usual $2n$ number of chromosomes plus one extra chromosome in the somatic cells. Monosomy is described numerically as $2n - 1$, which means that each somatic cell is missing one copy of a particular chromosome but has the usual two copies of every other chromosome. For example, human trisomy is $2n + 1 = 47$, and human monosomy is $2n - 1 = 45$. Aneuploidy is often designated by the chromosome that is extra or missing. For instance, most people with Down syndrome have **trisomy 21**, which

means that each somatic cell has three copies of chromosome 21 and two copies of every other chromosome.

Although trisomy and monosomy are the most common types of aneuploidy, there are other types. **Nullisomy** is the absence of both homologous copies of a chromosome that should normally be present, written numerically as $2n - 2$. **Double trisomy** is trisomy for two different chromosomes ($2n + 1 + 1$). **Tetrasomy** is the presence of two extra copies of the same chromosome ($2n + 2$).

Aneuploidy may arise in a number of ways, but the most common is nondisjunction during meiosis. **Nondisjunction** is the failure of a homologous chromosome pair to disjoin (separate) from one another and migrate to opposite poles during meiosis I, or the failure of sister chromatids to disjoin and migrate to opposite poles during meiosis II. Instead, both copies of a chromosome pair or chromatids move to the same pole. One of the daughter cells that arises from nondisjunction carries two copies of the chromosome, and the other daughter cell carries no copies of that chromosome. All of the gametes that develop after nondisjunction during meiosis I and half of the gametes that develop after nondisjunction during meiosis II are unbalanced, some with $n + 1$ and others with $n - 1$ chromosomes, as illustrated in Figure 17.2.

If a gamete with an extra chromosome ($n + 1$) unites with a normal, balanced gamete (n), the zygote is trisomic ($2n + 1$). If a gamete that lacks a chromosome ($n - 1$) unites with a normal, balanced gamete (n), the zygote is monosomic ($2n - 1$). Nondisjunction is reciprocal: for every nondisjunction event, equal numbers of $n + 1$ and $n - 1$ gametes are produced. Therefore, there should be as many trisomic as monosomic zygotes. However, in humans and most other animals, trisomy is much more frequent than monosomy at birth. Trisomies in humans are well-known, but monosomies for autosomal chromosomes are never observed because monosomic embryos do not survive gestation.

The effects of aneuploidy may vary significantly depending on the species. In diploid plants, $n - 1$ gametes usually abort before fertilization, but $n + 1$ gametes may survive and undergo fertilization. In animals, $n - 1$ and $n + 1$ gametes may undergo fertilization, but monosomic embryos usually abort very early during gestation. The phenotypic effects of aneuploidy are especially severe in humans and other mammals. Alterations in chromosome structure may also have severe phenotypic effects in humans. We will discuss examples of aneuploidy and alterations in chromosome structure in humans in section 17.4.

In many plant and a few animal species, trisomy may be transmitted from one generation to the next. Distinct phenotypes are often associated with each type of trisomy, as shown in Figure 17.3. During meiosis, trisomic chromosomes segregate from one another so that half the

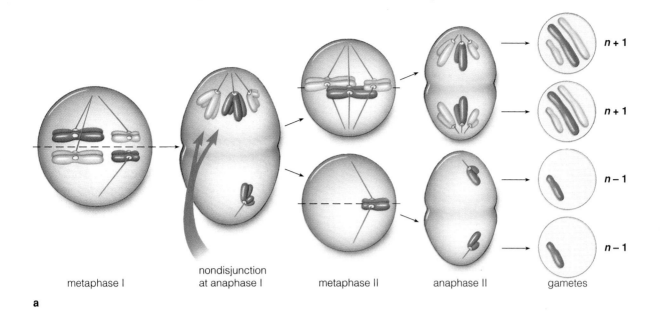

a

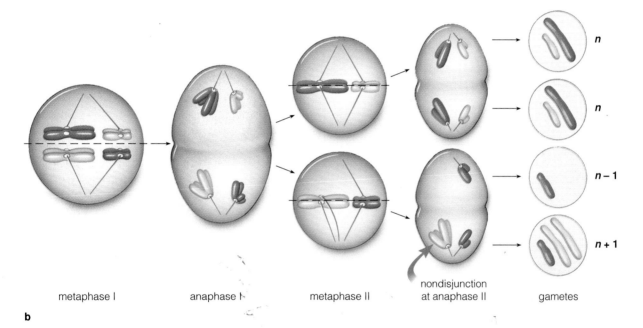

b

Figure 17.2 Nondisjunction **(a)** at anaphase I and **(b)** at anaphase II.

gametes have n chromosomes and half have $n + 1$ chromosomes, perpetuating trisomy in the next generation, as illustrated in the following example.

Example 17.1 Genetic segregation in a trisomic plant.

In 1920, Blakeslee et al. (*Science* 52:388–390) described the genetic and phenotypic effects of trisomy in the jimsonweed (*Datura stramonium*). Plants with the normal diploid number of chromo-

somes ($2n = 24$) have a normal phenotype, and plants that are trisomic for chromosome 12 ($2n + 1 = 25$) have a phenotype called poinsettia. The p locus that governs flower color is on chromosome 12. Plants that carry the dominant P allele have purple flowers, and plants that are homozygous for the recessive p allele have white flowers. The authors of this study examined haploid pollen mother cells (the products of meiosis) from a poinsettia plant and found that half of the cells had one full set of $n = 12$ chromosomes and the other half had $n + 1 = 13$ chromosomes, which is the expected

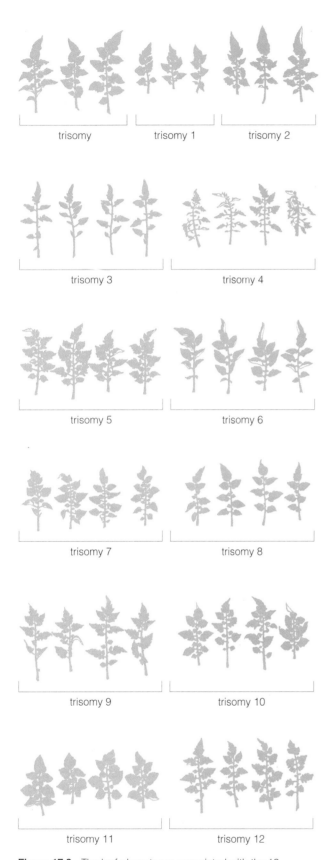

trisomy

trisomy 1

trisomy 2

trisomy 3

trisomy 4

trisomy 5

trisomy 6

trisomy 7

trisomy 8

trisomy 9

trisomy 10

trisomy 11

trisomy 12

Figure 17.3 The leaf phenotypes associated with the 12 trisomies of tomato. The numbers refer to each of the 12 chromosomes in tomato. (Adapted from Rick, C. M., and D. W. Barton. 1954. Cytological and genetical identification of the primary trisomics of the tomato. *Genetics* 39:640–666. Used by permission)

segregation in a trisomic plant. The authors also found that pollen grains with 13 chromosomes are inviable and cannot participate in fertilization, but pollen grains with $n = 12$ chromosomes are normal and fully capable of fertilization. Therefore, the extra chromosome cannot be transmitted by the male parent. On the other hand, ova (female gametes) bearing $n + 1 = 13$ chromosomes can be readily fertilized, as can ova with $n = 12$ chromosomes, so the extra chromosome can be transmitted through the female parent. In one experiment, they self-fertilized a poinsettia plant with purple flowers and recovered a ratio of 5 purple-flowered:4 white-flowered plants among the progeny that had the normal phenotype, and a ratio of 7 purple-flowered:2 white-flowered plants among the progeny that had the poinsettia phenotype.

Problem: Determine the genotype of the parent plant, and use a Punnett square to explain the observed ratios.

Solution: Because there are three copies of chromosome 12 in the trisomic poinsettia plants, there are four possible genotypes in the somatic cells: *PPP*, *PPp*, *Ppp*, and *ppp*. Only the *ppp* genotype produces white flowers; the other three genotypes produce purple flowers because *P* is dominant. The parent plant had purple flowers and the poinsettia phenotype, and it bore white-flowered offspring when self-pollinated, so it must have been trisomic, and it must have carried both the *P* and *p* alleles. Therefore, its genotype must have been *PPp* or *Ppp*. The gametes produced in a *PPp* plant are all possible meiotic segregations of the three chromosomes. In each segregation, half the gametes have two copies of chromosome 12 and half have one copy. If we number the three copies of chromosome 12 as 1, 2, and 3,

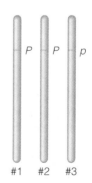

then there are three possible ways in which the chromosomes can segregate:

P and *Pp* (1 segregates from 2 and 3)

P and *Pp* (2 segregates from 1 and 3)

p and *PP* (3 segregates from 1 and 2)

female gametes

male gametes	2P	p	PP	2Pp
2P	4PP purple flowers normal	2Pp purple flowers normal	2PPP purple flowers poinsettia	4PPp purple flowers poinsettia
p	2Pp purple flowers normal	pp white flowers normal	PPp purple flowers poinsettia	2Ppp purple flowers poinsettia

a Punnett square for self-fertilization of a trisomic plant with the genotype *PPp*.

female gametes

male gametes	P	2p	2Pp	pp
P	PP purple flowers normal	2Pp purple flowers normal	2PPp purple flowers poinsettia	Ppp purple flowers poinsettia
2p	2Pp purple flowers normal	4pp white flowers normal	4Ppp purple flowers poinsettia	2ppp white flowers poinsettia

b Punnett square for self-fertilization of a trisomic plant with the genotype *Ppp*.

Figure 17.4   Solution to Example 17.1: Segregation of alleles in two trisomic genotypes.

The three segregation patterns are equally frequent, so the array of viable female gametes is $2P + p + PP + 2Pp$. The $n + 1$ male gametes are inviable, so the array of male gametes is $2P + p$. The Punnett square for self-fertilization is illustrated in Figure 17.4a. The expected phenotypic ratio is 8 purple-flowered plants:1 white-flowered plant among the normal plants and all purple-flowered plants among the poinsettia plants, which does not correspond with the observed results. The genotype of the parent plant must not be *PPp*. This leaves only *Ppp* as the possible genotype. The array of female gametes is $P + 2p + 2Pp + pp$, and the array of male gametes is $P + 2p$. The Punnett square for this genotype is illustrated in Figure 17.4b. The observed ratios match those predicted by this Punnett square. Therefore, the genotype of the parent plant is *Ppp*.

Aneuploidy is a deviation from euploid chromosome numbers. The most common forms of aneuploidy are trisomy, in which there is one extra chromosome, and monosomy, in which a single chromosome is missing.

17.2   POLYPLOIDY

Until now, we have focused on eukaryotes that are diploid (such as *Drosophila*, maize, mice, and humans) or haploid (such as *Neurospora* and *Saccharomyces*) during the major portion of their life cycles. However, about half of all flowering plant species and a few animal species are polyploid. **Polyploidy** is the presence of three or more complete chromosome sets in the somatic cells. Polyploidy in its various forms is responsible for much of the evolutionary history of flowering plants, a topic we will examine in Chapter 21. Polyploidy may appear occasionally in mammals as a result of errors in meiosis. However, polyploid mammals are nearly always aborted spontaneously before completing gestation. The rare polyploid individuals that survive birth are severely deformed and die shortly afterward.

Earlier in this chapter, we used the term n to describe the number of chromosomes in a gamete, which is the haploid number. Strictly speaking, the term *haploid* means the number of chromosomes in a gamete, not necessarily one copy of each chromosome. In discussing polyploidy, the term n retains its same definition: the haploid number of chromosomes in a normal gamete. However, in polyploids, n represents more than one set of chromosomes. For example, a tetraploid plant (one with four sets of chromosomes in each somatic cell) has diploid gametes. (Remember that meiosis halves the somatic number of chromosomes.) The number of chromosomes in one complete set is the **monoploid** number of chromosomes, and the haploid number is the number of chromosomes in a typical gamete. The term most often used to represent the monoploid number is x, and it can be applied to diploid as well as polyploid species. In diploid species, $n = x$, but in polyploids, n is a multiple of x. For example, in humans (a diploid species), $n = x = 23$,

Table 17.1 Examples of Polyploid Plants and Animals

Genus	Common Name	Ploidy
Plants		
Triticum	Wheat	2x, 4x, 6x
Avena	Oat	2x, 4x, 6x
Gossypium	Cotton	2x, 4x
Beta	Beet	2x, 4x, 6x
Nicotiana	Tobacco	2x, 4x
Solanum	Potato	2x, 4x, 6x
Musa	Banana	2x, 3x, 4x
Chrysanthemum	Chrysanthemum	2x, 4x, 6x, 8x, 10x
Rubus	Blackberry	2x, 3x, 4x, 5x, 6x, 7x, 8x, 9x, 10x, 12x
Atriplex	Saltbush, orache	2x, 4x, 6x, 8x, 10x, 12x, 14x, 20x
Animals		
Cnemidophorus	Whiptail lizards	2x, 3x
Hyla	Tree frogs	2x, 4x
Xenopus	Frogs	2x, 4x, 6x

and $2n = 2x = 46$. However, the somatic cells of a potato are tetraploid, $2n = 4x = 48$, so $x = 12$, and the gametes of a potato plant are diploid and have $n = 2x = 24$ chromosomes.

Ploidy levels are designated by numerical prefixes to indicate how many chromosome sets are in a cell:

x	monoploid	$7x$	heptaploid
$2x$	diploid	$8x$	octaploid
$3x$	triploid	$9x$	nonaploid
$4x$	tetraploid	$10x$	decaploid
$5x$	pentaploid	$11x$	undecaploid
$6x$	hexaploid	$12x$	dodecaploid

In nature, the most common polyploids are tetraploids, followed by hexaploids and octaploids. Levels of polyploidy rarely surpass dodecaploidy in nature, but a few plant species with ploidies as high as $20x$ have been observed. Table 17.1 provides examples of some polyploid plant and animal species. There are no polyploid mammalian species.

The Origin of Polyploidy

Polyploidy in plants usually arises from unreduced gametes. An **unreduced gamete** has the same number of chromosomes as the somatic cells. For example, a diploid plant may produce a few diploid gametes among its many normal monoploid gametes. If an unreduced diploid gamete unites with a normal haploid gamete, a triploid zygote results. Rarely, two unreduced gametes may unite to produce a tetraploid zygote.

Triploidy is a potential dead end for polyploidy because triploid plants are sterile. Meiosis halves the chromosome number, and it is not possible to divide an odd number of chromosome sets equally into complete euploid sets ($3x \div 2 = 1.5x$). So gametes that arise in a triploid plant are usually unbalanced and abort. Triploid plants are usually seedless or have undeveloped seeds. Bananas and seedless watermelons are common examples of seedless fruits from triploid plants. Plants with higher odd ploidy levels ($5x$, $7x$, $9x$, and so on) are also sterile because meiosis cannot halve the chromosome number to produce balanced gametes.

Although a triploid plant is sterile, a tetraploid one is not. Meiosis halves the number of chromosomes in tetraploid cells to produce balanced diploid gametes. Diploid gametes from a tetraploid plant participate normally in fertilization. Most polyploid species have even ploidies (tetraploid, hexaploid, octaploid, and so on). Plants with odd ploidies can survive but rarely reproduce because of sterility. Thus, odd ploidies are usually an aberration rather than the normal ploidy for a species.

There are a few exceptions, including some well-known triploid species. Most commercial types of bananas are triploid and have no seeds. There are some diploid and tetraploid types of bananas that contain

seeds. Banana growers propagate triploid banana plants asexually by placing cuttings (pieces of the live plant) into the ground instead of planting seeds. Roots sprout from the cutting, and a new plant grows. Cell division is entirely mitotic, and there is no opportunity for exchange of genetic material among plants. Plant breeders can create new triploid types of bananas by hybridizing diploid and tetraploid types.

Several species of whiptail lizards in the American Southwest are triploid. Some species consist entirely of females; in other species, males are very rare. Triploid females lay unfertilized eggs that arise from unreduced gametes, a process called **parthenogenesis**. The triploid eggs develop into triploid females. Again, there is no opportunity for genetic exchange through sexual reproduction.

Tetraploidy may arise in several ways. Two unreduced gametes from diploid plants may unite to produce a tetraploid zygote. Alternatively, a triploid plant may produce an occasional unreduced triploid gamete. When an unreduced triploid gamete from a triploid plant unites with a monoploid gamete from a diploid plant, a tetraploid zygote results. Occasionally, a somatic cell in a diploid plant may duplicate its chromosomes in preparation for mitosis but fail to complete mitosis, a process called **endoploidy**. An endoploid cell in a diploid plant has twice the number of chromosomes that it had before the chromosomes duplicated, so it is tetraploid. The tetraploid cell may duplicate its chromosomes again, then divide normally by mitosis to produce two tetraploid daughter cells. As the tetraploid cells continue to divide mitotically, they produce a sector on the plant that consists entirely of tetraploid cells. Flowers that develop on the tetraploid sector produce gametes that are all diploid (Figure 17.5). Self-fertilization in a tetraploid flower therefore produces all tetraploid seeds.

Higher levels of ploidy can arise in a similar manner. Unreduced tetraploid gametes from a tetraploid plant can unite with the normal reduced gametes from other tetraploid plants to produce a hexaploid zygote. Two unreduced tetraploid gametes can unite to produce an octaploid zygote. A triploid gamete from a hexaploid plant can unite with a diploid gamete from a tetraploid plant to produce a pentaploid zygote. Unreduced gametes on a pentaploid plant can unite with monoploid gametes to produce hexaploids. Among certain groups of plants, this sort of rampant polyploidization forms hybrid swarms of plants with various levels of polyploidy. In such a situation, the evolution of new polyploid species can proceed at an unusually rapid pace.

Autopolyploidy and Allopolyploidy

Polyploidy is often subdivided into autopolyploidy and allopolyploidy. These two terms describe differences in the origin of the polyploidy as well as the behavior of the

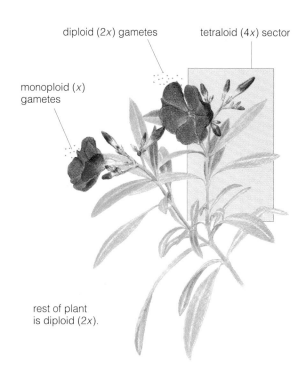

diploid (2x) gametes   tetraloid (4x) sector

monoploid (x) gametes

rest of plant is diploid (2x).

Figure 17.5 The production of diploid gametes in a flower on a tetraploid sector that arose through endoploidy. (Drawing by Marcus Alan Vincent.)

chromosomes during meiosis. **Autopolyploidy** is a situation in which all of the chromosomes in a polyploid species are from the same ancestral diploid species.

An autotetraploid has four homologous copies of each chromosome. Each of the four chromosomes is homologous to the other three, so any two homologous chromosomes can pair with each other during meiosis. Many autopolyploid species have some unusual types of chromosome pairing. Two paired chromosomes are called a **bivalent**, which is typical for diploids. In an autotetraploid, four homologous chromosomes may pair as two bivalents, as depicted in Figure 17.6a. Alternatively, two homologous chromosomes may initiate pairing at one point, while a third homologue simultaneously initiates pairing with one of the chromosomes at another point. When paired with one another, the three chromosomes are a **trivalent**. The fourth chromosome remains unpaired as a **univalent** (Figure 17.6b). Sometimes, all four homologous chromosomes pair with one another as a **quadrivalent** (Figure 17.6c).

Allopolyploidy is a situation in which the chromosomes of polyploid species are from different, but closely related, ancestral diploid species. When two species diverge from the same ancestral species, changes in the DNA and the chromosomes over time tend to separate the species genetically. Geographic barriers often prevent the two related species from exchanging genetic material, so the two species diverge genetically from one another (Figure 17.7a). When the two species encounter one another after being separated, they may hybridize. Each

a Two bivalents.

b One trivalent and one univalent.

c One quadrivalent.

Figure 17.6 Pairing possibilities for four homologous chromosomes in an autotetraploid.

of the diploid species produces monoploid gametes, so interspecific hybrids are diploid. However, if the chromosomes of the two species have diverged sufficiently, normal pairing during meiosis may not be possible in the diploid hybrid, so it is sterile. Chromosome 1 from species A is similar to chromosome 1 from species B even though the two chromosomes cannot pair properly. They are not homologous chromosomes because they cannot pair with each other, but they are like homologous chromosomes because of their similarity. They are called **homeologous chromosomes**, which means that there is some homology between them but that they do not normally pair with each other (Figure 17.7b).

If the cells in the sterile hybrid undergo endoploidy, the gametes that arise in flowers on the endoploid sector contain duplicated chromosomes that can pair normally with each other. Chromosome doubling in a diploid interspecific hybrid produces an **amphidiploid**, a plant that has a diploid set of chromosomes from each contributing species. If two species contribute a monoploid set of chromosomes to a zygote ($x + x$) and the chromosomes double so there is a diploid set from each of the two parents ($2x + 2x$), the resulting somatic cells are amphidiploid and **allotetraploid** because there are four monoploid sets of chromosomes, as shown in parts c and d of Figure 17.7.

Bread wheat is an example of an allohexaploid with three different genomes, named the A, B, and D genomes. Each of the genomes has 7 chromosomes in a monoploid set, so $x = 7$, and the somatic cells of wheat have $2n = 6x = 42$ chromosomes, $2x$ from the A genome, $2x$ from the B genome, and $2x$ from the D genome. Figure 17.8 shows a proposed model of how hexaploid bread wheat arose

from its three diploid ancestors. Two of the three diploid ancestors still grow as wild species in Mesopotamia. The third is apparently extinct.

Endoploidy is often an important event in the inception of polyploidy. Endoploidy can be induced artificially with a chemical called **colchicine**. Colchicine inhibits spindle fiber formation during mitosis, so chromosomes cannot separate to daughter cells and the cell fails to divide. After colchicine treatment, some of the treated cells have twice the number of chromosomes that they had before being treated. After the cell cycle is completed without a cell division, and the cells prepare to divide again, the colchicine has diffused and no longer has any effect. The result is artificially induced endoploidy. Flowers that develop on the endoploid sector of the plant produce unreduced gametes, which can give rise to polyploids.

This procedure has been used to make artificial polyploids of some garden plants. Autopolyploids of herbaceous (nonwoody) plants tend to be larger and more vigorous than their diploid counterparts, although seed production is often reduced. For most crop plants, the lack of seed production is a detriment to using autopolyploidy, although some artificial tetraploids, such as autotetraploid asparagus and autotetraploid apples, are grown commercially.

Plant breeders have induced allopolyploidy to create new species. One of the best examples is **triticale**, an induced allooctaploid derived from bread wheat and rye (Figure 17.9). Wheat is a hexaploid with three different genomes, and rye is a diploid. When wheat is pollinated by rye, a seed with a hybrid embryo forms. The plant derived from the hybrid embryo is tetraploid, but none of the chromosomes can pair properly during meiosis, so the plant is sterile. However, colchicine treatment produces a fertile allooctaploid with two copies of all the chromosomes in wheat and rye ($2n = 8x = 56$).

Rye is a very hardy plant and can thrive under adverse environmental conditions that are detrimental to wheat. However, wheat has certain qualities that make it superior to rye for bread making. The plant breeders who made triticale hoped to introduce the hardiness of rye into the new plant and maintain the favorable baking characteristics of wheat. Triticale is more hardy than wheat and is more productive than either wheat or rye. The first triticales produced had several problems, including a tendency for the grains to shrivel, but plant breeders have eliminated most of the problems through hybridization and selection. Triticale has recently started to attain commercial success.

Induced polyploidy has also been used to produce seedless fruits. The seedless watermelon is a good example (Figure 17.10). It is the triploid F_1 progeny of a cross between an induced autotetraploid and a normal diploid. Not all seedless fruits are triploid, however. Some, such as the navel orange, are seedless because mutant alleles prevent seed formation.

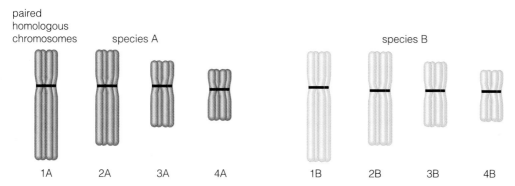

a Species A and species B are both diploids that diverged from a common diploid ancestor. Their chromosomes are similar.

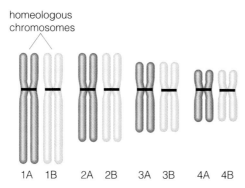

b A hybrid between species A and species B is also diploid. The chromosomes of species A and species B are homeologous; they have diverged to the point that they cannot pair in the diploid hybrid. The diploid hybrid is sterile.

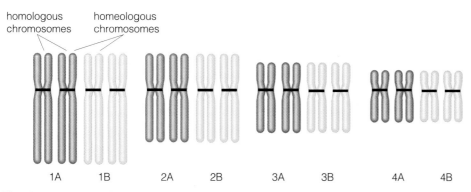

c The chromosomes double through endoploidy to form an amphidiploid that is an allotetraploid.

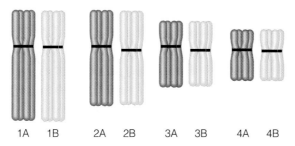

d Homologous chromosomes pair with one another. The allotetraploid is fertile.

Figure 17.7 Origin of an allotetraploid.

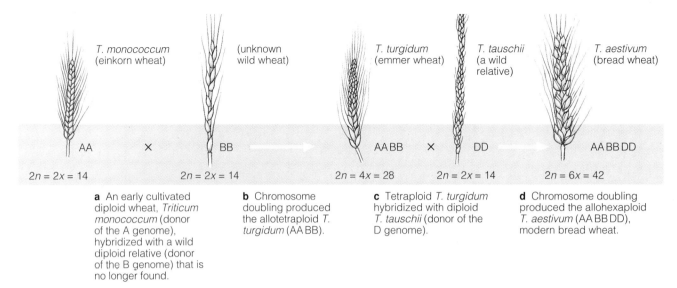

| T. monococcum (einkorn wheat) | | (unknown wild wheat) | | T. turgidum (emmer wheat) | | T. tauschii (a wild relative) | | T. aestivum (bread wheat) |

AA   ×   BB   →   AA BB   ×   DD   →   AA BB DD

$2n = 2x = 14$ $2n = 2x = 14$ $2n = 4x = 28$ $2n = 2x = 14$ $2n = 6x = 42$

a An early cultivated diploid wheat, *Triticum monococcum* (donor of the A genome), hybridized with a wild diploid relative (donor of the B genome) that is no longer found.

b Chromosome doubling produced the allotetraploid *T. turgidum* (AA BB).

c Tetraploid *T. turgidum* hybridized with diploid *T. tauschii* (donor of the D genome).

d Chromosome doubling produced the allohexaploid *T. aestivum* (AA BB DD), modern bread wheat.

Figure 17.8 Origin of hexaploid bread wheat.

In the following example, we examine polyploidy in interspecific hybrids produced in nature.

Example 17.2   Polyploidy and natural interspecific hybridization.

Perennial desert shrubs of the genus *Atriplex* are typically dioecious; they consist of separate male and female plants. Species within this genus also have varying levels of ploidy. Four-wing saltbush (*A. canescens*) and saltsage (*A. tridentata*) are widespread species throughout the American West. Both are fully fertile. In 1979, Stutz et al. (*American Journal of Botany* 66:1181–1193) reported the discovery of an isolated female four-wing saltbush plant surrounded by a large population of saltsage near the south shore of Great Salt Lake. Chromosome counts in somatic cells of the saltsage population consistently yielded $2n = 54$. Seed collected from the isolated female four-wing saltbush plant were grown in a garden. Most of the plants that arose from the seeds were sterile and had 45 chromosomes in the somatic cells.

Problem: **(a)** How many chromosomes are in the somatic cells of the four-wing saltbush female plant? **(b)** What are the ploidies of the four-wing saltbush plant, the saltsage plants, and the progeny plants grown in the garden?

Solution: **(a)** The four-wing saltbush plant is an isolated female, so there are no male four-wing saltbush plants to pollinate it. The pollen must have come from the surrounding saltsage plants. Saltsage has $2n = 54$ chromosomes, so its pollen cells have $n = 27$ chromosomes. The progeny plants had 45 chromosomes, 27 of which came from the saltsage pollen. The ova in the four-wing saltbush female plant, therefore, have $n = 45 - 27 = 18$ chromosomes, and the number of chromosomes in the somatic cells of the four-wing saltbush plant is $2n = 36$. **(b)** There are four hypotheses: (1) If four-wing saltbush is diploid, then $2n = 2x = 36$ and $n = x = 18$. Under this hypothesis, saltsage must be triploid ($3x = 54$), which would make it sterile. As saltsage is fertile, we rule out this hypothesis. (2) If four-wing saltbush is tetraploid, then $2n = 4x = 36$ and $x = 9$. Under this hypothesis, saltsage must be hexaploid ($2n = 6x = 54$), which is a plausible explanation. The progeny plants are pentaploid ($5x = 45$) and should be sterile, as observed. (3) If four-wing saltbush is hexaploid, then $2n = 6x = 36$ and $x = 6$. Under this hypothesis, saltsage must be nonaploid ($9x = 54$), which would make it sterile. As saltsage is fertile, we rule out this hypothesis. (4) Possibly, $x = 3$, which means that four-wing saltbush has $2n = 12x = 36$ chromosomes, and saltsage has $2n = 18x = 54$. Mathematically, this possibility is acceptable because both species have even ploidies. However, ploidies as high as $12x$ and $18x$ are very rare, as are genomes that consist of only 3 chromosomes. Situation 2 ($x = 9$) is the most plausible explanation. Microscopic examination of karyotypes has confirmed that indeed $x = 9$ in the genus *Atriplex*.

bread wheat
2n = 6x = 42

triticale
2n = 8x = 56

rye
2n = 2x = 14

Figure 17.9 Comparison of triticale with its parental species, bread wheat and rye.

Figure 17.10 A triploid seedless watermelon. Notice the undeveloped seeds that arise from chromosome imbalances in the seed embryos.

Polyploidy is common in plants but rare in animals. Most polyploid species have an even number of chromosome sets, which permits fertility. Odd numbers of chromosome sets cause sterility. Polyploid species may be autopolyploids or allopolyploids.

Genetics of Polyploidy

Polyploidy presents a different situation for inheritance when compared to diploidy. In allopolyploids, chromosomes pair, segregate, and assort as in diploids, so inheritance in allopolyploids follows typical Mendelian patterns. However, certain types of epistasis may be more

	AA	$4Aa$	aa
AA	$AAAA$	$4AAAa$	$AAaa$
$4Aa$	$4AAAa$	$16AAaa$	$4Aaaa$
aa	$AAaa$	$4Aaaa$	$aaaa$

Figure 17.11 Punnett square for self-fertilization of an autotetraploid plant with the duplex genotype $AAaa$.

common in allopolyploids than in diploids because genes on homeologous chromosomes are often identical. The example of duplicate gene action for spring and winter wheat given in Figure 13.19 is a result of allopolyploidy in wheat.

Inheritance in autopolyploids can be complicated, especially at high ploidy levels. Let's use the simplest case, an autotetraploid, to illustrate the phenomenon. In autotetraploids, there is no simple homozygosity or heterozygosity. For two alleles at a locus, one dominant allele and one recessive allele, there are five possible genotypes:

$aaaa$	nulliplex
$Aaaa$	simplex
$AAaa$	duplex
$AAAa$	triplex
$AAAA$	quadriplex

Suppose a nulliplex plant ($aaaa$) hybridizes with a quadriplex plant ($AAAA$). The F_1 offspring are all duplex:

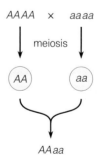

There are three possible ways for chromosomes to segregate during meiosis in a duplex $AAaa$ individual,

which results in a gametic array of 1 AA:4 Aa:1 aa and the F_2 progeny ratio in the Punnett square shown in Figure 17.11. If A is completely dominant over a, then the phenotypic segregation in the F_2 generation is 35:1.

Table 17.2 The Genotypic Arrays of Progeny from Self-Pollination for the Possible Combinations of 1, 2, 3, or 4 Alleles in an Autotetraploid Individual*

Parental Genotype	Gametic Array	Genotypic Array of Progeny from Self Pollination
$a^1a^1a^1a^1$ (nulliplex)	a^1a^1	All nulliplex
$a^1a^1a^1a^2$ (simplex)	$a^1a^1 + a^1a^2$	$\frac{1}{4}$ nulliplex + $\frac{1}{2}$ simplex + $\frac{1}{4}$ duplex
$a^1a^1a^2a^2$ (duplex)	$a^1a^1 + 4a^1a^2 + a^2a^2$	$\frac{1}{18}$ nulliplex + $\frac{4}{9}$ simplex + $\frac{1}{2}$ duplex
$a^1a^1a^2a^3$ (trigenic)	$a^1a^1 + 2a^1a^2 + 2a^1a^3 + a^2a^3$	$\frac{1}{36}$ nulliplex + $\frac{2}{9}$ simplex + $\frac{1}{4}$ duplex + $\frac{1}{2}$ trigenic
$a^1a^2a^3a^4$ (tetragenic)	$a^1a^2 + a^1a^3 + a^1a^4 + a^2a^3 + a^2a^4 + a^3a^4$	$\frac{1}{6}$ duplex + $\frac{2}{3}$ trigenic + $\frac{1}{6}$ tetragenic

*The genotypic array in the progeny is the square of the gametic array from the parent.

Not only are segregation patterns more complicated in autotetraploids than in diploids or allotetraploids, but there may also be as many as four different alleles at a single locus in one autotetraploid individual. The possible combinations of any two alleles form a **digenic** genotype that is simplex, duplex, or triplex. Three alleles combined, such as $a^1a^1a^2a^3$, is a **trigenic** genotype, and all four alleles combined, $a^1a^2a^3a^4$, is a **tetragenic** genotype. The progeny arrays of self-pollinated plants for the possible genotypes in an autotetraploid are given in Table 17.2.

Because genetic analysis in polyploids is considerably more complicated than in diploids, detection of linkage and chromosome mapping is not as straightforward as in a diploid. However, by using DNA markers and designing crosses appropriately, it is possible to construct genetic linkage maps in even the highest polyploids. The following example illustrates the procedure.

Example 17.3   Chromosome mapping in polyploids.

In 1996, Mudge et al. (*Crop Science* 36:1362–1366) published a genetic map of octaploid sugarcane ($2n = 8x = 80$) based on DNA markers. To construct the map, the researchers used the first-generation progeny of two genetically heterogeneous parents and examined hundreds of different DNA markers in the progeny. They found many markers that segregated in 1:1 ratios, as well as markers that segregated in ratios higher than 1:1. However, they used only those markers with 1:1 segregation ratios for mapping.

Problem: Why did the researchers search for markers that segregated in 1:1 ratios to construct a genetic map in a polyploid?

Solution: Suppose that a DNA marker locus has two alleles, *A* and *a*. Sugarcane is an octaploid, so the simplex genotype at the locus is *Aaaaaaaa*, and the nulliplex genotype is *aaaaaaaa*. Meiosis in the

simplex *Aaaaaaaa* plant produces two types of gametes, half that carry the *A* allele and half that do not:

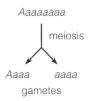

Hybridization of a simplex *Aaaaaaaa* parent with a nulliplex *aaaaaaaa* parent is the equivalent of a testcross. Under these circumstances, half of the first-generation progeny have the simplex genotype *Aaaaaaaa* and the other half have the nulliplex genotype *aaaaaaaa*. If the two alleles at a second marker locus also segregate in a 1:1 fashion (50% *Bbbbbbbb* and 50% *bbbbbbbb*) in the same cross, then the two marker loci are linked if their combined segregations deviate significantly from the 1:1:1:1 ratio expected for independent assortment. As long as one parent is simplex and the other is nulliplex for the DNA markers in question, calculation of map distances in polyploid testcrosses is done the same as in a diploid testcross. The genetic map distance in centimorgans is calculated as the number of recombinant progeny divided by the total number of progeny and multiplied by 100, which can be corrected with a mapping function if necessary. In this example, the researchers examined a large number of DNA markers in the progeny and used only those markers that fit the simplex × nulliplex situation. This type of mapping in polyploids is called single-dose marker mapping, because there is only one dose of the dominant allele from one of the two parents for each of the markers used.

Patterns of inheritance in allopolyploids are similar to those in diploids. Inheritance patterns in autopolyploids, however, are more complicated.

17.3 ALTERATIONS OF CHROMOSOME STRUCTURE

An alteration of a chromosome's structure is a type of mutation, but it is very different from the type of mutations we discussed in Chapter 5 that alter one or a few nucleotides. Alterations in chromosome structure delete, duplicate, or rearrange thousands to millions of nucleotides. Alterations in chromosome structure most often have their effect not by changing the gene structure, but by changing the number of genes or their chromosomal positions. There are six major types of chromosome structural alterations:

1. Deletion: loss of a chromosome segment.

2. Duplication: a repeat, often in tandem, of a chromosome segment.

3. Inversion: reverse orientation of a chromosome segment.

4. Translocation: transfer of a chromosome segment to a nonhomologous chromosome.

5. Fission: splitting of one chromosome into two.

6. Fusion: joining of two chromosomes into one.

Let's examine each of these types of alterations, starting with deletions.

Deletions

A **deletion** (also called a **deficiency**) is the loss of a chromosome segment. A **terminal deletion** arises when a chromosome arm breaks and the segment of the chromosome arm that is distal to the break fails to rejoin the chromosome (Figure 17.12a). An **interstitial deletion** arises when two breaks within a chromosome arm release a segment between the breaks, and the telomeric piece rejoins with the main chromosome (Figure 17.12b). A terminal deletion or an interstitial deletion produces an **acentric fragment**, a piece of chromosome with no centromere. A chromosome with a deletion still has its centromere, so it can behave normally in mitosis and meiosis. But an acentric fragment has no centromere and nothing to which spindle fibers can attach. An acentric fragment is not directed to the poles of the cell during mitosis or meiosis and usually does not become part of the nucleus of either daughter cell following cell division. Acentric fragments are usually lost in the cytoplasm outside of the two daughter nuclei, and are not replicated.

A small deletion may eliminate only one gene, or perhaps a few genes, from a chromosome. A large deletion may eliminate many genes. Some small deletions behave genetically like recessive alleles. The deleted genes are not present to produce a product, so the cell is dependent on the gene at the corresponding locus on the homologous chromosome to encode the product. Researchers

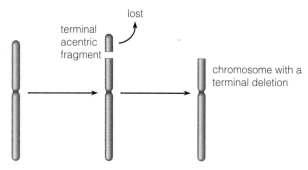

a A terminal deletion arises when a single break in a chromosome releases a terminal acentric fragment that is lost.

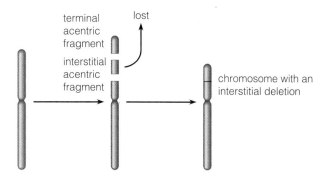

b An interstitial deletion arises when two breaks in a chromosome release an interstitial acentric fragment. The terminal acentric fragment reattaches to the chromosome.

Figure 17.12 Terminal and interstitial deletions.

rely on this phenomenon to map genes to their physical locations, as explained in section 15.7.

Some small deletions can persist when cells are heterozygous but may be lethal when they are homozygous. Large deletions often alter the phenotype even when cells are heterozygous. The phenotypic effect of large deletions depends on the species. Many plant species are tolerant of large deletions, but most animals—mammals in particular—are quite sensitive to deletions of any size.

Duplications

A **duplication** is a repeated segment of a chromosome. Duplications may arise in a number of ways, some of which are the result of translocations and inversions, which we will discuss shortly. For now, let's examine **tandem duplications**, which are repeated copies of a chromosome segment that lie adjacent to each other in the chromosome. Tandem duplications are not uncommon. They include segments of DNA that are repeated a few times, such as the genes of a gene family, and they include large numbers of repeats, such as the rRNA genes in the nucleolus organizer region (NOR). The number of repeats in a tandem duplication may increase or decrease through a process called **unequal crossing-over**.

a Different segments within the duplicated regions of homologous chromosomes pair with one another.

b When chromosome pairing is complete, unpaired segments loop out.

c A crossover takes place between paired duplicated segments.

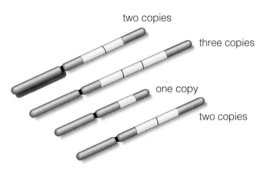

two copies

three copies

one copy

two copies

d The meiotic products include two chromosomes with two copies of the repeated segment, one with three copies, and one with one copy.

Figure 17.13 Unequal crossing-over between tandem duplications.

When two homologous chromosomes pair at the site of a tandem duplication, any one of the duplicated segments on one chromosome may pair with any one of the duplicated segments on the other. When the duplicated portions are not lined up equally, some of the duplicated segments may be left unpaired, as illustrated in parts a and b of Figure 17.13. A crossover between unequally paired segments increases the number of repeats in one chromatid and reduces the number in another, as diagrammed in parts c and d of Figure 17.13.

Duplications may or may not have a phenotypic effect. One of the best-studied cases of a duplication with a phenotypic effect is a tandem duplication at the *Bar* locus in *Drosophila melanogaster*. Alfred Sturtevant showed by genetic analysis that unequal crossing-over increased or decreased the copy number of the *Bar* locus. When the chromosomal segment containing the *Bar* locus is present in two tandem copies, it confers a dominant bar-shaped eye. As the number of tandem copies increases, the severity of the bar phenotype increases (Figure 17.14). Duplications may have detrimental phenotypic effects, but they are rarely as severe as the effects of deletions of comparable size.

~

Deletions and duplications often cause cells to be unbalanced for chromosomal segments. They may or may not have phenotypic effects. Many deletions are homozygous lethal.

Inversions

An **inversion** is a chromosomal segment that reversed its orientation in the chromosome, as shown in Figure 17.15. There is no loss or gain of genetic material, but the genetic material within the chromosome is rearranged. If one of the breaks is within a gene, then the gene's activity is lost, usually causing a recessive allele. However, the structures of the genes within the inversion remain unchanged.

Individuals may be homozygous or heterozygous for inversions. The chromosomes of inversion homozygotes pair and segregate normally. Crossovers within the inverted region have no effect because they perpetuate the

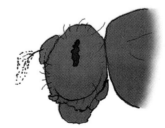

one copy of the *Bar* locus (wild-type)

two copies of the *Bar* locus

three copies of the *Bar* locus

Figure 17.14 Different phenotypes associated with the number of duplications of the *Bar* locus in *Drosophila melanogaster*.

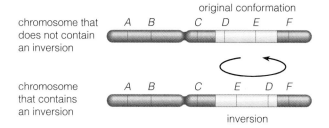

Figure 17.15 An inversion in a chromosome.

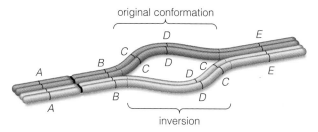

a The inverted region cannot pair normally with the corresponding region in a homologous chromosome that does not carry the inversion.

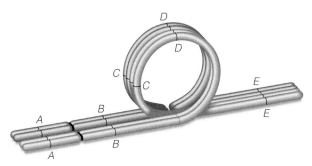

b The chromosomes can pair along their entire lengths when they loop in the inverted region.

Figure 17.16 Chromosome pairing during prophase I in an inversion heterozygote.

inversion. Inversions are often stable, especially when all individuals within a population are homozygous for the inversion. Inversions may persist in populations and may eventually become the most common form of a chromosome in a population, differentiating that population from others that do not carry the inversion.

Inversion heterozygotes have one chromosome with an inversion and a homologous chromosome without the inversion. Chromosome pairing is altered in inversion heterozygotes. In order to pair along their full lengths, the chromosomes must loop, as illustrated in Figure 17.16.

Looping has no effect on chromosome segregation in meiosis as long as there is no crossing-over within the looped region. However, crossovers may take place within the looped region, and when they do, there are profound consequences. In order to see what happens, we first need to distinguish between two types of inversions. If the centromere is located within the inverted region, the inversion is a **pericentric inversion**. If the inverted region does not include the centromere, the inversion is a **paracentric inversion**.

Figure 17.17 shows the consequences of a crossover within the looped region of paired chromosomes in a pericentric inversion heterozygote. In Figure 17.17a, starting at the telomere of the left arm of chromatid 2, trace with a pencil along the length of the chromatid. After you trace the crossover point within the loop, notice that you return to the left arm of chromatid 4. When this recombinant chromatid is partitioned to a gamete, it contains a deletion and a duplication (second chromosome from the top in Figure 17.17d). Now start at the right telomere of chromatid 2 and trace it. After tracing the crossover, you return to the right side of chromatid 4. When this recombinant chromatid is partitioned to a gamete, it also contains a deletion and a duplication (third chromosome from the top in Figure 17.17d). It is reciprocal to the first chromatid you traced. The two nonrecombinant chromatids have no duplications or deletions because they did not participate in the crossover. One carries the inversion; the other has the original conformation.

All four chromosomes that arise from this meiosis have a single centromere and are partitioned to the four cells normally, but two of the four cells are unbalanced,

with major deletions and duplications. Gametes that carry these deletions and duplications may abort, or they may transmit the deletions and duplications to the zygote. An embryo that inherits large duplications and deletions usually dies because it is unbalanced for chromosomal material.

Now let's look at a paracentric inversion in Figure 17.18. Start tracing at the telomere of the left arm of chromatid 2 in Figure 17.18a. Before reaching the inversion loop, you pass a centromere. After you trace the crossover point and complete the loop, you return to the left arm of chromatid 4, which also contains a centromere. The crossover creates a **dicentric chromatid**, a chromatid with two centromeres. When the centromeres migrate to opposite poles during anaphase I, a **chromatid bridge** forms between the two centromeres. The bridge stretches until it breaks at some point (Figure 17.18d). The resulting chromatids contain deletions and may contain duplications, depending on where the chromatid bridge breaks. Now start at the right telomere of chromatid 2 in Figure 17.18a and trace it through the crossover and back to the right telomere of chromatid 4. This is an acentric fragment with two telomeres. It has no centromere to direct its partitioning, so it is lost along with the genes it carries. Two of the four meiotic products that arise from this meiosis have large deletions and possibly duplications, depending on the site of the break (Figure 17.18e).

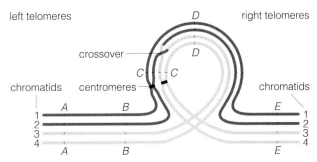

a Two-dimensional model of chromosome pairing in a pericentric inversion heterozygote with a crossover in the inversion loop.

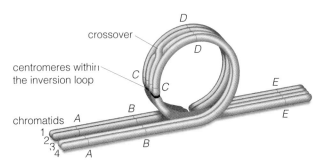

b Three-dimensional model of chromosome pairing.

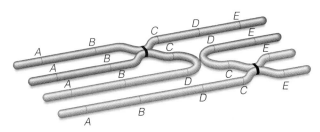

c Unraveling of chromosomes after a crossover in the inversion loop.

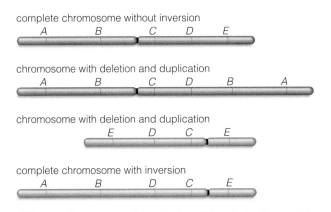

d The four chromosomes that arise from this meiosis. Two contain deletions and duplications.

Figure 17.17 Consequences of a crossover within the inversion loop in a pericentric inversion heterozygote.

Gametes or zygotes that carry deletions and duplications usually abort because they are unbalanced for chromosomal material. The frequency of crossing-over in an inversion loop is proportional to the size of the inversion. An inversion that spans 40 centimorgans should cause approximately 40% of the gametes to be unbalanced, because all crossover-type gametes contain deletions and possibly duplications. Thus, heterozygotes for large inversions may have significantly reduced fertility. However, inversion homozygotes have no fertility loss due to crossing-over within the inverted region, because there is no inversion loop and no duplications or deletions after crossing-over. This means that inversion heterozygotes are at a reproductive disadvantage when compared to inversion homozygotes or homozygotes with no inversion. Inversions can be important in the divergence of two species from a common ancestral species, a topic we will address in more detail in Chapter 21.

~

An inversion is a rearrangement, rather than a loss or gain, of genetic material. A crossover within the inverted region of the chromosomes in an inversion heterozygote produces crossover-type gametes that are unbalanced, with deletions and duplications.

Translocations

A break in a chromosome may liberate an acentric fragment, which can attach to another chromosome. In most cases, the other chromosomes have intact telomeres, which prevent the fragment from attaching. The fragment may reattach to its former site, or it may be lost as a deletion. However, under rare circumstances, the fragment may attach to the end of another chromosome, usually when the telomere of that chromosome has been broken. The result is a **simple terminal translocation**, in which a portion of one chromosome is translocated to another chromosome (Figure 17.19a).

The most common form of translocation is **reciprocal translocation**. A break in one chromosome releases an acentric fragment, and a break in a nonhomologous chromosome releases a second acentric fragment. The two acentric fragments exchange places, attaching to the break sites on nonhomologous chromosomes (Figure 17.19b). Some researchers have suggested that most, if not all, simple terminal translocations are actually reciprocal but may appear to be simple because the reciprocal product is a telomere, which cannot be readily detected. Cells that are homozygous or heterozygous for a translocation are balanced because there are no deletions or duplications, simply a rearrangement of genetic material.

Translocation homozygotes have no trouble retaining a correct balance during meiosis because the homologous chromosomes that carry a translocation pair normally as

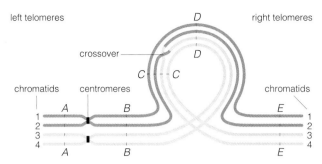

left telomeres

crossover

chromatids centromeres

a Two-dimensional model of chromosome pairing in a paracentric inversion heterozygote with a crossover in the inversion loop.

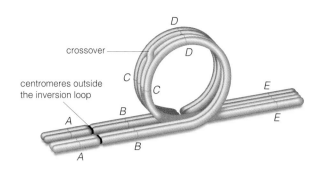

right telomeres

crossover

centromeres outside the inversion loop

chromatids

b Three-dimensional model of chromosome pairing.

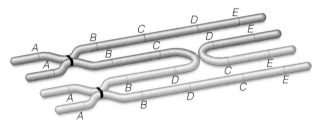

c Unraveling of chromosomes after a crossover in the inversion loop. One chromatid is attached to two centromeres.

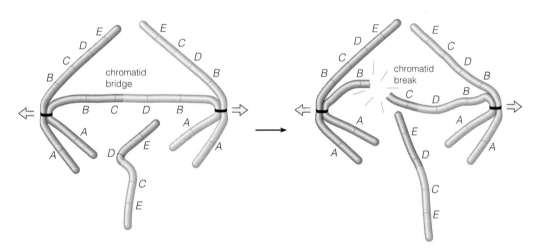

chromatid bridge

chromatid break

d During anaphase I, the two centromeres of the dicentric chromatid migrate to opposite poles, causing the chromatid to stretch until it breaks.

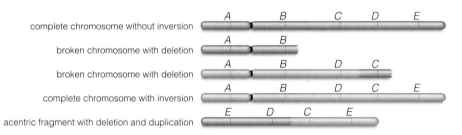

complete chromosome without inversion

broken chromosome with deletion

broken chromosome with deletion

complete chromosome with inversion

acentric fragment with deletion and duplication

e The four chromosomes and an acentric fragment that arise from this meiosis. Each of the chromosomes is partitioned to a cell. The acentric fragment is lost.

Figure 17.18 Consequences of a crossover within the inversion loop in a paracentric inversion heterozygote.

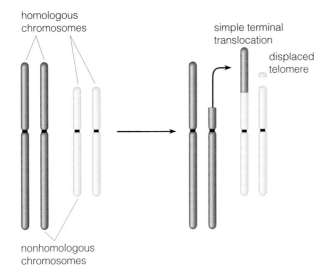

homologous chromosomes

nonhomologous chromosomes

simple terminal translocation

displaced telomere

a A simple terminal translocation. A segment of one chromosome is translocated to a nonhomologous chromosome, usually displacing the telomere on that chromosome.

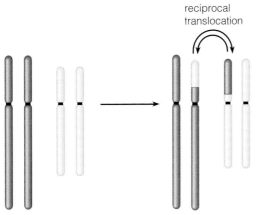

reciprocal translocation

b A reciprocal translocation. Segments on nonhomologous chromosomes exchange places.

Figure 17.19 Two types of translocation.

1   2   3   4

a Chromosomes in a reciprocal translocation heterozygote before chromosome pairing.

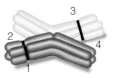

3
2
4
1

b Quadrivalent pairing during prophase I.

Figure 17.20 Chromosome pairing to form an X-shaped quadrivalent in a reciprocal translocation heterozygote.

bivalents and segregate normally. However, the gametes produced in translocation heterozygotes are often unbalanced following meiosis. The potential loss of chromosome balance in the gametes of translocation heterozygotes is the result of altered pairing during prophase I. In a reciprocal translocation heterozygote, the two nonhomologous chromosomes that carry the translocations and the two partners without translocations pair to form an X-shaped quadrivalent in which all chromosome segments are paired with their homologous counterparts (Figure 17.20).

The X-shaped quadrivalent contains four instead of two centromeres. Two of the centromeres are homologous to one another (labeled 1 and 2 in Figure 17.20), and the

other two centromeres are also homologous to one another (labeled 3 and 4 in Figure 17.20). Notice that the odd-numbered centromeres are in chromosomes that do not contain a translocation, and the even-numbered centromeres are in chromosomes that contain a translocation. At the end of prophase I, the chiasmata in the paired portions of the quadrivalent migrate to the telomeres. During the transition to metaphase I, spindle fibers attach to the sets of kinetochores at each of the four centromeres and direct the quadrivalent to the equator of the cell. Depending on how the spindle fibers attach, the quadrivalent assumes one of four possible orientations at metaphase I, each of which leads to a different type of segregation pattern: alternate-1 segregation, adjacent-1 segregation, alternate-2 segregation, and adjacent-2 segregation.

Alternate-1 segregation is diagrammed in Figure 17.21a. It begins with the quadrivalent initially aligned so that nonhomologous centromeres 1 and 4 face one pole and nonhomologous centromeres 2 and 3 face the other pole. The ring twists into a figure-eight conformation as spindle fibers attach and the cell enters metaphase I. The twisting causes the homologous centromeres to switch places, so centromeres 1 and 3 now face one pole, and centromeres 2 and 4 face the other pole. During anaphase I, the two chromosomes with translocations move to one pole, and the two chromosomes without translocations move to the other pole. All four meiotic products are balanced, two with a reciprocal translocation, and two with no translocation.

Adjacent-1 segregation is diagrammed in Figure 17.21b. It begins like alternate-1 segregation with nonhomologous centromeres 1 and 4 facing one pole and nonhomologous centromeres 2 and 3 facing the other

| | Number of Segregations | | | |
Genotype	Alternate-1	Adjacent-1	Alternate-2	Adjacent-2
T10-19	60	68	30	31
AG184	20	20	13	6

pole. Instead of twisting into a figure-eight shape, however, the quadrivalent retains its ring shape as the spindle fibers attach and the cell enters metaphase I. During anaphase I, centromeres 1 and 4 migrate to one pole, and centromeres 2 and 3 migrate to the other pole. Each of the four cells that arises from adjacent-1 segregation carries a duplication and a deletion and is therefore unbalanced.

Alternate-2 segregation is diagrammed in Figure 17.21c. It begins with the ring-shaped quadrivalent initially aligned so that homologous centromeres 1 and 2 face one pole and homologous centromeres 3 and 4 face the other pole. The ring twists into a figure-eight conformation as spindle fibers attach and the cell enters metaphase I. The twisting causes the nonhomologous centromeres to switch places, so that centromeres 1 and 3 now face one pole, and centromeres 2 and 4 face the other pole. The result is the same as in alternate-1 segregation: during anaphase I, the two chromosomes with translocations move to one pole, and the two chromosomes without translocations move to the other pole. All four meiotic products are balanced, two with a reciprocal translocation, and two with no translocation.

Adjacent-2 segregation is diagrammed in Figure 17.21d. It begins like alternate-2 segregation with homologous centromeres 1 and 2 facing one pole and homologous centromeres 3 and 4 facing the other pole. Instead of twisting into a figure-eight shape, the quadrivalent retains its ring shape as the spindle fibers attach and the cell enters metaphase I. During anaphase I, homologous centromeres 1 and 2 migrate to one pole, and homologous centromeres 3 and 4 migrate to the other pole. Each of the four cells that arises from adjacent-2 segregation carries a duplication and a deletion and is therefore unbalanced.

Figure 17.22 shows micrographs of chromosomes in prophase I and metaphase I in a rye plant that is heterozygous for a reciprocal translocation heterozygote. The plant has $2n = 2x = 14$ chromosomes that pair as five bivalents and one quadrivalent. The quadrivalent has an X shape in prophase I and a ring shape in metaphase I.

Alternate-1 and adjacent-1 segregations are equally frequent. Alternate-2 and adjacent-2 segregations are also equally frequent. However, the frequency of alternate-1 and adjacent-1 segregations is not necessarily equal to the frequency of alternate-2 and adjacent-2 segregations. Which types of segregations predominate depends on the position of the translocation break point in the chromosome. However, regardless of which type of segregation patterns predominate, the frequency of alternate segregations (alternate-1 and alternate-2) equals the frequency of adjacent segregations (adjacent-1 and adjacent-2). Thus, about 50% of the gametes in translocation heterozygotes arise from adjacent segregations and are unbalanced. The equality of alternate and adjacent segregations is illustrated in the following example.

Example 17.4 Alternate-1, adjacent-1, adjacent-2, and alternate-2 segregation frequencies in reciprocal translocation heterozygotes.

In 1974, Endrizzi (*Genetics* 77:55–60) reported his observations of the four segregation patterns in reciprocal translocation heterozygotes in cotton. His results are listed in Table 17.3.

Problem: (a) What do these results suggest about the relative frequencies of alternate and adjacent segregations? (b) Compare the frequencies of alternate-1 and adjacent-1 segregations and the frequencies of alternate-2 and adjacent-2 segregations.

Solution: (a) Let's test the hypothesis that about half of the segregations are alternate (alternate-1 and alternate-2 combined) and the other half are adjacent (adjacent-1 and adjacent-2 combined) with chi-square analysis. For T10-19, $\chi^2 = 2(4.5^2) \div 94.5 = 0.43$, with 1 degree of freedom. For AG184, $\chi^2 = 2(3.5^2) \div 29.5 = 0.83$, with 1 degree of freedom. In both cases, the segregation patterns do not deviate significantly from half alternate, half adjacent. (b) In both cases, alternate-1 segregation is more frequent than alternate-2 segregation, and adjacent-1 segregation is more frequent than adjacent-2 segregation.

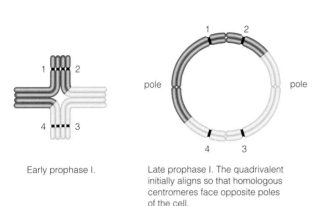

Early prophase I.

Late prophase I. The quadrivalent initially aligns so that homologous centromeres face opposite poles of the cell.

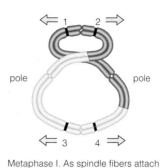

pole   pole

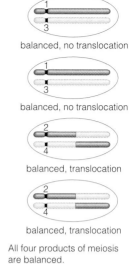

pole   pole

Metaphase I. As spindle fibers attach to the centromeres, the ring-shaped quadrivalent twists into a figure-eight shape. Homologous centromeres migrate to opposite poles.

balanced, no translocation

balanced, no translocation

balanced, translocation

balanced, translocation

All four products of meiosis are balanced.

a Alternate-1 segregation.

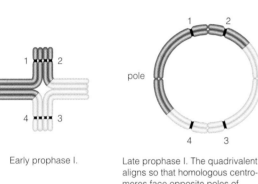

Early prophase I.

Late prophase I. The quadrivalent aligns so that homologous centromeres face opposite poles of the cell.

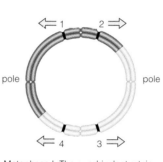

pole   pole

pole   pole

Metaphase I. The quadrivalent retains its ring shape as spindle fibers attach to the centromeres. Homologous centromeres migrate to opposite poles.

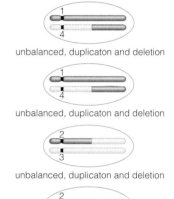

unbalanced, duplicaton and deletion

unbalanced, duplicaton and deletion

unbalanced, duplicaton and deletion

unbalanced, duplicaton and deletion

All four products of meiosis are unbalanced.

b Adjacent-1 segregation.

Figure 17.21 Segregation patterns in a reciprocal translocation heterozygote.

The loss of fertility in reciprocal translocation heterozygotes favors reproductive isolation of translocation homozygotes from homozygotes that do not carry a translocation. Translocations are therefore one of the most important genetic factors in speciation, a topic we will discuss in Chapter 21.

~

A reciprocal translocation, an exchange of chromosome segments between nonhomologous chromosomes, is the most common type of translocation. Alternate segregation patterns in reciprocal translocation heterozygotes produce balanced gametes; adjacent segregation patterns produce unbalanced gametes. About half of all gametes in recip-
rocal translocation heterozygotes are unbalanced and cause a reduction in fertility.

Chromosome Fission and Fusion

Chromosome fission is the splitting of one chromosome into two. Although it seems straightforward, it is rare. Usually, a chromosome break produces an acentric fragment that is either translocated or lost. In order for the acentric fragment to become a separate chromosome, it must gain a centromere. Centromere formation de novo probably does not happen very often, so for fission to be successful, it must come from a dicentric chromosome,

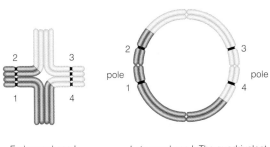

Early prophase I.

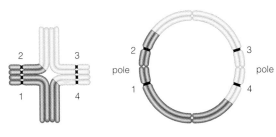

Late prophase I. The quadrivalent initially aligns so that homologous centromeres face the same poles.

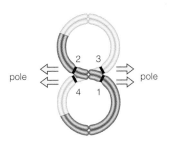

Metaphase I. As spindle fibers attach to the centromeres, the ring-shaped quadrivalent twists into a figure-eight shape. Homologous centromeres now face opposite poles and migrate to them.

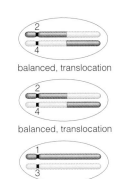

balanced, translocation

balanced, translocation

balanced, no translocation

balanced, no translocation

All four products of meiosis are balanced.

c Alternate-2 segregation.

Early prophase I.

Late prophase I. The quadrivalent align so that homologous centromeres face the same pole.

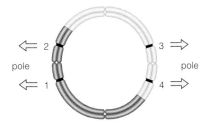

Metaphase I. The quadrivalent retains its ring shape as spindle fibers attach to the centromeres. Homologous centromeres migrate to the same pole.

unbalanced, duplication and deletion

unbalanced, duplication and deletion

unbalanced, duplication and deletion

unbalanced, duplication and deletion

All four products of meiosis are unbalanced.

d Adjacent-2 segregation.

or the acentric fragment must gain a centromere by translocation from another chromosome.

Chromosome fusions, however, are well-known. Chromosome fusion is usually part of a reciprocal translocation in which the long arms of two nonhomologous acrocentric chromosomes fuse to form a single metacentric chromosome called a Robertsonian translocation (Figure 17.23). The short arms of both chromosomes may also combine as the reciprocal product, but are often lost because they are small and contain mostly telomeric material and no essential genetic information.

An **isochromosome**, a chromosome with two identical arms, can arise in two ways: by fusion of homologous chromosome arms (Figure 17.24a), or by a crossover in the inversion loop in a cell that is heterozygous for a very small pericentric inversion (Figure 17.24b). The attached X chromosome in *Drosophila melanogaster* (see section 14.3) is an example of an isochromosome.

Position Effect

Deletions, duplications, inversions, and translocations alter the position of genes relative to one another. Most genes in an inversion or translocation are flanked by the genes they normally have near them. However, those genes near the break points of an inversion or translocation are

flanked on one side by a different set of genes. Does this alteration in position have any effect on expression of the gene? Extensive research in *Drosophila* has been conducted to answer this question. And the answer is sometimes yes and sometimes no, depending on the gene and the position it occupies. If a gene's expression is influenced by the position it occupies in the chromosome, then the gene is said to be subject to **position effect**.

When a gene that is subject to position effect is moved from a position near euchromatin to a position near heterochromatin, its expression may be modified. In some cases, genes located near the break point of an inversion or translocation show a variegated, or mosaic, type of expression after inversion of the chromosomal segment. For example, if a wild-type allele of the *white* locus in *Drosophila melanogaster* is moved to the opposite end of the chromosome by inversion, or to another chromosome by translocation, eye color often becomes variegated, as illustrated in Figure 17.25. If the gene is moved back to a site at or near its original location, it is expressed as usual. Thus, position, rather than a mutation within the coding sequence of the gene, is responsible for the variegated phenotype.

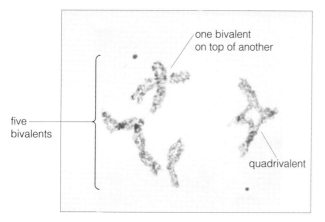

a Chromosome pairing during prophase I.

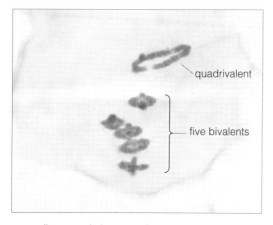

b Chromosome alignment during metaphase I.

Figure 17.22 Meiotic chromosome in a rye plant that is heterozygous for a reciprocal translocation. (Photos courtesy of H. C. Stutz.)

17.4 CHROMOSOME ALTERATIONS IN HUMANS

Humans and other mammals are especially sensitive to alterations in chromosome number and structure. About 15% of all recognized pregnancies end in spontaneous abortion (miscarriage), and major chromosomal abnormalities are often the cause. About half of all spontaneously aborted fetuses have recognizable alterations in chromosome number or structure. Of these, about two-thirds are aneuploid. The remaining third are polyploid or have another type of chromosomal imbalance. The frequency of chromosomal abnormalities in human conceptions is probably much higher than these frequencies suggest because some embryos with chromosomal abnormalities abort before pregnancy is recognized.

Only a few types of major chromosome abnormalities are found in infants born after a full-term gestation. The most common of these are deletions in the short arm of chromosome 5 (*cri du chat* syndrome), trisomy 21 (Down syndrome), reciprocal translocations, Robertsonian translocations, trisomy 18 (Edwards syndrome), trisomy

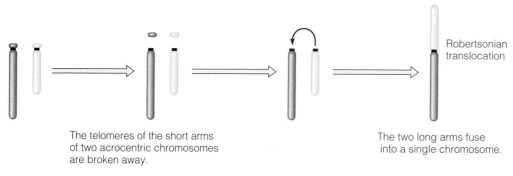

The telomeres of the short arms of two acrocentric chromosomes are broken away.

Robertsonian translocation

The two long arms fuse into a single chromosome.

Figure 17.23 Chromosome fusion to produce a Robertsonian translocation.

Table 17.4 Rules for Human Karyotype Nomenclature

1. The first symbol is the number of chromosomes in a somatic cell. It is followed by a comma. For example, a normal karyotype is 46 and a trisomic karyotype is 47.

2. The second set of symbols indicates the sex chromosome constitution (as XX, XY, or any deviations from this, such as XXY).

3. The third set of symbols designates deletions, duplications, translocations, monosomies, or trisomies. The chromosome with an alteration is designated by its number, and the chromosome arm is designated by *p* for the short arm and *q* for the long arm.

4. A "+" sign preceding a chromosome number indicates a trisomy for that chromosome. A "−" sign preceding a chromosome number indicates monosomy for that chromosome.

5. A "+" sign following a symbol indicates addition of chromosomal material to that chromosome. A "−" sign following a symbol indicates a deletion from that chromosome. For example, 5p− indicates a deletion in the small arm of chromosome 5, and 9q+ indicates addition of material (such as a simple translocation) to the long arm of chromosome 9.

6. A *t* followed by parentheses indicates a reciprocal translocation between the two chromosomes within the parentheses. For example, t(14q;21q) indicates a reciprocal translocation between the long arms of chromosomes 14 and 21.

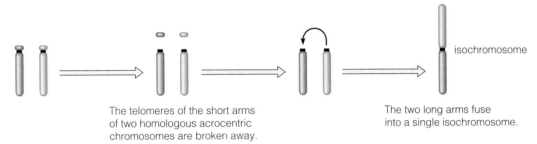

The telomeres of the short arms of two homologous acrocentric chromosomes are broken away.

The two long arms fuse into a single isochromosome.

isochromosome

a Fusion of the long arms of two homologous chromosomes.

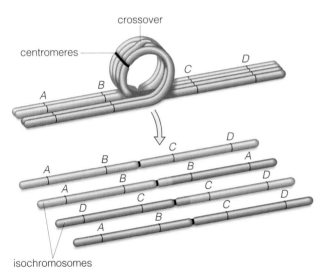

crossover

centromeres

isochromosomes

b A crossover in the inversion loop of a small pericentric inversion in a heterozygote.

Figure 17.24 Two ways by which an isochromosome can arise.

13 (Patau syndrome), and several aneuploidies for the X and Y chromosomes. All of the other aneuploidies nearly always cause spontaneous abortion. Although fetuses with trisomies for chromosomes 21, 18, and 13 sometimes reach birth, the majority of fetuses with these trisomies are also spontaneously aborted.

Human Karyotype Nomenclature

Before beginning our discussion of specific syndromes associated with chromosomal abnormalities in humans, we need to briefly review some of the basic rules for the symbols used in human karyotype nomenclature. Table 17.4 provides an abbreviated set of rules that describe most situations and explain the symbols used in our discussion. Let's now look at a few examples of human karyotype nomenclature and interpret them according to the rules given in Table 17.4.

47, XX, +21	Forty-seven chromosomes, XX female, trisomic for chromosome 21. This is the karyotype of most females with Down syndrome.
46, XY, 4p−	Forty-six chromosomes, XY male, deletion in the short arm of chromosome 4.
47, XXY	Forty-seven chromosomes, XXY trisomic male.
46, XX, t(9q;21q)	Forty-six chromosomes, XX female, reciprocal translocation between the long arm of chromosome 9 and the long arm of chromosome 21.

Figure 17.25 Phenotypic variegation caused by position effect for the *white* locus in *Drosophila melanogaster*.

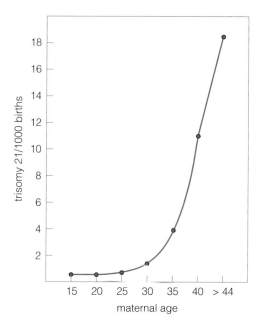

Figure 17.26 Relationship between the incidence of trisomy 21 and maternal age.

Reciprocal Translocations and Inversions

Reciprocal translocations are the most common type of chromosome abnormality in humans. Paracentric and pericentric inversions are also found in humans. Translocations and inversions are generally not associated with any phenotypic changes other than a loss of fertility as long as the chromosome complement is balanced, although there are exceptions to this. About 20% of all cases of hemophilia A are the result of an inversion with a break point in the *F8C* gene.

Many cases of subfertility in humans are the result of heterozygosity for reciprocal translocations. The most common translocation in humans is a Robertsonian translocation between the long arms of chromosomes 13 and 14, t(13q;14q). This translocation has arisen several times. Another Robertsonian translocation, t(14q;21q), can cause Down syndrome, as we will see shortly.

Cri du Chat Syndrome

A terminal deletion in the short arm of chromosome 5 (46, XY, 5p− or 46, XX, 5p−) causes a disorder called *cri du chat* **syndrome**. *Cri du chat* is a French expression that literally means "cat cry." The syndrome is so named because the crying sounds that infants with this disorder tend to make are similar to a cat's meow. Patients with *cri du chat* syndrome often die during the first years of childhood, although some patients have lived into adolescence. The syndrome is characterized by a small head, epicanthic folds over the eyes, multiple organ deformities, and mental retardation. Since children with this deletion usually die before their reproductive years, the deletion is almost never transmitted. Rarely, a reciprocal translocation between chromosome 5 and chromosome 15 balances the missing portion of chromosome 5, so no symptoms appear. Adjacent segregations produce gametes that carry a copy of chromosome 5 with an unbalanced deletion, and can be passed on to offspring, causing inherited *cri du chat* syndrome. The incidence of *cri du chat* syndrome is very low, about 1 in 50,000 births.

Down Syndrome

The most frequent syndrome associated with aneuploidy is **Down syndrome** (use of the possessive "Down's syndrome" is historically common, but no longer considered appropriate). Down syndrome is most often due to trisomy 21 (47, +21; see Figure 17.1 for the karyotype). Until recent years, the frequency of Down syndrome births was about 1 in 700 in the United States. The frequency has declined to about 1 in 1000 births because prenatal testing for Down syndrome is common and a significant number of women who discover that they are pregnant with a Down-syndrome fetus choose to end the pregnancy by elective abortion. Down syndrome is characterized by variable mental retardation, short hands with a simian crease, short stature, and an epicanthic fold over the eye, which is the reason that the inappropriate term *mongolism* has been used to describe Down syndrome. The severity of symptoms may vary among those who are affected.

Chromosome 21 is the smallest of human autosomes, which means that trisomy 21 is the least severe, in terms of chromosomal balance, of all autosomal trisomies. This may be the reason, at least in part, for the survivability of individuals with trisomy 21 when compared to other trisomies.

The frequency of trisomy-21 births has long been known to increase sharply when the mother's age exceeds 35, as depicted in Figure 17.26. Some researchers have speculated that the longer chromosomes are arrested in prophase I, the more likely they are to undergo nondisjunction. In human females, the chromosomes enter prophase I during fetal development and remain arrested

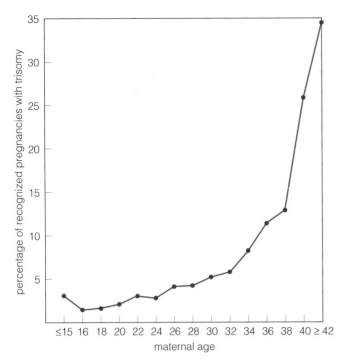

Figure 17.27 Relationship between the incidence of trisomies in recognized pregancies (including spontaneously aborted fetuses) and maternal age.

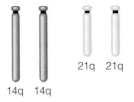

a Normal karyotype: two copies of 14q, two copies of 21q.

b Balanced translocation carrier: two copies of 14q, two copies of 21q.

c Translocation Down syndrome (trisomy 21q): two copies of 14q, three copies of 21q.

Figure 17.28 Chromosomal constitutions in **(a)** a normal karyotype, **(b)** a heterozygous carrier of the Robertsonian translocation t(14q;21q), and **(c)** a person who has translocation Down syndrome.

in prophase I for years until just before the onset of ovulation. This hypothesis suggests that the chromosomes in women over 35 years of age have remained in prophase I for so long that they are more likely to undergo nondisjunction than the chromosomes of a younger woman.

A second hypothesis recognizes that many trisomy-21 fetuses spontaneously abort and suggests that trisomy-21 fetuses may be less likely to abort spontaneously in older women. Under this hypothesis, it does not matter whether the extra chromosome comes from the female or the male parent. However, studies indicate that the rate of spontaneous abortion is not associated with age; thus, this hypothesis is probably not correct.

Although the graph in Figure 17.26 was not developed by measuring the father's age, it indirectly includes the father's age. With some exceptions, the age of the father is usually close to and often greater than the mother's age. After chromosome banding techniques became available, the extra chromosome in about 25% of Down syndrome cases was identified as paternal. However, more recent studies, based on DNA markers instead of chromosome banding, cast doubt on all chromosome banding studies for Down syndrome. Comparison of RFLP markers with chromosome banding shows that chromosome banding methods are prone to error and overestimate paternal contribution to aneuploidy. RFLP studies indicate that about 95% of Down syndrome cases arise from maternal nondisjunction. And so far, the number of paternal cases examined is too small

to determine whether or not there is a paternal age effect.

The increased frequency of Down syndrome with an increase in maternal age suggests that nondisjunction for all chromosomes may be correlated with maternal age. Studies of trisomic births and spontaneously aborted fetal tissue suggest that there is a general increase in nondisjunction for all chromosomes with increased maternal age, as diagrammed in Figure 17.27.

Down syndrome may also be caused by a Robertsonian translocation in which the long arm of chromosome 21 has fused to the long arm of chromosome 14. If there is only one copy of a normal chromosome 21, one copy of a normal chromosome 14, and the Robertsonian translocation (14q;21q) chromosome, the chromosomes are balanced, so there are no symptoms of Down syndrome (Figure 17.28b). In karyotype nomenclature, heterozygous female carriers of the 14q;21q translocation are

45, XX, t(14q;21q), and heterozygous males are 45, XY, t(14q;21q).

Although t(14q;21q) translocation heterozygotes do not have Down syndrome, their risk of having a child with Down syndrome is relatively high. When a gamete that carries one copy of the Robertsonian translocation and one copy of chromosome 21 unites with a normal balanced gamete, the zygote carries three copies of the long arm of chromosome 21, even though there are only 46 chromosomes (Figure 17.28c). The theoretical risk of having a Down syndrome child if one parent is heterozygous for a 14q;21q translocation is 1 in 3. However, the actual risk is lower, about 11% if the female parent is the translocation heterozygote, and about 2% if the male is the translocation heterozygote. The following example examines some of the data that demonstrate the actual risk.

Example 17.5   Inheritance of translocation Down syndrome.

In 1984, Boué and Gallano (*Prenatal Diagnosis* 4:45–68) reported the results of a collaborative study conducted at 71 European prenatal diagnosis centers. The researchers identified 188 pregnancies in which one parent was heterozygous for a 14q;21q Robertsonian translocation. Table 17.5 shows the fetal genotypes for these pregnancies.

Problem: **(a)** Use a Punnett square to show why the theoretical risk of having a Down syndrome child if one parent is heterozygous for a 14q;21q translocation is 1 in 3. **(b)** Compare the observed frequencies with the theoretical frequencies. Use chi-square analysis to determine whether or not the observed frequencies differ significantly from the theoretical frequencies.

Solution: **(a)** The t(14q;21q) chromosome pairs with chromosomes 14 and 21 to form a trivalent, as diagrammed in Figure 17.29. The trivalent may segregate in three possible ways, giving six possible gametes, which should appear in equal frequencies.

The Punnett square for the mating between a translocation carrier and a parent with a normal karyotype is shown in Figure 17.30. The frequencies of the six possible karyotypes in the Punnett square should be equal. However, three abort early during gestation because of monosomy or trisomy for the long arm of chromosome 14. Of the three karyotypes capable of survival, two are balanced, and the third is trisomic for the long arm of chromosome 21. **(b)** The observed frequency when the female parent is the translocation carrier is 21 ÷ 137 = 0.15. The observed frequency when the male is the carrier parent is 0 ÷ 51 = 0. The theoretical ratio for both parents is $\frac{1}{3}$ normal, $\frac{1}{3}$ heterozygous carrier, and $\frac{1}{3}$ trisomy 21. The chi-square value for female heterozygous carrier parents is $[(48 - 45.67)^2 + (68 - 45.67)^2 + (21 - 45.67)^2] \div 45.67 = 24.36$, with 2 degrees of freedom. The chi-square value for male heterozygous carrier parents is $[(20 - 17)^2 + (31 - 17)^2 + (0 - 17)^2] \div 17 = 29.06$, with 2 degrees of freedom. In both cases, the chi-square values exceed the 0.01 critical value of 9.21 for 2 degrees of freedom (from Table 12.5), indicating that the observed values deviate significantly from the theoretical values.

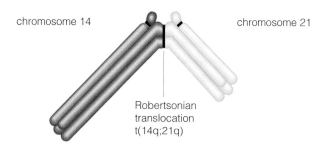

chromosome 14   chromosome 21

Robertsonian translocation t(14q;21q)

Figure 17.29   Information for Example 17.5: Trivalent pairing during prophase I in a cell that is heterozygous for the t(14q;21q) Robertsonian translocation.

Table 17.5   Information for Example 17.5: Fetal Genotypes for Which One Parent Is a Heterozygous Carrier of t(14q;21q)

Carrier Parent	Number of Pregnancies Examined	Fetal genotype		
		Normal	Balanced Translocation Heterozygote	Unbalanced Trisomy 21
Mother	137	48	68	21
Father	51	20	31	0
Totals	188	68	99	21

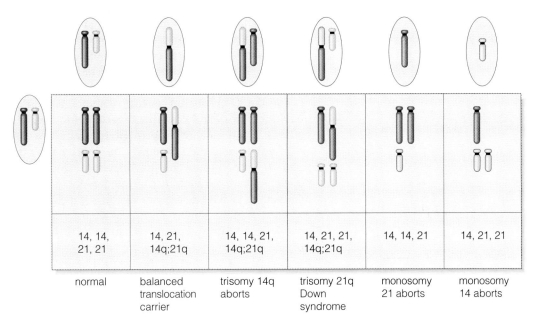

14, 14, 21, 21	14, 21, 14q;21q	14, 14, 21, 14q;21q	14, 21, 21, 14q;21q	14, 14, 21	14, 21, 21
normal	balanced translocation carrier	trisomy 14q aborts	trisomy 21q Down syndrome	monosomy 21 aborts	monosomy 14 aborts

KEY

= normal chromosome 14

= normal chromosome 21

= Robertsonian translocation 14q;21q

Figure 17.30 Information for Example 17.5: Punnett square that shows a theoretical probability of 1 in 3 births for translocation Down syndrome when one parent is a heterozygous carrier of the t(14q;21q) Robertsonian translocation.

Edwards Syndrome and Patau Syndrome

The two other autosomal trisomies found in humans cause severe syndromes. Trisomy 18 (47, +18) causes Edwards syndrome and trisomy 13 (47, +13) causes Patau syndrome. Both syndromes are characterized by severe multiple congenital defects. Most children with these disorders die within days to weeks of birth and rarely live longer than a few months. The incidences of both Edwards and Patau syndromes are less than that of Down syndrome, due to a higher rate of spontaneous abortion.

Aneuploidies for X and Y Chromosomes

Aneuploidies for the X chromosome tend to cause less severe effects than autosomal aneuploidies because of dosage compensation for the X chromosome. **Turner syndrome** is the only monosomy found in humans. The karyotype is 45, XO, which means there is one X chromosome in somatic cells and no corresponding Y chromosome. Turner syndrome is characterized by un-

developed ovaries, sterility, short stature, webbed neck, a broad chest, and a lack of secondary sex characteristics. There is no mental deficiency. Individuals can be treated with hormones to stimulate the development of secondary female sex characteristics. Individuals with 45, XO have no Barr bodies in their cells.

Males with a 47, XXY karyotype have **Klinefelter syndrome**. Klinefelter syndrome is characterized by a male phenotype with some breast development, little body hair, and tall stature, with little to no mental retardation. Males with XXY Klinefelter syndrome have a Barr body in their somatic cells. Males with 48, XXXY; 49, XXXXY; and 50, XXXXXY karyotypes also have Klinefelter syndrome, though with more severe symptoms, including mental retardation. The number of Barr bodies in somatic cells is one less than the number of X chromosomes in the karyotype.

The **triple-X karyotype**, 47, XXX, produces a female with no unusual phenotypic symptoms, so this trisomy is not associated with a syndrome. The **47, XYY karyotype** produces a male with no unusual phenotypic symptoms and also has no associated syndrome. Many

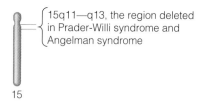

15q11—q13, the region deleted in Prader-Willi syndrome and Angelman syndrome

15

Figure 17.31 The 15q11–q13 region of human chromosome 15.

cases of XXX and XYY go undetected. Some authors have claimed that an extra Y chromosome encourages aggressive behavior. The hypothesis arose from a study that showed a slightly higher frequency of XYY males among men in prison when compared to unincarcerated men. However, there is little evidence to support this hypothesis. The vast majority of men with an XYY karyotype are not incarcerated and have no increased tendency to commit violent acts. Moreover, most XYY males who are incarcerated did not commit violent crimes.

~

Several alterations in chromosome number and structure are known in humans. Most are lethal in embryonic or fetal stages. Most of those that permit survival have characteristic syndromes associated with them.

Parental Imprinting

In diploid individuals, there are two copies of each gene, one on the maternal homologue and the other on the paternal homologue. In most cases, both genes are transcribed and translated unless a mutation disrupts transcription or translation. However, for some genes, either the maternal or the paternal copy is selectively inactivated, so only one of the genes is expressed, a phenomenon called **parental imprinting**. Parental imprinting usually has no phenotypic effect unless the gene that is not inactivated contains a mutation, or is missing because of a deletion.

The best-studied examples of a phenotypic effect associated with parental imprinting are two human genetic disorders called Prader-Willi syndrome and Angelman syndrome. **Prader-Willi syndrome** is characterized by mental retardation, obesity, a narrow face with almond-shaped eyes, lack of muscle tone, small feet and hands, and diminished growth. **Angelman syndrome** is characterized by mental retardation, lack of motor development, epilepsy, absence of speech, tongue extrusion, and excessive laughter. Except for mental retardation, the two disorders are phenotypically unrelated. However, both disorders are associated with interstitial deletions of a region near the centromere in the long arm of chromosome 15 called 15q11–q13 (Figure 17.31). The factor that de-

termines which disorder appears is the parental origin of the chromosome with the deletion. If the deletion comes from the father, then Prader-Willi syndrome appears. If it comes from the mother, then Angelman syndrome appears.

The 15q11–q13 region consists of about 400 million nucleotide pairs and contains several genes. Certain genes in this region are **imprinted**, meaning they are selectively inactivated on either the maternal or the paternal homologue. The mechanism of imprinting appears to be methylation; imprinted genes are highly methylated. When one copy of a gene is imprinted, the copy on the other homologue is actively transcribed, so only one copy of a gene subject to imprinting is expressed. If the copy that should be expressed is missing because of a deletion, the imprinted copy is not expressed, the gene product is not produced, and a mutant phenotype results.

Five genes are known to be imprinted in the 15q11–q13 region. Four of them, *ZNF127, NDN, SNRPN* and *IPW*(OMIM 176270, 602117, 182279, and 601491) are inactivated on the maternal homologue and expressed on the paternal homologue. Lack of expression of *ZNF127* and *IPW* is the cause of Prader-Willi syndrome. A single gene called *UBE3A* (OMIM 105830 and 601623) is inactivated on the paternal homologue and expressed on the maternal homologue. Lack of expression of this gene is probably the cause of Angelman syndrome.

Prader-Willi syndrome and Angelman syndrome are sometimes associated with rare cases of **uniparental disomy**, a situation in which both copies of a chromosome are inherited from the same parent. Uniparental disomy for chromosome 15 arises when nondisjunction during meiosis in one parent produces an $n + 1$ gamete with two copies of chromosome 15, a separate nondisjunction event in the other parent produces an $n - 1$ gamete with no copies of chromosome 15, and these two gametes unite to produce a $2n$ zygote with two copies of chromosome 15 from the same parent (Figure 17.32).

People with uniparental-maternal disomy for chromosome 15 have two maternal copies and no paternal copies of the chromosome. They have Prader-Willi syndrome because the *ZNF127* and *IPW* genes on both maternal copies are inactivated and there are no paternal copies of the genes. People with uniparental-paternal disomy for chromosome 15 have Angelman syndrome because the *UBE3A* gene on both paternal copies is inactivated and there are no maternal copies of the gene.

~

Parental imprinting is the selective deactivation of genes on either the paternal or maternal chromosome in a homologous pair. Genetic disorders that are associated with paternal imprinting usually appear when the nonimprinted gene is deleted or in cases of uniparental disomy.

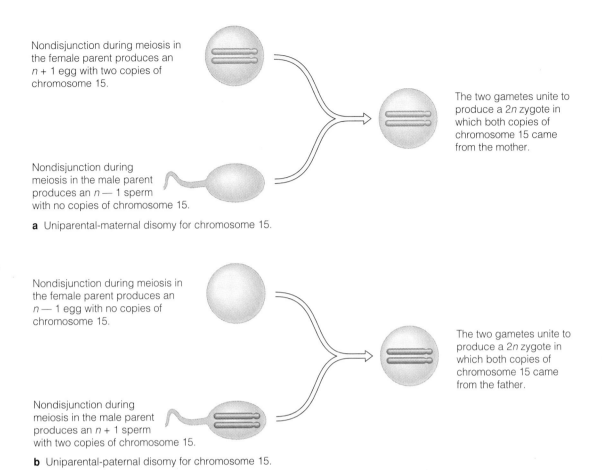

Nondisjunction during meiosis in the female parent produces an *n* + 1 egg with two copies of chromosome 15.

The two gametes unite to produce a 2*n* zygote in which both copies of chromosome 15 came from the mother.

Nondisjunction during meiosis in the male parent produces an *n* — 1 sperm with no copies of chromosome 15.

a Uniparental-maternal disomy for chromosome 15.

Nondisjunction during meiosis in the female parent produces an *n* — 1 egg with no copies of chromosome 15.

The two gametes unite to produce a 2*n* zygote in which both copies of chromosome 15 came from the father.

Nondisjunction during meiosis in the male parent produces an *n* + 1 sperm with two copies of chromosome 15.

b Uniparental-paternal disomy for chromosome 15.

Figure 17.32 The origin of uniparental disomy.

Chromosomal Alterations and Cancer

Chromosome number and structure can change in somatic cells, and the changes can be transmitted to the mitotic progeny of these cells. Cancerous cells are particularly notable for their number of chromosomal abnormalities. Certain types of cancer are caused by changes in gene expression induced by alterations in chromosome structure. One of the best-studied cases is the Philadelphia chromosome, named after the city in which it was discovered. The **Philadelphia chromosome** was initially described as a copy of chromosome 22 with a deletion in the long arm but has since been found to be part of a reciprocal translocation between the long arm of chromosome 22 and the long arm of chromosome 9 (Figure 17.33). The Philadelphia chromosome is found in the cells of patients with chronic myelogenous leukemia, a cancer of the bone marrow. The break point of the translocation on chromosome 9 is within the *ABL* gene, and the break point in chromosome 22 is within the *BCR* gene. The translocation creates a *BCR-ABL* fusion gene that is under the control of the *BCR* promoter and produces a fusion protein with the amino-terminal end of the

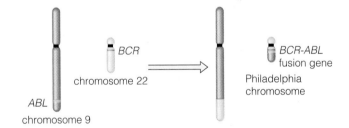

Figure 17.33 The reciprocal translocation (9q;22q) that forms the Philadelphia chromosome.

BCR protein attached to most of the ABL protein. The BCR-ABL fusion protein encourages development of chronic myelogenous leukemia.

Several other types of cancer, including Wilms tumor, Burkitt lymphoma, acute nonlymphocytic leukemia, acute myeloblastic leukemia, and ovarian cancer, are associated with translocations and inversions in specific regions. In nearly every case, one of the break points creates a mutation in or near a gene that promotes cancer. We will

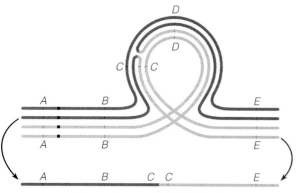

The region around *D* is deleted.

Figure 17.34 Information for Example 17.6: Crossover in the inversion loop that produces a chromosome with an interstitial deletion. (Adapted with permission from Sparkes, R. S., H. Muller, and I. Klisak. 1979. Retinoblastoma with 13q− chromosomal deletion associated with maternal paracentric inversion 13q. *Science* 203:1027–1029. Copyright 1979 American Association for the Advancement of Science.)

discuss cancer-causing genes and the mutations that affect them in detail in Chapter 24. In the following example, we discuss evidence of a chromosomal alteration associated with retinoblastoma, a cancer of the retina.

Example 17.6   Retinoblastoma caused by an inversion-induced deletion.

In 1979, Sparkes et al. (*Science* 203:1027–1029) found an interstitial deletion in a patient with retinoblastoma and mental retardation. The interstitial deletion was in the long arm of chromosome 13 (13q−), the same region in which deletions had been found in other patients with retinoblastoma. The patient's mother had a normal phenotype but was heterozygous for a paracentric inversion in chromosome 13. The mother's inversion spanned approximately the same region as the interstitial deletion in the patient, although the inversion appeared to be slightly larger than the deletion. The researchers examined the deleted chromosome carefully for evidence of a dicentric chromosome but found none.

Problem: **(a)** What evidence suggests a relationship between the inversion and the deletion? **(b)** Propose a model that explains how such a deletion might arise.

Solution: **(a)** The observation that the deletion and inversion are in the same region of chromosome 13 implies that the deletion arose from the inversion. **(b)** A crossover within the inversion loop of a paracentric inversion heterozygote results in a dicentric chromosome, which may then break during anaphase I due to the chromatid bridge. However,

such a break typically causes a terminal deletion rather than an interstitial deletion. A more plausible model suggests that chromatids within the inversion loop recombined as diagrammed in Figure 17.34. If the model in Figure 17.34 is correct, the interstitial deletion should be slightly smaller than the inversion, as was observed.

Certain cancers are associated with inversions or translocations. The break points of the inversions and translocations alter the activity of genes that promote cancer.

SUMMARY

1. Chromosome number is described using the term *ploidy*. Euploid individuals have complete sets of chromosomes with no missing or extra chromosomes. Aneuploid individuals have missing or extra chromosomes in their somatic cells. Polyploid individuals are euploid but have three or more sets of chromosomes in each somatic cell.

2. Aneuploidy is an imbalance in chromosome number. Monosomy and trisomy are the two most common forms of aneuploidy. Monosomy is the absence of one chromosome from a set, and trisomy is the presence of one additional chromosome.

3. Aneuploidy often arises from nondisjunction in meiosis. The end result is a gamete with one extra chromosome ($n + 1$) and a reciprocal gamete that is missing a chromosome ($n − 1$). The $n + 1$ gamete causes trisomy ($2n + 1$) when it unites with a normal haploid (n) gamete. The $n − 1$ gamete causes monosomy ($2n − 1$) when it unites with a normal haploid (n) gamete.

4. Polyploidy is rare in animals but common in plants. About half of all flowering plant species are polyploid.

5. Polyploids are designated by the multiple number of chromosome sets they contain. The symbol n is the number of chromosomes in a gamete, and x is the number of chromosomes in one complete set. In diploid individuals, $2n = 2x = $ the number of chromosomes in a somatic cell. In a tetraploid individual (four chromosome sets per somatic cell), $2n = 4x = $ the number of chromosomes in a somatic cell.

6. Polyploidy typically arises from unreduced gametes that unite with one another or with normal, reduced gametes.

7. Polyploid species may be either autopolyploids or allopolyploids. Autopolyploids arise from a single species. Allopolyploids arise from two or more different species.

8. Patterns of inheritance in allopolyploids resemble those in diploid species. Patterns of inheritance in autopolyploids are more complex.

9. Alterations in chromosome structure include deletions, duplications, translocations, inversions, fissions, and fusions.

10. Large deletions are often lethal, but small deletions may allow survival of heterozygotes. Most deletions are homozygous lethal.

11. Duplications may or may not have phenotypic effects. Tandem duplications may participate in unequal crossing-over, which serves to increase or reduce the number of tandem duplications.

12. Inversions are of two types: pericentric, in which the centromere is within the inversion, and paracentric, in which the centromere is outside of the inversion. Chromosomes loop within the inversion region to pair in inversion heterozygotes. Crossing-over within the inversion loop in inversion heterozygotes results in gametes that are unbalanced.

13. Translocations are typically reciprocal. Altered pairing in reciprocal translocation heterozygotes causes a quadrivalent to form. The chromosomes in the quadrivalent may segregate in one of four possible ways: alternate-1, adjacent-1, alternate-2, and adjacent-2 segregations. Alternate segregations result in balanced gametes, whereas adjacent segregations result in unbalanced gametes. Fertility is typically reduced by about 50% in translocation heterozygotes.

14. Chromosome fissions are rare; fusions are more common. A Robertsonian translocation is fusion that is a specialized type of reciprocal translocation in which the long arms of telocentric chromosomes fuse at their centromeres to form a single chromosome.

15. There are several known chromosome alterations in humans; reciprocal translocations are among the most common. Most chromosome imbalances (either for chromosomal segments or entire chromosomes) cause spontaneous abortion of embryos or fetuses that carry the aberration. The alterations that permit survival typically have medical syndromes associated with them.

16. *Cri du chat* syndrome is caused by a short terminal deletion in the short arm of chromosome 5.

17. Down syndrome is usually due to trisomy 21. Rarely, it is caused by trisomy for the long arm of chromosome 21 in a Robertsonian translocation.

18. Edwards syndrome is due to trisomy 18, and Patau syndrome is due to trisomy 13. Both are severe syndromes and usually cause death within days to weeks after birth.

19. Aneuploidies for X and Y chromosomes are associated with syndromes that are typically less severe than those associated with autosomal trisomies. The most common are Turner syndrome (45, XO), Klinefelter syndrome (47, XXY), and some karyotypes not associated with syndromes, such as 47, XYY and 47, XXX.

20. Certain genes may be selectively deactivated on the maternal or paternal homologue of a chromosome pair, a phenomenon known as parental imprinting.

21. Chromosomal aberrations in somatic cells may promote certain types of cancers.

QUESTIONS AND PROBLEMS

1. Monosomies for autosomes in humans are unknown and apparently abort very early during development. Yet monosomy for the X chromosome is known, and the phenotypic consequences are not particularly severe when compared with autosomal trisomies. Why is survival with monosomy for the X chromosome possible whereas survival is not possible for autosomal monosomies?

2. Using diagrams of chromosome pairing during prophase I, distinguish between the following alterations in chromosome structure: (a) interstitial duplication, (b) interstitial deletion, (c) pericentric inversion, (d) paracentric inversion, and (e) reciprocal terminal translocation.

3. According to G. Ledyard Stebbins, a famous evolutionary geneticist who taught for many years at the University of California at Davis, an autotetraploid is a polyploid in which the corresponding diploid is a fertile species. An allotetraploid contains the doubled genome of a sterile or semisterile hybrid. Explain this statement.

4. Suppose that all individuals in a wild population of *Drosophila melanogaster* are homozygous for an inversion in chromosome 3 that spans 8 cM. All individuals in a second population are homozygous or hemizygous for an inversion in the X chromosome that spans 14 cM. A female from the first population is mated with a male from the second. (a) What proportion of the gametes in the male progeny of these two flies should be unbalanced? (b) What proportion of the gametes in the female progeny should be unbalanced? (c) If the first-generation progeny intermate at random with each other, what proportion of the second-generation females should be heterozygous for both inversions?

5. Suppose that a man heterozygous for a t(14q;21q) Robertsonian translocation is married to a 40-year-old woman. If the woman becomes pregnant, what is the probability that she will bear a child with Down syndrome? (For this question, use the actual rather than the theoretical probabilities.)

6. In 1988, Eklund et al. (*Clinical Genetics* 33:83–86) reported the transmission of a t(13q;14q) Robertsonian translocation through nine generations in humans. One of the people in their study was a female who was homozygous for the translocation. She had six children, all of whom were heterozygous for the translocation. What is the most probable genotype of her husband, with respect to the translocation?

7. In 1984, Martinez-Castro et al. (*Cytogenetics and Cell Genetics* 38:310–312) published a pedigree of a family in which two first cousins had married. They had six children, three who were homozygous for t(13q;14q), two who were heterozygous for t(13q;14q), and one with an unknown genotype who was stillborn. **(a)** What are the genotypes of the parents? **(b)** Should there be a loss of fertility in the three homozygous offspring?

8. In 1978, Evans et al. (*Cytogenetics and Cell Genetics* 20:96–123) reported the results of a cytogenetic study of 14,069 newborn infants. Among the infants they examined, 86 had one parent who was heterozygous for one of several Robertsonian translocations. Of these 86 infants, 45 were heterozygous for a Robertsonian translocation, and 41 had normal karyotypes (no translocation), close to a ratio of 1:1. **(a)** Explain why a 1:1 ratio is expected. **(b)** Explain the absence of trisomic infants in these data.

9. In 1970, Jacobs et al. (*Annals of Human Genetics* 34:119–131) summarized data from several studies on translocations in humans. Among 95 children who had one parent who was heterozygous for a reciprocal translocation, 49 were heterozygous for a reciprocal translocation, and 46 had normal karyotypes (no translocation). **(a)** What are the theoretical frequencies of these two genotypes among live births? **(b)** Use chi-square analysis to test the hypothesis that these data conform to the theoretical frequencies.

10. In 1991, Antonarakis et al. (*New England Journal of Medicine* 324:872–876) published the results of a study to determine the parental origin of the extra copy of chromosome 21 in 200 children with trisomy 21. Through RFLP analysis, the researchers determined that 184 of the children inherited the extra chromosome from their mother, and 9 inherited it from their father. The parental origin of the extra chromosome could not be determined in 7 of the children. How do these results conflict with studies based on chromosome banding?

11. About 20% of all hemophilia A cases in humans are due to an inversion with a break point in intron 22 of the *F8C* gene. Apparently, this particular inversion has arisen independently many times. In 1993, Lakich et al. (*Nature Genetics* 5:236–241) characterized this inversion and proposed a model for how it could arise. Intron 22 is exceptionally large (32 kb) and contains a complete, small, intron-free gene within the intron. This gene, called *gene A*, has several functional copies upstream from the large *F8C* gene. The inversion has one break point within the copy of *gene A* in intron 22, and the other break point is within one of the copies of *gene A* that lies upstream from the *F8C* gene. Explain and diagram how this inversion may arise, and explain why this inversion may have arisen many times independently.

12. The Philadelphia chromosome arises through a reciprocal translocation between human chromosomes 9 and

22. The translocation break points are within the *ABL* gene on chromosome 9 and the *BCR* gene on chromosome 22. The translocation creates a *BCR-ABL* fusion gene that promotes the development of chronic myelogenous leukemia. In 1995, Chissoe et al. (*Genomics* 27:67–82) reported the results of experiments in which they examined the DNA sequences of several *BCR-ABL* fusion genes. They found that the break points varied from one fusion gene to the other and found no features that suggested how the break points arose or how the translocation occurred. How does this observation differ from the inversion situation described in the previous question?

13. Cotton was cultivated for millennia as a source of fiber in the Old World. The two most commonly cultivated species were *Gossypium herbacium* and *G. arboreum*. Both species have $2n = 26$ chromosomes. They differ from each other by a reciprocal translocation. Cotton was also cultivated by ancient civilizations in the New World. The two cultivated species were *G. hirsutum* and *G. barbadense*, each with $2n = 52$ chromosomes. A total of 26 wild species with $2n = 26$ chromosomes have been identified in Africa, South and Central America, Australia, and the Middle East. Two wild species with $2n = 52$ chromosomes have been identified, one with a limited range in northern Brazil and a second in Hawaii. **(a)** What is the basic chromosome number (x) of cotton? **(b)** What are the ploidies of the Old World cotton and the New World cotton?

14. The New World cultivated cotton species, *G. hirsutum* and *G. barbadense*, are characterized by strict bivalent chromosome pairing and genes that segregate in typical Mendelian patterns. Half of the chromosomes resemble the wild $2n = 26$ species found in the New World. The other half most closely resemble the cultivated $2n = 26$ species *G. herbacium*. **(a)** Is the New World cultivated cotton an auto- or allopolyploid? **(b)** What is the most probable origin of New World cultivated cotton?

15. Maize has a somatic chromosome number of $2n = 20$. Bread wheat has a somatic chromosome number of $2n = 42$. Gametes with 9 chromosomes abort in maize, but gametes with 20 chromosomes are viable and fully capable of fertilization in wheat. For this reason, monosomic wheat plants ($2n - 1 = 41$) may be grown and propagated readily, but monosomic maize ($2n - 1 = 19$) is very rare, and when it survives, it is sterile. Why is there such a difference for monosomy in these two species?

16. Most modern sugarcane varieties differ from each other in somatic cell chromosome numbers, generally ranging from 100 to about 125. Some are sterile, others are semisterile, and some are fully fertile. Prior to modern plant breeding, most sugarcane varieties (called noble canes) had $2n = 80$ chromosomes and were fully fertile. The higher chromosome numbers arose when noble canes (*Saccharum officinarum*) were crossed as females with a wild sugarcane (*S. spontaneum*) as the male parent.

Sugarcane may be hybridized with sorghum (*Sorghum bicolor*) and produce semisterile to sterile progeny. Sorghum has a somatic chromosome number of $2n = 20$. A single sterile hybrid plant of sugarcane and maize ($2n = 20$) has also been produced. **(a)** What is the most probable monoploid number (x) of sugarcane? **(b)** Provide a reasonable explanation of why modern cultivated varieties are able to tolerate such varied numbers of chromosomes and still retain fertility.

17. Most banana varieties are triploid and sterile. They produce fully sterile male flowers and partially sterile female flowers. Their wild diploid ancestors, however, have abundant fully developed seeds in them and are fully fertile. Edible and cultivated diploids are known. They are generally female-sterile (due to female-sterility alleles) but male-fertile, so their fruits produce no seeds even though they are diploid. Banana breeders are faced with a dilemma. Because the commonly cultivated triploid bananas are sterile, it is very difficult to introduce new genes into them. However, edible cultivated diploids are male-fertile, so it is possible to pollinate a triploid with an edible diploid and obtain a few viable seeds. The most common ploidy of the viable seeds from such a cross is tetraploidy. **(a)** How might a tetraploid seed arise in a triploid plant pollinated by a diploid? **b)** Once a tetraploid is obtained, how can the desirable triploid chromosome number be restored by further breeding?

18. In wheat, kernel color is governed in some crosses by three independently inherited genes with incomplete dominance at each of the three loci A, B, and C. The alleles A^1, B^1, and C^1 each give one additive dose of color to the kernel. The alleles A^2, B^2, and C^2 provide no doses of color to the kernel, so a plant with the genotype A^1A^1 B^1B^1 C^1C^1 has dark red kernels, and a plant with the genotype A^2A^2 B^2B^2 C^2C^2 has white kernels. A medium red plant A^1A^1 B^1B^1 C^2C^2 is crossed with a plant with light red kernels (A^2A^2 B^2B^2 C^1C^1). The F_1 has medium red kernels. **(a)** Predict the F_2 phenotypic ratios for kernel color in this cross, showing genotype and phenotype frequencies. **(b)** Bread wheat is an allohexaploid. Give a reasonable explanation for the existence of three loci that govern kernel color in bread wheat based on what you know about polyploidy.

19. In alfalfa, an autotetraploid, purple flowers are due to an allele (C_1) that shows complete dominance. White flowers appear when the plants are nulliplex ($c_1c_1c_1c_1$). A purple-flowered plant is crossed with a white-flowered plant. Half of the F_1 progeny are purple-flowered and half are white-flowered. **(a)** What is the genotype of the purple-flowered parent? **(b)** What is the genotype of the purple-flowered F_1 plants? **(c)** What proportion of purple-flowered progeny would be expected in the F_2 generation if one of the purple-flowered F_1 plants is self-pollinated?

20. In 1973, Bingham (*Crop Science* 13:393–394) reported the results of several crosses in alfalfa at both the diploid and tetraploid levels. He determined that flower color in his crosses was governed by two separate loci, named C_1 and C_2. (Note that this is a departure from standard genetic notation. In this case, C_1 and C_2 are different loci rather than different alleles at one locus.) C_1 is dominant over c_1 and causes purple flowers, and c_1 causes white flowers when no C_1 allele is present. C_2 is dominant over c_2 and causes purple flowers, and c_2 causes white flowers when no C_2 allele is present. Table 17.6 shows the results of several crosses with different genotypes for the C_1 and C_2 loci at the diploid and tetraploid levels. Assuming that the two loci are not linked, and remembering that tetraploid alfalfa is autotetraploid, determine the expected ratio of purple-flowered and white-flowered plants in each cross and a chi-square value for each cross.

21. An allotetraploid plant is heterozygous for a reciprocal translocation. Assume that (1) all gametes are viable and capable of fertilization, (2) 50% of the gametes arise from alternate-1 or alternate-2 segregations, (3) 40% of the gametes arise from adjacent-1 segregation, and (4) 10% of the gametes arise from adjacent-2 segregation. What proportion of the zygotes are unbalanced when the plant self-fertilizes?

22. An individual maize plant is heterozygous for genes a, b, c, and d and for an inversion indicated by the shaded area in the chromosomes shown in Figure 17.35. The lower homologue is divided into six regions as indicated. **(a)** Draw the pairing configuration of these homologous chromosomes in prophase I. **(b)** Assuming this individual is crossed with a plant homozygous for the lower chromosome, write the genotypes of expected progeny from a single crossover in each of the six regions. **(c)** Assume that the regions span the following genetic map distances:

Region 2	10 cM
Region 3	4 cM
Region 4	5 cM
Region 5	7 cM
Region 6	4 cM

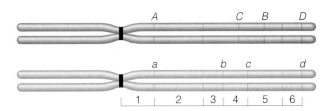

Figure 17.35 Information for Problem 22: Diagram of homologous chromosomes in an inversion heterozygote.

Table 17.6 Information for Problem 20: Crosses of Diploid and Tetraploid Alfalfa

Entry	Ploidy Level	Cross	Purple-Flowered Plants	White-Flowered Plants
1	2x	$C_1c_1\ C_2C_2$ self-fertilized	135	31
2	2x	$C_1C_1\ C_2c_2$ self-fertilized	245	80
3	2x	$C_1C_1\ c_2c_2 \times c_1c_1\ C_2C_2$	46	0
4	2x	$C_1c_1\ C_2c_2$ self-fertilized	33	28
5	4x	$C_1C_1c_1c_1\ C_2C_2C_2C_2$ self-fertilized	110	3
6	4x	$C_1C_1C_1C_1\ C_2C_2c_2c_2$ self-fertilized	770	21
7	4x	Cross of parents in 5 and 6	28	0
8	4x	$C_1C_1c_1c_1\ C_2C_2c_2c_2$ self-fertilized	122	6

Table 17.7 Information for Problem 23: Segregation Patterns and Pollen Abortion in a Reciprocal Translocation Heterozygote

Plant	Segregation Patterns (%)			% Pollen Abortion
	Alternate	Adjacent-1	Adjacent-2	
6913-2	48.7	24.7	26.6	51.1
6913-29	46.7	25.3	28.0	53.1
6913-7	60.7	34.6	4.7	41.0
6913-12	50.3	28.4	21.3	49.8
6913-14	53.1	22.6	24.3	47.1
6913-18	49.5	25.6	24.9	50.4

What proportion of meiotic cells are expected to have chromatid bridges at anaphase I? **(d)** Using the map distances given in part c, and assuming that unbalanced gametes during male gametogenesis result in aborted pollen grains, what is the expected frequency of aborted pollen grains?

23. Table 17.7 shows some of the data that Burnham (*Genetics* 35:446–481) reported in 1950 on translocation heterozygotes between chromosomes 5 and 6 in maize. What do these data suggest about the relationship between pollen abortion and segregation patterns?

24. In tomato, a gene designated *y* is located on chromosome 1. Assuming a plant is trisomic for chromosome 1 and the genotype is duplex for the dominant allele (*YYy*), write all possible gametes that may form, along with their expected frequencies. (Assume also that gametes with two copies of chromosome 1 develop normally.)

Table 17.8 Information for Problem 25: Progeny from a Cross in Which One of the Parents Carries an Inversion

Phenotype	Number of Progeny
Wild type	2214
Scarlet, stripe, ebony, rough, claret	2058
Stripe, ebony, rough, claret	219
Scarlet	238
Scarlet, stripe	4
Ebony, rough, claret	3
Scarlet, stripe, ebony	1
Ebony, rough	1
Total	4738

25. The genes *st* (*scarlet*), *sr* (*stripe*), *e* (*ebony*), *ro* (*rough*), and *ca* (*claret*) are linked on chromosome 3 of *Drosophila melanogaster*. The map distances are as follows:

st		sr	e		ro		ca
	18		8.7		20.4		9.6

In 1926, Sturtevant (*Biologisches Zentralblatt* 46:697–702) made the following cross:

$$+ + + + +/st\ sr\ e\ ro\ ca \times st\ sr\ e\ ro\ ca$$

The chromosome with the wild-type alleles contains an inversion. The progeny are listed in Table 17.8. Show how an inversion could produce these results, and explain how the last four classes of progeny could arise.

FOR FURTHER READING

An old but superb textbook that focuses on most of the topics covered in this chapter is **Burnham, C. R. 1962.** *Discussions in Cytogenetics.* **Minneapolis, Minnesota: Burgess.** An excellent book that discusses detailed medical and epidemiological aspects of Down syndrome is **Pueschel, S. M., and J. E. Rynders. 1982.** *Down Syndrome: Advances in Biomedicine and the Behavioral Sciences.* **Cambridge, Massachusetts: Ware Press.** Research showing the parental origin of the third chromosome in Down syndrome was presented by **Antonarakis, S. E., and the Down Syndrome Collaborative Group. 1991. Parental origin of the extra chromosome in trisomy 21 as indicated by analysis of DNA polymorphisms.** *New England Journal of Medicine* 324:872–876. Human chromosomal abnormalities and their consequences are discussed in **Epstein, C. J. 1986.** *The Consequences of Chromosome Imbalance: Principles, Mechanisms, and Models.* **Cambridge, England: Cambridge University Press.** The molecular basis of parental imprinting is explained in **Tilghman, S. M., T. Caspary, and R. S. Ingram. 1998. Competitive edge at the imprinted Prader-Willi/Angelman region?** *Nature Genetics* **18:206–208.** Genetic aspects of polyploidy and aneuploidy in plants are explained in detail in **Stebbins, G. L. 1971.** *Chromosome Evolution in Higher Plants.* **London: Edward Arnold.** Sturtevant's original work on duplications and unequal crossing-over is described in **Sturtevant, A. H. 1925. The effects of unequal crossing over at the** *Bar* **locus in** *Drosophila.* **Genetics 10:117–147.** Chromosomal deletions and translocations were first documented by Calvin Bridges in **Bridges, C. B. 1917. Deficiency.** *Genetics* **2:445–465;** and **Bridges, C. B. 1923. The translocation of a section of chromosome-II upon chromosome-III in** *Drosophila.* **Anatomical Record 24:426–427.** Inversions were first documented by **Sturtevant, A. H. 1926. A crossover reducer in** *Drosophila melanogaster* **due to inversion of a section of the third chromosome.** *Biologisches Zentralblatt* **46:697–702.**

For additional reading, go to InfoTrac College Edition, your online research library at: http: www.infotrac-college.com/brookscole

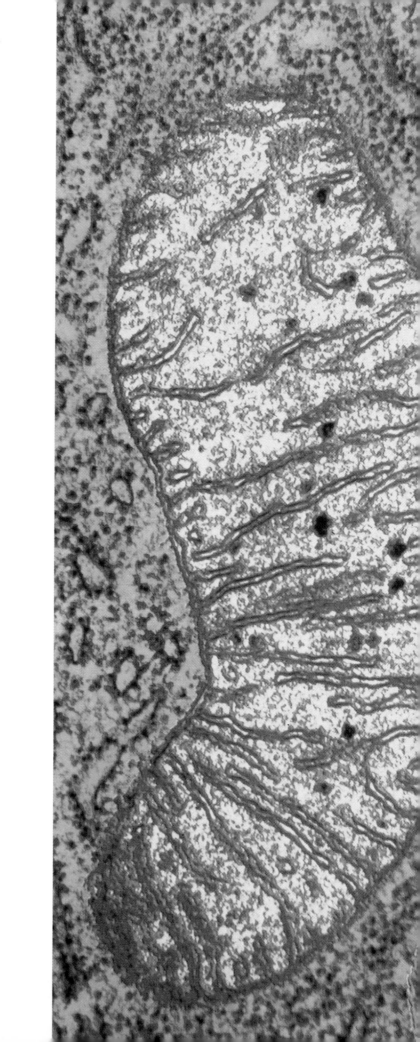

CHAPTER 18

EXTRANUCLEAR INHERITANCE

Within the cytoplasm of eukaryotic cells are **mitochondria**, specialized organelles that play a crucial role in energy conversion and are among the most essential components of eukaryotic cells. A single cell may have anywhere from one to thousands of mitochondria depending on the cell's function. Plants and other photosynthetic eukaryotes have **plastids**, a generic name for organelles that differentiate to perform specialized functions. The best known of the plastids is the **chloroplast**, which is the site of photosynthesis. Chloroplasts contain the pigment chlorophyll, which gives plants and other photosynthetic organisms their green color. Other types of plastids perform functions other than photosynthesis. They include **chromoplasts**, whose pigments give flowers and some other plant parts their colors, **amyloplasts**, which store starch in roots and tubers (such as potatoes), and **elaioplasts**, which store lipids. All plastids are derived from **proplastids**, undifferentiated plastids found in embryonic plant tissue. Figure 18.1 shows transmission electron micrographs of several of these organelles.

Geneticists are interested in mitochondria and plastids because these organelles contain DNA and have protein synthesis systems that are separate from the rest of the cell. Because mitochondria and plastids reside outside of the nucleus, mitosis and meiosis do not direct the

inheritance of their genes. Instead, the inheritance of organellar genes is described as **extranuclear inheritance**, because the genes reside outside of the nucleus. Extranuclear inheritance patterns differ substantially from those of nuclear genes.

In this chapter, we will focus on the genetics of organellar genomes. We begin with a brief discussion of the origins and cellular functions of mitochondria and chloroplasts as a background for understanding the structure and functions of their genomes. We will then examine the genomes of mitochondria and plastids at the molecular level. Next, we will study the inheritance of organellar genes. Lastly, we will look at some unusual types of nuclear inheritance that mimic organellar inheritance.

18.1 THE ORIGINS AND CELLULAR FUNCTIONS OF MITOCHONDRIA AND PLASTIDS

Mitochondria and plastids are like cells within a cell. In many respects, mitochondria and plastids look very much like prokaryotic cells trapped in the cytoplasm of a eukaryotic cell. The **endosymbiotic hypothesis** proposes that the remote ancestors of mitochondria and plastids were free-living prokaryotic cells that at some

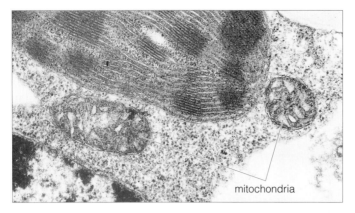

a Two mitochondria from a sugarcane leaf cell.

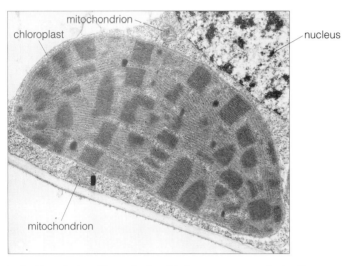

b A chloroplast and two mitochondria from a sugarcane leaf cell.

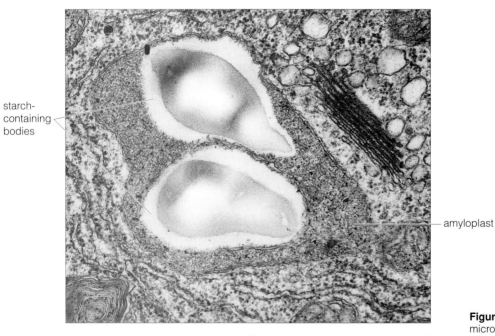

c An amyloplast from a wheat root cell. The white structures within the amyloplast are starch-containing bodies.

Figure 18.1 Transmission electron micrographs of mitochondria and plastids. (Photos courtesy of W. M. Hess.)

point entered primitive eukaryotic cells and remained captured within them. The prokaryotic cells within eukaryotic cells divided and grew within their hosts, and over time developed an obligatory symbiotic relationship. The eukaryotic cell could not live without the endosymbiont (the captured prokaryotic cell), nor could the endosymbiont live without its host (the eukaryotic cell).

There are several strong lines of evidence to support the endosymbiotic hypothesis. The genomes of mitochondria and plastids are circular and are not contained within a nucleus. They are not organized into chromosomes as are nuclear genomes. Their genes are similar in nucleotide sequence to bacterial genes, although organellar genes are much fewer in number than those in free-living bacteria because many of the physiological needs of organelles are now met by the products of nuclear genes.

Mitochondria and chloroplasts have their own systems of DNA synthesis, transcription, and translation, and they have their own ribosomes. The genes that encode the rRNAs of organellar ribosomes provide some of the most convincing evidence of the endosymbiotic hypothesis. The nucleotide sequences of mitochondrial and plastid rRNA genes are very different from the nucleotide sequences of eukaryotic rRNA genes, but they are similar to the sequences of rRNA genes of particular types of eubacteria. The rRNA genes of mitochondria are most similar to those of purple photosynthetic bacteria; the rRNA genes of plastids are most similar to those of cyanobacteria, a group of photosynthetic bacteria that contain chlorophyll.

Although mitochondria and plastids have probably descended from free-living bacteria, they have lost all but a few of their functions. However, the functions they have retained are among the most important in all of life: the acquisition and conversion of cellular energy. Like their cyanobacterial ancestors, chloroplasts carry out photosynthesis, converting energy from sunlight into chemical energy held in carbohydrates. Mitochondria take the energy derived from carbohydrates, proteins, and lipids and transfer it into ATP, which serves as a readily available source of energy in the cell. A detailed discussion of the complex cellular functions of mitochondria and chloroplasts is beyond the scope of this book. However, a brief review of their functions provides us with a framework for discussing the unique genetic aspects of these organelles. Let's look first at the chloroplast.

Chloroplastic Structure and Function

Figure 18.2 shows a three-dimensional representation of chloroplastic structure. The chloroplast is surrounded by two membranes, the **outer boundary membrane** and the **inner boundary membrane**. Between these two membranes is the **intermembrane compartment**. The inner boundary membrane encloses a second compart-

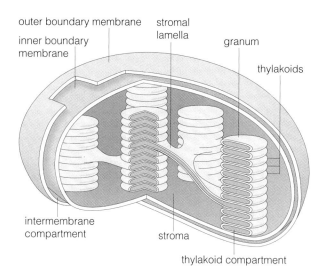

Figure 18.2 Major structural features of a chloroplast.

ment, called the **stroma**. Embedded within the stroma is a membrane system of flattened sacs called **thylakoids**, which are organized into stacked structures called **grana**. The grana are interconnected by extensions of the thylakoids called **stromal lamellae**. A third intermembrane compartment is enclosed by the thylakoid membrane and is called the **thylakoid compartment**. The stacked grana and the stromal lamellae can be seen in the electron micrograph of a chloroplast in Figure 18.1b.

Photosynthesis is the most important function of chloroplasts. It transfers energy from sunlight into carbohydrates by converting carbon dioxide and water into carbohydrates and oxygen. Photosynthesis is a complex process with many steps, but the overall process can be depicted as

$$nCO_2 + 2nH_2O \xrightarrow{\text{energy (from sunlight)}} (CH_2O)_n + nO_2$$

carbon dioxide   water   carbohydrates   oxygen

Photosynthesis takes place in the thylakoid membrane and within the stroma, as depicted in Figure 18.3. The reactions are divided into the **light-dependent reactions**, which occur in the thylakoid membrane, and the **light-independent reactions**, which occur in the stroma. As the names imply, the light-dependent reactions require light to proceed, whereas the light-independent reactions may proceed in the presence or absence of light.

The light-dependent reactions are partitioned among three major groups of proteins: **photosystem I**, **photosystem II**, and the **electron transport chains** that link the two photosystems and the light-dependent reactions with the light-independent reactions. Within the two photosystems are pigments called **chlorophylls**, substances that absorb light within the violet-blue and red portions of the spectrum. Most chlorophylls reflect light in the green

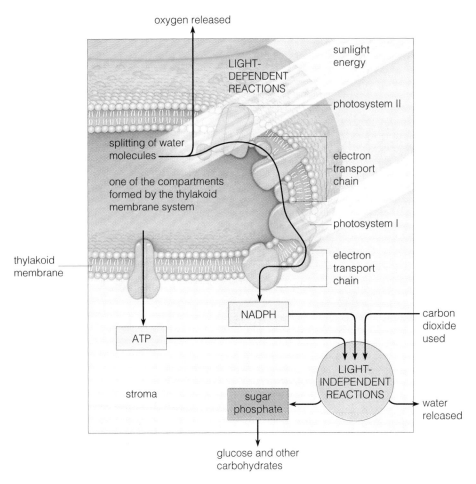

Figure 18.3 Locations of the photosynthesis reactions. The light-dependent reactions occur in the thylakoid membrane. The light-independent reactions occur in the stroma.

portion of the spectrum, giving plant leaves their green color.

Chlorophyll and other pigments in photosystem II absorb photons from sunlight. The photons excite electrons within the pigments to higher energy levels. The pigments transfer their excited electrons from one pigment molecule to another until the electrons reach a specialized chlorophyll called P680, which briefly holds the excited electrons. Simultaneously, several proteins of photosystem II carry out **photolysis**, the splitting of a water molecule to form an oxygen atom (which combines with another oxygen atom to form O_2), two hydrogen ions (H^+), and two electrons. The unexcited electrons from photolysis take the place of the excited electrons given up by the pigments that absorbed photons. The O_2 from photolysis is released into the atmosphere, and the hydrogen ions become part of a hydrogen ion reservoir in the thylakoid compartment.

The excited electrons in photosystem II are released from P680 and are ferried through an electron transport chain to photosystem I. As the electrons move through the electron transport chain, energy is released from the electrons, and that energy is used to transfer hydrogen ions across the thylakoid membrane from the stroma into

the thylakoid compartment. This creates a hydrogen ion concentration differential between the thylakoid compartment and the stroma.

The electrons coming out of the electron transport chain, now at a lower energy level, are accepted by pigments in photosystem II. As these pigments absorb photons from sunlight, the electrons are excited once again and transferred to a specialized chlorophyll called P700. The electrons, now at a higher energy level, are released from P700 through another electron transport chain to the acceptor molecule $NADP^+$, converting it to NADPH at the end of the light-dependent reactions.

The hydrogen ions that have accumulated in the thylakoid compartment move across the membrane back into the stroma, by virtue of the concentration differential, through channel proteins embedded in the thylakoid membrane. This process releases energy that is captured when ADP and inorganic phosphate combine to form ATP. Thus, the end products of the light-dependent reactions are ATP and NADPH, both of which hold the energy that came from sunlight.

The light-independent reactions transfer the energy contained in the ATP and NADPH synthesized by the light-dependent reactions into carbohydrates. The energy

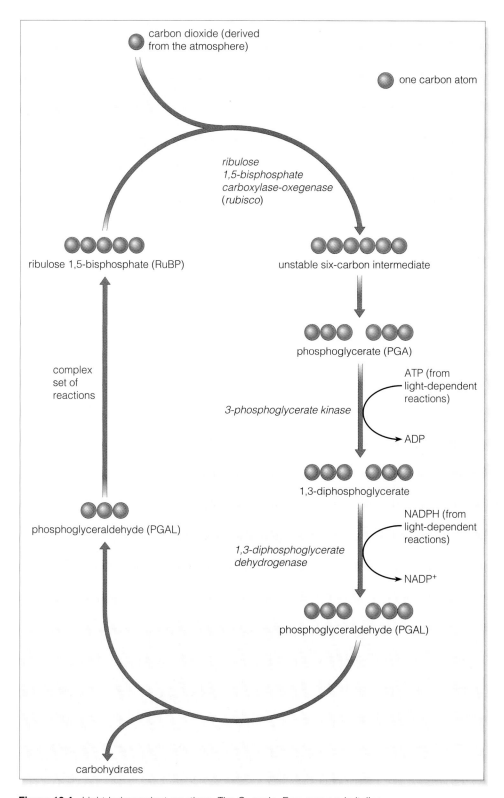

Figure 18.4 Light-independent reactions: The C_3 cycle. Enzymes are in italics.

acquired in the light-dependent reactions is transferred to carbohydrates through the **C_3 cycle**, also called the **Calvin-Benson cycle** (Figure 18.4). Carbon dioxide molecules are taken from the atmosphere and transferred to the stroma. Each CO_2 molecule combines with a five-carbon sugar called ribulose 1,5-bisphosphate (RuBP) to

form an unstable six-carbon intermediate. This intermediate is rapidly broken down into phosphoglycerate (PGA) molecules that each contain three carbon atoms. Each PGA molecule acquires a phosphate group from ATP and a hydrogen and electrons from NADPH to form phosphoglyceraldehyde (PGAL). This process transfers energy

from ATP and NADPH to PGAL. For every 12 molecules of PGAL produced, 2 are used to form sugar phosphates, which are converted into the sugars and starches found in plants. The remaining PGAL molecules are rearranged to form RuBP for continuation of the C_3 cycle.

The end products of photosynthesis are oxygen and carbohydrates that contain the energy acquired from sunlight. The carbohydrates can then be used as starting material to form other carbon-containing compounds, such as lipids and amino acids, needed by the plant.

Some plants do not depend entirely on the C_3 cycle for the light-independent reactions. The C_3 cycle is inefficient when temperatures are high. Many plants that are well adapted to warm climates utilize an alternative pathway known as C_4 photosynthesis to capture CO_2, a process that prevents much of the water loss suffered by plants that are obligated to use the C_3 cycle. C_4 plants tend to predominate in warm climates, whereas C_3 plants are more abundant in cooler climates.

Mitochondrial Structure and Function

Figure 18.5 shows a three-dimensional representation of the major structural features of a mitochondrion. Each mitochondrion has two major membranes: the **outer boundary membrane**, which surrounds the mitochondrion and forms its outer surface, and the **inner boundary membrane**, which is highly folded into flattened structures called **cristae** (singular, *crista*). The inner boundary membrane divides the mitochondrion into two compartments: the **intermembrane compartment**, between the outer and inner boundary membranes, and the **matrix**, which is enclosed by the inner boundary membrane.

The mitochondrion is responsible for most of the reactions of **cellular respiration**, the oxidation of energy-containing organic molecules, such as carbohydrates, proteins, and fats. Cellular respiration transfers the energy held in these molecules into ATP, a molecule that stores and transfers the energy so that it can be utilized elsewhere.

Figure 18.6 illustrates the pathways for the oxidation of proteins, fats, and carbohydrates during cellular respiration and the cellular locations of these reactions. Let's examine carbohydrate oxidation as an example of how energy is transferred to ATP in the mitochondrion.

Carbohydrate oxidation can be summarized as the reverse of photosynthesis:

$$(CH_2O)_n \;+\; nO_2 \;\longrightarrow\; nCO_2 \;+\; 2nH_2O$$
carbohydrates oxygen   carbon   water
                            dioxide

energy
(held in ATP)

It liberates energy that was captured by photosynthesis and produces carbon dioxide and water as by-products.

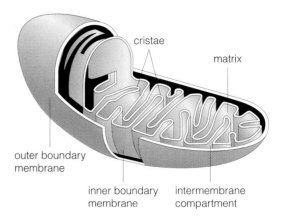

Figure 18.5 Major structural features of the mitochondrion.

Carbohydrate oxidation begins outside of the mitochondrion, where carbohydrates are converted to pyruvate through glycolysis. We will review glycolysis of glucose, although other carbohydrates may also be used. As illustrated in Figure 18.7, the initial steps that convert glucose into fructose-1,6-bisphosphate require energy from 2 ATP molecules. The fructose-1,6-bisphosphate is split into the three-carbon molecule phosphoglyceraldehyde (PGAL). The remaining steps convert PGAL into pyruvate and liberate energy through the synthesis of 4 ATP molecules, for a net gain of 2 ATP molecules.

Pyruvate enters the mitochondrion, where it undergoes further oxidation. The steps of pyruvate oxidation in the mitochondrial matrix are diagrammed in Figure 18.8. First, pyruvate is converted into acetyl coenzyme A (acetyl CoA), liberating carbon dioxide and transferring an electron to form NADH. The acetyl CoA then enters the **citric acid cycle**, also known as the **Krebs cycle**. In the first step, oxaloacetate combines with the acetyl CoA to form citrate while liberating coenzyme A. The citrate is then converted to isocitrate, which is converted to α-ketoglutarate following the liberation of carbon dioxide and the transfer of an electron to NAD^+ to form NADH. Coenzyme A now rejoins the α-ketoglutarate to form succinyl CoA while liberating an additional carbon dioxide and releasing an electron to form NADH. The conversion of succinyl CoA to succinate releases energy during the formation of ATP. The succinate is then converted into fumarate, coupled with the release of an electron, this time to form $FADH_2$. Fumarate is converted into malate and malate into oxaloacetate with the release of an additional electron to form NADH.

The Krebs cycle itself yields very little energy in the form of ATP, but it builds a reservoir of the electron carriers NADH and $FADH_2$. The electron carriers release electrons and hydrogen ions. They release the electrons into an electron transport chain in proteins that are embedded in the inner boundary membrane (Figure 18.9). The electron transport chain consists of four membrane-bound protein complexes, each with enzymatic activities. The electrons enter complex I from NADH produced in

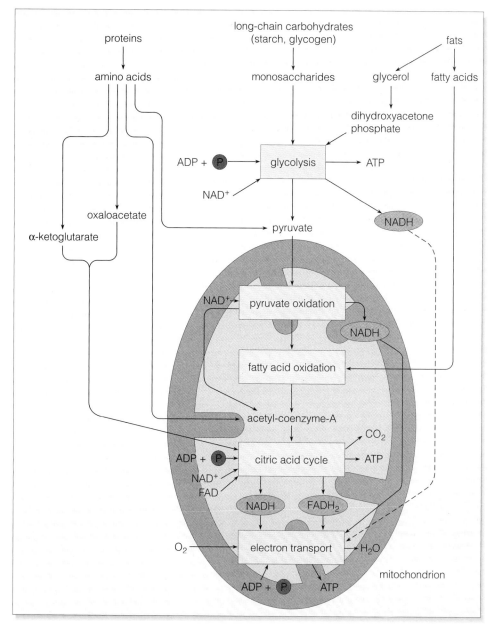

Figure 18.6 Pathways for the oxidation of proteins, fats, and carbohydrates.

the citric acid cycle. Electrons also enter complex II from FADH$_2$ produced in the conversion of succinate to fumarate in the citric acid cycle.

As the electrons pass through the electron transport chain, they drop to lower energy levels. The released energy is used to drive hydrogen ions across the inner boundary membrane from the mitochondrial matrix into the intermembrane compartment. This sets up a hydrogen ion concentration differential across the inner boundary membrane. The differential is relieved by passage of the hydrogen ions back into the matrix, which in turn releases energy. This energy is used to synthesize a substantial quantity of ATP, on average 32 ATP molecules for every six-carbon sugar molecule oxidized.

The electrons coming out of the electron transport chain are captured by oxygen, and the oxygen becomes a free radical, with unpaired electrons. The highly reactive oxygen radical is retained within complex IV, which ensures that the oxygen radical reacts only with hydrogen ions to form water. The overall process within the electron transport chain is called **oxidative phosphorylation** because oxygen is used as an electron and hydrogen ion acceptor at the end of the electron transport chain, and free phosphate is transferred to ADP to form ATP (Figure 18.9).

The end products of cellular respiration are carbon dioxide, water, and energy held in ATP. The ATP can then be transported out of the mitochondrion, where it can be used elsewhere as an energy source.

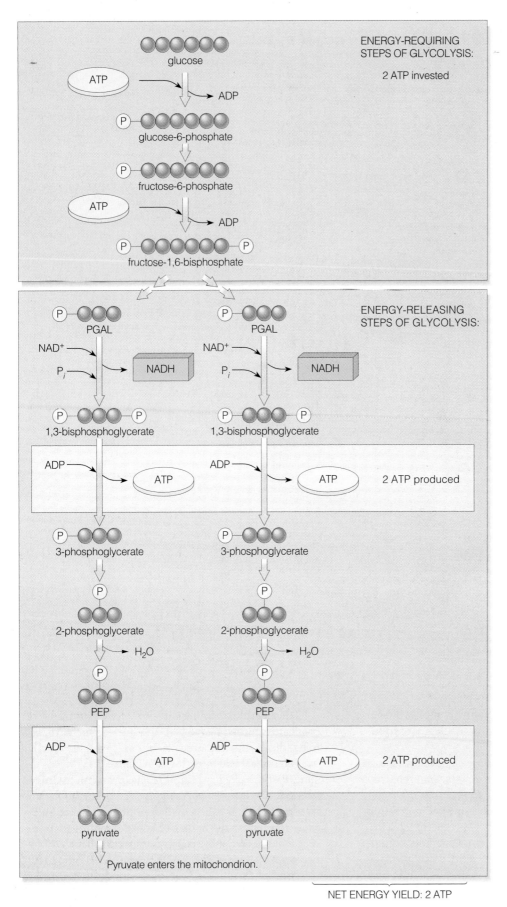

Figure 18.7 Glycolysis. Carbohydrates are broken down into pyruvate through glycolysis in the cytoplasm outside of the mitochondrion. The pyruvate then enters the mitochondrion.

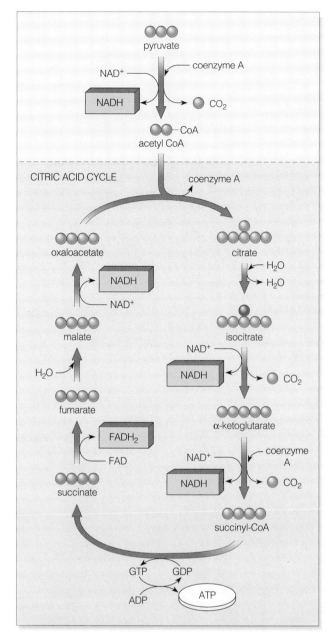

Figure 18.8 Pyruvate oxidation. Pyruvate oxidation yields only one molecule of ATP, but it also yields four molecules of NADH and one molecule of FADH$_2$, which are coenzymes that carry high-energy electrons.

Organellar Proteins

The biochemical reactions in mitochondria and chloroplasts require numerous proteins that function as enzymes and structural proteins. Some of the genes that encode these proteins are found in the circular genomes of the organelle. These genes are transcribed within the organelle, and the mRNAs are translated within the organelle on organellar ribosomes.

However, the organellar genomes encode only a fraction of the proteins needed within the organelle. For example, the human mitochondrial genome contains only 13 protein-encoding genes. These genes encode 13 of the approximately 80 proteins required for oxidative phosphorylation. None of the enzymes in the Krebs cycle are encoded in the mitochondrial genome. Most proteins required by mitochondria and plastids are encoded by nuclear genes and are synthesized on the cytoplasmic ribosomes outside of the organelle. Once synthesized, the proteins are imported into the organelle.

Protein-encoding genes in the organellar genomes typically do not encode all subunits of an enzyme. Instead, an organellar enzyme may include subunits encoded in both the organelle and the nucleus. Let's look at the mitochondrial enzyme cytochrome oxidase as an example. **Cytochrome oxidase** is part of complex IV of the mitochondrial electron transport chain. It is a multimeric, metal-containing protein that binds to oxygen and ensures that oxygen accepts electrons at the end of the electron transport chain during oxidative phoshorylation. It then directs the oxygen to react with hydrogen ions to form water (Figure 18.9). Cytochrome oxidase is composed of 10–13 protein subunits. Three of the subunits, numbered I, II, and III, are encoded by mitochondrial genes, and the remaining subunits are encoded by genes in the nucleus.

In both mitochondria and plastids, all proteins encoded by the organellar genome are utilized within the organelle in which they are synthesized. Many nuclear-encoded proteins are imported into the organelles, but no proteins are exported from the organelle.

The paucity of genes in organellar genomes raises an important question regarding the endosymbiotic hypothesis. Mitochondria and plastids contain far fewer genes than their free-living bacterial counterparts. Where are the additional genes that presumably were once found in the remote ancestors of mitochondria and chloroplasts? Apparently, many mitochondrial and plastid genes have been transferred to the nucleus throughout evolutionary time. For example, the enzyme ribulose-1,5-bisphosphate carboxylase-oxygenase, often referred to as rubisco, catalyzes the combination of carbon dioxide with RuBP in the first step of the C$_3$ cycle in photosynthesis. Rubisco is a very abundant protein in plant leaves and photosynthetic bacteria. In fact, it is probably the most abundant protein on earth. In cyanobacteria, the intact enzyme is composed of 16 subunits, eight identical large subunits and eight identical small subunits, which are encoded by two genes on the bacterial chromosome. Rubisco in vascular plants is quite similar to the bacterial form; it consists of eight identical large subunits and eight identical small subunits. However, only the large subunit is encoded by the plastid genome. The genes that encode the small subunit are located in nuclear chromosomes. The small subunit gene apparently moved from plastids to the nucleus during the early evolution of photosynthetic eukaryotic cells that were the ancestors of vascular plants. Furthermore, there is good evidence that such organellar-nuclear gene transfers are still under way.

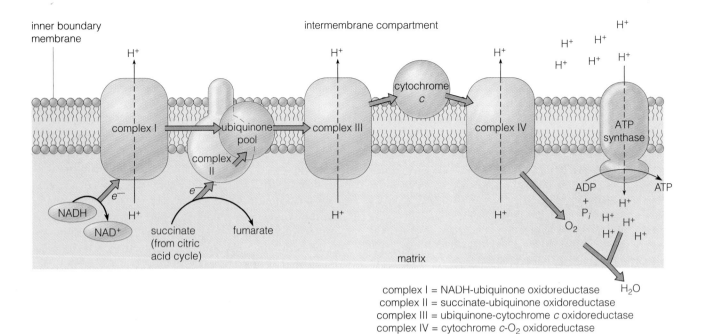

Figure 18.9 The electron transport chain of oxidative phosphorylation in the mitochondrion. The yellow arrows show the path of electrons.

complex I = NADH-ubiquinone oxidoreductase
complex II = succinate-ubiquinone oxidoreductase
complex III = ubiquinone-cytochrome *c* oxidoreductase
complex IV = cytochrome *c*-O$_2$ oxidoreductase

~
Mitochondria and plastids carry out energy conversions within the cell. Mitochondria and plastids contain their own genes, which are transcribed and translated within the organelle.

18.2 ORGANELLAR GENOMES

Organellar genomes contain only a small fraction of all of the genes in the cell. However, most of the proteins encoded by organellar genes are essential for the cell's survival. In this section, we examine organellar genes and their organization in the genomes of mitochondria and plastids. Let's begin with mitochondrial genomes.

Mitochondrial Genomes

The mitochondrial genome of each eukaryotic species is a circular molecule that is present in multiple copies within each mitochondrion. Mitochondrial genome size is highly variable among species (Table 18.1). In general, plants have the largest mitochondrial genomes and animals have the smallest. The mitochondrial genomes of animals are often an order of magnitude smaller than those of plants. Most cells have several mitochondria, each with multiple copies of the circular genome, so the mitochondrial genome is repeated anywhere from about 10 to hundreds or thousands of times in each cell (Figure 18.10).

Plant mitochondrial genomes are much more complicated than their animal and fungal counterparts. The

Table 18.1	Approximate Sizes of Mitochondrial Genomes
Species	**Genome Complexity (nucleotide pairs)**
Animals	**14,000–19,000**
Human	16,600
Drosophila	18,400
Fungi	**19,000–108,000**
Saccharomyces	74,000–85,000
Neurospora	62,000
Plants	**200,000–2,500,000**
Pea	240,000
Maize	320,000

DNA molecules in animal mitochondria are stable and fairly uniform within a species. On the other hand, plant mitochondrial DNA molecules have homologous regions that permit intramolecular recombination, so there may be a heterogeneous mixture of circular DNA molecules in plant mitochondria (Figure 18.11).

The DNA sequences of mitochondrial genomes in several species have been well studied. The complete DNA sequences of several mitochondrial genomes have been determined, and the sequencing has revealed some

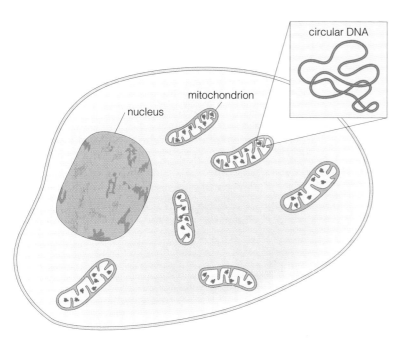

Figure 18.10 Repetition of mitochondrial genomes in a cell. Most cells contain several mitochondria, each of which contains several copies of the circular DNA molecule. Thus, there are several (often thousands of) copies of mitochondrial DNA molecules in each cell.

important characteristics about the organization of mitochondrial genes and their expression. Let's examine the human mitochondrial genome as an example.

Figure 18.12 shows the locations of genes in the circular human mitochondrial genome. There are 13 protein-encoding genes, 2 rRNA genes, and 22 tRNA genes. The genome consists of 16,569 nucleotide pairs and exemplifies remarkable efficiency of nucleotide utilization. With the exception of the origin of replication, nearly every nucleotide in the genome encodes part of an mRNA, tRNA, or rRNA. In fact, some genes overlap, and most termination codons in mRNAs are UAA codons in which only the U is encoded by the mitochondrial genome. The two A's in the termination codons are added as part of the poly (A) tail during polyadenylation of the mRNA.

Human mitochondrial genes have no promoters. Instead, one strand of the DNA is transcribed into one large RNA molecule called the **L strand**, and the other DNA strand is transcribed into a second large RNA molecule called the **H strand**. Only one protein-encoding gene (*NADH dehydrogenase 6*) and eight tRNA genes (*Pro, Glu, Ser, Tyr, Cys, Asn, Ala,* and *Gln*) are in the L strand. The remaining genes are in the H strand. After the two large RNAs are transcribed, they are cleaved into individual mRNAs, rRNAs, and tRNAs, and the mRNAs are polyadenylated.

In spite of their great divergence in size, the mitochondrial genomes of all eukaryotes have essentially the same genes, but the organization of these genes in the

genomes of different species is quite variable, suggesting that there has been substantial mitochondrial genomic reorganization over evolutionary time.

Plastid Genomes

Plastid genomes are circular DNA molecules that are highly similar in size and gene organization among all species that have plastids. They range from about 120 kb to almost 200 kb. The largest genomes are found in algae. There are over 120 genes in the plastid genome, compared to an average of about 40 in the mitochondrial genome, and the plastid genome encodes about 50 proteins, compared to less than 20 in mitochondrial genomes.

Plastid genes have prokaryotic-like promoters and in several cases are transcribed as multigenic units that resemble operons. Most of the protein products of plastid genes are subunits for enzymes. The plastid-encoded subunits combine with nuclear-encoded subunits to produce functional enzymes. Like the mitochondrion, plastids also synthesize many of their own rRNAs and tRNAs. However, unlike the mitochondrion, they also synthesize the subunits for their RNA polymerase, which are similar to the RNA polymerase subunits in *E. coli*.

Plastid genomes contain two inverted repeats that divide the genome into two segments of unique DNA: a short single-copy region, and a long single-copy region. The inverted repeats are 10,000–25,000 nucleotide pairs

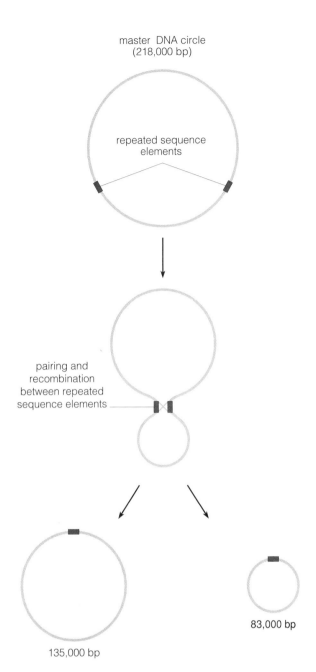

master DNA circle
(218,000 bp)

repeated sequence
elements

pairing and
recombination
between repeated
sequence elements

135,000 bp

83,000 bp

Figure 18.11 Formation of two smaller circular DNA molecules from one large "master" molecule in plant mitochondria.

long and contain several genes. The two repeated regions may recombine with one another, which inverts the short single-copy region relative to the long single-copy region without disrupting any of the genes (Figure 18.13). Most of the size differences in plastid genomes among species are due to differences in the size of the inverted repeats.

~

Animal mitochondrial genomes are smaller than plant mitochondrial genomes, and nearly all nucleotides in animal mitochondrial genomes are within genes. The arrangement of mitochondrial genes is highly variable among species. Plastid genome arrangement is similar among species that have plastids.

18.3 MITOCHONDRIAL INHERITANCE

Plastids and mitochondria reside in the cytoplasm outside of the nucleus. They reproduce within the cell through binary fission as bacteria do, and their reproduction is not necessarily synchronous with cell division. As a cell grows, the organelles grow and divide. When the cell divides mitotically, the cytoplasm with its organelles is partitioned to the two daughter cells. Except in a few single-celled eukaryotes, there is no mechanism to ensure equal partitioning of organelles. When gametes are formed, meiosis has no influence on organellar genomes. Organelles are inherited along with the cytoplasm that is transmitted to the gametes.

For all of these reasons, the inheritance of genes in mitochondrial and plastid DNA is very different from the inheritance of nuclear genes. Mendel's principles do not apply. Nor is partitioning of organellar genes equal and uniform in mitotic divisions as it is for nuclear genes. Therefore, extranuclear inheritance is in a class by itself for inheritance from one sexual generation to the next, and from one mitotic cell generation to the next. There are some fundamental differences in the formation of gametes and the transmission of cytoplasm among single-celled eukaryotes, animals, and vascular plants. For this reason, there are differences in the inheritance patterns for mitochondria both within and among these groups. Let's look first at yeast, one of the best-studied single-celled eukaryotes.

Mitochondrial Inheritance in Yeast

The manner in which organelles are inherited depends on the way the cytoplasm is transmitted at fertilization. There are two important features of cytoplasmic inheritance in yeast: (1) Yeast zygotes have **biparental inheritance** of mitochondria because the two haploid cells that unite to form the zygote both contribute cytoplasm to the zygote. (2) Yeast cells reproduce mitotically through **budding**, a process in which a small bud forms and receives an equal portion of nuclear material (due to the equal partitioning of chromosomes during mitosis) but may receive a smaller portion of cytoplasm from the parent cell (Figure 18.14).

In order to analyze mitochondrial inheritance, polymorphic genetic markers must be present in the mitochondrial genome. Some of the best examples of genetic markers for studying mitochondrial inheritance are antibiotic resistance alleles because there is no selection for or against them in the absence of the antibiotic, so their transmission at fertilization and their segregation during mitotic divisions represents the mechanisms of inheritance. At any point after mating, colonies can be transferred to a growth medium that contains the antibiotic to identify the phenotypes.

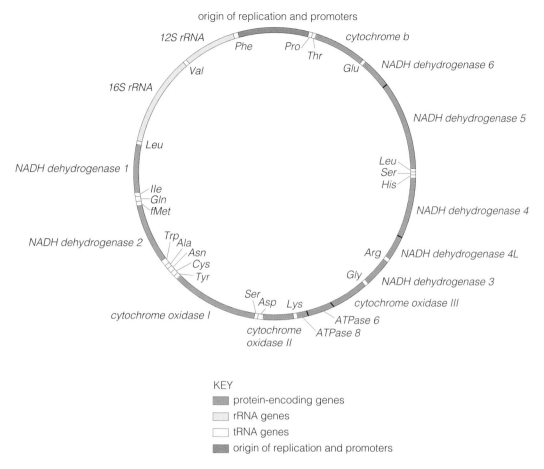

Figure 18.12 Locations of genes in the human mitochondrial genome.

KEY

■ protein-encoding genes
□ rRNA genes
□ tRNA genes
■ origin of replication and promoters

Several antibiotics interfere with the enzymes in the oxidative phosphorylation pathway. Susceptibility and resistance to these antibiotics is governed by mitochondrial genes that encode subunits of these enzymes. Antibiotic resistance genes known to reside on the yeast mitochondrial genome include genes for erythromycin resistance, chloramphenicol resistance, spiromycin resistance, paramomycin resistance, and oligomycin resistance.

If a erythromycin-resistant (ery^R) strain is crossed with a erythromycin-sensitive (ery^S) strain, the diploid hybrid zygote usually has both ery^R and ery^S types of mitochondria in its cytoplasm. If the diploid zygote is grown on erythromycin-free medium, it can divide mitotically by budding for several cell generations, each generation producing diploid cells. When a bud forms from a diploid cell, it receives some of the cytoplasm, and with it some of the mitochondria, from the parent cell. However, if the parental diploid cell has a 1:1 ratio of ery^R and ery^S mitochondria, because of sampling error the bud does not necessarily receive mitochondria in a 1:1 ratio.

Let's illustrate this phenomenon with a simple model. Suppose you have five black marbles and five white marbles in a cup, and you reach into the cup and remove four marbles at random. You could draw any one of five pos-

sible combinations: all black, three black and one white, two black and two white, one black and three white, or all white marbles. This range of outcomes is similar to what happens in the yeast cells by budding, and the smaller the number of mitochondria received by the bud, the greater the effect of sampling error. In some cases, a bud may receive only one type of mitochondrion. Once all of its mitochondria are of one type—say, ery^S—then all mitotic and meiotic progeny cells that descend from this cell have only ery^S mitochondria in their cytoplasms. A cell that has mitochondria that are all of one type in its cytoplasm cannot be called homozygous, because that term refers to the nuclear genetic component. Instead, it is **homoplasmic**. Conversely, a cell with two or more types of mitochondria in its cytoplasm is **heteroplasmic**.

As heteroplasmic cells divide through several mitotic generations, homoplasmic cells arise. With mitotic generation of cell division, the proportion of homoplasmic cells increases and the proportion of heteroplasmic cells decreases until eventually all cells are homoplasmic for one or the other type of mitochondria. In yeast, if a zygote that is heteroplasmic for ery^R and ery^S is allowed to proceed through several mitotic generations before meiosis, and after meiosis the ascospores from a single cell are cultured

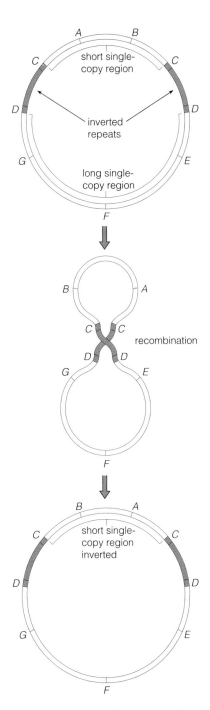

Figure 18.13 Intramolecular recombination between repeats in the plastid genome to form a molecule with one unique sequence inverted relative to the other.

individually to form colonies, each colony is homoplasmic *ery^R* or homoplasmic *ery^S*. In most experiments, all cells are homoplasmic after about 10–20 cell generations.

Statistical theory predicts such a trend toward homoplasmy. To illustrate the theory, let's simplify things by making a few assumptions that are not true in real life but make it easier to understand the principle. After we've looked at the simple (but unreal) situation, we'll then add the complications of real life to see what actually happens.

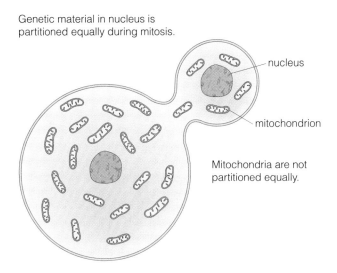

Figure 18.14 Budding in a yeast cell.

Using Figure 18.15 as a model, suppose there is a heteroplasmic cell with eight mitochondria, four that are antibiotic resistant and four that are antibiotic susceptible. Let's assume that each cell partitions mitochondria equally during mitosis, so each of the two daughter cells receives four mitochondria. Which progeny cell a mitochondrion is destined to enter is determined at random. Assume further that each progeny cell duplicates its mitochondria to eight so that the number of mitochondria per cell is the same in each cell generation.

In the first generation, the parent cell has 4 *S* mitochondria and 4 *R* mitochondria. The probability of homoplasmy in the daughter cells that arise from the first division is $2 \times \frac{4}{8} \times \frac{3}{7} \times \frac{2}{6} \times \frac{1}{5} = \frac{48}{1680}$, or 2.9%. Thus, the probability of heteroplasmy in the daughter cells that arise from the first division is 97.1%. Let's suppose that daughter cell 1 receives 3 *R* mitochondria and 1 *S* mitochondrion, so daughter cell 2 receives 1 *R* mitochondrion and 3 *S* mitochondria. After doubling their mitochondria, the ratios are 6*R*:2*S* (cell 1) and 2*R*:6*S* (cell 2). Now, the chance of homoplasmy for *R* arising from daughter cell 1 in the next generation is $\frac{6}{8} \times \frac{5}{7} \times \frac{4}{6} \times \frac{3}{5} = \frac{360}{1680}$, or 21.4%. After a cell is homoplasmic, all its mitotic progeny will be homoplasmic, so as the number of cells increases with mitotic divisions, a certain proportion become homoplasmic, reducing the frequency of heteroplasmic cells with each mitotic cell generation. After a number of generations, most cells are homoplasmic for resistant or susceptible mitochondria.

This model predicts that after complete homoplasmy, the frequency of cells homoplasmic for *R* mitochondria should equal the original frequency of *R* mitochondria in the zygote, and the frequency of cells homoplasmic for *S* mitochondria should equal the original frequency of *S* mitochondria in the zygote. Therefore, if one parent contributes most of the mitochondria, most of the progeny cells after several mitotic divisions become homoplasmic for the mitochondria of that parent. How many genera-

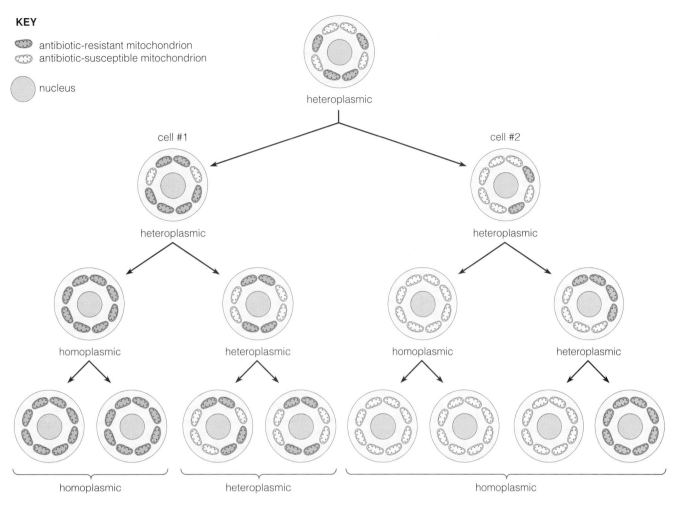

Figure 18.15 Model of somatic segregation of mitochondria.

tions are required for complete homoplasmy depends on the number of mitochondria in each cell. The greater the number, the longer it takes for complete homoplasmy to arise. This process is called **somatic segregation** or **sorting out** and is often observed in the mitotic progeny of heteroplasmic cells.

The theoretical model just described is sometimes called the **random-segregation model**. It can be tested under real situations to see if its predictions hold true. In real life, most of the assumptions of the random-segregation model are not met. This, however, does not nullify somatic segregation. On the contrary, the violations of the assumptions *enhance* the rate of somatic segregation. In yeast, complete homoplasmy arises earlier than expected on the basis of the random-segregation model. There are several good reasons for this. First, mitochondria are probably not mixed at random in the cell at each generation. When a mitochondrion divides within the cell, the two identical mitochondria from a single division are more likely to remain near each other. If those two mitochondria divide again before the cell divides, there are four genetically identical mitochondria next to each other. In other words, mitochondria may remain

spatially segregated within the cell before budding takes place (Figure 18.16a). Second, budding draws only a few mitochondria from the parent cell, so the probability of the bud being homoplasmic is increased (Figure 18.16b).

Experiments in all species in which homoplasmy has been detected show that complete homoplasmy eventually arises in the mitotic progeny of a heteroplasmic cell due to either random or nonrandom processes. Thus, heteroplasmy is usually a transient condition, which eventually disappears during mitotic cell generations.

In the following example, we examine some experiments in yeast that demonstrate somatic segregation of mitochondria.

Example 18.1   Mitochondrial inheritance in yeast.

In 1968, Linnane at al. (*Proceedings of the National Academy of Sciences, USA* 59:903–910) reported the results of tetrad analysis in colonies of *Saccharomyces cerevisiae* derived from crosses between erythromycin-resistant and erythromycin-susceptible strains. Some of their results are summarized in Table 18.2.

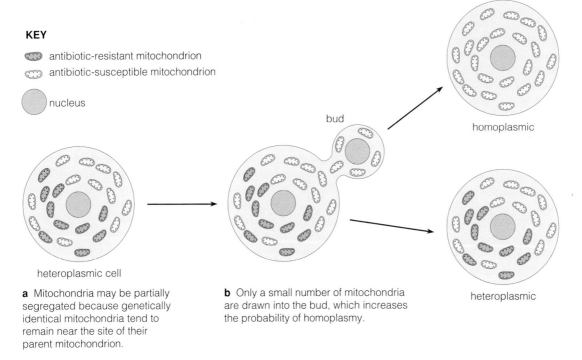

KEY

antibiotic-resistant mitochondrion

antibiotic-susceptible mitochondrion

nucleus

bud

homoplasmic

heteroplasmic cell

a Mitochondria may be partially segregated because genetically identical mitochondria tend to remain near the site of their parent mitochondrion.

b Only a small number of mitochondria are drawn into the bud, which increases the probability of homoplasmy.

heteroplasmic

Figure 18.16 Homoplasmy that arises from **(a)** spatial segregation and **(b)** unequal partitioning of mitochondria during budding.

Problem: **(a)** What aspects of the data in Table 18.2 suggest extranuclear rather than nuclear inheritance? **(b)** What evidence is there of biparental inheritance of mitochondria? **(c)** Is there any indication that there are heteroplasmic cells among those sampled? Describe how you arrived at your conclusion.

Solution: **(a)** If erythromycin resistance and susceptibility are governed by alleles at a nuclear locus, then a zygote from a cross between resistant and susceptible strains should be heterozygous for the two alleles. The diploid cells reproduced by budding, so they should remain heterozygous throughout all mitotic divisions. The spores should segregate 2:2 for resistant:susceptible in each ascus. The facts that all spore segregations are 4:0 or 0:4, and that their phenotypes correspond with their parent colonies, are good evidence of extranuclear rather than nuclear inheritance. **(b)** The observation of mixed colonies provides evidence of biparental inheritance because the traits of both parents are transmitted to progeny. **(c)** There is no evidence that any of the cells is still heteroplasmic. The mixed colonies provide evidence of heteroplasmy in the zygotes, but by the time the colonies were tested, all cells had segregated to full homoplasmy for either resistant or susceptible mitochondria. Evidence for this comes from the spore segregations during meiosis. All spores have the phenotype of their parent cell. If mitochondria were still undergoing somatic

segregation, some spores in the same ascus should have differed from the parent cell, but not necessarily in a 1:1 pattern typical of nuclear inheritance.

Somatic segregation produces homoplasmy in the progeny of a heteroplasmic cell after several mitotic cell generations.

Petite Mutants in Yeast

The first mitochondrial mutants in yeast were discovered by Boris Ephrussi in the 1950s and were called **petite mutants**, because yeast colonies that carried them were smaller due to slow growth. These mutants are often mentioned in textbooks as a classic example of cytoplasmic inheritance. The experiments were significant in that they were among the first to demonstrate non-Mendelian inheritance of an extranuclear gene, but they are not especially good examples of how extranuclear inheritance operates in yeast because selection favors one phenotype over the other and causes inheritance ratios to be skewed.

Ephrussi showed that petite mutants lacked aerobic (oxygen-requiring) respiration, a mitochondrial function. Yeast cells can live without aerobic respiration (unlike many other organisms) because they have the capability of deriving energy through anaerobic (oxygen-free) fermentation. The petite colonies grow using anaerobic fermentation even when they are in an aerobic environment.

Cross	Phenotype of Diploid Colony*	Number of Tetrads Analyzed	Segregation in Spores**	
			Resistant	Susceptible
1	R	2	2 (4:0)	0
2	R	8	8 (4:0)	0
3	R	8	8 (4:0)	0
4	S	2	0	2 (0:4)
5	M	14	2 (4:0)	12 (0:4)
6	M	4	1 (4:0)	3 (0:4)
7	M	7	3 (4:0)	4 (0:4)

*R = resistant, S = susceptible, and M = mixture of resistant and susceptible cells.
**The numbers in parentheses indicate the R:S segregation of spores within the asci.

Ephrussi crossed the petite colonies with normal colonies and found three different types of petite mutants. One was called a **segregational petite** because the petite and normal phenotypes segregated in a 1:1 ratio in the progeny, as expected for a nuclear gene. A second type, called a **neutral petite**, when crossed with the normal type always produced only normal progeny. When the normal progeny were crossed with the petite type, their progeny were also always normal. This was clearly not a case of typical nuclear inheritance, which should have produced a 1:1 ratio.

Later, the neutral petite mutants were found to have mitochondria that lacked mitochondrial DNA. When normal mitochondria are introduced into a cell through crossing, they are the only mitochondria with DNA and, therefore, the only mitochondria that transmit their genes. The normal mitochondria also outcompete the DNA-free mitochondria and are the only type transmitted to progeny.

The third type, called **suppressional petites**, produce both petite and normal types in variable ratios in the progeny. Suppressional petites arise from cells with mutant mitochondrial DNA, so somatic segregation, and perhaps some selection that favors normal mitochondria, account for the variable ratios.

Recombination of Mitochondrial Genes

Mitochondria in yeast may fuse with each other, allowing DNA of different mitochondrial types to come into contact and recombine in a heteroplasmic cell. On the surface, it appears that genetic mapping might be possible because of gene recombination. However, successful mapping is much more limited in mitochondria than in the nucleus. Mitochondrial genes do not recombine with a predicted frequency because fusion of different mito-

chondrial types in a heteroplasmic cell is not a predictable phenomenon. Also, heteroplasmic cells may vary in their ratio of heteroplasmy because of somatic segregation, so the likelihood of two different mitochondria fusing and their DNA recombining varies significantly from one cell to the next. Only linkages between very close genes can be detected because recombination between them is very rare. All mitochondrial genes are linked because they are on the same circular piece of DNA, but only genes that are very closely linked can be mapped using recombination.

Mitochondrial Inheritance in Animals

Animals have very different life cycles than yeast, particularly with regard to the cytoplasm. Animal sperm cells are much smaller than the ovum and carry very little cytoplasm (see Figure 11.20). The mitochondria present in sperm cytoplasm are few relative to those in the ovum and are there primarily to provide the energy needed to give the sperm cell motility. When the sperm cell unites with the ovum, its contribution of nuclear genetic material to the zygote is equal to the nuclear genetic material that the ovum contributes. However, the sperm cell contributes very little cytoplasm. The ovum, on the other hand, is large, with abundant cytoplasm that contains many mitochondria. The overwhelming majority of the mitochondria in the zygote are of maternal origin, and very few, if any, mitochondria are contributed by the sperm cell.

Thus, genes encoded by mitochondria conform to **uniparental-maternal inheritance** in animals—only the maternal parent contributes mitochondrial genes to all offspring. Several recent studies have shown that in reality a few paternal mitochondria may be transmitted to the zygote and inherited in progeny, as illustrated in

the following example. However, the paternal contribution of mitochondria in animals is so small that inheritance of mitochondrial genes is nearly always uniparental-maternal.

Example 18.2   Paternal inheritance of mitochondrial DNA in mouse.

For many years, mitochondrial inheritance in mouse (and all other animals) was thought to be strictly uniparental-maternal. Scientists knew that the sperm carried mitochondria, but no mitochondrial genes of paternal origin had been detected in mouse or any other mammal. Once PCR was available, detection of DNA present in very small concentrations became possible. In 1991, Gyllenstein et al. (*Nature* 352:255–257) reported a set of experiments designed to detect minute contributions of paternal mitochondrial DNA in mouse. A polymorphic DNA marker permitted mitochondrial DNA from the laboratory mouse (*Mus musculus domesticus*) to be distinguished from a closely related species called *M. spretus*. A female laboratory mouse (*M. musculus domesticus*) was mated with an *M. spretus* male. The F_1 females were fertile, and the F_1 males were sterile. An F_1 female was crossed with an *M. spretus* male, and one of the female progeny from this cross was crossed with an *M. spretus* male. The researchers continued this backcrossing procedure (crossing progeny females to *M. spretus* males) for 26 generations. Using PCR, the researchers found paternal (*M. spretus*) mitochondrial DNA in all progeny examined after 26 generations. They concluded that the fraction of mitochondrial DNA contributed by the paternal parent was about 1 paternal molecule per 100,000 maternal molecules per generation.

Problem: (a) Why did the authors backcross the mice through so many generations? **(b)** What do these results suggest about paternal transmission of mitochondrial DNA in mouse?

Solution: (a) If paternal transmission of mitochondria is very low, it might not be detected in the progeny of a single cross. If at least some of the paternal mitochondrial DNA remains in a heteroplasmic condition through each generation, then each backcrossing step should add a little more paternal mitochondrial DNA to the progeny. Eventually, the amount of accumulated paternal mitochondrial DNA may be sufficient to detect. **(b)** These results suggest that mitochondrial inheritance is almost entirely uniparental-maternal, but that the paternal parent transmits a very small amount of mitochondrial DNA to most, if not all, progeny.

The preponderance of uniparental-maternal inheritance of mitochondrial DNA in animals has some important consequences. First, there is very little chance for recombination of mitochondrial genes because animal zygotes are nearly homoplasmic. Heteroplasmy can arise only by mutation or by transmission of paternal mitochondria, and even then it may not be transmitted from one sexual generation to the next. If a gene mutates, it is transmitted to offspring only if it is in the germ line of a female. If it mutates in the male, or in the female's somatic cells, it is lost when the individual dies. Even if a gene mutates in the germ line of the female, the mutant gene is still not likely to be transmitted because each cell has several mitochondria, each with several copies of DNA. A mutation in a single copy of DNA is overwhelmed in number by all the other nonmutant molecules and may be quickly lost in most cells through somatic segregation.

However, when a mutation is established in a homoplasmic condition in the germ line, it is transmitted through the female lineage to all offspring. For this reason, mitochondrial mutations can spread rapidly in the offspring of a single female, and are excellent genetic markers for evolutionary analysis because they mark female lineages. The Y chromosome is inherited in a purely paternal fashion but is present only in males. It can be used to study paternal lineages in males. The mitochondrial genome is inherited in a uniparental-maternal fashion but is present in females and males. Analysis of mitochondrial DNA has provided important information about evolutionary patterns in animals, humans in particular, a topic we will examine in more detail in Chapter 21.

There is a rare and curious exception to uniparental-maternal inheritance of mitochondria in animals. In species of blue mussel (*Mytilus*), females inherit mitochondrial DNA only from the maternal parent, and males inherit mitochondria from both parents. The males then transmit only the paternal mitochondria to their male offspring.

~

Mitochondrial inheritance in animals is typically uniparental-maternal.

Mitochondrial Inheritance in Plants

Fertilization in plants is much different than it is in animals (see Figure 11.21). A pollen grain may contain organelles, and when it germinates on the stigma of a flower, the pollen tube may also carry organelles. This suggests that some organelles may be contributed by the paternal parent to the zygote, and indeed they are in some cases but not in others. In most flowering plant species studied, mitochondrial inheritance is uniparental-maternal, although some studies show evidence of biparental inheritance in a few species.

The best-studied mitochondrial trait in plants is **cytoplasmic male sterility**, a type of male sterility that is

governed by extranuclear genes. There has been considerable interest in this trait because it is used commercially to exclude self-pollination in the production of F₁ hybrid seed used on farms and in gardens. The best-studied species in this regard is maize, although similar genetic systems exist in other species. Cytoplasmic male sterility in maize is governed by genes on the mitochondrial genome that interact with genes on the nuclear genome, so a more appropriate name for it is **cytoplasmic-nuclear male sterility**.

The mitochondrial alleles are inherited in strict uniparental-maternal fashion. However, expression of the mitochondrial alleles depends on the nuclear genotype. A dominant nuclear restorer allele (*R*) causes male fertility when homozygous (*RR*) or heterozygous (*Rr*), even when the cells of the maize plant are homoplasmic for the mitochondrial male-sterility allele. When a plant is homozygous for the recessive allele (*rr*), and the plant is homoplasmic for the male-sterility allele in its mitochondria, then the plant is male sterile. If a plant is homozygous for the recessive allele (*rr*) but does not have the male-sterility allele in its mitochondria, the plant is male fertile. In other words, male sterility appears only when the plant has the mitochondrial male-sterility allele and is homozygous for the recessive *r* allele, as diagrammed in Figure 18.17.

This situation provides a convenience for commercial hybrid seed producers. Maize has a relatively high level of hybrid vigor when completely homozygous parents are crossed with each other to produce highly heterozygous F₁ hybrid progeny, so plants grown from F₁ hybrid seed are more productive than other types of maize plants. Maize is monoecious, with the male flowers located in the tassel at the top of the plant and the female flowers near the middle of the plant where the ear eventually grows. To produce all F₁ seed, the seed producers have to prevent plants from self-pollinating. To do this, they plant four rows of the genotype to be used as the female parent and two rows of the genotype to be used as the male parent in an alternating pattern. They then hire workers to go through the fields after the tassels have emerged but before the pollen is shed and cut off the tassels with a knife, an expensive and laborious procedure. However, if the female parent is male sterile, then there is no need to detassel the plants. The female parent is homozygous *rr* and has mitochondria with the male-sterility allele. The male parent is homozygous *RR*, so all the F₁ seeds sold to farmers are *Rr* and produce male-fertile plants, even though the plants inherit the mitochondrial male-sterility allele from their maternal parent (Figure 18.18). The nuclear restorer allele *R* is essential. A field of all male-sterile plants produces no seed because there is no functional pollen to fertilize the flowers and stimulate seed production.

KEY

mf mitochondrion with male-fertility gene

ms mitochondrion with male-sterility gene

RR and *Rr* genotypes restore male fertility in a cell with ms mitochondria

rr genotype fails to restore male fertility in a cell with ms mitochondria

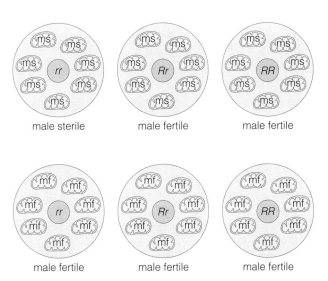

Figure 18.17 All possible nuclear-mitochondrial combinations for mitochondrial male-sterility genes and nuclear fertility-restoration genes. All genotypes are male fertile except one: the nuclear genotype *rr* in a cell with mitochondria that have the male-sterility gene.

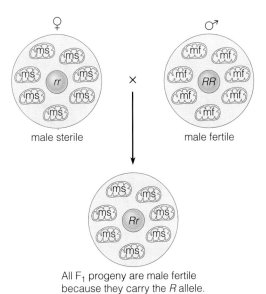

All F₁ progeny are male fertile because they carry the *R* allele.

Figure 18.18 The use of cytoplasmic-genetic male sterility in hybrid seed production. All of the F₁ hybrid progeny are heterozygous *Rr* and are male fertile even though they inherit the mitochondrial male-sterility gene from their maternal parent.

Cytoplasmic-nuclear male sterility was used for many years to produce hybrid maize seed. Eventually, nearly all hybrid maize seed sold in the United States was produced with the same mitochondrial male-sterility allele in a cytoplasmic-nuclear male sterility system. Then disaster struck. In 1970, much of the U.S. maize crop succumbed to southern corn leaf blight, a fungal disease that attacks the leaves and kills the plant before it produces seeds. The reason? Plants that carry the mitochondria with the male-sterility allele are susceptible to a particular race of the southern corn leaf blight fungus. DNA studies revealed that the allele responsible for male sterility also causes susceptibility to the fungus, an example of mitochondrial pleiotropy.

The fungal race (called race-T) is specific to a particular allele called the T-type male-sterility allele. The T designation stands for Texas, the state in which the male-sterility allele was first discovered. Two other types of male sterility have since been found: S-type and C-type. The S-type is interesting because the male-sterility allele is in a small plasmidlike segment of DNA in the mitochondrion. S-type and C-type male sterility genes do not cause susceptibility to race-T of the fungus, but fear among seed producers of another disaster (and possible lawsuits) has kept them from adopting cytoplasmic-nuclear male sterility again. Instead, they employ high school and college students to detassel the plants by hand, or they use a detasseling mower that cuts off the tassels a few feet above the ground.

~

Mitochondrial inheritance in plants is usually, though not always, uniparental-maternal. Mitochondrial genes may interact with nuclear genes to produce a particular phenotype.

18.4 MITOCHONDRIAL DNA AND GENETIC DISORDERS IN HUMANS

The human mitochondrial genome includes only 37 genes, 13 of which encode proteins. These genes are vitally important because they govern cellular respiration, one of the most essential cellular functions. The first conclusive association of a human genetic disorder with a mitochondrial mutation was found in 1988. Since then, researchers have identified several human disorders caused by mitochondrial mutations.

People who have cells that are heteroplasmic for a mitochondrial mutation may be unaffected phenotypically until the proportion of mutant DNA molecules in a cell line reaches a particular threshold through somatic segregation. For these and other reasons, genetic disorders associated with mitochondrial mutations often show partial penetrance, delayed onset, and a high degree of variation in phenotypic expressivity.

Let's examine a few of the human genetic disorders that are known to be associated with mutations in mitochondrial DNA.

Disorders Caused by Deletions in the Mitochondrial Genome

Deletions of mitochondrial DNA have been associated with progressive external ophthalmoplegia, Kearn-Sayre syndrome, and Pearson syndrome. Progressive external ophthalmoplegia is characterized by a gradual loss of the ability to control eye movement. Kearn-Sayre syndrome is characterized by a combination of symptoms that may include progressive external ophthalmoplegia, pigmentary disorders in the eye, heart disease, cerebellar dysfunction, high cerebrospinal fluid protein, muscle weakness, hearing loss, and diabetes. Pearson syndrome appears during childhood and is characterized by a combination of symptoms that may include anemia, reduction in the numbers of all blood cells, and dysfunction of the pancreas, liver, and kidneys.

The deletions in mitochondrial DNA that cause these disorders are not inherited but appear at some point during development and are replicated in somatic cell lineages. All individuals with mitochondrial DNA deletions have cells that are heteroplasmic for the deletion because cells that are homoplasmic for the deletion are unable to survive in the absence of cellular respiration. The phenotypic effects of a mitochondrial DNA deletion are highly variable and depend on the tissue in which the deletion is present and the proportion of mitochondrial DNA molecules that carry the deletion. The proportion of mutant mitochondrial DNA molecules may vary significantly among tissues because of somatic segregation.

The deletions responsible for these disorders vary in size from 1–9 kb and always include at least one tRNA gene. As cells acquire a greater proportion of mutant mitochondrial DNA molecules through somatic segregation, translation of all mitochondrial protein-encoding genes may be affected—which, in turn, reduces the ability of the cell to carry out oxidative phoshorylation.

Point Mutations in Protein-Encoding Genes

Point mutations in mitochondrial protein-encoding genes affect steps in the electron transport chain of oxidative phosphorylation. One of the best-studied examples of a genetic disorder caused by point mutations in a protein-encoding gene is Leber hereditary optic neuropathy (LHON), degeneration of the optic nerve that causes rapid onset of blindness, usually in men in their 20s. Most people with LHON are homoplasmic for a point mutation in one of the genes that encodes the subunits of complex I, which is responsible for the first step in the electron transport chain of oxidative phosphorylation. However, not all individuals who are homoplasmic for a mutation

have the LHON phenotype, indicating that the mutant alleles are partially penetrant. Also, LHON tends to be more penetrant in males than in females, as shown in the following example.

Example 18.3   Investigation of a possible nuclear-mitochondrial gene interaction in Leber hereditary optic neuropathy (LHON) in humans.

Leber hereditary optic neuropathy (LHON) is caused by several well-characterized mutations in mitochondrial DNA. Many people who are homoplasmic for a mutation do not have the LHON phenotype. Also, the disorder is about five times more prevalent in males than in females. This pattern of partial penetrance and preferential expression in males led several researchers to postulate that a recessive X-linked allele interacts with the mutant mitochondrial alleles. Under this assumption, males who are homoplasmic for mutant LHON mitochondrial allele and hemizygous for the X-linked susceptibility allele should have the LHON phenotype, and those who have the dominant allele at the X-linked locus should not have the LHON phenotype. Females who are homoplasmic for the mutant LHON mitochondrial allele could have the LHON phenotype if they are homozygous for the susceptibility allele or if X inactivation is skewed in such a way that it leaves the X chromosome with the susceptibility allele active in the neural tissues. The pattern of LHON in several pedigrees is consistent with this hypothesis. In 1996, Chalmers et al. (*American Journal of Human Genetics* 59:103–108) examined polymorphic X-linked DNA markers located throughout the X chromosome and polymorphic DNA markers located in the pseudoautosomal region of the X and Y chromosomes in LHON pedigrees of people from Great Britain and Italy. They found no association of any of the markers with penetrance of the LHON phenotype in the pedigrees. They also found no evidence of skewed X chromosome inactivation in females with the LHON phenotype.

Problem: (a) What observations suggest that a recessive X-linked susceptibility allele might explain incomplete penetrance for LHON? (b) What conclusions can be drawn from the results of the study by Chalmers et al.?

Solution: (a) The observation that penetrance is higher in males than in females suggests that a recessive allele on the X chromosome may affect phenotypic expression of LHON. Under this hypothesis, the LHON phenotype should appear more frequently in males but should not be excluded from females, particularly when skewed X inactivation is a possibility. (b) The results of Chalmers et al. provided evidence against the interaction of a recessive X-linked susceptibility allele with the mitochondrial mutations. Had there been a recessive X-linked allele that confers susceptibility to LHON, then polymorphic DNA markers that are closely linked to the susceptibility allele should have shown an association with the LHON phenotype.

The point mutations that cause LHON are inherited in a uniparental-maternal fashion and are nearly always homoplasmic in individuals who carry a mutation. All children of a mother who is homoplasmic for a mutation are also homoplasmic for it. But due to partial penetrance, many homoplasmic individuals do not have the LHON phenotype.

Point Mutations in tRNA and rRNA Genes

Point mutations in tRNA or rRNA genes affect protein synthesis in the mitochondrion. These are among the most common mitochondrial mutations in humans and are responsible for several neurological and muscular genetic disorders. The *tRNA^{leuUIR}* gene has a mutation hotspot and is the gene most often affected by point mutations.

Point mutations in tRNA and rRNA genes tend to have a widespread effect because they inhibit translation of all 13 proteins encoded in the mitochondrion. However, the proportion of mutant tRNA or rRNA molecules must be greater than 85% for the mutation to have a significant detrimental effect on mitochondrial translation.

Mutations in Mitochondrial DNA and Human Aging

There is a well-established relationship between the proportion of mitochondrial DNA molecules that contain deletions and the age of an individual. As humans age, deletions in mitochondrial DNA increase, especially in tissues where cells are no longer dividing. This observation suggests the possibility that many of the symptoms of aging, especially loss of neurological and muscular functions, may be associated with mitochondrial mutations that accumulate in somatic cells. Although the possibility that mitochondrial mutation may cause the symptoms of aging is an attractive hypothesis, current evidence is insufficient to show a clear causal relationship. Most mitochondrial genetic disorders require a threshold of at least 60% mutant mitochondrial DNA in an affected tissue for clinical symptoms to appear. The proportion of mutant mitochondrial DNA that appears with aging is rarely more than 0.1%. Nonetheless, the possible effects of mitochondrial

Figure 18.19 Variegation for green and white leaf tissue in a plant.

Figure 18.20 Flowers borne on a white sector of a variegated plant.

mutations on aging is an area of active research that should produce more conclusive results in the near future.

~

Several human genetic disorders, most of which affect the neural, muscular, and skeletal systems, are caused by mutations in mitochondrial DNA.

18.5 PLASTID INHERITANCE

Although mitochondrial inheritance in plants appears to be uniparental-maternal in most cases, plastid inheritance may be uniparental-maternal, biparental, or uniparental-paternal, depending on the species. Uniparental-maternal inheritance of a plastid-encoded trait was first reported in 1909 by Carl Correns, one of Mendel's rediscoverers, in the four o'clock plant. However, also in 1909, Erwin Baur discovered a case of non-Mendelian biparental inheritance of a plastid-encoded trait in geranium. Neither researcher knew that he had discovered plastid inheritance at the time, but later studies showed that the traits they studied were governed by plastid genes.

Much research on plastid inheritance has focused on variegation for green and white leaf tissue in plants (Figure 18.19). The tissues that are white consists of cells that are homoplasmic for a mutant chloroplast that cannot produce normal chlorophyll. Tissue sectors that are completely green with no nearby variegation are probably homoplasmic for the normal green chloroplasts. Sectors that are variegated still have heteroplasmic cells in them that are in the process of somatic segregation. The heteroplasmic tissue appears green, and each small sector of white tissue indicates that somatic segregation has produced a homoplas-

mic white cell that then produced only homoplasmic white progeny cells. A green sector may be homoplasmic for normal chloroplasts or heteroplasmic.

Inheritance can be studied using plants that are green or variegated. A plant that is entirely white survives only in the seedling stage, when it has sufficient nourishment from the energy stored in the seed. After the seedling stage, it dies because it cannot carry out photosynthesis, so it is not possible to use pure white plants as parents. However, a white sector may develop on a variegated plant and bear flowers (Figure 18.20). The sector survives because nutrients are transported from the normal green portions of the plant to the white tissue. The gametes produced in flowers on a white sector are homoplasmic for the mutant plastids that produce white tissue. By cross-pollinating flowers on a homoplasmic green sector with flowers on a homoplasmic white sector, researchers can study plastid inheritance.

Table 18.3 shows the results of all possible combinations of crosses between flowers borne on green, white, and variegated sectors in the four o'clock plant. Green plants arise from a zygote that is homoplasmic for normal plastids. White plants arise from a zygote that is homoplasmic for mutant chlorophyll-deficient plastids. Variegated plants arise from heteroplasmic zygotes. Plastid inheritance in the four o'clock plant is uniparental-maternal because the phenotypes of all progeny are like the maternal parent, with the exception of flowers on a variegated sector. The female gametes in such flowers may be homoplasmic green, homoplasmic white, or heteroplasmic, depending on how far somatic segregation has proceeded by the time the gamete is formed.

About two-thirds of all angiosperms (flowering plants) studied have uniparental-maternal inheritance of plastids. The remaining third of species studied have biparental inheritance of plastids. Examples of well-studied species with biparental inheritance of plastids are

geranium, evening primrose, and alfalfa. When flowers on a white sector are used to pollinate flowers on a green plant or green sector, the progeny may consist of a combination of purely white, variegated, and purely green individuals. This observation indicates that biparental inheritance is actually a combination of uniparental-paternal, uniparental-maternal, and biparental inheritance in the same cross. The following example highlights some of the data that led researchers to this conclusion.

Example 18.4   Biparental inheritance of plastids in alfalfa.

In 1989, Smith (*Journal of Heredity* 80:214–217) reported the data in Table 18.4 on plastid inheritance in alfalfa. He pollinated five genetically different green plants with flowers borne on chlorophyll-deficient (yellow-green) sectors. The yellow-green phenotype is caused by a mutation in chloroplastic DNA.

Problem: What do these data suggest about the type of plastid inheritance (paternal, maternal, or biparental) that predominates in alfalfa?

Solution: Plants that are pure yellow-green are homoplasmic for mutant plastids and inherit their plastids only from the paternal parent. Plants that are pure green are homoplasmic for normal plastids and inherit their plastids only from their maternal parent. Plants that are variegated inherit their plastids from both parents. The variegation is caused by somatic segregation. Pure green plants are the least frequent in the progeny of every cross except one (CUF-B × 301). Overall, they constitute only 5.5% of the progeny. Pure yellow-green progeny are the most frequent, although not in every cross. Overall, they constitute 63.4% of the progeny. Variegated progeny constitute 31.1% of the total. Plastid inheritance in these experiments is a combination of uniparental-paternal, uniparental-maternal, and biparental inheritance with predominance of uniparental-paternal inheritance.

There is no evident association between plants that have biparental inheritance and their evolutionary origin among flowering plants. Typically, species with biparental inheritance and species with uniparental-maternal inheritance of plastids are found in the same families. However, the situation may be different among gymnosperms, a group of plants that produce seeds but have no flowers. They include conifers (such as pine, spruce, and fir trees), ginkgoes, cycads, and gnetophytes. Most gymnosperms studied show predominantly

Table 18.3   Progeny Phenotypes from Crosses Between Flowers Borne on Green, White, and Variegated Sectors of Four-O'clock Plants

Cross*	First-Generation Offspring
Green × green	All green
Green × white	All green
Green × variegated	All green
White × green	All white
White × white	All white
White × variegated	All white
Variegated × green	Green, white, and variegated
Variegated × white	Green, white, and variegated
Variegated × variegated	Green, white, and variegated

*In each cross, the phenotype of the female parent is listed first.

uniparental-paternal inheritance of plastids, but depending on the species, mitochondrial inheritance may be uniparental-maternal, uniparental-paternal, or biparental.

A pattern of reverse uniparental inheritance for mitochondria and plastids is seen in the unicellular alga *Chlamydomonas reinhardtii*. Instead of male and female sexes, it has mt^+ and mt^- mating types among the haploid cells. An mt^+ cell unites with an mt^- cell to form the zygote. Uniparental inheritance predominates for both plastids and mitochondria, but plastids are inherited predominantly (> 90%) from the mt^+ mating type, and mitochondria are inherited in a uniparental fashion from the mt^- mating type.

Some of the mechanisms that enforce uniparental inheritance of plastids are now known. In *Chlamydomonas reinhardtii*, each haploid cell contains a single chloroplast. When the two haploid cells fuse to form a diploid zygote, the two chloroplasts also fuse. A nuclear gene closely linked to the mating-type locus encodes an enzyme that selectively degrades the DNA from the mt^- parent but not from the mt^+ parent. Mutations in this gene cause biparental inheritance of plastid DNA.

In some plants, plastids are abundant in the generative cells that eventually mature into pollen grains, but as the cells mature, the plastids are gradually eliminated. Once the mature pollen grain is formed, there are no more plastids present, so plastids are inherited in a uniparental-maternal fashion. In other cases, paternal plastids are transmitted by the pollen but are degraded at fertilization or shortly thereafter.

~

Plastid inheritance in plants varies among species. It may be uniparental-maternal, uniparental-paternal, or biparental.

Table 18.4 Information for Example 18.4: Plastid Inheritance in Alfalfa

Cross			Number of Progeny		
Maternal Parent (green)		Paternal Parent (chlorophyll-deficient)	Chlorophyll-Deficient (paternal)	Green (maternal)	Variegated (biparental)
MS5	×	301	28	5	27
MS5	×	272	25	0	3
MS5	×	262	13	0	2
CUF-B	×	301	76	3	2
CUF-B	×	272	57	2	4
CUF-B	×	7W	23	1	13
CUF-B	×	262	79	3	21
SAMS	×	301	44	12	33
SAMS	×	272	55	5	36
SAMS	×	7W	6	2	11
6–4	×	301	29	2	40
6–4	×	272	36	4	17
6–4	×	7W	5	4	11
6–4	×	262	7	1	7
33B	×	272	24	1	23
33B	×	262	12	0	4
Totals			519	45	254

18.6 OTHER TYPES OF EXTRANUCLEAR INHERITANCE

Not all traits that are inherited in an extranuclear fashion are governed by genes in mitochondria and plastids. Some cells of the protozoan *Paramecium aurelia* harbor a bacterial species, *Caedobacter taeniospiralis*, that lives symbiotically within the cytoplasm as if it were an organelle, a curious situation because it is a modern example of endosymbiosis. The bacteria in the cell are called **kappa particles**. A cell with kappa particles is called a killer cell because the kappa particles produce a toxic substance that the *Paramecium* cells release into the environment. The substance kills other *Paramecium* cells that do not have kappa particles. The killer phenotype is governed by genes in the nucleus and in the cytoplasmic bacterium. Kappa particles can only survive in a cell with a dominant allele *K*. Those cells that are recessive *kk* cannot harbor kappa particles and die when they are near a killer cell. The bacteria are transmitted from one *Paramecium* cell to another during conjugation between cells when cytoplasm is exchanged.

A similar situation is found in *Drosophila melanogaster*. Female flies with a phenotype called **maternal sex ratio** produce almost all female progeny when mated, and their female progeny pass the trait on to the next generation of female progeny. The rare male progeny do not transmit the trait, so inheritance is uniparental-maternal. Maternal sex ratio is encoded by a virus in bacterial cells that reside in the cytoplasm of a fly's cells. The virus kills most male embryos but does not affect the females.

In the fava bean (*Vicia faba*), one type of cytoplasmic male sterility is inherited in a uniparental-maternal fashion, as in other plants, but the trait is not governed by either plastid or mitochondrial genes. Instead, it is associated with double-stranded RNA molecules that reside within membrane-bound cytoplasmic bodies that are inherited like other organelles.

18.7 CORRECTLY RECOGNIZING EXTRANUCLEAR INHERITANCE

Extranuclear inheritance fails to adhere to Mendelian principles, including the principles of segregation, independent assortment, and parental equivalence. Typically, the first indication of extranuclear inheritance is the observation of uniparental-maternal inheritance in the progeny of reciprocal crosses.

However, what may appear to be uniparental-maternal inheritance cannot always be attributed to

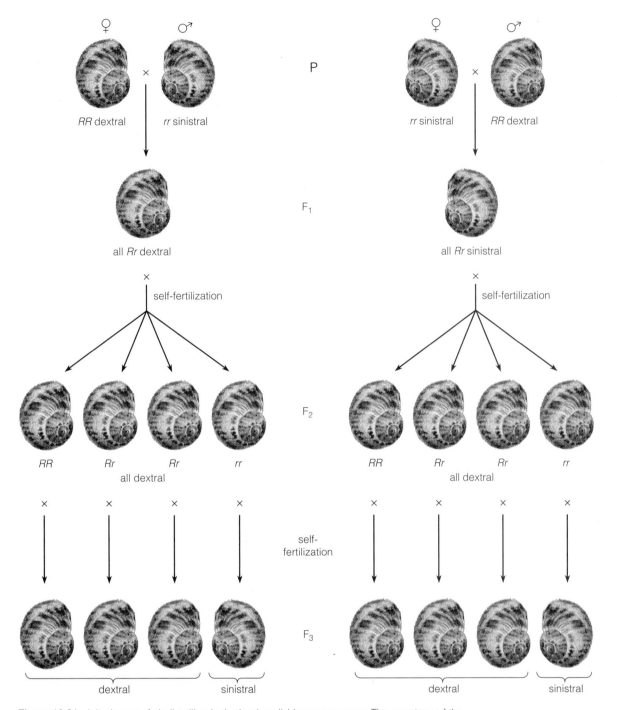

Figure 18.21 Inheritance of shell coiling in the land snail *Limnaea peregra*. The genotype of the female parent is expressed in the phenotypes of its first-generation offspring.

extranuclear inheritance. **Maternal effect** is a type of inheritance that appears to be uniparental-maternal but in reality is governed by nuclear genes. The classic example of maternal effect is shell coiling in *Limnaea peregra*, a hermaphroditic land snail that is capable of self- and cross-fertilization. Snail shells develop as a spiral that is either dextral (right-handed) or sinistral (left-handed). When a snail from a true-breeding dextral strain is mated as a female with a snail from a true-breeding sinistral strain, all the F$_1$ progeny are dextral, like the female par-

ent. The reciprocal cross (sinistral as the female parent, dextral as the male parent) produces all sinistral F$_1$ offspring—again, the same phenotype as the female parent. The results are those expected of uniparental-maternal inheritance of an allele in mitochondrial DNA. However, analysis of the progeny from subsequent generations reveals a pattern typical of nuclear genes. In the F$_2$ generations from both groups, all snails are dextral, a puzzling result that is not expected with either extranuclear inheritance or nuclear inheritance. The puzzle, however,

is easily resolved in the F_3 generation, in which both groups segregate in a 3 dextral:1 sinistral pattern, typical Mendelian inheritance of a single nuclear gene with two alleles. The phenotype is delayed one generation because each snail's phenotype is determined by the genotype of its maternal parent (Figure 18.21). The female parent's genotype determines the orientation of the spindle in mitosis for the first cell generations in the embryo. A right-handed orientation gives dextral coiling, and a left-handed orientation gives sinistral coiling. This delays the phenotypic expression of the nuclear genes by one generation.

Let's now analyze a genetic experiment with snail shell coiling to illustrate maternal effect.

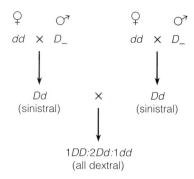

Figure 18.22 Solution to Example 18.5: Pedigree of two sinistral parents.

Example 18.5   Maternal effect in *Limnaea peregra*.

Problem: Two snails with sinistral coiling are mated in reciprocal crosses. All F_1 progeny from both crosses have dextral coiling. **(a)** What are the genotypes of the two sinistral parents? **(b)** Draw a pedigree showing how the two sinistral parents arose.

Solution: **(a)** In order to have a sinistral phenotype, the female parent of each of the sinistral snails must have had the genotype *dd*. This means each snail inherited a *d* allele from the *dd* parent. Both sinistral snails had all dextral progeny when used as a female parent, so both of them must have been heterozygous *Dd*. **(b)** The pedigree is shown in Figure 18.22. Both sinistral parents had a female parent with the genotype *dd*. Their male parents were *DD* or *Dd*.

Maternal effects may mimic extranuclear inheritance patterns but in reality are governed by nuclear genes.

SUMMARY

1. Mitochondria and plastids contain circular DNA genomes that contain genes.

2. The genes on organellar genomes are transcribed and translated within the organelle, and the products of these genes are utilized within the organelle.

3. Proteins encoded by genes in organellar genomes are usually subunits of enzymes that contain organellar and nuclear-encoded subunits.

4. The similarity of organellar genomes to prokaryotic genomes suggests that mitochondria and plastids have remote prokaryotic origins, a concept called the endosymbiotic hypothesis.

5. A cell with only one type of mitochondrion or plastid is homoplasmic. A cell with two or more types of mitochondria or plastids is heteroplasmic.

6. Mendel's principles of inheritance do not apply to extranuclear inheritance.

7. In yeast, mitochondria are inherited biparentally and undergo somatic segregation during mitotic cell generations.

8. In animals and vascular plants, mitochondria are usually inherited in a uniparental-maternal fashion.

9. Plastid inheritance in higher plants differs among species. It may be uniparental-maternal, uniparental-paternal, or biparental. When it is biparental, plastids undergo somatic segregation during mitotic cell generations.

10. There are a few examples of extranuclear inheritance of cytoplasmic bacteria or viruses.

11. Maternal effect mimics uniparental-maternal inheritance but is governed by nuclear genes.

QUESTIONS AND PROBLEMS

1. Digestion of human mitochondrial DNA with *Eco*RI produces three fragments, of 8050, 7366, and 1153 nucleotide pairs. **(a)** How many *Eco*RI restriction sites are there in human mitochondrial DNA? **(b)** According to these data, what is the exact size of the human mitochondrial genome in nucleotide pairs? **(c)** How does the human mitochondrial genome compare in size to mitochondrial genomes in plants?

2. In many plant species, about 10% of the DNA in leaf cells is chloroplastic DNA. Yet far less than 10% of inherited traits are governed by chloroplastic genes. Also, the number of genes on plastid DNA represents only a tiny fraction of the genes in the plant cell. Why is there so much chloroplastic DNA yet so few genes present in that DNA?

3. Mitochondrial and plastid DNA have been referred to as "populations within populations" (Birky. 1978. *Annual Review of Genetics* 12:471–512). Explain what this statement means.

4. One of the most important enzymes in photosynthesis is ribulose-1,5-bisphosphate carboxylase-oxygenase,

often referred to as rubisco. In cyanobacteria, the intact rubisco enzyme is composed of 16 subunits, eight identical large subunits and eight identical small subunits. In higher plants, rubisco is similar, with eight identical large subunits and eight identical small subunits. The large subunit is encoded by genes in the chloroplast, and the small subunit is encoded by genes in the nucleus. Under the assumptions of the endosymbiotic hypothesis, what does the presence of the small subunit gene in the nucleus and the large subunit gene in the chloroplast suggest?

5. Suppose a gene mutates in a single molecule of mitochondrial DNA in a single mitochondrion in the cytoplasm of a human female zygote. What is the most probable fate of the mutation?

6. Suppose a gene mutates in a single molecule of mitochondrial DNA in a single cell in the germ line of a human female embryo. What is the most probable fate of that mutation?

7. Mendel studied several sets of contrasting phenotypes expressed in pea seeds, including yellow and green seeds, round and wrinkled seeds, and opaque and transparent seed coats. If a plant from a pea variety that is true breeding for round seeds with opaque seed coats is pollinated by a plant from a variety that is true breeding for wrinkled seeds and transparent seed coats, all of the seeds harvested from flowers that are cross-pollinated are round and have opaque seed coats. When the reciprocal cross is made, all of the seeds harvested from flowers that are cross-pollinated are round and have transparent seed coats. **(a)** What is the reason for this difference in reciprocal crosses for one trait but not the other? Use Mendel's paper and information in Chapter 12 to help you answer this question correctly. **(b)** What principles (Mendelian and/or non-Mendelian) are portrayed by this example?

8. Suppose you discovered a female *Drosophila melanogaster* fly that was clearly smaller than usual and had already been fertilized by at least one of the wild-type males in the bottle. You isolate the female and allow her to lay eggs. All of her first-generation offspring are small like their maternal parent. Using the offspring as a starting point, design a set of experiments to help determine the type of inheritance for the trait. Indicate what results would be expected in each generation for each of the following types of inheritance: **(a)** autosomal dominant inheritance of the small-body phenotype, **(b)** X-linked dominant inheritance of the small-body phenotype, **(c)** mitochondrial inheritance of the small-body phenotype, and **(d)** maternal effect.

9. Suppose you discovered a male-sterile maize plant from hybrid corn seed grown in a garden. Design a set of experiments to determine whether or not the male sterility is cytoplasmic only (no effect by nuclear genes), nuclear only, or cytoplasmic-nuclear. As you design your experimental scheme, remember that a male-sterile plant cannot be used as a male parent in crosses. Also recognize that mitochondrial inheritance in maize is uniparental-maternal and that the male-sterile plant is a hybrid and is heterozygous for some nuclear alleles.

10. If T-type male-sterile maize plants are pollinated by male-fertile plants with the genotype *Rr*, **(a)** what proportion of the progeny should be male sterile? **(b)** If 1000 of the progeny plants are selected at random and grown together in a garden, approximately how many should fail to produce seeds?

11. Mitochondrial encephalopathy, lactic acidosis, and strokelike episodes (MELAS) syndrome is caused by an A→G transversion at position 3243 in the human mitochondrial genome. Nearly all individuals who have MELAS syndrome are heteroplasmic for the mutation. Also, there is substantial variation for clinical manifestations of the syndrome among people who inherit the mutation. In 1994, Matthews et al. (*Journal of Medical Genetics* 31:41–44) traced the mutation through four generations in a family. They found that the amount of mutant mitochondrial DNA in tissues varied within an individual and that the proportion of mutant mitochondrial DNA in a tissue was correlated with clinical manifestations of symptoms in that tissue. For example, a woman who died of cardiomyopathy (disease of the heart muscle) had a higher proportion of mutant mitochondrial DNA in her heart than in most other organs. Explain why the proportion of mutant DNA may be greater in one organ than another in the same individual. Also explain why heteroplasmy may cause partial penetrance and variable expressivity for mitochondrial genetic disorders.

12. The *ND4* gene in the human mitochondrial genome encodes a subunit of NADH dehydrogenase, an enzyme that is part of the oxidative phosphorylation electron transport chain in the mitochondrion. A particular mutation in the *ND4* gene that alters one amino acid causes Leber hereditary optic neuroretinopathy (LHON). The mutation can be readily detected as an RFLP because it eliminates an *Sfa*NI restriction endonuclease cleavage site in the mitochondrial DNA. In 1990, Vikki et al. (*American Journal of Human Genetics* 47:95–100) examined people from three generations of a human family and found that heteroplasmy was maintained in some individuals through all three generations. In one family, mitochondrial DNA samples from the blood of four siblings were examined. The four siblings had 0%, 15%, 30%, and > 95% mutant mitochondrial DNA in their blood. Only the sibling with > 95% mutant mitochondrial DNA had symptoms of LHON. All siblings had the same mother (who was deceased, so her mitochondrial DNA was not examined). No inheritance of paternal mitochondrial DNA was detected. **(a)** What do these results suggest about the rate of mitochondrial somatic segregation in humans? **(b)** Explain why only one of the four siblings had LHON even though two others carried the mutant allele.

13. In the same study highlighted in the previous question (Vikki et al. 1990. *American Journal of Human Genetics*

47:95–100), the sibling with > 95% mutant mitochondrial DNA was female and had two children, both males with > 95% mutant mitochondrial DNA. Given the wide variation for the proportion of mutant mitochondrial DNA among the siblings in the previous question, why do these children have proportions of mutant DNA that are similar to their mother's?

14. In *Limnaea peregra*, two snails with sinistral coils are crossed (single cross with no reciprocal cross). All progeny have dextral coiling. When these progeny are self-fertilized, half of them produce all dextral progeny, and the other half produce all sinistral progeny. What are the genotypes of the original male and female parents?

15. Plastid inheritance was long thought to be strictly uniparental-maternal in tobacco. In 1986, Medgyesy et al. (*Molecular and General Genetics* 204:195–198) pollinated a streptomycin-susceptible plant with a streptomycin-resistant plant. The streptomycin-resistance allele was known to be located on the plastid genome. A total of 6800 seedlings were tested, and all were susceptible to streptomycin. However, when cells from 1500 additional seedlings were cultured on medium that contained streptomycin, small clumps of streptomycin-resistant cells arose in the cultured cells from 44 of the seedlings. The researchers transferred some of the streptomycin-resistant cells to a medium that encouraged small plants to grow from the cultured cells, a process called plant regeneration. The regenerated plants were resistant to streptomycin and had plastid restriction fragment patterns that were identical to those from the paternal streptomycin-resistant parent. **(a)** What aspects of these data suggest biparental inheritance of plastids in tobacco? **(b)** What do these data suggest about the frequency of paternal transmission of plastids in tobacco? **(c)** Regenerated plants had a purely paternal restriction fragment pattern. What does this observation imply about the cells from which these plants were regenerated?

16. Cells from a T-type cytoplasmic male-sterile maize are sensitive to a toxin produced by the fungus responsible for southern corn leaf blight. When cells of T-type cytoplasmic male-sterile maize are cultured on medium supplemented with the toxin, they cannot survive. However, occasionally resistant cells that arise by mutation are recovered in cell cultures supplemented with the toxin. When plants are regenerated from the resistant cells, they are always male fertile. Several researchers (see Wise et al. *Plant Molecular Biology* 9:121–126; and Dixon et al. *Theoretical and Applied Genetics* 63:75–80 for examples and references) have studied the mitochondrial DNA from these mutant types. In each case, a mutation that disrupts a gene that encodes a 13 kD polypeptide was found. What do these results suggest about the relationship between male sterility and susceptibility to the toxin?

17. In 1989, Neale et al. (*Theoretical and Applied Genetics* 77:212–216) reported that hybrids of loblolly pine had paternal chloroplastic RFLP patterns but maternal mitochondrial RFLP patterns. Also in 1989, Neale et al. (*Proceedings of the National Academy of Sciences, USA* 86:9347–9349) reported that both chloroplastic and mitochondrial RFLP patterns in hybrids of coast redwood were identical to the paternal parent. **(a)** What taxonomic aspect of loblolly pine and coast redwood is consistent with the observation of uniparental-paternal inheritance of chloroplastic DNA? **(b)** What do these results suggest about mitochondrial inheritance in this taxonomic group?

18. There are two *Hae*III recognition sites in the mitochondrial DNA of *Drosophila mauritania*. One type of mitochondrial DNA molecule produces a 12.8 kb fragment and a 6.13 kb fragment when digested with *Hae*III. Another type produces one 12.8 kb fragment and one 5.63 kb fragment when digested with *Hae*III. In 1983, Solignac et al. (*Proceedings of the National Academy of Sciences, USA* 80:6942–6946) published the results of experiments in which they examined the inheritance of mitochondrial DNA in *D. mauritania*. They established a population of flies from a single founder female fly that had three mitochondrial DNA fragments, of 12.8, 6.13, and 5.63 kb. After 30 generations, about half of the flies in the population had two rather than three fragments. The remaining progeny had the three fragments, but the relative amounts of the two lower-molecular-weight fragments (6.13 and 5.63 kb) varied among the individuals. **(a)** Explain the presence of three instead of two restriction fragments in a single founder female. **(b)** Explain why after 30 generations the relative amounts of the two lower-molecular-weight fragments varied among individuals.

19. In a 1990 paper on *Drosophila*, Kondo et al. (*Genetics* 126:657–663) estimated that the proportion of paternal mitochondrial DNA in *Drosophila* zygotes is 0.1%. How does this compare with the estimated proportion of paternal mitochondrial DNA in mouse zygotes (see Example 18.2)?

20. In 1994, Reboud and Zeyl (*Heredity* 72:132–140) stated that "the lack of phenotypic markers and the use of RFLPs on small samples may have biased the prevailing [uniparental-maternal] view of organelle inheritance by underestimating the occurrence of low-frequency paternal transmission of organelles." In what way might methods of analysis have led to the general conclusion that organelle inheritance is strictly uniparental-maternal?

21. In 1988, Chiu et al. (*Current Genetics* 13:181–189) reported the results of reciprocal crosses of four plastid types in evening primrose in which the nuclear genetic background was uniform. Their objective in this study was to determine the effect of the plastid genome on transmission of maternal and paternal plastids. They used four different genetic types of normal green plastids, designated I, II, III, and IV. They then selected chlorophyll-deficient (white) mutant types from among these four normal types. The names given to the mutant

Cross*	F₁ Progeny (%)			Cross*	F₁ Progeny (%)		
	Green	White	Variegated		Green	White	Variegated
I × I-β	98.8	0.0	1.2	III × II-γ	96.0	0.0	4.0
I-β × I	0.0	59.1	40.9	II-γ × III	0.0	76.7	23.3
I × II-γ	100.0	0.0	0.0	III × III-γ	91.2	0.0	8.8
II-γ × I	0.0	73.7	26.3	III-γ × III	0.0	72.4	27.6
I × III-γ	92.2	0.0	7.8	III × IV-α	99.1	0.0	0.9
III-γ × I	0.0	78.3	21.7	IV-α × III	0.0	49.6	50.4
I × IV-α	100.0	0.0	0.0	IV × I-β	78.5	0.0	21.5
IV-α × I	0.0	60.0	40.0	I-β × IV	0.0	99.6	0.4
II × I-β	86.6	0.0	13.4	IV × I-ζ	67.4	0.0	32.6
I-β × II	0.0	94.0	6.0	I-ζ × IV	0.0	100.0	0.0
II × I-η	43.7	0.0	56.3	IV × I-η	47.9	0.0	52.1
I-η × II	0.0	98.0	2.0	I-η × IV	0.0	100.0	0.0
II × II-γ	100.0	0.0	0.0	IV × II-γ	81.6	0.0	18.4
II-γ × II	0.0	96.2	3.8	II-γ × IV	0.0	99.3	0.7
II × III-γ	68.4	0.0	31.6	IV × II-ε	85.0	0.0	15.0
III-γ × II	0.0	97.0	3.0	II-ε × IV	0.0	100.0	0.0
II × IV-α	98.8	0.0	1.2	IV × III-γ	44.0	0.0	56.0
IV-α × II	0.0	86.0	14.0	III-γ × IV	0.0	97.5	2.5
III × I-β	92.8	0.0	7.2	IV × IV-α	100.0	0.0	0.0
I-β × III	0.0	76.7	23.3	IV-α × IV	0.0	99.6	0.4
III × I-η	81.6	0.0	18.4				
I-η × III	0.0	82.0	18.0				

*In each cross, the female parent is listed first. Reciprocal crosses are paired.

types were the number of the green type in which the mutation was found (I, II, III, or IV) followed by a Greek letter to designate the mutation. The mutant types were maintained in a heteroplasmic condition in variegated plants, and crosses were made with flowers borne on homoplasmic white sectors. Table 18.5 lists some of the data reported in this paper. (a) Is inheritance predominantly maternal, paternal, or biparental? (b) Is it possible to conclude that the plastid genotype has an effect on the frequency of biparental inheritance of plastids? Justify your answer using examples from the data. (c) What differences are evident in reciprocal crosses?

FOR FURTHER READING

The structure and function of mitochondria and chloroplasts are covered in detail in Chapter 25 of **Lewin, B. 1994. *Genes V.* Oxford: Oxford University Press**; Chapters 9, 10, and 21 of **Wolfe, S. L. 1993. *Molecular and Cellular Biology.* Belmont, Calif.: Wadsworth**; and Chapter 14 of **Alberts, B., D. Bray, J. Lewis, M. Raff, K. Roberts, and J. D. Watson. 1994. *Molecular Biology of the Cell,* 3rd ed. New York: Garland**. Several excellent reviews of human genetic disorders associated with mitochondrial DNA mutations are available. They include **Zeviani, M., V. Tiranti, and C. Piantadosi. 1998. Mitochondrial disorders. *Medicine* 77:59–72; Lightowlers, R. N., P. F. Chinnery, D. M. Turnbull, and N. Howell. 1997. Mammalian mitochondrial genetics: Heredity, heteroplasmy and disease. *Trends in Genetics* 13:450–455; Grossman, L. I., and E. A. Shoubridge. 1996. Mitochondrial genetics and human disease. *BioEssays* 18:983–991**; and **Larsson, N. G., and D. A. Clayton. 1995. Molecular genetic aspects of human mitochondrial disorders. *Annual Review of Genetics* 29:151–178**. A review of mitochondrial inheritance that focuses on *Saccharomyces* and *Chlamydomonas* can be found in **Birky, W. C. 1978. Transmission genetics of mitochondria and chloroplasts. *Annual Review of Genetics* 12:471–512**. A detailed book on plastid structure, function, and inheritance is **Kirk, J. T. O., and R. E. A. Tilney-Bassett. 1978. *The Plastids: Their Chemistry, Structure, Growth and Inheritance,* 2nd ed. Amsterdam: Elsevier/North Holland**. Biparental inheritance of plastids in plants was reviewed by **Smith, S. E. 1988. Biparental inheritance of organelles and its implications in crop improvement. *Plant Breeding Reviews* 6:361–393**.

For additional reading, go to InfoTrac College Edition, your online research library at: http: www.infotrac-college.com/brookscole

PART IV

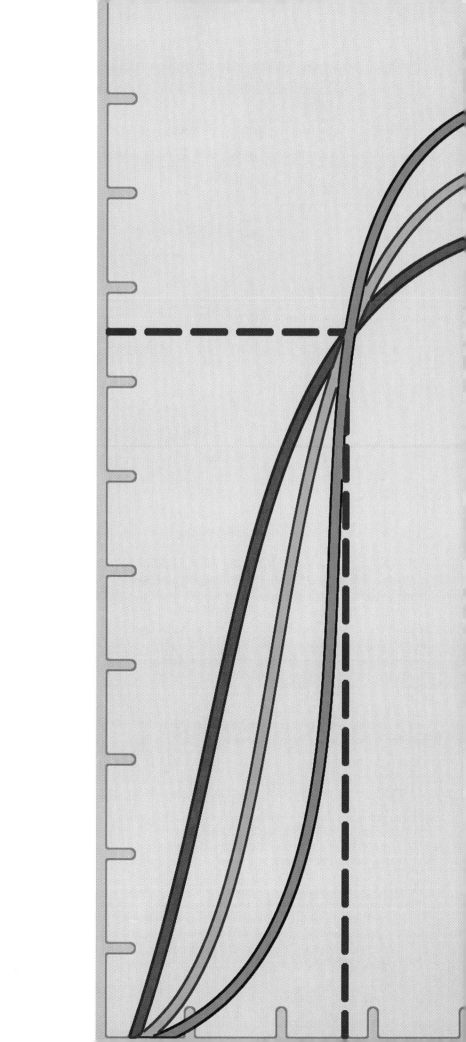

GENETICS OF POPULATIONS

CHAPTER 19

KEY CONCEPTS

Hardy-Weinberg equilibrium, which represents no change in allele frequencies in a population, serves as the foundation for a study of population genetics.

~

For Hardy-Weinberg equilibrium to apply, there must be no mutation, no migration, random mating with respect to genotype, no selection, and an infinitely large population.

~

The violations of these five assumptions are the factors that cause deviations from Hardy-Weinberg equilibrium in populations. These factors fall into five general categories: mutation, migration, nonrandom mating, selection, and drift.

~

It is possible to describe the effects of these factors mathematically.

POPULATION GENETICS

Geneticists often study the principles of inheritance by carefully designing controlled experiments and analyzing the results. In nature, however, an experimenter rarely controls mating, selection, population size, and migrations, so alleles in a population can be transmitted to offspring in many different ways. **Population genetics** is the study of how the principles of genetics function in real populations in their natural habitats. However, investigation of genetics in nature can often prove complicated because controls over experiments are much less rigid than they are for laboratory experiments.

One of the best approaches for studying genetics in nature is to establish a simplified, theoretical frame of reference to describe what should happen in an idealized population in which certain assumptions are met. After the theoretical framework has been established, researchers can then use it as a baseline to determine how the violations of theoretical assumptions in natural populations cause deviations from theory.

This approach has been successful for scientists who study how factors such as population size, nonrandom mating, natural and artificial selection, migration, and mutation interact to change or stabilize the genetic composition of natural populations. Population genetics is essential for understanding evolution, yet it is often misunderstood and misrepresented.

We begin this chapter with a discussion of Hardy-Weinberg equilibrium, the theoretical baseline for much of population genetics. We will then study the factors that cause populations to deviate from Hardy-Weinberg equilibrium.

19.1 HARDY-WEINBERG EQUILIBRIUM

Hardy-Weinberg equilibrium provides the theoretical framework for population genetics. Before discussing it, however, we need to first define two important terms: population and allele frequency. For the purposes of genetic analysis, a **population** is defined as a group of individuals of the same species that intermate with one another. **Allele frequency** is the proportion of a particular allele among all other alleles in a population. For example, let's suppose that in a population of 500 diploid individuals, there are two different alleles at a particular locus, a dominant allele A and a recessive allele a. Each individual carries two alleles (AA, Aa, or aa), so there are 1000 alleles in the population. Of the 1000 alleles, suppose that 300 are a, and 700 are A. The allele frequency of a is $300/1000 = 0.3$, and the allele frequency of A is $700/1000 = 0.7$.

Allele frequencies are usually represented by the letters p and q. These letters do not represent the alleles themselves, but rather their frequencies. For that reason, p and q always equal numbers between zero and one. Continuing the example above, let's call the frequency of the a allele q, and the frequency of the A allele p, so $q = 0.3$ and $p = 0.7$.

The values for p and q tell us only the frequencies of the alleles in a population, not how those alleles are distributed among individuals. For example, if 150 of the individuals are homozygous aa, and 350 are homozygous AA, then the allele frequencies are $q = 0.3$ and $p = 0.7$. On the other hand, if 45 individuals are homozygous aa, 210 are heterozygous Aa, and 245 are homozygous AA, the allele frequencies are still $q = 0.3$ and $p = 0.7$, but the distribution of alleles among individuals is different.

This leads us to the concept of **Hardy-Weinberg equilibrium**, a situation in which allele frequencies and genotype frequencies remain constant from one generation to the next when certain assumptions are met. The name Hardy-Weinberg refers to two scientists, Godfrey H. Hardy and Wilhelm Weinberg, who independently derived the equations for Hardy-Weinberg equilibrium in 1908. For populations that are in Hardy-Weinberg equilibrium, it is possible to determine not only allele frequencies but also **genotype frequencies**, the distribution of alleles among genotypes.

There are two basic equations that describe Hardy-Weinberg equilibrium. The first applies to all populations with two alleles at a locus, whether or not the population is in Hardy-Weinberg equilibrium:

$$p + q = 1 \qquad \textbf{[19.1]}$$

If there are only two alleles in a population, the sum of their frequencies must equal unity.

Equation 19.1 applies to all populations with only two alleles at a locus. But in real populations, there often are more than two alleles at a locus. In such instances, researchers may be interested in the frequency of just one allele, such as a mutant allele, compared to all wild-type alleles. In such an instance, equation 19.1 still suffices. The term q represents the frequency of the allele of interest, and p represents the combined frequencies of all other alleles.

The second equation applies to populations in Hardy-Weinberg equilibrium and provides the genotype frequencies:

$$p^2 + 2pq + q^2 = 1 \qquad \textbf{[19.2]}$$

where p^2 equals the frequency of homozygous AA individuals, $2pq$ equals the frequency of heterozygous Aa individuals, and q^2 equals the frequency of homozygous aa individuals.

Equation 19.2 is the square of equation 19.1. If individuals in a population mate at random, then the aver-

age probability that individuals in the population will inherit an A allele from one parent is equal to the frequency of that allele in the population. So the average probability of receiving an A allele from one parent is p, which in our example is 0.7. The probability of receiving the a allele from one parent is q, which is 0.3. Each individual receives two alleles, one from the maternal parent and one from the paternal parent. There is only one way to be homozygous AA, and that is to receive A from both parents. The average probability of receiving A from the maternal parent is p, and the average probability of receiving A from the paternal parent is also p, so the probability of receiving an A allele from both parents is p^2. Similarly, the only way to be homozygous aa is to receive the a allele from both parents, so the probability of receiving a from both parents is q^2. There are two ways for heterozygotes to arise. The first way is to receive A from the maternal parent and a from the paternal parent. The probability of that is pq. The other way is to receive a from the maternal parent and A from the paternal parent. The probability of that is also pq, so combining the two gives $2pq$. Equation 19.2 can be illustrated using a Punnett square (Figure 19.1).

Let's now look at an example of Hardy-Weinberg equilibrium in a human population.

Example 19.1 Hardy-Weinberg equilibrium for alleles at the *MN* locus in humans.

The *MN* locus (OMIM 111300) is located on chromosome 4 in humans. It encodes a polypeptide that is modified into a glycoprotein and expressed on the surface of red blood cells. Two alleles, *MN*M* and *MN*N*, are common in human populations. The M polypeptide differs from the N polypeptide by two amino acids. There are three possible genotypes for the two alleles: homozygous *MN*M/*M*, homozygous *MN*N/*N*, and heterozygous *MN*M/*N*. The MN blood type of an individual can be readily determined with an immunological assay similar to the test used to determine ABO blood type. The MN alleles are codominant, so three phenotypes can be readily distinguished: type M, type N, and type MN, which correspond to their respective genotypes. In 1980, Neel et al. (*Annals of Human Genetics* 44:37–54) reported the following observations of MN blood types in 1747 people from the Ticuna tribe, an isolated native tribe in the Amazon rain forest of Brazil:

Type M	1409
Type MN	310
Type N	28
Total	1747

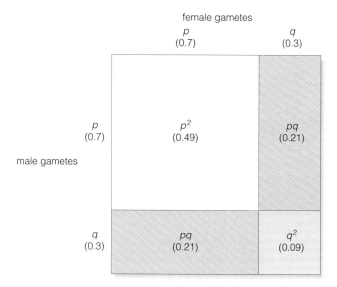

female gametes

	p (0.7)	q (0.3)
p (0.7)	p^2 (0.49)	pq (0.21)
q (0.3)	pq (0.21)	q^2 (0.09)

male gametes

Figure 19.1 A Punnett square illustrating the frequencies of gametes and genotypes in a population at Hardy-Weinberg equilibrium. The size of each box within the square represents the frequency of that genotype. The frequencies in this example are $p = 0.7$ and $q = 0.3$.

Problem: **(a)** Determine the frequencies of the MN^*M and MN^*N alleles in this sample. **(b)** Using the calculated allele frequencies, determine the expected allele frequencies for Hardy-Weinberg equilibrium. **(c)** Test the hypothesis that these data conform to Hardy-Weinberg equilibrium using chi-square analysis.

Solution: **(a)** Due to codominance, each genotype is represented by a separate phenotype. Thus, the exact allele frequencies in this sample can be computed directly from the data. Each individual carries two copies of MN alleles, so the total number of MN alleles in this population is $2 \times 1747 = 3494$. Each individual with type M blood is homozygous $MN^*M/^*M$ and, therefore, carries two copies of the MN^*M allele. The number of MN^*M alleles in the people with type M blood is $2 \times 1409 = 2818$. Each individual with type MN blood carries one copy of the MN^*M allele, so the number of MN^*M alleles in the people with type MN blood is 310. Therefore, the total number of MN^*M alleles in the entire sample is $2818 + 310 = 3128$. The total number of alleles at the MN locus in the entire sample is 3494, so the frequency of the MN^*M allele is $3128 \div 3494 = 0.8952$. If we let p equal the frequency of the MN^*M allele and q equal the frequency of the MN^*N allele, $p = 0.8952$ and $q = 0.1048$. Notice that q can be determined directly from the data using the same procedure used to calculate p. Alternatively, q can be calculated as $q = 1 - p$ (which we get by rearranging equation 19.1 to solve for q). **(b)** These values of p and q can be used to determine the

expected genotype frequencies if the population is in Hardy-Weinberg equilibrium. The expected frequency of individuals who are homozygous $MN^*M/^*M$ is $p^2 = (0.8952)^2 = 0.8014$. The observed frequency is $1409 \div 1747 = 0.8065$. The expected and observed frequencies for the other two genotypes can be calculated in the same way.

Genotype	Expected Frequency	Observed Frequency
$MN^*M/^*M$	$p^2 = 0.8014$	0.8065
$MN^*M/^*N$	$2pq = 0.1876$	0.1774
$MN^*N/^*N$	$q^2 = 0.0110$	0.0160

(c) In this example, chi-square analysis can be used to test the hypothesis that these data conform to Hardy-Weinberg equilibrium. Before doing so, however, let's look at two common pitfalls associated with this type of chi-square analysis. First, the frequencies in the table above cannot be used directly in chi-square analysis. The chi-square value is affected by the sample size, so expected and observed numbers of individuals, rather than frequencies, must be used. Second, the expected frequencies are determined from the observed values, not from a theoretical prediction. The people in this study are a sample of a population rather than the entire population, so both the expected and observed frequencies are subject to sampling error. In most applications of chi-square analysis, the expected frequencies are not subject to sampling error. In cases where expected frequencies are determined from observed data, the degrees of freedom do not equal one less than the number of phenotypic classes. In most cases, the degrees of freedom equal the number of phenotypic classes minus the number of alleles. In this instance, there are three phenotypic classes and two alleles, so $3 - 2 = 1$ degree of freedom. The chi-square analysis is $\chi^2 = [(1409 - 1400.95)^2 \div 1400.05] + [(310 - 327.74)^2 \div 327.74] + [(28 - 19.22)^2 \div 19.22] = 5.03$. This value exceeds the critical value of 3.84 at $P = 0.05$, but does not exceed the critical value of 6.64 at $P = 0.01$, with 1 degree of freedom, so the probability of observing a deviation as great as this due purely to sampling error (under the assumption that the population is in Hardy-Weinberg equilibrium) is less than 5% but greater than 1%. The appropriate conclusion from the chi-square analysis is that the observed values deviate significantly, but not highly significantly, from the expected values. At the $P = 0.05$ critical level, we reject the hypothesis that the observed values conform to Hardy-Weinberg equilibrium.

Hardy-Weinberg Equilibrium for Three Alleles

Hardy-Weinberg equilibrium also applies to multiple alleles in a population. Let's use the ABO blood group in humans as an example. Three different alleles can be distinguished, ABO*A, ABO*B, and ABO*O. The frequencies of the three alleles can be represented as p, q, and r. Let p equal the frequency of the ABO*A allele, q the frequency of the ABO*B allele, and r the frequency of the ABO*O allele. (The usual convention is to let the last alphabetic letter, r in this case, represent the recessive allele.) Then,

$$p + q + r = 1 \qquad \text{[19.3]}$$

and the equation that identifies the genotype frequencies when the population is in Hardy-Weinberg equilibrium is the square of $p + q + r$:

$$p^2 + 2pq + 2pr + q^2 + 2qr + r^2 = 1 \qquad \text{[19.4]}$$

The six possible genotypes and their corresponding frequencies are as follows:

Genotype	Frequency
ABO*A/*A	p^2
ABO*A/*B	$2pq$
ABO*A/*O	$2pr$
ABO*B/*B	q^2
ABO*B/*O	$2qr$
ABO*O/*O	r^2

The ABO*O allele is recessive, so it is masked whenever it is heterozygous with ABO*A or ABO*B. Thus, there are four possible phenotypes. Their corresponding frequencies are as follows:

Phenotype	Genotype(s)	Frequency
Type A	ABO*A/*A and ABO*A/*O	$p^2 + 2pr$
Type B	ABO*B/*B and ABO*B/*O	$q^2 + 2qr$
Type AB	ABO*A/*B	$2pq$
Type O	ABO*O/*O	r^2

Let's use some actual ABO blood group data to illustrate how equations 19.3 and 19.4 are used.

Example 19.2 Determining allele frequencies for multiple alleles.

In 1946, Dobson and Ikin (*Journal of Pathology and Bacteriology* 48:221–277) published phenotypic frequencies for types A, B, AB, and O blood from

---

190,187 people. Their results are summarized in Table 19.1.

Problem: (a) Estimate the allele frequencies for each of the three alleles in this population sample, and (b) test the hypothesis that the population is in Hardy-Weinberg equilibrium.

Solution: (a) In Example 19.1, it was possible to determine the exact allele frequencies in the sample because each phenotype represented a particular genotype. In cases where genotypes of all individuals cannot be determined, allele frequencies in the sample must be estimated rather than determined exactly. In this example, people with type A blood include two different genotypes (ABO*A/*A and ABO*A/*O). The same can be said for the type B phenotype. People with type O blood are homozygous ABO*O/*O, and the predicted frequency is r^2. The value for r can be estimated by taking the square root of the observed frequency for individuals with type O blood: $r = \sqrt{r^2} = \sqrt{0.46681} = 0.68324$.

One way to solve for p and q is to substitute the estimate of r into the frequency equations for the type A and type B phenotypes, then solve the quadratic equations. However, with a little algebraic juggling, it is possible to solve for p and q without having to solve a quadratic equation. Adding the type A and type O groups together, we get a combined frequency of $p^2 + 2pr + r^2$, which equals $(p + r)^2$. Rearranging equation 19.3, we get $1 - q = p + r$, so the combined frequency of the type A and type O phenotypes is $(1 - q)^2$. Therefore, substitution and algebraic rearrangement gives an equation that solves for q in terms of p and r:

$$q = 1 - \sqrt{(p^2 + 2pr + r^2)} \qquad \text{[19.5]}$$

The quantity $p^2 + 2pr$ is the phenotypic frequency of the type A phenotype, which is 0.41719, and r^2 is the frequency of the type O phenotype, which is 0.46681. So, using equation 19.5, we get $q = 1 - \sqrt{(0.41719 + 0.46681)} = 0.05979$.

We can solve for p in terms of q and r with the same type of algebraic manipulation:

$$p = 1 - \sqrt{(q^2 + 2qr + r^2)} \qquad \text{[19.6]}$$

which in this example is $p = 1 - \sqrt{(0.08560 + 0.46681)} = 0.25675$. In summary,

$$p = 0.25675$$
$$q = 0.05979$$
$$r = 0.68324$$

If these were actual rather than estimated frequencies, their sum would equal unity. The sum is close

Table 19.1 Information for Example 19.2: Frequencies of A, B, AB, and O Blood Types in a Sample of 190,187 People			
Phenotype	Number of Individuals	Observed Frequency*	Predicted Frequency
Type A	79,344	0.41719	$p^2 + 2pr$
Type B	16,280	0.08560	$q^2 + 2qr$
Type AB	5,781	0.03040	$2pq$
Type O	88,782	0.46681	r^2
Totals	190,187	1.00000	$p^2 + 2pq + 2pr + q^2 + 2qr + r^2$

*All frequencies in this example are rounded to five decimal places. However, all calculations are based on unrounded numbers.

to unity: $p + q + r = 0.25675 + 0.05979 + 0.68324 = 0.99978$. The values can be corrected so that they equal unity (or very close to it) using Bernstein correction formulas, which are as follows:

$$p_c = p(1 + \tfrac{1}{2}d) \qquad [19.7]$$
$$q_c = q(1 + \tfrac{1}{2}d) \qquad [19.8]$$
$$r_c = (r + \tfrac{1}{2}d)(1 + \tfrac{1}{2}d) \qquad [19.9]$$

where d is the deviation of the summed values from unity, and the subscript c denotes that the value is corrected. In our example, $d = 1 - 0.99978 = 0.00022$, so

$$p_c = p(1 + \tfrac{1}{2} \times 0.00022) = 0.25678$$
$$q_c = q(1 + \tfrac{1}{2} \times 0.00022) = 0.05979$$
$$r_c = (r + \tfrac{1}{2} \times 0.00022)(1 + \tfrac{1}{2} \times 0.00022) = 0.68343$$

which equal unity when summed. **(b)** It is possible now to test the hypothesis that the population is in Hardy-Weinberg equilibrium using chi-square analysis: $\chi^2 = [(79344 - 79292.97)^2 \div 79292.97] + [(16280 - 16223.29)^2 \div 16223.29] + [(5781 - 5840.09)^2 \div 5840.09] + [(88782 - 88830.64)^2 \div 88830.64] = 0.86$ with 1 degree of freedom $(4 - 3 = 1)$. This value is substantially less than the critical value of 3.84 for $P = 0.05$, so there is no evidence to reject the hypothesis that the population is in Hardy-Weinberg equilibrium.

Assumptions for Maintenance of Hardy-Weinberg Equilibrium

After Hardy-Weinberg equilibrium is established, five important assumptions need to be met in order to maintain it from one generation to the next:

1. *There must be no mutation.* There must be no mutation, because mutation can alter allele frequencies.

2. *There must be no migration.* Individuals must not migrate into and out of the population, because the infusion or loss of alleles in the migrants can alter allele frequencies in the population.

3. *Individuals must mate at random with respect to genotype.* Although nonrandom mating does not necessarily alter allele frequencies, it may tend to favor homozygotes over heterozygotes, or vice versa, thus disrupting the equilibrium.

4. *There must be no selection.* Selection that favors an allele may increase the frequency of that allele. Selection that disfavors an allele may decrease the frequency of that allele.

5. *The population must be infinitely large.* Sampling error affects allele frequencies when population sizes are finite. In very large populations, the effect of sampling error is usually negligible. In small populations, the effect of sampling error may be significant.

Several of these assumptions are never fully met in nature. Each gene naturally mutates at a low rate; migrants often move into and out of natural populations; no population is infinitely large; random mating does not always take place; and selection, be it natural or artificial, is common. If some of these assumptions are never met, why are some populations in Hardy-Weinberg equilibrium, such as humans for the ABO blood group alleles we just studied? Violation of any one assumption may cause a deviation from Hardy-Weinberg equilibrium, but the magnitude of the deviation is proportional to the degree of violation. In the example of human ABO blood groups, several assumptions were violated, but the degree of violation was negligible in each case. For example, the sample size (190,187 people) is finite, but it is large enough to render the effect of sampling error negligible.

Establishment of Hardy-Weinberg Equilibrium

Not all populations are in Hardy-Weinberg equilibrium. However, provided the violations of the assumptions are negligible, Hardy-Weinberg equilibrium can be established in a single generation. For example, suppose that

a dominant allele *B* gives rabbits a black coat, and the recessive allele *b* gives a brown coat when homozygous. If 4000 homozygous *BB* black rabbits are placed together with 6000 brown (*bb*) rabbits, and they mate at random with respect to genotype, Hardy-Weinberg equilibrium for the *B* and *b* alleles will be established in the first-generation offspring of the rabbits. The allele frequencies in the parental generation are $p = 0.4$ for the *B* allele and $q = 0.6$ for the *b* allele, but all the rabbits in the parental generation are homozygous, so they are not in Hardy-Weinberg equilibrium. However, the first-generation progeny are in Hardy-Weinberg equilibrium, so the frequency of homozygous *bb* individuals is $q^2 = 0.36$, the frequency of heterozygous *Bb* individuals is $2pq = 0.48$, and the frequency of homozygous *BB* individuals is $p^2 = 0.16$.

Also, if the violations of the five assumptions are negligible, each succeeding generation remains in Hardy-Weinberg equilibrium, and the allele frequencies, genotype frequencies, and phenotype frequencies remain constant from one generation to the next. Hardy-Weinberg equilibrium is the absence of genetic change in a population. Therefore, it represents the absence of evolution.

X-Linked Alleles and Z-Linked Alleles

There is a major exception to establishment of Hardy-Weinberg equilibrium in the first-generation progeny of individuals that mate at random. If the allele in question is located at a locus on an X or Z chromosome, the allele frequencies do not change from one generation to the next, but several generations are required to reach an equilibrium of allele frequencies between males and females. During establishment of equilibrium, the frequency of an allele on the X or Z chromosome oscillates between males and females. The magnitude of oscillation decreases with each generation until equilibrium is reached, as illustrated in Figure 19.2. After equilibrium is reached, the allele frequencies are as follows:

	Genotype	Frequency
Homogametic sex (XX or ZZ):	*AA*	p^2
	Aa	$2pq$
	aa	q^2
Heterogametic sex (XY or ZW)	*A*	p
	a	q

Even at equilibrium, the phenotype associated with a recessive X- or Z-linked allele is observed more often in the heterogametic sex because a recessive allele cannot be masked by a dominant allele in hemizygotes.

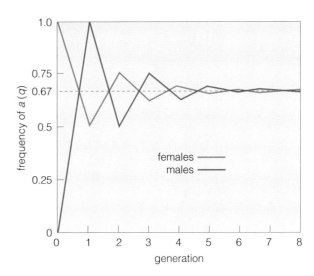

Figure 19.2 An example of the approach to Hardy-Weinberg equilibrium for alleles on the X chromosome. In this example, all females are homozygous *aa*, and all males are hemizygous *A* at generation 0. The frequency of the *a* allele is $q = 0.67$ in all generations. The allele frequencies in females and males are close to equilibrium after 6 generations of random mating.

~

When a population is in Hardy-Weinberg equilibrium, allele and genotype frequencies remain constant in each generation. Certain assumptions must be met to maintain Hardy-Weinberg equilibrium. Because there is no genetic change, Hardy-Weinberg equilibrium represents the absence of evolution.

19.2 GENETIC CHANGE IN POPULATIONS

If Hardy-Weinberg equilibrium represents no genetic change, of what value is it in populations that do change? First of all, it serves as a theoretical foundation to help us understand the factors that cause genetic change, which are violations of the assumptions that must be met for Hardy-Weinberg equilibrium. There are five assumptions, so there are five factors that disrupt the equilibrium. Defined briefly, those five factors are as follows:

1. *Mutation.* The definition of mutation we use here is any inherited alteration in the structure or arrangement of the genetic material. This definition is broad because it includes simple nucleotide alterations as well as chromosomal structural changes. As new mutations arise, they may cause a slow change in allele frequencies.

2. *Migration.* As individuals move into or out of a population, they may change allele frequencies by adding or removing alleles from the population.

3. *Nonrandom mating.* By itself, nonrandom mating does not alter allele frequencies, but it can cause

genotype and phenotype frequencies to deviate from their predicted values at equilibrium.

4. *Selection.* Selection is defined here as a reproductive advantage or disadvantage for particular genotypes. Alleles in genotypes that are favored by selection increase in frequency, and alleles that are in genotypes disfavored by selection decrease in frequency.

5. *Random genetic drift.* Random genetic drift is a change in allele frequency due to sampling error. Genetic drift is random in direction, but its magnitude is inversely proportional to population size. Drift is usually significant in small populations but negligible in very large ones.

Let's now take a detailed look at each of these five factors, paying particular attention to how they alter Hardy-Weinberg equilibrium.

19.3   MUTATION

Mutation rates tend to be very low in most populations, usually on the order of about 1 per million individuals at a single locus. This means that in the short term, mutation rate does little to alter allele frequencies. But in the long term over many generations, differences between the forward and reverse mutation rates for a locus may significantly alter allele frequencies.

There are, however, instances in which mutation has a significant effect in the short term, especially when the mutation rate interacts with other factors that modify allele frequencies, such as drift or selection. If populations are small, a single mutation may significantly alter the allele frequency of that population, because the total number of alleles is small. If selection favors a mutant allele, its frequency can increase very rapidly. Also, when selection disfavors a particular allele, a balance between mutation and selection may be reached. We will return to the balance between selection and mutation later when we discuss selection.

19.4   MIGRATION

If a few individuals migrate out of a large population and do so at random with respect to genotype, then migration has no effect. Likewise, if individuals who collectively have the same allele frequencies as a population migrate into it, then migration has no effect on altering allele frequencies. However, if migrants of a particular genotype leave a population preferentially, then the allele frequency is altered because there is a net flow of alleles out of the population. Likewise, if migrants into a population have allele frequencies different from the population they are entering, then the population's allele frequencies change. Provided the other assumptions of

Hardy-Weinberg equilibrium are met, the net effect of migration is to establish a new Hardy-Weinberg equilibrium with new allele frequencies.

~

Mutation is the only original source of genetic variation in populations. It typically alters allele frequencies slowly over many generations. Migration alters allele frequencies when there is a net flow of alleles into or out of a population.

19.5   NONRANDOM MATING

Nonrandom mating does not necessarily alter allele frequencies, but it disrupts Hardy-Weinberg equilibrium. It is usually one of two types, assortative mating or disassortative mating. **Assortative mating** is preferential mating between individuals of like genotypes. **Disassortative mating** is preferential mating of individuals with different genotypes. Let's examine assortative mating first.

Assortative Mating

Assortative mating is much more common than disassortative mating and tends to isolate alleles within certain subgroups. It increases the frequencies of homozygotes and decreases the frequency of heterozygotes when compared to Hardy-Weinberg equilibrium. An example of assortative mating in nature comes from mallard and pintail ducks. These two species of ducks can mate with one another and produce fully fertile hybrid offspring. Such matings, however, are rare, and the hybrid offspring usually do not mate. This behavior genetically isolates the two species, and their alleles.

Assortative mating is not always behavioral. Certain groups of plants are fully interfertile but fail to reproduce with one another because the time of flowering differs among them. Even though they may be in close proximity, hybrids rarely arise because flowers in one group may only be receptive to pollen after the pollination period of the other group has passed.

Inbreeding

The most extreme type of assortative mating is **inbreeding**, mating between individuals with common ancestry. Inbreeding increases homozygosity for all alleles in offspring, with a corresponding loss of heterozygosity. Some species of plants and animals inbreed routinely. Inbreeding is not uncommon in humans, particularly in cultures that promote it.

Mendel was the first to provide a mathematical explanation for the genetic consequences of inbreeding. His experimental organism was the pea plant, which naturally self-fertilizes. Self-fertilization is the most severe form of inbreeding because an individual's closest

relative is itself. Mendel allowed the F_1 plants from his monohybrid crosses to self-fertilize to form the F_2 progeny. He then allowed F_2 plants to self-fertilize to form the F_3 generation, and in some of his experiments he allowed plants to self-fertilize through the F_7 generation. Mendel predicted, and confirmed with his experimental observations, that the frequency of heterozygotes decreases by half with each generation of self-fertilization. When self-fertilization is continued for enough generations, every individual is homozygous at nearly all loci. Mendel's theory is illustrated in Figure 19.3. By the F_8 generation, over 99% of all individuals are homozygous. This theory explains why the pea plants that Mendel used as parents in his experiments were homozygous and true breeding—they had been propagated for many generations by natural self-fertilization.

Forms of inbreeding that are less severe than self-fertilization also increase homozygosity in populations, although at a slower rate. The ultimate result of any type of inbreeding is increased homozygosity for all alleles.

Inbreeding can be described quantitatively as the **inbreeding coefficient (F)**, which is the probability that any two alleles at a locus in an individual are alike by descent. The inbreeding coefficient can also be thought of as the proportion of loci that are homozygous in an individual due to inbreeding. When pedigree information is available, F can be calculated by tracing the paths from parents through common ancestors, then summing the coefficients for each path.

For example, the conventional way of drawing the pedigree of an individual whose parents are first cousins is illustrated in Figure 19.4a. A **consanguineous mating** (a mating between two individuals with common ancestry) is indicated by a double line. The pedigree in Figure 19.4a can be redrawn in path form for calculation of an inbreeding coefficient, as diagrammed in Figure 19.4b. The individual whose inbreeding coefficient we are calculating is represented as Z, and the individual's parents are represented as X and Y. Other individuals are represented by proceeding in reverse order through the alphabet. Each line between individuals in a path diagram represents transmission of genes from parent to progeny. Only those individuals in the pedigree who contribute to common ancestry need to be included in the path diagram. In the pedigree in Figure 19.4, V and W are two of Z's grandparents and are also full siblings, so X and Y are first cousins.

To identify a path of common ancestry, begin with individual Z in Figure 19.4b and follow all paths that return to Z without retracing any segment of the path. Any path that fails to return to Z is not a path of common ancestry and is not used to calculate F. In this example, there are two paths of common ancestry:

X V **T** W Y

X V **U** W Y

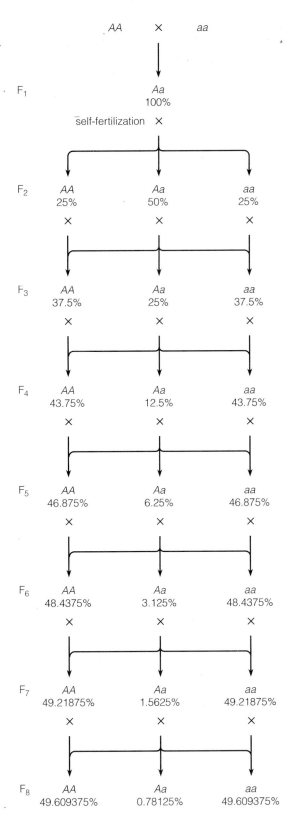

Figure 19.3 The effect of repeated self-fertilization. In each generation of self-fertilization, the proportion of heterozygotes decreases by 50%.

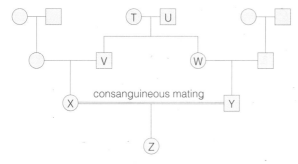

a Pedigree of a child (Z) whose parents (X and Y) are first cousins. The individuals denoted by letters are in paths of common ancestry.

b The pedigree redrawn in path form.

Figure 19.4 A conventional pedigree **(a)** and the same pedigree in path form **(b)** for calculating inbreeding coefficients.

The common ancestor in each path is identified in boldface. Notice that the paths always begin with one parent (X) and end with the other (Y). For purposes of calculating F, the individual whose inbreeding coefficient is being calculated (Z) is not included in the paths. The formula for calculating F is

$$F_Z = \Sigma(\tfrac{1}{2})^n(1 + F_A) \qquad [19.10]$$

where F_Z is the inbreeding coefficient of individual Z, n is the number of individuals in the path, and F_A is the inbreeding coefficient of the common ancestor. Unless there is information to indicate otherwise, each common ancestor in the pedigree is given an inbreeding coefficient of zero. When the common ancestor in each pathway has an inbreeding coefficient of zero, the final term $(1 + F_A)$ can be eliminated, which reduces equation 19.10 to

$$F_Z = \Sigma(\tfrac{1}{2})^n \qquad [19.11]$$

The inbreeding coefficient for each child from a marriage between first cousins is therefore calculated as follows:

Path	n	Contribution to F_Z
X V **T** W Y	5	$(\tfrac{1}{2})^5 = 0.03125$
X V **U** W Y	5	$(\tfrac{1}{2})^5 = 0.03125$
Total		$F_Z = 0.0625$

The probability that any two alleles at a locus are identical by descent in the children of first cousins is 6.25%, which is a relatively high inbreeding coefficient for humans. Some pedigrees can be quite complicated, but F_Z is determined in the same way, summing the contribution of each path to give the final value of F_Z, as the following example illustrates.

Example 19.3 Consanguinity in the British monarchy.

An example of inbreeding in humans is found in the pedigree of the British monarchy, shown in Figure 19.5a. In order to simplify the diagram, people in the complete pedigree who are not direct ancestors of Queen Elizabeth II and Prince Philip have been omitted, and the children of Queen Victoria and Prince Albert are not in birth order. The consanguineous marriages (indicated by double lines in the pedigree) are Queen Victoria and Prince Albert, who were first cousins; King George V and Queen Victoria Mary of Teck, who were second cousins once removed; and Queen Elizabeth II and Prince Philip, who are third cousins.

Problem: **(a)** Redraw the pedigree in path form, and use it to determine the inbreeding coefficient for the children of Queen Elizabeth II and Prince Philip. **(b)** Look at a more extensive pedigree showing intermarriages among European royal families in Figure 14.12. It has often been said that the recurrence of hemophilia in the descendants of Queen Victoria and Prince Albert was a result of inbreeding. Is there any evidence to support this assertion?

Solution: **(a)** Only individuals in a path of common ancestry need to be included in a pedigree for determining inbreeding coefficients. Each of those individuals is indicated by a letter. The pedigree in path form is illustrated in Figure 19.5b. The children of Queen Elizabeth II and Prince Philip are designated collectively as Z, and their inbreeding coefficient is calculated in Table 19.2. Even though there are six paths of common ancestry in this pedigree, the inbreeding coefficient for the children of Queen Elizabeth II and Prince Philip is very low (less than one-half of one percent). This analysis demonstrates that distant common ancestry does not have an appreciable effect on the inbreeding coefficient. **(b)** The more extensive pedigree in Figure 14.12 shows that each case of hemophilia affected a male whose mother was a heterozygous carrier of the recessive X-linked hemophilia allele. In no case was the allele inherited from both the father and the mother. Therefore, despite folk wisdom to the contrary,

inbreeding is not responsible for the recurrence of hemophilia among the descendants of Queen Victoria and Prince Albert.

In some plants and animals, including humans, inbreeding may uncover rare deleterious recessive alleles that tend to remain hidden in the heterozygous condition in the absence of inbreeding. For example, as diagrammed in Figure 19.6, two first cousins may each be heterozygous for the same rare recessive allele that causes a genetic disorder because they both inherited the allele from a common ancestor who is also heterozygous. Neither of the cousins has the disorder because the allele that confers the disorder is masked by a dominant allele.

Table 19.2	Solution to Example 19.3: Calculation of the Inbreeding Coefficient for Members of the British Monarchy		
Path		n	Contribution to F_Z
X V S P **N** Q T W Y		9	$(1/2)^9 = 0.001953125$
X V S P **O** Q T W Y		9	$(1/2)^9 = 0.001953125$
X V U R M **H** L O Q T W Y		12	$(1/2)^{12} = 0.00024414062$
X V U R M **I** L O Q T W Y		12	$(1/2)^{12} = 0.00024414062$
X V S P N J **F** K O Q T W Y		13	$(1/2)^{13} = 0.00012207031$
X V S P N J **G** K O Q T W Y		13	$(1/2)^{13} = 0.00012207031$
Total			$F_Z = 0.00463867186$

The recessive allele is rare in the general population, so the probability that either cousin will marry (in this context meaning "mate with") an unrelated person who is also heterozygous for the recessive allele is quite small. However, if the cousins marry each other, each of their children has a 25% chance of being homozygous for the recessive allele and having the genetic disorder. Marriages between close relatives are prohibited by law in many parts of the world to prevent the appearance of recessive genetic disorders.

Disassortative Mating

Disassortative mating is the reverse of assortative mating. It is a form of nonrandom mating in which individuals of dissimilar genotypes mate preferentially. It results in a higher frequency of heterozygotes in a population than expected with Hardy-Weinberg equilibrium. Disassortative mating is less common than assortative mating. An example of disassortative mating comes from plants with genotypic incompatibilities, such as apples. An apple tree cannot self-pollinate, nor can it be pollinated by another tree with the same or a similar genotype. For example, Jonathan apple trees cannot be self-pollinated or pollinated by another Jonathan tree's pollen. If a bee arrives carrying only Jonathan pollen and deposits that pollen in the flower of a Jonathan tree, the pollen fails to fertilize the flower. If the flower is not fertilized, an apple does not develop from that flower. However, if another bee that carries pollen from a Red Delicious apple tree visits the same flower, the pollen germinates, fertilizing the flower and causing a Jonathan apple to develop. Because of disassortative mating, the embryos in apple seeds are highly heterozygous.

~

Assortative mating is preferential mating between individuals of like genotypes. It promotes homozygosity and isolation of alleles. Inbreeding is a type of assortative mating in which individuals with common ancestry mate with one

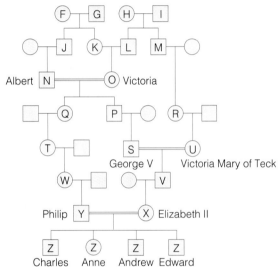

a Partial pedigree of the British monarchy. The individuals denoted by letters are in paths of common ancestry.

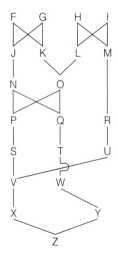

b A path diagram of the pedigree.

Figure 19.5 Information for Example 19.3: Pedigree of the British monarchy in **(a)** conventional and **(b)** path form.

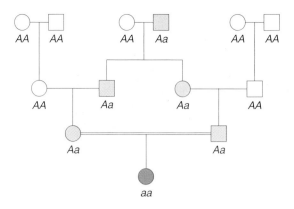

person affected by a recessive genetic disorder

heterozygous carrier of the recessive allele that causes the disorder

Figure 19.6 A human pedigree that illustrates a consequence of inbreeding. Two first cousins have a common ancestor (their grandfather) who carried a rare recessive allele that causes a genetic disorder when homozygous. Both cousins inherited the recessive allele, so there is a 25% chance that each child they have will be homozygous for the allele and therefore affected by the disorder.

another. **Disassortative mating is preferential mating between individuals of different genotypes. It promotes heterozygosity.**

19.6 SELECTION

Selection is a difference in the reproductive successes of different phenotypes. Selection may be either positive or negative. Any factor that favors the reproductive success of one phenotype relative to other phenotypes exerts **positive selection** on that phenotype, and any factor that disfavors the reproductive success of one phenotype relative to other phenotypes exerts **negative selection** on that phenotype. Selection acts directly on phenotypes and indirectly on the genotypes that confer the phenotypes. To the extent that a particular allele influences a favored or disfavored phenotype, selection may change the frequency of that allele. Positive selection increases the frequencies of alleles that confer a favored phenotype, and negative selection decreases the frequencies of alleles that confer a disfavored phenotype.

Directional Selection

Directional selection causes the frequencies of alleles affected by selection to move in a single direction for as long as the selection pressure is applied or until the frequency of the favored allele reaches unity. Directional selection may be natural or artificial. **Artificial selection** is intentional selection of plants and animals conducted by breeders to alter a population in a specific way, a prac-

a

b

Figure 19.7 Peppered moths on **(a)** light and **(b)** dark tree trunks. The green arrows point to the moths. (Photos courtesy of J. A. Bishop and L. M. Cook.)

tice that predates recorded history. Its systematic use is now an integral part of plant and animal breeding programs. **Natural selection** is selection in nature without the intentional aid of humans, although human activity can alter natural selection patterns.

One of the best examples of natural directional selection is found in a series of studies on wing color in the peppered moth (*Biston betularia*) in Great Britain. Peppered moths are nocturnal and spend most of the daylight hours resting on tree trunks. Prior to the industrial revolution, light-colored lichens were common on tree trunks. Peppered moths that were light colored were about the same color as the lichens and were camouflaged from predatory birds. Darker moths were not as well camouflaged and were easier prey for birds (Figure 19.7a). Although the dark moths were not completely eliminated from the population, their numbers were kept small by negative selection. As the industrial revolution began, however, air pollution from factories and vehicles increased. Lichens are very sensitive to airborne pollutants, so the light-colored lichens disappeared from many of the trees and the trees became darker. Soot deposition further darkened the tree trunks. The dark moths were now better camouflaged than the light moths

Table 19.3 Fitness Values for Different Degrees of Dominance

Degree of Dominance	Fitness Values*		
Complete dominance of A_1 over A_2	A_2A_2		A_1A_2 A_1A_1
	$1 - s$		1
Partial dominance of A_1 over A_2	A_2A_2	A_1A_2	A_1A_1
	$1 - s$	$1 - hs$	1
		$(\tfrac{1}{2} > h > 0)$	
No dominance	A_2A_2	A_1A_2	A_1A_1
	$1 - s$	$1 - hs$	1
		$(h = \tfrac{1}{2})$	
Partial dominance of A_2 over A_1	A_2A_2	A_1A_2	A_1A_1
	$1 - s$	$1 - hs$	1
		$(1 > h > \tfrac{1}{2})$	
Complete dominance of A_2 over A_1	A_1A_2 A_2A_2		A_1A_1
	$1 - s$		1

*On each scale, fitness increases from left to right.

(Figure 19.7b). The dark moths increased in number as selection favored them, and the numbers of light moths decreased as they fell prey to birds. Directional selection altered the frequencies of alleles that govern wing color.

Another example of directional selection that has repeated itself many times is selection for pesticide resistance. DDT was introduced as a pesticide in 1939 and was very effective—it killed nearly every insect it contacted. DDT was sprayed widely—on farms to control insect pests, in cities to control mosquitoes in the summer, and in forests to prevent insect damage to trees used for wood products. Its effectiveness eventually declined as insect populations rebounded. Susceptible individuals were killed before they could reproduce, leaving a few insects that were genetically resistant to the pesticide to mate with one another. The resistant insects mated and bore progeny that were resistant. After several generations of selection, most of the insects in populations were resistant, and DDT was no longer effective.

Directional selection can be described mathematically. The relative ability for reproductive success under selection pressure is called **fitness**. Fitness is often described by the **selection coefficient (s)**, which is a measure of the degree to which selection acts against a particular genotype. The higher the value of s, the more intense the selection. When $s = 1$ for a particular genotype, selection prevents all individuals with that genotype from reproducing. Fitness is measured by determining how suc-

cessfully genotypes in a population reproduce when compared to one another. Fitness is described mathematically by a relative value, w, called **relative fitness**. If we arbitrarily assign a relative fitness of 1.0 to the most successful genotype in a population, then the least successful genotype has a fitness of $w = 1 - s$, where s represents the reduction in fitness relative to the most successful genotype.

If $s = 1$, then the fitness of the least successful genotype is $w = 1 - s = 0$, and none of the individuals with that genotype reproduce. How effective selection is on altering allele frequencies depends not only on the value of s, but also on the nature of the alleles. If we assume that there are two alleles in a population, A_1 and A_2, with selection against the A_2 allele, then the effect of selection depends on the dominance relationships of the two alleles. The fitness values for different dominance relationships are given in Table 19.3.

The value h in Table 19.3 is a coefficient that must be factored in to calculate the relative fitness of heterozygotes when dominance is not complete. We assume a priori that the most favored genotype has a relative fitness of 1. That does not mean that all individuals with that genotype survive and reproduce, but simply that they are reproductively the most successful of all genotypes in the population. The success of other genotypes is then calculated relative to the most successful genotype. Let's look at the effects of selection under different situations

of dominance. We'll start with complete dominance and selection that disfavors recessive homozygotes.

Directional Selection Against Recessive Homozygotes

Among the most common situations in nature is selection against recessive mutant alleles that are expressed phenotypically only when homozygous. Many mutant alleles fail to encode a functional version of an enzyme that may be essential, or at least advantageous, for survival and reproduction. In such cases, the mutant allele is usually recessive, which means that selection acts against the allele only when the allele is in the homozygous state. If we let A_1 represent a favored allele that is completely dominant over A_2, the A_2A_2 genotype has a fitness of $1 - s$. Selection reduces the frequency (q) of the A_2 allele in each generation. The reduction of q in each generation is described mathematically as

$$q_1 = \frac{q - sq^2}{1 - sq^2} \qquad \text{[19.12]}$$

where q is the frequency of the A_2 allele in the parental generation and q_1 is the frequency of the A_2 allele in the progeny generation after selection. The difference between the allele frequency of the parental generation (q) and the allele frequency of the progeny generation (q_1) is $q_1 - q$, and is described as Δq, the change in q in one generation. We can subtract q from both sides of equation 19.12 to give

$$\Delta q = q_1 - q = \frac{q - sq^2}{1 - sq^2} - q$$

which, with some rearrangement, gives

$$\Delta q = \frac{-spq^2}{1 - sq^2} \qquad \text{[19.13]}$$

Because Δq is a negative number, there is a reduction in the value of q with each generation of selection. The greater the absolute value of Δq, the more effective selection is in reducing the allele frequency.

Let's use an example to illustrate how the effect of selection can be calculated with equation 19.12 or equation 19.13.

Example 19.4 Change in allele frequency in peppered moths due to natural selection.

In peppered moths, a dominant allele D confers the dark phenotype when homozygous or heterozygous. The light-colored moths are homozygous for a recessive allele d. In 1961, Kettlewell (*Annual Review of Entomology* 6:245–262) published his calculations of relative fitness values for light and dark moths

in an area of Great Britain with polluted air. He marked equal numbers of light and dark moths and released them, then later determined the proportions of both types when he recaptured the moths. He recaptured 53% of the dark-colored moths but only 25% of the light-colored moths.

Problem: Assuming that the rate of recapture represents the actual frequencies of marked moths, **(a)** determine the relative fitness values for the light type and the dark type, **(b)** calculate the value of Δq after one generation of selection when $q = 0.6$.

Solution: **(a)** The frequency of recapture for the dark type was higher than the frequency for the light type, so the dark type has a relative fitness of $w = 0.53 \div 0.53 = 1.0$. The light type then has a relative fitness of $w = 0.25 \div 0.53 = 0.47$, so $s = 1 - w = 0.53$. **(b)** We can use either equation 19.12 or equation 19.13 to calculate the value of Δq. Using equation 19.12, we get

$$q_1 = \frac{q - sq^2}{1 - sq^2} = \frac{0.6 - 0.53(0.6^2)}{1 - 0.53(0.6^2)} = 0.51$$

$$\Delta q = q_1 - q = 0.51 - 0.60 = -0.09$$

or with equation 19.13,

$$\Delta q = \frac{-spq^2}{1 - sq^2} = \frac{-0.53(0.4)(0.6^2)}{1 - 0.53(0.6^2)} = -0.09$$

After one generation of selection, the allele frequency of d is reduced from 0.6 to 0.51, a change of -0.09.

As illustrated in Figure 19.8, the effectiveness of selection against a recessive allele decreases as the value of q decreases. When $q < 0.1$, selection is very ineffective at reducing q, even when $s = 1$. As the value of q decreases due to selection, a greater proportion of individuals that carry A_2 are heterozygotes that are not disfavored by selection. For example, when $q = 0.8$, 64% of the individuals in the population are homozygous A_2A_2, and 32% are heterozygous A_1A_2. When $q = 0.1$, only 1% of the individuals in the population are homozygous A_2A_2, and the frequency of heterozygotes (A_1A_2) is 18%. In the first case, in which $q = 0.8$, only 20% of the A_2 alleles are in heterozygotes, where they are protected from selection. When $q = 0.1$, however, 90% of the A_2 alleles are in heterozygotes and are protected from selection.

In animal or plant breeding, it is often desirable to know how many generations are required to reduce an allele to a specific frequency by removing all homozygous recessive individuals from the population. If we let

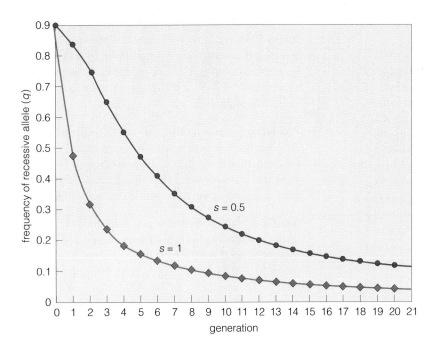

Figure 19.8 Selection against a recessive allele when $q = 0.9$ at generation 0.

t equal the number of generations, q_0 the initial allele frequency, and q_t the allele frequency after t generations,

$$q_t = \frac{q_0}{1 + tq_0}$$

which can be rearranged to give

$$t = \frac{1}{q_t} - \frac{1}{q_0} \qquad \textbf{[19.14]}$$

Let's work through an example to show how equation 19.14 can be used.

Example 19.5 Effect of selection against a recessive allele.

Problem: How many generations are required to reduce the frequency of albino rabbits in a population from 25% to 1% by removing all albino rabbits before they reproduce? Assume that the population is in Hardy-Weinberg equilibrium and that albino rabbits are homozygous for a recessive allele.

Solution: From the information above, $q_0^2 = 0.25$ so $q_0 = 0.5$, and $q_t^2 = 0.01$ so $q_t = 0.1$. Using equation 19.14,

$$t = \frac{1}{q_t} - \frac{1}{q_0} = \frac{1}{0.1} - \frac{1}{0.5} = 8 \text{ generations}$$

Directional Selection Against a Completely Dominant Allele
Directional selection against an allele that is completely dominant is more effective than selection against a re-

cessive allele because all genotypes that carry the allele are affected by selection. If the frequency of the dominant allele is p, then the change in its frequency due to selection is

$$\Delta p = \frac{-spq^2}{1 - s + sq^2} \qquad \textbf{[19.15]}$$

If $s = 1$, selection is 100% effective against a dominant allele. Substitution of 1 for s reduces equation 19.15 to $\Delta p = -p$, which means that the dominant allele is eliminated from the population after one generation by selection. For this reason, dominant alleles that are lethal before reproductive maturity are eliminated in a single generation.

Directional Selection with Partial Dominance As we saw in Table 19.3, an additional coefficient, called h, must be factored in when dominance is not complete to determine the relative fitness values for heterozygotes. For any value of h, the mathematical description of Δq is a little more complicated than in previous equations:

$$\Delta q = \frac{-spq[q + h(p - q)]}{1 - 2hspq - sq^2} \qquad \textbf{[19.16]}$$

Sometimes the phenotype of the heterozygote is halfway between the two homozygotes, in which case there is no dominance and $h = \frac{1}{2}$. Substituting $\frac{1}{2}$ for h in equation 19.16 and rearranging simplifies it to

$$\Delta q = \frac{-\frac{1}{2}spq}{1 - sq} \qquad \textbf{[19.17]}$$

The following example shows the results of an experiment in which a wild-type allele is partially dominant over a mutant allele.

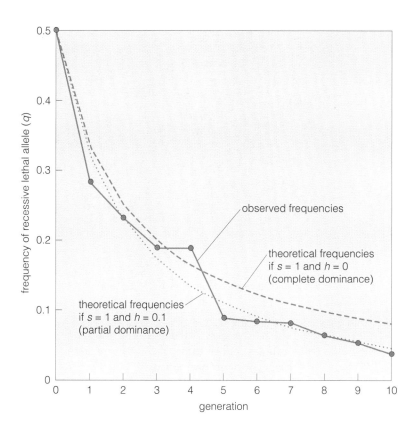

Figure 19.9 Selection against a recessive lethal allele in *Drosophila melanogaster* with a partial decrease in fitness for heterozygotes. (Adapted from Wallace, B. 1963. The elimination of an autosomal lethal from an experimental population of *Drosophila melanogaster*. *American Naturalist* 97:65–66.)

Example 19.6 Selection and partial dominance.

In 1963, Wallace (*American Naturalist* 97:65–66) reported an experiment on selection against a recessive lethal allele in *Drosophila melanogaster*. Flies that are homozygous for the recessive allele die before reproducing, and heterozygotes suffer a 10% decrease in fitness.

Problem: What are the relative fitness values for the genotypes, and what are the values of s and h?

Solution: All flies that are homozygous for the recessive lethal allele die before reproducing, so $s = 1$, and their relative fitness is $w = s - 1 = 0$. Heterozygotes suffer a 10% decrease in fitness, so $h = 0.1$, and their relative fitness is $w = 1 - hs = 1 - 0.1(1) = 0.9$. Figure 19.9 shows Wallace's observed results and theoretical curves for complete and partial dominance of the wild-type allele. The curve for partial dominance (in which $h = 0.1$) best matches the observations.

Stabilizing and Disruptive Selection

Selection operates on its own or in conjunction with other factors that alter allele frequencies to change allele frequencies in a predictable direction and magnitude. In some cases, selection can participate in genetic equilibria in which two or more selective pressures counter each other, or where selection is countered by another factor, to maintain constant allele frequencies in a population.

Stabilizing selection is the combined effect of positive and negative selection that maintains a stable phenotype in a population. Stabilizing selection is common in nature. A good illustration is a comparison of a domestic animal species with a wild counterpart. All coyotes are of the species *Canis latrans*, and the phenotype of the species is fairly simple to describe. An adult coyote is about half a meter tall and gray-colored, and has a bushy tail, a pointed nose, and erect ears. All domestic dogs are of the species *Canis familiaris*, but it is not as simple to provide a single descriptive phenotype of this species as it is for coyotes. The phenotypes of domestic dogs vary considerably because there are so many different breeds.

Natural stabilizing selection in coyotes has favored the phenotype that is most fit for the environment in which they live. The uniformity in phenotype is not necessarily an indication of genetic uniformity. Genetic variability is often present in populations with a stable phenotype because different genes may contribute to the same phenotypic trait. Extreme phenotypes, such as coyotes that are too large or too small, are less fit for their environment and are disfavored by selection.

On the other hand, humans have practiced disruptive selection in dogs to produce many different phenotypes. **Disruptive selection** is directional selection in more than

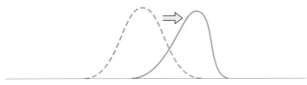

a Directional selection.

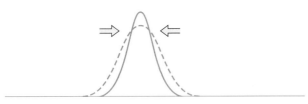

b Stabilizing selection.

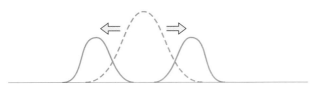

c Disruptive selection.

KEY

- - - - - population before selection
————— population after selection

Figure 19.10 A comparison of **(a)** directional, **(b)** stabilizing, and **(c)** disruptive selection.

one direction to produce separate populations with different phenotypes. In dogs, for example, the remote ancestors of Great Danes and Chihuahuas had an intermediate size, but humans have selectively bred Great Danes to be large and Chihuahuas to be small.

Underdominance is defined as a selective disadvantage for heterozygotes. When underdominance is present, disruptive selection favors two different homozygotes. Some alterations in chromosome structure, such as reciprocal translocations or inversions, provide examples of underdominance. Recall from Chapter 17 that inversion heterozygotes and reciprocal translocation heterozygotes are often at a selective disadvantage because they have reduced fertility. The loss of fertility in heterozygotes may cause reproductive separation of homozygous types and can eventually lead to speciation and full reproductive separation.

Figure 19.10 compares directional, stabilizing, and disruptive selection.

Heterozygote Advantage

When stabilizing selection operates, selective forces counter one another to set up an equilibrium that main-

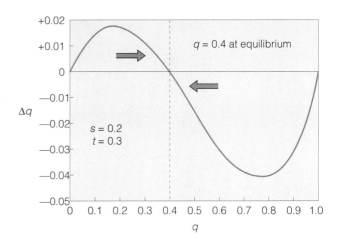

Figure 19.11 An example of the establishment of equilibrium when selection favors heterozygotes. The genotype frequencies differ from those in Hardy-Weinberg equilibrium.

tains constant allele frequencies. An example of two selective forces that counter one another is **heterozygote advantage**, or **overdominance**, in which two alleles are maintained at stable frequencies in a population because the heterozygotes are favored by selection over the two homozygous genotypes. In such a situation, the heterozygotes have a greater fitness than either homozygote, and the relative fitness values for homozygotes may be different. Thus, we assign two selection coefficients to the homozygous genotypes, s and t. Under these circumstances, the relative fitness values (w) are as follows:

A_2A_2	A_1A_1	A_1A_2
$1 - t$	$1 - s$	1.0

The change in allele frequency due to selection against both homozygotes is

$$\Delta q = \frac{pq(sp - tq)}{1 - sp^2 - tq^2} \qquad [19.18]$$

When selection favors the heterozygote over the two homozygotes, an equilibrium is reached when the numerator of equation 19.18 equals zero, and neither s nor t equals zero, so Δq equals zero. For example when $s = 0.2$ and $t = 0.3$, equilibrium is reached when $q = 0.4$, as is illustrated in Figure 19.11. This equilibrium is often called **balanced polymorphism**, a situation in which two different alleles are maintained in a population by selection that disfavors both homozygotes. The equilibrium reached in balanced polymorphism differs from Hardy-Weinberg equilibrium in that heterozygotes are more frequent than expected with Hardy-Weinberg equilibrium.

Some of the best-documented examples of balanced polymorphism are found in human populations in

Central Africa, the Middle East, and the Indian subcontinent, where the recessive allele for sickle-cell anemia in humans (HBB*S) is maintained at significantly higher frequencies than in nontropical regions of the world. Most HBB*S/*S homozygotes die from sickle-cell anemia before reproducing. However, the HBB*S allele has a pleiotropic effect in that it confers resistance to malaria in the homozygous or heterozygous condition. Thus, the HBB*S allele is recessive with respect to sickle-cell anemia but dominant with respect to resistance to malaria. In areas where the incidence of malaria is high, frequencies of the HBB*S allele are unusually high. Homozygotes for the HBB*S allele are disfavored by selection because they have sickle-cell anemia. Homozygotes for the HBB*A allele are disfavored by selection because they are susceptible to malaria. Heterozygotes (HBB*A/*S) have the highest fitness because they do not suffer from sickle-cell anemia and they are resistant to malaria.

The following example shows how balanced polymorphism for the sickle-cell anemia allele differs from Hardy-Weinberg equilibrium.

Example 19.7 Equilibrium with overdominance in a human population.

The Bedik people are native subsistence farmers in eastern Senegal, where malaria is common. They are the most ancient population in the area, and there is little intermarriage outside of the population. In 1975, Mauran-Sendrail and Bouloux (*Annals of Human Biology* 2:129–136) reported the results of a study of 875 Bedik adults that represented 60% of the entire Bedik population. They found that 626 were homozygous HBB*A/*A, 249 were heterozygous HBB*A/*S, and none were homozygous HBB*S/*S.

Problem: (a) Determine the frequency of the HBB*S and HBB*A alleles in this sample, and determine whether or not the population is in Hardy-Weinberg equilibrium. (b) Determine the values of s and t and the relative fitness values for the three genotypes under the assumption that the population is in a balanced polymorphism equilibrium.

Solution: (a) The frequencies can be determined directly from the data. The frequency of the HBB*S allele is $q = 249 \div [2(875)] = 0.1423$. The frequency of the HBB*A allele is $p = [2(626) + 249] \div [2(875)] = 0.8577$. The expected frequencies at Hardy-Weinberg equilibrium are $p^2 = 0.7357$, $2pq = 0.2441$, and $q^2 = 0.0202$.

The expected numbers of individuals for each genotype under the assumption of Hardy-Weinberg equilibrium and the observed numbers are shown

Table 19.4 Solution to Example 19.7: Observed and Expected Numbers of Individuals for Each of Three Genotypes at the *HBB* Locus

Genotype	Expected Number of Individuals	Observed Number of Individuals
HBB*A/*A	643.65	626
HBB*A/*S	213.59	249
HBB*S/*S	17.68	0

in Table 19.4. A chi-square test of the hypothesis that the population is in Hardy-Weinberg equilibrium produces a chi-square value of 24.08, which is substantially higher than the critical value of 6.64 at $P = 0.01$ with 1 degree of freedom. We reject the hypothesis that the population is in Hardy-Weinberg equilibrium. Heterozygotes are more frequent and both homozygotes are less frequent than expected under Hardy-Weinberg equilibrium, a result that is consistent with balanced polymorphism.

(b) The values of s and t can be calculated from these results. To determine the relative fitness of each genotype, we begin by dividing the observed frequency of each genotype by its expected frequency under Hardy-Weinberg equilibrium:

HBB*S/*S $0 \div 0.0202 = 0$

HBB*A/*S $0.2846 \div 0.2441 = 1.1659$

HBB*A/*A $0.7154 \div 0.7357 = 0.9725$

The heterozygote is the most fit of the three genotypes, so we arbitrarily assign it a relative fitness of $w = 1$. The relative fitness of the HBB*S/*S homozygote is $w = 0 \div 1.1659 = 0$, so $t = 1 - w = 1$. The relative fitness of the HBB*A/*A homozygote is $w = 0.9725 \div 1.1659 = 0.8341$, so $s = 1 - w = 0.1659$.

Balance Between Selection and Mutation

Mutations usually create deleterious recessive alleles, and selection is effective against such alleles only when they are homozygous. As we saw earlier, selection is very inefficient against a recessive allele that is present at a low frequency because most of the alleles are carried by heterozygotes. In spite of its inefficiency, selection should still continue to reduce the allele frequency of an infrequent recessive allele, albeit in very small increments with each generation. However, at some point the effect of

selection is negated by the effect of mutation. At low allele frequencies, the small effect of selection in decreasing an allele's frequency is offset by the small effect of forward mutation, which increases the allele's frequency. At a certain low allele frequency, an equilibrium between selection and mutation is attained, and the frequency of the recessive allele remains constant. If we let u equal the net forward mutation rate (forward mutation rate minus the reverse mutation rate), then we can describe the equilibrium between selection and mutation as

$$up = \frac{spq^2}{1 - sq^2} \qquad \text{[19.19]}$$

Without sacrificing much accuracy, we can simplify this equation with a few approximations. First, if q is very small, the denominator is very close to 1. Also, when q is small, p is close to 1. If we substitute 1 for the denominator and 1 for p, equation 19.19 reduces to

$$u = sq^2 \qquad \text{(approximate)} \qquad \text{[19.20]}$$

so at equilibrium between mutation and selection

$$q = \sqrt{(u \div s)} \qquad \text{(approximate)} \qquad \text{[19.21]}$$

When $s = 1$ for the recessive homozygote, as it does for many deleterious and lethal recessive alleles, then equation 19.21 becomes

$$q = \sqrt{u} \qquad \text{(approximate)} \qquad \text{[19.22]}$$

If we assume that the frequency of a recessive allele is at equilibrium, and that $s = 1$ for the homozygous recessive genotype, then we can use equation 19.22 to estimate the net forward mutation rate. For example, if a recessive disorder that is lethal before reproductive age or causes sterility is observed in 1 per 100,000 individuals, then q is equal to the square root of the phenotypic frequency, which is 0.003. Using equation 19.22, we find that the mutation rate equals q^2, which is the same as the phenotypic frequency of 1 per 100,000.

However, we must be wary about these assumptions. One might assume that the recessive allele that causes cystic fibrosis in humans is at a balance between selection and mutation. Until recently, people with cystic fibrosis rarely lived to reproduce, so the selection coefficient for homozygotes was $s = 1$. In humans of European descent, about 1 per 2000 people born have cystic fibrosis. The assumption of a balance between mutation and selection is probably incorrect because a mutation rate of 1 per 2000 is excessive. Another example is Tay-Sachs disease, which is lethal in early infancy and is observed in 1 in 3000 births among Ashkenazic Jews but at a much lower rate among most other ethnic groups. Again, it is unreason-

able to assume that a balance between mutation and selection has been reached. Because selection against a recessive allele is very inefficient when the allele frequency is low, it may take many generations for a balance between mutation and selection to be reached. Until the equilibrium is reached, a higher-than-expected allele frequency may be observed for certain alleles in a particular ethnic group.

~

Selection alters allele frequencies by favoring or disfavoring the reproductive success of phenotypes. The effect of selection depends on several factors, including the degree of dominance, allele frequency, and factors that interact with selection.

19.7 RANDOM GENETIC DRIFT

Of all the factors that affect allele frequencies in populations, random genetic drift is probably one of the most important in both natural and domestic species. Unfortunately, it is also among the least understood and most overlooked. **Random genetic drift** (often referred to simply as **drift**) is a random change in allele frequency due to sampling error. The three other factors (mutation, migration, and selection) that change allele frequencies are **systematic**—that is, they alter allele frequencies in a predictable direction and at a predictable magnitude. Drift, however, is **dispersive**—it changes allele frequencies in random directions, although it is predictable in magnitude. The factor that determines the magnitude of drift is population size. When the population size is small, the magnitude of drift is significant. As the size of a population becomes large, drift becomes negligible.

The effect of drift can be demonstrated experimentally by subdividing a large population into small populations, then maintaining the populations at a small size by randomly selecting the individuals that mate with one another to produce the next generation, as illustrated in the following example.

Example 19.8 Random genetic drift in *Tribolium castaneum*.

In 1979, Rich et al. (*Evolution* 33:579–584) published the results of a study on changes in allele frequencies in populations of the flour beetle *Tribolium castaneum*. They began their experiments with two true-breeding populations: one that was homozygous for a mutant allele called b, and another that was homozygous for the wild-type allele, b^+. Beetles that are homozygous b^+b^+ have a red body color, those that are homozygous bb have a black body color, and those that are heterozygous b^+b

have a brown body color. Thus, the researchers could easily distinguish all three genotypes on the basis of their phenotypes and thereby determine the allele frequencies of b^+ and b in a population. The researchers hybridized individuals from the two true-breeding populations to produce F_1 offspring that were all heterozygous b^+b. They then subdivided the heterozygous F_1 individuals into 48 populations: 12 populations maintained at 10 breeding individuals per generation, 12 populations maintained at 20 breeding individuals per generation, 12 populations maintained at 50 breeding individuals per generation, and 12 populations maintained at 100 breeding individuals per generation. Equal numbers of males and females were selected at random as breeding individuals for each generation in each population. The results of 20 generations are depicted in Figure 19.12.

Problem: Describe the evidence of random genetic drift and selection in the results of this study.

Solution: Random variations in allele frequencies among populations are evidence of drift. The magnitude of drift is indicated by the degree of random variation among populations. Drift is greatest in the set of populations with 10 breeding individuals and least in the set of populations with 100 breeding individuals. The magnitude of drift is inversely proportional to population size.

The average frequency of the b^+ allele increased in all four sets of populations. The average increase in the frequency of the b^+ allele is a consequence of natural directional selection that favors the b^+ allele.

Look again at Figure 19.12 (which was fully described in the preceding example). Notice in the set of populations with 10 breeding individuals that, after 20 generations, the frequency of the b^+ allele is $p = 1$ in six populations, and $p = 0$ in one population. After p reaches unity, the allele represented by q is lost, and the allele represented by p is said to be fixed in the population. **Fixation** is the establishment of a single allele in a homozygous state in all individuals in a population. All other alleles at that locus are lost from the population after fixation for one allele. After fixation, drift and selection can no longer affect the allele frequency until mutation or migration provides new alleles.

All populations are finite, so drift should cause all populations to eventually reach fixation for one of the alleles present in the population. In other words, drift causes homozygosity and loss of genetic diversity. The average number of generations required to reach fixation depends on population size. As is apparent in Figure 19.12, the smaller the population, the more rapid is the approach toward fixation.

If there are no factors other than drift that alter allele frequencies, then the allele frequency in the original population represents the probability that the population will become fixed for that allele. For example, if the allele frequency in a population is $q = 0.9$, then the probability that the population will eventually become fixed for the allele represented by q is 90%, and the probability that it will become fixed for the other allele is 10%. In Figure 19.12, the original frequency of the mutant allele was 0.5 in all populations, yet it is apparent that most populations will become fixed for the b^+ allele. In this example, selection and drift work simultaneously. Natural selection that favors the b^+ allele reduces the frequency of the b allele, and drift simultaneously causes variations in allele frequency equally in both directions.

The dispersive nature of drift and the tendency toward fixation are well illustrated in Figure 19.13. It shows the results of a study of 107 *Drosophila melanogaster* populations, each started with 16 flies that were heterozygous bw^1/bw^{75}. In the first generation, $p = 0.5$ and $q = 0.5$ in each population, where p represents the frequency of the bw^1 allele, and q represents the frequency of the bw^{75} allele. Each population was maintained at 16 breeding individuals for 19 generations by randomly selecting eight males and eight females to reproduce in each generation. The dispersive nature of drift over generations is evident as the allele frequencies among populations spread in the early generations. By generation 4, some populations have reached fixation, and by generation 19, 54% of the populations have become fixed for one or the other allele. Notice that in these experiments, the number of populations fixed for one allele is about the same as those fixed for the other allele, indicating that there is no selection. Had the original allele frequencies been $q = 0.9$ and $p = 0.1$, then the expected proportion of populations fixed for the bw^{75} allele would have been 0.9, and the expected proportion of those fixed for the bw^1 allele would have been 0.1.

Under some circumstances, selection can indirectly increase the effect of drift. For example, suppose that 2% of the individuals in an insect population are resistant to a pesticide. If the pesticide is applied and contacts every individual, only 2% of the individuals in the population survive. Directional selection increases the frequencies of alleles that confer resistance to the pesticide. However, while the population is small, drift has the opportunity to alter in random directions the frequencies of all other alleles not linked to the alleles that confer resistance. Under these circumstances, a few individuals are spared because of their resistance, and they become the founding parents for a new resistant population. They are called the founders of the population, and the allele frequencies at many loci in the new population may be different than those in the original population because of

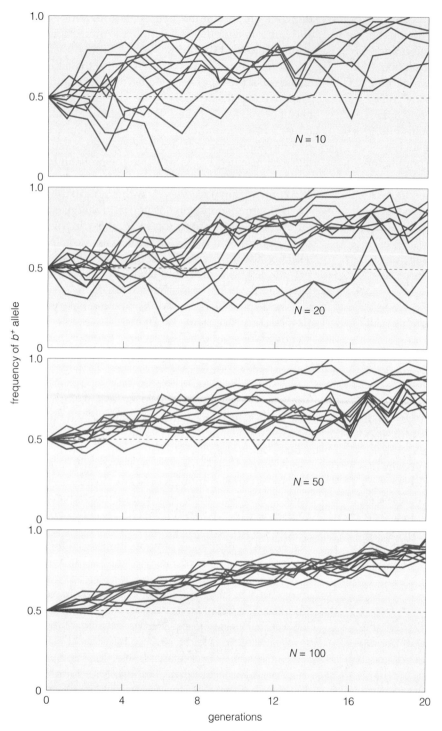

frequency of b^+ allele

generations

N = number of breeding individuals per generation

Figure 19.12   Information for Example 19.8: The effect of random genetic drift and directional selection in populations of *Tribolium castaneum*. (Adapted from Rich, S. S., A. E. Bell, and S. P. Wilson. 1979. Genetic drift in small populations of *Tribolium. Evolution* 33:579–584. Reprinted by permission of the publisher.)

drift. The degree of drift depends on how small the founder population is.

This phenomenon is called the **founder effect**, the effect of drift when a small number of individuals found a new population. The founder effect is significant whenever population size is reduced to a small number of individuals, whether or not the reduction is due to selection. When a small number of individuals migrate to a new location and become the founders of a new population, the new population also experiences founder effect.

The following example shows the consequences of the founder effect when coupled with selection.

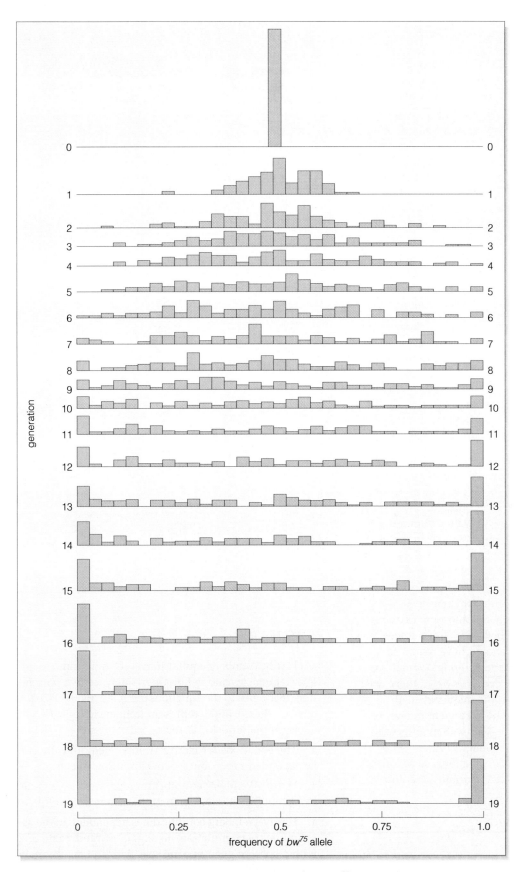

Figure 19.13 The dispersive nature of genetic drift for the *bw[1]* and *bw[75]* alleles in *Drosophila melanogaster*. The height of each bar on the graph represents the number of populations with a given allele frequency for the *bw[75]* allele. (Adapted with permission from Buri, P. 1956. Gene frequency in small populations of *Drosophila*. *Evolution* 10:367–402.)

Example 19.9   The consequences of founder effect.

In 1957, Dobzhansky and Pavlovsky (*Evolution* 11:311–319) reported the results of experiments in which they took flies of the species *Drosophila pseudoobscura* from two populations, each homozygous for inversions in chromosome 3. The two chromosome types were designated PP for "Pikes Peak" and AR for "Arrowhead." In one experiment, 20 populations were established that had initial frequencies of $p = 0.5$ for the AR chromosome and $q = 0.5$ for the PP chromosome. One set of 10 populations was initiated from 20 flies per population, and the other set from 4000 flies per population. From that point on, the populations were allowed to reproduce undisturbed for a period of 18 months. Table 19.5 shows the frequency of the PP chromosome in each of the populations after 18 months.

Problem: (a) Make two graphs of these data, one for set A and one for set B, in which the *x* axis represents time (18 months) and the *y* axis represents the frequency of the PP chromosome in each population. (b) Which set of populations was founded from 20 flies per population, and which set was founded from 4000 flies per population? (c) What evidence is there of selection in these data?

Solution: (a) Figure 19.14 shows graphs of the two sets of populations. (b) A smaller founding population size should result in a higher degree of drift for chromosome frequency among populations. The graphs in Figure 19.14 show that after 18 months the variation for set B was much greater than the variation for set A. Therefore, the populations in set B must have been founded from 20 flies each, and the populations in set A must have been founded from 4000 flies each. (c) Had drift alone been responsible for the changes in chromosome frequencies in the populations of both sets, the average chromosome frequencies should have remained unchanged. Only the variation among populations should change because of drift. However, there was a reduction in PP chromosome frequency in all populations of both sets. The average frequency of the PP chromosome for set A after 18 months was 0.2744, and the average frequency of PP for set B after 18 months was 0.3273. Directional selection reduced the frequencies of PP chromosomes in both sets.

Table 19.5   Information for Example 19.9: Change in Chromosome Frequency in *Drosophila pseudoobscura*

Subpopulation Number	Frequency of PP Chromosome After 18 Months
Set A	
145	0.317
146	0.290
147	0.347
148	0.340
149	0.227
150	0.203
151	0.320
152	0.223
153	0.257
154	0.220
Set B	
155	0.180
156	0.320
157	0.460
158	0.467
159	0.327
160	0.473
161	0.163
162	0.343
163	0.320
164	0.220

SUMMARY

1.   Hardy-Weinberg equilibrium is an equilibrium for allele frequencies and genotype frequencies in a population. It adheres to the equations $p + q = 1$ and $p^2 + 2pq + q^2 = 1$, where p and q represent frequencies of different alleles at the same locus.

2.   The equations for Hardy-Weinberg equilibrium can be expanded to account for more multiple alleles.

3.   Hardy-Weinberg equilibrium represents no genetic change and, therefore, represents the absence of evolution.

4.   Hardy-Weinberg equilibrium is dependent on the assumptions of no mutation, no migration, random mating with respect to genotype, no selection, and an infinitely large population.

5.   The violations of these five assumptions are the factors that disrupt Hardy-Weinberg equilibrium.

Random genetic drift is a change in allele frequency caused by sampling error. It is predictable in magnitude but unpredictable in direction. The magnitude of drift is inversely related to population size.

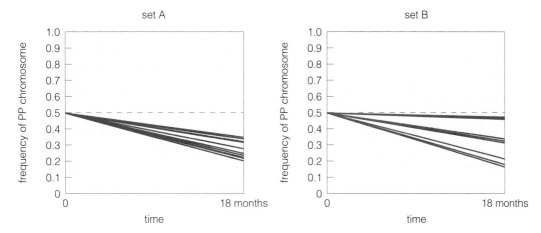

Figure 19.14 Solution to Example 19.9. Change in chromosome frequencies for two sets of populations. (Adapted from Dobzhansky, T., and O. Pavlovsky. 1957. An experimental study of interaction between genetic drift and natural selection. *Evolution* 11:311–319.)

6. Mutation is the only original source of genetic variation. It alters allele frequencies at slow rates over many generations.

7. Migration alters allele frequencies when there is a net flow of a particular allele into or out of a population.

8. Nonrandom mating includes assortative mating, inbreeding, and disassortative mating. Assortative mating and inbreeding increase homozygosity, whereas disassortative mating increases heterozygosity in populations.

9. Selection alters allele frequencies by favoring or disfavoring the reproduction of individuals with a particular phenotype.

10. The effectiveness of selection is dependent on the intensity of selection, the degree of dominance, and allele frequency.

11. Random genetic drift causes alterations in allele frequencies because of sampling error.

12. The magnitude of random genetic drift is inversely related to population size.

13. Random genetic drift is predictable in magnitude but unpredictable in direction.

QUESTIONS AND PROBLEMS

1. Albinism in rabbits is caused by homozygosity for a recessive allele called *c*. The allele that causes the agouti phenotype (which appears mottled gray) is called *C* and is completely dominant over *c*. If 16 albino rabbits (8 male and 8 female) are introduced into a population of 60 true-breeding agouti rabbits (30 male and 30 female), and all rabbits mate at random, what proportion of the first-generation offspring will be **(a)** albino, **(b)** gray, and **(c)** heterozygous *Cc*?

2. A population of agouti and albino rabbits has reached Hardy-Weinberg equilibrium, and there are four times as many gray rabbits as white rabbits. What is the frequency of the *c* allele?

3. If the frequency of albino rabbits is 10%, how many generations will be required to reduce the frequency to 1% if all albino rabbits are removed before reproducing in each generation?

4. If the frequency of albino rabbits is 10%, how many generations are required to increase the frequency to 20% if 30% of the gray rabbits are removed before reproducing in each generation?

5. Why is Hardy-Weinberg equilibrium attained in one generation for autosomal alleles, whereas several generations are required for Hardy-Weinberg equilibrium to be reached for X- or Z-linked alleles?

6. Explain why selection that favors an allele with partial dominance is more effective than selection that favors an allele with complete dominance. Use appropriate equations from this chapter in your explanation.

7. If the frequency of people in a human population with type O blood is 36%, and the frequency of people with type A blood is 45%, what are the frequencies of individuals with type AB and type B blood? Assume that the population is at Hardy-Weinberg equilibrium for ABO alleles.

8. Before World War II, it was common practice to encourage people with recessive genetic disorders to not have children in an attempt to reduce the frequencies of recessive genetic disorders in the population. Recessive genetic disorders are usually very rare. Suppose that the frequency of a particular autosomal recessive disorder is 1 per 5000 people. **(a)** How many generations would be required to reduce the frequency to 1 per 10,000 if all people with the disorder had no offspring? **(b)** Assuming an

average of 25 years per generation, how many years would be required to reduce the frequency from 1 per 5000 to 1 per 10,000? **(c)** Although there are ethical reasons for allowing all individuals the right to reproduce, what would be the efficacy of implementing restricted reproduction purely from a genetic perspective?

9. In the domestic cat, black coat color is caused by the *o* allele, and orange coat color is caused by the *O* allele at the same locus. The two alleles are codominant and produce a tortoiseshell phenotype in the heterozygous females due to differential X-chromosome inactivation. In 1964, Todd (*Heredity* 19:47–51) reported that among stray cats sampled in Boston, 102 females and 99 males were black, 4 females and 28 males were orange, and 48 females and no males were tortoiseshell. Determine whether or not the population is in Hardy-Weinberg equilibrium.

10. In 1956, Allison (*Annals of Human Genetics* 21:67–69) reported that among 287 infants in Tanzania, 189 had the genotype *HBB*A/*A*, 89 had the genotype *HBB*A/*S*, and 9 had the genotype *HBB*S/*S*. Determine the frequencies of the *HBB*A* and *HBB*S* alleles, and use chi-square analysis to test the hypothesis that the population is in Hardy-Weinberg equilibrium.

11. In 1962, Thompson (*British Medical Journal* 1:682–685) reported that among 840 young children in a population from Ghana, 701 had the genotype *HBB*A/*A*, 135 had the genotype *HBB*A/*S*, and 4 had the genotype *HBB*S/*S*. From these observations, determine the frequency of the *HBB*S* allele in this sample, and demonstrate using chi-square analysis that there is no evidence to reject the hypothesis that the population is in Hardy-Weinberg equilibrium.

12. The Niokholonko people are subsistence farmers in Eastern Senegal. In 1975, Mauran-Sendrail and Bouloux (*Annals of Human Biology* 2:129–136) reported that from a study of 599 Niokholonko people who represented 80% of the entire Niokholonko population, 483 had the genotype *HBB*A/*A*, 114 had the genotype *HBB*A/*S*, and 2 had the genotype *HBB*S/*S*. From these observations, determine the frequency of the *HBB*S* and *HBB*A* alleles in this sample and determine whether or not the population is in Hardy-Weinberg equilibrium.

13. When researchers examine infants in areas where malaria is common, the genotypic frequencies for the *HBB*A/*A*, *HBB*A/*S*, and *HBB*S/*S* genotypes usually correspond to expected frequencies for Hardy-Weinberg equilibrium. However, when researchers examine adult populations, the frequency of heterozygotes is usually higher than expected with Hardy-Weinberg equilibrium. Explain why such a discrepancy between infant and adult populations exists.

14. The M and N blood groups have been used for decades as a textbook example in population genetics.

In a study summarized by Mourant et al. (1976. *The Distribution of the Human Blood Groups and Other Polymorphisms*. Oxford: Oxford University Press), among 500 people of African descent who lived in New York City, 119 had type M blood, 242 had type MN blood, and 139 had type N blood. **(a)** What are the frequencies of the two alleles in this population? **(b)** Use chi-square analysis to test the hypothesis that the population is in Hardy-Weinberg equilibrium.

15. In 1958, Matsunaga and Itoh (*Annals of Human Genetics* 22:111–131) reported that among 2858 people in Japan, 810 had type M blood, 1424 had type MN blood, and 624 had type N blood. **(a)** What are the frequencies of the two alleles in this population? **(b)** Use chi-square analysis to test the hypothesis that the population is in Hardy-Weinberg equilibrium. **(c)** How do these allele frequencies compare with those determined in the previous question?

16. In 1939, Ikin et al. (*Annals of Eugenics* 9:409–411) reported that among 3459 people from southern England, 1546 had type A blood, 297 had type B blood, 113 had type AB blood, and 1503 had type O blood. **(a)** Determine the frequencies of each allele. **(b)** Use chi-square analysis to test the hypothesis that the population is in Hardy-Weinberg equilibrium.

17. In 1927, Waaler (*Zeitschrift für Induktive Abstammungs und Vererbungslehre* 45:279) reported the results of a study of red-green color blindness in Norwegian children. Among 18,121 children examined, 9049 were males, of whom 725 had red-green color blindness; and 9072 were females, of whom 40 had red-green color blindness. Test the hypothesis that this population is in Hardy-Weinberg equilibrium for red-green color blindness alleles.

18. Recall from Chapter 14 that there are two types of red-green color blindness, deuteranopia and protanopia, governed by two linked loci on the X chromosome. In Waaler's study cited in the previous question, of the 725 males with color blindness, 551 had deuteranopia and 174 had protanopia. Of the 40 females with color blindness, 37 had deuteranopia and 3 had protanopia. **(a)** Determine whether or not each mutant allele (independent of the other allele) is in Hardy-Weinberg equilibrium. **(b)** How do these data affect the conclusions drawn in the previous question?

19. One might assume that it is possible to estimate the mutation rate for a recessive allele when $s = 1$ for recessive homozygotes using equation 19.22. Why is it unwise to do so in most cases?

20. In 1979, Rich et al. (*Evolution* 33:579–584) published the results illustrated in Figure 19.12. Only one of 48 populations became fixed for the mutant *b* allele, whereas 10 became fixed for the wild-type b^+ allele. Given that selection favors the b^+ allele, explain how the mutant *b* allele could become fixed in one of the populations. In your answer, be sure to explain the effect of population size.

Table 19.6	Information for Problem 23: Chromosome Genotypes of Flies in Two Experiments		
	Number of Flies of Each Type		
Experiment	ST/ST	ST/CH	CH/CH
22	57	169	29
23	80	196	58

Table 19.7	Information for Problem 24: Frequencies of the ST Chromosome in *Drosophila pseudoobscura* Populations			
	Population			
Day	90	91	92	96
0	0.200	0.200	0.200	0.200
35	0.377	0.373	0.383	0.320
80	0.500	0.590	0.527	0.520
120	0.630	0.700	0.607	0.617
160	0.690	0.713	0.687	0.660
200	0.747	0.760	0.690	0.763
250	0.780	0.800	0.807	0.790
310	0.787	0.840	0.803	0.830
365	0.830	0.840	0.807	0.803

21. In the study highlighted in Figure 19.12, Rich et al. (1979. *Evolution* 33:579–584) determined that the average values of q after 20 generations were 0.852 for the populations in which $N = 10$, 0.781 for the populations in which $N = 20$, 0.811 for the populations in which $N = 50$, and 0.867 for the populations in which $N = 100$. Calculate the average value of Δq per generation in each group, and determine the value of s. Predict the value of q when $h = 0$ and $h = \frac{1}{2}$. Which value of h most closely corresponds to the observed data?

22. In Figure 19.13, 30 populations became fixed for the bw^1 allele, and 28 became fixed for the bw^{75} allele after 19 generations. The original allele frequencies were both 0.5. What do these observations suggest about the effect of selection in this experiment?

23. In 1947, Dobzhansky (*Genetics* 32:142–160) reported the results of a study on inversion frequencies in the third chromosome of *Drosophila pseudoobscura*. The genotype of individual flies could be identified by examining polytene chromosomes in their larval offspring. Flies that were homozygous for the standard chromosome conformation were designated ST/ST, and flies homozygous for an inversion called "Chiricahua" were designated CH/CH. Inversion heterozygotes were designated ST/CH. Flies raised in two cages were tested for their inversion genotypes. The results are listed in Table 19.6. **(a)** Calculate the frequencies of each chromosome in the two populations, and from these values determine the expected numbers of flies that have each genotype if the populations are in Hardy-Weinberg equilibrium. **(b)** How do the observed results deviate from Hardy-Weinberg equilibrium? **(c)** Calculate the relative fitness values and selection coefficients for each genotype. **(d)** What could account for these results?

24. Studies reported in 1953 on the ST and CH chromosomes in *Drosophila pseudoobscura* by Dobzhansky and Pavlovsky (*Evolution* 7:198–210) showed the effect of selection on these chromosomes. Table 19.7 provides the frequencies of the ST chromosome in four populations raised in cages over a period of 1 year. The initial frequency of the ST chromosome was 0.2 in each case. **(a)** On graph paper, plot these results in a graph with days on the x axis and the ST frequency on the y axis.

Connect the plotted data points for each population with lines. **(b)** Describe how these results are consistent with selection that favors heterozygotes. **(c)** What evidence of drift is present in the graph?

25. In 1944, Hartman and Lundevall (*Mat. Naturv. Klasse, no. 2*) reported that among 34,309 people in Norway, 10,354 had type M blood, 16,822 had type MN blood, and 7133 had type N blood. **(a)** Determine the frequencies of the MN^*M and MN^*N alleles in this sample. **(b)** Using the allele frequencies, determine the expected allele frequencies for Hardy-Weinberg equilibrium. **(c)** Test the hypothesis that these data conform to Hardy-Weinberg equilibrium using chi-square analysis.

FOR FURTHER READING

Excellent detailed explanations of the concepts discussed in this chapter can be found in **Crow, J. F. 1986. *Basic Concepts in Population, Quantitative, and Evolutionary Genetics*. New York: Freeman; Falconer, D. S. 1996. *Introduction to Quantitative Genetics*, 4th ed. Essex, England: Longman; and Hartl, D. 1997. *Principles of Population Genetics*, 3rd ed. Sunderland, Mass.: Sinauer Associates.**

For additional reading, go to InfoTrac College Edition, your online research library at: http: www.infotrac-college.com/brookscole

CHAPTER 20

QUANTITATIVE GENETICS

Geneticists usually describe phenotypes in qualitative or quantitative terms. A **qualitative description** is one that provides a simple, nonnumerical characterization of a trait. Such terms as "tall or short" and "heavy or light" are qualitative descriptions. A **quantitative description**, on the other hand, describes a phenotype numerically using some unit of measurement. For example, an individual may be 1.5 meters tall rather than just "tall."

Genes and their effects may be described qualitatively or quantitatively. Much of Mendelian genetics is based on qualitative assessments of phenotypes. Many phenotypes can be readily distinguished from one another qualitatively with no need to measure individuals to identify the difference. For example, Mendel classified pea plants as tall or dwarf without measuring the height of each plant. Quantitative characters, on the other hand, tend to exhibit **continuous variation**—meaning that individuals differ from one another over a wide range of phenotypes measured in small increments. To classify an individual by phenotype, researchers must use units of measurement instead of discrete qualitative classes.

Continuous variation is usually (though not always) a consequence of the combined effects of several genes on a single trait. The inheritance of each individual gene follows a Mendelian pattern. The combined inheritance of several genes that govern the same trait is called **poly-genic inheritance**. Continuous variation for quantitative traits may also be influenced by environmental (nongenetic) factors. A good example is height in humans. As Figure 20.1 illustrates, height among a population of adult humans usually shows a bell-shaped distribution. Much of the variation is influenced by inheritance. People who are taller than average usually have at least one parent who is taller than average. But height is also influenced by diet and childhood health. A child with a nutritious diet who is free of serious illnesses is likely to attain his or her full genetic potential for height. Conversely, a child who receives inadequate nutrition or who suffers from a serious illness is not likely to reach his or her full genetic potential for height. Both genetic and environmental factors act together to influence the ultimate phenotype, and both are responsible for the phenotypic variation in a population.

Quantitative genetics is the analysis of how genetic and environmental factors influence the inheritance and expression of quantitative traits. Geneticists have studied the quantitative genetics of human populations and natural populations of plants and animals. However, most research in quantitative genetics has been conducted as part of plant and animal breeding programs, and its applications have dramatically improved the quantity and quality of food and fiber worldwide.

1	4	8	10	16	16	15	15	14	13	13	11	9	8	8	5	1	2

number of individuals

60 (5 feet)	61	62	63	64	65	66	67	68	69	70	71	72 (6 feet)	73	74	75	76	77

height (inches)

Figure 20.1 Human height as an example of a quantitative trait. Quantitative traits often exhibit a bell-shaped distribution in a random sample from a population. The photo shows students from a general biology class at Brigham Young University lined up by height in increments of one inch.

20.1 STATISTICAL DESCRIPTIONS OF METRIC CHARACTERS

Before discussing the genetic aspects of quantitative variation, let's first take a brief look at how quantitative variation is measured and summarized. Any character that is distributed in a continuous manner in a population and can be measured in some way is a **metric character**. Examples include anatomical dimensions, such a body height, body weight, head size, and limb length; physical characteristics, such as the amount of pigment in flower petals, leaf tissue, hair, or skin, the amount of hair, and the number of bristles on insects; economic characters in animals and plants, such as grain yield, seed size, meat yield, feed efficiency, fat content, protein content, growth rates, and time to maturity; and, among the most controversial, intelligence in humans as measured by IQ scores.

Metric characters can be further classified as continuous or meristic traits. A **continuous trait** is one that varies from one extreme to another in a population with no discrete classes. Phenotypes are measured in incremental units, such as meters, kilograms, or percentages. Grain yield in wheat, body weight in cattle, and protein content of soybeans are examples of continuous traits.

A **meristic trait** is one for which the metric value is measured by counting rather than applying incremental units of measurement. Examples of meristic traits include the number of hairs or bristles per individual or the number of seeds per plant. Certain traits may be meristic

or continuous depending on how the trait is measured. For example, researchers may measure egg production in chickens by counting the number of eggs each hen lays in a given period, in which case egg production is measured as a meristic trait. Alternatively, researchers may determine the total weight of eggs produced by each hen in a given period, in which case egg production is measured as a continuous trait.

A **threshold trait** is one in which the measurement is qualitative but the underlying inheritance is polygenic. For example, susceptibility to certain types of cancer is inherited. An individual either has the cancer or does not, a qualitative assessment. But the appearance of cancer may be influenced by several genes that interact with environmental factors.

Distributions

When analyzing the inheritance of quantitative traits, rarely do geneticists study the effects of individual genes and their alleles in isolation. Instead, they study the combined effects of genetic and environmental factors in a population using statistical analyses. Often, a metric character in a population is represented as a two-dimensional **distribution**, in which the increments of measurement are portrayed on the horizontal (x) axis and the numbers or proportions of individuals within particular phenotypic ranges are plotted on the vertical (y) axis.

Figure 20.1 is a photographic example of a distribution for human height. The fact that we measure in set in-

crements (such as meters, feet and inches, grams, pounds, etc.) precludes a truly continuous representation of a distribution. Instead, we measure in chosen increments, then group individuals into classes of predefined ranges of increments. The portrayal of such groupings in a distribution is called a **histogram**. For example, in grouping individuals by height, we may measure them to the nearest inch, then group them into classes, each class with a 3 inch range. Individuals who are 5'0" to 5'2" are grouped as one class, those 5'3" to 5'5" fall into another class, and so on up the scale. Each of the classes is represented as a bar on the histogram. The classes may be few, as in this histogram:

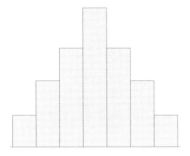

Or they may be many, as in this one:

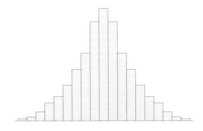

Sampling

Among the more important considerations in the analysis of a distribution is **sampling**, the selection of individuals from a population for measurement. Often, it is not practical to measure every individual in a population because the population may be very large. Instead, a sample of individuals from the population may be selected to represent the overall population. **Random sampling** is the selection of individuals at random from a population. The purpose of random sampling is to avoid biases that may influence the sample and prevent it from accurately representing the population. Alternatively, a sample derived from **systematic sampling**, the selection of individuals according to predetermined criteria, may better represent the overall population than a sample derived from random sampling. For example, researchers from reputable polling firms who conduct opinion surveys often systematically select individuals according to economic status, age group, or gender to ensure that the sample accurately represents the overall population.

To understand the difference between random and systematic sampling in genetics, let us suppose we wished to estimate the overall genetic diversity of a population using DNA markers. We could take a random sample of DNA markers—say, 100 markers—and determine the variation for those markers in the population. However, we might obtain a more accurate representation of diversity by systematically selecting 100 mapped molecular markers that represent incremental regions of chromosomes so that each chromosomal region is represented in the study. We could then analyze differences in variation for certain chromosomal regions to get a better understanding of the overall variation in the population.

Mean, Mode, Median, and Variance

Several statistical terms can be used to describe a distribution. We will discuss only a few of the most important ones. The **mean** is calculated by summing the phenotypic measurements for all individuals in the group, then dividing the sum by the number of individuals whose values were summed. In equation form,

$$x = M = \frac{\Sigma x_i}{n} = \frac{x_1 + x_2 + x_3 + \cdots + x_n}{n} \qquad \text{[20.1]}$$

where x and M are symbols that represent the mean, x_i represents the metric value of each individual, and n represents the number of individuals in the sample being studied.

The **mode** is the value at the peak of the distribution and usually corresponds to the most frequent class of individuals in the distribution. In a **symmetric distribution**, the mode and the mean are the same:

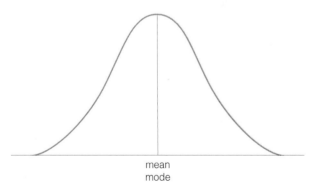

mean
mode

In an **asymmetric distribution**, the mode and the mean are different. Asymmetric distributions are often said to be **skewed** in one direction or another because the mode is not at the halfway point of the distribution:

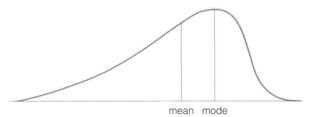

mean   mode

Most distributions have only a single mode, but some distributions have more than one mode, such as a **bimodal distribution**, with two modes:

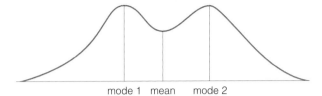

mode 1   mean   mode 2

Another descriptive character of a distribution is the **median**, the class that divides the distribution into equal halves by number of individuals.

Variance is a measurement of the degree to which individuals within the sample vary from one another. For example, two distributions may have the same mean and mode but may differ in their variances:

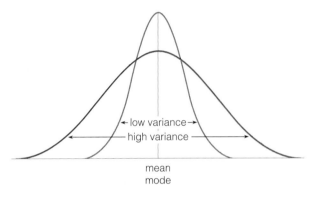

←low variance→
←———high variance———→

mean
mode

We calculate the variance by summing the squared deviations from the mean, then dividing this sum by the number of individuals minus one. In other words, the mean is subtracted from each individual's value to obtain that individual's deviation from the mean; the deviations for each individual are squared and summed; then the sum is divided by the number of individuals in the sample minus one:

$$s^2 = V = \frac{\Sigma(x_i - x)^2}{n - 1} \qquad [20.2]$$

where s^2 and V are terms that represent variance, and the other terms represent the same quantities as in equation 20.1.

Variance is often expressed as the **standard deviation** from the mean, which is simply the square root of the variance:

$$s = \sqrt{\frac{\Sigma(x_i - x)^2}{n - 1}} \qquad [20.3]$$

In discussing statistical terms, we must often distinguish between **theoretical terms**, which represent a hypothetical population (often one that is infinite in size), and **observed terms**, which are derived from observations in a sample and are intended as estimates of the

theoretical terms. The mean, variance, and standard deviation are often represented as either theoretical or observed terms, depending on the context of the discussion. Some of the symbols typically used to represent these terms are listed in Table 20.1.

The following example shows how a distribution can be characterized.

Table 20.1   Statistical Symbols

Term	Theoretical	Observed
Mean	μ	x or M
Variance	σ^2	s^2 or V
Standard deviation	σ	s

Example 20.1   Distribution of height in humans.

In the distribution portrayed in Figure 20.1, individual heights are measured in increments of 1 inch. The numbers of individuals in each height class are given below the photograph.

Problem: **(a)** Is height, as measured here, a continuous or meristic trait? **(b)** Determine the mean, mode, median, variance, and standard deviation for this sample of individuals. **(c)** Is this distribution skewed? If so, in which direction?

Solution: **(a)** Because height is measured in set increments (inches in this case), it is a continuous trait. **(b)** The mean is 67.799, the mode is 64.5 (halfway between 64 and 65, because these two classes have the most individuals and are adjacent to one another), the median is 67, the variance is 15.078, and the standard deviation is 3.883. **(c)** The distribution is skewed toward the lower values.

Normal Distributions

Distributions in nature, especially in genetics, often conform to what is known as a **normal distribution** (Figure 20.2), which has two key features:

1. Normal distributions are symmetrical. The mean and the mode are theoretically identical. In real situations the mean and the mode may differ slightly because of sampling error.

2. Sixty-eight percent of the population falls within one standard deviation of the mean, and 95% falls within 1.96 standard deviations of the mean. In real situations the deviations may be slightly different than this, again because of sampling error.

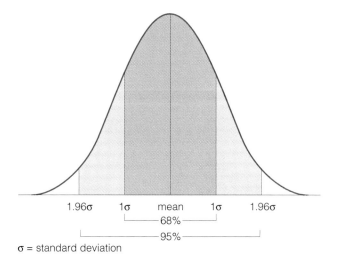

1.96σ   1σ   mean   1σ   1.96σ
|—— 68% ——|
|———— 95% ————|

σ = standard deviation

Figure 20.2 Characteristics of a normal distribution.

Because normal distributions are common in nature, many of the methods used for quantitative genetic analysis are based on a normal distribution. When populations do not conform to a normal distribution, researchers must sometimes make adjustments in their methods of analysis to account for the deviation from a normal distribution.

~

Quantitative traits are typically continuous or meristic. Samples from a population can be analyzed as distributions.

20.2 ADDITIVE GENE ACTION AND CONTINUOUS VARIATION

Many of the principles of quantitative genetics are based on polygenic inheritance. In nature, the several genes that determine the phenotype of an individual may have varying degrees of dominance, the genes may interact with one another in an epistatic fashion, and they may interact with environmental factors to produce the ultimate phenotype. Accounting for the contributions of all these variables to the phenotypic variation in a population is possible but not simple.

To understand how the combined effects of several genes may produce a particular phenotype, let's begin with the simplest of all possible situations: a group of genes with alleles that have equal and additive effects on the phenotype. Several loci affect a trait of interest, and the alleles at these loci show no dominance, so the effects of the alleles on the phenotype are purely additive, a situation known as **additive gene action**.

One of the best-documented examples of additive gene action is the inheritance of kernel color in wheat. This example was first described by Herman Nilsson-Ehle, who demonstrated that two loci, with alleles that have no dominance, governed kernel color. A third locus

that governs kernel color was later discovered. In our example, we will use all three loci. Let's call the loci A, B, and C, each with two alleles that have no dominance over one another. The alleles A^1, B^1, and C^1 each cause accumulation of a red pigment in the kernel. The alleles A^2, B^2, and C^2 add no red pigment. Because there is no dominance, the A^1, B^1, and C^1 alleles have equal and additive effects on kernel color, so each allele provides one equal dose of color to the kernel. The number of alleles with the superscript 1 determines how much red pigment accumulates in the kernel. The darkest red kernel has the genotype $A^1A^1 B^1B^1 C^1C^1$ (all six alleles add a dose of pigment), whereas a white kernel has the genotype $A^2A^2 B^2B^2 C^2C^2$ (none of the six alleles adds a dose of pigment). When three color-producing alleles are present, the kernel's color is halfway between the colors of the white and darkest red kernels. The segregation of F_2 progeny from the cross $A^1A^1 B^1B^1 C^1C^1 \times A^2A^2 B^2B^2 C^2C^2$ is illustrated in Figure 20.3. Notice that the two parental genotypes reappear as the least frequent classes of F_2 progeny, whereas the intermediate phenotypic class with three color-producing alleles is the most frequent. When plotted as a histogram, the distribution of phenotypes approximates a normal distribution (Figure 20.4).

In this example, each color-producing allele has an equal and additive effect on the phenotype, and environmental factors have no effect on wheat kernel color, so the phenotype of each individual is determined entirely by its genotype. The absence of dominance ensures that the effect of every allele appears in the phenotype, either by adding a dose of pigment or by failing to add a dose of pigment to the kernel. In quantitative genetics, this is the most straightforward of all possible situations, because there are no confounding effects of dominance, epistasis, or environmental factors on the phenotype. The effects of all alleles are equal and additive.

When certain assumptions are met, it is possible to estimate the number of genes that influence a trait by examining the variation for the trait in F_2 progeny. The formula for estimating the number of genes is

$$n = D^2 \div 8V_G \qquad [20.4]$$

where n is the number of genes, D is the difference of the means of the two parents, and V_G is the genetic variance of the F_2 progeny. For n to be a valid estimate of the number of genes, three assumptions must be met: First, alleles for the genes in question must have equal and additive effects. Second, the genes must all assort independently. Third, the original parents must be homozygous at all loci that influence the trait. Also, V_G is the genetic variance, so if there is any influence from environmental variation, it must be removed from the phenotypic variance to obtain an accurate estimate of genetic variance.

Let's now use an actual experimental situation with plants to illustrate how equation 20.4 is used.

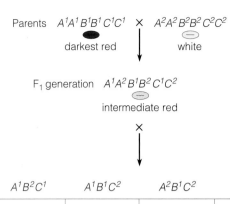

Parents $A^1A^1\,B^1B^1\,C^1C^1$ ✕ $A^2A^2\,B^2B^2\,C^2C^2$

darkest red   white

F_1 generation $A^1A^2\,B^1B^2\,C^1C^2$

intermediate red

✕

	$A^1B^1C^1$	$A^2B^1C^1$	$A^1B^2C^1$	$A^1B^1C^2$	$A^2B^1C^2$	$A^1B^2C^2$	$A^2B^2C^1$	$A^2B^2C^2$
$A^1B^1C^1$	$A^1A^1B^1B^1C^1C^1$	$A^1A^2B^1B^1C^1C^1$	$A^1A^1B^1B^2C^1C^1$	$A^1A^1B^1B^1C^1C^2$	$A^1A^2B^1B^1C^1C^2$	$A^1A^1B^1B^2C^1C^2$	$A^1A^2B^1B^2C^1C^1$	$A^1A^2B^1B^2C^1C^2$
$A^2B^1C^1$	$A^1A^2B^1B^1C^1C^1$	$A^2A^2B^1B^1C^1C^1$	$A^1A^2B^1B^2C^1C^1$	$A^1A^2B^1B^1C^1C^2$	$A^2A^2B^1B^1C^1C^2$	$A^1A^2B^1B^2C^1C^2$	$A^2A^2B^1B^2C^1C^1$	$A^2A^2B^1B^2C^1C^2$
$A^1B^2C^1$	$A^1A^1B^1B^2C^1C^1$	$A^1A^2B^1B^2C^1C^1$	$A^1A^1B^2B^2C^1C^1$	$A^1A^1B^1B^2C^1C^2$	$A^1A^2B^1B^2C^1C^2$	$A^1A^1B^2B^2C^1C^2$	$A^1A^2B^2B^2C^1C^1$	$A^1A^2B^2B^2C^1C^2$
$A^1B^1C^2$	$A^1A^1B^1B^1C^1C^2$	$A^1A^2B^1B^1C^1C^2$	$A^1A^1B^1B^2C^1C^2$	$A^1A^1B^1B^1C^2C^2$	$A^1A^2B^1B^1C^2C^2$	$A^1A^1B^1B^2C^2C^2$	$A^1A^2B^1B^2C^1C^2$	$A^1A^2B^1B^2C^2C^2$
$A^2B^1C^2$	$A^1A^2B^1B^1C^1C^2$	$A^2A^2B^1B^1C^1C^2$	$A^1A^2B^1B^2C^1C^2$	$A^1A^2B^1B^1C^2C^2$	$A^2A^2B^1B^1C^2C^2$	$A^1A^2B^1B^2C^2C^2$	$A^2A^2B^1B^2C^1C^2$	$A^2A^2B^1B^2C^2C^2$
$A^1B^2C^2$	$A^1A^1B^1B^2C^1C^2$	$A^1A^2B^1B^2C^1C^2$	$A^1A^1B^2B^2C^1C^2$	$A^1A^1B^1B^2C^2C^2$	$A^1A^2B^1B^2C^2C^2$	$A^1A^1B^2B^2C^2C^2$	$A^1A^2B^2B^2C^1C^2$	$A^1A^2B^2B^2C^2C^2$
$A^2B^2C^1$	$A^1A^2B^1B^2C^1C^1$	$A^2A^2B^1B^2C^1C^1$	$A^1A^2B^2B^2C^1C^1$	$A^1A^2B^1B^2C^1C^2$	$A^2A^2B^1B^2C^1C^2$	$A^1A^2B^2B^2C^1C^2$	$A^2A^2B^2B^2C^1C^1$	$A^2A^2B^2B^2C^1C^2$
$A^2B^2C^2$	$A^1A^2B^1B^2C^1C^2$	$A^2A^2B^1B^2C^1C^2$	$A^1A^2B^2B^2C^1C^2$	$A^1A^2B^1B^2C^2C^2$	$A^2A^2B^1B^2C^2C^2$	$A^1A^2B^2B^2C^2C^2$	$A^2A^2B^2B^2C^1C^2$	$A^2A^2B^2B^2C^2C^2$

F_2 generation

Figure 20.3 Additive gene action for kernel color in wheat.

Example 20.2 Estimating the number of genes that govern the inheritance of a quantitative character.

One of the first experiments designed to explain continuous variation from a Mendelian perspective was reported by East in 1916 (*Genetics* 1:164–176), who studied corolla length in *Nicotiana longiflora*, a wild species of tobacco. The corolla is the group of petals within a single flower. In *N. longiflora*, the corolla has a trumpet shape consisting of petals that are fused near the base and form a long tube before fanning out to form the open flower (Figure 20.5). East used two varieties as parents, one with short corollas (which we'll call P_1) and one with long corollas (which we'll call P_2). Both varieties bred true for short or long corolla length, indicating that the parents were fully homozygous for alleles that govern corolla length. East hybridized the two parent populations and measured the corolla lengths of individual plants in the two parent populations and

the F_1 and F_2 progeny. The results of his experiments are summarized in Figure 20.6.

Problem: **(a)** Determine the means and variances of the P_1, P_2, F_1, and F_2 groups. **(b)** Estimate the number of genes that govern corolla length in the experiment. **(c)** Are the results consistent with the hypothesis that genetic variation for corolla length is caused by additive gene action?

Solution: **(a)** The means and variances can be calculated using equations 20.1 and 20.2 and are given in Table 20.2. With large data sets such as this one, it is usually best to use a calculator or statistical software program that automatically calculates means and variances directly from the data.

(b) East intentionally chose two homozygous true-breeding parents because all plants within each of the two parental groups (P_1 and P_2) were genetically identical to one another. When hybridized, all of their F_1 progeny are genetically

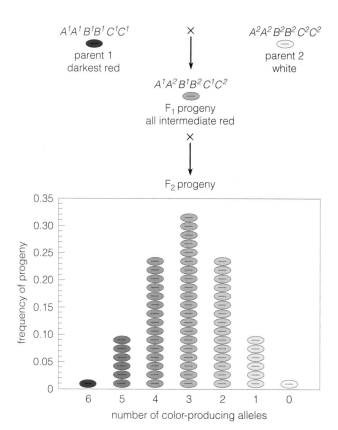

frequency of progeny

number of color-producing alleles

Figure 20.4 Histogram for kernel color in wheat in the F_2 generation of the cross $A^1A^1\ B^1B^1\ C^1C^1 \times A^2A^2\ B^2B^2\ C^2C^2$.

corolla

Figure 20.5 Information for Example 20.2: A tobacco flower.

Table 20.2 Partial Solution for Example 20.2: Means and Variances for Corolla Length in Tobacco

Group	Mean	Variance
P_1	40.37	3.03
P_2	93.11	5.64
F_1	63.53	8.62
F_2	68.76	42.37

and F_1 groups. The higher variance in the F_2 group is a result of the combined effects of genetic and environmental variation. Before estimating the number of genes that govern variation for corolla length, we must correct the phenotypic variance of the F_2 group to remove the effect of environmental variation. Because the variances of the P_1, P_2, and F_1 groups represent only environmental variation, we can use the mean of the variances of these groups to estimate the environmental variance in the entire experiment, which is 5.76. To estimate the genetic variance of the F_2 group, we subtract the estimate of environmental variance from the phenotypic variance of the F_2 group: $V_G = 42.37 - 5.76 = 36.61$. The difference between the two parental means is $D = 93.11 - 40.37 = 52.74$. These values can now be substituted into equation 20.4 to estimate the minimum number of genes, which is $n = 52.74^2 \div 8(36.61) = 9.497$. Variation for at least nine genes is required to produce the observed variation.

(c) If dominance and/or epistasis have a significant effect on the phenotypes, then the means of the F_1 and F_2 generations should deviate from the midpoint of the parental means. Also, the F_2 variation is likely to be skewed in one direction or the other. The midpoint between the parental means is 66.74, which is very close to the means of the F_1 and F_2 groups. Also, the distribution of the F_2 is not highly skewed in either direction. Therefore, the observed results are consistent with the hypothesis that genetic variation for corolla length is determined by additive gene action.

identical to one another and are heterozygous for all alleles that differed between the two parents. Thus, there is no genetic variation within each of the two parental groups or within the F_1 progeny. The F_2 progeny, however, differ from one another genetically. Because there is no genetic variation within each of the P_1, P_2, or F_1 groups, any variation within these three groups must be environmental.

Notice that the variances for the P_1, P_2, and F_1 groups are lower than the variance for the F_2 group because there is no genetic variation in the P_1, P_2,

In most situations, dominance, epistasis, and environmental factors have some role in determining phenotypic variation, so all of the assumptions for equation 20.4 are rarely met. However, even when some or all of the assumptions are violated, experiments to determine the number of genes can still be useful because equation 20.4 estimates the *minimum* number of genes that vary. If dominance or epistasis influences the characteristic under study, equation 20.4 may underestimate, but not overestimate, the number of genes that vary in the experiment.

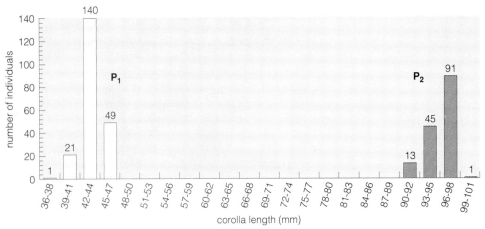

a Parental generation.

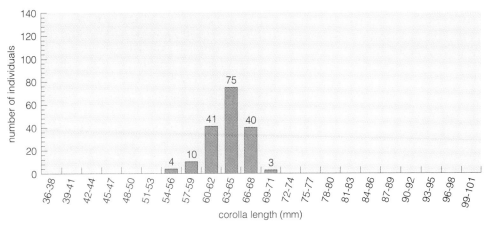

b F₁ generation.

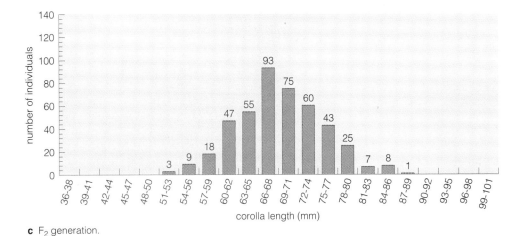

c F₂ generation.

Figure 20.6 Information for Example 20.2: A summary of the results of East's experiments with corolla length in tobacco. (Adapted from East, E. M. 1916. Studies on size inheritance in *Nicotiana*. *Genetics* 1:164–176.)

Under the assumption of additive gene action, two parents that are homozygous for alleles that affect a particular trait should produce F₁ offspring with a phenotype that is intermediate between that of the two parents. However, when the parents are heterozygous, some of the offspring may have phenotypes that exceed the phenotypes of both parents, such as a child who grows taller than both of her or his parents. The child's height may be due to the fortuitous combination of additive effects of several alleles, a phenomenon called **transgressive segregation**. A young woman may be taller than her parents because the combination of height-determining alleles re-

ceived from her parents happened to be the right combination to make her grow tall. She may have a sister who is not as tall as she is because her sister inherited a different combination of alleles from the same parents.

Transgressive segregation is typically observed for traits that show continuous variation. It often appears in the F_2 generation when some F_2 individuals exceed the range of parental phenotypes for a trait. A simple example of transgressive segregation is illustrated in Figure 20.7 using wheat kernel color, an extension of the example we used in Figures 20.3 and 20.4. In the cross A^2A^2 $B^1B^1 C^1C^1 \times A^1A^1 B^2B^2 C^2C^2$, neither parent's phenotype reaches the possible phenotypic extremes, even though both parents are homozygous at all three loci. One parent has four color-producing alleles that give the kernels a darker red color than the other parent, which has two color-producing alleles. The parents are homozygous at all three loci, so each breeds true for its phenotype when self-fertilized.

When the two parents are hybridized, the F_1 progeny are heterozygous at all three loci ($A^1A^2 B^1B^2 C^1C^2$) and have three doses of color-producing alleles. They have a kernel color that is between the two parental phenotypes. In the F_2 generation, $1/64$ of the kernels have six color-producing alleles, and $3/32$ have five color-producing alleles. Their phenotypes are darker than both parental phenotypes. On the other extreme, $1/64$ of the F_2 kernels have no color-producing alleles, and $3/32$ have only one color-producing allele. Their phenotypes are lighter than both parental phenotypes. Thus, about $7/32$ of the F_2 plants are **transgressive segregants**, whose phenotypes exceed the parental range of phenotypes at both extremes.

Plant and animal breeders rely on transgressive segregation for genetic improvement. Suppose a wheat breeder's objective is to produce a true-breeding variety of wheat with a high grain yield. The breeder may hybridize two high-yielding, true-breeding varieties and allow the progeny to self-pollinate through the F_6 generation to obtain homozygous, true-breeding lines. The breeder's hope is that a few of the lines will have a grain yield that exceeds the grain yields of both parental varieties because of transgressive segregation.

~

Additive gene action is the influence of alleles with no dominance or epistasis. Progeny means may sometimes exceed parental means because of transgressive segregation.

20.3 HETEROSIS AND INBREEDING DEPRESSION

In nature, most characteristics are influenced by alleles that have some degree of dominance associated with them. When dominance is present, there is the potential for two related phenomena called heterosis and inbreeding depression. In this section, we'll examine heterosis first, then inbreeding depression.

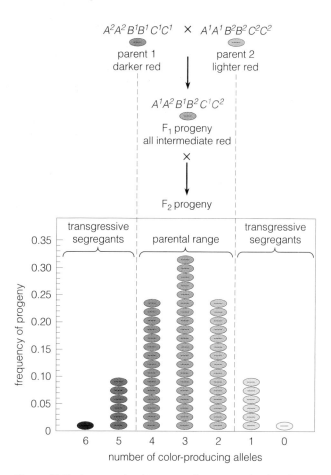

Figure 20.7 An example of transgressive segregation for wheat kernel color.

Heterosis

In some cases, the phenotype of a heterozygote may exceed the phenotypes of its two homozygous parents. For example, when two homozygous maize plants are hybridized, the F_1 plants are often much taller than their taller parent. They are also more vigorous in their growth and produce more biomass and seed than either parent. This phenomenon of increased vigor in heterozygotes is called **heterosis** or **hybrid vigor**.

Heterosis may be caused by **overdominance**, the interaction of two different alleles at a single locus that causes the heterozygote's phenotype to exceed the phenotypes of both parents. For example, there are several alleles at the R locus in maize that control kernel color. In certain combinations, the heterozygote has more pigment in the kernel than either homozygote. There are a few other isolated examples of overdominance.

Overdominance, however, is rare. Most cases of heterosis are due to heterozygosity for dominant and recessive alleles at many loci. Recall that transgressive segregation causes some progeny to exceed both parents in phenotype because of a fortuitous combination of alleles with additive gene action. Heterosis, on the other hand, is a direct result of heterozygosity for alleles with a high degree of dominance.

For example, assume that there are six loci that govern a quantitative trait and that alleles at the loci are fully dominant or recessive. Also assume that each dominant allele contributes to the expression of the trait, whereas the recessive alleles do not. A fully homozygous individual with the genotype *AA BB CC dd ee ff* carries dominant alleles at three of the six loci. Another homozygous individual with the genotype *aa bb cc DD EE FF* also carries dominant alleles at three loci, but the loci with dominant alleles differ from those in the first individual. If these two individuals hybridize, the F_1 progeny will have dominant alleles at all six loci, all in the heterozygous condition: *Aa Bb Cc Dd Ee Ff*. The fully heterozygous genotype *Aa Bb Cc Dd Ee Ff* is phenotypically equivalent to *AA BB CC DD EE FF* because each allele represented by an uppercase letter has complete dominance over its recessive counterpart.

Because the alleles are dominant, the F_1 progeny exceed the phenotypic range of both parents. Unlike the effect of transgressive segregation, the effect of heterosis is lost in subsequent generations of inbreeding as the level of heterozygosity decreases. The increased productivity of commercial hybrid varieties of agricultural plants and animals is a result of heterosis. Hybrid varieties are usually highly heterozygous F_1 progeny of two inbred homozygous parents. Farmers obtain high yields when they plant F_1 hybrid seed of certain crops such as maize. However, if the farmer collects seed from the hybrid plants to plant the next year, the plants that grow from the seed are F_2 plants and lose one-half of the heterozygosity present in their F_1 parents. The loss of heterosis in F_2 plants is so high that farmers must purchase new F_1 hybrid seed each year to maintain high yields, a profitable situation for companies that produce hybrid seed.

Inbreeding Depression

As we discussed in the previous chapter, inbreeding is the mating of individuals with common ancestry. The ultimate result of all forms of inbreeding is increased homozygosity for all alleles.

Most mutations that alter the product of a gene cause a recessive allele that is deleterious when homozygous. If a particular recessive allele has only a minor effect when homozygous, then selection is not very effective in removing the recessive allele from the population. Likewise, if the frequency of the recessive allele is low, selection is not effective because most individuals that have the recessive allele are heterozygous. As long as there is nothing to increase the frequency of homozygotes (such as drift, selection that favors the recessive alleles, or inbreeding), deleterious recessive alleles remain in the population and are shuffled among individuals with each generation. Thus, species that rely on cross-fertilization (including humans) often carry numerous deleterious recessive alleles at low frequencies. The proportion of deleterious recessive alleles in a population or an individual is called the **genetic load** of the population or individual.

Many recessive alleles have minor effects when homozygous, so any individual who is homozygous for only one or a few recessive alleles suffers little from their effects. Inbreeding increases the probability that recessive alleles will be together in the homozygous condition. When larger-than-usual numbers of deleterious recessive alleles are in the homozygous condition in an individual, they are expressed as a deleterious phenotype, a phenomenon called **inbreeding depression**. For example, humans who are highly inbred tend to be shorter than average and tend to have lower IQ scores. Also, recessive lethal alleles and recessive genetic disorders tend to be expressed more frequently in inbred individuals.

The susceptibility to inbreeding depression varies from one species to another and from one population to another, and is dependent on genetic load. The greater the genetic load, the greater the susceptibility to inbreeding depression. Some species are highly susceptible to inbreeding depression. Alfalfa, for example, suffers so severely from inbreeding depression that by the eighth generation of self-pollination, most plants are so weak that they die before reproducing.

On the other hand, many plant species rely entirely on self-pollination in nature, including a number of domesticated species such as peas, wheat, barley, soybean, and tomato. Why don't they suffer from inbreeding depression? In reality they do, but the levels of inbreeding depression are minor because the genetic load in these species is small. They self-pollinate in nature, so most individuals are homozygous at every locus. Thus, deleterious recessive alleles are expressed in the phenotype and can be removed efficiently by selection even when they are at low frequencies. Because the genetic load is low, there is usually little heterosis in a self-fertilized species.

Cross-fertilized species, on the other hand, may have large genetic loads and a high potential for heterosis. Heterosis has been studied and exploited commercially in maize more than in any other species; these studies provide a good example of how heterosis is used commercially. In a population of maize in which the plants mate at random, each individual suffers slightly from inbreeding depression because a few deleterious recessive alleles are homozygous in each individual. Open-pollinated populations are generally in Hardy-Weinberg equilibrium for each locus, so the frequency of homozygotes for each allele is the square of the allele frequency in the population (q^2). If the frequency of heterozygosity could be increased through nonrandom mating, some of the inbreeding depression present naturally in the population could be eliminated and the productivity of the plants could be increased.

Plant breeders exploit heterosis in maize by selecting individual plants from an open-pollinated population and self-pollinating them at least to the F_5 generation to achieve nearly complete homozygosity. With repeated

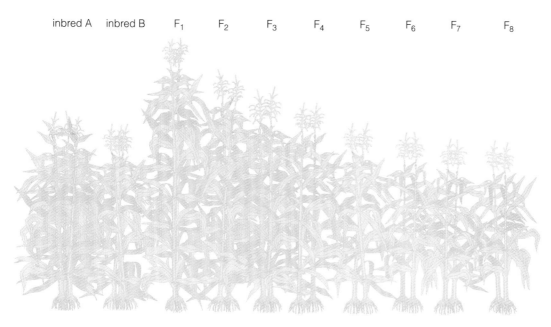

inbred A   inbred B F_1 F_2 F_3 F_4 F_5 F_6 F_7 F_8

Figure 20.8 Heterosis and inbreeding depression in maize. The two plants on the left (inbred A and inbred B) are two fully homozygous inbred lines. The third plant is from the F_1 progeny of the two inbred lines. From left to right, the remaining plants are from successive generations of self-pollination. (Adapted from Jones, D. F. 1924. The attainment of homozygosity in inbred strains of maize. *Genetics* 9:405–418.)

self-fertilization, inbreeding depression is readily observed in the plant's phenotype, as illustrated in Figure 20.8. After the F_5 generation, the increase in homozygosity is small with each generation of self-pollination, so inbreeding depression tends to level off in subsequent generations of self-fertilization, as is apparent in the F_5–F_8 generations in Figure 20.8. The self-fertilized progeny of a single highly inbred plant are collectively called an **inbred line**. Each plant within an inbred line is homozygous, true-breeding, and genetically identical to all the other plants in the line. The self-fertilized offspring are also genetically identical to their parents. This means that inbred lines can be maintained as a genetically stable, true-breeding genotype for an unlimited number of generations through self-pollination.

When two inbred lines are crossed with each other, their F_1 progeny are heterozygous at every locus for which the two parents differ. If two inbred lines are homozygous for different deleterious recessive alleles, then many of the recessive alleles that contribute to inbreeding depression are masked by dominant alleles in the F_1 progeny, and heterosis causes a dramatic increase in the plant's vigor and grain yield.

~

Inbreeding depression is the expression of deleterious recessive alleles in the homozygous condition that arises because of inbreeding. Heterosis is the genetic opposite of inbreeding depression and appears in individuals that are highly heterozygous such that dominant alleles mask deleterious recessive alleles.

20.4 ENVIRONMENTAL VARIATION

Environmental variation is the proportion of the phenotypic variation that is caused by environmental (nongenetic) factors. Most environmental factors are easy to understand. A field in which plants are growing may be more wet on one end than on the other, so the plants on the wet end receive more water than those on the dry end. The plants on the dry end may not be able to reach their full genetic potential for growth due to a lack of water, whereas those on the wet end may have adequate water at all times.

Causes of Environmental Variation

The causes of environmental variation may not always be so obvious. For example, a seed that has a high genetic potential for growth may come from a flower that develops late on the plant, so the seed may not be as large as other seeds that developed earlier. When this smaller seed germinates, its resources for seedling growth are not as abundant as those of other seedlings, so the seedling is off to a poor start compared to other seedlings. It may never overcome this poor start and may be out competed by the more vigorous seedlings in spite of its inherent genetic superiority.

The maternal environment of mammals can have a similar effect. A malnourished or ill female who is pregnant may not have a healthy gestation environment for her offspring. This environment may permanently

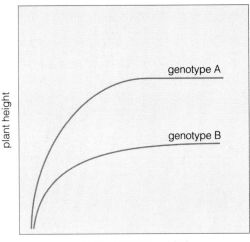

Figure 20.9 Norms of reaction for two genotypes for the response of plant height to increased water application. Over a range of environments, the two genotypes respond differently.

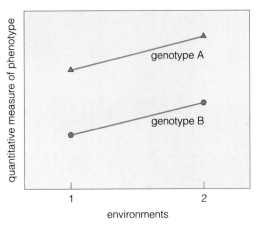

a No genotype-by-environment interaction. The norms of reaction for the two genotypes are additive, and the lines are parallel.

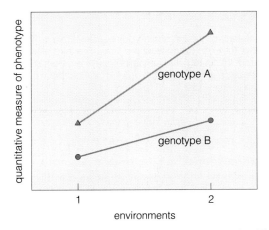

b One type of genotype-by-environment interaction. The genotypes retain their same rankings in both environments, but the response to environments is not additive and the lines are not parallel.

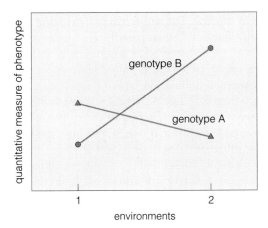

c Another type of genotype-by-environment interaction. The rankings of the genotypes change from one environment to the other, and the lines are not parallel.

Figure 20.10 Comparison of no genotype-by-environment interaction with two types of genotype-by-environment interaction.

damage the fetus before it is born regardless of its genetic potential.

Environmental factors can sometimes have predictable effects on phenotypic variation. Usually these take the form of a correlation between the environmental effect and the phenotype. For example, caloric intake is positively correlated with body weight in mammals. Likewise, fertilizer application is positively correlated with biomass production in plants. The same genotype may express a variety of phenotypes depending on the environment in which it lives. The genotype defines the potential range of phenotypes that are possible under a range of environmental conditions.

Let's use Mendel's experiments for plant height in pea plants as an example. The *Le* allele is dominant and causes plants to grow tall, and the *le* allele causes a dwarf plant when homozygous. A pea plant with the genotype *LeLe* may grow about 2 meters tall if it has enough water, is grown with plenty of light, is kept at the right temperature, and grows in fertile soil. However, if deprived of water during its growth, the plant cannot reach its full 2 meter potential height. How tall the plant grows depends on how much water it receives. The growth is positively correlated with water application until the amount of water allows the plant to reach its full genetic potential. The height of a pea plant with the genotype *le le* is also correlated with the amount of water it receives. However, when it receives adequate water, its maximum height is only about 0.5 meter.

The phenotypic response curve within a range of environments is defined by the genotype and is called the **norm of reaction** or **range of reaction**. As Figure 20.9

illustrates, the norm of reaction for one genotype may differ from the norm of reaction for another genotype.

Genotype-by-Environment Interaction

The norms of reactions for two genotypes may be additive or nonadditive for a quantitative trait. They are additive when the phenotypic response to environmental factors is additive for different genotypes across different environments. When the responses for two (or more) genotypes are graphed over different environments, additivity for norms of reaction produces parallel lines (Figure 20.10a). Even though the phenotypes vary between environments, the responses of the two genotypes to the different environments are additive. When the responses are additive, there is no interaction between environments and genotypes.

A genotype-by-environment interaction is a nonadditive response by two or more genotypes to different environments. It appears graphically as norms of reaction that are not parallel (Figure 20.10b and c). Genotype-by-environment interaction is one of the most important concepts in plant and animal breeding. When there is no genotype-by-environment interaction, the phenotypes vary in an additive way in response to variations in environment. A genotype that is superior in one environment is superior in all environments. However, when there is a genotype-by-environment interaction, conclusions about genotype performance apply only to specific environments and often cannot be applied generally. When a genotype-by-environment interaction is present, geneticists usually attempt to determine the causes of the interaction so they can make accurate predictions of genotype performance under a variety of environmental conditions.

The following example shows how genotype-by-environment interaction is detected and interpreted.

Example 20.3   Genotype-by-environment interaction in soybeans.

In the upper Midwest of the United States, soybeans are often affected by iron-deficiency chlorosis, a yellowing of the plant caused by the plant's inability to extract sufficient iron from soils that are calcareous (high in calcium carbonate). Soybean growers whose farms have calcareous soil can prevent yellowing by selecting a soybean variety that is genetically resistant to iron-deficiency chlorosis. There is substantial genetic variation for iron-deficiency chlorosis among soybean varieties. Soils from different locations also differ in their ability to produce iron-deficiency chlorosis. In 1987, Fairbanks et al. (*Crop Science* 27:953–957) presented data on the degree of iron-deficiency chlorosis observed among soybean varieties grown in different calcareous soils. Some of their results are summarized in Table 20.3.

Problem: (a) Graph these data with location of the soil source on the x axis and degree of chlorosis on the y axis. (b) What evidence is there in the graph of a genotype-by-environment interaction? (c) Interpret the genotype by environment interaction with recommendations regarding the best genotype to use to avoid iron-deficiency chlorosis in each environment.

Solution: (a) Figure 20.11 shows a graph of the data. (b) The lines in the graph are not parallel, so there is a genotype-by-environment interaction. (c) The degree of chlorosis is highest in the Bechyn soil and then in the Hanska soil when all varieties are considered. Although there is a significant genotype-by-environment interaction, varieties A7 and Dawson are consistently the least affected by iron-deficiency chlorosis in all soils, so they are the best varieties to use to avoid iron-deficiency chlorosis.

Table 20.3   Information for Example 20.3: Mean Chlorosis Ratings for Six Soybean Varieties Grown in Soil from Four Locations in Minnesota*

Variety	Location			
	Lamberton	Willmar	Hanska	Bechyn
A7	1.1	1.1	1.4	2.2
Corsoy 79	2.8	3.3	4.0	4.8
Dawson	1.2	1.2	2.1	3.3
Hodgson 78	1.3	2.4	2.8	4.7
Pride B-216	4.4	3.9	5.0	5.0
Simpson	2.9	1.6	3.3	4.7

The chlorosis rating denotes the degree of chlorosis: 1 = no chlorosis, and 5 = severe chlorosis.

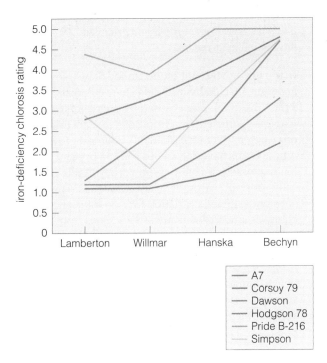

A7
Corsoy 79
Dawson
Hodgson 78
Pride B-216
Simpson

Figure 20.11 Information for Example 20.3: Iron-deficiency chlorosis ratings for different soybean genotypes grown in soils from different locations.

Genotype-by-environment interaction is a nonadditive response of phenotypes to varying environmental conditions.

20.5 BROAD-SENSE HERITABILITY

Geneticists usually analyze quantitative inheritance by analyzing variation and determining what proportion of the phenotypic variation is genetic. Variation is usually measured in terms of variance. The **phenotypic variance** is the variance that can be measured directly in a population. It can be partitioned into **components of variance**, which represent the genetic variance and the environmental variance, represented mathematically as

$$V_P = V_G + V_E \qquad [20.5]$$

where V_P is the phenotypic variance, V_G is the genetic variance, and V_E is the environmental variance.

There are two general types of heritability: broad-sense heritability and narrow-sense heritability. We'll discuss broad-sense heritability first and then explain narrow-sense heritability in the next section.

Mathematically, **broad-sense heritability** is expressed as the proportion of the phenotypic variance attributed to genetic variance:

$$H^2 = \frac{V_G}{V_P} \qquad [20.6]$$

or, substituting $V_G + V_E$ for V_P, we get

$$H^2 = \frac{V_G}{V_G + V_E} \qquad [20.7]$$

where H^2 represents broad-sense heritability.

Phenotypic variance is always greater than or equal to genotypic variance, so heritability is expressed as a fraction between zero and one. When $H^2 = 1$, all the phenotypic variation is due to genetic variance ($V_G = 1$) and there is no environmental variation ($V_E = 0$). When $H^2 = 0$, all the variation is due to environmental variance ($V_E = 1$) and there is no genetic variation ($V_G = 0$).

The term H^2 represents broad-sense heritability directly; there is no need to take the square root. The *square* designation in H^2 indicates that heritability is determined from variances, which are determined from squared deviations.

In plants and animals grown in controlled environments, broad-sense heritability can often be estimated from experiments like those illustrated in the following example.

Example 20.4 Estimating broad-sense heritability for mineral uptake in wheat.

In 1964, Rasmusson et al. (*Crop Science* 4:586–589) reported broad-sense heritabilities for strontium uptake in wheat and barley. In one wheat cross, two true-breeding, fully homozygous varieties of wheat (Kenya 117A and Kentana 52) were hybridized to obtain F$_1$ progeny. The F$_1$ progeny were allowed to self-fertilize to produce an F$_2$ generation. Strontium uptake was measured as disintegrations of radioactive strontium per second per gram of dry matter. All plants were grown under the same environmental conditions. The means and variances for strontium concentration in the parents and progeny are provided in Table 20.4.

Problem: Estimate the broad-sense heritability.

Solution: Three of the four groups listed in Table 20.4 consist of genetically identical plants. The two parents are true-breeding, homozygous varieties, so each plant is genetically identical to every other plant within the variety and all are homozygous for alleles at all loci. The F$_1$ plants are, therefore, genetically identical to one another and heterozygous for alleles that differ between the two parental varieties. There is no genetic variation within these three groups, so the phenotypic variances for each of these three groups must be equal to the environmental variances. In other words, if $V_G = 0$, then $V_E = V_P$. We can estimate the average environmental deviation for the entire experiment by taking the mean of the variances for the three groups that have no genetic variation: $V_E = (334 + 234 + 397) \div 3 =$

Table 20.4	Information for Example 20.4: Strontium Uptake in Wheat		
Group	Number of Plants	Mean Strontium Concentration*	Variance
Kenya 117A	28	161	334
Kentana 52	29	110	234
F_1	26	159	397
F_2	81	158	1083

*Measured as disintegrations of radioactive strontium per second per gram of dry matter.

321.67. This value is an estimate of V_E in the F_2 group because the F_2 plants were grown under the same environmental conditions. The phenotypic variance for the F_2 plants is the sum of the genetic variance and the environmental variance, according to equation 20.5. If we rearrange equation 20.5 to solve for the genetic variance, $V_G = V_P - V_E$, then for the F_2 group we get $V_G = 1083 - 321.67 = 761.33$. Using equation 20.6, we calculate the broad-sense heritability as $H^2 = 761.33 \div 1083 = 0.70$, which means that 70% of the phenotypic variance is genetic variance and 30% is environmental variance.

Broad-sense heritability can be easily calculated in this way in plants and animals that are selectively bred for the experiment. In cases where selective breeding is not an option, such as studies with humans and populations in nature, other methods must be applied to estimate broad-sense heritability. The best way to estimate environmental variance is to compare genetically identical individuals. In humans, identical twins are genetically identical, so any phenotypic differences must represent environmental variation. Numerous studies have compared identical twins with fraternal twins, genetic siblings with adopted siblings, or related people with unrelated people to estimate broad-sense heritability in humans. The major premise in such studies is that there is less genetic variation between related individuals than unrelated individuals, so measurements of variation can provide estimates of V_G and V_E and, therefore, heritability.

The meaning of heritability is often misinterpreted, especially in relation to human studies. For example, some studies on heritability in humans have found higher heritabilities among Americans of European descent than those of African descent for IQ (intelligence quotient). Questions about the validity of such studies and their interpretations have generated much controversy. Some people have erroneously concluded from such studies that people of African descent are genetically less intelligent than people of European descent.

However, heritability estimates refer to *variation* within a population rather than the magnitude of a quantitative trait. If heritability is lower in one group than in another for IQ scores, then a likely reason for the lower heritability is higher environmental variation in one group when compared to the other. For example, less uniform educational opportunities for one group than for another should cause differences in heritabilities for IQ. Different broad-sense heritability estimates for IQ among ethnic groups in the United States are probably influenced more by discrepancies for educational opportunities than differences in genetic variation for intelligence.

~

Broad-sense heritability is the proportion of the phenotypic variation that is caused by genetic variation.

20.6 NARROW-SENSE HERITABILITY AND SELECTION

Narrow-sense heritability predicts the effectiveness of selection. To define and understand narrow-sense heritability, we first need to discuss the factors that influence the effectiveness of selection. The most straightforward situation in quantitative genetics is complete additivity for genes with no dominance among alleles, no epistatic interactions among genes, and no environmental variation. Under these ideal circumstances, each genotype can be distinguished by its phenotype. Because selection (both natural and artificial) acts on the phenotype, selection is most effective when every genotype can be identified phenotypically. Dominance, epistasis, and environmental variation confound the correlation between genotype and phenotype and reduce the effectiveness of selection. Let's examine the effect of dominance first.

The Effect of Dominance on Selection

Suppose there is a single locus with two alleles, A_1 and A_2, and that each allele contributes differently to the expression of a trait in the phenotype. If there is no dominance, the phenotype of A_1A_2 is halfway between the phenotype of the two homozygotes, A_1A_1 and A_2A_2, as illustrated in Figure 20.12a. If there is partial dominance, then the heterozygote A_1A_2 resembles one of the homozygotes more than the other homozygote, as illustrated in Figure 20.12b. If the A_1 allele is completely dominant, then A_1A_2 is phenotypically identical to A_1A_1, as illustrated in Figure 20.12c. If selection favors the A_1 allele, then dominance of the A_1 allele reduces the ability of selection to remove the A_2 allele from the population because the A_2 allele remains hidden in heterozygotes.

Let's return to Mendel's pea plants for a simple example. The *Le* allele is completely dominant over *le* and causes a tall plant when homozygous or heterozygous. The recessive *le* allele causes a dwarf plant when homozygous. *Le* corresponds to the A_1 allele in Figure

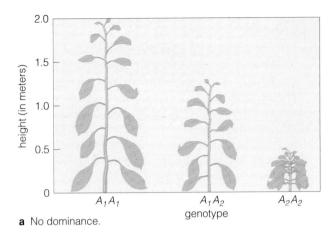

a No dominance.

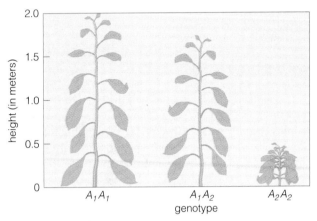

b Partial dominance.

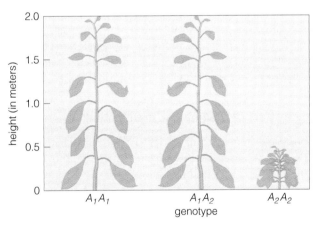

c Complete dominance.

Figure 20.12 Comparison of **(a)** no dominance, **(b)** partial dominance, and **(c)** complete dominance.

20.12c because it contributes more to plant height than *le*, which corresponds to the A_2 allele. *Le* is completely dominant, so the heterozygote *Lele* is phenotypically identical to the homozygote *LeLe*. Selection that favors tall plants preserves both the *LeLe* and *Lele* genotypes, and therefore fails to remove the *le* allele entirely from the population because the *le* allele persists in heterozygotes.

Had there been no dominance, the *Lele* heterozygotes could be distinguished from the *LeLe* homozygotes and selection could have effectively removed the *le* allele from the population.

If we extend this situation to multiple genes that affect the same trait, we find that any degree of dominance tends to mask recessive alleles, reduces the ability of selection to remove the recessive alleles, and allows them to persist in the population. In the context of quantitative genetics, **dominance** is defined as a nonadditive interaction of alleles at the same locus, and it reduces the effectiveness of selection.

The Effect of Epistasis on Selection

Just as dominance is a nonadditive interaction of alleles at the same locus, **epistasis** is a nonadditive interaction of alleles at different loci. Let's use an example of epistasis we discussed in Chapter 13 to illustrate this concept. In sweet pea, two loci called *p* and *c* interact in an epistatic way to govern flower color. If there is at least one dominant allele at both loci (*P_ C_*), the flowers are colored. If an individual is homozygous for a recessive allele at either locus (*P_ cc*, *pp C_*, or *pp cc*), the flowers are white. Thus, the interaction of the *p* and *c* loci is nonadditive. Had the effects been additive, then the phenotypes for *PP cc* and *pp CC* should have been halfway between the phenotypes of *PP CC* and *pp cc*. Instead, the phenotypes of *PP cc* and *pp CC* are equal to the phenotype of *pp cc* because all three genotypes produce white flowers. The effects of alleles at the two loci are not additive, so there is an epistatic interaction between the two loci.

Like dominance, epistasis reduces the effectiveness of selection. Using the flower color example we just discussed, suppose that selection favors plants with white flowers and disfavors plants with colored flowers. Selection for white flowers favors the *p* and *c* alleles and disfavors the *P* and *C* alleles. Plants with the genotypes *P_ cc* and *pp C_* have white flowers and are favored by selection. The *P* and *C* alleles persist in the population because selection fails to remove the *P_ cc* and *pp C_* genotypes. Had the interaction between the two loci been additive, then only plants with the genotype *pp cc* would have had white flowers, and selection would have eliminated the *P* and *C* alleles in a single generation.

The Effect of Environmental Variation on Selection

Like dominance and epistasis, environmental variation reduces the effectiveness of selection. Let's use pesticide resistance in insects as an example. As we discussed in Chapter 19, repeated use of certain pesticides, such as DDT, has caused directional selection to favor insects that are genetically resistant to the pesticides. DDT resistance has been studied in several insect species, and its inheri-

tance is usually polygenic. When an entire population of insects is exposed to uniform concentrations of DDT, those insects that are most resistant survive, reproduce, and transmit the alleles that confer resistance to their offspring. Alleles that confer susceptibility are removed from the population, so after several generations the entire population becomes resistant. The fact that all the insects in the population receive an equal dose of DDT ensures that all susceptible insects die before reproducing.

However, suppose that only a portion of the population is treated, as is often the case when pesticides are applied commercially. Only those susceptible insects that are treated fail to reproduce. Untreated insects that are susceptible survive, reproduce, and transmit their alleles that confer susceptibility to offspring. Selection reduces the frequencies of the alleles that cause susceptibility, but it fails to remove them entirely from the population because they persist in untreated individuals. Environmental variation for pesticide treatment thus reduces the effectiveness of selection.

Narrow-Sense Heritability

Having reviewed the effects of dominance, epistasis, and environmental variation on selection, we are prepared to discuss narrow-sense heritability. The genetic variance (V_G) in a population can be partitioned into the variance for additive effects of genes, the variance for dominance, and the variance for epistatic interactions:

$$V_G = V_A + V_D + V_I \qquad [20.8]$$

where V_A is the variance for additive effects, V_D is the variance for dominance, and V_I is the variance for epistatic interactions.

An allele may contribute variation to all three components. Every allele that influences a trait has an additive component, and variation for all alleles that influence a trait contributes to V_A. Alleles that vary only for additive effects (alleles that have no dominance or epistatic interactions) contribute only to V_A. Alleles that have any degree of dominance contribute to both V_A and V_D, and alleles that interact epistatically with alleles at other loci contribute to both V_A and V_I. The combined effects of all alleles that contribute to the variation of a particular trait in a population determine the actual values of V_A, V_D, and V_I in a population.

Narrow-sense heritability is defined as the proportion of the phenotypic variation that is caused by variation for the additive effects of genes and is expressed mathematically as

$$h^2 = \frac{V_A}{V_P} \qquad [20.9]$$

where h^2 is the narrow-sense heritability. If we partition the denominator of equation 20.9 into its components

of variance, then narrow-sense heritability is

$$h^2 = \frac{V_A}{V_A + V_D + V_I + V_E} \qquad [20.10]$$

Recall from our previous discussion that the efficiency of selection is reduced by variation for dominance, epistasis, and environmental influences. The terms that represent variation for these factors are V_D, V_I, and V_E, all of which appear only in the denominator.

Because narrow-sense heritability is based on variation for the additive components of genes, it provides the most accurate prediction of the effectiveness of selection. In order for selection to operate, there must be genetic variation for a trait. However, the only variation on which selection can effectively operate is the variation for the additive effects of genes. All other types of variation expressed in the phenotype, including variation for dominance, epistatic interactions, and environmental influences, decrease the effectiveness of selection. How efficiently selection alters the population in the direction of the favored phenotype depends on the value of narrow-sense heritability. When narrow-sense heritability is high, selection is effective in altering allele frequencies. When narrow-sense heritability is low, selection is ineffective.

Because it predicts the effectiveness of selection, narrow-sense heritability is an important concept in plant and animal breeding, and most examples of narrow-sense heritability come from studies in plant and animal breeding. But narrow-sense heritability is equally important for natural selection. Natural selection has a much greater effect on traits with high narrow-sense heritabilities than on traits with low narrow-sense heritabilities.

The relationship between heritability and selection can be illustrated with some simple diagrams. Suppose a population has a normal distribution for a particular trait (Figure 20.13a). The distribution has its mean (M) near the center of the distribution. A breeder who wishes to improve the population selects only individuals from one side of the distribution to mate and produce progeny for the next generation.

Conceivably, the breeder could take only the two best individuals from the population, based on their phenotypes, mate them, and eliminate all other individuals from the population. However, such a strategy is unwise for several reasons. First, some of the variation is probably environmental, so some of the desired alleles may be present in individuals whose phenotypes are not as good as the two best individuals. Also, if the number of individuals selected as parents for the next generation is small, then drift becomes significant and the population may become inbred. Genetic diversity is reduced, and further selection will have little effect because little genetic variation remains.

In terms of narrow-sense heritability, V_A is reduced to zero when genetic diversity is lost, so h^2 is also reduced

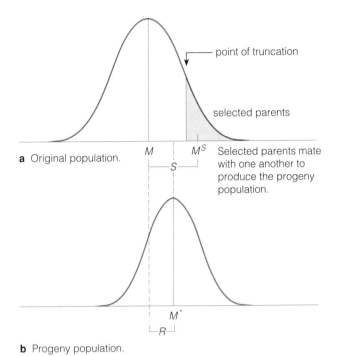

a Original population.

point of truncation

selected parents

M M^S Selected parents mate with one another to produce the progeny population.

S

M^*

R

b Progeny population.

Figure 20.13 Response to selection when $h^2 = 0.5$.

to zero. When $h^2 = 0$, further selection has no effect. For this reason, it is wise to select a relatively large number of individuals from the upper end of the distribution as parents for the next generation. Usually, researchers select a cutoff point called the **point of truncation**. All individuals whose phenotypes exceed the point of truncation are selected as parents for the next generation. All those whose phenotypes are below the point of truncation are not selected as parents for the next generation. Individuals selected as parents form a subpopulation with a mean M^S, which is higher than the overall population mean (M). The difference between the population mean (M) and the mean of the selected individuals (M^S) is called the **selection differential**, or S, as illustrated in Figure 20.13a.

The purpose of selection is to improve the mean phenotype of the progeny when compared to the original population. The efficiency of improvement depends on how much of the selection differential is gained by selection. Narrow-sense heritability provides a numerical estimate of what proportion of the selection differential will be gained by selection. The proportion of the selection differential gained by selection is often called the **response to selection**, or R, which can be represented mathematically as

$$R = h^2 S \qquad \text{[20.11]}$$

Figure 20.13b shows the response to selection (R) when $h^2 = 0.5$—50% of the selection differential (S) is gained by selection, thus creating a new mean for the

progeny population (M^*) that is halfway between M and M^s. The value of h^2 is an important one for plant and animal breeders whose objective is to genetically improve a population as quickly as possible with the most efficient use of resources. If h^2 is high, then improvement is efficient. If h^2 is low, then steps must be taken to improve h^2 so that selection will be efficient.

How can h^2 be improved? The answer comes from equation 20.9. The value of h^2 can be increased by increasing V_A or decreasing V_P. V_A can be increased by choosing parents that are as genetically diverse as possible. Selection of parents with diverse genotypes is routine in breeding programs. V_P can best be decreased by decreasing V_E. This can be done by raising plants or animals under as uniform and ideal environmental conditions as possible. In the case of plants, they should be grown in a field with soil that is uniform in texture, fertility, and water content so that plants are influenced as little as possible by environmental variation. For animals, each animal should be given an ideal diet and fed the same amount, and the animals should be well cared for and protected from adverse weather and overcrowding.

Quantitative Trait Loci and DNA Markers

In recent years, DNA markers have been used to improve narrow-sense heritability through a process called **marker-assisted selection**. DNA markers that have been mapped to their chromosomal locations are useful in tagging genes that contribute to the genotypic variation for a quantitative trait. Researchers determine which markers are present in individuals that have the best phenotypes, and eventually identify DNA marker loci that are linked to genes that govern a quantitative trait. The researchers can then select those individuals that carry favorable alleles by selecting those individuals that carry the DNA markers coupled to those alleles. A gene that influences a quantitative trait is called a **quantitative trait locus (QTL)**. QTLs that have significant phenotypic effects can often be identified by DNA markers, and the alleles can be tracked in progeny.

Variation for DNA markers is not influenced by environmental factors, so $V_E = 0$ for DNA markers. Also, most types of DNA markers are codominant and do not interact with one another, so V_D and V_I are also zero for the DNA markers. Thus, narrow-sense heritability for most types of DNA markers is 1.0, so the markers can be selected with 100% efficiency. Unfortunately, it is not usually possible to find a DNA marker linked to every gene that influences a trait, but it is often possible to find DNA markers linked to many such genes. Once the linkages are established, individuals in a breeding population can be tested for the DNA markers they carry, and the markers can then be used to aid in selection.

DNA markers that are linked to QTLs may also be used to identify the specific genes that function as QTLs.

One of the best examples is identification of alleles that confer susceptibility to alcoholism in humans and other mammals. Susceptibility to alcoholism is determined by both genetic and environmental factors. Broad-sense heritability estimates for susceptibility to alcoholism in humans often exceed 50%, and narrow-sense heritability estimates in several controlled experiments with mice are about 30%. In human and mouse studies, DNA markers that are associated with alcoholism identify chromosomal regions that contain genes that govern alcohol metabolism and brain reward systems. The following example highlights the results of two studies of a potential QTL associated with alcoholism in humans.

Example 20.5 Identification of a potential QTL associated with susceptibility to alcoholism in humans.

In 1990, Blum et al. (*Journal of the American Medical Association* 263:2055–2059) examined DNA from the brains of 35 deceased alcoholics and 35 deceased nonalcoholics for differences at the *DRD2* locus (*dopamine D2 receptor* locus, OMIM 126450). They examined the *DRD2* locus as a potential QTL for susceptibility to alcoholism because it encodes a protein that is part of the central dopaminergic nervous system, a brain reward system that influences addiction. The *A1* allele lacks a *Taq*I restriction endonuclease cleavage site that is found in other alleles, so the *A1* allele can be readily distinguished from other alleles through RFLP analysis. Blum et al. found that 24 of the alcoholics and 7 of the nonalcoholics were homozygous or heterozygous for the *A1* allele. They also found that 28 of the nonalcoholics and 11 of the alcoholics did not have the *A1* allele. Subsequent studies by other researchers, however, failed to find an association between the *A1* allele and alcoholism. In 1997, Kono et al. (*American Journal of Medical Genetics* 74:179–182) examined DNA from 100 alcoholics and 93 nonalcoholics and did not find a strong association between the *A1* allele and alcoholism. However, they found that among alcoholics there was an additive effect of the *A1* allele on the age of onset of alcoholism. *A1/A1* homozygotes were more susceptible to early onset alcoholism than were heterozygotes, and heterozygotes were more susceptible than people who did not have the *A1* allele.

Problem: (a) What type of trait is alcoholism, and how might it be influenced by environmental variation? (b) What do these results indicate about the association of the *A1* allele with alcoholism?

Solution: (a) Alcoholism is a threshold trait that probably has a polygenic basis. It is also influenced

by environmental variation because different cultural and social factors may influence alcohol consumption. (b) The studies highlighted in this example show that the *DRD2* locus is a potential QTL for alcoholism with an allele that increases susceptibility to early onset of alcoholism.

Realized Heritability

Narrow-sense heritability is often used to predict selection efficiency. Instead, it may be calculated after the results of selection have been observed, in which case it is called **realized heritability**. Realized heritability is calculated by rearranging equation 20.11 to solve for h^2:

$$h^2 = R/S \qquad \text{[20.12]}$$

R is determined by subtracting the mean of the original population (M) from the mean of the progeny population (M^*):

$$R = M^* - M \qquad \text{[20.13]}$$

Figure 20.13 illustrates the relationships among M, M^*, and R.

Of what value is realized heritability if selection has already been done? Realized heritability is easy to calculate whenever selection is practiced. When realized heritabilities have been determined in many populations and environments, it is often possible to determine how generally heritable a trait is. Heritability is dependent on both genetic and environmental variation, so a particular estimate of heritability applies to only one population in one environment. However, heritabilities for a particular trait are often similar across populations and environments. Some traits tend to be highly influenced by the environment, whereas others are not. A trait that is sensitive to environmental variation generally does not have high heritabilities except in a uniform environment. A trait that is not highly influenced by environmental variation has high heritabilities regardless of the environment.

For example, amino acid composition of seed proteins is an important trait for both human and animal nutrition. The amino acid composition of maize protein is not highly influenced by environment, so its heritabilities tend to be high no matter where the maize is grown. Grain yield of maize, however, is highly influenced by environmental variation and by dominance, so its heritabilities are generally low under most conditions.

~

Narrow-sense heritability is the proportion of the phenotypic variation attributed to variation for the additive effects of genes. Narrow-sense heritability predicts the efficiency of selection.

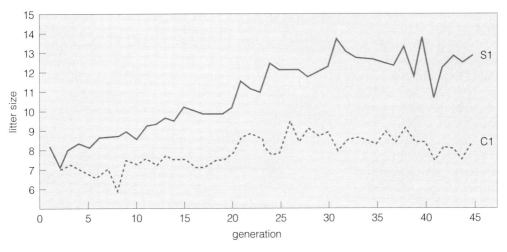

Figure 20.14 A selection plateau reached for upward selection for litter size in mice. S1 is the population subjected to selection for large litter size, and C1 is an unselected control population. Selection was effective until generation 31, when it reached a plateau. (Adapted from Eklund, J., and G. E. Bradford. 1977. Genetic analysis of a strain of mice plateaued for litter size. *Genetics* 85:529–542. Copyright Genetics Society of America. Used by permission.)

20.7 LIMITS ON SELECTION

Continued selection alters allele frequencies by increasing the frequencies of favored alleles. As selection proceeds, it eventually eliminates disfavored alleles as favored alleles become fixed in the population. Because it eliminates alleles, selection reduces genetic variation. As more alleles become fixed, the amount of genetic variation upon which selection can operate decreases, so selection becomes less effective. After a period of selection, a plateau beyond which selection is no longer effective should be reached.

Indeed, plateaus have been reached in many selection experiments. In some cases, they are the result of fixation for favored alleles and the exhaustion of genetic variation. However, plateaus may be reached for other reasons. Selection for large litter size in mice is a good example. Each female mouse is physically capable of carrying only a certain number of fetuses without causing physical harm to herself and the fetuses. When litter size approaches the physiological limits of the female, a second selection force operates to reduce litter size. With selection for high litter size opposed by a physiological limit to litter size, stabilizing selection causes an equilibrium for litter size. A plateau is reached even though genetic variation for litter size may still be present (Figure 20.14).

Selection in the other direction, for low litter size, has a numerical limit of zero—it is impossible to have a litter size of less than zero individuals. In fact, a litter size of zero automatically halts selection because there are no offspring to form the next generation. Even though there may still be variation for genes that govern litter size, the numerical limit prevents selection from continuing further.

True genetic plateaus, in which genetic variation has been exhausted, can be distinguished from physiological plateaus through reverse selection experiments. **Reverse selection** is a reversal of selection pressure to favor the trait that was previously disfavored. If genetic variation has been exhausted and alleles favored by selection are fixed in a homozygous state in the population, then reverse selection should be ineffective. However, if genetic variation remains, reverse selection should be successful in causing the mean phenotype of the population to change in the direction of the new selection pressure.

The effect of reverse selection is exemplified by long-term selection experiments for high and low oil content in maize kernels, as illustrated in Figure 20.15. These experiments have been under way since 1896. At generation 49, researchers began reverse selection for low oil content in a subset of the population that had been selected for high oil content. They also began reverse selection for high oil content in a subset of the population that had been selected for low oil content. Reverse selection was successful in both populations, indicating that genetic diversity had not been exhausted in either population, even after 49 generations of selection.

The following example shows how a physiological plateau may prevent continued progress from selection.

> **Example 20.6 Selection plateaus and reverse selection.**
>
> In 1977, Eklund and Bradford (*Genetics* 85:529–542) described a selection plateau in a population of mice selected for high litter size. Their results for 45 generations of selection are illustrated in Figure 20.14. In the first generations, the mean litter size

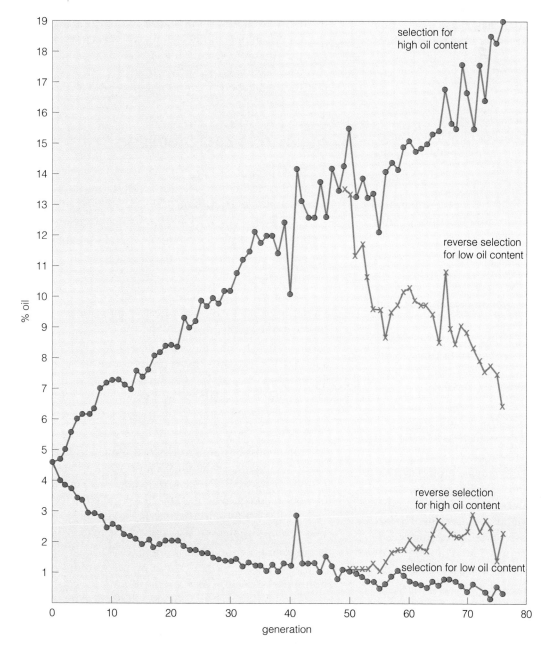

Figure 20.15 Selection experiments in maize for oil content in kernels at the University of Illinois. Selection was started in 1896 in an open-pollinated population, and no new genes have been introduced into the selected populations since that time. The graph shows 76 generations of selection. (Adapted from Dudley, J. W. 1977. 76 generations of selection for oil and protein percentage in maize. In Pollack, E., O. Kempthorne, and T. B. Bailey, Jr., eds. *Proceedings of the International Conference on Quantitative Genetics.* Ames, Ia.: Iowa State University Press, pp. 459–489.)

was about 8. By generation 31, the mean litter size was 13.6 and failed to increase thereafter. In generation 34, reverse selection for small litter size was initiated on a subgroup of the plateaued population, while selection for large litter size was continued with the remaining mice in the plateaued population. By generation 45, the mean litter size of the population selected for large litter size had not increased since generation 31. However, the mean litter size decreased to 10.3 by generation 45 in the population in which reverse selection for small litter size had been implemented.

Problem: What do these results indicate about genetic variation and the reason for the plateau?

Solution: Reverse selection was successful in reducing litter size in the plateaued population, so genetic variation had not been exhausted. Instead, a physiological limit on selection had been reached.

Exhaustion of genetic variation may cause a selection plateau, or physiological limits may prevent selection from exploiting genetic variation.

SUMMARY

1. Many inherited traits may be described in qualitative or quantitative terms. Quantitative descriptions use units of measurement to describe a trait.

2. Quantitative traits are often influenced by several genes acting together, both with each other and with environmental factors, to give the final phenotype.

3. Quantitative traits may be classified as continuous or meristic. Continuous traits vary from one extreme to another without discrete classes. Meristic traits are measured by counting rather than applying incremental units of measurement.

4. Researchers usually analyze quantitative inheritance with summary statistics rather than attempting to determine the individual effects of genes. Common statistical values are the mean, mode, and variance.

5. Additive gene action is the most straightforward situation in quantitative genetics. When additive gene action is present, the number of genes that vary for the trait in a population can be estimated.

6. Deleterious recessive alleles may be protected from selection in the heterozygous condition. Inbreeding causes an increase in homozygosity, allowing recessive alleles to be expressed at higher frequencies. The expression of deleterious recessive alleles due to inbreeding is called inbreeding depression. The degree of potential inbreeding depression is directly related to the degree of dominance in a population.

7. Heterosis is the opposite of inbreeding depression. It occurs when deleterious recessive alleles are masked in heterozygous individuals. Hybrid varieties of plants and animals have been developed to exploit the productivity conferred by heterosis.

8. A genotype-by-environment interaction is a nonadditive response of different genotypes to changes in the environment.

9. Heritability is among the most important concepts in quantitative genetics. Broad-sense heritability is the proportion of the phenotypic variation in a population attributed to genetic variation. Narrow-sense heritability is the proportion of the phenotypic variation attributed to variation for the additive effects of genes.

10. Narrow-sense heritability predicts the effectiveness of selection. Narrow-sense heritability identifies the proportion of the selection differential gained by selection.

11. Limits on selection may cause selection to lose its effectiveness. Limits may be due to exhaustion of variation or to environmental limits. The limits can be distinguished from one another by reverse selection experiments. If genetic variation has been exhausted, then reverse selection is ineffective. If reverse selection is effective, then genetic variation has not been exhausted.

QUESTIONS AND PROBLEMS

1. Some people assume that quantitative genetics is the study of genes with equal and additive effects. What is a more accurate description of quantitative genetics?

2. Under what circumstances does narrow-sense heritability equal broad-sense heritability for a population?

3. What is the difference between broad-sense and narrow-sense heritabilities?

4. Define V_P, V_G, V_D, V_I, and V_E, and describe what these values represent. Also describe how they are related to broad- and narrow-sense heritabilities.

5. In light of the previous question, why is it incorrect to say that V_A represents the variance for additive genes, V_D represents the variance for dominant genes, and V_I represents the variance for epistatic genes?

6. Explain why it is impossible for V_A to equal zero when V_D is not zero.

7. (a) Is it possible for narrow-sense heritability in a population to equal zero while broad-sense heritability is greater than zero in the same population? (b) Is it possible for broad-sense heritability in a population to equal zero while narrow-sense heritability is greater than zero in the same population? Explain your answers.

8. In plants, genetic loads tend to be least among self-pollinated species, greater among diploid cross-pollinated species, and greatest among polyploid cross-pollinated species. Provide a theoretical reason for this general observation.

9. Self-pollinated species of plants generally have very little variance for dominance. Cross-pollinated plants generally have relatively high levels of variance for dominance. Based on your answer to the previous question, why is this so?

10. Plant breeders often self-pollinate plants to obtain homozygous inbred lines, then cross the inbred lines to form hybrids, which they then self-pollinate to form new sets of homozygous inbred lines. Given the information in the previous question, what type of variety (inbred or hybrid) would usually be marketed in (a) a self-pollinated species and (b) a cross-pollinated species?

11. Describe two ways that a plant breeder can optimize heritabilities when selecting for high grain yield in maize.

12. In general, plant breeders say that heritabilities for grain yield in wheat are low, but heritabilities for resistance to stem rust (a fungal disease) are high. Herita-

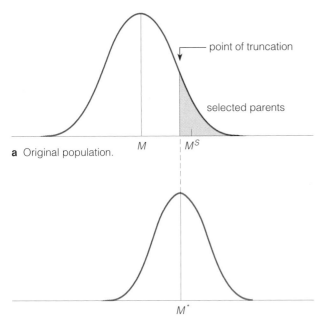

a Original population. M M^S

point of truncation

selected parents

M^*

b Progeny population.   b Progeny population.

Figure 20.16 Information for Problem 14: Response from selection.

Table 20.5	Information for Problem 15: Seeds per Pod in Pea Plants
Number of Seeds per Pod	**Number of Pods**
1	4
2	14
3	21
4	23
5	26
6	17
7	5
8	4

bilities differ from one environment to another, so why is it possible to make such a generalization?

13. Why does the use of DNA markers for marker-assisted selection have the potential to improve heritability?

14. From the diagram in Figure 20.16, determine the narrow-sense heritability. Be sure to provide your calculations.

15. Table 20.5 provides data on the number of seeds per pod in pea plants. **(a)** Is peas per pod a continuous or a meristic trait? **(b)** What are the mean, mode, median, variance, and standard deviation of this distribution? **(c)** How close is this distribution to a normal distribution? **(d)** Is this distribution skewed? If so, in which direction?

16. In Eklund and Bradford's paper on selection for litter size in mice (1977. *Genetics* 85:529–542) used for Figure 20.14 and Example 20.6, the average realized heritability for generations 0 to 31 was 0.16. After generation 31, the average realized heritability was zero, even though selection was still practiced for an additional 14 generations. Does this mean that the value for V_A equaled zero? Use information from Example 20.6 to explain your answer.

17. Using East's data from Example 20.2, determine the broad-sense heritability for corolla length in tobacco.

18. Hayes and Immer (1942. *Methods of Plant Breeding*. New York: McGraw-Hill, p. 340) reported data on the grain yields of five varieties of barley grown at four locations in Minnesota. Table 20.6 provides the mean yields of each variety at each of the four locations for the summer of 1932. **(a)** Graph these data with locations on

the *x* axis and grain yield on the *y* axis, then connect the points for each variety with lines as in Figure 20.11. **(b)** Is there evidence in the graph of a genotype-by-environment interaction? **(c)** What conclusions about the grain yield potential of these varieties can be derived from the data?

19. In 1924, Jones (*Genetics* 9:404–418) reported the data in Table 20.7 on heterosis and inbreeding in maize. **(a)** What evidence is there in this data of heterosis and inbreeding depression? **(b)** Determine the inbreeding coefficient for each generation beginning with the F_1 generation. (For self-fertilization, there is only one path of common ancestry, and the number of individuals in the path is one.) **(c)** For each of the traits, plot the inbreeding coefficient for each generation on the horizontal axis and the corresponding phenotypic values for each of the traits on the vertical axis. **(d)** What relationship is there between the inbreeding coefficient and the expression of each of these traits?

20. In an experiment for upward selection of 6 week weight in mice (Falconer. 1955. *Cold Spring Harbor Symposia on Quantitative Biology* 20:178–196), selection differentials averaged 1.39 g per generation over a period of 22 generations. Falconer determined that over these 22 generations, the realized heritability was fairly constant, averaging 0.175. What was the average gain from selection per generation?

21. In 1957, Robertson (*Journal of Genetics* 55:428–443) reported the variances for thorax length (in $\frac{1}{100}$ mm) and number of eggs laid in genetically uniform inbred lines of *Drosophila melanogaster* and the F_1 progeny of these lines, and in genetically diverse populations of *D. melanogaster* raised under the same conditions. The variance for thorax size was 0.186 in the genetically uniform lines and 0.366 in the genetically variable populations. The variance for number of eggs laid was 16.6 in the genetically uniform lines and 43.4 in the genetically variable populations.

Table 20.6 Information for Problem 18: Grain Yields (in bushels per acre) of Barley Varieties Grown at Four Locations in Minnesota

	Location			
Variety	University Farm	Waseca	Crookston	Grand Rapids
Manchuria	26.9	33.5	33.0	22.1
Glabron	36.8	37.7	26.2	14.4
Velvet	26.8	37.4	32.1	32.2
Barbless	38.0	36.0	25.2	20.7
Peatland	28.1	58.2	35.9	40.2

Table 20.7 Information for Problem 19: Phenotypic Characteristics of Maize Plants

Generation	Grain Yield (bushels per acre)	Ear Length (cm)	Daily Increase in Height (inches)
Inbred parent 1	19.5	8.4	0.86
Inbred parent 2	19.6	10.7	0.67
F_1	101.2	16.2	1.16
F_2	69.1	14.1	1.00
F_3	42.2	14.7	0.95
F_4	44.1	12.1	0.95
F_5	22.5	9.4	0.85
F_6	27.3	9.9	0.78
F_7	24.5	11.0	0.75
F_8	27.2	10.7	0.74

(a) Is thorax length a continuous or a meristic trait?

(b) Is number of eggs laid a continuous or a meristic trait?

(c) What is the broad-sense heritability for thorax size?

(d) What is the broad-sense heritability for number of eggs laid?

22. In the same study cited in the previous question, Robertson also calculated realized heritabilities. For thorax size, the selection differential was 5.72, and the response from selection was 2.7. For egg number, the selection differential was 1.174, and the response from selection was 0.108. Calculate the narrow-sense (realized) heritabilities.

23. Based on your answers for the previous two questions, what proportion of V_G is represented by V_A for (a) thorax size and (b) number of eggs laid? (c) Which of the two traits has more variance for dominance if variance for epistatic interactions is assumed to be negligible?

24. In 1975, Moreno-Gonzales, Dudley, and Lambert (*Crop Science* 15:840–843) hybridized maize plants that were selected for high oil content with plants that were selected for low oil content. They then intermated the progeny at random for six generations. The parent plants came from populations that had undergone 68 generations of selection. The authors estimated that $V_A = 0.812$, $V_D = 0.011$, and $V_P = 2.460$. Estimate broad- and narrow-sense heritabilities. Assume that $V_I = 0$.

25. In 1950, Powers, Lock, and Garrett (*USDA Technical Bulletin* 998:1–56) reported the following means for weight per locule (in grams) in the tomato cross Porter × Ponderosa. Assume that the two parents are inbred and fully homozygous.

Group	Mean Weight per Locule
Porter	10.2
Ponderosa	9.8
F_1	14.4
F_2	13.5

(a) Explain why the F_1 mean exceeds both parental means. **(b)** Explain why the F_2 mean exceeds both parental means but is less than the F_1 mean.

26. In the tomato plants from the study cited in the previous question, **(a)** is the distribution for the F_1 likely to be skewed? If so, in which direction? **(b)** Is the distribution for the F_2 likely to be skewed? If so, in which direction? **(c)** Please briefly explain your answers.

27. In 1944, Powers (*American Naturalist* 78:275–280) reported the following data for mean number of locules per fruit in tomato:

Group	Mean Number of Locules
4110	9.6
4101	11.7
F_1 (4110 × 4101)	7.3

Explain why these results are an example of heterosis even though the phenotypic values of both parents exceed the phenotypic value of the hybrid.

28. In 1995, Georges et al. (*Genetics* 139:907–920) identified DNA marker loci linked to QTLs that govern milk production in dairy cattle. They examined over 150,000 female progeny of 1518 sires from highly productive milking herds using 159 PCR-based microsatellite DNA markers. The researchers mapped the markers to linkage groups and assigned most of the linkage groups to specific chromosomes using somatic cell hybridization. They found five chromosomal regions, on chromosomes 1, 6, 9, 10, and 20, that were associated with variation for milk production. **(a)** What do these results imply about the number of genes that govern milk production? **(b)** How might the markers be used to increase narrow-sense heritability for milk production? **(c)** Why were PCR-based microsatellite markers a good choice for this particular study?

FOR FURTHER READING

Two specialized textbooks in population and quantitative genetics provide excellent explanations of the concepts discussed in this chapter. A brief review of quantitative genetics can be found in Chapter 5 of **Crow, J. F. 1986.** *Basic Concepts in Population, Quantitative, and Evolutionary Genetics.* **New York: Freeman.** Detailed discussions along with several examples of all the concepts discussed in this chapter can be found in **Falconer, D. S. 1996.** *Introduction to Quantitative Genetics,* **4th ed. London: Longman House.** Three back-to-back discussions of the controversy that surrounds heritability estimates in humans are **Stoltenberg, S. F. 1997. Coming to terms with heritability.** *Genetica* 99:89–96; Schonemann, P. H. 1997. On models and muddles of heritability. *Genetica* 99:97–108; and Kempthorne, O. 1997. Heritability: Uses and abuses. *Genetica* 99:109–112. Reviews on QTL identification and analysis with molecular markers include **Paterson, A. H. 1995. Molecular dissection of quantitative traits: Progress and prospects.** *Genome Research* 5:321–333; Martin, N., D. Boomsma, and G. Machin. 1997. A twin-pronged attack on complex traits. *Nature Genetics* 17:387–392; Haley, C. S. 1995. Livestock QTLs: Bringing home the bacon. *Trends in Genetics* 11:488–492; Kearsey, M. J., and A. G. Farquhar. 1998. QTL analysis in plants: Where are we now? *Heredity* 80:137–142.

For additional reading, go to InfoTrac College Edition, your online research library at: http: www.infotrac-college.com/brookscole

CHAPTER 21

KEY CONCEPTS

Speciation, the formation of new species, arises when populations are reproductively isolated from one another.

~

Evolution can be studied at the molecular level. The rate of nucleotide and amino acid substitution is relatively constant over evolutionary time.

~

Evolutionary forces may reduce or increase genetic diversity in a species depending on which forces predominate.

~

Evidence of genetic divergence can be used to reconstruct phylogenies, the evolutionary histories of groups of organisms.

EVOLUTIONARY GENETICS

Evolution, the process of genetic change within and among species, is one of the core principles of biology. It is based on genetic differentiation at the DNA and chromosomal levels, which is expressed as altered gene products, altered phenotypes, and altered reproductive abilities. At some point, genetic differentiation may preclude successful reproduction between individuals with common ancestry. After reproductive isolation, populations with common ancestry may eventually develop into separate species. The genetic variation responsible for evolution is closely intertwined with the principles of population and quantitative genetics discussed in the two preceding chapters, as well as the ecologies of populations, communities, and ecosystems.

Most early studies in evolutionary genetics were based on morphological similarity and dissimilarity. As cytogenetics (the study of chromosome structure and inheritance) developed, it was added to the body of morphological data to further refine our understanding of evolution. The discovery of DNA as the genetic material, together with the discovery of the genetic code, permitted the development of molecular evolution, the study of divergence in DNA and amino acid sequences within and among species. Scientists now utilize all of these techniques to piece together parts of the evolutionary puzzle.

In general, the results from these various areas of study correspond with one another and provide independent evidence of how life has diverged, how species arose in the past, and how the processes of evolution continue to function in the present. This chapter provides an introduction to the principles and some of the methods of evolutionary genetics. We begin with a discussion of speciation, the arisal of different species from common ancestors, and the mechanisms that promote it. We then examine molecular evolution with a focus on DNA and amino acid sequence diversity. We end the chapter with a discussion of phylogenetic analysis, the reconstruction of ancestral relationships and evolutionary divergence.

21.1 SPECIATION

The concept of species is one of the most important aspects of evolutionary genetics. The classic definition of a **species** is a group of interbreeding individuals that produce fully fertile offspring and are isolated reproductively from other such groups. Another way to think of speciation is in terms of gene pools. The **gene pool** of a species consists of all the genes that can be shared by the members of a species and cannot be shared by members of

other species. All members who share the gene pool or are capable of sharing the gene pool are of the same species.

However, nature does not always conform to definitions invented by humans, so although the definitions of species and gene pool are generally applicable, there are frequent and significant exceptions to them. Some species are clearly defined by their distinct genetic, phenotypic, and reproductive separation from even the most closely related species. Others are not so sharply defined. Genetic variation is present both within and among species, and speciation is under way today as it has been in the past. Speciation is often a gradual, ongoing process, so sometimes it is difficult to determine at which point two groups have diverged sufficiently to be called separate species.

Reproductive Isolation

Speciation is based on **reproductive isolation**, a reduction or complete exclusion of gene flow between two or more populations. The presence of reproductive isolation does not mean that speciation is complete, but some form of partial or complete reproductive isolation is required for speciation. After a form of reproductive isolation is present, the gene pools can evolve separately. Usually, reproductive isolation first produces separate **races**, groups that have distinct genetic and often phenotypic characteristics but are still fully interfertile if gene flow is restored. Speciation is well under way when hybrids between two races are less fertile than the members of each race.

Reproductive isolation is a result of a variety of **reproductive isolating mechanisms** that are not exclusive of one another. There are two basic classes of reproductive isolating mechanisms: **prezygotic reproductive isolating mechanisms**, which prevent or inhibit the formation of hybrid zygotes between groups, and **postzygotic reproductive isolating mechanisms**, which affect the viability or fertility of hybrids after a hybrid zygote has formed.

Prezygotic Reproductive Isolating Mechanisms Prezygotic reproductive isolating mechanisms produce their effect by preventing or inhibiting fertilization. In some cases they act on their own, and in other cases they are combined with postzygotic reproductive isolating mechanisms. The most obvious type of prezygotic isolation is **geographic isolation**, a geographic separation of two or more populations. If two populations are geographically isolated from one another, there can be no gene flow between them. Populations may be kept apart by land or water barriers that they cannot or do not cross. For example, populations of fish and other aquatic animals and plants may be isolated in lakes with no means of mating with similar populations in other lakes. As selection,

drift, and mutation alter these populations, the populations may diverge sufficiently to allow other types of prezygotic and postzygotic reproductive barriers to form—ultimately resulting in speciation. When speciation is a result of geographic isolation, it is called **allopatric speciation**.

Although geographic isolation contributes to speciation, it is not essential for speciation. **Sympatric speciation** is speciation of two or more groups with common ancestry that occupy the same area. **Parapatric speciation** is speciation in which there is no distinct separation between populations over a geographical range. Populations may be isolated at the extremes of the range but overlap and interbreed in intermediate areas of it, as diagrammed in Figure 21.1. All of the reproductive isolating mechanisms we will discuss in the following sections may contribute to allopatric, sympatric, or parapatric speciation.

Ecological isolation is the separation of populations into different habitats even though they may occupy the same territory. For example, two species of fish may live in the same lake, but one species may be a surface feeder and the other a bottom feeder. Or plants of related species may occupy the same territory, but one may grow on one soil type and the other on a different, nearby soil type.

Temporal isolation is a form of assortative mating in which populations are reproductively separated by time. In plants, closely related species that occupy the same area may flower at different times of the year and thus do not hybridize with one another. Animals may also have different seasonal mating periods that isolate them reproductively.

Behavioral or **ethological isolation** is a type of reproductive isolation found in animals in which males and females of different populations may not be attracted to one another because of innate differences in mating behavior. The development of elaborate mating rituals in some animals is evidence that behavioral isolation in animals is common. Behavioral isolation is a type of assortative mating. In some animals, especially insects, the release of pheromones into the environment by females stimulates the males to mate. The pheromones also act as an attractant, guiding the males to females of the same species. Without the proper pheromone, certain insects do not mate even though they may be in physical contact with one another.

Mechanical isolation is the prevention of fertilization because of physical differences in the structure and location of reproductive organs. An example in certain plant species is cleistogamy, the presence of flowers that fail to open until after the flowers have self-pollinated. Cleistogamous enforcement of self-fertilization effectively isolates each individual from all others. It also prevents any gene flow between populations, so populations may become genetically distinct.

KEY

species A

species B

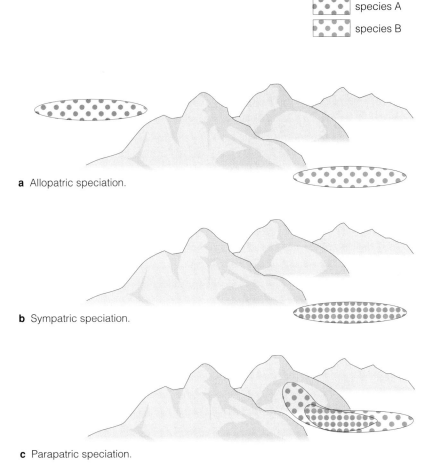

a Allopatric speciation.

b Sympatric speciation.

c Parapatric speciation.

Figure 21.1 Comparison of allopatric, sympatric, and parapatric speciation.

There are many examples of mechanical isolation between species that are closely related, especially in plants. As illustrated in Figure 21.2, mechanical isolation may include the insect (or other animal) that pollinates the plant species.

Gametic isolation is the incompatibility of male and female gametes that come into contact. Behavioral and chemical stimuli usually prevent animals of all but the same or very closely related species from mating. However, in the rare circumstances when gametes from different species do come together, biochemical incompatibilities may prevent the union of sperm and egg. In plants, gametic contact between diverse species is common. A bee, for instance, may visit flowers on different species of plants, so the pollen it carries is a mixture of pollen from all the species it has visited. When a bee loaded with mixed pollen visits a flower, the bee may deposit pollen from several different species on the flower. Biochemical triggers ensure that only pollen from the proper species fertilizes the ova.

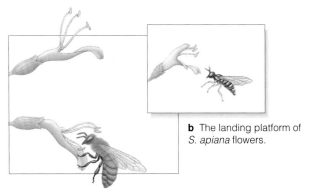

b The landing platform of *S. apiana* flowers.

a The landing platform of *S. mellifera* flowers.

Figure 21.2 Two species of sage (*Salvia*) are reproductively isolated by the ability of pollinators to pollinate the flowers. **(a)** *S. mellifera* has a small landing platform, whereas **(b)** *S. apiana* has a large landing platform. The flowers accommodate different pollinators, so the pollen of one species is isolated from the other. (Adapted from Grant, K. A. and V. Grant. 1964. Mechanical isolation of *Salvia apiana* and *Salvia mellifera* (Labiatae). *Evolution* 18:196–212.)

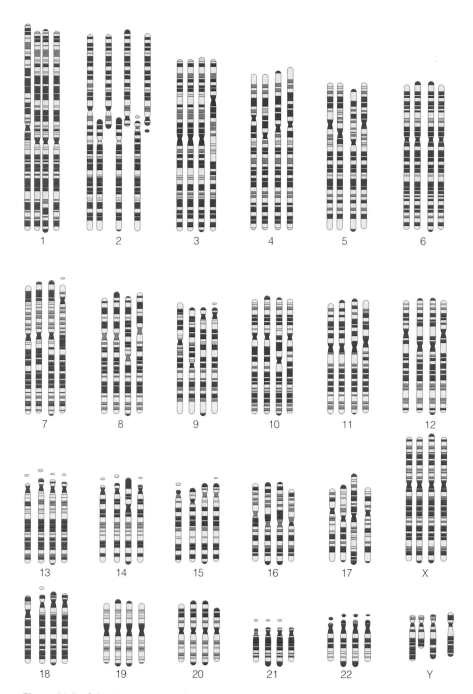

Figure 21.3 G-banded patterns of chromosomes from (left to right) human, chimpanzee, gorilla and orangutan. The numbers are those assigned to human chromosomes. (Adapted with permission from Yunis, J. J., and O. Prakash. 1982. Origin of man: A chromosomal pictoral legacy. *Science* 215:1525–1530. Copyright 1982 American Association for the Advancement of Science.)

Postzygotic Reproductive Isolating Mechanisms Postzygotic reproductive isolating mechanisms reduce the fertility and/or viability of hybrids after a zygote has formed. These mechanisms are especially important in the early stages of species divergence. Postzygotic reproductive isolation may be a consequence of prezygotic reproductive isolation that has been present for many generations, or it may precede prezygotic isolation.

Alterations in chromosome structure or number, such as inversions, translocations, fusions, fissions, or polyploidy, often cause postzygotic isolation. Let's review a hypothetical example to see how such isolation could arise, then we'll look at some actual examples.

Suppose that an impregnated female migrates from a large population to colonize a new area where her species is not present. Also suppose that she is heterozygous

for a reciprocal translocation that is generally absent from the large population. She transmits the translocation to some of her offspring. As the progeny population is small, drift can have a significant effect. If the population remains small, the translocation may eventually become fixed in a homozygous state in the population because of random genetic drift. Selection may speed up the rate of fixation if it favors the translocation homozygote. Translocation heterozygotes suffer from a loss in fertility (as described in section 17.3) and are disfavored by selection. If the new population becomes fixed for the translocation, then hybrids between individuals from this new population and the original population are less fertile because they are reciprocal translocation heterozygotes. If additional translocations or inversions become fixed in either of the populations, hybrids from matings between individuals of different populations may be completely sterile and perhaps inviable.

If translocations, inversions, and fusions or fissions that become fixed in populations are common postzygotic reproductive isolating mechanisms, then we should expect to find fixed inversions, translocations, and fusions or fissions when closely related species are compared. Among the best-documented cases is the differentiation between the chromosomes of humans and the great apes. Figure 21.3 shows the G-banded patterns of chromosomes from (left to right) human, chimpanzee, gorilla, and orangutan. The chromosomes of the four species are grouped side by side, and the numbers of the groups are those assigned to human chromosomes.

Although the chromosomes from the four species are quite similar, a large number of structural alterations are apparent. Chromosome 2 in humans arose from a fusion between two chromosomes that have remained separate in the great apes, so the great apes have $2n = 48$ chromosomes, whereas humans have $2n = 46$ chromosomes. Notice that the centromere in human chromosome 2 corresponds to the centromere of one of the chimpanzee chromosomes. In the orangutan and gorilla, this centromeric region is inverted. Several other prominent pericentric inversions are evident between human and chimpanzee in chromosome groups 4, 5, 9, 12, 16, and 17. There are large pericentric inversions in the orangutan chromosomes in groups 3 and 7 when compared to the other three species, and also in the gorilla chromosomes in groups 8, 10, and 14 when compared to the other three species. A large paracentric inversion is present in the orangutan chromosome of group 3. A reciprocal translocation is evident in the gorilla in chromosomes 5 and 17. Numerous other alterations in chromosome structure can be found, including deletions and insertions. However, of the nearly 1000 bands that can be identified, nearly all are shared by the four species, though many in rearranged positions.

In the following example, we examine one of the chromosome alterations in detail.

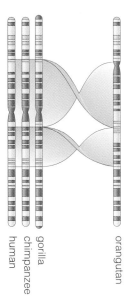

Figure 21.4 Solution to Example 21.1: Two inversions in an orangutan chromosome. (Adapted from Yunis, J. J., and O. Prakash. 1982. *Science* 215:1525–1530.)

Example 21.1 **Chromosomal evolution in primates.**

Compare the orangutan chromosome that corresponds to chromosome 3 in humans with the human, chimpanzee, and gorilla chromosomes.

Problem: (a) How many inversions are present in the orangutan chromosome? (b) Indicate approximately where the break points of the inversions are, and identify each inversion as a pericentric or paracentric inversion.

Solution: (a–b) As illustrated in Figure 21.4, there are two inversions in the orangutan chromosome, one pericentric inversion and one paracentric inversion. The inversions are not present in any of the other three species.

A female horse mated with a male donkey to produce a sterile mule is an example of postzygotic isolation of horses and donkeys. Horses have $2n = 64$ chromosomes, and donkeys have $2n = 62$ chromosomes. The hybrid mule with 63 chromosomes in its somatic cells is sterile because chromosome imbalances arise during meiosis. Horses and donkeys are still close enough to produce viable hybrid offspring, but the hybrids are sterile, so the horse and donkey gene pools are isolated from one another.

The genus *Drosophila* has also been studied extensively to determine the patterns of speciation. The common fruit fly, *Drosophila melanogaster*, is found on most of the land masses of the world. There are many other

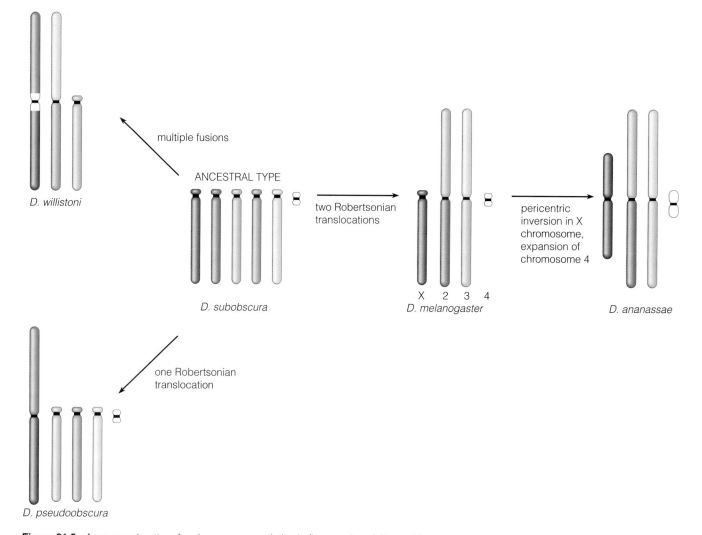

Figure 21.5 A proposed pattern for chromosome evolution in five species of *Drosophila*. *D. subobsura* is the ancestral type.

species of *Drosophila*, some with more limited ranges. When the chromosomes of the various *Drosophila* species are compared, patterns of speciation emerge. For example, four species (*D. melanogaster, D. pseudoobscura, D. ananassae,* and *D. willistoni*) have apparently evolved from a common type (*D. subobscura*) through chromosome fusions and a major inversion (Figure 21.5).

Chromosomal rearrangement may be the first step in speciation. A rearrangement, such as a fusion, inversion, or translocation, can become fixed in a small colony within a large population because of drift within the colony. Initially, the colony differs little, in terms of the alleles that it carries, from its parent population. The colony remains partially isolated because of the reduced fitness of the hybrids, and if the colony is favored by selection, it may expand its range and displace members of its parent population. Eventually, as the colony remains reproductively isolated, other genetic changes accumu-

late in a homozygous condition, and a new species arises.

An alteration in chromosome structure can quickly become fixed in the homozygous state when some form of inbreeding is present, especially when the number of interbreeding members of a colony is small. Thus, both inbreeding and random genetic drift operate to fix a chromosomal alteration in the population. Fixation may be rapid in plant populations that self-fertilize. Remember that self-fertilization is the most rapid way to approach complete homozygosity through inbreeding. Other forms of inbreeding likewise increase homozygosity, so fixation of a chromosomal alteration is more likely in an inbred population than in a random-mating one.

Inbreeding in animals increases when the animals are not highly mobile and congregate in small colonies. Indeed, fixation of chromosomal alterations is found among some less mobile, closely related animal species, such as flightless grasshoppers, mole rats, and pocket go-

phers. Often, the closely related species are indistinguishable by phenotype but can be distinguished by karyotypes, an indication that they still share the same genes even though they are isolated genetically.

Polyploidy is another type of postzygotic reproductive isolating mechanism. For example, a tetraploid hybridized with a diploid produces triploid offspring with greatly reduced fertility. The tetraploid and the diploid gene pools thus remain isolated from one another.

~

Speciation requires some form of reproductive isolation. Reproductive isolation may be either prezygotic or postzygotic.

Quantum Speciation

A single inversion or reciprocal translocation is usually insufficient to cause complete reproductive isolation and speciation by itself. The reduced fertility of inversion or translocation heterozygotes, however, causes partial reproductive isolation, which can lead to the accumulation of additional alterations in chromosome structure or other forms of reproductive isolation. After isolation is nearly complete, a genetic alteration that appears in one population is not readily transmitted to the other, and the two populations differentiate even more over time.

Speciation is often gradual—the first reproductive isolating mechanism that initially separates the populations leads to the accumulation of other types of isolating mechanisms that jointly cause complete speciation. Such a process often requires many generations sometimes spanning thousands to millions of years. However, speciation is sometimes very rapid—sometimes it requires only a single generation. Rapid speciation in one or a few generations is called **quantum speciation**.

In many plant and some animal species, quantum speciation is caused by polyploidy, an increase in the number of chromosome sets. There are several examples of recent quantum speciation in polyploid plants. Triticale (see Figure 17.9) is an example of a new allopolyploid species produced with human intervention. However, there are examples from nature as well. Under the right conditions, two related polyploid species may come into contact and form partially sterile hybrids. Some of these hybrids may produce fertile offspring that carry traits from both parent species. After several generations of cross-fertilization among the parental species and the hybrids, a **hybrid swarm**, which consists of plants with various ploidies and genes from both parent species, arises. A hybrid swarm is characterized by an explosion of genetic and phenotypic variation. If a particular phenotype in the hybrid swarm is well adapted to its environment, it may reproduce and spread. Natural selection may favor a particular phenotype among the individuals in the hybrid swarm and stabilize the phenotype. Eventually, the individuals with the favored phenotype may become re-

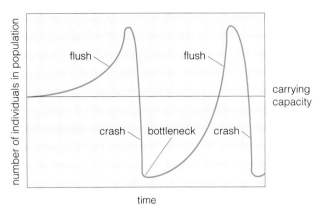

Figure 21.6 The flush-crash cycle.

productively isolated from their parent species and evolve into a new species with its own unique combination of genes derived from the two parent species.

Quantum speciation can arise by extremes of drift and selection. In nature, all species have the potential for exponential population growth. If the potential growth were actually realized, some species could overwhelm the entire earth by their numbers within a few years. Environmentally imposed limits on population growth prevent populations from reaching their growth potentials. Such limitations include availability of resources, such as food and water, predation, and parasitism. Some growth-limiting factors contribute to natural selection by curbing population growth selectively, favoring certain genotypes over others. If for any reason one type of growth-limiting factor is removed, the population may grow exponentially until most of the resources available to it are consumed. With few resources available, most of the individuals die, sharply reducing the number of individuals in the population.

For example, rabbits are capable of producing many offspring in a single year. Predators keep the rabbit population sizes fairly stable, so the population does not reach its potential for exponential growth. However, if for some reason the predator population decreases drastically in size, a limit on rabbit population growth is removed, and the rabbit population is free to increase. Eventually the large numbers of rabbits consume all the plants available to them, and most of the rabbits die from starvation. The rabbit population declines steeply, and food resources rebound. Without predatory control, the rabbit population could rebound from a few founders into a population that grows exponentially again until it overshoots the ability of its environment to sustain it. Once again, the population size can drop rapidly when all resources have been consumed. Such a cycle of rapid population growth followed by a steep drop in numbers is called the **flush-crash cycle** (Figure 21.6) and may be repeated several times in a population under the right circumstances.

During the flush, individuals may migrate in search of food and establish new populations that may be isolated geographically from the original population. If the number of migrants is small, a founder effect can take place in the new population. In the original population, random genetic drift can alter the genetic composition of the population while the number of individuals is small following the crash. The effect of drift in a population that is temporarily small is called a **bottleneck effect** (Figure 21.6). Selection can also act during the flush-crash cycle to alter the genetic structure of the population. As selection and random genetic drift rapidly alter the genetic structure of the population, the opportunities for reproductive isolation and speciation are greater than in a stable population.

The following example shows the arisal of reproductive isolation in populations subjected to flush-crash cycles.

Table 21.1 Information for Example 21.2: Isolation Indices for *Drosophila pseudo-obscura* Populations

Population	Isolation Index	Chi-Square Value
0	+0.09	5.00*
1	+0.31	36.16***
2	+0.13	9.58***
3	+0.12	6.88**
4	+0.23	29.14***
5	+0.12	9.96***
6	+0.20	15.41***
7	+0.06	3.11
8	+0.13	1.59

* = significant at $P < 0.05$
** = significant at $P < 0.01$
*** = significant at $P < 0.001$

Example 21.2 Reproductive isolation from flush-crash cycles in *Drosophila*.

To test the effect of the flush-crash cycle, Powell (*Evolution* 32:465–474) created a series of artificial flush-crash cycles in *Drosophila pseudoobscura* populations maintained in the laboratory. He began the experiments by intermating four populations collected from different locations to create a hybrid swarm. The hybrid swarm was allowed to flush (expand in numbers through reproduction) for 2 months. He then made single male-female pair matings (simulating a crash and subsequent bottleneck) to establish eight founder populations that were carried through four more flush-crash cycles. He then tested the populations for mating preference by placing 12 males and females from one population with 12 males and females from another population in all possible combinations. He found that all F_1 individuals were fully fertile regardless of the cross from which they came. He then calculated an isolation index by taking the number of heterogametic matings (matings between individuals of different populations) and subtracting this from the number of homogametic matings (matings of individuals from the same population), then dividing the result by the total number of matings. This calculation produces a number between -1 and $+1$, where $+1$ represents complete assortative mating (all homogametic matings), -1 represents complete disassortative mating (all heterogametic matings), and zero represents an equal number of homogametic and heterogametic matings, which is the expected value for random mating. The deviation of the isolation index from zero (random mating) can be tested for significance with chi-square analysis. Table 21.1 provides the isolation indices and the chi-square values for the sum of all matings

for a population mated with all other populations. The numbers 1 through 8 represent the eight founder populations, and 0 represents the original hybrid swarm.

Problem: **(a)** What type of reproductive isolation is indicated by the information given above? **(b)** What type of mating (assortative, disassortative, or random) predominates? **(c)** Suggest some appropriate control experiments that could verify the validity of the data.

Solution: **(a)** No postzygotic isolation is indicated by the data because all F_1 individuals were fully fertile. Apparently, the isolation is behavioral because there are no physical or genetic barriers to prevent individuals of one population from mating with those of another population. **(b)** All isolation index values are positive, so assortative mating predominates. In other words, individuals preferentially mate with individuals of their own population. The isolation indices and the chi-square values indicate that assortative mating is greater in some populations than in others, and that in some cases the deviation from random mating is not significant.

(c) Control experiments help researchers verify their conclusions by excluding possible alternative interpretations of the results. In this case, the effect of flush-crash cycles is being tested. Populations subjected to flush-crash cycles are also subjected to physical isolation and to inbreeding. Possibly, the physical isolation or the inbreeding, apart from the flush-crash cycles, could be the causes of the reproductive isolation observed. Appropriate control experiments should test for deviation from random

mating in populations that are physically isolated and inbred, but that had not undergone flush-crash cycles. The most appropriate population to test is the original hybrid swarm. It can be divided at random into several subpopulations. The subpopulations should be isolated from each other throughout the experiments and maintained at a large size to avoid flush-crash cycles, inbreeding, and drift. They should then be tested for reproductive isolation at the same time as the populations subjected to the flush-crash cycles. If the populations not subjected to flush-crash cycles show no significant reproductive isolation, then the reproductive isolation in other populations could not be attributed to physical isolation alone.

Powell conducted this control experiment along with other control experiments to test inbreeding. He isolated three large subpopulations of the original hybrid swarm and maintained them throughout the duration of the experiment. When flies from these populations were placed together, they showed no significant deviation from random mating, so physical isolation could not explain the reproductive isolation observed in the flush-crash experiments. He also developed several inbred populations from the original hybrid swarm and maintained them as isolated populations throughout the duration of the experiments. They also showed no significant deviations from random mating, so the effect of inbreeding was not sufficient to cause reproductive isolation in the flush-crash experiments. The results of Powell's experiments indicate that the assortative mating observed in the populations subjected to flush-crash cycles is probably a result of the flush-crash effect and not to physical isolation or inbreeding.

Drift and selection may operate to heightened degrees in small populations to reproductively isolate populations.

21.2 MOLECULAR EVOLUTION

After the isolation of incipient species is complete, the gene pools evolve independently. Because gene flow has been disrupted, mutations that arise in one species are not transmitted to the other. The degree of genetic divergence should be positively correlated with the amount of time that the two species have been reproductively isolated. The degree of divergence among species for a particular gene can be determined by examining the amino acid sequences of the same gene product in several species and identifying how many differences in the amino acid sequence there are. Likewise, the DNA sequences of the same gene can be compared

in several species to identify the number of nucleotide substitutions. The accumulation of only a few amino acid or nucleotide substitutions indicates close ancestry, whereas a large number of substitutions indicates distant ancestry.

Amino Acid Sequence Divergence in Proteins

The first studies on molecular evolution were done in the 1960s with amino acid sequences. Even though amino acid sequencing was laborious and time-consuming, DNA sequencing methods were not yet well developed. Among the first polypeptides sequenced and compared among species were the hemoglobin subunits. Hemoglobin proved to be a good choice because it can be easily purified from the blood of animals, the subunits have a relatively small number of amino acids (141 in the α subunit and 146 in the β subunit) so the subunits can be sequenced rapidly, and hemoglobin was present in a large number of animal species so comparisons can be made over a wide range of animal groups.

The results of hemoglobin sequencing confirmed the evolutionary relationships that had been hypothesized on the basis of comparative anatomy. Humans and chimpanzees had identical hemoglobins, whereas humans and chimpanzees differed from gorillas by only one amino acid in the α subunit and one amino acid in the β subunit. Humans differed from cattle by 17 amino acids in the α subunit and 24 in the β subunit. As various groups were compared, especially those with distant common ancestry, a clear pattern emerged. If comparisons were made between major groups, such as mammals and reptiles, or birds and reptiles, and the proportions of amino acids that differed were plotted against the time of divergence of the groups as determined by the fossil record, a straight line could be drawn on the graph, an indication that amino acid substitution for hemoglobin was fairly constant in evolutionary time (Figure 21.7).

The α and β subunits of hemoglobin are about 35% identical to one another, an observation that indicates that they were derived anciently from a single gene that diverged into two genes. If this indication is correct, then we expect the divergence of the α and β subunits within one group to be the same as that within another group when the two groups are compared. This is indeed the case—amino acid substitutions arise at the same rate in both subunits in major groups of species.

As other proteins were sequenced, a similar pattern of straight-line relationships between amino acid substitution and evolutionary time in the fossil record for each of the proteins emerged. Although the rate was constant for each protein (as implied by the straight-line correlation), the rates differed among proteins (Figure 21.7). Hemoglobin diverged at a rate of about 1% per 5.8 million years, and cytochrome c diverged at a much slower rate, about 1% per 20 million years. Fibrinopeptides (proteins found in structural fibers in animals) diverged very

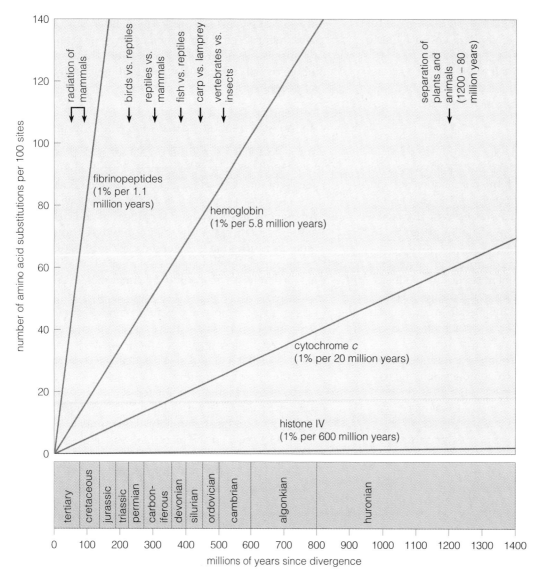

Figure 21.7 Comparison of amino acid substitutions and time of divergence based on the fossil record. (Adapted from an original drawing by Irving Geis published in Dickerson, R. E. 1972. The structure and history of an ancient protein. *Scientific American* 226 (Apr):58–72. Used by permission of the Estate of Irving Geis.)

rapidly, whereas histones hardly diverged at all, even when plants and animals were compared.

Differences in the rate of protein divergence can be readily explained when natural selection is taken into account. Point mutations that alter the functional activity of an essential protein are disfavored by selection. However, point mutations that alter amino acids that are not as essential for protein function are not disfavored by selection and can become fixed in the ancestors of the species by random genetic drift. Some mutations may be favored by selection and be rapidly fixed in a species.

Histones play extremely important roles in stabilizing the DNA molecules of all eukaryotes. Most amino acid substitutions in histones render them less functional, so mutations that alter histone structure are usually disfavored by selection. Selection ensures that the amino acid sequences of histones remain highly conserved over

evolutionary time. Proteins that play structural roles, such as fibrinopeptides, on the other hand, can undergo amino acid substitutions without significantly altering their function. Because selection does not remove many of the point mutations, amino acid substitutions can accumulate at a faster rate in fibrinopeptides than in other proteins. Some proteins, such as hemoglobin, fall in between these two extremes. Selection preserves amino acids that are essential for protein function, and amino acids that are not as essential may be changed by mutation, then become fixed in the population because selection does not disfavor their presence. The amino acids in the active sites of an enzyme can often be identified by their conservation among species.

Proteins that undergo amino acid substitution at a relatively high rate are useful for comparing species that are closely related. For example, fibrinopeptides are useful

for distinguishing relatedness among mammals. Proteins that change more slowly are useful for determining at which point distantly related groups diverged from common ancestry. Among the most useful proteins in this regard is cytochrome *c*. Because it is present in photosynthetic bacteria and in the mitochondria of all eukaryotes, it allows for comparison of amino acid sequence divergence among broad groups of organisms. It also has a relatively short amino acid sequence (an average of about 100 amino acids) and is easy to purify, making it a good candidate in all respects for analysis (Figure 21.8).

The following example uses cytochrome c to show how researchers compare and analyze amino acid sequences.

Example 21.3 Molecular evolution in cytochrome *c*.

At position 15 in cytochrome *c*, either a serine or an alanine is present in most species. In *Saccharomyces cerevisiae* (brewer's yeast), a glutamic acid is present in this position.

Problem: Using the genetic code and the information in Figure 21.8, determine which amino acid is most probably the ancestral type, and describe the mutations that must have caused the substitutions.

Solution: Most species in Figure 21.8 have either a serine or an alanine at position 15, so one of these amino acids is probably the ancestral type. All plants and fungi, as well as a majority of animals in diverse groups, have alanine at this site. Independent fixation of fewer mutant alleles is required to explain the information in Figure 21.8 if alanine is assumed to be the ancestral type. Alanine is encoded by codons with the sequence GCN, and serine is encoded by codons with the sequences UCN and AGPy. The simplest possible mutation for an alanine → serine substitution is a single transversion in the first nucleotide of an alanine codon: GCN → UCN. Further evidence that alanine is indeed ancestral comes from the observation that *S. cerevisiae* has glutamic acid at this position. Glutamic acid is encoded by codons with the sequence GAPu. A glutamic acid can arise by a transversion mutation of C → A in the second nucleotide of an alanine codon. For glutamic acid to arise by mutation of a serine codon, at least two mutations are required. Therefore, alanine is the most probable ancestral type.

Nucleotide Sequence Divergence in DNA

Substitutions in DNA sequence follow a pattern that is similar to amino acid substitutions. Those nucleotide substitutions that reduce or eliminate the function of an essential enzyme are often removed by selection, and the

DNA sequence at those positions is conserved, whereas mutations that do not significantly alter essential enzyme function are affected little by selection. The rate of change in DNA sequence depends on how highly the sequence is conserved by selection. As with proteins, sequences that are highly conserved are useful for comparing species whose ancestries diverged long ago, whereas sequences that are not highly conserved are useful for comparing species that are closely related. DNA sequence analysis often provides more information about evolution than protein sequence comparison because certain point mutations are **silent mutations**—they have no effect on the amino acid sequence of a protein and no effect on expression of the gene. Silent mutations may be same-sense mutations in the third (or sometimes first or second) position of a codon, mutations in the unessential nucleotide sequences of introns, and mutations in DNA sequences whose function is unessential or redundant, such as pseudogenes and highly repetitive DNA. Silent mutations should be completely unaffected by selection, so their presence or absence in a species should be due predominantly, if not entirely, to the effects of mutation and random genetic drift. Such mutations are **selectively neutral**—selection neither favors nor disfavors them—although, as illustrated in the following example, there is evidence that some same-sense mutations may not be entirely neutral.

Example 21.4 Codon usage bias in *Drosophila*.

The growing body of readily available DNA sequence information has made it possible to compare DNA sequences for many genes and identify patterns that emerge from the analysis. An example of this is the observation of codon bias, described in *Drosophila* by Powell and Moriyama (1997. *Proceedings of the National Academy of Sciences, USA* 94:7784–7790). Codon bias is defined as a higher frequency of a particular codon for an amino acid than expected if silent nucleotides (such as those in the third position) are truly neutral. Codon bias also appears to be related to the most abundant tRNA present. Table 21.2 lists the anticodons for the most abundant tRNAs and the most common codon found for 11 amino acids in *Drosophila* as identified by Powell and Moriyama.

Problem: **(a)** Which nucleotides are most frequent at the third position of the codon? **(b)** What relationship is there between the most frequent codons and the anticodons of the most abundant tRNA for each amino acid? **(c)** What do these results suggest about the possible cause of codon bias?

Solution: **(a)** All 11 most frequent codons have either C or G in the third position. **(b)** As discussed in section 4.4, wobble permits altered codon-anticodon pairing at the third position of the codon.

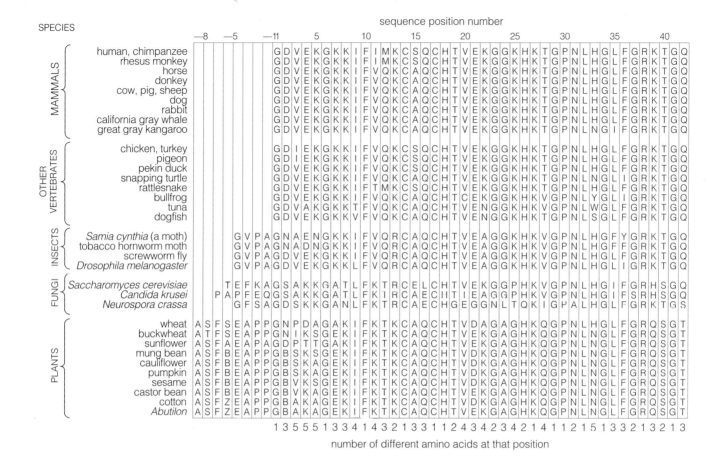

Figure 21.8 Amino acid sequences of cytochrome *c* in different species. (Adapted from an original drawing by Irving Geis published in Dickerson, R. E. 1972. The structure and history of an ancient protein. *Scientific American* 226 (Apr):58–72. Used by permission of the Estate of Irving Geis.)

The nucleotide present at the third position in the most frequent codon is usually the nucleotide that pairs with the corresponding nucleotide in the most abundant tRNA according to the standard base-pairing rules (i.e., C-G and U-A). **(c)** Even though wobble permits altered pairing, standard base pairs may be more effective than altered base pairs during translation. Possibly, selection has favored those mutations at the third position of codons that make translation most efficient.

Molecular Clocks

The observation that the rate of change in a given DNA or amino acid sequence is linear over time allows us to use amino acid or nucleotide sequence divergence to estimate how long ago two groups or species diverged. Any amino acid or DNA sequence used to determine time of divergence can be thought of as a **molecular clock**. As is evident in Figure 21.7, molecular clocks tick at different rates. Those sequences that are highly conserved by selection tick more slowly than those that are not. Sequences that are selectively neutral tick at the highest rate.

Molecular clocks can be compared to one another to confirm conclusions. For example, the graph of hemoglobin in Figure 21.7 can be used as a molecular clock to estimate that fish and reptiles diverged about 400 million years ago. That time period is also within the useful range of cytochrome *c*, which changes more slowly than hemoglobin. Cytochrome *c* also indicates that fish and reptiles diverged about 400 million years ago.

```
        45        50        55        60        65        70        75        80        85        90        95       100   104
A P G Y S Y T A A N K N K G I I W G E D T L M E Y L E N P K K Y I P G T K M I F V G I K K E E R A D L I A Y L K K A T N E
A P G Y S Y T A A N K N K G I I W G E D T L M E Y L E N P K K Y I P G T K M I F V G I K K E E R A D L I A Y L K K A T N E E
A P G F T Y T D A N K N K G I I T W K E E T L M E Y L E N P K K Y I P G T K M I F A G I K K K T E R E D L I A Y L K K A T N E E
A P G F S Y T D A N K N K G I T W K E E T L M E Y L E N P K K Y I P G T K M I F A G I K K K T E R E D L I A Y L K K A T N E E
A P G F S Y T D A N K N K G I T W G E E T L M E Y L E N P K K Y I P G T K M I F A G I K K K G E R E D L I A Y L K K A T N E E
A P G F S Y T D A N K N K G I T W G E D T L M E Y L E N P K K Y I P G T K M I F A G I K K K T G E R A D L I A Y L K K A T K E E
A V G F S Y T D A N K N K G I T W G E D T L M E Y L E N P K K Y I P G T K M I F A G I K K K D E R A D L I A Y L K K A T N E E
A V G F S Y T D A N K N K G I T W G E E T L M E Y L E N P K K Y I P G T K M I F A G I K K K G E R A D L I A Y L K K A T N E E
A P G F T Y T D A N K N K G I I W G E D T L M E Y L E N P K K Y I P G T K M I F A G I K K K G E R A D L I A Y L K K A T N E

A E G F S Y T D A N K N K G I T W G E D T L M E Y L E N P K K Y I P G T K M I F A G I K K K S E R V D L I A Y L K D A T S K
A E G F S Y T D A N K N K G I T W G E D T L M E Y L E N P K K Y I P G T K M I F A G I K K K A E R A D L I A Y L K Q A T A K
A E G F S Y T E A N K N K G I T W G E E T L M E Y L E N P K K Y I P G T K M I F A G I K K K A E R A D L I A Y L K D A T S K
A V G Y S Y T A A N K N K G I I W G D D T L M E Y L E N P K K Y I P G T K M V F T G L S K K K E R T N L I A Y L K E K T A A
A A G F S Y T D A N K N K G I T W G E D T L M E Y L E N P K K Y I P G T K M I F A G I K K K K G E R Q D L I A Y L K S A C S K
A E G Y S Y T D A N K S K G I V W N N D T L M E Y L E N P K K Y I P G T K M I F A G I K K K G E R Q D L V A Y L K S A T S S
A Q G F S Y T D A N K S K G I T W Q Q E T L R I Y L E N P K K Y I P G T K M I F A G L K K K S E R Q D L I A Y L K K T A A S

A P G F S Y S N A N K A K G I T W G D D T L F E Y L E N P K K Y I P G T K M V F A G L K K A N E R A D L I A Y L K E S T K —
A P G F S Y S N A N K A K G I T W Q D D T L F E Y L E N P K K Y I P G T K M V F A G L K K A N E R A D L I A Y L K Q A T K —
A A G F A Y T N A N K A K G I T W Q D D T L F E Y L E N P K K Y I P G T K M I F A G L K K P N E R G D L I A Y L K S A T K —
A A G F A Y T N A N K A K G I T W Q D D T L F E Y L E N P K K Y I P G T K M I F A G L K K P N E R G D L I A Y L K S A T K —

A Q G Y S Y T D A N I K K N V L W D E N N M S E Y L T N P X K Y I P G T K M A F G G L K K E K D R N D L I T Y L K K A C E —
A Q G Y S Y T D A N K R A G V E W A E P T M S D Y L E N P X K Y I P G T K M A F G G L K K A K D R N D L V T Y M L E A S K —
V D G Y A Y T D A N K Q K G I T W D E N T L F E Y L E N P X K Y I P G T K M A F G G L K K D K D R N D I I I T F M K E A T A

T A G Y S Y S A A N K N K A V E W E E N T L Y D Y L L N P X K Y I P G T K M V F P G L X K P Q D R A D L I A Y L K K A T S S
T A G Y S Y S A A N K N K A V T W G E D T L Y E Y L L N P X K Y I P G T K M V F P G L X K P Q E R A D L I A Y L K D S T E —
T A G Y S Y S A A N K N M A V I W E E N T L Y D Y L E N P X K Y I P G T K M V F P G L X K P Q E R A D L I A Y L K T S T A —
T A G Y S Y S T A N K N M A V I W E E K T L Y D Y L E N P X K Y I P G T K M V F P G L X K P Q D R A D L I A Y L K E S T A —
T A G Y S Y S A A N K N K A V E W E E K T L Y D Y L E N P X K Y I P G T K M V F P G L X K P Q D R A D L I A Y L K E A T A —
T P G Y S Y S A A N K N R A V I W E E K T L Y D Y L E N P X K Y I P G T K M V F P G L X K P Q D R A D L I A Y L K E A T A —
T A G Y S Y S A A N K N M A V Q W G E N T L Y A Y L E N P X K Y I P G T K M V F P G L X K P Q D R A D L I A Y L K E A T A —
T A G Y S Y S A A N K N M A V Q W G E N T L Y D Y L E N P X K Y I P G T K M V F P G L X K P Q D R A D L I A Y L K E S T A —
T P G Y S Y S A A N K N M A V N W G E N T L Y D Y L E N P X K Y I P G T K M V F P G L X K P Q D R A D L I A Y L K E S T A —
```

```
3 6 1 2 3 1 2 5 1 1 2 6 4 3 2 7 1 7 4 5 2 2 5 4 1 1 3 1 1 1 1 1 1 1 1 1 1 1 1 3 1 5 1 2 2 1 6 9 2 1 7 2 2 2 2 2 2 6 4 4 5 4
```

number of different amino acids at that position

Table 21.2 Information for Example 21.4: Comparison of the Most Frequent Codons and the Anticodon of the Most Abundant tRNA for 11 Amino Acids

Amino Acid	Most Frequent Codon	Anticodon of Most Abundant tRNA*
Valine	GUG	3' CAC 5'
Glycine	GGC	3' CCG 5'
Serine	UCC	3' AGI 5'
Tyrosine	UAC	3' AUG 5'
Histidine	CAC	3' GUG 5'
Asparagine	AAC	3' UUG 5'
Aspartic acid	GAC	3' GUC 5'
Lysine	AAG	3' UUC 5'
Glutamic acid	GAG	3' SUC 5'
Phenylalanine	UUC	3' AAG^m 5'
Arginine	CGC	3' GCI 5'

*G^m is methylguanine. S is 2-thiouridine, which pairs with A, G, and G^m.

Large-Scale DNA Sequence Comparisons

Examination of amino acid or nucleotide sequence divergence provides detailed information about the evolution of a single gene or portion of a gene. Ideally, it would be best to compare entire genomic DNA sequences so that we could have some idea how genomes, rather than a few genes, have evolved. This lofty goal is now a reality for several bacterial species and will be soon for a few eukaryotic species.

Even when the entire genome has not been sequenced, large-scale DNA sequence comparison is possible among those species for which substantial DNA sequence information is available. Also, DNA sequences discovered during a DNA sequencing project may be used to examine homologous sequences in other species by using the information to generate primers for PCR or DNA probes for hybridization analysis.

New DNA sequences are added to DNA sequence databases daily, and there is now a large body of DNA sequences from many species that can be compared. Rapid computerized comparison of DNA sequences is available gratis to all scientists worldwide (courtesy of the United States government) through the World Wide Web. The free and easy access to computerized DNA sequence analysis greatly facilitates research on molecular evolution.

Molecular-Marker Analysis

Although significant DNA sequence information is available for several model species, very little, if any, DNA sequence information is available for most other species. Genomes are so large that most studies on DNA and amino acid sequencing can provide information on just a small fraction of the genome. To obtain some idea of how entire genomes have evolved, it is possible to use molecular markers selected systematically or at random to serve as samples that represent the entire genome. The markers can then be compared both among and within species to determine the degree of marker similarity. The degree of marker similarity should represent the degree of genetic similarity, provided an adequate and appropriate sample of markers has been analyzed.

Molecular methods used in evolutionary genetic studies include isozymes, RFLPs, minisatellites, microsatellites, and DNA sequencing. These methods are described in Chapter 9 (sections 9.6–9.9) and Chapter 13 (section 13.4).

DNA Hybridization

DNA markers can be selected to represent the genome. Because DNA markers represent only a small portion of the genome, however, analyses based on marker similarity are subject to sampling error. It would be helpful to sample the entire genome, or at least a large fraction of it. Estimates of DNA sequence similarity can be obtained by determining how closely DNA from one species hybridizes with the DNA of another species. Double-stranded DNA molecules are most stable when all the nucleotides are paired with no mismatches. As the number of mismatches increases, so does the thermal instability of the molecule. A DNA molecule with many mismatches will dissociate into single strands at a lower temperature than a DNA molecule with few mismatches. DNA denaturation curves of DNA from one species hybridized with another can be compared to the DNA of a single species hybridized with itself to determine how different the two species are.

It is not possible to compare the entire genome in this manner. Highly repeated DNA hybridizes readily both within and between species and tends to obscure any differences that may be present in the single-copy fraction, which is usually the most informative fraction of the genome for evolutionary studies. Typically, the single-copy DNA is first isolated from one species, then labeled radioactively. The labeled DNA is then denatured and hybridized with excess unlabeled DNA of its own species and excess unlabeled DNA from other species that are to be compared to it. Because there is a surplus of unlabeled DNA, nearly all of the labeled DNA hybridizes with complementary sequences of unlabeled DNA.

After DNA hybridization is complete, the DNA is denatured by slowly increasing the temperature. This makes it possible to determine the denaturation curve for each sample. Recall that a DNA denaturation curve is S-shaped and can be described by a value called $t_{1/2}$, the temperature at which half the DNA is denatured (see section 10.1). The $t_{1/2}$ of the radioactively labeled DNA that has hybridized with the unlabeled DNA can be determined for each mixture. The difference between the $t_{1/2}$ of DNA hybridized with a different species and the DNA hybridized with itself, a value called $\Delta t_{1/2}$, is a measure of the genetic similarity of the two species. Each degree of $\Delta t_{1/2}$ corresponds to about 1% mismatched nucleotides, so a $\Delta t_{1/2}$ of 4°C between two species indicates that there is about a 4% difference in the nucleotide sequence of the single-copy DNA fraction. The less the value of $\Delta t_{1/2}$, the greater the genetic similarity between the two species.

A comparison of *Drosophila* species by DNA hybridization is illustrated in Figure 21.9. Labeled *D. melanogaster* DNA was hybridized with itself and with DNA from *D. funebris* and *D. simulans*. The thermal stability of the hybridized DNA was determined by measuring the amount of labeled DNA that was released as single strands as the temperature was increased. According to the graph, $\Delta t_{1/2} = 14°C$ between *D. funebris* and *D. melanogaster*, and $\Delta t_{1/2} = 3°C$ between *D. simulans* and *D. melanogaster*. These observations indicate that *D. simulans* is more closely related to *D. melanogaster* than *D. funebris* is.

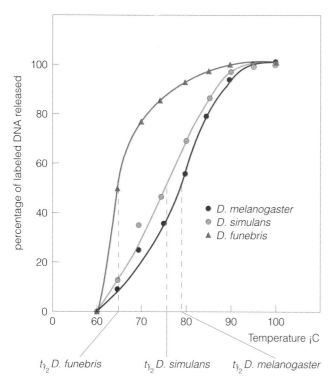

Figure 21.9 Comparison of *Drosophila funebris* and *D. simulans* with *D. melanogaster* using DNA hybridization. (Adapted from Laird, C. D., and B. J. McCarthy. 1968. Magnitude of interspecific nucleotide sequence variability in *Drosophila*. *Genetics* 60:303–322. Copyright Genetics Society of America. Used by permission.)

Gene Organization

Genes in eukaryotes are frequently organized in complex ways. Eukaryotic genes are often organized into gene families in which a cluster of genes may contain duplicated functional genes and pseudogenes. Because the complex organization of each gene family probably evolved from a single gene, differences in gene organization may provide clues about genetic divergence among species.

Gene duplication can have several consequences. Initially, it creates redundancy, so if a mutation alters the products of one gene, the other gene's functional product is still present. This reduces the amount of selection pressure against deleterious mutations in one of the genes. Eventually, as the duplicated genes diverge, one of them may take on a slightly different function. This appears to be the case with the α and β hemoglobins. They are about 35% similar in amino acid sequence, indicating a common origin, but have evolved into different subunits that combine to form the functional molecule. They are no longer redundant. Both the α and β forms are essential for survival in mammals. The α and β subunits are also encoded by several genes that are transcribed at different times during development (see

Figure 10.28). In addition, there are several pseudogenes that are no longer functional, apparently because of debilitating mutations that eliminated gene function. Most pseudogenes are no longer transcribed but remain as relics of evolution.

If we look at the organization of the globin gene family in other species and compare it to the mammalian organization, an evolutionary pattern emerges. Modern lampreys, parasitic fish that attach themselves to other fish, are very similar to their fossilized counterparts and have probably evolved relatively little from their common ancestor with bony fish. Lamprey skeletons are cartilaginous, and they (along with other cartilaginous fish, such as sharks) predate bony fish in the fossil record. Lamprey hemoglobin is a very primitive, monomeric type, consisting of a single protein molecule with identical subunits. Sharks, on the other hand, have a tetrameric (four-part) hemoglobin that consists of α and β subunits. Tetrameric hemoglobin apparently carries oxygen more readily than monomeric hemoglobin, so it should be favored by selection when improved oxygen transport contributes to a higher fitness, as is the case with sharks, who must expend much energy in swimming. Parasitic lampreys, on the other hand, attach themselves to fish and are simply carried along by their host, so improved oxygen transport is not essential.

Monomeric and tetrameric hemoglobin apparently diverged about 500 million years ago, before the appearance of bony fish in the fossil record. The β genes have since followed a pattern of divergence that has led to five active genes and one pseudogene in the mammalian globin gene family. The prenatal and adult β genes diverged about 200 million years ago, about the time when reptiles emerged, and the embryonic and fetal β genes diverged about 100 million years ago, when mammals first appeared. The emergence of fetal hemoglobin was particularly important for placental mammals because fetal hemoglobin (with two α subunits and two γ subunits) has a greater affinity for oxygen than adult hemoglobin (two α subunits and two β subunits). The higher oxygen affinity of fetal hemoglobin when compared to adult hemoglobin ensures that oxygen transport across the placenta has a net flow from mother to fetus, permitting the fetus to receive adequate oxygen during its development.

Ancient DNA

Most evolutionary studies with DNA are restricted to species currently alive. DNA is a very fragile and reactive molecule that is protected in living cells. When an organism dies, its DNA usually degrades rapidly. Thus, researchers could hardly expect DNA to remain intact in fossilized remains. However, several reports have surfaced in which putative ancient DNA has been recovered and sequenced. Among the most interesting is a case in

which a purported ancient bacterial species was revived (see problem 22 at the end of this chapter). In each case, special conditions (such as encasement of ancient insects in amber) may have permitted fragments of ancient DNA to escape complete degradation.

Much skepticism has been raised in response to reports of the recovery of ancient DNA. Ancient DNA, if present, has probably been degraded into small fragments that are very low in quantity. The polymerase chain reaction (PCR) has usually been used in attempts to recover ancient DNA sequences because PCR is effective in amplifying DNA fragments that are present in low quantities. However, this advantage is also a disadvantage because contaminating modern DNA that may be present in the sample or the surrounding environment might also be amplified. In attempts to recover true ancient DNA, most researchers take great care to avoid contamination with modern DNA, and conduct numerous control experiments to test for contamination.

Several scientists have published well-controlled and convincing studies that suggest the recovery of true ancient DNA. However, attempts to repeat these experiments often fail. The possible recovery of ancient DNA is intriguing and exciting. However, it must be viewed with some skepticism until additional research provides enough information to determine with a high degree of certainty whether or not the sequences recovered are truly from ancient DNA.

~

The forces that drive evolution have their ultimate effect at the DNA level. Evolution can be reconstructed by comparing DNA and protein polymorphisms.

21.3 MAINTENANCE OF GENETIC DIVERSITY

The classical Darwinian approach to evolution has centered on the hypothesized dominant role of natural selection as the driving force in evolution. This view has changed with the realization that forces other than selection can have a significant effect. In this chapter and in the previous two chapters, we have discussed factors that act in the absence of selection, or together with it, to determine the direction in which evolution proceeds. Most of the factors that drive evolution decrease diversity, working toward homozygosity and fixation of alleles. Directional selection, drift, inbreeding, and assortative mating eliminate some alleles and fix others in populations, with a subsequent loss of genetic diversity. Indeed, many alleles and chromosomal arrangements are fixed, or at least present in very high frequencies, within each species. Mutation, certain types of stabilizing selection, and infusion of new alleles by immigration all increase or maintain diversity. Many natural populations have very high levels of diversity and heterozygosity for cer-

tain genes. Natural populations of *Drosophila*, although highly uniform in phenotype, are actually quite diverse in their genetic constitution, with high levels of heterozygosity for DNA and enzyme markers. How can such high levels of diversity and heterozygosity be maintained when some of the most important evolutionary forces work toward uniformity and homozygosity?

There is no consensus among scientists to explain this situation. Apparently, there is no consensus in nature's methods either. Some species are highly heterozygous, and others are not. Cheetahs, for example, are highly homozygous and highly uniform, possibly due to a bottleneck effect that the entire species suffered at some point in its history. Populations of self-pollinating plants are also highly homozygous due to the inbreeding that their reproductive system forces upon them. Other plant and animal populations may be highly heterozygous for a number of reasons.

The classical approach to mutation assumes that each mutation has a different fitness. Nearly all mutant alleles are assumed to be deleterious, with reduced fitness compared to existing alleles, so the mutant alleles are removed, or maintained at very low frequencies, by selection. The few alleles that confer higher fitness than existing alleles are favored by selection and eventually become fixed in the population.

The **neutrality theory** recognizes that some mutations are deleterious and disfavored by selection, whereas a few are beneficial and favored by selection. The frequency and eventual fixation or loss of these alleles in a population is highly influenced by selection. Other mutations, however, are selectively neutral. These may include silent mutations and amino acid substitutions that do not significantly alter the function of a gene product. These mutations are unaffected by selection. Their frequency in the population is subject to the effects of random genetic drift. Because all populations are finite, random genetic drift can change the frequency of any neutral allele, driving it randomly toward fixation or loss. Mathematically, the probability of substitution of one neutral allele for another is equal to the rate at which that allele arises by mutation, regardless of population size. However, the rate of fixation for the substituted allele depends on population size. The larger the population, the slower the rate of fixation.

Random genetic drift can explain the presence of diversity and heterozygosity for neutral alleles. In large populations, the rate of fixation for an allele is slow, so neutral alleles that arise by mutation may be present at various stages of loss or fixation. At any point in time, the level of heterozygosity for neutral alleles should be close to $2pq$, as predicted by Hardy-Weinberg equilibrium, unless the population is very small. The smaller the population, the greater the possibility of significant deviations from Hardy-Weinberg equilibrium due to sampling error.

Selection theory proposes that there may be a selective advantage for heterozygosity at certain loci in the

population, and that selection maintains a high level of diversity and heterozygosity for the chromosomes that carry these loci. The often cited example of heterozygote advantage for the sickle-cell anemia (HBB*S) allele in humans shows how heterozygotes may be maintained by selection. Another well-studied example is the apparent selective advantage for individuals that are heterozygous for certain inversions in chromosome 3 in *Drosophila pseudoobscura* (see problems 23–25 in Chapter 19). Heterosis also provides a selective advantage for heterozygotes when genetic loads are high.

Selection and neutrality theories are not necessarily at odds with one another. There is good evidence that both theories are correct: while one species may have a high degree of heterozygosity due primarily to neutrality, another may be highly heterozygous due to selection. For the time being, it is not possible to determine the relative fitness levels of many alleles, particularly those represented by DNA or protein markers. With no estimate of relative fitness, it is difficult to determine whether selection or random genetic drift (or a combination of the two) is most responsible for the observed diversity.

Neutrality theory predicts a constant rate of change in regions of a gene that are selectively neutral. It also predicts that conserved nucleotide sequences within a gene change at a much slower rate than neutral mutations—slow enough to not significantly affect the overall rate of change. Beneficial mutations should be so rare that they also do not significantly affect the overall rate of change. The observation that amino acid substitutions in proteins are linear over evolutionary time indicates that some aspects of neutrality theory are correct, at least for certain regions of the proteins studied. At the same time, examples of heterozygote advantage in nature provide evidence of the effect of selection in maintaining heterozygosity.

~

Drift and selection, two of the most important forces in evolution, often reduce genetic diversity, eventually resulting in homozygosity and fixation of alleles. Neutrality and selection theories describe possible ways for maintaining heterozygosity and genetic diversity.

21.4 PHYLOGENETIC ANALYSIS

Much of evolutionary research is based on comparison of species, either living or extinct, for similarities and differences. A major assumption that underlies the comparisons is that genetic similarities imply common ancestry, so two species with recent common ancestry should be more similar genetically to one another than a third species with less recent common ancestry. The usual depiction of such an evolutionary history is in the form of a **phylogenetic tree**, a representation of the simplest path through which related races or species may have descended from common ancestry.

Construction of a Phylogenetic Tree

There are many methods for constructing phylogenetic trees, some of which are mathematically and statistically advanced. Most methods, however, are based on the premise that paths of common ancestry requiring the fewest mutations are most likely to be correct.

For example, let's use data from the amino acid sequence of cytochrome *c* in Figure 21.8 to compare five mammals: human, chimpanzee, rhesus monkey, horse, and donkey. The human and chimpanzee sequences are identical, and the rhesus monkey sequence differs from human and chimpanzee by one amino acid substitution (amino acid 102 is threonine in human and chimpanzee; it is alanine in rhesus monkey). The horse and donkey sequences also differ by only one amino acid substitution (amino acid 47 is serine in donkey and threonine in horse). All five species in question can be compared in all possible pairwise combinations, and the number of amino acid substitutions that differentiates each pair summarized, as shown in Table 21.3.

We can use the information in Table 21.3 to construct a phylogenetic tree. We'll let the horizontal axis of the tree represent the average number of amino acid substitutions that differentiate species in the tree. Let's begin with the two most similar sequences, which are chimpanzee and human. There is no difference in the cytochrome *c* amino acid sequences of these two species, so we connect them with a vertical line:

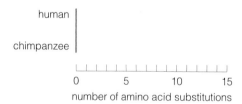

We can now add rhesus monkey, whose cytochrome *c* differs from the human and chimpanzee sequence by one amino acid substitution. These three species represent one group. Also, horse and donkey differ from one another by one amino acid substitution and can be added as a second group:

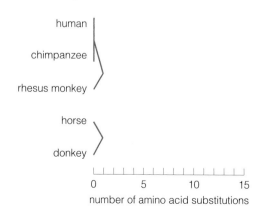

Table 21.3   All Pairwise Comparisons for Number of Amino Acid Substitutions in Cytochrome *c* for Five Mammalian Species

Species	Species				
	Human	Chimpanzee	Rhesus monkey	Horse	Donkey
Human	0	0	1	12	11
Chimpanzee		0	1	12	11
Rhesus monkey			0	13	12
Horse				0	1
Donkey					0

Table 21.4   Determination of Average Amino Acid Substitutions Between the Human/Chimpanzee/Rhesus Monkey Group and the Horse/Donkey Group

Pairwise Comparison	Number of Amino Acid Substitutions
Human–horse	12
Human–donkey	11
Chimpanzee–horse	12
Chimpanzee–donkey	11
Rhesus monkey–horse	13
Rhesus monkey–donkey	12
Average number of amino acid substitutions	11.83
Standard deviation	0.69

To connect these two groups, we must calculate the average number of amino acid substitutions that separates them. As shown in Table 21.4, there are six pairwise comparisons between the two groups. The number of substitutions in the six comparisons averages 11.83, so the two groups can be connected by representing the difference between them as 11.83 amino acid substitutions:

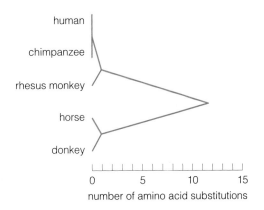

This is a simple example of how researchers construct phylogenetic trees. Most phylogenetic trees include many

entries, and constructing one can be quite complicated. Researchers usually rely on computer programs that make the necessary calculations rapidly. In most cases, alternative trees may arise from the analysis. Notice in Table 21.4 that a standard deviation was calculated along with the average amino acid substitutions for the branch point between the two groups. The best tree is usually the one that has the lowest set of standard deviations when averages at each branch point in the tree are calculated. Also, conversion and appropriate weighting of the data may be required to reduce the variation and select the best tree.

The following example shows how constructing a phylogenetic tree can reveal evolutionary relationships.

Example 21.5   Construction of a phylogenetic tree from amino acid sequence diversity in cytochrome *c*.

In 1967, Fitch and Margoliash (*Science* 155:279–284) published a phylogenetic tree based on amino acid sequences in cytochrome *c*. They converted each amino acid substitution for cytochrome *c* into the

minimum number of mutations required for that amino acid substitution for 20 species of animals and fungi. They then analyzed the differences among the 20 species and arrived at 40 alternative phylogenetic trees. Figure 21.10 shows the tree with the lowest standard deviations for average minimum number of mutational differences and thus represents the best tree given the available data.

Problem: How well does the phylogenetic tree correspond with traditional taxonomic groupings?

Solution: The groupings that emerged in this tree are based entirely on amino acid sequences and derived DNA sequences, with no consideration of traditional taxonomic classifications. Even so, the groupings correspond closely with traditional classifications. The most distant branch point on the tree separates the fungi (*Candida, Saccharomyces,* and *Neurospora*) from the animals. Among the animals, the most distant branch point separates the vertebrates from the invertebrates. Among the vertebrates, the birds are in one group, and all the mammals are in another group.

Comparison of Data from Different Sources in Phylogenetic Analysis

The patterns of evolution are manifest at many interrelated levels, including nucleotide sequences in DNA, amino acid sequences in proteins, gene and chromosomal organization, chromosome number, and the ultimate expression of genes in the external phenotype. Methods for studying evolution include comparative anatomy among both living and fossilized species, examination of differences in chromosome structure and number, identification of genome sequence and organizational divergence as revealed by amino acid or DNA sequencing, molecular marker analysis, and DNA hybridization. Data obtained from these various methods can be combined with information obtained from biogeographical and ecological studies to determine evolutionary histories. In most cases, information obtained by one method confirms the information obtained by another method, and provides independent evidence of how evolution proceeds.

In this section, we will look at our own species as an example of how scientific data gathered from a variety of sources have been compared in efforts to decipher evolutionary history. At the outset, it is important to point out that our understanding of human evolution is by no means complete. Conclusions drawn from several independent studies support one another, but some studies have produced conflicting data. The search to identify the ancient genetic heritage of humans remains very active and controversial.

Fossilized remains of species that have human characteristics have been studied for more than a century. At one time, such fossils were the principal source of information on human evolution. The fossil record has continued to grow as more fossils are discovered. In recent years, the fossil record has been augmented by studies with DNA markers and nucleotide sequences.

Most studies of the fossil record suggest an African origin for humans. Studies based on DNA markers and DNA sequences among modern humans also suggest an African origin. However, there is no general consensus about when and how humans spread to other parts of the world. There are two major theories: the multiple-origins theory and the single-origin theory. The **multiple-origins theory** is based on the hypothesis that modern humans evolved simultaneously from several *Homo erectus* populations in different parts of the world, aided by migration and gene flow among the developing human populations. *Homo erectus* is an extinct species that has numerous anatomical similarities to modern humans but is sufficiently different from humans to be classified as a separate species. The fossil record of *H. erectus* suggests migration from Africa about 1.8 million years ago. The fossil record also indicates that races of *H. erectus* persisted in the Far East as late as 200,000 years ago, which coincides with the dates of early human fossils outside of Africa. Some features of *H. erectus* fossils in the Far East and Australia are similar to features that are characteristic of indigenous humans in these areas, an observation suggesting that humans developed simultaneously in these areas from different *H. erectus* populations.

The **single-origin theory** (also known as the **"out of Africa" theory**) is based on the hypothesis that modern humans arose once in Africa and then migrated from there, displacing other humanlike species that had migrated earlier. Most studies of diversity for DNA markers and DNA sequences among modern humans support the single-origin theory. These studies are based on the assumption that the highest genetic diversity should exist in the region of origin and that less diversity should be present in populations that descended from founders who migrated to other regions. Studies of nuclear and mitochondrial DNA reveal that indigenous sub-Saharan African populations are much more genetically diverse than all non-African human populations combined.

Some of the most extensive studies of molecular evolution in humans are based on analysis of mitochondrial DNA. Their interpretations are based on the assumption that mitochondrial DNA is inherited in a strict uniparental-maternal fashion in humans, so there is no opportunity for recombination of alleles. Mitochondrial DNA diversity must, therefore, be due to mutation only, and the diversity must be transmitted only through maternal lineages. The initial interpretation of RFLP diversity in human mitochondrial DNA suggested that all mitochondrial genotypes in modern humans could be

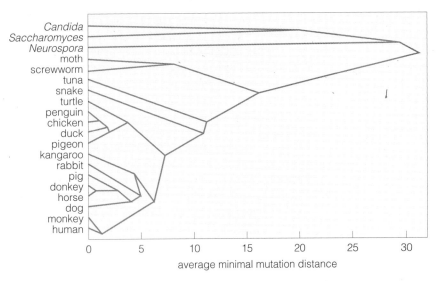

x-axis: average minimal mutation distance (0, 5, 10, 15, 20, 25, 30)

Labels (top to bottom): Candida, Saccharomyces, Neurospora, moth, screwworm, tuna, snake, turtle, penguin, chicken, duck, pigeon, kangaroo, rabbit, pig, donkey, horse, dog, monkey, human

Figure 21.10 A phylogenetic tree based on amino acid sequence similarities and differences in cytochrome *c*. (Adapted with permission from Fitch, W. M., and E. Margoliash. 1967. Construction of phylogenetic trees. *Science* 155:279–284. Copyright 1967 American Association for the Advancement of Science. Used by permission.)

traced phylogenetically to a single founder genotype from Africa. On the basis of the presumed mutation rate (which is higher in mitochondrial DNA than in nuclear DNA), the founder genotype was dated to a period between 100,000 and 300,000 years ago, which suggests that humans first migrated from Africa during or after this time. In popular and scientific articles, the founder genotype has been called the mitochondrial Eve.

Researchers conducted subsequent studies with larger samples and DNA sequencing of the D-loop region of mitochondrial DNA. Also, existing mitochondrial DNA data was reanalyzed using different assumptions, including the assumption that inheritance of mitochondrial DNA in humans may not be entirely uniparental-maternal. The subsequent studies also support an African origin for humans but differ substantially in the dating of the founder genotype with molecular clocks. Estimates range from as long as 800,000 years ago to as recent as 60,000 years ago.

Studies based on minisatellite markers and an *Alu* deletion allele derived from nuclear DNA also show high diversity among sub-Saharan African populations and much less diversity among humans throughout the rest of the world, again supporting the single-origin theory, as highlighted in the following example. Molecular clocks applied to data from nuclear DNA suggest dates of less than 150,000 years ago for the beginning of migration from Africa.

Example 21.6 Genetic diversity for coupled nuclear mutations in humans.

In 1996, Tishkoff et al. (*Science* 271:1380–1387) reported the results of experiments designed to measure DNA marker diversity for coupled mutations

at the *CD4* locus on chromosome 12 in humans. The two sites studied were located 9.8 kb apart (much less than 1 cM in map distance) in regions outside of the reading frame of the *CD4* locus. The first site is a microsatellite that contains a tandem pentanucleotide repeat (TTTTC). Twelve different alleles with 4–15 repeats were found at this site. The second site contains an *Alu* element (see section 10.1 for a description of *Alu* elements). Two different alleles are found in human populations at this site. One allele contains the intact *Alu* element; the other contains a deletion of 265 nucleotide pairs in the *Alu* element. The authors examined over 1600 individuals from 42 geographically diverse, native populations. The greatest diversity for the 12 microsatellite alleles was found in sub-Saharan African populations. In fact, all 12 possible alleles were found only in sub-Saharan African populations, and two of the alleles were exclusive to sub-Saharan African populations. The *Alu* deletion allele, *Alu*(−), was rare or absent in Asian, New World, and Pacific Island populations. Its frequency ranged from 25% to 30% among European, Middle Eastern, and northeast African populations, and from 7% to 28% in sub-Saharan African populations. When present, the *Alu*(−) allele was coupled almost exclusively with the six-repeat microsatellite alleles in all populations except the African populations. Among sub-Saharan African populations, the *Alu*(−) allele was coupled with 9 of the 12 microsatellite alleles.

Problem: (a) What advantages do these two mutation sites in the *CD4* locus provide for a study of nuclear DNA evolution? (b) From information pre-

sented in the article, summarize the authors' interpretations of their data.

Solution: (a) The two sites provide several advantages. First of all, they are in noncoding DNA, and the mutations should, therefore, create selectively neutral alleles. Second, the microsatellite site should undergo expansion and contraction mutations that create several different tandem repeat alleles, 12 alleles in this study. The deletion of 265 nucleotide pairs in the *Alu* sequence is a stable mutation that is unlikely to be repeated, so this particular mutation probably arose only once in human history. Third, the two sites are very closely linked, so recombination between them should be very rare. Thus, mutant alleles should nearly always remain coupled as they are transmitted from one generation to the next. **(b)** The authors of this study interpreted their data by concluding that at some time during the development of human populations in Africa, a deletion created the *Alu*(−) allele, which increased in frequency, probably due to random genetic drift. When the *Alu*(−) allele arose, it should have been coupled to only one allele at the microsatellite site (probably the allele with six repeats), but over time, through expansion and contraction mutations of the microsatellite and through rare recombination events, the *Alu*(−) allele became coupled to several different alleles at the microsatellite site, as was observed among the sub-Saharan African populations. People who migrated from Africa carried the chromosome with the *Alu*(−) allele coupled to the microsatellite allele with six repeats. This was apparently the only chromosome with the *Alu*(−) allele to leave Africa, an example of the founder effect. This chromosome was almost lost in Asian, New World, and Pacific Island populations (probably due to drift and secondary founder effects) but persisted in European and Middle Eastern populations. Given sufficient time, mutations at the microsatellite site and rare recombination events should produce chromosomes in which several microsatellite alleles are coupled to the *Alu*(−) allele in the European and Middle Eastern populations. The proportion of such chromosomes in European and Middle Eastern populations is small, indicative of a recent migration (probably less than 102,000 years ago, according to the authors' calculations) of the *Alu*(−) chromosome from Africa. The authors concluded that their results support the single-origin theory.

Information from morphological, geographical, biochemical, and molecular analyses can be synthesized to reconstruct possible phylogenies.

SUMMARY

1. A species is defined as a group of interbreeding individuals that produce fully fertile offspring and are isolated reproductively from other such groups. Some form of reproductive isolation is necessary for speciation.

2. There are several types of reproductive isolating mechanisms. They can be divided into two general categories: prezygotic and postzygotic.

3. Prezygotic mechanisms prevent or inhibit formation of zygotes. They typically prevent or inhibit mating. Examples include geographic isolation, mechanical isolation, and temporal isolation.

4. Postzygotic mechanisms do not prevent mating and zygote formation but cause a loss of viability or fertility in the hybrid zygote or embryo. Examples include chromosomal incompatibilities such as inversions, translocations, and differences in ploidy.

5. After speciation is complete, gene pools of different species evolve independently. Mutations may appear and become fixed in one group but not another. The genetic divergence of reproductively isolated species that share common ancestry is correlated with the amount of time that has passed since reproductive isolation. Thus, differences in DNA and amino acid sequence can be used to determine how long ago species, or major groups of species, diverged from common ancestry.

6. Nucleotide and amino acid substitutions appear at about the same rate for a particular gene or site in DNA. The number of substitutions can be used to estimate when two groups or species diverged, a concept known as a molecular clock.

7. Because of selection, genes may evolve at different rates, so molecular clocks for genes, and even for regions within genes, may tick at different rates.

8. Drift, selection, and inbreeding are mechanisms that cause fixation of alleles and loss of genetic diversity. Although many alleles are fixed in each species, other alleles may be maintained at high levels of heterozygosity. Neutrality and selection theories explain how genetic diversity can be maintained.

9. Information gathered from a variety of sources, such as analyses of morphological characteristics, chromosome organization, DNA sequences, and amino acid sequences, can be synthesized to identify phylogenetic relationships and reconstruct evolutionary histories.

QUESTIONS AND PROBLEMS

1. In Figure 21.3, what evidence and theoretical concepts suggest that human chromosome 2 arose from a fusion and that the corresponding chromosomes in the great apes did not arise by fission?

2. The most abundant proteins in plant seeds are seed storage proteins. When the seed germinates, these proteins are rapidly broken down and provide the amino acids for the developing seedling to synthesize proteins until it can synthesize its own amino acids from photosynthetic products. The amino acid sequences of seed storage proteins vary widely among plants. On the other hand, the amino acid sequence of cytochrome *c* is nearly the same in all plants. Why can so much divergence be present in seed storage proteins but not in cytochrome *c*?

3. In Figure 21.5, *Drosophila subobscura* is shown as ancestral to *D. melanogaster*, and *D. melanogaster* is shown as ancestral to *D. ananassae*. Alternatively, *D. ananassae* may be ancestral to *D. melanogaster*, and *D. melanogaster* may be ancestral to *D. subobscura*. Why is this alternative explanation an unlikely possibility?

4. In 1995, Dorit et al. (*Science* 268:1183–1185) reported results of their examination of a 729 nucleotide-pair intron sequence in the human Y chromosome among 38 males of different worldwide geographical origins. They found no nucleotide sequence diversity whatsoever. When the same region of DNA was examined in other primates, DNA sequence diversity was found. **(a)** What reasons could be given for choosing a DNA sequence from an intron for study instead of using the DNA sequence from a polypeptide-encoding region? **(b)** What possible interpretations can be made from the results of this study?

5. Comparison of mitochondrial DNA diversity in humans with mitochondrial DNA diversity in chimpanzees reveals that chimpanzee mitochondrial DNA is about four times as diverse as human DNA. Yet chimpanzees have a relatively small geographic distribution when compared to humans. What does this information suggest about the origin of humans?

6. When a female horse is mated with a male donkey, the resulting offspring is a mule. When a male horse is mated with a female donkey, the offspring is a hinny. Mules are far more valuable as draft animals than hinnies, and the two differ substantially in their phenotypes. Both mules and hinnies can be either male or female, and both mules and hinnies are sterile regardless of sex. **(a)** What genetic and nongenetic possibilities could explain the difference between mules and hinnies? **(b)** Why are both mules and hinnies sterile?

7. In 1975, Stutz et al. (*American Journal of Botany* 62:236–245) described a population of unusually large-statured four-wing saltbush (*Atriplex canescens*) that is diploid and grows only on a limited area of shifting sand dunes in central Utah. In close proximity just off of the sand dunes are populations of shorter-statured tetraploid four-wing saltbush. In some areas, the diploid and tetraploid populations overlap each other geographically. The diploid plants flower in July, whereas the tetraploid plants flower in late May to early June. Hybrids between plants of the two populations rarely form in nature and are sterile. **(a)** Name at least one prezygotic and one postzygotic mechanism that reproductively isolate the two types. **(b)** Describe why hybrids are rare and why they are sterile. **(c)** Describe why the diploid and tetraploid populations should or should not be classified as different species.

8. Cultivated soybean carries the Latin species name *Glycine max*. It originated in China and is now grown throughout the world. It is grown only under cultivated conditions and has never been found growing in the wild. *G. soja* grows wild throughout East Asia and differs substantially in phenotype from *G. max*. *G. max* × *G. soja* hybrids are never found in nature, but when hybridized artificially, the two species produce fully fertile F_1 offspring. Both species have $2n = 40$ chromosomes. *G. tabacina* and *G. tomentella* are species that grow wild in Australia, and both have $2n = 80$ chromosomes. When hybridized, they produce sterile F_1 hybrids. Neither of these two species produces viable hybrids (sterile or fertile) when hybridized with *G. max*. Another species, *G. wightii*, grows wild in Southeast Asia. It has two chromosome races, one with $2n = 22$ and one with $2n = 44$ chromosomes. When hybridized with *G. max*, it fails to produce viable hybrids. **(a)** Based on the information given for chromosome numbers, geographical distribution, and hybrid fertility and viability, propose a possible evolutionary scheme that describes chromosome evolution (how the observed chromosome numbers could have arisen) and geographical dispersal. **(b)** Describe the reproductive isolating mechanisms that are present among these species. **(c)** Identify the species that represents the most probable ancestral type, and describe why it is the best choice.

9. When *G. max* is compared with *G. soja*, there are several distinct phenotypic differences. The seeds of *G. max* are much larger and fewer in number per plant. The pods that contain the seeds of *G. max* tend to hold the seeds on the plant and do not open easily, whereas the pods of *G. soja* break open as soon as they are dry, dropping the seeds on the ground. The seeds of *G. max* germinate readily when planted, resulting in nearly 100% germination. The seeds of *G. soja* do not germinate as readily. Seeds that do not germinate one year may remain in the ground and germinate in subsequent years, however. These differences are highly heritable and have probably arisen by selection. Describe why selection would favor these differences for each species in its current habitat.

10. In 1968, McNeilly and Bradshaw (*Evolution* 22:108–118) described two adjacent populations of *Agrostis tenuis* (a wild grass), one that grows on normal pasture soils, and a second that grows on the polluted tailings of an abandoned mine. They found that the plants that grow on the mine tailings are genetically resistant to excess soil copper, whereas the plants that grow

on the adjacent pasture soil cannot grow in soil with elevated levels of copper. Further experiments showed that copper resistance is highly heritable. McNeilly and Antonovics (1968. *Heredity* 23:205–218) found that the population growing on the mine tailings flowered about a week earlier than the adjacent population on pasture soil. When plants from the two populations were grown in a uniform environment, they still flowered about a week apart. When the two populations are manually cross-fertilized, the hybrids that arise from them are fully fertile. **(a)** What types of reproductive isolating mechanisms are illustrated by this example? **(b)** Selection for copper resistance is expected for plants that grow on the mine tailings. However, there is a less obvious reason why differences in flowering time should also be favored by selection. Explain why differences for copper resistance and differences for flowering time are both affected by selection in these two populations. **(c)** Describe why speciation may be the eventual result of this situation.

11. Occasionally, a geographically isolated population is found in which all individuals have a trait that is not favorable for the environment in which the population exists. Other isolated populations of the same species in similar environments typically carry the more favorable trait. What genetic and evolutionary principles could explain this situation?

12. Among species in nature, most traits are governed by several genes, and the effects of dominance and epistasis for these traits are fairly low. There are good reasons for this observation from an evolutionary point of view. Using the information from this chapter and the previous two chapters, describe why this situation is expected.

13. There is no variation among species for the amino acid sequence of cytochrome *c* between residues 70 and 80 (see Figure 21.8). What does this imply about this segment of amino acids?

14. At position 12 in the amino acid sequence for cytochrome *c*, primates have methionine, whereas most other animals have glutamine (except the rattlesnake, which also has methionine). Fungi and plants have threonine at this site. **(a)** Using the genetic code and the information in Figure 21.8, determine which amino acid is most probably the ancestral type, and describe the mutations that must have arisen to cause the pattern in Figure 21.8. **(b)** Rattlesnake has the same amino acid at this site as primates. Does this observation indicate that it is phylogenetically closer to primates than to other reptiles? Please explain your answer.

15. Problems 23 and 24 in Chapter 19 discuss a well-studied case in which inversion heterozygosity in *Drosophila* is maintained in nature. Describe which theory, neutrality or selection, best explains this example of high inversion heterozygosity in *Drosophila*.

16. In 1994, Woodward et al. (*Science* 266:1229–1232) reported the recovery and sequencing of DNA from 80-million-year-old bone fragments recovered from a coal mine. They extracted several small bone samples under sterile conditions to prevent contamination, then amplified DNA from the *cytochrome b* gene using PCR. They also extracted small samples of the surrounding rock and subjected them to the same treatments and PCR reactions as the bone samples. The bone samples produced a DNA amplification product, but the rock samples did not. They then determined the DNA sequences of the amplified fragments. The DNA sequences did not match closely those of any known modern species, but they were most similar to birds, reptiles, and mammals, all with about 70% homology. The most distant sequence was a honeybee sequence, with about 38% homology. **(a)** What evidence suggests that the DNA sequences were truly obtained from ancient DNA? **(b)** Read their article and a skeptical response (Gibbons. *Science* 266:1159). What aspects of this work were challenged?

17. In 1974, Nevo et al. (*Evolution* 28:1–23) described genetic variation in different species of pocket gophers in the western United States. The gophers fill a subterranean niche whereby they rarely come above ground. For that reason, there is little migration among populations. The species are indistinguishable morphologically and do not differ substantially based on isozyme analysis. However, chromosome numbers among populations include $2n = 44$, $2n = 46$, $2n = 48$, and $2n = 60$. **(a)** What reproductive isolating mechanisms are present among these species? **(b)** What could explain the lack of isozyme and morphological diversity?

18. In 1995, Wills (*Evolution* 49:593–607) reported a reexamination of human mitochondrial DNA sequence data. Previous studies had been done under the assumption of constant mutation rates. Wills transformed the data to account for a higher mutation rate for transitions when compared to transversions and arrived at somewhat different conclusions than in previous studies. Why might such a transformation of data be appropriate when analyzing nucleotide sequences for phylogenetic analysis?

19. Look at chromosome 12 in Figure 21.3. Chimpanzee and gorilla chromosomes are similar to one another, and human and orangutan chromosomes are similar to one another. This pattern of similarity and diversity in chromosome 12 differs from the patterns for most other chromosomes. Explain how such a pattern for chromosome 12 may have arisen.

20. Construct a pairwise comparison table and a phylogenetic tree for chicken, turkey, pigeon, Pekin duck, snapping turtle, and rattlesnake using the amino acid sequences for cytochrome *c* in Figure 21.8. Do the groupings based on amino acid sequence substitutions correspond well with traditional taxonomic classifications of birds and reptiles?

21. Read the article by Powell and Moriyama (1997. *Proceedings of the National Academy of Sciences, USA* 94:7784–7790), then explain why preferential mutation toward C and G in *Drosophila* is not an adequate explanation of codon usage bias.

22. In 1995, Cano and Borucki (*Science* 268:1060–1064) reported the revival and growth of a purported ancient bacterial culture derived from within ancient bees fossilized in amber. Scientific journals often encourage scientific criticism of research published in the journal. In this case, *Science* published two responses criticizing aspects of the research, along with a response to the criticisms from one of the authors. In addition, several more recent studies designed to test the possibility that ancient DNA may have been preserved have been published. Read the article by Cano and Borucki (1995. *Science* 268:1060–1064) and its accompanying news release (Fishman. 1995. *Science* 268:977), the criticisms by Priest (1995. *Science* 270:2015) and Beckenbach (1995. *Science* 270:2015–2016), and the response by Cano to the criticisms (1995. *Science* 270:2016–2017). Also read the scientific news releases by Service (1996. *Science* 272:810) and Sykes (1997. *Nature* 386:764–765) that describe subsequent research. **(a)** What evidence provided by Cano and Borucki suggests that the bacteria they cultured were of ancient rather than modern origin? **(b)** What aspects of the research were criticized by the responding scientists, and what response did the authors provide? **(c)** Summarize the results of subsequent research highlighted in the news releases.

23. In 1996, Armour et al. (*Nature Genetics* 13:154–160) published a study of diversity in human populations at the *MS205* minisatellite locus. This locus has a polar mutational property, common in minisatellite loci, in which most repeat-number mutations arise at one end of the tandem-repeat sequence, whereas the other end mutates only rarely. Alleles could, therefore, be grouped according to similarities of sequences at the stable end of the locus. The authors identified 44 groups of alleles on this basis. Alleles identified in non-Africans were classified in 13 groups, and 93.4% of the alleles were classified in 6 groups that are similar to one another. Alleles identified in African populations were classified into 37 groups, with 42% of the alleles in the same 6 groups that predominated in non-African populations. **(a)** Compare these results with the results of Tishkoff et al. (1996. *Science* 271:1380–1387) highlighted in Example 21.6. **(b)** How might these results be interpreted in light of the multiple-origins and single-origin theories?

24. In the paper cited in the previous question (Armour et al. 1996. *Nature Genetics* 13:154–160), the authors discussed several alternative scenarios that could explain their data. In the end, they concluded that when their data were interpreted together with data from other experiments, the combined information was best explained by the single-origin theory. What alternative scenarios could explain the general observation that genetic diversity is much lower among non-Africans than among Africans? (You may wish to consult the paper by Armour et al. to review the alternatives they discussed.)

25. In 1994, Transue et al. (*Crop Science* 34:1385–1389) compared 282 randomly chosen polymorphic PCR-based DNA markers in three plant species of the same genus, *Amaranthus cruentus, A. caudatus,* and *A. hypochondriacus.* These three species are difficult to classify morphologically. The authors found that *A. hypochondriacus* and *A. caudatus* had 137 markers in common (i.e., 145 markers differed genetically when the two species were compared), *A. hypochondriacus* and *A. cruentus* had 73 markers in common, and *A. caudatus* and *A. cruentus* had 72 markers in common. According to this analysis, which two species are most closely related?

26. Gupta and Gudu, in 1991 (*Euphytica* 52:33–38) tested interfertility of the same three species of *Amaranthus* described in the previous question. They found that large numbers of F_1 hybrids between *A. hypochondriacus* and *A. cruentus* could be produced when the two species were hybridized. They also found that the hybrids had low pollen fertility. When they attempted to hybridize *A. caudatus* and *A. cruentus*, they recovered only a few F_1 plants, all of which died as seedlings. When they attempted to hybridize *A. hypochondriacus* and *A. caudatus*, they also recovered only a few F_1 plants, most of which died as seedlings. The few that survived were sterile. **(a)** According to this analysis, which two species are most closely related? **(b)** How does this information correspond with the information in the previous question? **(c)** What types of reproductive isolation are evident from this information?

27. The three species of *Amaranthus* described in the previous two questions were cultivated as food crops by ancient American civilizations (including the Aztec, Maya, and Inca civilizations). They do not grow well as wild plants. *A. hypochondriacus* was grown anciently from the southwestern United States into central Mexico. *A. cruentus* was grown anciently in northern and central Mexico (where its cultivation overlapped that of *A. hypochondriacus*) into Central America. Ancient cultivation of *A. caudatus* was confined to the Andes mountains of South America. What does the information from the previous two questions suggest about the geographic dispersal of these three species of *Amaranthus*?

FOR FURTHER READING

Most current genetics textbooks (including this one) devote no more than a single chapter to evolutionary genetics. Two genetics textbooks that are now out of print but still available in many libraries include several excellent and detailed chapters on evolutionary genetics. Chapters 16–22

of **Ayala, F. J., and J. A. Kiger. 1980.** *Modern Genetics.* **Menlo Park, Calif.: Benjamin/Cummings**, include information on evolutionary genetics with examples drawn from diverse species. Chapters 7–13 of **Bodmer, W. F., and L. L. Cavalli-Sforza. 1976.** *Genetics, Evolution, and Man.* **San Francisco: Freeman**, discuss evolutionary genetics with most examples drawn from studies involving humans. Papers presented in a 1997 colloquium on evolutionary genetics were published in 1997 in the *Proceedings of the National Academy of Sciences, USA* **94:7691–7806** as a collection. Several of those papers provide excellent reading in relation to this chapter. The molecular evolution of cytochrome *c* was reviewed by **Dickerson, R. E. 1972. The structure and history of an ancient protein.** *Scientific American* **226 (Apr):58–72.** Basic methods for constructing phylogenetic trees are given in **Fitch, W. M., and E. Margoliash. 1967. Construction of phylogenetic trees.** *Science* **155:279–284.** Two companion articles review the state of fossil and molecular studies on human evolution: the relationship of genetic studies to fossil history is reviewed by **Wood B. 1996. Human evolution.** *BioEssays* **18:945–953**; and DNA studies, with a focus on nuclear DNA, are reviewed by **Wills, C. 1996. Another nail in the coffin of the multiple-origins theory?** *BioEssays* **18:1017–1020.** Important human mitochondrial DNA studies were published by **Cann, R. L., M. Stoneking, and A. C. Wilson. 1987. Mitochondrial DNA and human evolution.** *Nature* **325:31–36**; Vigilant, L., M. Stoneking, H. Harpending, K. Hawkes, and A. C. Wilson. 1991. African populations and the evolution of human mitochondrial DNA.** *Science* **253:1503–1507**; and **Wills, C. 1995. When did Eve live? An evolutionary detective story.** *Evolution* **49:593–607.**

For additional reading, go to InfoTrac College Edition, your online research library at: http: www.infotrac-college.com/brookscole

PART V

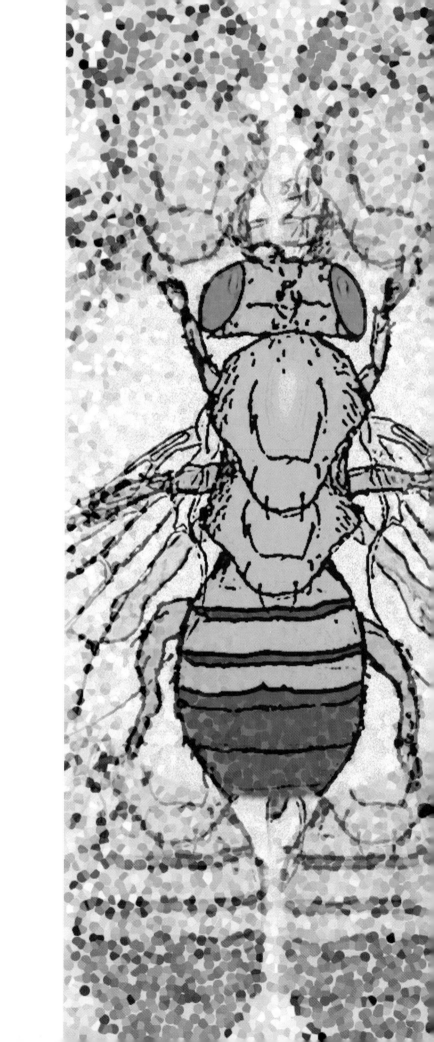

GENE EXPRESSION AND THE ORGANISM

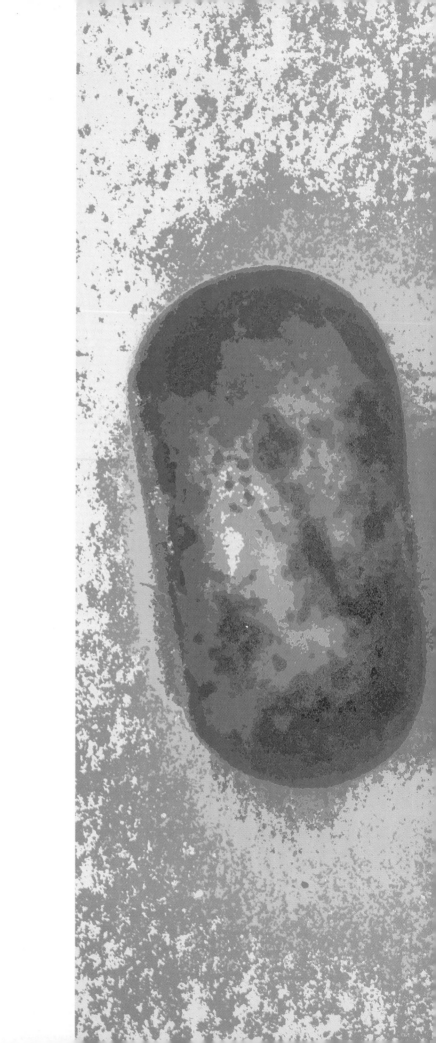

CHAPTER 22

TRANSPOSABLE ELEMENTS

Before the 1950s, most geneticists thought that the genomes of all organisms were stable entities, changed only by rare mutations and chromosomal structural alterations. Against this background, Barbara McClintock observed some unusual phenotypic and cytological phenomena in maize that led her to conclude that pieces of chromatin could move from one location to another in the genome. These mobile pieces of chromatin are called **transposable elements**. Many geneticists were at first reluctant to accept such an idea, in spite of the compelling evidence. However, in the decades that followed, geneticists found abundant evidence of transposable element activity in maize, *Drosophila*, and *E. coli*.

It was no coincidence that all three species are model organisms in genetics. Because they were the species that geneticists had studied most intensively, the inheritance patterns typical of transposable elements were first recognized in them. Transposable elements have since been found in many species, and are probably a part of the genetic constitution of all species. After years of traditional genetic research, many classes of transposable elements have been investigated at the molecular level. Like so many other aspects of genetics, the conclusions about the structure and behavior of transposable elements deduced solely from traditional genetic analysis were confirmed in detail by molecular analysis.

Transposable elements are significant in evolution and in human genetic disorders and cancers. In addition, transposable elements are now used routinely as a tool in biotechnology. For her pioneering work, McClintock received the 1984 Nobel prize in medicine.

Transposable elements have been called by many names. For the purposes of this chapter, we will discuss three types of transposable elements, with specific definitions:

1. **Transposons**—segments of DNA that move from one location of the genome to another and move as a DNA entity.

2. **Retrotransposons**—segments of DNA that resemble retroviruses and encode a reverse transcriptase. They transpose by means of an RNA intermediate that is transcribed from the retrotransposon. The RNA is then reverse-transcribed into DNA, which then inserts itself at a new position in the genome.

3. **Retroposons**—segments of DNA that are transcribed into an RNA intermediate, which is then reverse-transcribed into DNA and inserted into the genome. Unlike retrotransposons, they do not code for their own reverse transcriptase, nor do they resemble retroviruses.

The general mechanisms of transposition for transposable elements are illustrated in Figure 22.1. Transposons transpose through a process called **conservative transposition**, in which the DNA element is excised from the chromosomal DNA and moved to a new location in the chromosomal DNA. Conservative transposition does not increase the number of transposons in the genome; it simply moves them from one location to another. Retrotransposons and retroposons, on the other hand, transpose through a process called **replicative transposition**, in which the original element remains in its site and a copy of the element is inserted in another location. Replicative transposition increases the number of transposable elements in the genome with each transposition event.

Transposable elements have been studied most extensively in maize, *Drosophila,* and *E. coli.* The features of the elements in these species appear to be similar to those of transposable elements in other species. We will look first at eukaryotic transposons, focusing on examples from maize and *Drosophila,* then at eukaryotic retrotransposons and retroposons, including those in humans. Next, we will examine bacterial transposons, with examples from *E. coli.* We conclude with a discussion of the role of transposable elements in evolution and their applications as tools in recombinant DNA research.

22.1 EUKARYOTIC TRANSPOSONS

Transposons cause mutations in genes by inserting themselves into genes (Figure 22.2a). Once the element is present in the gene, a normal mRNA cannot be transcribed because the transposon interrupts the transcribed region of the gene (Figure 22.2b). A transposon-insertion mutation usually creates a recessive allele, because a functional gene product cannot be produced. Alternatively, a transposon can insert near a gene, but not within it, and disrupt the gene's regulatory sequences. Such mutations also usually create recessive alleles because the gene's product is no longer produced, or its production is reduced. Recessive alleles that arise from insertion of a transposon can often be distinguished from other types of mutations by their instability. A transposon is mobile, so it can be excised from the gene in which it resides. Excision of a transposon from a gene can restore the former function of the gene (Figure 22.2c). Recessive transposon-insertion alleles are often called **mutable alleles** because they revert to dominant alleles at rates that are higher than those for recessive alleles that contain other types of mutations.

Transposons were first discovered in maize, and later in *Drosophila.* Most of the maize and *Drosophila* transposons have been analyzed in detail at the molecular level. In this section, we will discuss the *Ac-Ds* transposons in maize and the P elements in *Drosophila* as

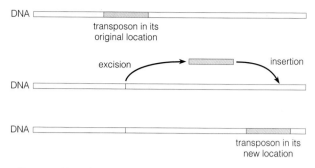

a Transposition of transposons.

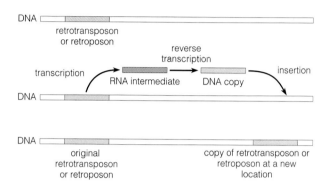

b Transposition of retrotransposons and retroposons.

Figure 22.1 Transposition of transposable elements.

examples of eukaryotic transposons. They are good examples of the general structure and function of DNA transposons in many species.

Detection of Transposable Element Activity in Maize

For centuries, people have observed dark spots or mottled areas on a light background in maize kernels (Figure 22.3). Such spotted kernels are common in the decorative maize often called Indian corn, and they led to the discovery of transposons. A light-colored (yellow or white) maize kernel is caused by homozygosity for recessive alleles at any of several loci that encode enzymes for the anthocyanin (purple pigment) synthesis pathway. When the pathway is blocked because of homozygosity for a recessive allele, the anthocyanin fails to form, and the kernel has a light color. The A locus is one of several loci that governs anthocyanin synthesis. A recessive allele, a_1, prevents anthocyanin formation when it is homozygous. This allele contains a transposon that is located within the A locus and interrupts the normal dominant A allele. The cells responsible for kernel pigmentation are located in the aleurone layer of the endosperm, which is triploid (even though the rest of the plant is diploid), because of double fertilization in plants (see section 11.4). Thus, the homozygous recessive genotype in the light-colored aleu-

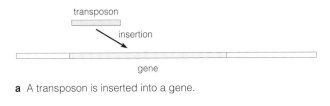

a A transposon is inserted into a gene.

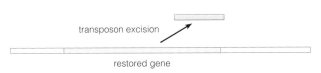

mutant allele with a transposon insertion

b The transposon interrupts the gene and creates a mutant allele.

transposon excision

restored gene

c When the transposon is excised from the gene, the gene s original function may be restored.

Figure 22.2 Mutation by insertion of a transposon, and reversion of the mutation by excision of the transposon.

Figure 22.3 A maize kernel with dark spots that arose from excision of transposons.

rone is $a_1a_1a_1$. If one of the alleles reverts to the dominant allele by excision of the transposon ($a_1a_1a_1 \rightarrow Aa_1a_1$), then the dominant phenotype is restored because the A allele encodes a functional enzyme.

The cell that contains the reversion mutation can now produce pigment. When it divides mitotically, the two daughter cells both have the genotype Aa_1a_1 and produce pigment. As mitosis proceeds, a group of cells that all have the Aa_1a_1 genotype arises. These cells appear as a dark spot on the kernel. Each spot on the kernel represents a transposon excision event in one cell. The timing of the excision can be estimated by the size of the spot. Excision early during development of the kernel produces a large spot because the cells divide several times before the kernel reaches maturity. Excision late during development of the kernel produces a small spot because only a few cell generations remain before the kernel reaches maturity (Figure 22.4).

The *Activator-Dissociator* (*Ac-Ds*) Transposon System in Maize

A mutable allele system in maize, called *Ac-Ds*, led Barbara McClintock to propose the existence of transposons as the most probable interpretation of her observations, much of which she deduced by studying the inheritance of spotted kernels. Several *Ac* and *Ds* transposons have been cloned and sequenced. They are good examples of how many transposons function.

The term *Ac* stands for *activator*, and *Ds* stands for *dissociator*. Both terms represent transposons that are part of the same system but function in different ways. The *Ds* element is often found within the coding sequence of a gene and causes a recessive allele. Mutant alleles that contain a *Ds* element are usually stable because the *Ds* element is incapable of excising itself from the allele. *Ac* elements, on the other hand, can excise themselves from the position where they reside and move elsewhere, causing the allele to revert to the dominant type after excision. The element moves to a new site, usually on the same chromosome. If that new site is within the transcribed region of a gene, the *Ac* element creates a mutation in the gene that is usually detected as a recessive allele.

Although *Ds* elements cannot excise themselves like an *Ac* element, they can be excised and move to a new location when an *Ac* element is present somewhere in the genome. In other words, an *Ac* element activates a *Ds* element. If no *Ac* element is present, a *Ds* element remains within the gene, and the allele behaves like any other stable recessive allele. However, if a plant that carries a *Ds* element hybridizes with a plant that carries an *Ac* element, and the *Ac* element is transmitted to the F_1 progeny, then the *Ds* element in the F_1 progeny can be excised from the gene where it resides and move to a new location. The gene where it resided reverts to a dominant allele. The *Ac* element may also transpose and cause the gene in which it resided to revert.

The *Ac* element is called an **autonomous transposon** because it is capable of self-transposition. The *Ds* element is called a **nonautonomous transposon** because it is incapable of self-transposition but can be transposed when the autonomous transposon (*Ac*) is present in the genome. The best-studied mutation is a recessive allele, *c-m1* (*colorless-mutable 1*), that contains a *Ds* element in the protein-encoding region of the *c1* locus on chromosome 9 in maize. The *c1* locus encodes an enzyme in the pathway for anthocyanin synthesis in the aleurone, so recessive alleles cause a white or yellow kernel (called colorless) when homozygous. The *c-m1* allele is a stable recessive allele as long as there is no *Ac* element in the

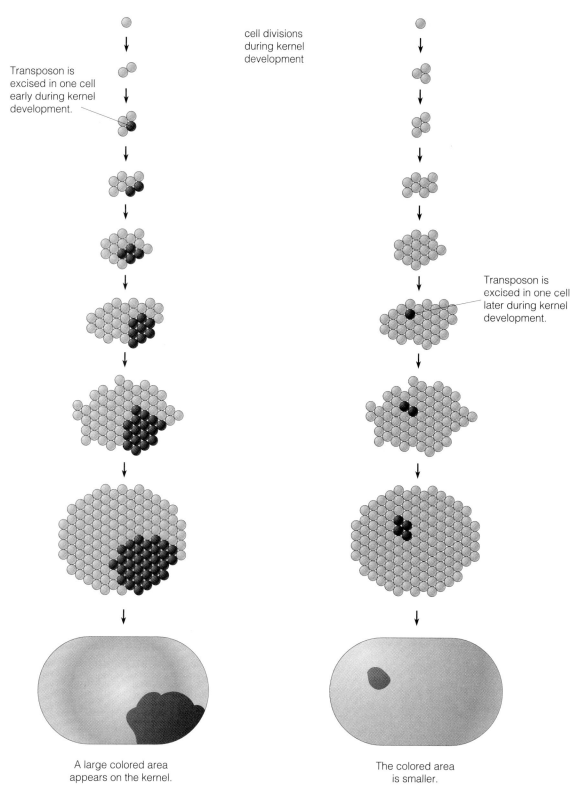

cell divisions
during kernel
development

Transposon is
excised in one cell
early during kernel
development.

Transposon is
excised in one cell
later during kernel
development.

A large colored area
appears on the kernel.

The colored area
is smaller.

Figure 22.4 The relationship between the developmental stage at which a transposon is excised and the size of the colored area on a maize kernel.

genome. However, if an *Ac* element is present somewhere in the genome, the *Ds* element can transpose. When the *Ds* element leaves its site in the *c1* locus, the functional sequence of the *c1* locus is restored, and pigment can be formed in the cell and in the cells that arise from this cell.

The result is a pigmented spot on the kernel. An *Ac* element can cause transposition of the *Ds* element in several cells, to produce several spots on a single kernel, one for each transposition event (Figure 22.5).

Ac and *Ds* elements transpose almost exclusively in

somatic cells. Rarely do they transpose in the germ line, so the elements are inherited as typical recessive alleles and can be analyzed genetically. The effect of an *Ac* element on a *Ds* element can be tested by hybridization, as illustrated in the following example.

Example 22.1 Genetic analysis with *Ac* and *Ds* elements.

In 1950, McClintock (*Proceedings of the National Academy of Sciences, USA* 36:344–355) described experiments in which she hybridized an inbred (homozygous) line of maize with colorless kernels that carried the *c-m1* allele but did not carry an *Ac* element, with a second inbred line with colored kernels that was homozygous for the *Ac* element.

Problem: Diagram the genotypes and phenotypes of the parent lines and the F_1 and F_2 generations. In the F_2 generation, describe the expected ratio of phenotypes.

Solution: If we let Ac^+ represent the presence of the *Ac* element on a chromosome and Ac^- the absence of the *Ac* element at the corresponding site on a homologous chromosome, then the parental genotypes and phenotypes are

$c\text{-}m1\ c\text{-}m1\ Ac^-Ac^-$ (colorless kernels) $\times$
$C1\ C1\ Ac^+Ac^+$ (colored kernels)

Both parents are homozygous, so the F_1 genotype and phenotype are

$C1\ c\text{-}m1\ Ac^+Ac^-$ (colored kernels)

The Punnett square in Figure 22.6 provides the possible genotypes and their ratios in the F_2 generation. The phenotypic ratio is

12   colored kernels ($C1\ \_\ \_\ \_$)

3   spotted kernels ($c\text{-}m1\ c\text{-}m1\ Ac^+\_$)

1   colorless ($c\text{-}m1\ c\text{-}m1\ Ac^-Ac^-$)

The presence of *Ac* and/or *Ds* in kernels of the dark-colored class has no phenotypic effect because the dominant *C1* allele already produces pigment in every cell. Any transposition event that reverts a *c-m1* allele to the dominant type cannot be observed because the cell already has pigment in it from the dominant *C1* allele.

In the early 1980s, Nina Fedoroff and her colleagues cloned several *Ac* and *Ds* elements. The nucleotide sequences of these cloned elements were determined, and

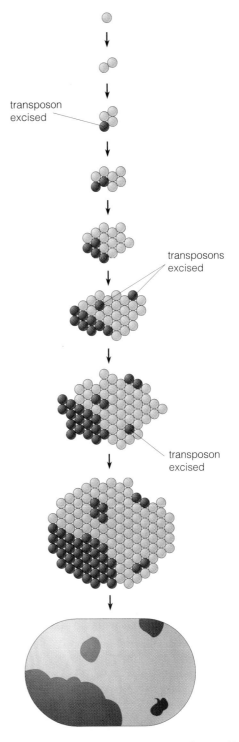

transposon excised

transposons excised

transposon excised

Figure 22.5 A phenotype that results from multiple transposon excision events at different stages of kernel development.

the sequences revealed two distinct features about the *Ac* element that are apparent in other autonomous elements in maize and other species:

1. Each element is bordered by imperfect inverted terminal repeats.

2. Each element contains a gene that encodes an enzyme called transposase.

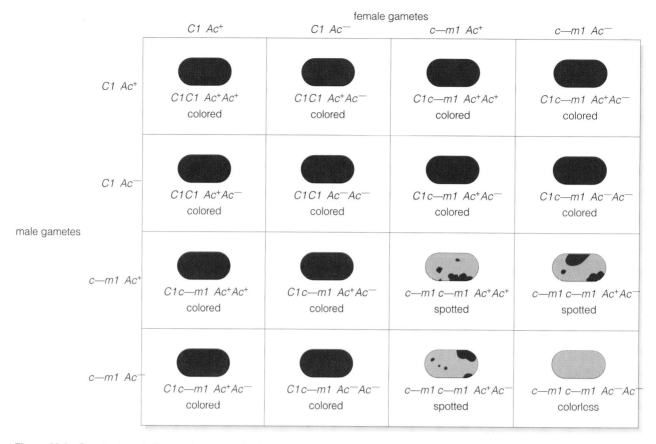

Figure 22.6 Genotypic and phenotypic segregation in the F₂ generation from the cross *c-m1 c-m1 Ac⁻Ac⁻* × *C1C1 Ac⁺Ac⁺*.

Figure 22.7 illustrates these features and how they function. All *Ac* elements that have been sequenced are nearly identical; they differ by only a few nucleotides. The *Ds* elements are like *Ac* elements, but they contain deletions, as diagrammed in Figure 22.8. In each case, the deletion includes part of the transposase gene, so each *Ds* element fails to encode functional transposase. The amount of deleted DNA varies considerably among the *Ds* elements that have been sequenced. In one *Ds* element (*Ds9*), only a small segment of the transposase gene is missing, whereas in another *Ds* element (*Ds6*), most of the element is missing; all that remains is a small segment of DNA bordered by the inverted terminal repeats.

The structural features and the mutations in *Ds* elements indicate how transposons function. Transposase is essential for transposition. It recognizes the DNA sequences at the inverted terminal repeats and excises the element at those sites. Only *Ac* produces functional transposase. *Ds* elements cannot produce functional transposase because all or some of their transposase gene has been deleted. *Ac* and *Ds* have the same inverted terminal repeats, and transposase recognizes the repeats and transposes both *Ac* and *Ds* elements. However, if there is no *Ac* element in the genome, no functional transposase is produced, and any *Ds* elements that are present cannot

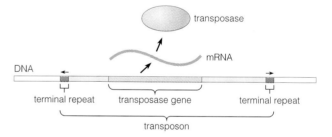

a An autonomous transposon is bordered by inverted terminal repeats and contains a transposase gene.

b Transposase acts on the inverted terminal repeats to excise the transposon and insert it elsewhere.

Figure 22.7 Structure and function of autonomous transposons.

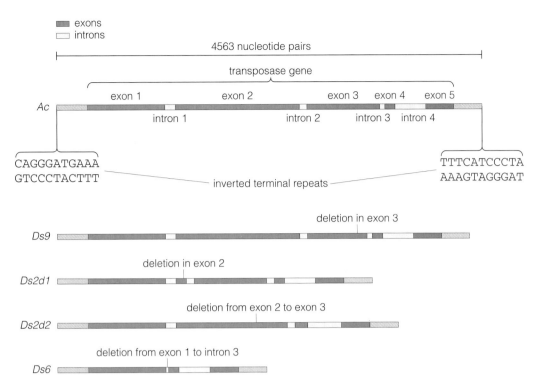

Figure 22.8 The structures of the *Ac* element and several *Ds* elements. The *Ds* elements contain deletion mutations in the transposase gene.

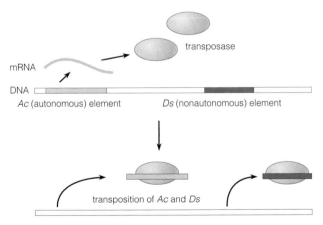

a When both *Ac* and *Ds* are present in the genome, the transposase encoded by a gene in *Ac* transposes both *Ac* and *Ds*.

b When only *Ds* is present in the genome, no transposase is produced, and *Ds* remains at the site it occupies.

Figure 22.9 The effect of autonomous and nonautonomous transposons.

transpose themselves, so they remain as stable entities where they reside.

The observation of deletion mutations in the transposase gene of *Ds* elements explains why *Ac* elements are autonomous and why *Ds* elements are nonautonomous. Autonomous elements encode functional transposase and are therefore capable of self-transposition. Nonautonomous elements do not encode functional transposase, so they transpose only when an autonomous element is present to produce transposase. The presence of transposase allows both autonomous and nonautonomous elements to transpose, as illustrated in Figure 22.9.

When transposase inserts an *Ac* or *Ds* element into a new site, the site is cut in a staggered fashion, with single-stranded portions of eight nucleotides on either end of the cut site (Figure 22.10a and b). A DNA polymerase adds nucleotides to fill in the gaps where the DNA is single stranded, creating short direct repeats of eight nucleotide pairs in the DNA sequences that flank the inverted repeats of the element (Figure 22.10c). When transposase excises the element, some of the repeated sequence is left behind as a "footprint" where the element once resided (Figure 22.10d).

If an excised element leaves a footprint behind, why does the gene revert to its functional form after the

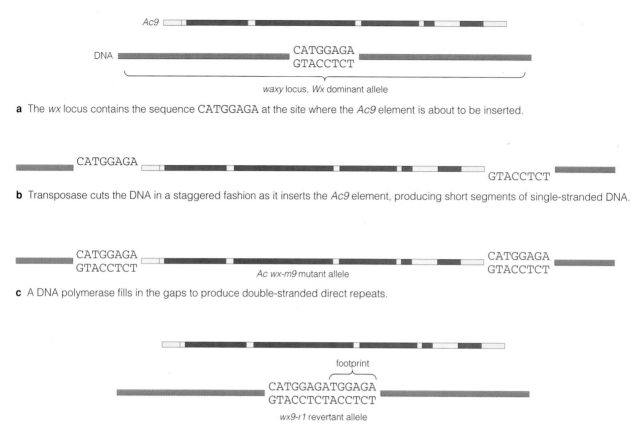

a The *wx* locus contains the sequence CATGGAGA at the site where the *Ac9* element is about to be inserted.

b Transposase cuts the DNA in a staggered fashion as it inserts the *Ac9* element, producing short segments of single-stranded DNA.

c A DNA polymerase fills in the gaps to produce double-stranded direct repeats.

d When transposase exises the element, it leaves part of the direct repeat behind, creating a footprint in the DNA where the *Ac9* element resided.

Figure 22.10 Creation of a footprint in DNA when a transposase excises a transposon.

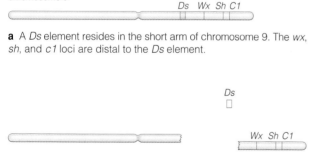

a A *Ds* element resides in the short arm of chromosome 9. The *wx*, *sh*, and *c1* loci are distal to the *Ds* element.

b When the *Ds* element is excised, the chromosome breaks at the site where the *Ds* element resided.

Figure 22.11 A chromosome-breaking *Ds* element.

element leaves? Many footprints are frameshift mutations, and there is no reversion to the dominant phenotype, so excision is not detected. Alleles that revert from the recessive to dominant expression are the only ones detected phenotypically. Several such revertant alleles have been sequenced, and they all have a similar pattern. When the element leaves the gene, the number of nucleotides in the footprint is a multiple of three, so the correct reading frame is maintained.

For example, Figure 22.10d shows the DNA sequence of a revertant allele (*wx9-r1*) that produces a functional product even though a footprint remains. Transposase duplicated eight nucleotides at the *waxy* locus when it inserted the *Ac9* element to create the *Ac wx-m9* mutant allele (Figure 22.10c). When the element transposed out of the *wx* locus, it left six of the eight duplicated nucleotides as a footprint in the revertant allele (Figure 22.10d). The *wx* protein translated from this allele has two additional amino acids because of the additional six nucleotides in the reading frame. Even though the protein is slightly longer than the normal form, it functions normally.

Ds Elements and Chromosome Breakage

Barbara McClintock observed chromosome breakage associated with some, but not all, *Ds* elements. She had a genotype of maize with a chromosome-breaking *Ds* element in the short arm of chromosome 9, and the loci for the *c1*, *wx*, and *sh* genes were all located distal to the *Ds* element. The genotype she used was homozygous for the dominant alleles at all three loci (Figure 22.11a). When this genotype was crossed with another line that carried *Ac* and was homozygous for recessive alleles at all three loci,

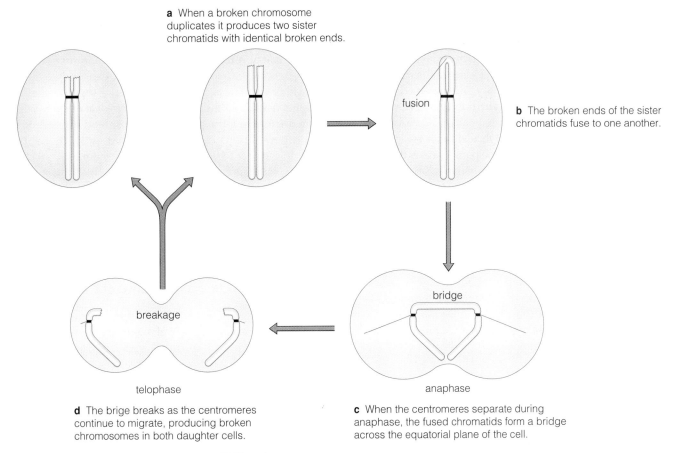

a When a broken chromosome duplicates it produces two sister chromatids with identical broken ends.

fusion

b The broken ends of the sister chromatids fuse to one another.

bridge

anaphase

c When the centromeres separate during anaphase, the fused chromatids form a bridge across the equatorial plane of the cell.

breakage

telophase

d The brige breaks as the centromeres continue to migrate, producing broken chromosomes in both daughter cells.

Figure 22.12 The breakage-fusion-bridge (BFB) cycle.

Ds transposed in cells of the F_1 progeny because *Ac* was present. The chromosome often broke as the *Ds* element was excised, so the part of the chromosome that contains the *c1*, *wx*, and *sh* loci was deleted (Figure 22.11b).

If transposition and breakage occurred before the chromosome was replicated, then the deletion was replicated, leaving two sister chromatids without a telomere on one end (Figure 22.12a). Broken chromosome ends tend to fuse with other broken sites. In this case, two broken sites are close to each other, one on each sister chromatid, so the broken sister chromatids often fuse with one another to create a dicentric chromosome (Figure 22.12b). As the fused chromosome proceeds through mitosis, the centromeres move toward opposite poles, causing a bridge to form across the equatorial plane (Figure 22.12c). Eventually, the tension on the bridge becomes so great that the chromosome breaks (Figure 22.12d). In the next cell generation, the broken chromosomes replicate once again, and the broken sites on sister chromatids can once again fuse, causing a dicentric chromosome. This again forms a bridge with breakage at a random site during anaphase. The process then repeats itself with each cell generation. McClintock called this the **breakage-fusion-bridge (BFB) cycle**, which was set in motion by *Ds* transposition. All the cells that participate in the BFB cycle lack

dominant alleles at the *c1*, *wx*, and *sh* loci, so the recessive phenotypes for these loci are expressed.

Only certain *Ds* elements are capable of breaking chromosomes. They can break the chromosome at the site where the element leaves the chromosome and at the site where it is inserted. The presence of chromosome breaks can lead not only to a BFB cycle but also to other types of chromosomal structural changes, including duplications, inversions, and translocations.

P Elements in *Drosophila*

In the 1940s, McClintock proposed that *Drosophila* must also have transposable elements that cause some of the variegated phenotypes that had been well studied at the time. After transposable elements were discovered in maize, researchers found several types of transposable elements in *Drosophila*, some with structural similarities to the maize elements. We will discuss P elements, one of the best-studied transposons, as an example of transposons in *Drosophila*.

In the 1960s and 1970s, several geneticists noticed that the offspring of certain matings between flies caught in the wild and flies from laboratory stocks were infertile or had substantially increased mutation rates. Researchers

observed infertility or increased mutation rates only when the wild-caught flies were used as male parents and the laboratory-stock flies as female parents. The reciprocal cross produced normal progeny with no reduction in fertility and only occasional slight increases in the mutation rate. This phenomenon of infertility and increased mutation in the offspring is called **hybrid dysgenesis**, which was later found to be due to transposition of a group of transposons called **P elements**. Hybrid dysgenesis appears in the offspring only when the male parent carries the P elements and the female does not. The name *P element* stands for "paternal element" because the element produces its effect when inherited from the paternal parent.

Flies without a P element are called M-cytotype flies (M for "maternal"), whereas flies that have a P element are called P-cytotype flies. Hybrid dysgenesis appears in offspring only when a P-cytotype male is mated with an M-cytotype female, a mating called a **dysgenic cross**. P-cytotype males mated with P-cytotype females, M-cytotype males mated with P-cytotype females, and M-cytotype males mated with M-cytotype females all do not produce hybrid dysgenesis in their offspring and are not dysgenic crosses.

Hybrid dysgenesis is the result of frequent transpositions of P elements in the offspring of a dysgenic cross. P elements encode a repressor that resides in the cytoplasm of a P-cytotype fly and prevents transposition. Because the cytoplasm is inherited only from the female, hybridization of a P-cytotype male with an M-cytotype female produces a zygote with P elements in the nucleus but a repressor-free cytoplasm inherited from the maternal M-type parent.

The results of a dysgenic cross depend on certain environmental factors. At higher temperatures (about 29°C), nearly all the F_1 offspring of a dysgenic cross are phenotypically normal but sterile. At lower temperatures (about 21°C), many of the F_1 flies are fully or partially fertile, but they transmit many mutations, including chromosome breaks and alterations in chromosome structure, to their progeny. In addition, crossovers are frequently observed in the male offspring of a dysgenic cross. In *Drosophila*, crossovers are usually restricted to females.

In order to have the P cytotype, a fly must have many copies of the P element. A P-cytotype fly typically carries about 30–50 P elements per genome. When a P-cytotype male is crossed with an M-cytotype female, the F_1 offspring are dysgenic but are not immediately transformed into P-cytotype flies. If dysgenic flies are mated with each other to form F_2 and subsequent generations, the progeny eventually acquire the P cytotype. Each generation of mating causes P elements to transpose and to accumulate. After the number of P elements reaches about 30 per genome, the P cytotype is established. The P cytotype tends to spread when introduced to M-cytotype flies, and eventually all the flies in a culture become P-cytotype flies after 20–80 generations.

Several scientists who have studied P elements arrived at a rather startling conclusion. When P elements were first discovered, they were present in nearly all wild-type populations and absent in laboratory stocks that descended from flies caught in the wild several decades ago. The accumulated evidence suggests that P elements entered the genome of *Drosophila melanogaster* somewhere in Latin America, then spread throughout wild populations worldwide very recently, probably within the past 50 years. Flies in many laboratory stocks lack P elements because the flies from which the stocks were founded were collected from populations that did not yet have P elements. P elements, therefore, represent a very recent, and highly significant, event in the evolution of *Drosophila*.

Molecular Structure and Function of P Elements

Several P elements have been cloned and sequenced. Like the transposons of maize, P elements have inverted terminal repeats and a transposase gene. Both autonomous and nonautonomous elements have been characterized, and the nonautonomous elements carry deletions in the transposase gene. The structure of an autonomous P element is shown in Figure 22.13. Most autonomous P elements are 2907 nucleotide pairs in length and have inverted terminal repeats, each with 31 nucleotide pairs. A second set of subterminal inverted repeats, each with 11 nucleotide pairs, is found 125 nucleotide pairs from each end of the element. The transposase gene occupies most of the element and has four exons. The transposase enzyme has been isolated and characterized. It contains DNA-binding proteins of the type discussed in section 8.5.

Like the *Ac-Ds* transposons in maize, a P element generates a direct repeat in the host DNA when transposase inserts the element into the DNA. The direct repeat is eight nucleotides long. P elements do not always excise perfectly. Sometimes, an internal portion of the element is excised, which may account for the high frequency of nonautonomous P elements that contain deletions. Also, a P element may be excised from a gene and carry a portion of the host gene's DNA along with it, an event that creates a deletion mutation within the gene. Other times, excision may be precise, removing not only the element but also one of the eight-nucleotide-pair direct repeats, restoring the exact gene sequence that was present before insertion. These latter excisions cause a gene to revert to its former function, whereas the other types of excision leave a stable mutant form of the gene behind.

The fact that transposition of P elements is suppressed in all but dysgenic flies suggests that P elements are regulated by a cytoplasmic repressor protein that is encoded by a gene in the P element. P elements in dysgenic flies usually transpose only in the germ line and not in somatic cells. In this sense, they differ from the

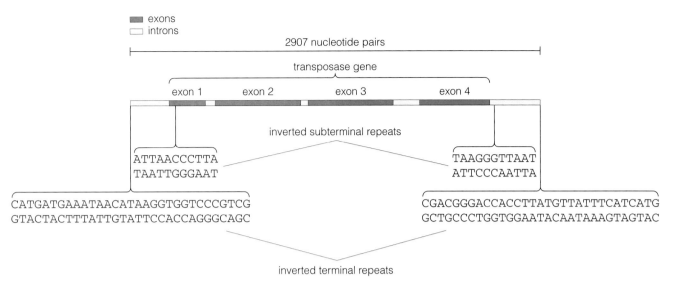

exons
introns

2907 nucleotide pairs

transposase gene

exon 1   exon 2   exon 3   exon 4

inverted subterminal repeats

ATTAACCCTTA   TAAGGGTTAAT
TAATTGGGAAT   ATTCCCAATTA

CATGATGAAATAACATAAGGTGGTCCCGTCG   CGACGGGACCACCTTATGTTATTTCATCATG
GTACTACTTTATTGTATTCCACCAGGGCAGC   GCTGCCCTGGTGGAATACAATAAAGTAGTAC

inverted terminal repeats

Figure 22.13   The structure of an autonomous P element in *Drosophila melanogaster*.

Ac-Ds transposons in maize, which transpose readily in somatic cells.

If the transposase gene occupies nearly all of an autonomous P element, where is the repressor gene? Apparently, the repressor protein is encoded by the same gene as the transposase, but the pattern for intron removal differs for the repressor. In somatic cells, the intron between exons 2 and 3 is not spliced out of the mRNA. It contains a termination codon in the reading frame, so the polypeptide encoded by this mRNA is a shortened version of the transposase, and it acts as the repressor. Germ-line cells in dysgenic flies produce an mRNA in which this intron is removed, so the polypeptide is longer and acts as a transposase rather than a repressor.

The following example highlights the relationship between the molecular structure of a P element and phenotypes associated with its transposition.

> **Example 22.2   The *sn^w* allele in *Drosophila melanogaster*.**
>
> The *singed* (*sn*) locus is on the X chromosome in *Drosophila melanogaster*. Mutant alleles in hemizygous males or homozygous females cause the bristles to appear like hairs that have been singed with a flame. The *sn^w* allele causes an intermediate singed phenotype that is stable in M-cytotype flies. When M-cytotype females with a stable *sn^w* allele are mated with P-cytotype males, the *sn^w* allele mutates very frequently to a more severe singed type (*sn^e*) or to the wild type *sn^+*. When DNA from the *sn^w* allele was analyzed (Roiha et al. 1988. *Genetics* 119:75–83), the allele was found to contain two P elements, one that was about 1150 nucleotide

pairs in length and the other about 950 nucleotide pairs in length.

Problem: **(a)** Based on the information above, are these elements both autonomous, both nonautonomous, or is one autonomous and the other nonautonomous? **(b)** What might explain the two phenotypes that arise from this allele after dysgenesis?

Solution: **(a)** The *sn^w* allele is stable in the M cytotype, so both elements within the allele must be nonautonomous. If one or both elements were autonomous, then the functional transposase gene would be transcribed, the allele would be unstable, and the M cytotype could not be maintained. Both elements are shorter than autonomous elements. A complete autonomous P element has 2907 nucleotide pairs, so both of these elements contain deletions. **(b)** The two phenotypes can be most readily explained if they arise from excision of different nonautonomous elements. Research has shown that the *sn^e* allele arises from excision of the smaller element, whereas the *sn^+* allele arises by excision of the larger element.

Transposons have characteristic common features. The most important are inverted terminal repeats and a gene that encodes a transposase. Transposons that encode functional transposase may transpose in the absence of other transposable elements and are said to be autonomous. Transposons with a mutant transposase gene that does not encode functional transposase are nonautonomous and can transpose only when an autonomous element is present in the genome.

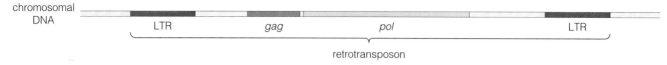

LTR   *gag*   *pol*   LTR

retrotransposon

Figure 22.14   General structure of retrotransposons.

22.2 RETROTRANSPOSONS AND RETROPOSONS

Retrotransposons and retroposons are DNA elements that transpose through a process called **retrotransposition**: an RNA is transcribed from the element's DNA, and the RNA is then reverse-transcribed into DNA, which is made double stranded and inserted elsewhere in the genome. The original DNA of the retrotransposon or retroposon remains at its site, and the new copy is inserted elsewhere. Thus, retrotransposons and retroposons increase in number over generations. Retrotransposition has been under way for so many generations in some species that a significant percentage of the genome consists of retrotransposons or retroposons. For example, the human genome contains approximately 500,000 copies of *Alu* elements, a type of retroposon we discussed in section 10.1. *Alu* elements constitute about 5% of the human genome. About 15% of the *Drosophila* genome consists of retrotransposons.

Although retrotransposons and retroposons both transpose via an RNA intermediate, there are some important differences between them. We'll discuss retrotransposons first.

Retrotransposons

Certain eukaryotic viruses are called retroviruses because their genetic information is encoded in RNA, which is transcribed into DNA through reverse transcription after the virus infects a cell. The DNA is then inserted into the genome of the host cell, where its genes are expressed using the cell's transcription and translation machinery. The human immunodeficiency virus (HIV) is an example of a retrovirus. Retrotransposons resemble retroviruses in many respects. Their structural organization is very similar, nearly identical in some cases, and the nucleotide sequences of retrotransposons and retroviruses are similar. These observations suggest that retrotransposons originated as retroviruses. Some retrotransposons encode structural proteins that form viruslike particles in the cells. The particles house the RNA transcribed from the retrotransposon and the enzymes translated from the genes in the RNA. Researchers have attempted to isolate retrotransposon particles and use them to infect other cells, with little success. The major difference between retrotransposons and retroviruses is the inability of retrotransposons to leave cells as a virus particle and infect other cells. Retrotransposons have lost the ability to lyse cells and infect other cells; they are transmitted genetically as part of the chromosomes in cells.

Most retrotransposons have a similar structure and mode of transposition. Among the best-studied retrotransposons are the *Ty* elements of yeast and the *copia* elements of *Drosophila*. The general structure of these elements is illustrated in Figure 22.14. They are bordered by direct terminal repeats, called **long terminal repeats (LTRs)** because the repeats are longer by about one order of magnitude than the inverted terminal repeats of transposons. They contain two genes, called *gag* and *pol,* that are highly similar to genes of the same name found in retroviruses.

The process of retrotransposition is illustrated in Figure 22.15. RNA polymerase II transcribes the retrotransposon DNA into a single-stranded RNA molecule, which is then polyadenylated to become an mRNA (Figure 22.15a). The proteins translated from the *gag* and *pol* genes are called **polyproteins** because they are destined to be cleaved into several individual proteins (Figure 22.15b and c). The polyprotein encoded by the *gag* gene is cleaved into three proteins, which form a viruslike capsid that surrounds the retrotransposon RNA. The *pol* gene encodes a polyprotein that is cleaved into four enzymes: a reverse transcriptase, an RNase, a protease, and an integrase. The protease cleaves the polyproteins into their individual proteins, and the proteins and RNA assemble into a particle that is very similar to a retrovirus particle (Figure 22.15d). Reverse transcriptase transcribes a single-stranded DNA molecule from the retrotransposon RNA template in the particle. The RNase then removes the RNA from the DNA. The DNA arranges itself into a circle, and a complementary DNA strand is synthesized to form a double-stranded, circular copy of the retrotransposon (Figure 22.15e). Integrase integrates this copy of the retrotransposon into a new site on the cellular DNA (Figure 22.15f).

Retroposons

Retroposons differ from retrotransposons in that they do not have LTRs, nor do they resemble retroviruses as retrotransposons do. For the most part, retroposons appear to be cellular mRNAs and snRNAs that have been reverse-transcribed into DNA and inserted into the genome. The presence of poly (A) segments in many retroposons is evidence that they arose from mRNAs with poly (A) tails. Most retroposons transpose only in the presence of re-

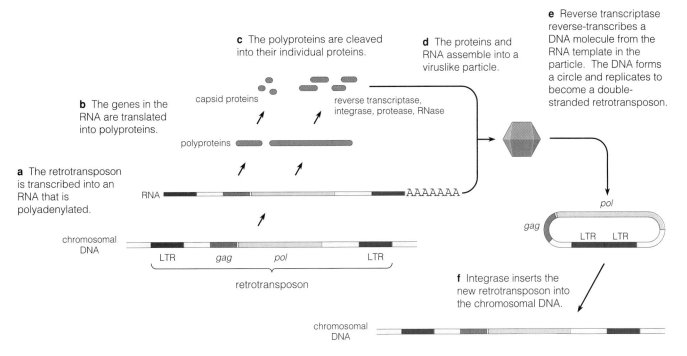

Figure 22.15 The process of retrotransposition.

verse transcriptases and integrases produced by retroviruses or retrotransposons.

The best-studied examples of retroposons are mammalian LINEs and SINEs, discussed in detail in section 10.1 and diagrammed in Figures 10.1 and 10.2. They constitute a significant portion of the human genome. For example, *Alu* elements, the most common SINE in humans, constitute about 5% of the human genome. Most LINEs and SINEs are transpositionally inactive and are inherited as part of the genomic DNA. However, cases of recent transposition have been identified, as highlighted in the following example.

Example 22.3   A new germ-line mutation from transposition of an *Alu* element.

Mutations in the human *NF1* gene (OMIM 162200) cause neurofibromatosis, a genetic disorder that is characterized by light brown spots on the skin, called café au lait spots, and by tumors on the peripheral nerves and skin. Nearly all mutations in this gene create a dominant allele that causes the disorder. Neurofibromatosis is a relatively common genetic disorder, affecting about 1 in 3000 people, and it is fully penetrant. In 1991, Wallace et al. (*Nature* 353: 864–866) described a case of neurofibromatosis that arose from insertion of an *Alu* element into the *NF1* gene. This study was the first documented case in humans of a recent transposable element–insertion mutation that caused a genetic disorder.

The authors examined the DNA of a male patient who had neurofibromatosis. The *NF1* gene in one of the patient's chromosomes contained an *Alu* element of 320 nucleotide pairs located in intron 5, 44 nucleotide pairs upstream from the intron's 3' end. Neither of the patient's parents had neurofibromatosis. The authors examined the DNA of the two parents and found that neither parent had an *Alu* element at this site in their *NF1* genes. DNA analysis confirmed that the patient's parents were his genetic parents and that the *Alu* element had been inserted into the *NF1* allele that the patient inherited from his father. When the authors examined the mRNA encoded by the mutant allele, they found that exon 6 was missing from the mRNA, and that all other exons were present and normally spliced. The missing exon caused a deletion and frameshift in the reading frame of the mRNA.

Problem: (a) The authors of this study concluded that the *Alu* element transposition was a recent mutation that arose in the germ line of the patient's father. What evidence led them to this conclusion? (b) Explain how the mutant allele could cause neurofibromatosis given that the *Alu* element was within an intron.

Solution: (a) The first indication that the mutation was recent is the observation that neither parent had the disorder. Neurofibromatosis is a fully penetrant dominant genetic disorder, so at least one of the parents should have had the disorder, unless a

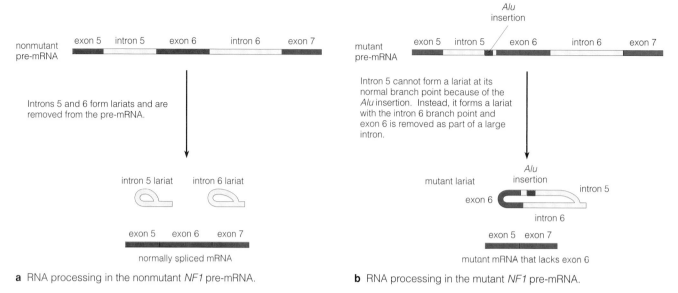

a RNA processing in the nonmutant *NF1* pre-mRNA.

b RNA processing in the mutant *NF1* pre-mRNA.

Figure 22.16 Solution to Example 22.3: Altered processing of a mutant pre-mRNA with an *Alu* insertion.

new mutation arose in the germ line of one of the parents. DNA analysis in the patient and his parents confirmed that the patient's cells had an *Alu*-insertion mutation in one of the *NF1* alleles, but the *NF1* alleles in both parents did not carry the insertion mutation. Thus, the insertion must have occurred in the germ line of one of the parents. DNA analysis showed that the insertion mutation occurred in the father's germ line. **(b)** Most mutations in an intron have no effect because the mutation is removed along with the intron during pre-mRNA processing. However, a mutation in an intron that prevents proper intron removal alters the mRNA and the protein it encodes. In this instance, the *Alu* insertion was 44 nucleotides upstream from the 3' end of the intron, which is the approximate location of the branch point for lariat formation during intron splicing (see section 3.7). If the 5' end of the intron cannot form a lariat at its proper site, then it may seek another site for lariat formation. If that site is in the next intron (intron 6 in this case), then a segment of RNA that contains intron 5, exon 6, and intron 6 should be removed as a single intron, as illustrated in Figure 22.16. When the authors examined the mRNA encoded by the mutant allele, they found that exon 6 was indeed missing from the mRNA, indicating that it had been removed during intron splicing.

Retrotransposons are similar to retroviruses and probably have a retroviral origin. They contain genes that facilitate retrotransposition. Retroposons do not contain genes for retrotransposition and apparently arose from cellular mRNAs or snRNAs.

22.3 PROKARYOTIC TRANSPOSONS

Several families of transposons have been identified in prokaryotes. Although nucleotide sequence similarities between prokaryotic and eukaryotic transposons are not common, they are structurally similar. Like eukaryotic transposons, prokaryotic transposons are bordered by inverted terminal repeats, they contain genes that encode transposases, and they generate short direct repeats at the site of insertion in the host DNA. Prokaryotic transposons may be simple transposons called **insertion sequences**, which carry only the information required for transposition, or they may be larger **composite transposons**, which consist of DNA that carries other types of genes. We'll discuss insertion sequences first, then the composite transposons derived from them.

Insertion Sequences

Insertion sequences are the simplest bacterial transposons. Like eukaryotic transposons, they are a segment of DNA that is bordered by inverted terminal repeats. The DNA typically contains one or two genes that encode a transposase or subunits of a transposase. More than 50 insertion sequences have been characterized. They are usually named by the prefix IS (for "insertion sequence") followed by a number. The features of one insertion sequence, IS*903*, are illustrated in Figure 22.17. Insertion sequences are relatively short, ranging from 768 nucleotide pairs to over 2500 nucleotide pairs long. The majority of their DNA is the gene or genes required for transposition.

Although insertion sequences were first identified in *E. coli*, they have since been found in many species of bacteria. They are found in both laboratory and native strains.

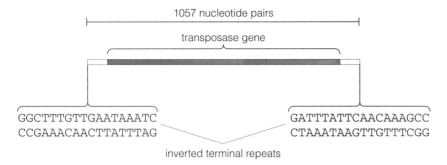

1057 nucleotide pairs

transposase gene

GGCTTTGTTGAATAAATC   GATTTATTCAACAAAGCC
CCGAAACAACTTATTTAG   CTAAATAAGTTGTTTCGG

inverted terminal repeats

Figure 22.17 The structure of the IS*903* insertion sequence.

Some insertion sequences are found in only one species of bacteria, whereas others may be present in several species. For example, IS*1*, one of the best-studied insertion sequences, was originally found in *E. coli* but is also found in the bacterial genera *Shigella, Klebsiella,* and *Serratia.*

Several copies of a particular insertion sequence may be present in a single cell. Usually, less than 10 copies are present, but some strains of bacteria may carry as many as 200 copies of a particular insertion sequence. Insertion sequences are found in the bacterial chromosome, in plasmids, and in bacteriophages. They can be transmitted among cells through transfer of plasmids or through bacteriophage infection.

Like eukaryotic transposons, insertion sequences create a short direct repeat on either side of the insertion site in the host DNA. Also like eukaryotic elements, they cause mutations when they reside within a gene or in its regulatory sequence. Excision of the insertion sequence may cause a gene to revert its former function, or a new stable mutation may result when a footprint or part of the insertion sequence is left behind after excision. Sometimes, excision creates a stable deletion in the DNA of the gene where the insertion sequence once resided.

Insertion sequences can transpose using conservative transposition or replicative transposition, although replicative transposition of insertion sequences does not require an RNA intermediate as it does for retrotransposons and retroposons. Certain insertion sequences, such as IS*10* and IS*50*, transpose only by conservative transposition. Others, such as IS*1*, IS*102*, and IS*903*, may transpose by either conservative or replicative transposition.

Composite Transposons

Sometimes two insertion sequences may be present in the same circular bacterial DNA molecule, as diagrammed in Figure 22.18. The two insertion sequences and the bacterial DNA that resides between them can transpose as a single unit, called a **composite transposon**. The two insertion sequences are called **terminal modules** and form the right and left arms of the composite transposon. The DNA between them is called the **central region**. The terminal modules may be inverted or direct repeats.

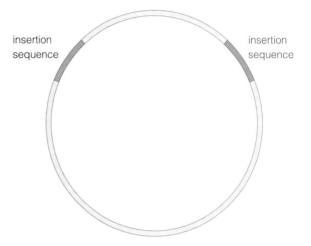

insertion sequence

insertion sequence

Figure 22.18 Two insertion sequences in a circular DNA molecule.

Composite transposons may transpose as a single entity, or the individual insertion sequences may transpose on their own. Circular molecules that have two insertion sequences contain two composite transposons, because any DNA bordered by two terminal modules is a composite transposon.

Depending on the orientation of the terminal modules, composite transposons may undergo several types of rearrangements (Figure 22.19). If the terminal modules are direct repeats, they may line up side by side and recombine to generate two circular molecules, each with one copy of the insertion sequence and one central region. If the terminal modules are inverted repeats, then recombination between the two generates an inversion.

Composite transposons are important because the DNA between the two insertion sequences often contains genes that can be readily transposed or repositioned in the genome. Many composite transposons have been identified by the antibiotic resistance genes they carry between the insertion sequence borders. In some cases, transposition and recombination of composite transposons produces plasmids with several antibiotic resistance genes in them. The plasmids can be transmitted from one bacterial strain to another. When antibiotics are

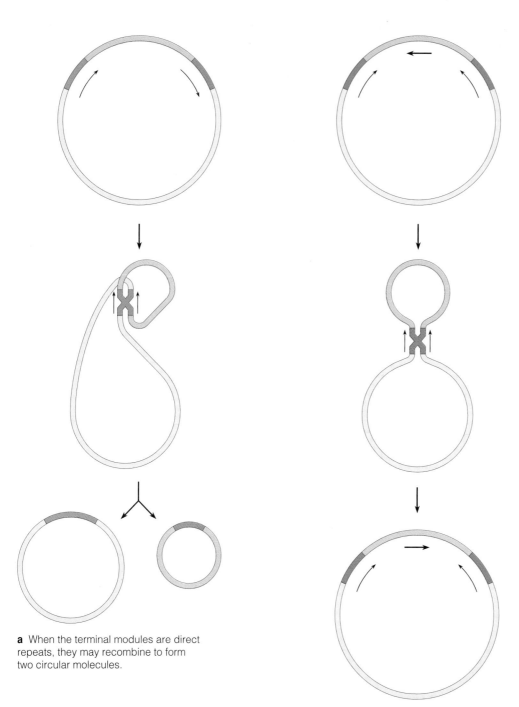

a When the terminal modules are direct repeats, they may recombine to form two circular molecules.

b When the terminal modules are inverted repeats, they may recombine to invert part of the molecule.

Figure 22.19 Effects of recombination in composite transposons.

in widespread use, natural selection can favor the formation of composite transposons with several resistance genes that confer multiple antibiotic resistance to the bacteria that carry the transposons. Several outbreaks of antibiotic-resistant *Salmonella* illness in humans have been attributed to composite transposons selected by antibiotic use in livestock and humans.

~
Insertion sequences are prokaryotic transposons with the same general features as eukaryotic transposons. Some insertion sequences utilize conservative transposition, and others utilize replicative transposition. Two insertion sequences may flank a segment of DNA to form a composite transposon, which transposes as a single unit.

22.4 USING TRANSPOSABLE ELEMENTS IN RESEARCH

The mutagenic effect of transposable elements implies that they can be used as a tool in both basic and applied research. Unknowingly, plant breeders have been using transposable elements for decades. Dark-colored kernels are not desirable in commercial maize, except for specialty items such as blue corn chips. To produce yellow or white corn, the plants must be homozygous for an allele that blocks purple-pigment synthesis in the kernel. One allele that has been used by plant breeders since the 1930s to produce yellow kernels contains a nonautonomous transposon.

With the discovery of transposable elements and their functions, it became apparent that transposable elements offered some intriguing possibilities to further basic genetic research and economically important research in biotechnology. The two most common applications of transposable element technology are called transposon tagging and transposon transformation.

Transposon Tagging

Researchers can most easily identify a gene at the molecular level or clone the gene if they have a DNA probe that is homologous to a portion of the gene (see section 9.4). In order to make a probe, at least a portion of the gene's DNA sequence needs to be available. If a probe for a gene is not available, transposons may be used to recover part of the gene to use as a probe, a process called **transposon tagging**. If a mutant allele of the gene that contains a transposon can be found, then the transposon can be used as a probe to isolate part of the gene's sequence. The transposon "tags" the gene, giving the researcher an identifying mark on the gene to distinguish it from other fragments of DNA. The following example highlights one of the first cases of transposon tagging.

Example 22.4   Cloning the *bronze* (*bz*) locus in maize by transposon tagging.

In 1984, Fedoroff et al. (*Proceedings of the National Academy of Sciences, USA* 81:3825–3829) reported cloning the *bronze* locus in maize by transposon tagging. Neither the DNA sequence of the gene nor the amino acid sequence encoded by the gene was known. No DNA probe for the *bronze* locus was available to assist with cloning. The authors selected a mutant allele, *Ac bz-m2*, known to be caused by insertion of an *Ac* element into the *bronze* locus. The *Ac* element had been previously cloned from another maize plant with an *Ac* element at the *waxy* locus, and was available as a probe. The authors used a lambda vector to create a genomic library of DNA fragments from a plant that was

---

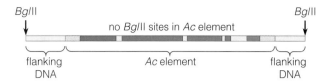

Figure 22.20   Information for Example 22.4: *Bgl*II sites in the DNA that flanks an *Ac* element.

homozygous for *Ac bz-m2*. When they hybridized clones from the library with the *Ac* probe, they found 25 clones with homology to the *Ac* probe. Restriction digestion of the clones revealed that only two had restriction sites that were identical to the cloned *Ac* element. The enzyme used for cloning (*Bgl*II) does not have a cleavage site within the *Ac* element, so cloned fragments with an *Ac* element in them must have host DNA on both sides of the *Ac* element, as illustrated in Figure 22.20.

If one of the clones had the *Ac* element at the *bronze* locus, then part of the *bronze* locus would be attached to the *Ac* element in the clone. The authors used one of the clones to probe genomic DNA from plants with different mutant *bronze* alleles and found different restriction patterns. They also attempted to hybridize the probe to a mutant genotype that had a deletion at the *bronze* locus. The probe failed to hybridize to DNA from this genotype. They then took the DNA that was outside the *Ac* element, cloned it into a new vector, and called this segment a subclone from the original clone. They created a new genomic library from a plant that was homozygous for the nonmutant *Bz* allele. Using the subclone as a probe, they found a clone from the new library that hybridized with the subclone. The new clone contained the intact *bronze* locus.

Problem: **(a)** Why did 25 clones (rather than just one) hybridize with the *Ac* probe? **(b)** What evidence suggests that the clone they selected for subcloning contained part of the *bronze* locus? **(c)** Why was it necessary to subclone the DNA that was outside the *Ac* element to use as a probe for identifying the clone with an intact *bronze* locus?

Solution: **(a)** There are three possible reasons for the presence of more than one clone that hybridized with the *Ac* probe. First, inactive *Ac*-like elements may be present in the genome at sites other than the *bronze* locus. Clones that contain these sequences should hybridize with the *Ac* probe. Indeed, there were several such clones. They could be identified and eliminated because their restriction patterns differed from that of the *Ac* element. Second, the same DNA sequence may be cloned

more than once, so several clones may be identical. Third, copies of the *Ac* element may have been present at sites other than the *bz* locus. Thus, the researchers had to conduct a series of experiments to eliminate all clones that did not contain part of the *bronze* locus. **(b)** The correlation of restriction patterns with mutations at the *bronze* locus is good evidence that the selected clone contained part of the *bronze* locus. However, the confirming evidence was the failure of the clone to hybridize with genomic DNA in which the *bronze* locus had been deleted. **(c)** The subcloning was necessary to eliminate the *Ac* element from the probe so that the probe contained only DNA from the *bz* locus. Had the *Ac* element (or a part of it) remained attached to *bz* DNA, then the probe would have hybridized with any *Ac*-like sequence in the genome. By eliminating the *Ac* DNA, the authors could be certain that their subcloned probe hybridized only with DNA from the *bz* locus.

Scientists have used transposon tagging to identify and clone genes from many different species using a variety of transposable elements as tags. If the target gene does not have a known allele that contains a transposon, it is a simple matter to make a cross in which transposons are active, and select the mutant phenotype. Complementation analysis can reveal if the new mutant is indeed in the target gene. After an allele that contains the transposon has been identified, the transposon can be used as a probe to clone a portion of the DNA at the locus.

Transposon-Mediated Transformation

Transformation is the process of integrating foreign DNA into a cell and is a natural process in bacteria. It is used routinely to introduce genetically engineered plasmids into bacteria. Although plasmid transformation in bacteria is straightforward, transformation of eukaryotes is no easy matter. A transformation vector that is capable of integrating the foreign DNA into the eukaryotic genome can be very helpful. Because of their ability to integrate into the genome, transposons are a good choice as vectors for transformation. If a transposon that carries a gene of interest is inserted into a cell, it can carry the gene with it as it integrates into the genome. The following example highlights transposon-mediated transformation in *Drosophila*.

Example 22.5   P element–mediated transformation of *Drosophila melanogaster*.

In 1982, Rubin and Spradling (*Science* 218:348–353) reported successful transformation of *Drosophila melanogaster* using a P element as the transformation vector. They inserted a wild-type *rosy* gene (*ry*$^+$) into

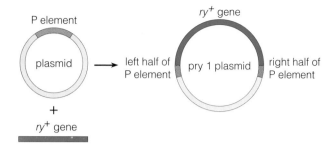

Figure 22.21   Information for Example 22.5: Insertion of the *ry* gene into a plasmid that contains a P element to create the pry 1 plasmid. (Adapted with permission from Rubin, G. M., and A. C. Spradling. 1982. Genetic transformation of *Drosophila* with transposable vectors. *Science* 218:348–353. Copyright 1982 American Association for the Advancement of Science. Reprinted by permission.)

the center of a cloned P element. The inserted element interrupted the transposase gene. The resulting plasmid (called the pry 1 plasmid) carried the *rosy* gene bordered on both sides by P element fragments that contained the inverted terminal repeat sequences, as diagrammed in Figure 22.21.

The authors conducted two separate transformation experiments. In the first experiment, they crossed P-cytotype males with M-cytotype females. Both male and female parents were homozygous for the recessive mutant *rosy* alleles. They then injected the pry 1 plasmid into the F$_1$ embryos from this cross. The F$_1$ flies that developed from the injected embryos were then mated with P-cytotype flies that were homozygous for recessive *rosy* alleles. In the second experiment, they used two plasmids, the pry 1 plasmid with the wild-type *rosy* gene and a plasmid called pπ25.7, which contained an intact P element but no *rosy* gene. They injected both plasmids into M-cytotype embryos that were homozygous for recessive *rosy* alleles. The flies that developed from these embryos were crossed with M-cytotype flies that were homozygous for recessive *rosy* alleles. All flies that developed from injected embryos in both experiments had the recessive rosy phenotype. Among their progeny, both wild-type and rosy flies appeared. The wild-type flies must have received their wild-type *rosy* allele by transformation. The authors used in situ hybridization to determine where the transformed *rosy* gene had been inserted in the chromosomes in the wild-type flies. In each case, the transformed gene was located near, but not at, the *rosy* locus on chromosome 3.

Problem: **(a)** The authors designed their experiments to provide transposase so that the wild-type *rosy* allele could be integrated into the chromosomes. What was the source of transposase in each of the experiments? **(b)** Why were no wild-type flies found among the flies that developed from the injected embryos? **(c)** What principle discussed in

previous chapters is illustrated by the observation that all wild-type flies examined had the transformed wild-type *rosy* allele near the site of the *rosy* locus on chromosome 3, instead of at random sites among all chromosomes?

Solution: **(a)** In both experiments, the pry 1 plasmid could not be the source of transposase because its transposase gene was interrupted by the wild-type *rosy* gene. However, the gene was bordered by both ends of the P element and, therefore, by the inverted terminal repeats. In essence, the pry 1 plasmid contained a nonautonomous P element that included the *rosy* gene. In order to transpose, this nonautonomous element needs a source of transposase. In the first experiment, P-cytotype males were crossed with M-cytotype females. This created hybrid dysgenesis in the progeny embryos. The source of transposase was the P elements donated by the P-cytotype male parents to the F$_1$ progeny. The transposase produced in dysgenic F$_1$ embryos could act on the nonautonomous P element with the *rosy* gene and transpose it into the fly chromosomes. In the second experiment, the embryos had no native transposase activity because they were M-cytotype embryos. The source of transposase was the transposase gene in the intact P element cloned in the pπ25.7 plasmid that was injected with the pry 1 plasmid into the embryos. This gene produced functional transposase that transposed the nonautonomous P element with the *rosy* gene into the embryo's chromosomes.

(b) P elements transpose only in the germ line. The injected elements transposed in the germ line of the injected embryos. The eyes of these embryos developed from somatic cells, so there was no P element activity in their cells, and the recessive alleles they inherited from their parents were expressed. However, their gametes developed from the germ line, in which transposition was active, so their progeny could express the dominant wild-type phenotype whenever transposition inserted an active *rosy* gene into the chromosomes.

(c) The observation that the wild-type flies all had the transformed rosy gene near the rosy locus on chromosome 3 is an example of position effect, a phenomenon described in section 17.3. Presumably, the transformed *rosy* gene was inserted at various sites in the chromosomes but was expressed only when near its proper location on chromosome 3. Its expression is probably due to its proximity to regulatory regions on chromosome 3 that are required for the gene to be transcribed.

P element–mediated transformation is now a routine method for introducing genes into *Drosophila*. Even human genes have been inserted and expressed in flies

using this method. Most protocols now call for injection of the engineered allele, using the *rosy* allele as a marker to identify transformed flies, and a P element that lacks one of the inverted repeats but has an intact transposase gene as a source of transposase. The intact transposase gene encodes transposase, but it cannot transpose the element in which it resides because one of the inverted repeats is missing. Instead, the transposase inserts only the engineered P element into the chromosome.

~

Transposable elements have several applications in research. Among the most important are transposon tagging and transposon-mediated transformation.

SUMMARY

1. Transposable elements may be divided into three general classes: transposons, retrotransposons, and retroposons.

2. Transposons are segments of DNA that can move from one location of the genome to another and move as a DNA entity.

3. Retrotransposons are segments of DNA that resemble retroviruses and code for a reverse transcriptase. They transpose by means of an RNA intermediate that is transcribed from the retrotransposon. The RNA is then reverse-transcribed into DNA, which is made double stranded and inserted into a new position in the genome.

4. Retroposons, like retrotransposons, are pieces of DNA that are transcribed into an RNA intermediate, which is then reverse-transcribed into DNA and inserted into the genome. Unlike retrotransposons, retroposons do not encode their own reverse transcriptase, nor do they resemble retroviruses.

5. Transposable elements were first found in maize, *Drosophila*, and *E. coli*, probably because these species were studied intensively in genetics. Transposable elements have since been found in many species and may be present in all species.

6. The *Ac-Ds* system is a good example of the general features of transposons. At the molecular level, *Ac* elements have inverted terminal repeats and contain a gene for a transposase. The transposase catalyzes transposition. *Ac* elements are said to be autonomous because the transposase produced by an *Ac* element can catalyze transposition of the element that encoded it.

7. *Ds* elements are *Ac* elements in which a deletion has inactivated the transposase gene. *Ds* elements cannot produce transposase, so they are incapable of transposing unless an *Ac* element is present to produce functional transposase. *Ds* elements are said to be nonautonomous because they are dependent on an autonomous element for transposition.

8. The best-studied transposons in *Drosophila* are the P elements. They have the same general features as maize

transposable elements, such as inverted terminal repeats and a transposase gene. They also exist in autonomous and nonautonomous forms.

9. Unlike maize transposons, which transpose in somatic cells, P elements transpose only in the germ line. They transpose in the progeny of a dysgenic cross, a cross between an M-cytotype female and a P-cytotype male. The high rate of transposition in the progeny of a dysgenic cross is called hybrid dysgenesis.

10. Retrotransposons are very similar to retroviruses in their mechanisms of transposition and their structural features, and are probably of retroviral origin. Retrotransposons are bordered by direct LTRs (long terminal repeats) and contain genes that encode enzymes for reverse transcription and for integration into the host genome.

11. The *Ty* element of yeast is an example of a retrotransposon that forms noninfective viruslike particles that encase the RNA prior to reverse transcription.

12. The *copia* element of *Drosophila melanogaster* is a well-studied example of a retrotransposon. Unlike *Ty*, it does not form viruslike particles. It comprises about 15% of the *Drosophila melanogaster* genome.

13. Retroposons also transpose by means of an RNA intermediate, but they have no reverse transcriptase gene. Instead, they rely on reverse transcriptase produced by retrotransposons or by retroviruses present in the cell.

14. Retroposons are present in humans and include LINEs and SINEs, retroposons are present in many copies in the human genome.

15. Transposons are also found in prokaryotes. The best-studied are insertion sequences, which have the same general features as maize and *Drosophila* transposons.

16. Two insertion sequences may border a segment of DNA and transpose as a single unit with the DNA between them. Such a unit of transposable DNA is called a composite transposon.

17. Transposable elements may be applied in genetic research for transposon tagging and transposon-mediated transformation.

QUESTIONS AND PROBLEMS

1. Distinguish between autonomous and nonautonomous transposable elements at the phenotypic and molecular levels, providing examples to illustrate the differences.

2. According to Battacharyya et al. (1990. *Cell* 60: 115–122), the recessive allele that caused wrinkled pea seeds when homozygous in Mendel's experiments contains a transposable element that disrupts a gene that encodes an enzyme that catalyzes the synthesis of branched starch. Apparently, the presence of this element did not alter Mendelian ratios, because Mendel conducted his experiments with plants that contained this allele. Explain why a transposable element does not necessarily alter Mendelian ratios.

3. McClintock's studies with maize transposable elements focused on reversion mutations. Transposons in maize can cause both forward mutations and reversions, so why did she focus her attention on reversions instead of forward mutations?

4. Many of the experiments with transposons in maize were conducted with mutable alleles that are expressed in the aleurone layer of the endosperm. The *c-m1* allele contains a *Ds* element inserted into the *c1* locus, and causes kernels to be light colored when homozygous. Any plant that is homozygous for a recessive allele at the *c1* locus can be tested for the presence or absence of *Ac* by crossing it with plants that are homozygous for the *c-m1* allele. What is the expected phenotype of the progeny kernels if a plant homozygous for an *Ac* element is crossed with a plant homozygous for *c-m1* if the parent that carries the *Ac* element is **(a)** homozygous for a stable recessive *colorless* mutant allele (*c1c1*), **(b)** homozygous for the dominant color-producing allele (*C1C1*)?

5. Suppose that a maize plant with yellow kernels is homozygous *c1c1* and is homozygous for an *Ac* element that is located 12 cM from the *c1* locus. This plant is crossed with a plant that is homozygous for *c-m1* and does not carry *Ac*. The F_1 kernels are spotted. The F_1 plants are then backcrossed with the parent line that is homozygous for *c-m1* but does not carry *Ac*. What is the expected phenotypic ratio of the backcross progeny kernels?

6. Recessive mutant alleles that contain transposable elements are often unstable and revert to alleles that cause the dominant phenotype. Occasionally, an unstable recessive allele changes into a stable recessive allele. What changes at the molecular level could account for such a change?

7. McClintock (*Carnegie Institution of Washington Year Book* 63:592–602 and 64:527–534) noted that active *Ac* elements can become inactive. In 1983, Fedoroff et al. (*Cell* 35:243–251) reported that there were several *Ac*-like sequences dispersed throughout the genome in a genotype with an active *Ac* element at only one site. They found that the additional *Ac* sequences were not cleaved by a methylation-sensitive restriction enzyme, but that the active *Ac* element was. Later, Chomet et al. (*EMBO Journal* 6:295–302) and Schwartz and Dennis (*Molecular and General Genetics* 205:476–482) showed that inactive *Ac* elements are highly methylated, whereas active *Ac* elements are not highly methylated. Why might methylation cause an *Ac* element to be inactive? (You may wish to consult Chapter 8 to help you answer this question.)

8. In 1987, Peschke et al. (*Science* 238:804–807) placed tissue from maize plants with no *Ac* activity into tissue cul-

ture where the cells grew rapidly in Petri plates. When whole plants were regenerated from the cultured cells, some of the plants had *Ac* activity. In light of the previous question, what could explain the *Ac* activity in these regenerated plants?

9. In 1938, Rhoades (*Genetics* 23:377–397) reported the first known study demonstrating the inheritance patterns associated with a transposon. He obtained an ear of corn from another researcher that had kernels in a 12 colored:3 spotted:1 colorless ratio. The mutation that caused the colorless phenotype was at the *a* locus. Describe the genotypes of the two parent plants for transposable element constitution and the types of alleles at the *a* locus.

10. When a chromosome-breaking *Ds* element transposes, it breaks the chromatid in which it resides and causes a single terminal deletion that initiates the BFB cycle. Both sister chromatids must be broken to initiate the BFB cycle, so how can a single break initiate the cycle?

11. In *Drosophila*, hybrid dysgenesis appears only in the progeny of an M-cytotype female hybridized with a P-cytotype male. It does not appear when a P-cytotype male is hybridized with a P-cytotype female. P-cytotype males and P-cytotype females both carry P elements, so why don't their progeny show hybrid dysgenesis?

12. Which type of inheritance described in Chapter 18 is illustrated by the mating of an M-cytotype female with a P-cytotype male?

13. Revertant alleles that arise when a transposon is excised often produce an enzyme that has reduced activity when compared to the enzyme encoded by the wild-type allele. What could explain the failure of such revertants to restore full enzyme activity?

14. In 1982, Schwartz and Echt (*Molecular and General Genetics* 187:410–413) reported that a *wx* revertant allele from which a *Ds* element had been excised produced a protein product with a lower electrophoretic mobility than the protein product of the usual dominant allele. Why is this result expected?

15. Nearly all nonautonomous P elements contain deletions. Why should deletions rather than point mutations be more common in nonautonomous P elements?

16. In 1983, O'Hare and Rubin (*Cell* 34:25–35) found that three of four P elements they examined at the *white* locus were inserted into exactly the same site. One of the three elements was in reverse orientation to the other two. When they examined other genes, they found similar patterns of P element insertion. What do these observations suggest about the DNA insertion sites for P elements?

17. The *sn*^w mutant allele described in Example 22.2 contains nonautonomous elements that are stable in the absence of an autonomous P element. It is possible to take the mutant *sn*^w allele in a P-cytotype fly and eventually recover the allele in M-cytotype progeny in which it is stable. Diagram how this could be done by traditional breeding without using any recombinant DNA methods. Be sure to identify which genotypes are used as males and females.

18. Antibiotic resistance genes in a composite bacterial transposon have a selective advantage when the corresponding antibiotic is in widespread use. Why would a composite transposon have a selective advantage over a simple antibiotic resistance gene in the bacterial chromosome?

19. In the insertion sequence IS*10*, an antisense RNA (an RNA transcribed from the sense strand in the DNA) is transcribed from the transposase gene when there are multiple copies of IS*10* in the cell. What possible role could this antisense RNA have?

20. When a clone of a transposable element is used as a probe, it is possible to clone genes when nothing is known about the gene's sequence or the amino acid sequence of the product it encodes. A similar procedure may be used to clone a transposable element when nothing is known about the DNA sequence of the transposable element or the amino acid sequence of its transposase. Describe a procedure by which such a transposable element may be cloned.

FOR FURTHER READING

An excellent book that contains chapters by different authors on all known types of transposable elements is **Berg, D. E., and M. M. Howe, eds. 1989. *Mobile DNA*. Washington, D.C.: American Society for Molecular Biology**. Chapters of particular interest are Fedoroff, N. Maize transposable elements (Chapter 14, pp. 375–412); Engels, W. R. P elements in *Drosophila melanogaster* (Chapter 16, pp. 437–484); Hutchison, C. A., et al. LINEs and related retrotransposons: Long interspersed repeated sequences in the eucaryotic genome (Chapter 26, pp. 593–618); and Deininger, P. I. SINEs: Short interspersed repeated DNA elements in higher eucaryotes (Chapter 27, pp. 619–636). Maize transposable elements were reviewed by **Fedoroff, N. 1984. *Scientific American* 250 (Jun):84–98**. P elements were reviewed by **Engels, W. R. 1997. Invasion of P elements. *Genetics* 145:11–15**; and **Engels, W. R. 1992. The origin of P elements in *Drosophila melanogaster*. *BioEssays* 14:681–686**.

For additional reading, go to InfoTrac College Edition, your online research library at: http: www.infotrac-college.com/brookscole

CHAPTER 23

KEY CONCEPTS

Genes govern the differentiation of somatic cells into specialized tissues and organs.

~

In the nematode *Caenorhabditis elegans*, the somatic lineage of every cell in the adult is known. Mutations in genes that regulate development have provided information about the function of those genes.

~

Much is known about gene expression and development in *Drosophila melanogaster*. Certain genes establish the number and position of body segments, whereas another set of genes determine segmental identity.

~

There are many similarities in DNA sequence, function, and organization between the genes that regulate development in *Drosophila* and those in mammals.

~

Genes that govern plant development are different from those that govern animal development.

DEVELOPMENTAL GENETICS

One of the most perplexing questions of biology is how a single-celled zygote can give rise to a complex, multicelled organism with specialized tissues and organs that interact with one another. Although the answers to this question are complicated and far from being fully answered, developments in recombinant DNA research have provided much information about the role of genes in development. We have already discussed how individual eukaryotic genes are regulated, in Chapter 8. The purpose of this chapter is to discuss how gene regulation is organized among cells and tissues to form complex multicellular organisms.

The diversity of life is readily manifest in development. The developmental patterns in a human differ substantially from those in an insect or a higher plant. Until recently, developmental genetics focused on just a few species that were relatively easy to study. Specifically, the nematode worm *Caenorhabditis elegans* and the fruit fly *Drosophila melanogaster* were, and still are, the model organisms for studies in developmental genetics. Both of these species are so different from a mammal that reluctance to apply the principles of development discovered in them to other species seemed well justified. However, although anatomical patterns of development are indeed

different among major taxonomic groups, there are some striking similarities at the molecular level in species as different as a fruit fly and a human. In this chapter, we will first review some basic aspects of developmental genetics. We'll then examine developmental genetics in animals, with a focus on *C. elegans* and *D. melanogaster*. We will also discuss research on developmental genetics in mammals. Finally, we examine the emerging field of plant developmental genetics and compare plant development genes with those of animals.

23.1 DIFFERENTIATION AND TOTIPOTENCY

Nearly all somatic cells in an organism contain the entire genetic complement necessary to form an entire multicelled individual. However, in adult multicelled organisms, most of the somatic cells have ceased dividing and have assumed the specialized tasks of the tissue in which they reside. These cells have undergone **differentiation**, the process of becoming specialized to perform a specific function. The number of genes that are expressed in a differentiated cell is relatively small,

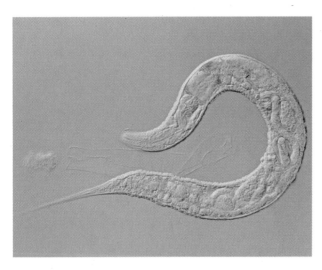

Figure 23.1 The nematode worm *Caenorhabditis elegans*. (Sinclair Stammers/Photo Researchers)

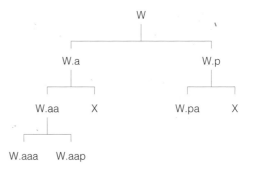

Figure 23.2 A diagram of mitotic lineages of nuclei in *C. elegans*.

and the genes expressed in a cell from one tissue are usually quite different from the genes expressed in a cell from another tissue. Often, somatic cells become so specialized that they completely lose the ability to develop into a new organism when the cell is separated from its tissue and induced to divide. A cell that is capable of giving rise to an entire organism is said to be **totipotent**. Totipotency in plant cells is common. It is much less common in animal cells, especially those of complex animals such as mammals. The specialization of cells during development is one of several factors that cause cells to lose totipotency.

In animals, differentiation begins soon after fertilization, and totipotency of most cells is usually lost early during development. In humans, cell differentiation is already well under way just 15 days after fertilization. At any stage of development, a nucleus can be extracted from a cell and inserted into a zygote whose nucleus has been removed or disrupted by radiation. If the introduced nucleus is totipotent, it can divide mitotically, and the zygote can develop into an embryo. Under the right conditions, such an embryo may eventually develop into a mature individual. Several totipotent nuclei obtained from the same individual and inserted into zygotic cells can produce cloned individuals. However, in animals, totipotency is usually lost very early. Some amphibian cells retain totipotency into the tadpole stage, whereas most mammalian cells lose totipotency during early stages of development, long before any anatomical structures are apparent in the embryo.

~

Differentiation is the process by which gene expression targets cells for specialized functions in a multicelled organism. Totipotency is the ability of a cell to produce progeny cells that can differentiate to form a whole organism. As cells differentiate into specialized tissues and organs, they usually lose their totipotency.

23.2 GENETIC CONTROL OF DEVELOPMENT IN *Caenorhabditis elegans*

At some point in development, two cells that came from the same original cell must diverge in their pattern of differentiation to produce different tissues. To trace the developmental fate of cells in a mammal that has trillions of cells is an overwhelming task. The nematode worm *Caenorhabditis elegans* (Figure 23.1), however, is microscopic and has relatively few cells. There are two sexes, males with an XO chromosome constitution, and hermaphrodites with an XX chromosome constitution. The mature male has 1031 nuclei, and the mature hermaphrodite has 959 nuclei. Researchers follow nuclei rather than cells because at some points during development, a single cell may have more than one nucleus. The exact mitotic lineage of every nucleus in both sexes is known, and the effects of mutations that alter cell development have been studied extensively. The studies conducted with *C. elegans* have provided a comprehensive view of how development proceeds in a single species.

Growth from zygote to adult passes through four distinct larval stages in a period of about 51 hours at 25°C. The end of each larval stage is marked by a brief resting period, during which the larva molts. The four larval stages are called L1, L2, L3, and L4. The L4 larva molts to become an adult. The hermaphroditic L1 larva has 558 nuclei, which eventually divide mitotically to produce the 959 nuclei of the adult. Some nuclei pass through more cell generations than others, and some nuclei are destined to die during development.

Mitotic nuclear lineages can be described using a tree diagram of the type illustrated in Figure 23.2. The letters in the diagram designate the nuclear lineage. In Figure 23.2, a nucleus, designated W, divides once in an anterior-posterior plane. The daughter nucleus at the anterior end is called W.a, and the posterior daughter nucleus is called W.p. The two nuclei divide again in an anterior-posterior plane, but the posterior nuclei W.ap and W.pp die, so they are designated X. The W.aa nucleus

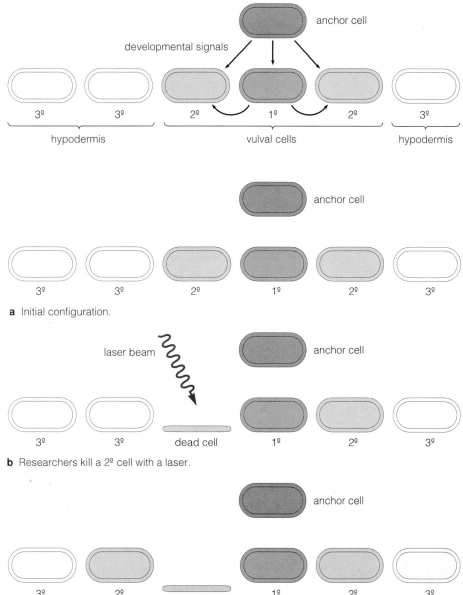

Figure 23.3 The pattern of intercellular signals in the single anchor cell and six vulval precursor cells of *C. elegans*. (Adapted from Sternberg, P. W. 1990. Genetic control of cell type and pattern formation in *Caenorhabditis elegans*. *Advances in Genetics* 27:63–116.)

a Initial configuration.

b Researchers kill a 2º cell with a laser.

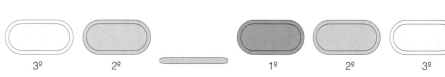

Figure 23.4 Evidence that vulval cell development is positional. When a laser beam is used to kill a 2º cell, the adjacent 3º cell becomes a 2º cell. (Adapted from Sternberg, P. W. 1990. Genetic control of cell type and pattern formation in *Caenorhabditis elegans*. *Advances in Genetics* 27:63–116.)

b The nearest 3º cell becomes a 2º cell.

divides once again in the anterior-posterior plane, so its daughter nuclei are designated W.aaa and W.aap. The lineage of any nucleus can be diagrammed in this manner.

The lineages of all nuclei in *C. elegans* are now known, so researchers can identify genes that govern development by analyzing mutations that alter normal development. By comparing the effects of mutant alleles with their nonmutant counterparts, researchers can determine how genes influence cell differentiation at different developmental stages. The work of Paul W. Sternberg and H. R. Horvitz on the genes that control vulval development in the hermaphrodite of *C. elegans* provides one of the best examples of how mutations can affect the development of an organ. The results of their studies illustrate how developmental regulation can be very complicated even in a relatively simple organism.

The vulva arises from seven cells, one called the anchor cell and the other six the vulval precursor cells. Each vulval precursor cell is destined to develop into one of three cell types, called 1º (primary), 2º (secondary), and 3º (tertiary). The 1º and 2º cells differentiate into vulval tissue, and the 3º cells develop into the hypodermis that surrounds the vulva. The distance of a precursor cell from the anchor cell determines its fate. Genes in the anchor cell produce a signal that affects genes in the precursor cells. Primary cells produce another signal that affects the adjacent 2º cells, as diagrammed in Figure 23.3.

If researchers kill a 2º cell with a laser beam, the 3º cell nearest the 1º cell becomes a 2º cell because of its newfound proximity to the 1º cell (Figure 23.4). This type of regulation, in which surrounding cells affect the development of a cell, is called **positional developmental**

regulation, because the cell's position relative to other cells determines its fate. On the other hand, in **autonomous developmental regulation**, the fate of the cell lineage is programmed within the cell and is unaffected by surrounding cells. Both positional and autonomous regulation of development are known in *C. elegans*, although in most animals, positional regulation tends to predominate.

Researchers have used mutations that alter the normal development of vulval cells to identify the developmental pathways and the genes that regulate them. These genes regulate development at four stages: (1) generation of the anchor cell and vulval precursor cells, (2) specification of the three precursor cell types (1º, 2º, and 3º), (3) execution of the three precursor cell types to produce progeny cells of each type, and (4) morphogenesis, the formation of the vulva's structure by interactions among the cell types and attachment of muscles and nerves. Mutations that disrupt any one stage prevent subsequent stages from proceeding.

At least five genes (and possibly more) control the first stage, differentiation of the anchor and vulval precursor cells. Geneticists identified these genes by analyzing mutations that disrupt this stage. For example, a mutation in the *lin-26* gene causes the cells that should be vulval cells to become neurons or neuroblasts, thus disrupting development of the vulva and preventing the development of subsequent stages. Other mutations, called **heterochronic mutations**, affect the timing of development. Mutations in the *lin-14* gene, for example, cause vulval development one stage earlier than in normal cells.

At least 15 genes regulate the second stage, specification of the cell types. Mutations in several of these genes prevent differentiation of the vulval precursor cells into 1º, 2º, and 3º cells. Instead, they all become 3º cells because the mutations either eliminate the signal from the anchor cell or prevent the synthesis of receptors in the vulval precursor cells.

The third stage is execution, the development of cell lineages from the three initial cell types. Mutations in genes that govern one cell type affect only that cell type. For example, mutations in the *lin-11*, *lin-17*, and *lin-18* genes cause abnormal development of the 2º lineage but have no effect on the 1º and 3º lineages.

Mutations in genes that regulate the fourth stage, morphogenesis, allow the normal number and the proper lineages of cells to develop but prevent normal formation of the vulva. One such mutation (*Egl*) causes defective egg laying because the vulva is malformed even though it has the normal number and types of cells.

Another class of genes studied in *C. elegans* are those that cause programmed cell death. Certain cells are destined to die in order for the organism to develop properly. For example, male development requires deactivation of the *egl-1* gene, which prevents selective death of the hermaphrodite-specific neuron. This cell dies during male development but survives during hermaphrodite development.

The following example shows how mutations in a gene can be used to determine the function of that gene in development.

Example 23.1   The role of *lin-12* in *C. elegans*.

The role of the *lin-12* gene in vulval development has been described in numerous publications (for examples, see Greenwald et al. 1983. *Cell* 34:435–444; Ferguson et al. 1987. *Nature* 326:259–267; and Newman et al. 1995. *Development* 121:263–271). One of two potential precursor cells, Z1.ppp or Z4.aaa, develops into the anchor cell. Each of the two cells has an equal chance of developing into the anchor cell. In normal development, as one cell develops into the anchor cell, the other cell develops into the ventral uterine precursor cell. Most recessive mutant alleles of the *lin-12* gene cause both cells to become anchor cells when the mutant allele is homozygous. Most dominant mutant alleles of the *lin-12* gene cause both cells to become ventral uterine precursor cells.

Problem: What do these results suggest about the role of *lin-12* in differentiating the anchor and ventral uterine precursor cells?

Solution: The recessive mutations are loss-of-function mutations that prevent synthesis of functional *lin-12* product in homozygotes. The dominant mutations are gain-of-function mutations that overproduce the *lin-12* product. The observations indicate that the *lin-12* product induces formation of the ventral uterine precursor cell, whereas the anchor cell forms in the absence of the *lin-12* product.

The somatic lineage of every cell in the adult forms of *C. elegans* is known. Mutations in genes that regulate cell differentiation have provided much information about the roles of these genes in development.

23.3   GENETIC CONTROL OF DEVELOPMENT IN *Drosophila melanogaster*

A fruit fly is much more complex than a microscopic worm. Nonetheless, much has been learned about the genetic regulation of development in *Drosophila*. Researchers have obtained much information by cloning and sequencing several genes that regulate development.

The major developmental stages of *Drosophila melanogaster* are diagrammed in Figure 23.5. Like many

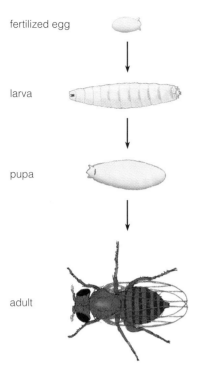

fertilized egg

larva

pupa

adult

Figure 23.5 The developmental stages of *Drosophila melanogaster*.

leg structures in place of antennae

a Antennapedia phenotype

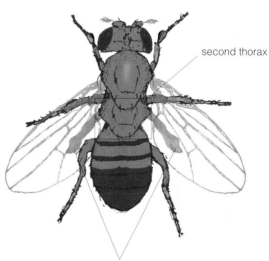

second thorax

undeveloped second set of wings

b Bithorax phenotype

Figure 23.6 Examples of mutant phenotypes in which development is altered. **(a)** In the mutant antennapedia phenotype, leg structures develop in place of the antennae. **(b)** In the mutant bithorax phenotype, a second thorax develops. (Part a: Oliver Meckes/Photo Researchers; part b: Adapted from Bridges, C., and T. H. Morgan. 1923. Carnegie Institution of Washington. Publication 327.)

other insects, *D. melanogaster* follows a pattern of metamorphosis, with egg, larval, pupal, and adult stages. Genes regulate development at each stage, and most of the important genes have been cloned and sequenced. The genes were first identified by mutations that cause altered development, such as the antennapedia and bithorax mutant phenotypes shown in Figure 23.6, and by mutations that cause defects in the developing larva. These latter mutations usually create recessive lethal alleles, but the mutant alleles can be maintained in heterozygotes. Cloning, transposon-mediated transformation, and DNA sequencing have added a wealth of information about how development proceeds.

P element–mediated transformation is an effective method for introducing foreign genes into *Drosophila* embryos. The method is explained in detail in section 22.4. P element–mediated transformation has provided information about how certain genes and their promoters behave. It has allowed researchers to genetically engineer genes outside of the fly embryo, then inject the engineered genes into embryos and study their effects.

The reconstructed genes are often called **fusion genes** because they consist of DNA from the regulatory region of a gene attached to the coding region of a **reporter gene**, a foreign gene whose product can readily be detected in the phenotype. In studies on gene regulation in *Drosophila*, the *lacZ* gene from *E. coli* has often been used as a reporter gene in fusion genes like the one diagrammed in Figure 23.7. Researchers inject the fusion

gene into embryos, and both the timing and site of the regulatory DNA expression can be followed by detection of β-galactosidase, the product of the *lacZ* gene. Because β-galactosidase is a bacterial enzyme not normally present in insects, and because it does not interfere with normal insect functions, researchers can use it as a marker to indicate how a particular promoter functions. Fly development may proceed normally as if the marker gene and its product were not there.

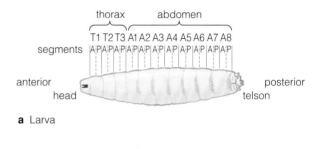

left end of   regulatory region   *lacZ* gene   right end of
P element   of developmental gene   P element

Figure 23.7 A genetically engineered P element that contains the regulatory region of a developmental gene positioned upstream from the *lacZ* gene of *E. coli.*

Another type of study that has contributed to our understanding of genes that regulate development is cloning and expression of *Drosophila* genes in *E. coli*, where gene products and their interactions with one another and with DNA can be studied more easily than in a developing fly. In a developing fly embryo, cells produce several regulatory proteins in very small quantities, making it difficult for researchers to isolate and study the proteins. When the genes that encode these proteins are cloned and expressed in bacteria, however, the proteins and their interactions with the DNA of other cloned genes can be studied directly.

Much research has focused on the early developmental stages in the embryo and larva that determine the development of body segments. In order to follow development, the segments have been numbered, as illustrated in Figure 23.8. Each segment has an anterior and a posterior portion (labeled A and P, respectively, in Figure 23.8). Mutations in genes that regulate development may alter the segment patterns.

Let's now examine in detail the sequential pattern of gene expression and how it regulates development in the embryo and larva.

Maternal Effects in Early Development

The first stages of development in the early embryo are controlled by maternal rather than zygotic genes. The maternal genes establish the initial patterns of development, particularly the embryo's **polarity**—which end will be anterior (head and thorax) and which will be posterior (abdomen and telson)—as diagrammed in Figure 23.8. When homozygous in the maternal parent, a mutation called **bicoid** (**bcd**) has no effect on the parent but causes failure of anterior development in all of the offspring. The developing larvae have two abdomens with a telson at each end and lack a head (Figure 23.9).

The fact that all progeny of a female fly that is homozygous for the *bicoid* allele have the bicoid phenotype, regardless of their genotype, indicates that the maternal genotype determines the larval phenotype. Additional evidence of this maternal effect was gleaned from the results of several experiments. When the anterior cytoplasm is removed from the developing embryo but the zygote is left in place, the embryos fail to form anterior structures. When anterior cytoplasm from a wild-type fly is added to a *bicoid* embryo, the embryo develops normally. All of these experiments suggest that a substance produced in the maternal parent determines anterior-posterior development in the embryo. Any substance that affects mor-

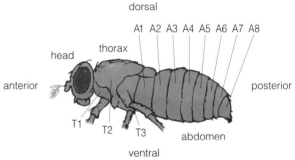

a Larva

Figure 23.8 Segment numbering in the larva and the location of these segments in the adult of *Drosophila melanogaster.*

phological development is called a **morphogen**. In this case, a maternal morphogen determines polarity.

Researchers found that the morphogen is a protein encoded by maternal mRNAs in the anterior pole of the embryo. When cloned and used as a probe, DNA from the *bicoid* gene hybridizes with mRNAs found only at the anterior pole. The mRNAs are translated into a protein that is concentrated at the anterior pole but diffuses throughout the anterior half of the embryo to form a concentration gradient from the anterior pole to the embryo's midline. The bicoid protein concentration gradient establishes anterior-posterior polarity. A portion of the protein contains a sequence of amino acids called a homeodomain that is homologous to homeodomains in other proteins that regulate development. A **homeodomain** is a DNA-binding region of the protein that includes a helix-turn-helix motif (see section 8.5). The *bicoid* gene product is a transcription factor that binds to DNA and regulates expression of zygotic genes.

At least 30 maternal genes govern the first stages of development. In addition to the *bicoid* gene, several others have been cloned and studied. Among the best-studied are the *nanos* (*nos*) and *dorsal* (*dl*) genes. The *nanos* gene regulates posterior development and formation of the abdomen. Its mRNAs are located exclusively at the posterior pole. The *dorsal* gene regulates the dorsal-

posterior   posterior

Figure 23.9 The mutant bicoid phenotype in a larva. Both ends of the larva have posterior structures.

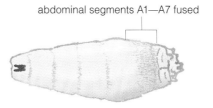

abdominal segments A1—A7 fused

Figure 23.10 The mutant knirps phenotype in a larva. Abdominal segments A1–A7 are fused into a single segment.

ventral orientation of the embryo. Its product is found throughout the embryo, but the gene itself is located in the nucleus of ventral cells, where it is active, and in the cytoplasm of all other cells, where it is inactive. A set of genes in the nuclei of the embryo direct the deposition of the *dorsal* protein in the cells. Mutations in any of these genes cause the location of the dorsal protein to change, resulting in ventral structures on the dorsal portion of the fly, or vice versa.

~

Maternal genes establish the initial anterior-posterior and dorsal-ventral orientations of the embryo. The products of these genes interact with gene products produced by the embryo's genes.

Segmental Gene Expression

With polarity established by maternal genes, the zygote's nucleus divides, and the progeny nuclei divide mitotically several times. Initially, the nuclei divide within the cytoplasm of a single cell, and none of the genes in the nuclei are expressed. Instead, the cell relies on maternal mRNAs for enzymes. Regulatory proteins encoded by maternal mRNAs in the embryo soon interact with the promoters of regulatory genes in the nuclei and activate transcription of some of the embryo's regulatory genes. From that point on, the embryo's own genes take over the task of developmental regulation. Several of the embryonic genes have been identified as **segmental genes** because their mutant phenotypes cause faulty formation of body segments in the larva.

Gap Gene Expression The first segmental genes to be expressed are collectively called **gap genes** because recessive mutations in them cause missing segments in the developing larva when the mutant alleles are homozygous. The missing segments are called gaps. Three well-studied gap genes are *hunchback*, *Krüppel*, and *knirps*, named after the phenotypes associated with mutations in the genes. For example, the mutant *knirps* phenotype is a short larva in which seven abdominal segments are fused into a single segment (Figure 23.10).

The gap genes control the initial development of segmentation in the larva. The product of the *bicoid* gene directly regulates some of the gap genes. For example, the bicoid protein suppresses transcription of the *knirps* gene. Thus, the *knirps* gene encodes a protein that is localized posterior to the midline of the embryo because the bicoid protein is absent in that location (Figure

23.11a). Transcription of the *hunchback* gene, on the other hand, is activated by the bicoid protein, so it is an anterior-specific protein (Figure 23.11b), although P maternal cells deposit some of their *hunchback* mRNAs in the extreme posterior portion of the embryo. A third gene, the *Krüppel* gene, is expressed near the midline of the embryo and is also under the influence of the *bicoid* gene product (Figure 23.11c). The gap genes encode transcription factors with zinc finger DNA-binding domains.

The following example shows how the bicoid protein interacts with the *hunchback* gene in *Drosophila*.

Example 23.2   Regulation of the *hunchback* gene by the bicoid protein.

In 1989, Driever et al. (*Nature* 340:363–367) described six sites where the bicoid protein binds to a region upstream from the *hunchback* gene. The sequences and their positions upstream from the startpoint for transcription are as follows:

−67 to −59	TCTAATCCC
−172 to −164	TCTAATCCA
−212 to −204	GATCATCCA
−225 to −217	CGCTAAGCT
−241 to −233	TGCTAAGCT
−282 to −274	CGTAATCCC

The consensus binding sequence for the bicoid protein is TCTAATCCC. The bicoid protein binds strongly to three of the above sites and weakly to the other three. The researchers constructed several fusion genes in which different fragments of regions upstream from the *hunchback* promoter were attached to a *lacZ* gene. The authors introduced the fusion genes into fly embryos using P element–mediated transformation and tested the progeny of these embryos for the presence of β-galactosidase. They found that deletion of any of the six binding sites reduced the expression of the *hunchback* gene.

Problem: **(a)** Given the consensus binding sequence for *bicoid*, which three of the binding sites should bind to the bicoid protein most strongly? **(b)** What

do the results of the fusion gene experiments indicate about the function of the bicoid protein?

Solution: **(a)** The three sequences that most closely resemble the consensus sequence are

−67 to −59	TCTAATCCC
−172 to −164	TCTAATCCA
−282 to −274	CGTAATCCC

These are the three sites that bind strongly to the bicoid protein. **(b)** The fusion gene experiments indicate that each of the upstream binding sites is important for full expression of the hunchback protein and that the bicoid protein is a positive regulator of *hunchback* at all of the binding sites.

Pair-Rule Gene Expression The next set of genes are called the **pair-rule genes** because they control the formation of adjacent paired segments in the developing embryo. Each gene is expressed in alternate segments, so most mutations cause a loss of alternate segments. Each gene is expressed as distinct regions in the embryo, called **stripes**, that correspond to alternate segments in the larva. The stripe locations for one gene may differ from those of another gene, as is apparent in the photograph in Figure 23.12. Several pair-rule genes have been characterized, including *hairy, runt, fushi tarazu, even-skipped, paired,* and *odd-skipped.*

Transcription factors encoded by the gap genes regulate expression of pair-rule genes. A pair-rule gene is expressed in several stripes. Each pair-rule gene has a complex regulatory region, with regions that regulate the expression of the gene in each stripe. For example, mutations within the coding region of the *hairy* gene eliminate its function in all stripes in which it is normally expressed. However, some mutant alleles in the *hairy* gene cause a loss of function in just one stripe, rather than in alternate stripes. These mutant alleles contain specific deletions in the regulatory region upstream from the *hairy* gene promoter, rather than in the coding region.

P element–mediated transformation of fusion genes has provided detailed insight as to how pair-rule genes function. Researchers linked various DNA fragments from the *hairy* gene's regulatory region to the *lacZ* gene in P element vectors and inserted the fusion genes into embryos. Detection of β-galactosidase made it possible to match regulatory regions with the stripe they affected. Among the best examples is the regulatory interaction of gene products that cause the *hairy* gene product to form in stripe 6. Using a transformed fusion gene with the center portion of the *hairy* gene's regulatory region attached to the *lacZ* gene, researchers discovered β-galactosidase formation in stripe 6. When this fusion gene is in embryos that are homozygous for a mutant *knirps* allele (a gap gene allele), no β-galactosidase appears in stripe 6. This

a knirps.

b hunchback.

c Krüppel.

Figure 23.11 Expression of **(a)** the *knirps* gene, **(b)** the *hunchback* gene, and **(c)** the *Krüppel* gene. The dark areas represent the concentration of the gene products. (From Hülskamp, M., and D. Tautz. 1991. Gap genes and gradients: The logic behind the gaps. *BioEssays* 13:261–268. Copyright John Wiley & Sons, Inc. Reprinted by permission.)

result suggests that the *knirps* gene product is a positive regulator of the *hairy* gene. When this fusion gene is present in embryos with a mutant allele at the *Krüppel* locus (another gap gene), expression of β-galactosidase is higher in stripe 6 than in wild-type embryos. Because embryos that are mutant for *Krüppel* lack the *Krüppel* gene product, this result suggests that the *Krüppel* gene product inhibits expression of the *hairy* gene.

When *Krüppel* and *knirps* genes are cloned and expressed in *E. coli*, both proteins bind to the central portion of the *hairy* gene's regulatory region that is specific for stripe 6. Stripe 6 is produced in a region of the embryo where the Krüppel protein is at a low concentration and the knirps protein is at a high concentration. This interaction of genes with regulatory products ensures that

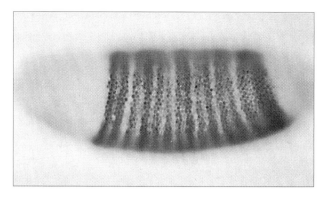

Figure 23.12 Alternate expression of the pair-rule genes *fushi tarazu* (blue) and *even-skipped* (red). (Lawrence, P. A. 1992. *The Making of a Fly: The Genetics of Animal Design.* Boston: Blackwell Scientific Publications.)

a Location of *Krüppel* gene expression in the embryo.

b Location of *knirps* gene expression.

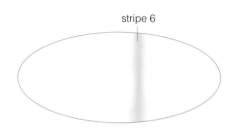

c Location of *hairy* gene expression.

Figure 23.13 Interaction of gene products to determine where the *hairy* gene is expressed.

stripe 6 develops where it should and nowhere else. It also illustrates the hierarchical nature of gene regulation. The bicoid product encoded by maternal mRNAs forms a concentration gradient from the anterior to the posterior end of the embryo. The *knirps* and *Krüppel* genes are both sensitive to bicoid product concentration in different ways. Thus, the bicoid product concentration gradient regulates where the *knirps* and *Krüppel* genes will and will not be expressed. At the location of stripe 6, the *knirps* gene is expressed at a high level, but the *Krüppel* gene is not, so the *hairy* gene is activated to produce its product in stripe 6 (Figure 23.13).

Segment Polarity Gene Expression   A third class of segmental genes are the **segment polarity genes**. Each segment consists of two parts, an anterior (A) compartment and a posterior (P) compartment (see Figure 23.8). Mutations in segment polarity genes cause a loss of one compartment (usually the P compartment), which is replaced by a mirror image of the remaining compartment (usually A), as illustrated in the gooseberry phenotype shown in Figure 23.14. Three segment polarity genes have been characterized in detail: *wingless, engrailed,* and *gooseberry*.

~

There are several types of segmental genes. These genes establish the number, position, and polarity of the segments in the larva and eventually the adult.

Homeotic Gene Expression

The maternal and segmental genes establish the number and position of segments, whereas **homeotic genes** are responsible for segmental identity. The term *homeotic* comes from *homeosis*, which William Bateson coined to describe unusual phenotypes in which body parts appear in the wrong places. Mutations in homeotic genes cause these phenotypes, in which one segment contains the structures and appearance of another body part, such as the antennapedia and bithorax phenotypes shown in Figure 23.6. The homeotic genes are arranged in two large complexes.

They are called the **antennapedia complex (ANT-C)** and the **bithorax complex (BX-C)**, named after the phenotypes they confer, and are located at separate sites on chromosome 3. Genes in these two complexes regulate the identity of each segment. Those in ANT-C regulate the development of the head, the first thoracic segment (T1), and the anterior compartment of the second thoracic segment (T2A). Genes in BX-C regulate development of the remaining segments, beginning with the posterior compartment of the second thoracic segment (T2P) and continuing to the end of the abdomen (Figure 23.15).

The homeotic gene complexes are large units that consist of protein-encoding regions and numerous regulatory regions. The BX-C region is about 300,000 nucleotide pairs long and contains four transcribed genes, three of which encode proteins. The fourth is transcribed but not translated and apparently serves no function. The complex can be divided into two domains, the ***Ultrabithorax* domain** and the ***Infraabdominal* domain**. One of the regions contains the *Ubx* (*Ultrabithorax*) gene and is in the *Ultrabithorax* domain. The gene that does not encode a protein is also within the *Ultrabithorax* domain. The two other protein-encoding genes are in the *Infraabdominal* domain and are called *abd-A* and *Abd-B*. All three proteins are DNA-binding proteins.

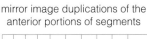

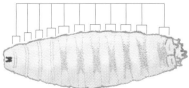

Figure 23.14 The mutant gooseberry phenotype in a larva. The posterior half of each segment is deleted, and the anterior half is duplicated in the opposite orientation.

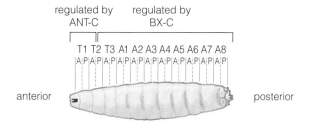

Figure 23.15 Targets of the antennapedia (ANT-C) and bithorax (BX-C) complexes in *Drosophila melanogaster*. ANT-C regulates development of the head and segments T1A–T2A. BX-C regulates development from segment T2P to the posterior end of the abdomen.

As mentioned earlier, the BX-C region regulates development of all segments from the posterior compartment of the second thoracic segment to the posterior end of the fly. Researchers have mapped the regions that regulate development of each segment by observing which segments are affected by mutations. The regions that regulate development of each segment are aligned in the BX-C region in a linear array that corresponds to the segmental order in the fly, as diagrammed in Figure 23.16.

Many of the mutations in BX-C that affect particular segments are in regulatory regions, rather than transcribed regions. Some of the regulatory regions are within introns in the transcribed regions, and many of the regulatory elements are enhancers. Others direct alternative splicing of the pre-mRNAs. The *Ubx* gene, for example, is transcribed into a single pre-mRNA, which can then undergo one of several splicing alternatives to yield different, but related, proteins. An amino acid sequence encoded by the first exon is present in all the *Ubx* proteins. An antibody that is specific for this sequence reacts with proteins in segments T2–A8. The reaction is strongest in T2P, T3A, T3P, and A1A, then grows progressively weaker throughout the remaining abdominal segments. This observation indicates that the effect of the Ubx proteins is greatest in the posterior portion of the thorax and the anterior portion of the abdomen. Indeed, mutations in the *Ultrabithorax* domain affect those parts of the fly. The proteins are typically found in nuclei and probably are transcription factors that interact with specific regulatory DNA sequences.

The *Infraabdominal* domain regulates development of the abdominal segments A3–A8. The two proteins encoded by the *abd-A* and *Abd-B* genes act together in each of the segments and are regulated in a sequential fashion by the regulatory regions diagrammed in Figure 23.16. Mutations in these regions (as well as in regulatory regions of the *Ultrabithorax* domain) cause the affected segment to take on an identity of a more anterior segment. There is an incremental type of regulation in which the development of one segment affects the development of segments posterior to it. This pattern is most evident when the entire BX-C region is deleted, which causes all

segments affected by BX-C to take on the form of T2, the most anterior of the segments.

The organization of genes in ANT-C is diagrammed in Figure 23.17. Contained within the complex are five homeotic genes (*Antp, pb, Scr, lab,* and *Dfd*) and three genes that are not homeotic but participate in development (*zen, bcd,* and *ftz*). As we have seen already, *bicoid* (*bcd*) is a gene with maternal influence that sets up the initial anterior-posterior gradient in the embryo, *fushi-tarazu* (*ftz*) is a pair-rule gene, and *zen* is a gene that affects dorsal-ventral differentiation. These three genes govern early development of segment location and number. The five homeotic genes then govern segment identity.

For the most part, mutations in four of the five homeotic genes of ANT-C (*pb, Scr, lab,* and *Dfd*) cause anterior structures to resemble structures posterior to them, the opposite effect of BX-C mutations. Deletion of the entire ANT-C region causes all the segments affected by ANT-C to be thoracic. In fact, deletion of both ANT-C and BX-C causes all segments in the entire embryo to develop as thoracic segments. The exception to this general rule is the *Antp* locus, in which loss-of-function mutations cause a more anterior type of development. Mutations that cause a loss of the *Antp* gene product cause development of head structures in thoracic segments. The *Antp* gene product is produced primarily in the anterior thoracic segments and not in the head. Thus, it represses head development and promotes thoracic development. Mutations in regulatory regions that cause the *Antp* gene product to be expressed in the head cause thoracic structures to form in the head, such as the antennapedia phenotype, in which legs form on the head in place of antennae (see Figure 23.6a).

Again, P element–mediated transformation has provided insights into the function of these genes. Certain genes in *Drosophila* (as well as most animals and plants) are called **heat-shock genes** because their protein products are produced at elevated temperatures. The promoters for these genes are temperature sensitive and activate transcription when temperatures are raised. Researchers fused a heat-shock promoter to the *Deformed* (*Dfr*) gene from ANT-C. When embryos that carry the

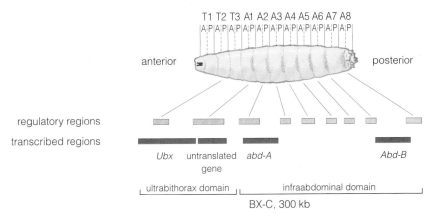

Figure 23.16 The molecular organization of BX-C and the larval segments affected by the regulatory regions of BX-C.

Figure 23.17 The locations of genes in ANT-C.

fusion gene are allowed to develop at lower temperatures, their development proceeds normally, but when the temperature is raised, head parts appear in thoracic segments. These observations indicate that the *Dfr* gene product is a positive regulator of head structure development.

The protein product of the *Dfr* gene contains a homeodomain that includes a helix-turn-helix DNA-binding site. When fusion genes in which the *Dfr* homeodomain is replaced by the *Ubx* homeodomain are present in developing larvae, the fusion protein causes the head region to develop into a thoracic segment. This observation indicates that the homeodomain is an important portion of the protein, not only for DNA binding but also for its effect on development.

In *Drosophila*, homeodomains are a common feature among regulatory proteins. They are found in maternal, segmentation, and homeotic genes and include helix-turn-helix DNA-binding sites. The homeodomains are encoded by a DNA sequence called a **homeobox**, which consists of about 180 nucleotide pairs typically located near the 3′ end of the region that encodes the pre-mRNA. Although the amino acid sequences in the protein homeodomains are fairly similar, the nucleotide sequences of *Drosophila* homeoboxes have diverged, mostly at the third nucleotide of several codons, so nucleotide sequence divergence leaves the amino acid sequence unaffected. The differences in nucleotide sequence allow researchers to compare homeoboxes and group them on the basis of their nucleotide sequence similarity. Homeoboxes of

genes in ANT-C are more similar to each other than to homeoboxes elsewhere, so an ANT-C homeobox can be defined based on its nucleotide sequence. The same is true for homeoboxes in the BX-C region.

~

Homeotic genes establish the identity of segments. These genes are arranged in two large complexes, called the antennapedia complex (ANT-C) and the bithorax complex (BX-C).

23.4 COMPARISON OF MAMMALIAN AND *Drosophila* DEVELOPMENTAL GENETICS

Mammals are substantially different from insects. We might expect that some of the basic developmental principles learned with model organisms, such as *C. elegans* and *D. melanogaster*, might apply to mammals as well, but few geneticists expected the genes themselves to be similar. Thus, geneticists were dismayed when probes that contain the homeobox sequences from *Drosophila* hybridized with the mammalian genes that govern development. Not only were the mammalian genes similar in DNA sequence to *Drosophila* genes (both within and, in some cases, outside of the homeoboxes), but they were also similar in their linear organization on the chromosome.

Human and mouse homeotic genes are localized in four similar complexes called **Hox clusters**, each located on a different chromosome. The Hox clusters are

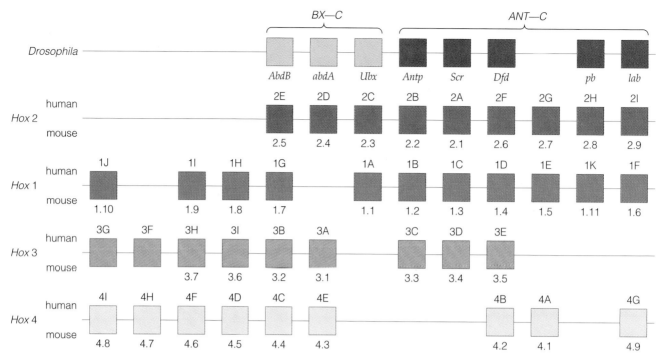

Figure 23.18 Illustration of the paralogous organization of the ANT-C and BX-C complexes of *Drosophila* with the Hox clusters of human and mouse. There is one set of BX-C and ANT-C genes in *Drosophila* that corresponds to four sets of Hox cluster genes in human and mouse. Genes that are aligned vertically are similar in DNA sequence. (Adapted from Krumlauf, R. 1992. Evolution of the vertebrate *Hox* homeobox genes. *BioEssays* 14:245–252.)

paralogous—that is, they are similar in both the nucleotide sequence of the genes and the linear organization of the genes in each cluster. The paralogous structure of Hox clusters in mouse extends to insects, in which the Hox clusters are similar in DNA sequence and gene organization to the ANT-C and BX-C gene clusters of *Drosophila*. Figure 23.18 illustrates the comparison. The similarity of homeoboxes within the ANT-C region extends to the Hox clusters as well. Five of the homeoboxes in the mouse Hox 1 and Hox 2 clusters correspond to the homeoboxes of the five ANT-C homeotic genes with about 70% homology. It is also possible to identify individual genes in the Hox clusters that correspond to specific genes in the ANT-C and BX-C clusters. For example, the homeodomains of the Hox 2.2 protein in mice and the Antp protein in flies differ in only 5 out of 60 amino acids. The homeodomains of the human Hox 4.2 protein and the Dfr protein in flies differ in 6 of 60 amino acids and retain 40% homology throughout the entire proteins.

In *Drosophila*, the ANT-C and BX-C clusters, although located on the same chromosome, are far from each other, whereas the Hox cluster genes are all in single clusters. The spatial separation of clusters in *Drosophila* is not conserved among all insects. When these same genes were mapped in flour beetles (*Tribolium castaneum*), researchers found that the ANT-C and BX-C clusters are in a single cluster with the same type of organization found in the mouse Hox clusters. Apparently, an alteration in

chromosome structure in the evolutionary history of *Drosophila* separated the ANT-C and BX-C clusters.

Not only is the spatial order of gene organization similar in insects and mice, but the order of gene expression is also similar. The eight homeotic genes in ANT-C and BX-C in *Drosophila* follow an anterior-to-posterior pattern of expression that corresponds to their linear array, beginning with *lab* and proceeding to *Antp* in the ANT-C cluster, then moving to *Ubx* and proceeding to *Abd-A* and *Abd-B* in the BX-C cluster. The same order of expression holds true in mouse. The farther genes are toward the right in Figure 23.18, the more anterior their expression in both *Drosophila* and mouse.

As in flies, transformation experiments in mouse have added to our understanding of how Hox cluster genes function. For example, a mutant *Hox 3.1* gene inserted into mice causes rudimentary ribs to form on the L1 vertebra, the first vertebra that does not have ribs. This tendency for a vertebra to take on a more anterior structure due to a mutant homeotic gene is similar to the tendency of *Drosophila* abdominal segments to take on more anterior structures due to mutant homeotic genes. Several mouse Hox cluster genes that were expressed in *Drosophila* via P element–mediated transformation, caused the same phenotypes as their corresponding genes in *Drosophila*. This observation further reinforced the conclusion that developmental genes in insects and mammals are similar in both structure and function.

The similarities in DNA sequence homology, organizational structure, and patterns of expression in homeotic genes extend beyond flies and mice to include much of the animal kingdom. This observation has led many geneticists to conclude that developmental genes in a broad spectrum of animals have a common evolutionary origin. Just as similarities of cell structure and function, inheritance patterns, and many other features have been preserved in diverse groups of animals, genetic organization and regulation of development are similar in spite of outward differences in developmental phenotypes.

The organizational similarities of mammalian and insect genes suggest that there is an evolutionary advantage to that organization. However, as the following example illustrates, the arrangement is not always essential for gene expression.

Example 23.3   Spatial placement of *pb* in ANT-C.

In 1991, Randazzo et al. (*Development* 113:257–271) reported studies on the homeotic gene *pb* (*proboscipedia*), which is located in ANT-C on chromosome 3 of *Drosophila melanogaster*. As illustrated in Figure 23.18, the *pb* gene is paralogous with the 2H and 1K Hox cluster genes in mouse and human. Flies that are homozygous for a mutant *pb* allele have leg structures in place of the proboscis on the head. When a genetically engineered fusion gene of all the coding regions of the *pb* gene plus some of the DNA upstream from the 5′ end of the gene was transformed into flies that were homozygous for a *pb* mutant allele using P element–mediated transformation, the wild-type phenotype was restored even when the introduced gene integrated into sites on the X chromosome and chromosome 2.

Problem: What do these results suggest about the spatial placement of *pb*?

Solution: The similarity of spatial placement of genes in the mammalian Hox clusters with ANT-C and BX-C in *Drosophila* suggests that there is some evolutionary advantage to maintaining the spatial organization of the genes in these clusters. If the genes within the clusters were somehow dependent on each other or on common regulatory elements within the clusters, the spatial conservation could be easily explained. However, the results of this study demonstrate that *pb* does not need to be located within or near ANT-C in order to carry out its function for normal development.

The organization and expression of many developmental genes appears to be conserved in much of the animal kingdom. Thus, studies on developmental genetics in flies can often be applied to other species, including mammals.

23.5   PLANT DEVELOPMENTAL GENETICS

Genetic control over development in plants has not been studied as extensively as in the animal species discussed in this chapter. Fortunately, the pace of research in plant developmental genetics has accelerated in recent years, resulting in a now large body of information. Much of this progress is tied to the genome sequencing project in *Arabidopsis thaliana*, a model organism for plant genetics. *Arabidopsis* is advantageous for studies of plant genetics (plant developmental genetics in particular) because of its short life cycle, rapid development, ease of cultivation, and relatively small genome (see Figure 1.6 for a photograph of *Arabidopsis*).

In this section, we will focus on the genetics of floral differentiation, one of the most intensively studied aspects of plant development. Then in the next section, we will compare animal and plant developmental genetics.

In plants, cell differentiation begins a few cell generations after the zygote divides, forming a tiny embryo with recognizable plant structures within the seed. However, in the embryo of a germinating seed, and later in the growing plant, small masses of undifferentiated cells, called **meristems**, divide rapidly at the growing point, as illustrated in Figure 23.19. Meristems are present at the tips of both roots and shoots. As the time for flowering approaches, cells within shoot meristems differentiate into inflorescence meristems. Within each inflorescence meristem, some of the cells differentiate as the germ line, which eventually forms the gametes in flowers. Due to this pattern of development, germ lines may develop several times in a single plant, and development of the germ line is much later than in animal development.

At some point during the plant's development, an external stimulus initiates a transition from vegetative to inflorescence meristems. The stimulus varies among plant species. In some species, changes in day length may be

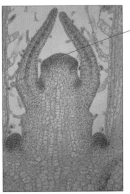

apical meristem

Figure 23.19 A cross section of a vegetative meristem in a plant of the genus *Coleus*. The dome-shaped meristem is embedded between leaf primordia that will develop into leaves. (Photo courtesy of H. B. Furniss.)

Figure 23.20 A wild-type *Arabidopsis thaliana* flower. (From Meyerowitz, E. M., J. L. Bowman, L. L. Brockman, G. N. Drews, T. Jack, L. E., Sieburth, and D. Weigel. 1991. A genetic and molecular model for flower development in *Arabidopsis thaliana*. *Development*, Supplement 1:157–167. (Copyright 1991 by the Company of Biologists Ltd. Used by permission.)

the stimulus, whereas in other species, a rise in temperature may stimulate floral development. After floral development is initiated, floral meristems form in the inflorescence meristem in place of leaf primordia found in vegetative meristems.

In *Arabidopsis*, there are two basic steps for floral development. The first step is **floral meristem identity**, the development of floral meristems from inflorescence meristems. Floral meristems develop into individual flowers. Several genes govern this first step, including *AP1, CAL, LFY, TFL, CLV1,* and *CLV2.*

The second step is **floral organ identity**, the development of the different organs of the flower. After the number of cells in the floral meristems has been established, genes that determine organ identity cause cells to differentiate into the different floral organs. There are four types of floral organs: the sepals, petals, stamens (male reproductive organs), and carpels (female reproductive organs). The wild-type *Arabidopsis* flower has four sepals, four petals, six stamens, and two fused carpels, as illustrated in Figure 23.20. These four organs develop from four concentric whorls in the floral meristems. The first whorl develops into sepals, the second into petals, the third into stamens, and the fourth into carpels.

The proteins of several genes that govern floral organ identity share a highly conserved region of 56 amino acids known as a **MADS box**. MADS stands for <u>M</u>CM1-<u>AG</u>-<u>DEF</u>-<u>SRF</u>, the symbols for the first four genes discovered that encode MADS boxes. *MCM1* is a yeast gene, *AG* and *DEF* are *Arabidopsis* genes, and *SRF* is a mammalian gene. The proteins encoded by MADS box genes in *Arabidopsis*

also share a second conserved region of 67 amino acids called the **K domain**. Over 20 MADS box genes are thought to exist in *Arabidopsis*, most of which govern floral development. The different MADS box genes probably arose from a single ancient gene that has duplicated several times. The duplicated genes then diverged to carry out distinct, but similar, developmental roles.

The MADS box is a DNA-binding site, and the K domain forms a coiled coil. All MADS box proteins probably act as transcription factors, although more research is needed to fully determine their molecular functions. Most MADS box genes in plants are homeotic in that mutations often cause a change in organ identity. Thus, MADS box genes perform a role that is analogous to that of the homeobox genes of animals, although there is no nucleotide sequence similarity between MADS box genes and homeobox genes. Also, unlike homeobox genes, which are clustered in animal genomes, MADS box genes are scattered throughout the *Arabidopsis* genome.

An example of a MADS box gene that governs floral organ identity is the *AGAMOUS* (*AG*) gene, which is expressed in the third and fourth whorls. The product of this gene limits cell division in the center of the flower and regulates development of the stamens and carpels. In the absence of the *AG* gene product (due to homozygosity for loss-of-function mutant alleles in the *AG* gene), the third whorl develops as petals instead of stamens, and the fourth whorl develops as another mutant flower. The resulting flower lacks stamens and carpels. In their place, petals and sepals proliferate in the flower (Figure 23.21).

After the organ identity of floral whorls has been determined, the *PERIANTHA* (*PAN*) gene influences the number and positioning of the organs. Mutations in the *PAN* gene produce alterations of sepal, petal, and stamen number. For example, flowers on a plant that is homozygous for the *pan-1* loss-of-function mutant allele have five sepals (instead of four), five petals (instead of four), and five stamens (instead of six). The number of carpels (two) is the same as in the wild-type flower (Figure 23.22).

MADS box genes have been studied most extensively for their roles in floral organ identity. However, as highlighted in the following example, some MADS box genes govern development in parts of the plant other than the flower.

Example 23.4   MADS box genes in *Arabidopsis thaliana*.

Genes that govern floral development in *Arabidopsis* were initially detected by the aberrant floral phenotypes associated with mutations in these genes. After a mutant phenotype was identified, the gene with the mutation was isolated, cloned, and sequenced. Comparison of DNA sequences revealed that most floral organ identity genes are MADS box genes. In 1995, Rounsley et al. (*The Plant*

Figure 23.21 Mutant phenotype of flowers on an *Arabidopsis* plant that is homozygous for a loss-of-function mutation at the *AG* locus. (Photo courtesy of J. L. Bowman and E. M. Meyerowitz.)

Figure 23.22 Mutant phenotype of flowers on an *Arabidopsis* plant that is homozygous for the *pan-1* loss-of-function mutation. (From Running, M. P. and E. M. Meyerowitz. 1996. Mutations in the *PERIANTHA* gene of *Arabidopsis* specifically alter floral organ numbers and initiation pattern. *Development* 122:1261–1269. Copyright 1996 by the Company of Biologists Ltd. Used by permission.)

Cell 7:1259–1269) reported the identification of six previously unknown MADS box genes in *Arabidopsis*. Before their study, researchers had identified 11 MADS box genes in *Arabidopsis* by searching for mutations that affected floral development. All 11 genes were expressed exclusively in the developing floral structures. Rounsley et al. took the reverse approach. They first identified genes with a MADS box, then searched for the plant structures in which these genes were expressed. Of the six genes they discovered, two were expressed in the ovules (part of the carpel in the flowers), one in the embryo, and three in the roots.

Problem: What do these results indicate about the function of MADS box genes in plant development?

Solution: Of the six genes they identified, four were expressed outside of the floral tissues. These results indicate that MADS box genes are not exclusively devoted to floral development but probably have developmental roles in other parts of the plant.

Plant developmental genetics is an emerging field that has produced much information in recent years. Among the most extensively studied genes are those that regulate floral development.

23.6 COMPARISON OF PLANT AND ANIMAL DEVELOPMENTAL GENETICS

The similarity of developmental genes in insects and mammals suggests that the patterns for animal development were determined early in evolutionary history. Might these similarities extend to both plants and animals? The available evidence suggests that plants and animals share the same types of genes that each group utilizes in development, but that the functions of these genes are very different in the two kingdoms.

An excellent example of this is comparison of the roles for homeobox genes and MADS box genes in animals and plants. Both plants and animals have homeobox genes. In animals, there are several homeobox genes, and they determine organ identity. In plants, there are only a few homeobox genes, and they encode proteins that serve as generalized activators of cell division. In plants, there are numerous MADS box genes, and they determine organ identity. MADS box genes are also found in animals, but the few MADS box genes in animals are transcription factors that govern generalized cell division, similar to the role played by homeobox proteins in plants.

In early evolutionary history, the ancestors of animals and plants probably diverged from one another while they were still single-celled eukaryotes. The arisal of multicelled plants and animals required genes that could regulate differentiation and development. It appears that different genes duplicated and diverged to fulfill developmental roles in plants and animals. Evidence that such genes were present in the unicellular ancestors comes from the observation that *Saccharomyces cerevisiae* (a unicellular eukaryote) has both homeobox and MADS box genes. As these genes duplicated and diverged from one another, groups of similar genes took on regulation of specific aspects of development.

~

The genes that govern animal development are very different from those that govern plant development. Plants and animals probably diverged from one another before the transition from single-celled to multicelled eukaryotes in evolutionary history.

SUMMARY

1. A cell that can divide and give rise to an entire multi-celled organism is said to be totipotent.

2. During somatic cell differentiation, totipotency is often lost as cells become specialized to carry out the function of the tissue in which they reside.

3. Developmental genetics is the study of genes that govern cell differentiation and organismal development.

4. The nematode worm *Caenorhabditis elegans* and the insect *Drosophila melanogaster* are the two best-studied species in developmental genetics.

5. In *C. elegans*, the developmental fate of every cell is known, and the effect of many genes that govern cell development has been determined through analysis of mutations in the genes.

6. Development in *Drosophila melanogaster* follows a hierarchical pattern, beginning with a gradient of maternal gene products that influences embryonic genes. The embryonic genes that were activated by maternal proteins activate another set of genes that then determine the pattern for development.

7. Mutations in homeotic genes may cause replacement of one structure with another body part. For example, certain homeotic mutations may cause legs to form in place of antennae.

8. The homeotic genes of *Drosophila* are located in two clusters on chromosome 3: the antennapedia complex (ANT-C), which determines structural development in the head and anterior portion of the thorax, and the bithorax complex (BX-C), which regulates structural development in the posterior portion of the thorax and the abdomen.

9. The DNA sequences and spatial arrangement of genes in ANT-C and BX-C in insects are similar to paralogous sequences and arrangements of homeotic genes in vertebrates. The similarity suggests a common evolutionary origin and arrangement of these genes.

10. The patterns of development and the genes that govern development in plants differ from those of animals. Among the best-studied plant development genes are the MADS box genes, which perform functions that are analogous to those of the homeobox genes in animals.

QUESTIONS AND PROBLEMS

1. In *Drosophila*, mutant alleles at the *bicoid* locus that are homozygous in a female fly are expressed in all the progeny of that fly. This is an example of maternal effect, described in section 18.7. The *bicoid* locus is on an autosome. How could the inheritance of mutant *bicoid* alleles be distinguished from uniparental-maternal inheritance of a mutant mitochondrial allele that affects development?

2. It is not uncommon to hear that each somatic cell of a plant or animal has the full genetic complement of an individual and that it may someday be possible to clone an individual from any somatic cell. Define the term *totipotency*, and describe why most somatic cells in animals and some plants are not suitable for cloning.

3. What characteristics of *Caenorhabditis elegans* have made it a model organism for studying the genetics of development?

4. What characteristics of *Drosophila melanogaster* have made it a model organism for studying the genetics of development, and how do some of these characteristics differ from those of *C. elegans*?

5. Define *positional* and *autonomous regulation* of genes that govern development. How might it be possible to distinguish positional from autonomous regulation experimentally?

6. Using the imaginal disc implantation experiments of Beadle and Ephrussi described in section 6.3, determine whether the genes they studied are regulated in an autonomous or positional manner. Briefly explain how you arrived at your conclusion.

7. Make a chart that shows the hierarchical nature of the major classes of genes that govern *Drosophila* embryo development.

8. Using a mitotic nuclear lineage diagram like the one shown in Figure 23.2, draw the nuclear lineage of a cell that is W.aappap.

9. The paralogous linear arrangement of genes in the *Drosophila* ANT-C and BX-C complexes is retained in the mouse and human Hox cluster gene complexes, as illustrated in Figure 23.18. Define what is meant by the term *paralogous*, and describe the evolutionary implications of this observation.

10. In relation to your answer to the previous question, what possible explanations could account for the results of the experiments highlighted in Example 23.3?

11. In 1994, Keith et al. (*The Plant Cell* 6:589–600) described a mutation called *fusca3* in *Arabidopsis thaliana* that altered the timing of late embryogenesis. Functions associated with late embryogenesis, such as deposition of seed storage proteins and establishment of dormancy, were reduced. The embryos appeared more like germinating embryos than the dormant embryos of a seed. Would this mutation be best classified as homeotic or heterochronic? Defend your answer on the basis of the information provided.

12. Most cell lineages are totipotent during the early stages of embryogenesis, then lose totipotency at some stage during cell differentiation. In plants and some animals, the retention of totipotency in some cells is evident when cells restore tissues or structures following an injury. In mammals, however, it appears that only one cell lineage remains totipotent in adults. Which cell lineage is this, and why is it essential that this lineage remain totipotent?

13. In 1995, Rosenberger (*BioEssays* 17:257–260) summarized a common phenomenon observed in cultured mammalian cells. When cultured mouse embryonic stem cells are prevented from differentiating, they may grow and divide through an unlimited number of cell generations. However, if the cells are allowed to differentiate in culture, they cease dividing after a predictable number of generations and eventually die. What does this suggest about the developmental genetic mechanisms involved in aging and death in mammals?

14. Rex and Scotting (1994. *Gene* 149:381–382) found through comparative DNA sequencing that the amino acid sequence of the *Hoxb-3* gene in chicken has 70% homology to the mouse *Hoxb-3* gene. They further found that the homeodomain regions show 100% identity between mice and chickens. Why is the similarity of these genes and the complete invariance of the homeodomains an expected result?

15. In 1990, Driever at al. (*Development* 109:811–820) published studies on the effect of altered positions of *bicoid* mRNAs. They injected mRNAs that encode a functional bicoid product into different locations of embryos with mutant *bicoid* alleles. Anterior structures developed at each site along the anterior-posterior axis where the bicoid mRNA was injected. What do these results suggest about the role of the bicoid product relative to other proteins that regulate development of anterior-posterior structures?

16. MyoD is a family of muscle-specific, helix-loop-helix transcription factors that regulate vertebrate muscle development. Similar DNA sequences appear in muscle development genes of many invertebrates, including *Drosophila melanogaster, Caenorhabditis elegans*, and sea urchins. The mammalian genes are not found in the Hox clusters, nor is the *Drosophila* gene found in the ANT-C or BX-C complexes. What do these observations suggest about the evolutionary relationships of the genes that regulate muscle development?

17. In 1995, Burchard et al. (*Journal of Cell Science* 108:1443–1454) found a P element–induced mutant gene they called *not enough muscles* (*nem*) because muscle development was substantially reduced in mutant embryos. They also found that development of the peripheral nervous system was altered in mutant embryos. Do these observations indicate that the *nem* gene product has two developmental roles, one in muscle development and the other in nerve development? Briefly explain your answer.

18. The *AG* gene of *Arabidopsis thaliana* is a MADS box organ identity gene that is expressed in the third and fourth whorls of floral meristems. Loss-of-function mutations in this gene typically cause the third whorl to develop into sepals and the fourth whorl to develop as a new mutant flower. In 1995, Sieburth et al. (*The Plant Cell* 7:1249–1258) described the phenotypic effects of two partial loss-of-function mutations in this gene. The purpose of their study was to determine whether the multiple phenotypic effects of *AG* could be separated. They examined two mutations, *ag-4* and *AG-Met205*. The *ag-4* mutation causes a partial deletion of the K domain at the C-terminal end of the protein. The phenotype in plants that are homozygous for *ag-4* is development of stamens from the third whorl but development of a new floral meristem from the fourth whorl. The *AG-Met205* mutation was induced artificially in isolated DNA and consisted of a single amino acid substitution of Arg → Met in the K domain. When the *AG-Met205* DNA was reintroduced, using recombinant DNA methods, into a plant that was homozygous for *ag-3* (a typical full loss-of-function mutation), all four floral organs were produced, but the fourth whorl developed into four carpels rather than two. What do these results suggest about the *AG* protein and the function of the K domain?

19. In 1997, Meyerowitz (*Genetics* 145:5–9) noted that both plants and animals have genes with MADS boxes, but that the number of MADS box genes discovered in plants is much greater than the number discovered in animals. The converse is true for homeobox genes; more homeobox genes have been discovered in animals than in plants. What might account for these discrepancies?

FOR FURTHER READING

The developmental genetics of *C. elegans* is reviewed by **Sternberg, P. W. 1990. Genetic control of cell type and pattern formation in *Caenorhabditis elegans*. *Advances in Genetics* 27:63–116**. Several reviews discuss various aspects of *Drosophila* development. A concise and well-written book is **Lawrence, P. A. 1992. *The Making of a Fly: The Genetics of Animal Design*. Boston: Blackwell Scientific Publications**. Other reviews include **Kaufman, T. C., M. A. Seeger, and G. Olsen. 1990. Molecular and genetic organization of the antennapedia gene complex of *Drosophila melanogaster*. *Advances in Genetics* 27:309–362**; and **Hülskamp, M., and D. Tautz. 1991. Gap genes and gradients: The logic behind the gaps. *BioEssays* 13:261–268**. The homeobox genes in vertebrates are reviewed by **Krumlauf, R. 1992. Evolution of the vertebrate *Hox* homeobox genes. *BioEssays* 14:245–252**. Plant developmental genetics is reviewed by **Meyerowitz, E. M. 1997. Genetic control of cell division patterns in developing plants. *Cell* 88:299–308**; and by **Coen, E. S., and E. M. Meyerowitz. 1991. The war of the whorls: Genetic interactions controlling flower development. *Nature* 353:31–37**. A review comparing developmental genetics in plants and animals is **Meyerowitz, E. M. 1997. Plants and the logic of development. *Genetics* 145:5–9**.

For additional reading, go to InfoTrac College Edition, your online research library at: http://www.infotrac-college.com/brookscole

CHAPTER 24

GENES AND CANCER

During the twentieth century, deaths from infectious disease have declined substantially in developed countries because of modern medicine. With the incidence of infectious disease reduced, cancer has become one of the leading causes of suffering and death. About one-third of all people in the United States suffer from some form of cancer during their lives. Research aimed at the prevention, early diagnosis, and treatment of cancer has been under way for decades. In recent years, great strides have been made in understanding the molecular basis of cancer. This information is currently being used to develop more effective methods for the prevention, detection, and treatment of cancer.

Cancer is a disease in which tumors arise, invade tissues, and spread throughout the body. A **tumor** is a mass of cells that have lost the normal constraints over growth and division. The changes in cells that cause tumors to form are called **oncogenesis**. Most tumors are **benign**, meaning that they remain localized to the tissue where they originated and do not invade other tissues. Moles on the skin are examples of benign tumors. In most cases, a benign tumor is harmless, unless it crowds out other organs or causes pain because of its size or location. A few benign tumors may grow as large as 20 cm (8 inches) in diameter, and people may live to old age with even a large benign tumor. Most benign tumors can be surgically removed if needed.

Sometimes a benign tumor becomes **malignant**, meaning that its cells penetrate tissues adjacent to those where the tumor originated. Malignant tumors are often dangerous and life threatening. Cells in a malignant tumor often lose their cell-to-cell adhesion and break free from the tumor. If these cells enter the circulatory or lymphatic system, they can migrate to another site in the body and establish a new malignant tumor, a process called **metastasis**. Tumors may release chemical signals that stimulate the growth of blood vessels around the tumor to supply it with nourishment, a process called **angiogenesis**. If left unchecked, malignant tumors can eventually destroy the organ in which they reside, causing severe illness and death. The progression from a benign to a malignant tumor in epithelial tissue is illustrated in Figure 24.1.

Cancer can be considered a genetic disease. Although cancer itself is usually not inherited, susceptibility to certain types of cancer often is. Also, cancer is a genetic disease in the sense that it is always caused by mutations in the cellular genome. Although there are many types of cancer, they all have one aspect in common: mutations cause cancerous cells to lose the normal controls over

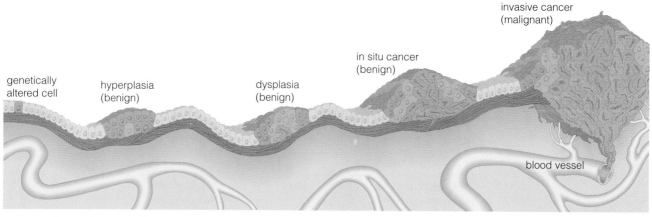

genetically altered cell

hyperplasia (benign)

dysplasia (benign)

in situ cancer (benign)

invasive cancer (malignant)

blood vessel

a A gene that regulates cell division mutates in one cell (orange).

b Because of the mutation, the cell and its progeny cells grow and divide more than they should. Another gene that regulates cell division mutates in one of the cells (pink).

c The cells with the second mutation (pink) continue to grow and divide. Their shapes and orientations are abnormal. Another gene mutates in one of the cells (purple).

d The cells with mutations in three genes grow and divide in an uncontrolled fashion and are abnormal in size and shape. The tumor increases in size but remains within the tissue of origin. A fourth gene mutates in one of the tumor cells (blue).

e Tumor cells invade surrounding tissues. The tumor is now malignant. Blood vessels form to supply the tumor with nutrients (angiogenesis). As the tumor cells acquire more mutations, cells may break away from the tumor and enter the bloodstream to spread to other parts of the body (metastasis).

Figure 24.1 Progression from a benign to a malignant tumor in epithelial tissue. (Adapted from an original drawing by Dana Burns-Pizer in Weinberg, R. A. 1996. How cancer arises. *Scientific American* 275 (Sep):62–70. Used by permission of the author.)

division and growth, so the cells grow and divide in an abnormal fashion. Mutations are inherited from one somatic cell generation to the next, so each cell that divides within a tumor passes the mutations that cause cell proliferation onto progeny cells.

Control of cell division is determined by genes that regulate the cell cycle. In the human body, most cells have become specialized to perform a particular function and will never divide again. Cancer results when mutations in genes that normally regulate cell division cause cells to divide in an uncontrolled fashion. There is no single gene or mutation that is responsible for cancer. Instead, cancer usually requires mutations in several genes.

We begin this chapter with a discussion of how genes regulate the cell cycle. We will then see how mutations in genes that regulate the cell cycle can cause cancer. We will examine some of those genes and their cancer-causing mutations in detail. Next, we will discuss the development of cancer and look at the role that heredity plays in conferring susceptibility to cancer. We conclude with a discussion of cancer prevention, detection, and treatment.

24.1 GENETIC CONTROL OF THE CELL CYCLE

Research on the causes of cancer has revealed much about genetic control of the cell cycle because the genes that regulate the cell cycle are often the same genes that are mu-

tated in tumor cells. Before reading about genetic control of the cell cycle, you may wish to review the basic steps of the cell cycle, which are discussed in section 11.1.

The cell cycle is initially active in all cells of a developing embryo. However, as cells differentiate into various tissues and organs, cell division must be controlled, and eventually arrested in most cells. Intracellular and intercellular signals are required to control cell division. Cells that are in contact with one another secrete signals that bind to receptors on neighboring cells. The receptors then activate a chain of events within the cell that determines whether or not the cell proceeds through the cell cycle. Thus, cells within a tissue or organ act together to ensure that the tissue or organ develops and functions properly. The process that stimulates a cell to undergo division is illustrated in Figure 24.2.

Numerous genes in the cell nucleus regulate the cell cycle. The products they encode govern a complex series of regulatory steps, some that promote cell division, others that inhibit it. These genes act at the various stages of the cell cycle, starting with G_1 (first gap) and eventually ending with M (mitosis). Figure 24.3 shows some of the major proteins and steps of cell-cycle regulation in the nucleus. The proteins illustrated in green promote advancement through the cell cycle; those in red inhibit advancement through the cell cycle.

The concentrations of a group of proteins called **cyclins** rise and fall throughout the stages of the cell cycle and are among the most important proteins that regulate

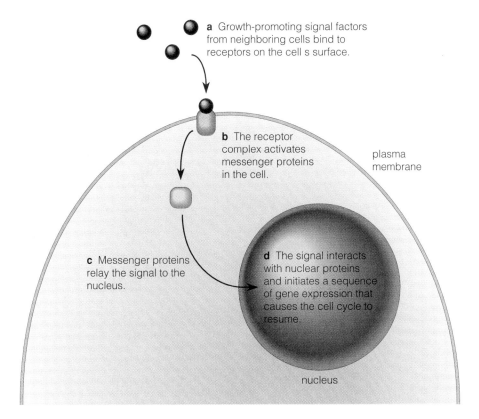

a Growth-promoting signal factors from neighboring cells bind to receptors on the cell s surface.

b The receptor complex activates messenger proteins in the cell.

plasma membrane

c Messenger proteins relay the signal to the nucleus.

d The signal interacts with nuclear proteins and initiates a sequence of gene expression that causes the cell cycle to resume.

nucleus

Figure 24.2 Transfer of an external signal to the cell nucleus stimulating the cell to initiate DNA synthesis and mitosis.

the cell cycle. Cyclins carry out their function by forming complexes with **cyclin-dependent kinases (CDKs)**. Kinases are enzymes that acquire phosphate groups (usually from ATP) and use them to phosphorylate other molecules. Cyclin-dependent kinases (as the name implies) are only able to carry out this function when in the presence of a cyclin.

During the early part of the G_1 stage of the cell cycle, levels of a cyclin known as **cyclin D** rise within the cell in response to signals received by receptors on the cell surface. Cyclin D then binds to **CDK4** or **CDK6**. The cyclin D–CDK4/6 complex transfers phosphate groups from ATP to a protein known as **pRB**. When pRB is underphosphorylated, it binds to certain transcription factors and prevents them from activating transcription of genes that encode cyclin A and cyclin E. Once fully phosphorylated, pRB releases the transcription factors, and they activate transcription of the genes that encode cyclin A and cyclin E. Thus, when underphosphorylated, pRB arrests the cell cycle in the G_1 stage at a point sometimes called the **restriction point** (abbreviated **R**). When the cell cycle proceeds beyond R, the cell is irreversibly committed to completing the cell cycle or dying. For this reason, pRB is sometimes called the **master brake** of the cell cycle.

After the cell has passed the restriction point, levels of cyclins A and E rise within the cell and bind to their respective CDKs. These cyclin-CDK complexes activate DNA synthesis, and the cell progresses through the S (DNA synthesis) stage of the cell cycle into the G_2 stage. The cyclin A–CDK1 complex also stimulates cyclin B to bind to CDK1. The cyclin B–CDK1 complex induces the cell to progress through G_2 and begin mitosis.

Cyclins, CDKs, and several other proteins are positive regulators of the cell cycle. When present, they promote progress through the cell cycle. Another group of proteins inhibit progress through the cell cycle. Thus, the cell has at its disposal a variety of positive and negative regulators of the cell cycle, which allow it to regulate any stage of the cycle.

The inhibitory proteins serve essential roles in regulating the cell cycle. We have already seen how pRB inhibits progress through the cell cycle by binding transcription factors. Several proteins, including p15, p27, and p21, inhibit activity of the cyclin D–CDK4/6 complex. Of these, **p21** is one of the most important. It acts as an inhibitor throughout the cell cycle, acting on several cyclin-CDK complexes. Its production is regulated by another very important protein called **p53**.

The p53 protein is present at very low levels in most cells, whether the cells are actively dividing or arrested in the cell cycle. The levels of p53 rise to a functional level only when the cell is in danger. Ionizing radiation, DNA damage, oxygen deprivation, and mutations in genes

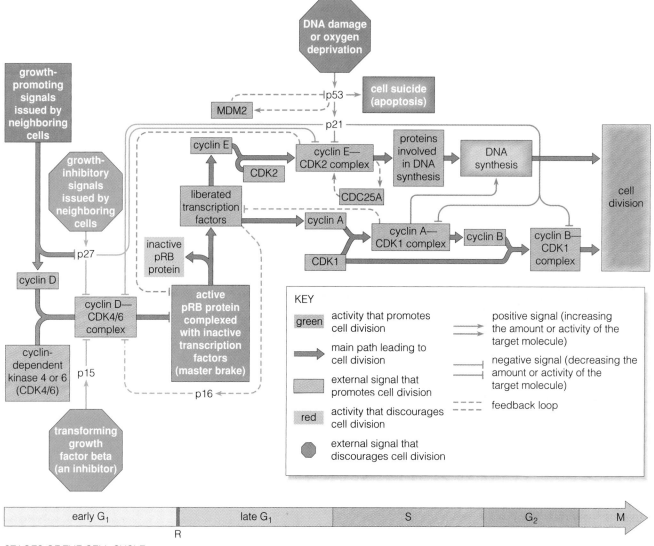

Figure 24.3 Interactions of some of the major nuclear proteins that regulate the cell cycle. (Adapted from an original drawing by Dimitry Schildlovsky in Weinberg, R. A. 1996. How cancer arises. *Scientific American* 275 (Sep):62–70. Used by permission of the author.)

responsible for the cell cycle are all signals that cause the cellular level of p53 to rise. The principal role of p53 is to recognize when something is wrong in the cell and arrest the cell cycle until the problem can be corrected. The p53 protein carries out this function by activating production of p21, which inhibits the activity of several cyclin-CDK complexes at all stages in the cell cycle. In response to cell damage, p53 activates p21, which then halts the cell cycle wherever the cell cycle happens to be. If the damage is repaired efficiently, then p53 ceases to stimulate production of p21, and the cell resumes its progress through the cell cycle.

The p53 protein also participates in determining whether or not a problem in the cell is correctable. If the problem appears to be beyond repair, then p53 stimulates the cell to undergo **apoptosis**, a genetically programmed cell suicide. Apoptosis can be thought of as a sacrifice of damaged cells to preserve the integrity of the tissue. Apoptosis is a powerful deterrent to cancer. When p53 is active, cells that lose control over the cell cycle usually undergo apoptosis and die, preventing them from developing into cancerous cells.

The following example illustrates the importance of p53 and pRB in human cancers.

Example 24.1 Cancer due to mutations in the genes that encode p53 and pRB.

According to Oliff et al. (1996. *Scientific American* 275 (Sep):144–149), about 40% of all human cancers contain mutations in the gene that encodes pRB, and about half of all cancers contain mutations in the gene that encodes p53.

Problem: Why should mutations in these genes be so common in human cancers?

Solution: Both proteins play very important roles in regulating the cell cycle. If pRB activity is lost, then the master brake that keeps the cell arrested in G_1 is lost, and cell division may proceed. If p53 activity is eliminated by mutation, then one of the major inhibitory controls over several steps in the cell cycle and a pathway to induce apoptosis when needed are lost.

Many genes participate in regulating the cell cycle. Some genes encode positive regulators, and others encode negative regulators that inhibit the cell cycle.

24.2 THE MOLECULAR BASIS OF CANCER

Cancer is caused by mutations in three classes of genes: proto-oncogenes, tumor suppressor genes, and DNA repair genes. **Proto-oncogenes** are the genes whose products stimulate normal progress through the cell cycle. Cyclins and CDKs are examples of products encoded by proto-oncogenes. **Tumor suppressor genes** are genes whose products prevent or inhibit progress through the cell cycle. The proteins p53, p21, and pRB are examples of products encoded by tumor suppressor genes. **DNA repair genes** encode enzymes that repair damage to DNA. Mutations in DNA repair genes may contribute to cancer by allowing DNA damage in proto-oncogenes and tumor suppressor genes to persist and cause mutations in those genes when the DNA replicates. **Oncogenes** are defined as genes whose products promote tumor formation. They are usually mutant alleles of proto-oncogenes.

Oncogenes typically take the form of dominant gain-of-function mutant alleles. A gain-of-function mutation causes a gene's product to be expressed when expression should be inhibited, or it may cause overexpression of the gene's product. Tumor suppressor genes, on the other hand, encode products that prevent or suppress tumor development by inhibiting cell division. Therefore, tumor-promoting mutations in these genes typically create loss-of-function alleles because the functional product must be eliminated to prevent its tumor-suppressing function. Oncogenes encode a variety of oncoproteins that stimulate the cell cycle, including cyclins, CDKs, DNA-binding proteins, protein-modifying enzymes, membrane receptor proteins, growth factors, transport proteins, and proteins that promote cell adhesion, so there is no consensus sequence of nucleotides that defines an oncogene. The definition is one of function only.

Although we now think of oncogenes as mutant genes that reside in the genome, oncogenes were first discovered in viruses. In 1911, Peyton Rous reported that chicken tumor extracts that had been filtered to remove the cells induced formation of new tumors when tumor-free chickens were inoculated with the extracts. Something other than the cells was causing the tumors. It was later shown that what Rous had discovered was a tumor virus, named the Rous sarcoma virus (RSV) in his honor. Many tumor-inducing viruses, including DNA viruses and RNA retroviruses, have since been discovered. What was it about these viruses that caused tumors to form, and could it be possible that all cases of tumor formation were due to viruses? If so, might cancer be an infectious disease?

Viruses can indeed cause cancer, but most cases of cancer are not caused by viruses. However, the discovery of cancer-causing genes in viruses led researchers to discover the genetic basis of cancer in cancer cells that are not infected with a virus. When a tumor-inducing virus infects a cell, the cell transcribes and translates the virus's genes. When the cell that carries a tumor-inducing virus changes into a tumor cell, the cell grows and replicates, replicating the viral genome in the process. Most viruses carry only those genes that are necessary for viral replication and assembly. Tumor-inducing viruses, however, have extra genes that are not directly related to virus replication and assembly. For example, the Rous sarcoma virus, an RNA retrovirus, has a gene called *src,* which is not essential for replication or viral assembly. When viruses with a deletion in this gene infect tumor-free chickens, no tumors form, an indication that *src* is essential for tumor formation. When the viral RNA was isolated and made into cDNA, the *src* gene cDNA hybridized with the DNA from uninfected cells. This observation indicated that a gene similar to *src* must be present in the chicken genome. Hybridization studies revealed that not only was the gene present in the chicken genome, but it was also present in the genomes of all vertebrates, suggesting that the gene played an important basic role.

Other tumor-inducing oncogenes were found in viruses, and in each case the story was the same. Similar genes were present in animal genomes, but in most cases the cellular genes contained introns, but the viral genes did not. Because viral oncogenes and cellular proto-oncogenes are similar, they usually carry the same names and symbols. To distinguish the two, the letter *v* precedes the symbols for genes of viral origin, and the letter *c* precedes the symbols for genes of cellular origin. For example, the viral *src* gene is called *v-src,* and the mouse cellular form is called *c-src.* Human proto-oncogenes are identified using standard human gene nomenclature, usually with all uppercase letters in the gene symbol (see the rules for human gene nomenclature in the introduction to Chapter 13). For example, the human homolog of *v-src* and *c-src* is named *SRC.*

So far, tumor-inducing viruses have been found in several species, including humans, meaning that cancer

can be an infectious disease. However, most cancers in humans are not infectious. They are caused by mutations in cellular proto-oncogenes and tumor suppressor genes. Studies of tumor-inducing viruses have proved valuable for the study of nonviral cancers. Nearly every viral oncogene has a counterpart in the mammalian genome. In the early years of cancer research, most oncogenes were first identified in tumor-inducing viruses and later found in mammalian cells by hybridization with a viral oncogene probe. Most oncogenes discovered in recent years were identified directly in tumor cells, and many of these genes do not have a viral counterpart.

Viral oncogenes have also provided much information about the mutations that change proto-oncogenes to oncogenes. Viral oncogenes are already mutant; when they are compared to their proto-oncogene counterparts, distinct differences in DNA sequence are observed. These differences provide clues about which mutations cause oncogenicity. If the same mutations found in oncogenes from tumor cells are present in the viral oncogenes, then the mutation is a likely candidate for an oncogenic mutation.

The following example shows how a mammalian proto-oncogene, identified because of its sequence similarity to a viral oncogene, serves an important cellular function.

Example 24.2   Vertebrate counterparts of the *v-src* oncogene.

Oncogenes were first discovered in viruses and then in vertebrate genomes. Did the virus transfer the oncogenes into vertebrate genomes, or did the virus acquire the oncogene from vertebrate genomes? Studies on the function of the *c-src* gene in mice provided the answer. In 1988, Azarnia et al. (*Science* 239:398–401) reported that the vertebrate *c-src* gene encodes a product that facilitates intercellular communication. Soriano et al. (1991. *Cell* 64:693–702) found that *c-src* is essential for normal bone formation and that mice homozygous for mutant *c-src* alleles usually died.

Problem: What do these results suggest about the evolutionary origin of the *v-src* oncogene?

Solution: Because a functional product of *c-src* is essential for normal development and intercellular communication in vertebrates, but is not necessary for viral function, the virus probably acquired the *v-src* oncogene from a vertebrate genome, possibly from a tumor in which the virus was present. It is highly unlikely that the gene originated in the virus.

Oncogenes were first discovered in tumor-inducing viruses. The viruses probably acquired the oncogenes from tumors in which they were present.

24.3   EXAMPLES OF GENES THAT INFLUENCE CANCER

Several genes that carry mutations in cancer cells have been well characterized at the molecular level. The number of well-characterized genes is too large to discuss each of them in detail. Instead, we will discuss a few examples to illustrate the types of genes that influence cancer and how the genes function. The genes encode many different types of proteins that regulate cell growth, cell division, and DNA replication. Table 24.1 lists several of these genes and the types of products they encode. Detailed descriptions of each of these genes can be found in the OMIM database. You can find a link to OMIM at the website for this book at www.brookscole.com/biology.

TP53

TP53 (OMIM 191170) encodes the tumor suppressor protein p53 (so named because its molecular weight is 53,000 daltons). Many cancers have at least one mutant *TP53* allele, so it is one of the most important genes in cancer. It is also among the best-characterized of all genes that influence cancer. The p53 protein has three major roles. First, it activates transcription of several genes, one of which is the *CDKN1A* gene that encodes p21. The p21 protein suppresses cyclin-CDK complexes throughout the cell cycle and may also inhibit certain DNA polymerases that synthesize DNA during the S stage of the cell cycle. The p53 protein may also directly inhibit proteins that stimulate progression from G_1 to S in the cell cycle. Thus, p53 inhibits progress through the cell cycle at many points.

The second role of p53 is initiation of apoptosis. The p53 protein is not required for apoptosis, but cells cannot induce apoptosis as readily in the absence of p53 as they can in its presence. DNA damage, cellular oxygen deficiency, deficiency of ribonucleotides for RNA synthesis, and activation of oncogenes are all types of cellular distress that raise p53 levels in the cell and may induce apoptosis.

Apparently, p53 has a third role in tumor suppression. Several proteins affect angiogenesis, the proliferation of blood vessels that supply nutrients and oxygen to tumors. Some of these proteins stimulate angiogenesis and others inhibit it. The p53 protein activates transcription of at least one gene that encodes an angiogenesis inhibitor. Therefore, p53 may also play a role in tumor suppression by preventing development of a blood supply for tumors.

The p53 protein has been very well characterized. The functional protein is a tetramer with four identical p53 subunits. Each subunit consists of 393 amino acids that have four functional domains. The first 42 amino acids are a domain that functions as a transcriptional activator by making contact with TFIID, the transcription factor

Table 24.1 Examples of Genes Associated with Cancer

Gene	Normal Gene Product Function	OMIM	Type of Cancer
ABL1	Tyrosine-specific protein kinase	189980	Chronic myelogenous leukemia
APC	Cell adhesion protein	175100	Adenomatous polyopsis of the colon
BCR	Serine/threonine kinase, GTPase activator	151410	Chronic myelogenous leukemia
BRCA1	Calcium-binding protein	113705	Breast cancer
BRCA2	Unknown	600185	Breast cancer
CCD1	Cyclin	168461	Parathyroid adenomatosis
CDK4	Cyclin-dependent kinase	123829	Melanoma
CDKN1A (p21)	Cyclin-dependent kinase inhibitor	116899	Wide variety of cancers
CDKN2B (p15)	Cyclin-dependent kinase inhibitor	600431	Wide variety of cancers
DCC	Cell adhesion molecule	120470	Colorectal carcinoma
E2F1	Transcription factor	189971	Wide variety of tumors
ERBA	Hormone receptor	190120	Leukemia
FOS	Transcription factor component	164810	Osteosarcoma
HRAS2	G-protein	190020	Bladder cancer
KRAS2	G-protein	190070	Wide variety of tumors
MYCN	Transcription factor	164840	Neuroblastoma
NRAS	G-protein	164790	Leukemias, colorectal cancer
PDGFB	Cell growth factor	190040	Glioma
RB1	Transcription factor–binding protein	180200	Retinoblastoma and many other cancers
TP53 (p53)	DNA-binding protein	191170	Wide variety of cancers

complex that binds to the TATA box and is part of the basal eukaryotic transcription complex (see section 3.3). Amino acid residues 102–292 are a DNA-binding domain that binds to specific sites in the promoters of genes that respond to p53. Over 90% of all *TP53* mutations in cancer cells are within the DNA that encodes this domain. Amino acids 324–355 constitute a domain that participates in tetramer formation. The final 26 amino acids are a regulatory domain. The cell regulates p53 levels by (1) controlling the rate of p53 degradation and (2) converting p53 from an inactive to an active form. The fourth domain is the portion of the protein affected during conversion of p53 from its inactive to its active form.

Because p53 is a tumor suppressor protein, most mutations that affect cancer are loss-of-function mutations. Some of the loss-of-function mutations, however, show some degree of dominance. We examined dominant loss-of-function mutant alleles in the *TP53* gene in section 13.1. All four subunits in a p53 tetramer bind to DNA. If any one of the four subunits in a tetramer lacks its DNA-binding ability, then the tetramer loses at least some of its function. As a consequence, a mutant *TP53* allele may contribute to cancer even though there is still a nonmutant *TP53* allele on the homologous chromosome. People who are heterozygous carriers of a mutant *TP53* allele have a high risk of developing some form of cancer, usually in childhood.

RB1

The pRB protein serves as the cell cycle's master brake, arresting the cell cycle at the G_1 stage by binding specific transcription factors and preventing the transcription factors from activating transcription of the proto-oncogenes they recognize. The *RB1* gene (OMIM 180200) encodes pRB. Because pRB plays such an important role in the cell cycle, it is no surprise that several types of cancers are associated with mutations in the *RB1* gene.

The designation *RB* stands for re<u>t</u>ino<u>b</u>lastoma, because mutations in the *RB1* gene were first found in cancer cells of patients with **retinoblastoma**, a cancer of the eye that often appears in children who inherit a mutant *RB1* allele. Typically, children who inherit a mutant *RB1* allele develop bilateral retinoblastoma (cancer in both eyes). Cases of retinoblastoma that appear in people who do not carry a mutant *RB1* allele are usually not bilateral. Mutations in *RB1* are also associated with several other types of cancer.

Because pRB inhibits the cell cycle, *RB1* is usually classified as a tumor suppressor gene. The gene is very large, over 180 kb of genomic DNA, most of it within introns. The gene contains 27 exons that encode a 4.7 kb mRNA, which is translated into the pRB phosphoprotein with 928 amino acids. *RB1* is expressed constantly in nearly all cells, even those that are actively dividing. Thus, regulation of pRB activity is not transcriptional, like that of most genes. Instead, cyclin D–CDK4/6 complexes determine the number of phosphate groups that are attached to pRB; the number of phosphate groups, in turn, determines its activity.

Loss-of-function mutations in *RB1* promote cancer. In order for a cell to lose complete pRB function, loss-of-function mutations must be present in the *RB1* gene on both homologous chromosomes. Many different types of *RB1* mutations have been identified in tumor cells, including deletions and many types of point mutations. Most tumor cells are compound heterozygotes, having different mutations in the *RB1* gene on the two homologous chromosomes.

The chromosomal location of the *RB1* gene on the long arm of chromosome 13 was first identified by the discovery of deletions there in patients with retinoblastoma who had a family history of the disease. In nearly all cases, tumor cells with a deletion also had a second mutation (often a point mutation) in the *RB1* gene on the other copy of chromosome 13. Because most mutations in *RB1* that influence cancer are recessive loss-of-function mutant alleles, a person may be a heterozygous carrier of a recessive mutant *RB1* allele and not have cancer. However, heterozygous carriers of a mutant *RB1* allele are highly susceptible to childhood retinoblastoma, because only one mutation is required (rather than two) to eliminate pRB function in a cell.

Many different cancers are associated with loss-of-function mutations in the *RB1* gene. However, children who inherit a single mutant allele of the gene (and are, therefore, heterozygous for the mutant allele) are especially susceptible to childhood onset of retinoblastoma, often in both eyes. It is puzzling why an inherited mutation in this gene is associated so often with retinoblastoma rather than any of the other different types of cancer. Other mammals may inherit *RB1* mutations, but inherited susceptibility to retinoblastoma is unique to humans.

The following example shows how comparison of molecular analysis of the *RB1* allele can resolve questions about inherited susceptibility to retinoblastoma.

Example 24.3 Penetrance of mutant *RB1* alleles.

People who inherit a single mutant allele at the *RB1* locus are at high risk for retinoblastoma, especially during childhood. Pedigree analysis has revealed that mutant *RB1* alleles may sometimes be nonpen-

etrant because not all individuals who apparently carry the mutant allele (according to the pedigree analysis) suffer from retinoblastoma. In 1995, Bia and Cowell (*Oncogene* 11:977–979) examined the mutant DNA at the *RB1* locus of two half first cousins who had bilateral retinoblastoma. Their mothers were half sisters, but only one of the mothers had retinoblastoma. When DNA samples of the two cousins were examined, one child had a C → T substitution that changed a CGA arginine codon to a TGA termination codon. This child's mother carried the same mutant allele and also had retinoblastoma. The other child carried a deletion mutation of eight nucleotide pairs in the coding region of the *RB1* gene; this deletion created a frameshift and a premature termination codon. This child's mother did not carry this mutant allele and did not have retinoblastoma.

Problem: **(a)** How does pedigree analysis suggest nonpenetrance of a mutant *RB1* allele in this example? **(b)** How can this apparent case of nonpenetrance be explained at the molecular level?

Solution: **(a)** Three related people (one mother and two half first cousins) had the same mutant phenotype (bilateral retinoblastoma). With no additional information, this observation can be most easily explained if all three inherited the same mutant allele from a common ancestor. Under this assumption, both mothers (the two half sisters) should have been heterozygous carriers of the mutant allele. However, one mother did not express the phenotype associated with the mutant allele, an apparent case of nonpenetrance. **(b)** DNA analysis revealed that the two half first cousins carried different mutant alleles and thus did not inherit the mutant alleles from a common ancestor. Each individual with a mutant allele expressed the mutant phenotype, a case of full penetrance of the mutant alleles among these four people.

E2F1

A set of similar genes encodes a set of transcription factors known as E2F factors. The most important of these genes is *E2F1* (OMIM 189971). The pRB protein binds to E2F transcription factors and releases them when it is fully phosphorylated. E2F-binding sites are present in the promoter regions of several genes that govern the transition from G_1 to S in the cell cycle, including the genes that encode cyclins A, D, and E and the genes that encode several proteins for DNA replication.

Overexpression of *E2F1* leads to tumor development, indicating that *E2F1* is a proto-oncogene. However, studies in mice show that tumors may also develop when

E2F1 is eliminated, suggesting that *E2F1* also may be a tumor suppressor gene.

RAS

The *v-ras* oncogene in murine (mouse) sarcoma virus was one of the first viral oncogenes identified. There are three similar *RAS* genes in humans, *HRAS2* (OMIM 190020), *KRAS2* (OMIM 190070), and *NRAS* (OMIM 164790), all of which encode G-proteins. **G-proteins** are part of the cascade of reactions initiated when a growth factor binds to a membrane receptor protein in a cell. G-proteins are part of a class of proteins called **signal transduction proteins** because they pass on the signal received by the receptor protein when the growth factor is bound to it (see Figure 24.2). They are called G-proteins because they bind GDP and convert it to GTP. A generalized mode of action for G-proteins is diagrammed in Figure 24.4. Typically, G-proteins consist of three subunits: α, β, and γ. The α subunit is bound to a GDP molecule, maintaining it in an inactive state. GTP can displace the GDP and separate the α subunit from the rest of the molecule, converting the molecule to an active form. The α subunit then reacts with its target molecule and passes the signal on to the next protein. The α subunit also has a GTPase activity that converts the GTP to GDP, thereby inactivating the α subunit and allowing it to bind again with the β and γ subunits. Thus, cellular levels of GTP regulate the activity of G-proteins.

Mutant RAS proteins usually contain α subunits that have lost their GTPase activity, so they remain activated constantly. The mutations that eliminate the GTPase activity are dominant loss-of-function mutations. Researchers have sequenced several normal and mutant *RAS* alleles and have compared them to see how mutations affect the protein products. The normal α subunits encoded by the *RAS* genes contain two GDP/GTP-binding domains. Most mutations are in the codons for the GTP-binding domains, particularly in codons 12, 13, 59, 61, and 63. Mutations in these codons alter the amino acid sequence and eliminate the protein's ability to convert GTP to GDP, causing the protein to remain constantly active.

APC

APC (for <u>a</u>denomatous <u>p</u>olyopsis of the <u>c</u>olon, OMIM 175100) was one of the first genes discovered that conferred an inherited susceptibility to cancer. Mutations in this gene are most often associated with colorectal cancer. This type of cancer begins with the appearance in the colon and rectum of **polyps**, benign tumors that are relatively common, especially in the elderly (Figure 24.5). Some of the polyps may become malignant, especially if a large number of benign polyps are present. The chromosomal location of *APC* on the long arm of chromosome 5 was first identified by the presence of deletions in this region in cancerous cells from patients with colorectal cancer.

Researchers have cloned and sequenced the *APC* gene. Its protein product promotes cell adhesion. When mutations cause a loss of APC function, cell adhesion in polyps is lost, and the cells metastasize. Mutations in *APC* also contribute to the initial formation of benign polyps, which may eventually become malignant. Because the protein encoded by *APC* inhibits metastasis, *APC* is considered a tumor suppressor gene.

Inherited susceptibility to colorectal cancer is nearly always associated with a mutant *APC* allele. Some mutant alleles have the characteristics of a dominant mutation in that all individuals who inherit the allele develop numerous benign colorectal polyps, some of which inevitably become malignant if left untreated. Thus, certain mutant alleles of *APC* are fully penetrant for eventual onset of colorectal cancer. Anyone who inherits one of these mutant alleles is destined to develop colorectal cancer unless preventative treatment is initiated.

Inherited colorectal cancer was first identified in a family by Eldon Gardner, who noted that the pattern of inheritance in the pedigree was typical of a fully penetrant dominant allele. He also noticed that patients with colorectal polyps often had tumors in other parts of the body, especially the teeth, bones, and adrenal glands. This association of other tumors with colorectal polyps became known as **Gardner syndrome**. Other cases of inherited colorectal cancer without associated tumors elsewhere in the body were identified and named **familial adenomatous polyopsis (FAP)**. Soon after *APC* was identified, it became apparent that both Gardner syndrome and FAP are caused by mutations in the *APC* gene and, therefore, represented different phenotypic expression of the same genetic disorder.

~

Many genes that influence cancer have been well characterized. They encode many different types of proteins, most of which govern the cell cycle.

24.4 THE ETIOLOGY OF CANCER

The cellular network that regulates cell growth and division is so pervasive that a single mutation in a proto-oncogene or tumor suppressor gene rarely causes cancer. In most cancers, at least two and usually four to six genes must mutate before a malignant tumor develops. The development of colorectal cancer is an excellent example. Mutations in several genes are required for a benign polyp to become a malignant tumor, as diagrammed in Figure 24.6. In the sequence of mutations illustrated in this figure, loss-of-function mutations in the *APC* gene on the long arm of chromosome 5 cause cells in the colon epithelium to proliferate to form an early adenoma (or class I adenoma), a slow-growing benign polyp. A mutation in the *KRAS2* gene on the short arm of chromosome 12 results in an intermediate (class II) adenoma, a

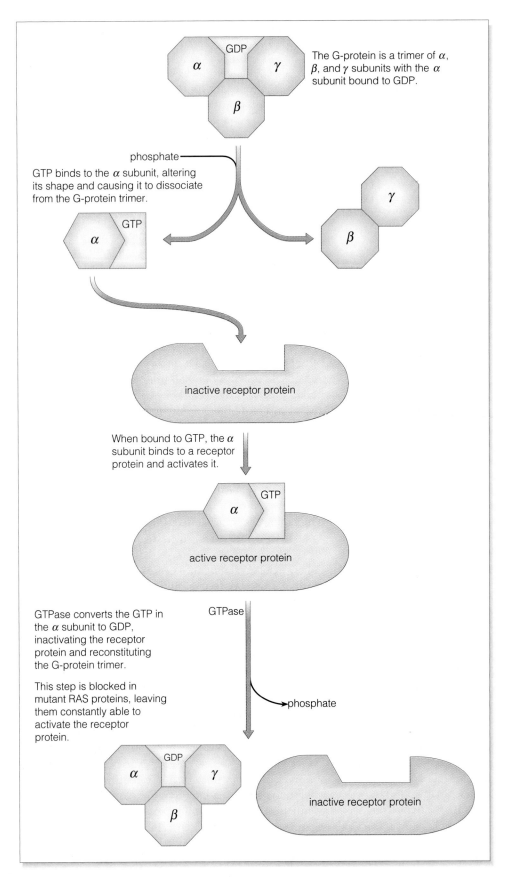

The G-protein is a trimer of α, β, and γ subunits with the α subunit bound to GDP.

phosphate

GTP binds to the α subunit, altering its shape and causing it to dissociate from the G-protein trimer.

inactive receptor protein

When bound to GTP, the α subunit binds to a receptor protein and activates it.

active receptor protein

GTPase converts the GTP in the α subunit to GDP, inactivating the receptor protein and reconstituting the G-protein trimer.

This step is blocked in mutant RAS proteins, leaving them constantly able to activate the receptor protein.

GTPase

phosphate

inactive receptor protein

Figure 24.4 Function of G-proteins.

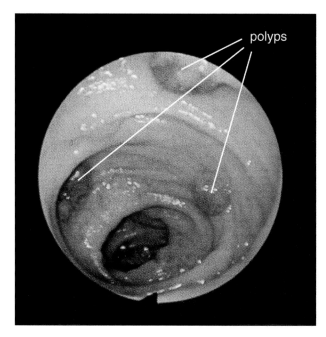

Figure 24.5 Colon polyps, benign tumors in the colon that can develop into malignant tumors. (I. Associates/Custom Medical Stock Photo.)

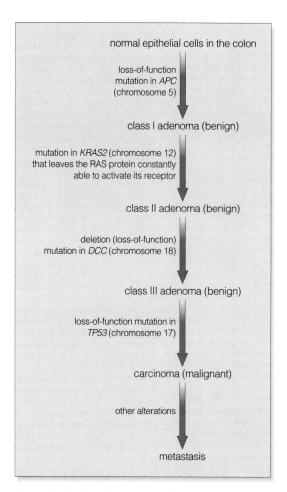

Figure 24.6 One of several possible sequences of mutations that can lead to malignant colorectal cancer and metastasis.

proliferative polyp that is also benign. A deletion in the *DCC* gene on chromosome 18 causes the adenoma to reach the advanced (class III) stage, which is still benign. A fourth mutation, in *TP53* on chromosome 17, converts the advanced adenoma into a malignant carcinoma. At that point, additional mutations may accumulate, and the tumor may metastasize.

Most cancerous cells contain mutations in one or more proto-oncogenes and at least one tumor suppressor gene. Cells have proteins on their surface that cause the cells to adhere to one another and to a proteinaceous extracellular matrix. Cellular adhesion ensures that cells stay at their correct positions within the tissue to which they belong. For tumor cells to metastasize, they must lose their adhesion, and this loss is usually a result of mutations in the genes that encode proteins that promote adhesion.

Cancer cells also gain telomerase activity. **Telomerase** is an enzyme that maintains telomere length and integrity during DNA replication (see section 2.4 for a detailed discussion of telomerase). Telomerase is usually active only in the germ line, so the telomeres in chromosomes of somatic cells are gradually degraded during mitotic cell divisions. This limits the number of cell divisions for cells that lack telomerase. Human somatic cells can divide only 50–60 times, after which the cells become **senescent**—that is, they are incapable of further divisions. Cancerous cells, however, have active telomerase and are **immortal**—meaning that, given the proper nutrients, the number of cell divisions is unlimited. Research is actively under way to find a treatment that blocks telomerase

activity with the hope that the treatment will cause cancer cells to become senescent.

Mutations That Promote Cancer

Tumor cells often carry many mutations, including point mutations as well as chromosome deletions and rearrangements. This is not surprising. In many tumors, cells grow and divide rapidly, and there is inadequate time for DNA repair and normal partitioning of chromosomes during mitosis, so mutations and alterations in chromosome structure and number accumulate in cancerous cells. Some mutations are the result of uncontrolled cell division rather than the cause of it. However, certain mutations are characteristic of particular types of cancer. These mutations usually activate an oncogene or the function of a tumor suppressor gene. Several mutations that are consistently associated with certain types of cancer have been well characterized. In this section, we will review a few of these mutations to illustrate how mutations can alter the normal controls over cell growth and division.

Point mutations in particular regions of proto-oncogenes and tumor suppressor genes are commonly

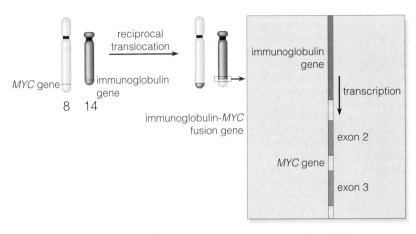

Figure 24.7 A reciprocal translocation between chromosomes 8 and 14 that creates a fusion gene by placing exons 2 and 3 of the *MYC* proto-oncogene under control of an immunoglobulin gene.

associated with cancer. Although transition mutations are more common than transversions, several common carcinogens, particularly those in tobacco smoke, tend to cause transversions. Tobacco smoke contains polycyclic hydrocarbons, which are highly mutagenic. Polycyclic hydrocarbons attack guanines in the DNA and cause the damaged guanines to pair with adenine during DNA replication. After the DNA molecule replicates, thymine pairs with the misplaced adenine, resulting in a transversion (see Example 5.3).

Deletions of small segments of a gene are associated with the transformation of proto-oncogenes into oncogenes. For example, the *v-erbB* gene encodes a mutant growth factor receptor that remains active even when a growth factor is not bound to it. The gene lacks DNA sequences that encode segments on both ends of the polypeptide chain, including the normal growth factor–binding domain. The result is a shortened polypeptide that lacks the growth factor–binding domain, so it is active in the absence of the growth factor.

Chromosome deletions are also associated with loss of antioncogene activity. Several tumors that have lost pRB function have deletions in chromosome 13 where the *RB1* gene resides. A second mutation, either a point mutation or a deletion, in the homologous copy of the *RB1* gene causes a loss of pRB activity.

Translocations are common oncogenic mutations. For example, as explained in section 17.4, cancerous cells from patients with chronic myelogenous leukemia often have a Philadelphia chromosome, a chromosome 22 that has a small segment of chromosome 9 translocated to it to form a fusion oncogene. Another example is Burkitt lymphoma, a cancer of B lymphocytes, the cells that produce antibodies. Immunoglobulin genes (the genes that encode antibodies) are expressed at high levels in B lymphocytes. They are located on chromosomes 2, 14, and 22. The *MYC* proto-oncogene is located on the long arm of chromosome 8. If a translocation moves a particular segment of chromosome 8 to a position near an active im-

munoglobulin gene (usually by reciprocal translocation between chromosomes 8 and 14, as diagrammed in Figure 24.7), then the active immunoglobulin gene's promoter activates transcription of the *MYC* gene, which causes abnormally high production of the protein encoded by the *MYC* gene. Elevated expression of *MYC* is one of the steps that leads to Burkitt lymphoma.

Some cancers are associated with DNA insertion mutations. A few tumor-inducing viruses do not carry viral oncogenes. Instead, they may insert themselves in or near the coding sequence of a cellular proto-oncogene and bring the proto-oncogene under the control of viral genes that are expressed at high levels. Some cancers are also associated with insertion mutations of transposable elements, especially L1 and *Alu* elements.

Why Cancer Is So Common

Cancer requires multiple mutations, and cells have built-in protections against loss of control over the cell cycle, so it may seem unlikely that cancer could ever arise. Mutations appear naturally at very low rates, so the probability that the necessary mutations will appear in a single cell is very small. However, an adult human is composed of as many as 30 trillion cells. A mutation that arises once per million cells arises in tens of thousands of cells in one person. Humans also have relatively long life spans. After a mutation appears in a somatic cell, the mutation persists, so as people age, they accumulate mutations in their cells. The older people become, the more likely they are to have at least one cell that has all the necessary mutations to induce tumor growth and cancer. Consequently, cancer is most prevalent among the elderly.

In addition, people are often exposed to radiation and **carcinogens**, substances that increase the risk of cancer. Carcinogens and radiation are around us at low levels in the form of ionizing and nonionizing radiation and carcinogenic chemicals in food, water, and air. Many car-

Table 24.2 Some Forms of Radiation and Suspected or Known Carcinogens to Which Humans Are Exposed

Radiation or Carcinogen	Description
Ionizing Radiation	
Cosmic radiation	Includes several forms of ionizing radiation from space.
Radon gas	Found in buildings and homes that are poorly ventilated and are located near natural sources of radon gas.
X-rays	Used as a diagnostic tool and for certain forms of therapy in medicine.
Nuclear fallout	Present near sites of nuclear weapons use or open-air testing and when radiation escapes from nuclear reactors or waste sites.
Nonionizing Radiation	
Ultraviolet light	Present in sunlight and some artificial light sources.
Suspected or Known Carcinogens	
Tobacco smoke	Contains polycyclic hydrocarbons that are mutagenic.
Industrial solvents	Include benzene, ether, formaldehyde, and other organic solvents used in manufacturing.
Airborne particulates	Present in vehicle exhaust, industrial emissions, and smoke from wood or coal burning, these carry several carcinogens, among them polycyclic hydrocarbons.
Soots, tars, and oils	Contain hydrocarbons that increase the risk of cancer with repeated exposure.
Nitrosamines	May appear when foods with nitrites are heated to high temperatures.
Asbestos	Present in some insulation, fireproofing, building, and roofing materials in older buildings.
Vinyl chloride	Used in plastics manufacturing.

cinogens are natural, such as ionizing radiation from space and natural carcinogens in food. Others are a result of lifestyle, workplace exposure, or pollution. Table 24.2 provides a list of radiation types and suspected carcinogenic compounds to which humans are often exposed.

Exposure to carcinogens increases the risk of cancer because the mutation rate in exposed tissues is increased. For example, at least 75% of the lung cancer cases in women and 85% in men can be attributed to smoking. Because of the large number of people who smoke throughout the world, lung cancer is quite prevalent. Exposure to airborne particulates in polluted areas may also contribute to lung cancers.

The following example provides evidence that exposure to nonionizing radiation in sunlight is associated with skin cancer.

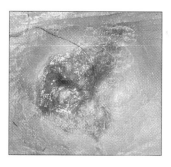

Figure 24.8 A squamous cell carcinoma, a common type of skin cancer. (Biophoto Associates/Science Source/Photo Researchers.)

Example 24.4 Mutations in the *TP53* gene in skin cancer.

In 1994, Ziegler et al. (*Nature* 372:773–776) reported the results of experiments that determined the type and timing of mutations in the *TP53* gene in human squamous cell carcinomas (SCCs), a type of skin cancer (Figure 24.8). They examined actinic keratoses, skin tumors that are in the early stages of SCC development. Of 45 tumors examined, they found that 27 (60%) had mutations in the *TP53*

gene, and that 8 of those had mutations in both copies of the *TP53* gene. Most mutations were C → T transitions, and 89% of the mutations affected adjacent pyrimidines. The results of their experiments indicated that over 80% of the *TP53* mutations examined were present in all cells of the tumor in which the mutation was found.

Problem: **(a)** What aspects of these results indicate that the mutations arose from exposure to sunlight? **(b)** The authors concluded that the *TP53* mutations were usually among the first mutations to arise in the tumor cells. What aspect of their results supports this conclusion?

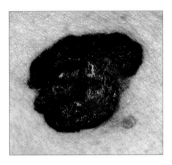

Figure 24.9 A malignant melanoma, a very dangerous type of skin cancer. (James Stevenson/SPL/Photo Researchers.)

Solution: **(a)** The mutagenic agent in sunlight is ultraviolet B radiation, a form of nonionizing radiation. Dimerization of adjacent pyrimidines is the most common DNA damage from nonionizing radiation (see section 5.3). If they are left unrepaired, pyrimidine dimers often cause transition mutations. The observation of transition mutations in adjacent pyrimidines is strong evidence that nonionizing radiation from sunlight is the mutagenic agent. **(b)** The observation that all cells in a tumor carried the mutation is evidence that the mutation appeared very early during tumor development. Had the mutation appeared later in tumor development, only a subset of the tumor cells should have carried the mutation.

Inherited Susceptibility to Cancer

Certain cancers tend to run in families because of inherited mutations in proto-oncogenes or tumor suppressor genes. Like other alleles, oncogenes and mutant tumor suppressor genes are present in every cell of a person when the genes are contributed by a parent's gamete. Although a single mutation in a proto-oncogene or tumor suppressor gene rarely causes cancer by itself, its inherited presence increases the likelihood of cancer because the first step toward cancer has already been taken in every cell of a person who inherits the mutant allele. There are several examples of inherited susceptibilities to cancer. We have already discussed several, including retinoblastoma and colorectal cancer earlier in this chapter, and xeroderma pigmentosum in section 5.4.

The association of certain mutant alleles with inherited susceptibility to certain types of cancer has great promise for early detection and treatment. For example, the success rate for overcoming breast cancer is very high if the tumors are detected and treated early. However, if breast cancer is not detected in time, the probability of successful treatment drops dramatically. The discovery that mutations in the genes *BRCA1* and *BRCA2* are as-

sociated with most cases of inherited susceptibility to breast cancer has made it possible for physicians to use DNA analysis to identify those people who carry the mutant alleles. The physicians can then closely monitor people who have an increased risk to detect tumors early and treat them while the probability of successful intervention is high.

~

Development of cancer usually requires mutations in several genes. Inherited mutations in proto-oncogenes and tumor suppressor genes render individuals more susceptible to cancer.

24.5 CANCER PREVENTION, DETECTION, AND TREATMENT

Our understanding of cancer has increased considerably over the past 20 years. Research into the causes and development of cancer has provided much information that we can use to prevent, detect, and treat cancer. In this section, we will review ways in which people and physicians can combat cancer and reduce its devastating effects.

Cancer Prevention

Without doubt, the best weapon against cancer is prevention. Not all cancers are preventable, but changes in lifestyle and public health efforts can prevent more than half of all cancers that now occur. In spite of all the information that has been gathered and disseminated, cancer rates in the United States rose by 6.3% in the 10-year period from 1973 to 1992, even after the rise in cancer due to increased longevity was factored out.* However, if increases in lung cancer are removed from the data, the rate of cancer declined by about 3.4% over the same period. About 30% of cancer fatalities can be attributed to smoking. Lung cancer caused by smoking is the single most preventable type of cancer. Unfortunately, declines in smoking among males in the United States have been offset by increases in smoking among females in the United States and among both males and females in the rest of the world.

Airborne particulate pollution from vehicle and industrial emissions, open burning, and even home barbecues that use wood or charcoal for fuel contains carcinogens that may contribute to lung cancer. Workers exposed to diesel exhaust fumes in bus, truck, and railroad operations have an increased risk of lung cancer. Many cities have undertaken efforts to reduce air pollution. However, increases in vehicular traffic have often offset reductions of industrial emissions. Reducing driving, use of public and nonmotorized transportation,

*Data in this section are from Rennie, J., and R. Rusting. 1996. *Scientific American* 275 (Sep):56–59.

and conversion of vehicles, industrial processes, and home heating systems to clean fuels such as natural gas can have a substantial impact on reducing airborne particulate pollution.

Changes in diet could also substantially reduce the number of cancer cases. About 30% of all cancers can be attributed to dietary factors. The relationship of diet to cancer is complex, but some associations are clear. Consumption of red meat and saturated fats is associated with colorectal cancer. Fat consumption is also associated with prostate cancer in men. High salt consumption may contribute to stomach cancer. Alcohol consumption, even at moderate levels, appears to be linked to several types of cancer, including liver, breast, and colorectal cancers. When combined with smoking, alcohol consumption increases the risk of respiratory and digestive tract cancers. A diet high in fruits and vegetables appears to provide significant protection against many types of cancer, due in part to natural compounds in plants that protect against DNA damage. Dietary planning to reduce meat and fat intake, avoid alcohol, and increase the proportion of vegetables and fruit in the diet can reduce the risk of many cancers.

Exposure to both ionizing and nonionizing radiation increases cancer risks. About 2% of all cancers can be attributed to radiation. Of these, exposure to nonionizing radiation in the form of ultraviolet radiation from the sun is the most common form of radiation exposure. Overexposure to ultraviolet B rays present in sunlight causes about 90% of all skin cancers. Avoiding intentional exposure to the sun, particularly when it causes sunburning, can do much to reduce skin cancers. In spite of public campaigns aimed at teaching people about the dangers of overexposure to sunlight, the incidence of skin melanomas (the most life-threatening type of skin cancer, shown in Figure 24.9) increased by 34.1% from 1973 to 1992.

Exposure to ionizing radiation is much less common than exposure to nonionizing radiation. Exposure to dangerous levels of ionizing radiation is usually localized to certain environmental or occupational situations. Radon is a gas that escapes from the earth in some areas. At one time, cancer due to radon exposure was common among miners who spent long periods of time underground. Most modern underground mines now have ventilation systems that significantly reduce radon exposure. Also, there is some concern about radon exposure in poorly ventilated basements of homes in high radon areas.

Exposure to ionizing radiation in nuclear fallout from nuclear explosions or nuclear reactor accidents is generally rare and localized. The rates of certain cancers are slightly higher among those exposed to the Hiroshima and Nagasaki nuclear explosions. In southern Nevada and Utah, the rate of leukemias and other cancers in both humans and livestock increased during the years that followed nuclear weapons tests in southern Nevada. Recently declassified U.S. government data confirm that ionizing radiation fallout reached dangerous levels in these areas following the tests. People who work with ionizing radiation, such as X-ray technicians in medical and dental clinics and researchers in molecular biology laboratories, are trained to take appropriate precautions to avoid exposure to ionizing radiation.

Some of the carcinogens listed in Table 24.2 are found mostly in certain occupational situations. They include benzene, asbestos, formaldehyde, ether, vinyl chloride, airborne particulates, ionizing and nonionizing radiation, and soot. People who work in situations where there is a potential long-term exposure to such compounds must be trained to appropriately protect themselves against potentially harmful exposure. Unfortunately, even when proper training has been given, some people choose to ignore their training and subject themselves unnecessarily to carcinogen exposure.

Clearly, the incidence of most types of cancer could be dramatically reduced by taking some straightforward preventative measures, such as dietary changes, avoiding overexposure to sunlight, and abstaining from tobacco and alcohol. The medical costs associated with treatment of preventable cancers are enormous. Both the economic and emotional costs of cancer could be substantially reduced with appropriate precautions to reduce or eliminate exposure to carcinogens and a willingness on the part of people to heed the warnings they receive.

Early Detection

Cancers that have been detected and treated during the early stages of tumor development can often be completely eliminated. On the other hand, advanced cancers that have invaded surrounding tissues or metastasized can usually be treated but not eliminated. Eventually, such cancers may elude treatment and cause severe disability and death. Efforts aimed at early detection of common cancers can significantly increase the effectiveness of treatments. Breast self-examination in all women and men (men can have breast cancer) and routine mammography in women over 40 are recommended for early detection of breast cancer. Pap smears can reveal cervical cancer in the early stages. Routine rectal examinations by a physician to detect rectal polyps and to examine the prostate gland in men can help in early detection of colorectal and prostate cancers.

These simple procedures facilitate early detection and successful treatment of many cancer cases. However, much current research is aimed at finding ways to rapidly detect cancer in the early stages, long before a patient experiences any symptoms, so that the cancer can be treated early by conventional means. Some tests are currently being developed to identify proteins or compounds present in the blood or urine that reveal the presence of

a tumor somewhere in the body. The tumor can then be sought out and treated.

Genetic susceptibility to certain cancers is inherited, often as a single mutant allele that is transmitted in a simple Mendelian fashion. DNA tests are now available for identifying inherited mutant alleles in several common proto-oncogenes and tumor suppressor genes. When there is a family history of a particular type of cancer, the mutant allele can often be readily identified and all carriers in the family identified. Physicians can then monitor carriers of a mutant allele to detect cancer and treat it early. Genetic testing for cancer susceptibility has raised much controversy over the legal and ethical issues associated with testing. We will discuss these issues in detail in Chapter 28.

Traditional Cancer Treatments

Most cancers are currently treated with surgery, radiation therapy, chemotherapy, or some combination of these. Surgery is often used to remove benign tumors before they become malignant and to remove malignant tumors before the cells metastasize. Surgery can be an effective treatment, especially if the cancer has not spread into vital parts of organs and has not metastasized. Unfortunately, microscopic portions of a malignant tumor may remain following surgery. If malignant cells remain, they can redevelop into a malignant tumor and may also metastasize. Other forms of treatment, such as radiation therapy and chemotherapy, are often combined with surgery to destroy any remaining cancer cells to reduce the possibility of recurrence.

Radiation therapy is based in part on the ability of ionizing radiation to damage DNA and induce expression of tumor suppressor genes such as *TP53* that halt cell growth and initiate apoptosis. Radiation is usually directed at the tumor so that exposure of noncancerous tissue to radiation is minimized. Unfortunately, tumor cells may become resistant to radiation therapy when gene function that is related to apoptosis is lost, as in cancer cells that have lost p53 activity.

Chemotherapy relies on drugs that are detrimental to actively dividing cells (often by inducting DNA damage that leads to apoptosis) but are less harmful to quiescent cells. Most medications used in chemotherapy inhibit DNA replication in some way. Some are mutagens, such as alkylating agents and base analogs that damage DNA (see section 5.3 for examples of these compounds). Others, such as topoisomerase inhibitors, inhibit the enzymes that participate in DNA replication. Compounds that inhibit DNA replication have their greatest effect on dividing tumor cells. DNA damage in quiescent cells is less consequential because these cells are not actively replicating their DNA. Thus the drugs administered in chemotherapy focus on cancer cells and allow quiescent cells to live and carry out their functions.

Some normal cells actively divide in adult humans, such as cells in the bone marrow, intestinal lining, and hair follicles. Chemotherapy usually has serious and sometimes painful side effects when it attacks the cells that divide normally. Some of the side effects include anemia and loss of immunity (due to a reduction of cell division in bone marrow), nausea and diarrhea (due to a reduction of cell division in the digestive tract lining), and hair loss (due to a reduction of cell division in the hair follicles). The serious nature of some of these side effects may limit the beneficial effects of chemotherapy when physicians must restrict dosages to minimize the harmful side effects.

Future Treatments

Discoveries of the molecular bases for initiation and development of many types of cancer have provided information that is now being used to design new treatments. Some future treatments of cancer may be modifications of conventional treatments to target cancer cells more specifically. Other treatments are based on recombinant DNA methodology. In this section, we will look briefly at some treatments that are currently under investigation.

Certain antibodies recognize and bind to substances found on tumor cells. When such an antibody is administered to a patient, the antibody seeks out the tumor cells and binds to them. Work is under way to genetically engineer antibodies and attach anticancer drugs to them so the administration of the drug is aimed specifically at the tumor cells to reduce some of the harmful side effects of chemotherapy. However, researchers still face some significant challenges. Most substances that antibodies seek out in tumors are present in normal cells, so the antibody may attack cells other than those in the tumor. Also, a person's immune system may attack the genetically engineered antibodies and render them ineffective.

Scientists are also studying medications that inhibit the activity of proteins required for cell division. For example, the E2F family of transcription factors are required for the cell cycle to proceed past G_1. Medications that specifically inhibit E2F transcription factors could be used to halt cell division and maintain tumor cells in a quiescent state. Likewise, telomerase inhibitors may cause tumor cells to eventually become senescent. Protein kinases are involved in the signal pathways that stimulate cell division. Protein kinase inhibitors are now in the clinical trial stages as a cancer treatment.

Among the most promising alternative treatments is the use of substances called **antiangiogenesis compounds**, chemicals that attack angiogenesis rather than attacking the tumor directly. Without an adequate blood supply, a malignant tumor ceases to develop, begins to shrink, and may eventually die. Antiangiogenesis drugs are currently being tested in clinical trials.

Gene therapy is the introduction of a functional gene into cells that have lost that gene's function. Cancers due to mutations in tumor suppressor genes could conceivably be controlled by gene therapy in which physicians introduce a functional version of a tumor suppressor gene into the cells of a tumor that has lost the function of that gene. Experimental treatment of lung cancer by introduction of functional *TP53* genes has shown some success, as highlighted in the following example.

Example 24.5 Successful treatment of lung tumors with gene therapy.

Gene therapy is the introduction of functional genes into cells with a mutant gene to restore lost gene function. In 1996, Roth et al. (*Nature Medicine* 2:985–991) reported the results of a clinical trial in which functional *TP53* genes were introduced into lung tumors in humans. Previous experiments with mice and with tumor cell cultures suggested that introducing a functional *TP53* gene into tumors that had lost p53 function could halt tumor growth. The researchers treated nine lung cancer patients with a genetically engineered retrovirus that contains a functional *TP53* gene spliced to a high-expression promoter. They injected the engineered virus directly into tumors in the lungs. Two of the patients could not be fully evaluated. The treatment eliminated viable tumors at the treated sites in three patients, one patient experienced tumor regression, and tumor growth ceased in two patients. One patient had a viable tumor before and after treatment. PCR tests of biopsied tissue revealed that the viral transfer to tumor cells was successful. Although most treated tumors responded to the viral injections, all nine patients died following the study from untreated metastatic tumors or from complications unrelated to the treatments.

Problem: (a) The tumors treated in this clinical trial were advanced, so cells in the tumors contained mutations in several proto-oncogenes and tumor suppressor genes. Why was it possible to treat the cancer by replacing the lost function of only a single gene? (b) What do these results suggest about the possibility of treating lung cancer with engineered viral vectors? (c) What limitations might be associated with such treatments?

Solution: (a) Functional p53 can halt cell division and induce apoptosis, even when there are mutations in other genes. Therefore, restoration of p53 function alone is often sufficient to halt tumor growth and cause tumor regression in tumor cells that have lost p53 function. (b) Six of the seven patients that were evaluated responded positively to the treatments. These results suggest that transfer of functional *TP53* genes using viral vectors may be an effective method for treating lung tumors that have lost p53 function. (c) Not all cancers are associated with a loss of p53 function. Therefore, the genetically engineered retrovirus should be effective in treating only lung cancers that have a loss of p53 function. Also, the human immune system naturally attacks viruses. Patients may develop immunity to the genetically engineered virus, which could render the treatment ineffective. Although gene therapy holds much promise, it is still in the early stages of testing.

Since some of these alternative treatments are now being investigated at the clinical level, several may soon be available for routine use. In many cases, however, significant obstacles have yet to be overcome. Nonetheless, recombinant DNA technology shows great promise in the eventual prevention and treatment of cancer without the harmful side effects inherent in some of our current methods of treatment.

~

Cancer has traditionally been treated with surgery, radiation, and chemotherapy. Alternative and potential treatments are currently being developed.

SUMMARY

1. Due to medical advances, deaths due to infectious diseases have declined in developed countries, resulting in longer life spans. As a consequence, cancer has become one of the leading causes of death in these countries.

2. Cancer can be considered a genetic disease because it is always the result of mutations in genes.

3. Mutations that influence cancer often convert proto-oncogenes into oncogenes. Oncogenes are mutant versions of genes that govern cell growth and division.

4. Mutations in more than one gene are required for a tumor to become malignant and to metastasize.

5. Tumor suppressor genes encode products that suppress cell division. Most cancers are due to mutations that create oncogenes as well as mutations in at least one tumor suppressor gene.

6. Because cancers are due to mutations, susceptibility to cancer may be inherited. A person who inherits a mutant oncogene has an elevated risk for cancer because the first step toward cancer has already been passed.

7. Proto-oncogenes are essential genes that govern the cell cycle. Their products include nuclear regulatory proteins, protein kinases, growth factors, growth factor

receptors, and G-proteins. These products may interact with the products of tumor suppressor genes.

8. Oncogenic mutations may include point mutations as well as deletions and chromosome rearrangements.

9. Radiation or carcinogenic substances may increase cancer rates by increasing mutation rates. Many cases of cancer can be prevented by eliminating or reducing exposure to harmful carcinogenic compounds and radiation.

10. Cancer is typically treated by surgery, radiation therapy, chemotherapy, or a combination of these. Several promising alternative treatments are currently being tested.

QUESTIONS AND PROBLEMS

1. What are the normal functions of proto-oncogenes?

2. Describe several types of proto-oncogenes and their roles in the cell.

3. Distinguish oncogenes, proto-oncogenes, and tumor suppressor genes from one another.

4. List some examples of proto-oncogenes, and describe their cellular functions.

5. Describe how a mutation in a proto-oncogene can cause a gain of function.

6. Describe the cellular role of the *TP53* gene and its product. Why do many cancers have a loss-of-function mutation in this gene?

7. Examples of frameshift, deletion, insertion, and chromosome-loss mutations for the *TP53* gene have been found in mammalian tumors. However, many cancers have a single-nucleotide substitution mutation in the *TP53* gene that causes a dominant loss-of-function allele. What aspect of the p53 protein makes its gene susceptible to dominant loss-of-function mutations? (Information from Chapter 13 may help you answer this question.)

8. Why is more than one mutation required for a tumor to become malignant?

9. Rarely are cancers themselves inherited. Yet the susceptibility to certain cancers is inherited. Why is this the case?

10. Define apoptosis. How is it utilized in cancer treatment?

11. Ionizing radiation and compounds used in chemotherapy are usually mutagenic and carcinogenic. Given that they cause cancer, why can they be effectively used to treat cancer?

12. Explain why p53 expression is elevated in cultured cells that are treated with ionizing radiation when compared to untreated cultured cells.

13. In 1994, Kharbanda et al. (*The Journal of Biological Chemistry* 269:20739–20734) reported that leukemia cells exposed to ionizing radiation induced expression of the *p56/p53^{lyn}* gene. The product of this gene is a protein kinase with an amino acid sequence similar to proteins encoded by *Src* genes. Another protein, $p34^{cdc2}$, governs the transition from the G_2 to the M stage of the cell cycle. When DNA is damaged, $p34^{cdc2}$ is phosphorylated, and it halts the G_2 to M transition to allow repair of DNA damage. The *p56/p53^{lyn}* gene product binds to $p34^{cdc2}$ following its induction by radiation. **(a)** Is *p56/p53^{lyn}* an oncogene or a tumor suppressor gene? **(b)** What is the probable role of *p56/p53^{lyn}*? **(c)** In what way might *p56/p53^{lyn}* affect the apoptotic response of cancer cells to radiation therapy?

14. When E2F is bound to DNA, it bends the DNA molecule at an angle of about 125°. In 1994, Huber et al. (*The Journal of Biological Chemistry* 269:6999–7005) reported that when pRB binds to E2F, DNA bending reduces to less than 80° and that the DNA-binding affinity of E2F increases. According to the information presented in this chapter, what is the role of pRB in binding E2F, and how might the observation of Huber et al. further define the role of pRB?

15. The *TP53* gene is usually not expressed in somatic cells unless there is some sort of cell stress or DNA damage. In 1993, Almon et al. (*Developmental Biology* 156:107–116) reported that p53 activity in mice was highest in the testes and was confined to the primary spermatocytes. They also reported that the level of p53 activity was related to the spermatogenesis cycle. Define a possible role for p53 activity in primary spermatocytes when there is no DNA damage.

16. Cancer rates are often correlated with environmental exposure to certain carcinogens or with the lifestyle of certain groups of individuals. Most people are aware of associations between smoking and lung cancer or high cancer rates associated with exposure to certain industrial chemicals or radiation. However, associations with certain career paths and cancer have also been observed. For example, in 1993, Rubin et al. (*American Journal of Public Health* 83:1311–1314) reported that breast cancer was higher among women in executive, managerial, administrative, and professional careers than in women with service, farming, transportation, or labor careers. Breast cancer was especially high in clergywomen, librarians, and teachers. With such strong associations of environmental influences, why can it still be said that all cancers have a genetic origin?

17. In 1995, Tallarida (*Life Sciences* 56(3):PL51–62) reported that cancer rates in various countries were highly correlated with the proportion of the population over 60 years of age. Among the 31 countries surveyed, those with the highest death rates from cancer were European countries, whereas those with the lowest rates were Egypt, Mexico, and Trinidad and Tobago. What is the most probable explanation for these differences?

18. In 1951, Gardner (*American Journal of Human Genetics* 3:167–176) described an inherited form of colorectal cancer in several generations of an extended family. Following his initial findings, the inherited presence of colorectal polyps associated with other tumors was named Gardner syndrome and was subsequently identified in additional families. Gardner found that Gardner syndrome appeared in about half of the offspring of an affected individual and in every generation. **(a)** What is the probable pattern of inheritance for Gardner syndrome? **(b)** What common aspect of many types of cancer is exemplified by the presence of tumors in different organs of the same person?

19. The compound 2-chlorodeoxyadenosine (CdA) has been used to effectively treat advanced chronic lymphocytic leukemia (according to Delannoy et al. 1994. *Nouvelle Revue Francaise d'Hematologie* 36:311–315). The enzyme deoxycitidine kinase phosphorylates CdA, and the phosphorylated form inhibits DNA repair, resulting in DNA fragmentation and apoptotic death of actively dividing cells. **(a)** To what class of compounds does CdA belong? **(b)** Based on the name, draw the structure of CdA. **(c)** Why does CdA selectively affect actively dividing cells rather than all living cells? (You may wish to use information from Chapter 5 to help you answer this question.)

20. Explain why radiation is effective in treating cancer and why some tumors may be resistant to radiation treatment.

21. The nucleotide sequences of the human telomerase gene and a yeast (*Saccharomyces pombe*) telomerase gene, as well as the amino acid sequences of the products encoded by these genes, were determined by Nakamura et al. (1997. *Science* 277:955–959). The human and yeast telomerases share amino acid homology with the reverse transcriptases of retroviruses and retrotransposons. Elimination of telomerase activity in yeast cells results in cell senescence. Explain the role of telomerase in cancer and describe why the discoveries reported in this article are significant for cancer research.

22. Thymidine kinase 1 (TK1) is an enzyme that phosphorylates thymidine to thymidine monophosphate, providing thymidine during DNA replication. In 1990, Robertson et al. (*British Journal of Cancer* 62:663–667) found that blood serum TK1 levels were elevated in individuals with breast cancer and that the more advanced the cancer, the higher the levels of TK1. How might TK1 detection be used for early detection of breast cancer?

23. In 1992, O'Neill et al. (*Journal of the National Cancer Institute* 84:1825–1828) measured TK1 levels in primary breast tumors of patients who underwent treatment for breast cancer. The authors found that the rate of cancer recurrence was higher for those patients with higher levels of TK1 in the primary tumor. What significance do these results have for treatment of breast cancer and prevention of cancer recurrence?

FOR FURTHER READING

The **September 1996 issue of** *Scientific American* is a special issue with 23 articles devoted to the major aspects of cancer **(pp. 56–167)**. These articles are excellent sources for detailed and well-presented information on the genetic basis of cancer. The cellular function of pRB and the role of pRB in cancer was reviewed by **Weinberg, R. A. 1995. The retinoblastoma protein and cell cycle control.** *Cell* **81:323–330.** The function of p53 was reviewed by **Levine, A. 1997. p53, the cellular gatekeeper for growth and division.** *Cell* **88:323–331.**

For additional reading, go to InfoTrac College Edition, your online research library at: http: www.infotrac-college.com/brookscole

CHAPTER 25

Cells of the immune system recognize foreign substances as antigens and target for destruction those particles or cells that display the antigen.

~

Antibodies are proteins that bind to specific antigens. Antibody genes are capable of a large number of rearrangements, which produce a wide variety of possible antibodies.

~

When an antibody binds to an antigen, it marks the antigen and the cell or particle to which the antigen-antibody complex is attached for destruction by the immune system.

~

T cells distinguish normal cells in the body from cells that have been infected by a virus and mark the infected cells for destruction.

~

Several human diseases result from immune system dysfunction.

~

Antibodies have been used extensively as tools in genetics research.

GENES AND IMMUNITY

Among the most important systems in the human body is the immune system. When a foreign object, such as a pathogenic virus or bacterium, enters the body, the body recognizes the object as foreign and mounts an immune response. After the immune system has successfully fought a first-time infection, the body retains the ability to mount a rapid secondary immune response to the same type of infection, should it ever happen again. Because of the secondary immune response, people suffer certain infectious diseases, such as chicken pox, only once in their lifetimes.

Immunity also makes it possible for health care providers to artificially immunize people or animals against certain diseases by introducing a vaccine that contains a weakened or dead pathogen into the body. The weakened or dead pathogen in the vaccine cannot produce a disease, but the body mounts an immune response to it. When a virulent strain of the pathogen enters an immunized body, the previously established immunity eliminates the pathogen before it can proliferate. Because of immunization, diseases that formerly took thousands of lives per year are no longer a serious threat. Successful immunization can drastically affect pathogen populations as well. At one time, smallpox and polio were serious diseases that spread in epidemics and took many lives. But because of widespread immunizations, the smallpox

virus has been eradicated and the polio virus is close to eradication. People are no longer immunized for smallpox because the pathogen no longer exists in nature.

The immune response is a joint effort among several cell types that each produce specialized molecules. Some of the most important of these molecules are **antibodies**, composed of particular types of protein molecules called **immunoglobulin** or **Ig proteins**. A specialized cell secretes antibodies into the blood in response to invasion of the bloodstream by foreign substances. Each antibody recognizes and binds to a specific **antigen**, a foreign substance such as a protein or polysaccharide on the surface of a virus or bacterium. The antibody bound to its antigen is a signal that marks both the antigen and the cell or particle that is attached to the antigen for destruction by other agents of the immune system.

Antibodies are proteins that are encoded by genes. Potential antigens number in the millions, and a different antibody recognizes each antigen. In order to encode all the necessary antibodies, millions of immunoglobulin genes would have to occupy the vast majority of the human genome. Immunoglobulin genes, however, represent only a tiny fraction of the genome. How can so few genes code for such a wide variety of antibodies? The answer eluded researchers for some time, but with the use of recombinant DNA, it became apparent that immunoglobulin genes are

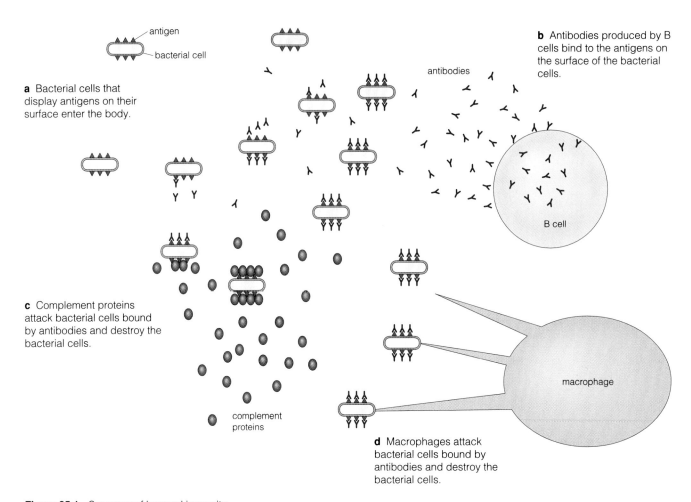

a Bacterial cells that display antigens on their surface enter the body.

b Antibodies produced by B cells bind to the antigens on the surface of the bacterial cells.

c Complement proteins attack bacterial cells bound by antibodies and destroy the bacterial cells.

d Macrophages attack bacterial cells bound by antibodies and destroy the bacterial cells.

antigen
bacterial cell
antibodies
B cell
complement proteins
macrophage

Figure 25.1 Summary of humoral immunity.

rearranged into millions of different combinations to code for the numerous antibodies needed. Researchers later discovered that rearranged genes encoded not only antibodies but also other proteins in the immune system. Only a few genes are rearranged to encode the millions of different immune system proteins.

In this chapter, we will first study the structure of antibodies and the genetic rearrangements that create their diversity. We will also look at the molecular diversity of other immune system molecules. Then we will consider scenarios of what can happen when things go wrong, including autoimmune disease, immune system deficiency, and infection by the human immunodeficiency virus (HIV, or the AIDS virus), which attacks part of the immune system and causes its ultimate failure. We conclude by discussing the use of monoclonal antibodies as a tool in genetics research.

25.1 IMMUNE RESPONSES

Several types of cells that act together to mount an immune response constitute the immune system. The immune response differs depending on the type of antigen

and its location in the body. For the purposes of this chapter, we are interested in the responses mediated by B cells and T cells, named after the tissues in which they mature. **B cells** mature in the bone marrow; **T cells** mature in the thymus. These cells participate in two basic types of immunity. **Humoral immunity** is the production of antibodies by B cells, the attachment of antibodies to antigens, and the destruction of particles attached to an antibody-antigen complex. **Cell-mediated immunity** is an immune response in which T cells attack body cells that have been infected by a pathogen and that express an antigen derived from the pathogen on their surface. In this section, we will discuss these two types of immunity.

Humoral Immunity

B cells produce antibodies through a process summarized in Figure 25.1. Each B cell contains specific arrangements of immunoglobulin genes, so each cell produces a specific antibody. B cells display their antibody molecules on their surface and secrete free antibodies into the blood.

When a foreign particle in the bloodstream (such as an infecting virus or bacterium) encounters an antibody that recognizes an antigen on the surface of the virus or

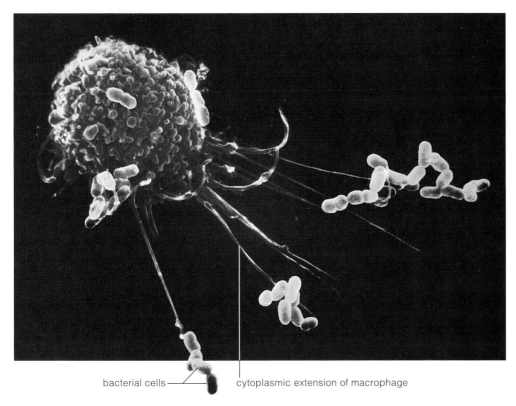

bacterial cells ——— cytoplasmic extension of macrophage

Figure 25.2 False-color scanning electron micrograph of a macrophage (red) attacking bacterial cells (green). (Lennart Nilsson © Boehringer Ingelheim International GmbH.)

bacterium, the antibody binds tightly to the antigen and marks both the antigen and the particle that carries it for destruction. **Macrophages**, white blood cells whose role is to engulf and digest foreign cells and particles, recognize the antibody-antigen complex and destroy it along with whatever is attached to it. Electron micrographs of this process often appear quite dramatic, as in Figure 25.2.

A second type of humoral immune response is the **complement system**, which consists of about 20 proteins that circulate in the blood. These proteins recognize the antibody-antigen complex and set off a chain of reactions that culminate in destruction of the virus or cell to which the antibody is attached. The reactions initiated by complement proteins also attract macrophages, and the macrophages engulf and destroy the antigen-bearing particle.

Cell-Mediated Immunity

T cells coordinate the cell-mediated immune response through a process summarized in Figure 25.3. One of the most important functions of T cells is to distinguish normal cells in the body from cells that have been infected by a foreign substance, usually a virus. They do this by recognizing two types of antigens in different ways. The first type of antigen consists of **histocompatibility antigens**, proteins produced by the body's cells to identify them as part of the body. The histocompatibility antigens are

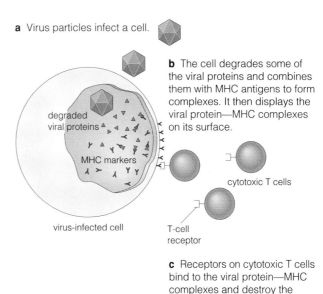

a Virus particles infect a cell.

b The cell degrades some of the viral proteins and combines them with MHC antigens to form complexes. It then displays the viral protein—MHC complexes on its surface.

degraded viral proteins

MHC markers

virus-infected cell

cytotoxic T cells

T-cell receptor

c Receptors on cytotoxic T cells bind to the viral protein—MHC complexes and destroy the infected cell.

Figure 25.3 Summary of cell-mediated immunity.

also called **MHC markers** (major histocompatibility markers) and are encoded by the *MHC* genes. These genes are highly variable among people, so the MHC markers displayed on one person's cells are different from those on the cells of other people. Even full siblings usually differ

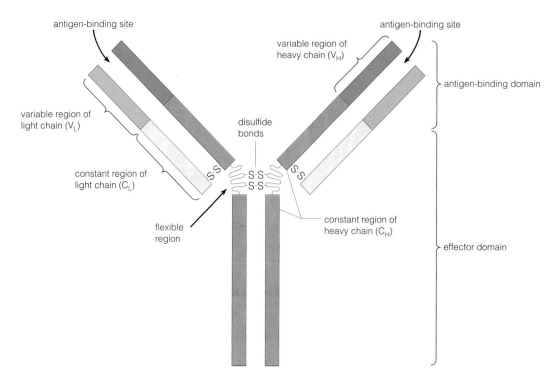

a Organization of light and heavy chains.

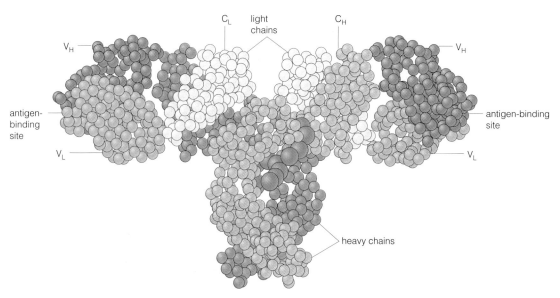

b Three-dimensional structure.

Figure 25.4 Antibody structure.

from one another in the MHC markers that their cells display. Only identical twins can be expected to have the same MHC markers on their cells. The second antigen recognized by T cells is a foreign antigen, like those recognized by antibodies. By recognizing these two antigens, T cells single out only those cells that belong to the body that are infected by foreign antigen-bearing particles.

When a virus infects a human cell, the cell degrades some of the viral proteins into fragments. These frag-

ments are antigens that combine with MHC markers and move to the surface of the cell, where they are displayed. The combination of viral protein fragments and MHC markers attracts **cytotoxic (killer) T cells**. These cells have T-cell receptors on their surface that recognize and bind to MHC-antigen complexes on the infected host cell. The cytotoxic T cells then attack and destroy the infected host cell along with the viruses in it, thus preventing the viruses from replicating further.

There are other classes of T cells and B cells. **Helper T cells** assist cytotoxic T cells with their attacks and assist B cells in antibody-antigen recognition. **Suppressor B cells** inhibit either T cells or B cells from responding to antigens.

~

B cells are part of the humoral immune response. They produce antibodies that bind to antigens. Other cells recognize the antibody-antigen complex and destroy it along with whatever is attached to it. T cells participate in cell-mediated immunity. They recognize histocompatibility antigens and foreign antigens on the surface of host cells from the body that have been infected by a pathogen. The infected cells are then destroyed, preventing further spread of the pathogen.

25.2 ANTIBODY STRUCTURE

Each antibody contains four polypeptide chains, two identical **heavy (H) chains** and two identical **light (L) chains**. Each light chain is about 220 amino acids long, and each heavy chain is about 440 amino acids long. The four polypeptides assemble into a Y-shaped structure, as illustrated in Figure 25.4. Four disulfide bonds connect the chains, all near the vertex of the Y structure. The two arms of the Y are connected to the stem by flexible regions that allow the arms to bend or rotate to facilitate antigen binding.

Each of the chains contains a constant (C) region and a variable (V) region. The **constant regions** are similar in all antibodies but may contain slight differences. The constant regions, particularly those of the heavy chains that form the stem of the Y, constitute the **effector domain**, a site that permits other agents of the immune system to recognize the antibody.

The **variable regions** differ substantially among antibodies and contain the **antigen-binding sites** (also known as **antigen-binding domains**). The variable regions give each antibody a unique structure that allows the antibody to bind only to a particular antigen (or, in some cases, to molecules that are very similar to the antigen). The antigen-binding site typically has a depression into which part of the antigen fits. The site is so small that only a small part of the antigen, called the **epitope**, is actually recognized. In protein antigens, the epitope generally consists of fewer than 16 amino acids. Each antibody has two identical antigen-binding sites, so it can bind two molecules of antigen, and can often cross-link the two antigen molecules to form an aggregate that a macrophage can recognize and destroy.

~

Antibodies are Y-shaped molecules that contain two identical light chains and two identical heavy chains. Each chain contains constant and variable regions. The variable regions vary among antibodies and allow them to recognize and bind different antigens.

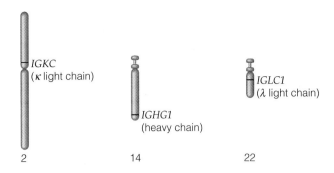

Figure 25.5 Chromosomal locations of genes that encode antibody chains in humans.

25.3 ANTIBODY GENES

The first indication that antibody genes could be rearranged came when researchers compared the DNA from an antibody gene isolated from cultured myeloma cells (cancerous B cells) with the corresponding DNA from cells that do not produce antibodies. The two genes had different arrangements of the antibody gene, indicating that the gene had been rearranged in the antibody-producing cells. Subsequent research has shown that programmed rearrangements are the rule for antibody genes.

Antibody genes are organized as large genes with many segments that must be rearranged into smaller genes before transcription. For each antibody gene, a B cell selects a few of the many gene segments that are available and links them together to form an intact antibody gene that encodes one subunit of the antibody. Differences in which segments are chosen and how they are combined account for the wide variety of antibodies in a single person.

Germ-line cells carry the entire array of antibody gene segments, with segments arranged in a constant order. The genes are transmitted in the same arrangement to the gametes and zygote. This initial arrangement remains intact during mitotic divisions and is passed unchanged in the germ line and the gametes that contribute genes to the next generation. Only in the B cells are antibody genes rearranged, so other cells in the body retain the original full array of antibody gene segments. Because antibody genes are not rearranged in germ-line cells, the rearrangements in B cells are not transmitted to offspring, so immunity to a particular pathogen must develop anew each generation.

Figure 25.5 shows the chromosomal locations of the three genes in humans that encode antibody chains. Two of the genes encode light chains, and one encodes the heavy chain. The light-chain gene on the short arm of chromosome 2 is called the *IGKC* gene (OMIM 147200), and it encodes the κ (kappa) light chain. The light-chain gene on the long arm of chromosome 22 is called the *IGLC1* gene (OMIM 147220), and it encodes the λ (lambda) light chain. The κ and λ light chains are very

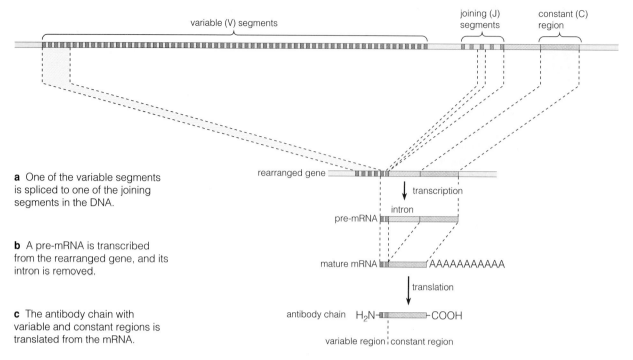

variable (V) segments

joining (J) segments

constant (C) region

a One of the variable segments is spliced to one of the joining segments in the DNA.

rearranged gene

transcription

intron

pre-mRNA

b A pre-mRNA is transcribed from the rearranged gene, and its intron is removed.

mature mRNA   AAAAAAAAAAA

translation

c The antibody chain with variable and constant regions is translated from the mRNA.

antibody chain   H₂N—   —COOH

variable region   constant region

Figure 25.6   Rearrangement of a gene that encodes an antibody chain.

similar but have slightly different constant regions. In humans, about 60% of antibodies have κ light chains, and about 40% have λ light chains. The heavy-chain gene is on the long arm of chromosome 14 and is called the *IGHG1* gene (OMIM 147100).

A generalized pattern of rearrangement in antibody genes is depicted in Figure 25.6. Each gene includes variable and constant regions that correspond to the variable and constant regions in the antibody chains. Segments of DNA in the variable and joining regions are selectively removed from the DNA and the remaining segments spliced to form the functional gene. After the gene has been rearranged, a pre-mRNA is transcribed and its intron removed to form the mature mRNA, which is then translated to produce the antibody chain. All mitotic progeny of a B cell that has a particular arrangement retain the same antibody gene arrangements and produce the same antibody.

Now let's take a detailed look at each of the three human antibody genes to see how rearrangements produce functional antibody genes.

The *IGKC* (κ Light-Chain) Gene

The *IGKC* gene and the steps for its rearrangement and expression are illustrated in Figure 25.7. The gene consists of three regions: a variable (V_κ) region, a joining (J_κ) region, and a constant (C_κ) region. The V_κ region is a cluster of about 300 different but related V_κ segments, each preceded by its own promoter and by a leader (L_κ) segment. Each L_κ-V_κ segment includes about 400 nucleotide

pairs and is separated from the next L_κ-V_κ segment by about 7000 nucleotide pairs. The entire variable region is over 2 million nucleotide pairs long. The V_κ region is followed by a spacer region of about 25,000 nucleotide pairs. This is followed by the J_κ region, which contains five J_κ segments, each 30–40 nucleotide pairs long. The J_κ segments are separated from each other by about 300–350 nucleotide-pair spacers. Following the J_κ region is another spacer, about 2500–4000 nucleotide pairs long, that encodes an intron in the pre-mRNA. Following the intron is a single C_κ segment, which encodes the constant region of the κ chain.

During rearrangement, a single L_κ-V_κ segment is selected and fused to a J_κ segment, and all intervening DNA is deleted. All V_κ segments to the left of the selected segment are part of the final gene, but only the selected V_κ segment is transcribed. The selected L_κ-V_κ-J_κ segment is joined to the intron that separates the J_κ region from the C_κ region. The final gene has the order L_κ-V_κ-J_κ-intron-C_κ.

The selected V_κ segment may have many other V_κ segments attached to it on the left, so there may be many promoters where transcription could conceivably begin. However, only the selected V_κ segment's promoter is active, so the transcript contains only one V_κ segment (Figure 25.7). The intron between the J_κ segment and the C_κ segment contains an enhancer sequence that is specific for V_κ promoters. In the rearranged gene, the enhancer stimulates transcription at the nearest promoter, which is the promoter in the selected V_κ segment. This stimulates transcription of the gene to form a pre-mRNA.

After transcription is complete, the enhancer in the

a In this example, the right end of the $V_{\kappa 2}$ segment is attached to the left end of the $J_{\kappa 3}$ segment, and the right end of the $J_{\kappa 3}$ segment is attached to the intron that contains the enhancer, to form the rearranged gene.

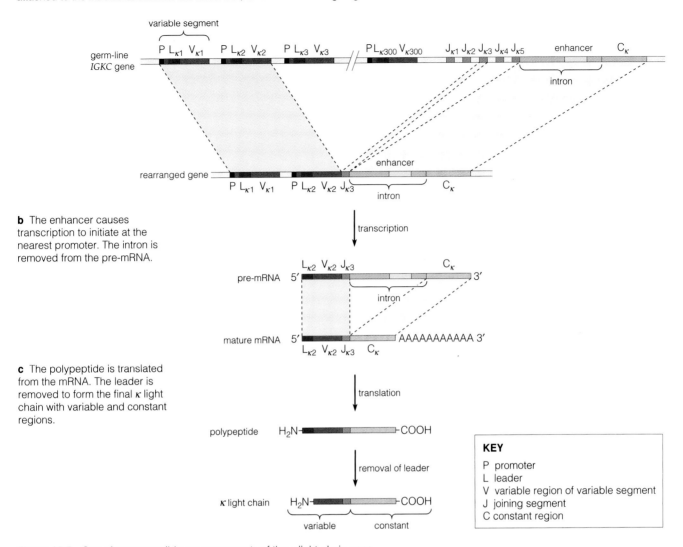

b The enhancer causes transcription to initiate at the nearest promoter. The intron is removed from the pre-mRNA.

c The polypeptide is translated from the mRNA. The leader is removed to form the final κ light chain with variable and constant regions.

KEY
P promoter
L leader
V variable region of variable segment
J joining segment
C constant region

Figure 25.7 One of many possible rearrangements of the κ light-chain gene.

pre-mRNA is no longer needed, so the intron is removed from the pre-mRNA to make a mature mRNA. The mRNA is translated to produce a polypeptide with a κ light chain that has a leader sequence 17–20 amino acids long, a variable sequence 108 amino acids long (95 amino acids encoded by the V_{κ} segment and 13 amino acids encoded by the J_{κ} segment), and a constant sequence of 112 amino acids.

After the entire antibody has formed, the leader sequence assists the antibody in moving through the plasma membrane. The leader sequence is then cleaved from the antibody.

The *IGLC1* (λ Light-Chain) Gene

The *IGLC1* gene is similar to the *IGKC* gene, with a few minor differences. The V_{λ} region is similar in size to the V_{κ} region in the *IGKC* gene. The *IGLC1* gene has more C_{λ} seg-

ments than the *IGKC* gene, although each rearranged *IGLC1* gene has only one C_{λ} segment. After rearrangement, the *IGLC1* gene has an arrangement similar to the rearranged *IGKC* gene: L_{λ}-V_{λ}-J_{λ}-intron-C_{λ}. The λ chains are similar to κ chains in size and in amino acid sequence.

The *IGHG1* (Heavy-Chain) Gene

In humans, a single gene, called the *IGHG1* gene, encodes the antibody heavy chains. The gene and its initial rearrangement are depicted in Figure 25.8. There is much greater diversity in heavy chains than in light chains, due in part to a greater number of combinations of variable and constant segments and the presence of an additional class of segments called diversity (D_H) segments. The gene is divided into four regions: the V_H region, the D_H region, the J_H region, and the C_H region. The V_H, J_H, and C_H segments contained in these regions perform the same

a In an immature B cell, the *IGHG1* gene is rearranged to join V_H, D_H, and J_H segments to the C_H region.

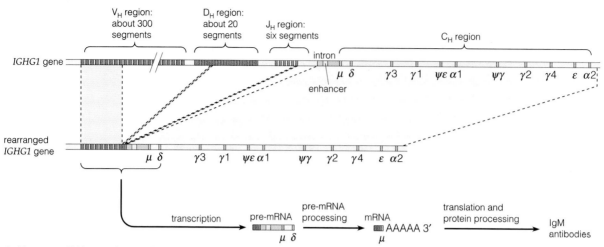

b The pre-mRNA contains one V_H segment, one D_H segment, one J_H segment, an intron with an enhancer, and the $C_{H\mu}$ and $C_{H\delta}$ segments. The pre-mRNA is processed to remove the intron and the $C_{H\delta}$ segment. The mRNA encodes only IgM antibodies, which are displayed on the surface of immature B cells.

Figure 25.8 Initial rearrangements of the *IGHG1* gene in immature B cells to produce IgM heavy chains. Those segments in the C_H region designated with a ψ are pseudogene segments that are not translated.

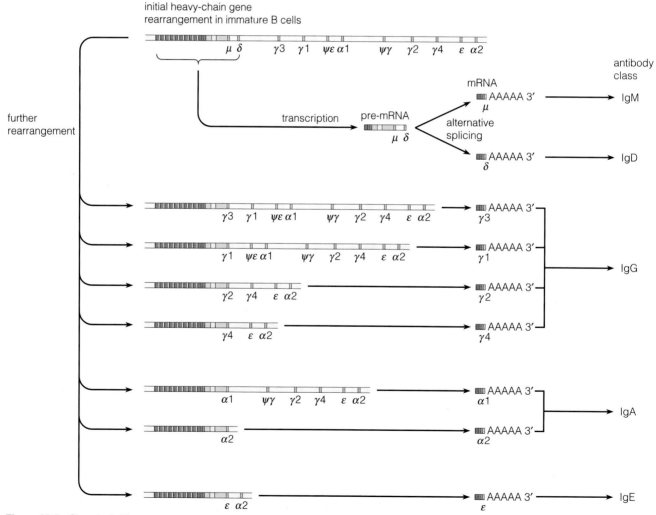

Figure 25.9 Class switching.

Table 25.1 Classes of Antibodies as Determined by the Heavy Chain They Carry, Their Tissue Localization, and the Gene Rearrangements That Encode the C_H Region of the Heavy Chain

Class	Function/Description	C_H Region of the Heavy-Chain Gene
IgM	Binds antigens on the surface of immature B cells and mature B lymphocytes	$C_{H\mu}$ segment
IgD	Binds antigens on the surface of B lymphocytes	$C_{H\delta}$ segment
IgG	Free antibodies in blood that bind antigens in blood during an infection	$C_{H\gamma1}$, $C_{H\gamma2}$, $C_{H\gamma3}$, and $C_{H\gamma4}$ segments
IgA	Antibodies present in tears, saliva, breast milk, mucus, and intestinal secretions	$C_{H\alpha1}$ and $C_{H\alpha2}$ segments
IgE	Stimulates histamine production	$C_{H\epsilon}$ segment

roles as their corresponding segments in the light-chain genes. The D_H segments provide additional diversity to the heavy-chain variable region in the antibody.

In humans, the V_H region consists of about 300 V_H segments, each preceded by a promoter and an L_H segment. A spacer separates the V_H region from the D_H region, which contains about 20 D_H segments, each an average of 13 nucleotide pairs in length. Following the D_H region is a spacer, followed by six J_H segments. These are followed by an intron that contains an enhancer, just as in the light-chain genes. Following the intron is a C_H region with several C_H segments. In humans, there are either 9 or 10 functional C_H segments (depending on the alleles a person inherits) and two nonfunctional pseudogene C_H segments.

Two deletion events rearrange segments from the V_H, D_H, and J_H regions. The intervening DNA between the selected V_H and D_H segments is deleted, and the intervening DNA between the selected D_H and J_H segments is deleted. This forms a gene that consists of an L_H-V_H-D_H-J_H segment in the **immature (or virgin) B cell**, a cell whose surface displays antibodies that have not contacted their antigen. In the immature B cells, the variable portion of the gene remains attached to the entire C_H region, which contains all the C_H segments. The heavy chains produced in these early cells, however, contain only the constant region encoded by the $C_{H\mu}$ segment, which is the first C_H segment in the gene.

The *IGHG1* gene arrangement in immature B cells encodes a heavy chain called an IgM chain. Pre-mRNAs encoded by the initial rearrangement of the heavy-chain gene have both $C_{H\mu}$ and $C_{H\delta}$ segments present in the transcript. The pre-mRNAs are processed to produce mature mRNAs with only the $C_{H\mu}$ segment. They encode IgM antibodies, which are the only class of antibodies displayed on the surface of immature B cells.

When an immature B cell contacts its antigen, it matures into a B lymphocyte. Mature B lymphocytes utilize alternative splicing of pre-mRNAs and further rearrangements of the *IGHG1* gene to produce several different classes of antibodies, a process called **class switch-ing**. As illustrated in Figure 25.9, the classes differ from one another only in the composition of the constant region of the heavy chain. The first class switch does not require further rearrangement of the *IGHG1* gene. Instead, the pre-mRNA is alternatively spliced to encode heavy chains with either the $C_{H\mu}$ or the $C_{H\delta}$ segments. The other class switches require further rearrangements in the C_H region that place different C_H segments after the L_H-V_H-D_H-J_H region of the final gene. Class switching produces five classes of heavy chains: **IgM** chains, encoded by the $C_{H\mu}$ segment; **IgD** chains, encoded by the $C_{H\delta}$ segment; **IgG** chains, encoded by the $C_{H\gamma1}$, $C_{H\gamma2}$, $C_{H\gamma3}$, and $C_{H\gamma4}$ segments; **IgA** chains, encoded by the $C_{H\alpha1}$ and $C_{H\alpha2}$ segments; and **IgE** chains, encoded by the $C_{H\epsilon}$ segment.

The different classes of heavy chains encoded in the progeny cells of a single immature B cell bind to the same antigen because their antigen-binding sites are identical. However, the function of each class of antibodies is different, and the different classes of antibodies are located in different parts of the body. IgM antibodies are found on the surface of immature B cells. They are also found on the surface of mature B lymphocytes, where they serve as a signal for macrophages and the complement system when bound to their antigen. IgA antibodies are present in tears, saliva, breast milk, mucus, and intestinal secretions. IgD antibodies remain on the surface of B lymphocytes, where they act as receptors. IgE antibodies stimulate histamine production during an allergic reaction. IgG antibodies are the principal class of antibodies in blood. They circulate as free antibodies that signal macrophages and the complement system when bound to an antigen. The specializations of the different classes of antibodies and the gene rearrangements that produce them are summarized in Table 25.1.

The variable region of antibodies contains sections that are called **hypervariable regions**. The amino acid sequences in hypervariable regions are more varied from one antibody to the next than other portions of the variable segments. The hypervariable regions are always found in the same positions. There are three hypervariable regions in the variable region of each light chain and

heptamer   12 nucleotide pairs   nonamer

```
CACAGTGNNNNNNNNNNNNNACAAAAACC
GTGTCACNNNNNNNNNNNNNTGTTTTTGG
```

or

heptamer   23 nucleotide pairs   nonamer

```
CACAGTGNNNNNNNNNNNNNNNNNNNNNNNNACAAAAACC
GTGTCACNNNNNNNNNNNNNNNNNNNNNNNNTGTTTTTGG
```

Figure 25.10 Consensus sequences that border segments in genes that encode antibody proteins. (*N* = any one of the four nucleotides.)

three in the variable region of each heavy chain. One hypervariable region of the heavy chain is encoded by the D segment. Other hypervariable regions are found at the splice junctions between the J and V segments in the light chains. The hypervariable regions contain most of the amino acids that bind directly to the antigen in the antigen-binding site. The other amino acids in the variable region form the structure of the antigen-binding site.

Segment Joining During Rearrangement

Each V and J segment in an antibody gene has a sequence of nucleotides called a **heptamer-nonamer sequence** on one end. Each heptamer-nonamer sequence has a conserved heptamer separated by a spacer sequence of either 12 or 23 nucleotide pairs, as shown in Figure 25.10. The heptamer and nonamer sequences are highly conserved, but the spacer sequences may be any sequence of 12 or 23 nucleotide pairs.

In the two light-chain genes (*IGKC* and *IGLC1*), each V segment is followed by a heptamer-nonamer sequence with a 12 nucleotide-pair spacer. Each J segment is preceded by a heptamer-nonamer sequence with a 23 nucleotide-pair spacer. The heptamer-nonamer sequences that precede the J segments are in reverse orientation to the heptamer-nonamer sequences that follow the V segments, as diagrammed in Figure 25.11.

Enzymes called **RAG1** and **RAG2** catalyze rearrangement of antibody genes. They are encoded by two genes, *RAG1* (OMIM 179615) and *RAG2* (OMIM 179616), that are located near one another on chromosome 11. As illustrated in Figure 25.12, the enzymes join the selected V segment to the selected J segment at the heptamers of the heptamer-nonamer sequences that are attached to the V and J segments. RAG1 and RAG2 can only catalyze recombination between a heptamer-nonamer sequence with a 12 nucleotide-pair spacer and a heptamer-nonamer sequence with a 23 nucleotide-pair spacer. Thus, the arrangement of heptamer-nonamer sequences shown in Figure 25.11 ensures that a V segment can join only to a J segment in the light-chain genes.

In the *IGHG1* gene, each VH segment is followed by a heptamer-nonamer sequence with a 23 nucleotide-pair spacer, each D_H segment is bordered on both sides by

heptamer-nonamer sequences with 12 nucleotide-pair spacers, and each J_H segment is preceded by a heptamer-nonamer sequence with a 23 nucleotide-pair spacer. Thus, RAG1 and RAG2 can join the segments in only one arrangement: V_H-D_H-J_H.

Productive and Nonproductive Arrangements

Recombination between heptamers during joining is not always precise. Often a few nucleotide pairs are lost during the recombination event. In fact, imprecise recombination accounts for some of the hypervariability found in the hypervariable regions of rearranged antibody genes. Although it may produce high variability, imprecise recombination also has its cost. When nucleotide pairs are lost during joining, the rearranged genes do not necessarily retain the correct reading frame. Deletions of any number of nucleotides other than multiples of three create frameshift mutations that alter the amino acid sequence downstream from the point of joining, including the constant portion of the antibody chain. Such frameshift mutations produce **nonproductive arrangements** because the altered polypeptide cannot function as part of an antibody. Only those deletions that retain the correct reading frame produce **productive arrangements**.

This seemingly wasteful process is actually beneficial. Each antibody gene is present in two copies in each cell, one on each of the two homologous chromosomes that bear the gene. However, only one of the two homologous genes is actively transcribed. When the two alleles for an antibody gene are compared in a cultured cell line derived from a single B cell, one is typically in a productive arrangement, and the other is in either a nonproductive arrangement or the original germ-line arrangement. Only the single productive arrangement is actively transcribed.

After a productive arrangement has formed, the *RAG* genes no longer produce their enzymes, and there are no further rearrangements. There are three possible fates for a B cell during rearrangement: (1) If the first gene to rearrange forms a productive arrangement, then it becomes the active gene, and the other gene is not rearranged. (2) If the first rearrangement is nonproductive, then the RAG enzymes rearrange the second copy of the gene. If this second rearrangement is productive, then it is transcribed, and the nonproductive arrangement is not. (3) If both copies have nonproductive arrangements, the enzymes continue to rearrange the genes until one gene assumes a productive arrangement.

Matching Specific Antigens

With so many antigens around, how can an antibody gene be rearranged to match a specific antigen? The answer is a simple one. Gene rearrangement is random and does not respond to the presence of antigens. Each of the millions of B cells produces its own unique arrangement, resulting in many different antibodies, all at very low con-

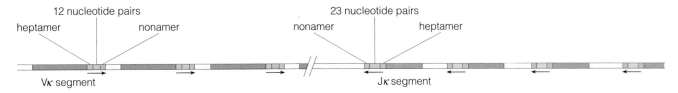

Figure 25.11 Arrangement of heptamer-nonamer sequences in relation to V_κ and J_κ segments in the *IGKC* gene.

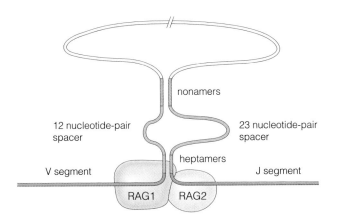

a Heptamers align at their corresponding sequences.

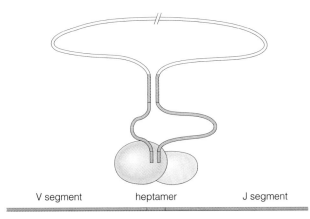

b The RAG1 and RAG2 enzymes delete all the DNA between the selected V and J segments and join the selected V and J segments.

Figure 25.12 V-J joining.

centrations. The alternative combinations of light and heavy chains, V, D, and J regions, and hypervariability in joining arrangements provide millions of possible productive arrangements. When an antigen enters the body, it circulates until it encounters an immature B cell that happens to have an antibody that can bind to it. Antigen binding stimulates the B cell to mature and divide to produce B lymphocytes that encode the same antibody. The dividing B lymphocytes produce many copies of their antibody to bind the antigen while the other B cells remain quiescent. In other words, the body protects itself by having large numbers of antibodies arranged at random, instead of selectively designing antibodies to match each antigen. The large number of different antibodies ensures

that at least one antibody can be found that is able to bind to each foreign antigen.

When a pathogen, such as a virus or bacterium, enters the body, a B cell with an antibody that can bind to an antigen on the surface of the pathogen divides several times to produce many identical B lymphocytes, a process called the **primary immune response**. The primary immune response produces an excess of B cells, and the excess B cells do not bind to the antigen. Instead, they remain in the bloodstream as **memory cells**, cells with the antibody still present on their surface. If the same antigen enters the body again, the memory cells mount a rapid secondary immune response, which quickly attacks the antigen-bearing pathogen and prevents it from reproducing and infecting the body. This is the basis of acquired immunity to diseases.

~

Three genes encode the polypeptides in antibodies, two that encode the light chains, called κ (kappa) and λ (lambda), and one that encodes the heavy chains. B cells rearrange the antibody genes in different ways to encode a wide variety of antibodies.

25.4 CELL-MEDIATED IMMUNITY GENES

At the beginning of this chapter, we discussed two types of immunity: (1) humoral immunity, the production of B cells and antibodies, and (2) cell-mediated immunity, the production of T cells, which seek out infected host cells and destroy them. So far, we've discussed only the genes of humoral immunity. In this section, we'll look at the genes that govern cell-mediated immunity.

On the surface of cytotoxic T cells are receptor proteins that recognize and bind to antigens that are displayed with MHC markers on the surface of a cell that has been infected by a pathogen (Figure 25.13). The genes that encode the receptor proteins are very similar to antibody genes. The genes include V, D, J, and C segments and utilize the same heptamer-nonamer sequences as antibody genes for rearrangement. Gene rearrangements produce a high degree of variation in receptor proteins, allowing T cells to recognize many different antigens on the surface of infected cells. There are four T-cell receptor genes in humans. The *TCRA* (OMIM 186880) and *TCRD* (OMIM 186810) genes overlap one another on

chromosome 14. The *TCRB* (OMIM 186930) and *TCRG* (OMIM 186970) genes are located at separate sites on chromosome 7.

The *TCRA* and *TCRD* genes form a large, complex locus (over a million nucleotide pairs) that rearranges to encode two different but related polypeptides, called α and δ polypeptides (Figure 25.14). Notice in Figure 25.14 that the *TCRD* gene is intact, but it divides the *TCRA* gene into two segments. During rearrangement, the δ segments join only with other δ segments to encode the δ polypeptide, and the α segments join only with other α segments to encode the α polypeptide. The *TCRD* gene is the first to rearrange. During early development of a T cell, the *TCRD* gene rearranges and produces a δ polypeptide. Its rearrangement leaves the *TCRA* gene unaffected. As the T cell matures, it rearranges the *TCRA* gene. The entire *TCRD* gene is between the V and J seg-

ments of the *TCRA* gene, so rearrangement of the *TCRA* gene deletes all of the *TCRD* gene. From that point on, the cell encodes an α polypeptide and no longer encodes a δ polypeptide.

The *TCRB* and *TCRG* genes encode polypeptides that are similar to antibody light chains. The structural organization of these genes is depicted in Figure 25.15. The *TCRG* gene is expressed only during early T-cell development. Expression shifts to the *TCRB* gene in mature T cells. Thus, mature T cells express the *TCRA* and *TCRB* genes, which encode the α and β polypeptides. As shown in Figure 25.13, T-cell receptors on mature T cells contain one α polypeptide and one β polypeptide.

Cytotoxic T cells recognize pathogen-infected cells that display antigens from the pathogen together with MHC markers on their surface. Like antibodies and T-cell receptors, MHC markers are quite diverse, so diverse that siblings usually have different MHC marker patterns. Unlike antibodies and T-cell receptors, MHC gene diversity is not provided by rearrangement. A large gene family encodes the MHC proteins. The genes are clustered at the **HLA locus** on chromosome 6 in humans (OMIM 142800, 142830, 142840, 142855, 142857, 142858, 142860, 142871, 142880, 142925, 142930, 143010, 143110, 146880, and 600629).

There are three classes of MHC genes. The **class I genes** encode **self-recognition histocompatibility antigens**, proteins that are present on the surface of most cells in the body. These genes vary substantially from one individual to another, allowing for unique combinations of proteins so that each individual has a unique set. The **class II genes** encode proteins found on the surface of B cells, T cells, and macrophages. These proteins participate in intercellular communication during immune responses. The **class III genes** encode many different proteins, including those in the complement system.

Cytotoxic T cells recognize and attack only those cells that have the host pattern of class I MHC markers and a foreign antigen on the surface. When a virus enters a host cell, proteases within the cell break down some of the viral proteins into polypeptides. The cell transports these foreign polypeptides to the cell surface, where they form a complex with the MHC markers. The T-cell receptors recognize only an antigen-MHC complex and bind to it, initiating the eventual lysis and destruction of the cell.

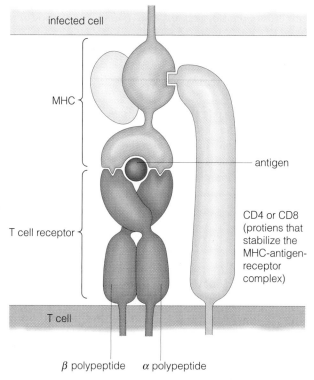

Figure 25.13 Interaction of an infected cell that displays a foreign antigen and MHC markers on its surface with a T-cell receptor.

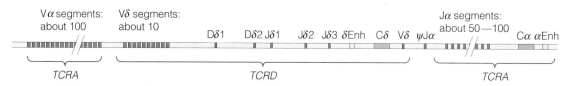

Figure 25.14 The human *TCRA* and *TCRD* genes on chromosome 14. The segment designated by ψ is a pseudogene segment.

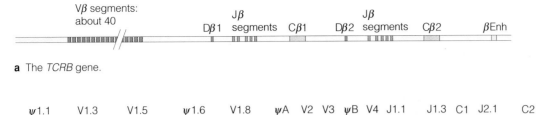

a The *TCRB* gene.

b The *TCRG* gene.

Figure 25.15 The human *TCRB* and *TCRG* genes on chromosome 7. The segments designated by ψ are pseudogene segments.

Cytotoxic T cells may also recognize tumor cells as foreign and bind to them to destroy the tumor. As highlighted in the following example, a tumor must overcome this immune response in order to develop into a malignant cancer.

Example 25.1 **Inhibition of transcription of MHC class I genes by the *JUN* oncogene product.**

MHC class I genes have an enhancer element upstream from the promoter that binds to the product of the proto-oncogene *JUN* (see section 24.2 for a description of proto-oncogenes). Howcroft et al. (1993. *EMBO Journal* 12:3163–3169) found that the *JUN* product reduced transcription of MHC class I genes when it was bound to the enhancer.

Problem: How might this property of the *JUN* gene product affect cancer development when the *JUN* proto-oncogene mutates to a form that is overexpressed?

Solution: The immune system may recognize and destroy tumor cells through the cell-mediated immune response. In order for a cancer to develop, cell-mediated immune recognition and destruction of the tumor cells must be suppressed. Because MHC markers are needed for cytotoxic T cells to recognize and destroy a target cell, reduction or elimination of these proteins may allow the cancerous cells to remain unaffected by cell-mediated immunity. Apparently, overexpression of a *JUN* oncogene may eliminate production of MHC proteins by tumor cells.

The cell-mediated immune response is dependent on genes that produce receptors on T cells that recognize foreign antigens and histocompatibility markers. Histocompatibility markers are present on the surface of most cells in the body and are encoded by highly variable MHC genes.

25.5 IMMUNE SYSTEM MALFUNCTION

Usually, the immune system works properly, attacking only particles that bear foreign antigens and cells that display foreign antigens complexed with MHC markers. However, immune system malfunction is the cause of many different diseases. For example, the immune system may overrespond to an antigen, producing an **allergic reaction**, such as hay fever, in which mast cells, stimulated by IgE-antigen complexes, overproduce a class of chemicals called histamines. Antihistamines are medicines that block histamine production and relieve the allergy symptoms.

In this section, we'll examine some serious immune system disorders, including autoimmune diseases, immune deficiency, and HIV infection.

Autoimmune Diseases

The immune system's ability to distinguish between substances that belong to the body and substances that are foreign is quite remarkable. However, occasionally the immune system incorrectly recognizes a substance that belongs to the body as a foreign antigen and attacks the cells that display the antigen. The attack can damage tissues or organs and produce one of many diseases called **autoimmune diseases**. Hyperthyroidism, rheumatoid arthritis, insulin-dependent juvenile onset diabetes, autism, multiple sclerosis, acquired hemolytic anemia, and pernicious anemia are examples of autoimmune diseases.

Among the best-understood autoimmune diseases is **insulin-dependent juvenile onset diabetes**, a severe form of diabetes that affects children, making them dependent on insulin treatments for life. The insulin-producing cells in the pancreas are called β cells. In certain individuals, the immune system recognizes a protein on the surface of β cells as a foreign antigen and destroys the β cells. After several years, most of the β cells have been destroyed, and the pancreas can no longer produce insulin. The disease is associated with specific alleles in the class II MHC proteins, so susceptibility is inherited.

Rheumatoid arthritis is another autoimmune disease that affects many people. The joints in fingers and toes become inflamed and swell, causing pain and discomfort. The disease may spread to other joints and, in severe cases, may affect every joint in the body. Rheumatoid arthritis can be distinguished from other types of arthritis by the presence of a protein called rheumatoid factor in the blood. This protein is an antibody that recognizes the body's own IgG antibodies as antigens and binds to them to form a rheumatoid factor–IgG complex. The complex is deposited on membranes in the joints, where it stimulates the complement system and macrophages to attack the membrane. The joint membranes become inflamed and often separate from the tissues that support them. Fluid accumulates in the intervening space, membrane cells divide to compensate, and the joint thickens and becomes stiff. Eventually, movement in the joint may be completely lost. In the most severe cases, patients become crippled and bedridden for life. Although the disease itself is not inherited, susceptibility to rheumatoid arthritis can be inherited. For unknown reasons, rheumatoid arthritis, as well as several other autoimmune diseases, tends to affect women more frequently and more severely than men.

Hyperthyroidism or **Graves disease** is among the most common autoimmune diseases. The thyroid gland regulates the overall metabolic rate of the body. Patients with hyperthyroidism have increased metabolic rates that cause nervousness, shakiness, heat intolerance, excessive perspiration, increased appetite, weight loss, and periodic loss of muscle function in some cases. Thyroid cells normally produce thyroid hormone when thyroid-stimulating hormone (TSH) produced in the pituitary gland binds to TSH receptors on thyroid cells. In autoimmune hyperthyroidism, an antibody called long-acting thyroid stimulator (LATS) recognizes the TSH receptors on the membrane of thyroid cells as an antigen. When the antibody binds to the TSH receptors, it stimulates them in the same way that TSH does, causing increased thyroid activity. Persistent hyperthyroidism is usually treated by surgically removing thyroid tissue or destroying thyroid tissue by treatment with radioactive iodine, which is concentrated in the thyroid gland, where the ionizing radiation kills some of the thyroid cells.

The biochemical mechanisms of most autoimmune diseases remain unknown, although certain factors are associated with some autoimmune diseases. Often, the onset of an autoimmune disease follows a viral infection. In its response to the virus, the body may produce an antibody that binds to an antigen on the pathogen but also happens to bind to one of the body's own proteins. Certain associations between viral infection and autoimmune response are known. For example, some antibodies that attack the hepatitis B and the Epstein-Barr viruses also attack the myelin basic protein and cause multiple sclerosis.

Immune Deficiencies

There are several inherited **immune deficiencies**, diseases in which part or all of the immune system fails to function properly. An example is **X-linked agammaglobulinemia**, which is caused by a recessive allele at the *BTK* locus (OMIM 300300) on the long arm of the X chromosome. Because the locus is on the X chromosome, the disease is more frequent in males; it usually affects infant boys at about 5–6 months of age. The *BTK* locus encodes a protein kinase that is part of the cellular signal transmission pathway that coordinates B-cell maturation. Affected people have immature B cells but no mature B cells and fail to produce antibodies in response to infections. The cell-mediated immune system, however, is functional. People with this disorder have high susceptibility to bacterial and viral infections, particularly infections that cause pneumonia.

Severe combined immunodeficiency syndrome (SCID) is characterized by failure of both the humoral and cell-mediated immune systems, resulting in no immunity. It typically allows an infectious disease, such as pneumonia or measles, to spread unabated until the person dies. The disorder usually affects lymphoblasts, the precursor cells of both B cells and T cells in the bone marrow. Patients have no T cells or B cells, although some have B cells that produce no antibodies. Bone marrow transplants from close relatives introduce functional lymphoblasts into the patient and are sometimes successful in treating the disease. This disease is famous because children who have it are kept in sterile plastic bubbles in hospitals, where they are isolated from infectious pathogens. One boy with SCID was isolated in such a bubble for the first 12 years of his life, until he died from complications associated with a bone marrow transplant (Figure 25.16).

Infectious Defeat or Destruction of Immunity

Several bacteria and viruses are dependent on infection of human or animal hosts for their reproduction and survival. The immune system is a major obstacle to their growth and reproduction, and some have developed strategies to overcome host immunity. Certain bacterial diseases, such as meningitis, rheumatic fever, African sleeping sickness, and gonorrhea, are caused by bacteria that alter their surface proteins to avoid attack by the immune system. After the immune system mounts an antibody response against a surface protein antigen, the bacterium may switch to another protein to avoid the immune response against the first protein. Some viruses alter their coat proteins by frequent mutation, so new primary immune responses are required to provide immunity.

The **human immunodeficiency virus (HIV)**, which is responsible for **acquired immune deficiency syndrome**

Figure 25.16 A boy who spent the first 12 years of his life in a protective enclosure because of SCID. (Photo courtesy of Baylor College of Medicine/Peter Arnold, Inc.)

(AIDS), is a virus that frequently mutates. It also has the unusual ability to attack the immune system directly, eventually causing immune system failure. Secondary infections cause most deaths associated with HIV because the weakened immune system cannot adequately fight the infection. HIV is a retrovirus with the same basic structure as most retroviruses, although its genes are more complicated and it encodes more protein products than most retroviruses. Its ability to attack the immune system comes from its ability to infect T4 cells, a helper T cell that is a positive regulator of the cell-mediated immune response. Normally, people have 800–1000 T4 cells per cubic millimeter of blood. A drop in the T4 count is directly related to the progress of HIV infection. The Federal Centers for Disease Control and Prevention define AIDS as a T-cell count of 200 or less.

When T4 cells are depleted, immune responses decline, and secondary infections can establish themselves. Among the most common secondary infections in AIDS cases is *Pneumocystis carinii*, a protozoan that causes a rare form of pneumonia called pneumocystic pneumonia. The immune system usually eliminates this protozoan before it can cause pneumonia. When AIDS first appeared, an increase in the incidence of pneumocystic pneumonia was one of the first indications that an acquired immune disorder was spreading.

HIV recognizes a protein called CD4 as a cell surface receptor. HIV preferentially infects cells that display CD4 on their surface. Helper T4 cells display the CD4 protein on their surface, but cytotoxic T cells do not, so the helper T4 cells are preferentially infected. The number of T4 cells declines, which allows secondary infections to proceed.

The following example shows one of the ways that HIV defeats the immune system.

Example 25.2   Inhibition of MHC gene expression by HIV.

Part of the cell-mediated immune response is the identification and selective destruction of cells that are infected by a virus. The human immunodeficiency virus type 1 (HIV-1) produces a protein called HIV-1 Tat. According to Howcroft et al. (1993. *Science* 260:1320–1322), HIV-1 Tat inhibits expression of MHC class I genes.

Problem: How might this situation favor progression of the viral infection?

Solution: The cell-mediated immune response recognizes virus-infected cells by the presence of complexes of MHC antigens and viral antigens on the surface of the infected cell. If a cell infected by HIV fails to produce enough MHC antigens, the cell-mediated immune system fails to attack the cell, and the viruses within it can multiply.

Autoimmune disorders appear when the immune system mounts an immune response against the body's own substances. Infectious immune deficiency appears when a pathogen attacks part of the immune system.

25.6   ANTIBODIES AS TOOLS IN GENETICS RESEARCH

All the antibodies produced by a B cell and its mitotic progeny have identical antigen-binding sites, so they are specific for a single antigen. If researchers isolate a single type of antibody and purify it, they can use it to identify and track an antigen. Ideally, an antibody-producing B cell could be isolated and cultured so that its antibody could be produced in unlimited amounts. Unfortunately, B cells do not grow in culture, but cancerous B cells, called **myeloma cells**, do. Isolated normal B cells can be fused with cultured myeloma cells that no longer produce antibodies. The fusion products are called **hybridomas**. Hybridomas grow and divide indefinitely while expressing the antibody from the B cell's rearranged antibody genes. Researchers can isolate a particular antibody by injecting a mouse with the desired antigen, isolating B cells from the spleen, and fusing these cells with cultured antibody-deficient mouse myeloma cells to form hybridomas (Figure 25.17). Each individual hybridoma cell produces mitotic progeny cells that all produce the same single antibody that was raised against the injected antigen. These antibodies are called **monoclonal antibodies**, because they are single antibodies produced in cloned cells.

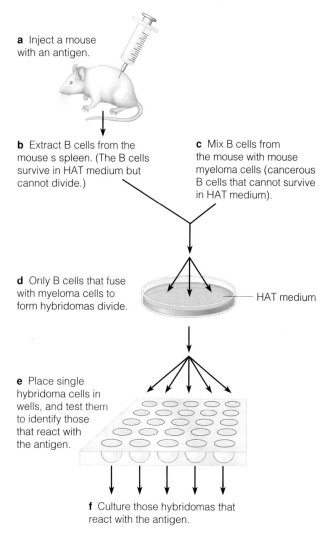

a Inject a mouse with an antigen.

b Extract B cells from the mouse s spleen. (The B cells survive in HAT medium but cannot divide.)

c Mix B cells from the mouse with mouse myeloma cells (cancerous B cells that cannot survive in HAT medium).

d Only B cells that fuse with myeloma cells to form hybridomas divide.

HAT medium

e Place single hybridoma cells in wells, and test them to identify those that react with the antigen.

f Culture those hybridomas that react with the antigen.

Figure 25.17 The method used to produce monoclonal antibodies. HAT medium (HAT = hypoxyxanthine + aminopterin + thymidine) is a medium that blocks synthesis of purines and pyrimidines in myeloma cells, but permits synthesis of purines and pyrimidines in B cells and hybridomas.

Monoclonal antibodies have many applications in clinical medicine. Home pregnancy tests, for example, rely on monoclonal antibodies that react with proteins present in the urine of a woman who is pregnant. The antibody-antigen complex can be detected either as a precipitate or, more commonly, as a reaction with other chemicals that causes a color to develop. Many other proteins, such as rheumatoid factor, can be detected clinically using a monoclonal antibody that reacts with the protein. Also, monoclonal antibodies that react with antigens on tumor cells are used to identify internal tumors in patients for directed radiation treatment.

Monoclonal antibodies have been particularly useful as a tool in recombinant DNA research. They can be used to detect specific proteins in much the same way as a DNA probe is used to detect DNA fragments, as described in section 9.8. Monoclonal antibodies also show

promise in the diagnosis and treatment of cancer, as discussed in section 24.5.

Using recombinant DNA methodology, it is possible to synthesize **bispecific antibodies**, antibodies that bind two different antigens. Each antibody has two antigen-binding sites, one on the end of each arm in the Y-shaped protein. A natural antibody has identical binding sites on both arms. However, using cloned genes in hybridomas, researchers can produce an antibody with different antigen-binding sites on the two arms (Figure 25.18). Such an antibody could be used to bind an antigen present on tumor cells with one arm, while delivering an antigen that is toxic to tumor cells bound on the other arm. This would permit the effective use of small quantities of toxin by bringing the toxin in immediate contact with the tumor cells while preventing other cells from being exposed to dangerous levels of the toxin.

The following example highlights one of many potential applications of antibodies.

Example 25.3 Medical and agricultural applications of transgenic monoclonal antibody genes.

Using genetic engineering methods, Hiatt et al. (1989. *Nature* 342:76–78) transferred rearranged antibody genes extracted from mouse B cells into tobacco plants. They transferred a gene for the light chain into one plant and the gene for the heavy chain into another plant. When the two plants were hybridized, those progeny that carried both light and heavy chains produced fully functional monoclonal antibodies.

Problem: How might this be beneficial from **(a)** a medical point of view and **(b)** an agricultural point of view?

Solution: **(a)** This method provides a way that is potentially less expensive than hybridoma culture for producing large quantities of monoclonal antibodies. Plants that produce the monoclonal antibody can be grown inexpensively on a farm instead of through expensive cell culture methods in a laboratory. **(b)** Plants do not have immune systems like those in vertebrate animals, so they do not raise antibodies against pathogens that may infect them. A monoclonal antibody may be raised against a plant pathogen by injecting the pathogen into an animal. The rearranged genes that encode the antibody can then be transferred into the chromosomes of the plant so that the plant produces the antibody. Because the rearranged antibody genes are in the chromosomes, they will be inherited like any other gene, and they may provide the plant with inherited immunity to the pathogen.

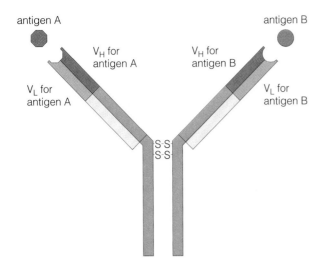

Figure 25.18 A bispecific antibody. Each arm binds a different antigen.

Antibodies are valuable tools in recombinant DNA research and diagnostic medicine. They have many current and potential applications.

SUMMARY

1. There are two major types of immune responses, humoral immunity and cell-mediated immunity.

2. In humoral immunity, B cells express genes that encode antibodies. Antibodies are proteins that recognize and bind to a wide variety of substances that are foreign to the body called antigens. An antibody bound to an antigen marks the antigen and anything attached to it for destruction.

3. Each B cell rearranges its antibody genes to encode a unique antibody. When an antibody encounters an antigen, the B cells that produce that antibody proliferate to mount an immune response.

4. Cell-mediated immunity relies on the association of MHC antigens and viral antigens on a cell surface. The associated antigens attract cytotoxic T cells, which destroy the infected cell.

5. Antibodies consist of two identical heavy chains and two identical light chains that associate to form a Y-shaped antibody molecule. Each of the chains contains a constant (C) region and a variable (V) region. The constant regions are similar in all antibodies.

6. The variable regions form the antigen-binding sites and give each antibody a particular structure that causes it to bind with its specific antigen.

7. Three genes encode antibodies: two light-chain genes called κ (kappa) and λ (lambda), and a heavy-chain gene. Each gene includes a variable region and a constant region.

8. The κ light-chain gene consists of three regions: a variable (V_κ) region, a joining (J_κ) region, and a constant (C_κ) region. The V_κ region consists of a cluster of about 300 V_κ segments, each preceded by its own promoter and by a leader (L_κ) segment. The V_κ region is followed by a spacer sequence, which is followed by the J_κ region, consisting of five J_κ segments. Following the J_κ region is another spacer, which encodes an intron, followed by a single C_κ segment, which encodes the constant region of the κ chain.

9. During rearrangement, a single L_κ-V_κ segment is selected and fused to a J_κ segment with all intervening DNA deleted. The selected L_κ-V_κ-J_κ segment is joined to the intron separating the J_κ region from the C_κ region. The final gene has the order L_κ-V_κ-J_κ-intron-C_κ. The mRNA is translated to produce a polypeptide with a κ light chain that has a leader sequence 17–20 amino acids long, a variable sequence 108 amino acids long (V_κ and J_κ segments), and a constant sequence of 112 amino acids.

10. The λ gene is similar to the κ gene, with a few differences. The human V_λ region is similar in size to the V_κ region, with about 300 V_λ segments. In humans, about 60% of antibodies have κ light chains, while about 40% have λ light chains.

11. There is much greater diversity in heavy chains than in light chains because there are a greater number of combinations of variable and constant segments, and the heavy-chain gene includes an additional class of segments called diversity (D_H) segments. The gene is divided into four regions: the V_H region, the D_H region, the J_H region, and the C_H region. The V_H, J_H, and C_H segments contained in these regions perform the same roles as their corresponding segments in the light-chain genes. The D_H segments add additional diversity to the heavy-chain variable region in the antibody.

12. Rearrangement of antibody genes is the responsibility of enzymes encoded by two linked genes, *RAG1* and *RAG2*. These genes encode enzymes that catalyze recombination.

13. There are millions of different antibody gene arrangements in B cells. When an antigen enters the body, it circulates until it encounters an immature B cell that happens to have an antibody that can bind to it. Antigen binding stimulates the B cell to divide several times to produce many B cells that encode the same antibody. This is called a primary immune response. The excess B cells produced during the primary immune response are called memory cells. If the same antigen enters the body again, the memory cells mount a rapid secondary immune response, providing acquired immunity to diseases.

14. On the surface of cytotoxic T cells are receptor proteins that recognize a complex of antigens and MHC markers. The genes that encode the receptor proteins are very similar to the antibody genes and include V, D, J, and C segments.

15. In the cell-mediated immune response, cytotoxic T cells recognize only antigens that present themselves along with MHC markers on the surface of an infected host cell. There are three classes of MHC genes. The class I genes encode the self-recognition histocompatibility antigens that are present on the surface of each cell. The class II genes encode proteins found on the surface of B cells, T cells, and macrophages. The class III genes encode a variety of proteins, including some involved in humoral immunity. T-cell receptors recognize only a foreign antigen–MHC complex and bind to it, initiating the eventual lysis and destruction of the cell.

16. Immune system malfunctions include autoimmune disorders, in which the immune system mounts an immune response against the body's own substances. Mutations in the genes that encode immune system proteins may cause inherited immune deficiency. Infectious immune deficiency arises when a pathogen successfully overcomes an immune response.

17. All the antibodies produced by a B cell and its mitotic progeny have identical antigen-binding sites, so they are specific for a single antigen. Isolated B cells can be fused with cultured myeloma cells to produce hybridomas. Each individual hybridoma cell produces a progeny cell line that produces one antibody, called a monoclonal antibody. Monoclonal antibodies have many current and potential applications.

QUESTIONS AND PROBLEMS

1. The one gene–one polypeptide hypothesis described in Chapter 6 has been taught for many years in genetics courses. Antibodies are made of protein and are encoded by genes. In what ways does the one gene–one polypeptide hypothesis apply or not apply to antibodies **(a)** at the cellular level and **(b)** at the organismal level?

2. Distinguish between humoral and cell-mediated immunity.

3. Viruses that have entered an individual may be outside the cells in capsular form, or they may have entered the cells. Describe how humoral and cell-mediated immune systems may fight viruses both inside and outside the cells.

4. Antibody gene rearrangements are specific to a particular individual. Explain why an antibody gene rearrangement is not transmitted to offspring.

5. When an individual has been immunized against a disease, that individual carries antibodies against a particular antigen in the pathogen that causes the disease. A second individual who is immunized against the same disease may carry a different antibody that likewise confers immunity to the same disease. Why may immunity to the same disease be due to different antibodies?

6. After rearrangement of a κ light-chain gene, many promoters may be present in the final gene, but only one promoter is utilized for transcription. What is the position of this promoter in the gene relative to the other promoters, and what structural arrangement ensures that only this promoter is utilized for transcription?

7. A large intron is transcribed into κ light-chain pre-mRNA. **(a)** What is the function of this intron? **(b)** The intron plays an important role, so why is it removed during pre-mRNA processing?

8. Explain the difference between productive and nonproductive arrangements of antibody genes. Why are nonproductive arrangements produced?

9. Describe some examples of autoimmune disorders.

10. Susceptibility to autoimmune disorders is often inherited. What genes are likely candidates for alleles that confer susceptibility to autoimmune disorders?

11. What are bispecific antibodies, and what possible uses might they have in research and medical applications?

12. Describe what a monoclonal antibody is and how monoclonal antibodies are usually produced for use in research and medicine.

13. HLA alleles are highly diverse in humans. They encode a variety of MHC class I proteins. One of these proteins, HLA-B53, binds to peptide antigens derived from the parasite *Plasmodium falciparum*, which causes malaria. According to McMichael (1993. *Science* 260:1771–1772), 25% of the human population of The Gambia (where malaria is prevalent) carry the allele that encodes HLA-B53, whereas only 1% of Europeans do. How might this observation help explain the high diversity for HLA alleles in humans?

14. MHC class I genes are expressed in nearly all nucleated cells in mammals. Transcription of these genes tends to increase when antigens on the cell are recognized during a cell-mediated immune response. What advantages would increased production of MHC proteins have during a cell-mediated immune response?

15. Shortly after infection by HIV, patients mount a cell-mediated immune response. The HIV-encoded protein gag produces peptides that are presented on the surface of infected cells by the MHC protein HLA-B8 (see McMichael. 1993. *Science* 260:1771–1772). However, HIV-positive patients who have mounted a cell-mediated response often have epitopes of gag peptides complexed with HLA-B8 that cytotoxic T cells fail to recognize. Given the nature of HIV, what is a possible origin of these epitopes?

16. B cells do not grow well in culture on their own. However, it is possible to fuse a B cell with a myeloma (cancerous) cell to form a hybridoma. Hybridoma cells grow well in culture, permitting large-scale production of antibodies. Describe some current and potential applications of this technology.

17. In 1993, Ehlich et al. (*Cell* 72:695–704) reported that rearrangements at the heavy-chain locus had no influence on rearrangements at the κ light-chain locus in the same B cell. Why is independence of arrangement in light- and heavy-chain genes possible?

18. The recombination-activating gene *RAG2* is expressed at high levels during rearrangement of antibody genes in chicken B cells. In 1992, Takeda at al. (*Proceedings of the National Academy of Sciences, USA* 89:4023–4027) found that when both copies of the *RAG2* gene are deleted in chicken B cells, the cells still rearrange antibody genes. What does this observation suggest about the role of *RAG2* in antibody gene rearrangement? Does this observation demonstrate that *RAG2* does not catalyze antibody gene rearrangement? Explain your answer.

19. Oettinger et al. (1990. *Science* 248:1517–1523) found that when the *RAG1* gene is artificially transferred into fibroblasts (a cell type that does not normally produce antibodies), the cells rearranged antibody genes. They found further that if both *RAG1* and *RAG2* genes were cotransfected into the cells, antibody rearrangement increased 1000-fold over rearrangement with *RAG1* alone. What do these results suggest about the possible roles of *RAG1* and *RAG2*? How does this observation help explain the results of Takeda et al. cited in the previous question?

20. In 1992, Mombaerts at al. (*Cell* 68:869–877) found that mice that are homozygous for a mutation in the *RAG1* gene fail to produce mature B and T lymphocytes and suffer from immunodeficiency. Also in 1992, Shinkai et al. (*Cell* 68:855–867) reported that mice that are homozygous for a deletion in the *RAG2* gene likewise fail to produce mature B and T lymphocytes and suffer from immunodeficiency. **(a)** What do these results suggest about the roles of *RAG1* and *RAG2*? **(b)** The results described in the two previous questions were obtained from studies with cultured cells that had been altered genetically while in culture, whereas the studies cited in this question were conducted with individual animals that had inherited the immunodeficiencies. How might these differences in methods explain the results obtained?

21. The *RAG1* and *RAG2* genes have been cloned, and their DNA sequences, as well as the amino acid sequences of the polypeptides they encode, are known. The polypeptides could directly catalyze gene rearrangement, or they could act indirectly as transcription factors that activate other genes, which, in turn, catalyze rearrangement, of antibody genes. Comparison of amino acid sequences of the RAG1 and RAG2 proteins to other proteins revealed that both proteins have amino acid sequences similar to those of transcription factors and topoisomerases that participate in DNA recombination. In 1993, Silver et al. (*Proceedings of the National Academy of Sciences, USA* 90:6100–6104) tested the ability of mutant versions of these polypeptides to cause rearrangements in antibody genes in fibroblasts. Their results indicate that

mutations in the regions similar to transcription factors reduce rearrangement but do not eliminate it. Mutations in the segments similar to topoisomerases eliminate rearrangement. What do these observations suggest about the most probable roles of these genes?

22. Certain alleles of the class II MHC genes have been considered likely candidates for inherited susceptibility to certain autoimmune diseases, including rheumatoid arthritis. In 1989, Wordsworth et al. (*Proceedings of the National Academy of Sciences, USA* 86:10049–10053) found an association of inherited susceptibility of rheumatoid arthritis with certain alleles of the gene that encodes the HLA protein DRβ. They found that alleles that alter the antigen-binding site are associated with susceptibility. Among 149 people with rheumatoid arthritis, 124 had a particular epitope of DRβ, whereas only 46 of the 100 individuals without rheumatoid arthritis in the control group had the epitope. **(a)** Using chi-square analysis, test the null hypothesis that there is no association between the epitope and rheumatoid arthritis. **(b)** Why might alleles in class II MHC genes be considered likely candidates for susceptibility to rheumatoid arthritis? **(c)** Why might alteration of antigen-binding sites affect susceptibility? **(d)** Genetic susceptibility to rheumatoid arthritis has long been thought to be conferred by several different genes. In what ways do these results support or contradict the multigene theory?

23. Selection that favors heterozygotes has been implied as a reason for the high degree of diversity for MHC alleles. Explain why heterozygotes may have an advantage over homozygotes given the nature of MHC alleles.

24. In 1993, Serwe and Sablitsky (*EMBO Journal* 12:2321–2327) reported that when the intron enhancer is deleted from the DNA in one of the two heavy-chain genes in mouse cells, rearrangement of the gene is reduced but not eliminated. **(a)** What is the usual role of the intron? **(b)** What do these results suggest about possible additional roles for the DNA within the intron?

FOR FURTHER READING

Details of the genetic basis for immunity can be found in Chapter 33 of **Lewin, B. 1997.** *Genes VI.* Oxford, England: **Oxford University Press**; Chapter 19 of **Wolfe, S. L. 1993.** *Molecular and Cellular Biology.* Belmont, Calif.: **Wadsworth**; **Riott, I. M. 1997.** *Essential Immunology,* **9th ed. Oxford, England: Blackwell Science**; and **Tonewaga, S. 1985. The molecules of the immune system.** *Scientific American* **253 (Oct):122–131.**

For additional reading, go to InfoTrac College Edition, your online research library at: http: www.infotrac-college.com/brookscole

PART VI

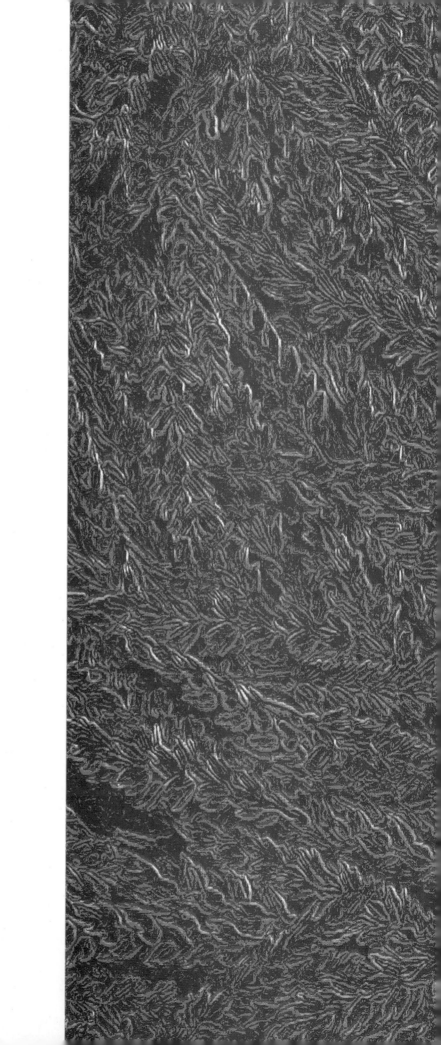

APPLICATIONS OF GENETICS

CHAPTER 26

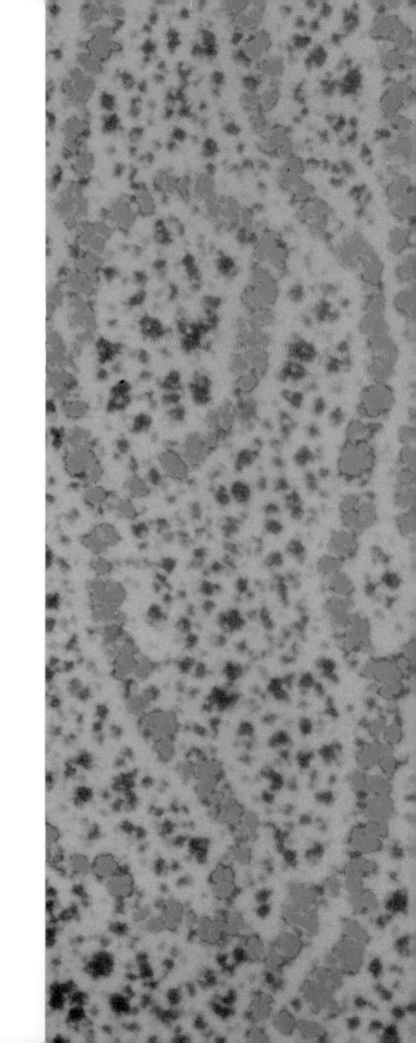

KEY CONCEPTS

The human genome project is an effort to map the human genome, identify human genes, and sequence the entire genome.

~

Genetic testing identifies alleles that confer genetic disorders.

~

DNA fingerprinting is valuable in forensics and paternity identification.

~

Genetic pharmacology, the production of proteins in bacterial, yeast, or mammalian cells, provides proteins that are useful in medicine.

~

Gene therapy, the genetic modification of cells to restore gene function, provides a means for treating some genetic disorders and cancer.

~

Clinical genetics is a medical specialty that includes diagnosis of genetic disorders, treatment of patients with genetic disorders, and genetic counseling of patients and their relatives.

GENETICS IN MEDICINE AND FORENSICS

Genetics has always been an important part of medicine. However, recent advances in human genetics have made possible many applications of genetics that were unavailable only a few years ago. Physicians and researchers have known for many years that certain medical disorders have a genetic basis, but only recently have researchers identified, cloned, and sequenced the genes responsible for these disorders. Discoveries of genetic influences on cancer and immune system disorders, which we discussed in the previous two chapters, have placed genetics at the forefront of research in these areas. Modern medicine has reduced the proportion of illnesses caused by infectious diseases. The result has been an increase in the proportion of diseases that have a strong genetic component. Even infectious diseases are often influenced by genetic factors in both the pathogen and the person who suffers from the disease.

Recombinant DNA technology has revolutionized many areas of medicine. As the previous three chapters illustrate, much of what we know about development, cancer, and immunity has come from advances in genetics research. Diagnosis of important diseases now often rely on a DNA or protein analysis component. DNA markers can be used to identify most genetic disorders in

a fetus early in pregnancy. Embryos fertilized in vitro can be selected to exclude certain genetic disorders. Some diseases, such as diabetes and pituitary dwarfism, are now routinely treated with genetically engineered pharmaceutical products. Gene therapy, a procedure in which a functional gene is added to a person's cells to correct the lack of gene function in those cells, is now in the final stages of clinical trials for several diseases. Genetic research and biotechnology have revolutionized medical forensics. People can be identified and paternity determined unambiguously with DNA markers derived from tiny amounts of tissue. Many applications are currently available for routine use, and many more are in developmental stages. These applications are the beginning of what promises to be a bright albeit uncertain future for medical biotechnology.

We have discussed numerous applications of genetics in medicine in previous chapters. In this chapter, we will focus on some important areas of genetics in medicine and forensics in the context of their practical applications. First, we will look at the human genome project, one of the most significant undertakings in the history of science and one that has already impacted the diagnosis and treatment of genetic disorders. We will

then look at genetic testing in medicine and forensics. Following this, we will examine applications of recombinant DNA in medicine, including genetic pharmacology and human gene therapy. Finally, we will discuss clinical genetics, a medical specialty that includes the diagnosis, treatment, and counseling of patients with genetic disorders and their relatives.

26.1 THE HUMAN GENOME PROJECT

Few scientific projects have generated as much excitement (and as much controversy) as the **Human Genome Project**, the coordinated effort of many laboratories to map the human genome, identify all of its genes, and eventually determine the entire genomic nucleotide sequence. Much misinformation about the human genome project has been disseminated in the popular press, largely due to misunderstandings. Among these misunderstandings is the question of whose genome should be sequenced. Many different scientists are collaborating to complete the sequencing effort, and they acquire their sequences from the DNA of many different people. Because many of the DNA sequences are repeated in different people, indications of conserved and variable nucleotide sequences will be a natural outcome of the project. There is no such thing as a single DNA sequence for the human genome, but the conserved sequences of genes and other regions can be identified by comparing homologous DNA sequences from many individuals.

Although DNA sequencing is a major part of the human genome project, the project also includes high-resolution mapping of human chromosomes using RFLPs and PCR-based genetic markers, cDNA cloning and sequencing, cloning large fragments with cosmid clones and YACs, mapping and sequencing genes in model organisms and comparing the results to human genes in order to speed the sequencing process, and managing the enormous amount of information generated by the project. Among the ultimate goals are to improve our understanding of the human genome and to apply that understanding for the benefit of humanity, primarily through medicine.

The ultimate goal of the project, to determine the entire nucleotide sequence of the human genome, was originally targeted for completion in 2005. In order to meet this goal, researchers had to develop the necessary technology for genetic mapping, DNA sequencing, and database management for the enormous quantity of data that would be generated. In spite of many obstacles and uncertainties, the technological advances were achieved, and in 1995 the project entered the final phase of massive DNA sequencing. Because of advancements in technology, the project directors revised the target date for completion of the entire sequence to 2001, 4 years earlier than the original target date.

Human Chromosome Mapping

One of the main efforts in the human genome project is high-resolution genetic mapping of the human chromosomes. Genetic map distances in centimorgans do not correspond in a linear fashion to numbers of nucleotide pairs, so the human genetic map is a product of the human genome project that stands separate from the DNA sequence but can be compared to it for understanding the relationship between genetic distances and physical distances in the genome. We discussed the mapping procedures in section 15.6. The original goal was to obtain a map with at least one marker every 2 cM along each of the chromosomes by 1996, a goal that had already been achieved by that date. By 1990, researchers had mapped 6726 markers, 1867 of them representing known genes. By 1994, just 4 years later, researchers had mapped over 60,000 markers. The potential benefits of mapping include identifying genetic markers that can be used for genetic testing and chromosome jumping and walking to locate genes, as described in section 15.8.

Mapping Cloned Fragments into Contigs

To develop the DNA sequence of the human genome, researchers had to find a way to identify overlapping fragments of DNA. Large DNA fragments have been cloned at random into YACs and cosmids. However, unless these fragments can be mapped in relation to one another, they are of little value. After a cloned fragment of DNA is assigned to a chromosome or a particular segment of a chromosome, it can be fingerprinted by mapping frequently repeated segments such as *Alu* elements, L1Hs elements, or poly (GT) repeats within the cloned fragment. These repeated segments form a particular pattern in a large clone. If part of the pattern in one clone matches part of the pattern in another clone known to be on the same chromosome or within the same chromosomal segment, there is a strong possibility that the two clones overlap along part of their sequences. Overlapping clones are called **contigs**, because they represent contiguous segments (Figure 26.1).

Contigs can be lined up with one another to form a linear map of overlapping contigs that represent large chromosomal regions. In some cases, entire chromosomes are now represented by a set of overlapping contigs. Once a set of contigs is identified, each contig can be sequenced to provide a set of overlapping DNA sequences that can be combined to provide the overall DNA sequence.

Clones can also be ordered into contigs using sets of specific sites that have been previously sequenced, called **sequence tagged sites (STSs)**. STSs are short DNA segments (200–500 nucleotide pairs) that have been sequenced and assigned to a chromosomal location. Each STS can be amplified by PCR using defined primers. The sequence of each primer for STS amplification is available

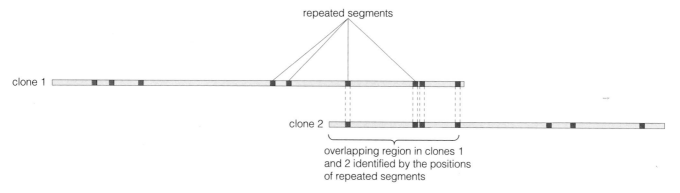

repeated segments

clone 1

clone 2

overlapping region in clones 1
and 2 identified by the positions
of repeated segments

Figure 26.1 Overlapping regions in cloned fragments of DNA. Repeated segments that are dispersed throughout the chromosomes can be used to identify clones with overlapping regions.

to researchers, so they can synthesize primers to amplify each STS. Geneticists can test large clones, such as YAC or cosmid clones, for the presence of known STSs to determine their location and identify overlapping clones for identification of contigs.

Geneticists can also identify contigs by aligning the overlapping DNA sequences of sequenced clones. After geneticists obtain the DNA sequence of a clone, they submit the sequence to a database, such as GenBank. DNA sequence databases contain the sequences of hundreds of thousands of cloned fragments of human DNA. Geneticists can take the sequence of any clone and use powerful computer programs that are available on-line to compare the clone's sequence with all other sequences in the database. Matches in DNA sequences among human DNA fragments identify potential contigs.

Expressed Sequence Tags (ESTs)

Much of the human genome consists of DNA sequences that do not encode proteins or mature RNAs. Although the entire sequence of a chromosome is useful for many types of research, the most beneficial portions for medical applications are the protein-encoding sequences contained within mature mRNAs. Researchers can rapidly obtain such sequences by isolating all the mRNAs from different tissues, producing cDNA clones of the mRNAs, and then determining the sequences of the cDNA clones. DNA sequences obtained in this way are called **expressed sequence tags (ESTs)**.

Ideally, a set of ESTs could be included as a large database of DNA sequences that researchers could access to match genes with an EST sequence. Such an EST database was initiated in 1991 but was compiled at a slow pace because an EST database was not originally envisioned as part of the human genome project. Consequently, scientists who worked with ESTs found it difficult to obtain public funding. Recognizing the commercial potential of EST databases, several private corporations initiated their own EST projects, in some cases maintaining rigid pro-

prietary control over the DNA sequence information. A major financial barrier was overcome when the pharmaceutical corporation Merck and Co. committed to sponsoring the sequencing of 200,000 human ESTs, all for free public release. A large EST database compiled from this project and a variety of other sources is now available gratis to all interested scientists worldwide via the Internet.

In many cases, known DNA sequences from other species have been used to identify the homologous human gene from the human EST database. For example, the cDNA sequence of the human telomerase gene was identified by comparing a yeast telomerase sequence with sequences in the human EST database, as highlighted in the following example.

Example 26.1 Identification of the cDNA sequence for human telomerase from the human EST database.

In 1997, Nakamura et al. (*Science* 277:955–959) reported identification of the human telomerase gene sequence by searching the human EST database for homology using known yeast (*Saccharomyces pombe*) and ciliate (*Euplotes aediculatus*) telomerase sequences. In subsequent experiments, they confirmed that the human EST sequence was the human telomerase gene (named *TRT* for telomere reverse transcriptase, OMIM 187270).

Problem: Of what significance is this discovery to medicine?

Solution: As described in section 24.4, telomerase is usually not expressed in somatic cells but is expressed in cancer cells. If telomerase could be inhibited, then cancer cells should become senescent, halting the tumor growth. Discovery of the gene that encodes human telomerase has facilitated the search for cancer treatments based on inhibition of telomerase activity.

Human Genome Sequencing and Data Management

The haploid human genome consists of about 3 billion nucleotide pairs. Robotic science has contributed greatly to speeding up the process of DNA sequencing. Human sequencing projects currently use automated DNA sequencers, which perform most of the sequencing steps automatically and provide the DNA sequence in computerized format. The sequences are entered into worldwide databases that researchers can access to compare a DNA sequence with others from different laboratories. The GenBank database, described in section 9.9, is one of several databases that exchange data on a daily basis to update sequences.

With many laboratories contributing large amounts of information, the task of information management is potentially overwhelming. One of the major efforts in the human genome project has been the development of interactive computer databases and analysis programs that researchers around the world can use in a coordinated effort. Such databases, as well as services for DNA sequence comparison, descriptions of genes, and integrated literature searching, are now available gratis to all interested scientists via the Internet.

Among the major efforts of the human genome project was the development of automated DNA sequencing. Sequencing is now much more rapid and inexpensive than it was just a few years ago. The human genome project is not the only beneficiary from improved sequencing methods. Scientists who conduct DNA analysis with any species can benefit from the improved efficiency of laboratory methods and information management. Major DNA sequence databases are not restricted to human DNA sequences but contain sequences submitted by researchers from around the world for many different species. The broad benefits of improved technology and information management are among the strongest justifications for the large expenditures required by the human genome project (an estimated $3 billion). Biotechnology is now taking its place as one of the most significant industries of our time. The investments made in the human genome project will likely be repaid many times over as they are applied in biotechnology aimed at improvements in medicine, agriculture, and industry.

~

The human genome project consists of mapping the human genome, identifying and sequencing human genes, and eventually determining a complete consensus sequence of the human genome. Although the project is very expensive, its benefits are many.

26.2 GENETIC TESTING

One of the major benefits of the human genome project is its contribution to the development of genetic tests.

Inherited genetic disorders are often caused by recessive alleles that remain phenotypically hidden in heterozygotes. Other genetic disorders, such as Huntington disease, may be conferred by a dominant allele but do not appear until later in life because of delayed onset. Some genetic disorders, such as phenylketonuria, can be treated successfully if diagnosed early, before symptoms appear. Advances in genetic testing have made it possible to identify people who are heterozygous carriers of recessive alleles or people who are at risk for later onset of a genetic disorder. Many genetic disorders can now be identified in a fetus during the early stages of pregnancy or even from a single cell extracted from a human embryo produced through in vitro fertilization.

Physicians now have at their disposal numerous genetic tests for identifying genetic disorders. These tests can provide physicians with information that they can use to develop effective prevention or treatment programs. For example, when a physician determines from the results of a genetic test that a young woman has an inherited susceptibility for breast cancer, the physician can monitor that woman more closely for the early onset of breast cancer and can treat the cancer early, when the probability of successful treatment is high. Under such circumstances, genetic testing is clearly beneficial. However, as genetic testing for certain disorders has become widespread, serious legal and ethical questions have arisen regarding the proper use of genetic testing and the information derived from it. In this section, we will describe some of the methods used for genetic testing and the disorders they diagnose. We will, however, postpone a discussion of the legal and ethical issues until Chapter 28.

Biochemical and Cytological Genetic Tests

Several genetic tests that are not based on DNA analysis have been available for many years. These tests are usually based on simple and inexpensive biochemical procedures or microscopic observations of chromosomes. About 50 genetic disorders can be diagnosed with such tests, several using fetal tissue.

Among the most widely administered genetic tests is newborn screening for phenylketonuria (PKU). Every state in the United States as well as many foreign countries require mandatory screening of newborns for PKU. PKU is a devastating illness that causes severe mental retardation if left untreated. It is caused by the absence of functional phenylalanine hydroxylase due to homozygosity for mutant alleles at the *PAH* locus, which encodes the enzyme phenylalanine hydroxylase. PKU is present in about 1 in 12,500 newborns. Clinicians detect it using a relatively simple test that costs only about $1.25 per person.

The test, known as the Guthrie test, requires a small amount of blood from an infant. The blood is placed on a filter paper and allowed to dry (Figure 26.2). The dried blood is then added to the growth medium of a bacte-

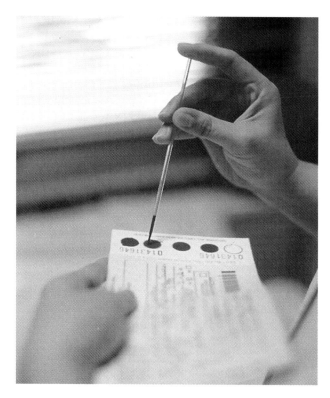

Figure 26.2 Placement of blood from a newborn on a filter paper for analysis with the Guthrie test, a test that detects phenylketonuria. (Custom Medical Stock Photo.)

rial strain of *Bacillus subtilis* that is auxotrophic for phenylalanine. High levels of phenylalanine in the blood cause rapid bacterial growth. When rapid growth is observed, the child's blood is then tested directly for phenylalanine and tyrosine levels to confirm the diagnosis. When evidence of PKU is present, physicians initiate extensive testing and treatment (if necessary) to prevent lifelong mental retardation in the affected children.

Sickle-cell trait can also be diagnosed in newborns, using a simple biochemical test that costs about $0.22 per person. Sickle-cell anemia is due to homozygosity for the recessive *HBB*S* allele. Sickle-cell trait is much less severe than sickle-cell anemia and appears in people who are heterozygous for the *HBB*S* allele. In general, the symptoms of sickle-cell trait are minor and do not have adverse consequences. However, one of the important symptoms of sickle-cell trait is increased susceptibility to streptococcal infection before 3 years of age, which currently causes unnecessary deaths among children with sickle-cell trait, particularly those who live in poverty and lack adequate medical care. Newborn screening and preventative treatment with antibiotics can avert potentially serious infections. For this reason, the National Institutes of Health (NIH) has recommended that sickle-cell trait testing be made available to all newborns in the United States.

Unfortunately, genetic testing for sickle-cell trait does not have a sterling history. Because selection against malaria has increased the frequency of the *HBB*S* allele in certain regions of the world (see section 19.5), sickle-cell trait is more common in certain ethnic groups, including those with ancestries originating in west Africa and the eastern Mediterranean. The frequency of heterozygous carriers of *HBB*S* among African Americans is quite high, about 1 in 12. The test for sickle-cell trait was developed in the 1960s and was made available to the public as a voluntary test in some states. However, in some states the test was mandatory for African American children before they could enter school. Misunderstandings regarding the test, breaches of confidentiality for test results, and racial discrimination in its application have raised serious questions about how it should be used.

Down syndrome and other aneuploidies can be detected by cytological examination of chromosomes prenatally in fetal cells and postnatally when symptoms in a newborn suggest the possibility of an aneuploid syndrome. Other currently incurable genetic disorders such as fragile-X syndrome and Huntington disease can be detected prenatally using recombinant DNA methods. Prenatal cytological tests are typically recommended by physicians for pregnant women who are 35 or older because of the increased probability of Down syndrome births (see section 17.4). Although many genetic disorders cannot be cured, early detection assists in planning for appropriate education and medical care of the affected child. Also, some parents elect voluntary abortion for fetuses that are diagnosed with aneuploidies or other genetic disorders detected early during pregnancy.

DNA-Based Genetic Tests

DNA tests have revolutionized genetic testing. The human genome initiative has produced saturated genetic maps in which many genes responsible for genetic disorders have been mapped in relation to DNA marker loci. A large number of these genes have now been cloned and sequenced, and the DNA sequences of numerous mutant alleles have also been determined as part of the human genome project. The alleles responsible for over 200 genetic disorders can now be detected using DNA analysis. DNA analysis can also identify people who are heterozygous for alleles that confer recessive disorders and people who are prone to delayed onset of a genetic disorder.

The first DNA tests for detection of mutant alleles were based on RFLP loci that were closely linked to the gene in question. However, the gene itself is a better marker than linked DNA marker loci for two reasons. First, because the gene itself is used, crossovers between the mutant allele and the DNA marker that is coupled to the allele are no longer a concern. Second, because several disorders are caused by multiple mutant alleles, a marker coupled to a mutant allele may only identify one of the several possible mutant alleles. Using the cloned gene often makes it possible to identify many, if not all, mutant alleles responsible for a disorder.

Huntington disease was among the first to be predicted using a linked RFLP locus. A specific RFLP marker

was about 95% accurate in predicting Huntington disease. The inaccuracies were due in part to rare crossovers between the RFLP marker locus and the *HD* locus. After the *HD* gene was cloned and sequenced, the RFLP test for Huntington disease became obsolete. The mutation that causes Huntington disease is a trinucleotide expansion that can be readily detected with a PCR test, eliminating the need to use linked marker loci (see Example 13.5).

Most human genetic disorders are due to recessive mutant alleles. Often, individuals with a recessive genetic disorder are compound heterozygotes for two different mutant alleles. This complicates genetic testing because each mutant allele may need to be identified with a separate DNA test, and certain mutant alleles may escape detection. In some disorders, such as fragile-X syndrome and Huntington disease, most mutant alleles are trinucleotide repeat expansions that can be readily detected with a simple PCR test. DNA testing for these genetic disorders is highly reliable. However, other genetic disorders, such as PKU and cystic fibrosis, may be caused by many different kinds of mutant alleles. Let's examine the cystic fibrosis gene as an example of how genetic tests are developed.

Genetic Testing for Mutant Alleles of the Cystic Fibrosis (*CF*) Gene

After the DNA of a gene that causes a genetic disorder has been identified and sequenced, much can be learned in a short period of time about the gene's function, whether or not there are multiple mutant alleles, and what variations in the disorder are due to differences in mutant alleles. Also, researchers can use sequence information to design highly reliable RFLP or PCR tests that directly identify specific mutant alleles. Among the most elusive yet important cases of this was cloning and sequencing the gene responsible for cystic fibrosis. Cystic fibrosis is one of the most common genetic disorders in humans of European descent, with a frequency of mutant alleles of about $q = 0.022$ and a frequency of heterozygous carriers of about 4% ($2pq = 0.043$). Approximately 1 in 2000 children are affected.

The nucleotide sequence of the *CF* gene (OMIM 219700) was reported in September 1989. After the gene was cloned and sequenced, information about cystic fibrosis came quickly. Researchers determined the amino acid sequence of the protein product from the DNA sequence. They identified the gene product as a large protein (1480 amino acids) that functions as a membrane transport protein for chloride ions. When no functional protein is produced, chloride ion transport across cell membranes is reduced, which explains the abnormally high salt accumulation in the perspiration of people who are afflicted with the disorder.

With the DNA sequence of the gene and the amino acid sequence of the gene product known, researchers could analyze the DNA sequences of the *CF* gene in people with cystic fibrosis to see what types of mutations caused the disorder. About 70% of people with cystic fibrosis have the same mutant allele (*CF**ΔF508), an allele with a deletion of three nucleotides (CTT) that encode phenylalanine at position 508 in the amino acid chain. Scientists have developed several DNA tests that detect this allele (Figure 26.3). The remaining 30% of cystic fibrosis cases are due to over 100 different mutant alleles, which include substitution, deletion, frameshift, nonsense, and intron-splicing mutations. The type of mutation is related to the severity of symptoms in some cases. The wide variety of mutant alleles that cause cystic fibrosis significantly complicates the detection process. However, after a mutant allele has been identified in a person, and a specific test is designed to detect that allele, family members can be tested for that specific allele to identify heterozygous carriers.

The following example shows how a DNA test for a specific allele can be used in clinical medicine.

Example 26.2   Genetic testing for preimplantation diagnosis of cystic fibrosis.

In 1992, Handyside et al. (*New England Journal of Medicine* 327:905–909) reported the results of genetic testing to identify the presence or absence of mutant *CF* alleles in human embryos that had been derived from in vitro fertilization. Three couples participated in this study. In each case, the wife and husband were heterozygous carriers of the *CF**ΔF508 allele. The authors extracted unfertilized oocytes from each of the women, and fertilized the oocytes in vitro with sperm from the respective husbands. For one woman, the authors recovered only one embryo that could be genotyped, and it was homozygous for the *CF**ΔF508 allele, so they did not implant that embryo. They implanted an embryo in the second woman, but she did not become pregnant. From the third woman, the authors recovered 11 oocytes, and six of these produced embryos whose cells divided normally following in vitro fertilization. They isolated DNA from single cells that they had removed from each of the embryos, and they amplified the DNA using a modified form of PCR that is effective for very small amounts of template DNA. They were able to characterize five of the six embryos genetically. Two were homozygous for the *CF**ΔF508 allele, one was heterozygous for the allele, and two were homozygous for the normal allele. They implanted one embryo that was homozygous for the normal allele and one that was heterozygous. The woman became pregnant from one of these embryos and bore a healthy girl at full term. Subsequent genetic testing

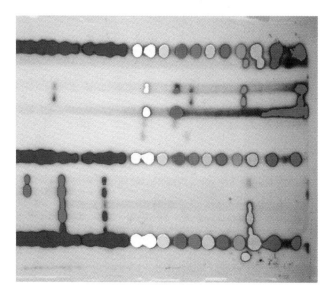

Figure 26.3 A DNA test for detection of the *CF*ΔF508* mutant allele. (Custom Medical Stock Photo.)

revealed that the girl was homozygous for the normal allele.

Problem: (a) Why was it possible to focus on a single *CF* allele for genetic testing in this study? (b) What is the purpose of genetic testing coupled with in vitro fertilization in a study such as this?

Solution: (a) All of the parents were known heterozygous carriers of the *CF*ΔF508* mutant allele. None of the parents had cystic fibrosis, so the allele on the homologous chromosome in each parent must have been normal. Therefore, there was no need to test for mutant alleles other than the *CF*ΔF508* allele. (b) Genetic testing on embryos derived from in vitro fertilization can be used to identify embryos whose genotypes indicate that they will not have cystic fibrosis. Implantation of only these embryos prevents pregnancy with a fetus that will develop cystic fibrosis.

Genetic testing is in widespread use for screening infants for phenylketonuria. Many other genetic tests are available for detection of genetic disorders, or for detection of heterozygous carriers of the alleles responsible for genetic disorders.

26.3 DNA FINGERPRINTING

DNA fingerprinting is the identification of individuals based on a pattern of DNA markers the scientists detect in the genomic DNA of an individual. In most cases, DNA fingerprinting is based on minisatellite or microsatellite markers (see section 13.4 for a description of these types of DNA markers). The patterns detected in DNA fingerprinting are so variable among people that only identical twins share the same DNA fingerprints. In this section we examine how DNA fingerprints are used in forensic science and paternity testing, and how DNA fingerprinting has fared as evidence in court.

DNA Fingerprinting in Forensic Science

In 1985, two rape-murders thought to be committed by the same criminal near Leicester, England, gained worldwide attention because of the way the culprit was identified. Alec Jeffreys, a molecular biologist at the University of Leicester, had recently developed minisatellite DNA analysis in humans that detected segments of DNA that are highly variable in human populations. The term DNA fingerprint was coined to describe these unique patterns because they can be used in a manner similar to that of a true fingerprint to identify an individual. Just as criminals tend to leave fingerprints at the scene of a crime, they may leave traces of their DNA behind in the form of skin, hair, blood, or semen, which provide a source of DNA for forensic scientists. In the Leicester cases, the police had a suspect in custody but little concrete evidence against him. They sought DNA fingerprints from all the men in three towns near the crime scenes and compared those to the DNA fingerprint obtained from semen found at the crime scenes. All were mismatches, including the DNA fingerprint that belonged to the suspect being held, thus clearing him of the crime. The actual criminal attempted to elude DNA fingerprinting efforts by trying to convince someone else to pose as him for the DNA test. These actions led police to him, and he confessed the crime. Since that time, the use of DNA fingerprinting in criminal identification has become routine in criminal investigations and has received much attention in the popular press.

DNA fingerprinting for criminal identification is straightforward. The DNA fingerprint of a criminal obtained from tissue, blood, or semen from a crime scene must exactly match the DNA fingerprint of a suspect to provide evidence against the suspect (Figure 26.4). A mismatch indicates that the DNA fingerprints belong to different individuals. Thus, DNA fingerprinting has tremendous power of exclusion. An innocent suspect can be cleared readily by a mismatch in DNA fingerprints. Exclusion is also possible on the basis of other types of evidence. For example, before DNA fingerprinting was available, exclusion of suspects was possible when blood types, or other types of biochemical markers, showed a mismatch.

Mismatches in blood type or other markers can be used only to exclude suspects, not to prove identity. For example, if a culprit's blood found at a crime scene is type

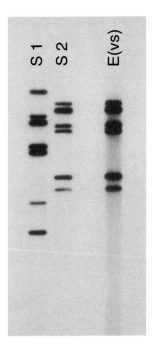

Figure 26.4 A DNA fingerprint used for criminal identification. S1 and S2 are DNA fingerprints from two suspects, and E(vs) is a fingerprint from DNA obtained from evidence at a crime scene. The S2 and E(vs) fingerprints match. (Leonard Lessin/Peter Arnold, Inc.)

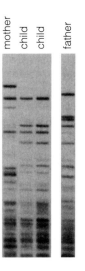

Figure 26.5 Information for Example 26.3: DNA fingerprints of a father, mother, and two children. (Reprinted with permission from Jeffreys, A. J., V. Wilson, and S. L. Thein. 1985. Individual-specific "fingerprints" of human DNA. *Nature* 316:76–79. Copyright 1985 Macmillan Magazines Limited.)

O, then a suspect with type O blood is not excluded but cannot be considered guilty on that basis because many people have type O blood. DNA fingerprinting, on the other hand, provides inclusive as well as exclusive evidence. The DNA fingerprint in Figure 26.4 is from a single probe and is sufficient to exclude one of the suspects as the criminal. However, for inclusive evidence, forensic scientists must generate DNA fingerprints using several different probes. When different probes are used to make several fingerprints, the likelihood that any two individuals chosen at random will have identical matches in all of them is extremely small (often less than 1 in 1 trillion).

DNA fingerprinting is also used in forensics to identify crime victims. In violent crimes, a criminal may remove the body of a victim from the crime scene, but remnants of blood or tissue from the victim may remain at the scene. DNA fingerprints from the crime scene can be matched to a body found elsewhere. Likewise, severely mutilated or decomposed bodies can be identified through DNA fingerprinting. Although the DNA fingerprint of a missing person may not be available to match to a body, forensic scientists can use DNA fingerprints from a person's parents, siblings, or offspring to determine if blood or tissue from a crime scene or a recovered body matches a missing person. Such identification is possible because each DNA marker in a DNA fingerprint is inherited. Just as relatives are each unique but share physical resemblances, DNA fingerprints are also unique among close relatives, but relatives have more fragments in common than people who are unrelated.

DNA Fingerprinting in Paternity Testing

Because the DNA markers in a DNA fingerprint are inherited, scientists can use DNA fingerprints for both exclusion and inclusion of paternity and maternity. In most cases of questioned parentage, the mother is known, but the identity of the father is in question. A child inherits about half of the fragments in his or her DNA fingerprint from the mother and half from the father. Any fragments present in the child's DNA fingerprint that are not in the mother's DNA fingerprint must be inherited from the father. Comparison of the child's DNA fingerprint with those of the mother and a suspected father can exclude or confirm the suspected father as the biological father. If the child has several fragments in his or her DNA fingerprint that are not present in the DNA fingerprint of either the mother or the suspected father, then the suspected father is not the child's genetic father. If, on the other hand, all fragments present in the child's DNA fingerprint that are not present in the mother's DNA fingerprint are present in the suspected father's DNA fingerprint, then the suspected father is the biological father.

The following example illustrates how DNA fingerprints are interpreted.

Example 26.3 DNA fingerprinting.

The DNA fingerprints in Figure 26.5 (from Jeffreys et al. 1985. *Nature* 316:76–79) are from a mother, a father, and their two children.

Problem: What do these fingerprints indicate about the genetic relationships of the parents and the children?

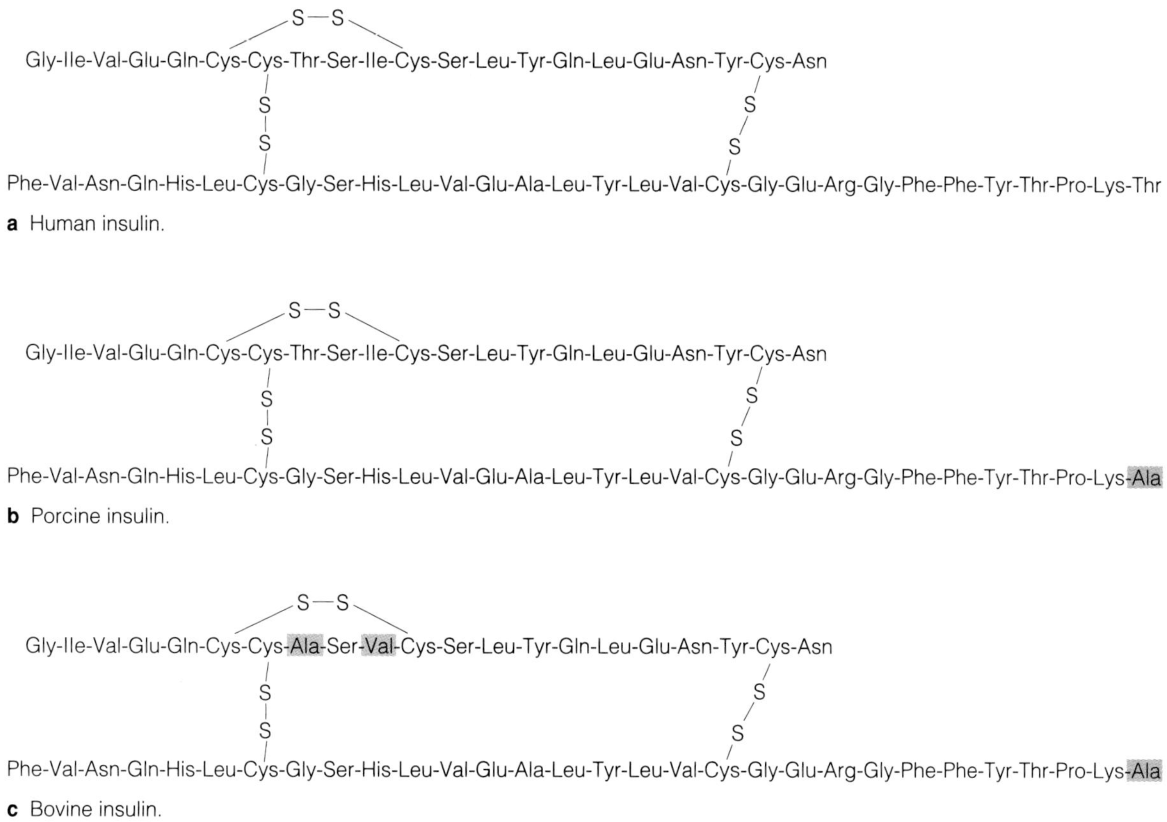

S—S

Gly-Ile-Val-Glu-Gln-Cys-Cys-Thr-Ser-Ile-Cys-Ser-Leu-Tyr-Gln-Leu-Glu-Asn-Tyr-Cys-Asn

S   S
S   S

Phe-Val-Asn-Gln-His-Leu-Cys-Gly-Ser-His-Leu-Val-Glu-Ala-Leu-Tyr-Leu-Val-Cys-Gly-Glu-Arg-Gly-Phe-Phe-Tyr-Thr-Pro-Lys-Thr

a Human insulin.

S—S

Gly-Ile-Val-Glu-Gln-Cys-Cys-Thr-Ser-Ile-Cys-Ser-Leu-Tyr-Gln-Leu-Glu-Asn-Tyr-Cys-Asn

S   S
S   S

Phe-Val-Asn-Gln-His-Leu-Cys-Gly-Ser-His-Leu-Val-Glu-Ala-Leu-Tyr-Leu-Val-Cys-Gly-Glu-Arg-Gly-Phe-Phe-Tyr-Thr-Pro-Lys-Ala

b Porcine insulin.

S—S

Gly-Ile-Val-Glu-Gln-Cys-Cys-Ala-Ser-Val-Cys-Ser-Leu-Tyr-Gln-Leu-Glu-Asn-Tyr-Cys-Asn

S   S
S   S

Phe-Val-Asn-Gln-His-Leu-Cys-Gly-Ser-His-Leu-Val-Glu-Ala-Leu-Tyr-Leu-Val-Cys-Gly-Glu-Arg-Gly-Phe-Phe-Tyr-Thr-Pro-Lys-Ala

c Bovine insulin.

Figure 26.6 Comparison of human, porcine, and bovine insulins. The amino acids in porcine and bovine insulin that differ from those in human insulin are indicated in red.

Solution: The two children are identical twins, as indicated by their identical DNA fingerprints. Each DNA fragment in the children is also found in at least one of the parents, indicating that both father and mother are the biological parents of these children.

Court Challenges of DNA Fingerprinting

Because of this tremendous power of both exclusion and inclusion, attorneys routinely use DNA fingerprinting as legal evidence. The ability of DNA fingerprinting to exclude individuals is rarely questioned. Inclusion, however, has been challenged at length. Laboratory techniques and assumptions of allele frequencies associated with DNA fingerprinting have been raised as reasons to exclude DNA fingerprinting as evidence. In recent years, however, the legal admissibility of DNA evidence for inclusion has gained widespread legal acceptance. In Chapter 28, we will discuss some of the legal aspects of DNA fingerprinting in detail.

~

DNA fingerprinting is used widely for paternity testing and in criminal forensics.

26.4 GENETIC PHARMACOLOGY

Genetic pharmacology is the production of a pharmaceutical product in a host organism or cell culture (usually bacteria) using recombinant DNA. Because the genetic code is the same in bacteria and humans, a bacterium can produce the same polypeptide as a human using the same gene, provided the gene has a bacterial promoter and is devoid of introns.

Unfortunately, not all human proteins can be properly encoded and assembled in a bacterium. Although bacterial and human cells use the same genetic code, they do not have the same apparatus for protein modification and processing. Although a bacterium can assemble the same linear array of amino acids as a human cell using the same gene, a bacterium cannot always produce a protein with the same secondary, tertiary, and quaternary structures as the native human protein. A bacterial cell also cannot always attach the proper additional compounds, such as phosphate groups, carbohydrates, and lipids, that a functional protein may need. These obstacles are formidable in some cases, but they are not always insurmountable. The case of **recombinant human insulin**, the first genetically engineered pharmaceutical to be licensed for human use (see Figure 1.2), is a good

preproinsulin

a During secretion, the *pre-* amino acids are cleaved from the preproinsulin molecule to produce proinsulin.

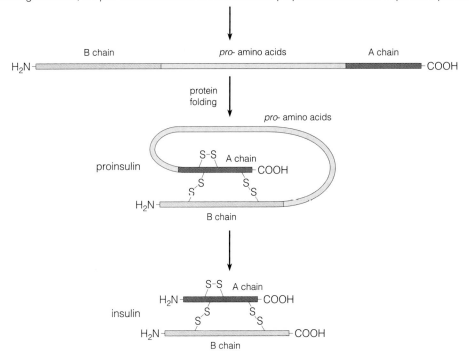

b After tertiary structure formation, the *pro-* amino acids are cleaved from the proinsulin molecule to produce insulin.

Figure 26.7 Normal cellular processing of insulin in humans.

example of how some innovative techniques can be used to overcome inherent problems and produce a genetically engineered product commercially.

Human Insulin Expression in Bacteria

Insulin is produced by β cells in the pancreas and is secreted into the bloodstream. It regulates sugar metabolism in the body. Insulin deficiency causes diabetes and often requires frequent hypodermic injections of insulin to maintain proper levels in the blood. Until recently, supplemental insulin was extracted from the pancreas tissue of cattle or pigs that had been killed for meat. Bovine (cattle) insulin differs from human insulin in three of the 51 amino acids, and porcine (pig) insulin differs by only one amino acid (Figure 26.6). In spite of the similarities of bovine and porcine insulin to human insulin, some patients mount an immune response to bovine or porcine insulin. In most cases, a switch to human insulin overcomes the problem. However, in a few cases, the immune response to animal insulin persists when the switch is made to human insulin. If a patient's treatments begin with human insulin, there is

no adverse immune response. Many people suffer from insulin-dependent diabetes, so there is a great need for human insulin.

Human insulin is a protein that is needed commercially in abundance, so it was a prime candidate for genetic pharmacology. However, fully processed human insulin in the proper conformation could not be made in bacteria directly from a human insulin gene. Figure 26.7 illustrates the normal cellular processing of insulin in humans. Insulin is encoded by a single gene that is transcribed and translated into a single polypeptide called **preproinsulin**. There are two segments in preproinsulin that are not present in the final form. The *pre-* amino acids are at the amino end of the polypeptide and facilitate secretion from the β cells. As the β cells secrete the preproinsulin, they cleave the *pre-* amino acids from the molecule to generate **proinsulin**. The *pro-* amino acids are located in the center of the polypeptide, in a segment called the C peptide, and help to form the proper tertiary and quaternary structures. After the proper structure is formed, the C peptide is cleaved from the molecule to form the final protein. The C peptide is present in the middle of the molecule, so its cleavage generates two

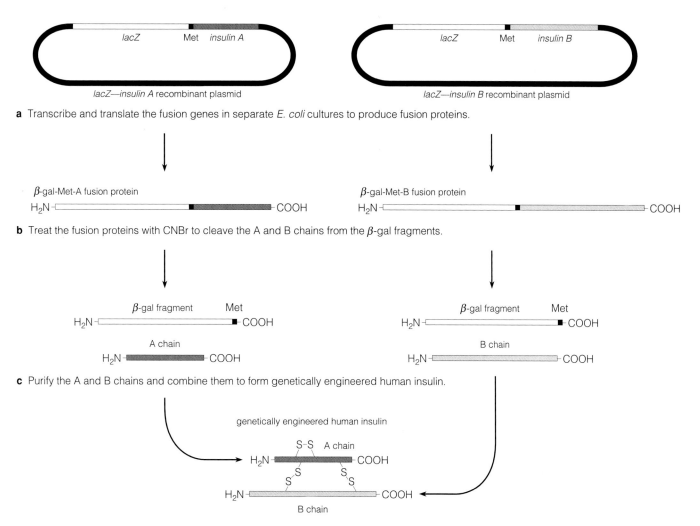

a Transcribe and translate the fusion genes in separate *E. coli* cultures to produce fusion proteins.

β-gal-Met-A fusion protein

H₂N—[   ]—COOH

β-gal-Met-B fusion protein

H₂N—[   ]—COOH

b Treat the fusion proteins with CNBr to cleave the A and B chains from the β-gal fragments.

β-gal fragment   Met

H₂N—[   ]■—COOH

A chain

H₂N—[   ]—COOH

β-gal fragment   Met

H₂N—[   ]■—COOH

B chain

H₂N—[   ]—COOH

c Purify the A and B chains and combine them to form genetically engineered human insulin.

genetically engineered human insulin

Figure 26.8 The two-chain method for production of recombinant human insulin.

polypeptides (A and B chains) in the final insulin molecule with three disulfide bonds, two that connect the A and B chains to each other, and one that stabilizes the tertiary structure of the A chain.

Human insulin produced in bacteria requires mature A and B chains that can be connected with the appropriate disulfide bonds. The process first used for commercial production of recombinant human insulin is diagrammed in Figure 26.8. Researchers synthesized two artificial genes, one that encodes the A chain and the other that encodes the B chain. They attached the genes to the *lacZ* gene in the *lac* operon in such a way that the insulin codons were in frame with the *lacZ* codons, with an extra methionine codon placed between the *lacZ* gene and the insulin gene. This arrangement created fusion genes in which the insulin genes are under the control of the *lac* operon promoter and operator. The researchers placed the fusion genes (*lacZ–insulin A* and *lacZ–insulin B*) into plasmids and transformed the recombinant plasmids into separate colonies of *E. coli*. When they grew the

cells in the absence of lactose, the insulin genes were not expressed, and the cells grew and divided.

When the bacterial cultures were transferred to a medium with IPTG (an inducer of the *lac* operon), the introduced genes were expressed. The product of each introduced gene is a fusion protein with the left portion of the β-galactosidase protein (called a β-gal fragment) and either an A chain or a B chain of insulin attached to it by a methionine. The β-gal-Met-A and β-gal-Met-B fusion proteins are large and precipitate into inclusion bodies (aggregates of protein) in the bacterial cells, where they are protected from bacterial enzyme destruction. If the small insulin chains were produced by themselves, bacterial enzymes would destroy them, so an attached polypeptide (such as the β-gal fragment) is needed. Researchers can then isolate the bacterial cells for further processing.

Cyanogen bromide (CNBr) cleaves polypeptide chains on the carboxyterminal end of methionine only. Treatment of the isolated fusion proteins with cyanogen bromide cleaves the insulin chains from the β-gal fragments

lacZ   Met   proinsulin

lacZ-proinsulin fusion gene

a Transcribe and translate the fusion gene in a single *E. coli* culture to produce a β-gal-proinsulin fusion polypeptide.

β-gal   Met   B chain   *pro-* amino acids   A chain

H₂N ⎯   ⎯ COOH

β-gal-proinsulin fusion polypeptide

b Remove the β-gal fragment and induce tertiary structure formation to produce proinsulin.

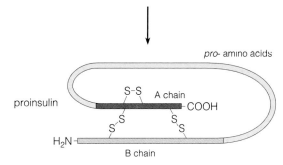

pro- amino acids

proinsulin

S-S   A chain
⎯ COOH

S   S   S   S

H₂N ⎯

B chain

c Remove the *pro-* amino acids from the proinsulin to produce genetically engineered human insulin.

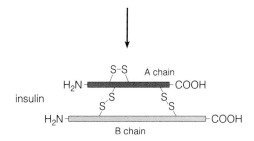

S-S   A chain
H₂N ⎯   ⎯ COOH

insulin   S   S   S   S

H₂N ⎯   ⎯ COOH

B chain

Figure 26.9 The single-chain method for production of recombinant human insulin.

because of the methionine between them. Neither of the insulin chains has methionine in it, so CNBr leaves the insulin chains intact. After researchers have purified the A and B chains, they mix them and treat them chemically so that the disulfide bonds form. The final product is genetically engineered human insulin.

The method that is now the preferred choice for commercial production of recombinant human insulin does not require two separate bacterial cultures. Instead, an engineered sequence that encodes proinsulin is attached to the β-gal fragment. In this situation, the entire insulin molecule can be made in a single bacterial culture. The proinsulin is purified and separated from the β-gal fragment either by treatment with CNBr or through enzymatic means. Enzymes then remove the *pro-* amino acids to produce intact human insulin (Figure 26.9).

Recombinant human growth hormone (HGH) was the second product of genetic pharmacology to be li-

censed for human use. The genetic engineering of recombinant HGH was much simpler than that for recombinant human insulin. The design of the expression vector for genetically engineered HGH is described in Example 9.5.

Foreign Protein Expression in Yeast

Bacteria are not eukaryotes, so the enzymes that they use for transcription and translation are different from those used in human cells. Although some eukaryotic proteins assume the correct structure when translated in a bacterial cell, there are other eukaryotic proteins that cannot be properly constructed in a bacterium. Yeast species are single-celled eukaryotes that grow much like bacteria. Some proteins that fail to assume the proper conformation in bacteria can be produced successfully in yeast.

The hepatitis B vaccine is among the first commercial successes with genetically engineered pharmaceuticals in yeast. The hepatitis B virus has a protein coat that certain human antibodies recognize as an antigen during a hepatitis B infection. Vaccines often contain weakened viruses or purified viral antigens that are then administered to the patient to elicit an immune response. However, in some cases, it is not possible to sufficiently weaken the virus or purify the antigen without contaminating the preparation with active viruses. If active virus particles contaminate a vaccine, then the vaccine can cause the disease that it is intended to prevent. In the case of hepatitis B, a purified coat protein without the viral DNA would make an ideal vaccine.

Researchers cloned the virus genes and chose the *HBsAg* gene because it encodes a protein that acts as an antigen on the virus's surface. They cloned the gene into a bacterial expression vector but could not use it to produce a functional antigen because of problems with protein structure. However, when they placed the gene into a yeast expression vector, the protein it encoded functioned as an antigen. Genetically engineered hepatitis B vaccine is now widely available, and many people have been immunized with it. In the coming years, many other safe and effective vaccines will be produced in a similar manner.

Human Proteins Produced in Cultured Mammalian Cells

Some functional human or viral proteins can be produced in bacteria with a little manipulation. Others can be produced in yeast. However, all of these proteins are relatively simple in their structure. Certain complex proteins that cannot be produced effectively in either bacteria or yeast are needed in abundance for medical purposes. These proteins must be produced in mammalian cells that can process the proteins properly and fold them into their correct conformations. Work is under way to produce such proteins in genetically engineered dairy cattle so that the proteins are secreted in the milk. The cow's cells could process human proteins effectively and produce them in large quantities. Because the cattle themselves are genetically altered, they should pass the engineered gene to their progeny so that unlimited quantities of the protein can be produced. Even simple proteins now produced in genetically engineered bacteria and yeast could potentially be produced at far greater savings in the milk of dairy cattle.

So far, however, production of proteins in cattle is still in the experimental stages and has not yet been developed sufficiently to use commercially. A readily available alternative is mammalian cell culture. Certain mammalian cells can be cultured and express the products of foreign genes. Mammalian cells, however, are used only when bacteria and yeast systems do not work, because mammalian cells grow more slowly and are very expensive to maintain. Thus, the protein must be highly valuable for a mammalian cell system to be used.

The first protein produced commercially in a mammalian cell system was tissue plasminogen activator, an enzyme that modifies plasminogen, a protein in the blood, into plasmin. Plasmin, in turn, digests another protein, fibrin, into smaller fragments. Fibrin is the protein that causes blood clots in coronary arteries that stop blood flow to the heart during a heart attack. When plasminogen activator protein is administered to a heart attack victim, it changes plasminogen into plasmin, which then digests the clot and restores blood flow to the heart.

Researchers placed a plasminogen activator gene derived from cDNA under the control of a strong promoter and added the gene into a mammalian cell culture. Attached to the gene is a second gene, called *DHFR*, that encodes the enzyme dihydrofolate reductase (DHFR). This enzyme provides resistance to the drug methotrexate, so methotrexate can be used to select cells with the plasminogen activator gene. As levels of methotrexate are increased, resistance only occurs when the *DHFR* gene is amplified into many copies. The cells that survive have multiple copies of the *DHFR* gene, which, in the process of amplification, multiplies the copy number of the closely linked plasminogen activator gene. The multiple copies of the plasminogen activator gene produce useful quantities of plasminogen activator protein.

~

Genetic pharmacology is the production of medically useful proteins in bacterial, yeast, or mammalian cells. Several products produced by genetic pharmacology are currently being marketed.

26.5 HUMAN GENE THERAPY

Currently, most genetic disorders are treated by focusing on alleviating symptoms rather than correcting the genetic cause of the disorder. Human gene therapy is focused on correcting the genetic cause of the disorder, which, in turn, alleviates the symptoms. Gene therapy can be divided into two types, somatic-cell gene therapy and germ-line gene therapy.

Somatic-cell gene therapy is based on the principle that a functional gene can be delivered to somatic cells in a tissue to correct a disorder in the cells. There are two types of somatic-cell gene therapy. The first is **in vivo therapy**, a procedure in which physicians introduce a functional gene into a virus and add the virus directly to the cells in the body so that the virus introduces the functional gene into the cells. Example 24.5 illustrates how this type of gene therapy has been used to treat lung cancer. The second type is **ex vivo therapy**, a procedure in which physicians extract somatic cells from a patient, insert a functional gene into the extracted cells, and return the cells to the tissue from which they were extracted. The

gene is expressed in the introduced cells and their somatic progeny cells. The inserted gene is not present in the germ line, so the gene functions only in specific somatic cells and is not transmitted to offspring.

Germ-line gene therapy is the alteration of the genetic composition of the germ line so that people who carry a mutant allele no longer transmit that allele to their progeny. Germ-line gene therapy has not yet been developed for humans, although germ-line alteration is now routine in many plant and animal species. Technical and ethical constraints have prevented the development of human germ-line gene therapy. Because germ-line gene therapy has not been developed for use in humans, all references hereafter in this chapter to gene therapy refer to somatic-cell gene therapy.

Gene therapy is still in its infancy, although progress in recent years has been impressive. The first federally approved clinical trials of gene therapy began in September 1990. By 1995, over 100 clinical trials of both in vivo and ex vivo gene therapy were under way. Some forms of gene therapy may be available for widespread clinical application soon. Unfortunately, the cost of gene therapy is very high. However, costs may decline as improvements are found.

Viral Vectors in Gene Therapy

For gene therapy to be successful, a functional copy of the gene must be delivered to living cells, and the gene must be expressed. Ideally, the gene will be incorporated into the chromosomal DNA, where it will be expressed as if it were a part of the original genome. There are many ways to introduce DNA into cells, but the methods used most often require viral vectors. A retrovirus is an RNA virus that is reverse-transcribed into DNA after it has infected a cell. The DNA is then inserted into the chromosomal DNA of the host cell, where many copies of the virus can be produced by transcription. Retroviruses can be engineered so that most of the viral genome is replaced by the functional human gene. These retroviral vectors can transfer an engineered gene into the chromosomes after the virus has infected the cell. Because the process utilizes infection and gene transfer, it is called **transfection**. Once inside the cell, the engineered virus is reverse-transcribed into DNA, and the engineered gene is incorporated into the chromosomal DNA.

Researchers construct viral vectors by deleting the viral genes that are necessary for replication, lysis, and infection of other cells. Consequently, the engineered virus cannot spread to cells other than those that are directly transfected. The vector also cannot be transmitted from one person to another like a wild-type virus.

Retroviruses can only infect cells that are actively dividing. This makes retroviruses an excellent choice as a vector for cancer gene therapy because the virus may target actively dividing tumor cells. On the other hand, this characteristic limits the use of retroviruses in other types of gene therapy because tissues that do not contain a large fraction of actively dividing cells cannot be successfully treated with retroviral vectors. Under these circumstances, adenovirus vectors are often used. Adenoviruses are double-stranded DNA viruses that insert themselves into the chromosomal DNA of the cells they infect. Unlike retroviruses, adenoviruses may infect cells that are not actively dividing.

Treatment of ADA Deficiency with Gene Therapy

One of the first examples of successful gene therapy was treatment of ADA deficiency. The *ADA* locus on chromosome 20 (OMIM 102700) encodes the enzyme adenosine deaminase (ADA), which degrades excess adenosine monophosphate (AMP) into uric acid. When functional ADA is absent, the excess AMP is converted into compounds that are toxic to lymphoblasts, a type of cell that is active in the immune system. ADA deficiency causes severe combined immune deficiency (SCID, see section 25.5), a disease in which the immune system fails to fight infections. People with ADA deficiency may die from infections that are usually harmless to most people. ADA deficiency has traditionally been treated with transplants of bone marrow that produces functional ADA.

ADA deficiency was a prime candidate for gene therapy for a number of reasons. First, it required the addition of a single enzyme encoded by a single gene. Second, it was known to be curable by introduction of foreign cells. Bone marrow transplants had been successful in the past, but their use was limited by the need to find an immunologically compatible donor. Lastly, ADA-deficient cells contain toxic levels of metabolites, which prevent them from growing and dividing. Cells with a functional *ADA* gene that have been introduced into a patient should have a selective advantage over the ADA-deficient cells.

The first experiments were conducted on a 4-year-old ADA-deficient girl. Some of her T cells were isolated and incubated with interleukin-2 to stimulate growth. Physicians inserted a functional *ADA* gene, together with a neomycin resistance marker gene, into the cells using a retroviral vector. They then reintroduced the engineered cells into the bone marrow and followed their progress by extracting blood samples and using PCR amplification of the neomycin marker as an indicator that engineered T cells were still present. The initial treatment has been followed by periodic maintenance infusions of engineered T cells. The treatment was successful, and the patient now lives a normal life with an immune system that successfully fights infections. A second ADA-deficient child has received gene therapy with similar success.

The following example shows how researchers have genetically engineered retroviral vectors for use in gene therapy of ADA deficiency.

Example 26.4 Genetic engineering of a retroviral vector for gene therapy.

In 1986, Kantoff et al. (*Proceedings of the National Academy of Sciences, USA* 83:6563–6567) described a genetically engineered retroviral vector used for introducing a functional *ADA* gene into bone marrow cells for treatment of ADA deficiency with ex vivo gene therapy. To construct the vector, Kantoff et al. attached the cloned cDNA sequence of the human *ADA* gene to a high-expression promoter from the SV40 virus. They placed the engineered gene into an *Xho*I restriction site in the cloned DNA sequence of an altered Maloney murine leukemia virus (a retrovirus). The viral structural genes had been deleted from the viral DNA and replaced with a bacterial neomycin resistance gene. They named the vector with the inserted gene SAX (for S̲V40 promoter, human A̲DA gene inserted into the *Xho*I site). To replicate the SAX vector, the authors introduced the plasmid containing the SAX virus into a genetically engineered cell line (called a helper cell line) that contained all the genes necessary for viral transcription and assembly. They then isolated intact retroviral particles from this cell line and used them to transfect cultured ADA-deficient human T and B cells.

Problem: **(a)** What precautions did the authors take to ensure that the retroviral vector could not replicate in human cells, spread among cells within the patient's body, or be transmitted from the patient to other people? **(b)** What was the purpose of the helper cell line? **(c)** Retroviruses are composed of RNA rather than DNA, so why was the engineered viral sequence cloned as DNA in a plasmid?

Solution: **(a)** The viral structural genes had been removed from the viral vector so it could not replicate nor infect other cells after it had been introduced into human cells. **(b)** Because the virus is incapable of replication in normal cell lines, the researchers needed a way to replicate and assemble the virus in the laboratory so that a sufficient number of viral particles were available for transfection of human cells. They used helper cells that were genetically engineered to contain all the genes necessary for viral replication and assembly. The viral vectors can only replicate and assemble in such a cell culture because the cultured cells contain all the functions for replication and assembly that the engineered viral vector lacks. After the viruses had been isolated from the cell culture, they could no longer replicate. **(c)** The vector DNA sequence can be maintained and replicated as part of a DNA plasmid in bacteria. When the assembled RNA viral

vectors are needed, researchers transfer the DNA to the helper cell line, where the helper cells transcribe the DNA into viral RNA molecules, which are then assembled into viral particles that can transfect cultured human cells.

Although the future of gene therapy is very promising, there are formidable obstacles. Gene therapy can currently be used to treat only a few genetic disorders. It is so expensive that only a small number of patients with these disorders can be treated. Its use has also raised serious ethical and legal questions, some of which we will discuss in Chapter 28.

~

Although still in the experimental stages, human gene therapy has proven successful in treating certain genetic disorders.

26.6 CLINICAL GENETICS

With the rapid emergence of medical genetic information and methods for diagnosis and treatment of medical disorders, clinical use of genetic technology has increased substantially over the past two decades. In 1991, the American Board of Medical Specialties recognized clinical genetics as a separate medical specialty. **Clinical genetics** is defined as the diagnosis of genetic disorders and the care of individuals with genetic disorders. The phenotypes associated with most genetic disorders are characterized as syndromes, a collection of symptoms that affect different organs and tissues. Thus, clinical geneticists often work closely with physicians and practitioners in many other specialties to ensure that patients with a genetic disorder receive appropriate treatments for the variety of symptoms that accompany each disorder.

Clinical genetics starts with the primary care physician, who must recognize the collection of symptoms that may be associated with a genetic disorder. Primary care physicians must therefore maintain a current understanding of advances in medical genetics. The primary care physician refers patients with a suspected genetic disorder to a clinical geneticist for an accurate diagnosis and development of a treatment plan. Clinical geneticists review the symptoms along with the family history to determine what genetic disorders may be possible. The geneticist then determines which diagnostic tests should be utilized to accurately determine which, if any, known genetic disorder is the cause of the symptoms. Advances in cytogenetics, biochemical genetics, genetic mapping, and gene cloning and sequencing have greatly improved the accuracy and range of diagnostic genetic tests currently available for clinical use. It is now possible for clinical geneticists to accurately diagnose many genetic disorders in a short period of time.

Although a family history of a disorder is strong evidence of a genetic component, a lack of family history does not exclude a genetic component. Most patients with a genetic disorder have no family history of the disorder, because many genetic disorders are caused by homozygosity for recessive alleles (or, in some cases, new mutations), which may not have appeared phenotypically in the patient's known ancestry.

After the geneticist has diagnosed a genetic disorder, patients and relatives often receive **genetic counseling**. The American Society of Human Genetics has provided this widely accepted definition of genetic counseling that includes the responsibilities of a genetic counselor:

> Genetic counseling is a communication process which deals with the human problems associated with the occurrence or risk of occurrence of a genetic disorder in a family. This process involves an attempt by one or more appropriately trained persons to help the individual or family to: (1) comprehend the medical facts including the diagnosis, the probable course of the disorder, and the available management, (2) appreciate the way heredity contributes to the disorder and the risks of recurrence in specified relatives, (3) understand the alternatives for dealing with the risk of recurrence, (4) choose a course of action which seems to them appropriate in their view of their risk, their family goals, and their ethical and religious standards and act in accordance with that decision, and (5) make the best possible adjustment to the disorder in an affected family member and/or to the risk of recurrence of the disorder.

In most medical situations, patients must make decisions about appropriate treatment. However, when faced with genetic disorders, people must also make decisions about future reproduction. Implicit in the definition of genetic counseling is the need to provide information about the risks associated with recurrence, so that parents can make informed decisions about reproduction. Explicit in the definition is the need to respect the ethical and religious standards of the family, which are likely to play a role in their decisions.

Alleles that confer genetic disorders are inherited, so relatives of a patient with a genetic disorder often receive genetic counseling and testing to determine what risks, if any, they may have for appearance of the genetic disorder in their offspring. In some cases, such counseling and testing may prevent serious consequences. For example, susceptibility to colorectal cancer is often inherited as a mutation in the *APC1* gene. When diagnosed as an inherited disorder in a family, the mutant allele that confers inherited susceptibility can be identified, and relatives can be contacted and tested so that physicians can monitor those who are at risk. Early detection and treatment of cancer under these circumstances can save the lives of many people in an extended family.

Because most people who meet with genetic counselors are unfamiliar with the principles of genetics, ge-

netic counselors must explain the circumstances in a way that is understandable, so that people can make informed choices. Genetics counselors typically receive training in clinical genetic medicine, psychotherapy and counseling, ethics, and legal issues related to genetics. Genetics counselors are trained in specialized programs and are certified by the American Board of Genetic Counseling. Most are employed at large metropolitan hospitals. With recent advances in genetic testing, the need for genetic counseling has increased.

~

Clinical genetics is a specialty in medicine that includes the diagnosis of genetic disorders and the care of patients with genetic disorders. Genetic counseling is available to families whose members are at risk of inheriting a genetic disorder.

SUMMARY

1. Applications of genetics in medicine have been important for many years. With improvements in genetic technology, applications of genetics in medicine are widespread and increasing in importance.

2. The human genome project includes mapping the human genome with DNA markers, identification and characterization of previously unmapped or unknown genes, studies on model organisms with applications to human genetics, and determination of a consensus nucleotide sequence for the entire human genome.

3. Genetic testing is now available for detecting the mutant alleles that cause many genetic disorders. Genetic testing may be based on biochemical, cytological, or DNA tests. In many cases, genetic testing may be used prenatally.

4. Certain biochemical genetic tests are inexpensive and may be used in widespread screening programs. Mandatory testing of newborns for PKU in the United States and other countries is an example of inexpensive genetic screening.

5. DNA fingerprinting is valuable for identity exclusion and inclusion in criminal cases and paternity testing.

6. Genetic pharmacology is the production of a pharmaceutical product in a host organism or cell culture (usually bacteria) using recombinant DNA. Currently, several genetically engineered pharmaceuticals, such as human insulin, are in widespread use.

7. Not all human proteins can be successfully produced and processed in bacteria. Yeast and animal cells are alternative hosts that may produce and process the gene products properly.

8. Human gene therapy is the introduction of genetically engineered somatic cells into a patient to provide a genetic function that is missing in the patient. Gene therapy for ADA deficiency is an example of successful

gene therapy. Currently, most applications of gene therapy are still in the experimental stages.

9. Clinical genetics is a medical specialty that focuses on the diagnosis and treatment of genetic disorders. Trained genetic counselors provide genetic counseling to families whose members are at risk for a genetic disorder. With the added importance of genetics in medicine, the need for genetic counseling has increased.

QUESTIONS AND PROBLEMS

1. Name four reasons why a human insulin gene isolated from a genomic library could not produce functional insulin in bacteria without modification.

2. What circumstances would make a protein a good candidate for genetic pharmacology using bacterial hosts? Provide both biological and economic considerations in your answer.

3. What types of proteins could not be effectively produced using genetic pharmacology?

4. Most types of hemophilia are caused by deficiency for protein-clotting factors. Would these clotting factors be good candidates for genetic pharmacology? Why or why not?

5. Why has genetic testing for PKU become widespread whereas genetic testing for sickle-cell trait has not, even though the frequency of sickle-cell trait is higher in the United States than that of PKU?

6. In constructing the plasmids for genetically engineered insulin, genetic engineers attach the DNA sequences that encode the two insulin chains to the *lacZ* gene via a methionine bridge. Provide three reasons why genetic engineers constructed the genes in this way instead of simply placing a cDNA copy of the insulin gene in an expression vector by itself.

7. ADA deficiency has been treated with gene therapy. Seemingly, it would be much simpler to use the cloned gene in genetic pharmacology to produce the functional product then administer the product, rather than go through the elaborate and expensive procedures required for gene therapy. Why is it not possible to use genetic pharmacology to treat ADA deficiency?

8. Research is under way to provide gene therapy for people who suffer from cystic fibrosis. One of the most severe symptoms of the disorder is excessive mucus production in the lungs. Researchers have proposed delivery of the functional *CF* gene to the lungs in a virus that normally causes respiratory disease. Considering that it is often difficult to achieve proper protein conformation and activity in genes cloned into bacteria that infect humans (such as *E. coli*), why should it be possible to achieve proper protein conformation and activity from a gene cloned in a virus that infects humans?

9. What aspects of DNA fingerprinting make it more valuable than blood typing for identification of people?

10. A woman whose age is 22 and whose husband's age is 24 bears a child with Down syndrome. The child is their first. What genetic tests would be appropriate under these circumstances, and why? Use information from this chapter and Chapter 17 to help you answer this question.

11. A single DNA test is capable of detecting many carriers of mutant *CF* alleles. However, a single test is insufficient for detecting all carriers. Why is this so?

12. A genetic test for fragile-X syndrome usually includes PCR amplification of the region where expansion mutations are found, a test that detects the level of methylation present in the 5′ region of the gene, and a mitotic karyotype from cultured cells. (a) Using information from this chapter and from Chapters 5 and 14, describe the purpose of each of these tests. (b) What types of mutations that can cause fragile-X syndrome would not be detected with these tests?

13. The perception of most people is that the human genome project is an effort to determine the DNA sequence of the human genome. What is a more complete and accurate definition of the human genome initiative?

14. Each person (except for identical twins) has a unique DNA sequence in their genome. Given this observation, how is it possible to define the sequence of the human genome?

15. Highly repetitive DNA sequences appear frequently throughout the human genome. Explain how these sequences are used as a tool in genome sequencing.

16. What are sequence tagged sites, and how are they used in mapping the human genome?

17. The hepatitis B virus produces a coat protein (HBsAg) that is glycosylated (attached to a lipid) in its final form. This protein is highly antigenic in humans and is now used as a vaccine against the virus. When the cloned gene that encodes HBsAg is expressed in *E. coli*, the entire protein is produced, but it is not glycosylated. MacKay et al. (1981. *Proceedings of the National Academy of Sciences, USA* 78:4510–4514) tested the protein produced in *E. coli* with monoclonal antibodies that had been raised against the native HBsAg glycosylated product and found that the antibodies failed to react with the protein produced in *E. coli*. However, when Ammerer and Hall (1982. *Nature* 298:347–350) expressed the cloned gene in yeast and tested it, all properties, including the immunological response, were identical to those of the native protein. What do these results indicate about differences in processing of this protein in yeast and *E. coli* cells and the differences in antigenic reactions to the products?

18. Fragile-X syndrome and phenylketonuria are genetic disorders that can both be detected with DNA-based tests and tests that do not directly utilize DNA. Both disorders are characterized by mental retardation. Of the two, fragile-X syndrome is more common, about 1 in 4000 in

mixed ethnic populations. Until the mid-1990s, the most common genetic test for fragile-X syndrome was cytological examination of metaphase chromosomes to detect a secondary restriction in the X chromosome. This test has been supplemented by PCR or RFLP DNA analysis. Only children with symptoms of fragile-X syndrome or a family history of fragile-X syndrome are tested. PKU affects about 1 in 12,500 newborns in the United States. Newborns with PKU have elevated levels of serum phenylalanine at birth, which can be detected using a biochemical test of phenylalanine levels in blood drawn from each infant at birth. All newborns in the United States, and in many other countries as well, are routinely screened for elevated phenylalanine. DNA tests for detection of many of the mutant alleles at the *PAH* locus have been developed but are used only when there is a positive indication of elevated phenylalanine or a family history of the disorder. Why is PKU screening routine among all newborns, whereas fragile-X syndrome screening is not, even though fragile-X syndrome is more common? Use information from this chapter and other chapters to answer this question.

19. One of the symptoms associated with cystic fibrosis is excessive production of mucus in the lungs. This leads to bacterial lung infections, which are the most common cause of death in patients with cystic fibrosis. The identification and cloning of the *CF* gene made possible the development of vectors that could potentially deliver functional copies of the *CF* gene to cells in the lung to restore *CF* function and prevent excess mucus formation. Since viruses infect lung cells during respiratory infections, an engineered respiratory virus could potentially be a good vector for delivering functional *CF* genes to lung cells. Adenovirus vectors with functional *CF* genes were developed, and initial in vitro experiments suggested that the gene could be delivered to lung cells and *CF* function in those cells restored. In 1996, Boucher (*Trends in Genetics* 12:81–84) reviewed the progress of *CF* gene therapy in clinical trials with humans. The lung epithelium consists of a surface layer of columnar cells in which the *CF* gene is normally expressed in humans. Below the layer of columnar cells is a layer of basal cells in which the *CF* gene is not normally expressed. Patients with cystic fibrosis lack *CF* expression in both the columnar and basal cells. Gene therapy administered in vivo was not successful in clinical trials in spite of success from in vitro studies. The lack of correlation between these studies was resolved when it was discovered that columnar cells are resistant to infection by adenoviruses, but basal cells are not. The tissues used in the in vitro studies contained basal cells that were exposed to the virus, but basal cells are not exposed to the virus in vivo. An additional problem was the discovery that adenovirus vectors may induce apoptosis in the cells they infect. Given these limitations, what steps might be taken to improve vectors for use in gene therapy for cystic fibrosis?

20. Hypothyroidism is a common disorder in humans that is often caused by deficiency for thyroid hormone. It is usually treated by oral administration of an effective and inexpensive synthetically produced thyroid hormone substitute. The chemical structure of native human thyroid hormone is shown below.

(a) To which general class of molecules does thyroid hormone belong? **(b)** Might it be advisable to attempt construction of genetically engineered bacteria to produce this molecule in bacteria using genetic pharmacology? If so, how could those strains be designed? If not, why not?

21. T4 endonuclease VII is an enzyme that cleaves DNA at the site of a single nucleotide-pair mismatch. In 1995, Youil et al. (*Proceedings of the National Academy of Sciences, USA* 92:87–91) reported the results of experiments to test the ability of T4 endonuclease VII to detect mutations. They amplified mutant and normal DNA of several genes, denatured the DNA, and allowed heteroduplex normal-mutant DNA molecules to form when the DNA hybridized. They then treated the heteroduplex DNA with T4 endonuclease and detected cleavage by electrophoresis. They examined 26 mutations, 22 of which were single nucleotide-pair substitutions and 4 of which were deletions. All but 2 of the 22 substitution mutations and 3 of the 4 deletion mutations were detected using this method. What advantage might a mutation detection method such as this provide in genetic testing?

22. Proinsulin consists of the A, B, and C chains with the *pre-* amino acids removed. Currently, recombinant human insulin is produced as proinsulin in a single bacterial culture, and the purified proinsulin is cleaved with enzymes to remove the C peptide, producing mature human insulin. The C peptide of insulin is removed from proinsulin both in the body and in processing recombinant human insulin. However, there is some evidence that the free C peptide plays an important biological role and should be administered to some patients with type I diabetes along with insulin. In 1996, Jonasson et al. (*European Journal of Biochemistry* 236:656–661) described a vector with a redesigned recombinant human insulin gene. The gene encodes a fusion protein that consists of two IgG-binding domains (called ZZ) on the amino end of a proinsulin polypeptide. The ZZ-proinsulin fusion protein can be easily purified from other proteins because of the binding affinity of the ZZ domains for IgG antibodies. The purified protein can then be induced to fold and form the disulfide bonds between the A and B chains. The enzyme trypsin cleaves polypeptide at the C-terminal end of basic amino acids. There are three trypsin cleavage sites: (1) between the ZZ domains and the B

chain, (2) between the B chain and the C chain, and (3) between the C chain and the A chain. When the purified ZZ-proinsulin is treated with trypsin, mature recombinant human insulin and free C peptide are released from the ZZ domains. What aspects of this design are advantageous for commercial processing of recombinant human insulin?

FOR FURTHER READING

An excellent textbook that focuses on medical applications of genetics is **Jorde, L. B., J. C. Carey, and R. L. White. 1997.** *Medical Genetics.* **St. Louis, Mo.: Mosby**. A detailed review of the history of the human genome project is **Cook-Deegan, R. 1994.** *The Gene Wars: Science, Politics, and the Human Genome.* **New York: Norton**. A recent book that explains current methods of DNA fingerprinting and its use in forensic analysis is **Inman, K. 1997.** *An Introduction to Forensic DNA Analysis.* **Boca Raton, Fla.: CRC Press**. The status of somatic-cell gene therapy was reviewed by **Bank, A. 1996. Human somatic cell gene therapy.** *BioEssays* **18:999–1007;**

Hanania, E. G., J. Kananagh, G. Hortobagyi, R. E. Giles, R. Champlin, and A. B. Deisseroth. 1995. Recent advances in the application of gene therapy to human disease. *The American Journal of Medicine* 99:537–552; Crystal, R. G. 1995. Transfer of genes to humans: Early lessons and obstacles to success. *Science* 270:404–410; and Marshall, E. 1995. Gene therapy's growing pains. *Science* 269:1050–1055. A book that explains methods and medical applications of gene therapy is **Culver, K. W. 1996.** *Gene Therapy: A Primer for Physicians,* **2nd ed. New York: Mary Ann Liebert, Inc**. The development of genetic pharmacology is explained in **McKelvey, M. D. 1996.** *Evolutionary Innovations: The Business of Biotechnology.* **Oxford, England: Oxford University Press**.

For additional reading, go to InfoTrac College Edition, your online research library at: http: www.infotrac-college.com/brookscole

CHAPTER 27

The origins and locations of genetic diversity for most domesticated plants and animals are found in less developed parts of the world.

~

Breeding and biotechnology of domesticated plants and animals have contributed substantially to increases in agricultural production and will continue to be a major part of agricultural improvement.

~

There are numerous applications of genetics in industry. And with the development of recombinant DNA technology, industrial applications of genetics are increasing.

GENETICS IN AGRICULTURE AND INDUSTRY

For millennia, people have practiced hybridization and selection to genetically improve plants and animals. We benefit today from plants and animals that our remote ancestors domesticated. With the discovery of Mendelian genetics and the subsequent development of population and quantitative genetics, plant and animal breeders developed systematic methods of hybridization, testing, and selection that contributed to rapid genetic improvement of domestic plants and animals. In the past 50 years, plant and animal production has increased severalfold. We now produce more food than at any other time in the history of the world. We also produce more food per capita than at any other time in the history of the world, at a time when the world's human population is the largest it has ever been.

The fact that we produce more food now than ever before does not mean that everyone on the planet has enough to eat. Although we currently produce enough food to provide an adequate diet for every person on the planet, tragically, over 1 billion people suffer from severe hunger and malnutrition. Due to advances in agricultural production and improvements in political and economic stability, the proportion of people who suffer from severe malnutrition and hunger worldwide has declined somewhat in recent years.

Some of the increases in food production can be attributed to environmental modifications. Improvements in irrigation, weed control, fertilization, animal feed regimes, and disease and pest control, along with many other environmental modifications, have contributed to increased production. However, genetic improvement through modern plant and animal breeding has contributed more than any other factor to increased production in nearly every domesticated species. For many plant species, it is possible to measure the effect of genetic improvement. Plant varieties that were used decades ago are still available. They can be grown side by side with new varieties using modern agricultural practices. Because they are grown in the same environment, any differences in productivity between the old and new varieties can be attributed to genetic improvement. The results of experiments in several important crop species indicate that over 50% of the improvements in productivity are due to genetic improvement. In maize, for example, approximately 89% of the yield gains in the United States from 1930 to 1980 are due to genetic improvement.

Animal breeders have achieved similar genetic gains in domestic animals. Milk production per cow has doubled in 25 years. The time it takes to raise chickens to market size

is half of what it was 50 years ago. With the demand for leaner meat, breeders have reduced overall fat in pork by 31% and saturated fat by 29% in only 10 years.

The development of recombinant DNA methods has added to the potential for improvement through genetic engineering. The first genetically engineered plants and animals are just now entering the market. In spite of periodic downturns in farm economies, agriculture remains one of the largest industries in the United States and most other countries. Sales of seed and animal stock are among the most lucrative of enterprises. Even small genetic improvements in plants and animals can bring enormous profits to the companies that achieve them.

Some people, who were uninformed about the applications of genetics in agriculture, assumed that with the advent of genetic engineering, traditional plant and animal breeding would become outdated as modern biotechnology took the place of breeding programs. This notion was, and still is, mistaken. With the promise of quick profits, investors provided the funds to initiate agricultural biotechnology companies during the 1980s. However, many of those companies went bankrupt within a few years after their inception. And those that succeeded were usually tied to large traditional breeding programs. Rather than replace traditional plant and animal breeding, biotechnology has created an even greater need for it. The products of the biotechnology laboratory generally focus on a single trait. Traits altered by biotechnology must then be refined and incorporated into plants and animals with other desirable traits through a breeding program.

Most people are surprised to learn that one of the greatest contributions biotechnology has made in agriculture has nothing to do with the genetic engineering of plants and animals. Plant and animal breeders use molecular markers to tag important genes and increase heritabilities, thus increasing the efficiency of selection in modern breeding programs. Although no new genes have been added, recombinant DNA methods used to generate DNA markers have improved the effectiveness of breeding.

Industry has also benefited from applications of genetics. The food-processing industry has utilized improved microbes for fermentation, and plant and animal products have been altered to meet processing demands. Foods have been genetically improved for easy packing and longer shelf lives. Other types of industries have likewise benefited. Enzymes used in detergents have been improved through selected alterations in the genes that encode them. Improved strains of bacteria and other microbes that process sewage and industrial waste are now available. Native plant species have been bred and selected to revegetate mine spoils and other degraded lands in an overall effort by industry to reclaim the lands they have used and to comply with regulations designed to minimize ecological degradation.

Figure 27.1 Genetic diversity in potato. The potatoes pictured here are representative of the genetic diversity found in potatoes grown in Andean countries of South America. (Visuals Unlimited)

In this chapter, we will review modern methods of plant and animal breeding and the ways in which biotechnology has contributed and will contribute in the future to agriculture. We will also look at several of the applications of genetic improvement in industry.

27.1 GENETIC DIVERSITY

Many domesticated plant and animal species have spread throughout the world from the area where they were domesticated. Few of the plants and animals used for food in the United States are North American natives. Let's use as an example the most American of meals, a Thanksgiving dinner of roast turkey, stuffing, cranberry sauce, mashed potatoes, yams, peas and carrots, and a green salad of lettuce, red cabbage, tomatoes, cucumbers, and radishes. The turkey and tomatoes originated in Mexico and Central America. Although there are wild turkeys in the United States, the domesticated type used in modern Thanksgiving dinners originated in Mexico. Potatoes came from South America, in the Andes mountains of Bolivia, Peru, and Ecuador. The yams are African. Peas, radishes, and the wheat in the stuffing bread originated in the Near East. The carrots and lettuce come from the Mediterranean region. Cabbage is European; cucumbers and chicken eggs (used in the stuffing) are from India. The only food of North American origin in the meal is the cranberries.

The **center of origin** (the region where a species was domesticated) is usually the same as the **center of diversity** (the region in which the greatest genetic diversity for the species is found). The usual pattern is substantial genetic diversity in the area where a plant or animal species was domesticated, and very little genetic diver-

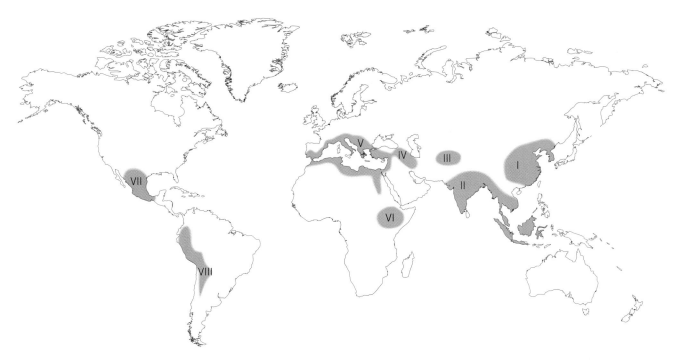

Figure 27.2 The eight centers of origin identified by Nicolai Vavilov.

sity for that species throughout the rest of the world. Potatoes are a good example. In a U.S. supermarket, there are typically three types: white, red, and russet (the brown-skinned type used for baking). In a Peruvian street market, on the other hand, it is not unusual to find an enormous variety of potatoes: potatoes with purple, black, red, brown, orange, or white skins; long and spindly ones, small spherical potatoes that look like marbles, as well as the types that are common in the United States (Figure 27.1).

There are several major centers of origin in the world, where most plant and animal species were domesticated and where most of the genetic diversity for these species exists. The Russian geneticist Nicolai Vavilov identified eight primary centers of origin for plants, which are also often called centers of diversity (Figure 27.2). These centers also apply to the origin and diversity of most domesticated animals. Agriculture is a prerequisite for a civilization to rise; thus, it is no surprise that the centers of diversity are located in regions where advanced ancient civilizations flourished. For example, the two American centers of diversity are located in central Mexico, where the ancient Aztecan and Mayan civilizations arose, and in the central Andes mountains of South America, where the ancient Aymaran and Incan civilizations thrived. Although we often admire ancient civilizations for their arts, literature, and culture, perhaps the greatest legacy they left us is the genetic heritage of plants and animals that we use for food. Table 27.1 provides a list of the places where many of our domesticated plants and animals originated.

Genetic Erosion

Genetic diversity in agricultural species is important because it provides the raw material for genetic improvement of those species. Selection, be it natural or artificial, is ineffective unless there is genetic variation. Unfortunately, many of our domesticated plant and animal species have little diversity outside of their center of origin. Soybean is a good example. It was domesticated in China and is a relative newcomer to the rest of the world. Currently, over 90% of soybeans are produced outside of China, mostly in the United States and Brazil. Modern plant breeders developed the soybean varieties grown in the United States and Brazil. The pedigrees of those varieties, for the most part, trace back to only 20 homozygous true-breeding varieties introduced to the United States in the early 1900s. Throughout east Asia, the number of genetically distinct varieties numbers in the hundreds of thousands. Thus, the genetic base for over 90% of the soybeans grown worldwide is extremely narrow. Other domesticated plant and animal species face a similar situation.

Ironically, the success of modern plant and animal breeding threatens the very diversity on which it depends. The improved varieties and breeds released from

Species	Center of Origin	Vavilov's Center*
Plants		
Alfalfa	Middle East	IV
Apple	Central Asia	III
Apricot	Central Asia	III
Avocado	Middle America	VII
Banana	South Asia	II
Barley	Middle East	IV
Bean	Middle and South America	VII, VIII
Beet	Mediterranean	V
Blueberry	North America	None
Broccoli	Mediterranean	V
Brussels sprouts	Mediterranean	V
Cabbage	Mediterranean	V
Cantaloupe	East Africa	VI
Carrot	Mediterranean	V
Cauliflower	Mediterranean	V
Celery	Mediterranean	V
Cherry	Central Asia	III
Coffee	East Africa	VI
Cotton	Middle East, South America	IV, VIII
Cranberry	North America	None
Cucumber	South Asia	II
Date	South Asia	II
Eggplant	South Asia	II
Fig	Middle East	IV
Flax	Mediterranean to South Asia	II through V
Garlic	Central Asia	III
Grape	Middle East	IV
Grapefruit	West Indies	None
Guava	West Indies	None
Jerusalem artichoke (sunflower root)	North America	None
Lemon	South Asia	II
Lettuce	Mediterranean	V
Lime	South Asia, China	I, II
Maize	Middle America	VII
Mango	South Asia	II
Millet	East Africa	VI
Oats	Northwestern Europe	None
Olive	Mediterranean	V
Onion	Central Asia	III
Orange	South Asia	II

*See Figure 27.2.

Species	Center of Origin	Vavilov's Center*
Plants (continued)		
Orange	South Asia	II
Papaya	Middle America	VII
Pea	Middle East	IV
Peach	China	I
Peanut	South America	VIII
Pear	Central Asia	III
Pecan	North and middle America	VII
Peppers (bell, chili)	Middle America	VII
Pineapple	Brazil	None
Plum	Central Asia	III
Potato	South America	VIII
Pumpkin	North and middle America	VII
Radish	Mediterranean	V
Rice	Southeast Asia	II
Rye	Middle East	IV
Sorghum	East Africa	VI
Soybean	China	I
Spinach	Middle East	IV
Squashes	Middle America	VII
Sugarcane	South Asia	II
Sunflower	North America	None
Sweet potato	Middle America	VII
Tangerine	Southeast Asia	II
Tobacco	South America	VIII
Tomato	Middle America	VII
Watermelon	East Africa	VI
Wheat	Middle East	IV
Yam	South-Southeast Asia, Africa	II, VI
Animals		
Alpaca	South America	VIII
Camel	Middle East, North Africa	IV, V
Chicken	South Asia	II
Cow	Middle East	IV
Goat	Middle East	IV
Horse	Middle East	IV
Llama	South America	VIII
Pig	Middle East	IV
Rabbit	Europe	V
Sheep	Middle East	IV
Turkey	Middle America	VII

*See Figure 27.2.

plant and animal breeding programs are usually much more productive than the native types from which they were developed. These improvements make new varieties attractive throughout the world. The spread of modern varieties and agricultural practices throughout the world began in the 1940s and has more than tripled food production worldwide since that time, a phenomenon known as the **green revolution**.

Although the spread of genetically improved plants and animals has contributed to increased food production worldwide, it has two inherent problems. First, most plant varieties and animal breeds are more genetically uniform than their unimproved counterparts. Genetic uniformity provides the potential for pest and disease epidemics. In 1970, the U.S. maize crop suffered an epidemic of southern corn leaf blight because of mitochondrial genetic uniformity (see section 18.3). Green revolution varieties of Mexican wheat suffered a serious fungal infection epidemic in 1976, also due to genetic uniformity. Similar epidemics have appeared sporadically in other parts of the world. When one plant is genetically susceptible to a disease, and all other plants are genetically similar or identical, they are all susceptible. Fortunately, where active breeding programs exist, plant breeders usually become aware of a new disease or pest before it spreads and are able to breed resistant varieties before a widespread outbreak. To do this, though, they must often rely on alleles that confer resistance derived from the pool of genetic diversity that came from the center of origin.

The second problem is irreplaceable loss of genetic diversity. When improved plant varieties or animal breeds are introduced into a center of diversity, farmers often choose the uniform improved varieties and breeds over the diverse plants and animals that their families have raised for generations. The genetically improved plants and animals thus displace the older, more diverse types. As the older plant varieties and animal breeds disappear, the alleles they carry are lost forever, unless efforts are made beforehand to preserve the diversity. This pattern of displacement is called **genetic erosion**, and it is now under way in every center of diversity throughout the world.

Conserving Genetic Resources

Recognizing the potential loss of the valuable genetic diversity in these older varieties, governments and international organizations have established **gene banks**, where plants, animals, and seeds of many different native types have been collected and are maintained in central facilities. One of the most moving events in the history of genetics took place in one of the oldest and largest gene banks, the Vavilov Institute in St. Petersburg, Russia. During the winter of 1941–1942, the city was under siege from Hitler's army. Tens of thousands of Russians died of starvation because food was in such short supply.

Housed within the Institute were thousands of pounds of seeds and tubers, representing a vast store of genetic diversity of rice, wheat, corn, peanuts, potatoes, and peas. As the scientists who worked in the Institute suffered from hunger, they refused to eat the genetic resources. One by one, nine of the scientists who worked there died of starvation, surrounded by seeds and potato tubers, rather than deprive future generations of the genetic diversity housed there. The Vavilov Institute remains one of the largest gene banks in the world, although the genetic diversity it holds is once again threatened, this time by economic hardship.

Organizations throughout the world coordinate efforts to conserve genetic resources. In the United States, the Department of Agriculture and state agricultural research centers operate several central facilities for maintaining genetic resources. Worldwide efforts are coordinated by the Food and Agriculture Organization (FAO) of the United Nations. However, as important as the preservation of genetic diversity is to the future of agriculture, genetic conservation is severely underfunded because few people are aware of its necessity, making it a low priority on political and scientific policy agendas.

Genetic erosion and its relationship to the rights of a nation to control its genetic resources have become issues of international legal debate, which we will discuss in the next chapter.

With inadequate funding for preserving genetic resources, scientists have sought ways to preserve the maximum amount of diversity with available funding. The following example shows how the development of core collections improves the efficiency of genetic resource conservation.

Example 27.1 Development of core collections in gene banks.

Plant varieties maintained in gene banks are much more useful in plant breeding if they have been well characterized for important agricultural traits. For example, if plant breeders suddenly face a new disease for which there is no genetic resistance among existing varieties, they search for alleles that confer resistance in the material held in gene banks. If the material in the gene bank has already been characterized for resistance to the disease, then plant breeders do not need to conduct large-scale screening experiments to identify the resistant genotypes. However, it is very difficult and expensive to maintain and characterize large numbers of varieties in gene banks. In some cases, varieties maintained in a gene bank may number in the zhundreds of thousands per species. In 1989, Brown et al. (*The Use of Plant Genetic Resources*. Cambridge, England: Cambridge University Press, pp. 136–156) recommended that a core collection, consisting of

about 10% of the large collection, be selected to represent the species. The core collection could then be well characterized for important agricultural traits so that plant breeders could have ready access to needed materials. Brown indicated that a core collection could be selected either at random or systematically.

Problem: Describe why systematic selection is better than random selection, and suggest how systematic selection of a core collection should be conducted.

Solution: Varieties collected from different locations may not uniformly represent the genetic diversity of a species. Suppose that 90% of the varieties collected have a common genetic base and are genetically similar, although perhaps not identical. The remaining 10% may be substantially diverse. Ideally, the core collection should represent the diversity of the species. A core collection selected at random could include a large proportion of similar varieties, and some of the more diverse varieties might be excluded. Through systematic selection, most of the varieties in the core collection would be selected from the 10% that are diverse. In systematic selection, it is necessary to identify which varieties are most diverse by examining geographic origins, morphological traits, and diversity revealed by DNA analysis. After characterization, the core collection can be selected to represent the highest possible degree of genetic diversity.

Most of the genetic diversity for domesticated species of animals and plants comes from centers of diversity that are located in less developed regions of the world. Much of the genetic diversity in domesticated plant and animal species is being lost through genetic erosion. Gene banks are used to preserve genes that might otherwise be lost through genetic erosion.

27.2 PLANT BREEDING

Most plant breeders utilize established breeding methods that are based on the principles of quantitative genetics. Also important in the design of a breeding program is the plant's reproductive system, the level of inbreeding depression that is typical of the species, and the expenses of large-scale seed production. The type of variety sold for commercial production depends on all these factors. Plant-breeding programs typically include some form of hybridization, followed by testing and selection. For those species that can be self-pollinated without severe effects of inbreeding depression, breeders often self-pollinate plants for several generations between the hybridization and testing steps to achieve homozygosity.

A ✕ B

a Two-parent cross: Two parents contribute equally to the offspring.

(A ✕ B) ✕ C

b Three-parent cross: Parent C contributes half of the alleles, and parents A and B each contribute one-quarter of the alleles to the offspring.

(A ✕ B) ✕ (C ✕ D)

c Four-parent cross: Four parents contribute equally to the offspring.

A ✕ B
A ✕ C   B ✕ C
A ✕ D   B ✕ D   C ✕ D
A ✕ E   B ✕ E   C ✕ E   D ✕ E
A ✕ F   B ✕ F   C ✕ F   D ✕ F   E ✕ F

d Diallel cross: Parents are hybridized in all pairwise combinations, excluding self-fertilization and reciprocal crosses.

e Circular cross: Each parent is hybridized with two other parents.

Figure 27.3 Examples of hybridization schemes in plant breeding.

Hybridization

In order for selection to be effective, genetic variation must be present, and hybridization is a way of generating genetic diversity in progeny. Also, breeders often use hybridization to combine the favorable traits found in two or more varieties into a single variety. Hybridization is the best way to create a population of genetically diverse individuals and is usually the first step in a breeding program. The form of hybridization and the type of population developed depend on the objectives of the breeding program.

There are several systematic hybridization schemes, some of which are illustrated in Figure 27.3. The simplest is a **two-parent cross**, in which the plant breeder hybridizes one genotype with another. The F_1 progeny are then self-pollinated to form an F_2 population. **Three-parent** and **four-parent crosses** are also common and require two generations of hybridization. After completing the hybridization scheme, the plant breeder usually self-pollinates the progeny.

A variation of the two-parent cross is **backcrossing**. Two parents are crossed with each other, and the F_1 progeny are backcrossed with one of the original parents to form a backcross population. Breeders often use repeated backcrosses to introduce a new trait into an existing variety while maintaining all other characteristics of that variety. For example, suppose that a well-accepted wheat variety has many favorable characteristics but is susceptible to a fungal disease. Suppose further that a dominant

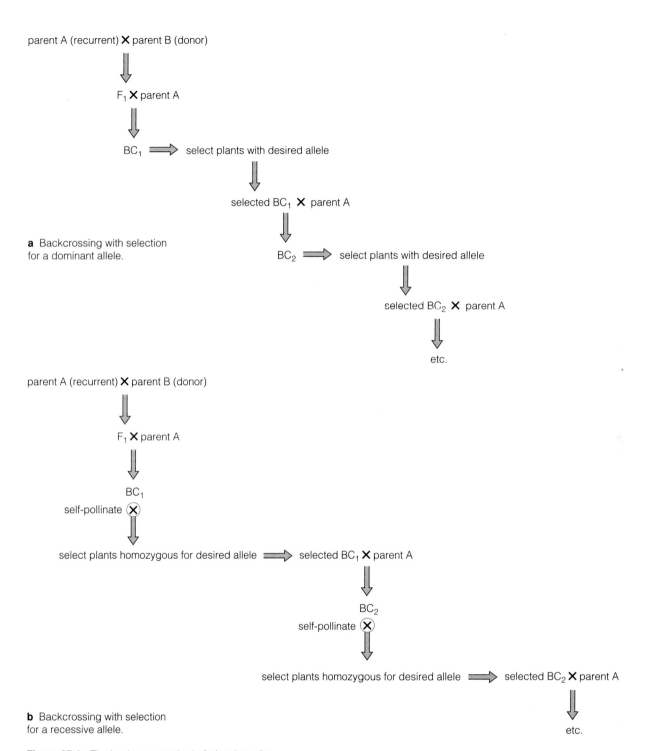

parent A (recurrent) ✕ parent B (donor)

F₁ ✕ parent A

BC₁ ⟹ select plants with desired allele

selected BC₁ ✕ parent A

BC₂ ⟹ select plants with desired allele

selected BC₂ ✕ parent A

etc.

a Backcrossing with selection for a dominant allele.

parent A (recurrent) ✕ parent B (donor)

F₁ ✕ parent A

BC₁

self-pollinate ✕

select plants homozygous for desired allele ⟹ selected BC₁ ✕ parent A

BC₂

self-pollinate ✕

select plants homozygous for desired allele ⟹ selected BC₂ ✕ parent A

etc.

b Backcrossing with selection for a recessive allele.

Figure 27.4 The backcross method of plant breeding.

allele confers resistance to the disease but is found only in a variety that lacks most of the desirable characteristics of the first variety. The breeder introduces the disease resistance allele with hybridization, then reconstitutes the favorable variety through backcrossing.

The procedure is illustrated in Figure 27.4a. First, the two parents are hybridized. The parent with the disease resistance allele is called the **donor parent**, because it donates a particular allele. The variety with the desirable characteristics that the breeder wishes to reconstitute is called the **recurrent parent**. Assuming that the parents are fully homozygous (which most wheat plants are), all the F_1 progeny will be resistant because the allele from the donor parent is dominant. Also, half of all the alleles throughout the genome are from the donor parent and half are from the recurrent parent. The F_1 plants are then backcrossed with the recurrent parent. The dominant resistance allele should segregate in a 1:1 ratio among the

progeny, called the BC_1 generation. On average, 75% of all alleles in the BC_1 generation are from the recurrent parent, and 25% from the donor parent. The plant breeder selects those progeny that are resistant and backcrosses them with the recurrent parent. Once again, the resistance allele should segregate in a 1:1 ratio in the progeny of this cross (called the BC_2 generation), although by this generation 87.5% of the alleles are from the recurrent parent. The breeder repeats the backcrossing procedure for several generations, reducing the proportion of alleles from the donor parent by half with each generation. After five or six generations of backcrossing, the genotype of the recurrent parent is essentially restored, but it now contains the disease resistance allele.

When the allele contributed by the donor parent is recessive, a generation of self-pollination along with each backcross is required to generate homozygous plants that express the introduced allele so that they can be selected as parents in the next backcross (Figure 27.4b). This additional generation of self-pollination can be eliminated with the use of **marker-assisted selection**, a procedure in which breeders use a DNA marker that identifies the recessive resistance allele to select plants that are heterozygous for the allele.

When traits from multiple parents are desired, breeders use some form of **multiple-parent cross** to form a highly diverse population called a **complex population**. The hybridization may be systematic or random. An example of systematic hybridization is a **diallel cross**, in which the breeder hybridizes each parent with every other parent. For example, a diallel cross with 6 parents (with self-fertilization and reciprocal crosses excluded) requires 15 single crosses (Figure 27.3d). As the number of parents increases, the number of single crosses required in a diallel cross increases exponentially. For this reason, breeders often use other forms of systematic crosses, such as the **circular cross**, which is diagrammed in Figure 27.3e.

Sometimes breeders create a complex population by selecting the parent plants and allowing them to cross-pollinate one another at random. A randomly pollinated population derived from selected parents is called a **synthetic population**. In some species, synthetic populations are tested, and those that perform well are released as **synthetic varieties** for commercial sale and production. Plants that require genetic diversity in the field due to their use in varied environments (such as rangeland or pasture grasses) and species that are highly sensitive to inbreeding depression, such as alfalfa, are usually developed as synthetic varieties.

Sometimes, breeders practice selection during the development of a complex population. Among the most common procedures is **recurrent selection**, in which breeders test the population after each generation of hybridization and select the best plants as the parents for the next generation (Figure 27.5).

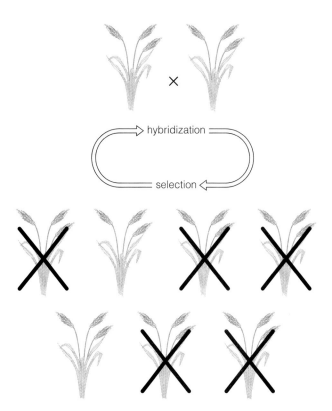

Figure 27.5 Recurrent selection. After each generation of hybridization, plants are tested for their performance (or for the performance of their progeny), and the best performers are selected as parents for the next generation of hybridization.

Self-Pollination to Achieve Homozygosity

In many species, plant breeders self-pollinate the progeny from hybridization for several generations to obtain complete homozygosity. This is especially true of naturally self-pollinated species such as wheat, barley, soybean, pea, and oats. Breeders usually allow plants to self pollinate through the F_5 or F_6 generation, after which each plant is homozygous at nearly every locus. All of the seed harvested from a single F_5 or F_6 plant is nearly genetically identical and constitutes a **true-breeding line**. Breeders can test true-breeding lines in replicated trials and in different environments simultaneously because each plant within a line is genetically identical to every other plant in the same line.

The following example highlights one of several methods that plant breeders use to develop true-breeding lines.

Example 27.2 Single-seed descent for development of true-breeding lines in self-pollinating plants.

In 1966, Brim (*Crop Science* 6:220) described a method for developing true-breeding lines that is now called single-seed descent. Breeders allow F_1 plants from a two-parent cross, or plants derived

from more complex hybridization schemes, to self-pollinate to form an F_2 generation. Then the breeder takes a single seed from each F_2 plant (with no selection) and plants the seeds to form the F_3 generation. He or she repeats the same procedure with each subsequent generation until the F_5 or F_6 generation.

Problem: What advantage is there to taking a single seed from each plant rather than planting all the progeny in the next generation?

Solution: If a plant breeder planted every seed from every plant in each generation of self-pollination, the number of plants would increase exponentially to the point that the number of plants would be overwhelming. By taking only one seed per plant, the breeder maintains populations at the same size in each generation until true-breeding lines are attained.

Late-Generation Testing

After they have developed true-breeding lines, breeders test them for desirable characteristics and select the best lines for further propagation. Line selection is much more effective than single-plant selection because many of the most important characteristics, such as seed yield, are highly influenced by the environment. The effect of environmental variation can be reduced when breeders plant several replications of a true-breeding line in at least two different environments. The results of testing in replicated tests over two or more environments provides a much more reliable estimate of genetic differences among lines, so breeders can effectively select the best genotypes for further testing. With each generation of testing, breeders save only the best lines. By the time lines reach the F_{10} to F_{12} generations, only a few lines from each cross remain. Breeders then compare the performance of these lines against the performance of existing commercial varieties. They release as new varieties only those lines that perform better than the existing varieties. In a commercial breeding program, a breeder may test as many as 10,000–20,000 F_6 lines in a single year, but may end up releasing only one or two of those lines as a commercial variety.

For naturally self-pollinated species, seed companies usually sell **pure-line varieties**, completely homozygous, genetically uniform, true-breeding lines. Every plant within the variety is genetically identical to every other plant, and each generation is genetically identical to the previous generation. The seed the farmer harvests is genetically the same as the seed purchased from the seed supplier. For this reason, farmers often save a small por-

tion of the harvested seed from one generation for planting in the next. This practice has caused some legal concerns regarding the rights of a seed company to control the use of seed of a pure-line variety developed by the company's breeders, a topic we will address in the next chapter. You may recall that the pea plants Mendel used as parents in his crosses were homozygous because they came from pure-line varieties.

F_1 Hybrid Varieties

As explained in section 20.3, most naturally cross-pollinated plant species are sensitive to inbreeding depression and show high levels of heterosis when highly heterozygous. Maize, sorghum, and sugar beets are examples. Breeders self pollinate plants of these species over several generations to produce true-breeding inbred lines, using the same methods as those used to develop homozygous true-breeding lines in self-pollinated species. Inbred lines of cross-pollinated species tend to be weak and unproductive because of inbreeding depression. However, the degree of inbreeding depression when plants are homozygous is inversely related to the degree of heterosis when they are highly heterozygous. Breeders achieve maximum heterosis by crossing two inbred lines to obtain an **F_1 hybrid variety**. Maize is one of the most important crops worldwide, and most commercial varieties are F_1 hybrids. Many seed companies make the bulk of their profits from sales of F_1 hybrid maize seed, which is planted extensively throughout the United States, Europe, Brazil, Asia, and Australia.

F_1 hybrid varieties are advantageous for seed companies because seed from the variety is F_2 seed, which has half the heterotic potential of F_1 seed. Breeders have expended much effort in developing F_1 hybrid varieties in species other than maize. For hybrid varieties to be valuable economically, several criteria must be met: (1) Breeders must be able to make inbred lines, (2) hybridization must be simple and efficient to produce large quantities of hybrid seed, (3) the level of heterosis must be sufficient for F_1 hybrid varieties to significantly outperform other types of varieties, and (4) the economic value of the crop must be sufficient to justify the extra cost of F_1 hybrid seed. In some species, F_1 hybrid varieties are technically possible, but at least one of those criteria is not met, so companies do not produce F_1 hybrid varieties.

Asexually Propagated Varieties

Farmers propagate certain plant species asexually through cuttings. Asexually propagated species include potatoes, sugarcane, and most long-lived perennial plants

Figure 27.6 Potato "eyes," small embryos that can grow into vegetative clones of the original potato plant.

potato "eyes"

includes some sort of hybridization scheme, like the ones diagrammed in Figure 27.3, followed by asexual propagation of the first-generation offspring to form vegetative clones. For example, potato breeders usually hybridize potato plants, then grow the seed into hybrid plants. They harvest the potatoes from the hybrid plants and cut them into sections that each have at least one embryonic plant embryo, called an "eye" (Figure 27.6). They plant the sections in the ground, and each embryo germinates into a new plant that is genetically identical to every other plant derived from the same parent plant.

A collection of cloned plants derived from the same hybrid is called a **clonal line**. Breeders propagate clonal lines to produce many plants that are genetically identical. They then test the cloned plants in replicated trials, much as homozygous true-breeding lines are tested. They release those lines that perform best as commercial **clonal varieties**. The different varieties of potatoes, apples, grapes, pears, and onions found in supermarkets are examples of clonal varieties.

Table 27.2 provides examples of different plant species, their reproductive system, their degree of inbreeding depression, and the type of varieties that are sold commercially for these species. Notice how the type of variety sold is related to the reproductive system and the degree of inbreeding depression.

~

Plant breeding procedures vary depending on the reproductive system, the degree of inbreeding depression, and the cost of commercial seed production of the species.

such as fruit trees, shrubs, and many ornamental species. Most of these plants produce seeds and cross-pollinate naturally, but asexual propagation from cuttings is more efficient and productive. Because they naturally cross-pollinate, these species are usually subject to inbreeding depression and heterosis. A breeding program usually

Table 27.2 Examples of Plant Species, Their Reproductive System, Degree of Inbreeding Depression, and Type of Variety Typically Sold

Species	Reproductive System	Degree of Inbreeding Depression	Type of Variety Sold
Barley (*Hordeum vulgare*)	Self-pollinated	Mild	Pure-line
Soybean (*Glycine max*)	Self-pollinated	Mild	Pure-line
Pea (*Pisum sativum*)	Self-pollinated	Mild	Pure-line
Wheat (*Triticum aestivum*)	Self-pollinated	Mild	Pure-line
Maize (*Zea mays*)	Cross-pollinated	Moderate	F_1 hybrid
Sorghum (*Sorghum bicolor*)	Cross-pollinated	Moderate	F_1 hybrid
Alfalfa (*Medicago sativa*)	Cross-pollinated	Severe	Synthetic
Potato (*Solanum tuberosum*)	Cross-pollinated, asexual reproduction	Severe	Clonal
Sugarcane (*Saccharum spp.*)	Cross-pollinated, asexual reproduction	Severe	Clonal
Apple (*Malus spp.*)	Cross-pollinated, asexual reproduction	Severe	Clonal
Grape (*Vitis spp.*)	Cross-pollinated, asexual reproduction	Severe	Clonal

The term **biotechnology** is now a popular one, used for many purposes. In this chapter, we define it as the use of in vitro genetic manipulation and recombinant DNA methods to genetically alter plants, animals, and microbes. The first commercial products of plant biotechnology have entered the market, and many will soon follow. The hope for improving agricultural plants through biotechnology has led some people to believe that biotechnology will replace traditional plant breeding. That view still persists in some circles, usually among those who do not work with agricultural plants, and has jeopardized funding for traditional breeding research. Now as the products of plant biotechnology are reaching the market, plant breeding and biotechnology have become an integrated process in which both are dependent on one another. Rather than replacing traditional breeding, plant biotechnology has created an even greater need for it.

Plant biotechnology can be divided into three general areas: (1) plant cell and tissue culture, (2) plant transformation, and (3) genetic marker technology. Let's take a brief look at each of them.

Plant Cell and Tissue Culture

Cells at a wound site in plants form a mass of rapidly dividing cells, called a **callus**, that repairs the wound. Callus cells can grow in culture much like bacterial cells. Generally, plant geneticists cut a piece of tissue from a plant, called an **explant**, and place the explant on agar-solidified medium that contains a carbon source (usually sucrose), minerals and vitamins needed for plant cell growth, and hormones. Callus cells grow from the wound sites in the explant into masses of callus on the surface of the medium (Figure 27.7). As long as hormone concentrations are kept at the correct levels, callus cells may grow and divide indefinitely.

When the callus is transferred to a medium that has no hormones or an altered hormone composition, small plant embryos, called **somatic embryos**, may form on the surface of the callus (Figure 27.8). The somatic embryos grow into small plant shoots, which researchers can excise from the callus and place in a rooting medium, where roots form at the base of the shoot. After the new plantlet is established, the researchers can transplant it into soil, where it will grow into a mature plant.

Somaclonal Variation

Cells in a callus divide mitotically, so all the plants regenerated from a single source of callus should be genetically identical clones of each other. However, for a variety of reasons, callus culture tends to be highly mutagenic; plants regenerated from callus culture often have point mutations, deletions, duplications, inversions, translocations, aneuploidies, polyploidies, and new transposable element activity when compared to the original explant source. The variation that arises because of mutation in culture is called **somaclonal variation**.

Somaclonal variation can be either detrimental or beneficial depending on the researcher's objectives. Sometimes callus culture is used to clonally propagate plants that should be genetically identical to one another. In these cases, somaclonal variation can be detrimental because it creates variation among the regenerated plants.

On the other hand, somaclonal variation can be advantageous for **cell selection**, the selection of desirable genetic traits in callus culture, followed by regeneration

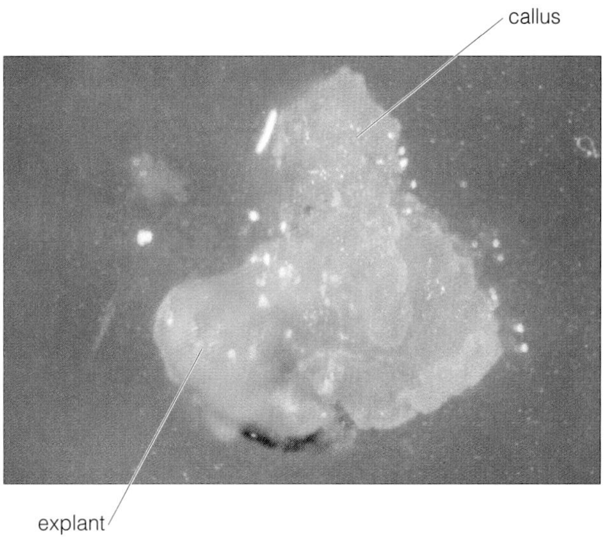

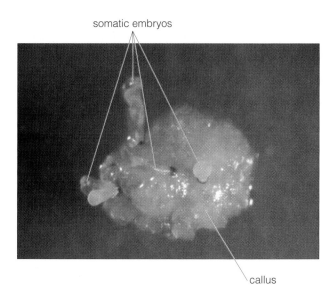

callus

explant

Figure 27.7 Callus cells growing from an explant in plant tissue culture.

somatic embryos

callus

Figure 27.8 Somatic embryos on the surface of cultured callus cells.

of plants from the selected cells. As in whole plants, genetic variation among the cells in culture is essential for selection to be effective. The high mutation rate in culture provides this variation. For example, plant breeders can culture cells of a variety that is susceptible to a particular disease in a medium that contains the toxin produced by the disease-causing organism. Only those cells that have a new mutation that confers resistance are able to divide and grow. Breeders can then regenerate whole plants that carry a new mutation for disease resistance. New mutations that confer resistance to several plant diseases have been found in this way. Once the new mutations are in a regenerated plant, they can be included in a breeding program.

Cell selection is effective only for those traits that are expressed in cell culture the same way as they are in the whole plant. Resistance to disease toxins is an example of such a trait. Another example is lysine-plus-threonine resistance, described in section 8.6. At one time, researchers thought it might be possible to select for salt tolerance in callus culture and recover mutations that could make plants resistant to salinity in soils. Researchers added callus to medium with low salt concentrations and selected the callus that grew well. They transferred the salt-tolerant callus to incrementally higher levels of salt with each cycle of selection. After several cycles of selection, salt-tolerant callus had developed that could grow on medium with salt concentrations that were sufficient to kill most cells. However, when plants were regenerated from the salt-tolerant callus, in nearly every case there was no increase in the salt tolerance of the whole plants.

How could cells be genetically salt tolerant in culture but not express the trait in a whole plant? The most probable explanation has to do with the physiology of salt tolerance in whole plants. Salt tolerance is a complex process in which specialized root cells exclude or excrete salt from the cells, or transfer the salt to other cells that excrete the salt from the plant. Different tissues in the root work cooperatively to resist the toxic effects of salt. Cells in callus are not differentiated into specialized root tissues, so the mechanism of salt tolerance is probably quite different and may be governed by genes that are effective only under culture conditions.

Plant Transformation

Plant transformation is the introduction of foreign genes into plants, where the genes are expressed and inherited as if they were a natural part of the plant's genome. After breeders have introduced foreign genes into a species, they can transfer those genes into new varieties through traditional breeding. Plant breeders use several methods for plant transformation. We'll highlight *Agrobacterium*-mediated transformation, one of the most widely used methods.

Agrobacterium-mediated transformation is the transfer of genes from one species to another using a bacterial plant pathogen called *Agrobacterium tumefaciens*. *A. tumefaciens* causes crown-gall disease, an economically important disease in species such as grapes and roses. *A. tumefaciens* is a natural genetic engineer. It can grow and divide in the soil like many bacteria, but it reproduces most rapidly after infecting a plant, where it relies on the plant's genetic system to express some of its genes.

A. tumefaciens carries a large plasmid called the **Ti plasmid** (for "tumor-inducing plasmid"). This plasmid carries a segment of DNA called **T-DNA**, which contains genes whose products cause tumorous galls to form on the plant. The plasmid also carries a set of genes called *vir* **genes**, which encode the enzymes necessary for infection (Figure 27.9). When the bacteria come in contact with a wound, they enter the site, and the *vir* genes produce enzymes that transfer the T-DNA into callus cells at the wound site and insert the T-DNA into the plant cell chromosomes (Figure 27.10). The plant cell expresses the genes in the DNA as if they were plant genes. In fact, the T-DNA genes are eukaryotic genes with plant-specific promoters and are never expressed in the bacterial cell.

Genes in the T-DNA encode enzymes that produce plant hormones designed to cause the tumor cells to grow. Other genes in the T-DNA cause unusual compounds to form that the bacterium, but not the plant, can use as an energy source. The bacteria then grow and divide inside the gall, relying on the plant to supply them with the resources they need. Later, the bacterial cells can leave the gall to infect other plants.

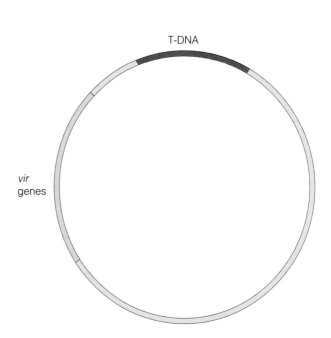

Figure 27.9 The Ti plasmid of *Agrobacterium tumefaciens*.

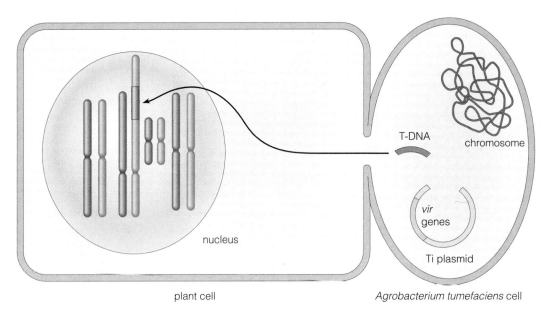

Figure 27.10 T-DNA transfer during infection by *Agrobacterium tumefaciens*. During infection of a plant cell, *A. tumefaciens* inserts its T-DNA into a chromosome in the plant cell. The plant cell expresses the genes in the T-DNA.

Plant geneticists have exploited the ability of *A. tumefaciens* to transfer genes to insert genetically engineered genes into plant cells. The process of engineering Ti plasmids for plant transformation is diagrammed in Figure 27.11. Researchers isolate the Ti plasmid and place engineered genes into the T-DNA, where they can be carried into the plant cells. The process of inserting engineered genes into the T-DNA is usually done in such a way that the engineered genes displace the natural tumor-inducing genes. With the natural genes missing, the bacterium cannot induce crown-gall disease. Instead, the bacterium inserts the engineered genes into the plant cell's chromosomes, where they are expressed in the plant cell. Researchers then culture the transformed plant cells as callus and regenerate whole plants from the transformed callus cells. Each plant carries the genetically engineered gene in its chromosomes. Breeders can then use the gene in a breeding program.

Breeders now have at their disposal several commercially important genes that have been transferred in this way, have been field-tested, and are entering the market. One of the most interesting is a gene that confers resistance to moth and beetle larvae. These larvae are among the most agriculturally destructive pests. A bacterium called *Bacillus thuringensis* is a natural disease-causing organism of these insects. It produces a crystallized protein that paralyzes the mouthparts of the larvae when they ingest the bacteria. *B. thuringensis* cultures have been used as a biological pesticide for decades. However, because the toxin is a protein encoded by a single gene, researchers were able to clone the gene from the bacteria, place it in a Ti plasmid, and transfer it into plants using *Agrobacterium*-mediated transformation. The trans-

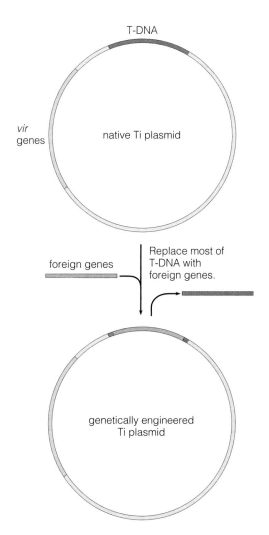

Figure 27.11 Construction of a genetically engineered Ti plasmid.

formed plant cells produce the bacterial toxin, so when a larva begins to feed, its mouthparts become paralyzed, and it soon dies. Humans and other animals are unaffected by the toxin; in fact, we are exposed to it naturally by *B. thuringensis* that is airborne or present on our food. So plants that express the toxin are harmless to humans and animals that eat them.

Other genes that have been successfully incorporated into agricultural species using *Agrobacterium*-mediated transformation include genes that confer resistance to plant viral diseases and to herbicides. For example, scientists transferred a gene that confers resistance to glyphosate (an herbicide sold under the trade name Roundup) into soybeans. The genetically engineered soybeans are called Roundup-ready soybeans. Glyphosate is a broad-spectrum herbicide that kills most plants, including most soybeans. Now farmers can plant Roundup-ready soybeans and spray the soybeans with glyphosate. All of the weeds die, but the soybeans are unaffected.

Agrobacterium-mediated transformation has one drawback, however. Not all plants are susceptible to *Agrobacterium* infection. *Agrobacterium* naturally infects only broadleaf plants. Plants of the grass family (which includes the grains, such as maize, wheat, rice, barley, and oats) are not susceptible and cannot be readily transformed using *Agrobacterium*, although scientists have recently reported some success in the development of methods for *Agrobacterium*-mediated transformation of grass species.

Marker-Assisted Selection

Marker-assisted selection is now used to tag agriculturally important genes in plants and animals and select the genes by virtue of their close linkage to the markers. Earlier in this chapter, we discussed how DNA markers can be used to improve selection in backcrossing programs. DNA markers can be used not only for tagging single genes, but also to tag **quantitative trait loci (QTLs)**, the loci that influence quantitatively inherited traits. We discussed the theoretical rationale for marker-assisted selection of QTLs in section 20.5.

Plant and animal breeders have successfully incorporated marker-assisted selection for QTLs into breeding programs, as illustrated in the following example.

Example 27.3 Marker-assisted identification of QTLs.

In 1996, Tanksley et al. (*Theoretical and Applied Genetics* 92:213–224) described a procedure in which QTLs from plant genetic resources held in gene banks could be identified and incorporated into modern plant varieties through marker-assisted selection. Among the examples they cited

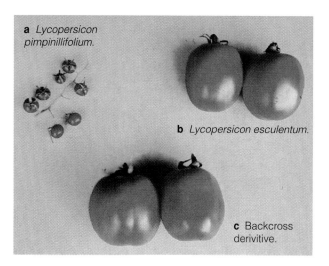

a *Lycopersicon pimpinillifolium.*

b *Lycopersicon esculentum.*

c Backcross derivitive.

Figure 27.12   Differences in fruit size for **(a)** *Lycopersicon pimpinellifolium* (a wild relative of tomato), **(b)** *L. esculentum* (cultivated tomato), and **(c)** a backcross derivative that was selected for a QTL derived from *L. pimpinellifolium* that conferred increased fruit size. (Reprinted with permission from Tanksley, S. D., and S. R. McCouch. 1997. Seed banks and molecular maps: Unlocking genetic potential from the wild. *Science* 277: 1063–1066. Copyright 1997 American Association for the Advancement of Science.)

was breeding for increased fruit weight in cultivated tomato (*Lycopersicon esculentum*). In this example, they hybridized *L. esculentum* with a wild relative, *L. pimpinellifolium*, of Peruvian origin. *L. esculentum* has relatively large fruits, whereas *L. pimpinellifolium* has small, berry-type fruits (Figure 27.12a and b). They subjected the plants to a backcrossing program with *L. pimpinellifolium* as the donor parent and *L. esculentum* as the recurrent parent. In the second backcross (BC$_2$) generation, alleles from eight QTLs donated by the *L. pimpinellifolium* parent were found to be associated with fruit weight. Seven of the QTL alleles caused smaller fruit weight. The eighth QTL allele, located on chromosome 9, caused an increase in fruit weight of 14% when compared to the recurrent parent (Figure 27.12c).

Problem: **(a)** How could a plant with very small fruits contribute to increased fruit size in progeny when crossed with a large-fruited type? **(b)** What implications does this research have regarding the use of plant genetic resources in a breeding program?

Solution: **(a)** The small-fruit phenotype of *L. pimpinellifolium* is probably governed by genes that mask the effects of other genes that could potentially contribute to larger fruit size. In a backcross generation, a gene that contributes to larger fruit size can be separated from those that confer smaller

fruits, and the gene for larger fruit size can be identified by its association with DNA markers. Once identified, such a gene can be incorporated through backcrossing into many different varieties to increase fruit size. **(b)** Perhaps the most important implication is that the phenotype of an unimproved variety in a plant genetic resource collection does not always reflect the genes that the variety may carry. Marker-assisted selection provides the opportunity to identify beneficial QTLs that may not be expressed in the phenotype of a parent.

Plant biotechnology augments breeding programs with plant tissue culture, plant transformation, and marker-assisted selection.

27.4 ANIMAL BREEDING AND BIOTECHNOLOGY

Animal breeding and biotechnology typically focus on the improvement of **breeds**, which are the counterpart to varieties in plants. An animal breed is a defined group of animals that have particular genetic characteristics that are maintained from one generation to the next. For example, the holstein breed in cattle is characterized by high production of excellent milk. Holsteins can be readily identified by the black and white patches on their bodies (Figure 27.13a). Angus cattle, on the other hand, are used for beef production and are rarely raised as dairy cattle. Black Angus cattle are solid black in color (Figure 27.13b). Most efforts in animal breeding and biotechnology focus on improving existing breeds rather than developing new ones.

Animal Breeding Methods

The nature of animal breeding is much different from that of plant breeding. Plant breeding is typically done by professionals with an advanced degree in plant breeding who are employed by seed companies or universities. Farmers rarely participate in plant breeding. Instead, they purchase the genetically improved seed from a seed company. Farmers and ranchers, on the other hand, often practice animal breeding together with professional breeders. Animal breeding is usually less systematic than plant breeding. The differences between plant and animal breeding are due largely to reproductive differences between plants and animals. Many domestic animals require several years to reach sexual maturity, and domestic animals cannot be self-fertilized to produce fully homozygous true-breeding lines. In poultry breeding, generation times are much shorter than in livestock, so full-sibling matings are sometimes used to intentionally

a

b

Figure 27.13 Cattle breeds. **(a)** Holstein, a dairy breed. **(b)** Black Angus, a beef breed.

produce inbred parents, which can then be hybridized to produce highly productive F_1 hybrid offspring. Animal breeders also utilize backcrossing methods that are similar to those used by plant breeders.

Most animal-breeding programs are confined to methods that intentionally balance the negative effects of inbreeding depression with the positive effects of maintaining desirable genes together in a herd or flock. **Closebreeding** is a type of breeding in which the number of new individuals brought into a breeding population is few, in order to maintain the frequencies of desirable alleles fairly constant from one generation to the next in the population. By bringing in a few new sires (male parents) each generation, a breeder can negate the effects of inbreeding depression. Often, breeders keep elaborate pedigrees and plan matings to ensure that inbreeding within the population is kept at a minimum. An inbreeding coefficient of 6% is generally considered the highest acceptable level of inbreeding for any particular mating (see section 19.4 for an explanation of inbreeding coefficients).

Figure 27.14 Two sets of identical twin Holstein calves derived from embryo splitting. The two calves on the left came from the same embryo, and the two calves on the right came from another embryo. The people in the photograph are university students who worked on the embryo-splitting project. (Photo courtesy of College of Biology and Agriculture, Brigham Young University.)

In order to conduct such a breeding program, breeders identify individuals with high genetic potential either by their own phenotypic characteristics or, more importantly, by the phenotypic characteristics of their offspring. A prize bull is one whose offspring consistently perform well. If the offspring of one parent perform well regardless of which animal is the other parent, then that parent is said to have a high **general combining ability**. If, on the other hand, two particular parents produce outstanding offspring when mated with one another, they are said to have high **specific combining ability**. Combining ability, either general or specific, is often the result of heterosis.

In large animal species, such as cattle, horses, sheep, or pigs, a male with a high general combining ability is valuable for his semen. Breeders may use semen from a single prize bull to fertilize many cows through artificial insemination, and the bull's owner may sell the semen for a high price. A bull produces poor-quality beef (and can produce no milk), so cattle producers castrate male cattle when they are calves. A castrated male calf grows into a steer, which produces the highest-quality beef. Only those males that are identified as having high genetic potential are left uncastrated and allowed to grow into bulls. Male castration is also practiced in horses and sheep. A castrated male horse (called a geld-ing) is much more tame and makes a better riding horse than a stallion (an uncastrated male horse). Male castration also allows breeders to prevent unwanted matings in a herd or flock.

Animal Biotechnology

Until biotechnology became available, specific combining ability was not of great value because two parents could only produce a limited number of offspring. With biotechnology, it is now possible for two parents to have hundreds to thousands of genetic offspring. In dairy cattle, for example, a bull and a cow whose female offspring consistently produce high quantities of milk can be the genetic parents of many calves. Technicians collect the bull's semen and store it cryogenically. They treat the cow with hormones that cause her to **superovulate** (produce several eggs at the same time) and collect the eggs. They then fertilize the eggs in a test tube with the bull's sperm, a process called **in vitro fertilization**. They then implant the resulting embryos into surrogate mothers, cows whose genetic potential is not high but who can bear the genetic offspring of another selected cow. The selected cow serves as a source of valuable eggs for in vitro fertilization, much as a prize bull provides high-priced semen.

Aside from the increased genetic potential to produce milk though breeding, dairy producers may inject dairy cows with genetically engineered bovine growth hormone. The bovine growth hormone gene has been cloned into *E. coli*, where its product is expressed and can be purified. Milk cows injected with the hormone produce more milk than cows who are not injected. Milk produced from injected cows has been declared safe by the U.S. Food and Drug Administration (FDA) and has entered the market.

Ironically, all these advances in milk production efficiency have been developed at a time when there are more dairy products available than the market demands. Dairy product prices are depressed because of oversupply, and genetic improvements have only exacerbated the situation. The U.S. government has instituted buyout programs to encourage dairy farmers to sell their cows as beef and go out of the dairy business, in order to reduce the number of milk-producing dairy cattle and stabilize the market for dairy products.

Animal Cloning

In 1997, scientists at the Roslin Institute in Scotland reported successful cloning of sheep by transfer of a nucleus from a somatic cell of an adult ewe into a zygote, which they then implanted into a surrogate mother, a process called **somatic-cell nuclear transfer**. This event received much attention from the press and the public, and instigated numerous discussions about the possibility of cloning humans. In response, the U.S. government, through executive order, placed a ban on government funding of research aimed at human cloning using similar methods.

Most of the discussions failed to note that animal cloning has been routine for over a decade. Cattle embryos that have been fertilized in vitro are quite valuable. Scientists may split these embryos in two to produce two identical twin embryos and thus increase the number of genetically superior embryos. They then implant the embryos into surrogate mothers, where the embryos develop into genetically identical calves that are clones of one another (Figure 27.14). The 1997 report of somatic-cell nuclear transfer in sheep was novel in that it was the first case of successful cloning from a somatic cell of an adult mammal, rather than cloning from embryonic somatic cells.

In a practical sense, cloning from adult cells is significant because an adult animal that has already demonstrated its genetic superiority can potentially be cloned and its genotype maintained indefinitely. Cloned animals derived from embryonic cells represent genotypes that have yet to be tested for their productivity, so there is an advantage for cloning from adult cells. Whether or not animal cloning from adult cells becomes commercially feasible remains to be seen.

The following example highlights the Scottish study in which researchers cloned sheep from an adult somatic cell.

Example 27.4 Somatic cell nuclear transfer in sheep.

In 1997, Wilmut et al. (*Nature* 385:810–813) reported the birth of a lamb that arose from a somatic nucleus taken from an adult ewe. The authors isolated cells from the mammary gland of a pregnant 6-year-old ewe (hereafter called the donor ewe). They cultured the cells from the donor ewe and induced them to enter the G_0 phase of the cell cycle. The cells ceased dividing and became quiescent. The authors then isolated oocytes from ewes that had been treated with gonadotropin-releasing hormone to cause them to superovulate. The authors enucleated the oocytes (eliminated their nuclei). They then transferred nuclei from the cultured cells of the donor ewe into the enucleated oocytes and cultured the cells. They implanted those cells that developed into embryos into surrogate ewes. The surrogate ewes were from a breed that differed phenotypically from the donor ewe's breed. One of the surrogate ewes bore a healthy female lamb at full term. This lamb displayed the phenotypic characteristics of the donor ewe's breed and not those of her surrogate mother's breed. In addition, DNA fingerprints from the lamb matched those of the donor ewe and did not match those of the surrogate ewe.

Problem: **(a)** What novel aspect of this research brought popular and scientific attention? **(b)** Which experiments could be considered control experiments, and what conclusions can be drawn from the results of the control experiments?

Solution: **(a)** In the authors' words, this is the "first mammal to develop from a cell derived from an adult tissue." The attention arose because of the possibility that such a technique might be adapted for use in humans. **(b)** An inadvertent mating of the surrogate ewe could produce a lamb that did not arise through somatic-cell nuclear transfer. Two control experiments excluded this possibility. First, the authors chose a surrogate ewe from a different breed than the donor ewe. Had the mating been inadvertent, then the lamb should have had some characteristics of the surrogate ewe. Second, the authors compared DNA fingerprints of the donor ewe, the surrogate ewe, and the lamb. The lamb had the phenotype of the donor ewe's breed. It also

had a DNA fingerprint that matched the donor ewe but differed from the surrogate ewe. These observations indicate that the lamb did not arise from an inadvertent mating.

Transgenic Animals

Techniques to introduce foreign genes into laboratory mice through microinjection of genes into zygotes are now routine. Similar methods can be applied to livestock, although the procedures are more complicated, and the proportion of successfully introduced genes is much lower than in mice. Researchers inject the genes to be transferred into a zygote, then allow the zygote to grow into an embryo in vitro. They use PCR to identify those embryos that have been transformed by amplifying the inserted gene from a single embryo cell. They then transfer the successfully transformed embryos into surrogate mothers. Growth-hormone genes have been successfully transferred to livestock, causing the animals to grow to a larger size and at a more rapid rate. However, refinements are still necessary before such animals can enter the market.

~

Animal breeding focuses on genetic improvement of existing breeds and the avoidance of inbreeding. Animal biotechnology has already made significant impacts through in vitro fertilization, surrogate motherhood, and genetic transformation.

27.5 APPLICATIONS OF GENETICS IN INDUSTRY

Industry is a relative newcomer to the world of genetics when compared to medicine and agriculture. Genetics has indirectly impacted industry though the improvement of biologically derived raw materials, such as food and fiber products. However, in recent years, there has been much public concern about environmental degradation caused by industrial processes. Scientists have utilized genetics to help overcome these environmental problems. Also, scientists have improved some industrially produced products through the applications of genetics.

Land Reclamation

Some industrial activities are particularly destructive to native ecosystems. Oil well drilling, forest clear-cutting, and mining are a few examples. Strip-mining is among the most destructive. Mining companies use strip-mining to extract coal, which often lies in strata that are hundreds of feet below the ground surface. Using heavy machinery, workers dig a strip of ground about 100 feet wide and several miles long down to the coal stratum. After the coal has been removed, they dig a second strip next to the first, using the material from the second strip to fill in the first strip. They continue the process strip by strip throughout the coal field, disturbing hundreds to thousands of acres of land.

In the past, mining companies abandoned strip-mined land and left it to erode after extracting the coal. Now, laws are in effect that require mining companies to post large bonds to guarantee that they will reclaim the land and restore an appropriate ecosystem on the mine spoils. Mine workers save the topsoil and replace it on the backfilled strips so that plants can grow. Even so, the native soil structure is disturbed, and some native plants that are very sensitive to microenvironments cannot be restored. The replanted material often dies, which is one of the most serious problems in reclamation. To overcome this problem, plant breeders have developed native populations of plants that are highly diverse genetically so that they can adapt to a wide variety of sites. As opposed to agricultural breeding, in which uniformity is desirable, breeding of native plants focuses on high diversity in a source of seed, often through interspecific hybridization. In response to an increased demand for seed used in reclamation, the breeding of native wildland plants and associated seed production programs are now under way.

Genetically Modified Bacteria for Hazardous-Waste Cleanup

One of the legacies of the industrial and agricultural revolutions is pollution with hazardous waste and unintentional leaks or spills of hazardous material. Scientists have selected bacteria to combat certain types of hazardous waste. The first patent issued on a microbe was for bacteria that degrade polymers in crude oil spills into compounds that are less harmful and more easily degraded. Most bacteria used in such treatments are not genetically engineered but have been selected through traditional breeding. However, recombinant DNA plays an important role in this process. Genetically altered strains of *E. coli* are capable of reducing phosphate pollution. Phosphates are used in detergents and as fertilizer on agricultural lands. They often make their way into waterways and groundwater through wastewater discharge and agricultural runoff. When they do, they act as fertilizer for algae and small aquatic plants. These organisms grow prolifically in the phosphate-polluted water, choking out the other species that normally live in the ecosystem, a process called **eutrophication**. *E. coli* and some other bacterial species naturally utilize inorganic phosphate and convert it to a polyphosphate (polyP) form that does not cause eutrophication. These bacteria are used routinely in wastewater treatment to remove phosphates, but typically they remove only about 40% of

the phosphates. Researchers have cloned the genes that affect phosphate polymerization and have engineered the genes to overexpress their products for more efficient phosphate removal.

Biosynthesis of Fuels and Industrial Products

The energy crisis of the 1970s stimulated research into alternative fuel technology. Among the most promising is biosynthesis of fuels using biomass as a resource. Industries can synthesize low-sulfur diesel fuel substitutes from algae that grow in saline water, methane (natural gas) though anaerobic digestion of biomass, and ethanol from sugar, starch, or cellulose for use as an automobile fuel. There are many advantages to using biosynthetic fuels. They come from renewable resources, unlike petroleum, which is nonrenewable. They also tend to produce less pollution when compared to petroleum-based fuels. Dependence on foreign petroleum can be reduced significantly by development of biosynthetic fuels.

Some industries are researching possible applications of genetic technology to improve the efficiency of biosynthetic fuel production. For example, yeast cultures produce ethanol from maize starch by breaking down the starch enzymatically into glucose then converting the glucose to ethanol through fermentation. Scientists can improve each of these steps genetically. They can develop maize with high starch contents and select yeast for efficient fermentation.

The major limitation to biosynthetic fuel production is economic competition from petroleum products. When inflation is taken into account, petroleum is now less expensive than it has ever been. The drop in petroleum prices has eliminated several biosynthetic fuel research programs, and government funding of research has declined. In order for biosynthetic fuels to be successful, they must be able to compete economically with their petroleum counterparts, or governments must provide economic incentives for the use and development of biosynthetic fuels. There are examples of successful biosynthetic fuel programs. Federal laws require that ethanol and its derivatives be blended with gasoline (at about 8% ethanol) in some of the most polluted cities of the United States to reduce air pollution. Through government-sponsored programs, Brazil has developed the most extensive ethanol fuel production program in the world. About 80% of Brazil's automobiles operate on ethanol derived from genetically improved sugarcane. Air pollution in Brazil's major cities has dropped considerably since ethanol has been adopted as a fuel, and Brazil's dependence on foreign petroleum has declined.

Many industries use bacteria to synthesize products. One of the most common is the enzyme subtilisin, which is used in most laundry detergents. This enzyme is a natural bacterial protease that cleaves proteins into smaller peptides, which are more easily dissolved by the deter-

gent. The natural enzyme, however, is deactivated by bleach, making commercial enzyme detergents less effective when bleach is used. The bleach inactivation is due to a methionine at position 222 in the polypeptide. Using recombinant DNA methods, researchers have substituted different codons at this position and tested the enzyme for its sensitivity to bleach. When alanine replaces methionine at this site, the bleach sensitivity is lost.

Another industrial product produced commercially in bacteria is biodegradable plastic consisting of polyhydroxybutyrate (PHB) or polyhydroxyalkanoate (PHA). These compounds are found naturally in many bacterial species, but often the species cannot be efficiently grown on a large scale, or they do not yield sufficient quantities of the products for commercial production. Researchers have cloned the genes that govern production of these polymers and have transferred the genes to bacterial species (such as *E. coli*) that can be cultured on a large scale.

The following example shows how researchers are working to produce PHB in plants.

Example 27.5   Genetic modification of plants for PHB production.

In 1992, Poirier et al. (*Science* 256:520–523) described transformed plants that produce PHB. The authors designed two separate recombinant Ti plasmids, one that contained the bacterial gene for acetoacetyl-CoA-reductase (*phbB*) and a second that contained the bacterial gene for PHB synthase (*phbC*), genes that encode enzymes necessary for PHB synthesis in plants. Using *Agrobacterium*-mediated transformation, they introduced each of the genes into different cultures of *Arabidopsis thaliana* cells and regenerated transformed plants from each culture. They then self-fertilized the transformed plants to obtain progeny that were homozygous for the transformed genes. They hybridized plants that were homozygous for the *phbB* gene with plants that were homozygous for the *phbC* gene. The F_1 hybrids produced PHB in the leaves, but the PHB-producing plants were stunted and produced fewer seeds than plants that did not produce PHB.

Problem: (a) What aspects of the experimental design ensured that the researchers could obtain plants with both the *phbB* and *phbC* genes? **(b)** *Arabidopsis thaliana* is a small plant with no agricultural potential. Why might researchers choose such a plant for these experiments?

Solution: (a) Plants that regenerate from tissue culture following transformation of callus are heterozygous for the introduced gene. By transforming

two genes into separate sets of plants, the researchers could select for those transformed individuals that were homozygous in the self-fertilized progeny of the transformed plants. When they hybridized plants that were homozygous for *phbB* with plants that were homozygous for *phbC*, the first-generation hybrids each carried one copy of both genes.

(b) *Arabidopsis thaliana* is a model organism in genetics. Among its characteristics are short generation time, large numbers of progeny, susceptibility to *Agrobacterium* infection, and ease in cultivation in the laboratory. Although it has no commercial potential for PHB production, it is a good choice for initial experiments because researchers can grow and manipulate the plants easily and quickly, making development of procedures for PHB production more efficient. After they have developed and tested the procedures in *Arabidopsis*, they can apply them to other species that are important agriculturally but may not be as easy to work with in a laboratory setting.

Industry has benefited from applications of genetics in breeding and biotechnology through the production of biosynthetic fuels, restoration of environments, and improvement of food and industrial products.

SUMMARY

1. Worldwide, agricultural production has increased severalfold in the past 50 years. Although many factors are responsible for the increases, genetic improvement has contributed more than any other factor.

2. Continued genetic improvement is dependent on genetic diversity. Most genetic diversity for agricultural plants and animals is found in centers of diversity, which correspond to the centers of origin.

3. Most varieties and breeds of agricultural species grown throughout the world have narrow genetic bases.

4. As genetically improved plants and animals replace their more diverse counterparts in the centers of diversity, genetic diversity is lost through genetic erosion.

5. Scientists maintain genetic diversity in gene banks, where seeds, plants, and animals with diverse genotypes are held for future use in breeding. Most gene banks are underfunded.

6. Plant-breeding methods depend on the reproductive system of the species, the level of inbreeding depression, and the cost of commercial seed production.

7. Plant biotechnology includes applications of cell and tissue culture, plant transformation, and marker-assisted selection. Plant biotechnology has created a greater need for traditional plant breeding.

8. Animal breeding focuses on genetic improvement of existing breeds through selective breeding and minimization of inbreeding depression.

9. Animal biotechnology has significantly impacted animal genetic improvement through applications such as in vitro fertilization, surrogate motherhood, and animal transformation.

10. There are many industrial applications of genetics, including breeding plants to reclaim mined lands, treatment of wastewater with genetically improved microorganisms, and biosynthesis of industrial and pharmaceutical compounds.

QUESTIONS AND PROBLEMS

1. In the 1950s, U.S. farmers used genetically improved plant varieties derived from a very narrow genetic base, yet genetic erosion was not nearly as serious as it is today. Why are concerns over genetic erosion so prevalent now?

2. In 1970, southern corn leaf blight caused widespread losses in the U.S. maize crop because of genetic uniformity. Using information from this chapter and Chapter 18, describe what portion of the genome was uniform, why it was uniform, and how the uniformity is related to genetic erosion.

3. Until this century, farmers in centers of diversity maintained plant and animal diversity with no systematic management. Now, scientists maintain much of the genetic diversity in gene banks with management systems that are based on research. Yet the former system was much more successful in maintaining genetic diversity. Even when large numbers of varieties are collected, genetic diversity decreases over time when gene banks are utilized. What factors could account for the loss of genetic diversity in gene banks when compared to maintenance of diversity in farmers' fields?

4. Most animal and plant breeders recognize the serious nature of narrow genetic bases for agriculturally important species. However, rarely are the genetically diverse native breeds or varieties used as parents in breeding programs, even though these diverse animals and plants are often freely available to breeders. What factors might be responsible for this paradox?

5. Both maize and alfalfa suffer from inbreeding depression, and heterosis may be exploited in both species. Maize varieties are typically sold as F_1 hybrids, whereas alfalfa varieties are synthetic. Why are different types of varieties used for these two species?

6. Maize cross-pollinates naturally and suffers from inbreeding depression, yet plant breeders often intentionally

self-pollinate maize plants in breeding programs. What is the purpose of self-pollination in a cross-pollinated species such as maize?

7. An alternative method to single-seed descent (described in Example 27.2) is the bulk method. In this method, the breeder harvests all the seed from the F_2 plants and plants a random sample of the seed equivalent to one seed per F_2 plant for the next generation. The breeder repeats this procedure until the F_5 or F_6 generation. What advantages and disadvantages does this method have over single-seed descent?

8. In 1991, Michelmore et al. (*Proceedings of the National Academy of Sciences, USA* 88:9828–9832) described a method of marker-assisted selection called bulk segregant selection. The objective of their research was to find DNA markers that are closely coupled to alleles that confer resistance to downy mildew (a fungal disease) in lettuce. They took F_2 plants from a single cross between a true-breeding susceptible parent and a true-breeding resistant parent and self-pollinated them for several generations to obtain true-breeding inbred lines. They then tested the lines for resistance and susceptibility and extracted DNA from each of the lines tested. They bulked the DNA from the susceptible plants into a single sample and bulked the DNA from the resistant plants into a second sample. The two bulked samples were then tested with a series of RFLP and PCR-based DNA markers. They found several markers that differed between the two samples. When tested in the DNA from individual plants, most of these marker loci were linked at less than 5 cM to the locus that governs resistance to downy mildew. **(a)** Why is bulking DNA into two samples an efficient method for identifying linked DNA-marker loci? **(b)** Of what practical value is a DNA marker closely coupled to a disease resistance allele in plants?

9. Recurrent selection is often used for population improvement in maize. After the population has gone through several rounds of selection, plants are self-fertilized for four or five generations to develop inbred lines that can be used for F_1 hybrid production. During the recurrent selection cycles, plants are selected based on the performance of their progeny, rather than on their own performance. Why is progeny performance a better indicator for selection than the plant's own performance?

10. In recurrent selection in maize, plants in the population under selection are often tested by crossing them with an inbred line that has already been proven in hybrid production (a procedure called topcrossing), then measuring the performance of the progeny. What is the purpose of topcrossing?

11. Sometimes maize breeders use a method called reciprocal recurrent selection for population improvement. Two genetically different populations are maintained as separate populations. Selection is based on the performance of progeny when plants of one population are crossed with random samples of plants from the other population. The plants whose progeny perform well are then used as parents for the next cycle of selection. The progeny of these crosses are used only to provide information. They do not become parents in the recurrent selection scheme, thus preventing any gene flow between the two populations. What is the purpose of reciprocal recurrent selection?

12. Barley breeders sometimes rely on a fortuitous phenomenon to make breeding programs more efficient. The species of cultivated barley is *Hordeum vulgare*, which is naturally self-pollinated and is typically grown as pure-line varieties. A related wild species, *H. bulbosum*, is also available. The breeders make their hybridizations in *H. vulgare* as they would in a typical breeding program, then take the F_1 plants and hybridize them with *H. bulbosum*. *H. vulgare* and *H. bulbosum* form an interspecific hybrid zygote that develops into a small embryo. If left to itself, the small embryo dies because of the genetic incompatibility between the two species. However, the small embryo may be removed from the plant and cultured on an agar-solidified nutrient medium in a test tube, where it develops into a plant. As the embryo develops into a plant, all of the *H. bulbosum* chromosomes are lost, so the plant recovered from this process is haploid, with only *H. vulgare* chromosomes. If allowed to grow, the plant is sterile, because haploid plants cannot proceed through normal meiosis. However, breeders treat the haploid plants with colchicine to double the chromosome numbers so that the full diploid complement of chromosomes is present before the plant flowers. The plant is then fertile. What advantage does this method have over traditional breeding methods?

13. When *Agrobacterium tumefaciens* is used to introduce foreign genes into plants, no crown galls develop even though the bacteria have infected plant cells. In nature, infection by *A. tumefaciens* usually causes crown galls. Describe what prevents crown galls from forming when breeders use genetically engineered *A. tumefaciens*.

14. In what ways do breeding methods for domesticated plants differ from the methods for domesticated animals, and what are the reasons for these differences?

15. General combining ability in cattle breeding has historically been of greater importance than specific combining ability. However, specific combining ability has taken on added importance in recent years. **(a)** Distinguish between general and specific combining ability. **(b)** Why is specific combining ability more important now than it was in the past?

16. The zona pellucida is a protective membrane that surrounds the zygote and the cells that develop from the zygote during early embryonic development in mammals. Damage to the zona pellucida is usually lethal to embryos

of large mammalian species. In 1979, Willadsen (*Nature* 277:298–300) described a procedure in which cells from a sheep embryo at the two-cell stage could be removed from the zona pellucida and each cell transferred to damaged zona pellucida from which the native embryonic cells had been removed. These cells, contained within a foreign zona pellucida, were then embedded in agar and the agar transferred to a surrogate ewe's fallopian tube, which had been ligated at the junction of the fallopian tube and uterus. The embryos, still embedded in the agar, could develop within the fallopian tube. After the embryos had developed to the late morula or early blastocyst stage, they were removed and transferred to the uterus of a surrogate ewe. Of 16 surrogates, 5 bore single lambs, and 5 bore monozygotic twins. **(a)** What was the purpose of embedding the embryos in agar? **(b)** Of what significance is this research in large-animal breeding?

17. Modern agriculture typically requires phenotypic uniformity among plants so that the quality and harvestability of commercial crops can be assured. For these (and other) reasons, plant-breeding methods are designed to produce uniformity. For example, all of the plants in a pure-line variety or an F_1 hybrid variety are genetically identical. On the other hand, plants bred to reclaim disturbed lands (such as strip-mine spoils) are often developed to be as genetically diverse as possible. It is not uncommon for breeders to cross two genetically compatible species, then plant the segregating F_2 seed for reclamation. Why is diversity, rather than uniformity, an objective in breeding plants for reclamation?

18. *E. coli* strain MV1184 is capable of removing inorganic phosphates from wastewater. In 1993, Kato et al. (*Applied and Environmental Microbiology* 59:3744–3749) used genetic engineering to enhance expression of genes that promote inorganic phosphate metabolism. They found that the modified bacteria increased phosphate removal as much as threefold when compared to unmodified bacteria. Of what significance is this work in the use of genetic manipulation in industry?

19. Why are domesticated transgenic plants and animals considered better than bacteria for producing pharmaceutical or industrial products? Include in your answer both economic and biological considerations.

20. Administrators in academia and industry have sometimes increased investment in biotechnology while decreasing investment in breeding with the idea that biotechnology can replace traditional breeding methods. Why can it be justifiably said that plant and animal biotechnology will increase the need for traditional breeding rather than replace it?

FOR FURTHER READING

The current status and future of genetic gains in agricultural research were discussed by **Mann, C. 1997. Reseeding the green revolution.** *Science* 277:1038–1043. Plant genetic resources and their conservation are reviewed in **Brown A. D. H., O. H. Frankel, D. R. Marshall, and J. T. Williams, eds.** *The Use of Plant Genetic Resources.* **Cambridge, England: Cambridge University Press.** The use of molecular markers to incorporate genes from genetic resources was explained by **Tanksley, S. D., and S. R. McCouch. 1997. Seed banks and molecular maps: Unlocking genetic potential from the wild.** *Science* 277:1063–1066; and **Mann, C. 1997. Cashing in on seed banks' novel genes.** *Science* 277:1042. Details of plant-breeding methods can be found in **Poehlman, J. M., and D. A. Sleper. 1995.** *Breeding Field Crops,* **4th ed. Ames, Ia.: Iowa State University Press.** Details of animal breeding and biotechnology can be found in **Bearden, H. J. 1997.** *Applied Animal Reproduction,* **4th ed. Upper Saddle River, N.J.: Prentice-Hall.** The relationship between biotechnology and environmental protection is addressed in **Krimsky, S. 1996.** *Agricultural Biotechnology and the Environment: Science, Policy, and Social Issues.* **Urbana, Ill.: University of Illinois Press.**

For additional reading, go to InfoTrac College Edition, your online research library at: http: www.infotrac-college.com/brookscole

CHAPTER 28

KEY CONCEPTS

Eugenics, the intentional hereditary improvement of the human population through selection, was popular at one time but is usually shunned today.

~

Issues of informed consent and confidentiality are among the most important considerations in genetic testing and the treatment of genetic disorders.

~

The legal admissibility of DNA fingerprinting has been challenged.

~

To ensure public safety, institutional review and approval are required for many applications of recombinant DNA.

~

To protect investments in research, scientists and corporations often patent a wide range of genetically altered organisms, the genes they carry, and the techniques used to produce them.

~

At a time when progress in genetics is moving rapidly, appropriate science education is essential.

LEGAL AND ETHICAL ISSUES IN GENETICS

Hope and fear are often a part of genetics. The hope of developing effective treatments for genetic disorders and cancer drives efforts to fund research in medical genetics. Humanitarian foundations and governments sponsor large-scale agricultural breeding and biotechnology programs in less developed countries with the hope that genetically improved plants and animals can provide food to people who suffer from hunger. There are many other examples of how scientists have applied the principles of genetics to improve the well-being of humanity or to reap financial profits. Regardless of its purpose, each application of genetics raises concerns and fears among the public, as well as legal and ethical questions that are often not easy to resolve.

The practical applications of genetics create significant financial concerns. Biotechnology companies invest enormous sums of money to develop microbes, cell lines, plants, and animals with particular genetic characteristics. Most companies seek legal protection to prevent others, who have not made the investment, from reaping undeserved profits. The very nature of genetic traits makes it difficult for a company to legally protect a genotype or a gene. Because organisms transmit their genes from one generation to the next, and multiply them in the process, it is often a simple matter for someone to prop-

agate a genetically altered organism and use it for profit without the permission of the company that developed the genotype.

For example, a farmer can purchase a small amount of seed of an improved plant variety with little money, then propagate the seed year after year without any additional payment to the company that developed the variety. Companies have sought ways to protect their investments in genetic modifications by patenting genotypes, genes, and the methods used to develop them. Although the promise of financial protection stimulates private investment in research, protection sometimes inhibits cooperative progress because business executives choose to keep proprietary information secret or place legal restrictions on methods that others might use in research. Patented products may benefit thousands of people, especially those who suffer from poverty, ill health, and malnutrition. However, all too often, the products of genetic improvement carry such a high price that they do not reach the people who need them most.

The current pace of genetics research is unprecedented. Now is a time of excitement and hope as major breakthroughs appear often. The pace of genetics research is a source of concern, however, as the constantly changing face of genetics can make legal decisions and

ethical guidelines outdated before they are instituted. Courts often rely on legal precedents established over a period of years to interpret laws and legal guidelines. However, as the procedures and products of genetics research rapidly change, historical precedent can become outdated. The same is true of ethical guidelines. Guidelines may be established for specific questions that are no longer an issue after the guidelines are in place. Even general guidelines must be reinterpreted as new procedures become available.

Those who must make legal and ethical decisions are often unable to understand the increasingly complex and sophisticated nature of genetics research. Few lawyers have significant training in the sciences. Those who do have a distinct advantage. Judges and juries must often make decisions on the basis of conflicting expert testimony. Because court decisions often conflict, it is difficult for the legal system to produce a consensus for many issues in genetics. The scientific community also has difficulty reaching consensus about certain legal issues. Some legal experts have lamented that if scientists cannot come to a consensus regarding scientific issues, how can the legal system be expected to do so?

Caught in the midst of this is the general public. Perceptions of genetic technology are varied, with mingled feelings of hope and skepticism. Many people look to biotechnology as a way to make our world better, but these same people may harbor deep fears about what scientists can or might do. Some would prefer that we establish laws to ban many aspects of genetics research. Books, movies, and television often portray fictional and outlandish consequences of genetics research as if they were plausible, often diverting public concern from important issues to meaningless ones. News reports may follow the same line. The result is an uninformed, or sometimes misinformed, public. The consequences are confusion and mistrust.

In this chapter, we will examine several of the legal and ethical issues related to genetics. First we will look at the history of eugenics practice and legislation. We will then examine the legal and ethical aspects of genetic testing and screening. Next, we will discuss protection rights for the products of genetic research. We then will examine the development of safety guidelines for research with recombinant DNA. To conclude, we will see how science education can bring the most important issues to the attention of the general public.

28.1 EUGENICS

The concept of genetic purity in humans has existed throughout recorded history. Beliefs of racial or hereditary birthrights have been a part of nearly every human culture, and are evident in historical literature. Some cul-

tures, political regimes, and religions have encouraged (or, in extreme cases, required) people to marry within their own ethnic or familial background in order to preserve perceived racial or ethnic purity, often with the notion that certain races or ethnic groups are superior and should not be contaminated by those that are inferior. Such ideas have persisted in one form or another to the present day.

The idea that people can improve humans through selection in the same way that breeders improve animals and plants has been voiced for centuries, but began to receive serious attention from scientists in the late-nineteenth century. Sir Francis Galton, a cousin to Charles Darwin, was among the most outspoken proponents of improving the human population by discouraging reproduction among those people that he (and many others) considered to be the inferior members of society. He coined the term **eugenics** in 1883 to describe the intentional hereditary improvement of the human population through selection. During the early part of the twentieth century, up until the 1940s, eugenics was popular (although not unanimously) among philanthropists, scientists, politicians, lawyers, judges, journalists, and the better-educated and wealthier members of society.

During this period, the idea that one of the most lofty goals of genetics is to improve the human race appeared in countless scientific papers and usually was the concluding subject of genetics textbooks. Before long, some applications of eugenics made their way into law. Some people attributed traits such as imbecility, feeblemindedness, criminal behavior, lack of intelligence, antisocial behavior, insanity, idiocy, epilepsy, chronic alcoholism, and many others to heredity (often with simplistic single-gene inheritance) and proposed that society could rid itself of those traits by preventing people who had them from reproducing.

As early as 1896, some state legislatures passed laws forbidding the marriage of a woman and a man when both were considered imbeciles, feebleminded, or epileptic. Mandatory sterilization laws began in 1907. Governments in the United States and Europe passed laws requiring that criminals or people in mental institutions be sterilized to prevent transmission of their behavioral characteristics to subsequent generations. In a famous 1927 case challenging a Virginia sterilization law before the U.S. Supreme Court, Chief Justice Oliver Wendell Holmes, Jr., writing for an 8-to-1 majority upholding the law, stated,

> It is better for all the world, if instead of waiting to execute degenerate offspring for a crime, or to let them starve for their imbecility, society can prevent those who are manifestly unfit from continuing their kind. The principle that sustains compulsory vaccination is broad enough to cover the cutting of fallopian tubes.

Between 1907 and 1960, more than 60,000 people in prisons and mental institutions in the United States were involuntarily sterilized on the presumption that they carried detrimental genetic defects. State legislatures have since rescinded most mandatory sterilization laws.

Eugenics reached its height during the Nazi regime in Europe, when the purportedly superior Aryan race was to have improved itself through selective marriages and elimination of undesirable people. This philosophy led to measures far worse than involuntary sterilization. During the Holocaust, 6 million people were killed in mass executions because their genetic heritage was considered inferior. Such events in history are tragic reminders of how ideas about eugenics can be taken to extremes. The Nazi Holocaust is not that far removed from us in time; some of its survivors are alive today. However, we need not look as far back as World War II for examples of travesties committed in the name of genetic purity. Executions, incarceration, genocide, and rape have been justified during recent conflicts as a means of preserving ethnic superiority and racial purity.

The Effect of Eugenic Selection

Proponents of eugenics justify its practice on the assumption that preventing people with a genetic disorder from reproducing should eliminate, or significantly reduce, the frequency of an undesired trait in human populations. However, many people consider the right to reproduce to be a basic human right, which is violated by involuntary sterilization or laws that forbid certain people from reproducing. The United Nations' *Universal Declaration of Human Rights*, *The International Covenant of Political and Civil Rights*, and the *European Convention on Human Rights* are just a few of the more prominent agreements on basic universal human rights that include the right to reproduce.

The human rights argument is an important one. However, even if we do not consider the issue from this perspective, there is little theoretical justification for laws that selectively prevent reproduction to reduce the incidence of genetic disorders. Most alleles that cause genetic disorders are recessive and have low allele frequencies. As discussed in section 19.5, when the frequency of a recessive allele in a population is low, selection against the allele is very inefficient. The following example illustrates this principle as it applies to eugenics.

Example 28.1   Recognition of the ineffectiveness of eugenic selection against recessive genetic disorders.

Feeblemindedness is a catchall term that refers to any type of subnormal intelligence. As research on men-

tal retardation and its causes developed, the term *feeblemindedness* fell into disuse and is now considered inappropriate. During the early part of the twentieth century, however, most scientists accepted the idea that feeblemindedness was a genetic disorder caused by homozygosity for a recessive allele at a single locus (see East, E. M. 1917. *Journal of Heredity* 8:215–217). On the basis of this assumption, eugenicists proposed that a society could eliminate feeblemindedness from the human gene pool in a few generations by preventing feebleminded people from reproducing. Reginald C. Punnett, like many geneticists of his day, supported some aspects of the eugenics movement. However, he countered the idea that feeblemindedness could be easily eliminated in a 1917 article entitled "Elimination of Feeblemindedness" (*Journal of Heredity* 8:464–465). Punnett argued that eugenic selection against recessive traits should be very ineffective. Citing data from East's study that the frequency of feeblemindedness was 3 in 1000 in the United States, Punnett (with the assistance of Godfrey H. Hardy) concluded that over 250 generations would be required to reduce the frequency to 1 in 100,000, under the assumption that feeblemindedness was a recessive characteristic due to single-gene inheritance.

Problem: (a) What does this article illustrate about some of the assumptions upon which the eugenics movement was based? (b) How accurate was Punnett's calculation for reducing the frequency of a recessive allele by selection against homozygotes?

Solution: (a) Some of the assumptions upon which the eugenics movement was based were faulty. At the time Punnett wrote this article, quantitative genetics had only recently emerged as an important field. Many scientists incorrectly assumed that variation for most traits was governed by simple single-gene inheritance, although, at the time, they did not have diagnostic methods that permitted them to conclude otherwise. Currently, medical specialists can distinguish many different types of mental disorders (all of which were lumped under the designation of feeblemindedness). Medical geneticists have identified a hereditary basis for some of the disorders. In many cases, mental disorders are complex traits influenced by both genetic and nongenetic factors. Characterization of feeblemindedness as a trait governed by a recessive allele at a single locus was an oversimplification. (b) Under Punnett's assumption of single-gene inheritance, the initial frequency of the recessive allele is $q_0 = \sqrt{(3/1000)} = 0.05477$. The allele frequency following selection is $q_t = \sqrt{(1/100,000)} = 0.00316$. Using

equation 19.14, $t = (1/0.00316) - (1/0.05477) = 298$ generations to reduce the phenotypic frequency from 3 in 1000 to 1 in 100,000. Punnett's estimate of over 250 generations was theoretically accurate. This article is significant in that published information regarding the ineffectiveness of eugenic selection against recessive traits was available as early as 1917. Even so, involuntary sterilization of people considered genetically inferior (including those diagnosed with feeblemindedness) continued for decades after this date.

Although the effect of selection is usually negligible in altering the frequency of recessive genetic disorders, laws that forbid marriages between close relatives can have an appreciable effect in preventing recessive genetic disorders. When inbreeding coefficients are high because of matings between close relatives, the frequency of recessive disorders increases considerably. Laws that prohibit marriages between close relatives are far more effective than mandatory sterilization in preventing recessive genetic disorders. Most states in the United States, as well as many foreign countries, prohibit marriages of first cousins and more closely related people for this reason. Also, many religions discourage marriages between close relatives.

Contemporary Selection in Humans

Most people consider the eugenics of the first half of the twentieth century deplorable. Nonetheless, we continue to practice certain forms of selection today. Like eugenic measures of the past, our current practices have little effect on overall allele frequencies. Unlike the eugenics of the past, the aim of modern practices is personal in nature, rather than an attempt to improve society. Some of our current practices include technologies that provide prenatal testing, donor screening in sperm banks, and sex selection.

Prenatal testing, the testing of a fetus to determine whether or not it has a genetic disorder, provides couples with information that permits them to choose an abortion rather than have a child with a genetic disorder. Voluntary abortion of fetuses with genetic disorders has decreased the frequencies of children born with certain genetic disorders, such as Down syndrome and Tay-Sachs disease. Abortion of fetuses that have a genetic disorder is a form of selection against the alleles that confer genetic disorders, but its effect in reducing allele frequencies is negligible.

Donor screening in sperm banks is the selection of semen donors to exclude alleles that may cause genetic disorders. Many couples cannot have children because of male infertility. For a variety of reasons, a woman may undergo artificial insemination to conceive a child. Some hospitals and clinics maintain sperm banks in which they store semen samples provided by donors, usually med-

ical students and physicians, who have been screened for family histories of genetic disorders. Sperm banks usually maintain anonymity for donors, although they keep genetic records on file in case questions arise later. With DNA tests now available, some hospitals and clinics screen donors to identify heterozygous carriers of recessive alleles that cause genetic disorders. Such screening is a mild form of selection that will have a negligible effect on the frequency of genetic disorders. Much popular attention has focused on commercial clinics that recruit male Nobel laureates and male celebrities to voluntarily donate sperm.

Sex selection, a practice that selectively biases the probability of bearing a child with a particular sex, is now available for humans. Animal breeders were the first to develop sex selection technologies. For example, some dairy operations use sex selection technologies to increase the frequency of female calves. Medical researchers have adapted the procedures for use in humans. Medical technicians can sort sperm cells by centrifugation (X-bearing sperm sediment at a different rate than Y-bearing sperm), by separation in a flow cytometer (a machine that runs sperm cells one by one through a capillary tube and directs X-bearing sperm into one container and Y-bearing sperm into another using an electrical field), or by allowing sperm to move through a column of albumin (in which X- and Y-bearing sperm move at different rates). None of the methods ensures that a child of a particular sex will be born. A few Y-bearing sperm may contaminate a sample of X-bearing sperm and vice versa, but the ratios are substantially biased toward the desired sperm-cell type. The prospective mother is artificially inseminated with the chosen sperm sample, and the probability of her bearing a child of the desired sex is greatly increased.

Much attention has focused on the potential imbalance between sexes should sex selection become widespread. Currently, sex selection is expensive, and very few people use it. Consequently, its effect on the overall sex ratio is negligible. Should it become a widespread practice, concerns over alteration in the human sex ratio may be justified.

~

The idea that humanity can be improved genetically has been used to justify mandatory sterilization and some of the most tragic cases of genocide in history. There are no theoretical or empirical grounds for most eugenic practices. The effects of eugenic practices and modern technologies in altering allele frequencies through selection in humans are usually negligible.

28.2 GENETIC TESTING AND SCREENING

Physicians can successfully treat certain genetic disorders if they diagnose the disorder early in the patient's life.

Phenylketonuria (PKU) is a good example. Most states test all newborns for PKU to identify the disorder so that a physician can treat it before serious symptoms develop. The state usually funds PKU testing from tax revenues. There is a clear ethical obligation to treat a child with PKU and prevent mental retardation. There is also a financial benefit to doing so. People with mental retardation require expensive care for education and medical treatment. Often, the state bears much of the cost. Expenditures that the state makes to detect PKU are repaid by prevention of mental retardation.

For this reason, most states have enacted genetic screening laws when mandatory or voluntary screening can detect treatable genetic disorders. States have enacted laws for widespread screening of PKU, sickle-cell anemia, sickle-cell trait, and Tay-Sachs disease. In 1976, a federal law consolidated previous federal legislation into the National Genetics Diseases Act, which provided federal funding for genetic screening, counseling, research, and education. Congress rescinded the act in 1981 and diverted funding to the states.

The success of genetic screening has varied. PKU screening is an example of genetic screening that has gained widespread success and acceptance. Tay-Sachs disease is most common among people of Eastern European Jewish descent, and Jewish political groups have provided strong support of screening programs. Screening for sickle-cell trait, on the other hand, has met much opposition because it has focused on certain ethnic groups and provided a basis for racial discrimination.

Occupational Genetic Testing

A few genetic disorders that are not particularly difficult to deal with in daily life might be dangerous in certain occupational situations. Among the best-studied examples is **glucose-6-phosphate dehydrogenase (G6PD) deficiency**, caused by lower-than-normal levels of the enzyme glucose-6-phosphate dehydrogenase. When they are exposed to certain chemical agents, particularly oxidizing compounds, people with this disorder may have an increased tendency to suffer from anemia, although this conclusion has not been adequately demonstrated.

Several recessive X-linked alleles at the *G6PD* locus (OMIM 305900) cause G6PD deficiency, so it appears more often in men than in women. Some mutant alleles provide resistance to malaria when heterozygous in women, so the frequencies of these alleles are higher in people with ancestry from tropical regions where malaria was common. Because some oxidizing chemicals may pose a more severe health threat to people with G6PD deficiency than to others, industrial employers may consider people with G6PD deficiency a greater health risk than those without the disorder, although conclusive data on the severity of occupational hazards for this trait are not available.

A simple and inexpensive biochemical test can detect G6PD deficiency, so a number of companies have screened workers for G6PD deficiency as part of the hiring process when the work may entail exposure to oxidizing agents. Because G6PD deficiency is more common in men and among certain ethnic groups, this screening practice may be considered a type of gender, ethnic, or racial discrimination, and in this sense could be challenged legally. On the other hand, companies justify screening programs by arguing that they are fulfilling their legal and moral obligation to prevent injuries and damage to worker health.

In the 1970s, several companies in the United States conducted genetic screening for G6PD deficiency, sickle-cell trait, and α-antitrypsin deficiency among potential and current employees. In some cases, they used the results to deny employment or to alter worker responsibilities. As the practice of genetic screening received adverse public attention, some companies terminated their screening programs. However, a survey sponsored by the U.S. Congress found that 10% of responding companies conducted some form of genetic screening. Other researchers have criticized the report, citing their own data that the practice is much more widespread than the report suggests.

Existing laws that regulate discrimination on the basis of ethnic origin and disabilities could conceivably restrict the extent to which employers can use occupational genetic testing. In particular, the 1990 Americans with Disabilities Act (ADA) forbids discrimination against workers with disabilities that do not seriously inhibit their performance on the job. To what extent this law applies to actual or potential genetic disorders will probably be decided in the courts. As the number of genetic tests available to employers increases, the legal questions may become more difficult to resolve.

Genetic Testing and Health Insurance

Some insurance companies use genetic information to deny coverage or increase premiums for health insurance to people considered at risk for a genetic disorder. Genetic risks identified by insurance companies include information gleaned from genetic testing as well as examination of medical and family histories. Most employers pay for a large proportion of the insurance coverage for their employees and thus have an added incentive to work with insurance companies to identify employees or employee dependents who are at risk for a genetic disorder. In some cases, they may exclude or reduce health and life insurance benefits on the basis of predisposition for expensive medical care or early death.

With an increase in the number of genetic tests available, the potential for insurance companies to use genetic testing and analysis to determine health care risks is growing. In response, many scientists and physicians

have recommended legislation to prevent insurance companies and employers from using genetic information to deny or alter health care coverage. The Working Group on Ethical, Legal, and Social Implications of the Human Genome Project and the National Action Plan on Breast Cancer, with an endorsement from the National Advisory Council for Human Genome Research, have proposed the following guidelines to prevent insurance providers from discriminating on the basis of genetic information:

1. Insurance providers should be prohibited from using genetic information or an individual's request for genetic services, to deny or limit any coverage or establish eligibility, continuation, enrollment, or contribution requirements.

2. Insurance providers should be prohibited from establishing differential rates or premium payments based on genetic information or an individual's request for genetic services.

3. Insurance providers should be prohibited from requesting or requiring collection or disclosure of genetic information.

4. Insurance providers and other holders of genetic information should be prohibited from releasing genetic information without prior written authorization of the individual. Written authorization should be required for each disclosure and include to whom the disclosure would be made.

Several states have implemented laws that restrict the ability of insurance providers to discriminate on the basis of genetic information, and federal legislation is under consideration.

Legal Challenges of DNA Testing

As we saw in Chapter 26, DNA fingerprinting is a powerful forensic tool, which law enforcement officials can use to identify criminals and exonerate innocent suspects. Likewise, it is used routinely to identify paternity (and, less often, maternity). In spite of its value, attorneys have challenged its accuracy and admissibility as evidence in court.

Those who challenge the admissibility of DNA fingerprinting usually base their arguments on the possibility of faulty technique or potential inaccuracy of statistical assumptions. There are several types of potential technical errors. The purity and identity of the DNA sample being tested must not be in question. Sources of DNA that belong to a criminal may come from sources at the scene of a crime, such as tissue under the fingernails of a victim or from semen in a rape case. Under such circumstances, the identity of the DNA source may be certain. The identity of blood at the scene of a crime,

however, may be less certain. Also, the DNA must not be contaminated with DNA from another human. If it is, then DNA fingerprints of two or more people may be superimposed on one another. Any alterations in the methods used for electrophoresis, blotting, hybridization, probe detection, and PCR can potentially cause artifacts that may lead to false conclusions.

Most commercial, university, and government laboratories that provide DNA testing services avoid technical error through the use of rigorous controls. They identify possible contamination by including DNA from the suspect, the victim, and laboratory workers in the analysis. They prevent technical problems by providing extensive training to technicians and by using high-quality equipment and chemicals. They often replicate experiments to ensure that the results are the same each time. In recent years, DNA fingerprinting has established itself as a reliable method for identification. Legal challenges of DNA fingerprinting on the basis of laboratory errors are now rare.

Courts usually determine the admissibility of scientific evidence by the degree of consensus among experts in the scientific community. If most experts in the field accept a technique or assumption as reliable, then it is admissible in court. Significant disagreement among experts about the reliability of a technique or assumption may be grounds for inadmissibility of evidence that is based on the technique or assumption. DNA fingerprinting has suffered both of these fates in court. Numerous convictions have been made when DNA fingerprinting was the principal evidence. In other cases, courts have dismissed DNA fingerprints as evidence due to questions about their reliability.

In order to develop a consensus on DNA fingerprinting, the National Research Council (NRC), an agency of the National Academy of Sciences, developed a report entitled *DNA Technology in Forensic Science* in 1992 to establish the standards by which DNA fingerprinting and other DNA methods used in forensics could be judged. The report provided standards for technical procedures and for estimating allele frequencies that the NRC considered to be reliable. The report also recommended accreditation of laboratories to ensure adherence to standards. The objective of the report was to provide standards that the scientific community, as well as the legal community, would consider sufficiently reliable to form a general consensus about the admissibility of DNA evidence in court.

Respected scientists have debated aspects of the report, jeopardizing the consensus that the NRC had hoped to develop. Some attorneys have used conflicting scientific opinions in attempts to convince judges and juries that DNA fingerprinting may be unreliable. However, in recent years, DNA fingerprinting has established itself in the legal system as reliable and conclusive evidence in criminal and pa-

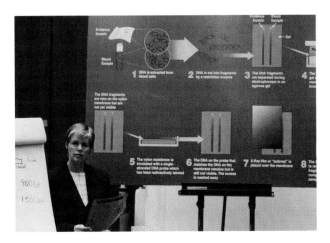

Figure 28.1 An expert witness explaining DNA fingerprinting to a jury. (Todd Bigelow/Black Star)

ternity cases. Figure 28.1 shows an attorney explaining the process of DNA fingerprinting to a jury.

Informed Consent and Confidentiality

Genetic testing is not new. Clinical geneticists and physicians have relied on blood typing, biochemical tests, cytological tests, and pedigree analysis for decades. However, the number of DNA tests that are now available is numerous and grows significantly each year. With so many ways to test people for genetic disorders and identify people with DNA fingerprints, issues of informed consent and confidentiality are of great importance. **Informed consent** for a genetic test is the voluntary submission to the test only after the person receives counseling about the test, what it will and will not identify, what the risks are, and how the information can be used once it is obtained. **Confidentiality** is a person's right to be certain that those who obtain information will guard it from disclosure to unauthorized individuals or groups. Sharing information with potential employers, insurance companies, law enforcement officials, physicians, researchers, and other interested parties may violate a person's legal right to privacy.

Confidentiality has often been breached in genetic testing. Results from tests identifying sickle-cell trait, XYY males, G6PD deficiency, and several other genetic traits have not always been held in strict confidence, causing legal, social, and financial problems for the people identified with these traits. Some insurance companies and employers have required genetic testing when a family member was known to have an identifiable genetic disorder.

The issues of confidentiality extend to genetic researchers as well. Human geneticists who possess records on familial genetic disorders have sometimes denied other researchers access to the records in order to protect a family's right to privacy. In some cases, questions re-

garding access to records for the purpose of research are resolved only after a legal dispute. In such cases, the need for advancement in medical and scientific research must be weighed against the right to privacy and laws designed to protect that right.

Many legal questions regarding informed consent and confidentiality are yet to be resolved. As the number of genetic tests increases and as technical questions surrounding the reliability of these tests are raised, the legal questions may become even more difficult to answer.

Informed consent and confidentiality are important issues for people who undergo any medical procedure. However, as highlighted in the following example, some aspects of informed consent and confidentiality are unique to genetic testing.

Example 28.2 Informed consent and confidentiality in genetic testing.

The legal definition of informed consent has evolved throughout this century and was not well established until the 1970s (see Frielander, W. J. 1995. *Perspectives in Biology and Medicine* 38: 498–510). In the early part of the twentieth century, physicians, rather than patients, usually chose whether or not a patient should undergo a medical procedure or treatment. The legal implementation of informed consent provides patients with the right to know enough about the risks and benefits associated with a medical treatment for them to choose whether the treatment is necessary and appropriate under the circumstances. The principles of informed consent and confidentiality apply to genetic testing as well. In other words, the patient usually has the right to accept or reject a genetic test when a physician recommends it. However, the ethics of informed consent and confidentiality can be more complicated with genetic testing than with medical treatments.

Problem: Why is this so, especially when nearly all genetic tests pose little medical risk to the patient?

Solution: Unlike most medical treatments, the results of genetic testing often apply not only to the person tested, but also to that person's relatives. A person may consent to a genetic test and to disclosure of results; however, relatives may oppose testing on the grounds that they have not provided informed consent regarding medical testing that may have an influence on them.

Many genetic tests are now available. Legal issues include questions about informed consent and confidentiality, the reliability of the tests, and how tests should be funded and administered.

28.3 LEGAL PROTECTION AND INTELLECTUAL PROPERTY RIGHTS IN GENETICS

Corporations, stockholders, foundations, and governments invest large sums of money in the development of medical and agricultural applications in genetics. Because genotypes and genes can often be propagated from a small number of individuals, private biotechnology companies and public institutions (such as universities and government research centers) seek legal protection to prevent others from exploiting the products of their investments. Companies also invest much in developing techniques that are unique and profitable and seek protection of their investment. In many cases, the protection allows the company to require that others who use the techniques pay royalties for their use.

The Constitution of the United States as well as numerous laws provide means for limited protection of inventions. Protection usually takes the form of a **utility patent**, which gives the patent holder a period of 17 years to control the exclusive rights to market and develop the invention. In many cases, genotypes, genes, and methods meet the legal definition of an invention and as such may be covered by utility patents.

Patenting Microbes

In 1980, the U.S. Supreme Court determined that genetically altered microbes designed to digest crude oil could be patented. Since that time, numerous microbes, including those that produce genetic pharmaceuticals, have been patented. In fact, for a time most patents on life-forms were for microbes, although the numbers of patents that cover aspects of human, animal, and plant genetics are rapidly catching up.

Patenting Genes

Genes can be patented when someone clones a gene or discovers its sequence. For example, the genetically engineered genes that produce human insulin in bacteria have been patented and have been the subject of a drawn-out and expensive court battle (see Example 28.3). Even natural human genes can be patented. The long search for the gene responsible for cystic fibrosis (*CF*) culminated in a rush to patent the gene. If an inventor discloses an invention in an article before patenting the invention, the invention becomes part of the public domain and cannot receive patent protection. In the case of the *CF* gene, the discoverers had submitted the articles for publication, but the articles were not scheduled to appear in print until after the patent applications had been filed. However, leaks to the press produced a flurry of questions about

the gene, forcing an earlier-than-expected disclosure of the gene and its sequence. The University of Michigan and the Hospital for Sick Children in Toronto, the two cooperating institutions where the gene was cloned and sequenced, had to quickly prepare patent applications before disclosure.

Many genes have been patented. Whereas patenting was once in the realm of private industry, now public universities, nonprofit corporations, and even government-funded research institutions patent many of their genetic discoveries to reap the financial benefits that can come from legal protection. Major universities have technology transfer offices that coordinate the legal protection of potentially lucrative discoveries made in university research.

Patenting Traits

Genetic traits themselves may be patented, provided the development or use of a trait is sufficiently novel. For example, high-lysine maize derived from tissue culture and the use of cytoplasmic male sterility in seed production are genetic traits in plants that have been patented. In these examples, the phenotype and its use, rather than the gene itself, is covered by the patent.

Patenting Techniques

Development of new techniques is important for advancement in research. For example, much effort and funding has been invested in the human genome project to find more efficient methods of DNA sequencing. With large investments being made in development of techniques, patent protection of new techniques has been sought and awarded. For example, Hoffmann-LaRouche, Inc., and F. Hoffmann-LaRouche Ltd. hold patents on the polymerase chain reaction (PCR) and on the use of *Taq* DNA polymerase in PCR. Any *Taq* polymerase sold for PCR can be done only under license.

Plant Genetic Protection Rights

Before 1930, there was no legal protection for plant genotypes. The first plant genetic protection law was the Plant Patent Act, passed in 1930. It extended plant patent protection for a period of 17 years, but only to asexually propagated varieties, such as fruit trees, grapes, and ornamental species. The protection did not extend to seed-propagated varieties. Universities and companies produced seed-propagated plant varieties for decades with no legal protection, and farmers often sold seed without paying any royalties to the company that developed the variety. For a time, state-owned universities developed most pure-line varieties for self-pollinated species, such as wheat, barley, soybean, and oats. Agricultural development of the state, rather than profit

from the sale of seed, was the principal motive. Private companies, however, produced most F_1 hybrid varieties. Because F_2 seed loses its heterotic potential (see section 20.3), a farmer gains nothing from saving seed. The genetic system of the plant ensures that a company's seed is protected, as long as the inbred lines used to produce F_1 seed are guarded.

This situation changed, however, when private companies began to market pure-line varieties and considered the genotype proprietary. The first attempt by Congress to provide protection of pure-line varieties and inbred lines was the **Plant Variety Protection (PVP) Act** of 1970, which gave protection to the genotype of a pure-line variety or inbred line. A PVP certificate provides 18 years of protection to a variety or inbred line. During the time of protection, the PVP certificate owner has exclusive rights to propagate and market the variety or to use the inbred line in hybrid seed production.

However, there were two aspects of the PVP Act that some seed companies found objectionable. First, a farmer could purchase a protected variety then save some of the seed for planting in subsequent seasons without paying royalties to the company. Second, breeders could use the protected variety as a parent in hybridization schemes to produce their own varieties without paying royalties to the PVP certificate owner when the new variety was marketed. Because of these objections, seed companies sought utility patents on pure-line varieties in order to obtain protection for any type of commercial propagation and for the right to use the variety as a parent in the development of new commercial varieties. The U.S. Patent Office denied such patents on the grounds that pure-line varieties were covered by the PVP Act. However, in 1985, under the *Ex Parte Hibberd* decision, the Board of Patent Appeals and Inferences reversed this policy and granted the right to patent plant varieties produced from seed.

As companies seek protection of newly developed plant varieties, countries that provided the original genetic resources from which these varieties were developed have raised questions about their rights. For example, a plant breeder may utilize an old potato variety from Peru that has resistance to a specified pathogen as a source of genes for development of a new variety in the United States. The breeder can then receive a patent on the new variety, but the old variety remains in the public domain, and anyone can propagate it freely. When a Peruvian company wishes to market the new variety or use it as a parent in a breeding program, the company may be legally unable to proceed unless it pays royalties. Yet no one paid royalties for the use of the older variety as a parent in breeding programs.

Some countries have passed legislation that prohibits transfer of their genetic resources outside of the country without compensation to the country. Unfortunately, many of these same countries fail to provide the financial backing necessary to adequately preserve the genetic resources they have legally restricted, causing irreplaceable loss of their protected resource. In addition, genetic resource protection laws are often not enforced, so genetic material may spread to other countries by contraband. In the meantime, legitimate scientists may be unable to obtain the material that they need through legal means.

How to best regulate the exchange of plant genetic resources is a difficult question. Some legal experts have suggested that plant genetic resources be treated much like minerals or other tangible resources that are exported. The country of origin would receive royalties on new varieties developed from their older varieties. In many cases, it would be difficult to ascertain how much of a new variety's genotype was derived from the older variety and how much the royalties should be. The costs of litigation in such a situation may be so prohibitive that breeders may not wish to use protected resources to expand the genetic base of their varieties. Such a scenario is not good because the genetic foundations of most crop species are narrow and should be expanded.

Many scientists have suggested that the rights to use a plant variety as a parent in a breeding program should be freely available to anyone in the world, but the commercial propagation and sale of a variety should be protected. In other words, a genotype can be protected, but its genes cannot. Seed companies usually oppose that policy because patents provide protection of a variety's use as a parent. In addition, some opponents of that policy believe that it discriminates against less developed countries because many of those countries rarely produce varieties that are used commercially. Such a policy could divert most of the benefits of protection to the economically developed countries.

This problem is so important that it became one of the most debated questions in the 1992 Earth Summit held in Rio de Janeiro. The United States did not sign the Convention on Biological Diversity, which was finalized at the Summit, in large part due to pressure from seed and biotechnology companies that opposed the free exchange of genetic resources.

Animal Genetic Protection Rights

A 1987 ruling by the U.S. Patent Office provided protection to animals that carry altered genes inserted into their genomes through genetic engineering. The genotypes of animals produced by traditional breeding cannot be protected, in part because each individual's genotype is different. However, an altered gene carried by an animal may be protected. Any animal carrying a gene altered by genetic engineering can be patented, and the patent extends to progeny that carry the gene. Also patented are inbred laboratory strains of mice that carry genes important to biomedical research.

Implications of Protecting Genetic Materials, Traits, and Techniques

Often the single most important question raised by legal protection in genetics is whether legal protection promotes or inhibits the overall progress of research. On the one hand, those in favor of protection argue that products must be protected to allow investors to recover their investments. Without the promise of legal protection, many companies would not risk the investment necessary to develop new products and techniques, and progress in private research would be seriously jeopardized.

On the other hand, legal restrictions on the use of a product or technique may inhibit beneficial research. Scientists have suppressed or delayed publication of significant results in order to obtain patent rights. Products such as patented enzymes or test animals for use in research can carry such a high price that researchers may need to divert limited research funds from other uses in order to purchase the patented product. Patented procedures may carry restrictions or requirements that royalties be paid if the procedures are employed in experimental research with commercial applications, inhibiting the widespread use of such procedures.

Striking a balance between the need to protect investment in research and the need to encourage cooperative research is not easy. Fortunately, some companies that own patents allow exemptions from some patent restrictions for nonprofit research institutions, requiring royalties only when products that have been developed using a patented product are marketed. Even so, the race to patent products and procedures in genetics research has accelerated. As the value of biotechnological products grows, so does litigation and the costs associated with it.

Even universities and nonprofit or governmental institutions have found patenting to be a way of raising needed funds. At one time, many of the products and techniques used in genetics research were developed at universities and nonprofit or governmental research institutes. Disclosure through publication in scientific journals was considered the primary goal. As the cost of research has risen at a time when public funding is difficult to obtain, universities have found legal protection through patents to be a means of raising funds to continue research. The university patents any product developed in its laboratories, then licenses the product to a private company, which pays royalties to the university for the right to market the product. In this way, the university can reap some of the financial benefits of the product without jeopardizing its nonprofit status by directly involving itself in marketing.

Perhaps the greatest beneficiary of biotechnology patents is the legal profession. Law firms that specialize in biotechnology law are growing in number, and the need for lawyers who understand the intricacies of genetics and patent law is on the rise.

The following example illustrates some of the concepts we have just discussed by highlighting a legal dispute over genetically engineered human insulin and the high costs of litigation.

Example 28.3 A patent dispute over recombinant human insulin.

In 1997, Marshall (*Science* 277:1028–1030) reviewed the issues surrounding a high-profile legal dispute over patents on recombinant human insulin. In 1977, a team of researchers from the University of California (UC) discovered the DNA sequence of the rat insulin gene. They applied for a patent on the sequence in May 1977 and received the patent (U.S. Patent Number 4,652,525) in 1987. At the time they conducted this work, rules on the safety of recombinant DNA prohibited work with human DNA. After the rules were modified to permit recombinant DNA research on human DNA, UC researchers determined the DNA sequence of the human insulin gene, and in 1979 applied for a patent on vectors containing the human preproinsulin gene and the human proinsulin gene, and on the expression of those genes in microorganisms. The patent (U.S. Patent Number 4,431,740) was awarded in 1984. In the meantime, researchers at Genentech, a private biotechnology company, developed the two-chain recombinant insulin method (described in section 26.4) for producing human insulin in *E. coli*. In 1982, under an agreement with Genentech, Eli Lilly and Company began producing and marketing recombinant human insulin synthesized by the two-chain method. (In 1996, Lilly switched to the single-chain proinsulin method, also described in section 26.4, which is more efficient than the two-chain method.) UC sued Lilly for royalties in 1990, claiming that Lilly's refusal to pay royalties to UC violated both its patents. In 1995, the court ruled in favor of Lilly, concluding that Lilly had infringed neither patent because the rat sequence differed from the human DNA sequence, and the procedures used by Lilly differed sufficiently from the procedures described in the patents. The judge also ruled that the UC scientists obtained their patents through "inequitable conduct." The judge based this conclusion on claims by attorneys for Lilly that the UC scientists did not adhere strictly to recombinant DNA safety regulations in effect at the time and, therefore, had an unfair advantage in obtaining one of the patents. UC appealed the decision, and in 1997 the appeals court dismissed the inequitable conduct claim but allowed the ruling that Lilly had not violated the patents to stand. The legal costs of this case for Lilly

were about $18.5 million, while those for UC were about $12 million.

Problem: In what way did the progress of genetic research influence the rulings in this case?

Solution: The technologies used in genetics research and in the biotechnology industry are frequently modified as new information becomes available. Patents typically cover technologies currently in use at the time of development. In patent disputes, courts are often faced with the challenge of determining when a technology is sufficiently different from a patented technology for the patent to no longer apply. In this example, the courts determined that the differences in technologies were sufficient to preclude infringement of the patents.

In recent years, legal protection of property rights associated with genetic discoveries has become one of the most significant legal and ethical issues in genetics. Genotypes, genes, traits, and techniques are now patented to protect the financial interests of the patent holder.

28.4 RECOMBINANT DNA AND SAFETY

Few applications of science are more frightening to the public than recombinant DNA research. Movies, television, and popular literature paint pictures of genetically engineered beasts that escape from the laboratory, leaving a path of destruction wherever they go. Children watch cartoons that depict genetically altered humans and animals with supernatural powers. All too often, such images create misguided fears. While people invest much hope in the potential of biotechnology to produce cures for disease, food for a hungry world, and products that are less harmful to the environment, they may also harbor fears about the possible misuse of genetic technology. Are such fears justified?

In 1974, as recombinant DNA research was in its infancy, several of the scientists who most closely worked on its development urged that a moratorium on recombinant DNA research was needed until its potential harm could be assessed. This urging brought about the Asilomar Conference in February of 1975, when over 100 scientists met to decide how recombinant DNA research should proceed. The participants adopted strict recommendations, which included (1) prohibiting the release of organisms carrying recombinant DNA outside of laboratory confines and (2) requiring the use of genetically altered bacteria that could not survive and reproduce outside of the laboratory for all recombinant DNA experiments with bacteria. Compliance was voluntary, but scientists throughout the world agreed to follow the guidelines.

In 1976, the National Institutes of Health Recombinant DNA Advisory Committee (NIH RAC) established guidelines that applied to all NIH-funded research. Institutions that failed to comply with the guidelines jeopardized their NIH funding. Soon, other government funding agencies developed similar guidelines that required compliance by all federally funded researchers. Privately funded researchers also complied, although their compliance was still voluntary.

Most universities and research institutions established institutional biosafety committees to ensure that researchers adhered to the guidelines. These committees continue to oversee recombinant DNA research in their institutions. As agricultural and industrial applications of biotechnology grew imminent, the U.S. Environmental Protection Agency, the U.S. Department of Agriculture, and the Food and Drug Administration established their own sets of guidelines.

At first, all guidelines prohibited any environmental release of genetically engineered organisms. However, it soon became clear that field tests would be necessary to complete product development. Confusion arose over the interpretation of complicated guidelines. In some cases, researchers were uncertain about which government agency had jurisdiction over their projects. Because of the confusion, some researchers conducted field tests with the understanding that they were in compliance with guidelines, only to find out later that an agency considered them in violation. In some cases, researchers undertook unusually strict precautions to ensure that they were not in violation of any written or perceived safety requirement.

As recombinant DNA research has matured, scientists have collected sufficient data to indicate that most applications pose very little hazard to human health and the environment. Revised guidelines permit field-testing of genetically engineered organisms subject to prior review and approval. Some recombinant DNA products have been field-tested and are now on the market. Even so, strict safety and environmental guidelines remain in place in research laboratories and in field tests. Each project is reviewed before it is initiated and before field tests can begin.

~

The advent of recombinant DNA raised concerns about its safety. Regulations currently require prior approval and review of any field-testing and of the release of organisms that contain recombinant DNA.

28.5 GENETICS AND SCIENCE EDUCATION

Many periods in the history of genetics have generated excitement. However, at no time in the history of genetics has the pace of discovery been so fast, and the applications so broad, as now. Unfortunately, we also live

in a time when many people are only vaguely aware of advances in genetics and have avoided even a basic education in science.

People often perceive science as being too boring or too difficult to understand. Surveys of nonscience students indicate that the students rarely have an aversion toward science but also fail to see science as having any relevance to their lives. Ironically, this attitude comes at a time when science, genetics in particular, is more relevant than ever.

Problems of ignorance are not confined to nonscientists. At a time when genetic tests are widely available, and the genetic aspects of cancer, congenital abnormalities, immunity, viral infection, psychiatric disorders, and many other medical problems are being revealed, lack of a current understanding of medical genetics is apparently widespread among physicians who are not specialists in genetics. A patient's informed consent can only happen if the health care provider is sufficiently informed. The following example highlights a study that documented the lack of understanding of medical genetics among physicians.

Example 28.4 Physicians' understanding of medical genetics.

In 1993, Hofman et al. (*Academic Medicine* 68:625–632) reported the results of a survey that tested understanding of medical genetics among physicians. The authors selected physicians from the medical society memberships of five specialties: family practice, internal medicine, pediatrics, obstetrics-gynecology, and psychiatry. They selected these five specialties to represent areas in which some genetic counseling by the physician may be required. They sent the questionnaire to 1795 physicians, of whom 1140 (63.5%) responded. For validation purposes, they also sent the questionnaire to 360 medical geneticists and genetics counselors, of whom 280 (77.8%) responded. The average score on the questionnaire for medical geneticists and genetic counselors was 94.6%, with a standard deviation of 4.2%, indicating that the questionnaire was valid. The physicians from the other five specialties averaged 73.9%, with a standard deviation of 13.9%. Several factors were associated with a higher performance on the questionnaire. The most significant was recency of graduation from medical school: those who had graduated more recently tended to have higher scores. The second most significant was the degree of exposure to medical genetics in actual practice, which was positively correlated with higher scores. Some physicians were not familiar with certain genetic tests, and some were unable to correctly analyze pedigree information. Surprisingly, those who had taken a required genetics course during medical school scored only slightly higher than those who did not.

Problem: (a) What do these results suggest about the level of physicians' understanding of medical genetics? (b) What might be done to improve physicians' understanding of medical genetics?

Solution: (a) These results reveal that the average understanding of medical genetics by physicians in specialties that may require them to provide some genetic counseling is inadequate, especially given the increased number of genetic tests that are now available. (b) The answer to this question is well summarized in the words of the authors of this study: "Although a medical school course in genetics may improve knowledge, it is not sufficient. Greater emphasis is needed at all levels of medical education to reduce the chance of physician error as more genetic tests become available."

Among the many ethical imperatives that professionals in genetics face, few are more important than appropriate education. Clearly, those who utilize and provide genetic technologies must be adequately trained. However, perhaps the most important task is appropriate public education. Primary and secondary schools can accomplish much by ensuring that science teachers receive adequate training, salaries, and resources. Programs designed to include social responsibility as a part of primary and secondary science curricula have been initiated. Public and educational television networks have developed outstanding programming designed to spark interest in science and highlight recent advances in ways that are both entertaining and informative. Science books written for nonscientists are now more common than ever, many in the field of genetics.

Many universities are providing increased support of general undergraduate science education. Just a few years ago, research prowess in the sciences was considered the primary barometer of success at many universities. There were few significant rewards for teaching excellence, and rarely was anyone dismissed or reprimanded for being a poor teacher, as long as success in research was evident. That situation still persists in some cases, but in general it is changing. In the past few years, universities have placed more emphasis on undergraduate science education, for both majors and nonmajors. The National Science Foundation has provided new emphasis (and funding) toward development of undergraduate education as well as primary and secondary school education.

Not everyone agrees on how to best improve science education. But there is plenty of constructive dialogue,

a sign that things are moving in the right direction. The ultimate hope is that more students will choose careers in science, especially in science teaching. Equally important in a democratic society is the hope that an informed public will participate in making rational decisions about the many ethical and legal issues that are already upon us and will only increase in the future.

~

As the applications of genetics become increasingly relevant to society, appropriate education of scientists, legal professionals, health care providers, and the public is essential.

SUMMARY

1. As genetics has assumed an increasingly important role in medicine and agriculture, legal and ethical issues have become both more common and more complicated.

2. Eugenics is the intentional improvement of the human race through selective mating or prevention of reproduction. Although eugenics was once popular, its application has declined significantly. Most applications of eugenics violate human rights and have little effect from a genetic standpoint.

3. Mandatory screening of some genetic disorders, such as phenylketonuria, helps to prevent the adverse consequences of these disorders. DNA tests are available for detecting many mutant alleles in homozygotes and heterozygotes.

4. Genetic testing has been used for occupational screening and for determination and possible denial of health insurance. Governments have passed laws in some areas to prevent discrimination or denial of insurance benefits on the basis of genetic information.

5. DNA fingerprinting has been challenged in court. It currently has gained widespread legal acceptance for both exclusion and inclusion, provided the purity and identity of the sample are not in question.

6. Informed consent and confidentiality are essential parts of genetic testing, to protect the rights of individuals. These issues are especially important in genetic testing because the results of genetic tests may affect relatives, as well as the person being tested.

7. Genetic discoveries and applications may be protected legally. In most cases, patent protection is sought.

8. Rules governing the safety of recombinant DNA research are in place and are updated as needed.

9. Genetics is now more relevant to the public than ever before, underscoring the need for appropriate public education in science.

QUESTIONS AND PROBLEMS

1. State-funded genetic testing for PKU has widespread support. Yet proposals for state-funded genetic testing for sickle-cell trait and G6PD deficiency and prenatal testing for Down syndrome and Tay-Sachs disease have generated many criticisms. **(a)** What factors associated with these tests may be responsible for the criticisms? **(b)** Why hasn't PKU testing received some of the same criticisms?

2. PKU testing of newborns is mandatory throughout most of the United States and in much of the world. Rarely are parents informed about the nature of the test unless the results indicate that the child is at risk. Why is informed consent not an important issue with PKU testing?

3. Why is it possible to patent a pure-line or clonal plant variety but not an animal breed?

4. What can be patented in domestic animals?

5. What is different about current practices that implement some form of selection in humans when compared to the eugenics of 70 years ago?

6. In what ways does intellectual property right protection promote progress in research? In what ways does it inhibit progress in research?

7. Why are restrictions placed on recombinant DNA research less stringent than they were two decades ago, when scientists recommended a moratorium on recombinant DNA research?

8. In 1995, Baird (*Perspectives in Biology and Medicine* 38:159–166) pointed out that epidemiological studies of public health that include human DNA testing require DNA sampling from large numbers of people, a procedure that could potentially conflict with informed consent and confidentiality. How might issues of informed consent and confidentiality be addressed under such circumstances?

9. In the article cited in the previous question, Baird also pointed out that researchers may store DNA samples for long periods of time. When a person consents to DNA extraction, a large amount of potential information about that individual's genetic constitution is present in the sample and may be acquired many years after the sample is taken. Legal precedents have established that DNA is part of a person's body and that each person has the right to control his or her own body. What does this imply about future testing using DNA samples that have been stored?

10. Why are laws that prohibit consanguineous marriages more effective in controlling the frequency of genetic disorders than mandatory sterilization laws?

11. Sometimes people who have a recessive autosomal genetic disorder are discouraged from having offspring so that they will not transmit the gene to future generations. Evaluate the appropriateness of such advice. In your evaluation, consider both the effect on the human population as a whole if such advice were followed and the effect on the person's descendants.

12. In what ways would the situation described in the previous question be the same or different for an X-linked recessive allele?

13. Review the recommendations of the Working Group on Ethical, Legal, and Social Implications of the Human Genome Project and the National Action Plan on Breast Cancer, which are listed in section 28.2. What consequences might the implementation of these recommendations have on (a) genetic testing, (b) people with a genetic predisposition for a particular disease, and (c) the insurance industry?

14. Under what conditions might employment discrimination on the basis of certain types of genetic information be a form of racial, ethnic, or gender discrimination? Provide examples of specific genetic disorders in your answer.

15. As legal and ethical issues become more numerous and technical, why is it important to provide appropriate public education on these issues when it is difficult for nonscientists to fully understand the principles and techniques that underlie these issues?

16. What technical difficulties with DNA fingerprinting have been used to challenge the use of DNA fingerprinting in court? How can researchers overcome these difficulties or compensate for them?

17. The human genome project (see section 26.2) has been questioned on ethical grounds because of its cost and the potential uses of the information it seeks. What aspects must people consider when judging the value of the human genome project and applying information gained from it in medical genetics?

18. Why has protection of intellectual property rights become a frequent practice in governmental, university, and other nonprofit research organizations?

FOR FURTHER READING

Several important ethical issues and their relationship to genetics are discussed in **Suzuki, D. T. 1989.** *Genethics: The Clash Between the New Genetics and Human Values.* **Cambridge, Mass.: Harvard University Press**. The history of eugenics is covered in **Paul, D. B. 1995.** *Controlling Human Heredity, 1865 to the Present.* **Atlantic Highlands, N.J.: Humanities Press**. Legal and ethical issues related to human genetics are discussed in Chapter 19 of **Cummings, M. R. 1997.** *Human Heredity: Principles and Issues,* **4th ed. Belmont, Calif.: Wadsworth**; and in Chapters 11 and 12 of **Jorde, L. B., J. C. Carey, and R. L. White. 1997.** *Medical Genetics.* **St. Louis, Mo.: Mosby**. Issues related to patenting genes, materials, or information from living organisms were reviewed by **Byrne, N. 1993. Patents for genes, other biological materials, and animals.** *Trends in Biotechnology* **11:409–411.**

For additional reading, go to InfoTrac College Edition, your online research library at: http: www.infotrac-college.com/brookscole

ANSWERS TO SELECTED QUESTIONS AND PROBLEMS

Chapter 1

1. Genetics is among the oldest of the sciences because plant and animal breeding have been practiced for millennia. It is a twentieth-century science because the fundamental principles of inheritance became a part of mainstream science in the year 1900. **6.** Humans have relatively long generation times and do not produce large numbers of progeny. It can also be very expensive to conduct genetic research with humans. DNA analysis compensates for the limitations of traditional genetic analysis by making possible large-scale studies on human families. **12.** Genetics is an analytical science and cannot be adequately studied as a body of theories and facts. Instead, it is best studied by examining how genetic experiments are conducted and their results interpreted.

Chapter 2

2. (a) Three. **(b)** $C_{39}H_{56}N_{15}P_4O_{32}$. **(c)** The relative proportions of carbon, hydrogen, nitrogen, oxygen and phosphorus atoms correspond closely to Miescher's basic empirical formula. **(d)** A particular DNA molecule may vary in the proportions of the four nucleotides from another DNA molecule, so we expect the relative proportions of the atoms to vary slightly. **3.** After becoming incorporated into the living R cells, the transforming principle transformed them to the S type. This change was genetic because all of the offspring of each succeeding generation were the S type.

6. In such a DNA double helix, there is no relationship between C and G or A and T as Chargaff's rules predict. In addition, pyrimidines (C and T) are smaller than purines, so C-C and T-T pairs would be markedly less wide than A-A and G-G pairs, which is contrary to the observations obtained by X-ray diffraction analysis. **7.** T = 30%, C = 20%, and G = 20%. **8.** The viral DNA is single stranded. **10.** The end marked "b" is 3′, "a," "c," and "d" are 5′. **15.** DNA polymerase II synthesizes DNA during repair of DNA damage and is not an essential enzyme for synthesis of undamaged DNA. If left unrepaired, DNA damage can cause mutations. In the absence of DNA polymerase II activity, DNA damage cannot be efficiently repaired, so mutations are more frequent. **17.** Less energy is required to separate AT-rich strands of the double helix. Regions rich in A-T pairs at origin of replication sites facilitate separation of the DNA strands for initiation of replication. **18.** Yes. Unidirectional replication is the migration of a replication fork in one direction from an origin of replication. Each replication fork has a leading and a lagging strand. The lagging strand contains Okazaki fragments. **20.** If replication were in the $3′ \rightarrow 5′$ direction, the last added nucleotide on a DNA strand would have to carry the triphosphate group to furnish energy for the bond. If this last nucleotide were to lose one or more of the phosphate units, there would not be enough energy to bond to the 3′ carbon of the new incoming nucleotide. This loss would block DNA replication.

25. (a) Semiconservative, **(b)** unidirectional, **(c)** and semi-discontinuous. **26.** If the 3′ end dissociated from the circle, the RNA primer that bound to that end could not be replaced with DNA because there would be no 3′ end on the newly synthesized strand to which new nucleotides could be added.

Chapter 3

2. A gene is a segment of DNA from which an RNA molecule is transcribed. **4.** A DNA sequence that encodes an RNA molecule is a gene. A gene does not necessarily encode a protein. **8.** There are several RNAs that are not among the three major types (mRNA, tRNA, and rRNA). These include primers which participate in DNA synthesis; telomerase RNA, which provides a template for synthesis of telomeric DNA; snRNAs, which assist in intron removal; small RNAs that direct secretory proteins to receptors in the cell membranes; and ribozymes, which catalyze biochemical reactions. **10.** TFIIS causes a stalled RNA polymerase II to back up, and removes a few nucleotides from the 3′ end of the RNA, permitting RNA polymerase to attempt elongation again. **12.** The ribosomes block sites where transcription may terminate in the reading frame, and thus prevent premature termination of transcription. **13. (c)** mRNAs. The genes that encode tRNAs and rRNAs are highly conserved among prokaryotes because they encode molecules that are similar in structure. **17.** Both sequences are high in A-T base pairs, and the two strands separate readily at these sites for initiation of transcription. **19.** Staring from the 5′ end of the molecule, the −35 site begins at the 11th nucleotide and has the sequence TTGACA, the −10 site begins at the 34th nucleotide and has the sequence TTTAAT, and the transcription startpoint is the 46th nucleotide (a G). The −35 site is identical to its consensus sequence, and the −10 site differs from its consensus sequence by one nucleotide. **22.** It causes RNA polymerase to stall. **25.** Ribosomes that are translating the mRNA block formation of any potential hairpins in the reading frame.

27.

```
        C
    T       T
  A           C
  G           C
  A           T
  A           T
  G           C
  G           C
  C           G
  C           G
  C           G
  C           G
  G           C
  A           T
  A           T
  A           T
5' AACGCATGAG     ATATAATTAGCGCGGTTGAT 3'
```

30. The position of the intron is indicated below by parentheses and the sequences at the splice sites that correspond to consensus sequences are underlined:

TTGGTGGTGAGGCCCTGGGCAG (<u>GTTGGT</u>ATCAAGGTTACAA-
GACA- GGTTTAAGGAGACCAATGAAACTGGCATGTGGAGACA-
GAGAAGACT-
CTTGGGTTTCTGATAGGCACTGACTCTCTCTGCCTATTGGTC-
TATT- TTC<u>CCACCCTTAG</u>) GCTGCTGGTGGTCTAC

The underlined sequences on the 5′ and 3′ ends of the intron differ from their consensus sequences by two nucleotides each.

Chapter 4

1. (a) The upper strand. **(b)** Leucine. **(c)** G. **4.** The AUG codon is nested within the sequence CCUUCUGCC<u>AUG</u>G, which differs by four nucleotides from the consensus sequence GCCGCCPuCC<u>AUG</u>G. The G following the AUG codon and the purine three nucleotides before the AUG codon are fully conserved. **7.** Prokaryotic mRNAs can carry several separate reading frames because the small ribosomal subunit binds to a Shine-Dalgarno sequence within the mRNA, and there can be several Shine-Dalgarno sequences in a single mRNA, one for each reading frame. In eukaryotes, all small ribosomal subunits must attach at the 5′ end of the eukaryotic mRNA, and, therefore, they must all begin translation at the first initiation codon. There is no mechanism in eukaryotes for attachment of ribosomes downstream from the 5′ end of mRNA. **9.** SerValSerGln-SerValSerGln . . . in a repeating sequence. **11.** No, because the genes are repeated many times. **12.** 31. **14.** AAG can pair with the anticodons 3′ UUU 5′ and 3′ UUC 5′. AUG can only pair with the anticodon 3′ UAC 5′, UUU can pair with the anticodons 3′ AAA 5′ and 3′ AAG 5′. CUA can pair with the anticodons 3′ GAU 5′ and 3′ GAI 5′. GUU can pair with the anticodons 3′ CAA 5′, 3′ CAG 5′, and 3′ CAI 5′. And CAG can pair with the anticodons 3′ GUU 5′ and 3′ GUC 5′. Base pairings that the genetic code does not allow are not included. **15.** The anticodon 3′ ACI 5′ would pair with the termination codon UGA, which would allow the UGA, codon to encode cysteine instead of termination of tranlation, which is a violation of the genetic code. **16.** 3′ AUG 5′. **20.** Each mutation that substituted a single nucleotide within a gene substituted a single amino acid in the protein specified by that gene. If the code were overlapping, a single nucleotide substitution within the gene would affect three adjacent amino acids. **21.** The mutation is just upstream from the initiation codon and eliminates part of the recognition sequence for initiation of translation, preventing initiation of transcription at the correct initiation codon. **23.** The anticodon in the tRNA, not the amino acid attached to the tRNA, determines whether or not the amino acid is incorporated. **26.** With the exception of A-I pairs, purine-purine or pyrimidine-pyrimidine pairs do not form at the wobble position. **31.** The edited sequence is GAGAUU<u>U</u>GUGGAACUAAUCAUGCCUUUACGCCU-AUC. The edited nucleotide is underlined. RNA editing changes the C in CGU to a U to make a UGU codon, which codes for cysteine.

Chapter 5

4. Because many insertions and deletions cause frameshift mutations, they affect a larger proportion of the amino acids and tend to have a greater deleterious effect than substitutions. **7.** Within the DNA sequence are three copies of the tetranucleotide repeat CTGG. A slippage mutation during DNA replication that *adds* one copy of the CTGG repeat creates a mutant sequence that encodes the first mutant polypeptide. A slippage mutation during DNA replication that *deletes* one copy of the CTGG repeat creates a mutant sequence that encodes the second mutant polypeptide. This is a mutation hotspot because it is a repeated sequence that is prone to slippage mutations. **10.** All mutagenic agents listed in this question promote transition mutations

during DNA replication. Of the amino acid substitutions listed, only three result from transition mutations: Val → Ile, Ala → Thr, and Trp → Ter all arise from a G → A mutation in the sense strand (which corresponds to a C → T mutation in the antisense strand). Each of the other amino acid substitutions listed requires at least one transversion mutation. Of the mutagens listed, nitrous acid, bisulfite compounds, and hydroxylamine can cause G → A and C → T transition mutations **17. (a)** All of the mutations listed arise from transitions except Pro → Thr, which arises from a C → A transversion mutation in the sense strand. **(b)** 96.875%. **(c)** Transition mutations are much more frequent than transversion mutations in RNA treated with nitrous acid. **18. (a)** The deletion is outside of the reading frame, so it is not a frameshift mutation. **(b)** The deletion is within the recognition sequence for initiation of translation, so transcription and RNA processing are unaffected. The ribosomes cannot correctly translate the mRNA because they fail to recognize the correct initiation codon.   **20.** These mutations are all loss-of-function mutations that eliminate the normal function of the FMR1 protein. When FMR1 function is lost, fragile X syndrome develops. **26. (a)** Transition. **(b)** The mutation is not in the reading frame. **(c)** The mutation changes the −35 sequence of this gene so that it more closely resembles its consensus sequence.   **30.** The first four mutations are not within the reading frame. The four remaining mutations are same-sense mutations.

Chapter 6

5. (a) A T → C transition mutation in the first position of the codon is the most probable cause of the Ser → Pro substitution in the polypeptide. **(b)** Proline cannot participate in the α helix, so it alters the protein's secondary and tertiary structure and eliminates the activity of the enzyme. **7.** Chain-terminator mutations produce abnormally shortened β-globin polypeptides that are completely nonfunctional. Defective promoter mutations reduce transcription but do not alter the product produced by the reduced transcription. Defective mRNA processing mutations that affect intron removal eliminate protein activity, whereas defective mRNA processing mutations that affect poly (A) tail addition or capping on the 5′ end do not alter the reading frame but reduce translation.   **11.** Anthranilic acid → indole → tryptophan.   **14.** The mutation is at the site where the signal peptide is cleaved and prevents cleavage of the signal peptide.

Chapter 7

1. (a) The high temperature sterilizes the medium (kills all microorganisms). **(b)** Substances that are altered by the high temperature in an autoclave are sterilized by filtration. **5.** Mutation.   **6.** They grew the triple-mutant strains in pure culture but observed only single-gene alterations in the pure cultures, presumably due to mutation. Only in mixed culture did they observe multiple-gene alterations. **8.** Davis's U-tube experiments showed that cell contact was required, excluding transformation and transduction. **11.** When bacteria from the cultures in which conjugation occurred are cultured in the presence of the antibiotic, the antibiotic eliminates all Hfr cells so that they do not interfere with detection of recombinant types.   **13.** The *thr*, *leu*, and

str genes allowed the researchers to eliminate the parental cells by growing the cells in medium that contained streptomycin and lacked threonine and leucine.
15. (a)

(b) 34.5.   **18.** Zinder and Lederberg observed generalized transduction in *S. typhimurium*, whereas Morse, Lederberg, and Lederberg observed specialized transduction in *E. coli*. **22. (a)** The genetic position of the two genes was the same, mutations in them caused similar phenotypes, and there were no alternative genes. **(b)** Complementation analysis.

Chapter 8

3. (a) The lactose operon would be transcribed, rather than repressed, in the absence of lactose, because the repressor would not be present. **(b)** The lactose operon would be repressed when lactose is absent, as it should be.   **4.** This mutation is polar because it eliminates β-galatosidase, which converts lactose into the inducer, allolactose. In the absence of the inducer, the operon remains repressed, even in the presence of lactose, and the entire operon is not transcribed. Other operons are not necessarily repressed when there is a mutation in the first gene, so the other genes can be transcribed and translated.   **7.** Yes. There is no functional repressor protein, and the constitutive operator is in the chromosome, so the *lac* operon will be constantly transcribed regardless of the concentrations of lactose and glucose.   **9.** Attenuation relies on the concentration of an amino acid encoded in the reading frame of the leader peptide. Only a low supply of an amino acid can directly inhibit ribosomes as they translate the leader peptide. **11. (a)** MetLysAlaIlePheValLeuLysArgLeuValAlaHisPheLeuLysArgAlaValTyrSerProCysValLysGlnSerValThrGlnProAla. **(b)** There are no tryptophan codons in the mutant leader peptide, so translation will continue to the first termination codon, which is within the sequence for hairpin 3. Transcription of the operon will continue regardless of the level of tryptophan.   **16.** The DNA that contains an enhancer, whether upstream or downstream from the promoter, bends to bring the enhancer and the proteins bound to it into contact with the basal transcription apparatus. **17.** β-galactosidase can convert neolactose into glucose, but it cannot convert neolactose into allolactose, the inducer of the lac operon. A constitutive operator mutation is not dependent on the presence of the inducer to allow transcription of the operon, so transcription can proceed in the presence of neolactose.   **23.** The region encoded by codons 1–60 is the DNA-binding domain of the protein. The region that binds subunits into dimers for tetramer formation is downstream from codon 60.   **25.** Dimer formation is essential for DNA binding.   **27.** In mouse, DNA methylation is important in regulating gene expression. DNA methylation is probably an important mechanism that cells use to prevent certain genes from being transcribed.

Chapter 9

3. Type II enzymes cleave DNA at a recognition site and do not require ATP. Type I and type III enzymes do not cleave

DNA at the recognition site and require ATP. Thus, they are not as useful as type II enzymes. **7. (a)** Because one primer is present in excess relative to the other primer, the number of strands initiated from it exceeds the number of those initiated from the other primer. **(b)** To produce many copies of the target DNA strand, the template strand must also be present in multiple copies. The other primer initiates synthesis of the template strands. **10. (a)** Reverse transcriptase synthesizes a DNA strand from the RNA of RNA viruses. The DNA strand then serves as a template for transcription of many copies of the viral RNA. **(b)** Researchers use reverse transcriptase to synthesize cDNA from mRNAs isolated from cells that are actively transcribing genes. After the cDNAs are made double stranded, they are cloned into vectors to produce a cDNA library. **14.** The *Bam*HI site is located within the *tet^r* gene, so insertion of a fragment into this site disrupts the gene. **15.** pUC19 is not an expression vector and thus does not have a promoter and other necessary sequences for expression of cloned genes. **18.** ddNTPs terminate DNA synthesis because they do not have a 3' hydroxyl group to which new nucleotides can be added. **21.** The fragment's sizes are approximately 600, 430, and 220 nucleotide pairs. **24.** This procedure placed the Shine-Dalgarno sequence at exactly the same distance from the initiation codon as in the *lacZ* gene of the *lac* operon.

Chapter 10

1. (a) 1,000,000 nucleotide pairs; **(b)** 2,100,005 nucleotide pairs; **(c)** 2,000,000 nucleotide pairs. **3. (a)** T_m would be lower. **(b)** T_m would be higher. **6. (a)** 4348. **(b)** Linear DNA molecule with no nucleosomes: 3,400,000 Å; 10 nm fiber: 485,714–566,667 Å ($\frac{1}{7}$ to $\frac{1}{6}$ the original length); 30 nm fiber: 85,000 Å ($\frac{1}{40}$ the original length). **10. (a)** 46, **(b)** 92, and **(c)** 92. **11.** Heterochromatin. **13. (a)** Duplication of a gene followed by mutation. **(b)** Reverse transcription of an mRNA followed by insertion of the DNA into the chromosome. **16.** rRNA genes are transcribed at very high levels. **18.** The genes are present in many copies, so the loss of one copy due to mutation is not likely to cause serious consequences and, thus, the psuedogene is not likely to be eliminated by natural selection. **19.** The $\psi\alpha$-globin pseudogene arose before divergence of the ancestors of modern humans and chimpanzees. **22. (a)** They bind to the negatively charged phosphate groups of DNA. **(b)** This histone binds to DNA at two places and thus links different parts of the DNA molecule together. **24.** In the constitutive heterochromatin of centromeres and telomeres. **29.** Duplication of chromosome segments, followed by divergence.

Chapter 11

3. (a) In meiosis I, duplicated homologous chromosomes pair with and then separate from one another. In mitosis, homologous chromosomes do not pair, and sister chromatids separate from one another. **(b)** In meiosis II, the chromatids are not identical (because of crossing-over) as they are in mitosis. Also, there is no duplication of chromosomes immediately before meiosis II as there is just before mitosis. **4.** Mitosis does not reduce the chromosome number as meiosis does, so mitosis can operate at any ploidy level. Meiosis halves the chromosome number. Haploid and triploid cells have odd chromosome numbers, so their

chromosome numbers cannot be halved and maintain full sets of chromosomes in the cell. **7.** Two-thirds. **8. (b)** Primary oocyte. Primary oocytes form during gestation in human females, then remain arrested in prophase I until just before ovulation. **11.** No. The chromatids consist of maternal and paternal segments because of crossing over. **13.** They hold the homologous chromosomes together until anaphase I. **17.** Telomerase is not active in most human somatic cells

Chapter 12

5. The chi-square values for the experiments in the order listed in Table 12.1 are 0.2629, 0.0150, 0.3907, 0.0635, 0.4506, 0.3497, and 0.6065, each with 1 degree of freedom. All but the last one (0.6065) have associated probabilities that are greater than 0.5. In a set of independent repetitions of an experiment, we expect about half of the associated probabilities to exceed 0.5. In this set of Mendel's experiments, six of seven experiments have associated probabilities that exceed 0.5, so the results are indeed close to expected **8.** The *su* allele is a mutant allele that fails to encode a functional enzyme in the pathway for starch synthesis. In an *Su su* heterozygote, the *Su* allele encodes functional enzyme, so starch is produced. Only in the *su su* homozygote is there no functional enzyme. **10.** They do not alter the polypeptide product. **14.** It fails to encode an enzyme that is a part of the melanin-synthesis pathway. **15.** Neither. There is a dominant allele at each locus that encodes a functional enzyme, so the pathway is not blocked at either step. **17. (a)** Purple-flowered, tall parent: *Aa Bb*; purple-flowered, dwarf parent: *Aa bb*. **(b)** The chi-square value is 1.629 with 3 degrees of freedom, which is not significant at the 0.05 probability level. There is no evidence to reject the hypothesis that the genotypes are correct. **19.** 0.578125. **22.** The *A* locus does not encode chalcone synthase. **26.** Purple-flowered, tall parent: *Aa Bb*; white-flowered, dwarf parent: *aa bb*; purple-flowered, tall progeny: *Aa Bb*; white-flowered, tall progeny: *aa Bb*; purple-flowered, dwarf progeny: *Aa bb*; white-flowered, dwarf progeny: *aa bb*. **29. (a)** The chi-square value for the hypothesis that these results are not in excess of expectation is $(115 - 72.25)^2 \div 115 + (174 - 216.75)^2 \div 216.75 = 33.727$ with 1 degree of freedom, which has an associated probability of < 0.01. Bateson's conclusion is justified statistically. **(b)** Those who gave Bateson the pedigrees may not have known of or counted all the children in the families. **39.** The 3:1 ratio appears when both parents are heterozygous. In natural populations, all genotypes may exist among the individuals in the population, so some matings will not be between two heterozygotes.

Chapter 13

2. The Bombay phenotype is an example of nonpenetrance for ABO blood types. **3. (a)** The wild-type allele w^+ is dominant over w^{bf}, which is dominant over w^l. **(b)** The w^+ allele encodes a functional enzyme in the biosynthesis pathway for eye pigments that when heterozygous with either the w^{bf} or the w^l allele produces wild-type pigmentation. The w^{bf} allele is a leaky recessive mutant allele that encodes a partially functional enzyme that produces reduced amounts of pigment in w^{bf}/w^{bf} homozygotes or in w^{bf}/w^l heterozygotes. The w^l allele is a mutant recessive allele that fails to encode a functional product. In the absence of functional

product in w^l/w^l homozygotes, the eyes have no pigment and are white. **4.** The mutant *PAH* alleles that do not cause PKU when homozygous probably encode phenylalanine hydroxylase (PAH) that has slightly reduced function when compared to the gene product of the normal allele. When homozygous, such alleles probably jointly produce enough PAH to cause a normal phenotype. However, when heterozygous with a mutant allele that fails to encode funtional PAH, the level of functional enzyme is reduced to the point that the non-PKU HPA phenotype appears. **7. (a)** codominance, **(b)** codominance, **(c)** dominance. **12.** The A^y allele is a fusion gene caused by a large deletion between the *Merc* and *A* loci. The absence of the *Merc* gene product is the cause of lethality. The other alleles contain mutations within the *A* locus that do not affect the *Merc* locus. **14. (a)** Two. **(b)** Call the two loci *A* and *B*. *A_B_* is black, *A_bb* is chocolate, *aaB_* is blue, and *aabb* is silver-faun. **(c)** First cross: Expected ratio is 9:3:3:1; chi-square value is 22.3520 with 3 degrees of freedom. Second cross: Expected ratio is 9:3:3:1; chi-square value is 0.8564 with 3 degrees of freedom. Third cross: Expected ratio is 3:1; chi-square value is 0.1323 with 1 degree of freedom. The deviation from expected values is significant only in the first cross. The hypothesis is probably correct. **18.** The genotypes *CCPP*, *CCpp*, *Ccpp*, *ccPP*, *ccPp*, and *ccpp* breed true. They constitute half of the F_2 progeny. **19.** Line 1: *ccPP* (given); line 2: *CCpp*; line 3: *ccpp*; line 4: *CCPP*. **22. (a)** All walnut. **(b)** 1/4 walnut, 1/4 rose, 1/4 pea, 1/4 single. **(c)** 3/8 walnut, 3/8 rose, 1/8 pea, 1/8 single. **(d)** 3/8 walnut, 1/8 rose, 3/8 pea, 1/8 single. **26.** 13/16 white, 3/16 colored. **28. (a)** F_1: all wild type; F_2: 9/16 wild type, 7/16 black. **(b)** In both crosses, half the progeny should be wild type and half black. **33.** Line 1: *aaRR cc*; line 2: *AA RR cc*; line 3: *aa rr CC*. **36.** The subcloned DNA reveals a maximum of two fragments, whereas the other probe reveals at least five fragments, four of which are uninformative with respect to sickle-cell anemia.

Chapter 14

2. Heteromorphic chromosomes are typically distinguishable by their different sizes and centromere locations. Also, had there been a pair of heteromorphic chromosomes, the authors should have found more than a single male-specific marker if the markers were sampled at random. Their results suggest that a small heteromorphic region on otherwise homologous chromosomes, perhaps just a single gene, is responsible for sex determination. **7.** In most insects, sex phenotype is determined at the level of individual cells. In mammals, sex phenotype is determined in the entire body by hormones that circulate throughout the body. **9.** The association of altered sex phenotype with a new mutation in *SRY*, coupled with the association of *SRY* with the male phenotype in sex-reversed individuals, is strong evidence that *SRY* is the male-determining gene. **12.** X-linked disorders are expressed in males who are hemizygous for the allele. There is no dominant allele on a homologous chromosome to mask the effect of a recessive X-linked allele in males. **16.** Queen Elizabeth's lineage to Queen Victoria is through her father, George VI, who did not have hemophilia. Therefore, Queen Elizabeth did not inherit the allele from her father and is not likely to be a heterozygous carrier. Prince Philip did not have hemophilia, so he did not inherit the allele. Thus, none of their children inherited the allele. **19.** Although several copies of the gene may be present, only one copy is expressed. **21. (a)** It is caused by a recessive X-linked allele. **(b)** The mother is probably a heterozygous carrier of the allele; the father's G6PD deficiency phenotype is coincidental and unrelated to the child's phenotype. **23.** The gold phenotype is caused by a recessive Z-linked allele, and the silver phenotype is caused by a dominant Z-linked allele at the same locus. Remember that chickens have a ZZ-ZW form of sex determination. **27.** Two. **28.** No. There are no Barr bodies in *Drosophila*. **30. (a)** Black females and a tortoiseshell male. **(b)** The tortoiseshell male probably had an XXY genotype. The black females were probably mosaics, or may have had XO genotypes (the O indicates that they had one X chromosome and no corresponding X or Y chromosome). **33. (a)** Sex-influenced trait. **(b)** The allele for red-spotting is dominant in females and recessive in males. The allele for mahogany-spotting is dominant in males and recessive in females. **36.** The female parent has an attached X chromosome.

Chapter 15

1. 14. **3.** They cause researchers to underestimate the actual map distance unless they adjust the results to account for undetected crossovers. **7. (a)** Crossover-type gametes will be less frequent. **(b)** Map distances will be underestimated. **11. (a)** Repulsion. **(b)** 0.0857. **14. (a)** No, *ca* is in repulsion to *ro* and *e*. **(b)** *ca*—10.7—*ro*——19.3——*e*.**(c)** 0.91. **16. (a)** The three loci are so close to one another that there are no double-crossover types. **(b)** No, *D* is in repulsion to *sp* and *sr*. **(c)** *D*—05.5—*sr*—03.4—*sp*.**(d)** 1. **23. (c)** *clcl Wxwx* knob and *Clcl wxwx* knobless. **(b)** The distance between the knob and *cl* is 15.9 cM, and the distance between *cl* and *wx* is 11.1 cM. **(c)** All double- and triple-crossover types are missing from the data. Single-crossover types between the knob and *cl* are also missing from the data. **(d)** The genotypes *clcl wxwx* knobless translocation and *Clcl Wx_* knob no translocation arose from single crossovers between *cl* and *wx*. The genotype *cl cl WxWx* knobless translocation arose from a single crossover between *wx* and the translocation. The genotype *clcl Wxwx* knobless translocation arose from a crossover between *cl* and *wx* or between *wx* and the translocation. All other genotypes are noncrossover types. **24. (a)** 1.3 cM. **(b)** All F_1 females are heterozygous at both loci, and all F_1 males are hemizygous for both recessive alleles. Therefore, all crossovers are in the F_1 females, and the intermating of F_1 individuals is the equivalent of a testcross. **27. (a)** Females: 36.75 cM; males: 38.66 cM. **(b)** Both male and female progeny inherit their recombinant X chromosomes from the same female parents, so there should be no difference. The slight observed difference is due to sampling error. **(c)** The design of the cross makes it the equivalent of a testcross.

Chapter 16

4. In a 5:3 segregation, two different alleles segregate from one another during the single mitotic division that takes place after meiosis is completed, so the segregation is postmeiotic. **7.** Gene conversion is best detected by examining all products of a single meiotic event. It is difficult, and often impossible, to recover the products of the same meiotic event in multicellular eukaryotes, whereas in ascomycete fungi, each ascus contains all of the products of a single

meiotic event.   **9.** Gene conversion transfers to a chromosome a strand of DNA from the homologous chromosome. If the homologuos chromosome contains the wild-type sequence at the site of a mutant allele, then the mutant allele can be converted to a wild-type allele. If both homologous chromosomes contain the same mutant allele, then the mutant allele cannot be changed to a wild-type allele through gene conversion because there is no wild-type sequence available for the conversion. Only a mutation can make such a change.   **10.** 4.12 cM.   **11.** C17: 7.59 cM; C22: 8.80 cM; C29: 6.00 cM.   **17. (a)** Recombination between flanking markers was detected in each "wild-type for lozenge" fly, indicating that the wild-type phenotype is associated with crossing-over. **(b)** A crossover is not expected with all gene conversion events. In this experiment, each "wild-type for lozenge" fly had a crossover between flanking markers. **(c)** About half of the flies with crossover-type chromosomes should be female, but the "like lozenge-spectacled" phenotype can be detected only in males, so the females with the crossover-type chromosome are not detected. **(d)** ct-sn-lz^{BS}-lz^{46}-lz^g-ras-v.   **18.** Many scientists envisioned genes as indivisible particles, much like beads on a string, rather than divisible segments of DNA.   **23. (a)** The researchers crossed each genotype with itself, and in each case there were no black spores in the progeny. Had reversion mutation been responsible for the black spores, then some black spores should have been observed in the progeny of these crosses. **(b)** Gene conversion. **(c)** Mutation B is probably a deletion mutation. **(d)** Mutation 63 lies outside of the region deleted in mutation B, whereas the other mutations are within this region.

Chapter 17

1. Human somatic cells function with only one active X chromosome. In females, each cell has two chromosomes, but one is inactivated as a Barr body. Females with an XO chromosome constitution have a single X chromosome in their cells and no Barr body.   **4. (a)** None; there is no crossing-over in male *Drosophila*. **(b)** 0.2088. **(c)** 0.25.   **6.** He probably does not carry the translocation.   **9. (a)** The theoretical frequencies are 0.5 heterozygous for a reciprocal translocation and 0.5 with a normal karyotype. **(b)** The chi-square value is 0.095 with 1 degree of freedom. There is no evidence to reject the hypothesis that the data conform to the theoretical frequencies.   **12.** There is no homologous region between the two genes in which recombination could repeatedly create the same translocation.   **15.** Maize is diploid, and bread wheat is hexaploid. Monosomic maize gametes are missing proportionally much more genetic material than monosomic wheat gametes.   **17. (a)** An unreduced triploid gamete from a triploid plant unites with a reduced monoploid gamete from a diploid plant to produce a tetraploid zygote. **(b)** The tetraploid can be hybridized with a diploid to produce all triploid offspring. **19. (a)** $C_1c_1c_1c_1$. **(b)** $C_1c_1c_1c_1$. **(c)** 0.75.   **23.** The proportion of pollen abortion is about the same as the proportion of adjacent segregations (adjacent-1 and adjacent-2 segregations combined). Aborted pollen grains probably arose from adjacent segregations.   **25.** The sr, e, ro, and ca loci are within the inversion. The st locus is outside of the inversion. The last four classes of progeny probably arose from double crossovers within the inversion.

Chapter 18

1. (a) Three. **(b)** 16,569 nucleotide pairs. **(c)** The human mitochondrial genome is much smaller than plant mitochondrial genomes.   **4.** At some point in the evolutionary history of plants, the small subunit gene was transferred from the plastid to the nucleus.   **7. (a)** The seed coat is composed of maternal tissue and, therefore, displays the phenotype of the maternal parent. The cotyledons (internal parts of the seed) are composed of embryonic tissue and, therefore, display the phenotype of the embryo. **(b)** Seed coat color, seed color, and seed shape are all inherited in a Mendelian fashion.   **10. (a)** 0.5. **(b)** None, because the plants that produce pollen pollinate those that do not.   **11.** Mitochondrial somatic segregation is a somewhat random process, so the relative proportions of two mitochondrial types may vary among tissues and organs. The degree of heteroplasmy varies among heteroplasmic cells, and changes from one cell generation to the next. The number of mitochondria of a particular type in a cell can determine to what degree characteristics influenced by mitochondrial genes are expressed. **13.** The proportion of mitochondria present in the cytoplasm of an ovum is determined largely by the proportion in the cytoplasm of the cells in the germ line. A mother with > 95% mutant mitochondrial DNA is likely to transmit a similar proportion of mutant mitochondrial DNA in the ova she produces.   **17. (a)** Loblolly pine and coast redwood are both conifers and thus classified as gymnosperms. **(b)** Mitochondrial inheritance patterns may vary among species of gymnosperms.   **21. (a)** Inheritance is predominantly maternal, but there is significant biparental inheritance in many of the crosses. **(b)** There is substantial evidence that plastid genotype has an effect on inheritance throughout the entire data set. For example, in the first four sets of reciprocal crosses, plastidial inheritance is predominantly uniparental-maternal when the green type is the maternal parent, but significant proportions of the progeny display biparental inheritance when the mutant types are the maternal parent.

Chapter 19

1. (a) 0.044, **(b)** 0.956, and **(c)** 0.332.   **3.** Seven generations ($t = 6.84$).   **7.** Type AB: 6%; type B: 13% ($p = 0.3$, $q = 0.1$, $r = 0.6$).   **12.** The frequency of the $HBB*S$ allele is $q = 0.0985$, and the frequency of the $HBB*A$ allele is $p = 0.9015$. The chi-square value is 3.074 with 1 degree of freedom, which is not significant. There is no evidence to reject the hypothesis that the population is in Hardy-Weinberg equilibrium. **13.** With random mating, Hardy-Weinberg equilibrium is established in one generation. Thus, regardless of the allele distribution in an adult population, the infants in the next generation should be in Hardy-Weinberg equilibrium if there is random mating with respect to genotype. Among adults, however, selection has had a chance to have an effect in favoring heterozygotes and disfavoring homozygotes. **17.** Under the assumption of a single allele, the allele frequency can be determined directly from the males. It is $725/9049 = 0.0801$. Now we can test the females for Hardy-Weinberg equilibrium. The expected number of females with red-green color blindness is $9072(0.0801^2) = 58.23$. The expected number of females *without* red-green color blindness is 9013.77. The chi-square value is 5.744 with 1 degree

of freedom, which is significant. We reject the hypothesis that the population is in Hardy-Weinberg equilibrium. **22.** Selection has little, if any, effect on allele frequency.

Chapter 20

2. When all genetic variation is additive. **5.** All genes that influence variation for a trait contribute to V_A. **7. (a)** No. **(b)** No. If there is any genetic variation, then both V_A and V_G must be greater than zero and both narrow- and broad-sense heritabilities must be greater than zero. **10. (a)** Inbred; **(b)** hybrid. **13.** DNA markers are not influenced by environmental factors, and many DNA markers are not influenced by dominance or epistatic interactions. **16.** No, the population had reached a physiological selection plateau. **17.** 0.86. **20.** 0.24325 g. **21. (a)** Continuous. **(b)** Meristic. **(c)** 0.49. **(d)** 0.62. **22.** For thorax size $h^2 = 0.47$. For egg number $h^2 = 0.09$. **25. (a)** The F_1 plants exhibit maximum heterosis. **(b)** Half the heterozygosity that contributed to heterosis in the F_1 generation is expected to be lost in the F_2 generation because of inbreeding. **28. (a)** Variation for at least five genes is responsible for the genetic variation. **(b)** Researchers can use the markers to help reduce the influence of environment, dominance, and epistasis on selection efficiency. **(c)** They are highly variable and permit the use of large numbers of markers at a relatively low cost.

Chapter 21

3. The alternative requires chromosome fissions, which are less likely than fusions. **6. (a)** The major genetic difference between the two is in the mitochondrial DNA. The gestation environment in female horses and donkeys may be quite different and have a significant effect on the hybrid offspring. **(b)** The chromosome numbers differ, and chromosome imbalances arise during meiosis. **10. (a)** Ecological and temporal isolation. **(b)** Copper resistance allows those plants that are most resistant to survive on the high copper soil and reproduce. Copper-resistant plants that flower at a different time mate only with one another, and their progeny are more likely to inherit copper-resistance alleles than the progeny of plants that mate with the copper-susceptible plants. **(c)** The populations are reproductively isolated, so genetic changes in one population are not transferred to the other population, and the two populations may change independently. **13.** It includes a part of the enzyme that is essential for its function and may be at least part of the enzyme's active site. **18.** Transition mutations are typically more frequent than transversion mutations, so it is appropriate to apply different mutation rates to them to refine phylogenetic analysis. **23. (a)** The results of Armour et al. are similar to those of Tishkoff et al. in that both show a higher degree of allele diversity among Africans when compared to non-Africans, and that most alleles in non-African populations are also found in African populations, albeit at lower frequencies. **(b)** These results are best explained by the single-origin theory.

Chapter 22

2. Although transposition is often more frequent than other types of mutation, it usually is not so frequent that it disrupts Mendelian ratios. Also, some species have mecha-

nisms that inhibit transposition. **6.** The footprint left behind when the element is excised may be a stable mutation that continues to disrupt the gene. **9.** The mutant allele at the *a* locus probably contained a nonautonomous element. An autonomous element was located on another chromosome. If the two parents were heterozygous for the mutant *a* allele and the autonomous element, the pattern that Rhoades observed should appear in their progeny. **10.** When an unreplicated broken chromosome replicates, both sister chromatids contain the terminal deletion. **16.** P-element insertion is not random; the elements are probably preferentially inserted at particular DNA sequences.

Chapter 23

6. The genes are regulated in an autonomous manner because transplanted eye disks developed into eye tissue in another part of the body. **9.** Genes that are paralogous have similar DNA sequences and similar linear organizations on the same DNA molecule in two or more species. The paralogous organization may be important for development to proceed properly and is thus preserved by natural selection. **13.** Cell differentiation initiates a genetically programmed process for cessation of cell division and eventual cell death. **15.** The bicoid protein initiates the expression of genes that are responsible for anterior development. **19.** Animals and plants have common evolutionary origins, and their common ancestors probably had small numbers of both homeobox and MADS box genes. In the evolutionary history of animals, homeobox genes duplicated and diverged to fulfill developmental roles. The same is true for the MADS box genes in plants.

Chapter 24

5. Mutations in the regulatory region of a gene can eliminate binding of proteins that inhibit transcription. In the absence of such proteins bound to the regulatory region, the genes are transcribed at abnormally high levels. Also, translocations can bring a gene under the control of a different regulatory region, which causes abnormally high transcription. **9.** For most cancers to develop, mutations in several genes are required. Rarely does an individual inherit all the mutations. Instead, the mutations occur over a period of time in somatic cells. In those individuals who inherit one of the mutations, every cell is already beyond the first step toward cancer, so accumulation of the necessary mutations for cancer is more likely. **13. (a)** A tumor suppressor gene. **(b)** As a protein kinase, it probably phosphorylates p34^{cdc2}, which then halts the G_2 to M transition. **(c)** With the cell cycle halted, the cell may initiate apoptosis if damage due to radiation is sufficient. **18. (a)** It shows the inheritance pattern of a dominant allele. **(b)** Because the mutant allele is inherited in every cell, all organs are predisposed to cancer. **22.** Because TK1 levels are correlated with the development of breast cancer, higher-than-normal TK1 levels may indicate that a cancer is in its early stages. Measurement of serum TK1 levels may be a diagnostic tool to help physicians determine whether or not further tests for detection of cancer are warranted.

Chapter 25

1. (a) At the cellular level, each rearranged antibody gene encodes one polypeptide. **(b)** At the organismal level, one

individual carries many different rearrangements of each antibody gene. Thus, a single gene rearranges into many genes to encode many different, but related, polypeptides. **6.** Only the promoter that is closest to the joined portion of the gene is activated. It is activated because it is closest to the enhancer contained within the gene's intron.   **13.** Natural selection has favored those HLA alleles that recognize antigens from those pathogens that are found in the immediate environment. Because pathogens may differ throughout the world, selection probably favors different HLA alleles in different parts of the world.   **17.** The products of the genes interact with one another, but they do not interact until after gene rearrangement, transcription, and translation. Therefore, rearrangements of the two genes are independent even though the gene products interact. **20. (a)** These results suggest that the roles of *RAG1* and *RAG2* may extend beyond antibody rearrangement to include a role in the development of B and T cells. **(b)** The studies highlighted in questions 18 and 19 did not require development of the B and T cells in an organism, so the developmental roles of *RAG1* and *RAG2* were not tested in these studies.

Chapter 26

1. (1) The gene contains introns that the bacterial cell cannot remove. (2) The gene does not have a bacterial promoter. (3) The gene does not have the appropriate recognition sequences for translation. (4) The cell has no mechanism for creating the A and B chains and connecting them correctly. **7.** The gene must be delivered into the cells of the tissue where it normally functions (in the bone marrow). Oral or intravenous administration of genetically engineered ADA fails to deliver the enzyme to the cells that require ADA. **12. (a)** PCR amplification detects trinucleotide expansion mutations; a methylation test determines the degree of methylation, which is related to the gene's transcription; and the karyotype detects the fragile site in the chromosome. **(b)** None of these tests detect single-nucleotide substitution mutations.   **18.** Infants who are diagnosed with PKU can be successfully treated to prevent mental retardation if the treatment is initiated in infancy. There is no medical treatment for fragile X syndrome, so early diagnosis has no effect in preventing the onset of symptoms. Also, the PKU test is inexpensive and easy to administer, whereas fragile X testing is expensive.

Chapter 27

4. The genetically diverse varieties and breeds have not been developed for use in modern agriculture. When used as parents in a breeding program, they introduce many characteristics that are disadvantageous in modern agricultural production.   **8. (a)** DNA marker alleles that are closely coupled to desirable alleles are present in one bulked sample and are usually absent from the other sample. Markers at unlinked marker loci are randomly distributed between the bulked samples and thus usually appear in both samples. **(b)** Rather than inoculate the plant with a pathogen, breeders can indirectly identify those plants that carry a disease-resistance allele by testing for a coupled DNA marker.   **16. (a)** The agar compensated for the damage to the zona pellucida, allowing the embryos to develop. **(b)** The two cells derived from the original embryo develop into clones of one another. In this way, scientists can increase the number of genetically desirable offspring from a single cross.   **20.** Agricultural biotechnology usually focuses on the improvement of single traits under laboratory conditions, whereas breeding focuses on the simultaneous improvement of all traits under field conditions. The direct products of biotechnology usually must undergo field-testing and further improvement through breeding before they are ready for commercial use.

Chapter 28

2. The test poses virtually no health threat to the infant, so there is no need to inform parents about the risk of an infant taking the test. Informed consent is only an issue when the test indicates that the infant may have PKU. At that point, the parents are fully informed about the nature of subsequent tests that will determine whether or not the infant actually has PKU.   **10.** Most recessive alleles that are present at low frequencies in a population are in heterozygotes and are not expressed phenotypically. The probability of a mating between two heterozygotes in the population as a whole is low. However, the probability of a mating between two heterozygotes among close relatives is much higher. **14.** Certain genetic disorders, such as sickle-cell anemia, cystic fibrosis, or G6PD deficiency, are more frequent in particular ethnic groups when compared to others. X-linked recessive genetic disorders are more frequent among males than females. Discrimination on the basis of a genetic disorder indirectly discriminates against the group in which that disorder is more frequent.

GLOSSARY

A See *adenine*.

acceptor arm The portion of a tRNA molecule that accepts or binds with an amino acid.

acentric fragment A piece of chromosome with no centromere, usually caused by a terminal or interstitial deletion.

acrocentric chromosome A chromosome containing a centromere located near one of the ends of the chromosome. Also called *subtelocentric chromosome*.

activator A transcription factor that, when bound to an enhancer, stimulates transcription of the gene under the control of the enhancer. In transcription, a binding protein that binds typically in the enhancer region.

active site The specific structural site found on an enzyme where the reaction changes a substrate to a product.

additive gene action A situation in which several loci affect a trait of interest and the alleles at these loci show no dominance so the effects of the alleles on the phenotype are purely additive.

adenine (A) One of the four different bases in nucleotides composing the DNA molecule. Adenine is a purine that pairs with thymine (T).

adjacent-1 segregation Unbalanced segregation of a translocation quadrivalent in which homologous centromeres migrate to opposite poles.

adjacent-2 segregation Unbalanced segregation of a translocation quadrivalent in which homologous centromeres migrate to the same pole.

A-DNA A form of DNA in which the bases are tilted at about 20° from being perpendicular to the axis of the molecule. The helix is wound less tightly than in B-DNA, with a 32.7° rotation per base pair.

alkylating agent Chemical mutagen that adds alkyl groups to the nitrogenous bases of DNA.

alkyltransferases A group of enzymes that remove alkyl groups from the bases of DNA, countering the mutagenicity of alkylating agents.

allele A specific DNA sequence found at a gene locus. Each different DNA sequence at a locus is a different allele.

allele frequency The proportion of a particular allele among all other alleles in a population.

allelic Reference to different alleles at the same locus.

allergic reaction Overresponse of the immune system to an antigen.

allolactose A lactose derivative that is the inducer of the *lac* operon.

allopatric speciation Speciation as a result of geographic isolation.

allopolyploidy A situation in which the chromosomes of a polyploid species are from different, but closely related, ancestral species.

allosteric site A site on an enzyme to which regulatory molecules may bind.

allosteric feedback inhibition A posttranslational regulation mechanism in which the end product of the biochemical pathway regulates the enzyme's activity.

allotetraploid Four monoploid sets of chromosomes derived from two different species in the same cell.

α helix A helical portion of a polypeptide held together by hydrogen bonds.

alternate-1 segregation Balanced segregation of a translocation quadrivalent in which homologous centromeres migrate to opposite poles.

alternate-2 segregation Balanced segregation of a translocation quadrivalent in which homologous centromeres migrate to opposite poles. The orientation of the quadrivalent distinguishes alternate-1 and alternate-2 segregation.

amino acid A molecule consisting of a central carbon atom bound to a carboxyl group, an amino group, a hydrogen atom, and one of 20 side groups (R groups).

amino acid attachment site The position on the acceptor arm of a tRNA where the amino acid attaches.

aminoacylated tRNA See *charged tRNA*.

A site One of the two sites within a ribosome that can hold tRNAs. This site receives the arriving tRNA molecule that is not attached to the polypeptide chain. Also called *aminoacyl site*.

amphidiploid A plant that has a diploid set of chromosomes from each contributing species in an interspecific hybrid.

anaphase The third stage of mitosis during which chromosomes migrate toward opposite poles.

anaphase I The stage of meiosis during which homologous pairs of chromosomes segregate from each other and migrate to opposite poles.

anaphase II The second anaphase in meiosis. The sister chromatids separate and migrate to opposite poles.

aneuploidy A situation in which each cell has at least one extra chromosome or is missing at least one chromosome.

angiogenesis A process in which tumors release chemical signals that stimulate the growth of blood vessels around the tumor to supply it with nourishment.

antennapedia complex (ANT-C) Cluster of homeotic genes that regulate the development of the head, the first thoracic segment (T1) and the anterior compartment of the second thoracic segment (T2A) in *Drosophila*.

antibiotic-resistant strain Bacterial cells that grow and divide in the presence of an antibiotic, a substance that normally prevents bacterial growth.

antibody A protein produced by the immune system of an animal. Each antibody binds tightly to specific foreign substances, called antigens.

anticoding strand See *antisense strand*.

anticodon Three nucleotides in a tRNA molecule that pair with the three nucleotides that make up a codon in an mRNA molecule.

anticodon arm The portion of the tRNA that contains the anticodon.

antigen A foreign substance to which an antibody binds.

antigen-binding site The site in the antibody to which the antigen binds.

antiparallel conformation The conformation of a double-stranded nucleic acid in which the $5' \rightarrow 3'$ orientation of one strand is reversed relative to the other strand.

antisense strand The strand of a double-stranded DNA molecule that serves as the template for transcription and contains the complementary nucleotide sequence of the mRNA. Also called the *anticoding strand* or *template strand*.

AP endonuclease An enzyme of the excision repair system that recognizes AP sites and initiates their removal.

apoptosis Genetically programmed cell suicide.

AP site A site where the base is missing from a nucleotide.

artificial selection The intentional selection practiced by humans to alter a population in a specific way.

ascus The sac that contains the spores of an ascomycete.

asexual reproduction Organismal reproduction with no union of gametes.

assortative mating Preferential mating between individuals of like genotypes.

asters Microtubules that radiate out from the centrioles.

asymmetric distribution A distribution in which the mode and mean are different.

asymmetric strand-transfer model See *Meselson-Radding model*.

attenuation A gene expression control mechanism that terminates transcription at a terminator sequence called the attenuator.

attenuator A transcription termination sequence that ends transcription prematurely and represses gene expression.

autogenous regulation Regulation of gene expression by the product encoded by that gene.

autoimmune disease A disease caused by the attack of an antigen on a substance that belongs to the body, damaging tissues or organs.

automated DNA sequencing A modified version of Sanger dideoxy sequencing that uses fluorescent dyes to label the ddNTPs.

autonomous developmental regulation The fate of the cell lineage is programmed within the cell and is unaffected by surrounding cells.

autonomous transposon A transposable element that encodes a transposase and is capable of self-transposition.

autopolyploidy A situation in which all of the chromosomes in a polyploid species are from the same ancestral species.

autosomes All chromosomes in a cell not designated as sex chromosomes.

auxotrophic mutation A mutation in an organism that requires supplementation of the minimal medium with a specific nutrient for the organism to survive.

bacterial artificial chromosomes (BACs) Genetically engineered F factors that carry very large segments of foreign DNA.

bacterial chromosome The large circular DNA molecule found in each bacterial cell.

bacteriophage A bacterial virus. Also called a phage.

balanced The condition of cells that have the correct number of chromosomes and the correct amount of genetic material in those chromosomes.

balanced polymorphism A situation in which two different alleles are maintained in a population by selection that disfavors both homozygotes.

Barr body A dark-stained body found in the nuclei of females. It represents the inactivated X chromosome that has been condensed into facultative heterochromatin, so nearly all of its genes are inactive.

basal eukaryotic transcription complex The process in eukaryotic transcription where RNA polymerase binds with a transcription factor to form a complex that is essential in the initiation of transcription.

base analogs Nitrogenous bases that are structurally similar to the four bases of DNA and may be incorporated into DNA during replication in place of the usual nucleotides.

B cells An antigen that matures in the bone marrow.

B-DNA See *B form*.

behavioral isolation A type of reproductive isolation found in animals in which males and females of different populations may not be attracted to one another because of innate differences in mating behavior.

benign Tumors that remain localized to the tissue where they originated and do not invade other tissues.

β-barrel A secondary protein structure in the form of a cylindrical β-pleated sheet.

β-galactosidase A bacterial enzyme encoded by the *lacZ* gene that breaks down lactose into glucose and galactose.

β-galactoside permease An enzyme encoded by the *lacY* gene that allows lactose to permeate the cell wall and membrane and enter the cell.

β-galactoside transacetylase An enzyme encoded by the *lacA* gene that acetylates lactose or molecules that resemble lactose.

β globin A subunit of adult hemoglobin.

β-pleated sheet A common secondary protein structure; formed when two or more β strands line up side by side and hydrogen bonds form between the carboxyl and amino groups of the adjacent strands.

β strand A common feature of protein secondary structure; a linear strand of amino acids.

β-thalassemia A genetic disorder characterized by expansion of bone marrow, skeletal changes and fractures.

β turns An irregular conformation of amino acids within a polypeptide, which serves as a bridge between two regular secondary structural conformations.

BFB cycle See *breakage-fusion-bridge cycle*.

B form (B-DNA) A form of DNA in which the bases are stacked nearly perpendicular to the axis of the molecule at about a 36° angle of rotation in a right-handed double helix.

bidirectional replication DNA replication that proceeds in both directions from the origin of replication.

bimodal distribution A distribution with more than one mode.

binary fission The process of cell division in bacteria.

binomial distribution A distribution that describes all possible combinations of two phenotypes or genotypes and their expected frequencies.

biochemical pathway The series of enzyme-catalyzed steps to produce a final product.

biotechnology The use of in vitro genetic manipulation and recombinant DNA methods.

biparental organellar inheritance Organellar inheritance in which both maternal and paternal gametes contribute cytoplasm to the zygote.

bispecific antibodies Antibodies that bind two different antigens.

bithorax complex (BX-C) Genes in *Drosophila* that regulate development of the posterior compartment of the second thoracic segment (T2P) to the end of the abdomen.

bivalent A pair of homologous chromosomes.

blue-white screening A procedure that relies on color to distinguish recombinant from nonrecombinant cloning vectors.

Bombay phenotype People with the Bombay phenotype have neither the A nor B antigens, yet their blood type is distinctly different from type O.

bottleneck effect The effect of drift in a population that is temporarily small.

breakage-fusion-bridge (BFB) cycle Breakage and fusion of a dicentric chromosome in the cell cycle.

breeds A defined group of animals that have particular genetic characteristics that are maintained from one generation to the next.

broad-sense heritability Heritability expressed as the proportion of the phenotypic variance attributed to genetic variance.

budding A yeast cell reproduction process in which a small bud forms and receives an equal portion of nuclear material due to mitosis but may receive a smaller portion of cytoplasm from the parent cell.

BX-C See *bithorax complex*.

C See *cytosine*.

C See *coefficient of coincidence*.

CAAT box A conserved sequence, found in eukaryotic promoters, generally located at about −75 to −80 with the consensus sequence GGCCAATCT.

cancer A disease in which tumors arise, invade tissues, and spread throughout the body.

CAP See *catabolite activator protein*.

capsid The protein capsule that encases the nucleic acid of a virus.

carbon source mutation A mutation in bacteria that renders a bacterial cell unable to utilize a particular carbon source.

carcinogen A substance that increases the risk of cancer.

catabolite activator protein (CAP) A regulator protein that must bind to the DNA in order for transcription to occur in the *lac* operon.

cDNA "Complementary DNA"; a DNA molecule that is synthesized from an mRNA template.

cDNA library A collection of cDNA clones.

cell cycle The cycle of events that occurs in a cell as it undergoes mitosis and cell division.

cell-mediated immunity An immune response in which T cells attack body cells that have been infected by a pathogen and express an antigen derived from the pathogen on their surface.

center of diversity The region in which the greatest genetic diversity for a species is found.

center of origin The region in which a species was domesticated, usually the same as the center of diversity.

centimorgan (cM) A unit of measurement used in gene mapping to represent the frequency of crossovers as a percentage value.

central dogma A gene composed of DNA is transcribed into RNA, which is then translated into a polypeptide that is processed to become a protein.

centrioles The organelles in eukaryotic cells to which spindle fibers attach.

centromere The constricted site on a chromosome where spindle fiber attachment occurs. Also called a primary constriction.

charged (aminoacylated) tRNA A tRNA with its amino acid attached.

chiasma An X-shaped structure that forms between paired nonsister chromatids during meiosis.

chi-square analysis A statistical measure that compares observed values with hypothesized expected values and determines the probability of observing that outcome under the assumption that the hypothesis is correct.

chloroplast The organelle in which photosynthesis occurs in plants.

chromatid The long and short arms of a chromosome after the DNA in the chromosome has been duplicated and identical duplicated portions of the chromosome are present.

chromatid bridge A dicentric chromatid that stretches across the equatorial plane of a cell during anaphase of mitosis or meiosis.

chromatin The DNA and its associated proteins in the nucleus of the cell.

chromosome arms The portions of the chromosome that are measured when the centromere is used as a reference point to divide the chromosome.

chromosome bands The specific patterns that are visible on a stained metaphase chromosome.

chromosome fission The splitting of one chromosome into two.

chromosome fusion Fusion of two chromosome arms.

chromosome jumping The detection of linked segments of DNA that do not overlap and are separated by up to tens of thousands of nucleotide pairs.

chromosome mapping A process used to determine the positions of genes on a chromosome.

chromosome satellite A small portion of the chromosome, usually near one end of the chromosome, extending beyond a secondary constriction.

chromosome walking The reconstruction of a long DNA sequence from many cloned segments of DNA by identifying overlapping clones.

cis-acting element Any gene or DNA sequence that affects the expression of a gene located in its same DNA molecule.

cistron A term used to describe a complementation group identified by a series of cis-trans tests.

class switching The process by which mature B lymphocytes utilize alternative splicing of pre-mRNAs and further rearrangements of the *IGHG1* gene to produce several different classes of antibodies.

clinical genetics The diagnosis of genetic disorders and the care of individuals with genetic disorders.

cloned DNA A fragment of recombinant DNA that can be replicated in bacterial cells.

cloning vector Bacterial plasmid or phage DNA that is used to carry recombinant DNA molecules.

cM See *centimorgans*.

coactivator In transcription, a protein bound to the basal eukaryotic transcription complex that interacts with an activator in the enhancer region through DNA bending.

coding strand See *sense strand*.

codominance The distinct expression of two phenotypes in heterozygotes, not an intermediate of the two.

codon A combination of three adjacent nucleotides in an mRNA molecule that encodes a specific amino acid or the termination of translation.

coefficient of coincidence (C) The observed number of double crossovers divided by the expected number.

cognate tRNAs Different tRNAs that carry the same amino acid and are therefore recognized by the same aminoacyl synthetase.

cohesive ends (sticky ends) The short sequences of single-stranded DNA that are left after a restriction endonuclease has cut the DNA molecule.

colchicine A chemical that inhibits spindle fiber formation during mitosis and can artificially induce endoploidy.

colinear In translation, the process by which the linear sequence of nucleotides in a portion of the mRNA corresponds directly to the linear sequence of amino acids in the polypeptide.

colony A clump of genetically identical cells that arose from a single cell grown on a solidified medium.

competent cells Cells that are treated in a manner that increases the rate of transformation.

complementary gene action A type of epistasis in which homozygosity for recessive alleles at either of two loci causes the same phenotype, characterized by a 9:7 phenotypic ratio in F_2 progeny.

complementation analysis A procedure in which two mutant alleles that confer the same phenotype are placed into the same cell to see if the two alleles belong to the same gene.

complement system About 20 proteins that circulate in the blood and are part of the humoral immune system.

complexity The length, usually expressed in nucleotide pairs, of different DNA sequences found in the genome.

components of variance Partitioned phenotypic variances that represent the components of genetic variance and the environmental variance.

composite transposons Prokaryotic transposons that consist of two insertion sequences and intervening DNA.

compound heterozygotes Individuals who are heterozygous for different alleles that confer a mutant phenotype.

conjugation Direct transfer of DNA from one bacterial cell to another.

conjugation bridge The structure that forms between two bacterial cells during conjugation.

consanguineous mating Mating between two individuals with common ancestry.

consensus sequence The most frequent sequence of a conserved sequence of nucleotides.

conservative DNA replication A DNA replication model in which the original double-stranded molecule is conserved, and both strands of the replicated molecule are new.

conservative transposition Transposition in which a transposable element is excised from the chromosomal DNA and moved to a new location in the chromosomal DNA.

conserved sequence Regions of DNA that are identical or similar when the nucleotide sequences of many genes are compared.

constant-length strand In the polymerase chain reaction, any strand synthesized from DNA other than the original genomic DNA, having a specific length defined by the primer-binding sites.

constant regions Regions in the heavy and light chains that are similar in all antibodies.

constitutive mutation A mutation that permits constant transcription of a gene under all conditions.

constitutive heterochromatin A region of the chromosome consisting of highly repetitive DNA that is permanently condensed into heterochromatin.

constitutive promoter A promoter that causes a gene to be transcribed constantly.

continuous variation Variation in which individuals in a population differ from each other by various degrees, resulting in a wide range of phenotypes and measured in small increments.

core DNA The DNA that is found within the nucleosome.

core enzyme A part of the DNA polymerase holoenzyme in prokaryotes that synthesizes DNA.

corepressor A molecule that complexes with a repressor protein to prevent transcription.

cosmid vector A cloning vector consisting of plasmid and phage elements that can carry large DNA insertions.

cos sites Short single-stranded segments at the ends of linear λ DNA.

Cot curve A graph depicting the renaturization of single-stranded DNA fragments as charted by time.

cotransformation The introduction of two genes on two different DNA fragments during transformation.

coupled transcription and translation The process in prokaryotes in which translation of an mRNA molecule begins before transcription of the mRNA has terminated.

coupling conformation A situation in which two alleles are coupled to one another on the same chromosome. Also called *cis conformation*.

cri du chat syndrome A genetic disorder caused by the terminal deletion of the short arm of chromosome 5.

crisscross pattern of inheritance An inheritance pattern in which an allele is passed from an affected male to carrier females, who then pass it to their male offspring.

crossing-over Exchange of chromosomal segments between two nonsister chromatids. Also called *nonsister chromatid exchange*.

cryptic splice site A cleavage site used during intron removal when cleavage at the normal splice site is prevented.

C value A value that refers to the total amount of DNA in a single (haploid) genome of a species.

C-value paradox The discrepancy between C value and the number of genes in an organism.

cyclin-dependent kinases (CDKs) Kinases that are only able to carry out their function when in the presence of a cyclin.

cyclins A group of proteins that rise and fall throughout the stages of the cell cycle and are among the most important proteins that regulate the cell cycle.

cystic fibrosis A genetic disorder in humans that disrupts the normal functioning of the exocrine system.

cytochrome oxidase A mitochondrial enzyme that binds to oxygen and ensures that oxygen accepts electrons at the end of the electron transport chain during oxidative phosphorylation.

cytokinesis The division of the cytoplasm during mitosis or meiosis.

cytological markers Distinct chromosomal differences that can be observed microscopically.

cytoplasmic male sterility A type of male sterility in plants that is governed by extranuclear genes.

cytoplasmic-nuclear male sterility Cytoplasmic male sterility that is governed by extranuclear genes that interact with nuclear genes.

cytosine (C) One of the four different bases in nucleotides composing the DNA molecule. Cytosine is a pyrimidine that pairs with guanine (G).

cytosine deamination A spontaneous lesion where the amino group on the number 2 carbon of C is removed and a double-bonded oxygen is left in its place, resulting in conversion of C to U.

cytotoxic (killer) T cells Cells that have T-cell receptors on their surface that recognize and bind to MHC-antigen complexes on the infected host cell.

ddNTP See *dideoxynucleotide triphosphate*.

deaminating agent Chemical mutagens that alter existing DNA bases by deamination.

degeneracy The redundancy in the genetic code; a single amino acid may be coded for by more than one codon.

deletion The loss of a chromosome segment.

deletion mapping The identification of genes located in the deleted region of a deletion heterozygote.

deletion mutation A mutation in which one or more nucleotides are deleted from a DNA sequence.

deoxyribonucleic acid (DNA) The primary carrier of genetic information; a polymer composed of deoxy nucleotide subunits.

depurination A spontaneous lesion in which the purine base is released from the DNA backbone, leaving a hydroxyl group in its place.

depyrimidination A spontaneous lesion in which a pyrimidine base is released from the DNA backbone, leaving a hydroxyl group in its place.

deuteranopia A common type of color blindness that is due to the lack of functional pigment involved in green perception, causing difficulty distinguishing red from green. A type of *red-green color blindness*.

developmental genetics The study of the relationships between gene regulation and cell differentiation during development.

diakinesis A stage of prophase I of meiosis in which the chromosomes condense and chiasmata terminalize.

diallel cross An example of systematic hybridization in which the breeder hybridizes each parent to every other parent.

dicentric chromatid A chromatid with two centromeres.

dideoxynucleotide triphosphate (ddNTP) A molecule that lacks a hydroxyl group on both the 2′ and 3′ carbons.

differentiation The process of cells becoming specialized to perform specific functions.

digenic The genotype that is possible for different combinations of any two alleles, it can be simplex, duplex, or triplex.

dihybrid experiment Experiment in which two true-breeding parents differ in two traits but are the same for all other traits.

dikaryotic cell A single cell with two separate haploid nuclei.

dioecy A situation in which members of a species are either male or female.

diploidy A situation in which the somatic cells of an organism contain two complete sets of chromosomes (2*n*).

diplonema A stage of prophase I of meiosis where the chromosomes decondense and paired chromosomes begin to separate, except where a chiasma is present.

directional selection The frequencies of alleles affected by selection change in a single direction.

disassortative mating Preferential mating of individuals with different genotypes.

discontinuous variation Phenotypic variation in which differing traits fall into discrete classes.

dispersive DNA replication A DNA replication model in which each strand on both daughter molecules consists of portions of both original and newly synthesized DNA.

disruptive selection Directional selection in more than one direction to produce separate populations with different phenotypes.

distribution A two-dimensional representation of a metric character in which the increments of measurement are portrayed on the horizontal axis and the number of individuals or proportion of individuals are plotted on the vertical axis.

disulfide bond The bond formed between the R groups of two cysteines, which stabilizes the tertiary structure of a polypeptide.

displacement loop (D-loop) During heteroduplex DNA formation the loop that is formed as the displaced strand loops out as a single-stranded DNA molecule.

D-loop See *displacement loop*.

D-loop replication A mode of replication in which two DNA strands in a circular molecule separate at an origin of replication and DNA synthesis begins. As synthesis progresses, the newly synthesized strand displaces an old strand and creates a displacement loop (D-loop).

DNA See *deoxyribonucleic acid*.

DNA-binding site The site in a protein that binds to the DNA.

DNA clone DNA from any species that researchers introduce into bacterial cells where the introduced DNA replicates as the bacterial cells grow and divide.

DNA ligase An enzyme that catalyzes the formation of a bond between single-stranded free ends of DNA.

DNA markers Fragments of DNA that can be distinguished from one another because of differences in their nucleotide sequences.

DNA polymerase An enzyme that catalyzes the synthesis of DNA.

DNA polymerase I An enzyme found in *E. coli* that catalyzes DNA synthesis and has the ability to accomplish both primer removal and DNA synthesis simultaneously.

DNA polymerase II An enzyme in *E. coli* that synthesizes DNA during repair of DNA damage.

DNA polymerase III An enzyme in *E. coli* that catalyzes most of the DNA synthesis.

DNA probe A segment of DNA that is homologous to at least a portion of a DNA fragment of interest.

DNA repair genes Genes that encode enzymes that repair damage to DNA.

DNA-RNA hybrid A double-stranded nucleic acid with one strand of RNA and one of DNA.

dominance The ability of one allele to phenotypically mask the effect of another allele in a heterozygote.

dominance series A series of differing alleles that dominate over certain alleles while being recessive to other alleles, depending on the genotypic arrangement.

dominant epistasis A form of epistasis that is called dominant because one allele may mask the effect of other alleles at two loci, characterized by a 12:3:1 ratio in F_2 progeny.

dominant suppression A form of epistasis in which a dominant allele at one locus suppresses the effect of a dominant allele at another locus, characterized by a 13:3 ratio in F_2 progeny

donor parent A parent that donates a particular allele to its progeny.

donor screening The selection of semen donors to exclude alleles that may cause genetic disorders.

dosage compensation Mechanism that compensates for the differences in chromosomal constitution of males and females.

double crossover The simultaneous occurrence of two crossovers between linked genes.

double fertilization A process in the fertilization of flowering plants in which two fertilizations take place, one that forms the zygote and one that forms the endosperm.

double-strand-break repair model A model of DNA recombination that requires breaks in both DNA strands.

double trisomy Trisomy for two different chromosomes.

downstream The position of sites on the DNA molecule relative to the direction of transcription. Downstream is in the direction of transcription.

Down syndrome A syndrome that includes mental retardation and is most frequently due to trisomy 21.

drift See *random genetic drift*.

duplicate gene action An epistatic pattern of inheritance in which a dominant allele at either of two loci produces the dominant phenotype, characterized by a 15:1 ratio in F_2 progeny.

duplication A repeat, often in tandem, of a chromosome segment.

dysgenic cross A mating in *Drosophila* where a P-cytotype male is mated with an M-cytotype female.

ecological isolation The separation of populations into different habitats even though they may occupy the same territory.

effector domain A site in constant regions that permits other agents of the immune system to recognize the antibody.

electrophoresis A procedure that separates DNA, RNA, or protein molecules according to size and/or charge.

endogenote The chromosomal portion of DNA in a merozygote.

endoploidy Doubling of chromosome number in a somatic cell.

endosymbiotic hypothesis A hypothesis that proposes that the remote ancestors of mitochondria and plastids were free-living prokaryotic cells that at some point entered primitive eukaryotic cells and remained there.

enhancer A DNA sequence element that, when bound by an activator, stimulates transcription of a gene.

environmental variation The proportion of the phenotypic variation that is caused by environmental (nongenetic) factors.

enzyme A protein that catalyzes a specific biochemical reaction in a living system.

episome A plasmid that has been integrated into the chromosomal DNA of a prokaryotic cell.

epistasis A phenomenon in which one gene influences the phenotypic expression of another gene. A nonadditive interaction of alleles at different loci.

epitope The part of the antigen that is recognized by the antigen-binding site in an antibody.

equational division A cell division in which sister chromatids separate from one another. Describes mitosis and meiosis II.

ESTs See *expressed sequence tags*.

essential amino acids Amino acids that cannot be synthesized and must be included in the diet.

ethological isolation See *behavioral isolation*.

euchromatin A region of the chromosome that is less condensed relative to other portions of the chromatin and contains most genes.

eugenics The intentional hereditary improvement of the human population through selection.

eukaryote An organism whose cells have a cell nucleus.

euploidy The situation in which each cell contains complete sets of chromosomes, with no missing or extra chromosomes.

evolution The process of genetic change within and among species.

evolutionary genetics The study of all levels of genetics as they affect evolutionary processes.

excision repair A group of DNA repair mechanisms that excise damaged nucleotides from the DNA molecule.

exinuclease An enzyme that excises damaged nucleotides in excision repair.

exogenote The introduced portion of DNA in a merozygote.

exon The portion of the RNA molecule in eukaryotes that remains after the processing and removal of introns.

exonuclease activity The ability of certain enzymes to remove nucleotides from a DNA molecule.

expressed sequence tags (ESTs) Nucleotide sequences of cDNA clones.

expressivity The degree of gene expression in a phenotype.

extranuclear inheritance The inheritance of genes that reside outside of the nucleus.

ex vivo therapy A procedure in which physicians extract somatic cells from a patient, insert a functional gene into the extracted cells, and return the cells to the tissue from which they were extracted.

F See *inbreeding coefficient*.

F₁ generation The first filial generation; the first generation produced after mating between parents that are homozygous for different alleles.

F₂ generation The second filial generation; the generation produced by self fertilization or intermating of F₁ individuals.

facultative heterochromatin A type of chromatin found in chromosomes that are highly condensed.

familial adenomatous polyposis (FAP) Inherited colorectal cancer without associated tumors elsewhere in the body.

F-duction See *sexduction*.

F factor A DNA molecule capable of existing as a plasmid or an episome that contains genes that control conjugation and transfer of DNA from one cell to another in *E coli*.

F′ factor An F factor containing a segment of chromosomal DNA.

first-division segregation The segregation of alleles in the first meiotic division.

fission Splitting of one chromosome into two chromosomes.

fitness The relative ability for reproductive success under selection pressure.

5′ cap A methylated guanine nucleotide attached by three phosphates to the 5′ end of an mRNA.

fixation The establishment of a single allele in a homozygous state in all individuals in a population.

fluorescence in situ hybridization (FISH) An in situ hybridization technique that uses a fluorescent probe.

flush-crash cycle A cycle of rapid population growth followed by a steep drop in numbers.

footprinting An experimental technique used to identify a specific DNA sequence. This technique depends on the ability of proteins to protect a desired sequence of DNA from deoxyribonucleases.

forward mutation A mutation in which there is a change from the original, usually functional, form of a gene to a mutant form.

founder effect The effect of drift when a small number of individuals found a new population.

fragile site A secondary constriction that appears at the sites of expanded trinucleotide repeats in the chromosomes of cultured cells.

fragile X syndrome (Martin-Bell syndrome) A genetic disorder due to expansion of a trinucleotide (CGG) repeat region of the *FMR1* gene.

frameshift mutation A mutation that alters the reading frame for translation from the point of the mutation.

free radical A highly reactive molecule containing an unpaired electron.

fusion The joining of two chromosomes into one chromosome.

fusion gene A gene that consists of parts of two or more genes.

G See *guanine*.

G₁ The first gap in the cell cycle, between mitosis and DNA synthesis.

G₂ The second gap in the cell cycle, between DNA synthesis and mitosis.

gain-of-function mutation A mutation that causes a gene to gain function relative to its former state.

galactoside permease A bacterial enzyme encoded by the *lacY* gene that causes lactose outside of the cell to enter the cell by permeating the cell wall and membrane.

gametic isolation Incompatibility of male and female gametes.

gap genes The first segmental genes in *Drosophila* to be expressed.

Gardner syndrome A type of familial colorectal cancer associated with the appearance of non-colorectal tumors.

GC box A conserved sequence found in eukaryotic promoters upstream from the CAAT box. It is named for its consensus sequence GGGCGG.

Genbank The largest DNA sequence database in the United States.

gene A segment of DNA that encodes an RNA molecule.

gene bank A collection of plants, animals, and seeds of many different native types in central facilities.

gene conversion The process by which one allele in a heterozygote is converted into the corresponding allele on the homologous chromosome.

gene expression The process by which a gene produces its product and the product carries out its function.

gene family A series of identical or similar genes that may be clustered together in a single chromosomal region or in clusters on two or more chromosomes.

gene mapping The process of determining the relative locations of genes in a segment of DNA.

gene pool All the genes that can be shared by the members of a species and cannot be shared by members of other species.

generalized transducing phage A phage with chromosomal DNA in place of phage DNA.

generalized transduction Infection of bacterial cells by phage particles that contain DNA from any part of the chromosomal DNA.

gene regulation The process by which a cell determines when and to what extent a specific gene is expressed.

genetic anticipation A pattern of inheritance in which the severity or frequency of an inherited disorder increases with each generation of transmission.

genetic background All genes in an individual except the gene under study.

genetic counseling The counseling that a patient and relatives receive after a clinical geneticist has diagnosed a genetic disorder.

genetic erosion The loss of genes of older plant varieties and animal breeds due to the introduction of genetically improved plants and animals.

genetic load The proportion of deleterious recessive alleles in a population or an individual.

genetic recombination An inheritance phenomenon in which genetic material is recombined in progeny relative to their parents.

genome The entire genetic complement of an organism.

genomic DNA library A collection of cloned DNA fragments from an entire genome.

genotype The genetic makeup of an individual organism.

genotype-by-environment interaction A nonadditive response by two or more genotypes to different environments.

genotype frequencies The distribution of alleles among genotypes in a population.

germ line The cells that become destined to eventually undergo meiosis and produce gametes in multicellular organisms.

germ-line gene therapy The alteration of the genetic composition of the germ line so that people who carry a mutant allele no longer transmit that allele to their progeny.

geographic isolation Geographic separation of two or more populations.

guanine (G) One of the four different bases in nucleotides composing the DNA molecule. Guanine is a purine that pairs with cytosine.

gynandromorph A mosaic phenomenon in *Drosophila* in which the same individual has some XX cells that produce a female phenotype and some XO cells that produce a male phenotype.

hairpin structure Base pairing of complementary nucleotides in a single-stranded RNA molecule that forms a double-stranded segment with a single-stranded loop at the end.

Haldane function A mapping function, derived by J. B. S. Haldane in 1919.

haploid chromosome number (*n*) The number of chromosomes in a gamete; designated by the term *n*.

Hardy-Weinberg equilibrium A situation in which allele frequencies and genotype frequencies remain constant from one generation to the next when certain assumptions are met.

H chains See *heavy chains*.

Hb A See *hemoglobin A*

Hb S See *hemoglobin S*.

H-DNA A triple-helical form of DNA.

heat-shock genes Certain genes whose protein products are produced at elevated temperatures.

heavy (H) chains The largest polypeptide chain found in antibodies, about 440 amino acids long.

helicases A group of enzymes that unwind the DNA molecule during replication.

helix-turn-helix domain A structure that consists of two α-helices within the DNA binding domain of a protein.

helper T cells A class of T cells that assist cytotoxic T cells with their attacks and assist B cells in antibody-antigen recognition.

heme group A chemical group consisting of an iron atom surrounded by a protoporphyrin ring that is found within each of the four subunits of a hemoglobin molecule.

heme pocket The pocket in each of the four subunits of the hemoglobin molecule into which oxygen is drawn.

hemizygous A condition of being neither homozygous nor heterozygous for an X-linked allele, as in males with only one X chromosome.

hemoglobin A tetrameric protein located in erythrocytes that is essential for the cells of the body to carry out aerobic respiration. It is responsible for binding oxygen and carrying it from the lungs to various tissues of the body.

hemoglobin A (Hb A) The structural form of hemoglobin found under normal conditions.

hemoglobin S (Hb S) The structural form of hemoglobin found under sickle-cell anemia conditions.

hemophilia A A genetic disorder that prevents normal blood clotting when blood vessels are ruptured, due to a lack of a protein called factor VIII.

hermaphrodite An individual with both male and female reproductive organs.

heterochromatin A region of the chromosome that is classified as being highly condensed relative to other portions of the chromatin.

heterochronic mutations A mutation that affects the timing of cell development.

heteroduplex DNA A segment of a double-stranded DNA molecule in which one strand is from a different DNA molecule.

heterogametic sex The sex whose gametes differ in chromosomal constitution (XY or XO males and ZW females).

heteromorphic chromosomes Chromosomes that pair as homologous chromosomes, but differ in size and structure.

heteroplasmy Two or more types of mitochondria or plastids in a cell's cytoplasm.

heterosis An increase of vigor in heterozygotes compared to either parent.

heterozygote advantage A situation in which heterozygotes are favored by selection over the two homozygous genotypes.

heterozygous A condition in which the alleles at the same locus on homologous chromosomes are different.

Hfr cell Bacterial cells that have an F factor integrated into the chromosome; "high-frequency recombination." The integrated F factor increases the frequency of recombination.

histocompatibility antigens Proteins produced by the body's cells to identify them as part of the body.

histogram The portrayal of classes (grouped individuals) in a distribution.

histone A class of proteins that stabilizes eukaryotic DNA molecules. The DNA winds around the individual histone cores.

HIV See *human immunodeficiency virus*.

Hogness box See *TATA box*.

Holliday junction A junction at the point of strand exchange between two DNA molecules that can migrate along the two molecules, forming heteroduplex DNA in both molecules as the strands are exchanged.

Holliday model A model that explains crossovers on the basis of a Holliday junction and accounts for gene conversion, first described by Robin Holliday in 1964.

homeobox A DNA sequence in genes that regulate development that encodes homeodomains consisting of about 180 nucleotide pairs.

homeodomain A DNA-binding region of a protein that regulates development that includes a helix-turn-helix motif.

homeologous chromosomes Chromosomes that have some homology between them but that do not normally pair with each other during meiosis.

homeotic genes Genes responsible for segmental identity in animals.

homogametic sex The sex whose gametes have the same chromosomal constitution (XX females and ZZ males).

homologous chromosomes Chromosomes that are identical in size and structure, are similar in nucleotide sequence, and pair with one another during meiosis.

homologues (homologs) Two chromosomes of a homologous pair.

homology Similarity, but not necessarily identity, between two DNA molecules.

homoplasmy A situation in which mitochondria, or plastids, in a cell are all the same type.

homozygous A condition in which the two alleles on the homologous chromosomes are identical.

H strand One of the large RNA molecules encoded by mitochondrial DNA.

Human Genome Project The coordinated effort of many laboratories to map the human genome, identify all of its genes, and eventually determine the entire genomic nucleotide sequence.

human immunodeficiency virus (HIV) A virus that is reponsible for acquired immune deficiency syndrome (AIDS) that frequently mutates. It has the unusual ability to attack the immune system directly, eventually causing immune system failure.

humoral immunity The production of antibodies by B cells, the attachment of those antibodies to antigens, and the destruction of particles attached to an antibody-antigen complex.

Huntington disease A genetic disorder characterized by progressive neurological degeneration. It is due to expansion of a trinucleotide repeat region of DNA.

hybrid dysgenesis The phenomenon of infertility and increased mutation in the offspring of a dysgenic cross.

hybridomas Fusion products of isolated normal B cells fused with cultured myeloma cells that no longer produce antibodies.

hybrid swarm Plants with various ploidies and genes from both parent species occurring after several generations of cross-fertilization among the parental species and the hybrids; characterized by an explosion of genetic and phenotypic variation.

hybrid vigor See *heterosis*.

hydroxylating agents Chemical mutagens that add a hydroxyl group to the bases of nucleotides.

hypervariable regions Sections in the variable region of antibodies that have varied amino acid sequences.

IEF See *isoelectric focusing*.

Ig protein See *immunoglobulin*.

imaginal disks Masses of embryonic tissue that are destined to develop into adult organs in *Drosophila*.

immature (or virgin) B cell A cell whose surface displays antibodies that have not contacted their antigen.

immune deficiencies A disease in which part or all of the immune system fails to function properly.

immunoglobulin A protein molecule that makes up antibodies. Also called *Ig protein*.

inbred line A group of self-fertilizing or interbreeding individuals that are homozygous and nearly genetically identical.

inbreeding Mating between individuals with common ancestry.

inbreeding coefficient (F) The probability that any two alleles at a locus in an individual are alike by descent.

inbreeding depression The appearance of deleterious phenotypes because of inbreeding.

incomplete dominance A situation in which the heterozygote has a phenotype that is intermediate between the phenotypes of the homozygotes.

independent assortment The process in meiosis I by which nonhomologous chromosomes assort independently of one another into two daughter cells.

induced mutation Mutation that is caused by known mutagens.

inducer A small molecule that induces gene transcription by binding to a repressor protein, causing the protein to leave the DNA.

inducer-binding site A site where the inducer binds to the repressor protein.

inducible operon An operon that is induced by the presence of a specific substrate or repressed in the absence of the same substrate.

induction The process that allows a prophage to enter the lytic pathway by excising itself from the chromosome.

informed consent The voluntary submission to a genetic test only after the person receives appropriate counseling about the test.

Infraabdominal **domain** One of two domains of the homeotic gene complex in *Drosophila* that has two protein-encoding genes, *abd-A* and *abd-B*.

initiation codon (start codon) A codon found near the 5′ end of an mRNA molecule that initiates translation and establishes the reading frame for translation. This codon is normally AUG, less frequently GUG or UUG.

insertion mutation A point mutation in which one or more nucleotides are inserted into the DNA sequence.

insertion sequences A specific class of prokaryotic transposons that carry only the information required for transposition.

in situ hybridization "On site"; a technique that makes it possible to detect specific fragments of DNA by probe hybridization directly to chromosomes visualized microscopically.

in vivo gene therapy A type of somatic-cell gene therapy in which physicians introduce a functional gene into a virus and add the virus directly to the cells in the body, where the virus introduces the functional gene into the cells.

integrase An enzyme encoded by phage DNA that causes phage DNA to integrate into chromosomal DNA, establishing lysogeny.

intercalating agent Mutagenic chemical compounds, similar to the nitrogenous bases of DNA, which insert themselves between bases in DNA.

interference The presence of one crossover that interferes with initiation of a second crossover in its immediate vicinity.

interphase The period of the cell cycle consisting of the G_1, G_2, and S phases.

interrupted mating A research technique used to map bacterial genes in which F′ cells are mixed with an excess of Hfr cells to rapidly establish conjugation. After the conjugation proceeds for a certain period of time, the researchers then disrupt conjugation by agitating the cells. The cells are then used to determine recombination rates.

interstitial deletion A deletion of a segment within a chromosome.

intragenic complementation Complementation of two alleles of the same gene that may mislead researchers to conclude that the alleles belong to different genes.

intragenic recombination A phenomenon in compound heterozygotes in which a crossover occurs between the mutant sites of two different alleles of the same gene.

intrinsic terminator A nucleotide sequence at the transcription termination site that contains a conserved sequence with the consensus sequence UUUUUUA in the RNA and a hairpin structure in the RNA just upstream from this conserved sequence.

intron Transcribed sequences of DNA in eukaryotes that do not appear in the final mRNA product because they are removed during the processing of the pre-mRNA molecule.

inversion Reverse orientation of a chromosome segment.

in vitro fertilization The fertilization of eggs in a test tube.

ionizing radiation High energy radiation that can strip electrons from atoms.

isochromosome A chromosome with two identical arms.

isoelectric focusing (IEF) Electrophoresis that separates proteins by net charge only.

isogenic An occurrence in which individuals in a population differ at one place in the genome and are genetically identical elsewhere in the genome.

karyotype The chromosomal makeup of a eukaryotic cell viewed during metaphase.

kilobase (kb) One thousand nucleotide pairs in double-stranded DNA.

kinetochores Bodies composed of protein found on either side of the centromere that are important in separation of the replicated chromosome during mitosis and meiosis.

Klinefelter syndrome A condition in humans due to an XXY karyotype.

knob Conspicuous thickenings of chromosomes at certain places during metaphase.

Kosambi function A mapping function derived by D. D. Kosambi in 1944 used to correct observed recombination frequencies.

lac **repressor protein** A protein that recognizes and binds to the operators in the *lac* operon and blocks transcription.

lagging strand The strand that is synthesized discontinuously during DNA replication.

lariat structure The structure formed by the intron through a 5′ to 2′ bond during the intron removal process.

lawn A layer of bacterial cells in a petri plate.

L chains See *light chain*.

leader peptide A small peptide that is encoded in operons that utilize attenuation.

leading gene The first gene to be transferred during conjugation of bacterial cells.

leading strand The strand that is synthesized continuously during DNA replication.

leaky recessive alleles A recessive allele that encodes a mutant enzyme that has some, but not all, of its function.

leptonema The first stage of prophase I of meiosis, during which the chromosomes begin to condense and homologous chromosomes begin to pair with one another.

lesion A chemically modified nucleotide in DNA that can cause mutations.

leucine zipper A segment of amino acids in a DNA-binding protein in which every seventh amino acid is a leucine.

life cycle The span of an organism's life from fertilization (or asexual generation) to the time it reproduces.

light (L) chain A polypeptide chain found in antibodies that is about 220 amino acids long.

LINE See *long interspersed nuclear element*.

linkage group A chromosomal segment that contains two or more linked markers.

linked genes Genes that are so close to one another on a chromosome that they do not assort independently.

linker DNA 1. An oligonucleotide that contains a specific restriction site and can be inserted into a DNA molecule, allowing recombination to occur. 2. The DNA that is found within the nucleosome.

linker histone A histone that typically has two binding sites for DNA for linking the DNA coils in the nucleosome.

locus A gene's location on a chromosome or within a DNA molecule.

long arm The chromosome arm that is longer relative to the other chromosome arm when the chromosome is divided using the centromere as the reference point.

long interspersed nuclear element (LINE) A highly repeated transposable element (retroposon) found mainly in mammals and a few other species.

loss-of-function mutation A mutation that causes an alteration in the amino acid sequence of a polypeptide causing it to lose all or some of its function.

Lyon hypothesis A hypothesis developed by Mary Lyon that proposes random inactivation of one X chromosome in cell lineages during development of mammalian females.

lysis Cell rupture.

lysogeny Integration of a temperate phage into the chromosomal DNA of a bacterium.

lytic cycle The cycle of phage infection followed by lysis.

macrophages White blood cells whose role is to engulf and digest foreign cells and particles, recognize the antibody-antigen complex, and destroy it along with whatever is attached to it.

MADS box A highly conserved region consisting of 56 amino acids shared with proteins of several genes that govern floral organ identity in plants.

major groove The larger of the two grooves in the DNA double helix.

malignant tumor A tumor whose cells penetrate adjacent tissues.

mapping functions Equations that can correct observed recombination frequencies to approximate true map distances.

marker-assisted selection A form of selection in which genetic markers are used to tag genes of importance.

Martin-Bell syndrome See *fragile X syndrome*.

maternal effect A type of inheritance that appears to be uniparental-maternal but in reality is governed by nuclear genes.

mean A statistical term calculated by summing the phenotypic measurements for all individuals in a group then dividing the sum by the number of individuals.

mechanical isolation Prevention of fertilization by physical differences in the structure and location of reproductive organs.

median The class that divides the distribution into equal halves by number of individuals.

meiocyte A cell in fungi that undergoes meiotic divisions to produce haploid cells.

meiosis The process by which gametes arise through cell divisions.

memory cells B cells from a previous primary immune response to a pathogen.

meristems Small masses of undifferentiated cells in the embryo of a germinating seed and, later, of a growing plant.

meristic trait A trait for which the metric value is measured by counting rather than applying incremental units of measurement.

merozygote A recipient cell that has received an F′ factor.

Meselson-Radding model A modified version of the Holliday model proposed in 1975. It differs from the Holliday model in that it asymmetrically transfers one DNA strand. Also called *asymmetric strand-transfer model*.

messenger RNA (mRNA) The RNA that contains the information that will be translated into a polypeptide.

metacentric chromosome A chromosome whose centromere is near the center of the chromosome.

metaphase The second stage of mitosis that is reached once all the chromosomes are oriented at the spindle equator.

metaphase I The stage of meiosis in which homologous chromosomes remain paired and align on the spindle equator.

metaphase II The second metaphase in meiosis. The duplicated chromosomes align along the spindle equator.

metastasis The process by which malignant tumor cells enter the circulatory or lymphatic system and then migrate to other sites in the body, where they establish new malignant tumors.

methylase An enzyme that methylates DNA.

methylation The addition of methyl groups to DNA or RNA.

metric character Any character that is distributed in a continuous manner in a population and can be measured in some way.

MHC markers (Major histocompatibility markers) Histocompatibility antigens that are encoded by the *MHC* genes.

microsatellite DNA segment that contains a tandem repeat region that consists of short repeats, usually mono-, di-, tri-, or tetranucleotide repeats. Also called *short tandem repeat*.

minimal medium A growth medium that contains only those substances required for cell growth, usually water, a simple carbon source, and inorganic salts.

minisatellite A region of DNA that contains segments of about 10–100 nucleotide pairs that are repeated several times in tandem. Also called *variable-number tandem repeat*.

minor groove The smaller of the two grooves in the DNA double helix.

−10 sequence See *Pribnow box*.

−35 sequence A conserved sequence in prokaryotes, found about 35 nucleotides upstream from the transcription startpoint with the consensus sequence TTGACA in the sense strand.

mismatch repair A mechanism in which mismatched nucleotides that have escaped correction by proofreading are recognized and replaced with correct ones.

missense mutation A mutation that causes a change in the amino acid sequence of a polypeptide.

mitochondria Specialized organelles in the cytoplasm of cells in which ATP is synthesized.

mitosis The process of chromosome replication and partitioning in somatic cells.

mobile genetic element See *transposable element*.

mode The largest class of individuals in a distribution.

model organism Organisms with characteristics that permit efficient genetic analysis.

moderately repetitive DNA sequences Sequences that are repeated a moderate number (usually tens to thousands) of times.

modifier genes Genes that have minor effects on a phenotype determined by a major gene.

molecular clock Any amino acid or DNA sequence used to determine the time of species divergence.

molecular genetics The study of the principal molecules of heredity (DNA, RNA, and protein), how they function in the cell.

monoclonal antibodies Single antibodies produced in cloned cells.

monoecy A situation in which a plant has male and female flowers that are physically separated to encourage cross-fertilization.

monohybrid experiments Experiment in which two true-breeding parents differ for only one trait.

monoploidy The number of chromosomes in one complete chromosome set.

monosomy A situation in which one chromosome is missing from a chromosome set.

morphogen A substance that affects morphological development.

mosaicism A condition in which some somatic cells have one genotype while other somatic cells in the same individual have another genotype.

mRNA See *messenger RNA*.

mRNA editing The process by which the nucleotide sequence in an mRNA molecule is changed after transcription.

multiple alleles More than two alleles for a single locus in a population.

multiple-origins theory A theory proposing that modern humans evolved simultaneously from several *Homo erectus* populations in different parts of the world, aided by migration and gene flow among the developing populations.

mutable allele An allele that mutates frequently because it contains a transposable element.

mutagens Agents that enter the cell and cause mutations.

mutant allele An allele that contains a mutation.

mutation Any change in the nucleotide sequence of cellular DNA.

mutational hotspots Sites in DNA that tend to mutate frequently.

myeloma cells Cancerous B cells.

n The symbol for the haploid number of chromosomes for a species.

native protein gel electrophoresis A method that permits most enzymes to retain their full enzymatic activity during electrophoresis.

natural selection Selection in nature without the intentional aid of humans.

negative regulation Gene regulation in which a repressor protein bound to DNA prevents transcription.

negative selection Any factor that disfavors the reproductive success of one phenotype relative to other phenotypes.

neutrality theory A theory proposing that selection does not differentially influence some alleles.

N-formyl methionine The amino acid that is encoded by the initiation codon, AUG, in prokaryotic translation, used only for initiation of transcription.

nitrous acid A deaminating mutagenic agent that has the ability to deaminate cytosine to uracil, adenine to hypoxanthine, and guanine to xanthine.

nonallelic Reference to alleles located at different loci.

nonautonomous transposon A transposable element that is incapable of self-transposition because it lacks a functional transposase gene.

nondisjunction The failure of a homologous chromosome pair to separate from one another and migrate to opposite poles during meiosis I, or the failure of sister chromatids to separate and migrate to opposite poles during meiosis II.

nonhomologous chromosomes Chromosomes that differ in size, structure, and nucleotide sequence from one another and do not pair with each other during meiosis.

nonionizing radiation Low-energy radiation that does not strip electrons from atoms.

nonparental ditype Tetrad that has two types of spores, both of which are different from the parental types.

nonpenetrance The inability of a genotype to produce the phenotype that is usually associated with the genotype.

nonproductive arrangement Arrangement of an antibody gene that encodes a nonfunctional polypeptide.

nonrecombinant chromosomes Chromosomes that arise when there is no crossover between two linked genes.

nonrecombinant plasmid Plasmid cloning vectors that do not contain foreign DNA.

nonselective medium A medium in which the mutant strain of bacteria cannot be distinguished from the normal or nonmutant strain.

nonsense mutation A mutation that creates a premature termination codon in the reading frame of a polypeptide.

nonsister chromatid exchange See *crossing over*.

nontemplate strand See *sense strand*.

NOR See *nucleolus organizer region*.

normal distribution Distributions in which the mean and mode are theoretically identical and 68% of the population falls within one standard deviation of the mean and 95% falls within 1.96 standard deviations of the mean.

norm of reaction The phenotypic response of a genotype to a range of environments.

Northern blotting A modification of Southern blotting with RNA instead of DNA.

nucleic acid A nucleotide polymer (DNA or RNA).

nucleoid A central region of a prokaryotic cell that contains the chromosome.

nucleolus A specific region of the cell nucleus where rRNA is transcribed.

nucleolus organizer region (NOR) The portion of cellular DNA that contains the tandemly repeated rRNA precursor genes and is contained within the nucleolus.

nucleoplasm The portion of the nucleus outside of the nucleolus.

nucleosome A beadlike structure that consists of DNA wound around a histone core.

nucleotide A molecule that consists of a five-carbon sugar (ribose or deoxyribose) with a nitrogenous base attached to the 1' carbon and one or more phosphate groups attached to the 5' carbon.

nullisomy The absence of both homologous copies of a chromosome that should normally be present.

octad An ascus with eight spores.

Okazaki fragments The fragments of DNA formed by discontinuous DNA synthesis.

oligonucleotide A short nucleotide sequence.

oncogenes Genes whose products promote tumor formation.

oncogenesis Changes in cells that cause the formation of tumors.

oogenesis Generation of an ovum.

operator Site in an operon where regulatory proteins bind.

operon A cluster of different genes that are transcribed together as a single mRNA.

ordered spores Spores whose spatial arrangement in the ascus is determined by meiotic divisions.

origins of replication Fixed sites along the DNA where synthesis begins.

"out of Africa" theory See *single-origin theory*.

overdominance See *heterozygote advantage*.

pachynema The third stage of prophase I of meiosis in which synapsis extends from one telomere to the other throughout the entire chromosome.

PAHs See *polycyclic aromatic hydrocarbons*.

pair-rule genes Genes that control the formation of adjacent paired segments in the developing *Drosophila* embryo.

palindrome A nucleotide sequence that is the same in both strands of a double-stranded segment when each strand is read in the 5'→3' direction.

paracentric inversion An inversion that does not include the centromere.

paralogous regions DNA regions that are similar in both the nucleotide sequence of the genes and the linear organization of the genes.

parapatric speciation Speciation in which there is no distinct separation between populations over a differing geographical range.

parental ditype A tetrad that has two types of spores, both of which are identical to the two parental types.

p arm The short arm of a human chromosome.

parthenogenesis The process in which females produce offspring without fertilization or any genetic contribution from a male.

partial diploid A bacterial cell that has received the F' factor and therefore contains two copies of a segment of chromosomal DNA.

Pascal's triangle A table of binomial coefficient factorials in which each value in the triangle is the sum of the two values immediately above it.

PCR See *polymerase chain reaction*.

P elements A class of transposable elements in *Drosophila*.

peptide bond The covalent bond formed between the carboxyl group of one amino acid and the amino group of another amino acid.

peptidyl site (P site) One of the two sites within a ribosome that can hold tRNAs. This site holds the tRNA molecule to which the polypeptide chain is attached.

peptidyl transferase The ribosomal enzyme that catalyzes the formation of a peptide bond.

pericentric inversion An inversion that contains a centromere.

phage Bacterial virus.

phage particle An intact phage consisting of DNA or RNA enclosed in a capsid.

phenotype The observable outward appearance of an organism, which is controlled by the genotype and its interaction with the environment.

phenotypic variance The variance that can be measured directly in a population.

phenotypic variation The differences in observed characteristics among individuals.

Philadelphia chromosome A chromosome that contains a reciprocal translocation between the long arm of chromosome 22 and the long arm of chromosome 9 and is associated with chronic myelogenous leukemia.

phosphodiester bond A bond formed when two ester linkages are formed on either side of a phosphorus atom. In DNA and RNA, this bond is formed by attachment of the 5' phosphate group of one nucleotide to the 3' carbon of another nucleotide.

photoreactivation A light-dependent repair system in which photoreactivating enzyme (PRE) recognizes pyrimidine dimers in DNA and cleaves the covalent bonds that form the dimers.

phylogenetic tree A representation of a path through which related groups or species may have descended from common ancestry.

pilus A protrusion formed on the surface of a bacterial cell when an F factor is contained within the cell.

plaque The area on a bacterial lawn that has been lysed by bacteriophages and appears as a transparent spot in the otherwise opaque lawn.

plasmid A small, typically circular extrachromosomal piece of DNA that may carry a few genes found in some prokaryotic cells.

plasmid cloning vector Artificial plasmids into which a fragment of DNA can be inserted.

plastid A generic name for an organelle in plants and photosynthetic eukaryotes that differentiates to perform particular functions such as photosynthesis and starch storage.

pleiotropy A phenomenon in which a single gene influences more than one phenotypic trait.

ploidy The number of chromosome sets a cell or an individual carries.

point mutation A change in a single nucleotide or in a few adjacent nucleotides of the nucleotide sequence of cellular DNA.

point of truncation A phenotypic cutoff point for selecting the parents of the next generation in artificial selection.

polar mutation A mutation that affects the genes downstream from it as well as the gene that contains the mutation.

polyadenylation The addition of a poly (A) tail on the 3' end of a eukaryotic mRNA molecule.

polycyclic aromatic hydrocarbons (PAHs) Alkylating agents found in smoke and other common substances.

polygenic inheritance The inheritance of multiple genes that govern the same trait.

polylinker A segment of DNA in a cloning vector that contains unique restriction sites for several enzymes.

polymerase chain reaction (PCR) An in vitro procedure that rapidly and efficiently replicates a fragment of DNA from genomic DNA.

polymorphic markers Genetic markers that are distinguishable as different alleles of a single locus.

polypeptide A linear chain of amino acids connected by peptide bonds.

polyploidy The presence of three or more complete chromosome sets in somatic cells.

polyps Benign tumors found in the colon and rectum.

polysome Several ribosomes attached in sequence onto the same mRNA molecule.

polytene chromosome Large *Drosophila* chromosomes consisting of DNA molecules that are replicated several times side by side and that, instead of forming new chromatids, remain parallel to one another in a single chromosome unit.

population A group of individuals of the same species that intermate with one another.

population genetics The study of inheritance in populations and how it is affected by external forces such as natural selection.

positional developmental regulation A type of regulation in which surrounding cells affect the development of a cell.

position effect Expression of a gene influenced by the position it occupies in the chromosome.

positive regulation Gene regulation in which a regulator protein bound to DNA stimulates transcription.

positive selection Any factor favoring the reproductive success of one phenotype relative to other phenotypes.

posttranscriptional regulation The control mechanisms that regulate the processing, transport and longevity of mRNAs.

posttranslational regulation The control mechanisms that regulate the modification of polypeptides after translation, the activity of functional enzymes, and the assembly of subunits to form functional proteins.

postzygotic reproductive isolating mechanism A type of reproductive isolating mechanism that affects the viability or fertility of hybrids after a hybrid zygote has formed.

potential terminators Sites in a gene that have termination sequences, but do not terminate transcription.

pre-mRNA The unprocessed form of the mRNA molecule.

prezygotic reproductive isolating mechanisms A type of reproductive isolating mechanism that prevents or inhibits the formation of hybrid zygotes.

Pribnow box Conserved sequence in prokaryotes that consists of a six-nucleotide sequence whose center is usually located about 10 nucleotide pairs upstream from the transcription startpoint.

primary constriction See *centromere*.

primary immune response A process whereby a B cell with an antibody that can bind to an antigen on the surface of a pathogen divides several times to produce many identical B lymphocytes.

primary protein structure The linear sequence of amino acids attached to each other by peptide bonds.

primase An RNA polymerase that synthesizes a primer.

primer A short segment of RNA (or DNA) with a terminal $3'-$OH group onto which DNA polymerases can add new deoxyribonucleotides.

processed pseudogene A pseudogene that lacks introns and often has a long sequence of A-T pairs on one end.

productive arrangement Antibody gene arrangement that encodes a functional polypeptide.

promoter A nucleotide sequence found in double-stranded DNA molecules to which RNA polymerase binds to initiate transcription.

proofreading The ability of DNA replication enzymes to correct base mispairing during replication.

prophage A temperate phage DNA molecule that has been integrated into chromosomal DNA.

prophase The first stage of mitosis that begins when chromosomes condense.

prophase II The second prophase in meiosis. The chromosomes condense again, and the nuclear membrane dissipates.

protanopia A form of red-green color blindness caused by a lack of a functional pigment for red perception.

protein A macromolecule containing one or more chains of polypeptides.

proto-oncogenes Genes whose products normally stimulate progress through the cell cycle.

prototrophy The ability of an organism to synthesize all of its needed requirements from minimal medium.

pseudoautosomal inheritance Inheritance of alleles located at any locus within the portion of the Y chromosome that is homologous to a portion of the X chromosome.

pseudogene A sequence of DNA that has the structural elements of a gene but is not expressed as a gene.

P site See *peptidyl site*.

Punnett square A graphic model created by Reginald C. Punnett in which the phenotypic and genotypic proportions of zygotes can be predicted.

purine A double-ring base in a nucleotide (A and G).

pyrimidine A single-ring base in a nucleotide (C and T).

pyrimidine dimer Adjacent pyrimidines, whose bases are covalently bound to one another.

q arm The long arm of a human chromosome.

QTL See *quantitative trait locus*.

quadrivalent Four chromosomes that have paired with one another.

quantitative trait A trait that is characterized with numerical measurements.

quantitative genetics The analysis of how genetic and environmental factors influence the inheritance and expression of quantitative traits.

quantitative trait locus (QTL) A locus that influences a quantitative trait.

quantum speciation Rapid speciation in one or a few generations.

quaternary protein structure The final protein structure formed when two or more protein subunits bond together.

R See *response to selection*.

races Groups that have distinct genetic and often phenotypic characteristics but are still fully interfertile if gene flow is restored.

random amplified polymorphic DNA (RAPD) A PCR-based method in which a short arbitrary primer of about 10 nucleotides is added to the DNA and the bound primers amplify the DNA between them.

random genetic drift A random change in allele frequency due to sampling error.

random sampling The selection of individuals at random from a population.

range of reaction See *norm of reaction*.

RAPD See *random amplified polymorphic DNA*.

rare cutting restriction enzyme An enzyme that cuts DNA rarely, usually at an eight-nucleotide target sequence.

reading frame The uninterrupted, non-overlapping series of three-nucleotide codons in an mRNA that encodes a polypeptide.

realized heritability The calculation of narrow-sense heritability after the response from selection has been observed.

recessive allele An allele in an allelic pair that is not phenotypically expressed when the other, dominant allele is present.

recessive epistasis A type of epistasis in which the homozygous recessive genotype at one locus masks the effect of alleles at another locus, characterized by a 9:3:4 ratio in F_2 progeny.

recessive phenotype The phenotype that prevails when an individual is homozygous for a recessive allele.

reciprocal cross A set of two crosses in which, for one cross, a parent with one genotype is the female and the parent with the other genotype is the male, and the situation is reversed in the second cross: $♀AA \times ♂aa$ and $♀aa \times ♂AA$.

reciprocal translocation A translocation in which two nonhomologous chromosomes exchange acentric fragments.

recombinant chromosomes Also known as crossover-type chromosomes, these chromosomes arise from a crossover between two genes.

recombinant DNA DNA that has been recombined artificially, usually in cloning vectors.

recombinant plasmid A plasmid cloning vector that is cut by a restriction enzyme then recombined with a desired DNA fragment.

recombination The processes of bringing together new combinations of genes in a cell.

recombination frequency A calculation used to find the frequency of recombination between two genes, determined by dividing the number of recombinant progeny by the total number of progeny in a testcross.

recombination repair A postreplication repair mechanism based on recombination between newly replicated DNA molecules.

recurrent selection Selection in which breeders test the population after each generation of hybridization and select the best individuals as the parents for the next generation.

red-green colorblindness A human genetic disorder (usually X-linked) characterized by the inability to distinguish red from green.

relative fitness (w) A mathematical value of fitness measured by determining how successfully genotypes in a population reproduce when compared to one another.

repetitive DNA sequence A DNA sequence that is repeated anywhere in the genome.

replica plating A procedure used to recover mutant colonies that cannot survive on selective media and transfer them to nonselective media, where they are then grown.

replication fork The place at which two strands of a DNA molecule separate in the replication process.

replicative transposition A process by which retrotransposons and retroposons transpose whereby the original element remains in its site and a copy of the element is inserted in another location.

replisome In prokaryotic cells, a group of 20–30 proteins that collectively facilitate the replication process.

reporter gene A foreign gene whose product can readily be detected in the phenotype.

repressible operon An operon that is repressed by the presence of a specific substrate or induced by the absence of this substrate.

repressor A protein that, when bound to DNA, reduces or prevents transcription.

reproductive isolation A reduction or complete exclusion of gene flow between two or more populations.

repulsion conformation A conformation in which the two dominant alleles are not coupled to one another; one dominant allele is on each of the homologous chromosomes.

response to selection (R) The proportion of the selection differential gained by selection.

restriction endonuclease (restriction enzyme) An enzyme that has the ability to cut DNA at specific nucleotide sequences.

restriction fragment length polymorphism (RFLP) DNA fragments created by restriction endonuclease cleavage that differ in length because of mutations.

restriction site A particular DNA sequence that is recognized by a restriction endonuclease.

retinoblastoma A cancer of the eye that often appears in children who inherit a mutant *RB1* allele.

retroposon Segments of DNA that are transcribed into an RNA intermediate that is then reverse-transcribed into DNA that can be inserted into the genome.

retrotransposition Transposition through an RNA intermediate.

retrotransposons Segments of DNA that resemble retroviruses and encode a reverse transcriptase, transposing by means of an RNA intermediate.

reverse selection A reversal of selection pressure to favor the trait that was previously disfavored.

reverse transcriptase An enzyme that synthesizes a DNA molecule from a single-stranded RNA template.

reversion A mutation that restores the function of a mutant allele.

RFLPs See *restriction fragment length polymorphisms.*

rho-dependent termination One of the two modes of termination occurring in prokaryotes. This mode requires an ancillary protein known as rho to determine where and when termination should occur.

ribonucleic acid (RNA) A nucleic acid composed of ribonucleotides that is similar to DNA.

ribosome A cellular structure composed of rRNA molecules and proteins; its function is to translate the information in mRNA into a protein.

ribozyme An RNA molecule that catalyzes biochemical reactions.

RNA See *ribonucleic acid.*

RNA polymerase An enzyme that catalyzes the synthesis of RNA from a DNA template in transcription.

RNA polymerase I An enzyme that transcribes most genes that code for rRNA in eukaryotes.

RNA polymerase II An enzyme that transcribes genes that code for mRNAs.

RNA polymerase III An enzyme that transcribes short genes that code for tRNA and small rRNAs.

rolling circle (σ-mode) replication A type of DNA replication in which multiple tandem linear copies of a circular molecule are synthesized.

S See *selection differential.*

s See *selection coefficient.*

same-sense mutation (silent mutation) A mutation that has no effect on the amino acid sequence of a polypeptide.

sampling error The deviation from expected ratios due to random variation.

Sanger dideoxy sequencing A method, developed by Frederick Sanger, that permits DNA sequencing by utilizing dideoxynucleotide triphosphates.

scanning The process by which the small subunit–Met–tRNA–GTP complex in eukaryotic initiation of translation proceeds along an mRNA molecule until it encounters an appropriate AUG initiation codon.

SDS polyacrylamide gel electrophoresis (SDS-PAGE) An electrophoretic procedure that uses sodium dodecyl sulfate (SDS) to denature proteins and a polyacrylamide gel to separate the proteins.

secondary constriction A second constriction in a chromosome that is away from the primary constriction or centromere.

secondary protein structure The arrangement of a polypeptide chain into α helices, β strands, β turns, and random coils.

second-division segregation The segregation of alleles in the second meiotic division.

segment polarity gene Gene that regulates development of *Drosophila* body segments into anterior and posterior compartments.

segregation The separation of corresponding alleles on homologous chromosomes during meiosis.

segregational division A synonym of meiosis I, during which paired homologous chromosomes segregate from one another.

selectable markers Genes that allow researchers to distinguish among different cell genotypes.

selection A difference in the reproductive successes of different phenotypes.

selection coefficient (s) A measure of the degree to which selection acts against a particular genotype.

selection differential (S) The difference between the population mean and the mean of the selected individuals.

selection theory A theory that proposes that there may be a selective advantage for heterozygosity at certain loci in the population and that selection maintains a high level of heterozygosity for the chromosomes that carry these loci.

selective neutrality A situation in which selection neither favors nor disfavors an allele relative to other alleles.

selective medium A medium in which a mutant strain of bacteria can be distinguished from a nonmutant strain.

semiconservative DNA replication A DNA replication model in which two daughter molecules each contain one original strand and one newly synthesized strand.

semidiscontinuous DNA replication The process by which one strand is synthesized continuously and the other discontinuously.

senescence Permanent cessation of cell division. Cells are incapable of further divisions.

sense strand The DNA strand in a gene that contains the same nucleotide sequence as the mRNA molecule. Also called *nontemplate strand* or *coding strand.*

sequence tagged sites (STSs) Short DNA segments that have been sequenced and assigned to a chromosomal location.

sex-determining region Y (SRY) The gene on the Y chromosome in mammals that is responsible for sex determination.

sexduction (F-duction) The transfer of an F′ factor through conjugation in bacteria.

sex-influenced trait A trait governed by genes on autosomal chromosomes that appears in both sexes but is more frequent in one sex.

sex-limited trait A trait governed by autosomal genes that is limited to one sex or the other.

sexual reproduction Reproduction by gametic union to form a new organismal generation.

Shine-Dalgarno sequence A conserved sequence found in the translation initiation sequence of prokaryotes with the consensus sequence AG-GAGG.

short arm The chromosome arm that is shorter relative to the other chromosome arm when the chromosome is divided using the centromere as the reference point.

short interspersed nuclear element (SINE) A highly repeated transposable element found mainly in mammals and a few other species.

short tandem repeat See *microsatellite.*

shotgun cloning Cutting the entire genome of an organism with a restriction enzyme, then cloning all of the fragments into vectors.

sickle-cell anemia A genetic disorder characterized by erythrocytes that have a sicklelike appearance.

σ-mode replication See *rolling circle replication.*

signal peptide An amino acid sequence found on the end of a polypeptide that directs the polypeptide into the lumen of the endoplasmic reticulum.

silent mutation See *same-sense mutation.*

SINE See *short interspersed nuclear element.*

single-origin theory A theory proposing that modern humans arose once in Africa and then migrated from Africa, displacing other humanlike species that had migrated earlier. Also called *"out of Africa" theory.*

single-strand displacement The process by which a single strand of introduced DNA displaces its homologous counterpart on the chromosomal DNA.

single-stranded binding proteins (SSBs) Proteins that bind to and stabilize newly formed single strands of DNA resulting from DNA unwinding in replication.

sister chromatid One of the two chromatids of a duplicated chromosome.

sister chromatid cohesion The phenomenon that causes sister chromatids to adhere to one another during mitosis and meiosis until the chromatids are ready to separate.

small nuclear ribonucleoproteins (snRNPs) Small protein molecules found in the nucleus that bind with snRNAs to form the spliceosome.

small nuclear RNAs (snRNAs) Small RNA molecules found in the nucleus of eukaryotes that usually range in size from 100 to 250 nucleotides.

solenoid The helical coil that is formed when nucleosomes attach to each other and the total packing becomes about $\frac{1}{40}$th the original length of the DNA.

somaclonal variation Variation that arises because of mutation in cultured cells.

somatic cell A cell in an organism that arises by mitosis.

somatic-cell gene therapy Therapy based on delivery of a functional gene to somatic cells in a tissue to correct a disorder in the tissue.

somatic segregation The process of heteroplasmic cells becoming homoplasmic after several mitotic divisions. Also called *sorting out.*

sorting out See *somatic segregation.*

SOS mutagenesis The phenomenon occurring in the induced state of the SOS response that increases mutation.

SOS response A bacterial distress mechanism that stimulates error-prone repair mechanisms.

Southern blotting A procedure, developed by E. M. Southern, that allows the DNA in an electrophoretic gel to be transferred to a filter while keeping the DNA in the same position on the filter as it occupied in the gel.

specialized transduction Infection of bacterial cells by phage particles that contain a small specific fragment of chromosomal DNA attached to the phage DNA.

species A group of interbreeding individuals that produce fully fertile offspring and are isolated reproductively from other such groups.

spermatogenesis Formation of sperm cells.

S phase The synthesis phase of the cell cycle where nearly all DNA replication occurs.

spindle equator The central plane of the cell during mitosis.

spindle fibers A collection of microtubules that direct the chromosomes to either side of the cell during mitosis and meiosis. They begin to form in the cytoplasm during prophase.

spliceosome A nuclear macromolecule, composed of RNA and protein that removes introns from the pre-mRNA molecule and splices exons.

splicing The process of intron removal and exon joining to form a eukaryotic mRNA.

spontaneous lesion A spontaneous chemical alteration of DNA.

spontaneous mutation A mutation that is not caused by an obvious external agent.

spores The haploid cells in fungal species, which are the products of meiosis.

SSBs See *single-stranded binding proteins.*

stabilizing selection The effect of selection that maintains a stable phenotype in a population.

standard deviation The square root of the variance.

start codon See *initiation codon.*

sticky ends See *cohesive ends.*

stop codon See *termination codon.*

STSs See *sequence tagged sites.*

stuffer fragment A segment of DNA in a phage cloning vector that is cut out and replaced by foreign DNA during the cloning procedure.

submetacentric chromosome A chromosome containing a centromere located somewhat off center.

substitution mutation A point mutation in which a single nucleotide is substituted for another.

subtelocentric chromosome See *acrocentric chromosome.*

supercoils Secondary coils in DNA caused by tension.

suppressor mutation A mutation that nullifies the effect of a second mutation.

symmetric distribution A distribution in which the mode and the mean are the same.

sympatric speciation Speciation of two or more groups with common ancestry that occupy the same area.

synapsis The close association of paired homologous chromosomes during prophase I of meiosis.

synaptonemal complex A highly organized structure that maintains tight association between the nonsister chromatids of homologous chromosomes during meiosis I.

synonomous codon Codons that differ in nucleotide sequence but specify the same amino acid.

syntenic genes Genes located on the same chromosome.

systematic sampling The selection of individuals according to predetermined criteria.

T See *thymine.*

tandem duplications Repeated copies of a chromosome segment that lie adjacent to each other in the chromosome.

TATA box Conserved sequence in eukaryotic promoters located about –25 to –30 with the consensus sequence TATAAAA. Also called the Hogness box.

tautomeric shift A temporary shift of a base to a less stable chemical structure.

T cell A specialized immune-system cell that matures in the thymus.

TDF See *sex-determining region Y.*

T-DNA A segment of DNA in the Ti plasmid of *Agrobacterium tumefaciens* that contains genes whose products cause tumorous galls to form on plants.

telocentric chromosome A chromosome whose centromere is located at its end.

telomerase An enzyme that adds short segments of DNA, using its own RNA template, to the ends of eukaryotic chromosomes, maintaining telomere length during DNA replication.

telomere The structure at the end of a eukaryotic chromosome.

telophase The fourth stage of mitosis, which is reached once the centromeres have reached the poles.

telophase I The first telophase in meiosis. Duplicated chromosomes arrive at the poles.

telophase II The second telophase in meiosis. The centromeres reach the poles and nuclear envelopes form around the haploid sets of chromosomes.

temperate phage A phage that is capable of lysogeny.

temperature-sensitive mutation A mutation that prevents growth and cell division at elevated temperatures.

template A single-stranded nucleic acid that serves as a pattern for synthesis of a complementary strand.

template strand See *antisense strand.*

temporal isolation A form of assortative mating in which populations are reproductively separated by time.

10 nm fiber A linear string of stacked nucleosomes that is 10 nm in diameter.

terminal deletion Loss of a chromosomal segment distal to a single break.

terminalization In meiosis, the process in which the chiasmata migrate toward the telomeres.

termination codon (stop codon) One of three codons found on an mRNA molecule that does not specify an amino acid (UAA, UAG, UGA) but does signal for the termination of translation.

tertiary protein structure The formation of protein structure by interactions between the amino acid residues contained in secondary structures.

testicular feminization A condition in humans in which an XY embryo forms testes that produce normal hormone levels, but a recessive allele on the X chromosome prevents cells from forming the hormone receptors, and the individual develops into a phenotypic female.

testis-determining factor (TDF) See *sex-determining region Y.*

tetrasomy The presence of two extra copies of the same chromosome.

tetratype An unordered tetrad that has four different types of spores.

θ-mode replication When replication proceeds bidirectionally in a circular DNA molecule to form a θ-shaped intermediate when replication is partially completed.

30 nm fiber A coil of a solenoid that is 30 nm in diameter.

threshold effect The effect in which a particular threshold of gene product function must be attained for a phenotype to appear.

threshold trait A trait in which the measurement is qualitative but the underlying inheritance is polygenic.

thymine (T) One of the four different bases in nucleotides composing the DNA molecule. Thymine is a pyrimidine that pairs with adenine (A).

thymine dimer The formation of covalent bonds between adjacent thymine bases.

time-of-entry gene mapping A method used to map bacterial genes by measuring the time it takes the gene to enter a cell during conjugation.

Ti plasmid "Tumor-inducing plasmid." A large plasmid carried by *Agrobacterium tumefaciens.*

topoisomerases A group of enzymes that relieve the positive tension created by the unwinding of the DNA molecule during replication.

totipotent cell A cell that is capable of giving rise to an entire organism.

trans-acting element Any gene or DNA sequence that affects genes on a different DNA molecule than the one on which it is located.

transcription The process by which the genetic information is transferred from DNA to RNA.

transcriptional regulation The mechanisms that determine if, when, and to what extent a gene will be transcribed.

transcription factor An ancillary protein in eukaryotes that assists RNA polymerase in binding to DNA to initiate transcription.

transcription-repair coupling The repair of DNA during transcription.

transcription startpoint The site on the DNA molecule where RNA polymerase adds the first ribonucleotide to begin transcription.

transduction The process of bacterial DNA being packaged in a phage particle, in which case it can be transmitted from one bacterial cell to another through phage infection.

transfer RNA (tRNA) A small ribonucleic acid molecule that transfers amino acids to the ribosome. Each molecule contains an anticodon that pairs with the codon of an mRNA molecule.

transformation The uptake of DNA from a cell's environment.

transgressive segregation A situation in which the phenotypes of progeny exceed the parental range of phenotypes.

transition A substitution mutation in which a purine is substituted for a different purine, or a pyrimidine is substituted for a different pyrimidine.

translation Polypeptide synthesis in which the amino acid sequence is encoded by a specific RNA sequence.

translational regulation The mechanisms that determine whether or not and to what extent an mRNA will be translated.

transmission genetics The study of how genetic material and the traits encoded by that material are transmitted from parents to offspring.

transposable element (mobile genetic element) An element of DNA that can move from one location in the genome to another.

transposon A segment of DNA that moves from one location of the genome to another and moves as a DNA entity.

transposon tagging A process by which transposons are used to recover part of the gene to use as a probe, used when a probe for a gene is not available.

transversion A substitution mutation in which a purine is substituted for a pyrimidine or a pyrimidine is substituted for a purine.

trihybrid experiment An experiment in which two true-breeding parents differ in three traits.

trinucleotide repeat expansion A class of mutations that expand the number of normal tandem three-nucleotide repeats found in a gene.

trisomy The most common form of aneuploidy, in which there is one extra chromosome.

trisomy 21 A condition in which each somatic cell has three copies of chromosome 21 and two copies of every other chromosome. Its phenotypic consequence is Down syndrome.

triticale An allooctaploid derived from bread wheat and rye.

trivalent Three chromosomes that are paired with one another.

tRNA See *transfer RNA.*

true-breeding individual An individual that is homozygous and produces identical homozygous offspring when self-fertilized or hybridized with an individual of like genotype.

tumor A mass of cells that has lost the normal constraints over growth and division.

tumor suppressor gene A gene whose product prevents or inhibits progress through the cell cycle.

turbid plaque A plaque that appears cloudy because it contains a mixture of lysed cells and lysogenic cells.

Turner syndrome A condition in humans in which the heteromorphic chromosomal makeup of an individual is XO.

two-dimensional polyacrylamide gel electrophoresis (2D-PAGE) A method that combines IEF and SDS-PAGE to separate proteins by net charge in one dimension and size in the other.

two-factor linkage analysis The analysis of two linked genes.

2*n* See *diploid.*

type I error The erroneous rejection of a correct hypothesis.

U See *uracil*.

***Ultrabithorax* domain** One of two domains of the homeotic gene complex in *Drosophila* that contains the *Ubx* gene.

unbalanced cell A cell that has extra or missing genetic material.

underdominance A lower relative fitness for heterozygotes when compared to homozygotes.

unequal crossing-over A crossover between tandem repeats in homologous chromosomes that did not line up equally.

unidirectional replication Migration of a replication fork in one direction only from an origin of replication.

uniparental-maternal organellar inheritance Only the maternal parent contributes organellar genes to the offspring.

uniparental-paternal organellar inheritance Only the paternal parent contributes organellar genes to the offspring.

unique (or nonrepetitive) DNA sequence A DNA sequence that is not repeated within itself nor anywhere else in the genome.

univalent An unpaired chromosome.

unordered spores Asci with spores that do not correspond spatially to the meiotic divisions.

unprocessed pseudogene A pseudogene that typically has all the features of an active gene, including its own promoter and introns.

unreduced gamete A gamete that has the same number of chromosomes as the somatic cells.

upstream The position of sites on the DNA molecule relative to the direction of transcription. Upstream is opposite or against the direction of transcription.

uracil (U) One of the four bases in the nucleotides of RNA. Uracil is a pyrimidine that pairs with adenine (A).

variable-number tandem repeats See *minisatellites*.

variable regions Region of the heavy and light chains that differ substantially among antibodies and contain the antigen-binding sites.

variance A measurement of the degree to which individuals within a sample vary from one another.

***vir* genes** A set of genes carried by the Ti plasmid that encodes the enzymes necessary for infection.

virgin B cell See *immature B cell*.

virulent phage A phage that relies entirely on the lytic cycle for reproduction.

w See *relative fitness*.

Western blotting A modification of Southern blotting that utilizes protein instead of DNA or RNA.

wild-type allele The allele that is represented most frequently at a locus in the genome of a species.

wobble hypothesis A hypothesis explaining the pattern of degeneracy in the genetic code, based on a specialized set of pairing rules for the third nucleotide in the codon with its corresponding nucleotide in the anticodon.

X-chromosome inactivation A type of mammalian dosage compensation in which one of the X chromosomes is inactivated and remains nonfunctional in female cells.

X-linked gene Any gene located on the X chromosome.

yeast artificial chromosome (YAC) A linear cloning vector created from yeast chromosomal DNA.

Y-linked (holandric) gene Any gene located on the Y chromosome.

Z-DNA A structure in which DNA is wound in a left-handed conformation, where the molecule is narrow, consisting of more base pairs per helical turn than in B-DNA. The sugar phosphate backbones of Z-DNA form a zigzag pattern.

zinc fingers Structures found on some DNA-binding proteins that consist of two β strands and an α helix held together in four places by a zinc ion bound to cysteine and histidine residues in the protein molecule.

zinc finger Y A gene normally located on the Y chromosome that contains a zinc finger domain. Also called *ZFY*.

zygonema The second stage of prophase I of meiosis during which synapsis begins and chromosomes continue to condense.

INDEX

Some Type II Restriction Endonucleases with Their Restriction Sites, Shown After Cleavage

Restriction Endonuclease	Source	Restriction Site	Type of Cut
AluI	Arthobacter luteus	5' AG CT 3' 3' TC GA 5'	Blunt
BalI	Brevibacterium albidum	5' TGG CCA 3' 3' ACC GGT 5'	Blunt
BamHI	Bacillus amyloliquifaciens	5' G GATCC 3' 3' CCTAG G 5'	Staggered
BglII	Bacillus globigii	5' A GATCT 3' 3' TCTAG A 5'	Staggered
CfoI	Clostridium formoaceticum	5' GCG C 3' 3' C GCG 5'	Staggered
DraI	Deinococcus radiophilus	5' TTT AAA 3' 3' AAA TTT 5'	Blunt
EcoRI	Escherichia coli	5' G AATTC 3' 3' CTTAA G 5'	Staggered
EcoRV	Escherichia coli	5' GAT ATC 3' 3' CTA TAG 5'	Blunt
HaeIII	Haemophilus aegyptius	5' GG CC 3' 3' CC GG 5'	Blunt
HindIII	Haemophilus influenzae	5' A AGCTT 3' 3' TTCGA A 5'	Staggered
PstI	Providencia stuarti	5' CTGCA G 3' 3' G ACGTC 5'	Staggered
SacI	Streptomyces achromogenes	5' GAGCT C 3' 3' C TCGAG 5'	Staggered
SalI	Streptomyces alba	5' G TCGAC 3' 3' CAGCT G 5'	Staggered
TaqI	Thermus aquaticus	5' T CGA 3' 3' AGC T 5'	Staggered
XhoI	Xanthomonas holica	5' C TCGAG 3' 3' GAGCT C 5'	Staggered